Control Systems

Theory and Applications

Second Edition

Smarajit Ghosh
Professor and Head
Department of Electrical and Instrumentation Engineering
Thapar University, Patiala
Punjab, India

Pearson

Editor—Acquisitions: Sameer Gupta
Associate Editor—Production: M. Balakrishnan

ISBN 978-81-317-5837-3

First Impression, 2012
Twelfth Impression, 2023
Thirteenth Impression, 2024

Head Office: 1st Floor, Berger Tower, Plot No. C-001A/2, Sector 16B, Noida – 201 301,Uttar Pradesh, India.

Registered Office: Featherlite, 'The Address' 5th Floor, Survey No 203/10B, 200 Ft MMRD Road,Zamin Pallavaram, Chennai - 600044
Website: in.pearson.com; Email:companysecretary.india@pearson.com

Digitally printed in India by Trinity Academy for Corporate Training Ltd, New Delhi in the year of 2024.

To

my father

Late Saileswar Ghosh

to whom I am indebted for what I am

Brief Contents

Contents

Preface

Control Systems: Theory and Applications has been prepared for undergraduate students in electrical, electronics and communication, mechanical, instrumentation and computer science courses of all engineering colleges in India and abroad. The book covers conventional control, modern control and digital control. Control Systems has wide application in all branches of engineering and power systems. Spread across 19 chapters, the book presents a unique self-study material on Control Systems with excellent pedagogical features. Additional solved numericals in different patterns have been given in each chapter for practice. Significant points, Short questions with answers, Exercises, and Test yourself (MCQ's) have been provided at the end of each chapter. An overview of the topics covered in each chapter is shown below:

Chapter 1 presents the fundamental and basic concepts of control systems.

Chapter 2 deals with the Laplace transform and matrix algebra, without which it is not possible to discuss the concepts of Control Systems.

Chapter 3 presents the transfer function. It characterizes the input–output relationship of the system.

Chapter 4 covers control system components.

Chapter 5 deals with mathematical modelling of physical systems, force–voltage and force–current analogy of systems.

Chapter 6 discusses block diagram representation of systems, which simplifies the input and output relationship of any complicated system.

Chapter 7 discusses signal flow graph. It represents the system in another pictorial form and is regarded as a simplified version of a block diagram. Mason's gain formula and its applications have also been discussed in this chapter.

Chapter 8 deals with the time response. It is imperative for design and analysis of control systems.

Chapter 9 gives feedback characteristics of control systems.

Chapter 10 presents the stability of control systems and discusses Routh–Hurwitz Criteria in detail to get absolute stability of a system.

Chapter 11 presents the root locus method. The root locus technique provides knowledge of movement of poles by graphical method in *s*-plane.

Chapter 12 deals with frequency domain analysis, which is an analytical method that can be used for analysis and design.

Chapter 13 is on Bode plot. It is the plot of magnitude as well as phase angle versus frequency representing a sinusoidal transfer function. It shows both high frequency and low frequency characteristic of transfer function in a single diagram.

Chapter 14 deals with Polar plot. It is possible to plot the magnitude and phase angle of a sinusoidal function on polar co-ordinates varying ω from 0 to ∞.

Chapter 15 presents Nyquist Criteria by which we can find the absolute stability as well as relative stability of any system.

Chapter 16 gives an introduction to compensators.

Chapter 17 is on state variable approach. State variable analysis automatically takes care of initial conditions and it is also possible to analyze time varying or time-invariant, linear or non-linear, single or multiple input-output systems. In conventional control theory, the initial conditions are assumed to be zero and it is applicable to linear time invariant system having single input and output. It also covers the concepts of controllability and observability.

Chapter 18 is on digital control systems.

Chapter 19 is on non-linear control systems.

Appendix A caters the Fuzzy Logic which will be useful to the students for its applications in control systems.

Students having no prior knowledge in MATLAB will feel free to learn it because it has been introduced step by step, beginning from Chapter 1.

At the end of the book 400 MCQs, 100 Extra Solved Problems have been appended. MATLAB has been discussed in detail and 48 extra problems solved through MATLAB have been added.

This book will also be beneficial to students preparing for AMIE, IETE, UPSC, GATE, IES and other competitive examinations. Constructive criticism for further improvement of the book will be appreciated.

Acknowledgements

My sincere thanks goes to Dr Abhijit Mukherjee, Director of Thapar University, Patiala, for his encouragement to bring the second edition of the book. I am also thankful to Dr Abhijit Chakrabarti, Professor, Department of Electrical Engineering, BESU, Howrah, and Dr Samarjit Sengupta, Professor, Department of Applied Physics, University College of Science and Technology, Calcutta, for their inspiration to bring this edition of the book.

I am indebted to Thomas Rajesh Mathew, Sameer Gupta and M. Balakrishnan of Pearson Education for their role in bringing this book.

I am obliged to my colleague Moon Inder Singh and my students Gagan, Arminder Singh and Kultardeep Singh for their help in editing the contents for this book.

Finally, I wish to express my sincere gratitude to my mother Smt. Beena Ghosh, and my wife Dr Kuntal Ghosh, my sons Saurojit Ghosh (Manpreet), Shubhojit Ghosh (Harpreet) and my daughter Sneha Ghosh (Gurpreet), for their patience during the preparation of materials and editing of the manuscript.

SMARAJIT GHOSH

About the Author

Smarajit Ghosh is Professor and Head, Department of Electrical and Instrumentation Engineering, Thapar University, Patiala, Punjab. He did his B.Tech. and M.Tech. in electrical machines and power systems from Calcutta University and was awarded a doctorate in electrical power and distribution systems by IIT Kharagpur. He has contributed several research papers in international and national journals. His areas of research are load flow study, application of fuzzy logic, artificial neural networks and differential evolution in electric power distribution systems.

He is a recipient of the Bharat Jyoti Award. He had also served at BITS Pilani and Sikkim Manipal University as Assistant Professor and Professor, respectively. He has guided a number of students to their Ph.D. degrees.

Chapter 1

Fundamentals of Control Systems

Control systems play a vital role in our day-to-day life. The notion of automatic control system is basic and well mixed in every facet of our life. Automatic control systems play an important role in the advancement and improvement of engineering skills. There is a series of unending implementations of control systems, from a simple bread toaster to a complex power plant, and also in the working of satellites, guided missiles, etc. This chapter introduces the basic control systems.

1.1 Basic Definitions

System: An arrangement or combination of different physical components that are connected or related together to form an entire unit to achieve a certain objective is called a system.

Actually, each physical object can be considered a system. A classroom is the combination of benches, blackboard, fans, lights, etc. and it acts as an elementary system. A lamp is another example of a system because it is made up of glass, filament, etc. A kite, likewise, is an example of a system because it is made up of paper and sticks.

A system can be of any type i.e., physical, ecological, biological, etc.

Control: The meaning of control is to regulate, direct or command a system so that a desired objective is obtained.

Lecturing of a person in a classroom is an example of a control system, because the person not only regulates the harmony of the entire class but also disseminates knowledge to the students so that the objective is achieved. This is the input of good knowledge to the students.

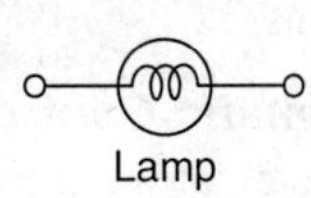

Fig. 1.1 Physical system

Switching "on" and "off" a lamp using a switch is an example of a control system because the entire system becomes a control system. Figure 1.1 shows a physical system, whereas Fig. 1.2 shows a control system.

When a person plays with a kite, the string is used to control the direction and flight of the kite. The entire system is, hence, considered as a control system.

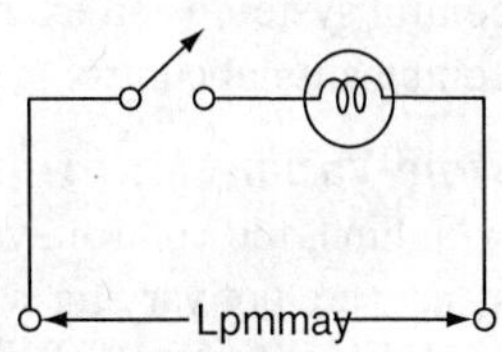

Fig. 1.2 Control system

A control system is the interconnection of the physical components to obtain a desired function, which involves some kind of controlling action on it.

***Plant*:** It is defined as the portion of a system, which is to be controlled or regulated. It is also called a process.

***Controller*:** It is the element of the system itself, or may be external to the system. It controls the plant or the process.

***Input*:** The applied signal or excitation signal that is applied to a control system to obtain a specified output is called input.

***Output*:** The actual response that is obtained from a control system due to the application of the input is termed as output.

***Disturbances*:** The signal that has some adverse effect on the value of the output of a system is called disturbance. If a disturbance is produced within the system, it is termed as an internal disturbance. Otherwise, it is known as an external disturbance.

***Control Systems*:** It is an amalgamation of different physical elements linked in such a manner so as to regulate, direct or command itself to obtain a certain objective. A control system must have (1) input, (2) output, (3) ways to achieve input and output objectives and (4) control action.

Figure 1.3 shows the cause-and-effect relationship between the input and the output of a plant.

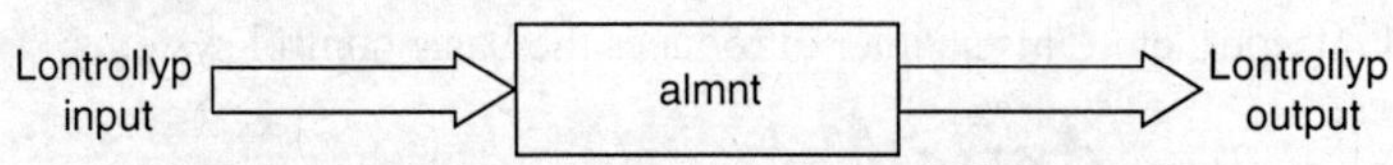

Fig. 1.3 Cause and effect relationship between the input and the output of a plant

1.2 Classification of Control Systems

***Natural Control Systems*:** The systems inside a human being or a biological system are known as natural control systems.

***Man-made Control Systems*:** The various control systems that are designed and developed by man are known as man-made control systems. An automobile system is an example of man-made control systems.

***Automatic Control Systems*:** Automatic control system is the theoretical base for mechanization and automation. It uses the methods from mathematics and engineering. Automatic control systems are the composition of the following three components: (i) sensors, (ii) responders, and (iii) actuator(s).

***Combinational Control Systems*:** The combination of a natural control system and a man-made control system is an example of combinational control systems. Driver driving a car is an example of combinational control systems.

***Time-Variant and Time-Invariant Control Systems*:** If the parameters of a control system vary with time, the control system is termed as time-varying control system. If the parameters of a control system are not varying with time, it is termed as time-invariant control system. A space vehicle leaving earth is an example of time-variant system. The elements of an electrical network such as resistance, inductance and capacitance are not time varying; this is an example of the time-invariant system.

Linear and Non-linear Systems: A control system is known as linear if it satisfies the additive property as well as the homogeneous property.

(a) Additive Property: If x and y belong to the domain of the function f, we have

$$f(x+y) = f(x) + f(y) \tag{1.1}$$

(b) Homogeneous Property: For any x belonging to the domain of the function f and for any scalar constant β, we have

$$f(\beta x) = \beta f(x) \tag{1.2}$$

The principle of superposition is a combination of the above two properties. If a function satisfies the above two properties, it is said to be linear in nature.

If $f(x) = x^3$, it is obvious that $f(x+y) = (x+y)^3 \neq x^3 + y^3$ and $f(\alpha x) = (\alpha x)^3 = \alpha^3 x^3 \neq \alpha(x)^3$. Therefore the function $f(x) = x^3$ is non-linear.

All physical systems are non-linear in nature. If the presence of certain non-linearity does not affect the performance of the system much, the presence of non-linearity can be neglected and the system can be treated as a linear system.

Continuous-Time and Discrete-Time Control Systems: If all the system variables of a control system are functions of time, it is termed as a continuous-time control system. If one or more system variables of a control system are known at a certain discrete time, it is termed as a discrete-time control system. The speed control of a DC motor with tacho-generator feedback is an example of continuous-time control systems. The microprocessor-based or computer-based system is an example of discrete-time control systems.

Deterministic-Control and Stochastic-Control Systems: If the response to input, and to external disturbances, of a control system is predictable and repetitive, the control system is known as a deterministic system. Any control system is called stochastic if such a response is unpredictable.

Lumped-Parameter and Distributed-Parameter Control Systems: If a control system can be represented by ordinary differential equations, such a control system is called lumped-parameter control system. In the case of electrical networks, parameters such as resistance, inductance, etc. are lumped-parameter systems.

If a control system can be described by partial differential equations, such a control system is known as distributed-parameter control system. In a transmission line, its parameters such as resistance and inductance are totally distributed along it. Therefore, transmission line characteristics are always described by partial differential equations.

Single-Input–Single-Output and Multiple-Input–Multiple-Output Systems: If a control system has one input and one output, it is termed as single-input–single output (SISO) system. If a control system has multiple input and multiple output, it is known as multiple-input–multiple output (MIMO) system. MIMO systems are also known as multivariable systems.

Open-loop and Closed-loop Systems: This is another classification of control systems. Both types are discussed in details in the following section.

1.3 Open-loop and Closed-loop Systems

Any control system can be classified as (a) open-loop system or (b) closed-loop system depending on the relation of control of action with output.

1.3.1 Open-loop System

A system in which control action does not depend on output is known as open-loop system. Figure 1.4 shows an open-loop system.

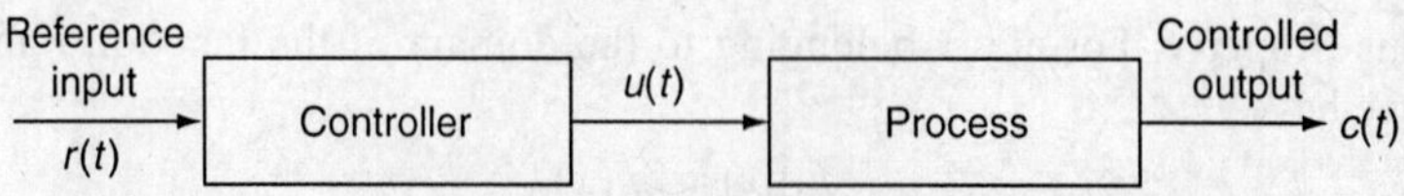

Fig. 1.4 Open-loop system

Examples of open-loop systems are automatic washing machines, bread toaster, electric hand drier, automatic milk server, automatic coffee server, electric lift, traffic signals and sprinkler etc.

An electric switch is a good example of an open-loop system because output is light and switch is controller of the lamp. There is no effect on the ON–OFF position of the switch due to any change of light.

The output of any automatic washing machine is the degree of its cleanliness of clothes. Its controlling action will not be affected due to any change in this output.

In an automatic bread toaster system, the quality of toast depends on the time for which toast is heated. Bread is simply heated in this system and it depends on the time setting. The toast quality is judged by the user and has no effect on the inputs. Figure 1.5 shows the block diagram of the function of a bread toaster.

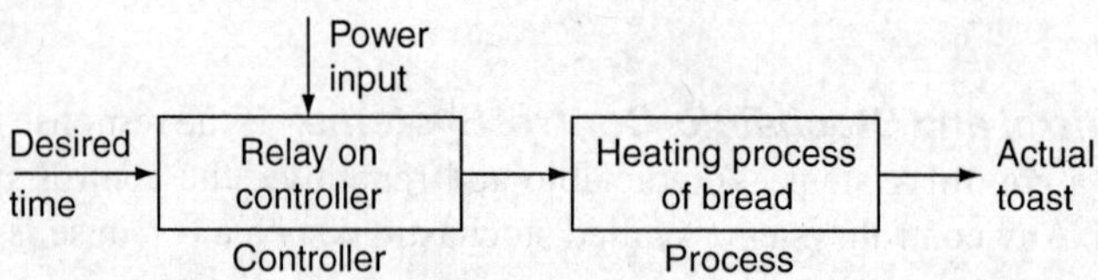

Fig. 1.5 Automatic bread toaster

Any traffic flow control system used on the road is an example of an open-loop control system, which is time dependent. The traffic on the road can be either mobile or stationary and it depends on the duration and sequence of lamp glow, which is time dependent and is controlled by relays. These are predetermined and do not depend on the volume of the traffic on the road. Figure 1.6 shows the block diagram of a traffic light controller.

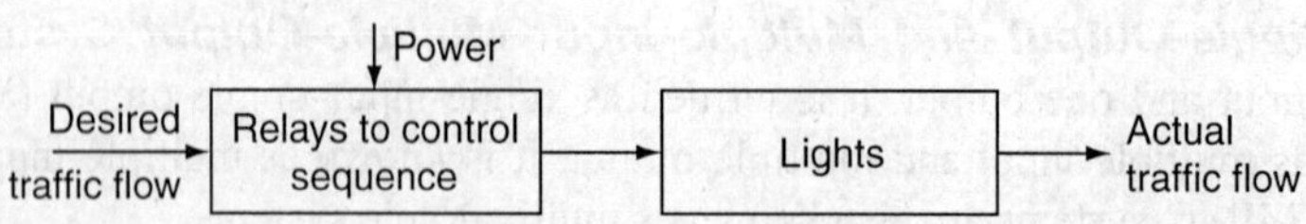

Fig. 1.6 Traffic light controller

1.3.1.1 Advantages and Disadvantages of Open-loop System

Advantages

- These systems are simple in construction and design.
- These systems are economic.
- These systems are easy from maintenance point of view.

- Usually these systems are not much troubled with problems of stability.
- These systems are convenient to use when output is difficult to measure.

Disadvantages

- These systems are not accurate and reliable because their accuracy is dependent on the accuracy of calibration.
- In these systems, inaccurate results are obtained with parameter variations, i.e., internal disturbances.
- Recalibration of the controller is required from time to time for maintaining quality and accuracy.

To overcome the above disadvantages, closed-loop control system is used.

1.3.2 Closed-loop System

If controlling actions of a system are somehow dependent on output or changes in output, system is called closed-loop system.

The property of the system which permits comparison of output with the input so that appropriate controlling action can be taken is called the feedback of the system. The part of the output is fed to input for comparison with the input signal in such systems.

Figure 1.7 shows a closed-loop control system.

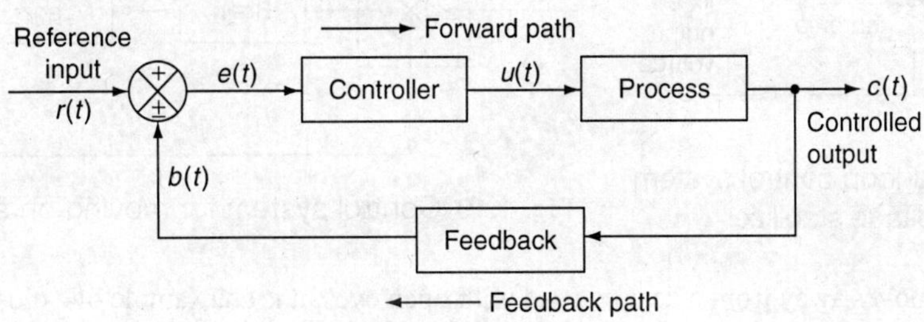

Fig. 1.7 Closed-loop system

In Fig. 1.7, $r(t)$ is the reference input, $e(t)$ is the error signal, $u(t)$ is the actuating signal, $c(t)$ is the controlled output, and $b(t)$ is the feedback signal.

For a positive feedback signal, we have

$$e(t) = r(t) + b(t) \tag{1.3}$$

For a negative feedback signal, we have

$$e(t) = r(t) - b(t) \tag{1.4}$$

This error $e(t)$ is applied to the controller. The controller gives the activating signal $u(t)$ for the process. The controlled output will be driven by the action of the controller in such a manner that $e(t)$ will be reduced to zero. Practical examples of closed-loop systems are automatic electric iron, railway reservation status display, display motor speed controlled by tacho-generator feedback, servo voltage stabilizer, water level controller, missile launching system, temperature control system, voltage stabilizer, etc.

Figure 1.8 depicts the closed-loop control of automatic electric irons. An automatic iron regulates the temperature of iron in such a way that the temperature for a specific cloth remains in a specified range.

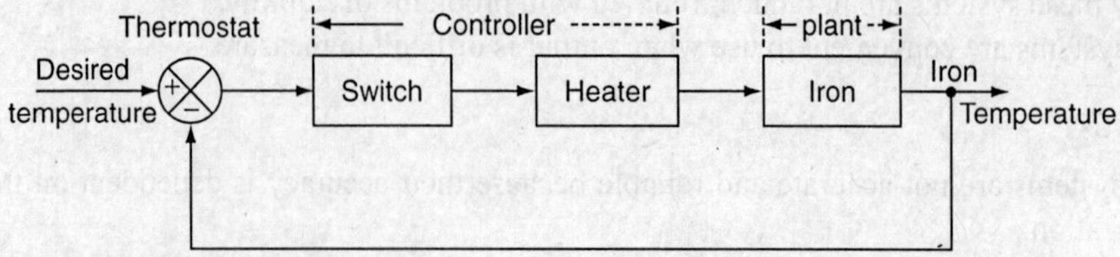

Fig. 1.8 Closed-loop control system of an automatic electric iron

Figure 1.9 depicts the closed-loop control of a voltage stabilizer. It is expected that the output voltage will be 230 V for a single-phase system. When input decreases, the output switch will be connected above (M). On the other hand, if input increases, the output switch will be connected below (M).

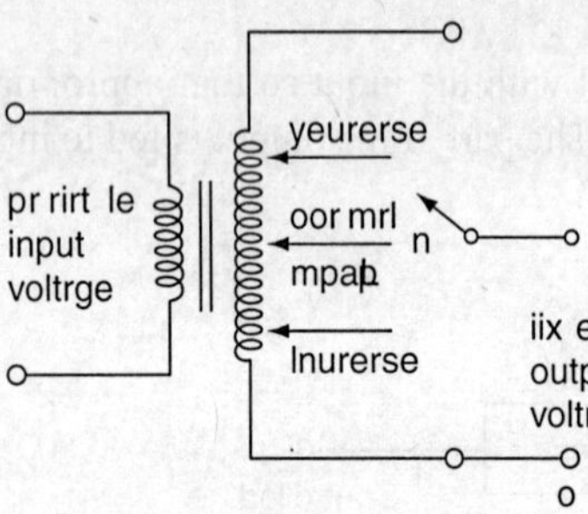

Fig. 1.9 Closed-loop control system of a voltage stabilizer

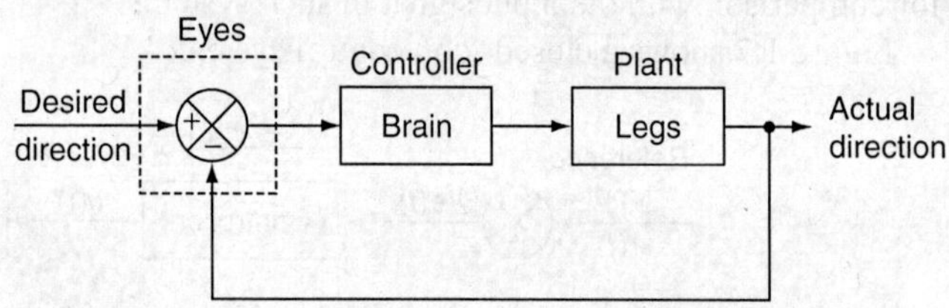

Fig. 1.10 Control system for moving on a desired road

If anybody moves away from a desired road with open eyes, it is an example of a closed-loop system as shown in Fig. 1.10. It will be an open-loop system if eyes are closed.

Figure 1.11 shows a sun seeker solar system, which is an example of a closed-loop system.

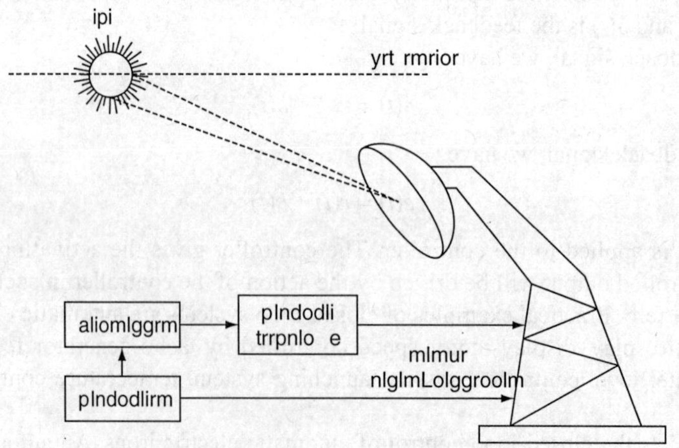

Fig. 1.11 Control system for a sun seeker solar system

Figure 1.12 shows a missile launcher, which is an example of a closed-loop system.

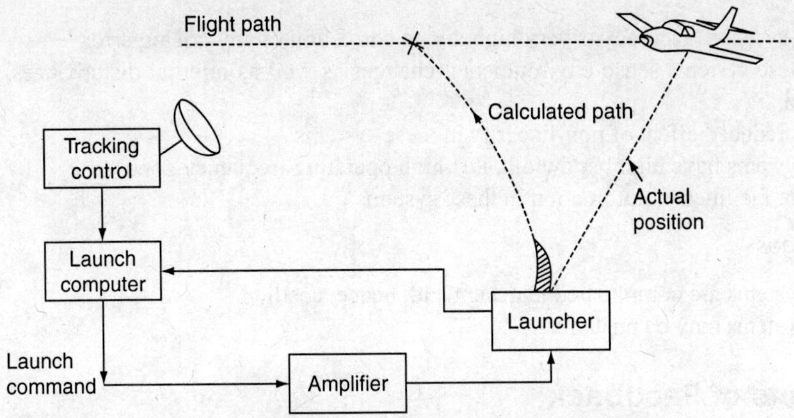

Fig. 1.12 Control system for a missile launcher

Figure 1.13 shows the speed control of a direct current (DC) motor where the DC shunt motor is used and the field current is kept constant. The armature voltage is favourably varied so that the desired speed is obtained. The tachometer is used to get the feedback, which generates voltage proportion to speed that is compared with the voltage required to obtain the desired speed. This difference is used to change the input to the controller, which cumulatively changes the speed of the motor as required. Figure 1.14 shows the block diagram for the speed control of a DC shunt motor.

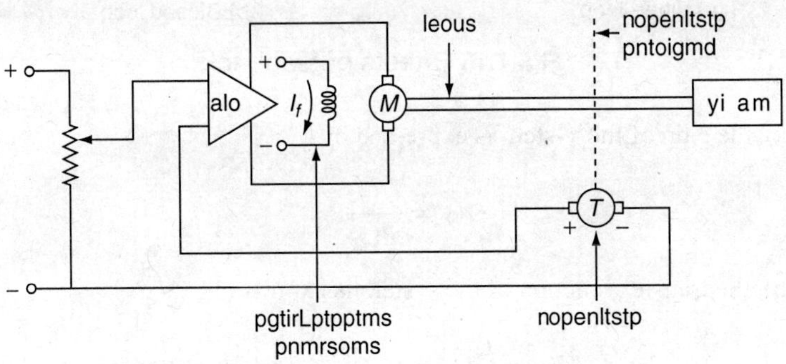

Fig. 1.13 Speed control of a DC shunt motor

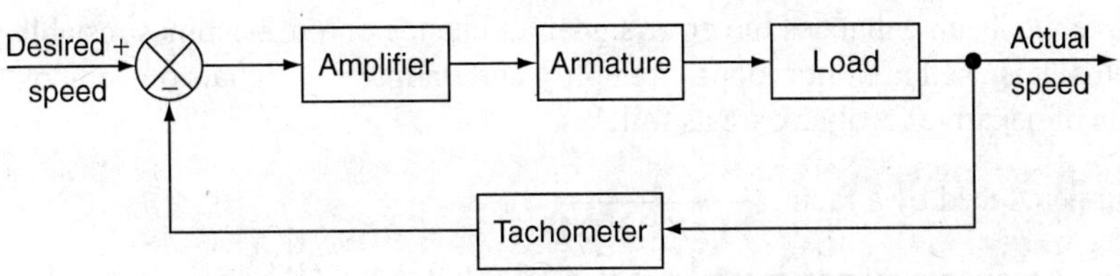

Fig. 1.14 Block diagram for speed control of a DC shunt motor

1.3.2.1 Advantages and Disadvantages of Closed-Loop System

Advantages

- In these systems, accuracy is very high due to correction of any arising error.
- Since these systems sense environmental changes as well as internal disturbances, the errors are modified.
- There is reduced effect of non-linearity in these systems.
- These systems have high bandwidth, i.e., high operating frequency zone.
- There are facilities of automation in these systems.

Disadvantages

- These systems are complicated in design and, hence, costlier.
- These systems may be unstable.

1.4 Effects of Feedback

The error between the system input and output can be reduced by using a feedback system as shown in Fig. 1.15.

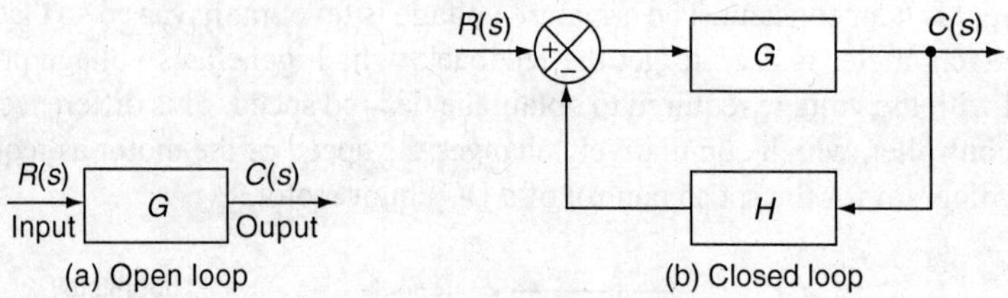

Fig. 1.15 Effects of feedback

From Fig. 1.15(a), the gain of the system is expressed by

$$G(s) = \frac{C(s)}{R(s)} \tag{1.5}$$

From Fig. 1.15(b), the transfer function of the system is expressed by

$$T(s) = \frac{C(s)}{R(s)} = \frac{G(s)}{1 \pm G(s)H(s)} \tag{1.6}$$

where the +ve and –ve signs in the denominator denote negative and positive feedback, respectively.

The positive feedback enhances the error signal and hence output becomes instable. Sometimes, in control systems, it is used in minor loops to amplify certain internal signals on components.

The effects of negative feedback are as follows:

(i) Gain is reduced by a factor $\dfrac{1}{1+G(s)H(s)}$.

(ii) There is reduction of parameter variation by a factor $1 + G(s)H(s)$.

(iii) There is improvement in sensitivity.

(iv) There may be reduction of stability.
(v) Accuracy in tracking steady-state value.
(vi) Rejection of disturbance signal.

The disadvantages of reduction of gain and reduction of stability can be overcome by gain amplification and good design, respectively.

1.5 Comparison of Open-loop and Closed-loop Control Systems

Table 1.1 shows the comparison of open-loop and closed-loop control systems.

Table 1.1 Comparison of open-loop and closed-loop control systems

Open loop	**Closed loop**
Feedback does not exist	*Feedback exists*
Output measurement is not necessary	*Output measurement is necessary*
Error detector is not present	*Error detector is present*
Not accurate and reliable	*Highly accurate and reliable*
Highly sensitive to disturbance	*Less sensitive to disturbance*
Very sensitive to environmental changes	*Less sensitive to environmental changes*
Bandwidth is small	*Bandwidth is large*
Design is simple and cheap	*Design is complicated and costly*
Usually stable in nature	*During designing, stability is the major consideration*
Highly affected by non-linearities	*Reduced effect of non-linearities*

1.6 Standard Test Signals

Standard test signals that are physically realizable are shown below:

(i) Step function
(ii) Ramp function
(iii) Parabolic function

1.6.1 Step Function

Figure 1.16 shows a step function. The value of the function is zero for $t < 0$, and its value is A for $t \geq 0$.

If $A = 1$, the function $u(t) = 1$, and it is known as unit step function.

Mathematically, it is expressed by

$$\left.\begin{array}{ll} u(t) = 0 & \text{for } t < 0 \\ u(t) = A & \text{for } t \geq 0 \end{array}\right\} \quad (1.7)$$

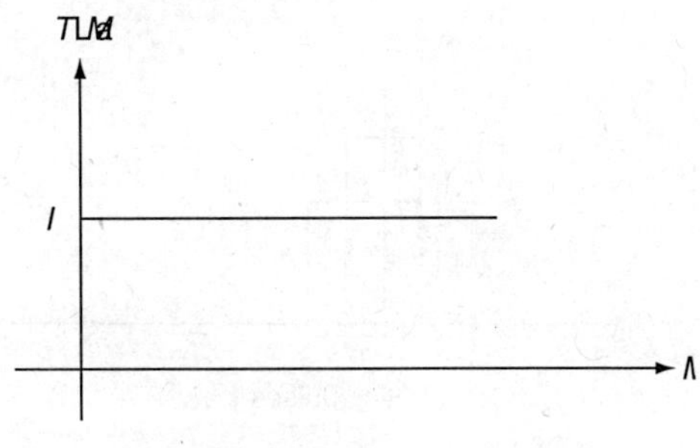

Fig. 1.16 Step function

1.6.2 Ramp Function

Figure 1.17 depicts a ramp function. The value of ramp function is zero for $t < 0$, and after $t \geq 0$, it linearly increases with time.

Mathematically, it is expressed by

$$\left.\begin{aligned} r(t) &= 0 \quad \text{for } t < 0 \\ r(t) &= At \quad \text{for } t \geq 0 \end{aligned}\right\} \tag{1.8}$$

1.6.3 Parabolic Function

Figure 1.18 depicts a parabolic function. The value of a parabolic function is zero for $t < 0$, and it is equal to $\frac{At^2}{2}$ for $t \geq 0$.

Mathematically, it is expressed by

$$\left.\begin{aligned} a(t) &= 0 \quad \text{for } t < 0 \\ a(t) &= \frac{At^2}{2} \quad \text{for } t \geq 0 \end{aligned}\right\} \tag{1.9}$$

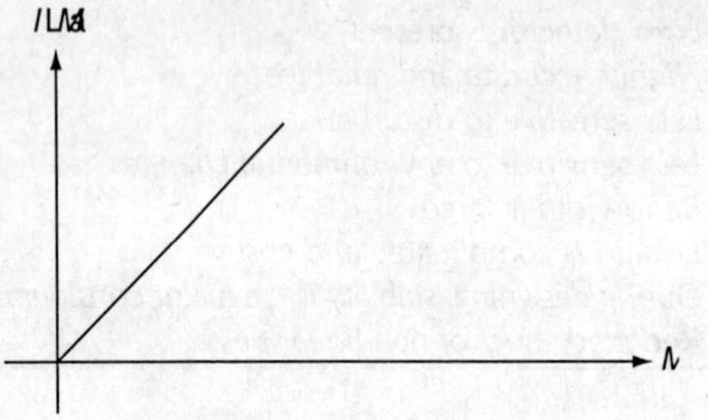

Fig. 1.17 Ramp function

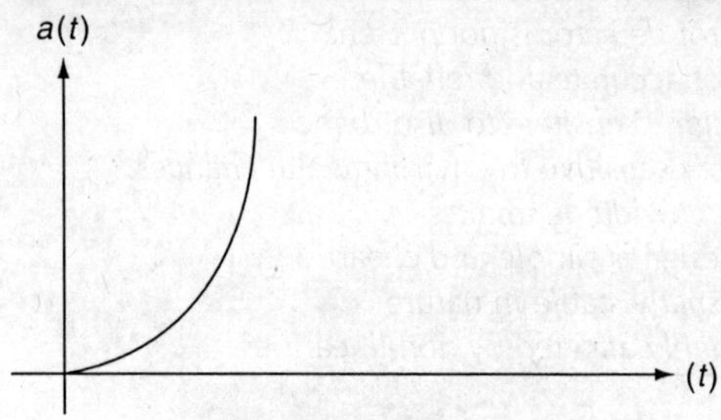

Fig. 1.18 Parabolic function

1.7 Impulse Function [δ(t)]

The impulse function can be defined as follows. A unit impulse function $\delta(t)$ can be obtained from Fig. 1.19. The width of the function is T and its height is $1/T$. If width is reduced, its height $1/T$ will increase. In the limit $t \to 0$, $T \to 0$ and $1/T \to \infty$. The pulse will be of infinite magnitude and it is termed as an impulse. It is denoted by $\delta(t)$ and is shown in Fig. 1.19.

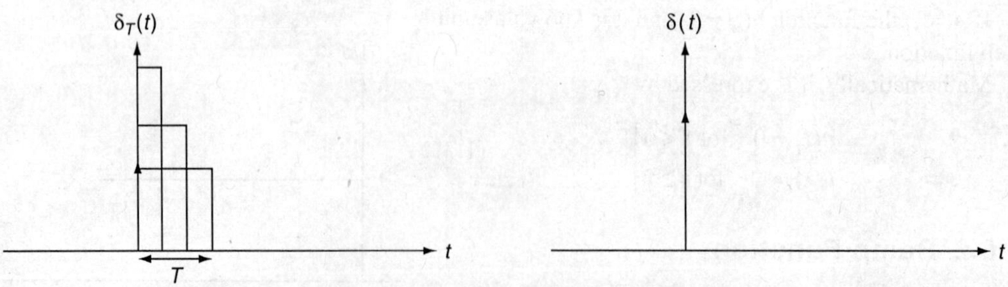

(a) Unit pulses (b) Impulse function

Fig. 1.19 Unit pulses and Impulse function

Mathematically, the Diracdelta (impulse) function is denoted as follows:

$$\left.\begin{aligned}\delta(t) &= \text{undefined}, & t = 0\\ \delta(t) &= 0, & t \neq 0\end{aligned}\right\} \tag{1.10}$$

With the requirement

$$\int_{-\infty}^{+\infty} \delta(t)\,dt = 1 \tag{1.11}$$

A unit pulse function $\delta_T(t)$ has been shown in Fig. 1.19(a) which is a brief rectangular pulse function of duration. $\delta_T(t)$ has been defined to have a constant amplitude $1/T$ over its extent. The area under the pulse becomes unity $[T \times (1/T) = 1]$. We have

$$\delta_T(t) = \begin{cases} 0 & \text{for } t < 0 \\ 1/T & 0 < t \leq T \\ 0 & \text{for } t > 0 \end{cases} \tag{1.12}$$

1.8 Feedback and Feedforward System

The output of any system is affected adversely by the disturbance. An output is measured and compared with the reference input to generate an error, which is fed to the controller. The controller successively operates the system and hence corrects the output.

In the feedback systems, the effect of disturbance must show up in the error before the controller can take proper corrective action.

For measurable disturbance, the signal can be added to the controller output so that the actuating signal is modified. A corrective action can be initiated without waiting for the effect of disturbance to show up in the output. Therefore, the undesirable effects of the measurable disturbances are possibly nullified before the output is affected. Whereas in normal feedback system, the output has been affected. In the feedforward systems, the corrective action is taken before disturbance affects the output. Figure 1.20 shows block diagram of a feedback and feedforward system.

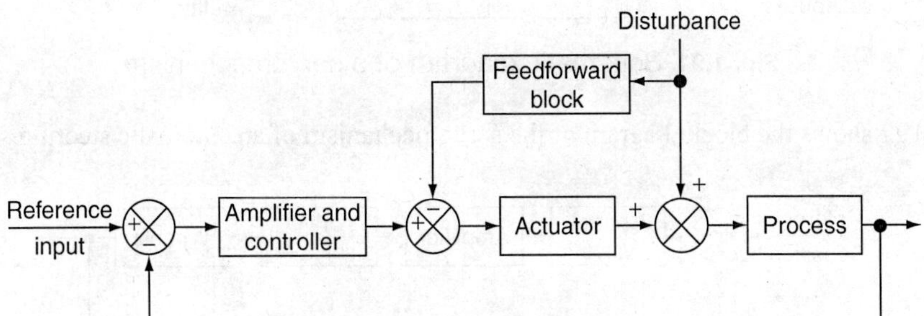

Fig. 1.20 Feedback and feedforward system

The following two constraints are associated with a feedforward system.

(i) The disturbance may not be measurable in some systems.

(ii) The feedforward compensation is an open-loop system. Without knowing the transfer function of the actuator accurately, the compensation cannot be realized.

1.9 Servomechanism

A servomechanism is a power-amplifying feedback control system in which the controlled variable or the output is a mechanical position or its time derivatives such as velocity or acceleration. A positional control system is a simple example of servomechanism. A load is considered here which requires a constant position in its application where the position is detected and converted to voltage using feedback potentiometer. This is compared with input potentiometer voltage to generate the error signal, which is amplified and fed into the controller. The controller now regulates the voltage applicable to the motor due to which it changes its position. Figure 1.21 shows a schematic diagram of servomechanism.

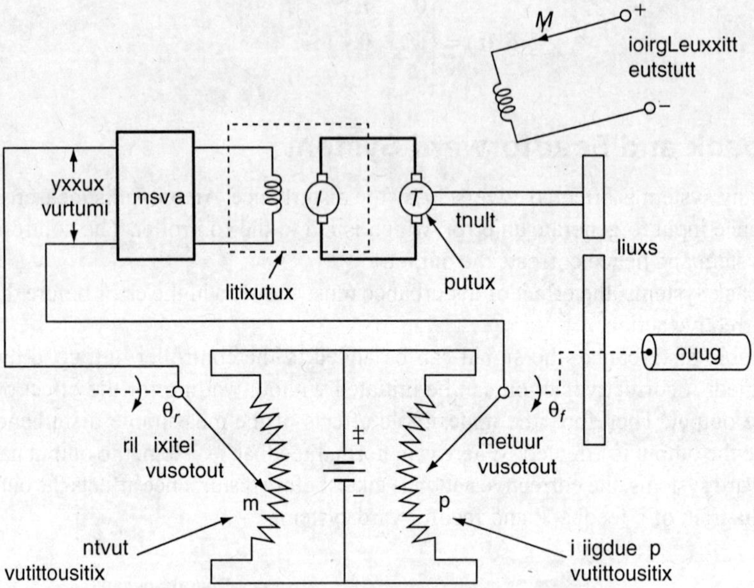

Fig. 1.21 Schematic diagram of a servomechanism

Figure 1.22 shows the block diagram of the servomechanism of an automatic steering system.

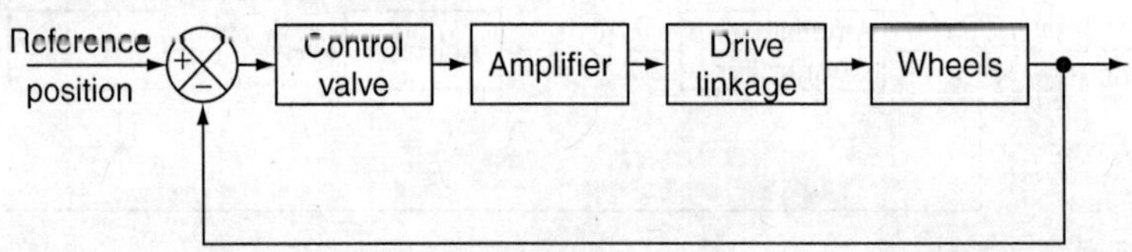

Fig. 1.22 Servomechanism of an automatic steering system

The following are the other examples of servomechanisms:

(i) Machine tool position control
(ii) Power-steering apparatus for an automobile
(iii) Missile launchers and
(iv) Roll stabilization of ships.

1.10 Regulating Systems (Regulators)

If the output is kept constant at its desired value for a preset value of the reference input in a feedback control system, the system is known as a regulating system or regulators.

The reference input is kept constant for a long time in such systems. The reference input on the desired output is kept constant on slowly varying with time during most of the time. The function of the regulator is mainly to maintain a constant output for a fixed input, whereas the function of a servomechanism is mostly to effect the output. Servo stabilizer is a simple example of such a regulator. In voltage stabilizer, the top or secondary is adjusted by using relay controls. The entire secondary can be smoothly tapped using a servomotor drive instead of a fixed tap. The shaft is driven by the servomotor and hence the position of the tap on the secondary is controlled as per the control signal. Figure 1.23 shows the actual scheme of regulating systems.

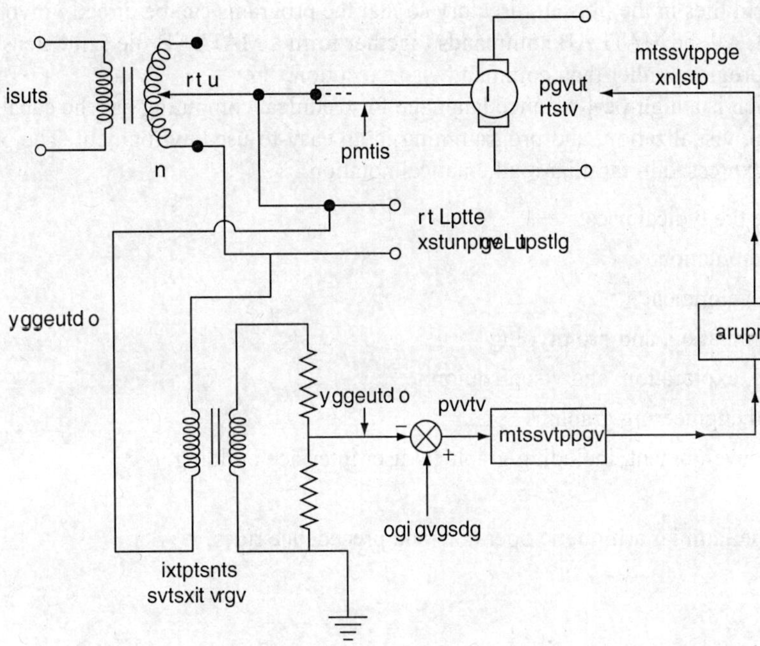

Fig. 1.23 Actual scheme of regulating systems

Figure 1.24 shows the block diagram of the regulating systems.

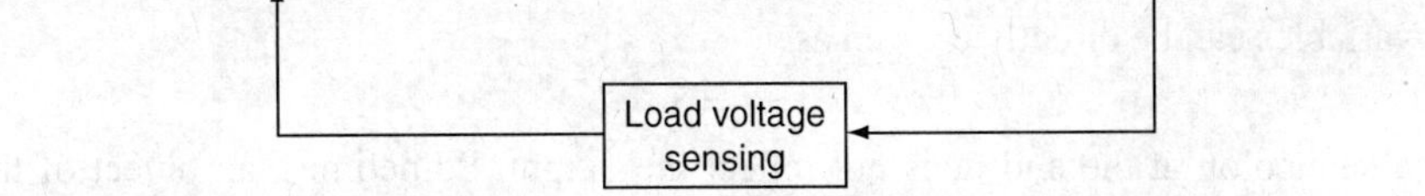

Fig. 1.24 Block diagram of regulating systems

The following are the other examples of a regulating system:

(i) temperature regulators,
(ii) frequency regulators, and
(iii) speed governors.

1.11 Introduction to MATLAB

Double-click on the ***MATLAB*** ® icon on the desktop, or navigate through the program menu to open the MATLAB window. After the program is initialized, we will get a command window where the MATLAB commands can be written. It is the command prompt of MATLAB. There will be pull-down menu where the current directory will be mentioned.

Current directory:c:\MATLABpl\work.

Store the present files in the present directory so that the programs can be directly invoked from the command prompt. All the MATLAB commands together form a MATLAB file (.m extension) and can be executed as a program rather than commandwise execution.

MATLAB which is a high-performance language for technical computing has the capability to integrate computation, visualization, and programming in an easy-to-use environment. The problems and solutions can be expressed in familiar mathematical notation.

The following are the typical uses:

- Math and computation
- Algorithm development
- Modeling, simulation, and prototyping
- Data analysis, exploration, and visualization
- Scientific and engineering graphics
- Application development, including graphical user interface building

operators

Expressions use familiar arithmetic operators and precedence rules.

+ Addition
– Subtraction
* Multiplication
/ Division
\ Left division
^ Power
' Complex conjugate transpose
( ) Specify evaluation order

1.12 MATLAB Variables

MATLAB variables can be directly created as

$$>>A = 5$$

Putting a semicolon at the end does not reflect the output. Sometimes, an object of the symbol is required during MATLAB operation (for example, Fourier transform). It will be dealt with separately.

SIGNIFICANT POINTS

System: An arrangement or combination of different physical components that are connected or related together to form an entire unit to achieve a certain objective is called a system.

Control: The meaning of control is to regulate, direct or command a system so that a desired objective is obtained.

Plant: It is defined as the portion of a system, which is to be controlled or regulated. It is also called process.

Controller: It is an element within the system itself, or external to the system, and it controls the plant or the process.

Input: The applied signal or excitation signal that is applied to a control system to get a specified output is called input.

Output: The actual response that is obtained from a control system due to the application of input is termed as output.

Disturbances: The signal that has some adverse effect on the value of the output of a system is called disturbance. If a disturbance is produced within the system, it is termed as an internal disturbance. Otherwise, it is known as an external disturbance.

Control Systems: It is an amalgamation of different physical elements linked in such a manner to regulate, direct or command itself to obtain a certain objective. A control system must have (1) input, (2) output, (3) ways to achieve input and output objectives and (4) control action. A control system may have more than one input and output. Such a system is called multi-input–multi-output system.

Time-Varying Control System: If the parameters of a control system vary with time, such a control system is termed as time-varying control system. These do not depend on the nature of inputs and output, i.e., these may or may not be functions of time.

Time-Invariant Control System: If the parameters of a control system do not vary with time, it is termed as time-invariant control system.

Linear System: A control system is called linear if it satisfies the additive property as well as the homogeneous property.

Non-linear System: A control system is called non-linear if it does not satisfy both the additive property and the homogeneous property.

Continuous-Time Control Systems: If all the system variables of a control system are functions of time, it is termed as continuous-time control system.

Discrete-Time Control Systems: If one or more system variables of a control system are known at a certain discrete time, it is termed as discrete-time control system.

Deterministic-Control Systems: If the response to input and to external disturbances of a control system is predictable and repeatable, the control system is said to be deterministic.

Stochastic-Control Systems: If the response to input and to external disturbances of a control system is not predictable and repetitive, the control system is said to be stochastic.

Lumped-Parameter Control System: If a control system can be represented by ordinary differential equations, it is called lumped-parameter control system.

Distributed-Parameter Control System: If a control system can be described by partial differential equations, it is known as distributed-parameter control system.

Single-Input–Single-Output Systems: If a control system has one input and one output, it is termed as single-input–single-output (SISO) system.

Multiple-Input–Multiple-Output Systems: If a control system has multiple input and multiple output, it is known as multiple-input–multiple-output (MIMO) system.

Open-Loop Systems: A system in which control action does not depend on output is known as open-loop system.

Closed-Loop Systems: If controlling actions of a system are somehow dependent on output or changes in output, the system is called a closed-loop system.

SHORT QUESTIONS WITH ANSWERS

1. Define the following:
   (i) System, (ii) Plant, (iii) Controller, (iv) Input, (v) Output, (vi) Control system and (vii) Disturbance.

   Ans. (i) An arrangement or a combination of different physical components, which are connected or related together to form an entire unit to achieve a certain objective, is called a system.
   (ii) It is defined as the portion of a system, which is to be controlled or regulated. It is also called process.
   (iii) It is the element of the system itself, or may be external to the system. It controls the plant or the process.
   (iv) The applied signal or excitation signal that is applied to a control system to get a specified output is called input.
   (v) The actual response that is obtained from a control system due to the application of input is termed as output.
   (vi) It is an amalgamation of different physical elements linked in a manner as to regulate, direct or command the process itself to obtain a certain objective. A control system must have (1) input, (2) output, (3) ways to achieve input and output objectives and (4) control action. A control system may have more than one input and output. It is multi-input and multi-output system.
   (vii) The signal that has some adverse effect on the value of the output of a system is called disturbance. If a disturbance is produced within the system, it is termed as an internal disturbance. Otherwise, it is known as an external disturbance.

2. Define the following control systems:
   (i) Time-invariant, (ii) Time-variant, (iii) Continuous, (iv) Discrete, (v) Deterministic and (vi) Stochastic.

   Ans. (i) If the parameters of a control system do not vary with lime, it is termed as time-invariant control system.
   (ii) If the parameters of a control system vary with time, this control system is termed as time-varying control system. Such systems do not depend on the nature of input(s) and output(s), i.e., they are functions of time or not.
   (iii) If all the system variables of a control system are functions of time, it is termed as continuous-time control system.
   (iv) If one or more system variables of a control system are known at certain discrete time, it is termed as discrete-time control system.
   (v) If the response to input and to external disturbances, of a control system is predictable and repetitive, the control system is known as deterministic.
   (vi) If the response to input, and to external disturbances, of a control system is not predictable and repetitive, the control system is said to be stochastic.

3. Define linear and non-linear control systems.

   Ans. Linear system: A control system is called linear if it satisfies the additive property as well as the homogeneous property.

   Non-linear system: A control system is called non-linear if it does not satisfy both the additive property and the homogeneous property.

4. Define open-loop and closed-loop control systems.

   Ans. Open-loop systems: A system in which control action does not depend on output is known as open-loop system.

   Closed-loop systems: If controlling actions of a system are somehow dependent on output or changes in output, this system is called closed-loop system.

5. State the effects of feedback.

   Ans. The effects of feedback are as follows:

   (i) Gain is reduced by a factor $\frac{1}{[1+G(s)H(s)]}$.

   (ii) There is reduction of parameter variation by a factor $1 + G(s)H(s)$.

   (iii) There is improvement in sensitivity.

   (iv) There may be reduction in stability.

6. What do you mean by servomechanism?

   Ans. A servomechanism is a power amplifying feedback control system in which the controlled variable or the output is a mechanical position or its time derivatives such as velocity or acceleration.

7. State the applications of open-loop and closed-loop control systems.

   Ans. The applications of open-loop control systems are in automatic washing machines, bread toaster, electric hand drier, automatic milk server, automatic coffee server, electric lift, traffic signals, sprinkler, etc. The applications of closed-loop control systems are in automatic electric iron, railway reservation status, display motor speed controlled by tacho-generator feedback, servo voltage stabilizer, water-level controller missile launching system, temperature control system, voltage stabilizer, etc.

EXERCISES

Questions for Practice

1. Define the following:
   (i) Man-made Control System
   (ii) Automatic Control System
   (iii) Deterministic System
2. What do you mean by open-loop control system? Give examples of open-loop control system. Discuss the advantages and disadvantages of open-loop control systems.
3. Give the difference between linear and non-linear control systems.
4. What do you mean by closed-loop control system? Give examples of closed-loop control system. What are the main advantages of closed-loop systems?

TEST YOURSELF

1. The principle of homogeneity and superposition is applied to
   (a) Linear time-variant system
   (b) Non-linear time-variant system
   (c) Linear time-invariant system
   (d) Non-linear time invariant system
2. In an open-loop control system
   (a) system variables affect the output signal
   (b) output signal has no control on the input signal
   (c) none of the variables has any effect on the input signal
   (d) none of these
3. Sprinkler is an example of
   (a) a single feedback control system
   (b) a multi-feedback control system

(c) an open-loop control system
(d) none of these

4. Human system is
(a) a multi-variable feedback control system
(b) an open-loop control system
(c) a single-variable control system
(d) a complex control system

5. An automatic iron is an example of
(a) an open-loop control system
(b) a single-feedback control system
(c) a multi-feedback control system
(d) none of these

6. Pick out the linear systems:
(i) $\frac{d^2y(t)}{dt^2}+8\frac{dy(t)}{dt}+y(t)=u(t)$

(ii) $y\frac{d^2y(t)}{dt^2}+7\frac{dy(t)}{dt}=8u(t)$

(iii) $3\frac{d^2y(t)}{dt^2}+2t\frac{dy(t)}{dt}+3t^2y(t)=6$

(a) (i) and (iii) (b) (ii) only (c) (i) only (d) (ii) and (iii)

7. Pick out the time-invariant systems:
(i) $\frac{dy(t)}{dt}+8y(t)=2u(t)$

(ii) $\frac{d^2y(t)}{dt^2}+6\frac{dy(t)}{dt}+y(t)=8$

(iii) $t\left(\frac{d^2y(t)}{dt^2}\right)+2\frac{dy(t)}{dt}+3y(t)=2u(t)$

(a) (i) and (ii) (b) (ii) and (iii) (c) (i) only (d) (iii) only (e) (i) and (iii)

8. Pick out the non-linear systems:
(i) $\frac{dy(t)}{dt}+8y(t)=2u(t)$

(ii) $\frac{d^2y(t)}{dt^2}+6\frac{dy(t)}{dt}+8y(t)=3u(t)$

(iii) $t\left[\frac{d^2y(t)}{dt^2}\right]+2\frac{dy(t)}{dt}+3y(t)=2u(t)$

(a) (i) and (ii) (b) (ii) only (c) (i) only (d) (iii) only (e) (i) and (iii)

9. Consider the following statements regarding a linear system $y=f(x)$:
1. $f(x_1+x_2)=f(x_1)+f(x_2)$
2. $f[x(t+T)]=f[x(t)+x(T)]$
3. $f(Kx)=Kf(x)$.

Of these statements

(a) 1, 2 and 3 are correct (b) 1 and 2 are correct (c) 3 alone is correct (d) 1 and 3 are correct

10. The input-output relationship of a linear system is given by

(a) $y = a_0 x^2 + a_1 x + a_0$ (b) $y = a_1 x + a_0$ (c) $y = a_1 x$ (d) $y = a_0$

ANSWERS

1 (c) 2 (c) 3 (c) 4 (a) 5 (b) 6 (a) 7 (a) 8 (e) 9 (d) 10 (c)

Chapter 2

Laplace Transform and Matrix Algebra

The method of transforming a function from time domain into s domain is known as Laplace transform, where 's' is a complex operator denoted by $s = \sigma + j\omega$. In other words, the Laplace transform will change a function from time domain to s domain. In a control system, setting up of system's equations and their solutions are required. Laplace transform is very handy to solve the differential equations because it automatically includes initial conditions compared to the classical methods. Matrix algebra has its wide application in modern control systems. The aim of this chapter is to explore Laplace transform and matrix algebra.

2.1 Laplace Transform

The Laplace transform of the function $f(t)$ is represented by $F(s)$, and it is expressed by

$$F(s) = LT[f(t)] = \int_0^\infty f(t)e^{-st}\,dt \tag{2.1}$$

where 's' in Eq. (2.1) is the intermediate or transformed variable, and it is a part of the exponential function, LT indicates Laplace transform.

2.2 Inverse Laplace Transform

Inverse Laplace transform permits to go back from s domain to t domain, and it is expressed by

$$f(t) = \mathrm{LT}^{-1}[F(s)] = \frac{1}{2\pi j}\int_{\sigma-j\infty}^{\sigma+j\infty} F(s)\,e^{st}\,ds \tag{2.2}$$

2.3 Laplace Transform of Standard Input Signals

2.3.1 Unit Step Function

It is the most common input signal and is denoted by $u(t)$.

$$\therefore \qquad \mathrm{LT}[u(t)] = \int_0^\infty u(t)e^{-st}\,dt = \int_0^\infty e^{-st}\,dt = \frac{1}{s}\left[e^{-st}\right]_0^\infty$$

i.e.,
$$\text{LT}[u(t)] = \frac{1}{s} \tag{2.3}$$

Similarly,
$$\text{LT}[Cu(t)] = \frac{C}{s} \tag{2.4}$$

2.3.2 Ramp Function

It is denoted by $r(t) = At$, where A is the amplitude of $r(t)$.

$$\therefore \quad \text{LT}[r(t)] = \int_0^\infty r(t)e^{-st}\,dt = \int_0^\infty Ate^{-st}\,dt$$

$$= \left[\frac{-tAe^{-st}}{s} - \frac{Ae^{-st}}{s^2}\right]_0^\infty = \frac{A}{s^2} \tag{2.5}$$

2.3.3 Parabolic Function

It is denoted by $a(t) = At^2$.

$$\therefore \quad \text{LT}[a(t)] = \int_0^\infty a(t)e^{-st}\,dt = \int_0^\infty At^2e^{-st}\,dt = \left[-\frac{At^2}{s}e^{-st} - \frac{2At}{s^2}e^{-st} + \frac{2A}{s^3}e^{-st}\right]_0^\infty$$

i.e.,
$$\text{LT}[a(t)] = \frac{2A}{s^3} \tag{2.6}$$

2.4 Exponential Function

Let
$$f(t) = e^{-at}$$

$$\therefore \quad \text{LT}[f(t)] = \int_0^\infty e^{-at}e^{-st}\,dt = \int_0^\infty e^{-(s+a)t}\,dt = \frac{1}{s+a}$$

i.e.,
$$\text{LT}[e^{-at}] = \frac{1}{s+a} \tag{2.7}$$

Similarly,
$$\text{LT}[e^{at}] = \frac{1}{s-a} \tag{2.8}$$

2.5 Sine Wave

Here
$$f(t) = \sin\omega t$$

$$\therefore \quad \text{LT}[f(t)] = \int_0^\infty f(t)e^{-st}\,dt = \int_0^\infty \sin\omega t\, e^{-st}\,dt$$

Now,
$$\sin\omega t = \frac{1}{2j}(e^{j\omega t} - e^{-j\omega t})$$

$$\therefore \quad \text{LT}[f(t)] = \frac{1}{2j}\int_0^\infty [e^{-(s-j\omega)t} - e^{-(s+j\omega)t}]\,dt = \frac{1}{2j}\left[-\frac{e^{-(s-j\omega)t}}{s-j\omega} + \frac{e^{-(s+j\omega)t}}{s+j\omega}\right]_0^\infty$$

$$= \frac{1}{2j}\left[\frac{1}{s-j\omega} - \frac{1}{s+j\omega}\right]$$

i.e.,
$$\text{LT}[\sin\omega t] = \frac{\omega}{s^2+\omega^2} \tag{2.9}$$

2.6 Cosine Wave

$$f(t) = \cos\omega t$$

$\therefore$
$$f(t) = \cos\omega t = \frac{1}{2}(e^{j\omega t} + e^{-j\omega t})$$

$\therefore$
$$\text{LT}[f(t)] = \int_0^\infty f(t)e^{-st}dt = \frac{1}{2}\int_0^\infty e^{-(s-j\omega)t}dt + \frac{1}{2}\int_0^\infty e^{-(s+j\omega)t}dt$$

$$= \frac{1}{2}\left[-\frac{e^{-(s-j\omega)t}}{s-j\omega} - \frac{e^{-(s+j\omega)t}}{s+j\omega}\right]_0^\infty = \frac{1}{2}\left[\frac{1}{s-j\omega} + \frac{1}{s+j\omega}\right]$$

i.e.,
$$\text{LT}[\cos\omega t] = \frac{s}{s^2+\omega^2} \tag{2.10}$$

2.7 Hyperbolic Sine Wave

Here
$$f(t) = \sinh\omega t = \frac{e^{\omega t} - e^{-\omega t}}{2}$$

$\therefore$
$$\text{LT}[f(t)] = \int_0^\infty f(t)e^{-st}dt = \int_0^\infty \left(\frac{e^{\omega t} - e^{-\omega t}}{2}\right)e^{-st}dt$$

$$= \frac{1}{2}\int_0^\infty \left[e^{-(s-\omega)t}dt - \int_0^\infty e^{-(s+\omega)t}dt\right] = \frac{1}{2}\left[\frac{-e^{-(s-j\omega)t}}{s-\omega} + \frac{e^{-(s+j\omega)t}}{s+\omega}\right]_0^\infty$$

$$= \frac{1}{2}\left[\frac{1}{s-\omega} - \frac{1}{s+\omega}\right]$$

$\therefore$
$$\text{LT}[\sinh\omega t] = \frac{\omega}{s^2-\omega^2} \tag{2.11}$$

2.8 Hyperbolic Cosine Wave

Here
$$f(t) = \cosh\omega t = \frac{e^{\omega t} + e^{-\omega t}}{2}$$

$\therefore$
$$\text{LT}[f(t)] = \int_0^\infty f(t)e^{-st}dt = \int_0^\infty \left(\frac{e^{\omega t} + e^{-\omega t}}{2}\right)e^{-st}dt$$

$$= \left[\frac{1}{2}\int_0^\infty e^{-(s-\omega)t} + \int_0^\infty e^{-(s+\omega)t}\,dt\right]$$

$$= \frac{1}{2}\left[\frac{-e^{-(s-\omega)t}}{s-\omega} - \frac{e^{-(s+\omega)t}}{s+\omega}\right]_0^\infty = \frac{1}{2}\left[\frac{1}{s-\omega} + \frac{1}{s+\omega}\right] = \frac{s}{s^2-\omega^2}$$

i.e., $$\text{LT}[\cosh\omega t] = \frac{s}{s^2-\omega^2} \tag{2.12}$$

2.9 Laplace Transform of a Derivative [*df*(*t*)/*dt*]

It is known that $\int u\,dv = uv - \int v\,du$. Let $u = f(t)$ and $dv = e^{-st}dt$. Therefore,

$$du = f'(t)$$

and $$v = \frac{-1}{s}e^{-st}$$

$$\therefore \quad \int f(t)e^{-st}dt = \left[-\frac{f(t)}{s}e^{-st}\right]_0^\infty - \int\left(-\frac{1}{s}\right)e^{-st}f'(t)dt$$

i.e., $$F(s) = \frac{f(0)}{s} + \frac{1}{s}\int_0^\infty f'(t)e^{-st}dt$$

i.e., $$\int_0^\infty f'(t)e^{-st}dt = sF(s) - f(0) \tag{2.13}$$

2.10 Laplace Transform of an Intergral t^n

Here $f(t) = t^n$

$$\therefore \quad F(s) = \int_0^\infty t^n e^{-st}dt = \left[-\frac{t^n}{s}e^{-st}\right]_0^\infty - \int_0^\infty\left(-\frac{1}{s}\right)e^{-st}nt^{n-1}dt$$

$$= 0 + \frac{n}{s}\int_0^\infty t^{n-1}e^{-st}dt = \frac{n}{s}\text{LT}[t^{n-1}]$$

$$= \left(\frac{n}{s}\right)\left(\frac{n-1}{s}\right)\left(\frac{n-2}{s}\right)\cdots\left(\frac{3}{s}\right)\left(\frac{2}{s}\right)\left(\frac{1}{s}\right)\text{LT}[t^0] = \frac{n!}{s^n}\{\text{LT}(t^0)\}$$

Since $$\text{LT}[t^0] = \frac{1}{s} \tag{2.14}$$

$$\therefore \quad \text{LT}[t^n] = \frac{n!}{s^{n+1}} \tag{2.15}$$

2.11 Laplace Transform of $e^{-\alpha t}$ sinω*t*

Here $$f(t) = e^{-\alpha t}\sin\omega t = e^{-\alpha t}\left[\frac{e^{j\omega t} - e^{-j\omega t}}{2j}\right] = \frac{e^{-(\alpha - j\omega)t} - e^{-(\alpha + j\omega)t}}{2j}$$

$$\therefore \quad \text{LT}[e^{-\alpha t}\sin\omega t] = \int_0^\infty \left[\frac{e^{-(\alpha - j\omega)t} - e^{-(\alpha + j\omega)t}}{2j}\right] e^{-st}\,dt$$

$$= \frac{1}{2j}\left[\frac{1}{s + (\alpha - j\omega)} - \frac{1}{s + (\alpha + j\omega)}\right] = \frac{\omega}{(s+\alpha)^2 + \omega^2} \tag{2.16}$$

2.12 Laplace Transform of $e^{-\alpha t}$ cosω*t*

Here $$f(t) = e^{-\alpha t}\cos\omega t = \frac{1}{2}[e^{-(\alpha - j\omega)t} + e^{-(\alpha + j\omega)t}]$$

$$\therefore \quad LT[e^{-\alpha t}\cos\omega t] = \frac{1}{2}\int_0^\infty [e^{-(\alpha - j\omega)t} + e^{-(\alpha + j\omega)t}]e^{-st}\,dt = \frac{s+\alpha}{(s+\alpha)^2 + \omega^2} \tag{2.17}$$

2.13 Laplace Transform of $e^{-\alpha t}$ sinhω*t*

Here $$f(t) = e^{-\alpha t}\sinh\omega t = \frac{1}{2}[e^{-(\alpha - \omega)t} - e^{-(\alpha + \omega)t}]$$

$$\therefore \quad \text{LT}[e^{-\alpha t}\sinh\omega t] = \frac{1}{2}\int_0^\infty [e^{-(\alpha - \omega)t} - e^{-[\alpha + \omega]t}]e^{-st}\,dt = \frac{\omega}{(s+\alpha)^2 - \omega^2} \tag{2.18}$$

2.14 Laplace Transform of $e^{-\alpha t}$ coshω*t*

Here $$f(t) = e^{-\alpha t}\cosh\omega t = \frac{1}{2}[e^{-(\alpha - \omega)t} + e^{-(\alpha + \omega)t}]$$

$$\therefore \quad \text{LT}[e^{-\alpha t}\cosh\omega t] = \frac{1}{2}\int_0^\infty [e^{-(\alpha - \omega)t} + e^{-(\alpha + \omega)t}]e^{-st}\,dt = \frac{s+\alpha}{(s+\alpha)^2 - \omega^2} \tag{2.19}$$

2.15 Laplace Transform of δ(*t*)

Here $$f(t) = \delta(t) = \frac{du(t)}{dt}$$

$$\therefore \quad F(s) = \text{LT}[f(t)] = \int_0^\infty \delta(t)e^{-st}\,dt = \int_0^\infty \frac{du(t)}{dt} = sF(s) = s \times \left(\frac{1}{s}\right)$$

i.e., $$\text{LT}[\delta(t)] = 1 \tag{2.20}$$

Table 2.1 shows the Laplace transform of different functions.

Table 2.1 Laplace transform of different functions

$f(t)$	$F(s)$
$u(t)$	$\frac{1}{s}$
C	$\frac{C}{s}$
t	$\frac{1}{s^2}$
t^n	$\frac{n!}{s^{n+1}}$
e^{-at}	$\frac{1}{s+a}$
e^{at}	$\frac{1}{s-a}$
$\sin\omega t$	$\frac{\omega}{s^2+\omega^2}$
$\cos\omega t$	$\frac{s}{s^2+\omega^2}$
$\sinh\omega t$	$\frac{\omega}{s^2-\omega^2}$
$\cosh\omega t$	$\frac{s}{s^2-\omega^2}$
$e^{-at}\sin\omega t$	$\frac{\omega}{(s+a)^2+\omega^2}$
$e^{-at}\cos\omega t$	$\frac{s+a}{(s+a)^2+\omega^2}$
$e^{-at}\sinh\omega t$	$\frac{\omega}{(s+a)^2-\omega^2}$
$e^{-at}\cosh\omega t$	$\frac{s+a}{(s+a)^2-\omega^2}$
$\frac{df(t)}{dt}$	$sF(s)-f(0)$
$\int f(t)dt$	$\frac{F(s)}{s}+\frac{1}{s}\int f(t)dt\Big\vert_0$
$e^{-at}t^n$	$\frac{n!}{(s+a)^{n+1}}$
$\delta(t)$	1

Example 2.1 Determine the Laplace transforms of the following functions:

(i) $f(t)=13$,

(ii) $f(t)=4+5e^{-3t}$ and

(iii) $f(t)=2t-3e^{-t}$

Solution

(i) $f(t)=13$

$$\therefore \quad \mathrm{LT}[f(t)]=\int_0^\infty f(t)e^{-st}dt=13\int_0^\infty e^{-st}dt=\frac{13}{s}$$

(ii) $f(t)=4+5e^{-3t}$

$$\therefore \quad \mathrm{LT}[f(t)]=\int_0^\infty f(t)e^{-st}dt=\int_0^\infty (4+5e^{-3t})e^{-st}dt=4\int_0^\infty e^{-st}dt+5\int_0^\infty e^{-(s+3)t}dt=\frac{4}{s}+\frac{5}{s+3}$$

(iii) $f(t)=2t-3e^{-t}$

$$\mathrm{LT}[f(t)]=\int_0^\infty f(t)e^{-st}dt\int_0^\infty (2t-3e^{-t})e^{-st}dt=2\int_0^\infty te^{-st}dt-3\int_0^\infty e^{-(s+1)t}dt=\frac{2}{s^2}-\frac{3}{s+1}$$

Example 2.2 Find the Laplace transform of $y(t)$ in the equation

$$\frac{d^2y(t)}{dt^2}+6\frac{dy(t)}{dt}+2y(t)=8$$

Assume initial conditions to be zero.

Solution

Here
$$\frac{d^2y(t)}{dt^2}+6\frac{dy(t)}{dt}+2y(t)=8$$

Taking Laplace transform, we have after putting initial conditions

$$s^2Y(s)+6sY(s)+2Y(s)=\frac{8}{s}$$

i.e.,
$$(s^2+6s+2)Y(s)=\frac{8}{s}$$

i.e.,
$$Y(s)=\frac{8}{s(s^2+6s+2)}$$

2.16 Initial Value and Final Value Theorems

We know that $\mathrm{LT}[df(t)/dt]=sF(s)-f(0)$

$$\therefore \quad \underset{s\to\infty}{\mathrm{Lt}}\left\{\mathrm{LT}\left(\frac{df(t)}{dt}\right)\right\}=\underset{s\to\infty}{\mathrm{Lt}}[sF(s)-f(0)] \tag{2.21}$$

Now, $$\text{LT}\left[\frac{df(t)}{dt}\right]=\int_0^{\infty}\frac{df(t)}{dt}e^{-st}dt \tag{2.22}$$

$\therefore$ $$\underset{s\to\infty}{\text{Lt}}\left\{\text{LT}\left(\frac{df(t)}{dt}\right)\right\}=\int_0^{\infty}\left[\underset{s\to\infty}{\text{Lt}}\frac{df(t)}{dt}e^{-st}dt\right]=0 \tag{2.23}$$

From Eq. (2.21) and Eq. (2.23), we have

$$\underset{s\to\infty}{\text{Lt}}[sF(s)-f(0)]=0$$

i.e., $$f(0)=\underset{s\to\infty}{\text{Lt}}\,sF(s)$$

i.e., $$\underset{t\to 0}{\text{Lt}}\,f(t)=\underset{s\to\infty}{\text{Lt}}\,sF(s) \tag{2.24}$$

Equation (2.24) represents the initial value theorem.

Again $$\underset{s\to 0}{\text{Lt}}\left\{\text{LT}\left(\frac{df(t)}{dt}\right)\right\}=\underset{s\to 0}{\text{Lt}}[sF(s)-f(0)]=\underset{s\to 0}{\text{Lt}}[sF(s)-f(0)] \tag{2.25}$$

and also $$\underset{s\to 0}{\text{Lt}}\left\{\text{LT}\left(\frac{df(t)}{dt}\right)\right\}=\int_0^{\infty}\left[\underset{s\to 0}{\text{Lt}}\left\{\frac{df(t)}{dt}e^{-st}dt\right\}\right]=f(\infty)-f(0) \tag{2.26}$$

From Eq. (2.25) and Eq. (2.26), we have

$$f(\infty)-f(0)=\underset{s\to\infty}{\text{Lt}}\,sF(s)-f(0)$$

i.e., $$f(\infty)=\underset{s\to 0}{\text{Lt}}\,sF(s)$$

i.e., $$\underset{t\to\infty}{\text{Lt}}\,f(t)=\underset{s\to 0}{\text{Lt}}\,sF(s) \tag{2.27}$$

Equation (2.27) represents the final value theorem.

Example 2.3 Find the initial and final values of the function having Laplace transform

$$F(s)=\frac{s+9}{s^2+7s+3}$$

Solution

Here $$F(s)=\frac{s+9}{s^2+7s+3}$$

$\therefore$ $$f(0)=\underset{s\to\infty}{\text{Lt}}\,sF(s)=\underset{s\to\infty}{\text{Lt}}\,s\left(\frac{s+9}{s^2+7s+3}\right)=\underset{s\to\infty}{\text{Lt}}\left(\frac{1+\frac{9}{s}}{1+\frac{7}{s}+\frac{3}{s^2}}\right)=1$$

$\therefore$ $$f(\infty)=\underset{s\to 0}{\text{Lt}}\,sF(s)=\underset{s\to 0}{\text{Lt}}\,s\left(\frac{s+9}{s^2+7s+3}\right)=0$$

Therefore, the initial and final values are 1 and 0 respectively.

2.17 Displacement Theorem

If the function $f(t)$ be the Laplace transformable, we can write by displacement theorem as follows:

$$\text{LT}[f(t-T)] = e^{-sT}F(s) \tag{2.28}$$

2.18 Convolution

If $f_1(t)$ and $f_2(t)$ be two real functions, the Laplace transform of convolution of $f_1(t)$ and $f_2(t)$ is the multiplication of $F_1(s)$ and $F_2(s)$ where $F_1(s)$ and $F_2(s)$ are the Laplace transform of $f_1(t)$ and $f_2(t)$ respectively.

$$\text{LT}[f_1(t) * f_2(t)] = F_1(s)F_2(s) \tag{2.29}$$

where $$F_1(s) = \text{LT}[f_1(t)] \tag{2.30}$$

and $$F_2(s) = \text{LT}[f_2(t)] \tag{2.31}$$

The meaning of the word **convolve** is to revolve continuously, and the functions $f_1(t)$ and $f_2(t)$ are multiplied in such a manner that one is continually moving with time, say τ with respect to the other. Therefore, we can write

$$f_1(t) * f_2(t) = \int_0^t f_1(t-\tau) f_2(t) d\tau \tag{2.32}$$

Let $F(s) = \text{LT}[f_1(t) * f_2(t)] = \int_{t=0}^{\infty} [f_1(t) * f_2(t)] e^{-st} dt$

$$= \int_{t=0}^{\infty} \left[\int_{\tau=0}^{\infty} f_1(t-\tau) f_2(\tau) d\tau \right] e^{-st} dt = \int_{t=0}^{\infty} \left[\int_{\tau=0}^{\infty} f_1(t-\tau) f_2(\tau) e^{-st} d\tau \right] dt \text{ [since } u(t-\tau) = 0 \text{ for } t < \tau]$$

$$= \int_{\tau=0}^{\infty} f_2(\tau) \left[\int_{t=0}^{\infty} [f_1(t-\tau)] e^{-st} dt \right] d\tau = \int_{\tau=0}^{\infty} f_2(\tau) e^{-s\tau} F_1(s) d\tau = F_1(s) F_2(s) \tag{2.33}$$

Therefore, the convolution theorem is proved.

2.19 Partial Fraction Expansion of *F*(*s*)

Let $$F(s) = \frac{P(s)}{Q(s)} = \frac{P(s)}{(s-a_1)(s-a_2)\ldots(s-a_n)} \tag{2.34}$$

where $Q(s)$ is of higher degree than $P(s)$.

Different cases are discussed below depending upon the roots of $Q(s)$.

2.19.1 General Case

This method can be implemented in all cases irrespective of the roots of denominator.

Let $$F(s) = \frac{P(s)}{Q(s)} = \frac{s}{s^2 + 6s + 5} \tag{2.35}$$

$\therefore$ $$F(s) = \frac{s}{s^2 + 6s + 5} = \frac{s}{(s+1)(s+5)} = \frac{A_1}{(s+1)} + \frac{A_2}{(s+5)} \tag{2.36}$$

i.e., $$s = A_1(s+5) + A_2(s+1)$$

i.e., $$s = s(A_1 + A_2) + (5A_1 + A_2) \tag{2.37}$$

Equating the terms of both sides of Eq. (2.37) corresponding to s and the constant term, we have

$$A_1 + A_2 = 1 \tag{2.38}$$

and $$5A_1 + A_2 = 0 \tag{2.39}$$

Solving Eq. (2.38) and Eq. (2.39), we have

$$A_1 = -\frac{1}{4} \quad \text{and} \quad A_2 = \frac{5}{4}$$

$$F(s) = \frac{-\frac{1}{4}}{s+1} + \frac{\frac{5}{4}}{s+5} \tag{2.40}$$

Taking inverse Laplace transform of Eq. (2.40), we have

$$f(t) = -\frac{1}{4}e^{-t} + \frac{5}{4}e^{-5t} \tag{2.41}$$

This method is very simple, but it is very cumbersome.

2.19.2 The Roots of *Q*(*s*) are Real and Unequal

Equation (2.36) is rewritten below.

$$F(s) = \frac{s}{s^2 + 6s + 5} = \frac{s}{(s+1)(s+5)} = \frac{A_1}{(s+1)} + \frac{A_2}{(s+5)} \tag{2.42}$$

Now $$A_1 = [(s+1)F(s)]_{s=-1} = \left[\frac{s}{s+5}\right]_{s=-1} = \frac{-1}{-1+5} = -\frac{1}{4}$$

and $$A_2 = [(s+5)F(s)]_{s=-5} = \left[\frac{s}{s+1}\right]_{s=-5} = \frac{-5}{-5+1} = \frac{5}{4}$$

From Eq. (2.42), we have

$$F(s) = \frac{-\frac{1}{4}}{s+1} + \frac{\frac{5}{4}}{s+5} \tag{2.43}$$

Taking inverse Laplace transform of Eq. (2.43), we have

$$f(t) = -\frac{1}{4}e^{-t} + \frac{5}{4}e^{-5t} \tag{2.44}$$

2.19.3 The Roots of *Q*(*s*) are Real and Repeated

To determine the coefficients in this case, the general formula is

$$a_{nm} = \left[\frac{1}{(m-1)!}\frac{d^{m-1}}{ds^{m-1}}\left\{(s-s_0)^m \frac{P(s)}{Q(s)}\right\}\right]_{s=s_0} \tag{2.45}$$

and for other non-repeated roots the procedure is same as above.

$$F(s) = \frac{s}{(s+1)^2(s+3)} \tag{2.46}$$

By partial fraction expansion of Eq. (2.46), we have

$$F(s) = \frac{A_{11}}{(s+1)^2} + \frac{A_{12}}{s+1} + \frac{A_2}{s+3} \tag{2.47}$$

Now, $A_{11} = [(s+1)^2 F(s)]_{s=-1} = \left[(s+1)^2 \times \frac{s}{(s+1)^2(s+3)}\right]_{s=-1} = \frac{-1}{-1+3} = -\frac{1}{2}$

$$A_{12} = \left[\frac{d}{ds}\left\{(s+1)^2 \times \frac{s}{(s+1)^2(s+3)}\right\}\right]_{s=-1} = \left[\frac{d}{ds}\left\{\frac{s}{(s+3)}\right\}\right]_{s=-1} = \left.\frac{3}{(s+3)^2}\right|_{s=-1} = \frac{3}{4}$$

and $A_3 = [(s+3)F(s)]_{s=-3} = \left[(s+3) \times \frac{s}{(s+1)^2(s+3)}\right]_{s=-3} = \left[\frac{s}{(s+1)^2}\right]_{s=-3}$

$$= \frac{-3}{(-3+1)^2} = -\frac{3}{4}$$

Now $$F(s) = \frac{-\frac{1}{2}}{(s+1)^2} + \frac{\frac{3}{4}}{s+1} + \frac{-\frac{3}{4}}{s+3} \tag{2.48}$$

Taking inverse Laplace transform of Eq. (2.48), we have

$$f(t) = -\frac{1}{2}te^{-t} + \frac{3}{4}e^{-t} - \frac{3}{4}e^{-3t} \tag{2.49}$$

2.19.4 The Roots of *Q*(*s*) are Complex

Let $$F(s) = \frac{1}{(s+1+j)(s+1-j)} = \frac{A_1}{s+1+j} + \frac{A_2}{s+1-j} \tag{2.50}$$

$\therefore$ $A_1 = [(s+1+j)F(s)]_{s=-1-j} = \left[\frac{1}{s+1-j}\right]_{s=-1-j} = \frac{1}{-1-j+1-j} = \frac{1}{-2j} = \frac{1}{2}j$

$\therefore$ $A_2 = [(s+1-j)F(s)]_{s=-1+j} = \left[\frac{1}{s+1+j}\right]_{s=-1+j} = \frac{1}{-1+j+1+j} = \frac{1}{2j} = -\frac{1}{2}j$

$\therefore$ $A_2 = A_1^*$

Now $$F(s) = \frac{\frac{1}{2}j}{s+1+j} + \frac{-\frac{1}{2}j}{s+1-j} \tag{2.51}$$

Taking inverse Laplace transform of Eq. (2.51), we have

$$f(t) = \frac{1}{2}j[e^{-(1+j)t} - e^{-(1-j)t}] = \frac{1}{2}je^{-t}(-2j)\sin t = e^{-t}\sin t \tag{2.52}$$

All the methods except (i) were originally derived by Oliver Heaviside. Hence these methods are termed as Heaviside expansion theorems.

Example 2.4 Find $f(t)$ if $F(s)=\dfrac{(s+3)}{(s+1)(s+2)(s+4)}$.

Solution

Here $$F(s)=\frac{(s+3)}{(s+1)(s+2)(s+4)}=\frac{A}{s+1}+\frac{B}{s+2}+\frac{C}{s+4}$$

Now $$A=[(s+1)F(s)]_{s=-1}=\left[\frac{s+3}{(s+2)(s+4)}\right]_{s=-1}=\frac{(-1+3)}{(-1+2)(-1+4)}=\frac{2}{3}$$

$$B=[(s+2)F(s)]_{s=-2}=\left[\frac{s+3}{(s+1)(s+4)}\right]_{s=-2}=\frac{(-2+3)}{(-2+1)(-2+4)}=-\frac{1}{2}$$

$$C=[(s+4)F(s)]_{s=-4}=\left[\frac{s+3}{(s+1)(s+2)}\right]_{s=-4}=\frac{(-4+3)}{(-4+1)(-4+2)}=-\frac{1}{6}$$

$\therefore$ $$F(s)=\frac{\frac{2}{3}}{s+1}+\frac{-\frac{1}{2}}{s+2}+\frac{-\frac{1}{6}}{s+4} \tag{1}$$

Taking inverse Laplace transform of Eq. (1), we have

$$f(t)=\frac{2}{3}e^{-t}-\frac{1}{2}e^{-2t}-\frac{1}{6}e^{-4t}$$

2.20 Matrix

A rectangular array of elements, which may be real numbers, complex numbers, functions or operators, is called a matrix. Let us consider a matrix A as follows:

$$A=\begin{bmatrix} a_{11} & a_{12} & \dots & a_{1n} \\ a_{21} & a_{22} & \dots & a_{2n} \\ \vdots & \vdots & \dots & \vdots \\ a_{m1} & a_{m2} & \dots & a_{mn} \end{bmatrix}_{m\times n}$$

The order of the matrix is $m \times n$ since it has m rows and n columns. The matrix A can also be written as

$$A=[a_{ij}]_{m\times n}$$

2.20.1 Equality of Two Matrices

Two matrices are called equal if and only if their corresponding elements are equal. Therefore, equal matrices should have equal number of rows and columns.

2.20.2 Vector

A matrix having only one column is known as a column vector. If the column vector has n elements, it is known as n vector or n-dimensional vector. A column vector is shown below:

$$B = \begin{bmatrix} a_1 \\ a_2 \\ \vdots \\ a_n \end{bmatrix}_{n\times 1}$$

If the matrix has only one row, it is known as row vector such as $\begin{bmatrix} a_1 & a_2 & \dots & a_n \end{bmatrix}_{1\times n}$

2.20.3 Square Matrix

If the number of rows of any matrix is equal to its number of columns, the matrix is called a square matrix.

2.20.4 Diagonal Matrix

If the other elements except the main diagonal elements in a square matrix are zero, the matrix is called a diagonal matrix.

$$A = \begin{bmatrix} a_{11} & 0 & \dots & 0 \\ 0 & a_{22} & \dots & 0 \\ \vdots & \vdots & \dots & \vdots \\ 0 & 0 & \dots & a_{mn} \end{bmatrix} = [a_{ij}\delta_{ij}]$$

Where $\delta_{ij} = 1 \quad \text{if } i = j$

$= 0 \quad \text{if } i \neq j$

and δ_{ij} is known as Kronecker delta.

2.20.5 Identity Matrix or Unity Matrix

If the main diagonal elements of a matrix are unity and other elements are zero, the matrix is known as identity or unity matrix.

2.20.6 Zero Matrix

A matrix whose elements are all zero is known as zero matrix.

2.20.7 Determinant of a Matrix

There exists a determinant for every matrix if the matrix is square. The properties of determinant are shown below:

1. The determinant changes its sign if any two constructive rows or columns are interchanged.
2. The value of the determinant is zero if any row or column consists of only zero elements.
3. The value of the determinant is also zero if the elements of any row or any column are exactly k times those of another row or column.

4. If any determinant is multiplied by a constant, then only one row or column is multiplied by that constant.
5. If A and B are two square matrices and both are $n \times n$, the following relation holds good:

$$|A \times B| = |A|\,|B|$$

2.20.8 Singular Matrix

A square matrix is known as singular if the associated determinant is zero.

2.20.9 Transpose of a Matrix

If the rows and columns of an $m \times n$ matrix are interchanged, the resulting matrix is known as the transpose of the matrix. The transpose of a matrix A is denoted by A^T.

If
$$A = \begin{bmatrix} a_{11} & a_{12} & \cdots & a_{1n} \\ a_{21} & a_{22} & \cdots & a_{2n} \\ \vdots & \vdots & \cdots & \vdots \\ a_{m1} & a_{m2} & \cdots & a_{mn} \end{bmatrix}_{m\times n}$$

then
$$A^T = \begin{bmatrix} a_{11} & a_{21} & \cdots & a_{m1} \\ a_{12} & a_{22} & \cdots & a_{m2} \\ \vdots & \vdots & \cdots & \vdots \\ a_{1n} & a_{2n} & \cdots & a_{mn} \end{bmatrix}_{n\times m}$$

Again $(A^T)^T = A$

2.20.10 Symmetric Matrix

If a square matrix A is equal to its transpose, then the matrix A is known as a symmetric matrix. For symmetric matrix $A = A^T$. Again, if $A = -A^T$, the matrix A is termed as skew matrix.

2.20.11 Conjugate of a Matrix

If the elements of a matrix A are complex numbers, the elements of the conjugate of matrix A are the conjugates of the elements of the matrix A.

2.20.12 Conjugate Transpose

The conjugate of the transpose of a matrix A is known as the conjugate transpose.

2.20.13 Hermitian Matrix

If the elements of a matrix A are complex numbers and $A = A^*$ or $a_{ij} = \overline{a_{ji}}$, where $\overline{a_{ji}}$ is complex conjugate of a_{ij}, A is called Hermitian matrix. If $A = -A^*$, A is known as skew Hermitian matrix.

2.20.14 Addition of Matrices

If $A = [a_{ij}]_{m\times n}$ and $B = [b_{ij}]_{m\times n}$, then $A + B = [a_{ij} + b_{ij}]_{m\times n}$

2.20.15 Multiplication of a Matrix by a Scalar

If a matrix $A=[a_{ij}]_{m\times n}$ is multiplied by a scalar k, we have

$$kA = k[a_{ij}]_{m\times n}.$$

2.20.16 Multiplication of a Matrix by a Matrix

If $A=[a_{ij}]_{m\times n}$ and $B=[b_{ij}]_{n\times m}$, then $AB = C = \sum_{l=1}^{k} a_{il}b_{lj}$,

where $i = 1, 2, \ldots, m$ and $j = 1, 2, \ldots, n$.

For multiplication, the following properties hold good:

1. $(AB)C = A(BC)$
2. $(A+B)C = AC+BC$
3. $C(A+B) = CA+CB$

If $AB = BA$, then A and B are known as commutative. Generally, AB may not be equal to BA.

2.20.17 Power of a Matrix

The nth power of a square matrix A is defined by

$$A^n = \underbrace{A\cdot A\cdot A\cdot A\cdot A\cdots}_{n}$$

2.20.18 Rank of a Matrix

A matrix (A) has a rank m if there exists an $m\times m$ submatrix B of A such that its determinant is non-zero and the determinant of every $k\times k$ submatrix, where $k \geq m+1$, of A is zero.

2.20.19 Minor of a Matrix

If the ith row and jth column are deleted from an $n \times n$ matrix, then there results a matrix of order, $(n-1)\times(n-1)$, and the determinant of this matrix is called the minor (M_{ij}) of the matrix A.

2.20.20 Co-factor

The co-factor A_{ij} of the element a_{ij} of a matrix is defined by

$$A_{ij} = (-1)^{i+j} M_{ij}$$

2.20.21 Adjoint of a Matrix

Adjoint of matrix A is defined as a matrix where the element of the ith row and the jth column equals to A_{ji}.

$$Adj \text{ of } A = \begin{bmatrix} A_{11} & A_{21} & \cdots & A_{n1} \\ A_{12} & A_{22} & \cdots & A_{n2} \\ \vdots & \vdots & \cdots & \vdots \\ A_{1n} & A_{2n} & \cdots & A_{nn} \end{bmatrix}_{n\times n}$$

2.20.22 Inverse of a Matrix

For a square matrix A, a matrix B exists if and only if $AB = BA = I$ and B is denoted by A^{-1} and it is called inverse of A. Therefore $AA^{-1} = A^{-1}A = I$, where I is the identity matrix. If A is non-singular and $AB = C$, then we can write $B = A^{-1}C$. Again, $(AB)^{-1} = B^{-1}A^{-1}$ and $(AB)(A^{-1}B^{-1}) = I$.
The following relations hold good:

$$(A^{-1})^{-1} = A$$
$$(A^{-1})^T = (A^T)^{-1}$$
$$(A^{-1})^* = (A^*)^{-1}$$

By definition, the inverse of the matrix $[A]_{m\times n}$, where $m = n$ is given by

$$A^{-1} = \frac{AdjA}{|A|} \text{ provided } |A| \neq 0.$$

2.20.23 More Properties of Matrices

$$(A+B)^T = A^T + B^T$$
$$(AB)^T = B^T A^T$$
$$(A+B)^* = A^* + B^*$$
$$(AB)^* = B^* A^*$$

2.21 MATLAB Commands for Matrix

The best way to learn MATLAB is to learn matrix manipulation in MATALB. There are several ways of defining a matrix-one of which is the explicit declaration. The following rules are to be followed while entering the matrix:

- Separate the elements of the row by spaces or commas.
- Separate the end of the rows by semi colons.
- Surround the entire list by square brackets.

For example,

$$A = [1\ 2\ 3\ 4\ ;\ 5\ 6\ 7\ 8]$$

The matrix displayed will be

```
A=
   1   2   3   4
   5   6   7   8
```

sum (A) will give the sum of the column as a row vector as given below:

```
   6   8   10   12
```

A' gives the transpose of the matrix

```
   1   5
   2   6
   3   7
   4   8
```

It is also possible to transfer the matrix elements by subscripts.

For example $A(1,1)$ responds 1 as the output.

$A(2,3)$ responds 7 as output

2.21.1 Matrix Multiplication

Two compatible matrices can be multiplied using '*' operator. For example,

$$B = [1\ 2\ 3\ 4]$$
$$C = [1;\ 2;\ 3;\ 4]$$
$$D = B * C$$

Ans. 30

Similarly, other operations, such as addition and subtraction, can be performed on two compatible matrices.

2.21.2 Inverse of the Matrix

The inverse of the matrix can be calculated using the command **inv**.

For example,

```
>>A = [1 0 0;0 1 0;0 0 1];
>> det(A)
  1
>> inv(A)
  1  0  0
  0  1  0
  0  0  1
```

2.21.3 Eigenvalue

The eigenvalue can be calculated using **eig** function.

```
>> eig(A)
  1
  1
  1
```

There are some other matrix functions which cater to different needs.

SIGNIFICANT POINTS

Laplace Transform: The Laplace transform of the function $f(t)$ is represented by $F(s)$, and it is given by

$$F(s) = \text{LT}\,[f(t)] = \int_0^\infty f(t)e^{-st}\,dt$$

Laplace Transform of Different Functions

***f*(*t*)**	***F*(*s*)**
$u(t)$	$\frac{1}{s}$

(*Continued*)

$f(t)$	$F(s)$
C	$\frac{C}{s}$
t	$\frac{1}{s^2}$
t^n	$\frac{n!}{s^{n+1}}$
e^{-at}	$\frac{1}{s+a}$
e^{at}	$\frac{1}{s-a}$
$\sin\omega t$	$\frac{\omega}{s^2+\omega^2}$
$\cos\omega t$	$\frac{s}{s^2+\omega^2}$
$\sinh\omega t$	$\frac{\omega}{s^2-\omega^2}$
$\cosh\omega t$	$\frac{s}{s^2-\omega^2}$
$e^{-at}\sin\omega t$	$\frac{\omega}{(s+a)^2+\omega^2}$
$e^{-at}\cos\omega t$	$\frac{s+a}{(s+a)^2+\omega^2}$
$e^{-at}\sinh\omega t$	$\frac{\omega}{(s+a)^2-\omega^2}$
$e^{-at}\cosh\omega t$	$\frac{s+a}{(s+a)^2-\omega^2}$
$\frac{df(t)}{dt}$	$sF(s)-f(0)$
$\int f(t)dt$	$\frac{F(s)}{s}+\frac{1}{s}\int f(t)dt\,\vert_0$
$e^{-at}t^n$	$\frac{n!}{(s+a)^{n+1}}$
$\delta(t)$	1

ADDITIONAL SOLVED EXAMPLES

Example 2.5 Determine the Laplace transform of the following functions:

(i) $f(t) = e^{-t} + 5t + 6e^{-3t}$

(ii) $f(t) = 5 + 6t^2 + 3e^{-2t}$

Solution

(i) $f(t) = e^{-t} + 5t + 6e^{-3t}$

$$\therefore \qquad \text{LT}[f(t)] = \int_0^{\infty} f(t)e^{-st}dt = \int_0^{\infty} (e^{-t} + 5t + 6e^{-3t})e^{-st}dt = \frac{1}{s+1} + \frac{5}{s^2} + \frac{6}{s+3}$$

(ii) $f(t) = 5 + 6t^2 + 3e^{-2t}$

$$\therefore \qquad \text{LT}[f(t)] = \int_0^{\infty} f(t)e^{-st}dt = \int_0^{\infty} (5 + 6t^2 + 3e^{-2t})e^{-st}dt = \frac{5}{s} + \frac{12}{s^3} + \frac{3}{s+2}$$

Example 2.6 Determine the Laplace transformation of $f(t) = 3 - 2e^{-3t}$

Solution

$$\text{LT}[f(t)] = F(s) = \int_0^{\infty} f(t)e^{-st}dt$$

$$= \int_0^{\infty} (3 - 2e^{-3t})e^{-st}dt$$

$$= 3\int_0^{\infty} e^{-st}dt - 2\int_0^{\infty} e^{-(s+3)}dt$$

$$= \frac{3}{s} - \frac{2}{s+3} = \frac{3s + 9 - 2s}{s(s+3)} = \frac{s+9}{s(s+3)}$$

Example 2.7 Determine the Laplace transformation of $f(t) = 7\cos\omega t + 5\sin\omega t$.

Solution

$$\text{LT}[f(t)] = F(s)$$

$$= 7\,\text{LT}[\cos\omega t] + 5\,\text{LT}[\sin\omega t]$$

$$= \frac{7s}{s^2 + \omega^2} + \frac{5\omega}{s^2 + \omega^2} = \frac{7s + 5\omega}{s^2 + \omega^2}$$

Example 2.8 Obtain the Laplace transform of the gate function shown in Fig. E.2.1

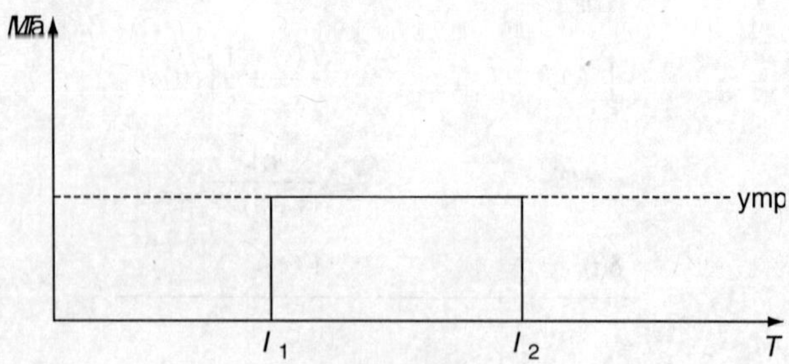

Fig. E.2.1

Solution Using delayed unit step function, Fig. E.2.1 can be constituted from Fig. E.2.1(a).

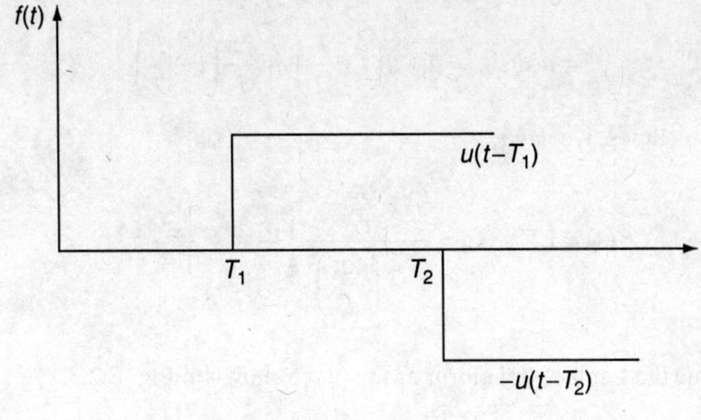

Fig. E.2.1(a)

$$\therefore \quad f(t-T) = u(t-T_1) - u(t-T_2)$$

$$\therefore \quad \text{LT}[f(t-T)] = \text{LT}\left[u(t-T_1)\right] - \text{LT}[u(t-T_2)]$$

$$= \frac{e^{-sT_1}}{s} - \frac{e^{-sT_2}}{s}$$

Example 2.9 Determine the Laplace transform of the wave shown in Fig. E.2.2.

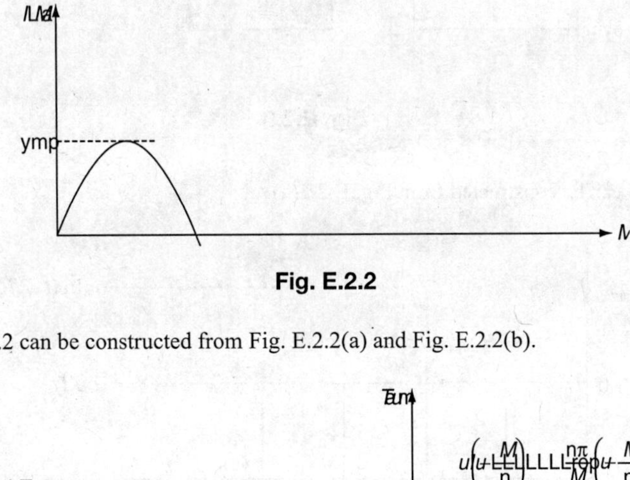

Fig. E.2.2

Solution Figure E.2.2 can be constructed from Fig. E.2.2(a) and Fig. E.2.2(b).

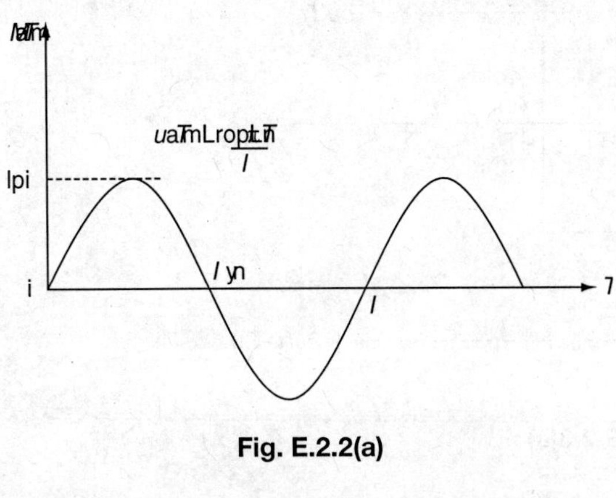

Fig. E.2.2(a)

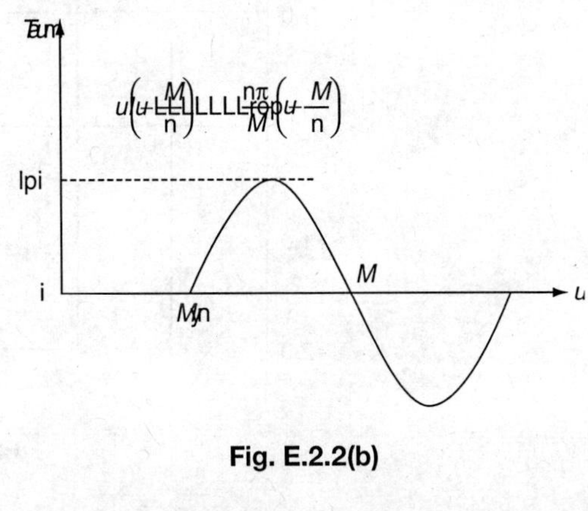

Fig. E.2.2(b)

$$f(t) = f_1(t) + f_2(t)$$

$$= u(t)\sin\frac{2\pi}{T}t + u\left(t - \frac{T}{2}\right)\sin\frac{2\pi}{T}\left(t - \frac{T}{2}\right)$$

Using Laplace transform methods, we have

$$F(s) = \text{LT}[f(t)] = \frac{\frac{2\pi}{T}}{s^2 + \left(\frac{2\pi}{T}\right)^2}\left[1 + e^{-\frac{T}{2}s}\right]$$

Example 2.10 Determine the Laplace transform of the wave shown in Fig. E.2.3.

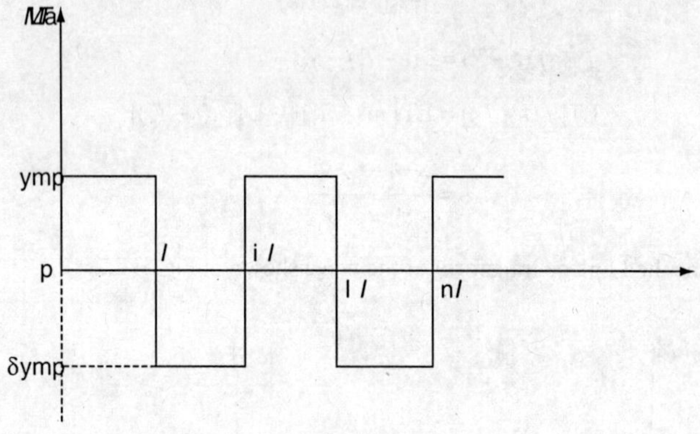

Fig. E.2.3

Solution Figure E.2.3 can be constructed from Fig. E.2.3(a)

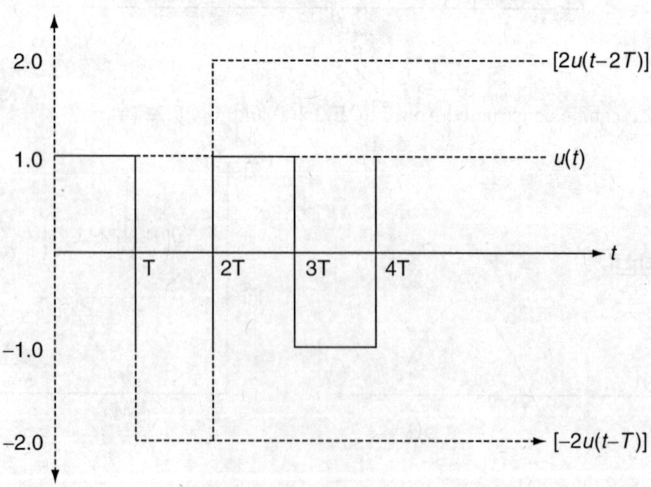

Fig. E.2.3(a)

$\therefore \quad f(t) = u(t) - 2u(t-T) + 2u(t-2T) - \ldots$

$\therefore \quad F(s) = \text{LT}[f(t)] = \dfrac{1}{s} - \dfrac{2e^{-Ts}}{s} + \dfrac{2e^{-2Ts}}{s} - \ldots$

$$= \frac{1}{s}[1 - 2e^{-Ts}(1 - e^{-Ts} + e^{-2Ts} - e^{-3Ts} + e^{-4Ts} - \ldots)]$$

Example 2.11 Find the initial and final values of $F(s) = \dfrac{0.38}{s(s^2 + 2.48s + 0.543)}$

Solution

Here $\quad F(s) = \dfrac{0.38}{s(s^2 + 2.48s + 0.543)}$

$\therefore \quad f(0) = \underset{s\to\infty}{\text{Lt}}\ sF(s) = \underset{s\to\infty}{\text{Lt}}\ s\left[\dfrac{0.38}{s(s^2 + 2.48s + 0.543)}\right] = 0$

$\therefore \quad f(\infty) = \underset{s\to 0}{\text{Lt}}\ sF(s) = \underset{s\to 0}{\text{Lt}}\ s\left[\dfrac{0.38}{s(s^2 + 2.48s + 0.543)}\right] = \dfrac{0.38}{0.543} = 0.699$

Therefore, the initial and final values are 0 and 0.699, respectively.

Example 2.12 Find $f(t)$ if $F(s) = \dfrac{(s+3)}{s(s+1)(s+2)}$.

Solution

Here $\quad F(s) = \dfrac{(s+3)}{s(s+1)(s+2)} = \dfrac{A}{s} + \dfrac{B}{s+1} + \dfrac{C}{s+2}$

Now, $\quad A = \dfrac{3}{2}, B = -2 \text{ and } C = \dfrac{1}{2}$

$\therefore \quad F(s) = \dfrac{\frac{3}{2}}{s} + \dfrac{-2}{s+1} + \dfrac{\frac{1}{2}}{s+2}$

Taking inverse Laplace transform, we have

$$f(t) = \frac{3}{2} - 2e^{-t} + \frac{1}{2}e^{-2t}.$$

Example 2.13 Determine $f(t)$ for a system whose $F(s) = \dfrac{s+3}{s(s+1)^2(s+2)}$. Use the Heaviside theorem

Solution $\quad F(s) = \dfrac{s+3}{s(s+1)^2(s+2)}$

$$= \frac{a_1}{s} + \frac{a_{21}}{(s+1)^2} + \frac{a_{22}}{s+1} + \frac{a_3}{s+2}$$

$\therefore \quad a_1 = \left[s \times \dfrac{(s+3)}{s(s+1)^2(s+2)}\right]_{s=0} = \left[\dfrac{s+3}{(s+1)^2(s+2)}\right]_{s=0} = \dfrac{3}{2}$

$$\therefore \quad a_{21} = \left[(s+1)^2 \times \frac{(s+3)}{s(s+1)^2(s+2)}\right]_{s=-1} = \left[\frac{s+3}{s(s+2)}\right]_{s=-1}$$

$$= \frac{-1+3}{(-1)(-1+2)} = \frac{2}{(-1)(1)} = -2$$

$$\therefore \quad a_{22} = \left[\frac{d}{ds}\left\{(s+1)^2 \times \frac{(s+3)}{s(s+1)^2(s+2)}\right\}\right]_{s=1} = \left[\frac{d}{ds}\left\{\frac{(s+3)}{s(s+2)}\right\}\right]_{s=1}$$

$$= \left[\frac{(s^2+2s)-(s+3)(2s+2)}{(s^2+2s)^2}\right]_{s=-1} = \frac{(1-2)-(-1+3)(-2+2)}{(1-2)^2}$$

$$= \frac{-1}{1} = -1$$

$$\therefore \quad a_3 = \left[(s+2) \times \frac{(s+3)}{s(s+1)^2(s+2)}\right]_{s=-2} = \left[\frac{s+3}{s(s+1)^2}\right]_{s=-2} = \frac{-2+3}{(-2)(-2+1)^2}$$

$$= -\frac{1}{2}$$

$$F(s) = \frac{3/2}{s} + \frac{-2}{(s+1)^2} + \frac{-1}{(s+1)} + \frac{-1/2}{s+2}$$

Taking inverse Laplace transform, we have

$$f(t) = \frac{3}{2} - 2te^{-t} - e^{-t} - \frac{1}{2}e^{-2t}$$

Program 1 If $A = \begin{bmatrix} 1 & 4 \\ 2 & 6 \end{bmatrix}$ and $B = \begin{bmatrix} 1 & 4 \\ 3 & 9 \end{bmatrix}$, using suitable MATLAB program obtain the matrix A and B with display and without display, addition of A and B, multiplication of A and B, determinant of A and B, inverse of A and B, eigen values of A and B, multiplication of inverse of A twice and thrice, multiplication of A two and four times, rank of A and B, transpose of A and B.

Solution

```
>> % Display the matrix A
>> A = [ 1 4; 2 6]
A =
   1   4
   2   6
>> % Do not display the matrix A
>> A = [ 1 4; 2 6];
>> % Display the matrix B
>> B = [ 1 4; 3 9]
B =
   1   4
   3   9
```

```
>> % Do not display the matrix B
>> B = [ 1 4; 3 9];
>> % Add matrix A and matrix B and display the result
>> C = A + B
C =
   2   8
   5  15
>> % Multiply A and B and display the result
>> D = A*B
D =
  13  40
  20  62
>> % Find determinant of matrix A
>> det(A)
ans =
  -2
>> % Find determinant of matrix B
>> det(B)
ans =
  -3
>> % Find inverse of matrix A
>> inv(A)
ans =
  -3.0000    2.0000
   1.0000   -0.5000
>> % Find inverse of matrix B
>> inv(B)
ans =
  -3.0000    1.3333
   1.0000   -0.3333
>> % Eigen value of A
>> eig(A)
ans =
  -0.2749
   7.2749
>> % Eigen value of B
>> eig(B)
ans =
  -0.2915
  10.2915
>> % A^(-2) effectively multiplies inv(A) by itself 2 – 1 times
>> A^(-2)
ans =
  11.0000   -7.0000
  -3.5000    2.2500
>> % A^(-3) effectively multiplies inv(A) by itself 3 – 1 times
>> A^(-3)
ans =
 -40.0000  25.5000
  12.7500  -8.1250
```

```
>> % Multiplication of A by one time i.e., 2 – 1 = 1 times
>> A^2
ans =
   9  28
  14  44
>> % Multiplication of A by four times i.e., 5 – 1 = 4 times
>> A^5
ans =
      3441     10796
      5398     16936
>> rank(A)
ans =
   2
>> rank(B)
ans =
   2
>> A'
ans =
   1   2
   4   6
>> B'
ans =
   1   3
   4   9
>>
```

Program 2 Obtain suitable MATLAB program for the matrix $A = \begin{bmatrix} 1 & 4 & 9 \\ 2 & 8 & 10 \\ 3 & 6 & 9 \end{bmatrix}$ to obtain the following:

(i) display it, (ii) determinant of the matrix A, (iii) inverse of A, (iv) transpose of A, (v) multiplication of A twice, (vi) multiplication of A six times, (vii) multiplication of inverse of A twice, (viii) multiplication of inverse of A thrice and (ix) rank of A.

Solution

```
>> A = [1  4  9; 2  8  10; 3  6  9]
A =
   1   4   9
   2   8  10
   3   6   9
>> det(A)
ans =
  -48
>> inv(A)
ans =
  -0.2500  -0.3750   0.6667
  -0.2500   0.3750  -0.1667
   0.2500  -0.1250   0
>> A'
```

```
ans =
   1    2   3
   4    8   6
   9   10   9
>> A^2
ans =
  36    90  130
  48   132  188
  42   114  168
>> A^7
ans =
  66562056  180049104  261252504
  95977296  259616736  376705584
  84438504  228404448  331416360
>> A^(−2)
ans =
   0.3229  −0.1302  −0.1042
  −0.0729   0.2552  −0.2292
  −0.0313  −0.1406   0.1875
>> A^(−4)
ans =
   0.1170  −0.0606  −0.0233
  −0.0350   0.1069  −0.0939
  −0.0057  −0.0582   0.0706
>> rank(A)
ans =
   3
>>
```

Program 3 Find the Laplace transform of $g(t) = e^{-3t}(1 - \sin t)$. Find the transfer function of the system.

Solution

```
>> syms t
>> a = exp(−3*t)*(1 − sin(t))
a =
−(sin(t) − 1)/exp(3*t)
>> laplace(a)
ans =
1/(s + 3) − 1/((s + 3)^2 + 1)
>>
```

Program 4 Use MATLAB for partial fraction expression expansion for the following transfer function

$$F(s) = \frac{s^2 + 3s + 2}{s^3 + 12s^2 + 47s + 60}.$$

Solution

```
>> num = [1  3  2]
num =
   1   3   2
```

```
>> den = [1 12 47 60]
den =
   1  12  47  60
>> [r, p, k] = residue(num, den)
r =
   6.0000
  -6.0000
   1.0000
p =
  -5.0000
  -4.0000
  -3.0000
k =
   [ ]
```

Program 5 Use MATLAB for partial fraction expression expansion for the following transfer function $F(s) = \dfrac{s^2+3s+4}{s^3+4s^2+5s+2}$.

Solution

```
>> num = [1 3 4]
num =
   1  3  4
>> den = [1 4 5 2]
den =
   1  4  5  2
>> [r, p, k] = residue(num, den)
r =
   2.0000
  -1.0000
   2.0000
p =
  -2.0000
  -1.0000
  -1.0000
k =
   [ ]
```

Program 6 Use MATLAB for partial fraction expression expansion for the following transfer function $F(s) = \dfrac{s^2+4s+6}{s^4+8s^2+23s^2+28s+12}$.

Solution

```
>> num = [1 4 6]
num =
   1  4  6
>> den = [ 1 8 23 28 12]
den =
   1  8  23  28  12
```

```
>> [r, p, k] = residue(num, den)
r =
   -1.5000
    0.0000
   -2.0000
    1.5000
p =
   -3.0000
   -2.0000
   -2.0000
   -1.0000
k =
   []
```

Program 7 Use MATLAB for partial fraction expression expansion for the following transfer function

$$F(s) = \frac{2s^4 + 2s^3 + 3s^2 + 5s + 10}{s^4 + 11s^2 + 41s + 61s + 30}.$$

Solution

```
>> num = [2 2 3 5 10]
num =
   2   2   3   5   10
>> den = [1 11 41 61 30]
den =
   1   11   41   61   30
>> [r, p, k] = residue(num, den)
r =
  -44.1667
   32.5000
   -9.3333
    1.0000
p =
   -5.0000
   -3.0000
   -2.0000
   -1.0000
k =
   2
```

Program 8 Use MATLAB for partial fraction expression expansion for the following transfer function

$$F(s) = \frac{s^2 + 3s + 5}{s^3 + 6s^2 + 12s + 8}.$$

Solution

```
>> num = [1  2  5]
num =
   1   2   5
>> den = [1  6  12  8]
```

```
den =
    1   6   12   8
>> [r, p, k] = residue(num, den)
r =
    1.0000
   -2.0000
    5.0000
p =
   -2.0000
   -2.0000
   -2.0000
k =
    []
```

Program 9 Use MATLAB for partial fraction expression expansion for the following transfer function $F(s) = \dfrac{2}{s^4 + 3s^3 + 9s^2}$.

Solution

```
>> num = [0 2]
num =
    0   2
>> den = [1 3 9 0 0 ]
den =
    1   3   9   0   0
>> [r, p, k]=residue(num, den)
r =
   0.0370 + 0.0214i
   0.0370 - 0.0214i
  -0.0741
   0.2222
p =
  -1.5000 + 2.5981i
  -1.5000 - 2.5981i
       0
       0
k =
    []
```

Program 10 Use MATLAB for partial fraction expression expansion for the following transfer function $F(s) = \dfrac{2s+1}{s^4 + 3s^3 + 9s^2}$.

Solution

```
>> num = [2 1]
num =
    2   1
>> den = [1 3 9 0 0 0]
```

```
den =
   1   3   9   0   0   0
>> [r, p, k]=residue(num, den)
r =
   0.0370 + 0.0143i
   0.0370 – 0.0143i
  –0.0741
   0.1852
   0.1111
p =
  –1.5000 + 2.5981i
  –1.5000 – 2.5981i
       0
       0
       0
k =
   []
```

EXERCISES

QUESTIONS FOR PRACTICE

1. Find the Laplace transform of the following:

   (i) $5+e^{-2t}+te^{-3t}$ **Ans.** $\dfrac{5}{s}+\dfrac{1}{s+2}+\dfrac{1}{(s+3)^2}$

   (ii) $5t+3e^{-2t}+e^{-2t}\sin t$ **Ans.** $\dfrac{5}{s^2}+\dfrac{3}{s+2}+\dfrac{1}{(s+2)^2+1}$

   (iii) $e^{-2t}\cos t+3e^{-5t}+2t$ **Ans.** $\dfrac{s+2}{(s+2)^2+1}+\dfrac{3}{s+5}+\dfrac{2}{s^2}$

2. Find the initial and final values of $F(s)=0.39/s(s^2+2.52s+1)$ **Ans.** 0,0.39

3. Find $f(t)$, where

   (i) $F(s)=\dfrac{s}{(s+1)(s+2)}$ **Ans.** $-e^{-t}+2e^{-2t}$

   (ii) $F(s)=\dfrac{(s+1)}{s(s+2)^2}$ **Ans.** $\dfrac{1}{4}+\dfrac{1}{2}t\,e^{-2t}-\dfrac{1}{4}e^{-2t}$

   (iii) $F(s)=\dfrac{(s+2)}{s(s+1)(s+3)}$ **Ans.** $\dfrac{2}{3}-\dfrac{1}{2}e^{-t}-\dfrac{1}{6}e^{-3t}$

4. Find $Y(s)$ if

   (i) $\dfrac{d^2y}{dt^2}+8\dfrac{dy}{dt}+3y=11$ **Ans.** $\dfrac{11}{s(s^2+8s+3)}$

   (ii) $\dfrac{d^3y}{dt^3}+9\dfrac{dy}{dt}+5y=13$ **Ans.** $\dfrac{13}{s(s^2+9s+5)}$

TEST YOURSELF

1. The Laplace transform of a positive sided function $f(t)$ is given by

   (a) $\int_0^\infty f(t)e^{st}dt$ (b) $\int_0^\infty f(t)e^{-st}dt$

   (c) $\int_{-\infty}^0 f(t)e^{-st}dt$ (d) $\int_{-\infty}^0 f(t)e^{-st}dt$

2. The Laplace transform of a unit step function is

   (a) $\frac{1}{s}$ (b) 1 (c) s (d) $\frac{1}{s^2}$

3. The Laplace transform of shifted unit step funiton $u(t-a)$ is

   (a) $\frac{1}{s^2}$ (b) $\frac{1}{s}e^{at}$ (c) $\frac{1}{s}e^{-at}$ (d) none of these

4. The Laplace transform of a unit impulse function is

   (a) 1 (b) s^2 (c) $\frac{1}{s}$ (d) $\frac{1}{s^2}$

5. The advantage(s) of Laplace transform is (are):

   (a) it gives total solution more systematically
   (b) it gives solution in frequency domain only
   (c) initial conditions are incorporated in the very first step.
   (d) none of these

6. The Laplace transform of $df(t)/dt$ is

   (a) $sF(s)-f(0)$ (b) $sF(s)$ (c) $sF(s)+f(0)$ (d) $F(s)-f(0)$

7. The Laplace transform of $f(t) = Ae^{\alpha t}$ is

   (a) $\frac{A}{s-\alpha}$ (b) $\frac{A}{s}$ (c) $\frac{A+\alpha}{s}$ (d) $\frac{A}{s-\alpha}$

8. The Laplace transform of $f(t) = 5 \sin 3t$ is equal to

   (a) $\frac{5}{s^2-4}$ (b) $\frac{4}{s^2+1}$ (c) $\frac{5}{s^2+9}$ (d) none of these

9. Laplace transform and Fourier integrals are related through

   (a) time domain
   (b) frequency domain
   (c) both frequency and time domains
   (d) none of these

10. The Laplace transform of $e^{\alpha t}$ is

    (a) $\frac{1}{s-\alpha}$ (b) $\frac{1}{s+\alpha}$ (c) $\frac{1}{s}e^{\alpha t}$ (d) none of these

11. The Laplace transform of $te^{\alpha t}$ is

    (a) $\frac{1}{(s-\alpha)^2}$ (b) $\frac{1}{s^2}e^{\alpha t}$ (c) $\frac{1}{(s+\alpha)^2}$ (d) none of these

12. The Laplace transform of $f(t)e^{-at}$ is

(a) $F(s)\,e^{at}$ (b) $F(s)\,e^{-at}$ (c) $F(s+a)$ (d) $\dfrac{F(s)}{s+a}$

13. The Laplace transform of $t^n e^{at}$ is

(a) $\dfrac{n!}{(s-a)^{n+1}}$ (b) $\dfrac{n!}{(s+a)^{n+1}}$ (c) $\dfrac{n!}{s^n}$ (d) $\dfrac{n!}{s^{n+1}}$

14. The Laplace transform of $\sin\beta t$ is

(a) $\dfrac{s}{s^2+\beta^2}$ (b) $\dfrac{1}{s^2+\beta^2}$ (c) $\dfrac{s}{s^2-\beta^2}$ (d) none of these

15. The Laplace transform of $\cos\beta t$ is

(a) $\dfrac{s}{s^2+\beta^2}$ (b) $\dfrac{\beta}{s^2+\beta^2}$ (c) $\dfrac{1}{s^2+\beta^2}$ (d) $\dfrac{\beta^2}{s^2+\beta^2}$

16. The Laplace transform of $e^{-at}\sin\omega t$ is

(a) $\dfrac{1}{(s+a)^2+\omega^2}$ (b) $\dfrac{\omega}{(s+a)^2+\omega^2}$ (c) $\dfrac{s}{(s+a)^2+\omega^2}$ (d) none of these

17. The inverse Laplace transform of $6/(s^2+6s+9)$ is

(a) $6e^{-2t}$ (b) $8e^{-3t}$ (c) $9te^{-4t}$ (d) $6te^{-3t}$

18. The inverse Laplace transform of $2/(s^2+5s+6)$ is

(a) $e^{-2t}-e^{-3t}$ (b) $\dfrac{1}{2}(e^{-2t}-e^{-3t})$ (c) $2(e^{-2t}-e^{-3t})$ (d) $2(e^{-3t}-e^{-2t})$

19. The initial value of $F(s)=1/[s(s^2+3s+2)]$ is given by

(a) 0.5 (b) 1 (c) 0.8 (d) 0.2

20. The final value of $F(s)=s^2/[s(s^2+3s+2)]$ is given by

(a) 1 (b) 0 (c) 2 (d) 0.5

ANSWERS

1 (b)	2 (a)	3 (c)	4 (a)	5 (d)	6 (a)	7 (d)	8 (d)	9 (c)	10 (a)
11 (a)	12 (c)	13 (a)	14 (d)	15 (a)	16 (b)	17 (d)	18 (c)	19 (a)	20 (a)

Chapter 3

Transfer Function

Due to the existence of cause-and-effect relationship between input and output of a system, it is possible to define the transfer function of a given system mathematically. Generally, it characterizes the input-output relationship of the system. The aim of this chapter is to explore transfer function.

3.1 Transfer Function

It is defined as the ratio of the Laplace transform of output response to the Laplace transform of input (excitation) assuming all the initial conditions to be zero. Figure 3.1(a) depicts the system in time domain whereas Fig. 3.1(b) shows the system in Laplace domain.

(a) System in time domain (b) System in laplace domain

Fig. 3.1 Transfer function of a system

If $G(s)$ be the transfer function of the system, we can write mathematically

$$G(s) = \left.\frac{\text{Laplace transform of output}}{\text{Laplace transform of input}}\right|_{\text{all initial conditions are zero}}$$

$$= \left.\frac{C(s)}{R(s)}\right|_{\text{all initial conditions are zero.}} \tag{3.1}$$

Example 3.1 Determine the transfer function of Fig. E3.1 where $v_i(t)$ is the input to the system and $v_o(t)$ is the output of the system.

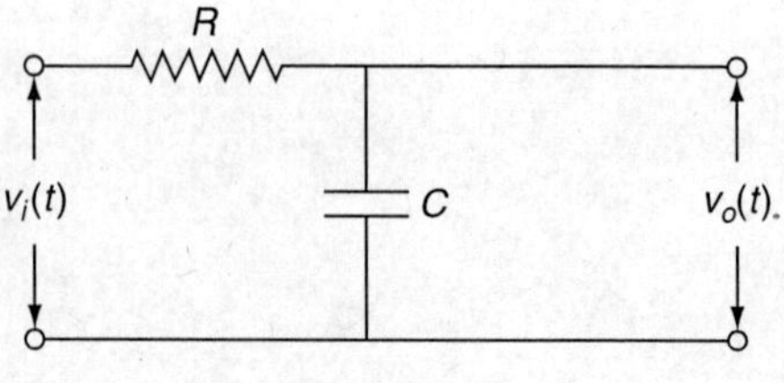

Fig. E3.1

Solution Let $i(t)$ be the current in the circuit. Using KVL (Kirchhoff's Voltage Law), we have

$$v_i(t) = Ri(t) + \frac{1}{C}\int i(t)dt \tag{1}$$

and $$v_o(t) = \frac{1}{C}\int i(t)dt \tag{2}$$

Taking Laplace transform of Eq. (1) and Eq. (2), we have

$$V_i(s) = RI(s) + \frac{1}{sC}I(s) \tag{3}$$

and $$V_o(s) = \frac{1}{sC}I(s) \tag{4}$$

It has been assumed that the system is initially relaxed. From Eq. (4), we have

$$I(s) = sCV_o(s) \tag{5}$$

From Eq. (3), we have

$$V_i(s) = \left(R + \frac{1}{sC}\right)I(s) = \left(\frac{sCR+1}{sC}\right)sCV_o(s) = (sCR+1)V_o(s) \quad \text{[using Eq. (4)]}$$

i.e., $$\frac{V_o(s)}{V_i(s)} = \frac{1}{1+sCR}$$

$\therefore$ Transfer function $= G(s) = \dfrac{V_o(s)}{V_i(s)} = \dfrac{1}{1+sCR}$

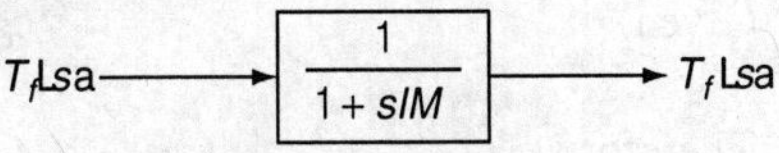

Fig. E3.1(a)

The system of Fig. E3.1 is shown in Fig. E3.1(a)

3.2 Impulse Response and Transfer Function

The impulse function has already been discussed in Chapter 1. In Chapter 2, it has been proved that the Laplace transform of an impulse function [$\delta(t)$] is unity. From Fig. 3.1(b), we can write Eq. (3.1) as follows:

$$G(s) = \frac{C(s)}{R(s)}$$

i.e., $$C(s) = G(s)R(s) \tag{3.2}$$

The response $C(s)$ can be found in terms of transfer function from Eq. (3.2). Here $G(s)$ and $R(s)$ are Laplace transforms of $g(t)$ and $r(t)$ respectively. Therefore, the response of any input $r(t)$, can be found out. Equation (3.2) represents a general equation and it is valid for any input. Let us take the impulse input

$$r(t) = \delta(t)$$

$\therefore$ $$R(s) = \text{LT}[\delta(t)] = 1 \tag{3.3}$$

Using Eq. (3.3), Eq. (3.2) can be written as follows

$$C(s) = G(s)$$

$\therefore$ $$c(t) = \text{LT}^{-1}[G(s)] = g(t) \tag{3.4}$$

Equation (3.4) suggests that for an impulse input to system, impulse response $C(s)$ is equal to the transfer function $G(s)$ of the system. For a linear time-invariant system, the transfer function of this system is the Laplace transform of the impulse response, assuming all the initial conditions to be zero.

Example 3.2 Find the transfer function of a system if its impulse response is e^{-3t}.

Solution

$$\text{LT}[e^{-3t}] = \frac{1}{s+3}$$

Since impulse response of a system gives its transfer function, we have

$$\text{Transfer function} = [G(s)] = \frac{1}{s+3}$$

Example 3.3 If the transfer function of a system and applied input to it are e^{-3t} and e^{-4t} respectively, obtain the response of the system.

Solution

Given $\quad g(t) = e^{-3t}$ and $r(t) = e^{-4t}$

Therefore $\quad G(s) = \text{LT}[g(t)] = \dfrac{1}{s+3}$

and $\quad R(s) = \text{LT}[r(t)] = \dfrac{1}{s+4}$

Since $\quad C(s) = G(s)R(s)$

$$C(s) = \frac{1}{(s+3)(s+4)}$$

$$\therefore \quad C(s) = \frac{1}{s+3} - \frac{1}{s+4}$$

Taking inverse Laplace transform, we have

$$c(t) = e^{-3t} - e^{-4t}$$

Note: $c(t) \neq g(t)r(t)$

3.3 Properties of Transfer Function (TF)

The properties of the transfer function are shown below:

- The ratio of the Laplace transform of output to input with all initial conditions to be zero is known as transfer function of a system.
- The transfer function of a system is the Laplace transform of its impulse response under assumption of zero initial conditions.

- Replacing 's' variable with linear operation $D \equiv d/dt$ in transfer function of a system, the differential equation of the system can be obtained.
- The transfer function of a system does not depend on the input(s) to the system.
- The system poles and zeros can be determined from its transfer function.
- Stability can be found from characteristic equation.
- Transfer function cannot be defined for non-linear systems. It can be defined for linear systems only.

3.4 Advantages and Disadvantages of Transfer Function

Advantages

- Transfer function is a mathematical model and it gives the gain of the system.
- Since Laplace transform is used, the terms are simple algebraic expressions and differential terms are not present.
- If transfer function of a system is known, the response of the system to any input can be determined very easily.
- Poles and zeros of a system can be determined from the knowledge of the transfer function of the system. Both poles and zeros have a vital role in the system's response.
- Transfer function helps in the study of stability analysis of the system.

Disadvantages

- Transfer function can be defined for linear systems only.
- Initial conditions lose their importance since transfer function does not take into account the initial condition.
- No inferences can be drawn about the physical structure of a system from its transfer function.

3.5 Poles and Zeros of a Transfer Function

From Eq. (3.1), the transfer function can be written as

$$G(s) = \left.\frac{C(s)}{R(s)}\right|_{\text{assuming zero initial conditions}} \tag{3.5}$$

$C(s)$ and $R(s)$ both are polynomial in s.

Let $$C(s) = b_m s^m + b_{m-1} s^{m-1} + \cdots + b_0$$

and $$R(s) = a_n s^n + a_{n-1} s^{n-1} + \cdots + a_0$$

$\therefore$ $$G(s) = \frac{b_m s^m + b_{m-1} s^{m-1} + \cdots + b_0}{a_n s^n + a_{n-1} s^{n-1} + \cdots + a_0} \tag{3.6}$$

$$= \frac{K(s-b_1)(s-b_2)(s-b_3)\ldots(s-b_m)}{(s-a_1)(s-a_2)(s-a_3)\ldots(s-a_n)} \tag{3.7}$$

where K is the gain factor.

Zeros of a transfer function are defined as the values of s for which the magnitude of the transfer function becomes zero. If the zeros of a transfer function are not repeated, they are known as simple zeros. If the zeros of a transfer function are repeated, they are termed as multiple zeros.

Poles of a transfer function are defined as the values of s for which the magnitude of the transfer function becomes infinity. If the poles of a transfer function are not repeated, they are known as simple poles. If the poles of a transfer function are repeated, they are termed as multiple poles.

3.5.1 Representation of Pole and Zero on the *s*-plane

There are two axes of a s-plane known as the real axis and the imaginary axis. Again $s = \sigma + j\omega$. Therefore, the σ-axis stands for the real axis and the $j\omega$-axis stands for the imaginary axis. The values of σ are plotted along the real axis as the values of $j\omega$ are plotted along the imaginary axis. Simple pole and zero are represented by X and O, respectively. Repeated poles and zeros are represented by repeated X and O, respectively.

Figure 3.2 shows the pole-zero plot of the transfer function $G(s)$ shown below

$$G(s) = \frac{K(s+b_1)(s+b_2)^2}{(s+a_1)(s+a_2)^3} \tag{3.8}$$

In Eq. (3.8), K is the gain function and $0 > -a_2 > -a_1 > -b_1 > -b_2$.

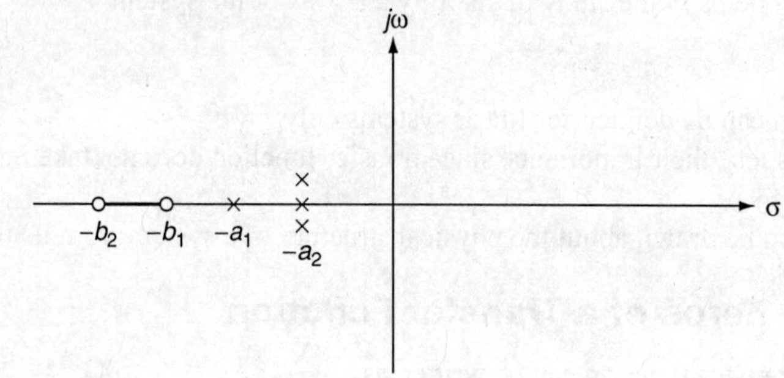

Fig. 3.2 Pole zero plot of $G(s)$ of Eq. (3.8)

3.5.2 Characteristic Equation

The denominator polynomial in terms of s of a transfer function is known as characteristic polynomial. If this polynomial is equated to zero, characteristic equation will be obtained. The characteristic polynomial of the transfer function $G(s)$ shown in Eq. (3.1) is $a_n s^n + a_{n-1}s^{n-1} + \cdots + a_0$ and the characteristic equation is given by $a_n s^n + a_{n-1}s^{n-1} + a_{n-2}s^{n-2} + \cdots + a_0 = 0$. Solving characteristic equation of a transfer function, we obtain poles of the transfer function.

Example 3.4 Find the pole(s) and zero(s) of the transfer function given by

$$G(s) = \frac{4(s+2)(s+2.5)^3}{(s+6)(s+4)^2}$$

Solution Poles are at $s = -6$ and $s = -4$. Pole at $s = -6$ is a simple pole, whereas the pole at $s = -4$ is multiple pole of order 2, i.e., it has second-order multiplicity. Zeros are at $s = -2$ and $s = -2.5$. Zero at $s = -2$ is simple, whereas the zero at $s = -2.5$ is multiple zero of order 3, i.e., it has third-order multiplicity.

3.6 MATLAB Command for Transfer Function and Roots

Transfer functions can be defined in MATLAB using the numerator and denominator matrices. For example, we have to define a transfer function $\frac{5}{s^2+2s+10}$.

The sequence of commands will be like this

>> $A = [0\ 5]$;

>> $B = [1\ 2\ 10]$;

>> tf(A, B)

Transfer function:

The coefficient in the numerator and denominator of the matrix should be in the increasing order of the powers, towards the left of the matrix. For example,

>> num = [1 2 10];

>> den = [1];

>> tf(A, B)

Transfer function:

s^2+2s+10

roots

The function roots(C) computes the polynomial roots whose elements are vector C.

>> num = [2 1 10];

>> roots(num)

$-0.25 + 2.2220i$

$-0.25 + 2.2220i$

The polynomial is defined only as a coefficient matrix vector which is a row vector. Inversely, given the roots of the polynomial, the polynomial can be constructed using **poly** command.

>> poly (res)

Ans. 1.000 0.500 5.000

SIGNIFICANT POINTS

Transfer Function: It is defined as the ratio of Laplace transform of output response to Laplace transform of input (excitation) assuming all the initial conditions to be zero.

Impulse Response and Transfer Function: If the transfer function of a system is not given but its impulse response is known, the impulse response will represent the transfer function of the system. *Therefore, for a linear time-invariant system, the transfer function of this system is the Laplace transform of the impulse response assuming all the initial conditions to be zero.*

Characteristic Equation: The denominator polynomial in terms of s of a transfer function is known as characteristic polynomial. If this polynomial is equated to zero, characteristic equation is obtained.

ADDITIONAL SOLVED EXAMPLES

Example 3.5 Calculate the transfer function of the electrical network shown in Fig. E3.2.

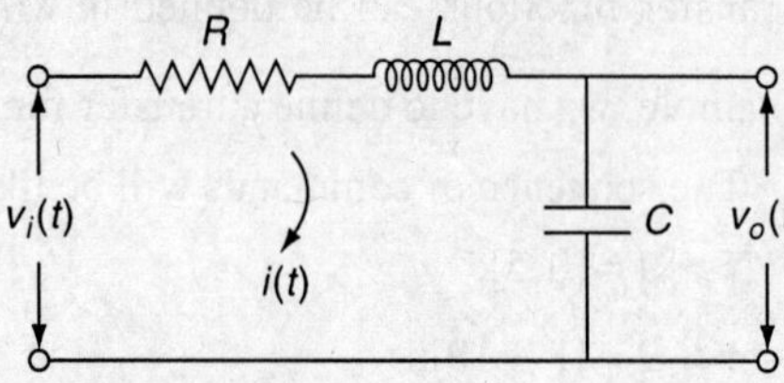

Fig. E3.2

Solution Using KVL, we have

$$v_i(t) = Ri(t) + L\frac{di(t)}{dt} + \frac{1}{C}\int i(t)\,dt \tag{1}$$

and $$v_o(t) = \frac{1}{C}\int i(t)\,dt \tag{2}$$

Taking Laplace transform of Eq. (1) and Eq. (2), we have

$$V_i(s) = RI(s) + sLI(s) + \frac{I(s)}{sC} \tag{3}$$

and $$V_o(s) = \frac{I(s)}{sC} \tag{4}$$

All the initial conditions have been put to zero.
From Eq. (4), we have

$$I(s) = sCV_o(s) \tag{5}$$

From Eq. (3), we have

$$V_i(s) = \left(R + sL + \frac{1}{sC}\right)I(s) = \left(R + sL + \frac{1}{sC}\right)sCV_o(s) = (s^2LC + sCR + 1)V_o(s)$$

i.e., $$\frac{V_o(s)}{V_i(s)} = \frac{1}{s^2LC + sCR + 1}$$

Therefore, the transfer function of the electrical network shown in Fig. E3.2 is given by

$$G(s) = \frac{1}{s^2LC + sCR + 1}$$

Example 3.6 Derive the $V_i(s)/I(s)$ and $V_o(s)/I(s)$ of the circuit shown in Fig. E3.3.

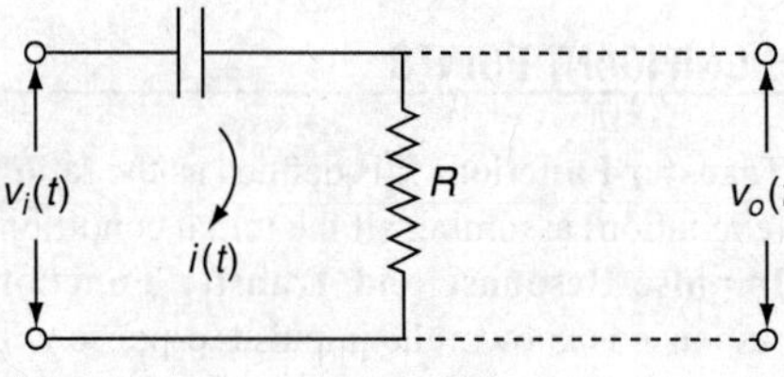

Fig. E3.3

Solution

$$v_i(t) = Ri(t) + \frac{1}{C}\int i(t)dt$$

Taking Laplace transform of Eq. (1), we have

$$V_i(s) = RI(s) + \frac{I(s)}{sC}, \text{ i.e., } V_i(s) = I(s)\left(\frac{RsC + 1}{sC}\right) \tag{1}$$

$$\therefore \qquad \frac{I(s)}{V_i(s)} = \frac{sC}{RsC+1}$$

Now
$$v_o(t) = Ri(t)$$

$$\therefore \qquad V_o(s) = RI(s)$$

i.e.,
$$I(s) = \frac{V_o(s)}{R} \tag{2}$$

Using Eq. (2), from Eq. (1), we have

$$V_i(s) = \frac{V_o(s)}{R}\left(\frac{sRC+1}{sC}\right) = \left(\frac{sRC+1}{sCR}\right)V_o(s)$$

i.e.,
$$\frac{V_o(s)}{V_i(s)} = \frac{sRC}{1+sRC}$$

The transfer function of the electrical network shown in Fig. E3.3 is $\dfrac{sRC}{1+sRC}$.

Example 3.7 Derive the transfer function of the network shown in Fig. E3.4.

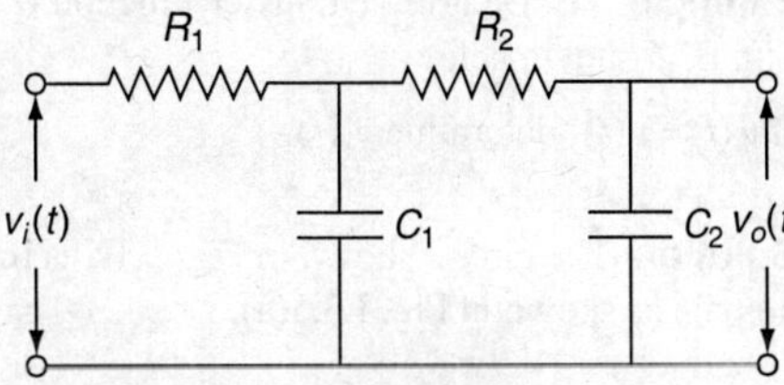

Fig. E3.4

Solution Figure E3.4 is redrawn in Fig. E3.4(a) in Laplace domain

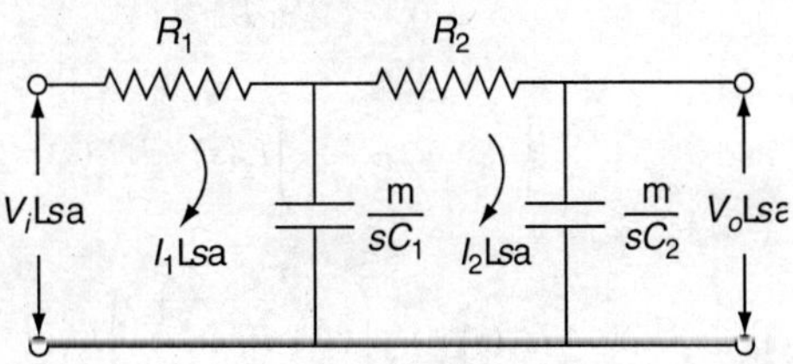

Fig. E3.4(a)

From Fig. E3.4(a), we have

$$V_i(s) = \left(R_1 + \frac{1}{sC_1}\right)I_1(s) - \frac{1}{sC_1}I_2(s) \tag{1}$$

and
$$\left(R_2 + \frac{1}{sC_1} + \frac{1}{sC_2}\right)I_2(s) - \frac{1}{sC_1}I_1(s) = 0 \tag{2}$$

From Eq. (2), we have

$$\frac{1}{sC_1}I_1(s) = \left(R_2 + \frac{1}{sC_1} + \frac{1}{sC_2}\right)I_2(s)$$

i.e.,
$$I_1(s) = sC_1\left(R_2 + \frac{1}{sC_1} + \frac{1}{sC_2}\right)I_2(s) \tag{3}$$

Using Eq. (3), Eq. (1) can be written as

$$V_i(s) = \left(R_1 + \frac{1}{sC_1}\right) \times sC_1 \times \left(R_2 + \frac{1}{sC_1} + \frac{1}{sC_2}\right)I_2(s) - \frac{1}{sC_1}I_2(s)$$

$$\therefore \qquad V_i(s) = \left[\frac{(sC_1R_1+1)(sR_2C_1C_2+C_1+C_2)-C_2}{sC_1C_2}\right]I_2(s)$$

Again,
$$V_o(s) = \frac{I_2(s)}{sC_2} \tag{4}$$

Substituting Eq. (4) in Eq.(3), we have

$$V_i(s) = \left[\frac{(sC_1R_1+1)(sR_2C_1C_2+C_1+C_2)-C_2}{C_1}\right]V_o(s)$$

$$\therefore \quad \frac{V_o(s)}{V_i(s)} = \frac{C_1}{(sC_1R_1+1)(sR_2C_1C_2+C_1+C_2)-C_2}$$

$$= \frac{C_1}{s^2C_1^2R_1R_2C_2 + sC_1^2R_1 + sR_1C_1C_2 + sR_2C_1C_2 + C_1 + C_2 - C_2}$$

$$= \frac{1}{s^2C_1R_1R_2C_2 + sC_1R_1 + sR_1C_2 + sR_2C_2 + 1}$$

$$= \frac{1}{s^2C_1R_1R_2C_2 + s(R_1C_1 + R_1C_2 + R_2C_2) + 1}$$

Example 3.8 Determine transfer function $V_o(s)/V_i(s)$ of Fig. E3.5 shown below.

If $v_i(t) = u(t)$, determine $v_o(t)$.

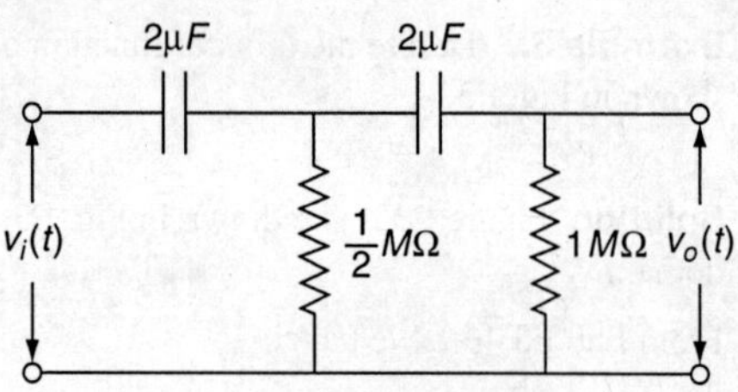

Fig. E3.5

Solution The circuit shown in Fig. E3.5 is redrawn in Laplace domain as shown in Fig. E3.5(a).

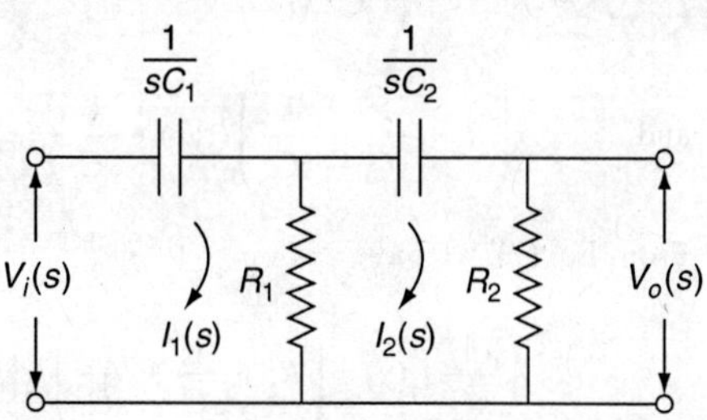

Fig. E3.5(a)

$$\therefore \quad V_i(s) = \left(R_1 + \frac{1}{sC_1}\right)I_1(s) - R_1I_2(s) \qquad (1)$$

and

$$\left(R_1 + R_2 + \frac{1}{sC_2}\right)I_2(s) - R_1I_1(s) = 0$$

i.e.,

$$I_1(s) = \frac{1}{R_1}\left(R_1 + R_2 + \frac{1}{sC_2}\right)I_2(s) \qquad (2)$$

Using Eq. (2), we have from Eq. (1)

$$V_i(s) = \left(R_1 + \frac{1}{sC_1}\right) \times \frac{1}{R_1} \times \left(R_1 + R_2 + \frac{1}{sC_2}\right)I_2(s) - R_1I_2(s)$$

$$= \frac{(1+sC_1R_1)[1+sC_2(R_1+R_2)]}{s^2R_1C_1C_2}I_2(s) - R_1I_2(s)$$

$$= I_2(s)\left[\frac{1 + sC_2(R_1+R_2) + sC_1R_1 + s^2C_1R_1C_2(R_1+R_2) - s^2C_1C_2R_1^2}{s^2R_1C_1C_2}\right]$$

$$= \left[\frac{s^2 C_1 C_2 R_1 R_2 + s(R_1 C_1 + R_2 C_2 + R_1 C_2) + 1}{s^2 R_1 C_1 C_2} \right] I_2(s) \tag{3}$$

Again, $V_o(s) = R_2 I_2(s)$

$$\therefore \quad I_2(s) = \frac{V_o(s)}{R_2} \tag{4}$$

Using Eq. (4), Eq. (3) becomes

$$V_i(s) = \left[\frac{s^2 C_1 C_2 R_1 R_2 + s(R_1 C_1 + R_2 C_2 + R_1 C_2) + 1}{s^2 R_1 R_2 C_1 C_2} \right] V_o(s)$$

i.e.,

$$\frac{V_o(s)}{V_i(s)} = \left[\frac{s^2 R_1 R_2 C_1 C_2}{s^2 R_1 R_2 C_1 C_2 + s(R_1 C_1 + R_2 C_2 + R_1 C_2) + 1} \right] \tag{5}$$

Equation (5) represents the transfer function of the system.

Now $v_i(t) = u(t)$

$$\therefore \quad V_i(s) = \frac{1}{s} \tag{6}$$

Using Eq. (6), we have from Eq. (5)

$$V_o(s) = \frac{s R_1 R_2 C_1 C_2}{s^2 R_1 R_2 C_1 C_2 + s(R_1 C_1 + R_2 C_2 + R_1 C_2) + 1}$$

$$= \frac{s \times \frac{1}{2} \times 1 \times 2 \times 2}{s^2 \times \frac{1}{2} \times 1 \times 2 \times 2 + s\left(\frac{1}{2} \times 2 + 1 \times 2 + \frac{1}{2} \times 2 \right) + 1} = \frac{2s}{2s^2 + 4s + 1}$$

$$= \frac{2s}{s^2 + 2s + 0.5} = \frac{2s}{(s + 1.707)(s + 0.293)} = \frac{A}{s + 1.707} + \frac{B}{s + 0.293}$$

$$\therefore \quad A = \left. \frac{s}{s + 0.293} \right|_{s = -1.707} = \frac{-1.707}{-1.707 + 0.293} = 1.2072$$

$$\therefore \quad B = \left. \frac{s}{s + 1.707} \right|_{s = -0.293} = \frac{-0.293}{-0.293 + 1.707} = -0.2072$$

$$\therefore \quad V_o(s) = \frac{1.2072}{s + 1.707} - \frac{0.2072}{s + 0.293}$$

$$\therefore \quad V_o(t) = 1.2072 e^{-1.707t} - 0.2072 e^{-0.293t}$$

Example 3.9 Derive the transfer function of the network shown in Fig. E3.6.

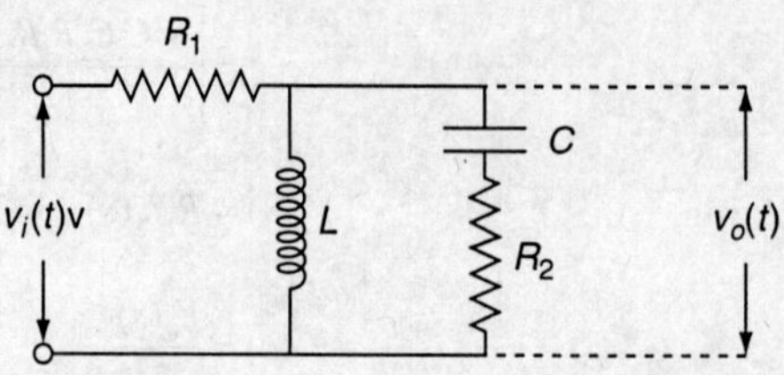

Fig. E3.6

Solution The circuit shown in Fig. E3.6 is redrawn in Laplace domain as shown in Fig. E3.6(a).

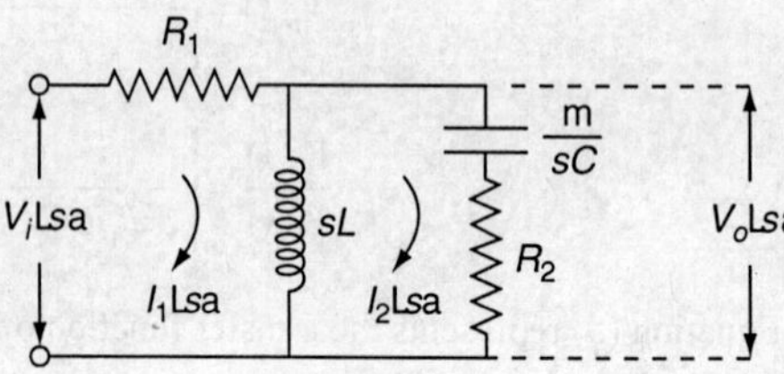

Fig. E3.6(a)

From Fig. E3.6(a), we have

$$V_i(s) = (R_1 + sL)I_1(s) - sLI_2(s) \tag{1}$$

and

$$\left(sL + \frac{1}{sC} + R_2\right)I_2(s) - sLI_1(s) = 0$$

i.e.,

$$I_1(s) = \frac{1}{sL}\left(sL + \frac{1}{sC} + R_2\right)I_2(s) \tag{2}$$

Using Eq. (2), we have from Eq. (1)

$$V_i(s) = \left[\frac{(R_1 + sL)\left(sL + \frac{1}{sC} + R_2\right)}{sL} - sL\right]I_2(s)$$

$$= \left[\frac{s^2R_1LC + sR_1R_2C + R_1 + s^3L^2C + s^2LCR_2 + sL - s^3L^2C}{s^2LC}\right]I_2(s)$$

$\therefore$

$$V_i(s) = \left(\frac{s^2R_1LC + s^2LCR_2 + sL + sR_1R_2C + R_1}{s^2LC}\right)I_2(s) \tag{3}$$

Again,

$$V_o(s) = \left(R_2 + \frac{1}{sC}\right)I_2(s)$$

$$I_2(s) = \left(\frac{sC}{1 + sR_2C}\right)V_o(s) \tag{4}$$

Substituting Eq. (4) in Eq. (3), we have

$$V_i(s) = \left[\frac{s^2R_1LC + s^2LCR_2 + sL + sR_1R_2C + R_1}{s^2LC}\right]\left(\frac{sC}{1 + sR_2C}\right)V_o(s)$$

$$= \left[\frac{s^2LC(R_1 + R_2) + s(L + R_1R_2C) + R_1}{sL(1 + sR_2C)}\right]V_o(s)$$

$\therefore$

$$\frac{V_o(s)}{V_i(s)} = \frac{sL(1 + sR_2C)}{s^2LC(R_1 + R_2) + s(L + R_1R_2C) + R_1}$$

Example 3.10 Determine the transfer function of the electric network shown in Fig. E3.7.

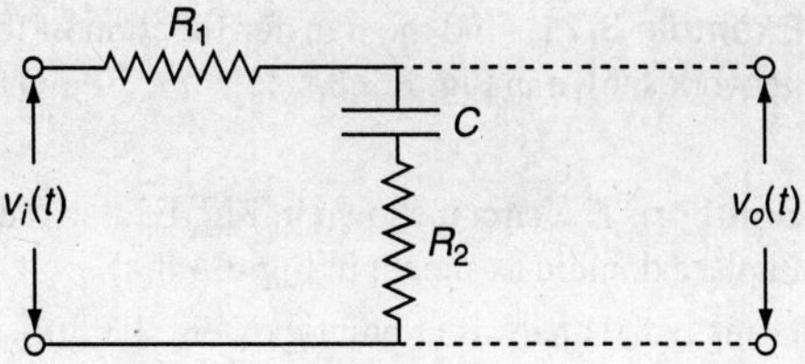

Fig. E3.7

Solution Using KVL, we have

$$v_i(t) = R_1 i(t) + \frac{1}{C}\int i(t)dt + R_2 i(t) \qquad (1)$$

and $$v_o(t) = R_2 i(t) + \frac{1}{C}\int i(t)dt \qquad (2)$$

Taking Laplace transform of Eq. (1) and Eq. (2), we have

$$V_i(s) = R_1 I(s) + \frac{1}{sC} I(s) + R_2 I(s) \qquad (3)$$

and $$V_o(s) = R_2 I(s) + \frac{1}{sC} I(s) \qquad (4)$$

It has been assumed that all initial conditions are zero. From Eq. (4), we have

$$V_o(s) = \left(R_2 + \frac{1}{sC}\right) I(s) = \left[\frac{sCR_2 + 1}{sC}\right] I(s)$$

$\therefore$ $$I(s) = \left(\frac{sC}{1 + sCR_2}\right) V_o(s) \qquad (5)$$

$\therefore$ $$V_i(s) = \left[\frac{(R_1 + R_2)sC + 1}{sC}\right] I(s) = \left[\frac{(R_1 + R_2)sC + 1}{sC}\right]\left(\frac{sC}{1 + sCR_2}\right) V_o(s) \quad \text{[using Eq. (5)]}$$

The transfer function is given by

$$G(s) = \frac{V_o(s)}{V_i(s)} = \frac{1 + sCR_2}{(R_1 + R_2)sC + 1}$$

$$= \frac{CR_2\left(s + \dfrac{1}{CR_2}\right)}{\left[s + \dfrac{1}{(R_1 + R_2)C}\right](R_1 + R_2)C} = \frac{\dfrac{s + \dfrac{1}{CR_2}}{(R_1 + R_2)C}}{\dfrac{s + \dfrac{1}{(R_1 + R_2)C}}{CR_2}} = \frac{a(s + b)}{b(s + a)}$$

where $a = \dfrac{1}{(R_1 + R_2)C}$ and $b = 1/CR_2$

This network is known as lag compensator.

Example 3.11 Find the transfer function of the electrical network shown in Fig. E.3.8.

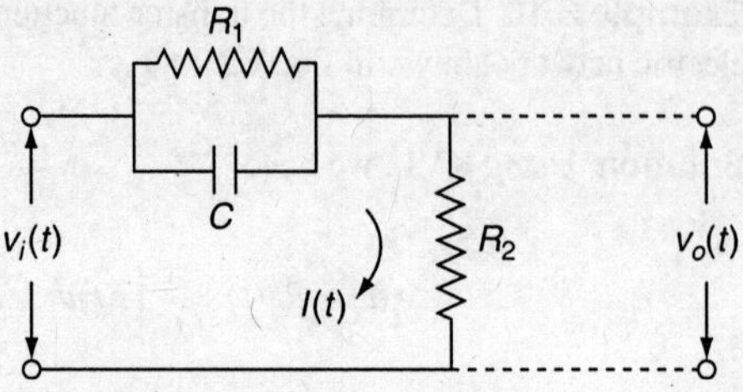

Fig. E3.8

Solution The circuit shown in Fig. E3.8 is redrawn in Laplace domain as shown in Fig. E3.8(a).

Using nodal analysis at point a of Fig. E3.8(a), we have

$$\frac{V_i(s)-V_o(s)}{R_1}+\frac{V_i(s)-V_o(s)}{\frac{1}{sC}}=\frac{V_o(s)}{R_2}$$

i.e.,
$$\left(\frac{1}{R_1}+sC\right)V_i(s)=\left[\frac{1}{R_1}+\frac{1}{R_2}+sC\right]V_o(s)$$

i.e.,
$$\frac{V_o(s)}{V_i(s)}=\frac{sC+(1/R_1)}{sC+\frac{1}{R_1}+\frac{1}{R_2}}=\frac{s+(1/CR_1)}{s+\frac{1}{R_1C}+\frac{1}{R_2C}}$$

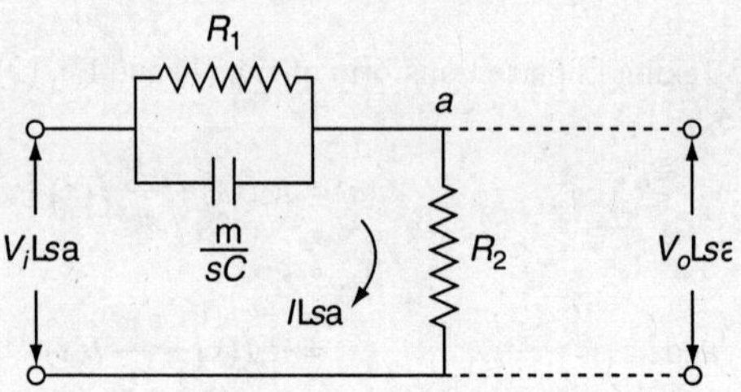

Fig. E3.8(a)

The transfer function of the system is given by

$$G(s)=\frac{V_o(s)}{V_i(s)}=\frac{s+a}{s+b}$$

where $a=\frac{1}{CR_1}$, $b=\frac{1}{R_1C}+\frac{1}{R_2C}$

and $a<b$. This network is known as lead network.

Example 3.12 A system having input $x(t)$ and output $y(t)$ is represented by

$$\frac{dy(t)}{dt}+3y(t)=\frac{dx(t)}{dt}+2x(t)$$

Find the transfer function of the system.

Solution The system is represented by

$$\frac{dy(t)}{dt}+3y(t)=\frac{dx(t)}{dt}+2x(t) \quad (1)$$

Taking Laplace transform of Eq. (1), we have

$$sY(s)+3Y(s)=sX(s)+2X(s)$$

Initial conditions of the system are assumed to be zero.

$$\therefore \quad (s+3)Y(s)=(s+2)X(s)$$

i.e.,
$$\frac{Y(s)}{X(s)}=\frac{s+2}{s+3}$$

Therefore, the transfer function of the system is $G(s)=Y(s)/X(s)=\frac{s+2}{s+3}$.

Example 3.13 Determine the transfer function of a system represented by

$$\frac{d^3y(t)}{dt^3}+3\frac{d^2y(t)}{dt^2}+2\frac{dy(t)}{dt}+y(t)=\frac{d^2x(t)}{dt^2}+2\frac{dx(t)}{dt}+3x(t)$$

Solution Taking Laplace transform of the given equation and assuming initial condition to be zero, we have

$$s^3Y(s)+3s^2Y(s)+2sY(s)+Y(s)=s^2X(s)+2sX(s)+3X(s)$$

i.e.,

$$(s^3+3s^2+2s+1)Y(s)=(s^2+2s+3)X(s)$$

i.e.,

$$\frac{Y(s)}{X(s)}=\frac{s^2+2s+3}{s^3+3s^2+2s+1}$$

Therefore, the transfer function of the system is

$$G(s)=\frac{s^2+2s+3}{s^3+3s^2+2s+1}$$

Example 3.14 The transfer function of a system is given by

$$G(s)=\frac{3s+2}{s^2+2s+3}$$

Find the differential equation of the system having input $x(t)$ and output $y(t)$.

Solution

Here

$$G(s)=\frac{Y(s)}{X(s)}=\frac{3s+2}{s^2+2s+3}$$

$$s^2Y(s)+2sY(s)+3Y(s)=3sX(s)+2X(s)$$

Taking inverse Laplace transform, we have

$$\frac{d^2y(t)}{dt^2}+2\frac{dy(t)}{dt}+3y(t)=3\frac{dx(t)}{dt}+2x(t)$$

This is the required differential equation of the system.

Example 3.15 Find the transfer function of a system if its impulse response is $e^{-3t}\sin 2t$.

Solution The impulse response of the system is given by

$$c(t)=e^{-3t}\sin 2t$$

$\therefore$

$$\text{LT}[c(t)]=\text{LT}[e^{-3t}\sin 2t]=\frac{2}{(s+3)^2+4}=\frac{2}{(s^2+6s+13)}$$

Since the Laplace transform of the impulse response of a system represents its transfer function, the transfer function of the system is

$$G(s)=2/(s^2+6s+13).$$

Example 3.16 The step response of a system is given by

$$y(t) = 6 - \frac{4}{5}e^{-5t} + e^{-3t}.$$

Find the transfer function of the system.

Solution Since the derivative of a step response gives impulse response, the impulse response (IR) is given by

$$(\text{IR}) = \frac{dy(t)}{dt} = 4e^{-5t} - 3e^{-3t}$$

$$\therefore \qquad \text{LT[IR]} = \frac{4}{s+5} - \frac{3}{s+3} = \frac{4(s+3) - 3(s+5)}{(s+3)(s+5)} = \frac{4s+12-3s-15}{(s+3)(s+5)} = \frac{s-3}{(s+3)(s+5)}$$

Since Laplace transform of impulse response of a system gives its transfer function, the transfer function is represented by

$$G(s) = \frac{s-3}{(s+3)(s+5)}$$

Example 3.17 The pole zero plot of a transfer function is shown in Fig. E3.9. Find the transfer function of the system.

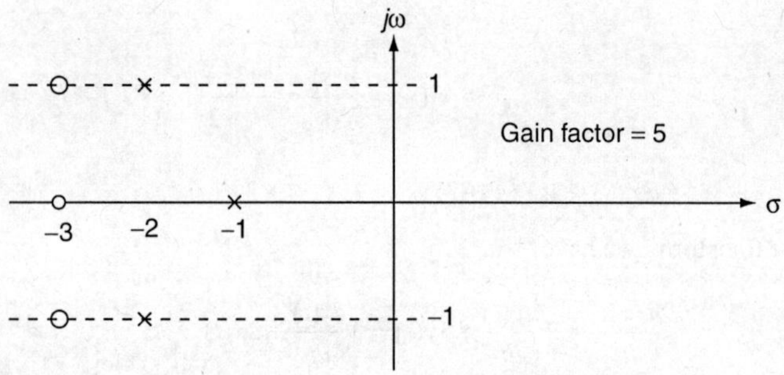

Fig. E3.9

Solution The poles are at $s = -1$ and $s = -2 \pm j$. The zeros are at $s = -3 \pm j$. The gain factor is 5. The transfer function of the system is given by

$$G(s) = \frac{5(s+3-j)(s+3+j)}{(s+1)(s+2-j)(s+2+j)} = \frac{5\{(s+3)^2 - j^2\}}{(s+1)\{(s+2)^2 - j^2\}}$$

$$\therefore \qquad G(s) = \frac{5(s^2+6s+10)}{(s+1)(s^2+4s+5)}$$

Example 3.18 Determine the characteristic polynomial and its roots for a system having transfer function as follows:

$$G(s) = \frac{4(s+2)}{(s+3)(s+4)}$$

Solution The denominator of the transfer function is

$$(s+3)(s+4) = s^2 + 7s + 12$$

The characteristic polynomial is $s^2 + 7s + 12$. The characteristic equation is given by

$$s^2 + 7s + 12 = 0$$

i.e.,

$$(s+3)(s+4) = 0$$

$\therefore$

$$s = \text{either} -3 \text{ or} -4$$

These are the poles of the system.

Example 3.19 The pole and zero of a system are at $s = -3$ and at $s = -2$, respectively. The value of gain factor is 5. Determine the response if unit step function is applied as input.

Solution The system transfer function is given by

$$G(s) = \frac{5(s+2)}{(s+3)}$$

Now input is $r(t) = u(t)$. Therefore

$$R(s) = \frac{1}{s}$$

$$C(s) = G(s)R(s) = \frac{5(s+2)}{s(s+3)} = 5\left[\frac{2/3}{s} + \frac{1/3}{s+3}\right] = \frac{5}{3}\left(\frac{2}{s} + \frac{1}{s+3}\right)$$

Taking inverse Laplace transform, we have

$$c(t) = \frac{5}{3}(2 + e^{-3t})$$

Example 3.20 Determine the transfer function of the system represented by

$$5\frac{dy(t)}{dt} + 3y(t) = 2x(t-T)$$

Solution The Laplace transform of the given equation becomes

$$5sY(s) + 3Y(s) = 2e^{-sT}X(s)$$

i.e.,

$$(5s+3)Y(s) = 2e^{-sT}X(s)$$

i.e.,

$$\frac{Y(s)}{X(s)} = \frac{2e^{-sT}}{5s+3}$$

Therefore, the transfer function of the system is given by

$$G(s) = \frac{2e^{-sT}}{5s+3}$$

Example 3.21 Determine V_2/V_1 for the circuit shown in Fig. E3.10.

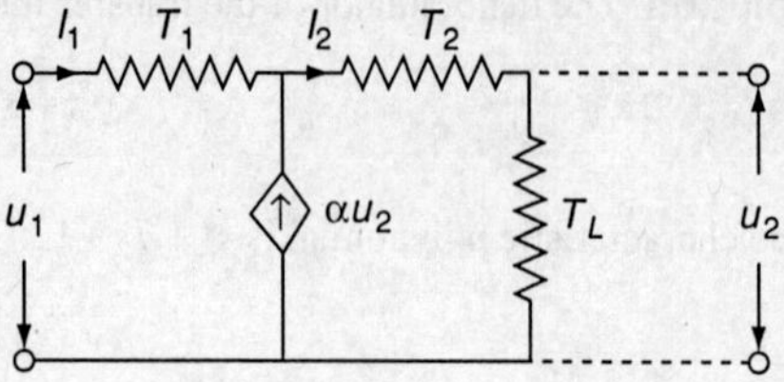

Fig. E3.10

Solution Using Kirchhoff's voltage law, we have

$$V_1 = I_1R_1 + I_2(R_2 + R_L) \quad (1)$$

and $$V_2 = I_2R_L \quad (2)$$

Again, by Kirchhoff's current law, we have

$$I_2 = I_1 + \alpha V_2 \quad (3)$$

From Eq. (2) and Eq. (3), we have

$$V_2 = (I_1 + \alpha V_2)R_L$$

i.e., $$(1 - \alpha R_L)V_2 = I_1R_L$$

i.e., $$I_1 = \left(\frac{1 - \alpha R_L}{R_L}\right)V_2 \quad (4)$$

Using Eq. (4), we have from Eq. (1)

$$V_1 = \left(\frac{1 - \alpha R_L}{R_L}\right)V_2R_1 + \frac{V_2}{R_L}(R_2 + R_L) \qquad \left[\therefore I_2 = \frac{V_2}{R_L}\right]$$

$$= \frac{V_2}{R_L}[(1 - \alpha R_L)R_1 + (R_2 + R_L)]$$

i.e., $$\frac{V_2}{V_1} = \frac{R_L}{(1 - \alpha R_L)R_1 + (R_2 + R_L)}$$

This is the transfer function of Fig. E3.10.

Example 3.22 Find V_2/V_1 of Fig. E3.11, where OPAMP is ideal.

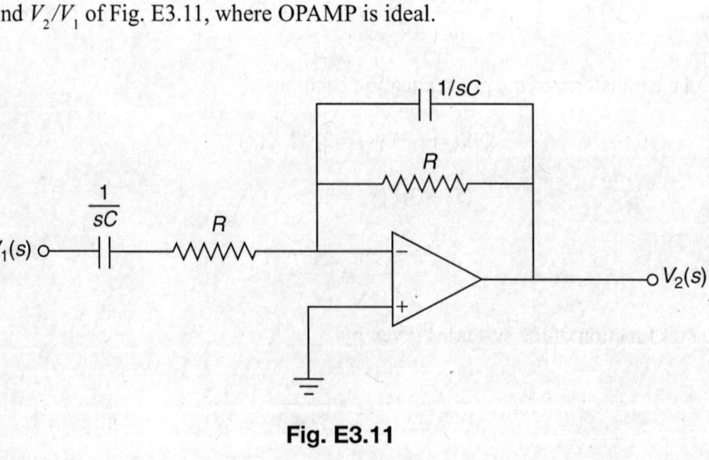

Fig. E3.11

Solution Figure E3.11 is redrawn and shown in Fig. E3.11(a)

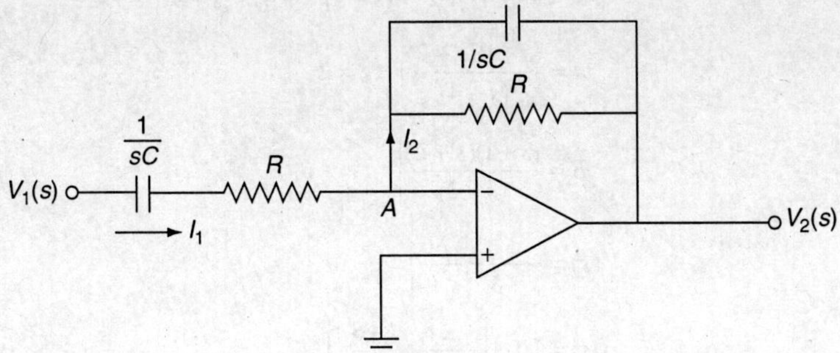

Fig. E3.11(a)

Using KCL at note A, we have

$$I_1 = I_2$$

i.e.,
$$\frac{V_1(s)-0}{R+\dfrac{1}{Cs}} = \frac{0-V_2(s)}{\dfrac{R}{1+sCR}}$$

i.e.,
$$\left[\frac{V_1(s)}{\dfrac{1+sCR}{sC}}\right]\left(\frac{R}{1+sCR}\right) = -V_2(s)$$

i.e.,
$$\frac{V_2(s)}{V_1(s)} = -\frac{sCR}{(1+sCR)^2}$$

Example 3.23 Draw poles and zeros for $V(s) = (s^2+3s+2)/(s^2+7s+12)$ and evaluate $v(t)$.

Solution

$$V(s) = \frac{s^2+3s+2}{s^2+7s+12} = \frac{(s+1)(s+2)}{(s+3)(s+4)}$$

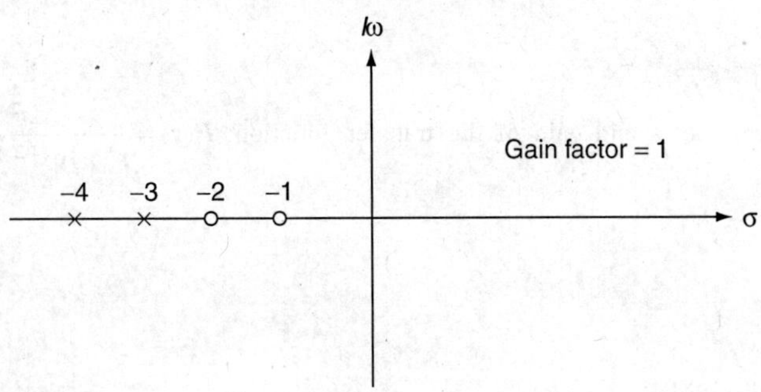

Fig. E3.12

$$V(s) = \frac{s^2+3s+2}{s^2+7s+2} = \frac{(s+1)(s+2)}{(s+3)(s+4)} = \frac{A}{s+3} + \frac{B}{s+4}$$

$$A = \left.\frac{(s+1)(s+2)}{(s+4)}\right|_{s=-3} = 2$$

and $$B = \left.\frac{(s+1)(s+2)}{(s+3)}\right|_{s=-4} = -6$$

∴ $$V(s) = \frac{2}{s+3} + \frac{-6}{s+4}$$

∴ $$v(t) = \text{LT}^{-1}\left[\frac{2}{s+3} + \frac{-6}{s+4}\right]$$

$$= 2e^{-3t} - 6e^{-4t}.$$

MATLAB Programs

Program 1 Find zeros, poles and gain of the transfer function $F(s) = \dfrac{s^2+5s+4}{s^4+16s^3+91s^2+216s+180}$ using MATLAB.

Solution

```
>> num = [1  5  4]
num =
   1   5   4
>> den = [1  16  91  216  180]
den =
   1   16   91   216   180
>> [z, p, k] = tf2zp(num, den)
z =
   -4
   -1
p =
   -6.0000
   -5.0000
   -3.0000
   -2.0000
k =
   1
```

Program 2 Find zeros, poles and gain of the transfer function $F(s) = \dfrac{5s^2+30s+25}{s^4+16s^3+89s^2+206s+168}$ using MATLAB.

Solution

```
>> num = [5  30  25]
num =
   5   30   25
>> den = [ 1  16  89  206  168]
den =
   1   16   89   206   168
```

```
>> [z, p, k] = tf2zp(num, den)
z =
   -5
   -1
p =
   -7.0000
   -4.0000
   -3.0000
   -2.0000
k =
   5
```

Program 3 If the zeros and poles of the system having gain 6 are at −1, −4 and −2, −3, −5 , −7 respectively, find the transfer function of the system using MATLAB.

Solution

```
>> z = [-1; -4]
z =
   -1
   -4
>> p = [-2; -3; -5; -7]
p =
   -2
   -3
   -5
   -7
>> k = 6
k =
   6
>> [num, den] = zp2tf(z, p, k)
num =
   0   0   6   30   24
den =
   1   17   101   247   210
>> printsys(num,den,'s')
num/den =
           6 s^2 + 30 s + 24
   -----------------------------------
   s^4 + 17 s^3 + 101 s^2 + 247 s + 210
```

Program 4 If the zeros and poles of the system having gain 5 are at −1, −3 and −2, −4, −5 , −7 respectively, find the transfer function of the system using MATLAB.

Solution

```
>> z = [-1; -3]
z =
   -1
   -3
```

```
>> p = [-2; -4; -5; -7]
p =
   -2
   -4
   -5
   -7
>> k = 5
k =
   5
>> [num, den] = zp2tf(z, p, k)
num =
   0  0  5  20  15
den =
   1  18  115  306  280
>> printsys(num,den,'s')
num/den =
          5 s^2 + 20 s + 15
   ----------------------------------
   s^4 + 18 s^3 + 115 s^2 + 306 s + 280
```

Program 5 If the zeros and poles of the system having gain 7 are at −1, −8 and −2, −3, −6 ,-7 respectively, find the transfer function of the system using MATLAB.

Solution

```
>> z = [-1; -8];
>> p = [-2; -3; -6; -7];
>> k = 7;
>> [num, den] = zp2tf(z, p, k);
>> printsys(num,den,'s')
num/den =
          7 s^2 + 63 s + 56
   ----------------------------------
   s^4 + 18 s^3 + 113 s^2 + 288 s + 252
```

Program 6 The transfer function of a system is shown below:

$$G(s) = \frac{5(s+3)(s+5)}{s(s+2)(s^2+6s+7)}.$$ Find the transfer function and pole-zero plot of the system using MATLAB.

```
>> p1 = [1 3];
>> p2 = [1 5];
>> num = conv(p1,p2);
>> num = 5*num
num =
   5  40  75
>> q1 = [1 0];
>> q2 = [1 2];
>> q3 = [1 6 7];
>> den = conv(q1,q2);
>> den = conv(den,q3)
den =
   1  8  19  14  0
```

```
>> % command tf
>> sys = tf(num,den)
Transfer function:
   5 s^2 + 40 s + 75
--------------------------
s^4 + 8 s^3 + 19 s^2 + 14 s

>> % command printsys
>> printsys(num,den)
num/den =
      5 s^2 + 40 s + 75
   --------------------------
   s^4 + 8 s^3 + 19 s^2 + 14 s

>> pzmap(sys);
>>
```

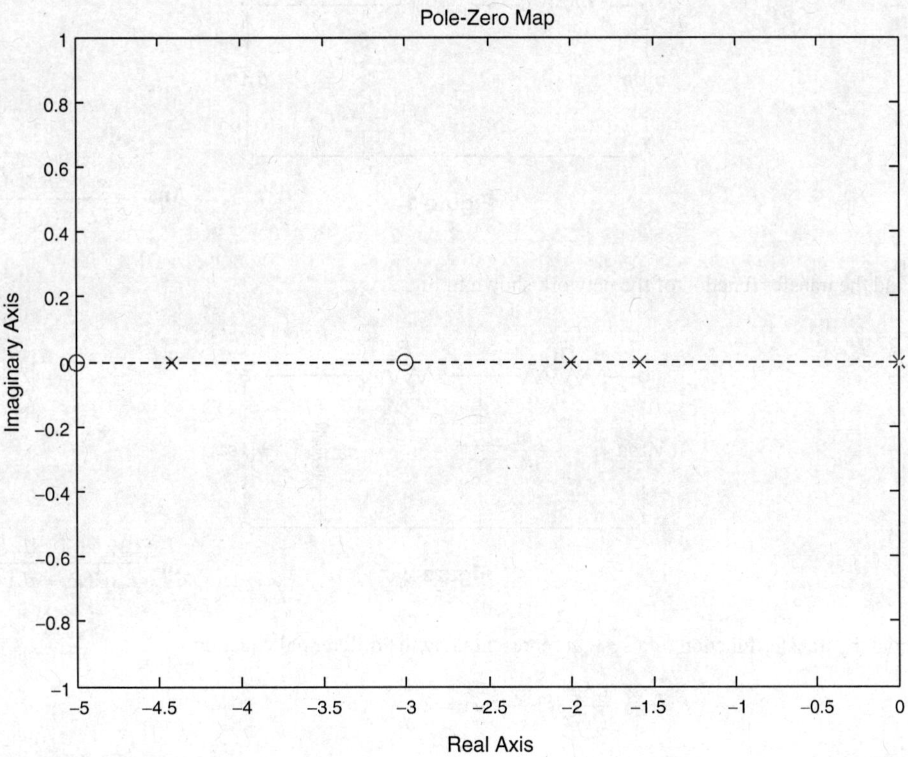

Fig. P6

SHORT QUESTIONS WITH ANSWERS

1. What do you mean by transfer function?

   Ans. It is defined by the ratio of Laplace transform of output response to Laplace transform of input (excitation), assuming all the initial conditions to be zero.

2. Relate impulse response and transfer function.

   Ans. If the transfer function of a system is not given but its impulse response is known, the impulse response will represent the transfer function of the system. Therefore, for a linear time-invariant system, the transfer function of this system is the Laplace transform of the impulse response, assuming all the initial conditions to be zero.

3. Define characteristic equation.

   Ans. The denominator polynomial in terms of s of a transfer function is known as characteristic polynomial. If this polynomial is equated to zero, characteristic equation is obtained.

EXERCISES

Questions for Practice

1. Find the transfer function of the network shown in Fig. 1.

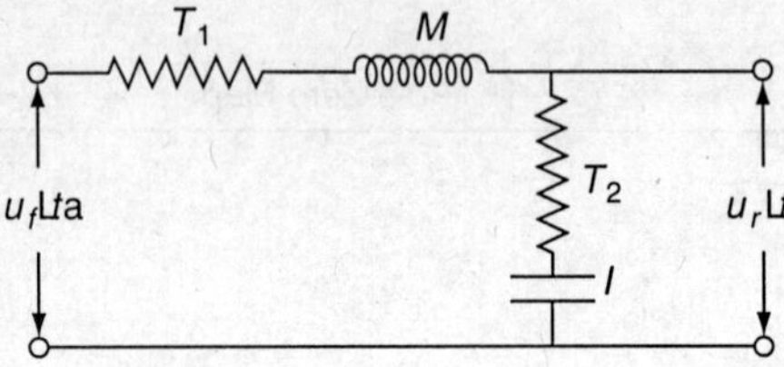

Figure 1

Ans. $\dfrac{sR_2C+1}{s^2LC+(R_1+R_2)sC+1}$

2. Find the transfer function of the network shown in Fig. 2.

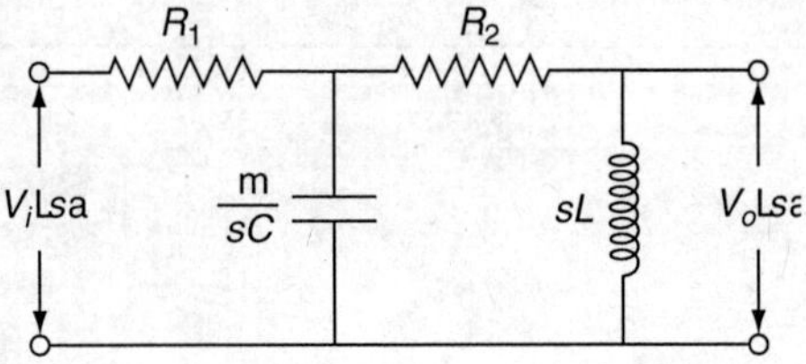

Figure 2

Ans. $\dfrac{sL}{R_1sC(R_2+sL)+R_2+sL}$

3. Find the transfer function of a system represented by the differential equation

$$\frac{d^2y}{dt^2}+3\frac{dy}{dt}+4y=2\frac{d^2x}{dt^2}+6x$$

Ans. $\dfrac{Y(s)}{X(s)}=\dfrac{2s^2+6}{s^2+3s+4}$

4. The dynamic behaviour of a system is described by the equation $dc/dt + 12c = 20r$, where r is the input and c is the output. Find the transfer function of the system.

Ans. $\dfrac{C(s)}{R(s)}=\dfrac{20}{s+12}$

5. The differential equation of a system is $3\frac{d^2c(t)}{dt^2}+5\frac{dc(t)}{dt}+c(t)=r(t)+3r(t-2)$ Find the transfer function of above system.

Ans. $\frac{C(s)}{R(s)}=\frac{1+3e^{-2s}}{3s^2+5s+1}$

6. The transfer function of a system is given by

$$G(s)=\frac{4s+1}{s^2+2s+3} \qquad \left[\frac{d^2y(t)}{dt^2}+2\frac{dy(t)}{dt}+3y(t)=4\frac{dx(t)}{dt}+x(t)\right]$$

Find the differential equation of the system having input x and output y.

7. The impulse response of a system is e^{-5t}. Determine the transfer function of the system.

Ans. $\frac{1}{s+5}$

8. The unit impulse response of system is $e^{-t}(1-\cos 3t)$. Determine its transfer function.

Ans. $\frac{9}{(s+1)(s^2+2s+10)}$

9. A system has poles at $s=-4$, $s=-2$ and zero at $s=-3$. Sketch the pole zero plot and find the transfer function of the system if the value of gain factor is 4.

Ans. $\frac{4(s+3)}{(s+2)(s+4)}$

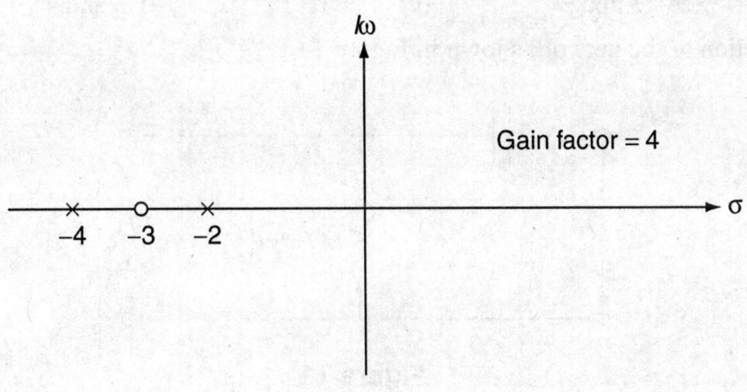

10. Find the pole zero plot of the following transfer function:

Ans. $\frac{20(s+3)}{(s+1)(s^2+7s+10)}$

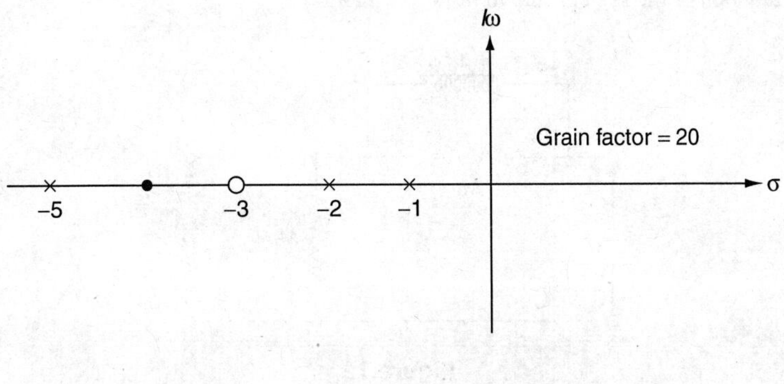

TEST YOURSELF

1. The transfer function is defined for
   (a) linear and time-variant system (b) linear and time-invariant system
   (c) non-linear and time-variant system (d) non-linear and time-invariant system
   (e) all of these.
2. The transfer function of a system is used to study its
   (a) steady state behaviour only (b) transient state behaviour only
   (c) transient and steady state (d) transient state and partly steady state
3. The transfer function of a system is defined as the ratio of its output to input in
   (a) Fourier transform (b) Laplace transform
   (c) Z transform (d) Both (a) and (b)
4. The steady state output of a unit feedback control system is
   (a) equal to reference input (b) more than reference input
   (c) not related to reference input (d) very near to reference input
5. The type of a transfer function denotes
   (a) the number of zeros at origin (b) the number of poles at origin
   (c) the number of poles at infinity (d) the number of zeros at infinity.
6. The transfer function of a system is $T(s) = K/[s^3(1 + sT)]$. The type and order of the system are
   (a) 2 and 3 (b) 3 and 2 (c) 3 and 3 (d) 3 and 4
7. The transfer function of the network shown in Figure T1 is

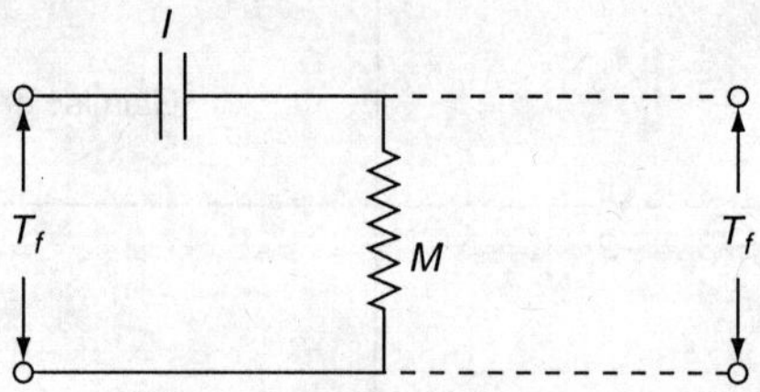

Figure T1

   (a) $\frac{1}{1+sRC}$ (b) $\frac{sRC}{1+sRC}$ (c) $\frac{RC}{1+sRC}$ (d) $\frac{1+sRC}{1-sRC}$

8. The transfer function of the system shown in Figure T2 is

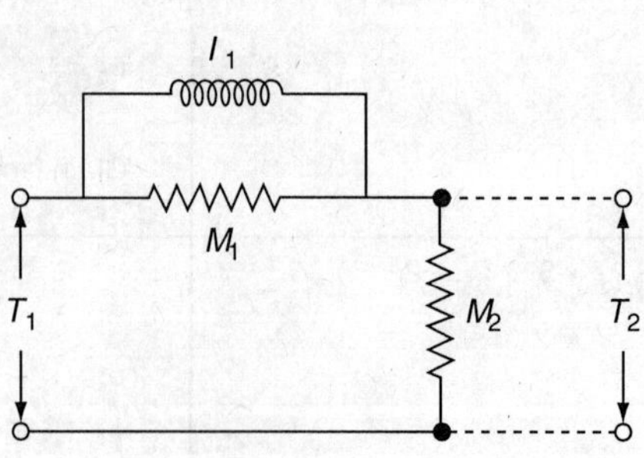

Figure T2

(a) $\dfrac{R_1 + sL}{sL\left(\dfrac{R_1 + R_2}{R_2}\right) + R_1}$ (b) $\dfrac{R_2}{\dfrac{1 + R_1 sL}{R_2} + sL}$ (c) $\dfrac{s - R_1/L}{s + \dfrac{R_1 R_2}{L}}$ (d) $\dfrac{s + L/R_2}{s + \dfrac{R_1 L}{R_2}}$

9. A system has the following transfer function:

$$G(s) = \frac{100(s+5)(s+50)}{s^4(s+10)(s^2+3s+10)}$$

The type and order of the system are, respectively,

(a) 4 and 9 (b) 4 and 7 (c) 5 and 7 (d) 7 and 5

10. For the function $X(s) = 1/s(s+1)^3(s+2)$, the residues associated with the simple poles at $s = 0$ and $s = -2$ are, respectively

(a) $\frac{1}{2}$ and $\frac{1}{2}$ (b) 1 and 1 (c) −1 and −1 (d) $-\frac{1}{2}$ and $-\frac{1}{2}$

11. The output of a linear system for a unit step input is given by $t^2 e^{-t}$. The transfer function is given by

(a) $\dfrac{s}{(s+1)^3}$ (b) $\dfrac{2s}{(s+1)^3}$ (c) $\dfrac{1}{s^2(s+1)}$ (d) $\dfrac{2}{s(s+1)^2}$

12. The starting point of the description of a linear system may be system's

(a) transfer function (b) dynamic equations

(c) differential equations (d) any one of (a), (b), and (c)

13.

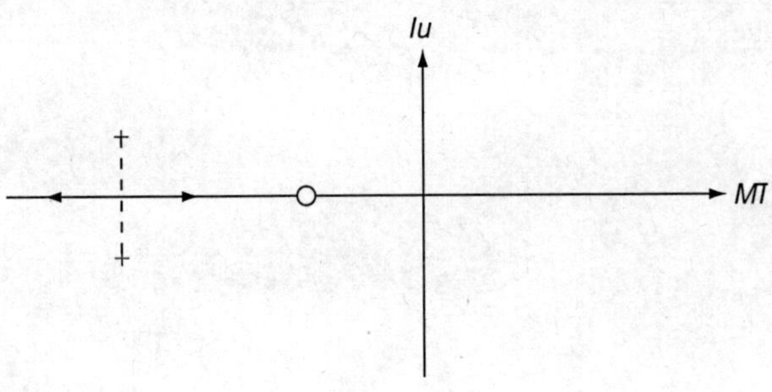

Figure T3

The open loop system has

(a) two complex zeros and a pole (b) two complex poles and a zero

(c) two real poles (d) two complex poles

14. The unit step response of a system function is given by $y(t) = 1 - e^{-8t}$. The transfer function of the system is given by

(a) $\dfrac{8}{s(s+8)}$ (b) $\dfrac{8s}{s+8}$ (c) $\dfrac{8}{s+8}$ (d) $\dfrac{8(s+8)}{s}$

15. The impulse response of a system $G(s) = 2/(s+1)(s+3)$ is given by

(a) $e^{-t} + e^{-2t}$ (b) $e^{-t} + e^{-3t}$ (c) $e^{-3t} + e^{-t}$ (d) $e^{-t} - e^{-2t}$

16. The input-output relationship of a system is given by

$$r(t) = \frac{d^2c(t)}{dt^2} + 3\frac{dc(t)}{dt} + 2c(t)$$

(a) $\frac{1}{s^2+s+2}$ (b) $\frac{1}{s^2+3s+2}$ (c) $\frac{2}{s^2+3s+2}$ (d) $\frac{1}{s^2+2s+2}$

ANSWERS

1 (b)	2 (c)	3 (b)	4 (d)	5 (b)	6 (d)	7 (b)	8 (a)
9 (b)	10 (a)	11 (b)	12 (d)	13 (b)	14 (c)	15 (d)	16 (b)

Chapter 4

Control System Components

Transducer, which is the first major component in a control system, is a device that senses the output in one form and converts it to another form. The sensing may be of temperature, pressure, position and conversion is generally into electrical. We shall consider some commonly used transducers. In this chapter we will introduce control system components.

4.1 Components of Automatic Control System

Error detector, amplifier and controller, actuator (power actuator), plant, and sensor of feedback system are the basic components of an automatic control system. Figure 4.1 shows the block diagram of an automatic control system.

The plant is an open-loop control system. Its output is controlled by closed-loop system. Automatic controller is the combined unit of error detector, amplifier, and controller. The system becomes open-loop system without automatic controller unit.

The reference input becomes an input signal proportional to the desired output in an automatic control system. The feedback signal is a signal, which is proportional to current output of the system.

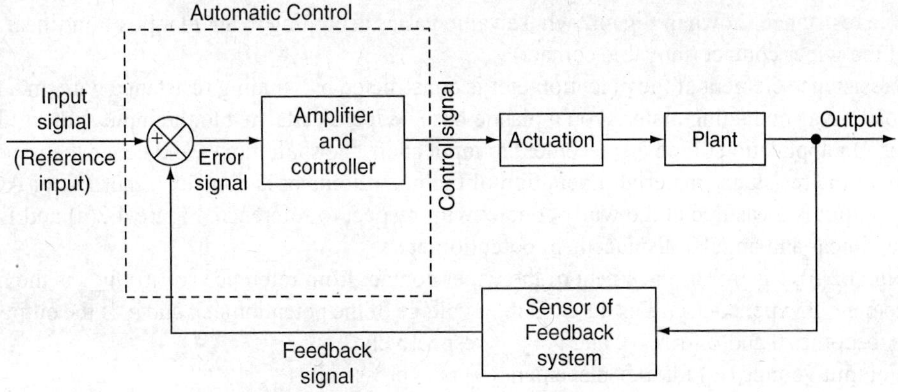

Fig. 4.1 Block diagram of an automatic control system

The function of the error detector is to compare the reference input and feedback signals. An error is produced by it if there is a difference between these two. The examples of error detector are potentiometer, LVDT, synchro, etc. Usually, the error signal is weak. The error signal can be amplified with the help of an amplifier. The controller modifies the error signal to make the control action better. The controller in most of the systems is designed in such a way that it itself amplifies the error signal and integrates or differentiates to produce a control signal. P, PI, PD, and PID controllers are the different types of controllers.

The function of the actuator is to amplify the controller output and to convert the required form of energy, which is acceptable for the plant. The output is dependent on the input of the plant. If there is difference between reference input and feedback signal, the process will be continued. When error signal is zero, the output settles at the desired value. The actuator may be pneumatic motor/valve, hydraulic motor or electric motor. DC servomotor, AC servomotor, and stepper motor are the examples of electric motors used as actuator.

The output is sampled by the feedback system so that a feedback signal is produced. This feedback signal is proportional to the current output. The feedback system also converts the output variable into another suitable variable (displacement, pressure, or voltage) for comparison with the reference input. Sensor and associated circuits/devices are usually in feedback system. Tachogenerators, transducers, etc. are also used as feedback systems. The transducers such as thermocouple photo electric cells generate direct voltages proportional to the quantity, which is to be measured. The DC signal is to be converted to an AC signal at first when the signal is to actuate an AC system. The device transfers the information available in a DC signal to AC signal known as modulator and the conversion process is known as modulation. The control system is superimposed on a high-frequency carrier during modulation. AC tachogenerators produces a modulated output signal. By demodulation, the information can be obtained. Demodulator is a device, which is used to extract information available on a high-frequency carrier and the process is known as demodulation. Power transmitters, levers, and gear trains are also employed sometimes in control system. The function of the power transmitter is to change the voltage and current level of power being transmitted. The function of lever and gear train is to alter the force and torque respectively of the power being transmitted.

4.2 Potentiometers

A device used to convert a linear or angular displacement into a voltage is known as potentiometer. It is a variable resistance shown in Fig. 4.2 whose value varies according to angular position/linear displacement of the wiper contact (movable contact).

The resistance element of the potentiometer is constructed by winding resistance wire on a former or by depositing a conducting material on a plastic base. Wiper is attached to the input shaft of the potentiometer. On application of the displacement to input shaft, the shaft moves and hence the wiper contact slides over the resistance material. Excitation of the potentiometer is done by either DC or AC voltage and the output is measured at the wiper contact with respect to reference. Figure 4.2(a) and Fig. 4.2(b) show the linear and angular displacement potentiometers.

In Fig. 4.2(a), y is the displacement of the wiper contact from reference ($y = 0$), y_{max} is the maximum displacement of wiper contact, E is the excitation voltage of the potentiometer and e_0 is the output voltage. Here, e_0 becomes 0 and E for $y = 0$ and $y = y_{max}$ respectively.

The output voltage (e_0) for a displacement (y) is expressed by

$$e_0 = \left(\frac{E}{y_{max}}\right) y = K_p y \tag{4.1}$$

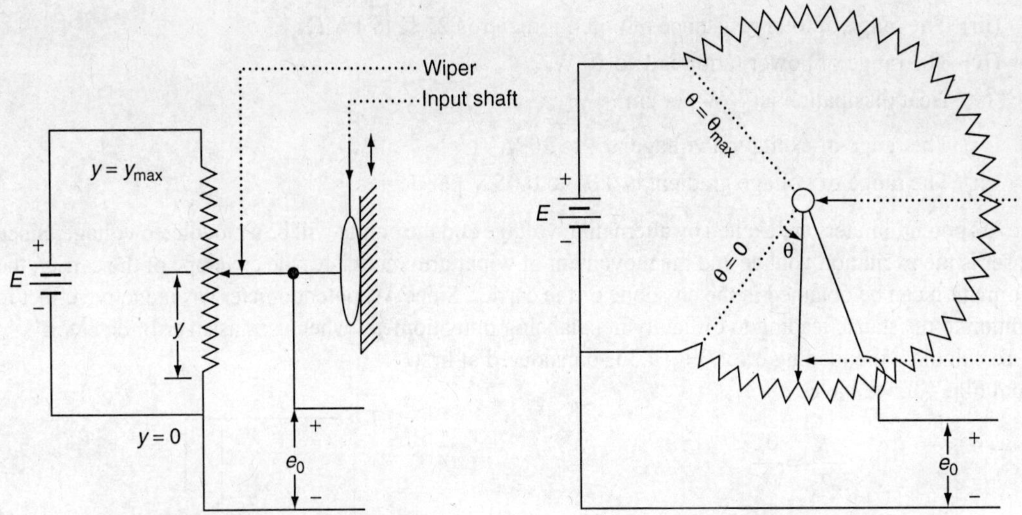

(a) Linear displacement potentiometer (b) Angular displacement potentiometer

Fig. 4.2 Potentiometers

In Eq. (4.1), $K_p = \dfrac{E}{y_{max}}$ is known as the sensitivity of the potentiometer in V/mm.

In Fig. 4.2(b), θ is the angular displacement of wiper arm from reference (θ = 0), θ_{max} is the maximum displacement of wiper arm, E is the excitation voltage of potentiometer, and e_0 is the output voltage. Here, e_0 becomes 0 and E for θ = 0 and θ = θ_{max}, respectively.

The output voltage (e_0) for a displacement (θ) is expressed by

$$e_0 = \left(\frac{E}{\theta_{max}}\right)\theta = k_p\theta \tag{4.2}$$

In Eq. (4.2), $K_p = \dfrac{E}{\theta_{max}}$ is known as the sensitivity of the potentiometer, which is in V/degree. K_p is also known as gain or gain constant of potentiometer.

The following are the characteristics of potentiometer:

(i) The linear variation of resistance is the ideal characteristics of a potentiometer, which can be realized by having very large radius, more number of turns, and high resistance elements.

(ii) The impedance of the device used to measure the output voltage of the potentiometer must be high, so that the loading error can be avoided.

(iii) As the wiper slides over the resistance, the simultaneous contact with adjacent turn is made and hence the discontinuity in output is avoided. The output becomes in the form of staircase steps. The output voltage per step is called resolution, which determines the minimum value of output voltage.

The following are the specifications of potentiometers used in control systems:

(i) The range of turns per unit length is 5 to 30.

(ii) The range of torque required for wiper movement is 1×10^{-3} kg-m to 1×10^{-2} kg-m.

(iii) The range of total resistance of potentiometer is 25 Ω to 1 MΩ.
(iv) The range of power rating is 1 to 10 W.
(v) Heat dissipation is $\frac{1}{2}$ W per cm^2.
(vi) The range of excitation voltage is 4 to 20 V.
(vii) The range of voltage gradient is 0.01 to 0.05 V per degree.

The AC potentiometers are excited by alternating voltage and the output will be a modulated voltage. Since carrier is the excitation voltage and the movement of wiper arm modulates the envelope of the carrier, the information can be obtained in the envelope of the carrier. Since AC potentiometer has inductive effect in addition to resistance leading to difficulty in balancing potentiometer when used as an error detector.

To calculate the loading error, Fig. 4.3 is considered at first.

From Fig. 4.3, we have

$$\frac{R}{R_T} = \frac{\theta_C}{\theta_T} \tag{4.3}$$

i.e.,

$$\frac{R}{R_T} = \frac{\theta_C}{\theta_T} = \frac{E_o}{E_{in}} \tag{4.4}$$

i.e.,

$$E_o = \left(\frac{\theta_C}{\theta_T}\right) E_{in} = \left(\frac{R}{R_T}\right) E_{in} = \text{gain} \times \text{input} \tag{4.5}$$

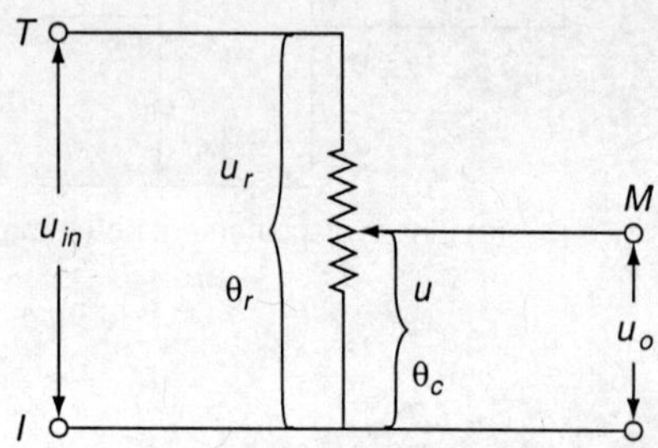

Fig. 4.3 Potentiometer

Generally, the potentiometers are circular resistance elements where one turn is 2π radians and for N turns it is $2\pi N$. The E_{in}/θ_T can be varied by varying θ_T for fixed E_{in}.

Here,

$$\theta_T = 2\pi N$$

$$\text{Gain} = \frac{E_{in}}{2\pi N} \text{ volt/radian} \tag{4.6}$$

The ideal characteristic for a potentiometer is linear variation of resistance with angular position. The per cent resolution is defined by

$$\text{Per cent resolution} = \frac{100}{\text{number of turns of wire}} \tag{4.7}$$

and it is an important factor in determining the minimum value of error signal. Figure 4.4 shows the loading effect of potentiometer.

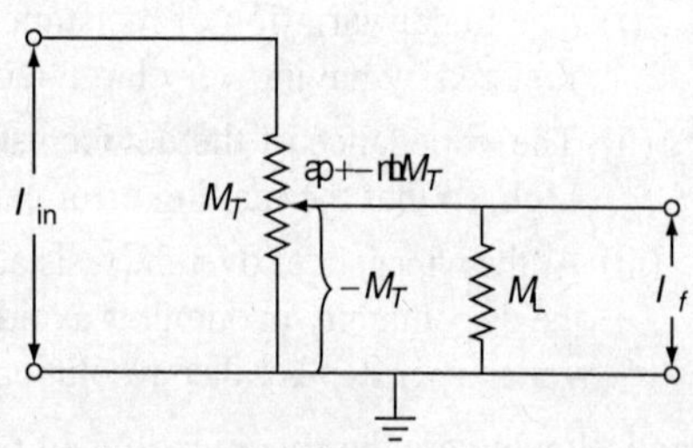

Fig. 4.4 Loading effect of potentiometer

The resistance due to load at an angular position is

$$R = (1-\alpha)R_T + \frac{\alpha R_T R_L}{R_L + \alpha R_T} \tag{4.8}$$

The output voltage is

$$E_o = \frac{\dfrac{\alpha R_T R_L}{R_L + \alpha R_T}}{(1-\alpha)R_T + \dfrac{\alpha R_T R_L}{R_L + \alpha R_T}} \times E_{in}$$

$$= \frac{\alpha R_L}{\alpha(1-\alpha)R_T + R_L - \alpha R_L + \alpha R_L} \times E_{in}$$

$$= \frac{\alpha E_{in}}{1 + \dfrac{\alpha(1-\alpha)R_T}{R_L}} \qquad (4.9)$$

$$\alpha E_{in} - \frac{\alpha E_{in}}{1+\dfrac{\alpha(1-\alpha)R_T}{R_L}} = \frac{\alpha + \alpha^2(1-\alpha)\dfrac{R_T}{R_L} - \alpha}{1+\dfrac{\alpha(1-\alpha)R_T}{R_L}} \times E_{in} = \frac{\alpha^2(1-\alpha)}{\alpha(1-\alpha)+\dfrac{R_L}{R_T}} \times E_{in}$$

Therefore, loading error $= \dfrac{\alpha^2(1-\alpha)}{\alpha(1-\alpha)+\dfrac{R_L}{R_T}}$ (4.10)

There are variations of the single ended potentiometer (Fig. 4.5 and Fig. 4.6). One variation is a double-ended arrangement with the mid-point earthed. Voltages of opposite polarities are applied to

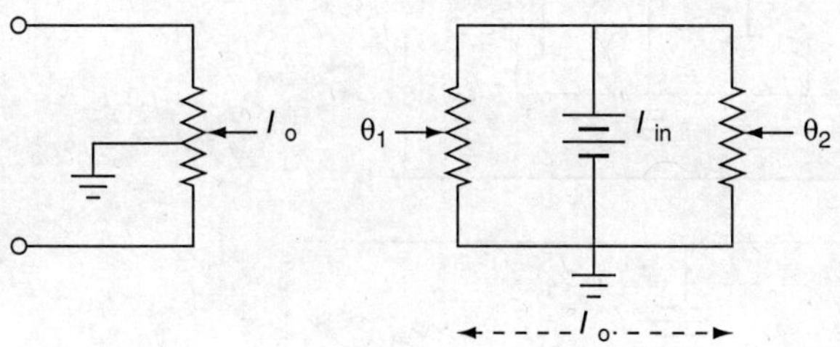

Fig. 4.5 Single-ended potentiometers

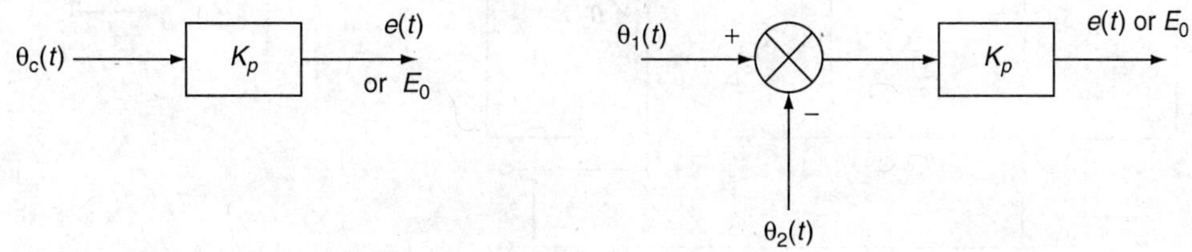

Fig. 4.6 Representation of Fig. 4.5

the ends. The double-ended arrangement can be considered as two single-ended potentiometers with their common point earthed.

$$E_o = \frac{E_{in}}{\theta_T} \times (\theta_1 - \theta_2) \tag{4.11}$$

These types of potentiometers are used in servomechanisms for error control. The potentiometer also acts as an attenuator so that $E_o \leq E_{in}$. For an attenuator, gain constant or potentiometer constant

$$(K_p) = \frac{E_{in}}{\theta_T} = \frac{\text{Input of excitation voltage}}{\text{Total angle of rotation}} \tag{4.12}$$

4.2.1 Applications of Potentiometers

It is possible to convert a mechanical motion proportional to voltage as an error signal with the help of potentiometers. Excitation of a single potentiometer can be made by direct voltage or alternating voltage, an output voltage proportional to displacement of the input shaft can be produced.

Figure 4.7 shows the requirement of two identical potentiometers, so that it can be used as an error detector. In Fig. 4.7, both potentiometers are excited by the same source and the potential is also

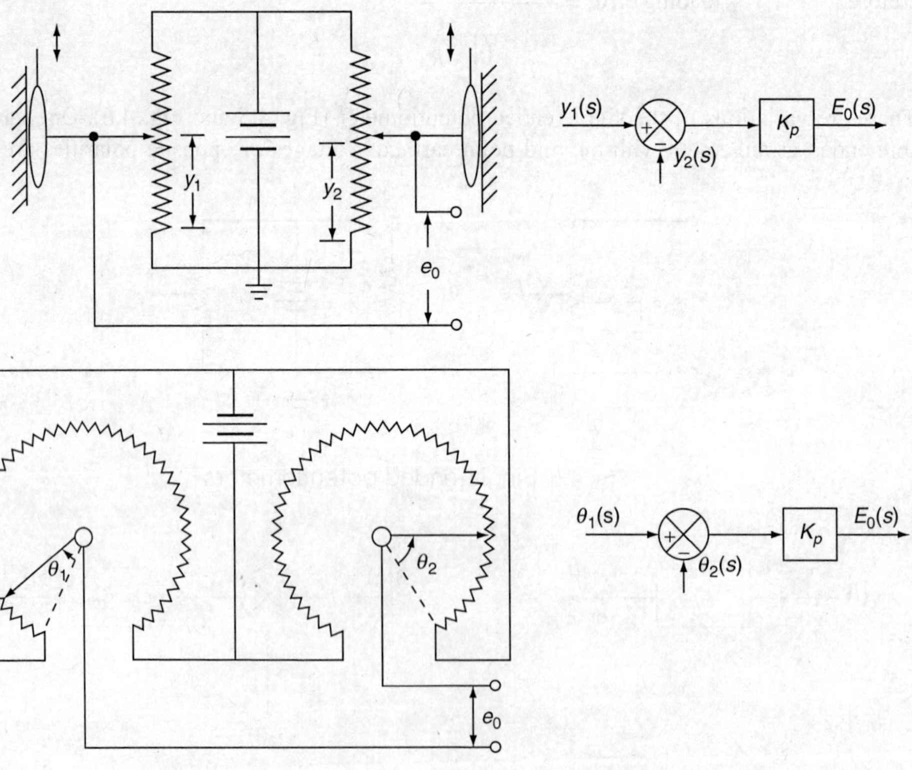

Fig. 4.7 Potentiometers as error detectors

maintained constant. The voltage between two wiper arms becomes zero when the wiper arm of both the potentiometers is in the same position. The position of one wiper arm is maintained as reference input whereas the displacement is applied to the wiper arm of another potentiometer, which is to be compared. The output voltage is measured between two wiper arms and it is proportional to the difference between the displacement of both the wiper arms.

Example 4.1 Find the output voltage of the 1 kΩ potentiometer shown in Fig. E4.1 when excitation voltage is 2 V and setting ratio is 0.4. Calculate the loading error and voltage error also.

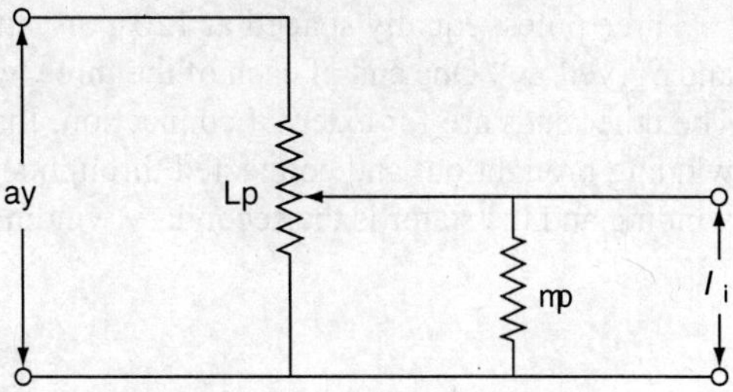

Fig. E4.1

Solution

$$E_o = \frac{\alpha E_{in}}{1+\dfrac{\alpha(1-\alpha)R_T}{R_L}} = \frac{0.4\times 2}{\dfrac{0.4\times(1-0.4)\times 10^3}{5\times 10^3}} = 0.775\text{ V}$$

$$\text{Loading error} = \frac{\alpha^2(1-\alpha)}{\alpha(1-\alpha)+\dfrac{R_L}{R_T}} = \frac{(0.4)^2\times(1-0.4)}{0.4\times(1-0.4)+5} = 0.018$$

Voltage error = 0.018 × 2 = 0.0036 V.

Example 4.2 A helical 5 turn potentiometer has a resistance of 10 kΩ and 9000 winding turns. Find the linearity when the measured resistance at its mid-point setting is 5050 ohm. Determine the resolution of the potentiometer in volts/winding turn and potentiometer constant in volts/radian if 90 V is impressed across it.

Solution

$$\text{Linearity} = \frac{\text{Deviation from nominal}}{\text{Total resistance}}\times 100$$

Deviation from nominal at mid-point is 5050 – 5000 = 50

$$\text{Linearity} = \frac{50}{10000}\times 100 = 0.5\%$$

$$\text{Resolution} = \frac{90}{9000} = 0.01\text{ volts/turn.}$$

$$\text{Potentiometer constant } (K_p) = \frac{E_{in}}{\theta_{max}} = \frac{90}{5\times 2\pi} = 2.865\text{ V/rad}$$

4.3 Synchros

Synchros are electromechanical devices, which produce an output voltage depending on angular position of the rotor and not on rotor speed, and it is different from a DC generator. The trade names for synchros are Selsyn, Antosyn, and Telesyn. The four basic types of synchros are transmitters, receivers, transformers, and differentials.

4.3.1 Transmitter

This is a mechanical to electrical transducer, which consists of a transformer core iron as stator shown in Fig. 4.8. The basic principle of the synchro is that the voltage induced in the secondary depends on the angle at which the magnetic lines of the flux cut the turns. This induced voltage will vary from a maximum of 1 when the two windings are in line or parallel to 0 when they are perpendicular.

Three poles, equally spaced at 120°, are attached to the stator, with windings S_1 (blue), S_2 (black) and S_3 (yellow). One end of each of the three windings is connected internally to each other to form a *Y*. The other ends are for external connection. Inside the core of the stator the rotor rotates with a single winding brought out and connected through slip rings for unlimited rotation. The rotor is the primary winding and the stator is the secondary winding.

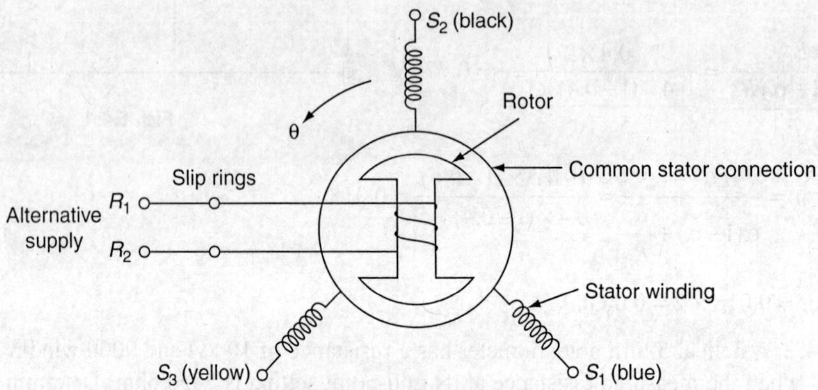

Fig. 4.8 Transmitter

Here the rotor and stator winding S_2 are in line and maximum voltage is induced in S_2 when an alternating voltage of 50 Hz or 400 Hz is fed to the rotor. The voltages induced in S_1 and S_3 will be less. As the rotor rotates, the voltage induced will be

$$E_{s_2} = KE_R \cos\theta \tag{4.13}$$

where E_{s_2} = rms voltage in stator coil 2,
E_R = rms voltage in rotor,
K = proportionality constant depending on magnetic coupling and turns ratio, and
θ = rotor angles in anti-clockwise direction.

The voltages induced in the other two windings S_3 and S_1 will be

$$E_{s_3} = KE_R \cos(\theta - 120°) \tag{4.14}$$

$$E_{s_1} = KE_R \cos(\theta + 120°) \tag{4.15}$$

The voltages between the respective terminals are

$$E_{s_{31}} = \sqrt{3}\,KE_R \sin\theta \tag{4.16}$$

$$E_{s_{12}} = \sqrt{3}\,KE_R \sin(\theta - 120°) \tag{4.17}$$

$$E_{s_{23}} = \sqrt{3}KE_R \sin(\theta + 120°) \tag{4.18}$$

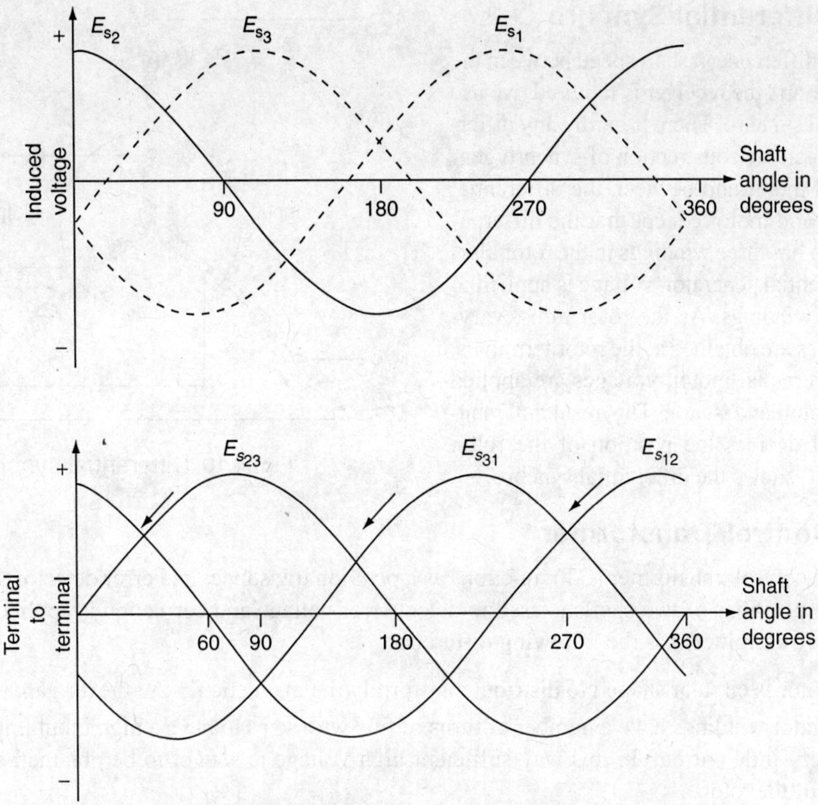

Fig. 4.9 Input as well as terminal to terminal voltage

Equation (4.16), Eq. (4.17), and Eq. (4.18) show the terminal voltages. From the measurement of terminal to terminal voltage, the shaft angle can be noted and for no two shaft positions the voltages are same. The accuracy of synchro is specified in terms of minutes of error and it is about 10. Figure 4.9 shows the input as well as the terminal to terminal voltage.

4.3.2 Synchro Receiver

In synchro transmitter the terminal to terminal voltage is governed by angular position of the rotor. In synchro receiver it is the reverse, that is, voltage to position transducer. The construction is the same except that a heavy flywheel is attached to the shaft of rotor to reduce oscillations. A single-phase voltage is applied to each of the three stator windings and AC excitation to the rotor winding and mutual repulsion gives the angle of the rotor shaft. Initially, the magnetic fields set by the stator and rotor windings cause the rotor to be aligned in line with the stator winding S_2. When the magnitudes of the stator voltages change, the stator current also changes. The resultant stator field changes direction, causing the rotor field to follow it. As the rotor field follows the resultant stator field, the rotor shaft starts to turn. Rotor motion is damped by the flywheel.

4.3.3 Differential Synchro

When the difference in shaft speed between the transmitter and the receiver is required, we use differential synchro. There is hardly any difference between the construction of synchro generator and motor and between the differential generator and motor except that the differential system has three windings in the rotor also. In a differential generator, voltage is applied to the stator windings. As the rotor turns, varying voltages are obtained at the rotor terminals. In the differential motor, voltages are applied to both rotor and stator. The resultant magnetic field decides the position of the rotor. Figure 4.10 shows the differential synchro.

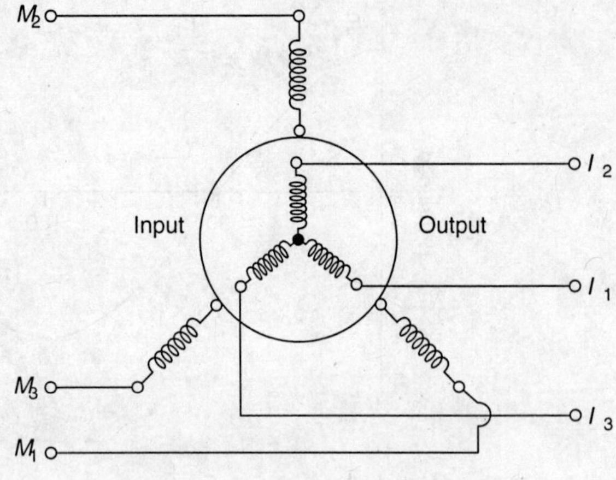

Fig. 4.10 Differential synchro

4.3.4 Control Transformer

A synchro control transformer (CT) functions as a position transducer and error detector. The difference in angular positions of two shafts is used to detect error voltage and correction is applied. The synchro generator (transmitter) has the following difference:

(1) The rotor is circular shaped to distribute uniformly the magnetic flux in the air gap.

(2) The stator windings have a number of turns of fine wire so as to have a high input impedance drawing very little current. In this way sufficient high voltage is sought to be obtained even for slight turns of the rotor.

Figure 4.11 shows a control transformer.

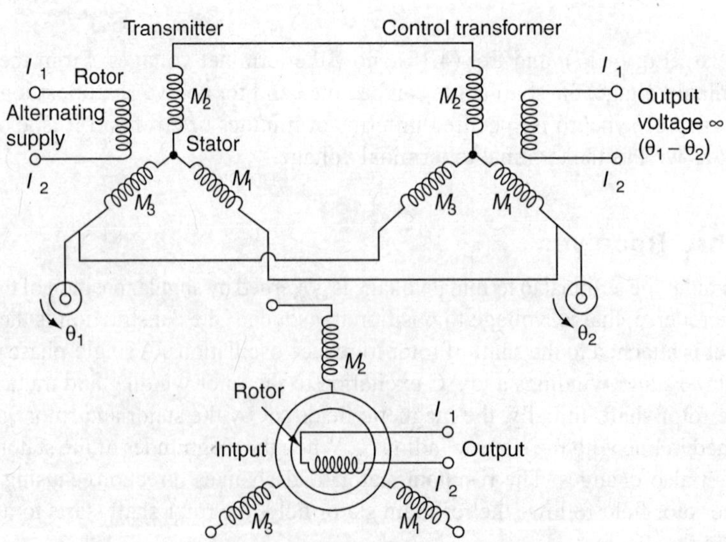

Fig. 4.11 Control transformer

The two inputs of the control transformer are the stator voltage from the synchro transmitter and the position of the two shafts, which are aligned. The stator voltage of the transmitter generated by the rotor of transmitter is also applied to the stator of the control transformer. This sets up a magnetic field within the device. The rotor of the CT is kept horizontal at right angles to the stator winding S_2 so that zero voltage is induced in the rotor. If the aligned shafts of the two synchros are disturbed and rotor moves counter-clockwise, magnetic flux lines are cut and voltage is induced in the rotor coil. The induced voltage will reach a maximum for rotor position of 90° from the horizontal. The output voltage of rotor winding will be sinusoidal and given by

$$E_R = E_{max} \sin(\theta_o - \theta_i) \tag{4.19}$$

where E_{max} is the maximum rotor voltage and $\theta_o - \theta_i$ is the difference in angle between CT and CX. As the output is sinusoidal, it is non-linear but can be considered linear for 20° and less.

Example 4.3 The voltage applied to the rotor of a synchro control transmitter is 28 V rms. The rotor shaft is moved 60° from the zero position. Determine the stator voltages w.r.t. the common stator connection for $k = 1$. Also determine the voltages between terminals S_1 and S_2, S_2 and S_3, S_3 and S_1. Take S_2 as the reference winding.

Solution Since S_2 is the reference stator winding, we have $E_{s_2} = KE_R \cos\theta$, where E_{S_2} and E_R are rms voltages.

Therefore,

$$E_{s_2} = 1 \times 28 \times \cos 60° = 14\,\text{V}$$

$$E_{s_1} = 1 \times 28 \times \cos(60° - 120°) = 14\,\text{V}$$

$$E_{s_3} = 1 \times 28 \times \cos(60° + 120°) = -28\,\text{V}$$

$$E_{s_{31}} = \sqrt{3}\, KE_R \sin\theta = 42\,\text{V}$$

$$E_{s_{12}} = \sqrt{3}\, KE_R \sin(\theta - 120°) = -42\,\text{V}$$

$$E_{s_{23}} = \sqrt{3}\, KE_R \sin(\theta + 120°) = 0\,\text{V}$$

4.4 Linear Transducers

Potentiometer and synchros are examples of rotary position transducers while the core in a transformer is moved up and down for linear variation of position. The core is enclosed by a non-magnetic hollow cylinder with two secondary coils and a single primary coil. In the central position of the magnetic core, there is the cylinder, no voltage is induced as the magnetic coupling to the two secondary windings is equal. When the core is moved up, there is more coupling to the upper secondary and less to the lower secondary with the result that the voltage difference appears in the upper coil. The secondary coils are normally connected in such a way that the voltages induced in them by the primary cancel each other.

4.5 Tachogenerator/Tachometer

An electromechanical device producing an output voltage proportional to its shaft speed is known as tachogenerator. A tachogenerator can be used as an analogue speed indicator, velocity feedback device, or signal integrator. DC and AC tachogenerators are two commonly used tachogenerators.

4.5.1 DC Tachogenerator

A DC tachogenerator is a miniature low-voltage DC generator having linear characteristics. It consists of a stator having a permanent magnetic field and an armature rotating along with a commutator and brush assembly. To measure the speed, the rotor is connected to the shaft. The output voltage of the tachogenerator is proportional to the angular velocity of the shaft. The direction of rotation of the shaft decides the polarity of the voltage. Figure 4.12 shows the schematic representation of a DC tachogenerator.

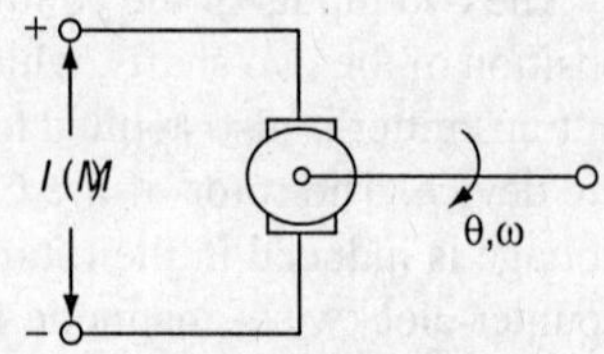

Fig. 4.12 DC tachogenerator

Let the output voltage of the tachogenerator in volts be $e(t)$, angular displacement in radian be θ, angular speed in rad/sec be $\frac{d\theta}{dt}$, and sensitivity of the generator in volts/rad/(rad/sec) be k_t.

The output voltage of the tachogenerator is expressed by

$$e(t) \propto \frac{d\theta}{dt}$$

$$\therefore \quad e(t) = k_t \frac{d\theta}{dt} \tag{4.20}$$

Laplace transform of Eq. (4.20) gives

$$E(s) = k_t\, s\theta(s)$$

$$\therefore \quad \frac{E(s)}{\theta(s)} = sk_t \tag{4.21}$$

Equation (4.21) shows the transfer function of the DC tachogenerator.

The output voltage of the DC tachogenerator varies linearly with the speed of rotation of armature. This is due to the fact that the flux ϕ is constant for the particular construction and this produced by permanent magnets. The generated output voltage is alternating in nature and it is converted to constant voltage by a split commutator. To reduce the generated harmonics, more segments in commutator are used. The tachogenerator is operated at the highest possible speed to get a larger output signal.

In servo system, the tachogenerator is used as a transducer to obtain signals proportional to the speed.

The following are the advantages of a DC tachometer:

(i) Generated voltages have no undesirable phase shifts or wave shapes.
(ii) Temperature is compensated.
(iii) No zero speed voltage error.
(iv) Linear characteristics.

The following are the disadvantages of a DC tachogenerator:

(i) Wear and tear of brushes; hence, it suffers from maintenance problems.
(ii) Commutation sparking gives noise generated voltages.
(iii) High-frequency ripple is generated by the commutator and brushes.
(iv) It needs more torque.

4.5.2 AC Tachogenerator

AC tachogenerator, which resembles a two-phase induction motor, is also known as drug cup tachogenerators. It consists of two-stator winding at right angles to each other shown in Fig. 4.13. It has a squirrel cage rotor. One rotor winding is excited by an alternating reference voltage of fixed amplitude and frequency. The output is taken from the other stator winding.

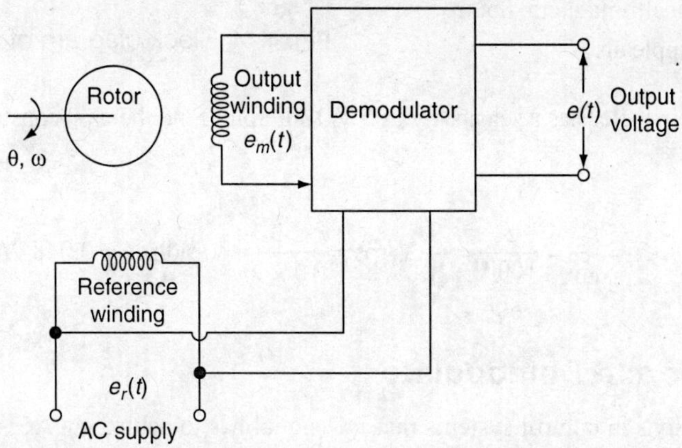

Fig. 4.13 Schematic diagram of AC tachogenerator

The rotor is in the form of a cup of non-magnetic material. When the cup rotates, eddy currents are induced, which generates rotor flux. The rotor flux is added with the reference flux. The direction of the total flux shifted, which results in an induced output voltage that is proportional to velocity. AC tachogenerators have good linearity and accuracy up to 0.1%. In practice output voltage is specified at a speed of 1000 rpm and hence values from 1 to 10 V/1000 rpm are typical for AC tachogenerator.

If a change in the direction of shaft rotation is made, it causes a 180°-phase shift in the output voltage. For an output voltage in phase with the reference, the direction of rotation becomes positive. For an output voltage 180° out of phase, the direction of rotation becomes negative. Therefore, the output voltage of the tachogenerator becomes in modulated form.

If $e_r(t) = E_r \sin \omega_c t$ be the voltage in reference winding, the output voltage can be expressed by $e_m(t) = e_r(t) \sin \omega_c t$, which is to be demodulated to obtain $e(t)$ because the envelope of $e_m(t)$ contains the information about shaft speed. The output voltage of AC tachogenerator is same to the output voltage of DC tachogenerator and it is expressed by

$$e(t) \propto \frac{d\theta}{dt}$$

$$\therefore \quad e(t) = K_T \frac{d\theta}{dt} \tag{4.22}$$

where K_T is the sensitivity of tachogenerator in volts (rad/sec)

The Laplace transform of Eq. (4.22) gives

$$E(s) = s K_T \theta(s)$$

i.e.,

$$\frac{E(s)}{\theta(s)} = s K_T \tag{4.23}$$

Equation (4.23) gives the transfer function of AC tachogenerator.

Figure 4.14 shows the block diagram of AC tachogenerator.

The advantages of AC tachogenerator are shown below.

(i) It is a brushless machine.
(ii) It provides speed feedback.
(iii) It reduces electromagnetic noise.
(iv) It reduces ripple also.

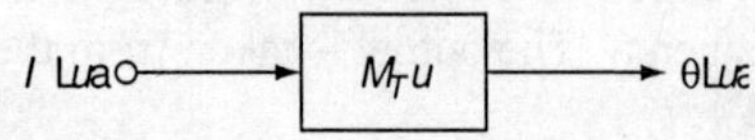

Fig. 4.14 Block diagram of AC tachogenerator

Example 4.4 A tachometer has a sensitivity of 5 V/1000 rpm. Find the gain constant of the tachometer.

Solution

Gain constant $$K_T = \frac{V_g}{\omega} = \frac{5}{\frac{1000}{60}} = \frac{3}{10}\text{ V/rps} = \frac{3}{10 \times 2\pi}\text{ V/rad/sec} = 0.048\text{ V/(rad/sec)}$$

4.6 Modulator and Demodulator

There may be situations in control systems that DC signal has to actuate an AC system and hence DC signal is to be converted to AC signal. The transducers such as thermocouple and photo electric cell produce direct voltages or slowly varying voltages, which are proportional to the quantity to be measured. Hence, these signals must be converted to AC signals before actuating an AC system.

A modulator is a device, which transfers the information available in DC or slowly varying signal to AC signal. This conversion process is known as modulation and in this process, the DC signal on control signal on low-frequency AC signal is superimposed on a high-frequency carrier.

The control system components such as AC tachogenerators and AC potentiometers produce a modulated output signal. The process by which the information on control signal is obtained is known as demodulation, which is the reverse process of modulation. During demodulation, the control signal is extracted from the carrier signal.

4.6.1 Synchronous Vibrator as Modulator

Figure 4.15 shows the schematic diagram of synchronous vibration, which consists of a driving coil, a vibrating reed, and a centre-tapped transformer.

A sinusoidal supply of frequency ω_c is needed to excite the driving coil of the vibrator. The magnetic field is produced by the driving coil. This makes the metal reed to vibrate between two fixed contacts A and B those are connected to the two ends of the primary winding of the transformer.

In Fig. 4.15, the DC error signal is applied between the centre tap of the primary of the transformer and the vibrating reed, which alternately touches A and B. During the contact of reed with A, the DC voltage circulates a current in one half of the primary winding of the transformer, whereas it is in other half of the primary winding of the transformer during the contact of reed with B. Therefore, the vibrating reed creates current reversals in the primary of the transformer where the frequency of the current reversals is equal to that of vibrating frequency. This periodical reversal of current through primary winding causes an alternating voltage, which is to be induced in the secondary. This vibrator is also called chopper because it converts DC voltage to AC voltage. The span of the time during which the vibrating reed rests on the fixed contact is known as dwell time.

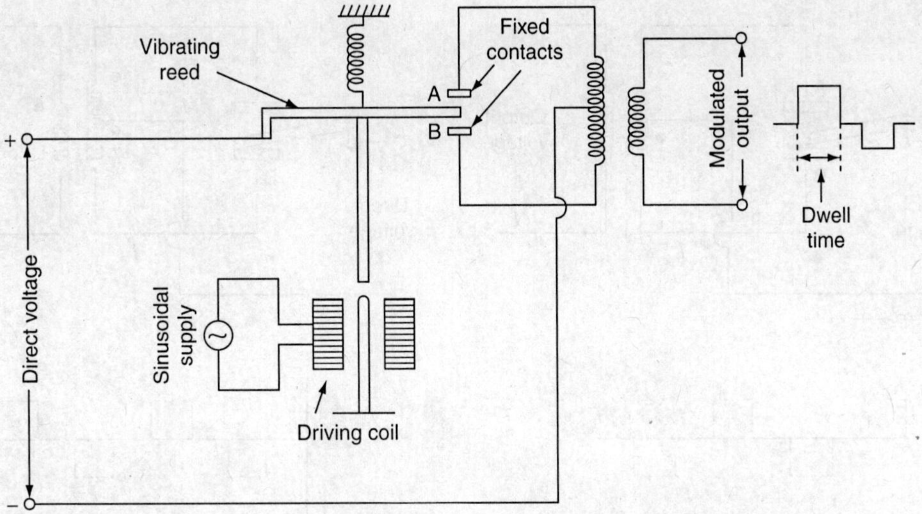

Fig. 4.15 Synchronous vibration as modulator

The time during which the reed moves from one contact to another there will be no primary current and hence output voltage will be zero.

Figure 4.16 shows the polarity of output voltages for the two positions of reed contacts. From Fig. 4.16, it is clear that the output waveform of the modulator is a square wave and the amplitude and phase of waveform are totally dependent on the amplitude and polarity of direct voltage. The output waveform has a phase shift of 180° during the reversal of the direct voltage.

The life time of the vibrator is low due to wear and tear of mechanical contacts. The noise is also produced but the noise level is low.

4.6.2 Diode Modulator

Figure 4.17 shows a diode modulation in which two diodes have been used. A sinusoidal supply of frequency ω_c was used here to excite the primary of the centre-tapped transformer. The centre point of the

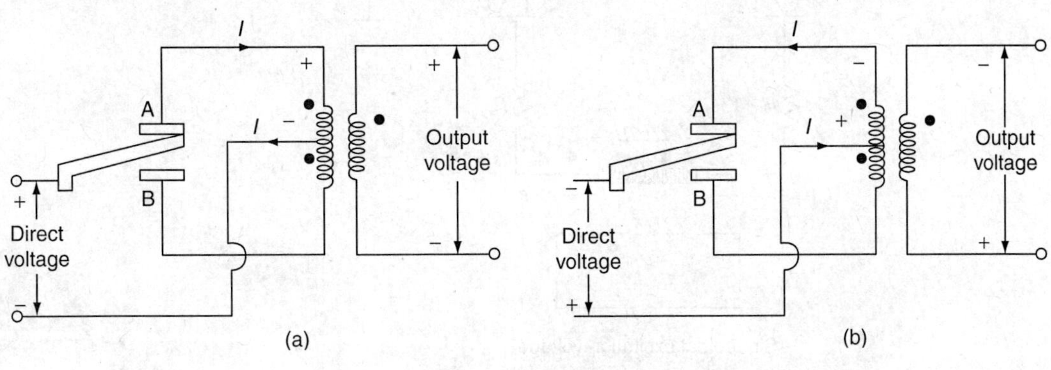

Fig. 4.16 Continued

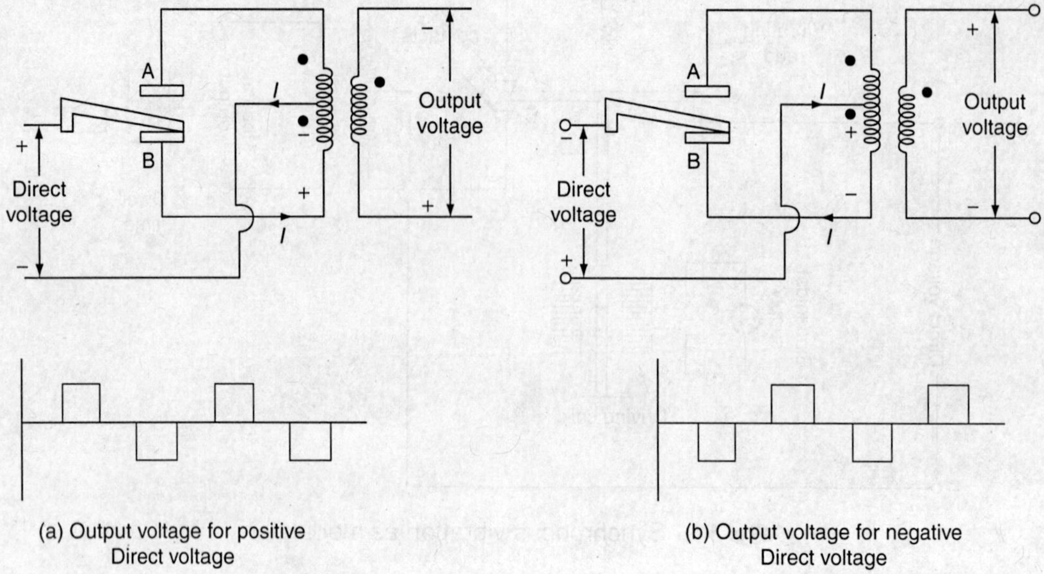

Fig. 4.16 Polarity of output voltages

secondary winding of the transformer is grounded and hence the phase shift of the emfs included in the two halves of secondary will be 180°.

The end s_2 is positive while the end s_1 is negative during the positive half cycle of the supply voltage, which leads the diode D_1 and D_2 to be forward biased. Hence, a current circulates in the secondary of transformer. The voltage between point B and ground becomes zero when the moving contact of

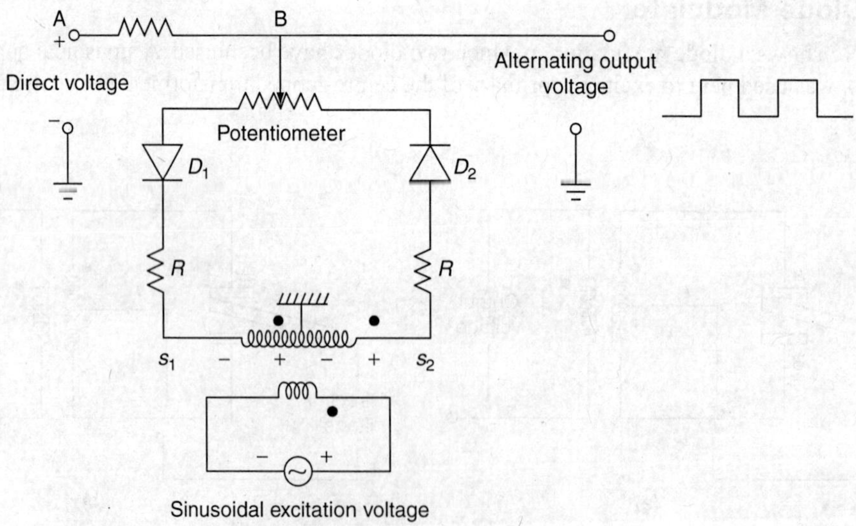

Fig. 4.17 Diode modulator

potentiometer is located exactly at the midpoint. This is valid strictly under the assumption that the diodes are identical and the resistances in series with the diode are identical. Therefore, the output voltage becomes zero.

The end s_2 becomes negative while the end s_1 becomes positive during the negative half cycle of the supply voltage. The diodes D_1 and D_2 become reverse biased. The point B is disconnected from the transformer and it is at the same potential of A. Therefore, the output voltage is equal to the DC voltage applied between point A and ground.

The output voltage is shown in Fig. 4.17, which oscillates between two voltage levels-ground potential and direct voltage and hence it becomes a square wave having frequency same as that of excitation voltage and an amplitude proportional to direct voltage applied at the input.

4.6.3 Diode Bridge Modulator

Figure 4.18 shows a bridge modulator, which consists of four diodes. These diodes are connected such that a bridge is formed. The circuit of bridge modulator consists of two centre-tapped transformers T_1 and T_2. The primary of the centre-tapped transformer T_1 is excited by the sinusoidal voltage having frequency ω_c. This voltage is known as reference voltage. The direct voltage, which needs to be modulated, is applied between the centre point of secondary of the transformer (T_1) and ground. On the other hand, the primary of the transformer (T_2) is centre tapped and the output is available from secondary side of T_2.

The polarity of emf induced in the secondary of T_1 during one half cycle is such that it forward biases the diodes D_1 and D_2. The diodes D_3 and D_4 are reverse biased. The point A becomes at the potential, which is equal to that of DC voltage. The DC voltage is available across the upper half of the transformer T_2 and hence the current flows from P_1 to ground in the primary of T_2, which induces an emf in the secondary of T_2.

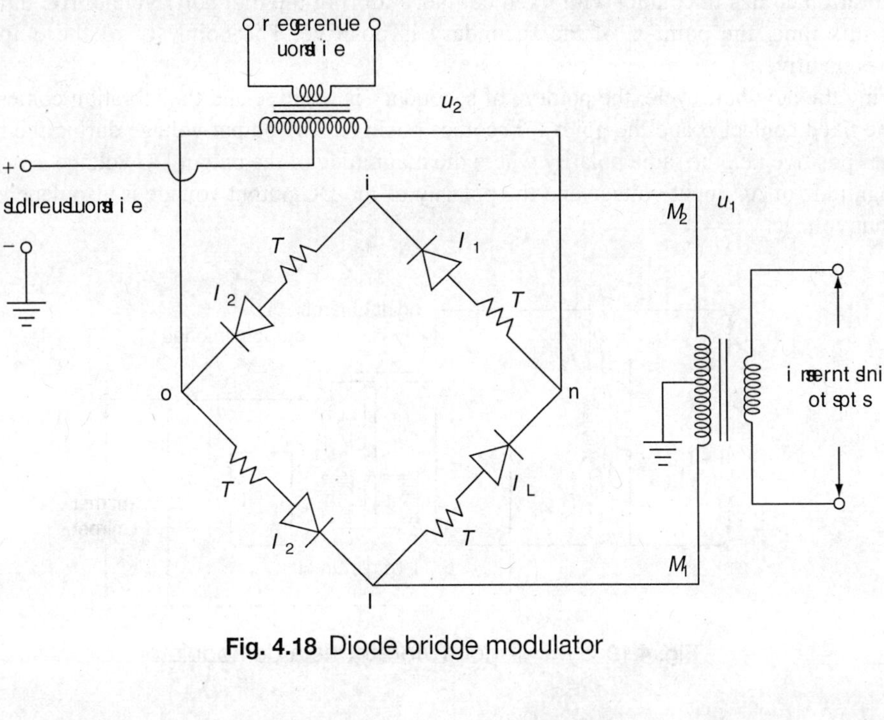

Fig. 4.18 Diode bridge modulator

The polarity of the emf induced in the secondary of T_1 during the next half cycle of reference voltage is such that it forward biases the diodes D_3 and D_4 and reverse biases the diodes D_1 and D_2. This leads to the case that the point B is at the potential same that of direct voltage. The direct voltage is available across the lower half primary winding of T_2. This causes a current to flow from P_2 to ground in the primary of T_2. The direction of flow current in this case is opposite to the previous cycle. This leads to induce an emf in the secondary of T_2 and the polarity of the induced emf is opposite to that of the emf induced during previous cycle.

From above discussion, it is clear that an alternating voltage is developed across the secondary of T_2 due to the circulation of current in the upper and lower halves of primary winding of T_2 in the alternate half cycles of reference voltage. This circulation of current is due to the direct input voltage. The direction of flow of currents is reversed in the alternate half cycles. The magnitude and polarity of alternating output voltage are totally dependent on the magnitude and polarity of DC output voltage, respectively. The frequency of alternating output voltage is same as that of reference voltage.

4.6.4 Synchronous Vibrator as a Demodulator

The reverse process of modulation is known as demodulation. If the modulated alternating voltage is used as input to the transformer, any synchronous vibration will function as demodulator. Figure 4.19 shows a synchronous vibration, which works as a demodulator. The driving coil of this synchronous vibration shown in Fig. 4.19 is excited by an AC signal having frequency same as that of the supply frequency. The secondary winding of the transformer is centre tapped and the DC output is obtained between the vibrator and the centre-tapped transformer, which has been shown in Fig. 4.19.

The vibrator, which is designed to operate in the range of 50–400 Hz, vibrates in synchronism with excitation voltage and the choice of excitation frequency is dependent on the signal frequency. The vibration comes in contact with fixed contact A during the first half cycle of AC input because during this time, the point s_1 of the secondary is positive. The point 'C' of the output voltage becomes positive.

During the next half cycle, the point s_2 of secondary is positive and the vibration comes in contact with the fixed contact B and the point C becomes positive. The output voltage during the both halves becomes positive i.e., the same polarity where the magnitude of the output DC voltage is dependent on the magnitude of AC input voltage and the polarity of the DC output voltage is also dependent on the AC input voltage.

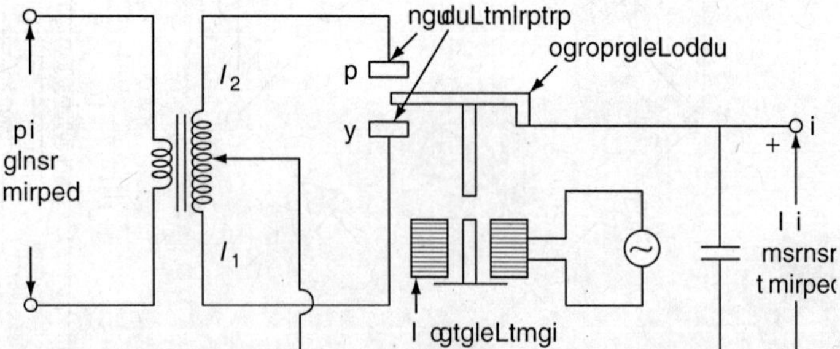

Fig. 4.19 Synchronous vibrator as a demodulator

4.6.5 Diode Demodulator

Figure 4.20 shows a diode modulator in which the excitation voltage is alternating having frequency same as that of input signal. The input signal has been applied to the primary of the transforms (T_1).

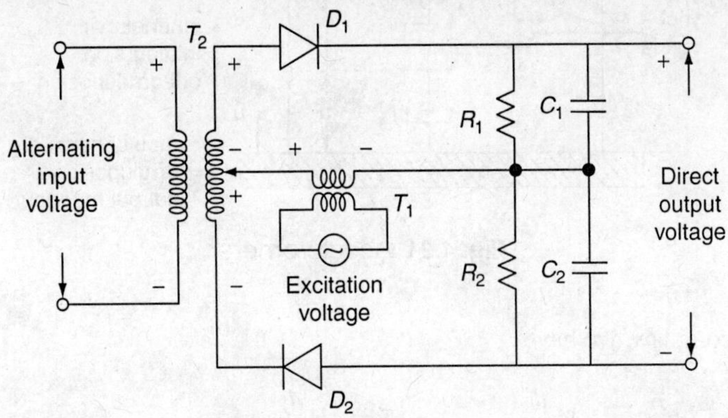

Fig. 4.20 Diode demodulator

The input voltage is assumed to be in phase with the excitation voltage. During positive half cycle, the diodes D_1 and D_2 will conduct due to the nature of the polarities of the induced emfs in the secondary of T_2 as shown in Fig. 4.20. The available voltage across D_1 is more than that of D_2 because there exists an additive secondary voltage in the closed path of D_1 and subtractive voltage in the closed path of D_2. Hence, the diode D_1 conducts more current than D_2 resulting in greater charge storage in C_1 compared to that of C_2. The resulting polarity has been shown in Fig. 4.20.

During the negative half cycle, the diodes D_1 and D_2 will be reverse biased because the polarity of the secondary emfs is exactly opposite to the polarity shown in Fig. 4.20. The capacitors C_1 and C_2 in this will function as filter and hence the direct output voltage will be maintained constant. The excitation voltage in the secondary of T_1 must be greater than the voltage induced in one half of secondary of T_1 to make the operation successful.

The diode D_2 conducts more current than the diode D_1 if the alternating voltage applied at the input is in phase opposition with the excitation voltage. In this case, the polarity of direct output voltage is opposite to the polarity shown in Fig. 4.20. The capacitors maintain the same direct voltage levels during the reactive half cycles.

4.7 Accelerometer

An accelerometer is also a transducer, which is used in high-speed aircrafts and space vehicles. The generated signal is proportional to linear acceleration along its sensitive axis, where the accelerometer is fixed to the object whose acceleration is to be measured and hence undergoes the same acceleration as the body. Mass M is attached by a spring to the case and the other end of the mass is connected to viscous dashpot by a shaft, which opposes the motion of the mass. If the body has acceleration along the sensing axis, the mass is displaced from its equilibrium position towards y, which is opposed by the dashpot and the spring. The mass tries to restore it to the original position. When the acceleration of the object and mass are equal, the displacement y is measured by a linear variable differential transformer (LVDT) or a linear potentiometer, which gives an output voltage e_0 proportional to displacement. Figure 4.21 shows an accelerometer.

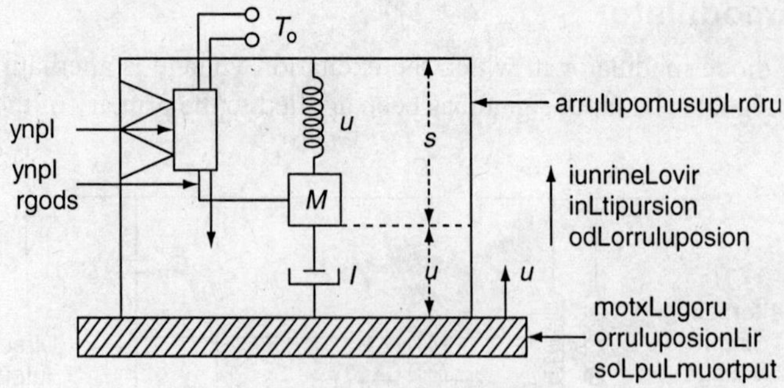

Fig. 4.21 Accelerometer

From Newton's second law, we have

$$M\frac{d^2(y-x)}{dt^2} = -Ky - D\frac{dy}{dt}$$

i.e.,
$$M\frac{d^2x}{dt^2} = M\frac{d^2y}{dt^2} + Ky + D\frac{dy}{dt}$$

i.e.,
$$Ma = M\frac{d^2y}{dt^2} + D\frac{dy}{dt} + Ky$$

i.e.,
$$a = \frac{d^2y}{dt^2} + \frac{D}{M}\frac{dy}{dt} + \frac{K}{M}y \tag{4.24}$$

where 'a' is the acceleration of the body and $(y - x)$ is the displacement from the initial position. To obtain the transfer function, we take Laplace transform of both sides of Eq. (4.24). We have

$$A(s) = s^2Y(s) + \frac{D}{M}sY(s) + \frac{K}{M}Y(s)$$

Input is acceleration, output is displacement.

$$G(s) = \frac{Y(s)}{A(s)} = \frac{1}{s^2 + \frac{D}{M}s + \frac{K}{M}} \tag{4.25}$$

This displacement is converted into a voltage E_o by the LVDT

$$E_o(s) = \frac{1}{K_l}Y(s) \tag{4.26}$$

where K_l is the proportionality constant of LVDT, Substituting Eq. (4.26) in Eq. (4.25), we have

$$G(s) = \frac{E_o(s)}{A(s)} = \frac{K_l}{s^2 + \frac{D}{M}s + \frac{K}{M}} \tag{4.27}$$

Figure 4.22 shows the block diagram of an accelerometer. Equation (4.27) is a second-order differential equation from which we have

$$2\xi\omega_n = \frac{D}{M}$$

and

$$\omega_n = \sqrt{\frac{K}{M}}$$

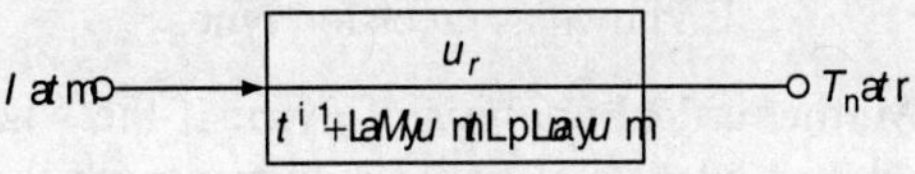

Fig. 4.22 Block diagram of accelerometer

4.8 Miscellaneous Transducers

4.8.1 Strain Gauge

It is used to measure force or weight conditions in machine tools, rolling units, pipes, aircraft structures, etc. It is based on the principle that when a wire stretches under strain, its resistance changes.
The gauge factor is defined below.

$$\text{Gauge factor} = \frac{\text{Resistance change}}{\text{Change in length}} = \frac{\frac{\Delta R}{R}}{\frac{\Delta L}{L}} \tag{4.28}$$

It is generally between 2 and 5.

4.8.2 Temperature Transducers

Temperature transducers convert temperature variations to an equivalent electrical signal. It is based on the principle that resistance changes with temperature represented by

$$R_t = R_0[1+\alpha(T_1 - 20)] \tag{4.29}$$

where 'α' is the temperature coefficient of resistance at 20°C, R_t is the resistance at T_1°C, and R_0 is resistance at 20°C.

There are two types of measuring instruments (1) resistance thermometer, (2) thermocouple. Materials used for resistance thermometer bulbs are platinum, nickel, and copper. Thermocouple junction materials are copper-constantan, iron-constantan, chromel-alumel, and platinum-rhodium.

4.8.3 Pressure Transducers

A pressure transducer converts the input pressure to voltage at the output. There are two types of pressure transducers: bourdon type and diaphragm type. Bourdon type is for pressures up to 10,000 lb/sq. in. and diaphragm type is for 30 lb/sq. in.

AC-type bourdon tube begins to straighten out on increase in pressure. The tube is connected to rotary shaft, which gives as output voltage. Bourdon tubes are used in aircrafts for measuring altitude as per barometric pressure.

4.8.4 Gyroscope

Gyroscope is an instrument used in space ships and aircrafts. The input is the angular velocity and the output is the angular displacement. The action of gyroscope is based on the following two physical principles:

(i) If no external torque acts on it, the spinning wheel maintains the direction of its spin axis in space, and this type of spinning is known as free gyro type.

(ii) If torque is applied to an axis inclined to the spin axis of a wheel, the wheel rotates about an axis at an angle 90° to both the spin axis as well as the input torque axis. This type of rotation is known as precision type.

Mathematical analysis of gyro: If the wheel spin with angular velocity ω_1 is aligned with the x-axis initially, the moment about the spin axis is given by

$$H = J_x \omega_1 \tag{4.30}$$

In Eq. (4.30), J_x is the moment of inertia about the spin axis. Let J_z and J_y be the moment of inertia about the input axis (z-axis) and the output axis (y-axis) respectively. If torque T be applied in the direction of the z-axis, it changes the momentum of the wheel by an amount ΔH in the direction of T in time Δt shown in Fig. 4.23 below.

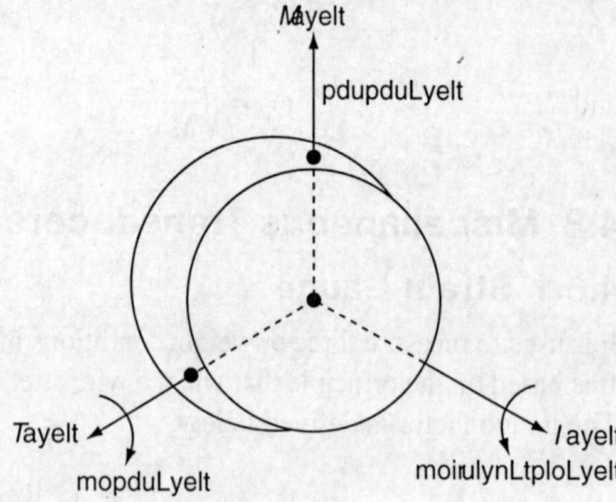

Fig. 4.23 Different Axes

If the spin axis turns by an angle $-\Delta\alpha$ in time Δt, we can write from Fig. 4.24

$$\tan(-\Delta\alpha) = -\Delta\alpha = \frac{\Delta H}{H} = \frac{T}{H}\Delta t \quad \text{[from Fig. 4.23]}$$

i.e., $$\frac{d\alpha}{dt} = \frac{-T}{H}$$

i.e., $$T = -H\dot{\alpha} \tag{4.31}$$

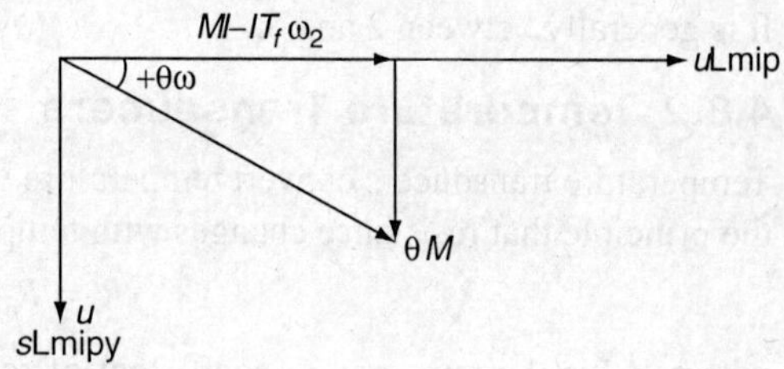

Fig. 4.24 Spin axis turned by an angle

The required torque to accelerate the wheel about the z-axis is expressed by

$$T = -H\dot{\alpha} + J_z \frac{d^2\beta}{dt^2} = -H\dot{\alpha} + J_z\ddot{\beta} \tag{4.32}$$

Here the wheel turns about the z-axis by an angle β. If an undesired disturbance torque is present at the y-axis, i.e., the output axis, the gyro response due to this torque can be determined by assuming the angular movement about the spin axis is very small. Therefore, we have

$$\Delta H = T_1 \Delta t$$

From Fig. 4.25, we have

$$\tan\Delta\beta = \Delta\beta = \frac{\Delta H}{H} = \frac{T_1}{H}\Delta t$$

i.e., $$\frac{d\beta}{dt} = \frac{T_1}{H}$$

i.e., $$H\dot{\beta} = T_1 \tag{4.33}$$

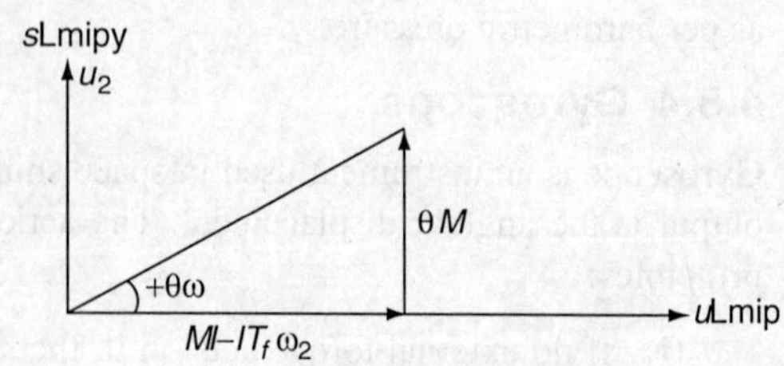

Fig. 4.25 z and x-axis

If we take into account the moment of inertia, the above equation becomes

$$T_1 = H\dot{\beta} + J_y \frac{d^2\alpha}{dt^2} = H\dot{\beta} + J_y\ddot{\alpha} \tag{4.34}$$

The wheel moves about the y-axis by an angle α.

Taking into account the effect of damping coefficient D and spring constant K, we have

$$T_1 = H\dot{\beta} + J_y\ddot{\alpha} + D\dot{\alpha} + K\alpha \tag{4.35}$$

The two final equations are

$$T = -H\dot{\alpha} + J_z\ddot{\beta} \tag{4.36}$$

$$T_1 = H\dot{\beta} + J_y\ddot{\alpha} + D\dot{\alpha} + K\alpha \tag{4.37}$$

Taking Laplace transform of Eq. (4.36) and Eq. (4.37), we have

$$T(s) = -sH\alpha(s) + s^2 J_z\beta(s) \tag{4.38}$$

i.e.,
$$HT(s) = -sH^2\alpha(s) + s^2 HJ_z\beta(s) \tag{4.39}$$

and
$$T_1(s) = sH\beta(s) + (J_y s^2 + Ds + K)\alpha(s) \tag{4.40}$$

i.e.,
$$sJ_z T_1(s) = s^2 J_z H\beta(s) + sJ_z(J_y s^2 + Ds + K)\alpha(s) \tag{4.41}$$

From Eq. (4.39) and Eq. (4.41), we have

$$-HT(s) + sJ_z T_1(s) = sJ_z(J_y s^2 + Ds + K)\alpha(s) + sH^2\alpha(s)$$

i.e.,
$$\alpha(s) = \frac{sJ_z T_1(s) - HT(s)}{sH^2 + sJ_z(J_y s^2 + Ds + K)} = \frac{sJ_z T_1(s) - HT(s)}{s[H^2 + J_z(J_y s^2 + Ds + K)]}$$

i.e.,
$$\alpha(s) = \frac{sT_1(s) - \dfrac{H}{J_z}T(s)}{s\left[J_y s^2 + Ds + K + \dfrac{H^2}{J_z}\right]} \tag{4.42}$$

Again
$$T(s) = -sH\alpha(s) + s^2 J_z\beta(s)$$

i.e.,
$$T(s) = -sH\left[\frac{sT_1(s) - \dfrac{H}{J_z}T(s)}{s\left(J_y s^2 + Ds + K + \dfrac{H^2}{J_z}\right)}\right] + s^2 J_z\beta(s)$$

i.e.,
$$s^2 J_z\beta(s) = \frac{T(s)\left[J_y s^2 + Ds + K + \dfrac{H^2}{J_z}\right] + sHT_1(s) - \dfrac{H^2}{J_z}T(s)}{\left[J_y s^2 + Ds + K + \dfrac{H^2}{J_z}\right]}$$

i.e., $$s^2 J_z \beta(s) = \frac{T(s)(J_y s^2 + Ds + K) + sHT_1(s)}{\left[J_y s^2 + Ds + K + \dfrac{H^2}{J_z}\right]}$$

i.e., $$\beta(s) = \frac{T(s)(J_y s^2 + Ds + K) + sHT_1(s)}{s^2 J_z \left[J_y s^2 + Ds + K + \dfrac{H^2}{J_z}\right]} = \frac{T(s)\left(\dfrac{J_y s^2 + Ds + K}{J_z}\right) + \dfrac{sH}{J_z} T_1(s)}{s^2 \left[J_y s^2 + Ds + K + \dfrac{H^2}{J_z}\right]} \tag{4.43}$$

If disturbing torque T_1 does not exist, we have

$$\alpha(s) = \frac{-\dfrac{H}{J_z} T(s)}{s\left[J_y s^2 + Ds + K + \dfrac{H^2}{J_z}\right]}$$

i.e., $$\frac{\alpha(s)}{T(s)} = \frac{\dfrac{-H}{J_z}}{s\left[J_y s^2 + Ds + K + \dfrac{H^2}{J_z}\right]} \tag{4.44}$$

Case 1: If J_y, D, K and $T_1(s)$ are zero, the gyro is termed a **free gyro**. From Eq. (4.42), we have

$$\alpha(s) = \frac{-\dfrac{H}{J_z} T(s)}{s\left(\dfrac{H^2}{J_z}\right)} = -\frac{T(s)}{sH}$$

i.e., $$s\alpha(s) = -\frac{T(s)}{H}$$

i.e., $$\frac{d\alpha}{dt} = -\frac{T}{H} \tag{4.45}$$

From Eq. (4.43), we can calculate the disturbance torque at the output y-axis assuming $T(s) = 0$ and J_y, D, and K become zero.

$$\beta(s) = \frac{\dfrac{sH}{J_z} T_1(s)}{s^2\left(\dfrac{H^2}{J_z}\right)} = \frac{T_1(s)}{sH}$$

i.e., $$s\beta(s) = \frac{T_1(s)}{H}$$

i.e., $$\frac{d\beta}{dt} = \frac{T_1}{H} \tag{4.46}$$

Case 2: The gyro system is termed as rate gyro if the disturbance torque is zero and the spring constant becomes very large in the output axis. From Eq. (4.40), We have

$$0 = sH\beta(s) + K\alpha(s)$$

i.e.,
$$-sH\beta(s) = K\alpha(s)$$

i.e.,
$$\alpha(s) = -\frac{H}{K}s\beta(s)$$

i.e.,
$$\alpha = -\left(\frac{H}{K}\right)\frac{d\beta}{dt} \tag{4.47}$$

Case 3: In restraint gyro there is single degree of freedom with respect to the frame and only y axis is free to move. For restraint gyro $T_1(s) = 0$ and K is negligible compared to D. From Eq. (4.40), we have

$$0 = sH\beta(s) + (J_y s^2 + Ds)\alpha(s)$$

i.e.,
$$\alpha(s) = -\frac{sH\beta(s)}{(J_y s^2 + Ds)} = -\frac{H\beta(s)}{D\left(\frac{J_y}{D}s + 1\right)} \tag{4.48}$$

Now $(J_y/D) \to 0$ because D is very large.

Therefore,
$$\alpha(s) = -\frac{H}{D}\beta(s)$$

i.e.,
$$\alpha = -\frac{H}{D}\beta \tag{4.49}$$

The free gyro has how accuracy while the rate gyro is very rugged and very much accurate.

The practical gyros are of two types such as free gyro and restrained gyro. There are no restraining forces in any direction of the free gyro. The gyro wheel is fixed in space. Hence, the angular motion of supporting frame is not transmitted to gyro wheel. Therefore, it measures the angular motions of the vehicle with respect to the gyro as a reference. The usage of free gyros is in automatic plotting, inertial navigational equipment, artificial horizons, stable platform, etc.

The restrained gyros are of two types, which are rate gyro or integrating gyro. Using restrained gyros, it is possible to provide either derivative on integral feedback from roll, pitch, or yaw to damp out vehicle oscillations about these axes.

Table 4.1 shows the summary of transducers.

Table 4.1 Summary of transducers

Variable	Transducer
Acceleration	Mass
Altitude	Gyroscope
Displacement	Potentiometer, Synchro, LVDT
Fluid flow	Photovoltaic cell, photoconductive cell, phototransistor
Liquid level	Float, resistance or capacitive probes, optical
Pressure	Bourdon tube, diaphragm, manometer strain gauge
Temperature	Resistance thermometer, thermocouple
Velocity	Tachometer

4.9 Power Actuators

Different types of power actuators are discussed below.

4.9.1 DC Generator

A generator, which is based on Faraday's Law of induction of voltage by a rotating conductor in a magnetic field, converts mechanical energy to electrical energy. The basic equation is

$$e_o = k_g Blv \quad \text{(if linear)}$$

and $$e_o = k_g Bl\omega \quad \text{(if angular)} \tag{4.50}$$

where 'k_g' is the generator constant. For a generator, B and l are constant so that the output voltage is proportional to angular velocity ω. Figure 4.26 shows a DC generator.

Generators are used as control elements in control systems rather than as transducers.

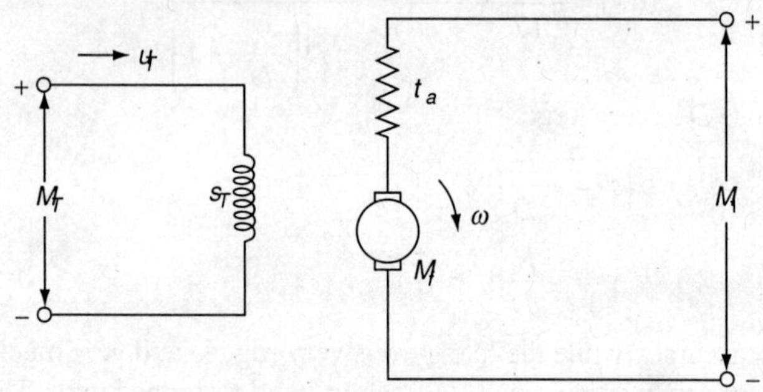

Fig. 4.26 DC generator

4.9.2 DC Motor

DC motor, which converts electrical energy to mechanical energy, is an important transducer. The DC motor is the opposite of a generator, though the DC motor can be used both as a generator and as a motor. An AC motor has the following disadvantages: (1) non-linear characteristic and (2) difficult to control its position through continuous speed or torque change. These disadvantages are absent in DC motors. The disadvantages of DC motors are as follows:

(i) Maximum torque is limited by commutation and not by heating as in other motors.

(ii) Contact between the brush and commutator is by arcing and high currents to armature to another segment of the commutator. This sparking, in course of time, causes pitting and has advanced a lot to reduce wear and tear and maintenance costs.

(iii) The brush and commutator requirement makes the DC motor expensive.

(iv) A basic requirement for a motor is high torque and low inertia (mass) or a high torque to inertia ratio. This is achieved by reducing the diameter and increasing the length. Again, modern technology has produced low-weight rotors and ironless rotors, i.e., electronic rotors.

4.9.3 Excitation of DC Motors

Figure 4.27 shows armature and field control of a DC motor. Table 4.2 compares armature and field control of a DC motor.

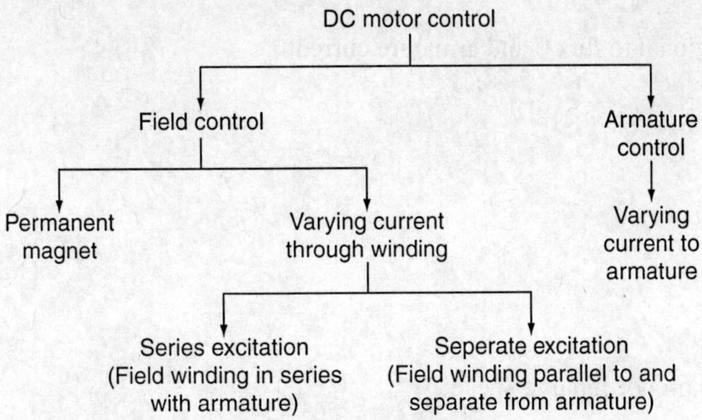

Fig. 4.27 Armature and field control of a DC motor

Table 4.2 Comparison of armature control and field control of a DC motor

Armature control	Field control
Large current is required as the source has to meet the full power requirements of the motor	Requirement of current is small
Back emf provides the damping	Damping has to be provided by the motor and load
Time constant on account of inductance of winding is less	Time constant will be more because large winding is required to produce the necessary flux
Speed of response of the motor to changing current is fast	Speed of response to changing current is slow
It is easy to provide constant field current	It is difficult to provide constant current to armature because of back emf
Efficiency is high	Efficiency is low

4.9.3.1 Armature-Controlled DC Motor

Figure 4.28 shows an armature-controlled DC motor. Control signal is applied to the armature terminals. Let

R_a be the armature resistance is ohms
L_a be the inductance of armature in henry,
i_a be the armature current in ampere,
i_f be the field current in ampere,
e_a be the applied voltage in volt,
e_b be the back emf in volt,
T be the torque in N-m,
J, f be the inertia in kg-m^2 and friction parameters in N-m/rad/sec of the motor and load, respectively.

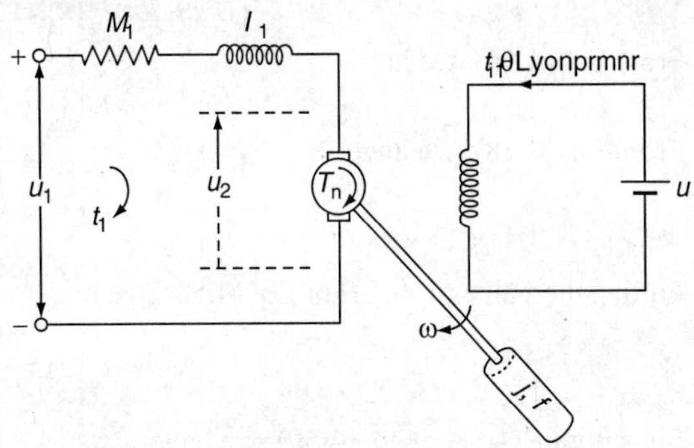

Fig. 4.28 Armature-controlled DC motor

Now,

(i) Air gap flux ϕ is proportional to i_f

$\therefore \quad \phi \propto i_f$

i.e., $\quad \phi = k_f i_f \qquad (4.51)$

(ii) Torque is proportional to flux ϕ and armature current i_a:

$\therefore \quad T \propto \phi i_a$

i.e., $\quad T \propto K_f i_f i_a$

i.e., $\quad T = K_l K_f i_f i_a$

$$\therefore \quad T = K_T i_a \qquad (4.52)$$

$$\text{where} \quad K_T = K_l K_f i_f \qquad (4.53)$$

(iii) Back emf (e_b) is proportional to speed:

$$\therefore \quad e_b \propto \frac{d\theta}{dt}$$

$$\text{i.e.,} \quad e_b = K_b \frac{d\theta}{dt} \qquad (4.54)$$

Applying KVL to the armature circuit, we have

$$e_a = i_a R_a + L_a \frac{di_a}{dt} + e_b \qquad (4.55)$$

The load torque equation is expressed by

$$J\frac{d^2\theta}{dt^2} + f\frac{d\theta}{dt} = T_m = K_T i_a \qquad (4.56)$$

To find $\theta(s)/E_a(s)$, let us take Laplace transform of Eq. (4.54), Eq. (4.55) and Eq. (4.56)

From Eq. (4.54), we have

$$E_b(s) = K_b s\theta(s) \qquad (4.57)$$

From Eq. (4.55), we have

$$(L_a s + R_a)I_a(s) = E_a(s) - E_b(s) \qquad (4.58)$$

From Eq. (1.56), wc have

$$(Js^2 + fs)\theta(s) = T_m(s) = K_T I_a(s) \qquad (4.59)$$

From Eq. (4.58), we have

$$I_a(s) = \frac{E_a(s) - E_b(s)}{sL_a + R_a} \qquad (4.60)$$

Putting the value of $E_b(s)$ in Eq. (4.58), we have

$$I_a(s) = \frac{E_a(s) - K_b s\theta(s)}{sL_a + R_a} \qquad (4.61)$$

Substituting $I_a(s)$ in eq. (4.59), we have

$$(Js^2 + fs)\theta(s) = \left[\frac{E_a(s) - K_b s\theta(s)}{sL_a + R_a}\right] K_T$$

i.e.,
$$\left[\frac{(sL_a + R_a)(Js^2 + fs)}{K_T} + K_b s\right]\theta(s) = E_a(s)$$

i.e.,
$$\left[\frac{(sL_a + R_a)(Js^2 + fs) + K_T K_b s}{K_T}\right]\theta(s) = E_a(s)$$

i.e.,
$$\frac{\theta(s)}{E_a(s)} = \frac{K_T}{[(sL_a + R_a)(Js + f) + K_T K_b]s} \tag{4.62}$$

From Eq. (4.62) we have
$$\frac{\theta(s)}{E_a(s)} = \left[\frac{\dfrac{K_T}{(sL_a + R_a)(Js + f)}}{1 + \dfrac{K_T K_b}{(sL_a + R_a)(Js + f)}}\right]\frac{1}{s} \tag{4.63}$$

The block diagram of Eq. (4.63) is shown in Fig. 4.29 below.

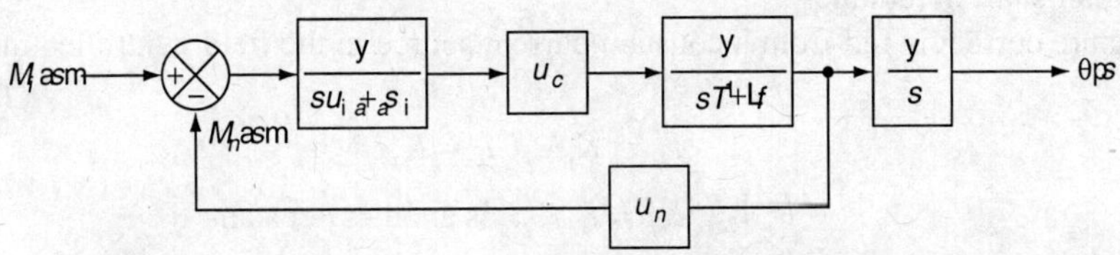

Fig. 4.29 Block diagram of an armature-controlled DC motor

In Eq. (4.62) if L_a is small, neglecting L_a we have
$$\frac{\theta(s)}{E_a(s)} = \frac{K_T}{R_a(Js^2 + sf) + K_T K_b s} = \frac{K_T/R_a}{Js^2 + s\left(f + \dfrac{K_T K_b}{R_a}\right)} = \frac{K_T/R_a}{Js^2 + f's}$$

i.e.,
$$\frac{\theta(s)}{E_a(s)} = \frac{\dfrac{K_T}{R_a}}{s(Js + f')} \tag{4.64}$$

where
$$f' = f + \frac{K_T K_b}{R_a} \tag{4.65}$$

Equation (4.64) can also be expressed as
$$\frac{\theta(s)}{E_a(s)} = \frac{K_m}{s(sT_m + 1)} \tag{4.66}$$

where $$K_m = \frac{K_T}{R_a f'} \tag{4.67}$$

and $$T_m = \frac{J}{f'} \tag{4.68}$$

4.9.3.2 Field-Controlled DC Motor

Figure 4.30 shows a field-controlled DC motor. In Fig. 4.30, R_f is the field winding's resistance in ohms, L_f is the field winding's inductance in henrys, e_f is the field-controlled voltage in volts, i_f is the field current in ampere, T_M is the torque developed by the motor in N-m, J is the equivalent moment of inertia of motor and load referred to the motor shaft in kg-m^2, f is the viscous coefficient of the motor and load referred to the motor shaft in N-m/rad/sec, and θ is the angular displacement of the motor shaft in radian.

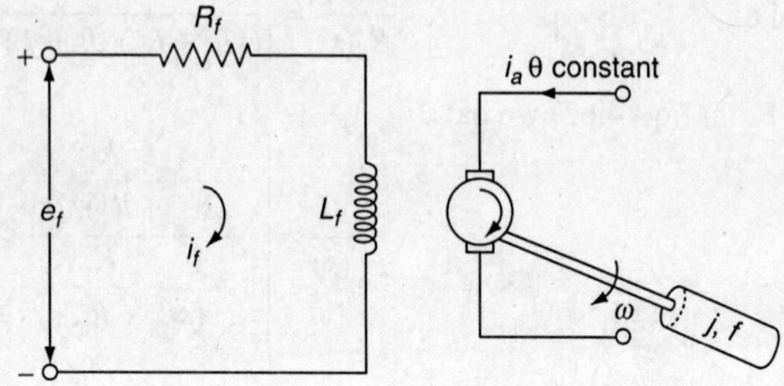

Fig. 4.30 Field-controlled DC motor

Since armature current is fed from a constant current source in the field-controlled motor, therefore we have

$$T_M = K_1 K_f i_f i_a = K'_T i_f \tag{4.69}$$

where $K'_T = K_1 K_f i_a$ is another constant. (4.70)

The equation in the field circuit is expressed by

$$L_f \frac{di_f}{dt} + R_f i_f = e_f \tag{4.71}$$

The torque equation is expressed by

$$J\frac{d^2\theta}{dt^2} + f\frac{d\theta}{dt} = T_M = K'_T i_f \tag{4.72}$$

Taking Laplace transform of Eq. (4.71) and Eq. (4.72) and assuming zero initial conditions, we have

$$(sL_f + R_f)I_f(s) = E_f(s) \tag{4.73}$$

and $$(Js^2 + fs)\theta(s) = T_M(s) = K'_T I_f(s) \tag{4.74}$$

From Eq. (4.73) and Eq. (4.74), we have

$$\frac{\theta(s)}{E_f(s)} = \frac{K'_T}{(L_f s + R_f)(Js^2 + fs)} \tag{4.75}$$

i.e., $$\frac{\theta(s)}{E_f(s)} = \frac{K_M}{s(T_f s + 1)(T_M s + 1)} \tag{4.76}$$

where $K_M = K'_T/(R_f f)$ is the motor gain constant, $T_f = L_f/R_f$ is the time constant of the field circuit, and $T_m = \frac{J}{f}$ is the mechanical time constant.

Figure 4.31 shows a block diagram of the field-controlled DC motor referred to Fig. 4.30.

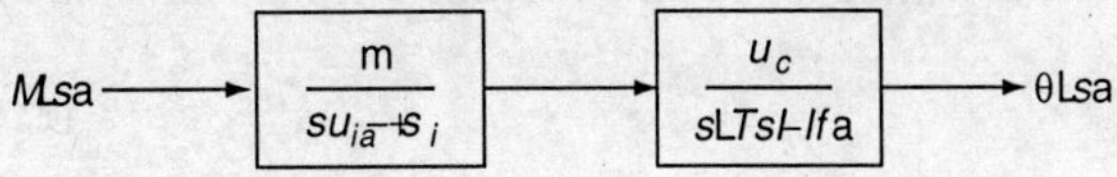

Fig. 4.31 Block diagram of a field-controlled DC motor

4.10 Servomotors

Servomotors are used in automatic control systems. The system is known as servomechanism when it is used to control the position of an object. The main purpose of the servomotors is to convert an electrical signal applied to them into an angular displacement of the shaft. The operation of the servomotor is either in a continuous duty or step duty, which depends on the construction.

A variety of servomotors are available for control system applications. The suitability of the motor for a particular operation is dependent on the characteristics of the system, the purpose of the system, and its operating conditions. The following features must be in any servomotor:

(i) There must be linear relationship between the speed and electric control signal.
(ii) Steady-state stability.
(iii) Wide range of speed control.
(iv) The mechanical characteristics must be linear throughout the entire speed range.
(v) Low mechanical and electrical inertias.
(vi) Fast rcsponse.

The servomotors are broadly classified as DC servomotors and AC servomotors based on the supply required to run the motor. The DC motors are costlier than AC motors. The DC servomotors have linear characteristics and hence their control is easier.

The following are the advantages of DC servomotors.

(i) Having higher output when compared to a motor of same size and frequency.
(ii) Linearity of characteristics can be achieved very easily.
(iii) It is possible to control the speed easily in both directions from zero speed to full speed.
(iv) The torque to inertia ratio is high which gives the quick response to control signals.
(v) Since the DC servomotors have light weight, low inertia, and inductance armature, there is quick response to command for change in position on speed.
(vi) Electrical time constants are low (0.1–6 ms) and mechanical time constants are also low (2.3–40 ms).
(vii) The capability of DC motors to deliver over three times their rated torque for a short time. The AC motors, on the other hand, have the capability to deliver 2 to 2.5 times their rated torque.

The usage of DC servomotors is usually for large power applications such as in machine tools and robotics.

On the other hand, the AC servomotors are cheaper, having higher efficiency, and less maintenance. The motors have non-linear characteristics and hence more difficult to control for position applications. These motors are best suited for low power applications such as instrument servo (control of pen in *X-Y* recorders) and also in computer-related equipment (disk drives, tape drives, printers, etc.).

4.10.1 DC Servomotors

The classifications of DC servomotors are shown in Fig. 4.32 below.

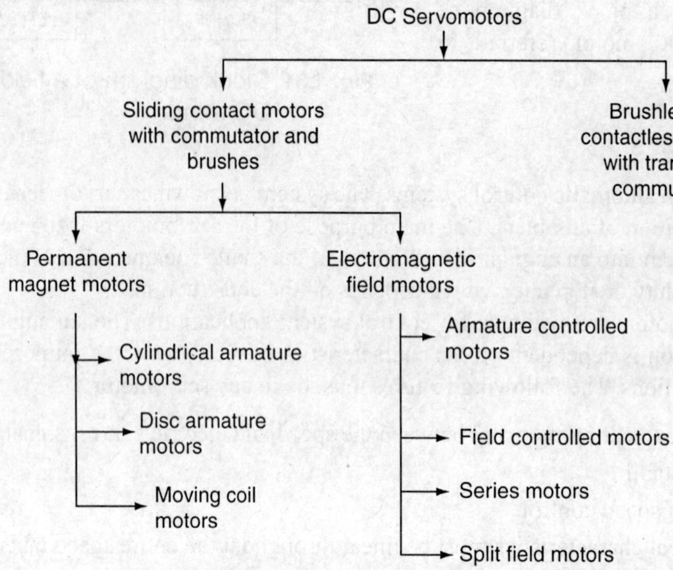

Fig. 4.32 Classifications of DC servomotors

4.10.1.1 Permanent Magnet DC Motors

Permanent magnet is used in these motors to replace the field winding to produce the required magnetic field. These types of motors are economical for power ratings up to a few kW. The advantages of permanent magnet motors are shown below.

(i) Since field power supply is not required, it is a simpler and reliable motor.
(ii) It has higher efficiency, as field loss is zero.
(iii) Due to the rise of temperature, field flux is less affected.
(iv) Since heating is less, it is possible to make it totally enclosed.
(v) There is no possibility of over speeding due to loss of field.
(vi) Torque-speed curve is more linear.
(vii) Higher power output at the same dimensions and temperature limitations.

The main disadvantage of permanent magnet motors is that the magnet deteriorates with time and hence demagnetized by large current transients. The high-grade magnetic materials such as ceramic magnets and rare earth magnets (samarium cobalt) are used to eliminate the above disadvantage. The cost of these materials is very high.

The armature is placed in rotor whereas permanent magnet poles are fixed to the stator in permanent magnet motors. The special type of construction is used to reduce the weight of rotor and hence inertia of the system. Cylindrical armature with small diameter and longer axial length, disc armature, and hollow armature are the special type of constructions.

4.10.1.2 Electro-Magnetic Field Motors

Usually, above 1 kW, these motors are very economical and these motors are very similar to conventional DC motors. The following are the features of these motors:

(i) The number of slots and commutator segments is made large to improve commutation.
(ii) To eliminate sparking, compoles and compensating windings are used.
(iii) To reduce inertia, the diameter to length ratio in kept low.
(iv) For withstanding the high torque stress, oversize shafts are used.
(v) The magnetic circuit is laminated and low-loss steel is used to reduce the eddy currents.

The armature current and/or field current are/is varied to control the torque and speed. One of these two is varied to control the torque and the other one is kept constant. The armature current is varied while the field current is kept constant to control the torque in armature-controlled mode of operation. On the other hand, the armature current is maintained constant while field current is varied to control the torque in field-controlled mode of operation.

4.10.1.3 Armature-Controlled DC Servomotor

The field of DC shunt motor, which is designed to satisfy the requirement of a servomotor, is excited by a constant DC supply. For constant field current i.e., excitation by a constant DC supply, speed is directly proportional to armature voltage and torque is directly proportional to armature current. Therefore, it is possible to control the torque and speed by armature voltage. To obtain reversible operation, armature voltage is reversed.

With the help of variable resistance, armature voltage is controlled in small motors. In large motors, armature voltage is controlled by thyristors, so that the power loss is also reduced. Figure 4.33 shows the operating characteristics of an armature-controlled DC servomotor.

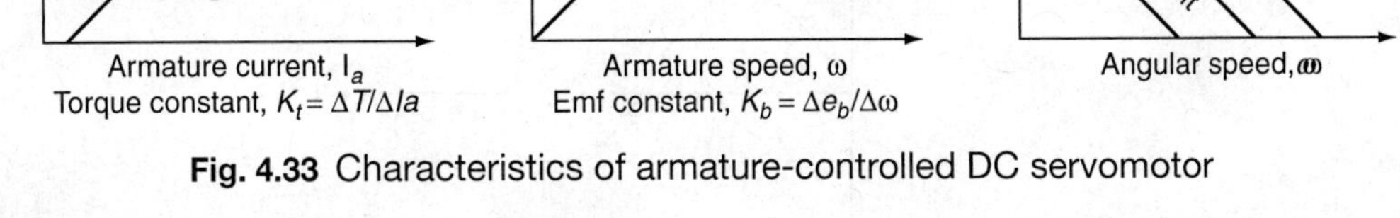

Fig. 4.33 Characteristics of armature-controlled DC servomotor

4.10.1.4 Field-Controlled DC Servomotor

The DC shunt motor is designed to satisfy the requirement of a servomotor in which armature is supplied with a constant current on voltage. The torque is directly proportional to field flux for constant armature voltage. Since field current is proportional to flux, the torque of the motor can be controlled by controlling the field current. If the field current is reversed, the reversible operation is possible. Due to field inductance, the response of field-controlled motor is slowed.

4.10.2 AC Servomotors

An AC servomotor is basically a two-phase induction motor. A two-phase servomotor differs from a normal induction motor as follows. The rotor of the servomotor is built with high resistance so that its *X/R* ratio is small and the torque-speed characteristic is nearly linear in contrast to the highly non-linear characteristic with large *X/R* ratio. A conventional induction motor has high *X/R* ratio. Because of the positive slope for part of the characteristic, the system using such a motor becomes unstable. In induction motor design, maximum torque is produced close to 4–5% slip. In an AC servomotor, the rotor has high resistance and it is usually squirrel-cage or drag-cup type having a small diameter to reduce inertia. Figure 4.34 shows a two-phase AC motor.

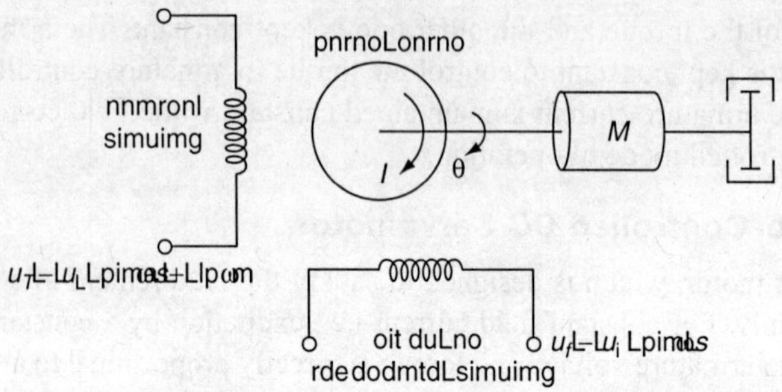

Fig. 4.34 Two-phase AC motor

The two-phase AC motor has two windings at right angles to each other, and the two windings are supplied with voltages with 90°-phase difference as through a capacitor. The rotor's diameter is small but long length to reduce inertia for faster motor response. The frequency is normally 50 Hz or 400 Hz. Voltage of varying amplitude but constant frequency is applied to the other winding. The flux will be rotating at the frequency of supply at the synchronous speed $(\omega_s) = 120\, f/P$. This rotating flux produces a varying flux and reacts with the stator flux, causing the rotor to rotate in a smooth manner. Figure 4.35

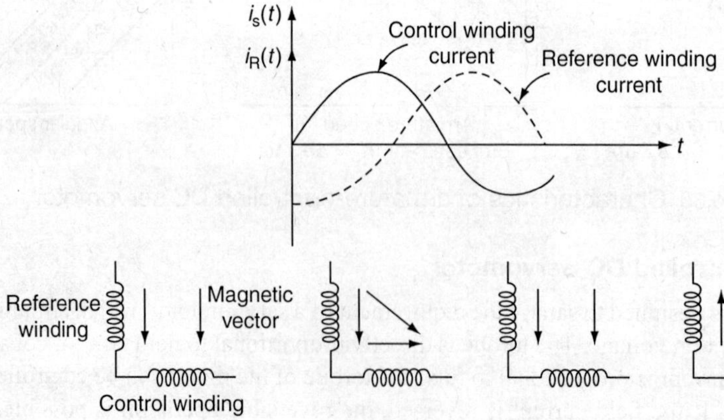

Fig. 4.35 Control winding and reference winding

shows the function of control winding and reference winding. Figure 4.36 shows the torque vs. speed characteristic. From the speed torque curve, the stall torque $(T_s) = K_m l_c$.

$$\text{No load speed } \omega_{nl} = \frac{K_m e_c}{D_m} \tag{4.77}$$

K_m is different for different e_c.

The ratio

$$\frac{\text{Stalled rotor torque}}{\text{No load speed}} = \frac{T_s}{\omega_{nl}} \tag{4.78}$$

is called m.

As in the case of DC motors,

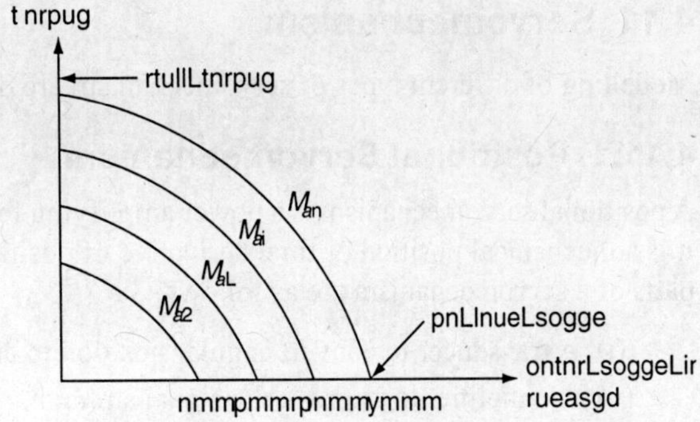

Fig. 4.36 Torque vs speed characteristic

$$T_D = K_m e_c - D_m \omega$$

$$T_D = J\frac{d^2\theta}{dt^2} + D\frac{d\theta}{dt} = K_m e_c$$

$$\text{Transfer function} \quad G(s) = \frac{\theta(s)}{E_c(s)} = \frac{K_m}{Js^2 + Ds} = \frac{K}{s(T_m s + 1)} \tag{4.79}$$

where $K = K_m/D$ is the motor gain constant; $T_m = J/D$ is the motor time constant.

The block diagram is shown in Fig. 4.37.

$\theta(s)$ is the viscous damping coefficient due to the motor and load. The damping is high if the slope of the T-ω curve is steep. For $|T_m s| << 1$, the servomotor acts as an integrator $(1/s)$.

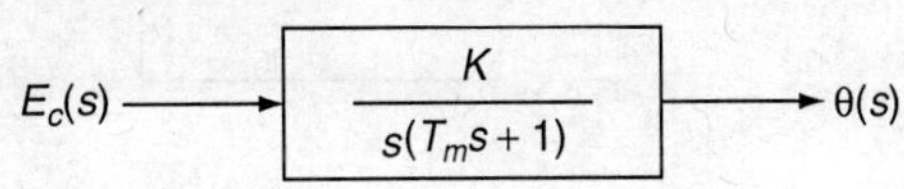

Fig. 4.37 Block diagram of two-phase AC motor

Example 4.5 An AC servomotor has both windings excited with 115V AC. It has a stall torque of 2 lb-ft and its coefficient of viscous friction is 0.2 lb-ft sec.

(i) Determine its no-load speed.

(ii) It is connected to a constant load of 0.8 lb ft and the coefficient of viscous friction is 0.05 lb-ft sec through a gear pass with a ratio of 4. Calculate the speed at which the motor will run.

Solution

(i) No load speed $\omega_{nl} = \frac{K_m V_m}{F_m} = \frac{2}{0.2} = 10$ rad/sec.

(ii) Total viscous friction $F_{\omega t} = F_m + N_1^2 F_1 = 0.2 + 4^2 \times 0.05 = 1$ lb ft sec.

$$T_e = K_m V_m - F_{\omega t}\,\omega$$

i.e., $$0.8 = 2 - \omega$$

i.e., $$\omega = 1.2 \text{ rad/sec.}$$

4.11 Servomechanism

Modelling of different types of servomechanism are discussed below.

4.11.1 Positional Servomechanism

A positional servomechanism is a power amplifying feedback control system. The controlled variable in it is a mechanical position or time derivative of position such as velocity or acceleration. The essential parts of a servomechanism are as follows:

(i) A transducer to convert angular position to an electrical signal,
(ii) an amplifier to improve the signal strength,
(iii) a device to develop torque such as motor and
(iv) devices to transmit torque to load such as gears.

In a positional servomechanism, the position of the load must be maintained both under steady-state and transient conditions.

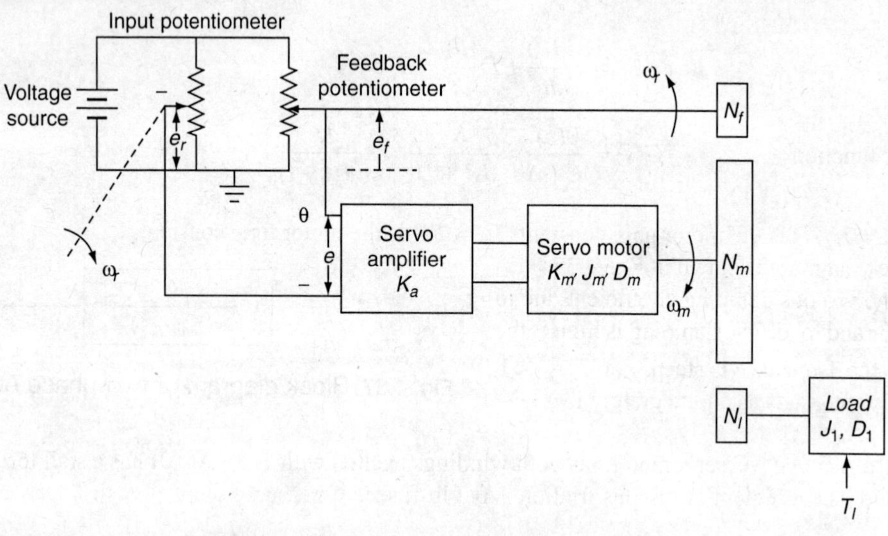

Fig. 4.38 Positional servomechanism

In Fig.4.38, $\theta_r(t)$ is reference angular position in radian, e_r is the input reference voltage at which the marked position of load is placed, e_f is the feedback voltage due to the feedback gear angular position, e is the error signal = $e_r - e_f$, K_a is the gain of the servo amplifier, K_m is the motor constant in lb-ft or Nm/V or A, J_m is the motor inertia in kg-m^2, D_m is damping friction of the motor in N-m/rad/sec, N_m is the gear tooth of the motor, θ_m is the motor shaft angular position in rad, N_l is the gear tooth of the load, J_l is the moment of inertia of the load in kg-m^2, D_l is the damping viscous friction of the load in N-m/rad/sec, T_l is the additional torque on the load, N_l is the gear tooth of feedback, and θ_f is the angular position of the feedback gear in rad.

The two potentiometers must be identical in their electrical characteristics and connected in parallel with their positions set at the initial marked position of the load. Figure 4.39 shows the block diagram of a servomotor.

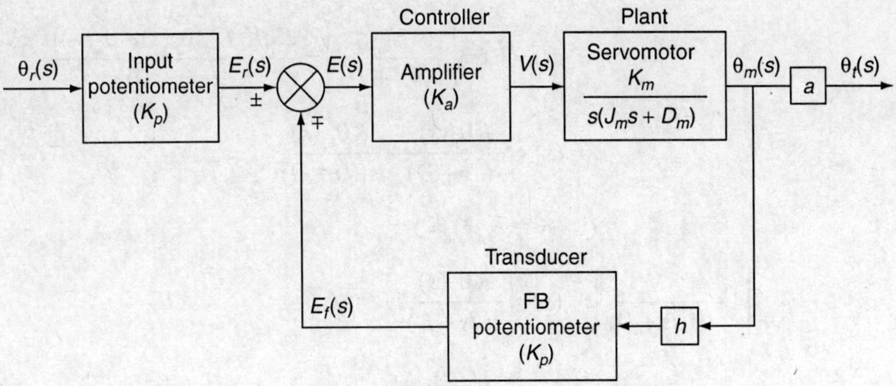

Fig. 4.39 Block diagram of a servomotor

$$e_r(t) = K_p\theta_r(t) \quad \text{and} \quad e_f(t) = K_p\theta_m(t)$$

$$E_r(s) = K_p\theta_r(s) \quad \text{and} \quad E_f(s) = K_p\theta_m(s)$$

$$e_r(t) - e_f(t) = e \ \text{ and } v = K_a e$$

$$E_r(s) - E_f(s) = E(s) \text{ and } V(s) = K_a E(s)$$

i.e.,

$$V(s) = K_a K_p[\theta(s) - \theta_m(s)]$$

Gear train to load $(a) = \dfrac{N_m}{N_l}$

Gear train to feedback $(h) = \dfrac{N_m}{N_f}$

It is possible to transfer the load due to the feedback gear to the motor as follows:

$$J = J_m + h^2 J_f$$

and

$$D = D_m + h^2 D_f$$

If we include the load,

$$J = J_m + h^2 J_f + a^2 J_l$$

and

$$D = D_m + h^2 D_f + a^2 D_l$$

Since we are interested in the feedback loop only, we have

$$J = J_m + h^2 J_f$$

and

$$D = D_m + h^2 D_f$$

$$\theta_m(s) = V(s)\left[\frac{K_m}{s(sJ+D)}\right] = \frac{K_p K_a K_m[\theta_r(s) - \theta_m(s)]}{s(Js+D)}$$

i.e.,
$$\theta_m(s) = \frac{K\theta_r(s)}{s(Js+D)} - \frac{K\theta_m(s)}{s(Js+D)}$$

where
$$K = K_p K_a K_m$$

i.e.,
$$\left[1+\frac{K}{s(Js+D)}\right]\theta_m(s) = \frac{K\theta_r(s)}{s(Js+D)}$$

i.e., Transfer function
$$G(s) = \frac{\theta_m(s)}{\theta_r(s)} = \frac{\dfrac{K}{s(Js+D)}}{1+\dfrac{K}{s(Js+D)}} = \frac{K}{s(Js+D)+K} = \frac{K}{Js^2+Ds+K}$$

i.e.,
$$G(s) = \frac{\dfrac{K}{J}}{s^2+\dfrac{D}{J}s+\dfrac{K}{J}} \tag{4.80}$$

The block diagram is reduced, and shown in Fig. 4.40 below.

The servomotor may be either a DC or two-phase AC servomotor. In a DC motor, the armature voltage is controlled by a DC amplifier and the field is provided by a permanent magnet. Figure 4.41 shows the wave-form diagram.

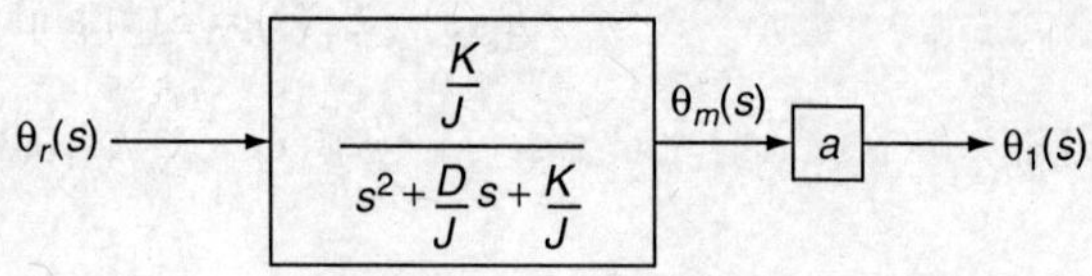

Fig. 4.40 Block diagram

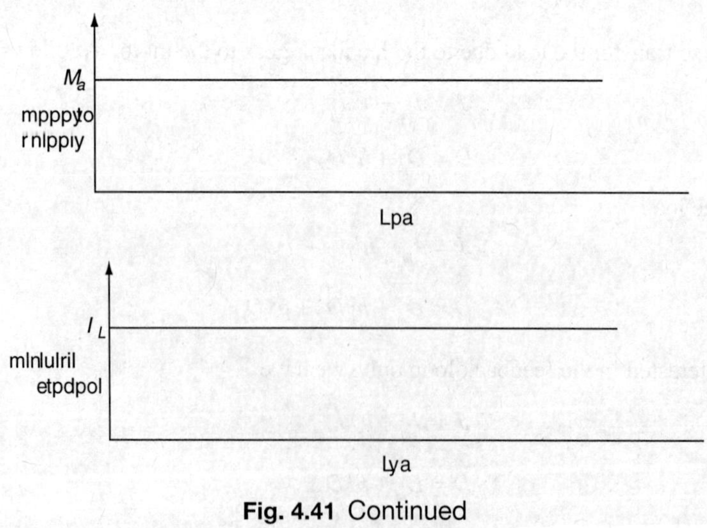

Fig. 4.41 Continued

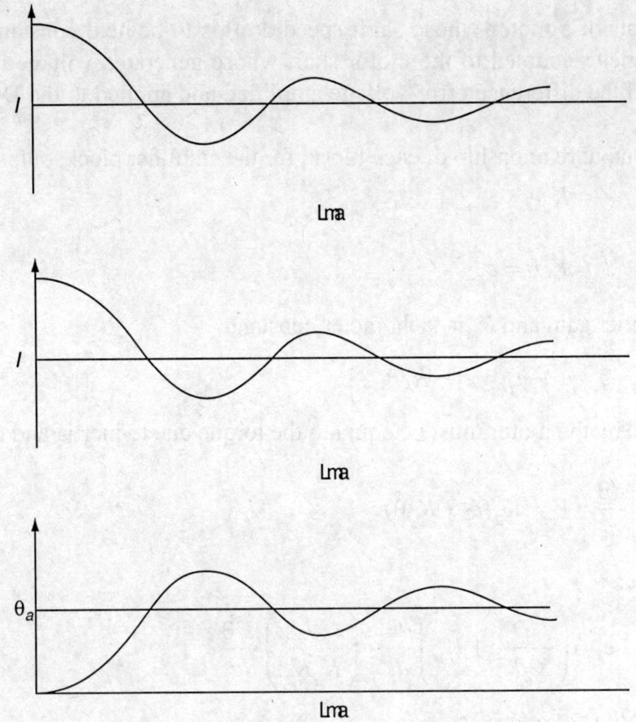

Fig. 4.41 Waveform diagram

In an AC system the reference voltage is AC and the amplifier is an AC amplifier. The motor demodulates the signal so that the feedback signal is DC. The final wave-form $\theta_l(s)$ will be the same as for DC.

4.11.2 Rate Servomechanism

The derivative of the position, i.e., velocity is monitored and kept constant here, with velocity as the controlled variable. Figure 4.42 shows the schematic diagram of a rate servomechanism.

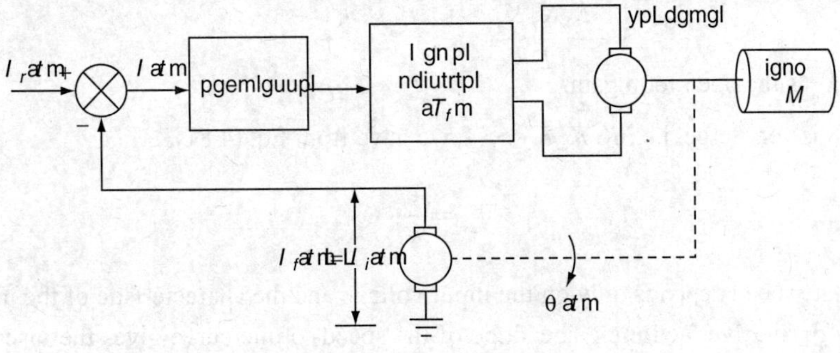

Fig. 4.42 Rate servomechanism

The system consists of a motor whose shaft speed $\dot{\theta}(\omega)$ is to be held constant. The load attached to the shaft, a tachometer coupled to the motor shaft whose generated voltage is fed along with the reference voltage and the difference error voltage amplified and applied to the DC motor to correct it to the desired speed.

Taking the input-output relationship of each block, for the amplifier block, output voltage is given by

$$V = K_a e$$

For the tachometer,

$$l_t = K_g \dot{\theta} = e_f$$

where K_a is the amplifier gain and K_g is tachometer constant:

$$e = e_r - e_f = e_r - K_g \dot{\theta}$$

The torque developed by the motor must be equal to the torque due to inertia and damping.

$$\left(\frac{J}{K_T}\right)\frac{d^2\theta}{dt^2} + \left(\frac{D}{K_T}\right)\frac{d\theta}{dt} = V = K_a(e_r - K_g\dot{\theta})$$

Solving for e_r, we have

$$e_r = \left(\frac{D}{K_a K_T} + K_g\right)\frac{d\theta}{dt} + \left(\frac{J}{K_a K_T}\right)\frac{d^2\theta}{dt^2}$$

In the steady state, $d^2\theta/dt^2 = 0$, therefore, we have

$$\dot{\theta}_{ss} = \frac{d\theta}{dt} = e_r\left(\frac{K_a K_T}{D + K_a K_T K_g}\right) = e_r\left(\frac{\frac{K_a K_T}{D}}{1 + \frac{K_a K_T K_g}{D}}\right) = \frac{e_r K_a K_m}{1 + K_a K_g K_m} \tag{4.81}$$

where $K_T/D = K_m$ is called the velocity gain constant, which relates the steady-state motor speed to the armature voltage and hence is a velocity gain. Its unit is (rad/sec)V. The time constant τ_c of the closed-loop system is defined by

$$\tau_c = \frac{\dot{\theta}}{\ddot{\theta}} = \frac{\frac{J}{D}}{1 + \frac{K_a K_T K_g}{D}} = \frac{\frac{J}{D}}{1 + K_a K_m K_g} \tag{4.82}$$

where $K_a K_m K_g$ is the open-loop gain.

If $K_a K_m K_g$ is very large, i.e., $K_a K_m K_g >> 1$, we have from Eq. (4.81),

$$\dot{\theta}_{ss} = \frac{e_r}{K_g} \tag{4.83}$$

The steady-state speed depends only on the input voltage and the characteristic of the tachometer.

Damping by derivative feedback The slope of the speed-torque curve gives the viscous damping or damping coefficient. The feedback introduced by the voltage generated by a tachometer is a derivative feedback as $V_g = K_g \omega$. Figure 4.33 shows the schematic diagram of damping by derivative feedback.

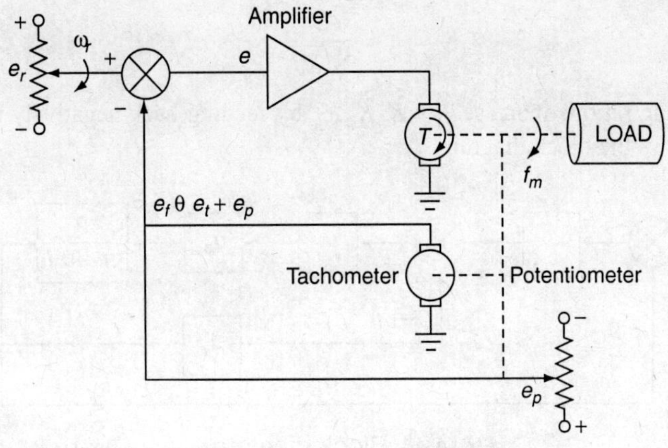

Fig. 4.43 Damping by derivative feedback

The error voltage is expressed by

$$e = e_r - (e_t + e_p) = K_p(\theta_r - \theta_m) - K_g\theta_m$$

Equating the torques, we have

$$J\ddot{\theta}_m + D\dot{\theta}_m = K_pK_aK_m(\theta_r - \theta_m) - K_gK_dK_m\dot{\theta}_m \tag{4.84}$$

where K_d is the amplifier gain of the voltage from the tachometer.
Solving θ_r,

$$\theta_r = \frac{J}{K}\ddot{\theta}_m + \left(\frac{D + K_gK_dK_m}{K}\right)\dot{\theta}_m + \theta_m \tag{4.85}$$

where $K = K_pK_aK_m$ is the open-loop gain.
The damping ratio from the second-order equation (4.85), after dividing throughout by J/K is

$$2\xi\omega_n = \frac{D + K_gK_dK_m}{J} \tag{4.86}$$

and
$$\omega_n = \sqrt{\frac{K}{J}} \tag{4.87}$$

∴
$$\xi = \frac{D + K_gK_dK_m}{2\sqrt{KJ}} \tag{4.88}$$

For potentiometer alone, we have from Eq. (4.86),

$$2\xi\omega_n = \frac{D}{J} \tag{4.89}$$

and
$$\omega_n = \sqrt{\frac{K}{J}} \tag{4.90}$$

$$\therefore \qquad \xi = \frac{D}{2\sqrt{KJ}} \tag{4.91}$$

Therefore, damping ratio is increased by $K_g K_d K_m$ by feeding back negatively the tachometer voltage. Figure 4.44 shows the block diagram.

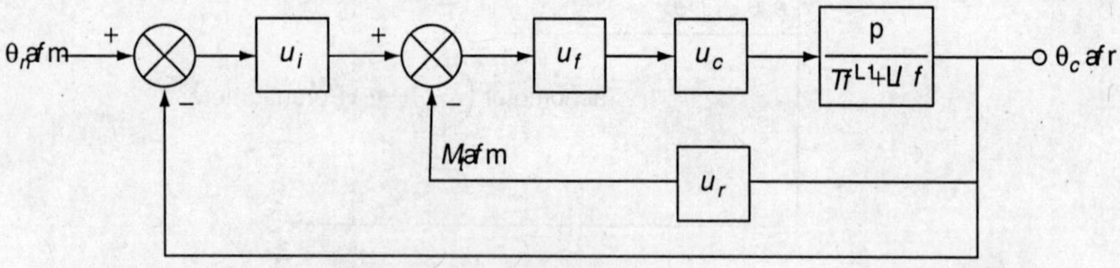

Fig. 4.44 Block diagram

If we take Laplace transform of Eq. (4.85), we have

$$\frac{\theta_m(s)}{\theta_r(s)} = \frac{\dfrac{K}{J}}{s^2 + \left(\dfrac{D + K_g K_d K_m}{J}\right)s + \dfrac{K}{J}}$$

Figure 4.45 shows the final block diagram of damping by derivative feedback.

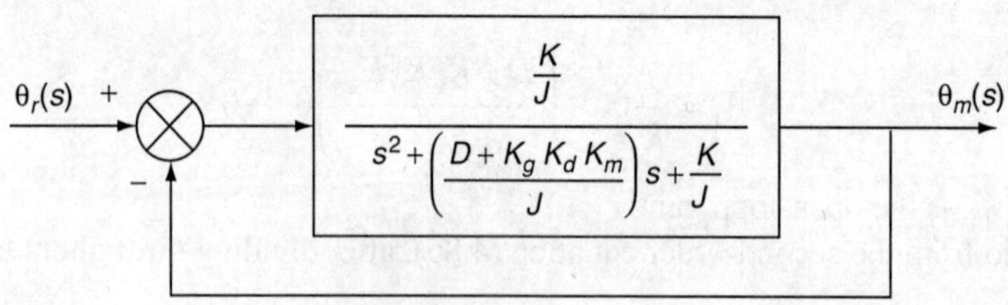

Fig. 4.45 Final block diagram

Example 4.6 Figure E4.2 shows a schematic diagram of an automatic control system where the angular position of an antenna is controlled.

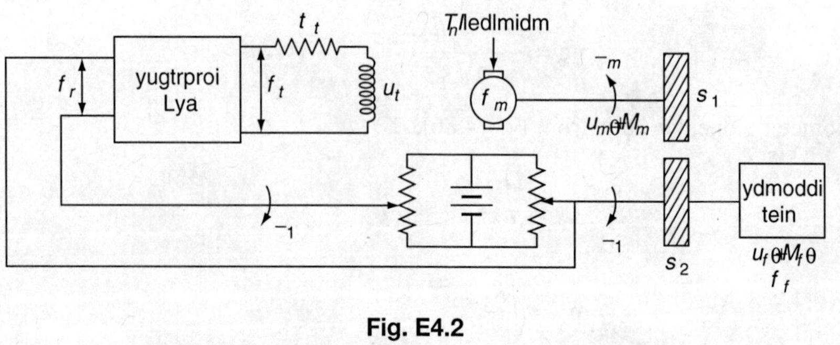

Fig. E4.2

The system has the following data:

The gear ratio $(N_1/N_2) = 0.1$, voltage gain of the amplifier (A) = 1000,
Motor torque constant = 0.25 Nm/A of field current
Field winding resistance $(R_f) = 20$ ohms
Self-inductance of winding $(L_f) = 1$H
Error detector sensitivity = 0.02 V/1 error
Moment of inertia of the load $(J_L) = 50 \times 10^{-6}$ kg-m^2
Moment of inertia of the motor $(J_L) = 05 \times 10^{-6}$ kg-m^2
Viscous friction coefficient of the load $(B_L) = 0.9\times10^{-5}$ N-m/rad/sec
Viscous friction coefficient of the motor $(B_m) = 0.1\times10^{-7}$ N-m/rad/sec
Determine the transfer function relating the input displacement θ_i and output displacement θ_0.

Solution

$$\frac{\theta_L}{\theta_m} = \frac{\dot{\theta}_L}{\dot{\theta}_m} = \frac{\ddot{\theta}_L}{\ddot{\theta}_m} = \frac{N_1}{N_2}$$

Equivalent moment of inertia referred to the motor shaft is

$$J = J_m + \left(\frac{N_1}{N_2}\right)^2 J_L = 1\times10^{-6}\,\text{kg-m}^2$$

Similarly, equivalent viscous friction referred to the motor shaft is

$$D = D_m + \left(\frac{N_1}{N_2}\right) B_L = 1\times10^{-7}\,\text{N-m/rad/sec}$$

Figure E4.2 (a) shows the block diagram of the system.

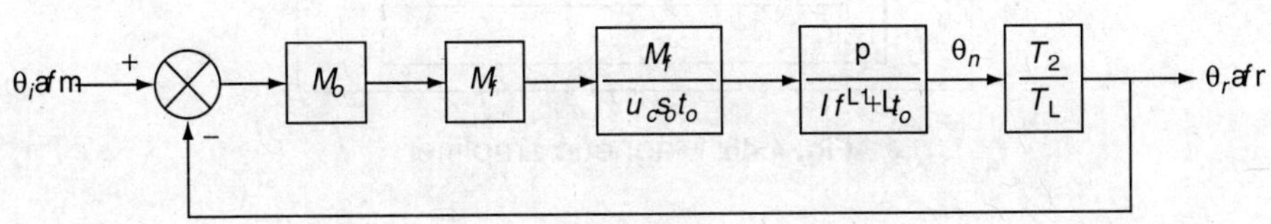

Fig. E4.2(a)

The open-loop transfer function is given by

$$G(s) = \frac{K_S A K_T \left(\frac{N_1}{N_2}\right)}{(R_f + sL_f)(Js^2 + Bs)} = \frac{0.2\times1000\times0.25\times0.1\frac{180}{\pi}}{(20+s)(s^2+0.1s)}$$

The closed-loop transfer function is given by

$$\frac{\theta_c(s)}{\theta_i(s)} = \frac{G(s)}{1+G(s)} = \frac{286.5\times10^6}{(s+20)(s^2+0.1s)+286.5\times10^6}$$

4.12 Magnetic Amplifier

These are simply another type of control valve and are electrically operated devices. These devices deliver power less than 1 kW and are used for higher power applications. These are used for frequency range of 50 Hz and 5 kHz. Figure 4.46 shows a basic magnetic amplifier, which has three limb core. The outer limb windings are identical and connected to an AC supply. The control winding is in the middle and it is connected to DC supply. If there is no direct current, there is no effective flux in the control limb. Hence there is induced voltage in the windings. The basic principle of magnetic amplifier is to control the current through one coil by passing a direct current through the another coil. The arrangement shown in Figure 4.46 is known as transductor or saturable reactor.

Flux is set up in the core if a direct current is passed through the central limb winding. The core flux set by the control winding current adjusts the impedance of two AC windings. The load current which is alternating in nature can be controlled by the direct current passing through the control winding. The input current polarity has no effect on the operation. It is possible to make less power requirement in control winding as compared to AC winding and the output by suitable selection of number of turns on the windings.

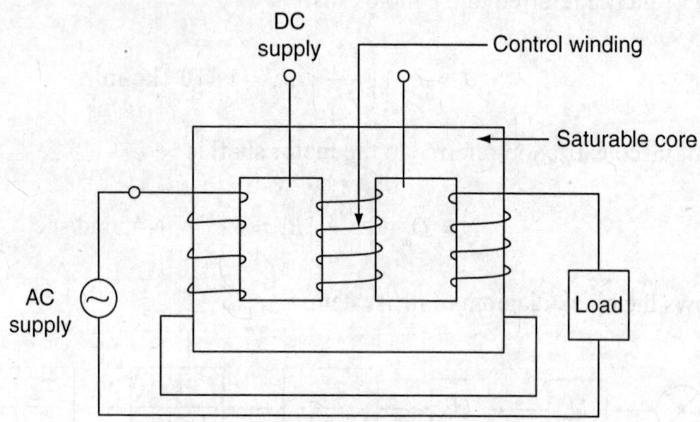

Fig. 4.46 Magnetic amplifier

4.13 Electronic Amplifier

An operational amplifier is called ideal if the following characteristics are satisfied:

(i) Infinite voltage gain
(ii) Zero output impedance

This type Op-Amp is also called as a voltage controlled voltage source (VCVS). Ideal Op-amps can be used to perform a large number of mathematical operations. It can also generate a number of circuit functions by applying passive feedback around the amplifier.

The symbol of Op-Amp in shown in Fig. 4.47 below:

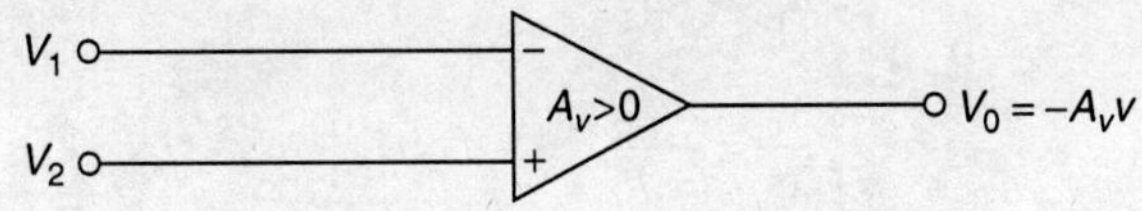

Fig. 4.47 Ideal operational amplifier

The ideal operational amplifiers have the following characteristics

1. The voltage gain $A_v \rightarrow \propto$. The output voltage v_0 is finite.
2. $R_i \rightarrow \propto$ (open circuited). Therefore, no current enters the input terminals and $i_1 = i_2 = 0$.
3. $R_0 = 0$.
4. When $v_1 = v_2$, $v_0 = 0$. This indicates zero offset.
5. The bandwidth is infinite. So the amplifiers gives response equally at all frequencies.

4.13.1 Ideal Inverting Op-Amp

The circuit symbol of an ideal inverting amplifier is shown in Fig. 4.48 below.

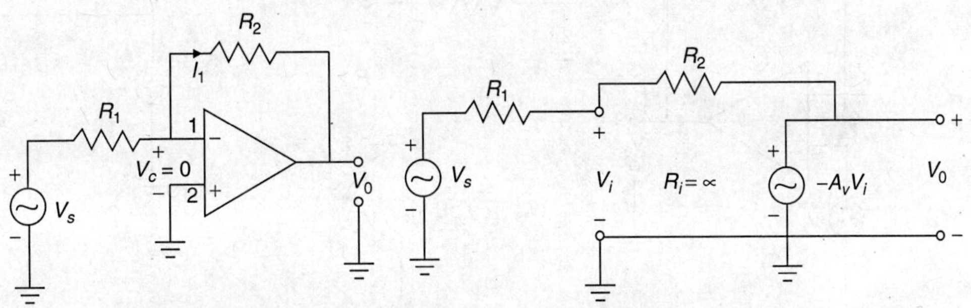

(a) Circuit symbol (b) Equivalent circuit

Fig. 4.48 Ideal op-amp in inverting mode

In inverting mode the positive terminal is being grounded. The input signal v_s is applied to negative terminal through a resistor R_1. In this circuit feedback is applied through a resistor R_2 from the output to input. For ideal Op-Amp the following two conditions are always satisfied:

(i) The current into each input terminal is zero.

(ii) Between the two input terminals the potential difference (v_i) is always zero.

Since the current into each input terminal is zero, the current passing through R_1 also passes through R_2. Again as v_i is zero, it can be written as

$$i = \frac{v_s - v_i}{R_1} = \frac{v_i - v_0}{R_2}$$

or $$\frac{v_s}{R_1} = -\frac{v_0}{R_2} \quad [\because v_i = 0]$$

or $$\frac{v_0}{v_s} = -\frac{R_2}{R_1} \tag{4.92}$$

$\frac{v_0}{v_s}$ is called the closed loop **voltage gain (A_v)**. In other words, it is the gain with feedback

$$\therefore \quad A_v = -R_2/R_1 \tag{4.93}$$

From Eq. (4.93) it is clear that the voltage gain is dependent on R_1 & R_2.

4.13.2 Ideal Non-Inverting Op-Amp

The ideal non-inverting Op-Amp is shown in Fig. 4.49(a) and Fig. 4.49(b) shows its equivalent circuit.

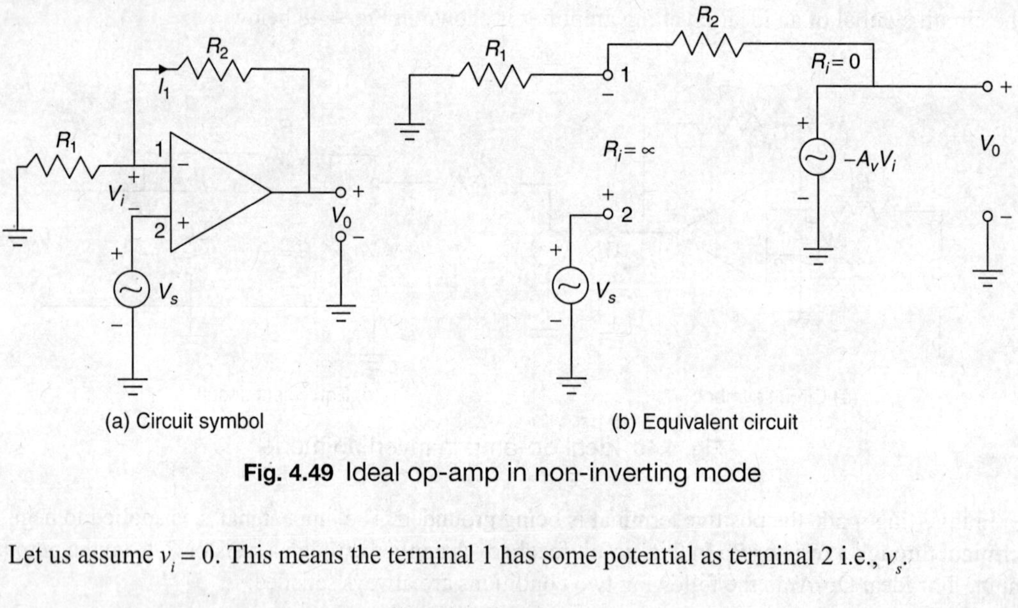

(a) Circuit symbol (b) Equivalent circuit

Fig. 4.49 Ideal op-amp in non-inverting mode

Let us assume $v_i = 0$. This means the terminal 1 has some potential as terminal 2 i.e., v_s.

$$\therefore \quad \frac{v_s}{R_1} + \frac{v_s - v_0}{R_2} = 0$$

or $$\frac{v_s - v_0}{v_s} = -\frac{R_2}{R_1}$$

or $$1 - \frac{v_0}{v_s} = -\frac{R_2}{R_1}$$

∴
$$\frac{v_0}{v_s} = 1 + \frac{R_2}{R_1}$$

∴
$$A_v = 1 + \frac{R_2}{R_1} \tag{4.94}$$

So closed loop gain upends on the ratio R_2/R_1.

Let us put $R_2 = 0$ in Fig. 4.49(a). A_v is equal to 1. In such case R_1 is not required. The final form of Fig. 4.49(a) is shown in Fig. 4.50. It is called as unity gain buffer or voltage follower as it has infinite (high) input resistance and unity gain.

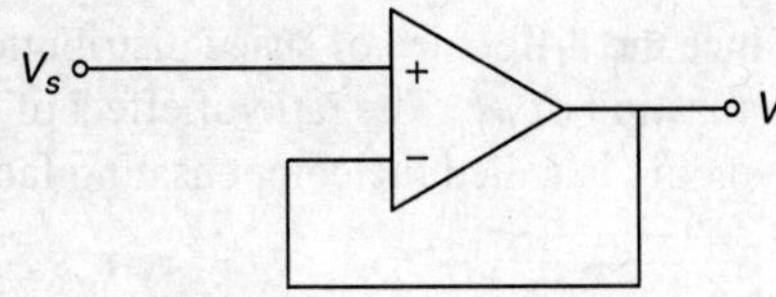

Fig. 4.50 Voltage follower

4.14 Cross-Field Machines

Cross field machines are also called rotating amplifiers, which are DC machines with additional brush set along α-axis. Figure 4.51 shows a DC generator. The output power is controlled by varying the field current of the generator. If P_2 is the output power and P_1 is the input power to the field winding, the power amplification becomes P_2/P_1. The input power is usually 1% of output power and hence $P_2/P_1 \cong 100$.There is a power amplification in DC generator. Since it is rotating, the rotating amplifier is derived. Depending on the degree of compensation, the cross field generators can be classified as follows:

(a) Metadyne

(b) Amplidyne

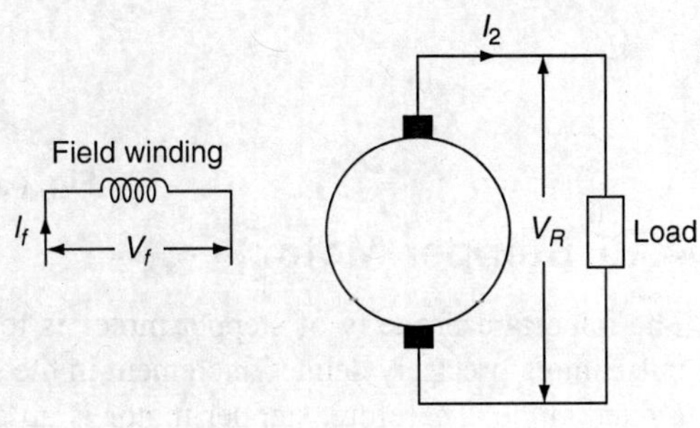

Fig. 4.51 DC generator

4.14.1 Metadyne

Figure 4.52 shows a metadyne generator which is obtained from an ordinary DC generator by providing an additional pair of brushes in the d-axis whereas the brushes in the q-axis is short circuited. The current (I_q) in the q-axis produces an mmf M_q along the q-axis, which is the first stage of amplification. The voltage induced across the q-axis due to controlled mmf M_f circulates I_q when the brushes are short circuited. In the second stage, due to rotation of armature in the mmf M_q, a voltage is developed across the d-axis across which load is connected.

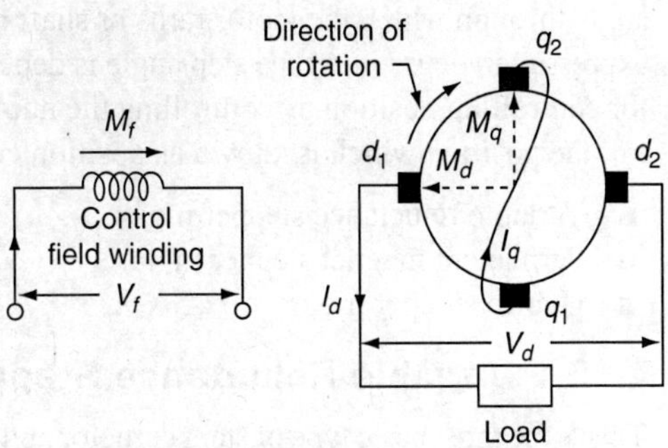

Fig. 4.52 Metadyne generator

Due to the current I_d, the armature mmf in the d-axis opposes the control mmf M_f as per Lenz's Law. I_q and I_d attains a steady state value for a given value of I_f and load resistance. If there is any increase in the value of I_d, it will increase M_d that neutralizes M_f. Hence the induced emf across the d-axis is reduced that prevents further increase of current keeping the output current of a metadyne substantially constant over a wide range of load variation. Therefore, the metadyne is also called constant current generator.

4.14.2 Amplidyne

Figure 4.53 shows an amplidyne, which is a metadyne having a compensation for a d-axis mmf M_d. In Fig. 4.53, a compensating winding is provided on the stator to produce an mmf M_c opposing M_d. Due to the flow of current through the compensating winding, the compensating mmf is produced and hence the negative feedback effect due to load current I_d is minimized.

The flux along the d-axis solely depends on M_f due to the presence of the compensating winding. Since the difference of space distribution of M_d and M_q are there, it is difficult to achieve complete neutralization of M_d. The ratio of effect of turns of compensating winding to effective turns of the armature winding is called the compensating factor.

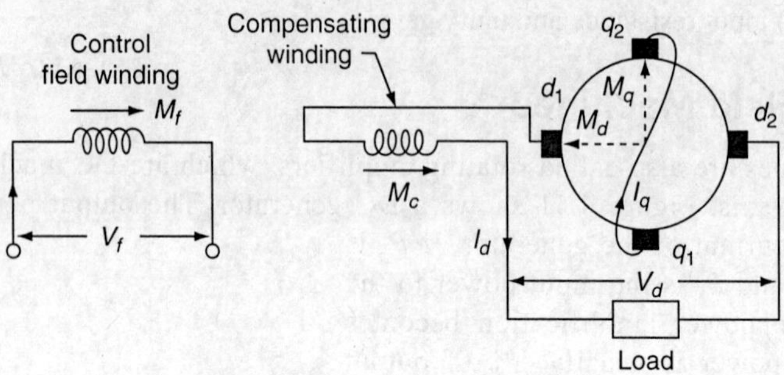

Fig. 4.53 Amplidyne

4.15 Stepper Motor

The important property of stepper motor is to convert a train of input pulses, that is, a square wave pulse into a precisely defined increment in the shaft position where each pulse moves the shaft through a fixed angle. Therefore, stepper motor is an electromechanical device which actuates a train of step movements of shaft in response to a train of input pulses where the step movement may be angular or linear and one-one relationship exists between the input pulse and step movement of the shaft. The angle through which the motor turns or shaft moves for each pulse is known as the *step angle*, which is expressed in degrees. As the step angle is dependent on the number of input pulses, the motor is suitable for controlling position by controlling the number of input pulses. Therefore, this system is used to control the position which is known as position control system. There are three types of stepper motors:

- Variable reluctance stepper motor
- Permanent magnet stepper motor
- Hybrid stepper motor

4.15.1 Variable Reluctance Stepper Motor

This is the most basic type of stepper motor and helps to explain the principle of operation of stepper motor. The stator of the motor is usually for three phases having six salient poles with concentrated exciting windings around each one of them. The stator has laminated construction and is assembled in a single stack. The number of poles of stator and rotor are different to secure the ability of the motor for bi-directional rotation and self-starting capability. The rotor is made of slotted steel laminations. Let N_s and N_r be the number of stator poles and rotor poles. For three-phase, the number of rotor poles in terms of stator poles is given by,

$$N_r = N_s \pm \left(\frac{N_s}{q}\right) \tag{4.95}$$

where q is the number of stator phases or stacks. Figure 4.54 shows the stepper motor having four salient-pole rotor.

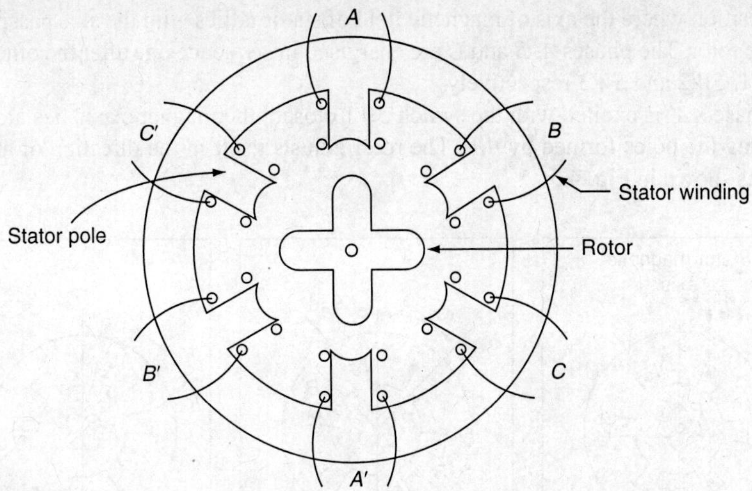

Fig. 4.54 Schematic arrangement of variable reluctance stepper motor

The basic diagram has been shown in Fig. 4.55 where the coils are wound in such a way that diametrically opposite poles are connected in series. The three phases are energised from a DC supply with the help of switches.

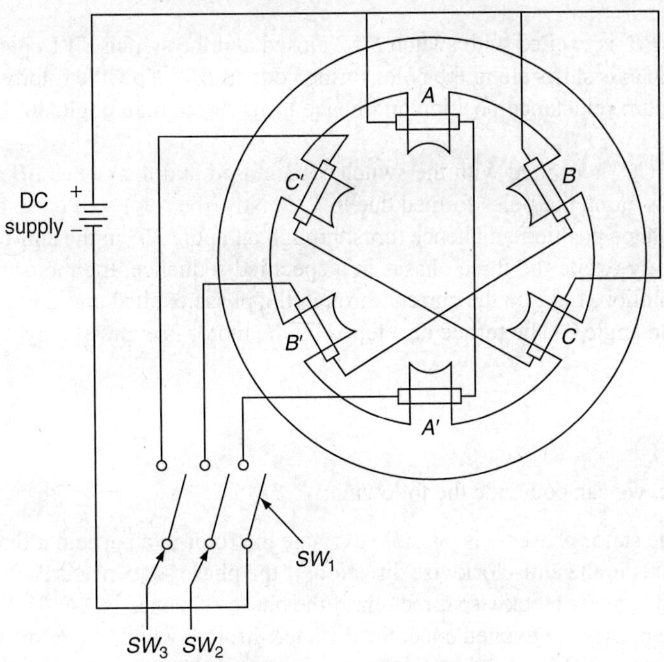

Fig. 4.55 Basic driving circuit

Operation The operation depends on various reluctance positions of rotor with respect to stator. If one phase of the stator is excited, it produces a magnetic field. Its axis lies along the poles, the phase around which it is excited. The rotor moves in the direction of minimum reluctance and such a position indicates the position where the axis of magnetic field of stator tallies with the axis passing through any two poles of the rotor. The phases *A, B* and *C* are energised in sequence one after the other with the help of switches *SW*1, *SW*2 and *SW*3 respectively.

When the phase *AA′* is excited with the switch *SW*1 closed, the magnetic axis lies along the vertical direction or along the poles formed by *AA′*. The rotor adjusts itself in the direction of minimum reluctance position as shown in Fig. 4.56.

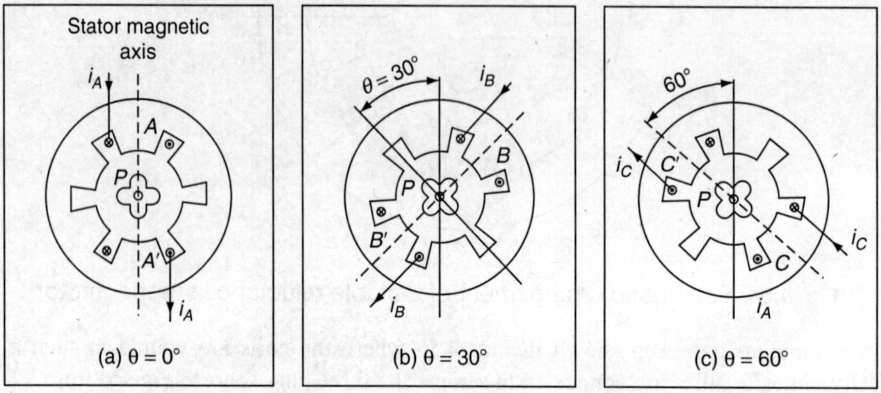

Fig. 4.56 Operation of variable reluctance stepper motor

Next the phase *BB′* is excited with switch *SW*2 closed and the switch *SW*1 open to de-energise *AA′*. The stator magnetic axis shifts along the poles formed due to *BB′*. Now the rotor will try to align in the direction of minimum reluctance position and hence turns through an angle 30° in the anti-clockwise direction.

Next the phase *CC′* is excited with the switch *SW*3 closed and the switch *SW*2 is open. The stator magnetic axis shifts along the poles formed due to *CC′*. Now the rotor will try to align in the direction of minimum reluctance position and hence turns through an angle 30° in the anti-clockwise direction.

If we successively excite the three phases in a specified sequence, the motor takes twelve steps to complete one revolution. Let I be the current through the phase excited and L be the inductance of the relevant phase at an angle θ. The torque developed by the motor is expressed by

$$T_m = \frac{1}{2} i^2 \frac{dL}{d\theta} \tag{4.96}$$

From the above, we can conclude the following:

- By exciting the stator phases it is possible to rotate the rotor in a particular direction.
- The rotor rotates in the anti-clockwise direction, if the phase sequence is A-B-C-A....
- The rotor rotates in the clockwise direction, if the phase sequence is A-C-B-A....
- If all the three phases are excited once, the distance through which the rotor is moved is known as rotor tooth pitch.

$$\text{Rotor tooth pitch} = \frac{360°}{N_r}$$

$$\text{The step angle}(\alpha_s) = \frac{360°}{qN_r} \tag{4.97}$$

In the above discussion we have seen that the rotor rotates through an angle 30° when each phase is excited simultaneously. Now if two phases are excited simultaneously, that is, *AA′* and *BB′*, the stator magnetic field axis shifts to a mid-position rather than along *BB′* resulting in the rotor being aligned along it, that is, through an angle 15°. Usually, the step angle is reduced by $\frac{1}{2}, \frac{1}{5}, \frac{1}{10}, \frac{1}{16}$ and $\frac{1}{32}$. This technique is known as *micro stepping*.

Multistack stepper motor: Figure 4.57 shows the cross-sectional view of multistack stepper motor. For a *m* stack motor, the motor is divided into a number of magnetically isolated sections known as stacks along its axial length. If *m* be the number of stacks, that is, the number of phases and N_r be the number of rotor poles, the step angle is given by,

$$\alpha_s = \frac{360°}{mN_r} \tag{4.98}$$

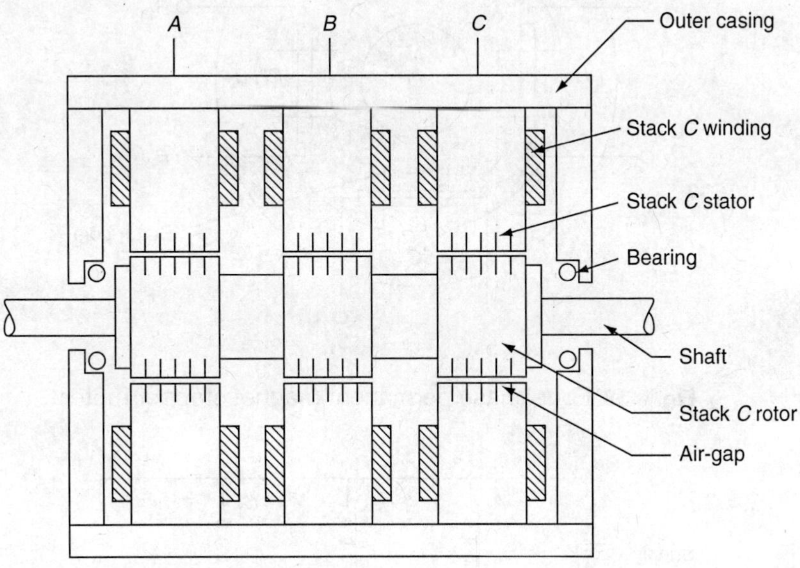

Fig. 4.57 Cross-sectional view of multistack stepper motor

Following are the advantages of variable reluctance induction motor:

- High torque to inertia ratio
- Fast dynamic response
- High rates of acceleration
- Simple and low cost

- Efficient cooling arrangement
- Robust rotor construction due to absence of brushes

4.15.2 Permanent Magnet Stepper Motor

This type of motor has multipolar stator, as shown in Fig. 4.58, having four stator poles. The exciting coils are wound around the poles. The number of slots per pole per phase is usually taken as one in such multipolar machines.

The rotor is generally taken as cylindrical pole type but it may be salient or smooth cylindrical type. The rotor is made of ferrite material, which is permanently magnetised and hence this motor is called permanent magnet stepper motor. Figure 4.59 shows the basic driving circuit for four-phase permanent magnet stepper motor to provide the voltage pulses to the stator winding of stepper motor shown in Fig. 4.58.

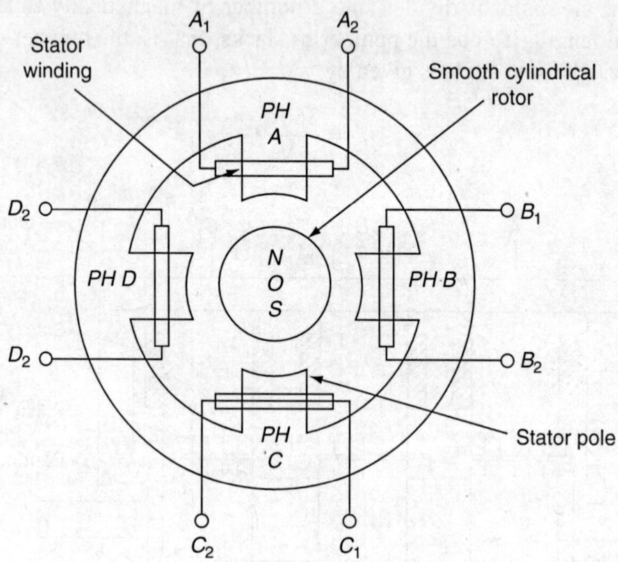

Fig. 4.58 Four-phase permanent magnet stepper motor

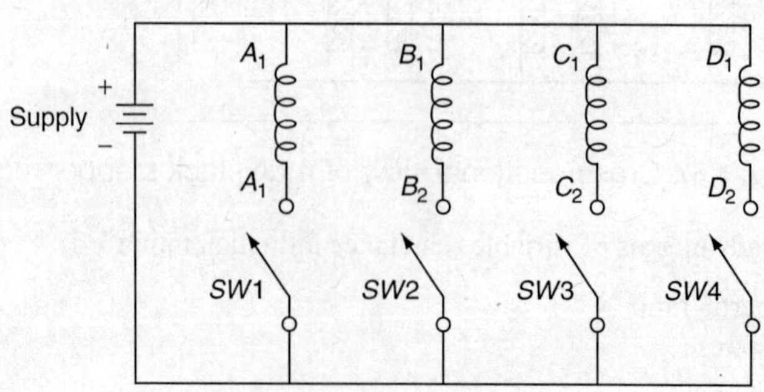

Fig. 4.59 Basic driving circuit for permanent magnet four-phase stepper motor

Just after the application of the voltage pulses to various phases with the help of driving circuit, the rotor starts rotating through a step for each input voltage pulse. After closing the switch *SW*1 to excite phase *A*, we get *N* pole in phase *A* and the rotor starts to rotate in such a way that the magnetic axis of the permanent magnet rotor adjusts with the magnetic axis of stator as depicted in Fig. 4.56(a). Next phase *B* is excited after closing the switch *SW*2 and disconnecting phase *A*. The rotor further adjusts its own magnetic axis with *N* poles of phase *B*. Therefore, it rotates through an angle 90° as shown in Fig. 4.56(b). Similarly, if the phases *C* and *D* are excited, the rotor rotates through an angle 90° in clockwise direction for excitation of each phase. Therefore, repetition of such sequence results in a step motion of a permanent magnet stepper motor. The major disadvantage of stepper motors is that they cannot he constructed with permanent magnet rotors with large number of poles and small steps are not possible. In order to overcome this, variable reluctance type stepper motor is used.

4.15.3 Hybrid Stepper Motor

The hybrid stepper motor is based on the principles of the permanent magnet and variable reluctance stepper motors. The rotor flux in hybrid stepper motor is generated by the permanent magnet and is directed by the rotor teeth to the appropriate parts of the air gap. The permanent magnet which is placed in the middle of the rotor is magnetised in the axial direction. Each pole of the magnet is surrounded by soft-toothed laminations. Figure 4.60 shows the construction of hybrid stepper motor.

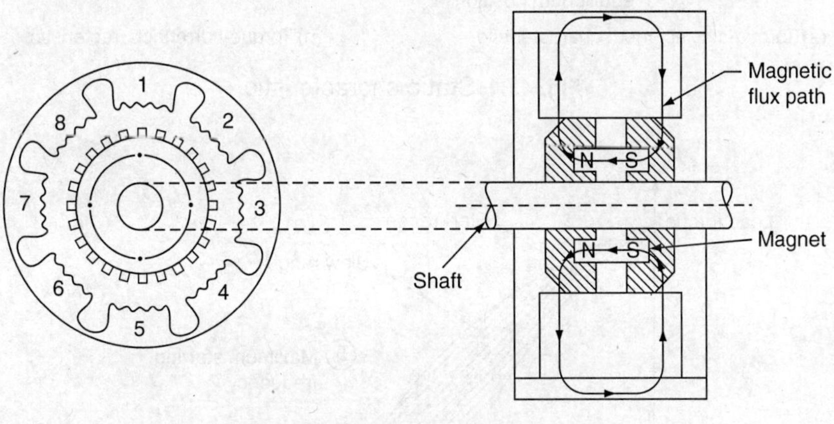

Fig. 4.60 Hybrid stepper motor

There are generally eight poles on the stator and each pole has between two to six teeth and there is two-phase winding. The coils on poles 1,3,5 and 7 are connected in series to form phase *A* whereas coils on poles 2,4,6 and 8 are connected in series to form phase *B*. If phase *A* carries positive current, stator poles 1 and 5 become South while 3 and 7 become North. Therefore, the rotor teeth with South and North polarity align with the teeth of stator poles 3 and 7 and 1 and 5 respectively. If phase *B* is excited and phase *A* is de-energised, the rotor will move by one-quarter of tooth pitch. Torque is produced in a hybrid motor due to interaction of stator and rotor-produced fluxes. Since the rotor field is produced by permanent magnet, it remains constant. Hence the motor torque is proportional to the phase current. Following are advantages of the hybrid stepper motor:

- Step angle is very small. It is 1.8°.
- Higher torque per unit volume. It is greater than that of reluctance motor.
- There is some detent torque due to the permanent magnet.

4.16 Characteristics of Stepper Motors

The characteristics of stepper motor are classified as follows:

- Static characteristic
- Dynamic characteristic

Torque displacement characteristic and torque current characteristic are known as static characteristic as shown in Fig. 4.61.

The torque stepping rate characteristic is known as dynamic characteristic as shown in Fig. 4.62.

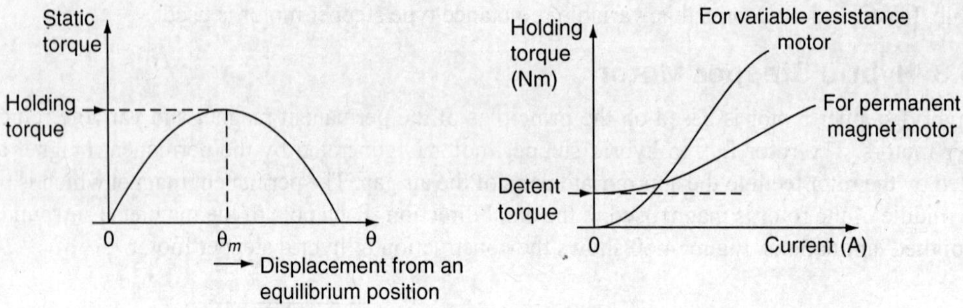

(a) Torque-displacement characteristic (b) Torque-current characteristic

Fig. 4.61 Static characteristic

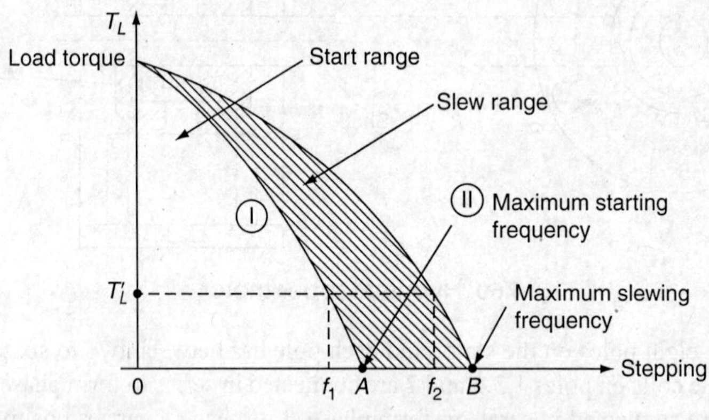

Fig. 4.62 Dynamic characteristic

4.17 Thermal Systems

In thermal plants, chemical processes and household appliances, one requires to control heating and cooling. In any thermal system, the flow of heat from one point to another is totally dependent on temperature rise (or drop) from point to point, and it is much in the same way as current is related to voltage. But there are certain important deviations in this analogy.

To control the temperature of a thermal system, a heating or cooling arrangement or a combination of both is required. Figure 4.63 shows a heating arrangement with heather as a source of heat.
The rate of generation of heat is expresses by

$$q = i^2 r = \frac{e^2}{r} W \tag{4.99}$$

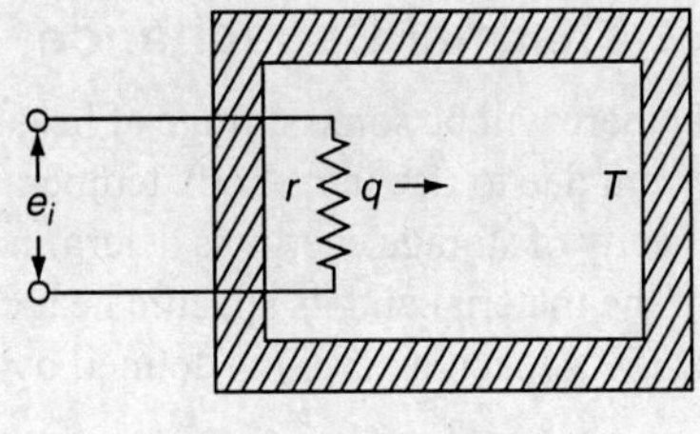

Fig. 4.63 Thermal system with electrical heating arrangement

Let us make the following assumptions to obtain the appropriate mathematical model:

(i) The heater has negligible inductance.
(ii) The resistance in the line is neglected.
(iii) The lead wire has negligible resistance.
(iv) The mass of metal in the heater resistance is small so that its thermal capacity will be neglected.
(v) The lead wire from the voltage source to the heather conducts negligible heat away from the coil.
(vi) The fluid is thoroughly mixed at all points.

4.17.1 Thermal Resistance

Figure 4.64 shows a metal rod subjected to temperature T_1 and T_2 at two ends. If heat storage within the thermal conductor is neglected, the flow of heat q is given by

$$q = \frac{T_1 - T_2}{R} \tag{4.100}$$

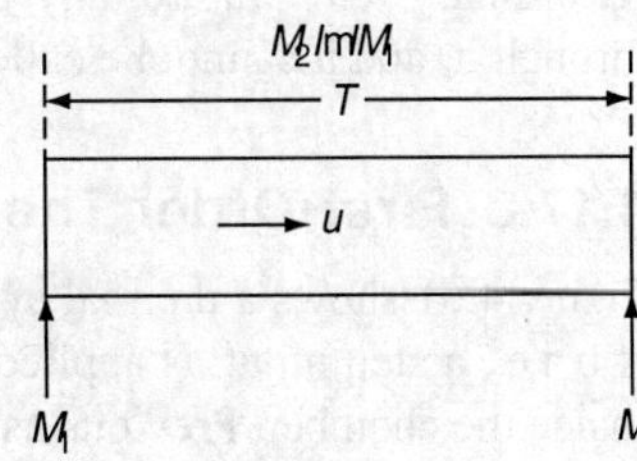

Fig. 4.64 Thermal resistor

where R is the thermal resistance and it is given by

$$R = \frac{l}{Ka} \tag{4.101}$$

This is defined as follows:

$$R = \frac{\text{Change in temperature difference, °C}}{\text{Change in heat flow rate, kcal/sec}}$$

In Eq. (4.101), K is the thermal conductivity of the material, 'l' is the length of the rod, and 'a' is the area through which heat flows.

The overall coefficient of heat transfer is given by

$$U = \frac{K}{l} \tag{4.102}$$

and thermal conductance is given by

$$G = \frac{1}{R} = Ua \tag{4.103}$$

4.17.2 Thermal Capacitance

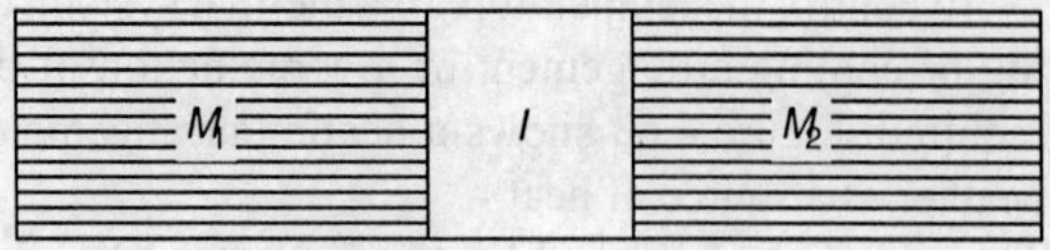

Fig. 4.65 Thermal capacitor

Usually there will be some storage of heat in a thermal system due to rise in its body temperature, and this capacity of storage energy is determined by the mass of the material and its specific heat capacity.

Thermal capacitance (C) is defined by

$$C = \frac{\text{Change in heat stored, kcal}}{\text{Change in temperature, °C}}$$

Figure 4.65 shows a highly idealized case of a thermal insulator having a thermal capacity (C) given by

$$C = m\sigma \tag{4.104}$$

where m is the mass (kg) and σ is the specific heat (kcal/kg°C) of the body. The heat stored is given by

$$Q_s = C(T_1 - T_2) \tag{4.105}$$

At ordinary temperature, any thermal element will not be able to store heat by virtue of heat flow through it, and it cannot be called a thermal inductance.

4.17.3 First-Order Thermal System

Figure 4.61 shows a thermal system where the ambient temperature outside the heating chamber is T_a. If e_i i.e., a step input, is applied to the thermal system, we will find the response of the temperature T inside the chamber. Previous assumptions (i)–(iv) are made here and $T(0) = T_a$. Let A be the area of the chamber surface through which heat leaks outside. U be the overall coefficient of heat transfer, and C be the thermal capacity of the chamber. The rate of heat leak (q_i) is given by

$$q_i = UA(T - T_a) = G(T - T_a) \tag{4.106}$$

If heat stored inside the chamber is $(q - q_i)dt$ in time dt, this will increase the temperature of the chamber by dT. We can write

$$CdT = (q - q_i)dt$$

$$\therefore \quad q - q_i = C\frac{dT}{dt}$$

i.e.,

$$q = C\frac{dT}{dt} + q_i$$

i.e.,

$$q = C\frac{dT}{dt} + G(T - T_a) \tag{4.107}$$

Let

$$\theta = T - T_a. \tag{4.108}$$

If we assume T_a remains constant,

$$\frac{d\theta}{dt} = \frac{dT}{dt} \tag{4.109}$$

and $\theta(0) = T(0) - T_a = 0$

Using Eq. (4.108) and (4.109), from Eq. (4.107) we have

$$q = C\frac{d\theta}{dt} + G\theta \tag{4.110}$$

Taking Laplace transform of Eq. (4.110), we have

$$q(s) = C[s\theta(s) - \theta(0)] + G\theta(s) = Cs\theta(s) + G\theta(s) = (Cs+G)\theta(s) \text{ [Putting } \theta(0) = 0]$$

$$\therefore \quad \theta(s) = \frac{q(s)}{Cs+G} = \frac{Q}{s(Cs+G)} = \frac{Q}{G}\left[\frac{1}{s} - \frac{1}{s+\frac{1}{\tau}}\right] \quad \left[\because q(s) = \frac{1}{s}\right] \tag{4.111}$$

where $\tau = C/G$.

Taking inverse Laplace transform, of Eq. (4.111), we have

$$\theta(t) = \frac{Q}{G}(1 - e^{-t/\tau}) \tag{4.112}$$

Figure 4.66 shows the plot of $\theta(t)$ versus time.

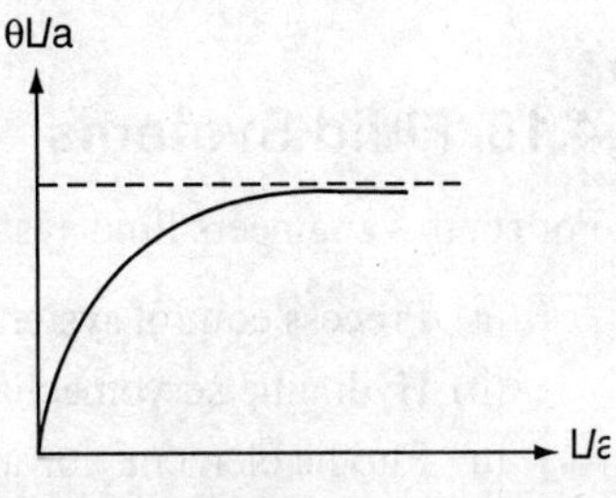

Fig. 4.66 Plot of $\theta(t)$ vs. time

4.17.4 Second-Order Thermal System

Figure 4.67 shows a second-order thermal system, which is the same as in Fig. 4.63. The heating element has considerable thermal capacity compared to that of the fluid being heated, and additional energy storage must be taken into account. Let C_1 and G_1 be the thermal capacity and thermal conductance of the heater, respectively.

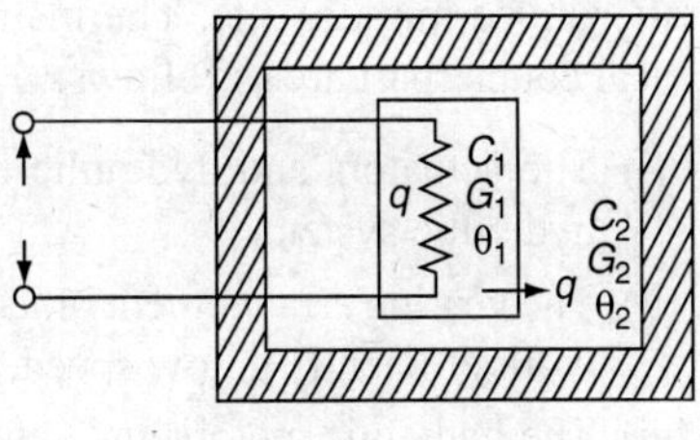

Fig. 4.67 Second-order thermal system

The transfer of heat from the heater to the surrounding fluid is given by

$$q_1 = G_1(\theta_1 - \theta_2) \tag{4.113}$$

where q_1 is the heat transferred from chamber 1 to chamber 2, θ_1 is the difference between the heater element temperature and the ambient temperature, and θ_2 is the difference between the temperature of the heating chamber and the ambient temperature.

We have for the thermal element for storage of energy,

$$q = q_1 + C_1\frac{d\theta_1}{dt}$$

i.e.,
$$q = C_1 \frac{d\theta_1}{dt} + G_1(\theta_1 - \theta_2) \tag{4.114}$$

where q is the total heat rate of the heating element embedded in the heater.

For the heating chamber, we have

$$G_1(\theta_1 - \theta_2) = G_2\theta_2 + C_2 \frac{d\theta_2}{dt} \tag{4.115}$$

Taking Laplace transform of Eq. (4.114) and Eq. (4.115), we have

$$(sC_1 + G_1)\theta_1(s) - G_1\theta_2(s) = q(s) + C_1\theta_1(0) \tag{4.116}$$

and
$$-G_1\theta_1(s) + [sC_2 + (G_1 + G_2)]\theta_2(s) = 0 + C_2\theta_2(0) \tag{4.117}$$

In matrix form, we have

$$\begin{bmatrix} sC_1 + G_1 & -G_1 \\ -G_1 & sC_2 + (G_1 + G_2) \end{bmatrix} \begin{bmatrix} \theta_1(s) \\ \theta_2(s) \end{bmatrix} = \begin{bmatrix} q(s) \\ 0 \end{bmatrix} + \begin{bmatrix} C_1\theta_1(0) \\ C_2\theta_2(0) \end{bmatrix} \tag{4.118}$$

4.18 Fluid Systems

For control engineers fluid systems are classified into three main types:

(a) Process control systems for controlling flow rates and fluid level.
(b) Hydraulic servomechanism for controlling forces and motion in machines.
(c) Fluidic elements for use with on-off and other types of controllers.

Hydraulic devices are extensively used in control systems, and these include power steering mechanism of large ships, control of large workshop machines power steering and braking in automobiles of aircraft, guided missiles, etc. The main advantages of the hydraulic servomechanism compared to its electrical counterpart are as follows:

(i) The actuators and hydraulic machines are relatively smaller due to high power density in a hydraulic system.
(ii) The gearing arrangement or its counterpart is relatively smaller because the hydraulic machines work efficiently at low speed.
(iii) The hydraulic motors and actuators are very rapid acting.
(iv) The hydraulic systems are very rugged and reliable.
(v) The hydraulic systems are less sensitive to noise pickup from shock and vibration.

4.18.1 Flow and Level as System Variable

Figure 4.68 shows laminar flow through a pipe section. The pressure difference $p_1 - p_2$ given by

$$q = C_1A(p_1 - p_2) = C_1A\rho(h_1 - h_2)g \tag{4.119}$$

where

p_1 is the upstream pressure for which the equivalent head is h_1,
p_2 is the downstream pressure for which the equivalent head is h_2,

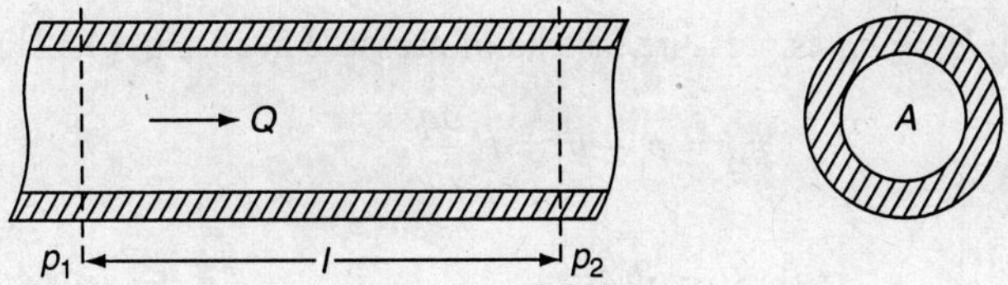

Fig. 4.68 Laminar flow through a pipe

A is the flow area,

C_1 is a constant for laminar flow and it depends on the length/between the upstream and downstream gauge points,

ρ is the density of the fluid and

g is the gravitational acceleration.

For turbulent flow, the corresponding expression for q is

$$q = CA(p_1 - p_2)^{\frac{1}{2}}$$

where C_1 is a constant of turbulent flow.

Fluid inductance is a measure of fluid inertia. Figure 4.69 shows a non-accelerating cylindrical pipe with negligible friction.

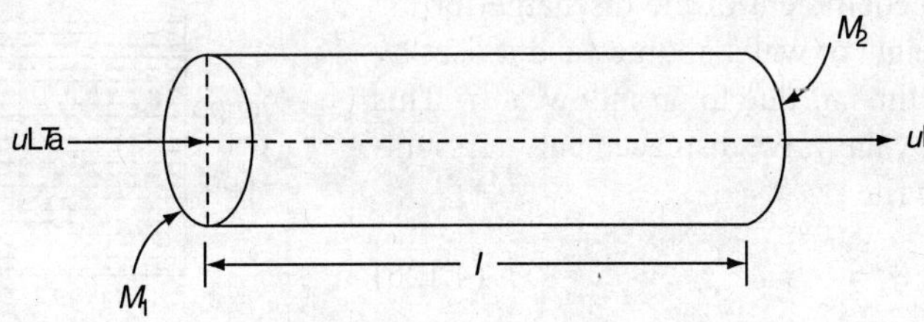

Fig. 4.69 Fluid inductance of a pipe

The fluid of mass ρlA in the tube experiences a force of $A_{p_{12}}$, provided the pressure differential $p_{12} = p_1 - p_2$ exists. The acceleration of fluid mass is given by

$$A_{p_{12}} = \rho lA \frac{dv}{dt} \tag{4.120}$$

where $v = q/A$ is the velocity of the fluid and $\rho gh = p$. From Eq. (4.119), we have

$$A\rho g(h_1 - h_2) = \rho lA \frac{d\left(\frac{q}{A}\right)}{dt}$$

Therefore,

$$h_{12} = h_1 - h_2 = \frac{l}{Ag}\frac{dq}{dt} = I\frac{dq}{dt} \tag{4.121}$$

where $I = l/Ag$. Taking $h_{12} = p_{12}$ as pressure instead of pressure head, Eq. (4.121) can be expresses as

$$p_{12} = p_1 - p_2 = I' \frac{dq}{dt} \tag{4.122}$$

where $$I' = \rho l/A. \tag{4.123}$$

If the friction is not negligible, the total head loss is due to the following factors:

(i) acceleration of the moving fluid in the pipe and
(ii) the loss due to friction.

If the resistance of the pipe is R, we have

$$h_{12} = I\frac{dq}{dt} + Rq \tag{4.124}$$

where $$R = 1/C_1 A\rho\, g \text{ [comparing with Eq. (4.119)]} \tag{4.125}$$

4.18.2 First-Order Fluid Level System

Figure 4.70 shows a first-order fluid level system where p and p_o are the pressures on the two sides of the orifice, respectively. C_l is the coefficient of the discharge for the orifice. Let h be the height of water at time t and it rises by an amount dh after a time dt due to an inflow of q_i. This inflow q_i is governed by the valve displacement x_i. Assuming linearity, we can write

$$q_i = C_v x_i \tag{4.126}$$

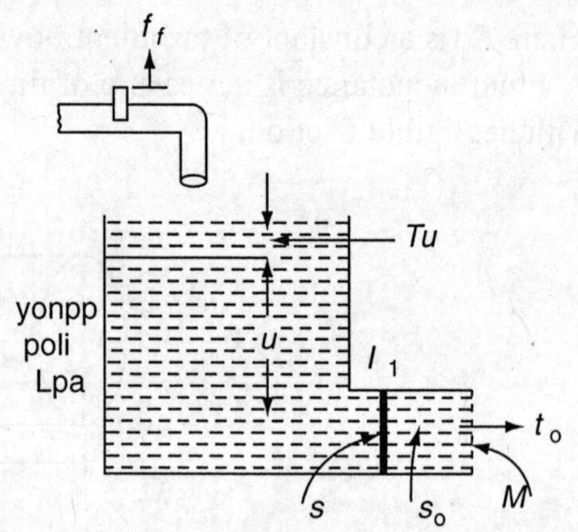

Fig. 4.70 First-order hydraulic system

where C_v is the valve constant.
We have the following quantities during the interval dt:

The net flow from source = $q_i dt$
The net storage in the tank = Adh
where ρ is the density of the fluid, g is the acceleration due to gravity, $p = \rho gh$ and a is the outlet area.

From conservation of mass,
Quantity of fluid flowing in = Additional storage + Quantity of fluid flowing out

$$q_i dt = A\,dh + C_1 a\rho gh\,dt - C_1 p_0\, a\,dt$$

Assuming $p_0 = 0$, we have

$$q_i = A\frac{dh}{dt} + C_1 a\rho gh \tag{4.127}$$

From Eqs. (4.126) and (4.127), we have

$$C_v x_i = A\frac{dh}{dt} + C_1 a\rho gh \tag{4.128}$$

Putting $A/C_v = C$ and $C_1\ a\rho gh\ h/C_v = G$, Eq. (4.128) becomes

$$x_i = C\frac{dh}{dt} + Gh \tag{4.129}$$

Equation (4.129) is a first-order differential equation and it gives a system of the same order. It relates fluid level h with valve opening x_i.

4.18.3 Second-Order Fluid Level System

Figure 4.71 shows a second-order system where two tanks are interconnected.

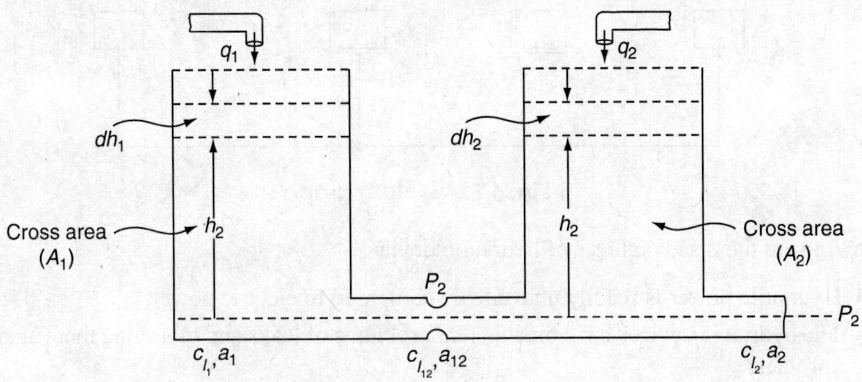

Fig. 4.71 Second-order hydraulic system

For tank 1:

Net fluid input = $q_1 dt$
Net output at $P_1 = C_{l_{12}}\rho g(h_1 - h_2)a_{12}dt$
Net storage = $A_1\ dh_1$
We can write

$$q_1 = A_1\frac{dh}{dt} + C_{l_{12}}\rho g(h_1 - h_2)a_{12} \tag{4.130}$$

For tank 2:

Net fluid input = $C_{l_{12}}\rho g(h_1 - h_2)a_{12}dt$
Net output at $P_2 = C_{l_2}\rho g\ h_2 a_2$
Net storage = $A_2\ dh_2$
We can write

$$q_1 = (C_{l_2}\rho g h_2 a_2 + C_{l_{12}}\rho g a_{12})h_2 + A_2\frac{dh_2}{dt} - C_{l_{12}}\rho g h_1 a_{12} \tag{4.131}$$

Putting $B_{12} = C_{l_{12}}\rho g a_{12}$ and $B_2 = C_{l_2}\rho g a_2$, from Eqs. (4.130) and (4.131), we have

$$q_1 = A_1\frac{dh_1}{dt} + B_{12}h_1 - B_{12}h_2 \tag{4.132}$$

and $$q_2 = -B_{12}h_1 + A_2\frac{dh_2}{dt} + (B_{12} + B_2)h_2 \tag{4.133}$$

Taking $B = 1/K$ where B designates the hydraulic constant. Figure 4.72 shows the system graph for the second-order system.

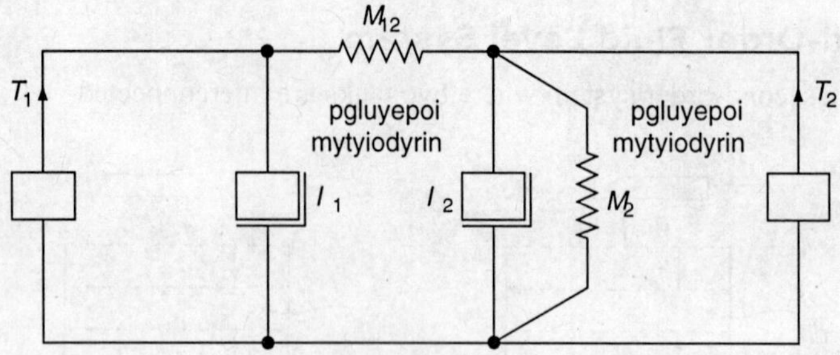

Fig. 4.72 System graph

The following are the disadvantages of hydraulic devices:

(i) Hydraulic power is readily unavailable compared to electric power.
(ii) The hydraulic devices have the inherent problems of leaks and of sealing them against foreign particles.
(iii) Operating noise is present.
(iv) Compared to electrical system, they are costly.
(v) At low temperature, the viscosity of fluid increases and hence these devices have a tendency to be sluggish.
(vi) Fire and explosion hazards exist.
(vii) Like electric cables, the hydraulic lines are not flexible.
(viii) There exists difficulty to design sophisticated hydraulic systems due to the involvement of non-linear and other complex characteristics.

4.18.4 Example of Hydraulic devices

Hydraulic motor pump set is the most frequently used hydraulic device in control system. Figure 4.73 shows a hydraulic motor pump set consisting of a variable stroke hydraulic pump and fixed stroke

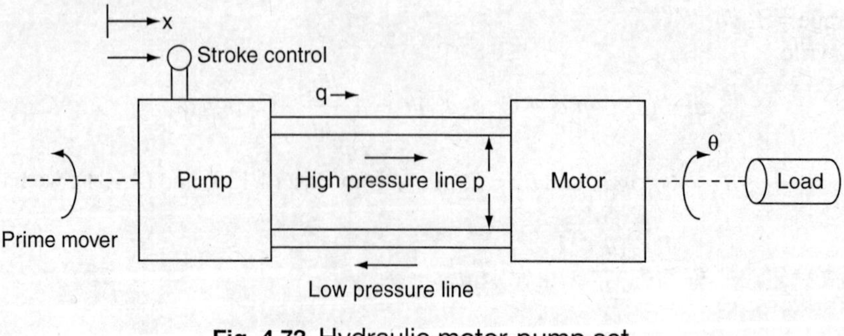

Fig. 4.73 Hydraulic motor-pump set

hydraulic motor. A linear displacement (stroke length) is used as input to this device and it delivers a large output torque.

The amount of oil delivered by the pump controls the hydraulic motor. The pump stroke is mechanically changed to control the oil delivered by the pump. There is no essential difference between hydraulic pump and motor like in DC generator and motor.

Let us take the following:

q_p = rate at which the oil flows from the pump,
q_m = oil flow rate through the motor,
q_i = leakage flow rate,
q_c = compressibility flow rate,
x = input stroke rate,
θ = output angular displacement of motor and
P = pressure drop across motor.

Since the rate of flow of oil from the pump is proportional to the stroke length, we have

$$q_p \propto x$$

i.e.,

$$q_p = K_p x \tag{4.134}$$

In Eq. (4.134) K_p is a constant and is equal to the ratio of rate of oil flow to unit stroke angle.

Since the rate of oil flow through the motor is proportional to motor speed, we have

$$q_m \propto \frac{d\theta}{dt}$$

i.e.,

$$q_m = K_m \frac{d\theta}{dt} \tag{4.135}$$

where K_m is known as motor displacement constant.

It is not possible to flow of all the oil through the motor in proper channel from the pump. Hence a portion of the ideal flow from the pump leaks back past the pistons off motor pump due to back pressure in the motor, which is built up by the hydraulic flow. This overcomes the resistance to free movement, which is offered by the load on the motor shaft.

The leakage flow is usually assumed to proportional to motor pressure. Therefore, we have

$$q_i \propto P$$

i.e.,

$$q_i = K_i P \tag{4.136}$$

In Eq. (4.136), K_i is the constant.

The developed back pressure also causes the oil in the lines to compress. Since volume compressibility flow is proportional to pressure, the rate of flow (i.e., compressibility flow rate) is proportional to the rate of change of pressure. Therefore, we have

$$q_c \propto \frac{dP}{dt}$$

i.e., $$q_c = K_c \frac{dP}{dt} \tag{4.137}$$

In Eq. (4.137), K_c is known as coefficient of compressibility.

Hence the sum of oil flow rate through the motor, leakage flow rate and compressibility flow rate gives the rate at which the oil flows from the pump.

$\therefore$ $$q_c = q_m + q_i + q_c \tag{4.138}$$

Substituting Eq. (4.134), Eq. (4.135), Eq. (4.136) and Eq. (4.137) in Eq. (4.138) we have

$$K_p x = K_m \frac{d\theta}{dt} + K_i P + K_c \frac{dP}{dt} \tag{4.139}$$

The torque developed by the motor is proportional to the pressure drop which also balances the load torque. Therefore, we have

$$T_m = K_t P \tag{4.140}$$

where K_t is motor torque constant.

The load is assumed to consist of moment of inertia J and viscous friction with coefficient D. Therefore, the expression for load torque becomes

$$T_l = J \frac{d^2\theta}{dt^2} + D \frac{d\theta}{dt} \tag{4.141}$$

$$\text{Hydraulic power input} = q_m P \tag{4.142}$$

Using Eq. (4.135), we have from Eq. (4.142)

$$\text{Hydraulic power input} = K_m P \frac{d\theta}{dt} \tag{4.143}$$

$$\text{Mechanical power output} = T_m \frac{d\theta}{dt} \tag{4.144}$$

Using Eq. (4.140), we have from Eq.(4.144)

$$\text{Mechanical power output} = K_t P \frac{d\theta}{dt} \tag{4.145}$$

Let the hydraulic motor losses be neglected or be included as a part of load. In this case the hydraulic motor input becomes equal to mechanical power output of hydraulic motor. Therefore, we have

$$K_m P \frac{d\theta}{dt} = K_t P \frac{d\theta}{dt}$$

i.e., $$K_m = K_t \tag{4.146}$$

Equation (4.140) becomes

$$T_m = K_t P = K_m P \tag{4.147}$$

Since the motor torque is equal to the load torque, we have

$$T_m = T_l$$

i.e.,
$$K_m P = J\frac{d^2\theta}{dt^2} + D\frac{d\theta}{dt}$$

i.e.,
$$P = \frac{J}{K_m}\frac{d^2\theta}{dt^2} + \frac{D}{K_m}\frac{d\theta}{dt}$$

i.e.,
$$P = \frac{J}{K_m}\frac{d^2\theta}{dt^2} + \frac{D}{K_m}\frac{d\theta}{dt} \tag{4.148}$$

i.e.,
$$\frac{dP}{dt} = \frac{J}{K_m}\frac{d^3\theta}{dt^3} + \frac{D}{K_m}\frac{d^2\theta}{dt^2} \tag{4.149}$$

Substituting P and $\frac{dP}{dt}$ from Eq. (4.148) and Eq. (4.149) in Eq. (4.139), we have

$$K_p x = K_m\frac{d\theta}{dt} + K_i\left(\frac{J}{K_m}\frac{d^2\theta}{dt^2} + \frac{D}{K_m}\frac{d\theta}{dt}\right) + K_c\left(\frac{J}{K_m}\frac{d^3\theta}{dt^3} + \frac{D}{K_m}\frac{d^2\theta}{dt^2}\right)$$

i.e.,
$$K_p x = \left(\frac{K_c J}{K_m}\right)\frac{d^3\theta}{dt^3} + \left(\frac{K_i J}{K_m} + \frac{K_c D}{K_m}\right)\frac{d^2\theta}{dt^2} + \left(K_m + \frac{K_i D}{K_m}\right)\frac{d\theta}{dt} \tag{4.150}$$

Laplace transform of Eq. (4.150) with zero initial conditions gives

$$K_p X(s) = \left(\frac{K_c J}{K_m}\right)s^3\theta(s) + \left(\frac{K_i J}{K_m} + \frac{K_c D}{K_m}\right)s^2\theta(s) + \left(K_m + \frac{K_i D}{K_m}\right)s\theta(s)$$

i.e.,
$$\frac{\theta(s)}{X(s)} = \frac{K_p}{\left(\frac{K_c J}{K_m}\right)s^3 + \left(\frac{K_i J}{K_m} + \frac{K_c D}{K_m}\right)s^2 + \left(K_m + \frac{K_i D}{K_m}\right)s}$$

i.e.,
$$\frac{\theta(s)}{X(s)} = \frac{K_p}{s\left[\frac{K_c J}{K_m}s^2 + \left(\frac{K_i J}{K_m} + \frac{K_c D}{K_m}\right)s + \frac{K_m^2 + K_i D}{K_m}\right]} \tag{4.151}$$

Usually $K_c << K_m$, therefore we have from Eq. (4.151) with $K_c = 0$

$$\frac{\theta(s)}{X(s)} = \frac{K_p}{s\left[\frac{K_i J}{K_m}s + \frac{K_m^2 + K_i D}{K_m}\right]}$$

i.e.,
$$\frac{\theta(s)}{X(s)} = \frac{\frac{K_p}{\frac{K_m^2 + K_i D}{K_m}}}{s\left[\frac{K_i J}{K_m^2 + K_i D}s + 1\right]}$$

i.e.,
$$\frac{\theta(s)}{X(s)}=\frac{K}{s(\tau s+1)} \tag{4.152}$$

In Eq. (4.152), $K=\dfrac{K_p}{\dfrac{K_m^2+K_i D}{K_m}}$ and $\tau=\dfrac{K_i J}{K_m^2+K_i D}$

Equation (4.152) shows the transfer function of the hydraulic motor pump set.

4.19 Pneumatic System

Compressible fluid is used as working medium in pneumatic system, where the compressibility effects of gas cannot be denied. Therefore, the dynamic equations will be resulted using conservation of mass. The change in the fluid's inertia energy and internal thermal energy are assumed to be negligible. The mass and volume flow rates are not readily interchangeable because pneumatic system uses compressible fluid as working fluid. Hence, the analysis of gas becomes more complicated.

The flow of gas in air through connected pipe lines and pressure vessels is involved in the pneumatic devices. The mass flow rate (q_m) is a through variable and it is analogous to current whereas the pressure is a cross variable and hence it is analogous to voltage.

The resistance and capacitance are the two basic elements of pneumatic system. The resistance to gas flow occurs due to the restrictions in the pipes and valves.

The gas flow resistance (R) is defined by

$$R=\frac{\text{Change in gas pressure difference, N/m}^2}{\text{Change in gas flow rate kg/sec.}}$$

The pneumatic capacitance (C) is defined by

$$C=\frac{\text{Change in gas stored, kg}}{\text{Change in pressure, N/m}^2}$$

which is defined for a pressure vessel and hence depends on the type of expansion process involved.

The pneumatic devices are used in air craft systems, guided missiles, automatic production machinery, and also in many other fields as automatic controllers.

The following are the advantages of pneumatic system:

(i) It offers safety from fire hazards due to the usage of non-inflammable air or gas.
(ii) The viscosity of air or gas is negligible compared to the high viscosity of hydraulic fluids.
(iii) Air can be let out at the end of defined work cycle and hence no return pipelines are required.

The response of the pneumatic is slower than that of hydraulic systems. This is due to the compressibility of the working fluid. This is the disadvantage of pneumatic system.

Figure 4.74 shows a simple pneumatic system, which consists of a pneumatic bellows consisting of a hollow chamber with thin pneumatic walls. The bellows have a corrugated side walls and the input and output surfaces are flat. If there is any increase in pressure within the bellows, it results in an increase in separation between the input and output surfaces. The following nomenclatures are used.

$\bar{P}_i$ = Steady-state value of input air pressure,
p_i = Increase in pressure of air–source,

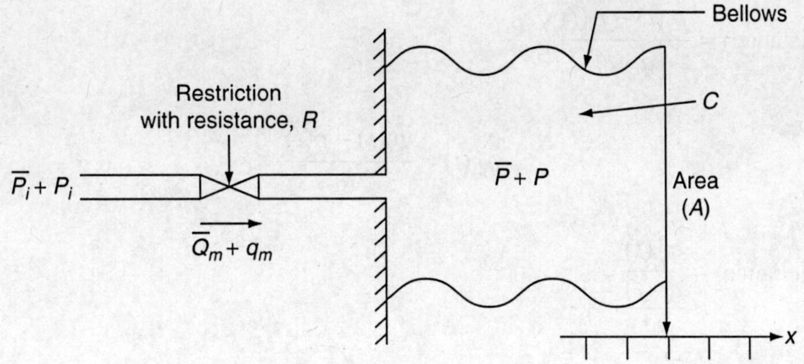

Fig. 4.74 Pneumatic system

$\overline{p}$ = Steady-state value of pressure inside the bellows,
p = Increase-in pressure inside the bellows,
$\overline{Q}_m$ = Steady-state value of air flow rate,
q_m = Increase in air flow rate,
A = Area of each flat surface of the bellows,
R = Resistance of the restrictions,
C = Capacitance of the bellows and
X = Displacement of the movable surface of the bellows.

Let P_i be the amount by which the pressure of air source is increased from its steady-state value. This causes an increase in air flow by q_m and hence there is increase in pressure inside the bellows, which results in the displacement of the movable surface of the bellows by the amount of x. The terms p_i, q_m, p, and x are all functions of time (t) and hence these can be expressed as $P_i(t)$, $q_m(t)$, $p(t)$, and $\lambda(t)$.

The force exerted on the surface of the bellows is directly proportional to the increase in pressure inside the bellows.

$$\therefore \quad f_b \propto p(t)$$

$$\therefore \quad f_b = Ap(t) \tag{4.153}$$

The force opposing the movement of the flat surface of bellow walls is expressed by

$$f_o \propto \lambda(t)$$

$$\therefore \quad \therefore f_o = K\lambda(t) \tag{4.154}$$

where K is the constant of proportionality and it represents the stiffness of the bellows.

During steady state, we have

$$f_b = f_o$$

$$\therefore \quad Ap(t) = K\lambda(t) \tag{4.155}$$

Here R (the resistance) $= \dfrac{p_i(t) - p(t)}{q_m(t)}$

i.e.,
$$q_m(t) = \frac{p_i(t) - p(t)}{R} \tag{4.156}$$

and C (the capacitance) $= \dfrac{q_m(t)}{\dfrac{dp(t)}{dt}}$

i.e.,
$$q_m(t) = C\frac{dp(t)}{dt} \tag{4.157}$$

From Eq. (4.156) and Eq. (4.157), we have

$$C\frac{dp(t)}{dt} = \frac{p_i(t) - p(t)}{R}$$

i.e.,
$$RC\frac{dp(t)}{dt} = p_i(t) - p(t)$$

i.e.,
$$RC\frac{dp(t)}{dt} + p(t) = p_i(t) \tag{4.158}$$

From Eq. (4.155), we have

$$p(t) = \frac{K}{A}\lambda(t) \tag{4.159}$$

i.e.,
$$\frac{dp(t)}{dt} = \frac{K}{A}\frac{d\lambda(t)}{dt} \tag{4.160}$$

Using Eq. (4.159) and Eq. (4.160), we have from Eq. (4.158)

$$RC\left[\frac{K}{A}\frac{d\lambda(t)}{dt}\right] + \frac{K}{A}\lambda(t) = p_i(t) \tag{4.161}$$

Laplace transform of Eq. (4.161) with zero initial condition gives

$$RC\left[\frac{K}{A}s\,\lambda(s)\right] + \frac{K}{A}\lambda(s) = p_i(s)$$

i.e.,
$$\frac{\lambda(s)}{p_i(s)} = \frac{A/K}{RCs+1} = \frac{A/K}{Ts+1} \tag{4.162}$$

where $T = RC$ = time constant of the system. Eq. (4.162) shows the transfer function of the system.

Table 4.3 shows the comparison between hydraulic and pneumatic devices.

Table 4.3 Comparison between hydraulic and pneumatic devices

Pneumatic	Hydraulic
Compressible working fluid	Incompressible wor king fluid
Lack of lubricating property in working fluid	Working fluid acts as lubricant
Lower operating pressure	Higher operating pressure
Less output power	More output power
Poor accuracy of actuator	Possible to make more accuracy
External leakage is permissible but internal leakage should be avoided	Internal leakage is permissible but external leakage should be avoided
Return pipes are not required	Return pipes are required
Insensitive to temperature changes	Sensitive to temperature changes
Fire and explosion proof	Neither fire proof nor explosion proof

4.20 Chemical Systems

The control temperature, flow rate, liquid level, etc. are required during the control of a chemical process. Figure 4.75 shows a simple chemical process where a solution of some chemical in water with a concentration of C_i is mixed with pure water to obtain an outflow stream concentration C_o. The solution flow rate is assumed to vary linearly with the controlling valve displacement x_v. Therefore, we can write

$$Q_i = K_v x_v \tag{4.163}$$

If Q_o be the output flow rate, we have

$$Q_0 = Q_i + Q_w \tag{4.164}$$

If mass of solute $Q_i\, C_i\, dt$ enters the tank while a mass of $Q_0 C_0\, dt$ leaves the tank, the change in the amount of solute dQ_o in the tank is given by

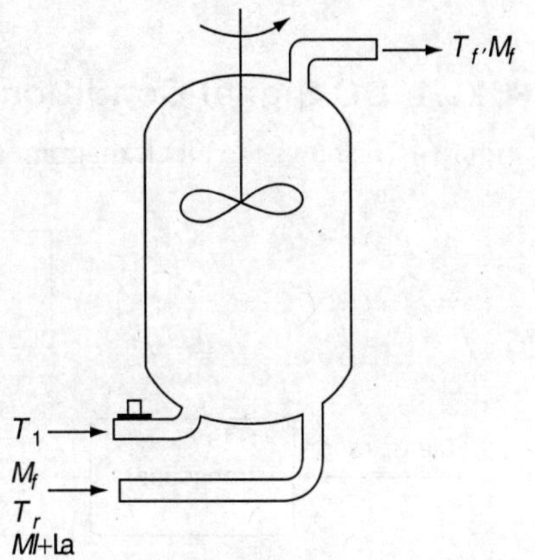

Fig. 4.75 Chemical process

$$dQ_o = \left(\frac{Q_i C_i - Q_o C_o}{V}\right) dt \tag{4.165}$$

Equation (4.165) gives

$$\frac{dQ_o}{dt} + \frac{Q_o}{V} C_o = \frac{Q_i}{V} C_i = \frac{C_i K_v}{V} x_v \tag{4.166}$$

If Q_o is assumed to be constant, from Eq. (4.166) we have

$$\frac{C_o(s)}{x_v} = \frac{K}{\tau s + 1}$$

where $\tau = V/Q_o$ is the time required to empty the tank at rate Q_o and $K = C_i K_v / Q_o$.

4.21 Signal Conditioning

Signal conditioning means modifying the transduced signal into the usable format for final stage of measurement systems. It involves faithful representation of analog or digital output of the transduced signal.

So the signal conditioning equipment may be required to do linear processes like attenuation, amplification, addition, subtraction, integration and differentiation. They are also required to do non linear processes like modulation, demodulation, filtering, sampling, clipping and clamping.

Signal condition equipment generally is an excitation amplification system for passive transducers and may be an amplification system for active transducers.

In both amplifications, the transducers output is increased to a sufficient level so that they can be processed, recorded.

Now depending upon the type of excitation used, signal conditioning system can be divided into two types:

(i) DC signal conditioning system and
(ii) AC signal conditioning system.

4.21.1 DC Signal Conditioning System

Figure 4.76 shows the block diagram of AC Signal Conditioning System.

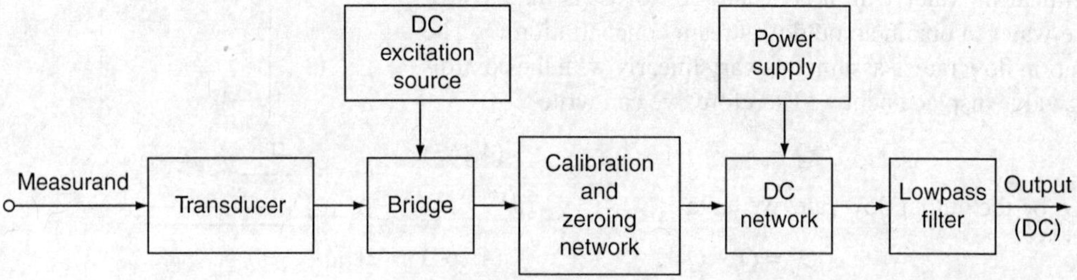

Fig. 4.76 DC signal conditioning system

Characteristic of DC amplifier:

(i) It may lead to high CMRR as it may read balanced differential input.
(ii) Extremely good thermal stability.

Advantages of DC amplifier:

(i) At low frequencies calibration is easier.
(ii) It is able to recover from overload condition.

Disadvantages of DC amplifier:

(i) It suffers from problem of drift. Due to drift low frequency spurious signals come out as data information. To overcome this, DC amplifier is followed by low pass filter which removes noise and high frequency component from data signal.

DC signal conditioning systems are generally used for common resistance transducers for example, potentiometers.

4.21.2 AC Signal Conditioning System

Figure 4.77 shows the block diagram of AC Signal Conditioning System. In AC signal conditioning system, variable resistance or variable inductances are used as transducers which works between carrier

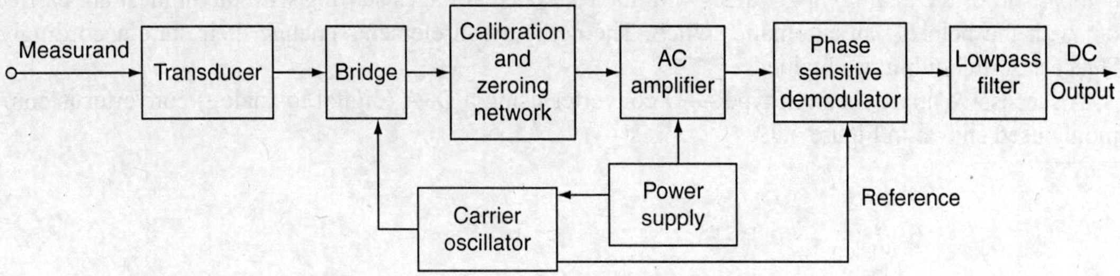

Fig. 4.77 AC signal conditioning system

frequencies 50 Hz to 200 kHz variation in transducer settings, modulate the carrier frequencies at the output and the waveform is amplified and demodulated. As demodulation is phase sensitive, polarity of DC output indicates the direction of parameter change.

AC signal conditioning systems are used where signals have to be transmitted via long cables to connect signal conditioning equipment to transducers and for variable reluctance transducers.

4.22 Analog to Digital (A–D) Converter

When the digital computer is used, the output form the electromechanical transducer, which is analog in nature, is required to be converted to a digital form. This is done with the help of analog to digital converter. Figure 4.78 shows a simplest type of an A–D converter, which is of potentiometric type. This

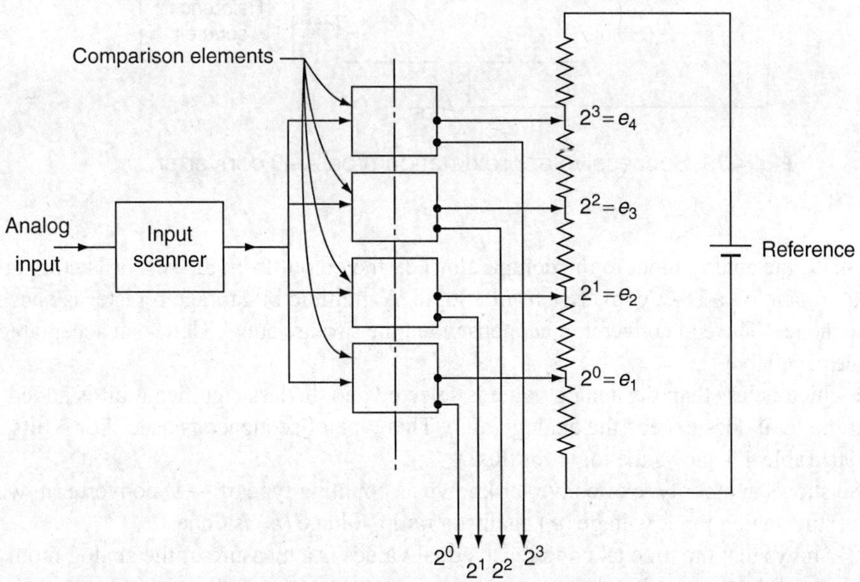

Fig. 4.78 Potentiometer type A–D converter

type of A–D converter uses comparison elements having in the form of semiconductor relays or magnetic cores. If the input signal is greater or less than a fraction of the reference voltage to which these elements are connected, these elements change their state (ON/OFF). The potentiometer resistances are weighted accordingly to get the reference potentials according to the binary values say 1,2,4,8 etc. The comparison of the analog input values with the reference voltages starting with the highest are carried out with the help of an automatic switch. The comparison elements change their state accordingly. Hence a digital output is obtained.

A successive approximation type A–D converter using a D–A (digital to analog) converter is commonly used shown in Figure 4.79.

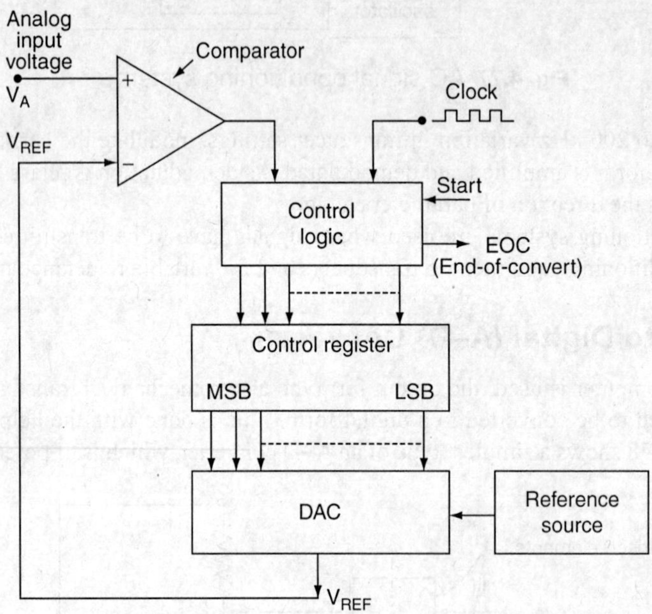

Fig. 4.79 Successive approximation type A–D converter

In Figure 4.79, the analog input to the comparator i.e., the output from the sensor is compared with the generated voltage by a D–A converter from a logic. A multiple bit storage register is successively in a certain sequence. The end conversion happens when the two are equal. Hence an acceptable digital output becomes available.

The MSB which is less than the analog value is selected and the less significant bit is added successively so that the total does exceed the analog value. This type of sequence is used. For 4 bits, we can start with 1000. Table 4.4 shows the logic for this.

Figure 4.80 shows another type of converter known as counting type of A–D converter in which the comparison of the analog input voltage (e_i) against a ramp voltage (e_r) is done.

Figure 4.81 shows that the time taken to reach equal values is a measure of the analog input voltage because the slope of the ramp voltage is constant.

Table 4.4 Logic for 4-bits

Cases	Serial nos. of successive trials	D–A converter input	D–A converter output	Comparator output	Resulting value
A	1	1000	8	Large	------
	2	0100	4	Small	------
	3	0110	6	Large	------
	4	0101	5	Nil	5 acceptable
B	1	1000	8	Large	------
	2	0100	4	Small	------
	3	0110	6	Large	------
	4	0111	7	Nil	7 acceptable
C	1	1000	8	Large	------
	2	0100	4	Small	------
	3	0010	2	Large	------
	4	0011	3	Nil	3 acceptable

An electronic gate is kept open so that the output from a frequency oscillator is fed to a counter. This ensures to get a digital output proportional to the time taken for the two voltages will be equal. As shown in Fig. 4.81 the gate is opened for the times t_1, t_2 and t_3. The time taken by e_i is equal to e_r. Due to the constant frequency of the oscillator the number of oscillations counted by the counter shows a

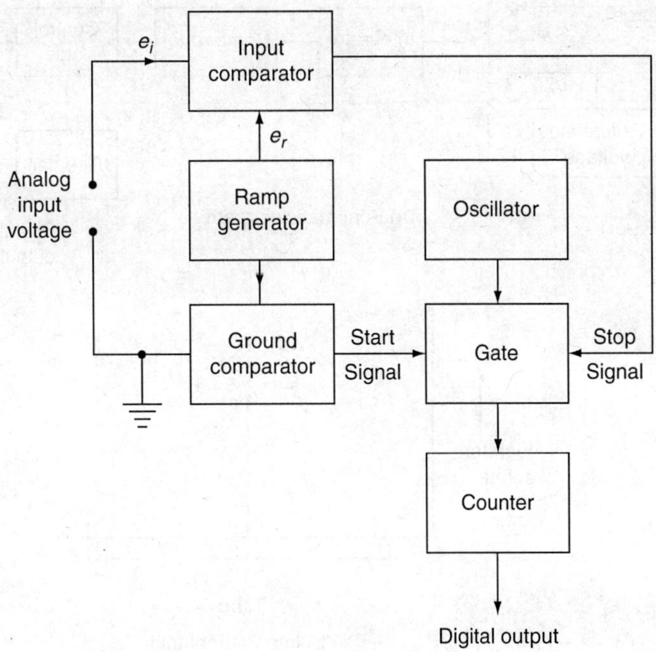

Fig. 4.80 Counting type of A–D converter

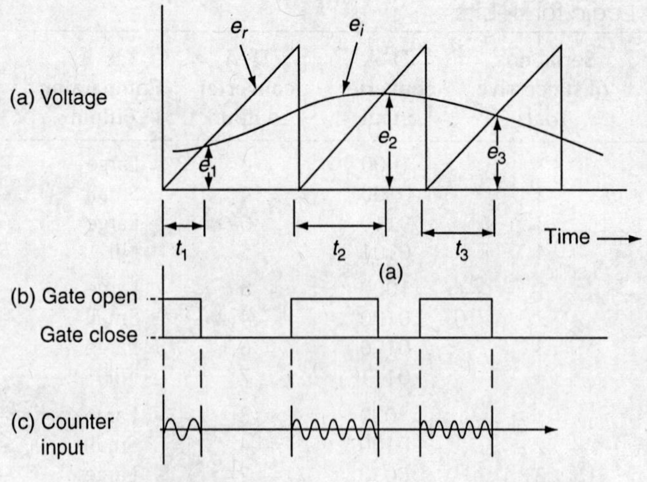

Fig. 4.81 Voltage, gate positions and counter inputs

digital measure of the analog input voltage (e_i). To achieve a binary output the counter can be arranged. The control is done in order to opening and closing the gate in the beginning of ramp and at $e_i = e_r$ as shown in Fig. 4.82.

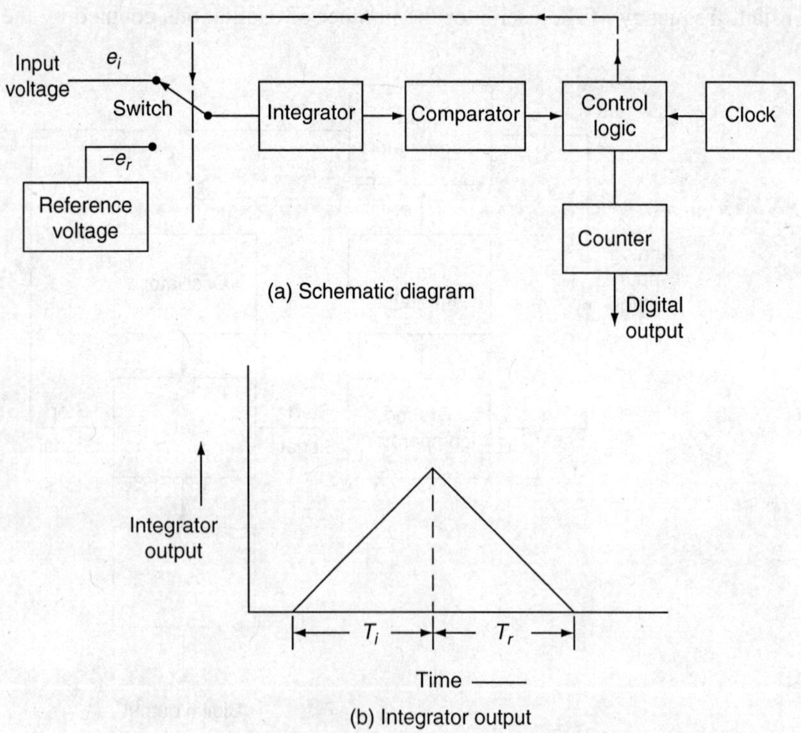

Fig. 4.82 Dual slope integrating type converter

Figure 4.82 shows another type of A–D converter known as dual slope integrating type converter. The performance of this type of converter is affected by the reference source. The accuracy and stability of this reference source should be ensured.

4.23 Digital to Analog (D–A) Converter

To convert a digital signal to analog signal, the value of digital signal defined by the switching is required to be changed to an equivalent value. One type of D–A converter is shown in Fig. 4.83. This performance of this converter is similar to the potentiometric A–D converter discussed in Art 4.22.

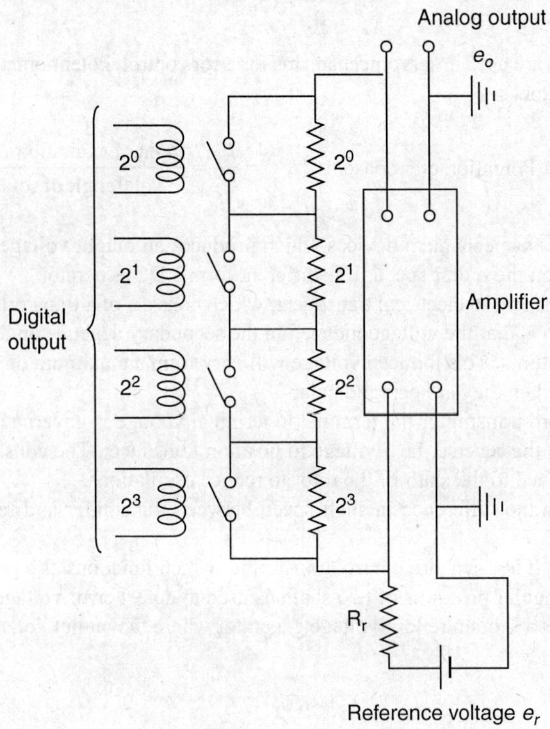

Fig. 4.83 D–A converter

SIGNIFICANT POINTS

Potentiometers: A potentiometer has a variable resistance. Its value varies according to the angular position of the wiper contact and slider contacts as in a rheostat are linear. The variation in resistance is directly proportional to the angular position.

$$\text{Gain} = \frac{E_i}{2\pi N} V/\text{rad}$$

$$\text{Per cent resolution } = \frac{100}{\text{Number of turns of wire}}$$

$$\text{Output voltage } (E_o) = \frac{\alpha E_{in}}{1 + \frac{\alpha(1-\alpha)R_T}{R_L}}$$

$$\text{Loading error } = \frac{\alpha^2(1-\alpha)}{\alpha(1-\alpha) + \frac{R_L}{R_T}}$$

The double-ended potentiometer can be considered as two single-ended potentiometers with their common point earthed:

$$E_o = \frac{E_i}{\theta_T} \times (\theta_1 - \theta_2)$$

These types of potentiometers are used in servomechanisms for error control. Potentiometers also act as an attenuator so that $E_o \leq E$. For an attenuator.

$$\text{Gain constant or Potentiometer constant } (K_p) = \frac{E_i}{\theta_T} = \frac{\text{Input of excitation voltage}}{\text{Total angle of rotation}}$$

Synchros: Synchros are electro-mechanical devices which produces an output voltage depending on the angular position of the rotor and not on the motor speed. It is different form a DC generator.
Transmitter: This is a mechanical to electrical transducer, which consists of a transformer iron core as stator. The basic principle of the synchro is that the voltage induced in the secondary depends on the angle at which the magnetic lines of the flux cut the turns. This induced voltage will vary from a maximum of 1V when the two windings are in line or parallel to 0 V when they are perpendicular.
Synchro Receiver: In synchro transmitter the terminal to terminal voltage is governed by the angular position of the rotor. In the receiver it is the reverse, i.e., voltage to position transducer. The construction is the same except that a heavy flywheel is attached to the shaft of the rotor to reduce oscillations.
Differential Synchro: When the difference in shaft speed between transmitter and receiver is required, we use differential synchro.
Control Transformer (CT): It is a synchro control transformer which functions as a position transducer and error detector. The difference in angular positions of two shafts is used to detect error voltage and correction is applied.
Tachometer: A tachometer is a miniature low-voltage generator where the output voltage of generator is given by the equation.

$$E_g = K\phi\omega$$

Accelerometer: An accelerometer is also a transducer, which is used in high-speed aircrafts and space vehicles.
Gyroscope: This is also an instrument used in spaceships and aircrafts. The input is the angular velocity and the output is the angular displacement.
AC Servomotors: An AC servo motor is basically a two-phase induction motor.
Positional Servomechanism: A servomechanism is a power amplifying feedback control system. The controlled variable in it is a mechanical position or time derivative of position such as velocity of acceleration.
The essential parts of servomechanism are as follows:

(i) a transducer to convert angular position to an electrical signal.
(ii) an amplifier to improve the signal strength,
(iii) a device to develop torque such as motor, and
(iv) devices to transmit torque to load such as gears.

Rate Servomechanism: The derivative of the position, i.e., velocity is monitored and kept constant here with velocity as the controlled variable.

ADDITIONAL SOLVED EXAMPLES

Example 4.7 A 1 kΩ potentiometer shown in Fig. E4.3 has setting ratio 0.4. If the potentiometer is a multi-turn unit having a shaft rotation of 1080°, find the positional error associated with the loading.

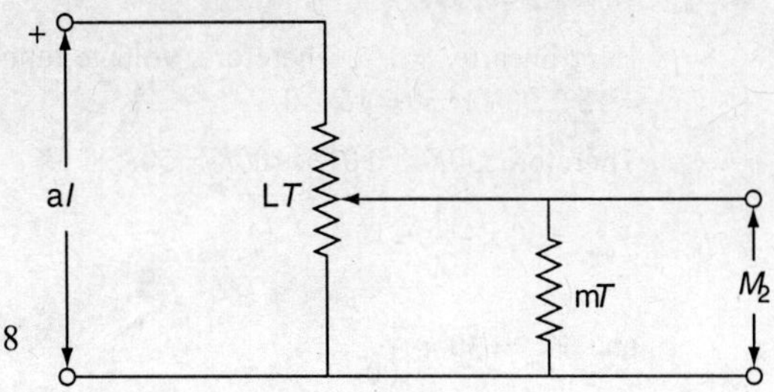

Fig. E4.3

Solution

$$\text{Loading error} = \frac{\alpha^2(1-\alpha)}{\alpha(1-\alpha)+\dfrac{R_L}{R_T}} = \frac{(0.4)^2(1-0.4)}{0.4\times(1-0.4)+5} = 0.018$$

Positional error = Maximum angle of rotation × loading error

$$= 1080° \times 0.018 = 19.44°$$

Example 4.8 Calculate the number of turns of wire needed to provide a potentiometer with a resolution of 0.05%.

Solution

$$\text{Per cent resolution} = \frac{100}{\text{Number of turns}} = \frac{100}{N}$$

i.e.,
$$0.05 = \frac{100}{N}$$

i.e.,
$$N = 2000 \text{ turns.}$$

Example 4.9 A 8-kΩ potentiometer is operated with a grounded centre tap as shown in Fig. E4.4. If the potentiometer has a rotation angle of 350°, find the gain constant expressed in units of V/rad.

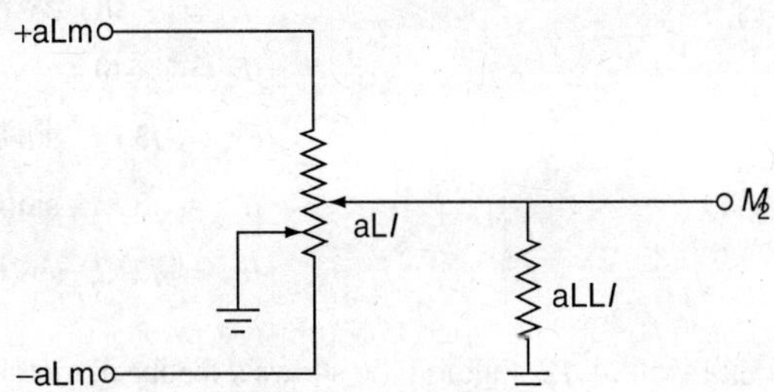

Fig. E4.4

Solution

$$\text{Gain constant } (K_p) = \frac{E_i}{\theta_{max}} = \frac{20}{350° \times \dfrac{2}{360°}} = 3.28 \text{ V/rad.}$$

Example 4.10 A 3 turn 100 kΩ potentiometer with 1% linearity uses a 30 V supply. Determine the following:

(i) the potentiometer constant in V/turn
(ii) range of voltages at the mid-point setting
(iii) the voltage at the mid-point if the potentiometer is loaded with 500 kΩ. Assume that the potentiometer is linear.

Solution

(i) Potentiometer constant $(K_p) = \dfrac{30}{3} = 10\,\text{V/turn}$

(ii) Mid-point setting 50*K*. Voltage at mid-point = $10 \times 1.5 = 15$V.

Here linearity = 1%. Therefore, voltage range is $= 15 \pm 0.01 \times 30 = 15 \pm 0.3$.

Therefore, $50K \pm 0.01 \times 100K = 50K \pm 1K$

$$V_{min} = 30 \times \frac{51}{100} = 15.3$$

and $V_{min} = 30 \times \frac{49}{100} = 14.7$

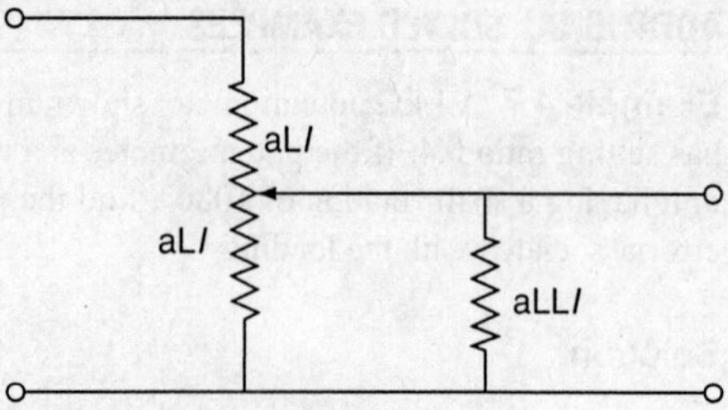

Fig. E4.5

(iii) At mid-way point 50*K* with perfect linearity.

$$\therefore \quad 500k \parallel 50k = 45.45k$$

Therefore, by voltage division we can write $\frac{45.45}{50 + 45.45} \times 30 = 14.3$V.

Example 4.11 The voltage applied to the rotor of a synchro control transmitter is 40V rms. The rotor shaft is moved 60° from the zero position. Determine the stator voltages w.r.t. the common stator connection for $k = 1$. Also determine the voltages between terminals S_1 and S_2, S_2 and S_3, S_3 and S_1. Take S_2 as the reference winding.

Solution Since S_2 is the reference stator winding, we have

$$E_{s2} = KE_R \cos\theta, \text{ where } E_{s2} \text{ and } E_R \text{ are rms voltages.}$$

$$E_{s2} = 1 \times 40 \times \cos 60° = 20\,\text{V}$$

$$\therefore \quad E_{s1} = 1 \times 40 \times \cos(60° - 120°) = 20\,\text{V}$$

$$E_{s3} = 1 \times 40 \times \cos(60° + 120°) = -40\,\text{V}$$

$$Es_{31} = \sqrt{3}\, KE_R \sin\theta = 60\,\text{V}$$

$$E_{s12} = \sqrt{3}\, KE_R \sin(\theta - 120°) = -60\,\text{V}$$

$$E_{s23} = \sqrt{3} KE_R \sin(\theta + 120°) = -0\,\text{V}$$

Example 4.12 Figure E4.6 shows a feedback motor system using a servomotor. The motor torque speed characteristic is shown in Fig. E4.7. Calculate the transient response to a 10° step displacement input. The inertia of the motor and load are 3×10^{-6} kg–m² and 3×10^{-4} kg–m² and the synchro sensitivity is 1.0 V/deg.

Solution The incremental torque balance equation for the motor is expressed by

$$\Delta T_M = k\Delta\Delta - f\Delta\Delta$$

Where $k = \frac{\partial T_M}{\partial E} = \frac{0.03}{25} = 1.2 \times 10^{-3}\,\text{N–m/V}$

and $f = \frac{\partial T_M}{\partial \theta} = \frac{0.03}{\frac{2\pi}{f_0} \times 2000} = 0.143 \times 10^{-3}\,\text{N–m/rad/sec}$

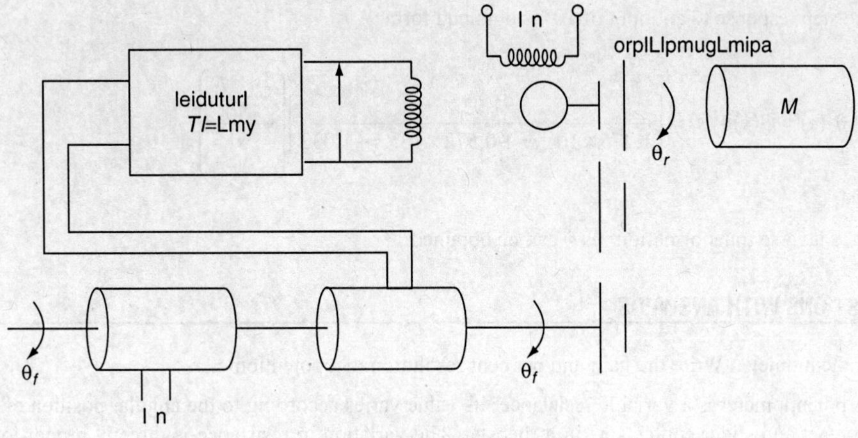

Fig. E4.6

The transfer function of the motor is expressed by

$$\frac{\omega(s)}{E(s)} = \frac{k}{Js + f}$$

where J is the total inertia reflected on the motor and

$$J = 3\times10^{-6} \times \left(\frac{1}{40}\right)^2 \times 3\times10^{-4} = 3.188\times10^{-6}\text{kg}-\text{m}^2$$

$$E = 15K_s(\theta_i - \theta_c) = 15\times1\times\frac{180}{\pi}\times(\theta_i - \theta_c)\text{ volts}$$

$$\frac{\theta_m}{E(s)} = \frac{k}{Js^2 + fs}$$

and $$\frac{\theta_c}{\theta_m} = \frac{1}{40}$$

i.e., $$\theta_m = 400_c$$

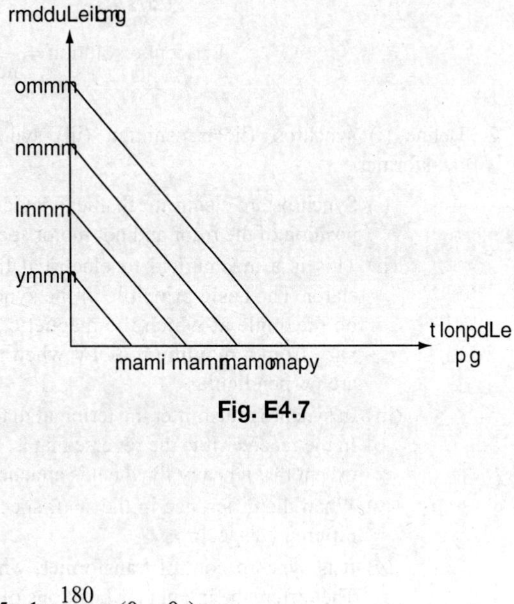

Fig. E4.7

The torque equation in the s domain is given by

$$(Js^2 + fs)\theta_m = k\times15\times1\times\frac{180}{\pi}\times(\theta_i - \theta_c)$$

i.e., $$[40(Js^2 + fs) + k\times860]\theta_c = 860\theta_i$$

$\therefore$ $$G(s) = \frac{\theta_c}{\theta_i} = \frac{k\times860}{40(Js^2 + fs) + k\times860} = \frac{1.032}{1.275\times10^{-4}s^2 + 0.572\times10^{-2}s + 1.032}$$

Therefore, the step response to an input of 10° is obtained form

$$\theta_c(s) = G(s)\theta_i(s) = \left[\frac{1.032}{1.275\times10^{-4}s^2 + 0.572\times10^{-2}s + 1.032}\right]\left(\frac{\frac{10\times\pi}{180}}{s}\right)$$

Taking Laplace inverse transformation, $\theta_c(t)$ can be obtained.

SHORT QUESTIONS WITH ANSWERS

1. Define potentiometer. Write the gain and per cent resolution of a potentiometer.

   Ans. A potentiometer is a variable resistance. Its value varies according to the angular position of the wiper contact and slider contacts as in a rheostat. The variation in resistance is directly proportional to the angular position.

$$\text{Gain } \frac{E_i}{2\pi N} \text{ V/rad}$$

$$\text{Per cent resolution} = \frac{100}{\text{number of turns of wire}}$$

2. Define (i) synchros, (ii) transmitter, (iii) synchro receiver, (iv) differential synchro, and (v) control transformer.

   Ans. (i) Synchros are electromechanical devices that produce an output voltage depending on the angular position of the rotor and not motor speed. It is different from a DC generator.
   (ii) This is a mechanical to electrical transducer, which consists of a transformer iron core as stator. The basic principle of the synchro is that the voltage induced in the secondary depends on the angle at which the magnetic lines of the flux cut the turns. This induced voltage will vary from a maximum of 1V when the two windings are in line or parallel to 0V when they are perpendicular.
   (iii) In synchro transmitter the terminal to terminal voltage is governed by angular position of the rotor. In the receiver it is the reverse that is, voltage to position transducer. The construction is the same except that a heavy flywheel is attached to the shaft of the rotor to reduce oscillations.
   (iv) When the difference in the shaft speed between the transmitter and receiver is required, we use differential synchro.
   (v) It is synchro control transformer, which functions as a position transducer and error detector. The difference in angular positions of two shafts in used to detect error voltage and correction is applied.

3. Define the following:
   (i) Tachometer, (ii) Accelerometer, and (iii) Gyroscope

   Ans. (i) A tachometer is a miniature low-voltage generator where the output voltage of a generator is given by $E_g = k_t\omega$.
   (ii) An accelerometer is also a transducer which is used in high-speed aircrafts and space vehicles.
   (iii) It is an instrument used in spaceships and aircrafts. The input is the angular velocity and the output is the angular displacement.

4. Define the following:
(i) Positional servomechanism (ii) Rate servomechanism.

Ans. (i) A positional servomechanism is a power amplifying feedback control system. The controlled variable in it is a mechanical position of time derivative of position such as velocity or acceleration.
(ii) The derivative of the position, i.e., velocity is monitored and kept constant here, with velocity as the controlled variable. Figure 4.29 shows the schematic diagram of a rate servomechanism.

EXERCISES

Questions for Practice

1. A 50k 10 turn potentiometer has a ± 12 V supply as shown in Fig. 1. Determine the following:
(a) The potentiometer constant K_p in V/Rad
(b) The output of the arm (unloaded) when the shaft is rotated 70° from the mid-point towards + 12 V.

Ans. 0.382 V/rad, 0.467V

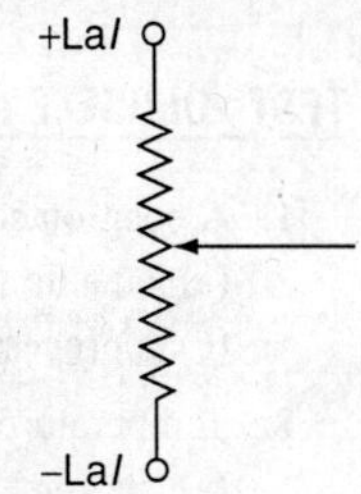

Figure 1

2. A tachometer has a gain of 0.05 V/(rad/sec). Determine (a) the output voltage when the shaft speed is 40 rad/sec, (b) the output voltage when the shaft speed is 20 rad/sec, (c) the shaft speed in rad sec when output voltage is 1.8 V.

Ans. (a) 0.6981317 rad/sec, (b) 1 V, (c) 36 rad/sec

3. A two-turn, 200 winding, 10 kW wire wound potentiometer has a resistance of 20 V impressed across its terminals. Assume that it turns full 72°. Find the potentiometer constant or gain in (i) volts per degree, (ii) volts per winding, (iii) volts per radian and (iv) volts per turn.

Ans. (i) 0.02777 V/deg, (ii) $\frac{1}{10}$ V/winding, (iii) 1.59 V/rad, (iv) 10 V/turn

4. An armature-controlled DC servomotor driving a system has following components:
(i) Potentiometer error sensor: Sensitivity (K_e) = 0.5 V/deg
(ii) Error amplifier; Gain (K_1) = 10.0
(iii) Motor: Armature resistance (R_e) = 10.0
Armature inductance (L_e) is negligible
Torque constant (K_T) = 1 N–m/A
Inertia (J_m) = 2 kg–m²
Friction (D_m) is negligible
(iv) Gear ratio $(n) = \frac{1}{10}$

Load: Inertia (J_L) = 0.2 kg–m²

Friction (D_L) = 0.2 N–m/rad/sec.

Find the overall transfer function of the system.

Ans. $\dfrac{5.7296\times 10^{-3}}{4s^2 + 3.1s + 5.7296\times 10^{-3}}$

5. Derive the expression for the transfer function of a servomotor having following data:

   Starting torque = 0.166 N–m^2

   Moment of inertia = 1×10^{-5} kg–m^2

   Supply voltage = 115 V

   No-load speed = 2.904 rpm

   Assume friction to be zero.

Ans. $\dfrac{2.64}{s(1+0.018s)}$

TEST YOURSELF

1. A position control is
   (a) an automatic regulating system (b) a servomechanism
   (c) a process control system (d) a stochastic control system
2. Temperature control system is known as
   (a) servomechanism (b) process control
   (c) cascade control system (d) none of these
3. The resolution of a potentiometer depends on
   (a) type of contact (b) wire material
   (c) size and width of the wire (d) none of these
4. The resolution of a potentiometer must be
   (a) infinite (b) very high (c) zero (d) medium
5. To minimize the loading effect of a potentiometer we use
   (a) Very high impedance (b) low resistance
   (c) low impedance (d) low inductance
6. LVDT is an
   (a) electromagnetic device (b) electrostatic device
   (c) electrodynamic device (d) electromechanical device
7. In comparison to a potentiometer, LVDT is preferred as an error detector because it
   (a) Produces stronger actuating signal (b) has high voltage sensitivity, low friction level, and infinite resolution
   (c) both (a) and (b) (d) uses capacitance to minimize loading error
8. Selsyn is the trade name of
   (a) spinning top (b) rotating transformer
   (c) rotating capacitor (d) synchros
9. Microsyn in based on the principle of
   (a) DC motor (b) resolver
   (c) saturable reactor (d) rotatory differential transformer
10. Find the correct device from the following to indicate the angular position of an aircraft
    (a) rotatory differential transformer (b) rotating transformer
    (c) rotating capacitor (d) none of these
11. To avoid the loading effect of a potentiometer we can use
    (a) constant current source (b) trimmer (c) battery (d) none of these

12. An error detector is a control system that
    (a) detects on error and signals out an alarm
    (b) detects the error of the system only
    (c) produces an error signal of difference between the actual output and the desired value of output
    (d) none of these
13. Output of a thermocouple is
    (a) direct voltage (c) alternating current
    (b) direct current (d) none of these
14. To measure pressure error, we use
    (a) differential bellows and strain gauge (b) pneumatic gauge
    (c) hydraulic actuator (d) selsyn
15. To convert the coordinates, the device used is
    (a) synchro resolver (b) synchro transmitter (c) microsyn (d) selsyn
16. Synchros are generally used as transmitters of
    (a) data logger (b) digital data (c) angular data (d) all of these
17. Match the following

System	Input	Output
(a) Thermocouple	a_1 temperature	a_2 watts
(b) Electric generator	b_1 acceleration	b_2 voltage
(c) Thermistor	c_1 temperature	c_2 ohms
(d) Accelerometer	d_1 rotating torque	d_2 velocity

18. The gyroscope is based on the principle of the law of
    (a) conservation of momentum (b) conservation of energy
    (c) action and reaction are equal and opposite (d) none of these
19. In servo applications, the motor used is
    (a) universal motor (b) single-phase shaded-pole motor
    (c) three-phase induction motor (d) synchronous motor
20. The term conformity is associated with
    (a) potentiometer (b) rate gyro (c) transformer (d) none of these
21. System analysis is carried out in
    (a) time domain (b) frequency domain
    (c) time as well as frequency domain (d) none of these
22. Identify the system that generally operates under the type-0 system.
    (a) servomechanism (b) regulator (c) both (a) & (b) (d) none of these
23. Pressure error can be measured by
    (a) strain gauge (b) differential bellows and strain
    (c) selsyn (d) strain gauge and potentiometer
24. Which of the following cannot be treated as an amplifier in control system?
    (a) amplitude (b) DC generator (c) DC motor (d) metadyne
25. Synchro generator and motor are
    (a) 3–ϕ DC devices (b) single-phase AC devices
    (c) DC devices (d) 2–ϕ DC devices

26. The characteristic equation of an armature-controlled DC motor is of
    (a) first-order equation (b) second-order equation
    (c) zero-order equation (d) third-order equation
27. A bimetallic thermostat is a
    (a) zero-term controller (b) two-term controller
    (c) one-term controller (d) on-off controller
28. The stator of synchro is made up of
    (a) silicon steel (b) cast iron (c) high-speed steel (d) high carbon steel
29. Total numbers of poles and zeros are not changed by use of a
    (a) tachometer (b) amplidyne (c) DC servomotor (d) DC servomotor
30 In a series DC servomotor, the following constitutes a drawback:
    (a) poor speed regulation (b) high starting torque (c) large current drawn (d) all of these
31. Match the following:

List A	List B
(a) AC servomotor	1. $\dfrac{s+2}{s+p}(2<p)$
(b) DC amplifier	2. $\dfrac{1+T_1 s}{1+T_2 s}(T_1<T_2)$
(c) lead network	3. $\dfrac{K}{1+T_s}$
(d) lag network	4. $\dfrac{K}{s(1+Ts)}$

32. Match the following:

(a) AC servomotor	(i) $\dfrac{K}{s(s+sT_m)}$
(b) field-controlled DC servomotor	(ii) $\dfrac{K}{s(1+sT_e)(1+sT_m)}$
(c) tachogenerator	(iii) Ks
(d) integrating gyro	(iv) $\dfrac{K}{1+sT}$

ANSWERS

1 (b)	2 (b)	3 (c)	4 (a)	5 (a)
6 (d)	7 (b)	8 (d)	9 (d)	10 (a)
11 (b)	12 (c)	13 (a)	14 (a)	15 (a)
16 (c)	17 (a)→(a_1,b_2) (b)→(d_1,a_2) (c)→(a_1,c_2) (d)→(b_1,d_2)	18 (a)	19 (c)	20 (a)
21 (c)	22 (b)	23 (b)	24 (c)	25 (b)
26 (d)	27 (d)	28 (a)	29 (a)	30 (d)
31 (a)→(iv) b→(iii) (c)→(i) (d)→(ii)	32 (a)→(i) (b)→(ii) (c)→(iii) (d)→(iv)			

Chapter 5

Mathematical Modelling of Physical Systems

A collection of physical objects that serves some objective is called a physical system. The study of equivalent representation of the physical system, which describes the various components of the system and their relations in the system is important. Transfer function has been introduced in Chapter 3. Block diagram will be introduced in Chapter 6. Before obtaining a block diagram, the set of mathematical equations describing the dynamic behaviour of the system must be known. This is known as mathematical modelling of a system. Since most control systems contain mechanical or both types of elements and components, it is helpful to convert such systems into mathematical models using transfer function approach for analysis. Models of mechanical and electrical components are exactly analogous to each other. For a given mechanical system, there always exists an analogous electrical network, and the converse is also true. The main aim of this chapter is to discuss the mathematical modelling of analogous systems.

5.1 MODELLING OF MECHANICAL SYSTEMS

Mechanical systems are classified into two types: (i) translational and (ii) rotational. The motion of the body during translational motion is along a straight line or a curved path, whereas during rotational motion, the motion of the body is about its own axis.

5.1.1 Translational Motion

There are three elements, which are dominantly involved in the analysis of translational motion. These are (i) mass, (ii) spring and (iii) damper.

Mass: In the physical model of a mass element, it is assumed that mass is concentrated at the centre of the mass element. Figure 5.1 shows that when a block ABCD of mass M is moved, all its particles undergo an equal displacement. Therefore, mass has only one displacement.

In Fig. 5.2, it has been shown that one displacement is enough for mass and $\mapsto x$ can be put anywhere of the mass as reference. In Fig. 5.2(a), a block of mass M is shown, whereas in Fig. 5.2, $\mapsto x$ represents the displacement x w.r.t. reference.

The relation between force f and mass M is given by

$$f = Ma \tag{5.1}$$

Fig. 5.1 Movement of a block of mass

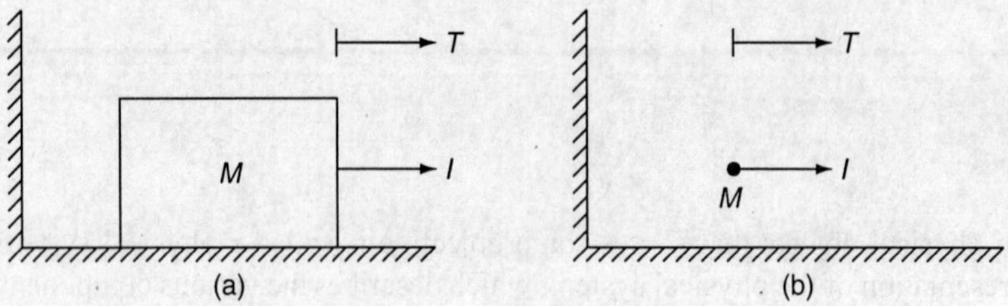

Fig. 5.2 Displacement for a mass (*M*)

$$= M\frac{du}{dt} \tag{5.2}$$

$$= M\frac{d^2x}{dt^2} \tag{5.3}$$

where 'a' is the acceleration, 'u' is the velocity, and 'x' is the displacement of the mass.

Spring: Figure 5.3 shows a spring of negligible mass. Let the linear spring constant for the spring be K. In this case the spring is subjected to force, and it undergoes elastic deformation. The relation is $f \propto$ displacement

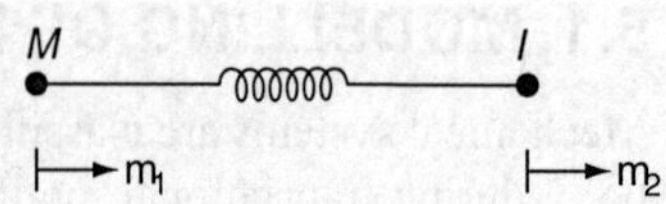

Fig. 5.3 Spring

$$\therefore \quad f = Kx \tag{5.4}$$

Here x_2 is the displacement of end A, whereas x_1 is the displacement of end B.
If force is applied at end B, $x = x_1 - x_2$, we have from Eq. (5.4),

$$f = K(x_1 - x_2) \tag{5.5}$$

If force is applied at end A, $x = x_2 - x_1$ We have from Eq. (5.4)

$$f = K(x_2 - x_1) \tag{5.6}$$

Damper: Motion is opposed by friction. Types of frictional forces are shown below.

(i) **Coulomb Frictional Force:** The sliding friction between dry surfaces is called coulomb frictional force.

(ii) **Viscous Friction:** The friction between moving surfaces by a viscous fluid or the friction between a solid body and a fluid medium is known as viscous friction. Viscous friction is proportional to velocity.

Although friction opposes motion, sometimes viscous friction is introduced intentionally as shown in Fig. 5.4. Since $f \propto$ velocity, we have

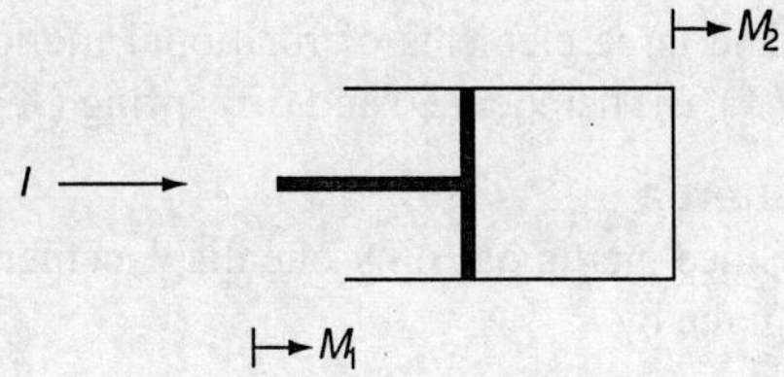

Fig. 5.4 A dashpot

$$f = D\frac{dx}{dt} = D\frac{d}{dt}(x_1 - x_2) \qquad (5.7)$$

where D is the damping coefficient

(iii) **Stiction:** The friction force that is required to initiate motion between two contact surfaces is known as stiction.

Example 5.1 Determine the mathematical model for the system shown in Fig. E5.1.

Solution The force f causes the displacement of mass M. This is opposed by spring as well as by damper. The forces Kx and $D\frac{dx}{dt}$ will oppose f.

$$\therefore \quad M\frac{d^2x}{dt^2} = f - D\frac{dx}{dt} - Kx$$

$$\therefore \quad M\frac{d^2x}{dt^2} + D\frac{dx}{dt} + Kx = f \qquad (1)$$

Equation (1) shows the mathematical model for the system shown in Fig. E5.1.

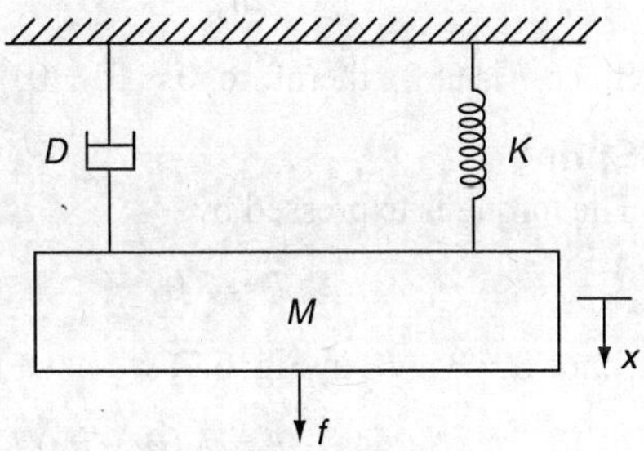

Fig. E5.1

5.1.2 Rotational Motion

There are many similarities between translational and rotational motion. Table 5.1 shows a list of translational and rotational counterparts.

Table 5.1 Counterparts of translational and rotational motion

Translational motion	Rotational motion
Mass (M)	Inertia (J)
Damper (D)	Damper (D)
Spring (K)	Spring (K)
Force (f)	Torque (T)
Displacement (x)	Angular displacement (θ)
Velocity (u) $= \frac{dx}{dt}$	Angular velocity (ω) $= \frac{d\theta}{dt}$

The three elements of rotational motion are (i) inertia (J), (ii) damper (D) and (iii) spring (K).

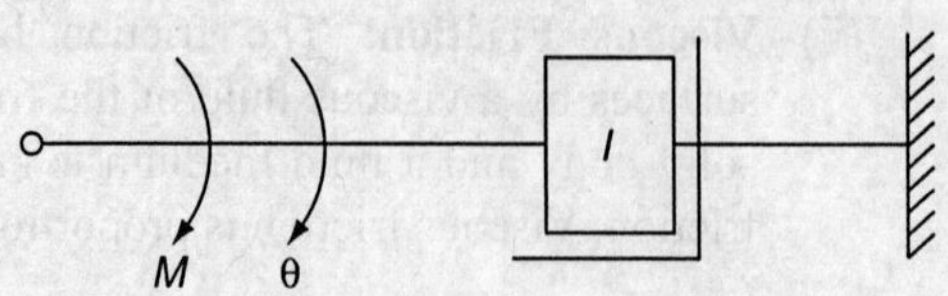

Fig. 5.5 Inertia

Inertia

Since inertia has only one displacement, the relation is given by

$$T = J\frac{d^2\theta}{dt^2} \tag{5.8}$$

$$T = J\frac{d\omega}{dt} \tag{5.9}$$

Damper

The torque is expressed by

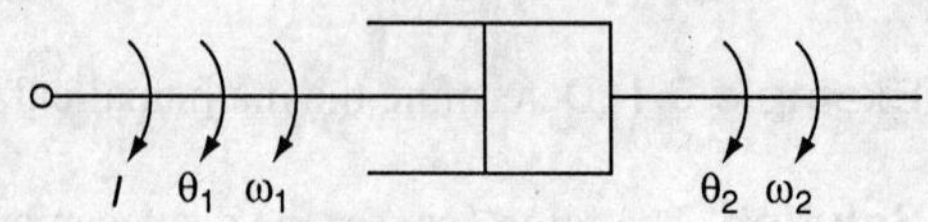

Fig. 5.6 Damper

$$T = D\frac{d\theta}{dt} \tag{5.10}$$

$$= D\frac{d}{dt}(\theta_1 - \theta_2) \tag{5.11}$$

Since T is at θ_1, therefore, $\theta = \theta_1 - \theta_2$

Spring

The torque is expressed by

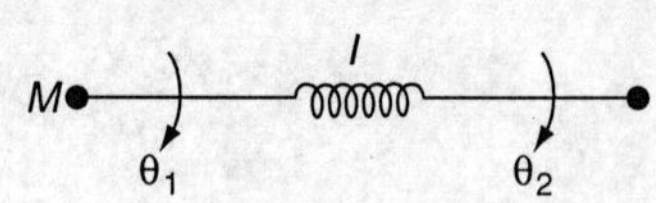

Fig. 5.7 Spring

$$T = K\theta \tag{5.12}$$

Since $\theta = \theta_1 - \theta_2$ in Fig. 5.7, we have from Eq. (5.12)

$$T = K(\theta_1 - \theta_2) \tag{5.13}$$

Example 5.2 Determine the torque equation of the system shown in Fig. E5.2

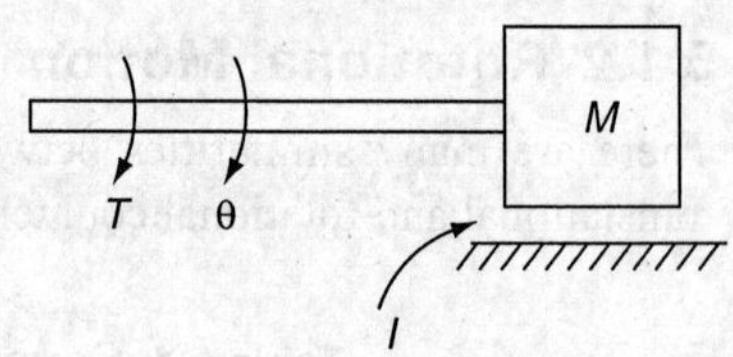

Fig. E5.2

Solution Let the displacement be θ.

The torque equation of the system is expressed by

$$T - D\frac{d\theta}{dt} - K\theta = J\frac{d^2\theta}{dt^2}$$

because $D\dfrac{d\theta}{dt}$ and $K\theta$ opposes T. We have

$$J\frac{d^2\theta}{dt^2} + D\frac{d\theta}{dt} + K\theta = T$$

Equation (1) shows the torque equation of the system shown in Fig. E5.2.

5.2 MODELLING OF ELECTRICAL SYSTEMS

The basic elements of an electrical system are (i) resistor, (ii) inductor and (iii) capacitor.

5.2.1 Resistor

From Fig. 5.8, we have

$$v = iR \tag{5.14}$$

where v is the voltage, i is the current and R is the resistance of the resistor.

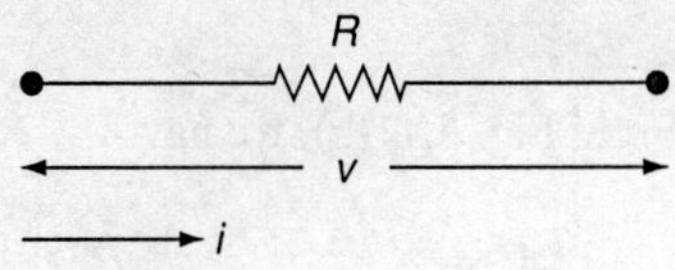

Fig. 5.8 Resistor

5.2.2 Inductor

From Fig. 5.9, we have

$$v = L\frac{di}{dt} \tag{5.15}$$

i.e.,

$$i = \frac{1}{L}\int_0^t V\,dt \tag{5.16}$$

where v is voltage, i is current and L is the inductance of the inductor.

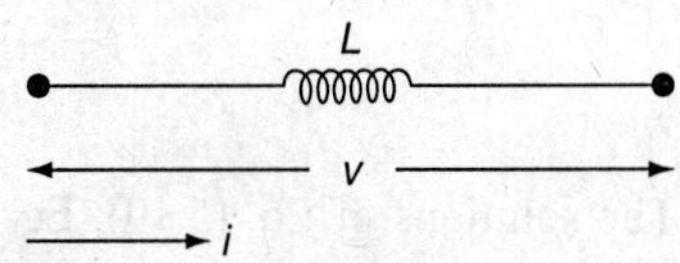

Fig. 5.9 Inductor

5.2.3 Capacitor

From Fig. 5.10, we have

$$v = \frac{1}{C}\int_0^t i\,dt \tag{5.17}$$

i.e.,

$$i = C\frac{dv}{dt} \tag{5.18}$$

where v is voltage, i is current and C is the capacitance of the capacitor.

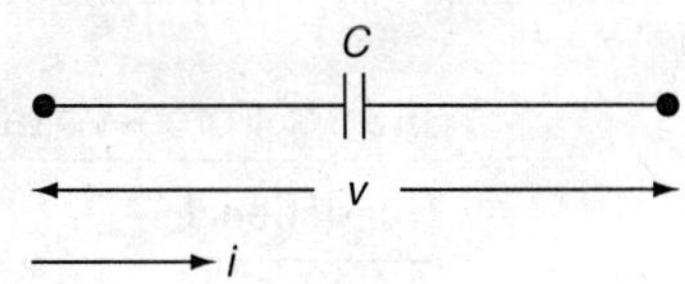

Fig. 5.10 Capacitor

5.3 ANALOGOUS SYSTEMS

Let us consider a mechanical (both translational and rotational) and an electrical system, as shown in Fig. 5.11.

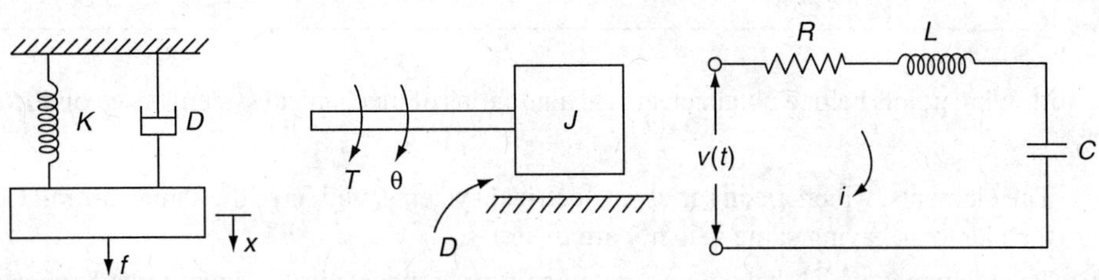

Fig. 5.11 Examples of analogous systems

From Fig. 5.11(a), we have

$$M\frac{d^2x}{dt^2}+D\frac{dx}{dt}+Kx=f \tag{5.19}$$

From Fig. 5.11(b), we have

$$J\frac{d^2\theta}{dt^2}+D\frac{d\theta}{dt}+K\theta=T \tag{5.20}$$

From Fig. 5.11(c), we have

$$L\frac{d^2q}{dt^2}+R\frac{dq}{dt}+\frac{1}{C}q=V(t) \tag{5.21}$$

where $q=\int idt.$

The solutions of Eq. (5.19), Eq. (5.20), and Eq. (5.21) are same. Hence they are analogous systems. There are two methods to get an analogous system. These are (i) force–voltage ($f-v$) analogy and (ii) force–current ($f-i$) analogy.

5.3.1 Force–Voltage Analogy

In this case, force is analogous to voltage. From Eq. (5.19), Eq. (5.20), and Eq. (5.21), the following analogies can be drawn as shown in Table 5.2.

Table 5.2 Force–voltage analogy

Translational	Electrical	Rotational
Force (f)	Voltage (v)	Torque (T)
Mass (M)	Inductance (L)	Inertia (J)
Damper (D)	Resistance (R)	Damper (D)
Spring (K)	Elastance $\left(\frac{1}{C}\right)$	Spring (K)
Displacement (x)	Charge (q)	Displacement (θ)
Velocity (u) $=\frac{dx}{dt}$	Current (i)	Angular velocity (ω) $=\frac{d\theta}{dt}$

The following points help to obtain electrical analogous of mechanical system based on force–voltage analogy:

(i) The elements, which are in series in electrical systems, will have the same current. Conversely, the elements having same velocity are in series.

(ii) The elements, which have the same velocity in a mechanical system, must have same analogous current in analogous electrical system.

(iii) In mechanical system, every node corresponds to a closed loop in the electrical system where a mass is considered as a node.

(iv) Since the number of meshes in electrical analogous is equal to the number of nodes (masses) in the mechanical system, the number of mesh currents and system equations will be equal to the number of velocities of nodes in the mechanical system.

(v) The mechanical driving source (i.e., force) and passive elements connected to the node (i.e., mass) in the mechanical system will be represented by analogous elements in a closed loop in analogous electrical system.

(vi) The element, which is connected between two masses, i.e., nodes, in the mechanical system becomes a common element between two meshes in electrical analogous system.

The following points are useful to obtain electrical analogous of mechanical rotational systems based on torque voltage analogy:

(i) The elements, which are in series in the electrical system, carry the same current. Conversely, the elements which have the same angular velocity are in series.

(ii) If the angular velocities of the elements in the mechanical system are same, the currents through the analogous elements in analogous electrical system are same.

(iii) Any node (moment of inertia of mass) in the mechanical system corresponds to a closed loop in the electrical system. A node in this case is the moment of inertia.

(iv) Since the number of meshes in analogous electrical system is equal to the number of nodes in the mechanical system, the number of mesh currents and system equations is equal to the number of angular velocities of nodes in the mechanical system.

(v) The mechanical driving sources (torque) and passive elements connected to the node in mechanical system are represented by analogous element in a closed analogous electrical system.

(vi) The element, which is connected between two nodes in mechanical system, is represented as a common element between two meshes in electrical analogous system.

5.3.2 Force–Current Analogy

Let us consider the electrical network shown in Fig. 5.12.

In this case, force is equal to current. From Fig. 5.12, we have

$$i = \frac{v}{R} + \frac{1}{L}\int v\,dt + C\frac{dv}{dt} \tag{5.22}$$

Since voltage

$$v = \frac{d\psi}{dt} \tag{5.23}$$

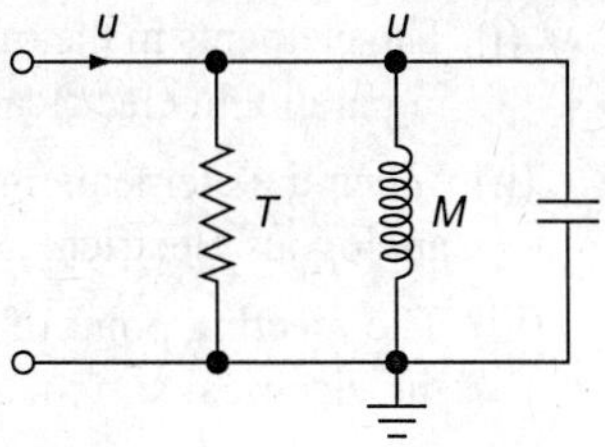

Fig. 5.12 Electrical system

From Eq. (5.22), we have

$$i = \frac{1}{R}\frac{d\psi}{dt} + \frac{\psi}{L} + C\frac{d^2\psi}{dt^2} \tag{5.24}$$

Comparing Eq. (5.19), Eq. (5.20), and Eq. (5.24), the following analogies can be drawn as shown in Table 2.3.

Table 5.3 Force–current analogy

Translational	Electrical	Rotational
Force (f)	Current (i)	Torque (T)
Mass (M)	Capacitance (C)	Inertia (J)
Damper (D)	Conductance $\left(\frac{1}{R}\right)$	Damper (D)
Spring (K)	Reciprocal of inductance $\left(\frac{1}{L}\right)$	Spring (K)
Displacement (x)	Flux linkage (ψ)	Displacement (θ)
Velocity (u) $= \frac{dx}{dt}$	Voltage (v) $= \frac{d\psi}{dt}$	Angular velocity (ω) $= \frac{d\theta}{dt}$

The following points are very useful to obtain electrical analogous of mechanical systems based on force–current analogy.

(i) The elements in mechanical systems having the same force are in parallel. Similarly, the elements in electrical systems will have the same voltage when they are in parallel.

(ii) The elements have the same velocity in the mechanical system must have same analogous voltage in analogous electrical system.

(iii) Any node, i.e., meeting point of elements, in mechanical system is also analogous to a node in electrical system. A mass is taken as a node.

(iv) The number of node in analogous electrical system is equal to the number of nodes in the mechanical system. Therefore, the number of node voltages and system equations is equal to the number of velocities of masses (nodes) in the mechanical system.

(v) Any element connected between two nodes (masses) in the mechanical system can be represented as a common element between two nodes in analogous electrical system.

The following points are useful to obtain electrical analogous of mechanical rotational systems based on torque–current analogy.

(i) The elements in the mechanical system in parallel will have same torque, whereas the elements in parallel in electrical system will have same voltage.

(ii) When the elements in the mechanical system have the same angular velocity, the elements in analogous electrical system will have the same voltage.

(iii) The meeting point of elements, i.e., node, in the mechanical system corresponds to a node in the electrical system. Here, the moment of inertia of mass is taken as a node.

(iv) The number of nodes in the mechanical system and that in analogous electrical system is equal. Therefore, the number of node voltages and system equations is equal to the number of angular velocities of nodes (moment of inertia of mass) in the mechanical system.

(v) The mechanical driving sources (torque) and the elements connected to a node in the mechanical system are represented by analogous elements, which are connected to a node in analogous electrical system.

(vi) The element, which is connected between two nodes in mechanical system, is represented as a common element connected between two nodes in analogous electrical system.

5.4 REPRESENTATION BY NODAL METHOD

This method is very simple to obtain mathematical models of both types of mechanical systems.

Let us consider a mechanical system shown in Fig. 5.13. Following are the various steps involved:

Step 1: The number of nodes is equal to the total number of displacements, i.e., 3.

One reference node is taken in addition, i.e., the total number of nodes becomes 4.

Step 2: M_1 is connected between x_1 and reference. M_2 is connected between x_2 and reference. M_3 is connected between x_3 and reference.

Step 3: K_1 is connected between x_1 and reference. K_{12} is connected between x_1 and x_2. K_2 is connected between x_2 and reference. K_{23} is connected between x_2 and x_3. K_3 is connected between x_3 and reference.

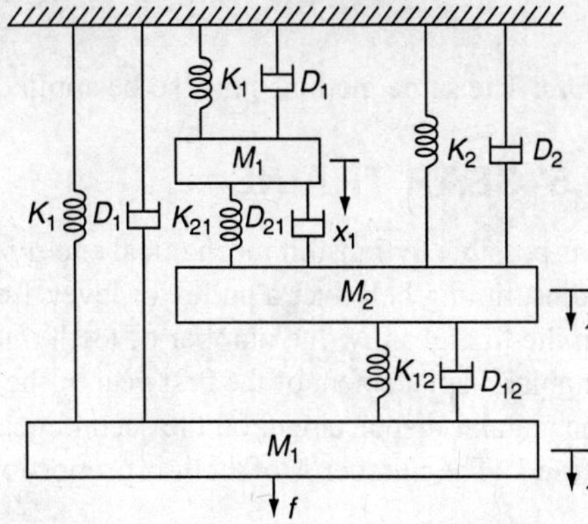

Fig. 5.13 A mechanical system

Step 4: D_1 is connected between x_1 and reference, D_{12} is connected between x_1 and x_2. D_2 is connected between x_2 and reference. D_{23} is connected between x_2 and x_3. D_3 is connected between x_3 and reference.

Step 5: Let us connect M's, K's, and D's, as discussed in steps 2, 3 and 4, respectively.

Step 6: The force f is connected between x_1 and reference because f is connected to M_1 in Fig. 5.13. Figure 5.14 shows the representation of Fig. 5.13 by nodal method.

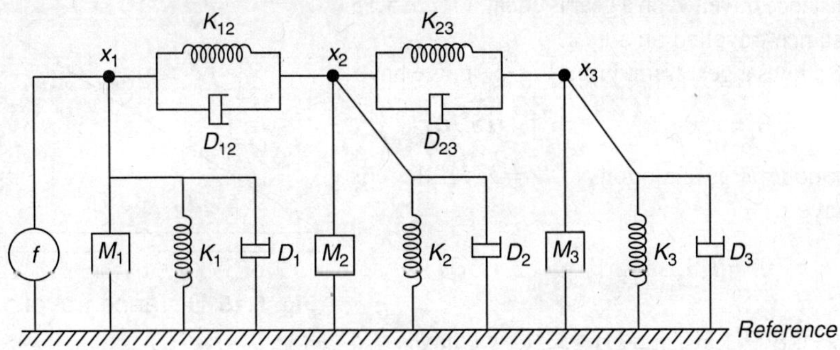

Fig. 5.14 Representation of Fig. 5.13 by nodal method

Step 7: Applying force equation at node x_1, we have from Fig. 5.14,

$$f(t) = M_1 \frac{d^2 x_1}{dt^2} + D_1 \frac{dx_1}{dt} + K_1 x_1 + K_{12}(x_1 - x_2) + D_{12} \frac{d}{dt}(x_1 - x_2) \tag{5.25}$$

At node x_2,

$$K_{12}(x_1 - x_2) + D_{12} \frac{d}{dt}(x_1 - x_2) = M_2 \frac{d^2 x_2}{dt^2} + D_2 \frac{dx_2}{dt} + K_2 x_2 + K_{23}(x_2 - x_3) + D_{23} \frac{d}{dt}(x_2 - x_3) \tag{5.26}$$

At node x_3,

$$D_{23}\frac{d}{dt}(x_2 - x_3) + K_{23}(x_2 - x_3) = M_3\frac{d^2x_3}{dt^2} + D_3\frac{dx_3}{dt} + K_3x_3 \tag{5.27}$$

***Note*:** The same method can also be applied to rotational systems.

5.5 GEAR TRAINS

It is possible to transmit mechanical energy from one part of a system to another via devices such as gear trains, timing belt over a pulley or lever. Let us consider a gear train with N_1 being the number of teeth on the first gear, N_2 the number of teeth on the second gear, T_1 the input torque on the first gear, θ_1 the angular displacement of the first gear, r_1 the radius of the first gear, T_2 input torque of the second gear, θ_2 the angular displacement on the second gear and r_2 radius of the second gear.

Rule 1: The number N of teeth is proportional to the radius r of a gear:

$$r \propto N$$

Therefore, $r_1 \propto N_1$ and $r_2 \propto N_2$

i.e.,
$$\frac{r_1}{r_2} = \frac{N_1}{N_2} \tag{5.28}$$

i.e.,
$$r_1N_2 = r_2N_1 \tag{5.29}$$

Rule 2: The distance travelled on a gear is equal. Figure 5.15 shows the distance travelled on a gear.

Figure 5.16 shows a gear train. From Fig. 5.16, we have

$$r_1\,\theta_1 = r_2\,\theta_2 \tag{5.30}$$

Rule 3: Work done $T\theta$ is analogous to $W = Fx$. If there is no loss, we have

$$T_1\theta_1 = T_2\theta_2 \tag{5. 31}$$

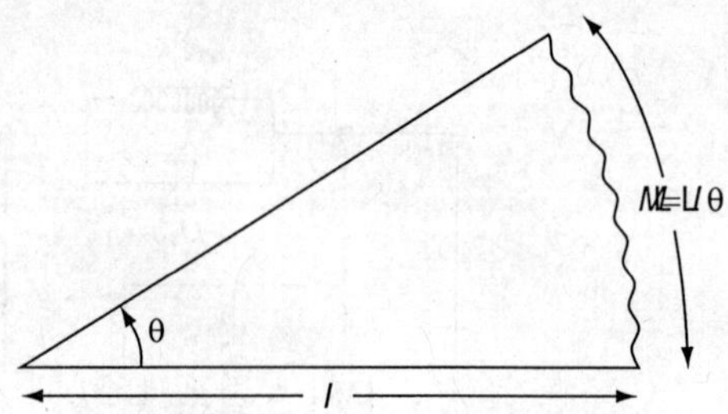

Fig. 5.15 Distance travelled on a gear

From Eq. (5.29), Eq. (5.30) and Eq. (5.31), we have

$$\frac{T_1}{T_2} = \frac{r_1}{r_2} = \frac{N_1}{N_2} = \frac{\theta_2}{\theta_1} \tag{5.32}$$

From Fig. 5.16, we have

$$T_M = J_1\frac{d^2\theta_1}{dt^2} + D_1\frac{d\theta_1}{dt} + T_1 \tag{5.33}$$

where T_1 is transmitted towards load. Assuming no loss in the gear, we have

$$T_2 = J_2\frac{d^2\theta_2}{dt^2} + D_2\frac{d\theta_2}{dt} + T_L \tag{5.34}$$

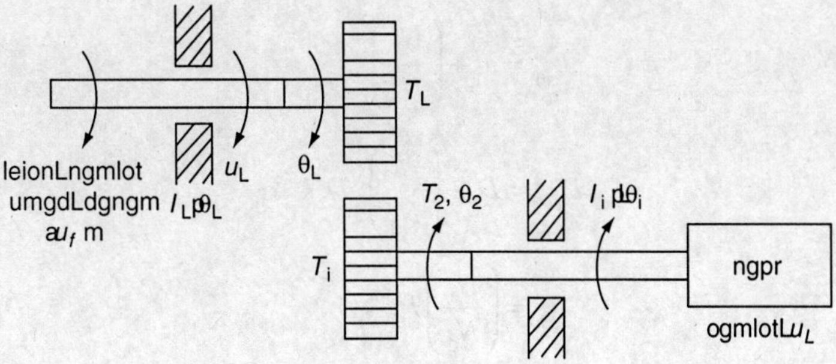

Fig. 5.16 Gear train

T_2 in Eq. (5.34) is the reflected torque on gear secondary and T_L is the load torque. The relation between T_1 and T_2 is expressed by

$$\frac{T_1}{T_2} = \frac{N_1}{N_2} \tag{5.35}$$

i.e.,
$$T_2 = \left(\frac{N_2}{N_1}\right) T_1 \tag{5.36}$$

Using Eq. (5.36), Eq. (5.34) can be expressed as

$$\left(\frac{N_2}{N_1}\right) T_1 = J_2 \frac{d^2\theta_2}{dt^2} + D_2 \frac{d\theta_2}{dt} + T_L$$

i.e.,
$$T_1 = \left(\frac{N_1}{N_2}\right)\left[J_2 \frac{d^2\theta_2}{dt^2} + D_2 \frac{d\theta_2}{dt} + T_L\right] \tag{5.37}$$

$$= \left(\frac{N_1}{N_2}\right)\left[J_2 \left(\frac{N_1}{N_2}\right)\frac{d^2\theta_1}{dt^2} + D_2 \left(\frac{N_1}{N_2}\right)\frac{d\theta_1}{dt} + T_L\right] \quad \text{[Using Eq. (5.30)]}$$

$$T_1 = J_2 \left(\frac{N_1}{N_2}\right)^2 \frac{d^2\theta_1}{dt^2} + D_2 \left(\frac{N_1}{N_2}\right)^2 \frac{d\theta_1}{dt} + \left(\frac{N_1}{N_2}\right) T_L \tag{5.38}$$

Using Eq. (5.38), we can write Eq. (5.33) as follows:

$$T_M = \left[J_1 + J_2 \left(\frac{N_1}{N_2}\right)^2\right]\frac{d^2\theta_1}{dt^2} + \left[D_1 + D_2 \left(\frac{N_1}{N_2}\right)^2\right]\frac{d\theta_1}{dt} + \left(\frac{N_1}{N_2}\right) T_L$$

i.e.,
$$T_M = J_{eq1} \frac{d^2\theta_1}{dt^2} + D_{eq1} \frac{d\theta_1}{dt} + T_L' \tag{5.39}$$

where
$$J_{eq1} = J_1 + \left(\frac{N_1}{N_2}\right)^2 J_2 \tag{5.40}$$

$$D_{eq1} = D_1 + \left(\frac{N_1}{N_2}\right)^2 D_2 \tag{5.41}$$

$$T_L' = \left(\frac{N_1}{N_2}\right) T_L \tag{5.42}$$

Equation (5.40), Eq. (5.41) and Eq. (5.42) show the terms referred to as primary. Similarly, from Eq. (5.32), we have

$$\frac{\theta_2}{\theta_1} = \frac{N_1}{N_2}$$

i.e.,
$$\theta_1 = \theta_2 \left(\frac{N_2}{N_1}\right) \tag{5.43}$$

∴
$$\frac{d\theta_1}{dt} = \frac{d\theta_2}{dt}\left(\frac{N_2}{N_1}\right) \tag{5.44}$$

and
$$\frac{d^2\theta_1}{dt^2} = \frac{d^2\theta_2}{dt^2}\left(\frac{N_2}{N_1}\right) \tag{5.45}$$

Once again from Eq. (5.32), we have

$$T_1 = T_2 \left(\frac{N_1}{N_2}\right) \tag{5.46}$$

Substituting $\frac{d^2\theta_1}{dt^2}, \frac{d\theta_1}{dt}$ and T_1 from Eq. (5.45), Eq. (5.44) and Eq. (5.46) in Eq. (5.33), we have

$$J_1 \frac{d^2\theta_2}{dt^2}\left(\frac{N_2}{N_1}\right) + D_1 \frac{d\theta_2}{dt}\left(\frac{N_2}{N_1}\right) + T_2\left(\frac{N_1}{N_2}\right) = T_M \tag{5.47}$$

Substituting T_2 from Eq. (5.34) in Eq. (5.47), we have

$$J_1 \frac{d^2\theta_2}{dt^2}\left(\frac{N_2}{N_1}\right) + D_1 \frac{d\theta_2}{dt}\left(\frac{N_2}{N_1}\right) + \left(\frac{N_1}{N_2}\right)\left(J_2 \frac{d^2\theta_2}{dt^2} + D_2 \frac{d\theta_2}{dt} + T_L\right) = T_M$$

i.e.,
$$\left[J_1\left(\frac{N_2}{N_1}\right)^2 + J_2\right]\frac{d^2\theta_2}{dt^2} + \left[D_1\left(\frac{N_2}{N_1}\right)^2 + D_2\right]\frac{d\theta_2}{dt} + T_L = T_M\left(\frac{N_2}{N_1}\right)$$

i.e.,
$$J_{eq2}\frac{d^2\theta_2}{dt^2} + D_{eq2}\frac{d\theta_2}{dt} + T_L = T_M\left(\frac{N_2}{N_1}\right) \tag{5.48}$$

where

$$J_{eq2} = J_1\left(\frac{N_2}{N_1}\right)^2 + J_2 = \text{ Total equivalent moment of inertia referred to load shaft} \quad (5.49)$$

$$D_{eq2} = D_1\left(\frac{N_2}{N_1}\right)^2 + D_2 = \text{ Total equivalent viscous friction co-efficient referred to load shaft.} \quad (5.50)$$

Equation (5.49) and Eq. (5.50) show the terms referred to secondary.

Example 5.3 Figure E5.3 shows a gear train system. Find the displacement of the load on the shaft (side 2) if $T_1 = 20e^{-2t}$ is applied at side 1.

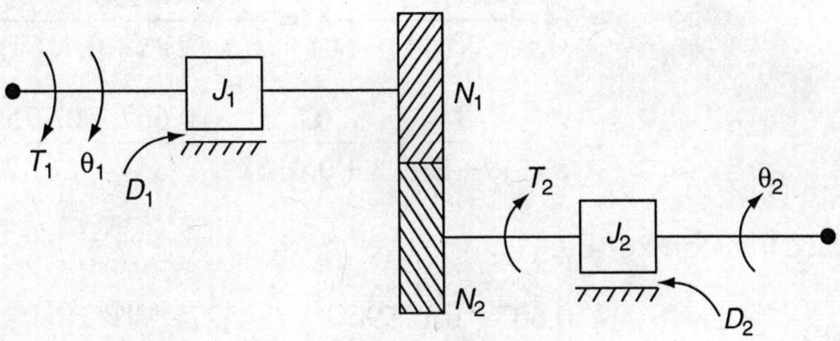

Fig. E5.3

From Fig. E5.3

$$J_1 = 6\,\text{kg-m}^2,\ J_2 = 48\ \text{kg-m}^2,\ D_1 = 3\,\text{N-m/rad/sec},\ D_2 = 12\text{N-m/rad/sec}, N_1 = 1 \text{ and } N_2 = 4.$$

Solution The equivalent J, D and T referring to side 2 are given by

$$J_{eq} = J_2 + J_1\left(\frac{N_2}{N_1}\right)^2,\ D_{eq} = D_2 + D_1\left(\frac{N_2}{N_1}\right),\ T_2 = T_1\left(\frac{N_2}{N_1}\right)$$

Again,

$$T_2 = J_{eq}\frac{d^2\theta_2}{dt^2} + D_{eq}\frac{d\theta_2}{dt}$$

Taking Laplace transform, we have

$$T_2(s) = s^2 J_{eq}\theta_2(s) + sD_{eq}\theta_2(s)$$

$$J_{eq} = J_2 + J_1\left(\frac{N_2}{N_1}\right)^2 = 48 + \left(\frac{4}{1}\right)^2 \times 6 = 144\ \text{kg-m}^2$$

$$D_{eq} = D_2 + D_1\left(\frac{N_2}{N_1}\right)^2 = 12 + 3 \times \left(\frac{4}{1}\right)^2 = 60\ \text{N-m/rad/sec}$$

and
$$T_2 = T_1\left(\frac{N_2}{N_1}\right) = \left(\frac{4}{1}\right)\times 20e^{-2t} = 80e^{-2t}$$

$\therefore$
$$T_2(s) = \frac{80}{s+2}$$

$\therefore$
$$\frac{80}{s+2} = s^2 J_{eq}\theta_2(s) + sD_{eq}\theta_2(s) = (s^2 J_{eq} + sD_{eq})\theta_2(s)$$

i.e.,
$$\theta_2(s) = \frac{80}{(s+2)(s^2 J_{eq} + sD_{eq})} = \frac{80}{(s+2)(144s^2+60s)}$$

$$= \frac{80}{s(s+2)(144s+60)} = \frac{0.556}{s(s+2)(s+0.4167)}$$

$$= \frac{A}{s} + \frac{B}{(s+2)} + \frac{C}{s+0.4167} = \frac{0.667}{s} + \frac{0.1759}{(s+2)} + \frac{0.8427}{s+0.4167} \quad (1)$$

Taking inverse Laplace transform of Eq. (1), we have

$$\theta_2(t) = 0.667 + 0.1759e^{-2t} - 0.8427e^{-0.4167t}$$

5.6 CHAIN DRIVER

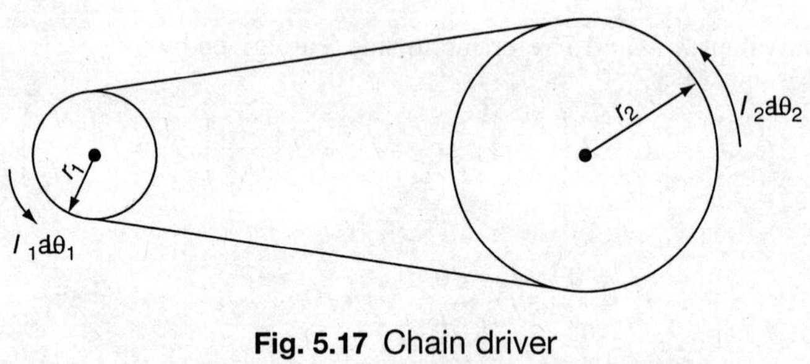

Fig. 5.17 Chain driver

Figure 5.17 shows a chain driver. Rule 2 and Rule 3 are applicable to chain drivers also.

$\therefore$
$$\frac{T_1}{T_2} = \frac{r_1}{r_2} = \frac{\theta_2}{\theta_1} \quad (5.51)$$

5.7 LEVER

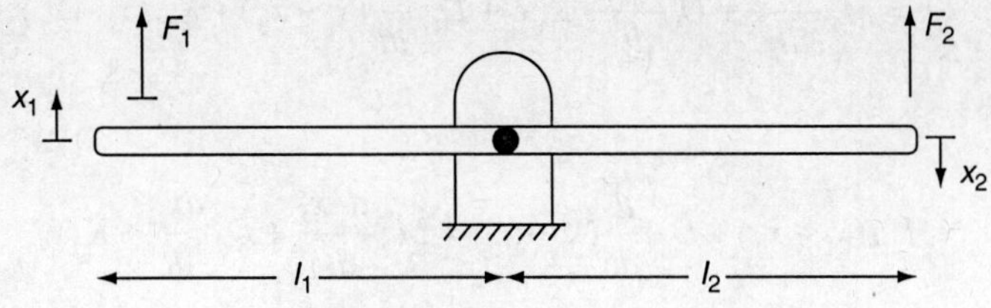

Fig. 5.18 Lever

Figure 5.18 shows a lever. Using the law of moments, we can write

$$F_1 l_1 = F_2 l_2 \tag{5.52}$$

This is Rule 5.

Example 5.4 Find the system equation of the mechanical system shown in Fig. E5.4. Also, find the force–voltage (f–v) and force–current (f–i) analogy.

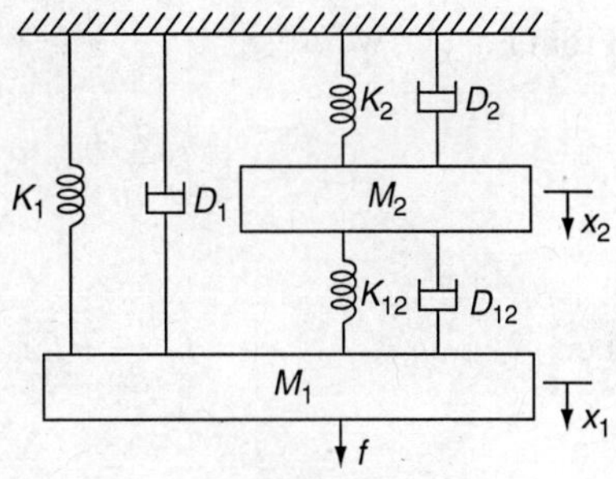

Fig. E5.4

Solution

Step 1: There are two nodes, x_1 and x_2 and an additional node known as reference node. Total number of nodes = 3.

Step 2: M_1 is connected between x_1 and reference, M_2 is connected between x_2 and reference.

Step 3: K_1 is connected between x_1 and reference. K_{12} is connected between x_1 and x_2. K_2 is connected between x_2 and reference.

Step 4: D_1 is connected between x_1 and reference. D_{12} is connected between x_1 and x_2. D_2 is connected between x_2 and reference.

Step 5: f is connected between x_1 and reference.

Step 6: Figure E5.4(a) shows the schematic diagram.

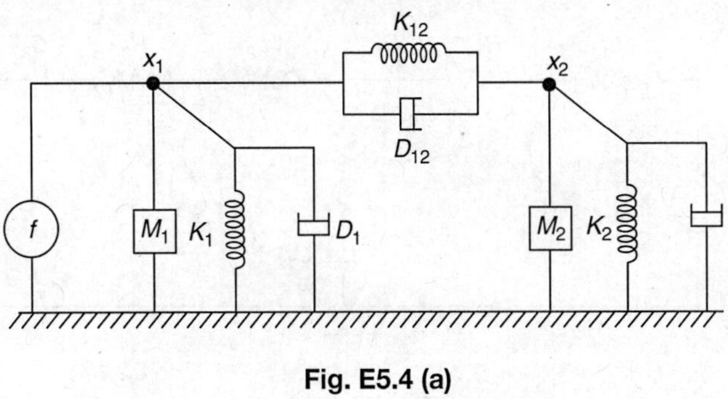

Fig. E5.4 (a)

Step 7: At node x_1,

$$f = M_1 \frac{d^2x_1}{dt^2} + D_1 \frac{dx_1}{dt} + K_1x_1 + D_{12}\frac{d}{dt}(x_1 - x_2) + K_{12}(x_1 - x_2) \tag{1}$$

At node x_2,

$$K_{12}(x_1 - x_2) + D_{12}\frac{d}{dt}(x_1 - x_2) = M_2 \frac{d^2x_2}{dt^2} + D_2 \frac{dx_2}{dt} + K_2x_2 \tag{2}$$

Force–voltage (f–v) analogy:

$$f \rightarrow v, M \rightarrow L,\ D \rightarrow R,\ K \rightarrow \frac{1}{C}, x \rightarrow q$$

From Eq. (1), we have,

$$v = L_1 \frac{d^2q_1}{dt^2} + R_1 \frac{dq_1}{dt} + \frac{1}{C_1}q_1 + R_{12}\frac{d}{dt}(q_1 - q_2) + \frac{1}{C_{12}}(q_1 - q_2)$$

i.e.,

$$v = L_1 \frac{di_1}{dt} + R_1i_1 + \frac{1}{C_1}\int i_1 dt + R_{12}(i_1 - i_2) + \frac{1}{C_{12}}\int (i_1 - i_2)dt \tag{3}$$

From Eq. (2), we have,

$$\frac{1}{C_{12}}(q_1 - q_2) + R_{12}\frac{d}{dt}(q_1 - q_2) = L_2 \frac{d^2q_2}{dt^2} + R_2 \frac{dq_2}{dt} + \frac{1}{C_2}q_2$$

i.e.,

$$\frac{1}{C_{12}}\int (i_1 - i_2)dt + R_{12}(i_1 - i_2) = L_2 \frac{di_2}{dt} + R_2i_2 + \frac{1}{C_2}\int i_2 dt$$

From Eq. (3) and Eq. (4), we can draw the f–v analogy as shown in Fig. E5.4(b)

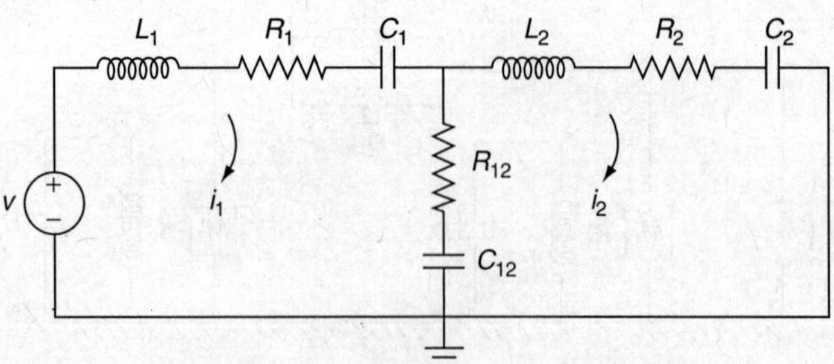

Fig. E5.4 (b)

Force–current (f–i) analogy:

$$f \to i,\ M \to C,\ D \to \frac{1}{R},\ K \to \frac{1}{L},\ x \to \psi$$

From Eq. (1), we have,

$$i = C_1 \frac{d^2\psi_1}{dt^2} + \frac{1}{R_1}\frac{d\psi_1}{dt} + \frac{1}{L_1}\psi_1 + \frac{1}{R_{12}}\frac{d}{dt}(\psi_1 - \psi_2) + \frac{1}{L_{12}}(\psi_1 - \psi_2)$$

i.e.,
$$i = C_1 \frac{dv_1}{dt} + \frac{1}{R_1}v_1 + \frac{1}{L_1}\int v_1 dt + \frac{v_1 - v_2}{R_{12}} + \frac{1}{L_{12}}\int (v_1 - v_2) dt \tag{5}$$

From Eq. (2), we have,

$$\frac{1}{L_{12}}(\psi_1 - \psi_2) + \frac{1}{R_{12}}\frac{d}{dt}(\psi_1 - \psi_2) = C_2 \frac{d^2\psi_2}{dt^2} + \frac{1}{R_2}\frac{d\psi_2}{dt} + \frac{1}{L_{12}}\psi_2$$

i.e.,
$$\frac{1}{L_{12}}\int (v_1 - v_2) dt + \frac{1}{R_{12}}(v_1 - v_2) = C_2 \frac{dv_2}{dt^2} + \frac{v_2}{R_2} + \frac{1}{L_{12}}\int v_2 dt \tag{6}$$

From Eq. (5) and Eq. (6), we can draw the force–current analogy circuit as shown in Fig. E5.4(c).

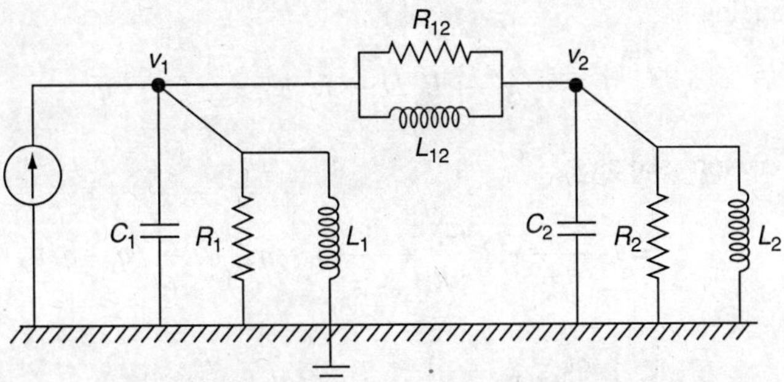

Fig. E5.4 (c)

Example 5.5 Determine equations of the system shown in Fig. E5.5. Show its f–v and f–i analogy.

Solution Using all steps as before, the system representation by the nodal method is shown in Fig. E5.5(a).

From Fig. E5.5(a)
At node x_1:

$$f = M_1 \frac{d^2x_1}{dt^2} + D_1 \frac{dx_1}{dt} + K_{12}(x_1 - x_2) + D\frac{d}{dt}(x_1 - x_3) \tag{1}$$

At node x_2:

$$K_{12}(x_1 - x_2) + K(x_3 - x_2) = M_2 \frac{d^2x_2}{dt^2} + D_2 \frac{dx_2}{dt} + K_2 x_2 \tag{2}$$

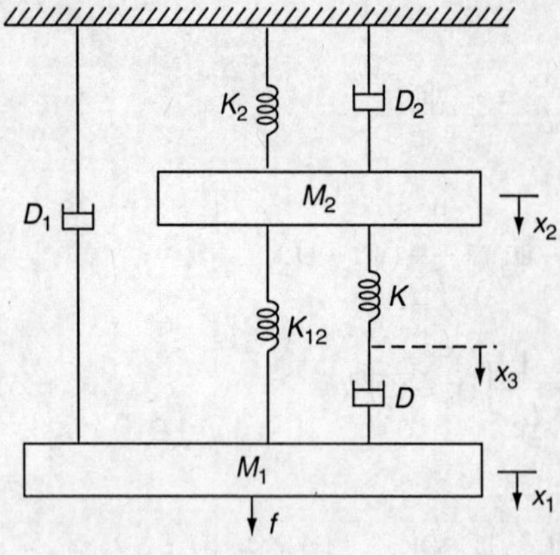

Fig. E5.5

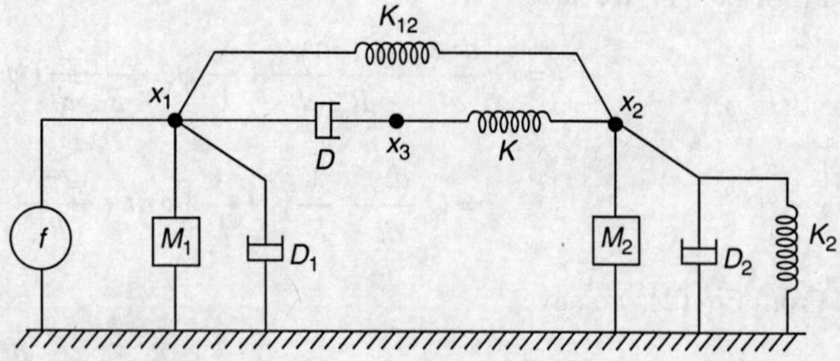

Fig. E5.5 (a)

At node x_3:

$$D\frac{d}{dt}(x_1 - x_3) = K(x_3 - x_2) \quad (3)$$

(i) Force–voltage analogy:

$$f \to v,\ M \to L,\ D \to R,\ K \to \frac{1}{C},\ x \to q$$

Equation (1) can be expressed as

$$v = L_1\frac{d^2q_1}{dt^2} + R_1\frac{dq_1}{dt} + \frac{1}{C_{12}}(q_1 - q_2) + R\frac{d}{dt}(q_1 - q_3)$$

i.e.,

$$v = L_1\frac{di_1}{dt} + R_1 i_1 + \frac{1}{C_{12}}\int (i_1 - i_2)\,dt + R(i_1 - i_3)$$

Equation (2) can be expressed as

$$\frac{1}{C_{12}}(q_1 - q_2) + \frac{1}{C}(q_3 - q_2) = L_2\frac{d^2q_2}{dt^2} + R_2\frac{dq_2}{dt} + \frac{1}{C}q_2$$

i.e.,

$$\frac{1}{C_{12}}\int (i_1 - i_2)\,dt + \frac{1}{C}\int (i_3 - i_2)dt = L_2\frac{di_2}{dt} + R_2 i_2 + \frac{1}{C}\int i_2 dt \quad (5)$$

Equation (3) can be expressed as

$$R\frac{d}{dt}(q_1 - q_3) = \frac{1}{C}(q_3 - q_2)$$

i.e., $$R(i_1 - i_3) = \frac{1}{C}\int (i_3 - i_2)dt \tag{6}$$

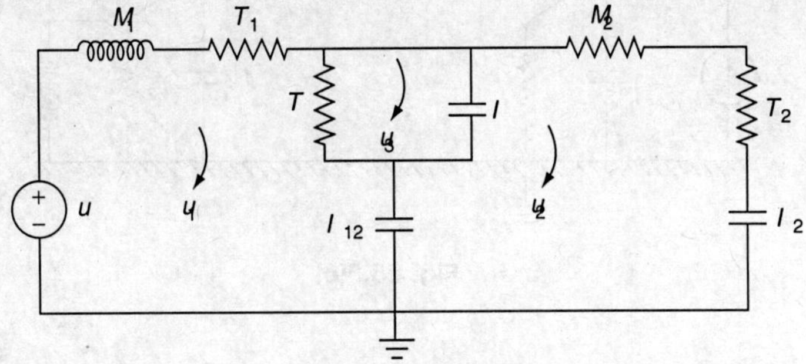

Fig. E5.5(b)

(ii) Force–current analogy:

$$f \to i,\ M \to C,\ D \to \frac{1}{R}, K \to \frac{1}{L},\ x \to \psi$$

From Eq. (1), we have

$$i = C_1\frac{d^2\psi_1}{dt^2} + \frac{1}{R_1}\frac{d\psi_1}{dt} + \frac{1}{L_{12}}(\psi_1 - \psi_2) + \frac{1}{R}\frac{d}{dt}(\psi_1 - \psi_3)$$

i.e., $$i = C_1\frac{dv_1}{dt} + \frac{v_1}{R_1} + \frac{1}{L_{12}}\int (v_1 - v_2)\ dt + \frac{v_1 - v_3}{R} \tag{7}$$

From Eq. (2), we have

$$\frac{1}{L_{12}}(\psi_1 - \psi_2) + \frac{1}{L}(\psi_3 - \psi_2) = C_2\frac{d^2\psi_2}{dt^2} + \frac{1}{R_2}\frac{d\psi_2}{dt} + \frac{1}{L_2}\psi_2$$

i.e., $$\frac{1}{L_{12}}\int (v_1 - v_2)dt + \frac{1}{L}\int (v_3 - v_2)dt = C_2\frac{dv_2}{dt} + \frac{v_2}{R_3} + \frac{1}{L_2}\int i_2 dt \tag{8}$$

From Eq. (3), we have

$$\frac{1}{R}\frac{d}{dt}(\psi_2 - \psi_3) = \frac{1}{L}(\psi_3 - \psi_2)$$

i.e., $$\frac{1}{R}(v_2 - v_3) = \frac{1}{L}\int (v_3 - v_2)\, dt$$

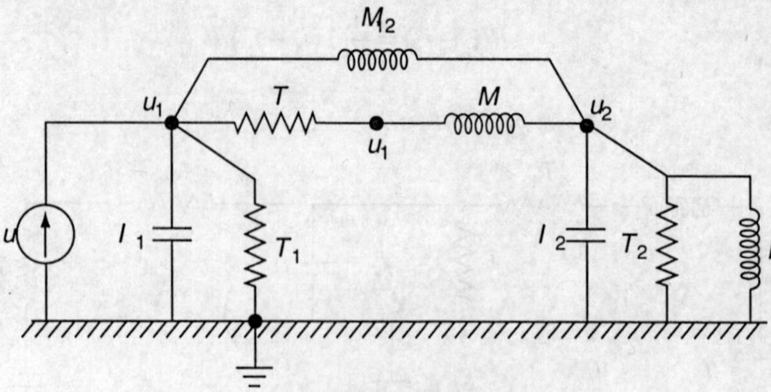

Fig. E5.5(c)

Example 5.6 Determine system equations of the system shown in Fig. E5.6. Draw *f*–*v* and *f*–*i* analogies.

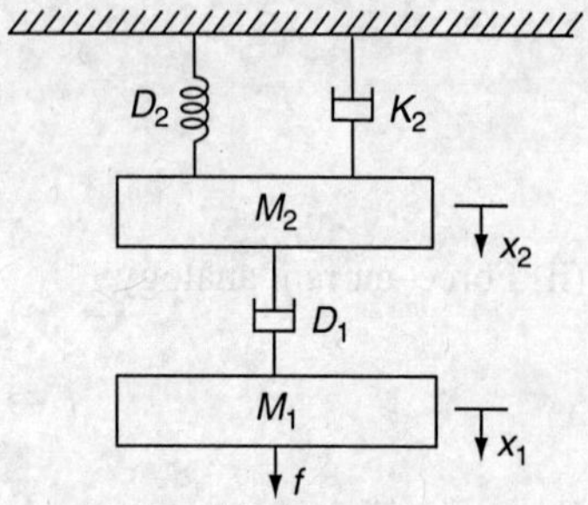

Fig. E5.6

Solution Using the step procedure as before, by nodal analysis, the system representation is shown in Fig. E5.6(a).

At node x_1:

$$f = M_1 \frac{d^2x_1}{dt^2} + D_1 \frac{d}{dt}(x_1 - x_2) \quad (1)$$

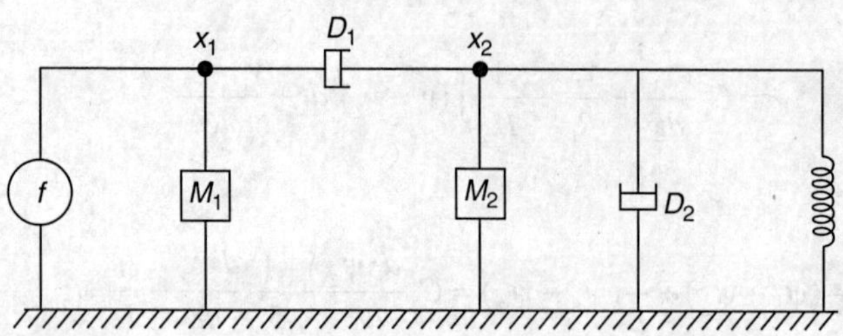

Fig. E5.6(a)

At node x_2:

$$D_1 \frac{d}{dt}(x_1 - x_2) = M_2 \frac{d^2x_2}{dt^2} + D_2 \frac{dx_2}{dt} + K_2 x_2 \quad (2)$$

(i) *f*–*v* analogy:

$$f \to v,\ M \to L,\ D \to R,\ K \to \frac{1}{C},\ x \to q$$

From Eq. (1), we have

$$v = L_1 \frac{d^2 q_1}{dt^2} + R_1 \frac{d}{dt}(q_1 - q_2)$$

i.e.,

$$v = L_1 \frac{di_1}{dt} + R_1 (i_1 - i_2) \tag{3}$$

From Eq. (2), we have

$$R_1 \frac{d}{dt}(q_1 - q_2) = L_2 \frac{d^2 q_2}{dt^2} + R_2 \frac{dq_2}{dt} + \frac{1}{C_2} q_2$$

i.e.,

$$R_1 (i_1 - i_2) = L_2 \frac{di_2}{dt} + R_2 i_2 + \frac{1}{C_2} \int i_2 dt \tag{4}$$

The *f-v* analogy is shown in Fig. E5.6(b)

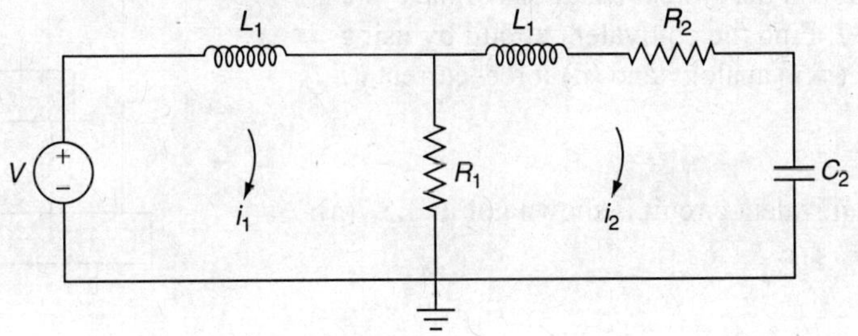

Fig. E5.6(b)

(ii) *f-i* analogy:

$$f \rightarrow i,\ M \rightarrow C,\ D \rightarrow \frac{1}{R}, K \rightarrow \frac{1}{L}, x \rightarrow \psi$$

From Eq. (1), we have

$$i = C_1 \frac{d^2 \psi_1}{dt^2} + \frac{1}{R_1} \frac{d}{dt}(\psi_1 - \psi_2)$$

i.e.,

$$i = C_1 \frac{dv_1}{dt} + \frac{v_1 - v_2}{R_1} \tag{5}$$

From Eq. (2), we have

$$\frac{1}{R_1} \frac{d}{dt}(\psi_1 - \psi_2) = C_2 \frac{d^2 \psi_2}{dt^2} + \frac{1}{R_2} \frac{d\psi_2}{dt} + \frac{1}{L_2} \psi_2$$

$$\frac{v_1 - v_2}{R_1} = C_2 \frac{dv_2}{dt} + \frac{v_2}{R_2} + \frac{1}{L_2}\int v_2 dt \tag{6}$$

f-i analogy is shown in Fig. E5.6(c).

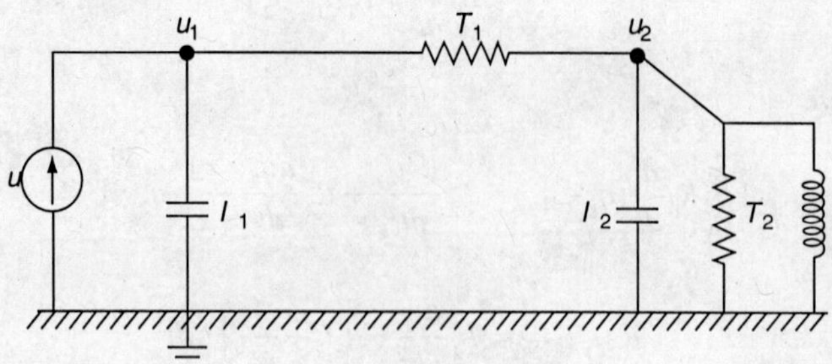

Fig. E5.6(c)

Example 5.7 Obtain the system equations of the system given in Fig. E5.7. Find the equivalent circuit by using (i) force–voltage (*f–v*) analogy and (ii) force–current (*f–i*) analogy.

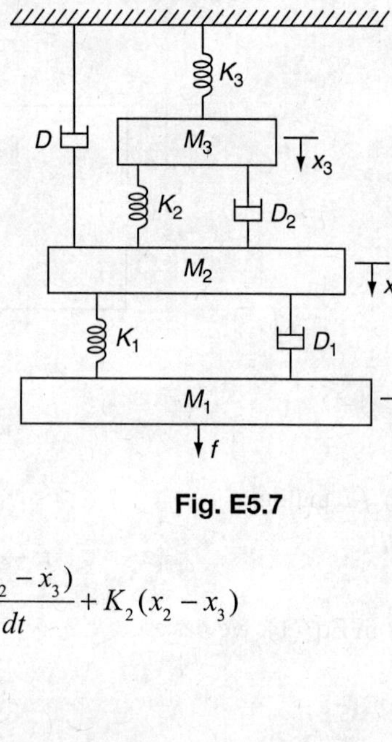

Fig. E5.7

Solution The equivalent circuit is shown in Fig. E5.7(a).

For node x_1,

$$f = M_1 \frac{d^2 x_1}{dt^2} + D_1 \frac{d(x_1 - x_2)}{dt} + K_1(x_1 - x_2) \tag{1}$$

For node x_2,

$$K_1(x_1 - x_2) + D_1 \frac{d(x_1 - x_2)}{dt} = M_2 \frac{d^2 x_2}{dt^2} + D\frac{dx_2}{dt} + D_2 \frac{d(x_2 - x_3)}{dt} + K_2(x_2 - x_3) \tag{2}$$

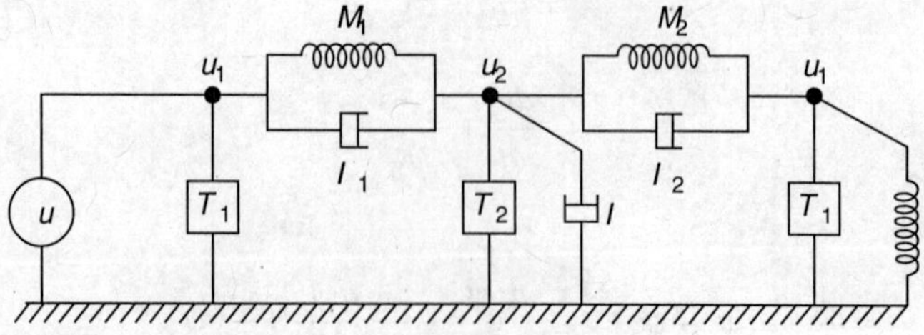

Fig. E5.7(a)

For node x_3,

$$D_2 \frac{d(x_2 - x_3)}{dt} + K_2(x_2 - x_3) = M_3 \frac{d^2 x_3}{dt^2} + K_3 x_3 \tag{3}$$

(i) In force–voltage analogy,

$$f \to v,\ M \to L,\ D \to R,\ K \to \frac{1}{C},\ x \to q$$

From Eq. (1), Eq. (2) and Eq. (3), we have

$$v = L_1 \frac{d^2 q_1}{dt^2} + \frac{1}{C_1}(q_1 - q_2) + R_1 \frac{d}{dt}(q_1 - q_2) \tag{4}$$

$$\frac{1}{C_1}(q_1 - q_2) + R_1 \frac{d}{dt}(q_1 - q_2) = L_2 \frac{d^2 q_2}{dt^2} + R \frac{dq_2}{dt} + R_2 \frac{d(q_2 - q_3)}{dt} + \frac{1}{C_2}(q_2 - q_3) \tag{5}$$

$$R_2 \frac{d(q_2 - q_3)}{dt} + \frac{1}{C_2}(q_2 - q_3) = L_3 \frac{d^2 q_3}{dt^2} + \frac{q_3}{C_3} \tag{6}$$

The force–voltage analogy is shown in Fig. E5.7(b).

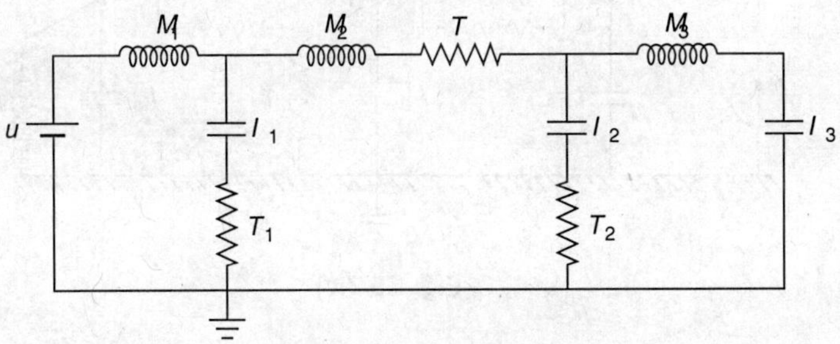

Fig. E5.7(b)

(ii) In force–current analogy,

$$f \to i,\ M \to C,\ D \to \frac{1}{R},\ K \to \frac{1}{L},\ x \to \psi$$

From Eq. (1), Eq. (2) and Eq. (3), we have

$$i = C_1 \frac{d^2 \psi_1}{dt^2} + \frac{1}{L_1}(\psi_1 - \psi_2) + \frac{1}{R_1}\frac{d}{dt}(\psi_1 - \psi_2) \tag{7}$$

$$\frac{1}{L_1}(\psi_1 - \psi_2) + \frac{1}{R_1}\frac{d}{dt}(\psi_1 - \psi_2) = C_2 \frac{d^2 \psi_2}{dt^2} + \frac{1}{R}\frac{d\psi_2}{dt} + \frac{1}{R_2}\frac{d(\psi_2 - \psi_3)}{dt} + \frac{1}{L_2}(\psi_2 - \psi_3) \tag{8}$$

$$\frac{1}{R_2}\frac{d(\psi_2-\psi_3)}{dt}+\frac{1}{L_2}(\psi_2-\psi_3)=C_3\frac{d^2\psi_3}{dt^2}+\frac{\psi_3}{L_3} \tag{9}$$

Since $v = d\psi/dt$, $\psi = \int v dt$

$$i = C_1\frac{dv_1}{dt}+\frac{1}{L_1}\int(v_1-v_2)dt+\frac{1}{R_1}(v_1-v_2) \tag{10}$$

$$\frac{1}{L_1}\int(v_1-v_2)\,dt+\frac{1}{R_1}(v_1-v_2)=C_2\frac{dv_2}{dt}+\frac{1}{R}v_2+\frac{1}{R_2}(v_2-v_3)+\frac{1}{L_2}\int(v_2-v_3)\,dt \tag{11}$$

$$\frac{1}{R_2}(v_2-v_3)+\frac{1}{L_2}\int(v_2-v_3)\,dt=C_3\frac{dv_3}{dt}+\frac{1}{L_3}\int v_3 dt \tag{12}$$

Figure E5.7(c) shows the force–current analogy of Fig. E5.7.

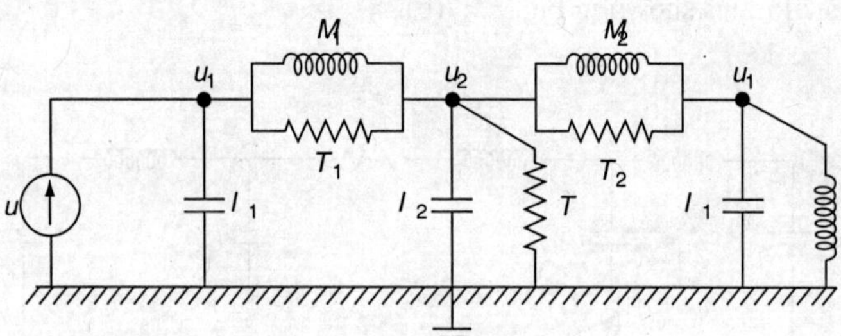

Fig. E5.7(c)

Example 5.8 Obtain the system equations for the system shown in Fig. E5.8. Determine (i) force–voltage (f–v) and force–current (f–i) analogy.

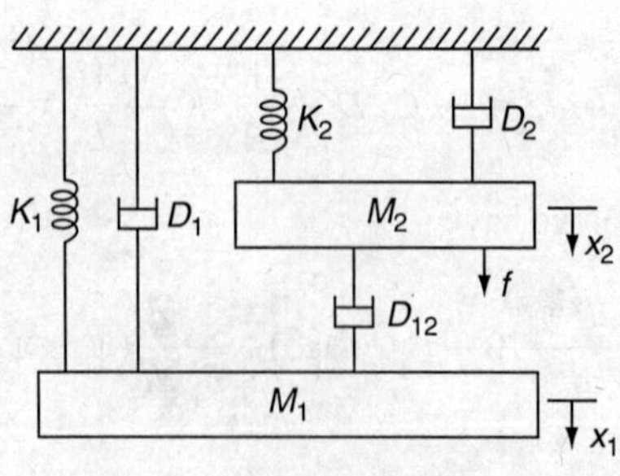

Fig. E5.8

Solution The equivalent circuit is shown in Fig. E5.8(a).

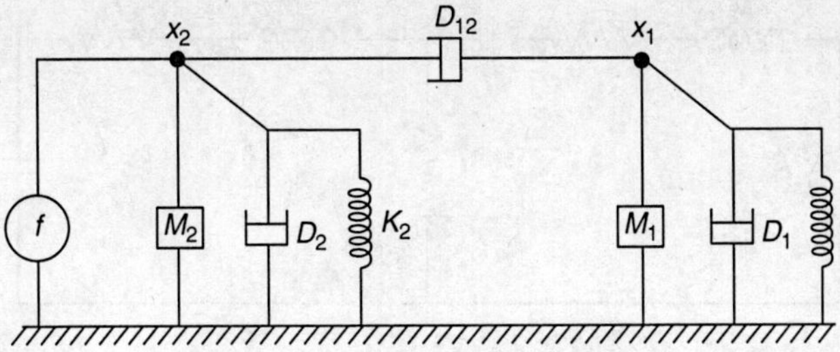

Fig. E5.8(a)

For node x_2, we have

$$f = M_2 \frac{d^2 x_2}{dt^2} + D_2 \frac{dx_2}{dt} + D_{12} \frac{d(x_2 - x_1)}{dt} + K_2 x_2 \tag{1}$$

For node x_1, we have

$$D_{12} \frac{d(x_2 - x_1)}{dt} = M_1 \frac{d^2 x_1}{dt^2} + D_1 \frac{dx_1}{dt} + K_1 x_1 \tag{2}$$

Equation (1) and Eq. (2) represent the system equations.
(i) In force–voltage analogy, we have

$$f \to v,\ M \to L,\ D \to R,\ K \to \frac{1}{C},\ x \to q$$

Equation (1) and Eq. (2) become

$$v = L_2 \frac{d^2 q_2}{dt^2} + R_2 \frac{dq_2}{dt} + \frac{q_2}{C_2} + R_{12} \frac{d}{dt}(q_2 - q_1)$$

i.e.,
$$v = L_2 \frac{di_2}{dt} + R_2 i_2 + \frac{1}{C_2} \int i_2 dt + R_{12}(i_2 - i_1) \tag{3}$$

$$R_{12} \frac{d}{dt}(q_2 - q_1) = L_1 \frac{d^2 q_1}{dt^2} + R_1 \frac{dq_1}{dt} + \frac{1}{C_1} q_1$$

i.e.,
$$R_{12}(i_2 - i_1) = L_1 \frac{di_1}{dt} + R_1 i_1 + \frac{1}{C_1} \int i_1 dt \tag{4}$$

The force–voltage analogy is shown in Fig. E5.8(b).

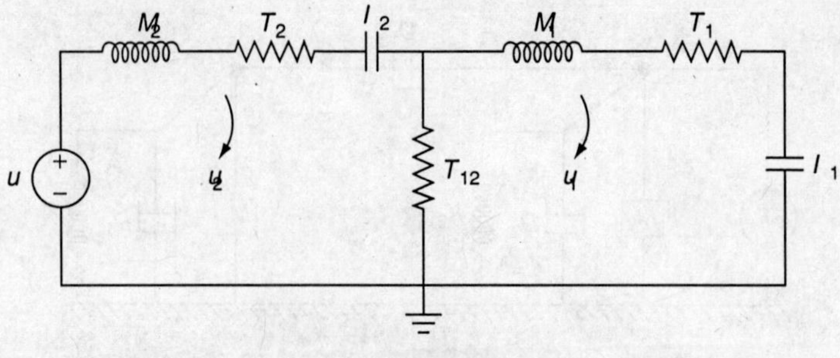

Fig. E5.8(b)

(ii) In force–current analogy,

$$f \to i,\ M \to C,\ D \to \frac{1}{R},\ K \to \frac{1}{L},\ x \to \psi$$

Equation (1) and Eq. (2) become

$$i = C_2 \frac{d^2\psi_2}{dt^2} + \frac{1}{R_2}\frac{d\psi_2}{dt} + \frac{1}{R_{12}}\frac{d}{dt}(\psi_2 - \psi_1) + \frac{\psi_2}{L_2} \tag{5}$$

$$= C_2 \frac{dv_2}{dt} + \frac{v_2}{R_2} + \frac{(v_2 - v_1)}{R_{12}} + \frac{1}{L_2}\int v_2 dt$$

and $$\frac{1}{R_{12}}\frac{d}{dt}(\psi_2 - \psi_1) = C_1 \frac{d^2\psi_1}{dt^2} + \frac{1}{R_1}\frac{d\psi_1}{dt} + \frac{\psi_1}{L_2}$$

i.e., $$\frac{(v_2 - v_1)}{R_{12}} = C_1 \frac{dv_1}{dt} + \frac{v_1}{R_1} + \frac{1}{L_1}\int v_1 dt \tag{6}$$

[Since $v = d\psi/dt$, $\psi = \int v dt$]

Figure E5.8(c) shows the force–current analogy of Fig. E5.8.

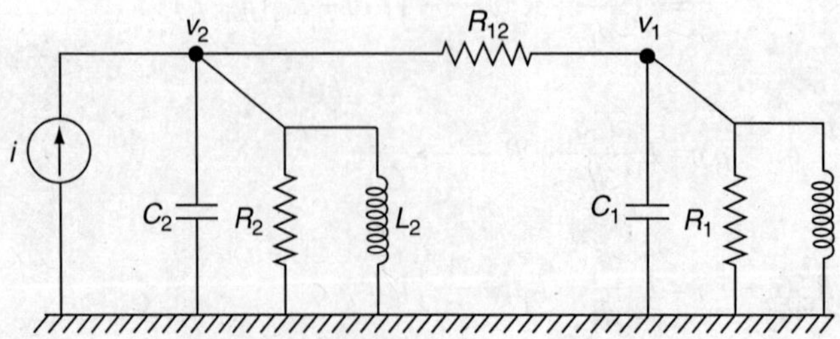

Fig. E5.8(c)

Example 5.9 Determine the system equations for the system shown in Fig. E5.9. Also find (a) torque–voltage (T–v) and (b) torque–current (T–i) analogies.

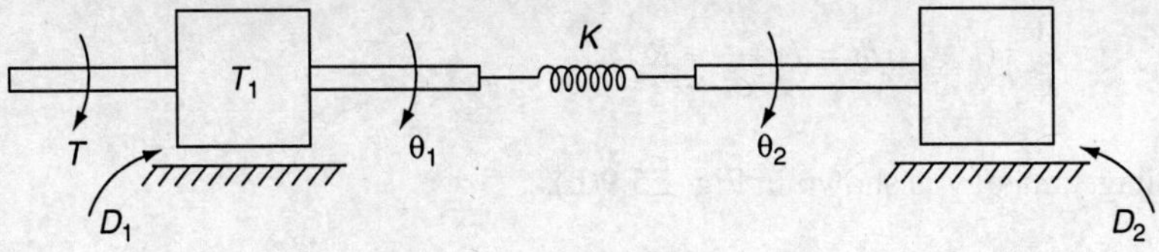

Fig. E5.9

Solution From Fig. E5.9, Fig. E5.9(a) is obtained.

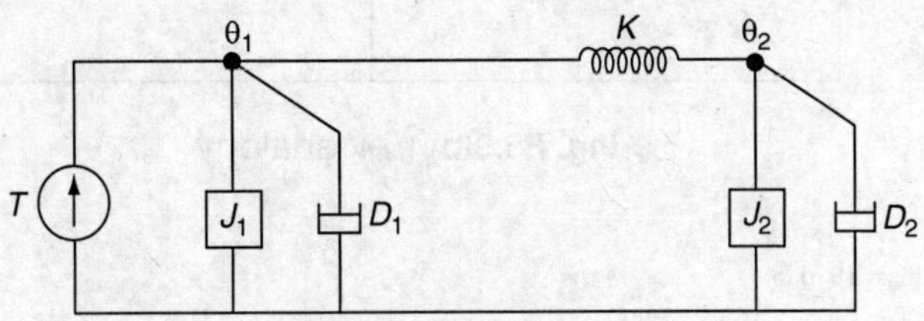

Fig. E5.9(a)

At node θ_1,

$$T = J_1 \frac{d^2\theta_1}{dt^2} + D_1 \frac{d\theta_1}{dt} + K(\theta_1 - \theta_2) \quad (1)$$

At node θ_2,

$$K(\theta_1 - \theta_2) = J_2 \frac{d^2\theta_2}{dt^2} + D_2 \frac{d\theta_2}{dt} \quad (2)$$

Equation (1) and Eq. (2) are the system equations.
(i) In torque–voltage analogy,

$$T \to v,\ J \to L,\ D \to R,\ K \to \frac{1}{C},\ \theta \to q$$

Equation (1) and Eq. (2) become

$$v = L_1 \frac{d^2q_1}{dt^2} + R_1 \frac{dq_1}{dt} + \frac{(q_1 - q_2)}{C}$$

$$= L_1 \frac{di_1}{dt} + R_1 i_1 + \frac{\int (i_1 - i_2)dt}{C} \quad (3)$$

and $$\frac{(q_1 - q_2)}{C} = L_2 \frac{d^2 q_2}{dt^2} + R_2 \frac{dq_2}{dt}$$

i.e., $$\frac{1}{C}\int (i_1 - i_2)\, dt = L_2 \frac{di_2}{dt} + R_2 i_2$$

The torque–voltage analogy is shown in Fig. E5.9(b).

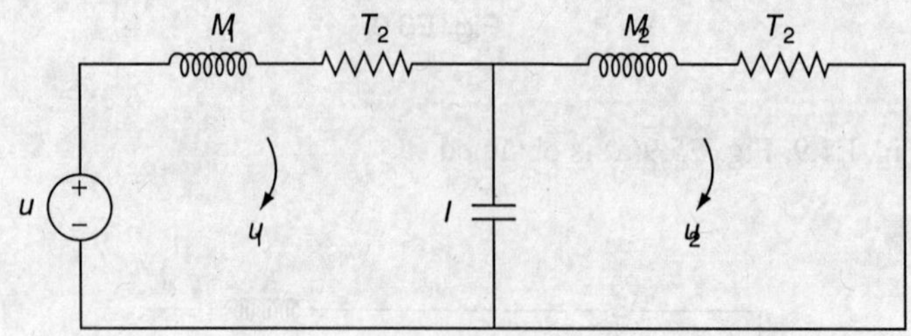

Fig. E5.9(b) *T–v* analogy

(ii) In torque–current analogy,

$$T \to i,\ J \to C,\ K \to \frac{1}{R},\ D \to \frac{1}{L},\ x \to \psi$$

Equation (1) and Eq. (2) become

$$i = C_1 \frac{d^2\psi_1}{dt^2} + \frac{1}{R_1}\frac{d\psi_1}{dt} + \frac{(\psi_2 - \psi_1)}{L}$$

$$= C_1 \frac{dv_1}{dt} + \frac{v_1}{R_1} + \frac{1}{L}\int (v_2 - v_1)\, dt$$

and $$\frac{(\psi_1 - \psi_2)}{L} = C_2 \frac{d^2\psi_2}{dt^2} + \frac{1}{R_2}\frac{d\psi_2}{dt}$$

i.e., $$\frac{1}{L}\int (v_2 - v_1) dt = C_2 \frac{dv_2}{dt} + \frac{v_2}{R_2}$$

$$\left[\text{Since } v = \frac{d\psi}{dt},\ \psi = \int v dt\right]$$

Figure E5.9(c) shows the torque–current (T–i) analogy of Fig. E5.9.

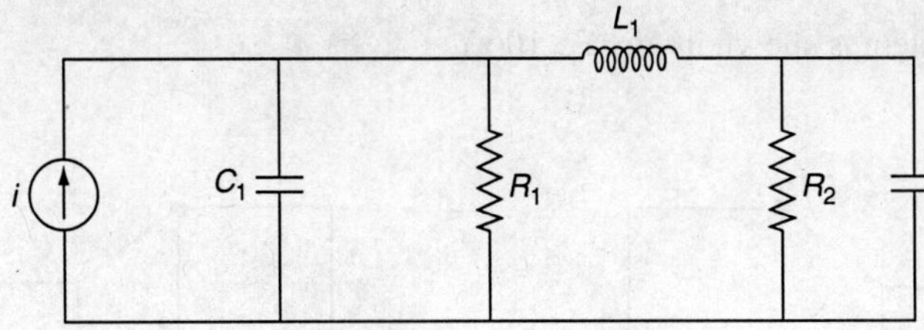

Fig. E5.9(c) T–i analogy

Example 5.10 Draw the force–voltage analogous network for the mechanical system shown in Fig. E5.10.

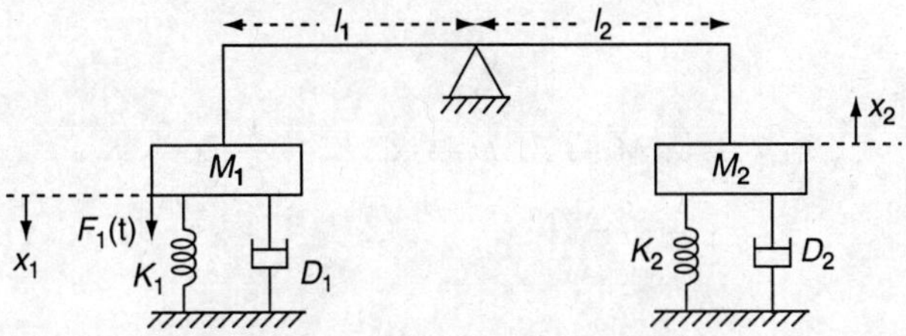

Fig. E5.10

Solution The displacement x_1 of mass M_1 is due to force $F_1(t)$.

$$F_1(t) = M_1 \frac{d^2 x_1}{dt^2} + D_1 \frac{dx_1}{dt} + K_1 x_1 \tag{1}$$

The force transmitted to mass M_2 is given by

$$F_1(t) \times l_1 = F_2(t) \times l_2$$

$$\therefore \quad F_2(t) = F_1(t) \times \left(\frac{l_1}{l_2}\right) \tag{2}$$

For mass M_2, the displacement x_2 is due to force $F_2(t)$.

$$F_2(t) = M_2 \frac{d^2 x_2}{dt^2} + D_2 \frac{dx_2}{dt} + K_2 x_2$$

$$\therefore \qquad F_1(t)\left(\frac{l_1}{l_2}\right) = M_2\frac{d^2x_2}{dt^2} + D_2\frac{dx_2}{dt} + K_2x_2 \qquad (3)$$

The equivalent system is shown in Fig. E5.10(a).

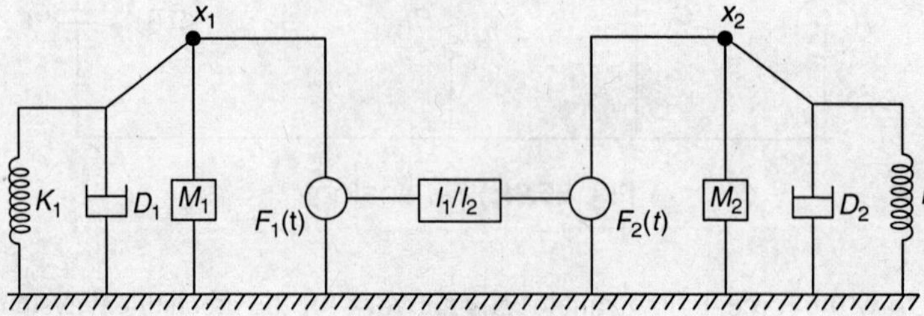

Fig. E5.10(a)

For f–v analogy,

$$f \to v,\ M \to L,\ D \to R,\ K \to \frac{1}{C},\ x \to q$$

Equation (1) becomes

$$v_1 = L_1\frac{d^2q_1}{dt^2} + R_1\frac{dq_1}{dt} + \frac{q_1}{C_1}$$

i.e.,
$$v_1 = L_1\frac{di_1}{dt} + R_1i_1 + \frac{1}{C_1}\int i_1dt \qquad (4)$$

Equation (3) becomes

$$v_2 = L_2\frac{d^2q_2}{dt^2} + R_2\frac{dq_2}{dt} + \frac{q_2}{C_2}$$

i.e.,
$$v_2 = L_2\frac{di_2}{dt} + R_2i_2 + \frac{1}{C_2}\int i_2dt \qquad (5)$$

where
$$v_2 = v_1\left(\frac{l_1}{l_2}\right) \qquad (6)$$

The f–v analogous network is shown in Fig. E5.10(b).

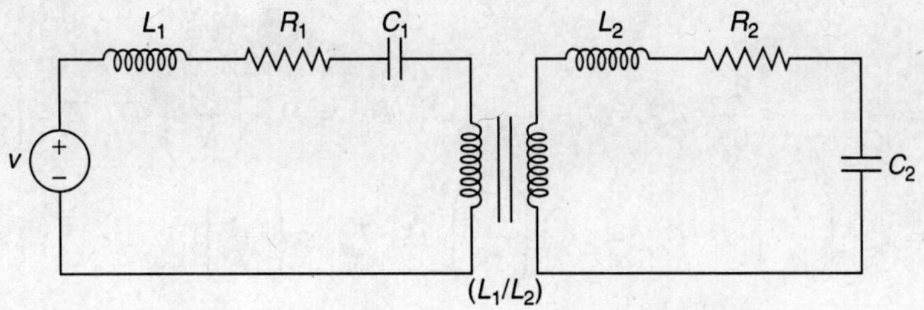

Fig. E5.10(b)

SIGNIFICANT POINTS

Counterparts of Translational and Rotational Motion

Translational motion	Rotational motion
Mass (M)	Inertia (J)
Damper (D)	Damper (D)
Spring (K)	Spring (K)
Force (f)	Torque (T)
Displacement (x)	Angular displacement (θ)
Velocity $(u) = \dfrac{dx}{dt}$	Angular velocity $\omega = \dfrac{d\theta}{dt}$

Force–voltage Analogy

Translational	Electrical	Rotational
Force (f)	Voltage (v)	Torque (T)
Mass (M)	Inductance (L)	Inertia (J)
Damper (D)	Resistance (R)	Damper (D)
Spring (K)	Elastance $\left(\dfrac{1}{C}\right)$	Spring (K)
Displacement (x)	Charge (q)	Displacement (θ)
Velocity $(u) = \dfrac{dx}{dt}$	Current (i)	Angular velocity $(\omega) = \dfrac{d\theta}{dt}$

Force–current Analogy

Translational	Electrical	Rotational
Force (f)	current (i)	Torque (T)
Mass (M)	Capacitance (C)	Inertia (J)
Damper (D)	Conductance $\left(\dfrac{1}{R}\right)$	Damper (D)
Spring (K)	Reciprocal of inductance $\left(\dfrac{1}{L}\right)$	Spring (K)
Displacement (x)	Flux linkage (ψ)	Displacement (θ)
Velocity $(u) = \dfrac{dx}{dt}$	Voltage $(v) = \dfrac{d\psi}{dt}$	Angular velocity $(\omega) = \dfrac{d\theta}{dt}$

For gear train

$$\frac{T_1}{T_2} = \frac{r_1}{r_2} = \frac{N_1}{N_2} = \frac{\theta_2}{\theta_1},\ J_{1eq} = J_1 + \left(\frac{N_1}{N_2}\right)^2 J_2,\ D_{1eq} = D_1 + \left(\frac{N_1}{N_2}\right)^2 D_2,\ T_L' = \left(\frac{N_1}{N_2}\right) T_L$$

and
$$J_{2eq} = J_2 + \left(\frac{N_2}{N_1}\right)^2 J_1,\ D_{2eq} = D_2 + \left(\frac{N_2}{N_1}\right)^2 D_1,\ T_{Meq} = \left(\frac{N_2}{N_1}\right) T_1$$

SHORT QUESTIONS WITH ANSWERS

1. Write down the counterparts of translational and rotational motion.

Ans.

Sl No.	Translational motion	Rotational motion
1	Mass (M)	Inertia (J)
2	Damper (D)	Damper (D)
3	Spring (K)	Spring (K)
4	Force (f)	Torque (T)
5	Displacement (x)	Angular displacement (θ)
6	Velocity $(u) = \frac{dx}{dt}$	Angular velocity $(\omega) = \frac{d\theta}{dt}$

2. Write down the force–voltage analogy.

Ans.

Translational	Electrical	Rotational
Force (f)	Voltage (v)	Torque (T)
Mass (M)	Inductance (L)	Inertia (J)
Spring (K)	Resistance (R)	Damper (D)
Damper (D)	Elastance $\left(\frac{1}{C}\right)$	Spring (K)
Displacement (x)	Charge (q)	Displacement (θ)
Velocity $(u) = \frac{dx}{dt}$	Current (i)	Angular velocity $(\omega) = \frac{d\theta}{dt}$

3. Write down the force–current Analogy.

Ans.

Translational	Electrical	Rotational
Force (f)	Current (i)	Torque (T)
Mass (M)	Inductance (L)	Inertia (J)
Damper (D)	Conductance $\left(\frac{1}{R}\right)$	Damper (D)
Spring (K)	Reciprocal of inductance $\left(\frac{1}{L}\right)$	Spring (K)
Displacement (x)	Flux linkage (ψ)	Displacement (θ)
Velocity $(u) = \frac{dx}{dt}$	Voltage $(v) = \frac{d\psi}{dt}$	Angular velocity $(w) = \frac{d\theta}{dt}$

EXERCISES

Questions for Practice

1. Using Fig. 1, answer the following questions.

(i) What is the ratio of diameters D_1/D_2?

(ii) If the gear 1 is to displace 35°, what is the displacement of gear 2?

(iii) If $\omega_1 = 20$ rad/sec, what is the value of ω_2?

(iv) If $\alpha_2 = 5$ rad/sec^2, what is the value of α_1, angular acceleration of gear 1?

Ans. (i) 2, (ii) 70°, (iii) 40 rad/sec, (iv) 2 rad/sec^2

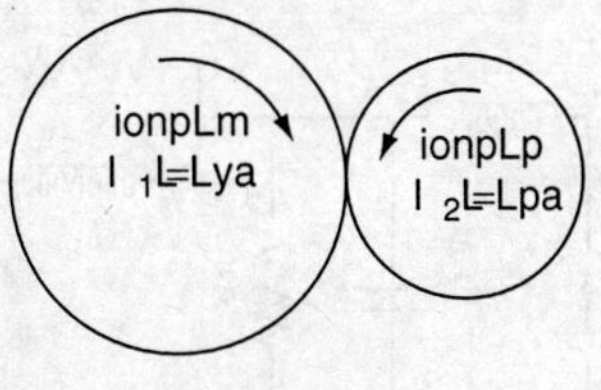

Figure 1

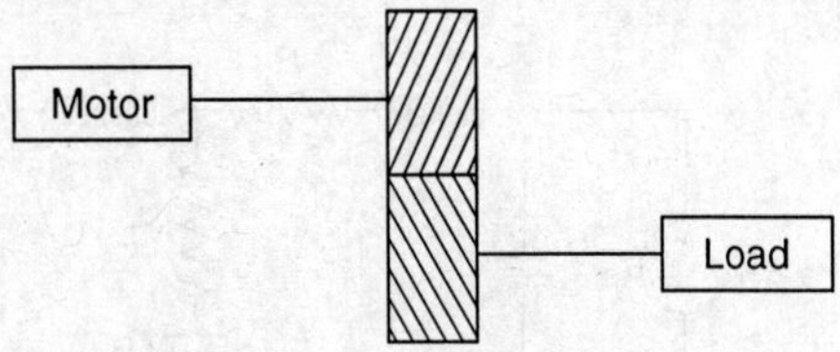

Figure 2

2. Using Fig. 2 and the values given below, answer the following questions.

$J_m = 8 \qquad F_m = 0.5 \qquad J_L = 4 \qquad F = 0.4$

(a) Calculate the total moment of inertia $J_{\omega t}$ experienced by the motor.

(b) Calculate the total coefficient of viscous friction (F_T) as experienced by the motor.

Ans. (a) 9 (b) 0.6

3. Obtain the analogous electrical network of Fig. 3 using (a) f–v and (b) f–i analogy.

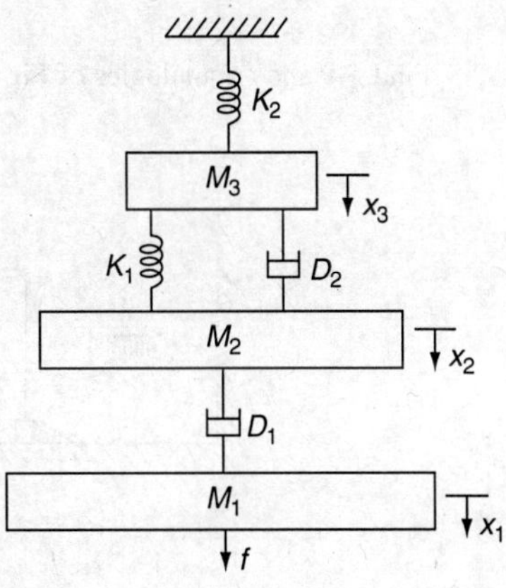

Figure 3a

Ans.

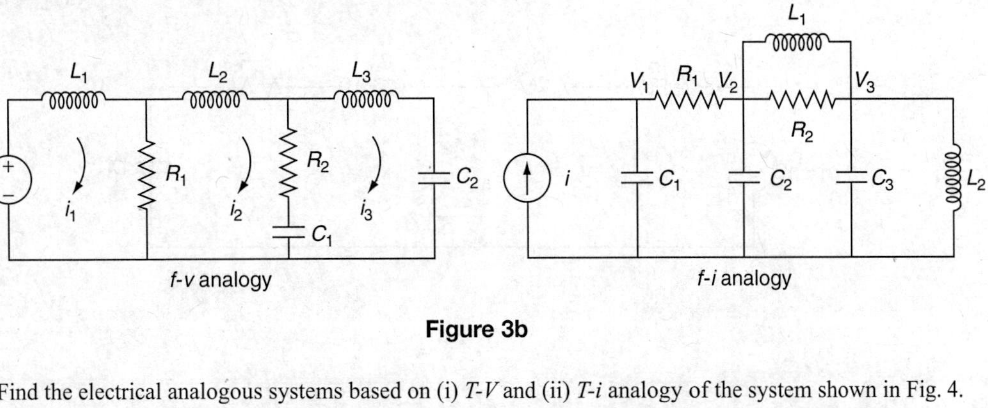

Figure 3b

4. Find the electrical analogous systems based on (i) T-V and (ii) T-i analogy of the system shown in Fig. 4.

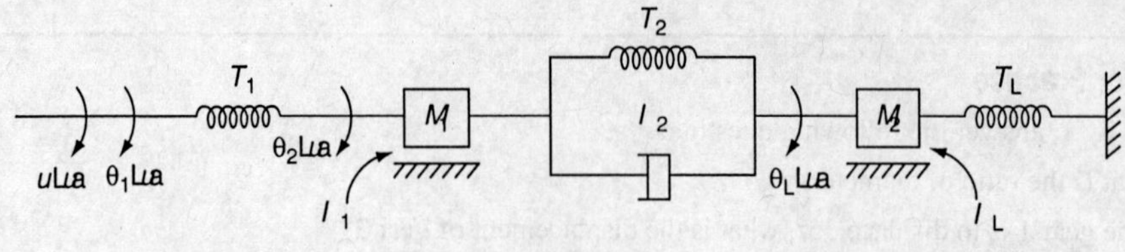

Figure 4

Ans.

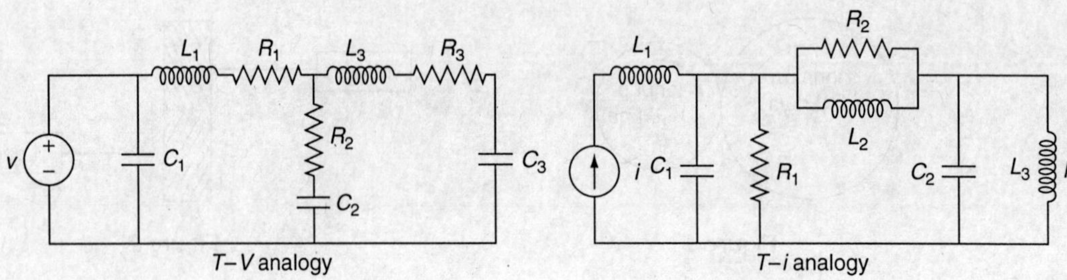

5. Find *f–υ* and *f–i* analogies of Fig 5.

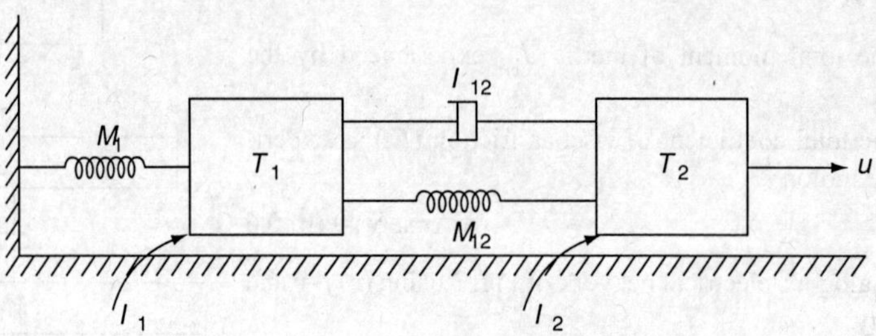

Figure 5

Ans. *f–v* analogy

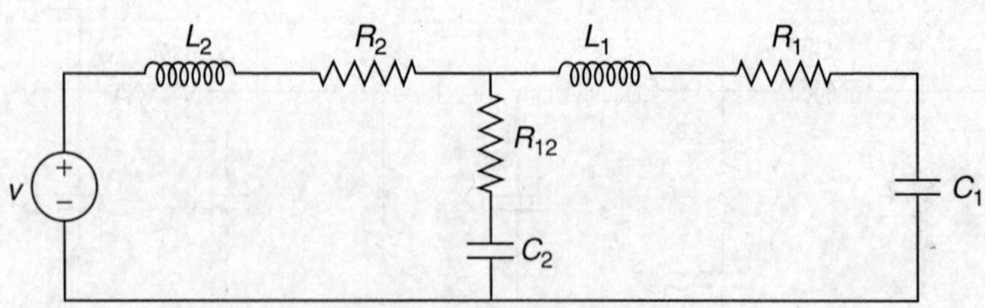

f–i analogy

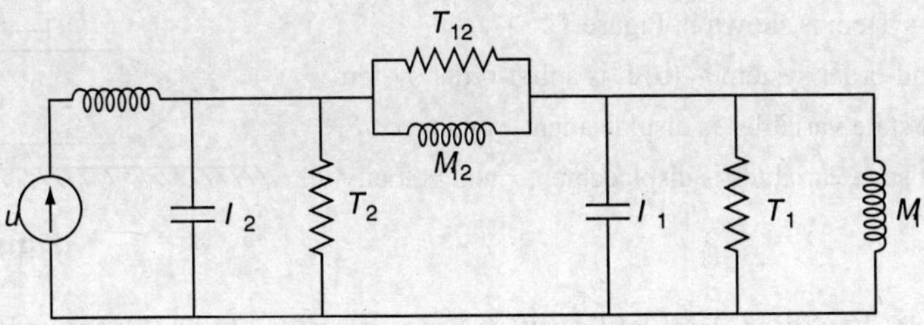

TEST YOURSELF

1. The force equation for the system shown in Figure T1 is given by

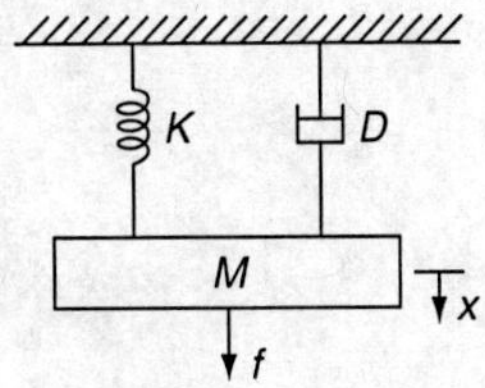

Figure T1

(a) $F = M\dfrac{d^2x}{dt^2} + D\dfrac{dx}{dt} + Kx$

(b) $F + M\dfrac{d^2x}{dt^2} = D\dfrac{dx}{dt} + Kx$

(c) $F + M\dfrac{d^2x}{dt^2} + Kx = D\dfrac{dx}{dt}$

(d) $F = M\dfrac{d^2x}{dt^2} + D\int x\,dt + Kx$

2. Match the following

(a) Force	(i) reciprocal of capacitance
(b) Velocity	(ii) charge
(c) Displacement	(iii) current
(d) Friction coefficient	(iv) resistance
(e) Spring constant	(v) inductance
(f) Mass	(vi) voltage

3. If two gears have the number of teeth N_1 and N_2, radii r_1 and r_2, angular displacements θ_1 and θ_2, and torques T_1 and T_2 (received on the shaft on the second gear), find the valid relation.

(a) $\dfrac{T_1}{T_2} = \dfrac{\theta_1}{\theta_2} = \dfrac{N_1}{N_2} = \dfrac{r_1}{r_2}$ (b) $\dfrac{T_1}{T_2} = \dfrac{\theta_2}{\theta_1} = \dfrac{N_1}{N_2} = \dfrac{r_2}{r_1}$

(c) $\dfrac{T_1}{T_2} = \dfrac{\theta_2}{\theta_1} = \dfrac{N_2}{N_1} = \dfrac{r_1}{r_2}$ (d) $\dfrac{T_1}{T_2} = \dfrac{\theta_1}{\theta_2} = \dfrac{N_1}{N_2} = \dfrac{r_2}{r_1}$

4. If torque T_1 is transferred from a gear with N_1 teeth to a gear with N_2 teeth, the value of the torque received at the shaft of second gear is

(a) $\left(\dfrac{N_1}{N_2}\right)T_1$ (b) $\left(\dfrac{N_2}{N_1}\right)T_1$ (c) N_1T_1 (d) $\left(\dfrac{N_2}{N_1}\right)^2 T_1$

5. A mass spring system is shown in Figure T2
   (a) It is a second-order system (b) It is a first-order system
   (c) A choice of state variables is displacement x and force F.
   (d) A choice of state variables is displacement x and velocity v.

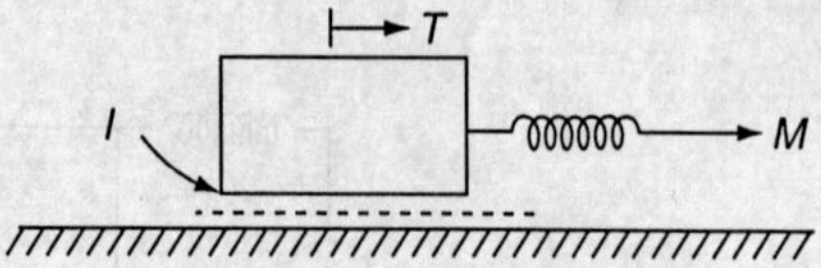

Figure T2

ANSWERS

1 (a);→2 (a)→(vi), (b)→(iii), (c)→(ii), (d)→(iv), (e)→(i), (f)→(v); 3 (c); 4 (b); 5 (a), (d)

Chapter 6

Block Diagram

The concept of transfer function has already been introduced in Chapter 3. For a complicated system, it is easy to find the transfer function of each and every element, and output of a certain block may act as an input to other block or blocks. Therefore, the knowledge of transfer function of each block is not sufficient in this case. The interrelation between the elements is required to find the overall transfer function of the system. There are two methods: (i) block diagram and (ii) signal flow graph. Block diagram is a pictorial representation of each system. Each element of a complicated system is represented by a block. Each block is a functional block. The main aim of this chapter is to introduce block diagram representation and the rules for block diagram reduction.

6.1 Definition of Basic Elements of a Block Diagram

Block diagram: The shorthand pictorial representation of the cause-and-effect relationship between the input and output of a physical system is known as a block diagram. Figure 6.1 shows the representation of a block diagram.

(a) A simple block

(b) Block diagram with gain (s-domain)

Fig. 6.1 Block diagram

Output: The value of input multiplied by the block gain is known as output.

From Fig. 6.1 (b)

$$C(s) = G(s)R(s) \tag{6.1}$$

Summing point: At summing point, two or more signals can be added or subtracted. Figure 6.2 shows a summing point.

In Fig. 6.2, $X(s)$, $Y(s)$, and $Z(s)$ are the inputs while $C(s)$ is the output. From Fig. 6.2, we have,

$$C(s) = X(s) - Y(s) + Z(s) \tag{6.2}$$

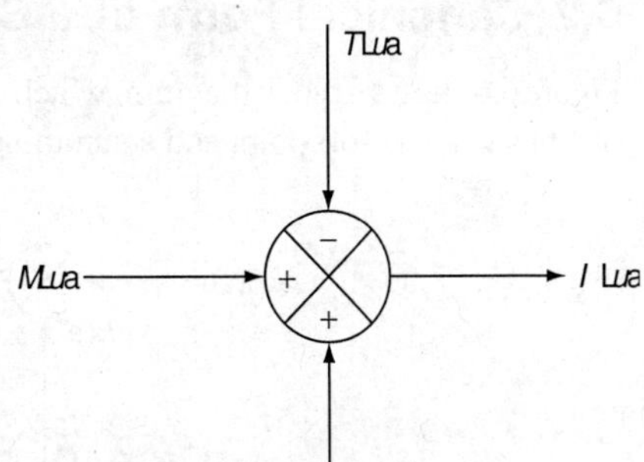

Fig. 6.2 A summing point

Take-off point: The point at which the output signal of any block can be applied to two or more points is known as take-off point. Figure 6.3 shows a take-off point. The output signal is analogous to voltage, not the current.

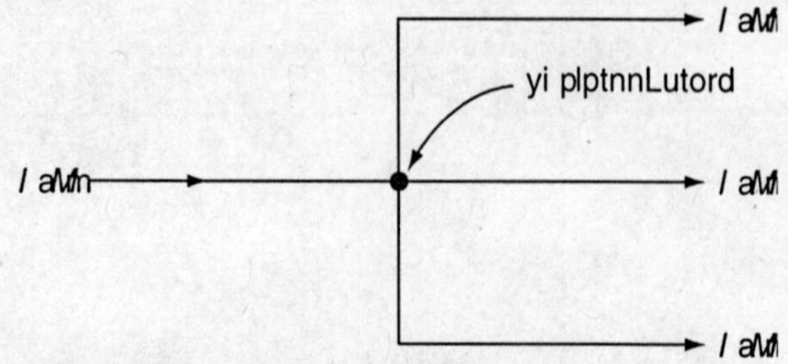

Fig. 6.3 Take-off point

Forward path: The direction of flow of signal from input to output is known as forward path. Figure 6.4 shows a forward path. Individual block gains are $G_1(s), G_2(s)$ and $G_3(s)$ in Fig. 6.4 from input to output of the system.

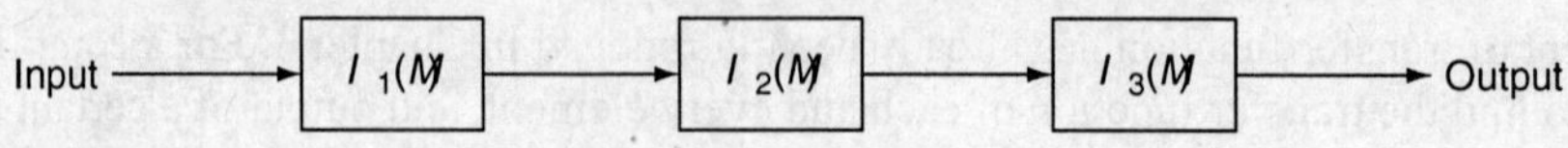

Fig. 6.4 Forward path

Feedback path: The direction of flow of signal from output to input is known as feedback path. Figure 6.5 shows the feedback path.

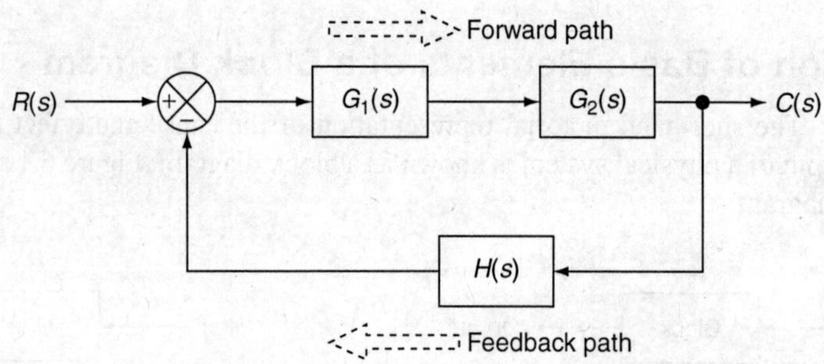

Fig. 6.5 Feedback path

6.2 Canonical Form of a Closed Loop System

Figure 6.6 shows a block diagram, which consists of a forward path having one block, a feedback path having one block, a take-off point and a summing point. It represents a canonical form of a closed-loop system.

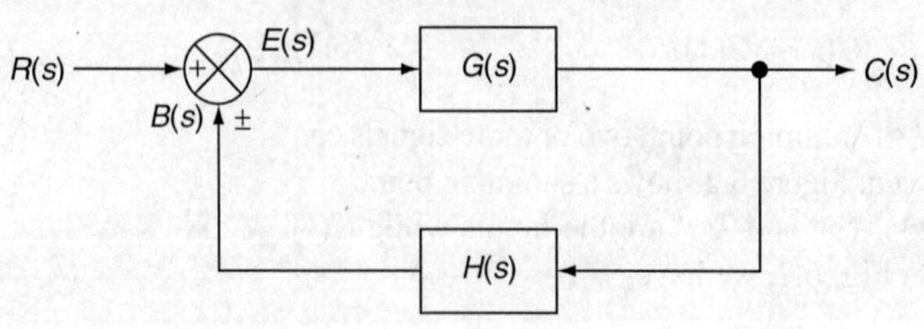

Fig. 6.6 Canonical form of a feedback control system

In Fig. 6.6, $R(s)$ is the Laplace transform of reference input $r(t)$; $C(s)$ is the Laplace transform controlled output $c(t)$; $E(s)$ is the Laplace transform of error signal $e(t)$; $B(s)$ is the Laplace transform of feedback signal $b(t)$; $G(s)$ is equivalent forward path transfer function; and $H(s)$ is equivalent feedback path transfer function.

$G(s)\,H(s)$ is the open loop transfer function

$\dfrac{C(s)}{R(s)}$ is the closed loop transfer function ≡ control ratio

$\dfrac{E(s)}{R(s)}$ is the actuating signal ratio ≡ error ratio

and $\dfrac{B(s)}{R(s)}$ is the primary feedback ratio.

6.3 Rules for Block Diagram Reduction

Any complicated system can be brought into simple form by the reduction of block diagram. The following rules are used in block diagram reduction.

Rule 1: Associative law

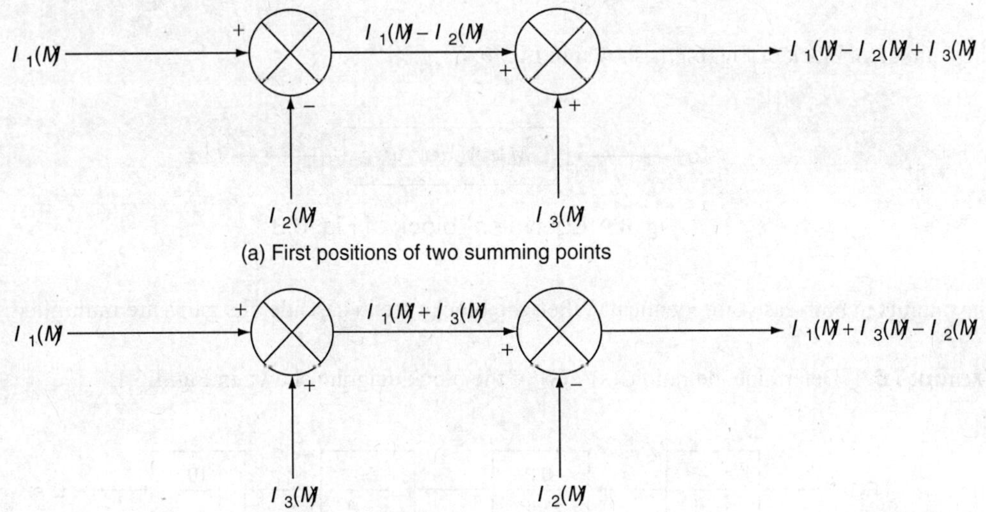

(a) First positions of two summing points

(b) Summing points are interchanged

Fig. 6.7 Associative law

In Fig. 6.7(a), two summing points have been taken into account. In the first case, the output is $R_1(s) - R_2(s) + R_3(s)$. In Fig. 6.7(b), the positions of summing points are interchanged. The output is $R_1(s) + R_3(s) - R_2(s)$. From Fig. 6.7(a) and Fig. 6.7(b), we have,

$$R_1(s) - R_2(s) + R_3(s) = R_1(s) + R_3(s) - R_2(s)$$

If any block is present in between the summing points, by interchanging the summing points, it can be shown that outputs will not be equal.

Rule 2: For blocks in cascade
Figure 6.8 shows that the three blocks are in cascade.

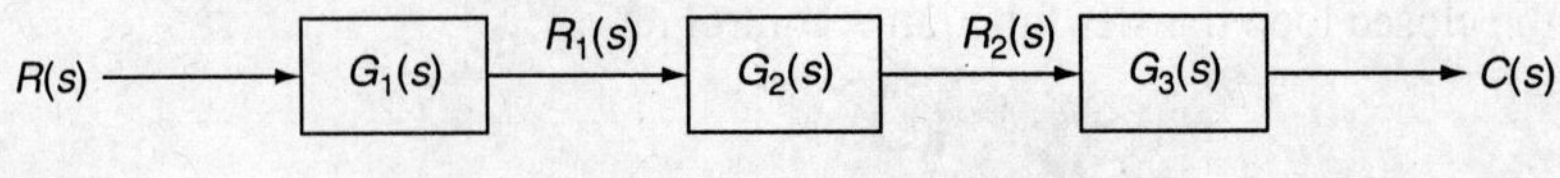

Fig. 6.8 Blocks in cascade

From Fig. 6.8, we have

$$R_1(s) = G_1(s)R(s)$$

$$R_2(s) = G_2(s)R_1(s) = G_1(s)G_2(s)R(s)$$

and $$C(s) = G_3(s)R_2(s) = G_1(s)G_2(s)G_3(s)R(s)$$

$\therefore$ Transfer function of the system $= \dfrac{C(s)}{R(s)} = G_1(s)G_2(s)G_3(s)$

The equivalent block of Fig. 6.8 is shown in Fig. 6.9.

Fig. 6.9 Equivalent block of Fig. 6.8

The outputs in both cases are identical. When the blocks are in cascade, the gains are multiplied.

Example 6.1 Determine the ratio $C(s)/R(s)$ of the block diagram shown in Fig. E6.1.

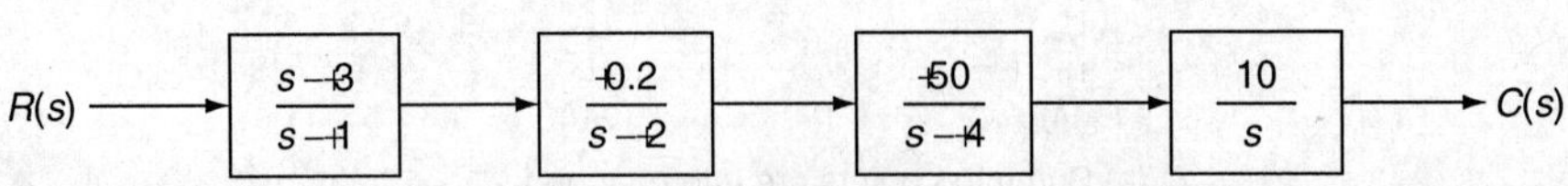

Fig. E6.1

Solution The blocks in Fig. E5.1 are connected in cascade. The overall transfer function is expressed by

$$\frac{C(s)}{R(s)} = \left(\frac{s+3}{s+1}\right)\left(\frac{0.2}{s+2}\right)\left(\frac{50}{s+4}\right)\left(\frac{10}{s}\right) = \frac{100(s+3)}{s(s+1)(s+2)(s+4)}$$

Rule 3: For blocks in parallel

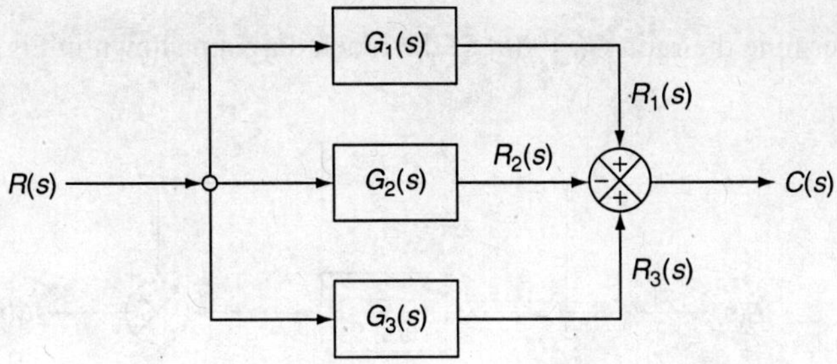

Fig. 6.10 Parallel blocks

The three blocks shown in Fig. 6.10 are in parallel.

$$\therefore \quad C(s) = R_1(s) - R_2(s) + R_3(s) = G_1(s)R(s) - G_2(s)R(s) + G_3(s)R(s)$$
$$= [G_1(s) - G_2(s) + G_3(s)]R(s)$$

Therefore, the transfer function of the system is given by,

$$\frac{C(s)}{R(s)} = G_1(s) - G_2(s) + G_3(s)$$

The equivalent block of Fig. 6.10 is shown in Fig. 6.11.

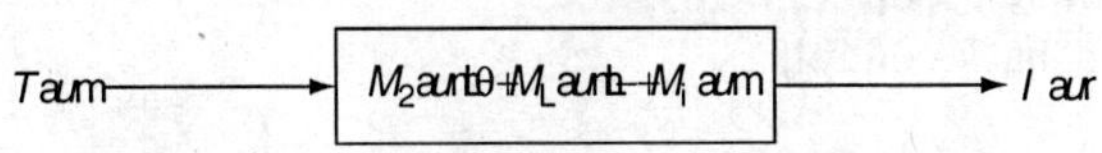

Fig. 6.11 Equivalent block of Fig. 6.10

When the blocks are in parallel, the gains are added algebraically. This rule cannot be applied directly if a take-off occurs as shown in Fig. 6.12.

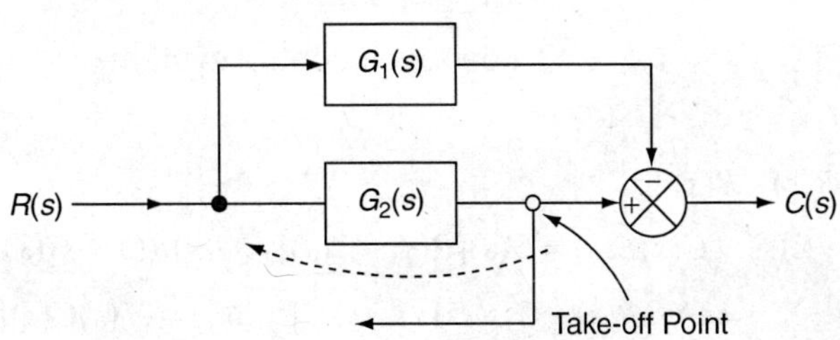

Fig. 6.12 Take-off point in between a block and a summing point

It is always advisable to shift the take-off point prior to the block $G_2(s)$ as shown by the dotted line arrow shown in Fig. 6.12. It is always better to avoid shifting the take-off point after the summing point.

Example 6.2 Determine the ratio $C(s)/R(s)$ of the block diagram shown in Fig. E6.2.

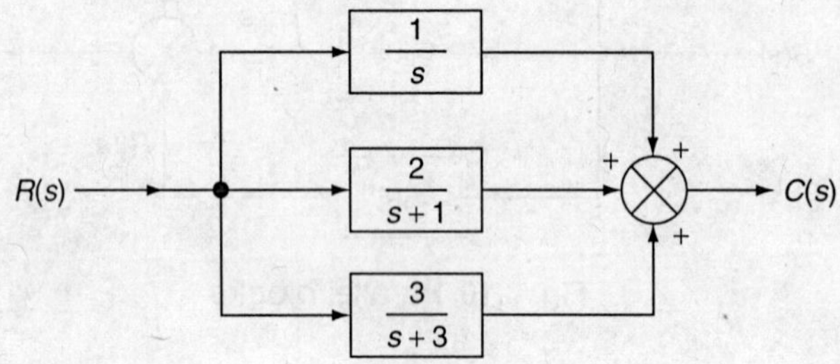

Fig. E6.2

Solution The blocks in Fig. E6.2 are connected in parallel. The overall transfer function is expressed by,

$$\frac{C(s)}{R(s)} = \frac{1}{s} + \frac{2}{s+1} + \frac{3}{s+3} = \frac{(s+1)(s+3) + 2s(s+3) + 3s(s+1)}{s(s+1)(s+3)}$$

$$= \frac{s^2 + 4s + 3 + 2s^2 + 6s + 3s^2 + 3s}{s(s+1)(s+3)} = \frac{6s^2 + 13s + 3}{s(s+1)(s+3)}$$

Rule 4: Eliminate feedback loop

Figure 6.13 shows a feedback control.

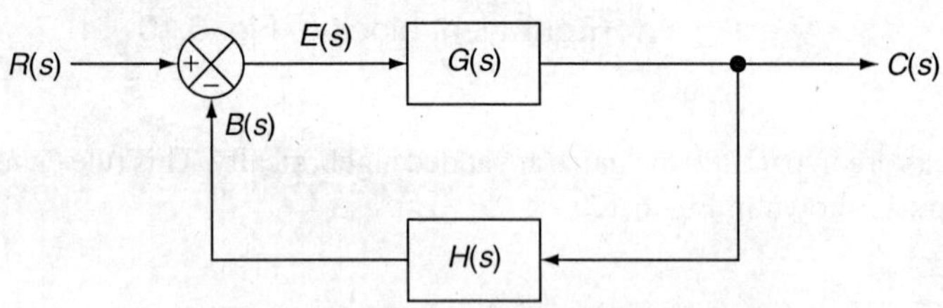

Fig. 6.13 Feedback control system

In Fig. 6.13, $E(s) = R(s) - B(s)$

$$C(s) = G(s)E(s) = G(s)[R(s) - B(s)] = G(s)R(s) - G(s)B(s)$$

$$= G(s)R(s) - G(s)H(s)C(s) \quad [\because B(s) = H(s)C(s)]$$

i.e., $$C(s)[1 + G(s)H(s)] = G(s)R(s)$$

i.e.,
$$C(s) = \frac{G(s)R(s)}{1+G(s)H(s)}$$

$$\therefore \quad \frac{C(s)}{R(s)} = \text{Transfer function of the system } = \frac{G(s)}{1+G(sH(s))} \tag{6.3}$$

Equation (6.3) is valid for negative feedback.
Similarly, for positive feedback, we have,

$$\frac{C(s)}{R(s)} = \frac{G(s)}{1-G(s)H(s)} \tag{6.4}$$

In general, we have,

$$\frac{C(s)}{R(s)} = \frac{G(s)}{1 \pm G(s)H(s)} \quad \begin{bmatrix} + \text{ for negative feedback and} \\ - \text{ for positive feedback} \end{bmatrix} \tag{6.5}$$

Example 6.3 Determine the ratio $C(s)/R(s)$ of the block diagram shown in Fig. E6.3. Also, find the characteristic equation.

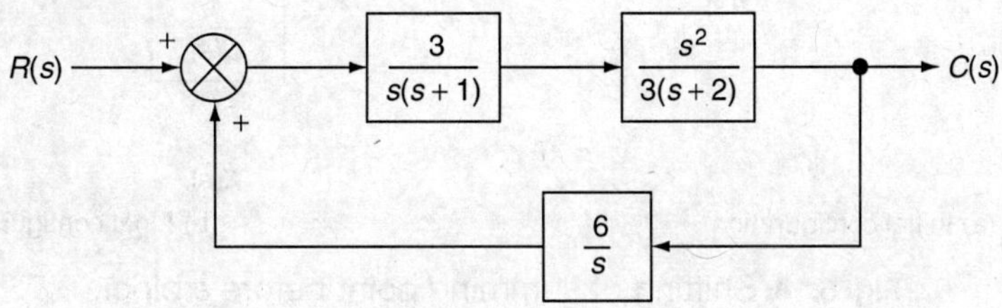

Fig. E6.3

Solution

Here
$$G(s) = \left[\frac{3}{s(s+1)}\right]\left[\frac{s^2}{3(s+2)}\right] = \frac{s}{(s+1)(s+2)}$$

and
$$H(s) = \frac{6}{s}$$

Now
$$G(s)H(s) = \frac{6}{(s+1)(s+2)}$$

Since the feedback is positive, the overall transfer function is expressed by,

$$\frac{C(s)}{R(s)} = \frac{G(s)}{1-G(s)H(s)} = \frac{\dfrac{s}{(s+1)(s+2)}}{1-\dfrac{6}{(s+1)(s+2)}} = \frac{s}{(s+1)(s+2)-6} = \frac{s}{s^2+3s-4}$$

The characteristic equation of Fig. E5.1 is expressed by,

$$1-G(s)H(s)=0$$

i.e.,
$$1-\frac{6}{(s+1)(s+2)}=0$$

i.e.,
$$(s+1)(s+2)-6=0$$

i.e.,
$$s^2+3s-4=0$$

Rule 5: Shifting a summing point before a block

Figure 6.14(a) shows a block diagram in which the summing point needs to be shifted before $G(s)$ as shown by the dotted line.

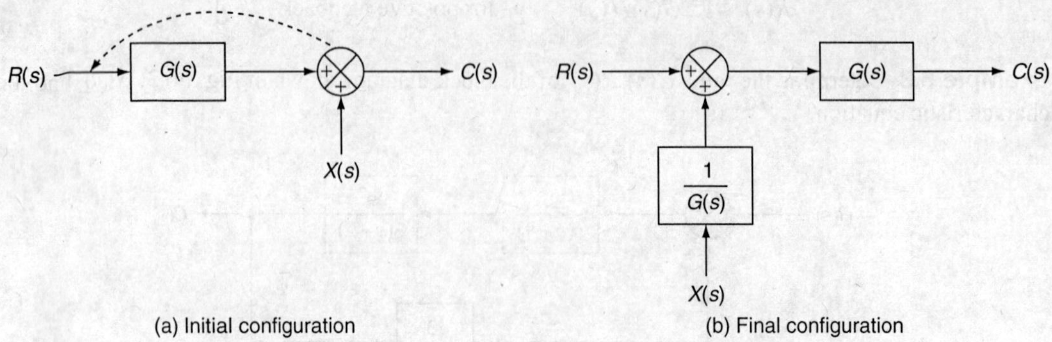

Fig. 6.14 Shifting of summing point before a block

From Fig. 6.14(a),

$$C(s)=G(s)R(s)+X(s)$$

From Fig. 6.14(b),

$$C(s)=G(s)\left[R(s)+\frac{X(s)}{G(s)}\right]=G(s)R(s)+X(s)$$

Therefore, the outputs in both cases are identical.

Rule 6: Shifting of summing point after a block

Figure 6.15(a) shows a block diagram in which the summing point needs to be shifted after $G(s)$ as shown by the dotted line.

From Fig. 6.15(a),

$$C(s)=G(s)[R(s)+X(s)]=G(s)R(s)+G(s)X(s)$$

From Fig. 6.15(b),

$$C(s)=G(s)R(s)+G(s)X(s)$$

Therefore, the outputs in both cases are same.

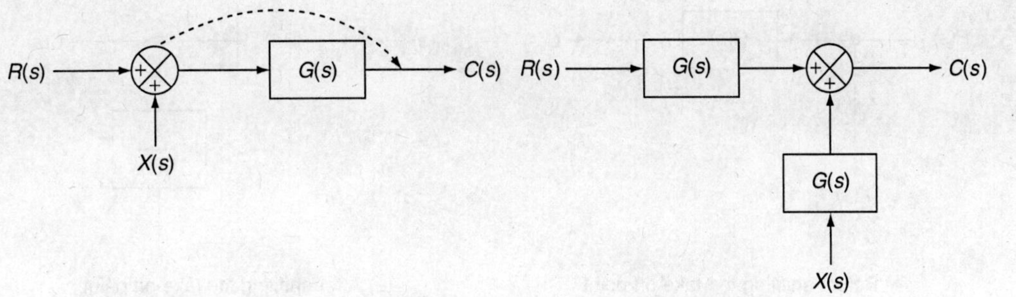

(a) Before shifiting the summing point (b) After shifiting the summing point

Fig. 6.15 Shifting of summing point after a block

Rule 7: Shifting of take-off point before a block
Figure 6.16(a) shows a block diagram in which the take-off point is to be shifted before the block as shown by the dotted line.

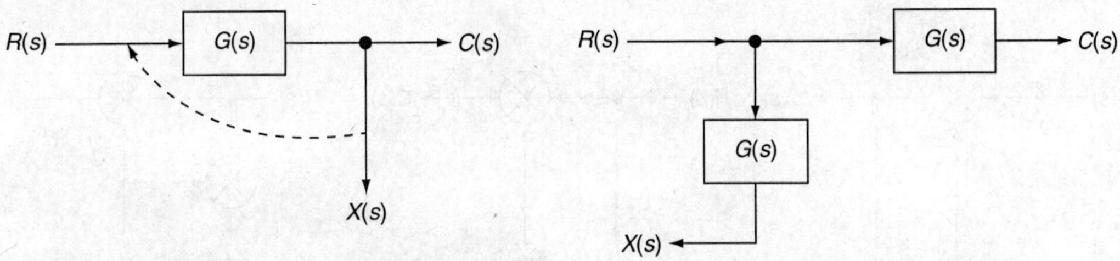

(a) Before shifiting the take-off point (b) After shifiting the take-off point

Fig. 6.16 Shifting of take-off point before a block

From Fig. 6.16(a),

$$X(s) = C(s) = G(s)R(s)$$

From Fig. 6.16(b),

$$X(s) = G(s)R(s)$$

Rule 8: Shifting of take-off point after a block
Figure 6.17(a) shows a block diagram in which the take-off point is to be shifted after the block as shown by the dotted line.
From Fig. 6.17(a),

$$X(s) = R(s)$$

From Fig. 6.17(b),

$$X(s) = \frac{1}{G(s)}C(s) = \frac{1}{G(s)}G(s)R(s) = R(s)$$

Therefore, $X(s)$ in both cases is the same.

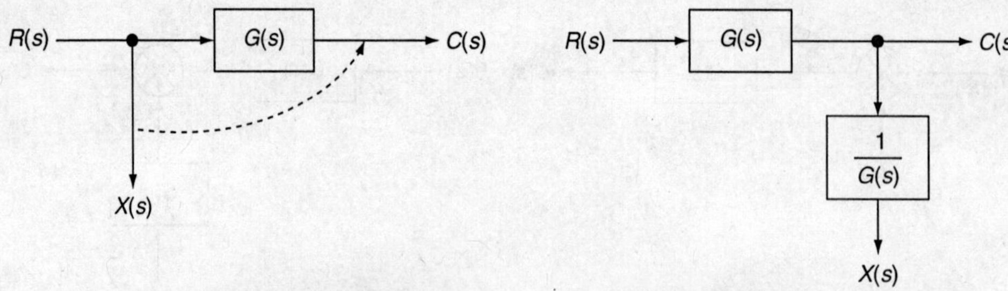

(a) Before shifiting the take-off point (b) After shifiting the take-off point

Fig. 6.17 Shifting of take-off point after a block

Rule 9: Shifting of take-off point before a summing block

Figure 6.18(a) shows the initial situation of the take-off point. In this case, $X(s) = R(s) \pm Y(s)$. The take-off point is to be shifted before the summing point as shown by the dotted line in Fig. 6.18(a). Figure 6.18(b) shows the intermediate configuration, i.e., before the corrected configuration.

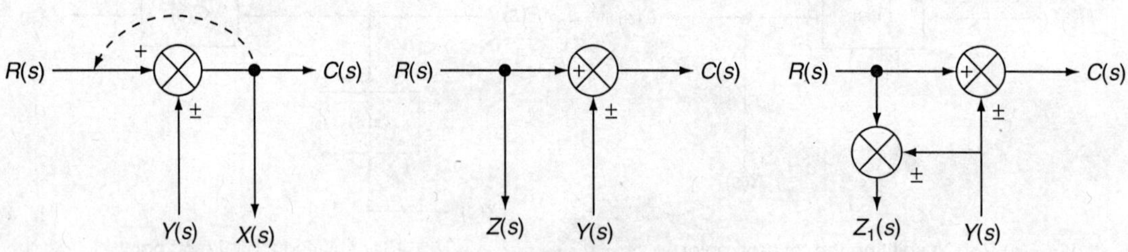

(a) Initial situation (b) After shifting the take-off point (c) Corrected configuration

Fig. 6.18 Shifting of take-off point before a summing block

From Fig. 6.18(b), $Z(s) = R(s)$. Therefore, $Z(s) \neq X(s)$. In Fig. 6.18(c), the signal $Y(s)$ is added to $Z(s)$ by the summing point with the same sign. In this case,

$$Z_1(s) = R(s) \pm Y(s)$$

$$\therefore \qquad X(s) = Z_1(s)$$

Therefore, Fig. 6.18(c) is equivalent to Fig. 6.18(a).

Rule 10: Shifting of take-off point after a summing block

Figure 6.19(a) shows the initial situation of the take-off point. In this case, $X(s) = R(s)$. The take-off point is to be shifted after the summing point as shown by the dotted line. Figure 6.19(b) shows the intermediate configuration i.e., before the corrected configuration. From Fig. 6.19(b), $X(s) = R(s) \pm Y(s)$. Therefore, $Z(s) \neq X(s)$. In Fig. 6.19(c), the signal $Y(s)$ is added to $Z(s)$ by the summing point with opposite sign.

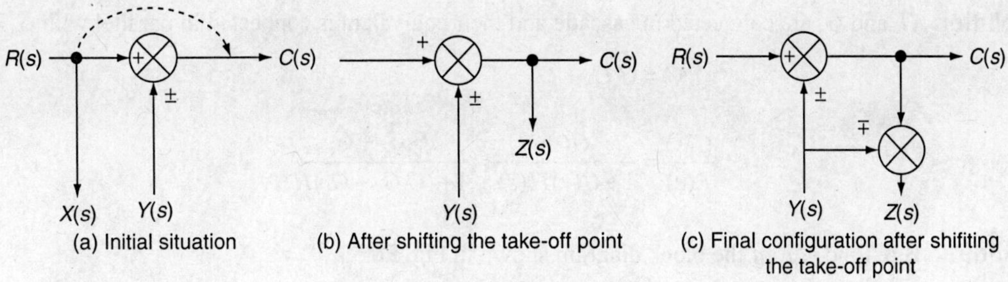

Fig. 6.19 Shifting of take-off point after a summing block

From Fig. 6.19 (a),

$$X(s) = R(s)$$

From Fig. 6.19 (c),

$$Z(s) = R(s) \pm Y(s) \pm Y(s) = R(s) = x(s)$$

Therefore, Fig. 6.19(c) is equivalent to Fig. 6.19(a).

6.4 Procedure for Reduction of Block Diagram

Step 1: Reduce the cascade blocks.
Step 2: Reduce the parallel blocks.
Step 3: Reducc thc internal feedback loops.
Step 4: It is advisable to shift take-off points towards right and summing points towards left. It is always better to avoid Rule 9 and Rule 10.
Step 5: Repeat Step 1 to Step 4 until the single block is obtained.
Step 6: Find transfer function of the overall system using formula $C(s)/R(s)$.

Procedure for Multiple Inputs

Step 1: Consider one input and put other inputs to zero. Find the output.
Step 2: Repeat step 1 until all inputs are covered.
Step 3: Find the resultant output by superposition.

Example 6.4 Determine the ratio $C(s)/R(s)$ of the block diagram shown in Fig. E5.4.

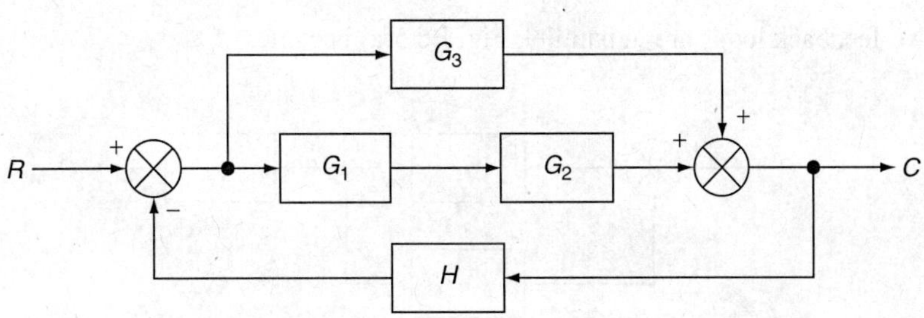

Fig. E6.4

Solution G_1 and G_2 are connected in cascade and their equivalent is connected in parallel with G_3.

$$\therefore \qquad G = G_1G_2 + G_3$$

$$\therefore \qquad \frac{C(s)}{R(s)} = \frac{G(s)}{1+G(s)H(s)} = \frac{G_1G_2+G_3}{1+(G_1G_2+G_3)H}$$

Example 6.5 Find C/R of the block diagram shown in Fig. E6.5.

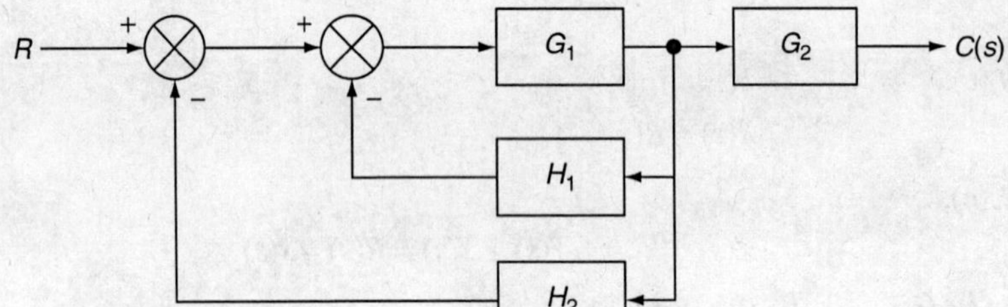

Fig. E6.5

Solution Figure E6.5 is redrawn and shown in Fig. E6.5(a).

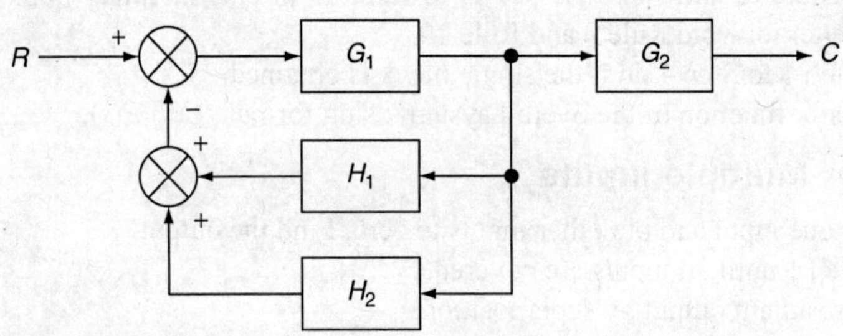

Fig. E6.5(a)

Since the two feedback loops are in parallel, Fig. E6.5(a) becomes

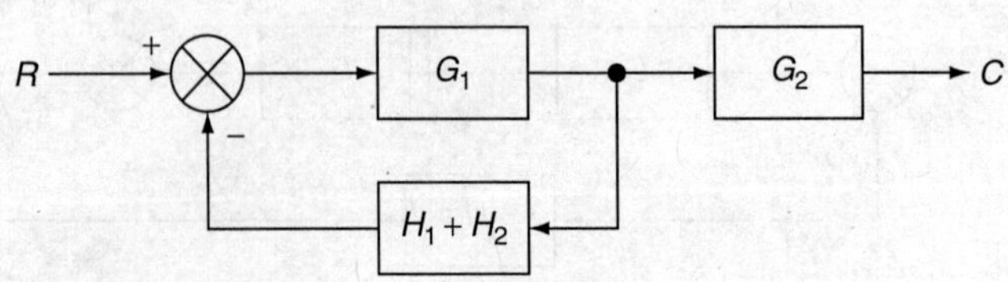

Fig. E6.5(b)

Replacing the feedback loop of Fig. E6.5(b) by its equivalent block, Fig. E6.5(b) becomes

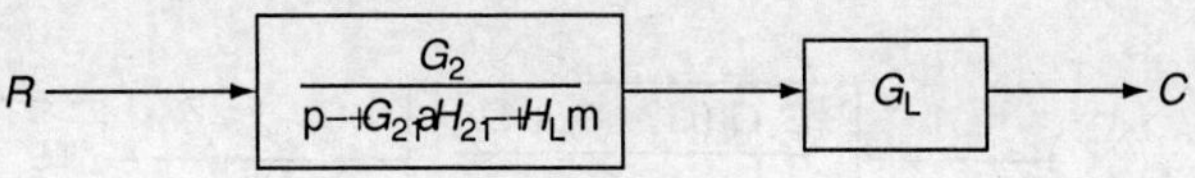

Fig. E6.5(c)

$$\therefore \qquad \frac{C(S)}{R(s)} = \frac{G_1G_2}{1+G_1(H_1+H_2)}$$

Example 6.6 Find the single block equivalent of Figure E6.6

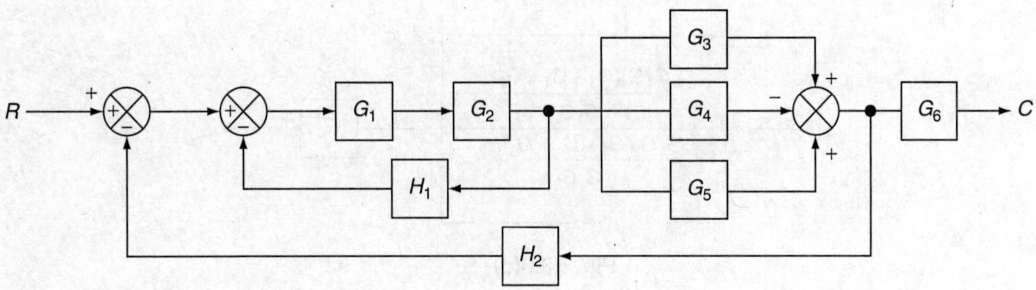

Fig. E6.6

Solution At first the cascade and parallel blocks of Fig. E6.6 are reduced as shown in Fig. E6.6(a).

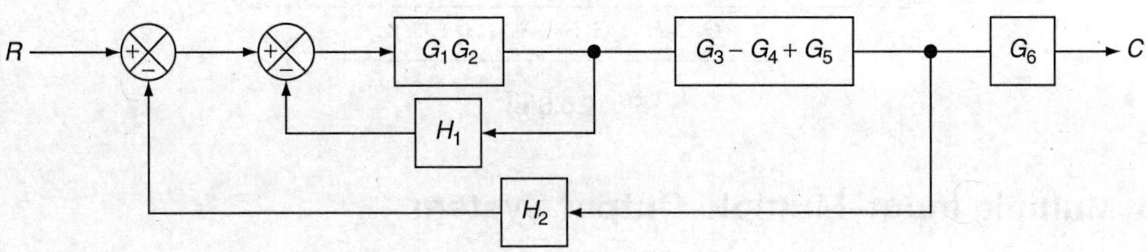

Fig. E6.6(a)

The first internal feedback loop of Fig. E6.6(a) is reduced by its equivalent block shown in Fig. E6.6(b).

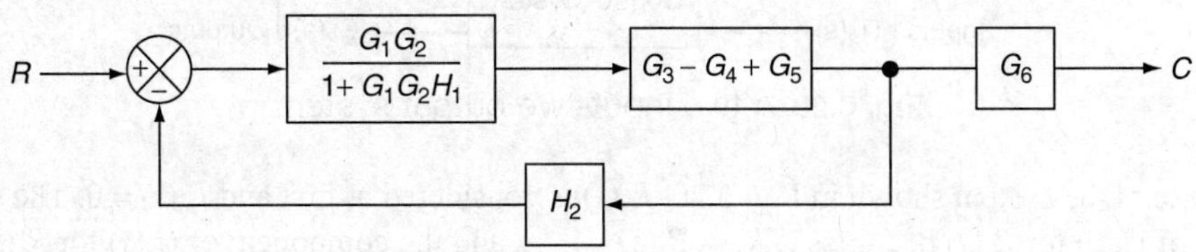

Fig. E6.6(b)

The cascade blocks of Fig. E6.6(b) are replaced by its equivalent block as shown in Fig. E6.6(c).

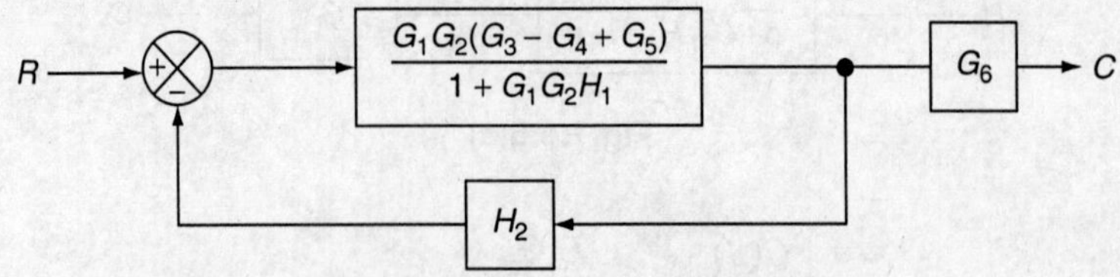

Fig. E6.6(c)

The feedback loops of Fig. E6.6(c) is replaced by its equivalent block shown in Fig. E6.6(d).

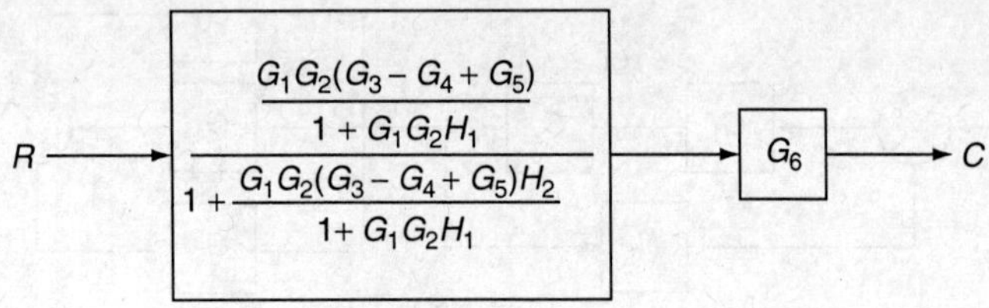

Fig. E6.6(d)

The two cascade blocks in Fig. E6.6(d) are replaced by equivalent block in Fig. E6.6(e).

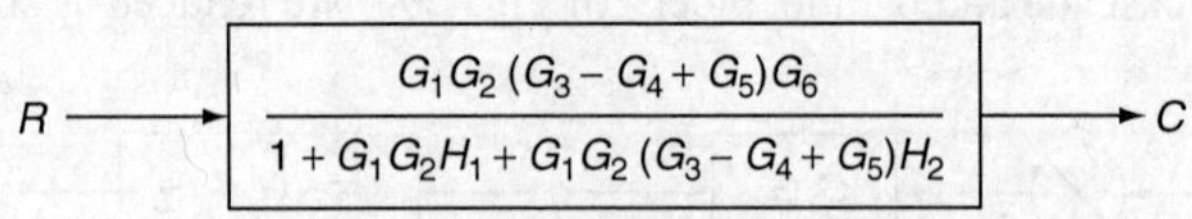

Fig. E6.6(e)

6.5 Multiple Input–Multiple Output System

In multiple input-multiple output system there are more than one input and more than one output. Figure 6.20 shows a two input two output system.

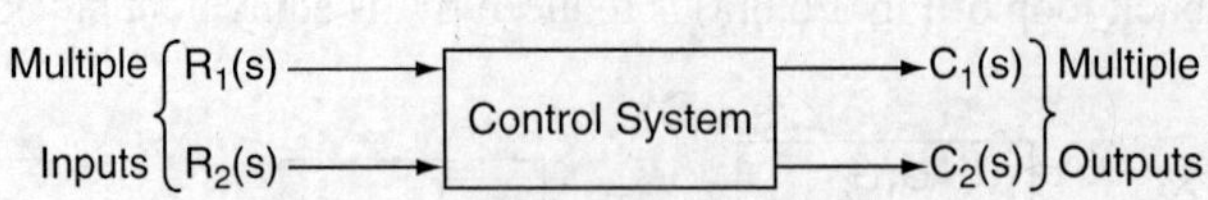

Fig. 6.20 A two input-two output system

For the case of the system shown in Fig. 6.20, $R_1(s)$ is considered at first and $R_2(s) = 0$. The component of output of $C_1(s)$ for $R_1(s)$ becomes $(TR_1) \times R_1(s)|_{R_2(s)=0}$ and the component of $C_2(s)$ for $R_1(s)$ becomes $(TR_2) \times R_1(s)|_{R_2(s)=0}$.

where $$TR_1 = \left.\frac{C_1(s)}{R_1(s)}\right|_{R_2(s)-0} \text{ and } TR_2 = \left.\frac{C_2(s)}{R_1(s)}\right|_{R_2(s)=0}$$

The component of output of $C_2(s)$ for $R_2(s)$ becomes $(TR_3)\times R_2(s)\big|_{R_1(s)=0}$ and the component of $C_2(s)$ for $R_1(s)$ becomes $(TR_4)\times R_2(s)\big|_{R_1(s)=0}$

where $$TR_3 = \left.\frac{C_1(s)}{R_2(s)}\right|_{R_1(s)=0} \text{ and } TR_4 = \left.\frac{C_2(s)}{R_2(s)}\right|_{R_1(s)=0}$$

$\therefore$ $$C_1(s) = (TR_1)\times R_1(s)\big|_{R_2(s)=0} + (TR_3)\times R_2(s)\big|_{R_1(s)=0}$$

and $$C_2(s) = (TR_2)\times R_1(s)\big|_{R_2(s)=0} + (TR_4)\times R_2(s)\big|_{R_1(s)=0}$$

To solve this type of problems, principle of superposition is applied.

Example 6.7 Find the output of the system shown in Fig. E6.7.

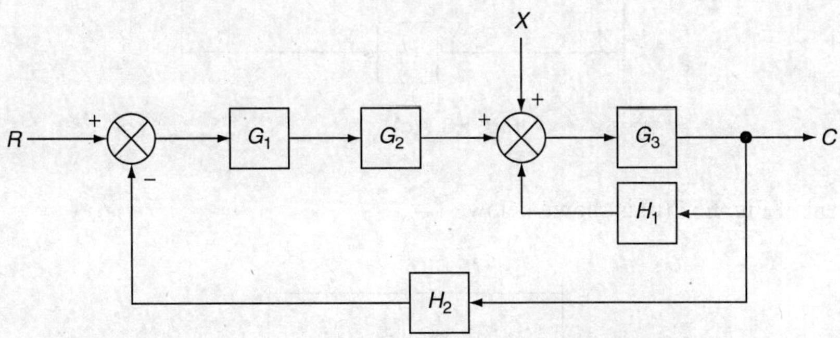

Fig. E6.7

Solution Putting $X = 0$, Fig. E6.7 becomes

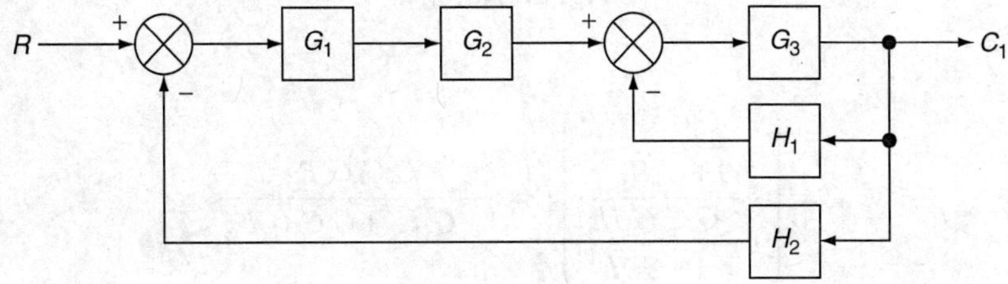

Fig. E6.7(a)

Removing the cascade blocks and internal feedback loop of Fig. E6.7(a), we have

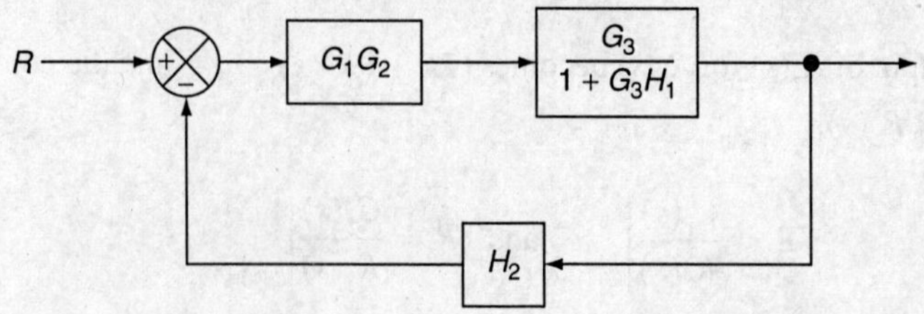

Fig. E6.7(b)

Removing the cascade blocks of Fig. E6.7(b), we have

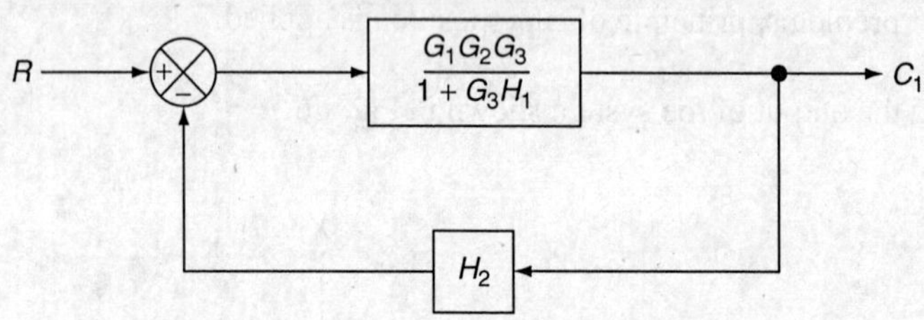

Fig. E6.7(c)

The equivalent of Fig. E6.7(c) is shown below

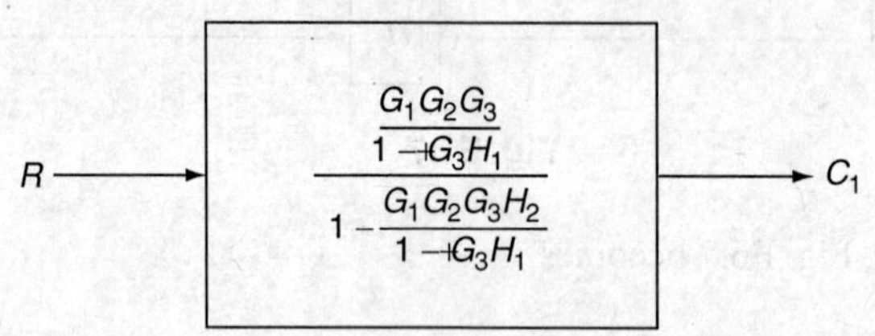

Fig. E6.7(d)

$$\therefore \qquad C_1 = \left(\frac{\dfrac{G_1G_2G_3}{1+G_3H_1}}{1+\dfrac{G_1G_2G_3H_2}{1+G_3H_1}} \right) R = \frac{G_1G_2G_3R}{1+G_3H_1+G_1G_2G_3H_2} \qquad (1)$$

Putting $R = 0$, Fig. E6.7 becomes

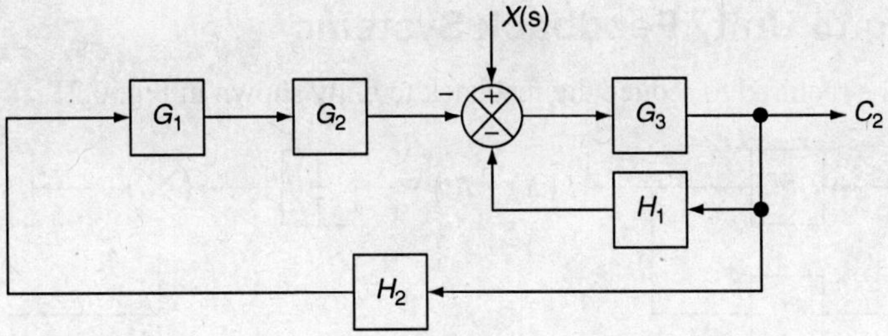

Fig. E6.7(e)

Removing internal feedback loop, we get

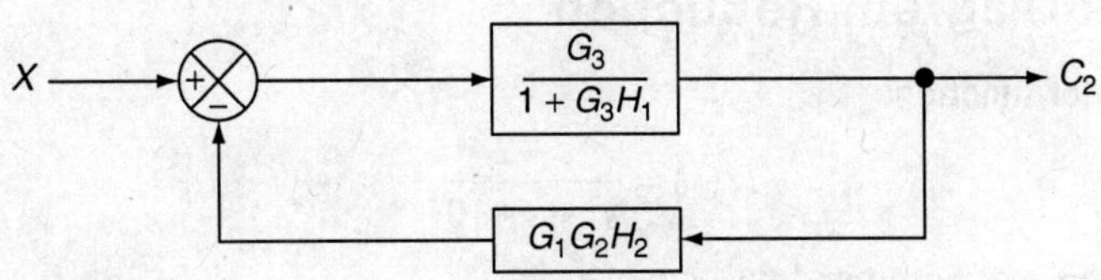

Fig. E6.7(f)

Removing the feedback block of Fig. E6.7(f), we get

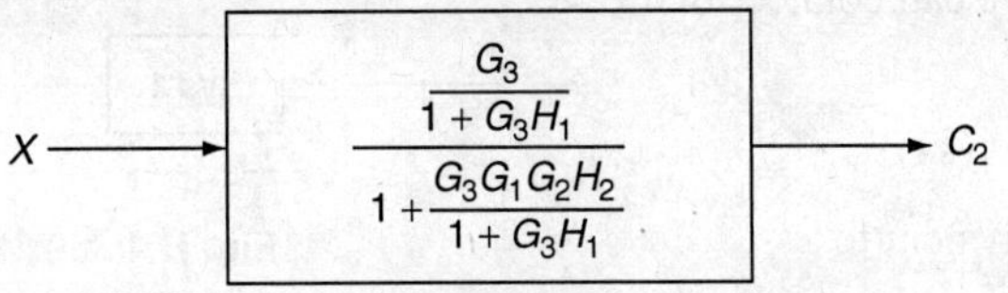

Fig. E6.7(f)

$$\therefore \qquad C_2 = \left(\frac{\dfrac{G_3}{1+G_3H_1}}{1+\dfrac{G_3G_1G_2H_2}{1+G_3H_1}} \right) X = \frac{G_3X}{1+G_3H_1+G_1G_2G_3H_2} \qquad (2)$$

From Eq. (1) and Eq. (2), we get the resultant output by superposition.

$$C = C_1 + C_2 = \frac{G_1G_2G_3R}{1+G_3H_1+G_1G_2G_3H_2} + \frac{G_3X}{1+G_3H_1+G_1G_2G_3H_2}$$

6.6 Reducing to Unity Feedback Systems

Sometime, it may be required to reduce the feedback to unity shown in Fig. 6.21.

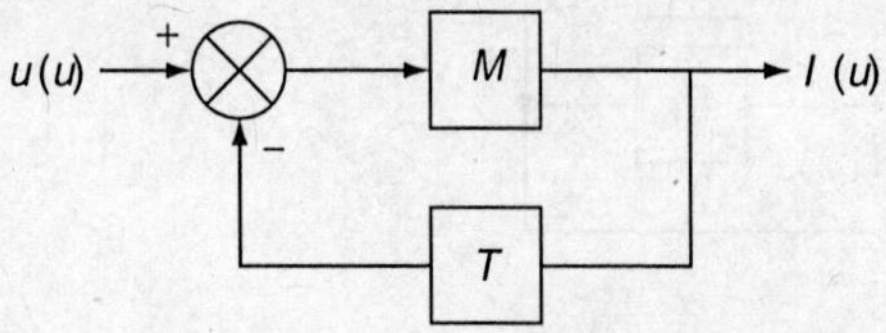

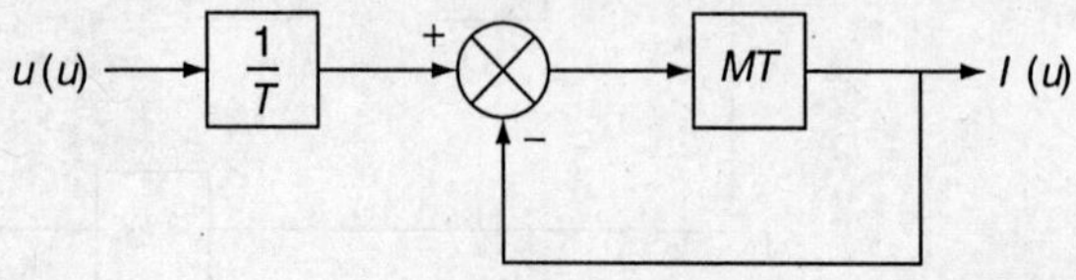

Fig. 6.21 Reducing feedback to unity

6.7 MATLAB Commands for Transfer Function and Block Diagram Reduction

Let us define a transfer function,

$$H(s) = \frac{s+2}{s^2+s+10}$$

The ***MATLAB*** sequence of command will be

num = [1 2]
den = [1 1 10]

Supposing, we want to add two transfer function, we use the '+' operator.

series

Using this function, we can connect two LTI models in series.

>> sys = series (sys1, sys2)

This command is equivalent to

sys = sys1 * sys2

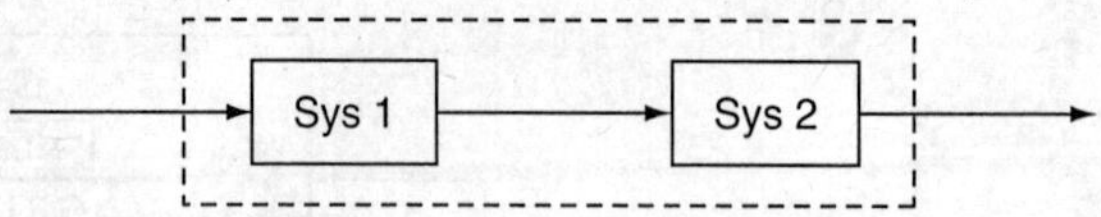

Fig. B.1 Series connection

parallel

This function performs the parallel combination of two LTI models

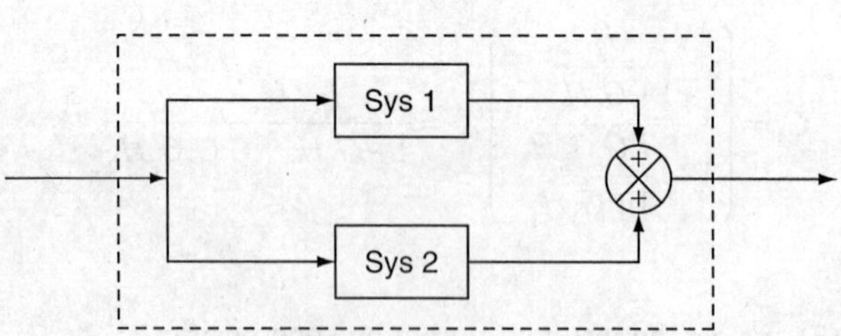

Fig. B.2 Parallel connection

>> sys = parallel (sys1, sys2)

This command is equivalent to direct addition

sys = sys1 + sys2

feedback

Fig. B.3 shows feedback connection of two LTI models

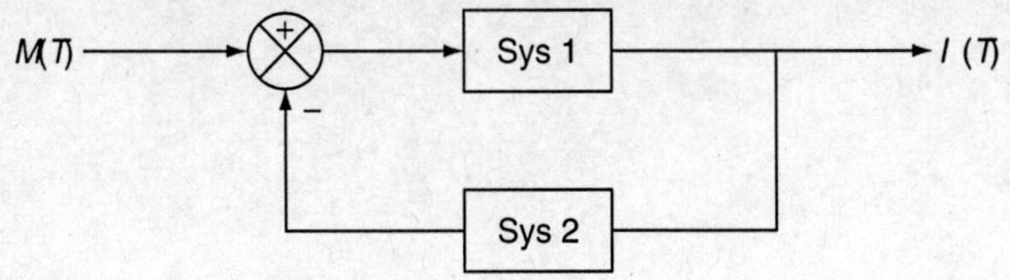

Fig. B.3 Feedback connection

```
>> sys = feedback (sys1, sys2)
```

In fact, feedback assumes positive sign. The exact command for negative feedback is

```
>> sys = feedback (sys1, sys2, −1)
```

dcgain

This function computes the low-frequency (*dc*) gain of the LTI system.

>> K = dcgain(sys)

pole

This function computes the poles of an LTI system.

>> p = pole(sys);

These are some of the functions that aid in the development and processing of the control system design and implementation. Other function and some other derivatives of these functions can be found in the MATLAB documentation and online help.

6.8 Block Diagram Reduction Example Using MATLAB

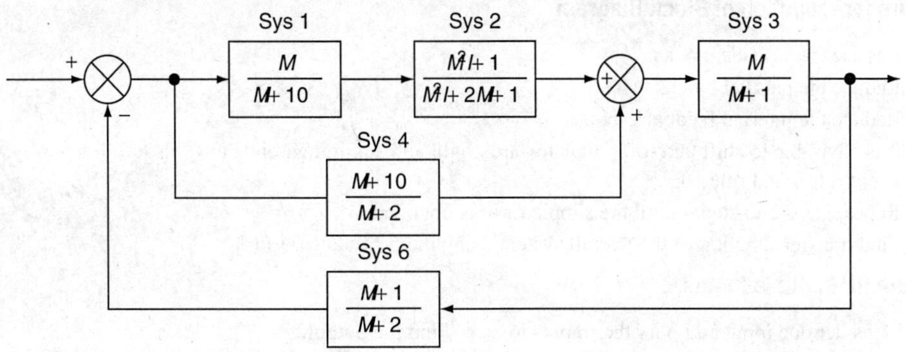

Fig. B.4

MATLAB sequence of commands

```
sys1num = [1 0];
sys1den = [1 10];
sys2num = [1 0 1];
sys2den = [1 2 1];
sys3num = [1 0];
sys3den = [1 1];
sys4num = [1 10];
sys4den = [1 2];
sys5num = [1 1];
sys5den = [1 2];
sys1 = tf (sys1num, sys1den);
sys2 = tf (sys2num, sys2den);
sys3 = tf (sys3num, sys3den);
sys4 = tf (sys4num, sys4den);
sys5 = tf (sys5num, sys5den);
sys6 = series (sys1, sys2);
sys7 = parallel (sys6, sys4);
sys8 = (series (sys7, sys8);
seriesfinal = feedback(sys8, sys5, –1);
```

For NEGATIVE FEEDBACK, negative sign must be present; otherwise MATLAB will consider this as POSITIVE FEEDBACK.

SIGNIFICANT POINTS

Block Diagram: The shorthand pictorial representation of the cause and-effect relationship between the input and output of a physical system is known as block diagram.
Output: The value of input multiplied by the block gain is known as output.
Summing Point: At summing point two or more signals can be added or subtracted.
Take-off Point: The point at which output signal of any block can be applied to two or more points is known as take-off point.
Forward Path: The direction of flow of signal from input to output is known as forward path.
Feedback path: The direction of flow of signal from output to input is known as feedback path.

Procedure for reduction of Block Diagram

Step 1: Reduce the cascade blocks.
Step 2: Reduce the parallel blocks.
Step 3: Reduce the internal feedback loops.
Step 4: It is advisable to shift take-off points towards right and summing points towards left. It is always better to avoid rule 9 and rule 10.
Step 5: Repeat steps 1 to step 4 until the simple form is obtained.
Step 6: Find transfer function of the overall system using the formula $C(s)/R(s)$.

Procedure for Multiple Inputs

Step 1: Consider one input and put other inputs to zero. Find the output.
Step 2: Repeat step 1 until all inputs are covered.
Step 3: Find the resultant output by superposition.

Summary of Block Diagram Rules

Sr. No.	Rule No.	Configuration	Equivalent	Name
1	Rule 1	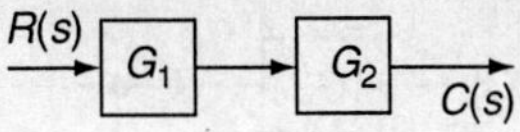	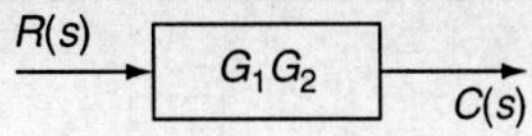	Series Cascade
2	Rule 2	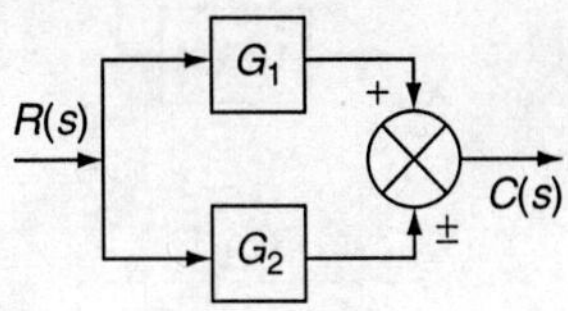	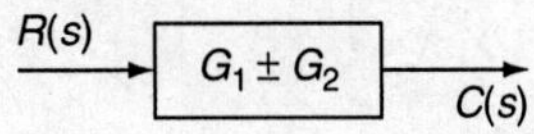	Parallel
3	Rule 3	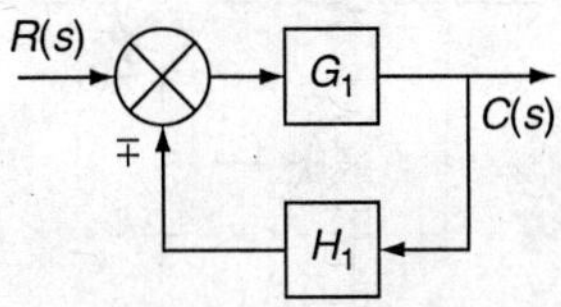	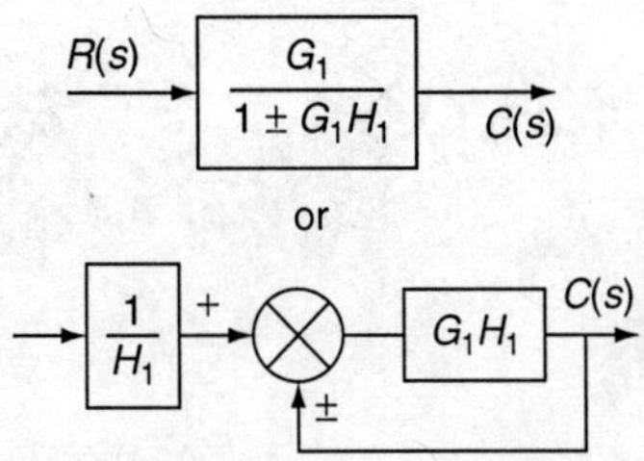	Loop
4	Rule 4	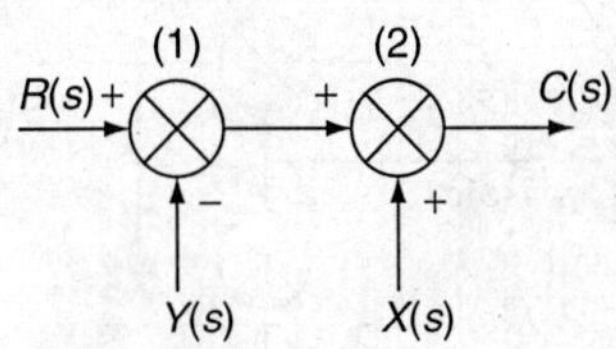	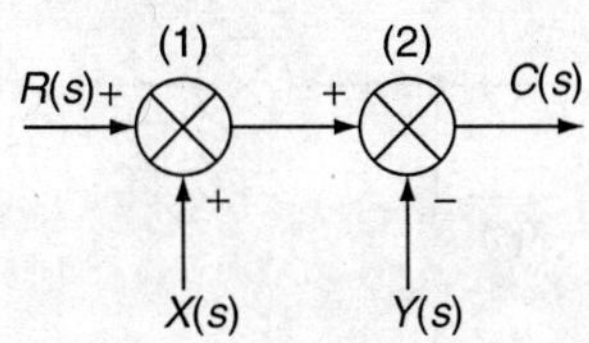	Associative Law
5	Rule 5	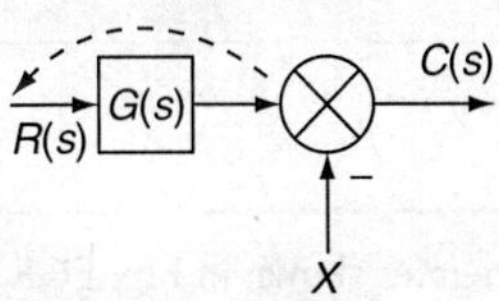	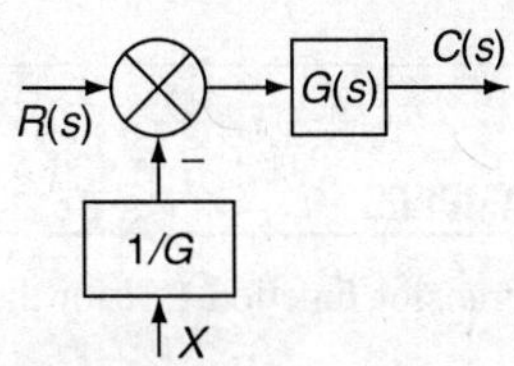	Move summing point before block
6	Rule 6	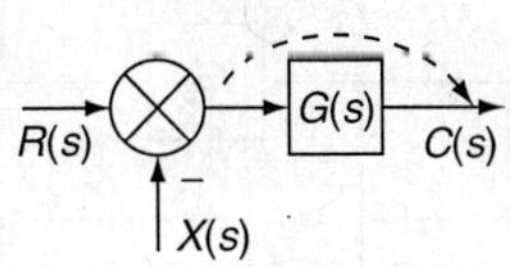	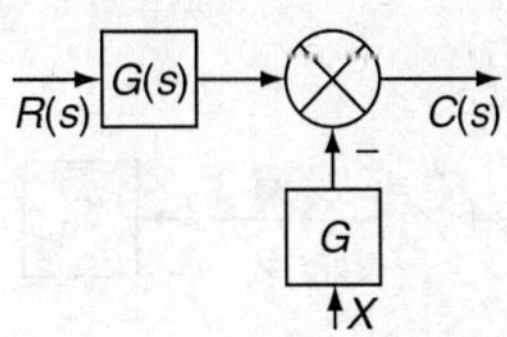	Move summing point after block
7	Rule 7	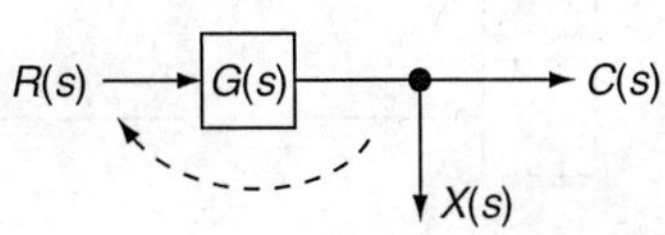	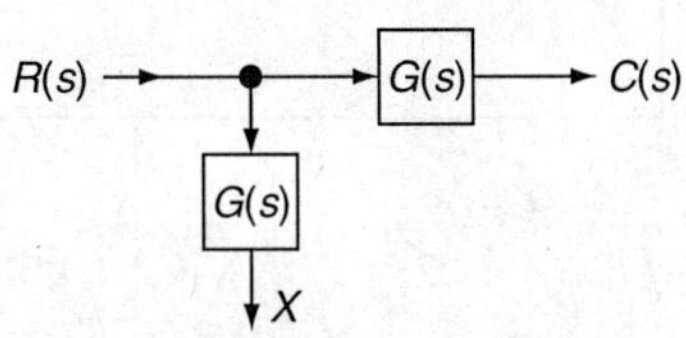	Move take-off point before a block

(Continued)

Summary of Block Diagram Rules (*Continued*)

Sr. No.	Rule No.	Configuration	Equivalent	Name
8	Rule 8	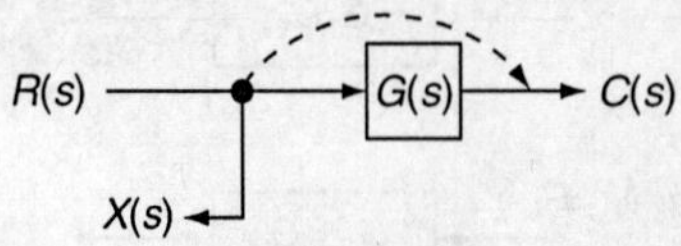	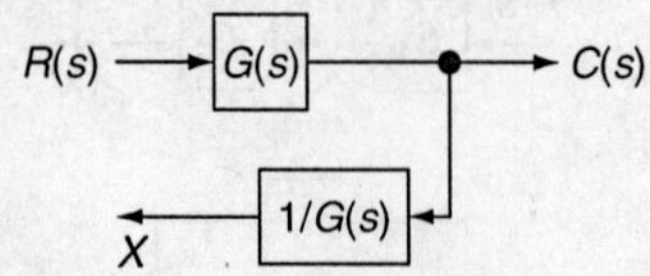	Move take-off point after a block
9	Rule 9	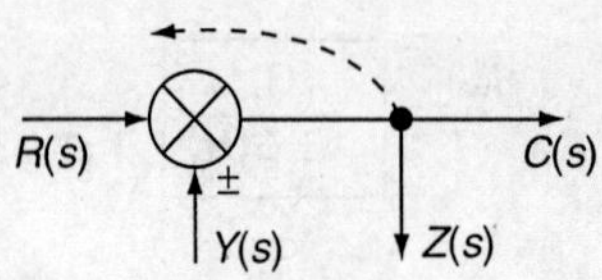	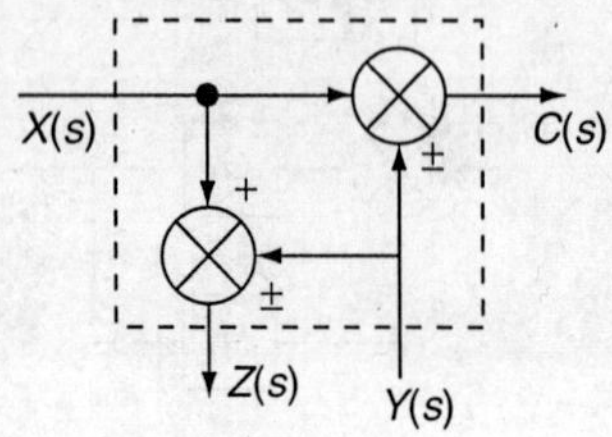	Move take-off point before a summing point
10	Rule 10	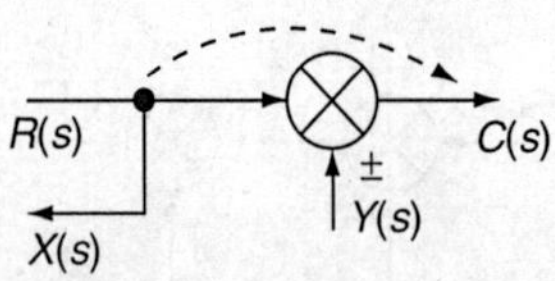	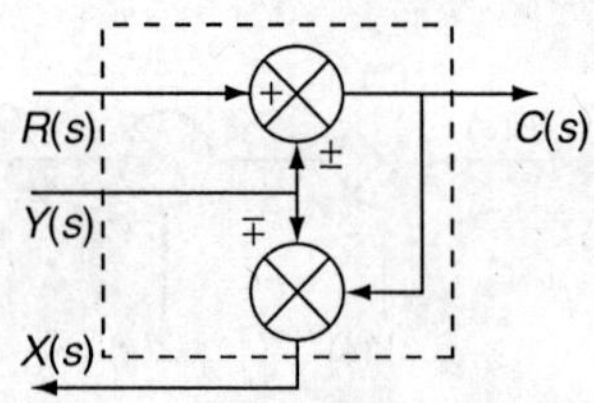	Move take-off point after a summing point

ADDITIONAL SOLVED EXAMPLES

Example 6.8 Obtain the transfer function C/R for the block diagram shown in Fig. E6.8

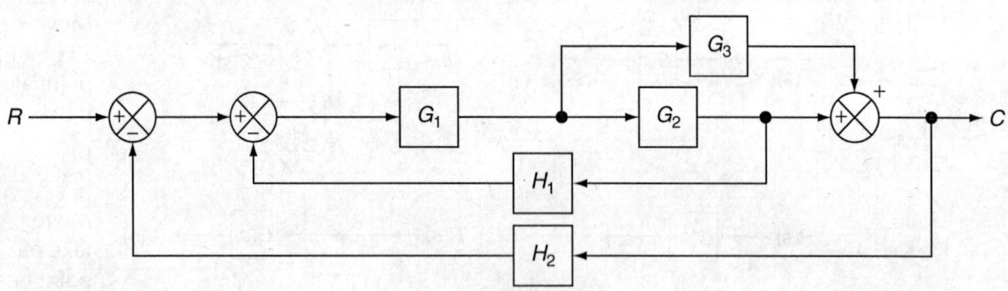

Fig. E6.8

Solution The take-off point is shifted after the block G_2. Fig. E6.8 becomes

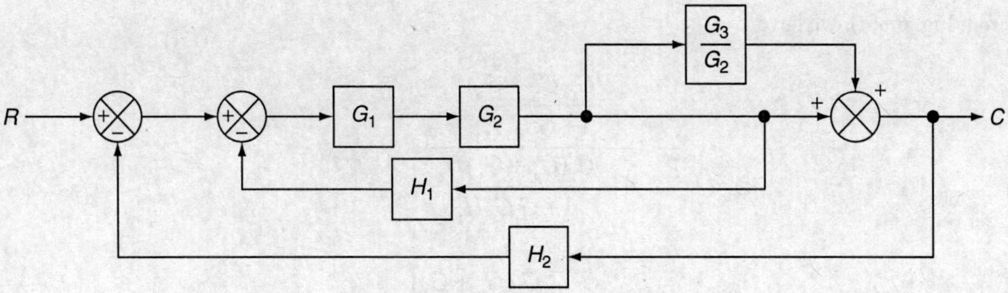

Fig. E6.8(a)

Reducing the cascade block and parallel block of Fig. E6.8(a), we have

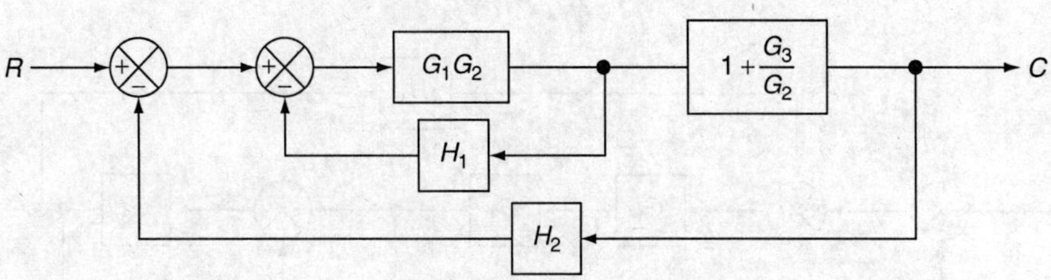

Fig. E6.8(b)

Replacing the internal feedback loop by its equivalent block, we get

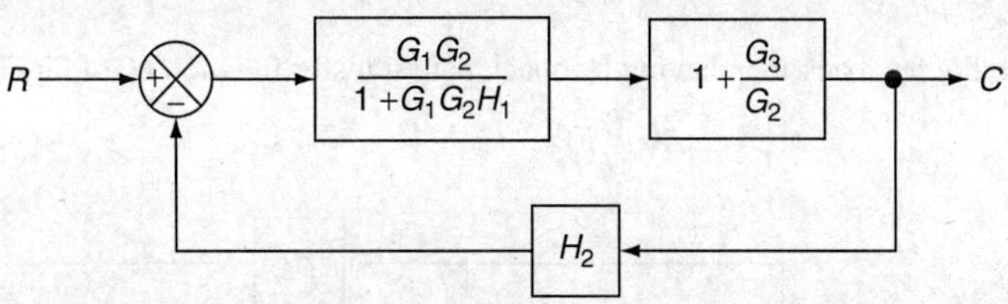

Fig. E6.8(c)

The equivalent block diagram of Fig. E6.8(c) becomes

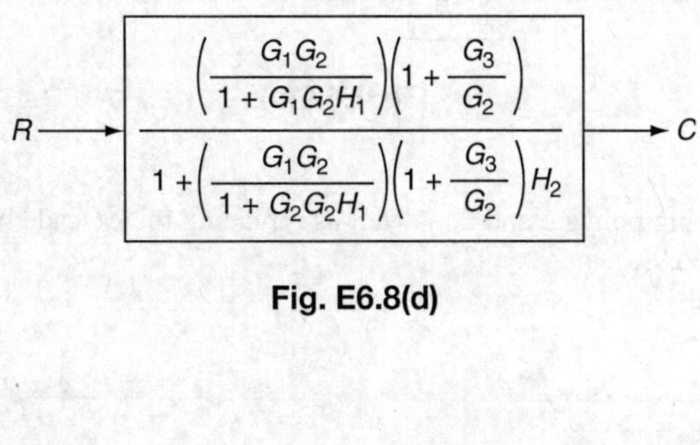

Fig. E6.8(d)

From Fig. 6.8(d), we have

$$\frac{C}{R}=\frac{\dfrac{G_1(G_2+G_3)}{1+G_1G_2H_1}}{1+\dfrac{G_1(G_2+G_3)H_2}{1+G_1G_2H_1}}$$

$$=\frac{G_1(G_2+G_3)}{1+G_1G_2(H_1+H_2)+G_1G_3H_2}$$

Example 6.9 Find the simplified block diagram of Fig. E6.9 and obtain the transfer function.

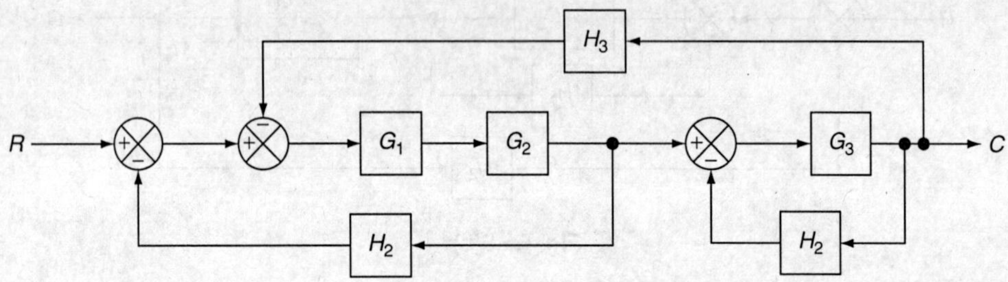

Fig. E6.9

Solution Removing the feedback loop having feedback path transfer function H_2 of Fig. E6.9, we get Fig. E6.9(a).

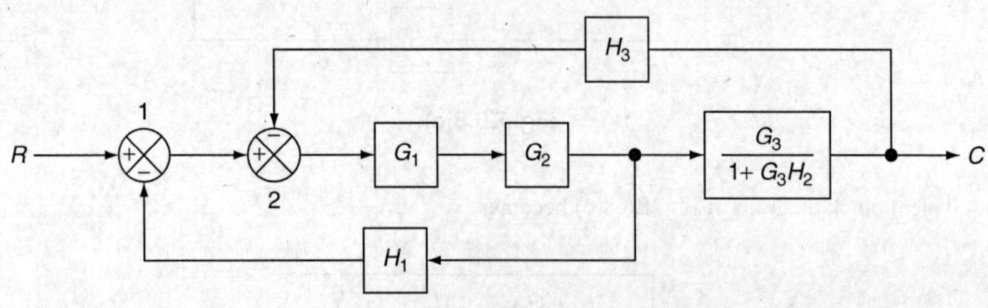

Fig. E6.9(a)

Interchanging the summing points 1 and 2, as well as replacing the cascade block by its equivalent block, we get Fig. E6.9(b) from Fig. E6.9(a),

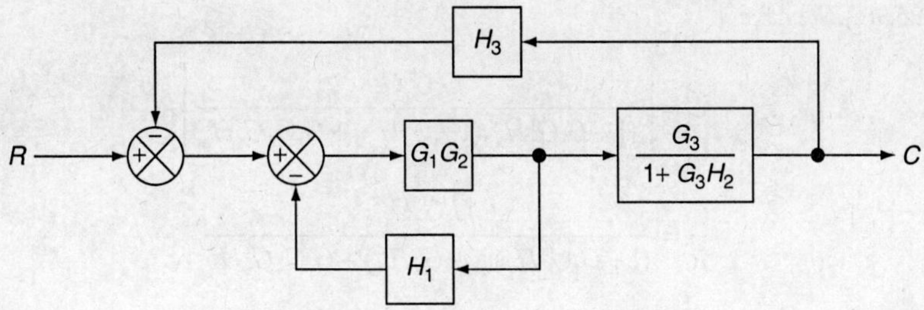

Fig. E6.9(b)

From Fig. E6.9(b),

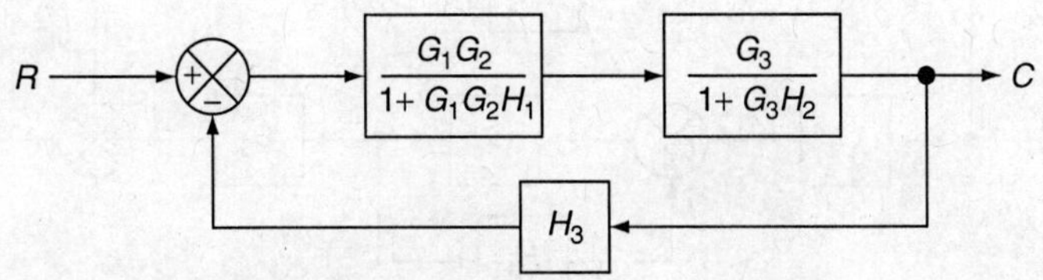

Fig. E6.9(c)

From Fig. E6.9(c), we have

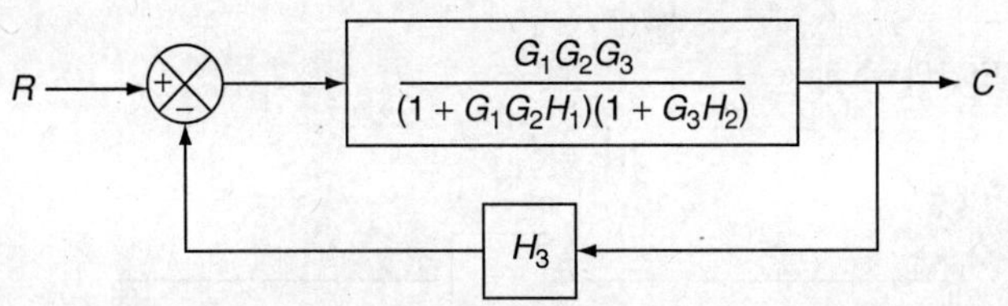

Fig. E6.9(d)

From Fig. E6.9(d), we have

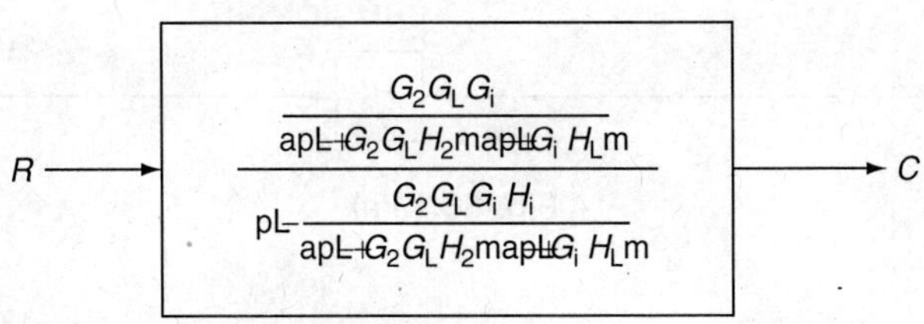

Fig. E6.9(e)

From Fig. E6.9(e), we have

$$C = \left[\frac{G_1 G_2 G_3}{(1 + G_1 G_2 H_1)(1 + G_3 H_2) + G_1 G_2 G_3 H_3} \right] R$$

$$\therefore \qquad \frac{C}{R} = \frac{G_1 G_2 G_3}{(1 + G_1 G_2 H_1)(1 + G_3 H_2) + G_1 G_2 G_3 H_3}$$

Example 6.10 Obtain transfer function of the block diagram shown in Fig. E6.10.

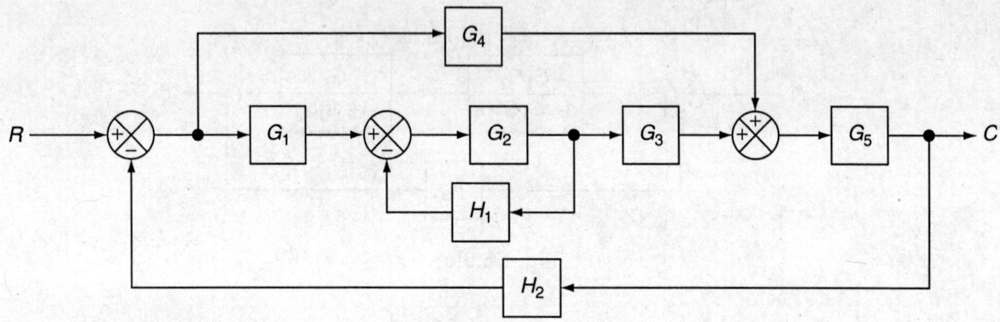

Fig. E6.10

Solution From Fig. E6.10, we have

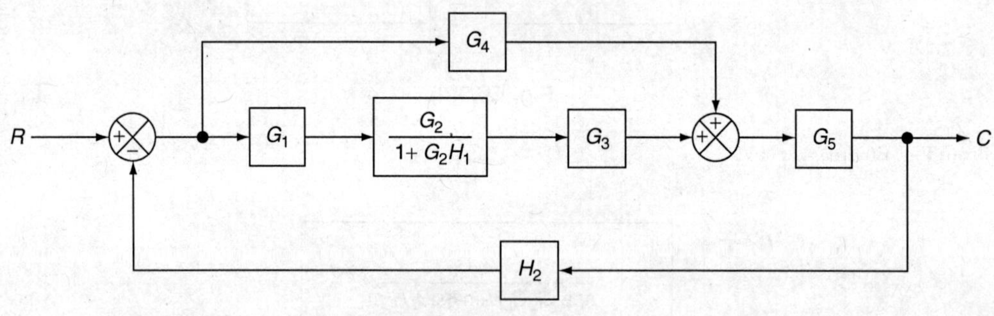

Fig. E6.10(a)

From Fig. E6.10(a),

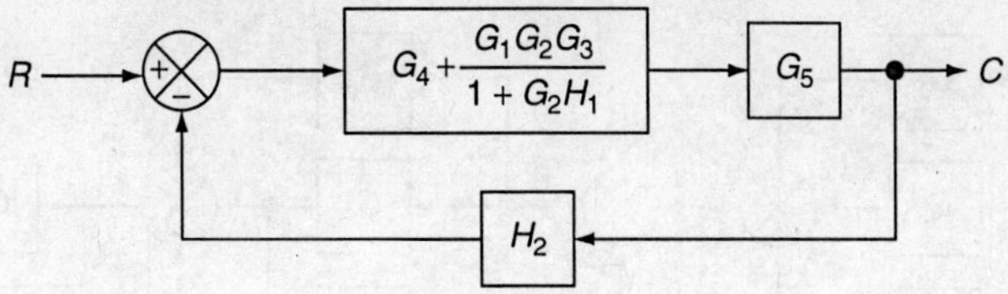

Fig. E6.10(b)

From Figure. E6.10(b), we have

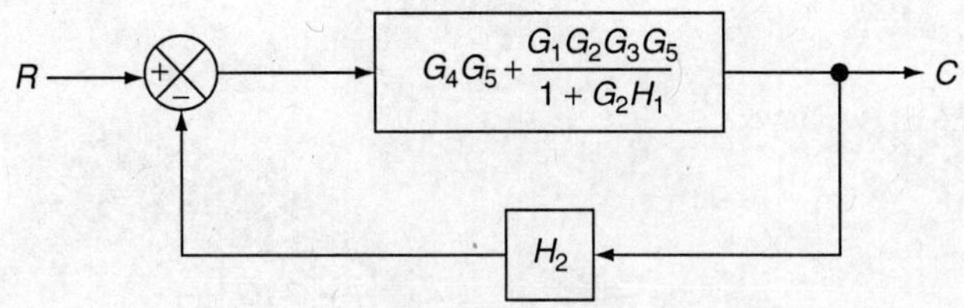

Fig. E6.10(c)

From Fig. E6.10(c), we have

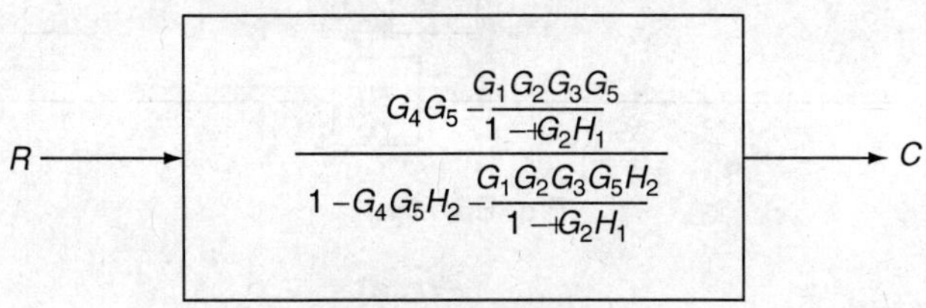

Fig. E6.10(d)

From Fig. E6.10(d), we have

$$C = \frac{(G_4G_5 + G_2G_4G_5H_1 + G_1G_2G_3G_5)R}{1 + G_2H_1 + G_4G_5H_2 + G_2G_4G_5H_1H_2 + G_1G_2G_3G_5H_2}$$

$$\therefore \quad \frac{C}{R} = \frac{G_4G_5 + G_2G_4G_5H_1 + G_1G_2G_3G_5}{1 + G_2H_1 + G_4G_5H_2 + G_2G_4G_5H_1H_2 + G_1G_2G_3G_5H_2}$$

This is the transfer function of Fig. E6.10.

Example 6.11 Obtain the simplified block diagram of Fig. E6.11.

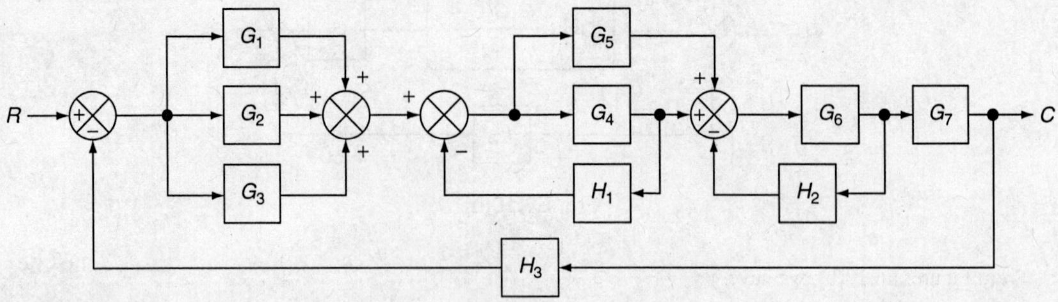

Fig. E6.11

Solution From Fig. E6.11, we have

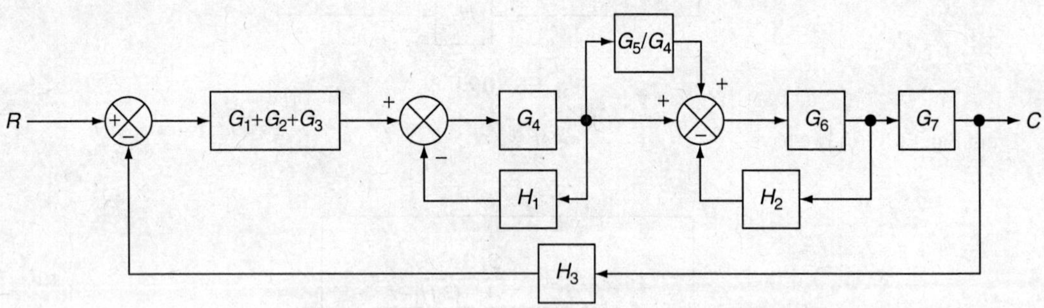

Fig. E6.11(a)

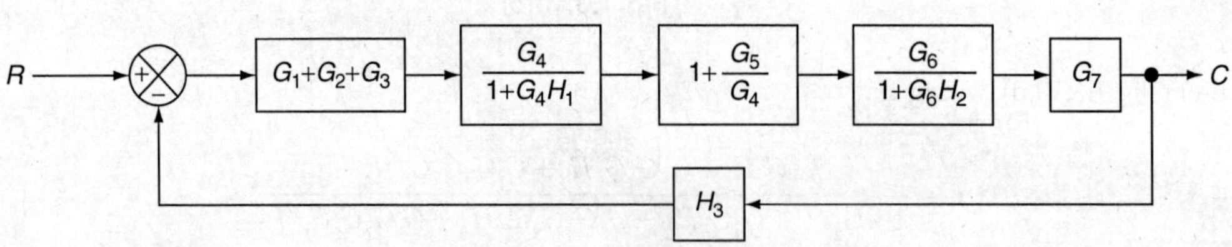

Fig. E6.11(b)

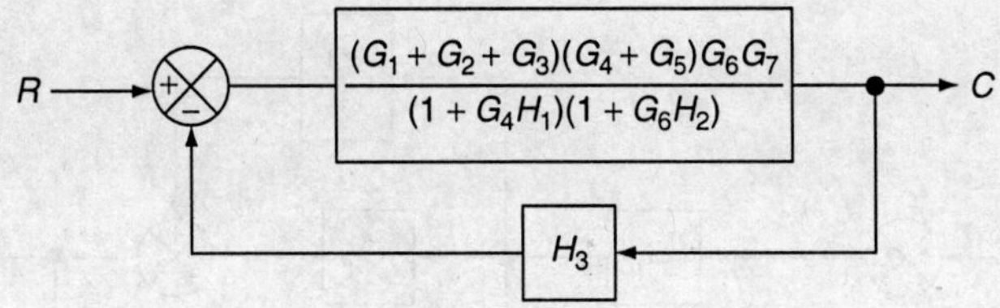

Fig. E6.11(c)

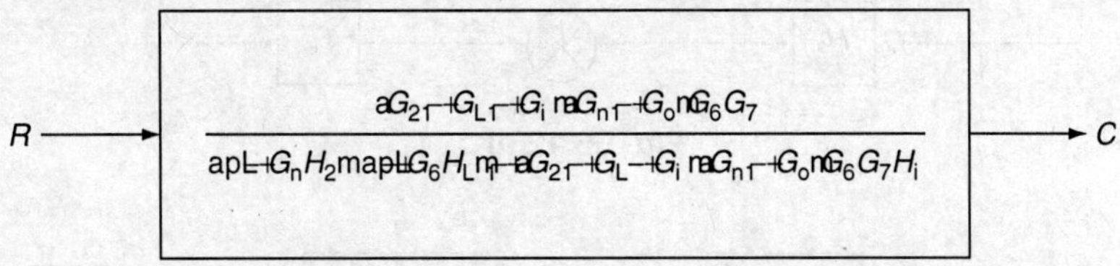

Fig. E6.11(d)

Figure E6.11(d) shows the simplified block diagram of Fig. E6.11.

Example 6.12 Obtain the simplified block diagram of Fig. E6.12.

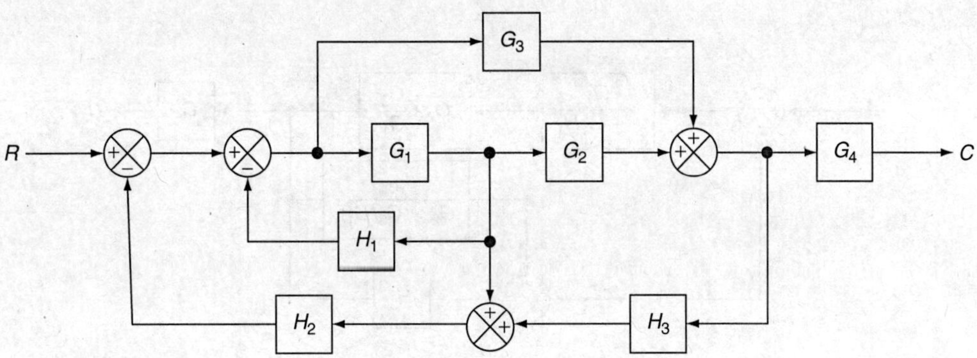

Fig. E6.12

Solution Shifting the take-off point after the block G_1, we have

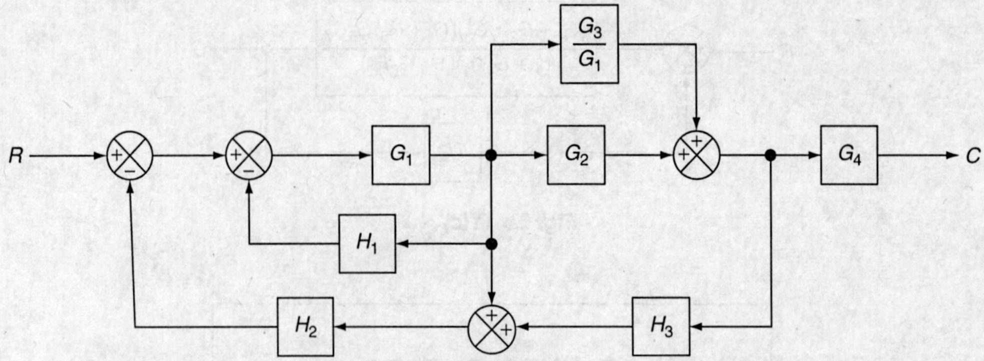

Fig. E6.12(a)

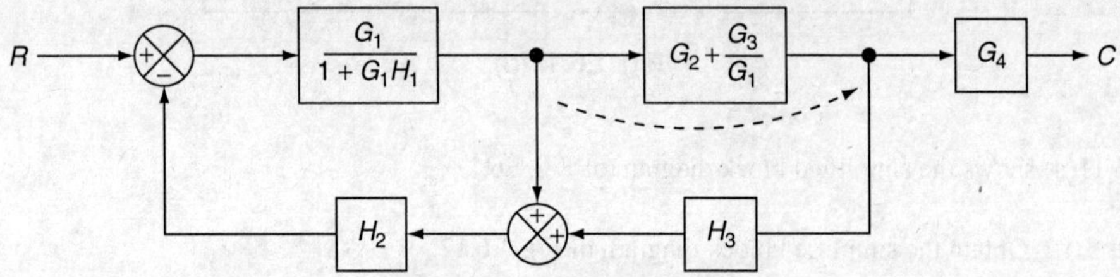

Fig. E6.12(b)

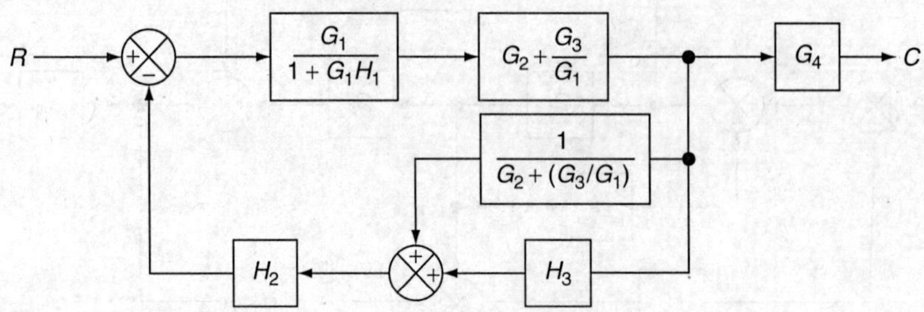

Fig. E6.12(c)

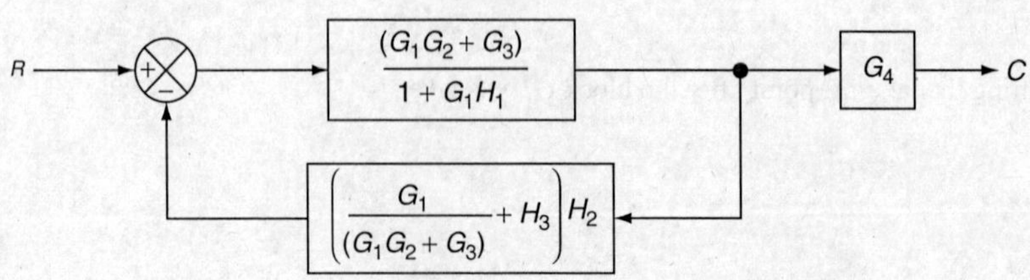

Fig. E6.12(d)

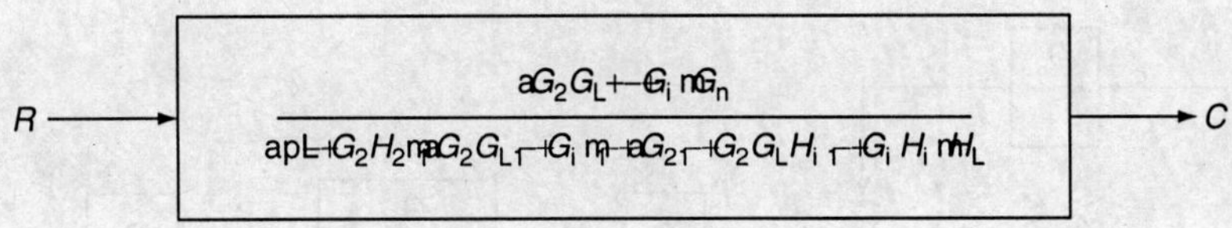

Fig. E6.12(e)

Figure E6.12(e) shows the simplified block diagram of Fig. E6.12.

Example 6.13 Find the simplified block diagram of Fig. E6.13.

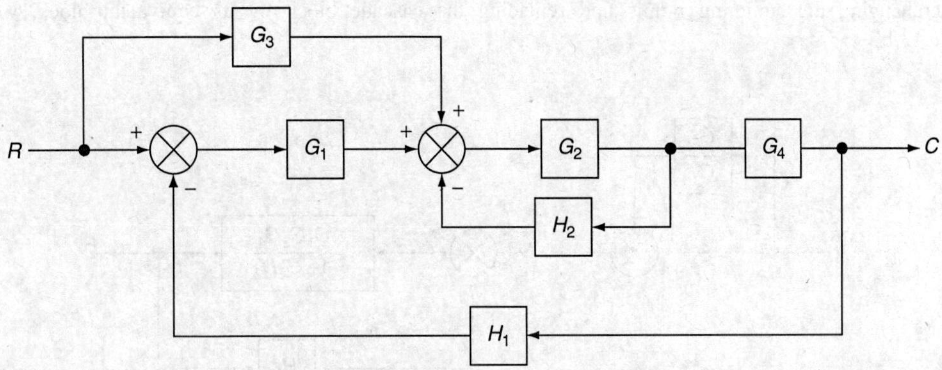

Fig. E6.13

Solution Reducing internal Feedback loops of Fig. E6.13, we have

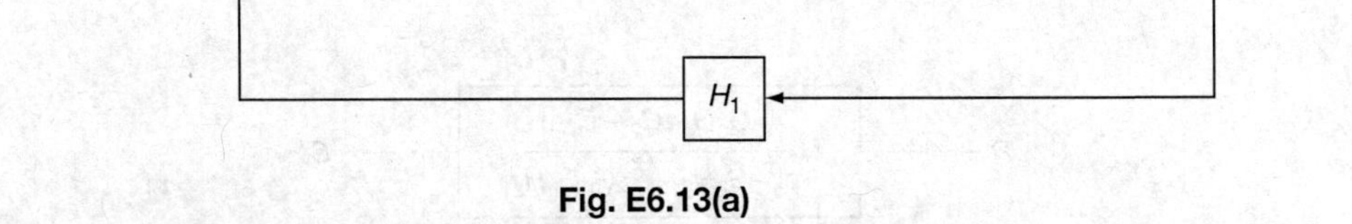

Fig. E6.13(a)

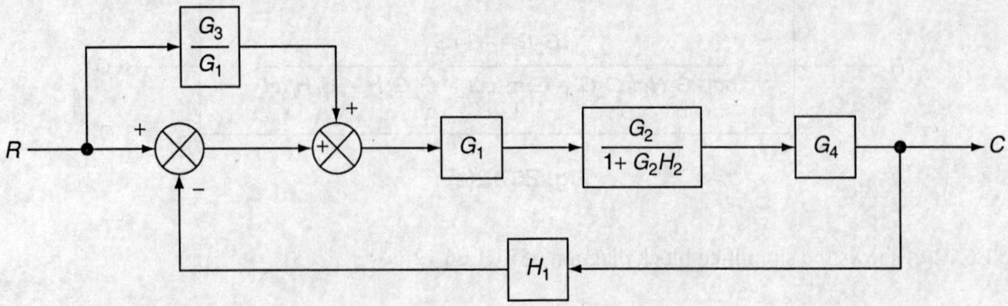

Fig. E6.13(b)

Interchanging the summing points and replacing the cascade block by its equivalent block, we have Fig. E6.13(b)

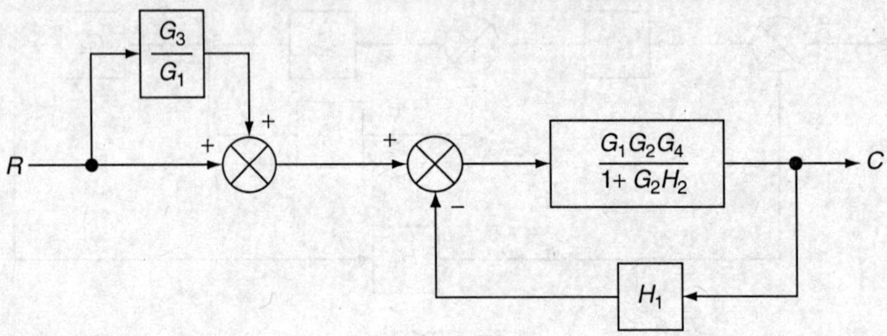

Fig. E6.13(c)

From Fig. E6.13(c)

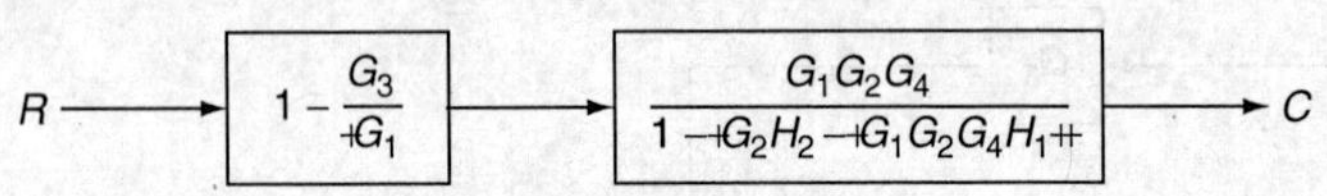

Fig. E6.13(d)

From Fig. E6.13(d),

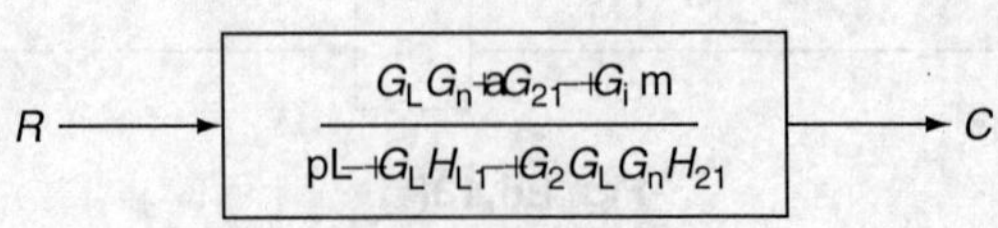

Fig. E6.13(e)

Fig. E6.13(e) is the simplified block diagram of Fig. E6.13.

Example 6.14 Obtain the transfer function of the block diagram shown in Fig. E6.14.

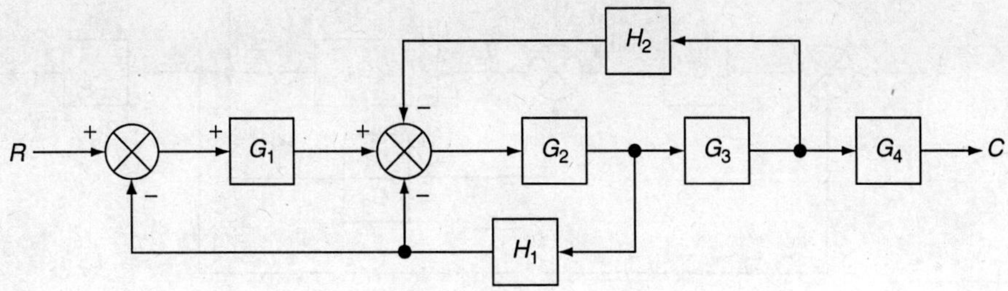

Fig. E6.14

Solution Shifting the take-off point after the block G_3, we have

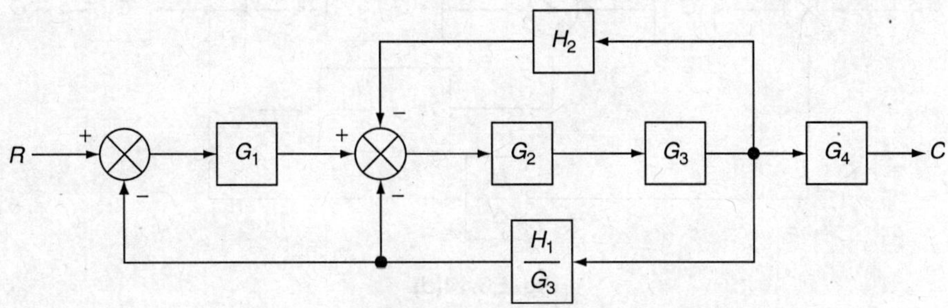

Fig. E6.14(a)

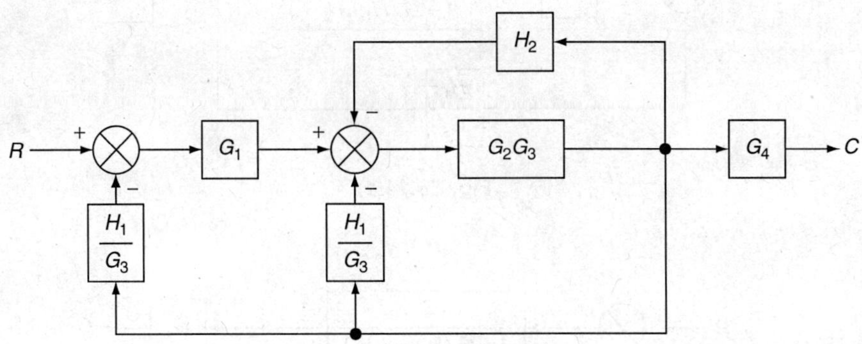

Fig. E6.14(b)

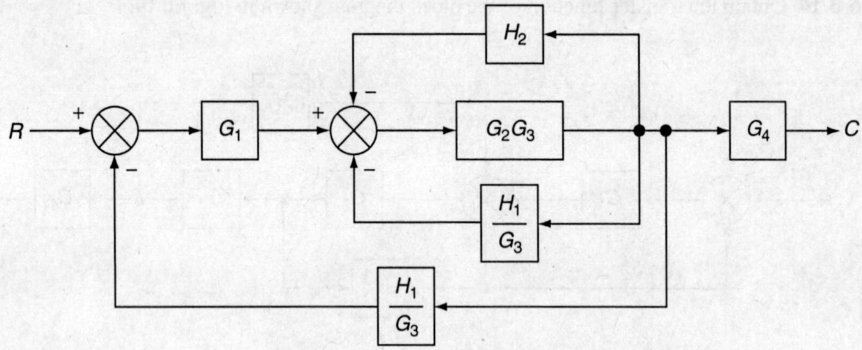

Fig. E6.14(c)

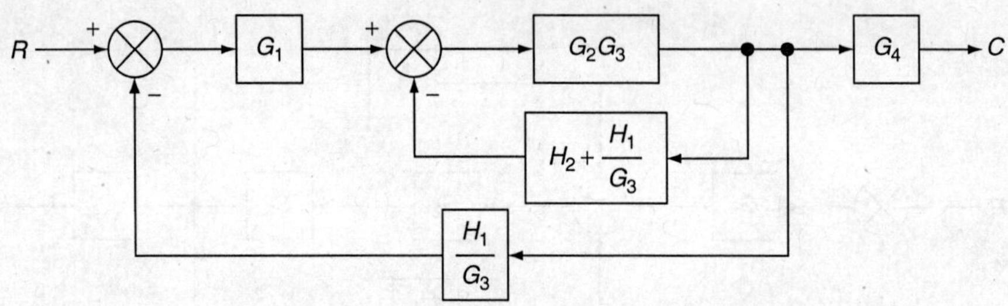

Fig. E6.14(d)

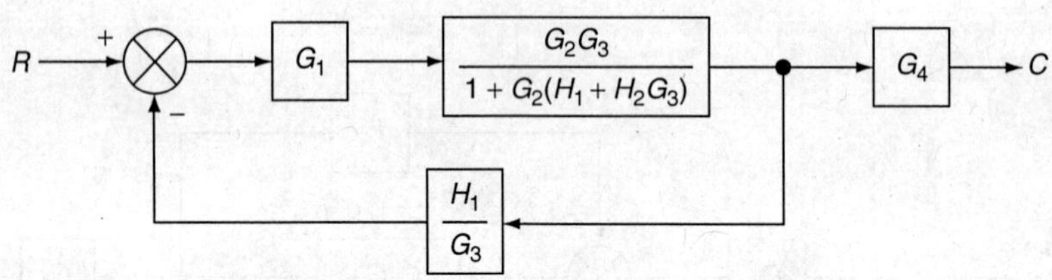

Fig. E6.14(e)

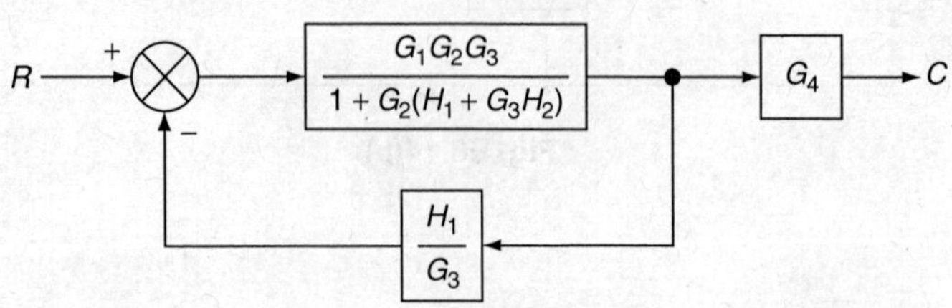

Fig. E6.14(f)

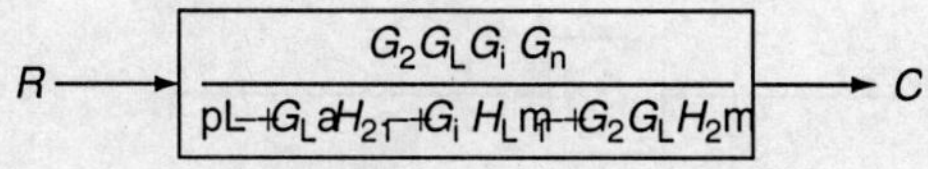

Fig. E6.14(g)

Fig. E6.14(g) is the simplified block diagram of Fig. E6.14.

Example 6.15 obtain the simplified block diagram of Fig. E6.15.

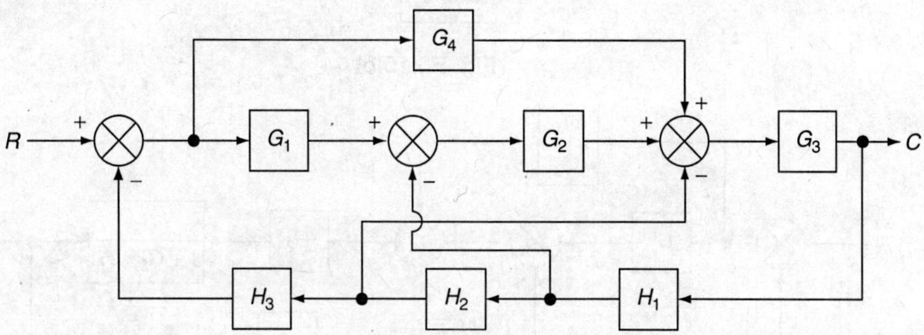

Fig. E6.15

Solution

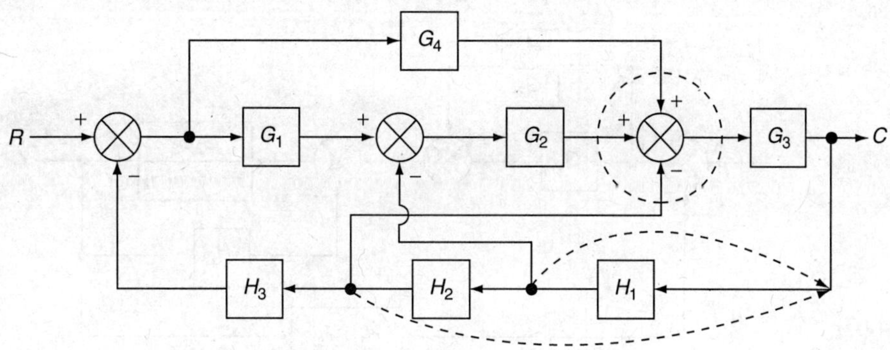

Fig. E6.15(a)

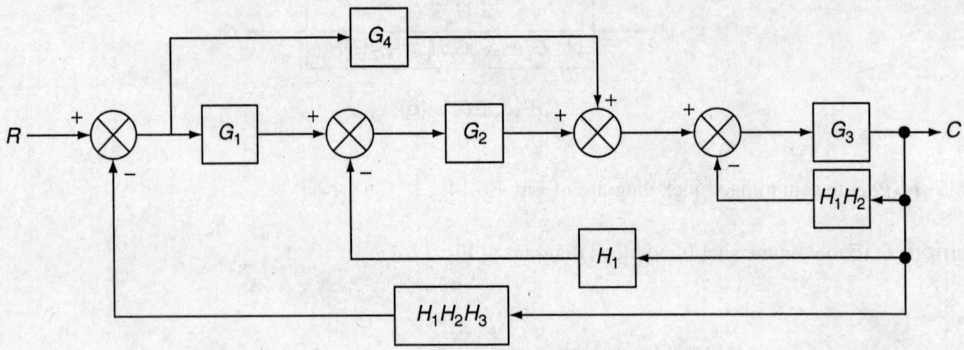

Fig. E6.15(b)

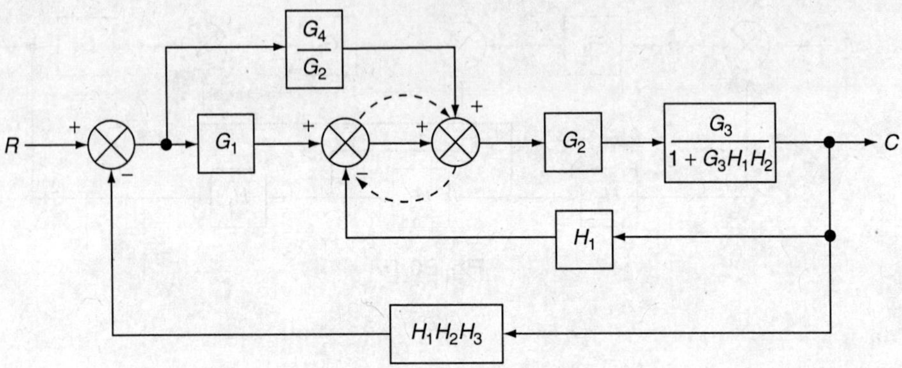

Fig. E6.15(c)

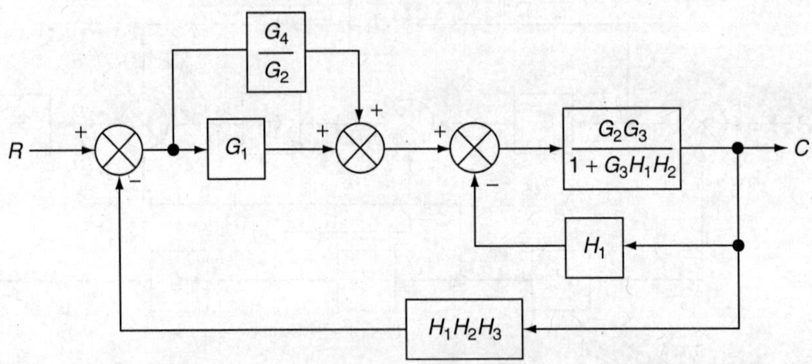

Fig. E6.15(d)

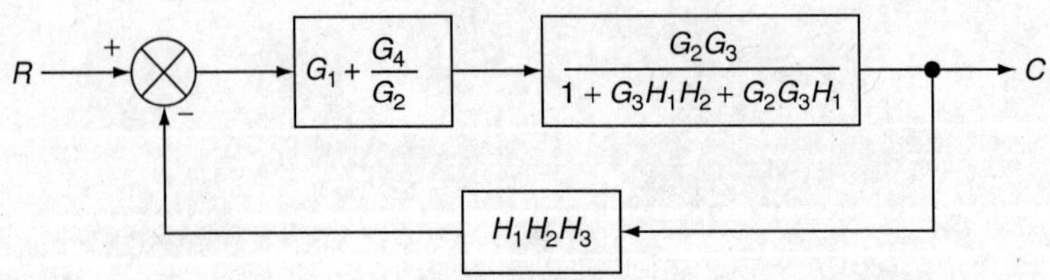

Fig. E6.15(e)

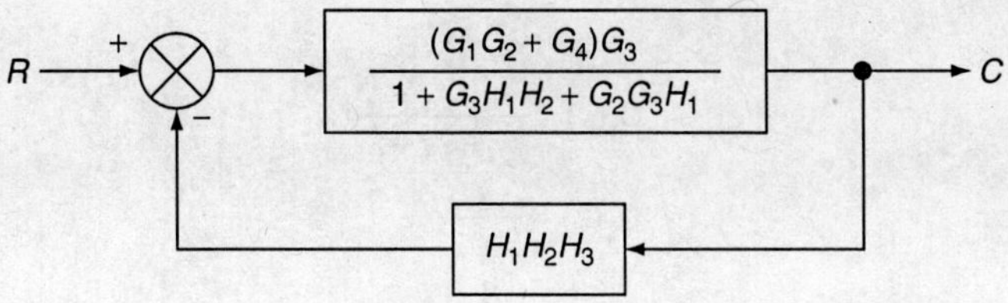

Fig. E6.15(f)

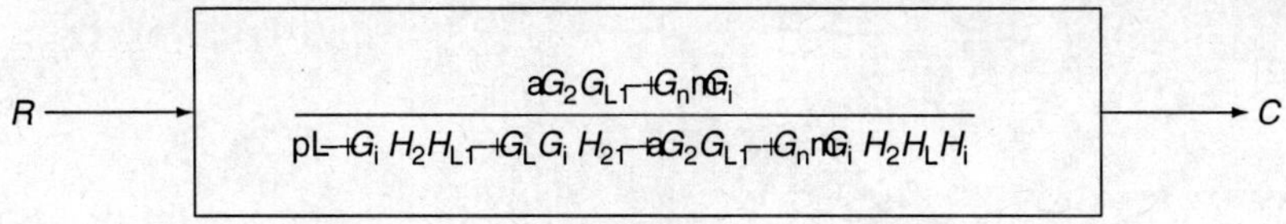

Fig. E6.15(g)

From Fig. E6.15(g) shows the simplified block diagram of Fig. E6.15.

MATLAB Program

Program 1 For Fig. P1, find the equivalent transfer function for each case.

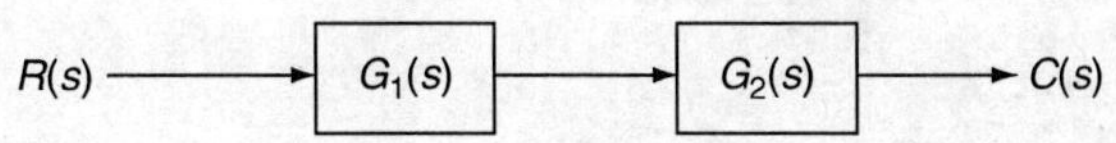

(a)

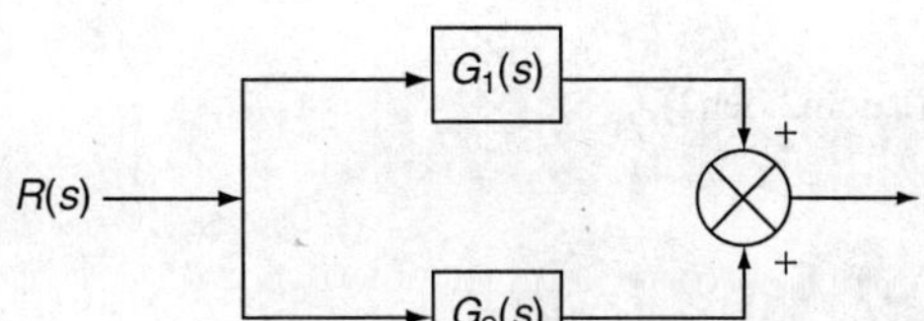

(b)

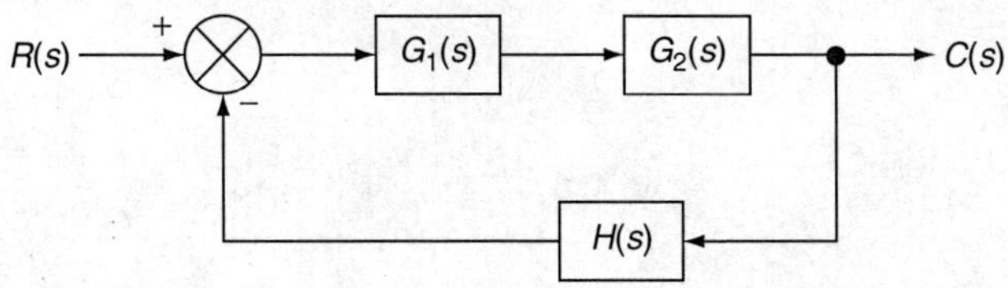

(c)

Fig. P1

In Fig. P1,

$$G_1(s) = \frac{2}{s^2 + 2s + 5}$$

$$G_2(s) = \frac{s}{s+2}$$

$$H(s) = \frac{1}{s+3}$$

Solution

```
>> num1 = [0 2]
num1 =
   0   2
>> den1 = [1 2 5]
den1 =
   1   2   5
>> num2 = [1 0]
num2 =
   1   0
>> den2 = [1 2]
den2 =
   1   2
>> % for Fig. P1(a)
>> [num,den]= series(num1,den1,num2,den2);
>> printsys(num,den)
num/den =
           2 s
  ----------------------
   s^3 + 4 s^2 + 9 s + 10
>> % for Fig. P1(b)
>> [num,den]= parallel(num1,den1,num2,den2);
>> printsys(num,den)
num/den =
   s^3 + 2 s^2 + 7 s + 4
  ----------------------
   s^3 + 4 s^2 + 9 s + 10
>> % for Fig. P1(c)
>> [num,den]= series(num1,den1,num2,den2);
>> num2 – [0 1]
num2 =
   0   1
>> den2 = [1 3]
den2 =
   1   3
>> [num,den]= feedback(num,den,num2,den2,–1)
num =
   0   0   2   6   0
```

```
den =
   1   7   21   39   30
>> printsys(num,den)
num/den =
           2 s^2 + 6 s
   -------------------------------
   s^4 + 7 s^3 + 21 s^2 + 39 s + 30
```

Program 2 Find the equivalent tranfer function of the block diagram shown in Fig. P2.

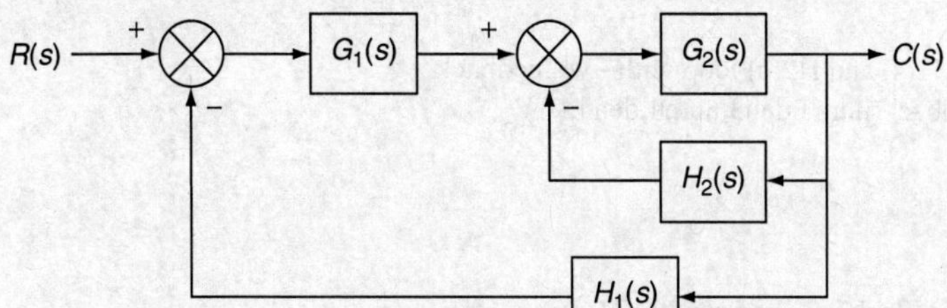

Fig. P2

In Fig. P2,

$$G_1(s) = \frac{2s}{s^2+3}, \quad G_2(s) = \frac{4}{s+6}$$
$$H_1(s) = \frac{2}{s+4}, \quad H_2(s) = \frac{5}{2s+1}$$

Solution

```
>> % for G1(s)
>> num1 = [2 0]
num1 =
   2   0
>> den1 = [ 1 0 3]
den1 =
   1   0   3
>> % for H1(s)
>> num2 = [0 2]
num2 =
   0   2
>> den2 = [1 4]
den2 =
   1   4
>> % for G2(s)
>> num3 = [ 0 4]
```

```
num3 =
   0   4
>> den3 = [ 1 6]
den3 =
   1   6
>> % for H2(s)
>> num4 = [0 5]
num4 =
   0   5
>> den4 = [2 1]
den4 =
   2   1
>> % Reduction of G2(s) and H2(s) loop with –ve feedback
>> [num,den] = feedback(num3,den3,num4,den4,–1)
num =
   0   8   4
den =
   2   13   26
>> % This equivalent block is in series with G1(s)
>> [num,den] = series(num1,den1,num,den)
num =
   0   0   16   8   0
den =
   2   17   62   117   130
>> %Reduction of this equivalent block and H1(s) with –ve feedback
>> [num,den] = feedback(num,den,num2,den2,–1)
num =
   0   0   16   72   32   0
den =
   2   25   130   397   614   520
>> printsys(num,den)
num/den =
              16 s^3 + 72 s^2 + 32 s
   ----------------------------------------------
   2 s^5 + 25 s^4 + 130 s^3 + 397 s^2 + 614 s + 520
```

Program 3 Draw the equivalent tranfer function of Fig. P3 shown below.

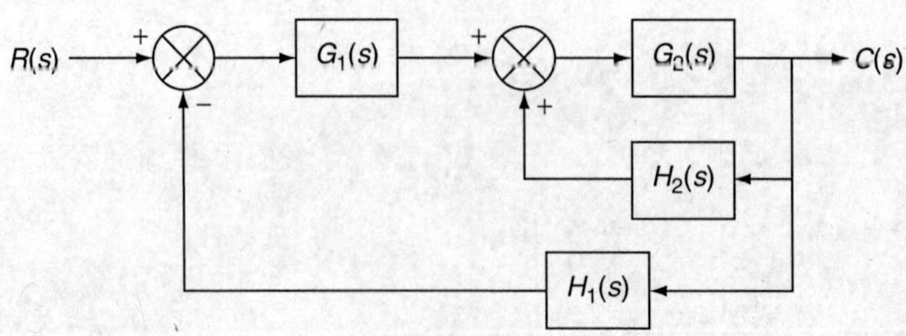

Fig. P3

In Fig. P3,

$$G_1(s) = \frac{2s}{s^2+3}, H_1(s) = \frac{2}{s+4}$$
$$G_2(s) = \frac{4}{s+6}, H_2(s) = \frac{5}{2s+4}$$

Solution

```
>> % for G1(s)
>> num1 = [2 0]
num1 =
   2   0
>> den1 = [ 1 0 3]
den1 =
   1   0   3
>> % for H1(s)
>> num2 = [0 2]
num2 =
   0   2
>> den2 = [1 4]
den2 =
   1   4
>> % for G2(s)
>> num3 = [ 0 4]
num3 =
   0   4
>> den3 = [ 1 6]
den3 =
   1   6
>> % for H2(s)
>> num4 = [0 5]
num4 =
   0   5
>> den4 = [2 1]
den4 =
   2   1
>> % Reduction of G2(s) and H2(s) loop with +ve feedback
>> [num,den] = feedback(num3,den3,num4,den4,+1)
num =
   0   8   4
den =
   2   13   –14
>> %this equivalent block is in series with G1(s)
>> [num,den] = series(num1,den1,num,den)
num =
   0   0   16   8   0
den =
   2   17   22   37   –70
>> %Reduction of this equivalent block and H1(s) with –ve feedback
>> [num,den] = feedback(num,den,num2,den2,–1)
```

```
num =
   0    0   16    72   32    0
den =
   2   25   90   157   94  -280
>> printsys(num,den)
num/den =
         16 s^3 + 72 s^2 + 32 s
   -----------------------------------------
   2 s^5 + 25 s^4 + 90 s^3 + 157 s^2 + 94 s - 280
```

SHORT QUESTIONS WITH ANSWERS

1. Define the following:
   (i) Block diagram, (ii) Summing point, (iii) Take-off point.

   Ans. (i) The shorthand pictorial representation of the cause-and-effect relationship between the input and output of a physical system is known as block diagram.
   (ii) The point at which two or more signals can be added or subtracted is known as summing point.
   (iii) The point at which output signal of any block can be applied to two or more points is known as take-off point.

2. State the procedures for reduction of block diagram.

   Ans. Step 1: Reduce the cascade blocks.
   Step 2: Reduce the parallel blocks.
   Step 3: Reduce the internal feedback loops.
   Step 4: It is advisable to shift take-off points towards right and summing points towards left.
   Step 5: Repeat steps 1 to step 4 until the simple form is obtained.
   Step 6: Find transfer function of the overall system using formula $C(s)/R(s)$.

Procedure for multiple inputs

Step 1: Consider one input and put other inputs to zero. Find the output.
Step 2: Repeat step 1 until all inputs are covered.
Step 3: Find the resultant output by superposition.

EXERCISES

Questions for Practice

1. Find the transfer function of the following:

   (i)

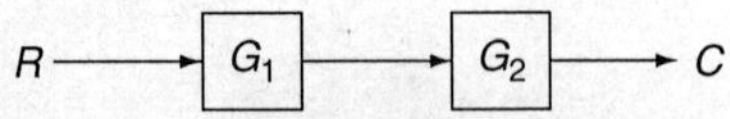

Fig. 1(a)

   (ii)

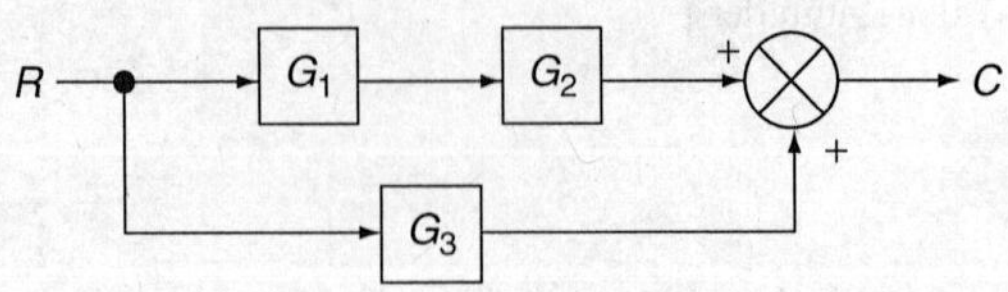

Fig. 1(b)

(iii)

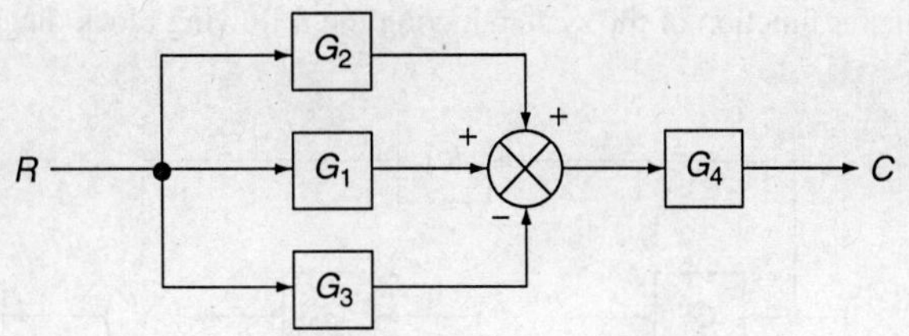

Fig. 1(c)

(iv)

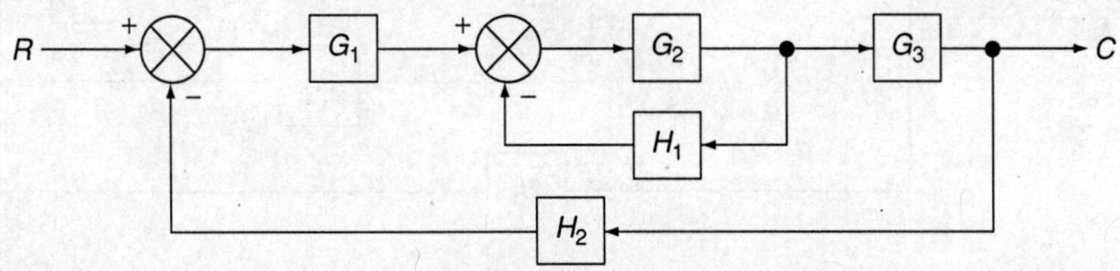

Fig. 1(d)

Ans. (i) G_1G_2, (ii) $G_1G_2 + G_3$, (iii) $(G_1 + G_2 - G_3)G_4$, (iv) $\dfrac{G_1G_2G_3}{1+G_2H_1+G_1G_2G_3H_2}$

2. Find the transfer function of the following block diagram:

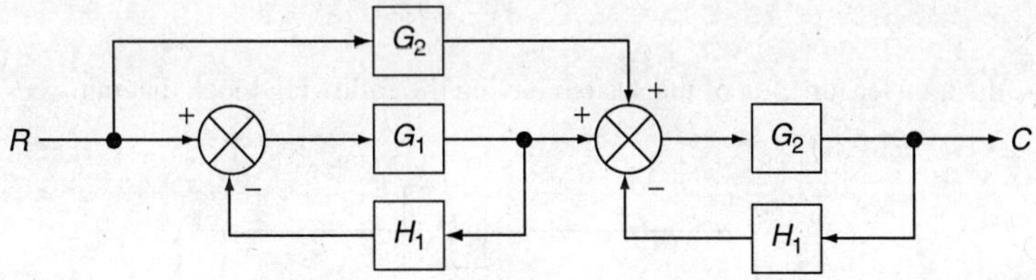

Fig. 2

Ans. $\dfrac{(G_1+G_2+G_1G_2H_1)G_3}{(1+G_1H_1)(1+G_3H_2)}$

3. Find the transfer function of the following block diagram:

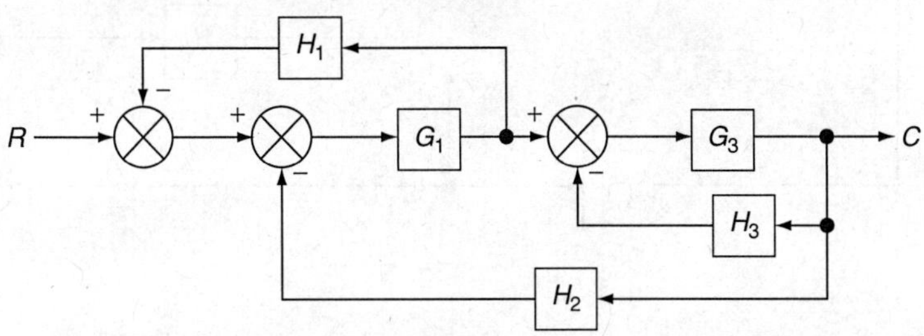

Fig. 3

Ans. $\dfrac{G_1G_3}{(1+G_1H_1)(1+G_3H_3)+G_1G_3H_2}$

4. Determine the transfer function of the system having the following block diagram.

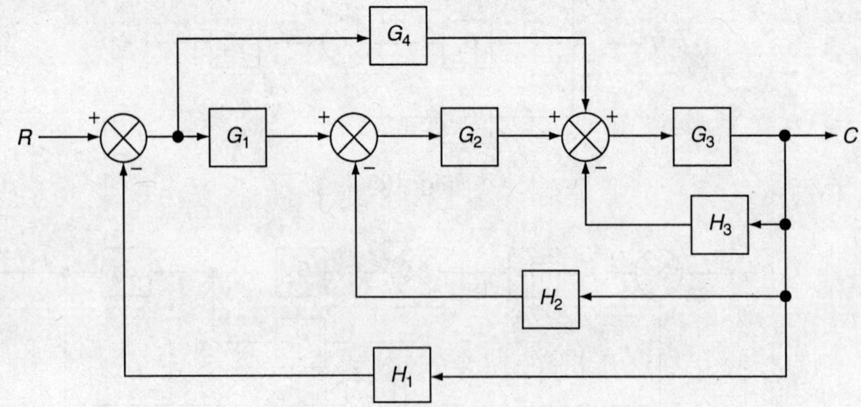

Fig. 4

Ans. $\dfrac{(G_1G_2 + G_4)G_3}{1 + G_1G_2G_3H_1 + G_4G_3H_1 + G_2G_3H_2 + G_3H_3}$

5. Determine the transfer function of the system having the following block diagram.

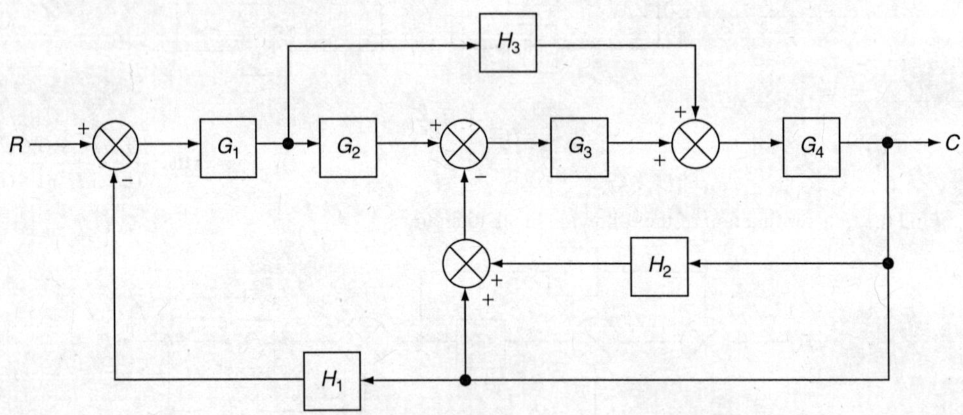

Fig. 5

Ans. $\dfrac{G_1G_2G_3G_4 + G_1G_4H_3}{1 + G_3G_4 + G_3G_4H_2 + G_1G_4H_1H_3 + G_1G_2G_3G_4H_1}$

6. Determine the transfer function of the system having the following block diagram.

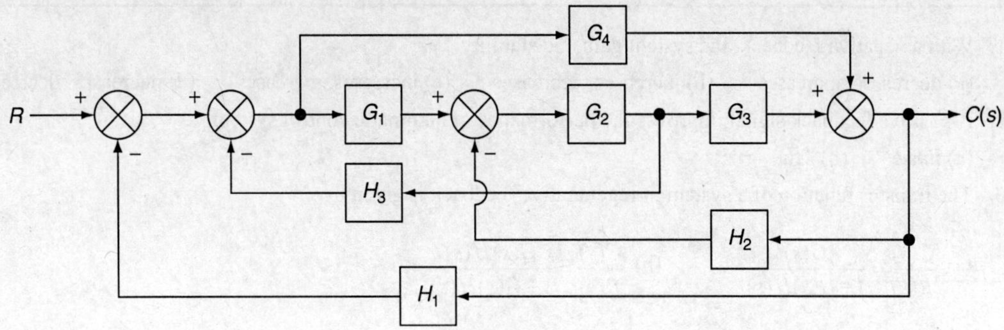

Fig. 6

Ans. $\dfrac{G_2G_3G_4 + G_4}{1 + G_1G_2G_3H_1 + G_1G_2H_3 + G_2G_3H_3 + G_4H_1 - G_2G_4H_2H_3}$

7. Determine the transfer function of the system having the following block diagram.

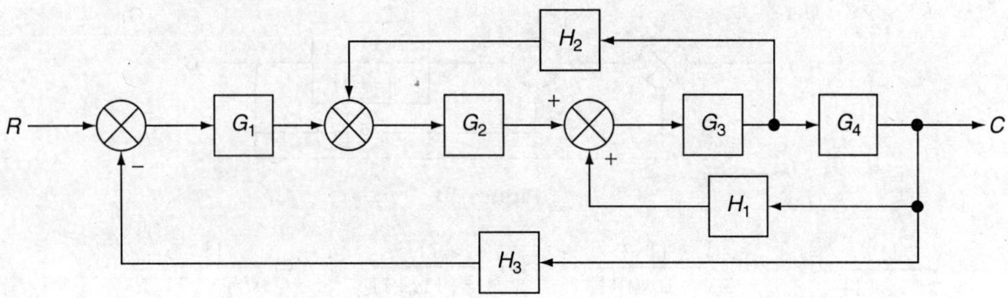

Fig. 7

Ans. $\dfrac{G_1G_2G_3G_4}{1 + H_2G_2G_3 - H_1G_3G_4 + G_1G_2G_3G_4H_3}$

8. Figure 5 shows the block diagram of a linear feedback system. Obtain a signal flow graph for the system and hence calculate the overall gain $C(s)/R(s)$ for the system.

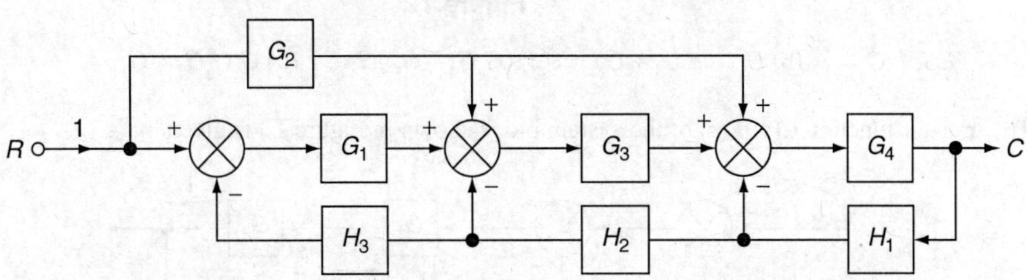

Fig. 8

Ans. $\dfrac{G_1G_3G_4 + G_2G_3G_4 + G_2G_4}{1 + G_4H_1 + G_3G_4H_1H_2 + G_1G_3G_4H_1H_2H_3}$

TEST YOURSELF

1. With a negative feedback, the system gain and stability

   (a) decreases, increases (b) increases, decreases (c) increases, increases (d) decreases, decreases

2. A positive feedback signal improves the performance of automatic control system.

   (a) False (b) True

3. The transfer function of a system in regenerative feedback is given by

   (a) $\frac{C(s)}{R(s)} = \frac{G(s)}{1+G(s)H(s)}$ (b) $\frac{C(s)}{R(s)} = \frac{G(s)H(s)}{1+G(s)H(s)}$

   (c) $\frac{C(s)}{R(s)} = \frac{G(s)}{1-G(s)H(s)}$ (d) $\frac{C(s)}{R(s)} = \frac{G(s)H(s)}{1+G(s)}$

4. It is given that $G(s) = K/[s^3(1 + sT)]$ and the system is operated in a closed loop with unit feedback. The order and type of the closed-loop system are

   (a) 2 and 3 (b) 3 and 4 (c) 3 and 0 (d) 7 and 0

5. The transfer function C/R of the system shown in Figure T1 is equal to

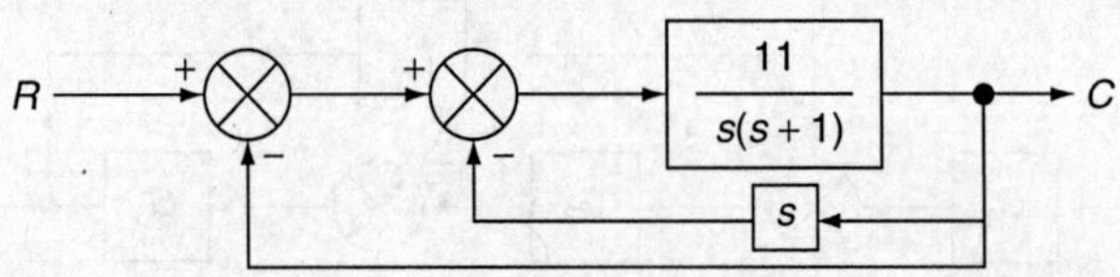

Figure T1

   (a) $\frac{11}{s^2+s+11}$ (b) $\frac{11}{s^2+10s+11}$ (c) $\frac{11}{s^2+11s+11}$ (d) $\frac{11}{s^2+s+11}$

6. The total gain of the system shown in Figure T2 is given by

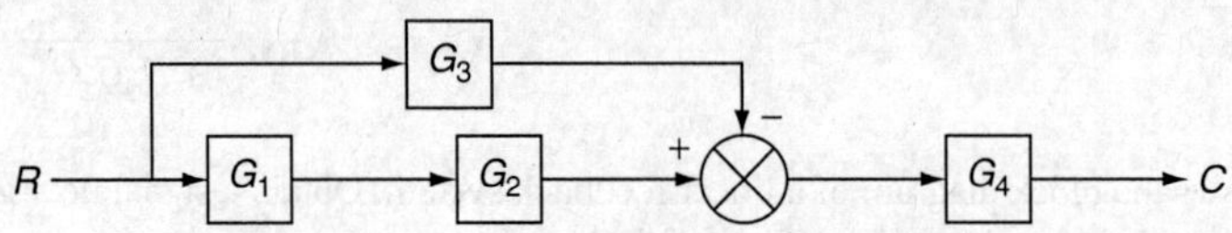

Figure T2

   (a) $G_1G_2G_3 + G_4$ (b) $G_1G_2 + G_3 + G_4$ (c) $(G_1G_2 - G_3)G_4$ (d) $G_1G_2G_4 + G_3$

7. The transfer function $C(s)/R(s)$ of the system block shown in Figure T3 is given by

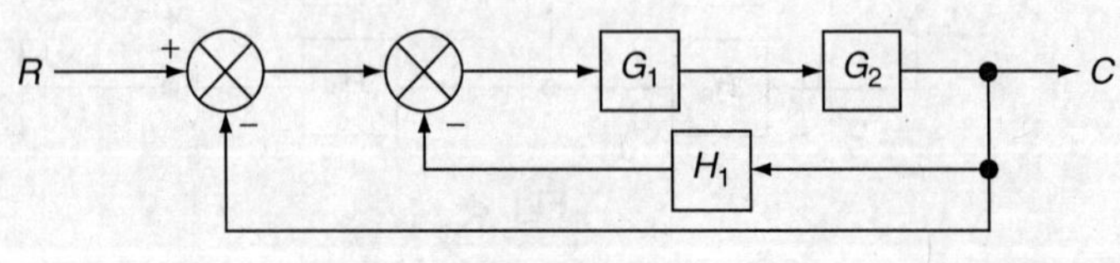

Figure T3

(a) $\dfrac{G_1 - G_2}{1+\dfrac{H_1G_1}{G_2}}$ (b) $\dfrac{G_1/G_2}{1+\dfrac{G_1+G_2}{H_2+1}}$ (c) $\dfrac{G_1G_2(1+H_1-G_1G_2)}{1+G_1G_2H_1+G_1G_2}$ (d) $\dfrac{G_1G_2}{1+G_1G_2H_1+G_1G_2}$

8. The given block diagram is equivalent to

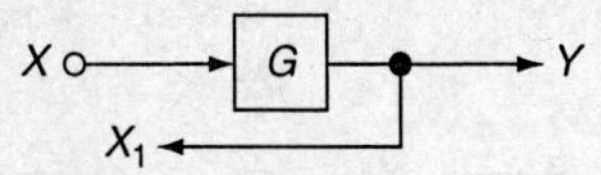

Figure T4

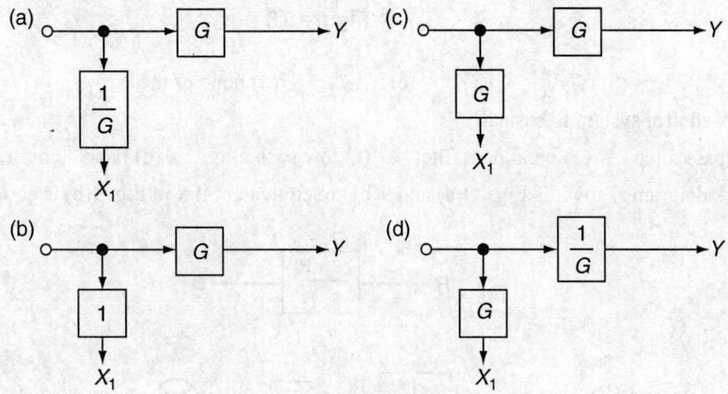

Figure T5

9. Regenerative feedback means the output is feedback with
(a) positive sign (b) negative sign (c) step input (d) oscillation

10. Two equivalent block diagrams are shown below. G_1 is equal to

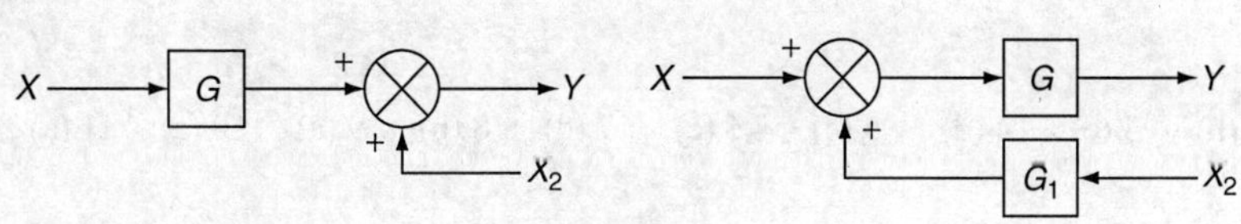

Figure T6

(a) $\dfrac{1}{G}$ (b) G (c) $G+1$ (d) $\dfrac{G}{2}$

11. The transfer function of the system shown in Fig. 7 is

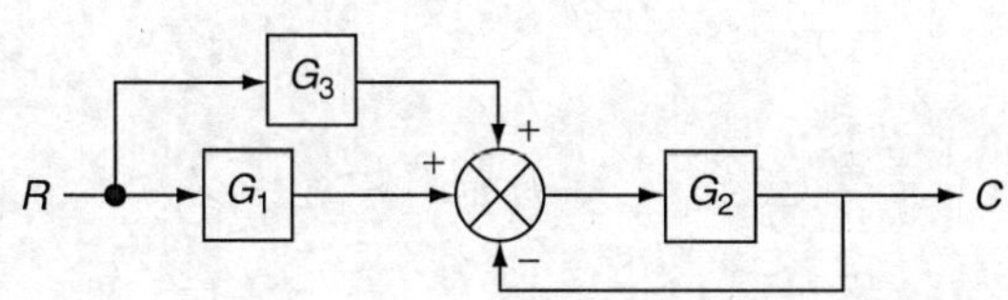

Figure T7

(a) $(G_1 + G_3)\left(\dfrac{G_2}{1+G_2}\right)$ (b) $\dfrac{G_1G_2 + G_3}{1+G_2}$ (c) $\dfrac{G_1 + G_3G_2}{1+G_2}$ (d) $\dfrac{G_1G_2G_3}{1+G_2}$

12. For the given block diagram, the output is equal to

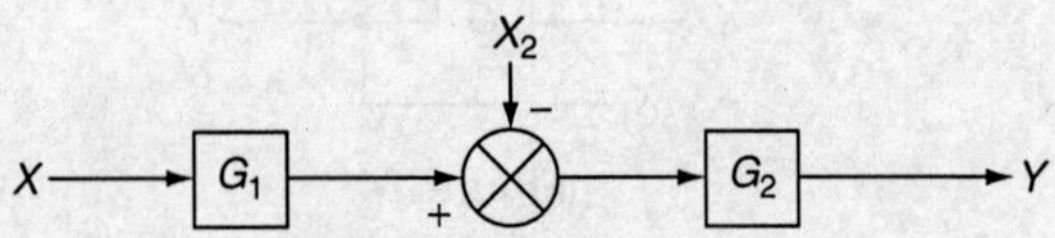

Figure T8

(a) $(X_1 + X_2)G_2$ (b) $(G_1X_1 - X_2)G_2$ (c) $X_1 - X_2$ (d) none of these

13. Feedback control system is basically

(a) high pass filter (b) band pass filter (c) low pass filter (d) band stop filter

14. The block diagrams shown is Figs. 9(a) and 9(b) are equivalent if x in Fig. 9(b) is equal to

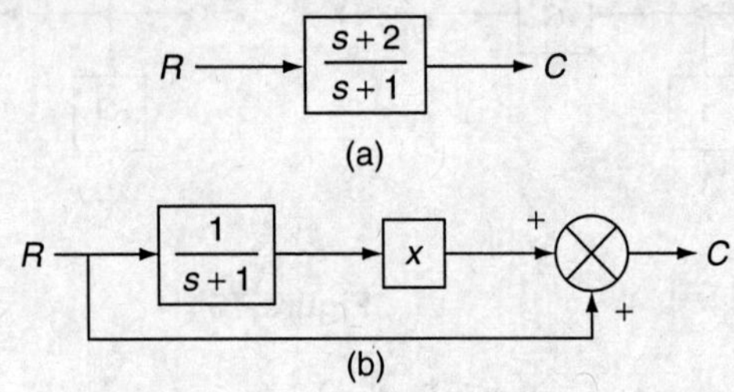

Figure T9

(a) 1 (b) 2 (c) $s + 1$ (d) $s + 2$

ANSWERS

1 (a) 2 (b) 3 (c) 4 (d) 5 (c) 6 (c) 7 (d) 8 (b) 9 (a) 10 (a) 11 (a) 12 (b) 13(c) 14 (a)

Chapter 7

Signal-Flow Graphs

If the set of equations describing a system is known, it is possible to represent the system in another pictorial form known as signal-flow graph. It may be regarded as a simplified version of a block diagram. Signal-flow graph is a pictorial representation of a system, which displays graphically the transmission of signals in it. S. J. Mason introduced the signal-flow graph for representing the cause-and-effect relationship of linear systems, which are modelled by algebraic equations. Although there is difference between physical appearances of the signal-flow graph (SFG) and block diagram, the SGF is constrained by more rigid mathematical rules whereas the usage of block diagram's notation is more liberal. In an SGF, all dependent and independent variables are represented by the nodes, and the lines joining the nodes are known as branches. A branch is associated with transmission gain and an arrow. Figure 7.1 shows the pictorial representation of a system having two variables where $x_2 = Ax_1$.

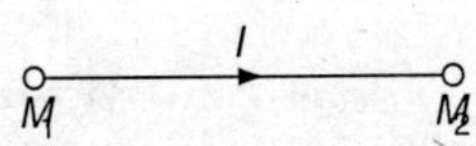

Fig. 7.1 Signal-flow graph of a system

In Fig. 7.1, A is the transmission gain and the arrow shows the flow of signal. The branches of any SFG are always unidirectional. All the branches of an SGF represent the cause-and-effect relationship existing between the various variables. The aim of this chapter is to introduce SFG in detail and its application in finding the transfer function of the complicated systems.

7.1 Basic Definitions in SFG

The terminologies used in SFG are discussed below.

***Input or source node*:** The node having only outgoing branches is called input or source node. In Fig. 7.2, x_1 is the input or source node.

***Output or sink node*:** The node having only incoming branches is called sink node or output node. In Fig. 7.2, x_9 is the sink or output node.

***Chain node*:** The node having both incoming and outgoing branches is known as chain node. In Fig. 7.2, x_2, x_3, x_4, x_5, x_6, x_7 and x_8 are all chain nodes.

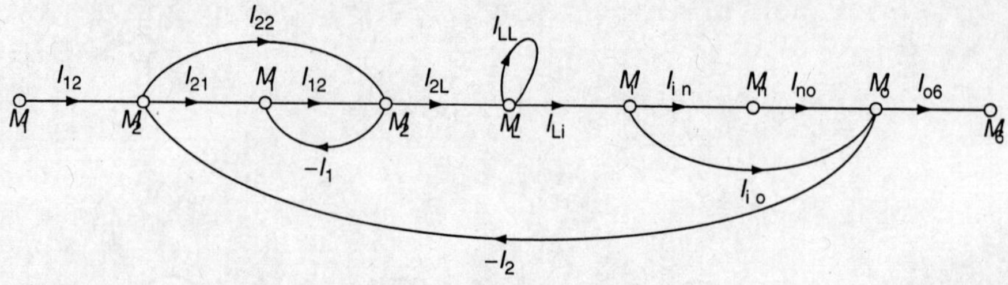

Fig. 7.2 Signal-flow graph

***Forward path*:** A path from input to output is known as forward path. In Fig. 7.2, there are four forward paths as follows:

$$x_1 - x_2 - x_3 - x_4 - x_5 - x_6 - x_7 - x_8 - x_9\text{: first forward path}$$

$$x_1 - x_2 - x_4 - x_5 - x_6 - x_7 - x_8 - x_9\text{: second forward path}$$

$$x_1 - x_2 - x_3 - x_4 - x_5 - x_6 - x_8 - x_9\text{: third forward path}$$

$$x_1 - x_2 - x_4 - x_5 - x_6 - x_8 - x_9\text{: fourth forward path}$$

In determining the forward path, any node should not be traded twice.

***Feedback loop/Feedback path*:** If a loop originates and terminates at the same node after trading through other nodes, it is known as feedback loop. In Fig. 7.2, there are three feedback loops or feedback paths as follows:

$$x_3 - x_4 - x_3\text{: first feedback loop}$$

$$x_2 - x_3 - x_4 - x_5 - x_6 - x_7 - x_8 - x_2\text{: second feedback loop}$$

$$x_2 - x_4 - x_5 - x_6 - x_7 - x_8 - x_2\text{: third loop}$$

$$x_2 - x_3 - x_4 - x_5 - x_6 - x_8 - x_2\text{: fourth loop}$$

$$x_2 - x_4 - x_5 - x_6 - x_8 - x_2\text{: fifth feedback loop.}$$

***Self-loop*:** A loop that consists of only one node is known as self-loop. In determining the forward path or feedback path, the self-loop should not be taken into account. In Fig. 7.2, there is the self-loop at x_5.

***Path gain*:** The product of gains going through a forward path is called path gain. In Fig. 7.2, the path gain for the first forward path is

$$P_1 = t_{12}t_{23}t_{34}t_{45}t_{56}t_{67}t_{78}t_{89}$$

Similarly, the path gains for the second, third and fourth paths are, respectively,

$$P_2 = t_{12}t_{24}t_{45}t_{56}t_{67}t_{78}t_{89},$$

$$P_3 = t_{12}t_{23}t_{34}t_{45}t_{56}t_{68}t_{89}\text{ and}$$

$$P_4 = t_{12}t_{24}t_{45}t_{56}t_{68}t_{89}$$

***Dummy node*:** If the incoming as well as outgoing branches exist at the first and the last node representing input and output variables, these nodes cannot be taken as input and output nodes as shown in Fig. 7.3(a). In such cases, separate input and output nodes are created by adding branches with gain 1 shown in Fig. 7.3(b). These nodes are known as dummy nodes. Figure 7.3 shows the dummy nodes.

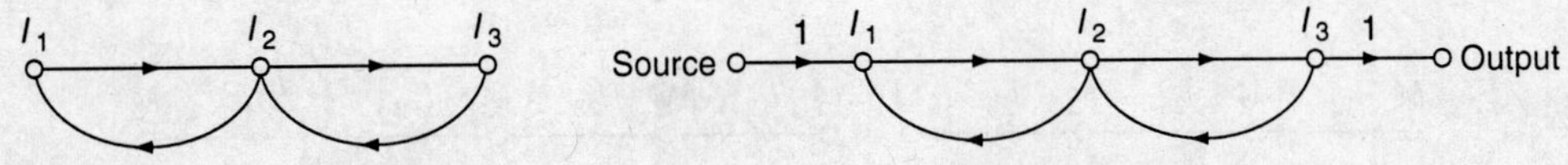

(a) Without input and output nodes

(b) Dummy nodes

Fig. 7.3 Dummy nodes

***Non-touching loops*:** If there is no common node between two or more loops, these loops are known as non-touching loops. In Fig. 7.4(a) and Fig. 7.4(b), there are two and three non-touching loops, respectively. In Fig. 7.4(c), there is no non-touching loop.

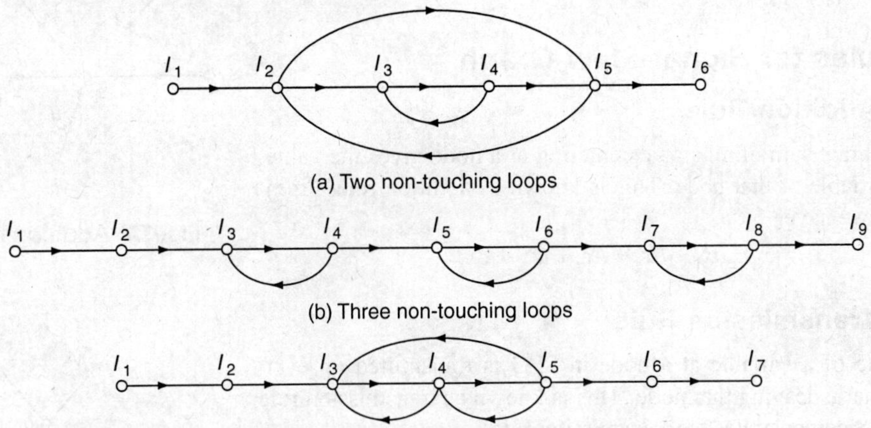

(a) Two non-touching loops

(b) Three non-touching loops

(c) No non-touching loops

Fig. 7.4 Existence of non-touching loop

Similarly, if a loop does not touch a forward path, i.e., no node is common between a forward path and a feedback loop, this loop is non-touching to that forward path. Figure 7.5 shows non-touching loops to the forward paths.

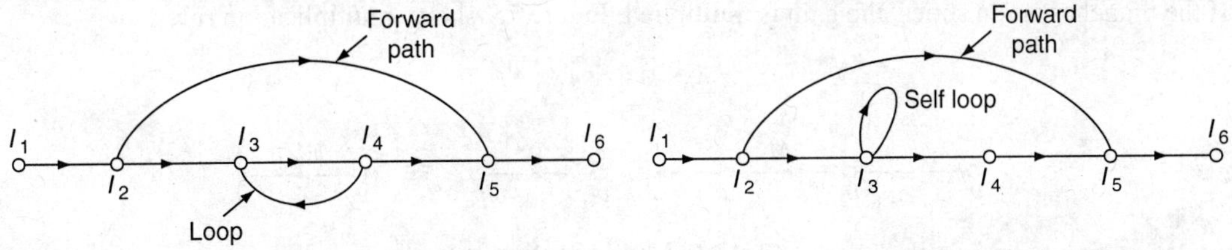

(a) Loop non-touching to forward path

(b) Self loop non-touching to the forward path

Fig. 7.5 Non-touching loop to a forward path

***Loop gain*:** It is defined as the product of all gains of the branches forming a loop. Figure 7.6 illustrates the loops.

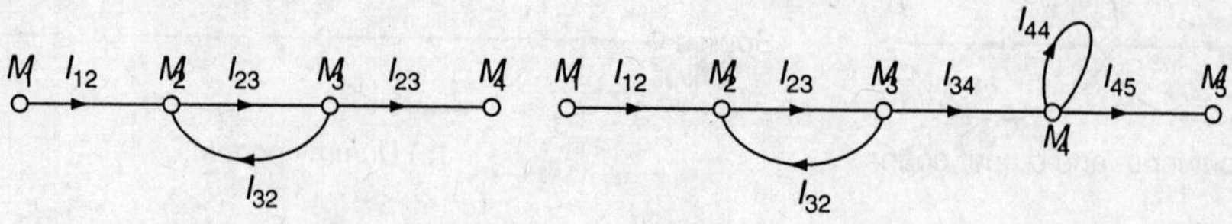

(a) One loop (b) Two loops

Fig. 7.6 Loop gain

In Fig. 7.6(a), the loop gain is $L = t_{23}t_{32}$. In Fig. 7.6(b), the loop gains are $L_1 = t_{23}t_{32}$ and $L_2 = t_{44}$ (self-loop).

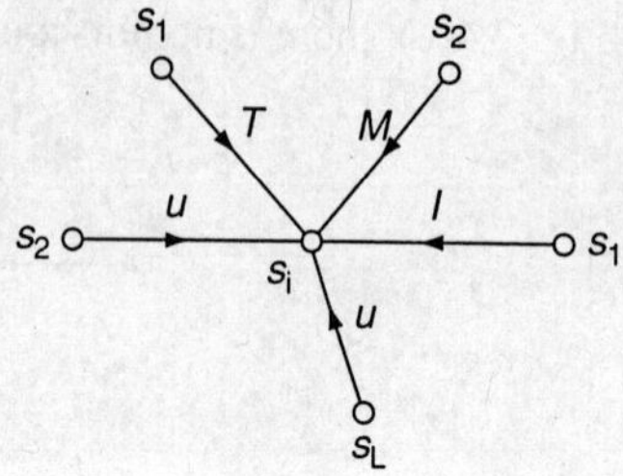

Fig. 7.7 Addition rule

7.2 Rules for Signal-Flow Graph

7.2.1 Addition Rule

The algebraic sum of all signals entering at a node gives the value of the variable of that node. This is known as addition rule. From Fig. 7.7, we have,

$$x_6 = ax_1 + bx_2 + cx_3 + dx_4 + ex_5$$

7.2.2 Transmission Rule

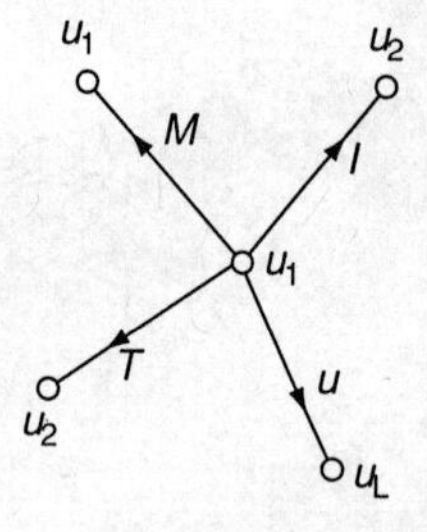

Fig. 7.8 Transmission rule

The value of a variable at a node in SFG is transmitted on every branch that is leaving that node. This is known as transmission rule. Figure 7.8 demonstrates the transmission rule.

Using transmission rule, we have from Fig. 7.8,

$$x_2 = ax_1$$
$$x_3 = bx_1$$
$$x_4 = cx_1$$
$$x_5 = dx_1$$

7.2.3 Multiplication Rule

If the branches are in series, the gain is multiplied. Figure 7.9 shows multiplication rules.

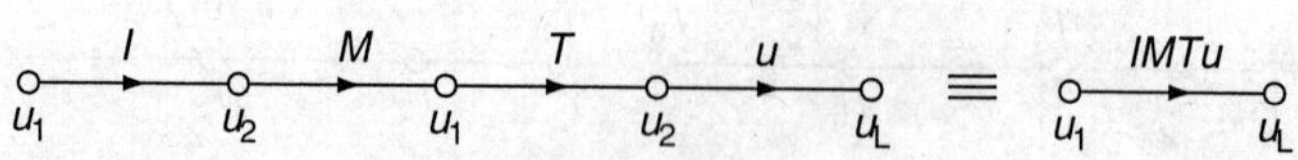

Fig. 7.9 Multiplication rule

7.3 Properties of Signal-Flow Graph

- It is applicable to linear systems.
- Arrow indicates signal.

- There is transmission of the value of a variable on each node, which leaves it.
- The algebraic sum of all signals entering a node gives the value of a variable at the node.
- Since different system equations can be written, the signal-flow graph of a system is not unique.
- Using Mason's gain formula, the overall gain of an SFG can be determined.
- An SFG can represent a block diagram, but the reverse is not true.

7.4 Mason's Gain Formula

To find the transfer function of a complicated system of block diagram, reduction technique is a cumbersome process because step-by-step reduction of block diagram is required. On the other hand, it is possible to obtain the transfer function very easily by using Mason's gain formula shown below:

$$\text{Transfer function} = \frac{C(s)}{R(s)} = \frac{\sum_{i=1}^{N} P_i \Delta_i}{\Delta} \tag{7.1}$$

where N = total number of forward paths

P_i = gain of the ith forward path

$\Delta = 1 - (\sum$ all individual feedback loop gains including self-loops)
$+ (\sum$ gain product of all possible combinations of two non-touching loops)
$- (\sum$ gain product of all possible combinations of three non-touching loops) + ...

$$\Delta = 1 - (L_{11} + L_{21} + \ldots) + (L_{12} + L_{22} + L_{32} + \ldots) - (L_{13} + L_{23} + L_{33} + L_{43} + \ldots) + \ldots.$$

Δ_i = value of Δ after eliminating all loops that touch the i th forward path.

7.5 Steps for Solving Signal-Flow Graph

Step 1: Find out all the possible forward paths, i.e., $P_1, P_2, P_3, \ldots, P_n$ and their gains. Here n forward paths are considered.

Step 2: Find out all single loops, i.e., $L_{11}, L_{21}, L_{31}, \ldots L_{m1}$ and their gains. Here m single loops are considered.

Step 3: Find out all two non-touching loops from Step 2, i.e., $L_{12}, L_{22}, L_{32}, \ldots, L_{p2}$ and their gains. Here p number two non-touching loops are considered.

Step 4: Find out all three non-touching loops from Step 2 if any, i.e., $L_{13}, L_{23}, L_{33}, \ldots, L_{q3}$ and their gains. Here q number three non-touching loops are taken. Similarly, other higher number(s) of non-touching loops if any is find out is (are) to be considered.

Step 5: The value of Δ is expressed by

$$\begin{aligned}\Delta = 1 &- (L_{11} + L_{21} + L_{31} + \ldots + L_{m1}) \\ &+ (L_{12} + L_{22} + L_{32} + \ldots + L_{p2}) \\ &- (L_{13} + L_{23} + L_{33} + \ldots + L_{q3}) + \ldots\end{aligned} \tag{7.2}$$

Step 6: For the forward path P_1, Δ_1 is found out by Eq. (7.2). The loops, which touch the forward path P_1, are excluded. Step 2 and Step 3 are followed for P_1 path to find Δ_1. Find out $P_1 \Delta_1$. This step is repeated for other forward paths, if any.

Step 7: The transfer function is found out by

$$T = \frac{P_1\Delta_1 + P_2\Delta_2 + \ldots}{\Delta}$$

Note that in two-subscript notation, the first letter stands for the serial number of loop and the second one stands for number of loops that are non-touching.

For example, $L_{11}, L_{21}, L_{31}, \ldots$ stands for 1st, 2nd, 3rd … of single loops.

$L_{12}, L_{22}, L_{32}, \ldots$ stands for 1st, 2nd, 3rd … of two non-touching loops.

$L_{13}, L_{23}, L_{33}, \ldots$ stands for 1st, 2nd, 3rd … of three non-touching loops.

Example 7.1 Find the transfer function of the system shown in Fig. E7.1 using Mason's gain formula.

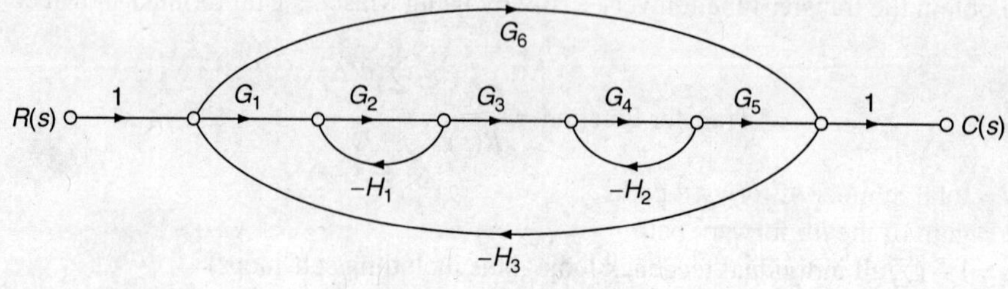

Fig. E7.1 SFG

Solution

Step 1: There are two forward paths in Fig. E7.1 as shown in Fig. E7.2.

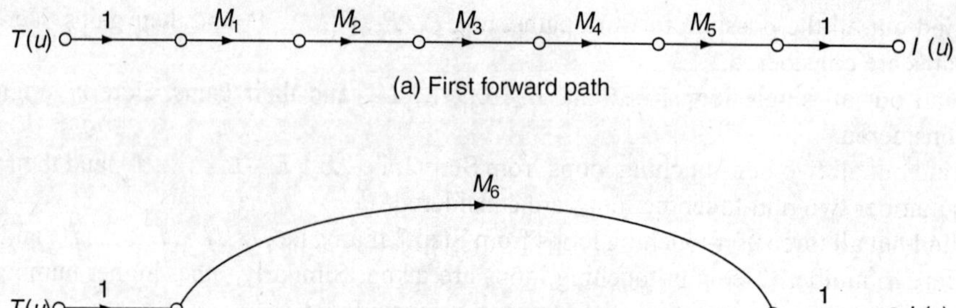

(a) First forward path

(b) Second forward path

Fig. E7.2 Forward paths in Fig. E7.1

Gain of first forward path $(P_1) = G_1G_2G_3G_4G_5$

Gain of second forward path $(P_2) = G_6$

Step 2: There are four loops as shown in Fig. E7.3.

Loop gain (L_{41}) of the loop shown in Fig. E7.3(a) $= -G_2H_1$

Loop gain (L_{32}) of the loop shown in Fig. E7.3(b) $= -G_4H_2$

Loop gain (L_{33}) of the loop shown in Fig. E7.3(c) $= -G_1G_2G_3G_4G_5H_3$

Loop gain (L_{41}) of the loop shown in Fig. E7.3(d) $= -G_6H_3$

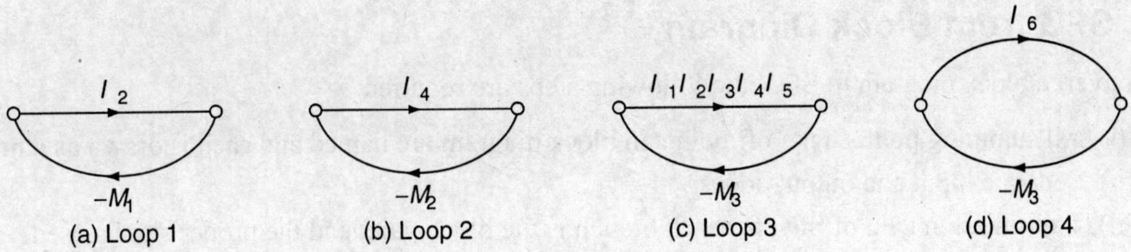

Fig. E7.3 The four loops

Step 3: Out of four loops, Loop 1, Loop 2 and Loop 4 are non-touching. The combinations of two non-touching loops are

(i) Loop 1, Loop 2: Loop gain $L_{12} = G_2G_4H_1H_2$
(ii) Loop 1, Loop 4: Loop gain $L_{22} = G_2G_6H_1H_3$
(iii) Loop 2, Loop 4: Loop gain $L_{32} = G_4G_6H_2H_3$

where L_{ij} indicates ith of j non-touching loops.

Step 4: Out of these four loops, Loop 1, Loop 2 and Loop 4 are possible combinations of three non-touching loops:

$$\therefore \qquad L_{13} = -G_2G_4G_6H_1H_2H_3$$

Step 5: There is no higher-order non-touching loops.

Step 6:

$$\begin{aligned}\Delta &= 1 - (L_{11} + L_{21} + L_{31} + L_{41}) + (L_{12} + L_{22} + L_{32}) - L_{13} \\ &= 1 + (G_2H_1 + G_4H_2 + G_1G_2G_3G_4G_5H_3 + G_6H_3) \\ &\quad + (G_2G_4H_1H_2 + G_2G_6H_1H_3 + G_4G_6H_2H_3) + G_2G_4G_6H_1H_2H_3\end{aligned}$$

Step 7:

(i) Loop 1, Loop 2, Loop 3 and Loop 4 touch P_1.

$$\therefore \qquad \Delta_1 = 1 - (0) = 1$$

(ii) Loop 1 and Loop 2 do not touch P_2.

$$\begin{aligned}\therefore \qquad \Delta_2 &= 1 - (-G_2H_1 - G_4H_2) + G_2G_4H_1H_2 \\ &= 1 + G_2H_1 + G_4H_2 + G_2G_4H_1H_2\end{aligned}$$

Step 8:

$$T = \frac{P_1\Delta_1 + P_2\Delta_2}{\Delta} = \frac{G_1G_2G_3G_4G_5 + G_6(1 + G_2H_1 + G_4H_2 + G_2G_4H_1H_2)}{\begin{aligned}1 &+ G_2H_1 + G_4H_2 + G_1G_2G_3G_4G_5H_3 + G_6H_3 \\ &+ G_2G_4H_1H_2 + G_2G_6H_1H_3 + G_4G_6H_2H_3 \\ &+ G_2G_4G_6H_1H_2H_3\end{aligned}}$$

7.6 SFG from Block Diagram

To convert a block diagram to SFG, the following steps are required:

(i) All summing points, take-off points in block diagram are named and each is drawn as a node.
(ii) Separate input and output nodes.
(iii) Transmission gain of SFG for each branch is the block gain, and the proper sign is used.
(iv) Mason's gain formula is used to obtain the transfer function of the SFG.

Example 7.2 Draw the equivalent SFG of the block diagram shown in Fig. E7.4. Find the transfer function of the system.

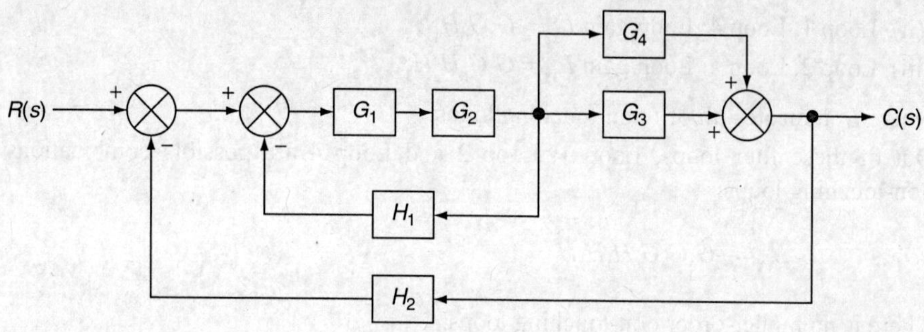

Fig. E7.4

Solution There are three summing points and two take-off points. The summing points are named as S_1, S_2 and S_3 while take-off points are named as T_1 and T_2 as shown in Fig. E7.5.

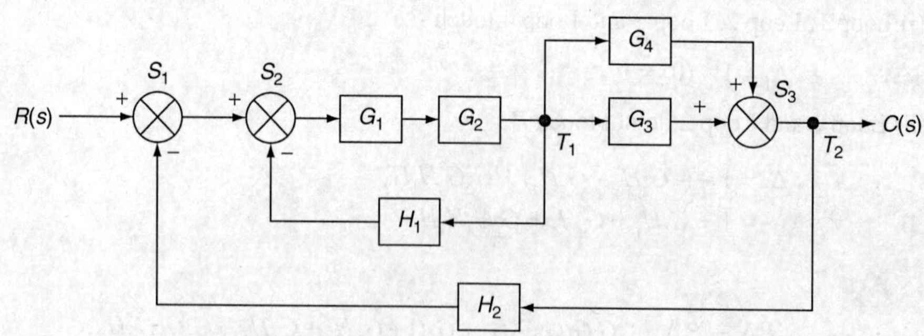

Fig. E7.5

Each summing point and each take-off point is drawn as a node and separate input and output nodes are also made as shown in Fig. E7.6.

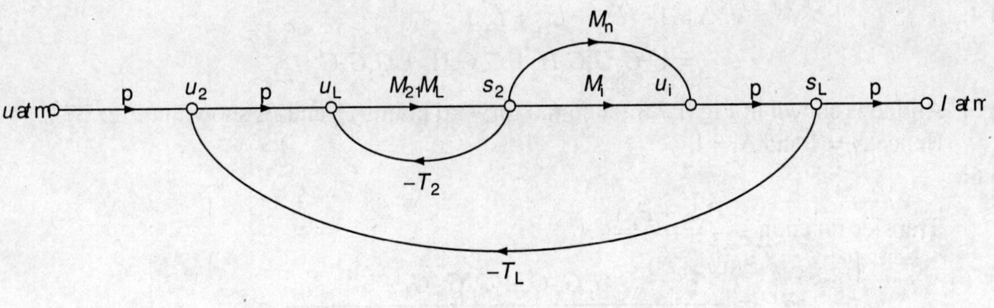

Fig. E7.6

To obtain the transfer function of the SFG shown in (Fig. E7.6) by using Mason's formula, the following steps are used:

Step 1: The forward paths of Fig. E7.6 are drawn as shown in Fig. 7.7(a) and Fig. 7.7(b).

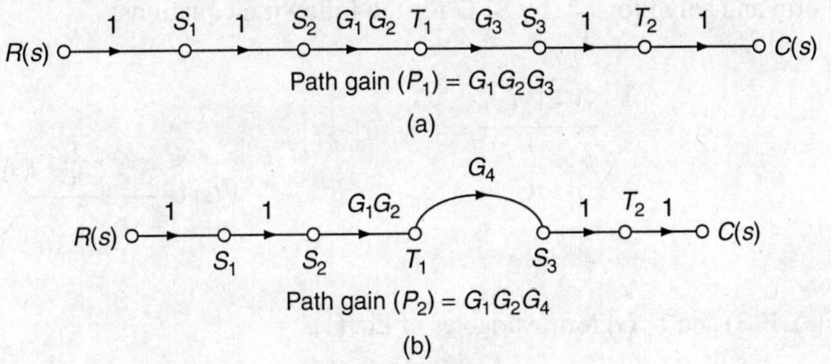

Fig. E7.7

Step 2: The individual single loops are shown in Fig. E7.8.

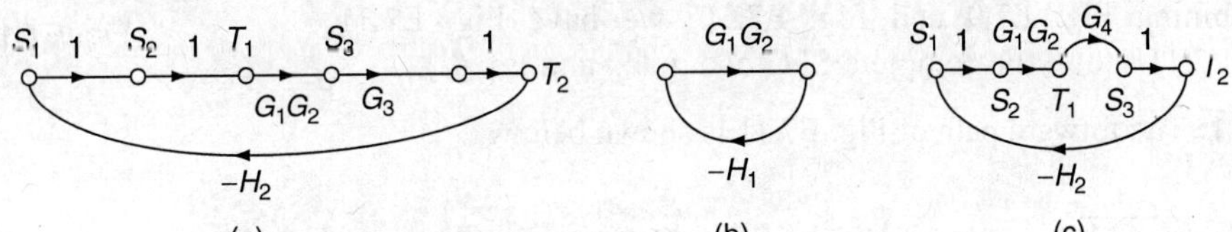

Fig. E7.8

Here $\quad L_{11} = -G_1G_2G_3H_2$

$L_{21} = -G_1G_2H_1$

and $\quad L_{31} = -G_1G_2G_4H_2$

Step 3: All single loops shown in Fig. E7.8 touch each other and hence there are no higher-order non-touching loops.

Step 4: $$\therefore \Delta = 1-(L_{11}+L_{21}+L_{31})$$
$$= 1+G_1G_2G_3H_2+G_1G_2H_1+G_1G_2G_4H_2$$

Step 5: All loops shown in Fig. E7.8 touch the forward points P_1 and P_2 shown in Fig. E7.7. Hence $\Delta_1 = 1$ and $\Delta_2 = 1$

Step 6:

$$\text{Transfer function} = \frac{P_1\Delta_1 + P_2\Delta_2}{\Delta}$$
$$= \frac{G_1G_2G_3 + G_1G_2G_4}{1+G_1G_2G_3H_2+G_1G_2H_1+G_1G_2G_4H_2}$$

7.7 Formation of SFG from Given Equation(s):

If the equations for an SFG are given, the variables form the nodes.

Example 7.3 Form and solve for $\frac{V_0}{V_i}$ by SFG for the following equations:

$$I(s) = \frac{V_i(s)-V_0(s)}{R+\frac{1}{sC}} \quad (1)$$

and
$$V_0(s) = sL\,I(s) \quad (2)$$

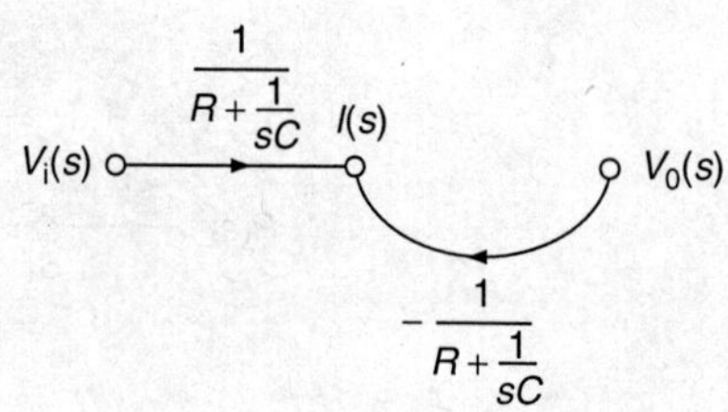

Fig. E7.9

Solution Here $I(s)$, $V_i(s)$ and $V_0(s)$ form variables of Eq. (1).

From Eq. (1), we have $$I(s) = \frac{V_i(s)}{R+\frac{1}{sC}} - \frac{V_0(s)}{R+\frac{1}{sC}} \quad (3)$$

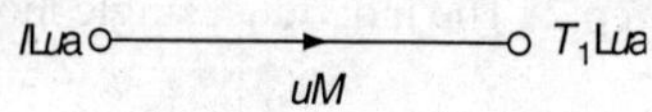

Fig. E7.10

Fig. E7.9 represents Eq. (3). From Eq. (2), we have Fig. E7.10. Combining Fig. E7.9 and Fig. E7.10, we have Fig. E7.11. Fig. E7.11 shows the complete SFG of Eq. (1) and Eq. (2).

Step 1: The forward path of Fig. E7.11 is shown below.

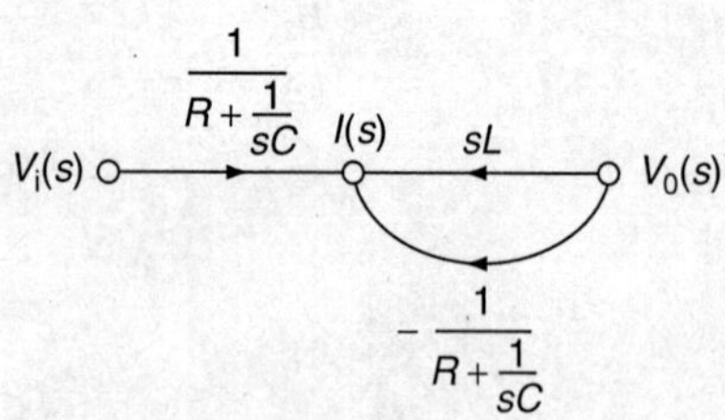

Fig. E7.11

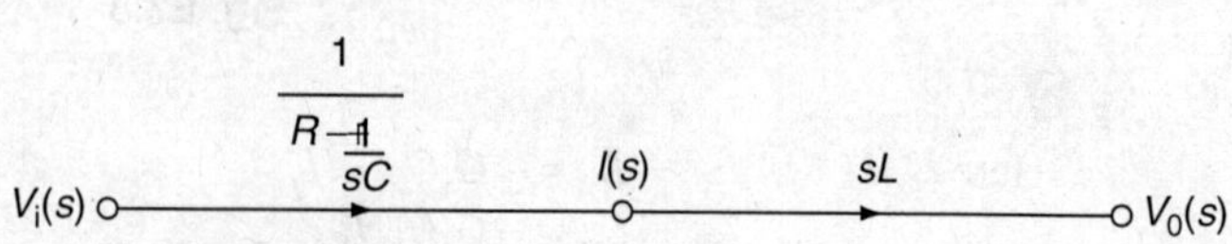

Fig. E7.12

$$\therefore \qquad P_1 = \frac{sL}{R+\frac{1}{sC}}$$

Step 2: The individual loop of Fig. E7.11 is shown below.

$$\therefore \quad L_{11} = -\frac{sL}{R+\frac{1}{sC}}$$

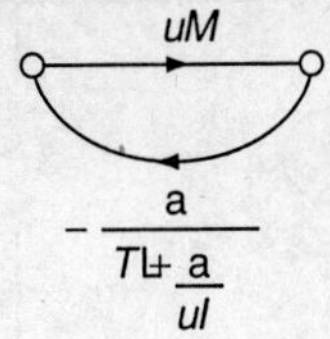

Fig. E7.13

Step 3: There are no higher-order non-touching loops.

Step 4: $\Delta = 1 - L_{11} = 1 + \frac{sL}{R+\frac{1}{sC}}$

Step 5: The loop touches the forward path.
Hence $\Delta_1 = 1$

Step 6: $TF = \frac{P_1\Delta_1}{\Delta} = \frac{\frac{sL}{R+\frac{1}{sC}}}{1+\frac{sL}{R+\frac{1}{sC}}}$

$$= \frac{sL}{R+\frac{1}{sC}+sL} = \frac{s^2LC}{s^2LC+sRC+1}$$

7.8 Signal-Flow Graph for Networks:

The steps shown below are followed to get the SFG of networks.

Step 1: Obtain Laplace domain representation of the network at first.
Step 2: Determine the equations for currents and voltages.
Step 3: From Step 2, draw the SFG of the network.
Step 4: Solve SFG using Mason's gain formula

Example 7.4 Obtain SFG of the network shown in Fig. E7.14. Find the TF using Mason's gain formula.

Solution

Step 1: The network shown in Fig. E7.14 is transferred to Laplace domain as shown in Fig. E7.15.

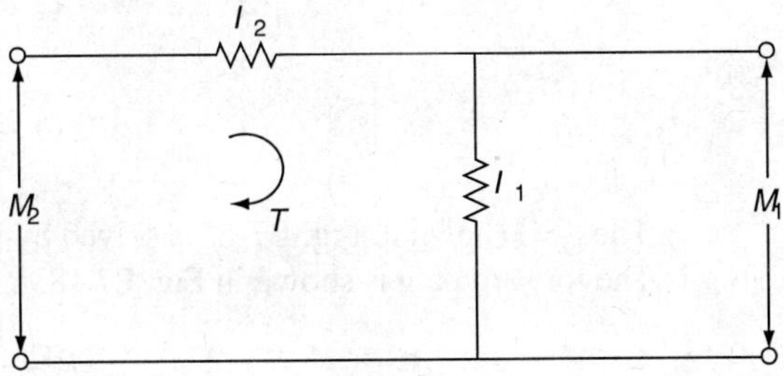

Fig. E7.14

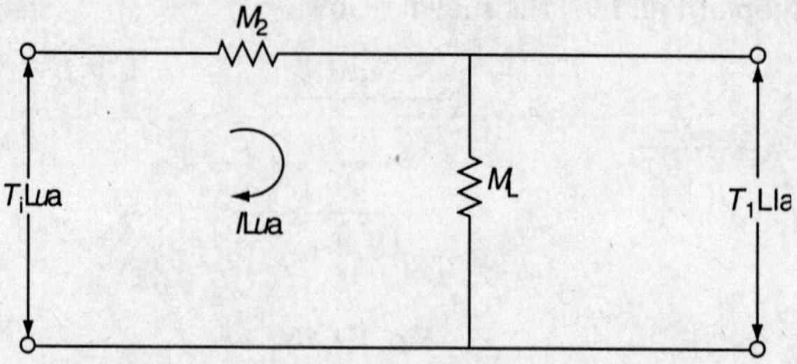

Fig. E7.15

Step 2:

$$I(s) = \frac{V_i(s) - V_0(s)}{R_1} = \frac{V_i(s)}{R_1} - \frac{V_0(s)}{R_1} \quad (1)$$

and

$$V_0(s) = R_2\, I(s) \quad (2)$$

Step 3: From Eq. (1), we have Fig. E7.16(a).

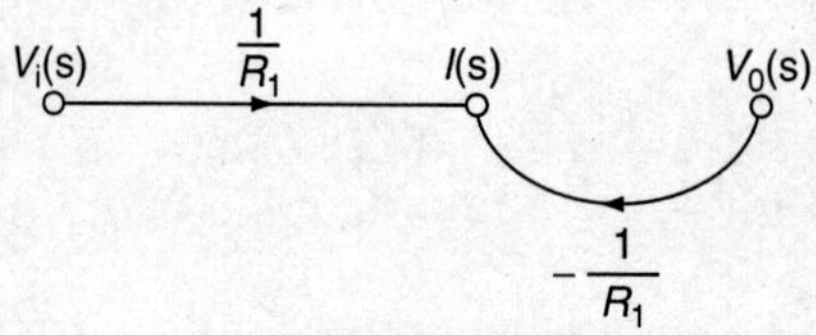

Fig. E7.16(a)

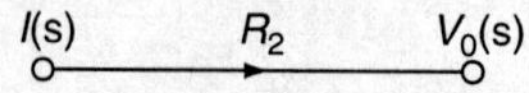

Fig. E7.16(b)

From Eq. (2) we have Fig. E7.16 (b)

Step 4: Combining Fig. E7.16(a) and Fig. E7.16(b), we have Fig. E7.17.

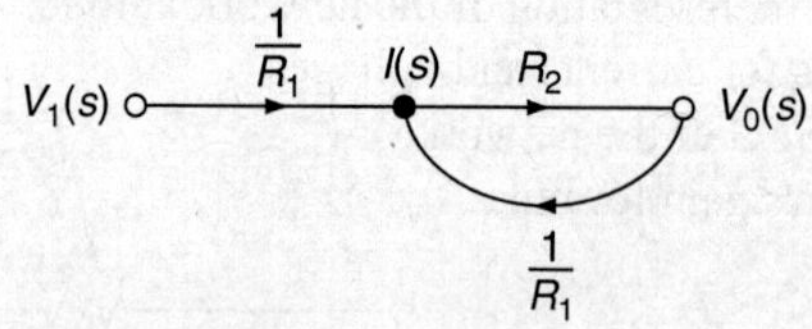

Fig. E7.17

The SFG shown in Fig. E7.17 is solved by Mason's gain formula.

Step 1: The forward path is shown in Fig. E7.18.

$$\therefore \qquad P_1 = \frac{R_2}{R_1}$$

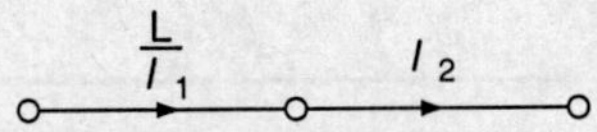

Fig. E7.18

Step 2: The individual loop is shown in Fig. E7.19.

$$\therefore \qquad L_{11} = -\frac{R_2}{R_1}$$

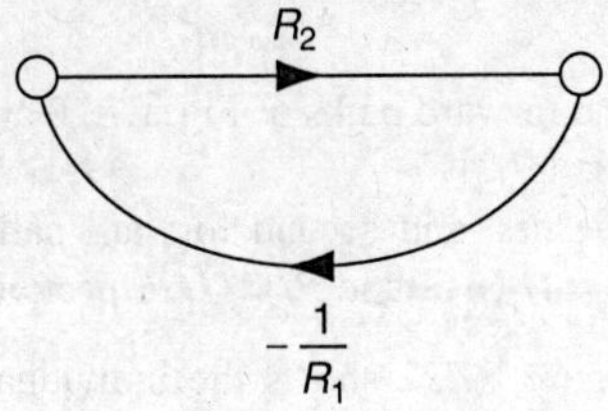

Fig. E7.19

Step 3: There are no higher-order non-touching loops.

Step 4: $\Delta = 1 - L_{11} = 1 + \frac{R_2}{R_1}$

Step 5: Since the loop touches the forward path, $\Delta_1 = 1$.

Step 6: $TF = \frac{P_1\Delta_1}{\Delta} = \frac{\frac{R_2}{R_1}}{1+\frac{R_2}{R_1}} = \frac{R_2}{R_1 + R_2}$

SIGNIFICANT POINTS

Signal-Flow Graph: Signal-flow graph is a pictorial representation of a system, which displays graphically the transmission of signals in it.

Input or Source Node: The node having only outgoing branches is called input or source node.

Sink Node: The node having only incoming branches is called sink node or output node.

Chain Node: The node having both incoming and outgoing branches is known as chain node.

Forward Path: A path from input to output is known as forward path.

Feedback Loop/Feedback Path: If a loop originates and terminates at the same node, it is known as feedback loop.

Self-Loop: A loop that consists of only one node is known as self-loop. In determining forward path or feedback path, the self-loop should not be taken into account.

Path Gain: The product of gains through a forward path is called path gain.

Dummy Node: If the incoming as well as outgoing branches exist at the first and last nodes representing input and output variables, these nodes cannot be taken as input and output nodes. In such cases, separate input and output nodes are created by adding branches with gains.

Mason's Gain Formula: To find the transfer function of a complicated system of block diagram, reduction technique is a cumbersome process because step-by-step reduction of block diagram is required. On the other hand, it is possible to obtain the transfer function very easily by using Mason's gain formula shown below:

$$\text{Transfer function} = \frac{C(s)}{R(s)} = \frac{\sum_{i=1}^{N} P_i\Delta_i}{\Delta}$$

where N = total number of forward paths

P_i = gain of the ith forward path

Δ = 1 – (Σ all individual feedback loop gains including self-loops)
+ (Σ gain product of all possible combinations of two non-touching loops)
– (Σ gain product of all possible combinations of three non-touching loops)
+ …

Δ_i = value of Δ after eliminating all loops that touch its forward path.

ADDITIONAL SOLVED EXAMPLES

Example 7.5 Find the transfer function of the system shown in Fig. E7.20 using Mason's gain formula.

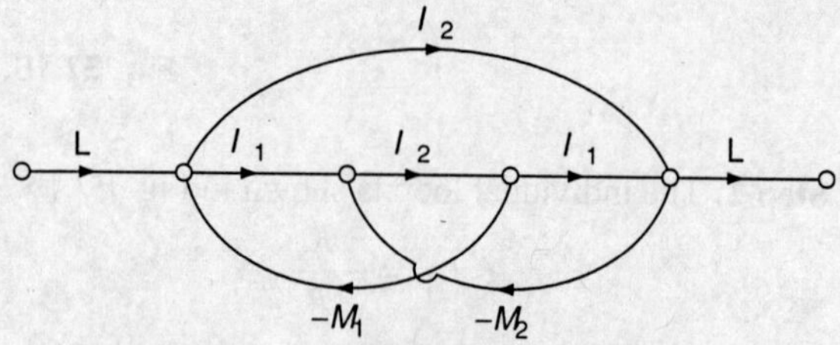

Fig. E7.20

Solution

Step 1: The forward paths of Fig. E7.20 are shown in Fig. E7.21.
The first and second forward path gains are $P_1 = G_1G_2G_3$ and $P_2 = G_4$ respectively.

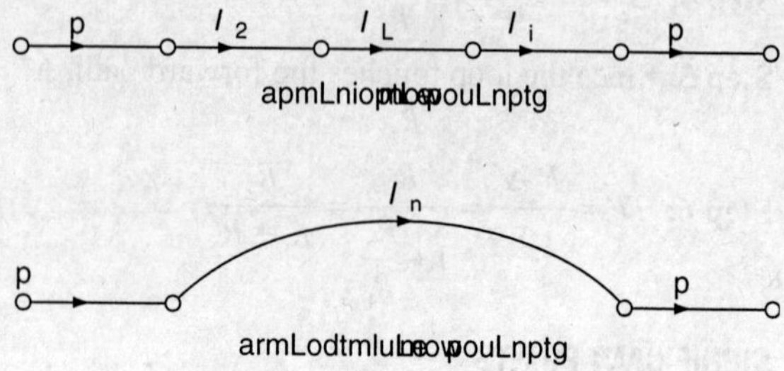

Fig. E7.21 Forward paths of Fig. E7.4

Step 2: Figure E7.22 shows the individual loops of Fig. E7.4.
The gains of Loop 1 and Loop 2 are, respectively,

$$L_{11} = -G_1G_2H_1$$

$$L_{21} = -G_2G_3H_2$$

Step 3: There are no higher-order non-touching loops.

Step 4: $\Delta = 1 - (L_{11} + L_{21}) = 1 + G_1G_2H_1 + G_2G_3H_2$

Step 5: Loop 1 and Loop 2 touch the first forward path and hence $\Delta_1 = 1$
Loop 1 and Loop 2 also touch the second forward path and hence $\Delta_2 = 1$.

Step 6:

$$\text{Transfer function} = \frac{P_1\Delta_1 + P_2\Delta_2}{\Delta} = \frac{G_1G_2G_3 + G_4}{1 + G_1G_2H_1 + G_2G_3H_2}$$

Example 7.6 The SFG of a system is shown in Fig. E7.23. Find the transfer function of the system.

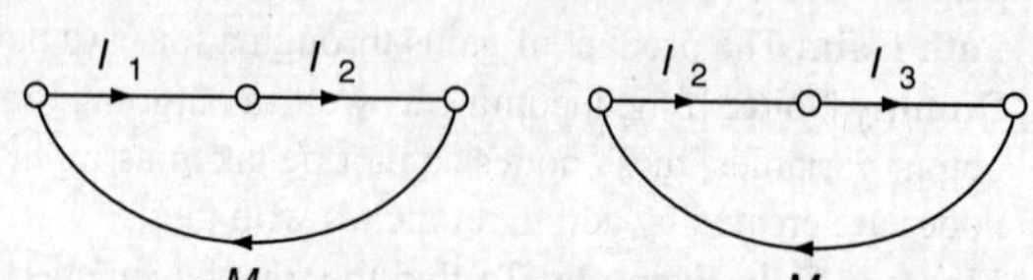

(a) Loop 1 (b) Loop 2

Fig. E7.22

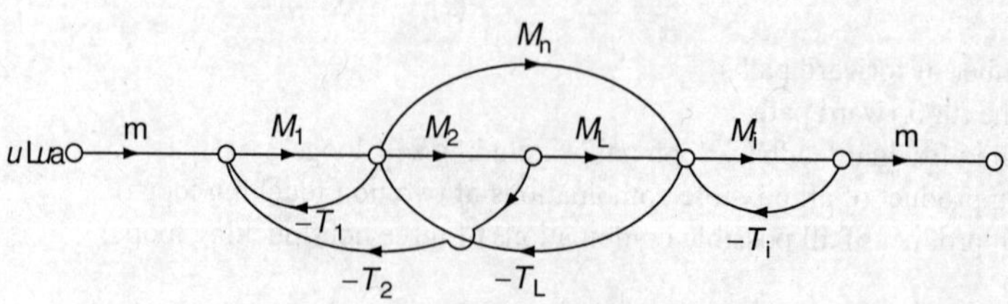

Fig. E7.23

Solution

Step 1: The forward paths of Fig. E7.23 are shown in Fig. E7.24.

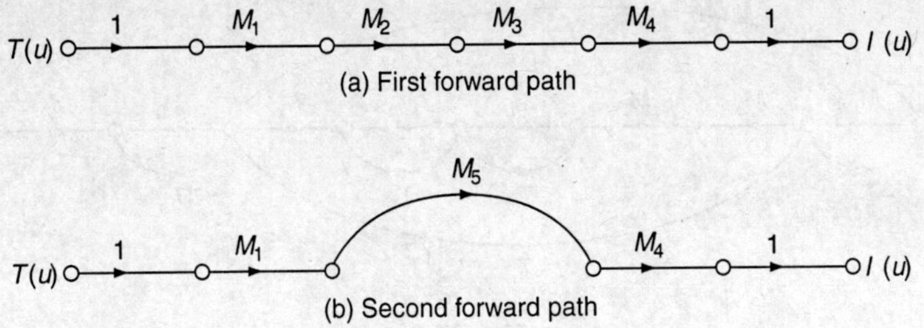

Fig. E7.24 Forward paths of Fig. E7.23

The gain of the first forward path and that of second forward path are $P_1 = G_1G_2G_3G_4$ and $P_2 = G_1G_4G_5$ respectively.

Step 2: The individual loops of Fig. E7.23 are shown in Fig. E7.25.

$\therefore \quad L_{11} = -G_1H_1, L_{21} = -G_1G_2H_2, L_{31} = -G_2G_3H_3, L_{41} = -G_4H_4$ and $L_{51} = -G_5H_3$

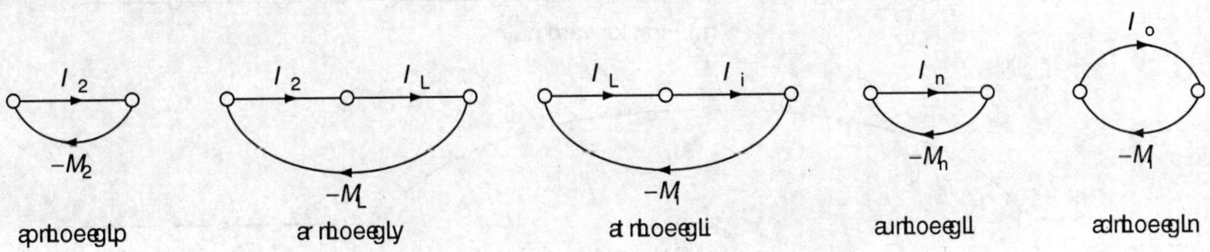

Fig. E7.25 Individual loops of Fig. E7.23

Step 3: Loop 1 and Loop 4, Loop 2 and Loop 4 are non-touching.

$\therefore \quad L_{12} =$ Loop 1 × Loop 4 $= G_1G_4H_1H_4$ and $L_{22} =$ Loop 2 × loop 4 $= G_1G_2G_4H_2H_4$

Step 4: There are no higher-order non-touching loops.

Step 5: $\Delta = 1 - (L_{11} + L_{21} + L_{31} + L_{41} + L_{51}) + (L_{12} + L_{22})$

$= 1 + (G_1H_1 + G_1G_2H_2 + G_2G_3H_3 + G_4H_4 + G_5H_3) + G_1G_4H_1H_4 + G_1G_2G_4H_2H_4$

Step 6: Since all the loops touch the first and second forward path, $\Delta_1 = 1$ and $\Delta_2 = 1$.

Step 7:

$$\text{Transfer function} = \frac{C(s)}{R(s)} = \frac{P_1\Delta_1 + P_2\Delta_2}{\Delta}$$

$$= \frac{G_1G_2G_3G_4 + G_1G_4G_5}{1 + G_1H_1 + G_1G_2H_2 + G_2G_3H_3 + G_4H_4 + G_5H_3 + G_1G_4H_1H_4 + G_1G_2G_4H_2H_4}$$

Example 7.7 Find the transfer function of the system shown in Fig. E7.26 using Mason's gain formula.

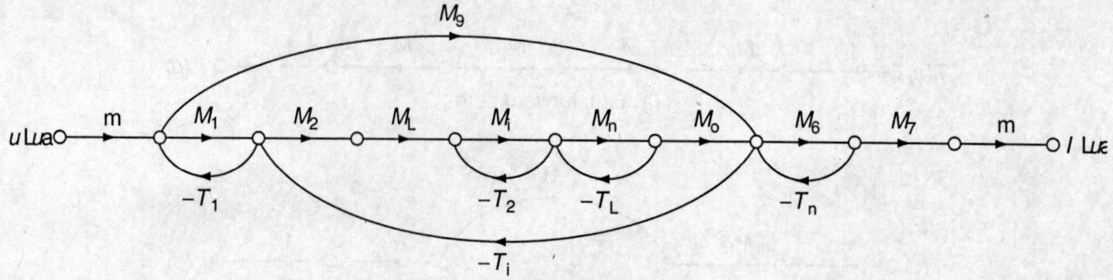

Fig. E7.26

Solution

Step 1: The SFG of Fig. E7.26 has the following forward paths as shown in Fig. E7.27.

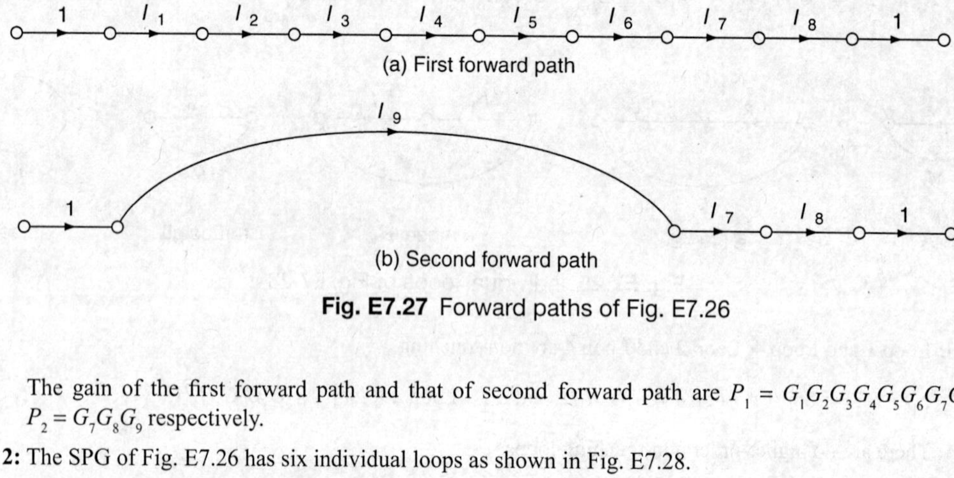

(a) First forward path

(b) Second forward path

Fig. E7.27 Forward paths of Fig. E7.26

The gain of the first forward path and that of second forward path are $P_1 = G_1G_2G_3G_4G_5G_6G_7G_8$ and $P_2 = G_7G_8G_9$ respectively.

Step 2: The SPG of Fig. E7.26 has six individual loops as shown in Fig. E7.28.

$$\therefore L_{11} = -G_1H_1, \quad L_{21} = -G_2G_3G_4G_5G_6H_4, \quad L_{31} = -G_4H_2$$

$$L_{41} = -G_5H_3, \quad L_{51} = -G_7H_5, \quad \text{and} \quad L_{61} = G_9H_1H_4$$

Step 3: Gains of two non-touching loops are shown below:

L_{12} = Gain of Loop 1 × Gain of Loop 3 = $G_1G_4H_1H_2$

L_{22} = Gain of Loop 1 × Gain of Loop 4 = $G_1G_5H_1H_3$

L_{32} = Gain of Loop 1 × Gain of Loop 5 = $G_1G_7H_1H_5$

L_{42} = Gain of Loop 3 × Gain of Loop 5 = $G_4G_7H_2H_5$

L_{52} = Gain of Loop 4 × Gain of Loop 5 = $G_5G_7H_3H_5$

L_{62} = Gain of Loop 3 × Gain of Loop 6 = $-G_4G_9H_1H_2H_4$

L_{72} = Gain of Loop 4 × Gain of Loop 6 = $-G_5G_9H_1H_3H_4$

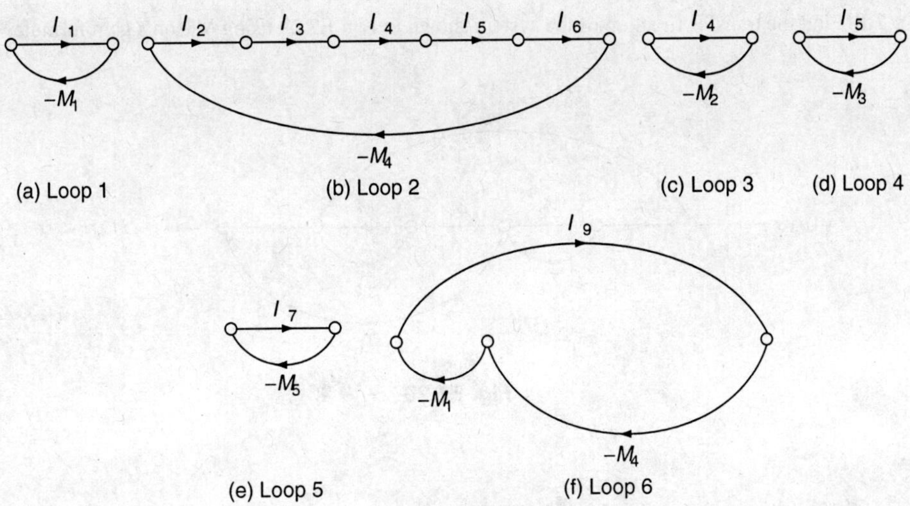

Fig. E7.28

Step 4: Loop 1, Loop 3, Loop 5 as well as Loop 1, Loop 4, Loop 5 are three non-touching loops. The gains of these combinations are

$$L_{13} = -G_1G_4G_7H_1H_2H_5$$
$$L_{23} = -G_1G_5G_7H_1H_3H_5$$

Step 5: There are no higher-order non-touching loops.

Step 6:

$$\begin{aligned}\Delta &= 1-(L_{11}+L_{21}+L_{31}+L_{41}+L_{51}+L_{61})+(L_{12}+L_{22}+L_{32}+L_{42}+L_{52}+L_{62}+L_{72})-(L_{13}+L_{23})\\ &= 1+G_1H_1+G_2G_3G_4G_5G_6H_4+G_4H_2+G_5H_3+G_7H_5-G_9H_1H_4\\ &\quad +G_1G_4H_1H_2+G_1G_5H_1H_3+G_1G_7H_1H_5+G_4G_7H_2H_5+G_5G_7H_3H_5\\ &\quad -G_4G_9H_1H_2H_4-G_5G_9H_1H_3H_4+G_1G_4G_7H_1H_2H_5+G_1G_5G_7H_1H_3H_5\end{aligned}$$

Step 7: All individual loops touch the first forward path and hence $\Delta_1 = 1$. Loop 3 and loop 4 do not touch the second forward path. These two loops touch each other.

$$\therefore \quad \Delta_2 = 1 + G_4H_2 + G_5H_3.$$

Step 8: Transfer function $= \dfrac{P_1\Delta_1 + P_2\Delta_2}{\Delta}$

$$= \frac{G_1G_2G_3G_4G_5G_6G_7G_8 + G_7G_8G_9(1+G_4H_2+G_5H_3)}{\begin{aligned}&1+G_1H_1+G_2G_3G_4G_5G_6H_4+G_4H_2+G_5H_3+G_7H_5-G_9H_1H_4\\ &+G_1G_4H_1H_2+G_1G_5H_1H_3+G_1G_7H_1H_5+G_4G_7H_2H_5\\ &+G_5G_7H_3H_5-G_4G_9H_1H_2H_4-G_5G_9H_1H_3H_4+G_1G_4G_7H_1H_2H_5\\ &+G_1G_5G_7H_1H_3H_5\end{aligned}}$$

Example 7.8 Find the transfer function of the system shown in Fig. E7.29 using Mason's gain formula.

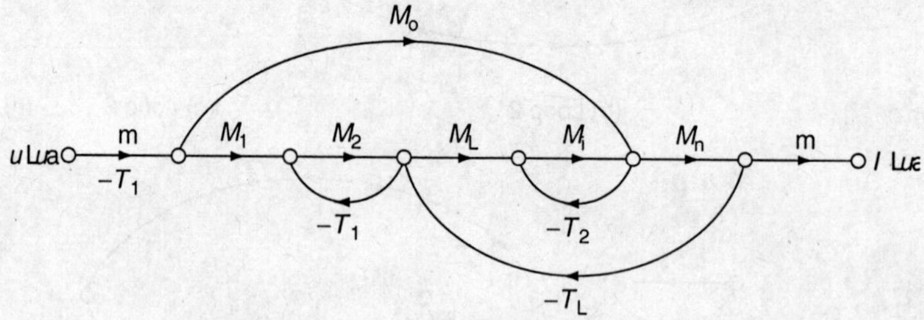

Fig. E7.29

Solution

Step 1: Fig. E7.30 shows the forward paths of Fig. E7.29.

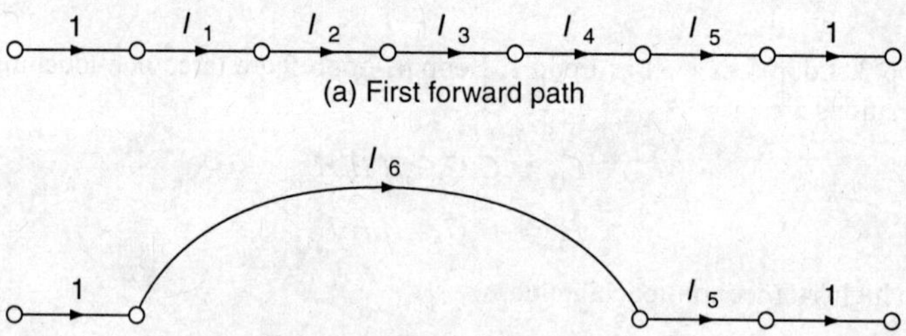

(a) First forward path

(b) Second forward path

Fig. E7.30 Forward paths of Fig. E7.29

$\therefore \quad P_1 = G_1G_2G_3G_4G_5 \quad$ and $\quad P_2 = G_5G_6$

Step 2: The individual loops of Fig. E7.29 are shown in Fig. E7.31.

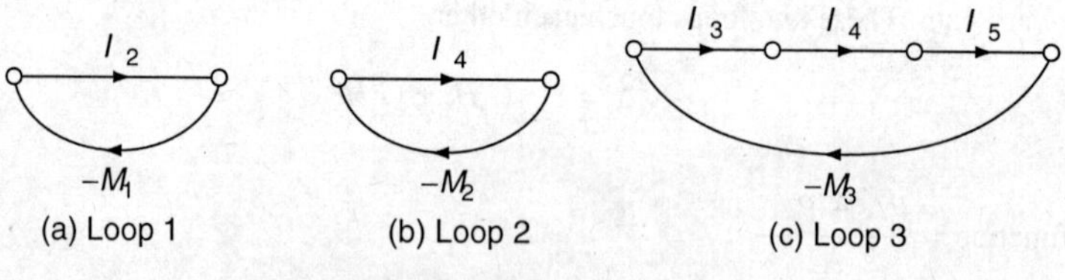

(a) Loop 1 (b) Loop 2 (c) Loop 3

Fig. E7.31 Loops of Fig. E7.29

$$\therefore \qquad L_{11} = -G_2H_1, L_{21} = -G_4H_2, L_{31} = -G_3G_4G_5H_3$$

Step 3: Loop 1 and Loop 2 are non-touching. Therefore,

$$L_{12} = G_2G_4H_1H_2$$

Step 4: There are no higher-order non-touching loops.
Step 5: $\Delta = 1 - (L_{11} + L_{21} + L_{31}) + L_{12} = 1 + (G_2H_1 + G_4H_2 + G_3G_4G_5H_3) + G_2G_4H_1H_2$
Step 6: Since all the loops touch the first forward path, $\Delta_1 = 1$. The first loop does not touch the second forward path; therefore, $\Delta_2 = 1 - (-G_2H_1) = 1 + G_2H_1$
Step 7:

$$\text{Transfer function} = \frac{G(s)}{R(s)} = \frac{P_1\Delta_1 + P_2\Delta_2}{\Delta} = \frac{G_1G_2G_3G_4G_5 + G_5G_6(1 + G_2H_1)}{1 + G_2H_1 + G_4H_2 + G_3G_4G_5H_3 + G_2G_4H_1H_2}$$

Example 7.9 Find the transfer function of the system shown in Fig. E7.32 using Mason's gain formula.

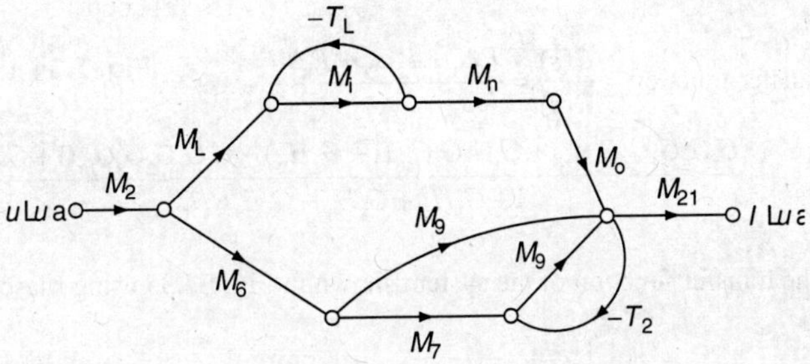

Fig. E7.32

Solution

Step 1: The forward paths of Fig. E7.32 are shown in Fig. E7.33.
The gain of the first forward path, second forward path and third forward path are, respectively,

$$P_1 = G_1G_2G_3G_4G_5G_{10}$$
$$P_2 = G_1G_6G_9G_{10}$$
$$P_3 = G_1G_6G_7G_8G_{10}$$

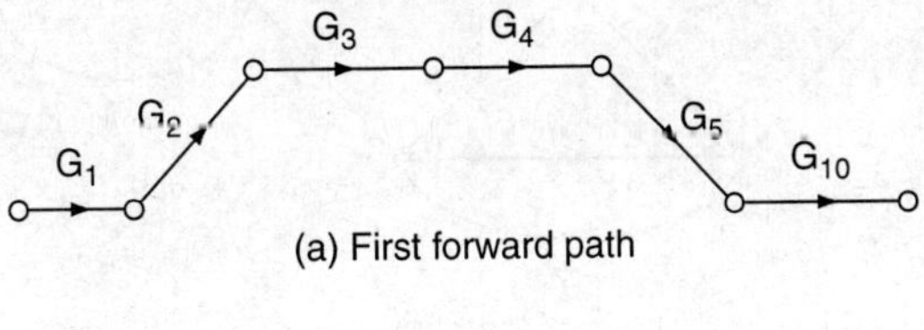

(a) First forward path

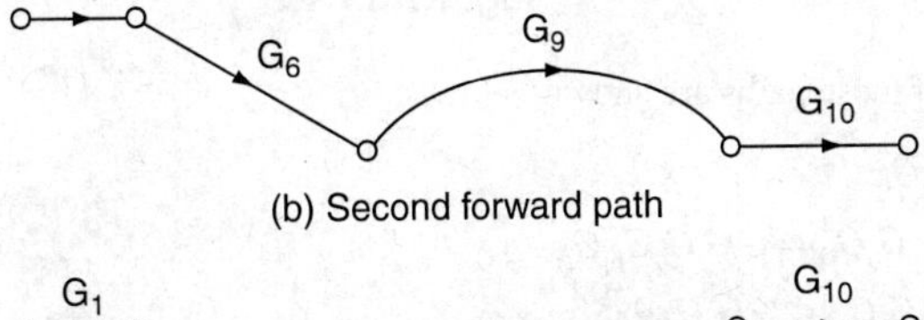

(b) Second forward path

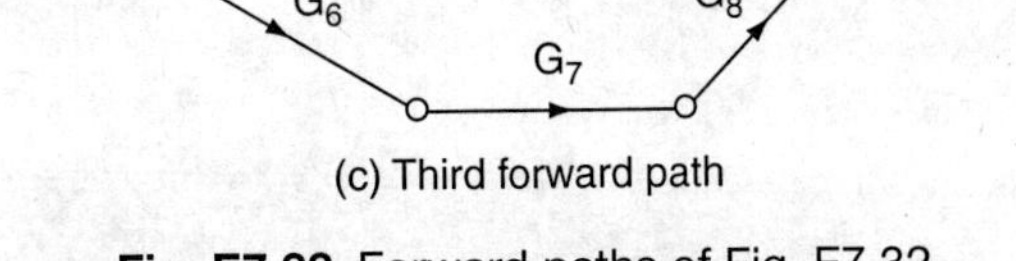

(c) Third forward path

Fig. E7.33 Forward paths of Fig. E7.32

Step 2: The individual loops of Fig. E7.32 are shown in Fig. E7.34.

Here, $L_{11} = -G_3H_2$ and $L_{21} = -G_8H_1$

Step 3: Loop 1 and Loop 2 are non-touching. Therefore,

$L_{12} = G_3G_8H_1H_2$

Step 4: There are no higher-order non-touching loops.

Step 5: $\Delta = 1 - (-G_3H_2 - G_8H_1) + G_3G_8H_1H_2 = 1 + G_3H_2 + G_8H_1 + G_3G_8H_1H_2$

Step 6: For first forward path, $\Delta_1 = 1$

For second forward path, $\Delta_2 = 1 - (-G_3H_2) = 1 + G_3H_2$

For third forward path, $\Delta_3 = 1 - (-G_3H_2) = 1 + G_3H_2$

(a) Loop 1 (b) Loop 2

Fig. 7.34 Loops of Fig. E7.32

Step 7:

$$\text{Transfer function} = \frac{C(s)}{R(s)} = \frac{P_1\Delta_1 + P_2\Delta_2 + P_3\Delta_3}{\Delta}$$

$$\frac{G_1G_2G_3G_4G_5G_{10} + G_1G_6G_9G_{10}(1+G_3H_2) + G_1G_6G_7G_8G_{10}(1+G_3H_2)}{1+G_3H_2+G_8H_1+G_3G_8H_1H_2}$$

Example 7.10 Find the transfer function of the system shown in Fig. E7.35 using Mason's gain formula.

Solution

Step 1: The forward paths of Fig. E7.35 are shown in Fig. E7.36.

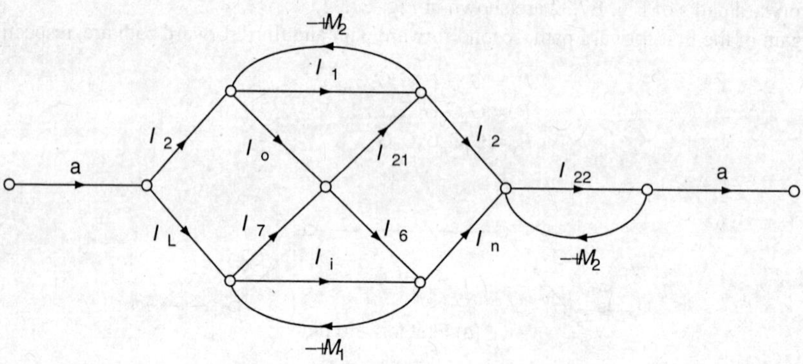

Fig. E7.35

The gains of each of these paths are shown below.

$P_1 = G_1G_2G_3G_{11}$

$P_2 = G_1G_7G_{10}G_3G_{11} = G_1G_3G_7G_{10}G_{11}$

$P_3 = G_1G_7G_8G_6G_{11} = G_1G_6G_7G_8G_{11}$

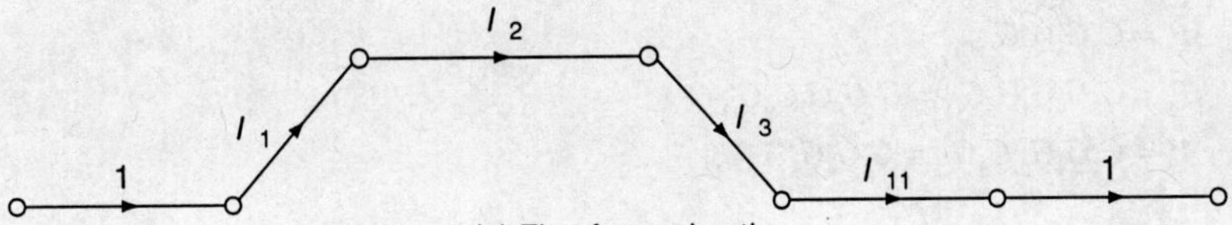

(a) First forward path

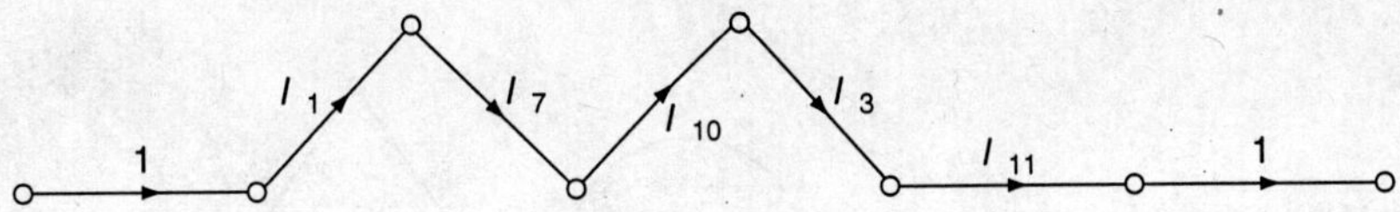

(b) Second forward path

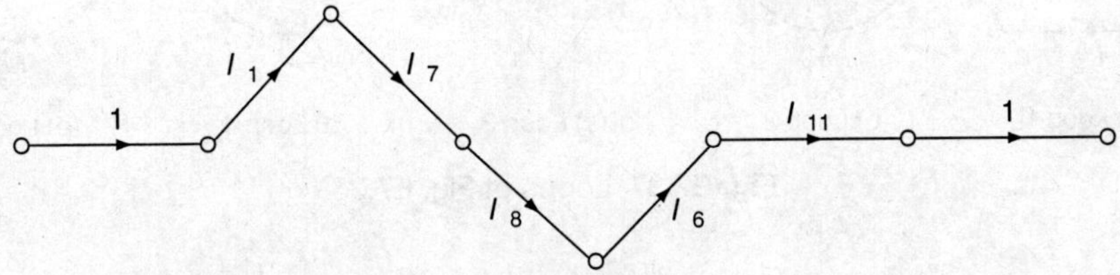

(c) Third forward path

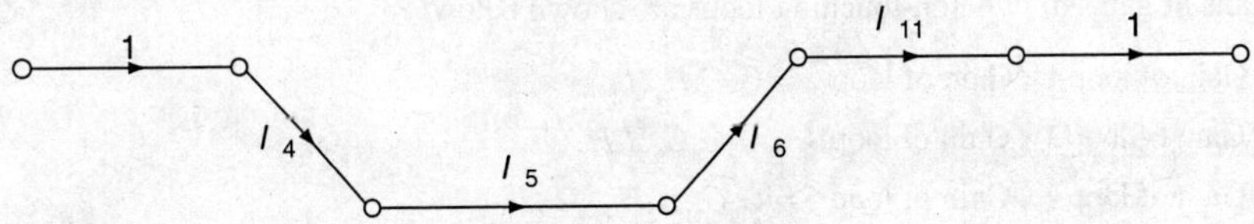

(d) Fourth forward path

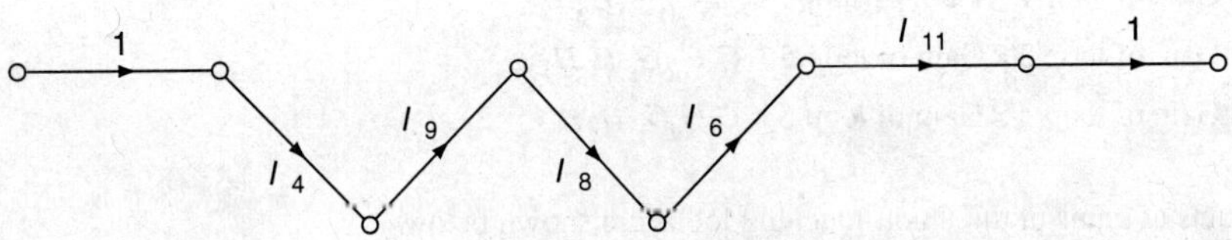

(e) Fifth forward path

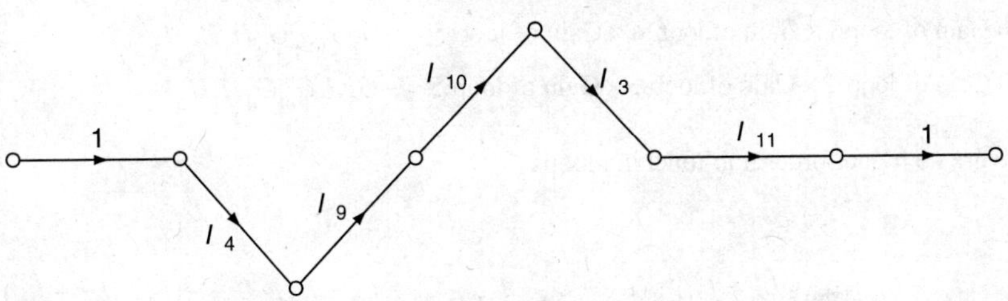

(f) Sixth forward path

Fig. E7.36 Forward paths of Fig. E7.35

$$P_4 = G_4G_5G_6G_{11}$$
$$P_5 = G_4G_9G_8G_6G_{11} = G_4G_6G_8G_9G_{11}$$
$$P_6 = G_4G_9G_{10}G_3G_{11} = G_3G_4G_9G_{10}G_{11}$$

Step 2: The individual loops of Fig. E7.35 are shown in Fig. E7.37.

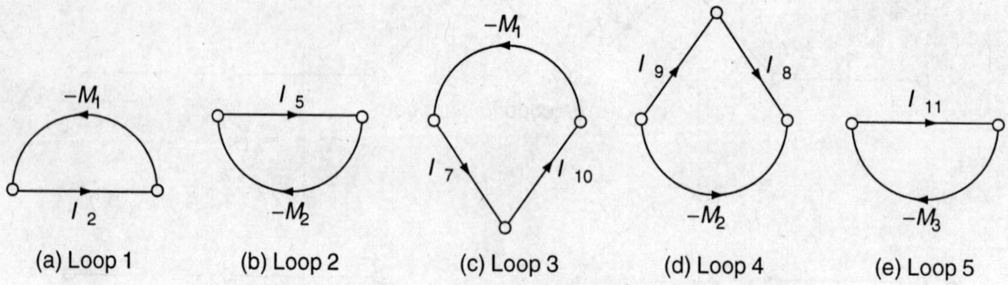

Fig. E7.37 Loops of Fig. E7.35

$$\therefore L_{11} = -G_2H_1,\ L_{21} = -G_5H_2,\ L_{31} = -G_7G_{10}H_1, L_{41} = -G_8G_9H_2 \text{ and } L_{51} = -G_{11}H_3$$

Step 3: Products of gains of two non-touching loops are shown below.

$$L_{12} = \text{Gain of loop } 1 \times \text{Gain of loop } 2 = G_2G_5H_1H_2$$
$$L_{22} = \text{Gain of loop } 1 \times \text{Gain of loop } 4 = G_2G_8G_9H_1H_2$$
$$L_{32} = \text{Gain of loop } 1 \times \text{Gain of loop } 5 = G_2G_{11}H_1H_3$$
$$L_{42} = \text{Gain of loop } 2 \times \text{Gain of loop } 3 = G_5G_7G_{10}H_1H_2$$
$$L_{52} = \text{Gain of loop } 2 \times \text{Gain of loop } 5 = G_5G_{11}H_2H_3$$
$$L_{62} = \text{Gain of loop } 3 \times \text{Gain of loop } 5 = G_7G_{10}G_{11}H_1H_3$$
$$L_{72} = \text{Gain of loop } 4 \times \text{Gain of loop } 5 = G_8G_9G_{11}H_2H_3$$

Step 4: Products of gains of three non-touching loops are shown below.

$$L_{13} = \text{Gain of loop } 1 \times \text{Gain of loop } 2 \times \text{Gain of loop } 5 = -G_2G_5G_{11}H_1H_2H_3$$
$$L_{23} = \text{Gain of loop } 1 \times \text{Gain of loop } 4 \times \text{Gain of loop } 5 = -G_2G_8G_9G_{11}H_1H_2H_3$$
$$L_{33} = \text{Gain of loop } 2 \times \text{Gain of loop } 3 \times \text{Gain of loop } 5 = -G_5G_7G_{10}G_{11}H_1H_2H_3$$

Step 5: There are no higher-order non-touching loops
Step 6:

$$\begin{aligned}\Delta &= 1-(L_{11}+L_{21}+L_{31}+L_{41}+L_{51})+(L_{12}+L_{22}+L_{32}+L_{42}+L_{52}+L_{62}+L_{72})-(L_{13}+L_{23}+L_{33})\\ &= 1+(G_2H_1+G_5H_2+G_7G_{10}H_1+G_8G_9H_2+G_{11}H_3)\\ &\quad +(G_2G_5H_1H_2+G_2G_8G_9H_1H_2+G_2G_{11}H_1H_3+G_5G_{11}H_2H_3+G_5G_7G_{10}H_1H_2+G_7G_{10}G_{11}H_1H_3\\ &\quad +G_8G_9G_{11}H_2H_3)+(G_2G_5G_{11}H_1H_2H_3+G_2G_8G_9G_{11}H_1H_2H_3+G_5G_7G_{10}G_{11}H_1H_2H_3)\end{aligned}$$

Step 7:

For first forward path $\Delta_1 = 1-(-G_5H_2 - G_8G_9H_2) = 1+(G_5H_2 + G_8G_9H_2)$

For second forward path $\Delta_2 = 1-(-G_5H_2) = 1+G_5H_2$

For third forward path $\Delta_3 = 1$

For fourth forward path $\Delta_4 = 1-(-G_2H_1 - G_7G_{10}H_1) = 1+G_2H_1 + G_7G_{10}H_1$

For fifth forward path $\Delta_5 = 1-(-G_2H_1) = 1+G_2H_1$

For sixth forward path $\Delta_6 = 1$

Step 8:

$$\text{Transfer function } = \frac{P_1\Delta_1 + P_2\Delta_2 + P_3\Delta_3 + P_4\Delta_4 + P_5\Delta_5 + P_6\Delta_6}{\Delta}$$

$$= \frac{\begin{gathered}G_1G_2G_3G_{11}(1+G_5H_2+G_8G_9H_2)+G_1G_3G_7G_{10}G_{11}(1+G_5H_2)\\ +G_1G_6G_7G_8G_{11}+G_4G_5G_6G_{11}(1+G_2H_1+G_7G_{10}H_1)\\ +G_4G_6G_8G_9G_{11}(1+G_2H_1)+G_3G_4G_9G_{10}G_{11}\end{gathered}}{\begin{gathered}1+G_2H_1+G_5H_2+G_7G_{10}H_1+G_8G_9H_2+G_{11}H_3+G_2G_5H_1H_2+G_2G_8G_9H_1H_2\\ +G_2G_{11}H_1H_3+G_5G_7G_{10}H_1H_2+G_5G_{11}H_2H_3+G_7G_{10}G_{11}H_1H_3\\ +G_8G_9G_{11}H_2H_3+G_2G_5G_{11}H_1H_2H_3+G_2G_8G_9G_{11}H_1H_2H_3+G_5G_7G_{10}G_{11}H_1H_2H_3\end{gathered}}$$

Example 7.11 Find the transfer function of the system shown in Fig. E7.38 using Mason's gain formula.

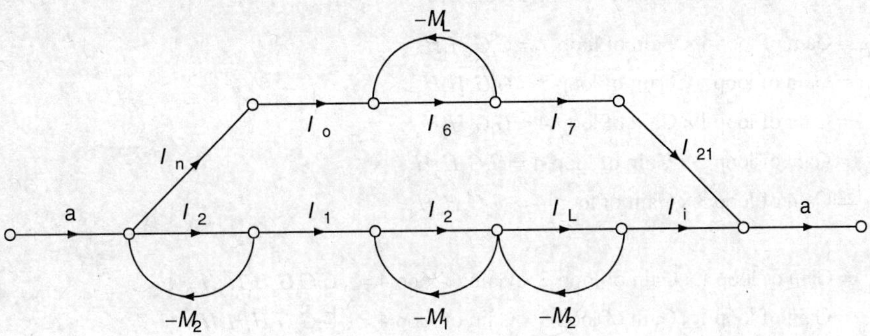

Fig. E7.38

Solution

Step 1: The forward paths of the given SFG shown in Fig. E7.38 are shown in Fig. E7.39.
The gain of the first forward path and that of the second forward path are $P_1 = G_1G_2G_3G_4G_5$ and $P_2 = G_6G_7G_8G_9G_{10}$ respectively.

Step 2: The individual loops of Fig. E7.38 are shown in Fig. E7.40.
The gains of each individual loop are

$L_{11} = -G_1H_1, L_{21} = -G_3H_2, L_{31} = -G_4H_3$ and $L_{41} = -G_8H_4$

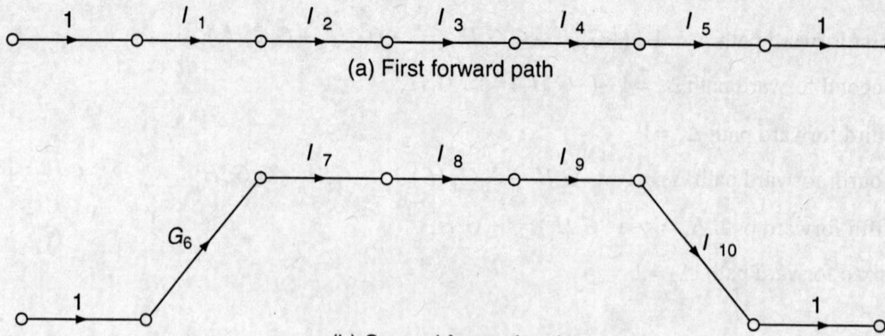

Fig. E7.39 Forward paths of Fig. E7.38

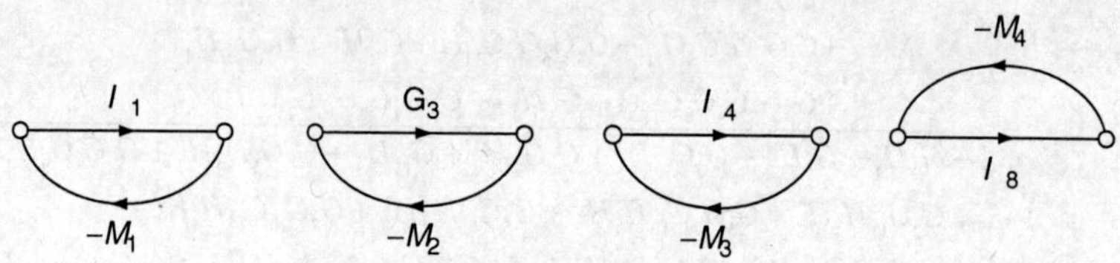

Fig. E7.40 Loops of Fig. E7.38

Step 3: Products of gains of two non-touching loops are

$$L_{12} = \text{Gain of loop } 1 \times \text{Gain of loop } 2 = G_1G_3H_1H_2$$
$$L_{22} = \text{Gain of loop } 1 \times \text{Gain of loop } 3 = G_1G_4H_1H_3$$
$$L_{32} = \text{Gain of loop } 1 \times \text{Gain of loop } 4 = G_1G_8H_1H_4$$
$$L_{42} = \text{Gain of loop } 2 \times \text{Gain of loop } 4 = G_3G_8H_2H_4$$
$$L_{52} = \text{Gain of loop } 3 \times \text{Gain of loop } 4 = G_4G_8H_3H_4$$

Step 4:

$$L_{13} = \text{Gain of loop } 1 \times \text{Gain of loop } 2 \times \text{Gain of loop } 4 = -G_1G_3G_8H_1H_2H_4$$
$$L_{23} = \text{Gain of loop } 1 \times \text{Gain of loop } 3 \times \text{Gain of loop } 4 = -G_1G_4G_8H_1H_3H_4$$

Step 5: There are no higher-order non-touching loops.

Step 6:

$$\begin{aligned}\Delta &= 1-(L_{11}+L_{21}+L_{31}+L_{41})+(L_{12}+L_{22}+L_{32}+L_{42}+L_{52})-(L_{13}+L_{23})\\ &= 1+(G_1H_1+G_3H_2+G_4H_3+G_8H_4)+(G_1G_3H_1H_2+G_1G_4H_1H_3+G_1G_8H_1H_4+G_3G_8H_2H_4\\ &\quad +G_4G_8H_3H_4)+(G_1G_3G_8H_1H_2H_4+G_1G_4G_8H_1H_3H_4)\end{aligned}$$

Step 8: For first forward path $\Delta_1 = 1 + G_8 H_4$

For second forward path $\Delta_2 = 1 + G_3 H_2 + G_4 H_3$

Step 9:

$$\text{Transfer function} = \frac{C(s)}{R(s)} = \frac{P_1\Delta_1 + P_2\Delta_2}{\Delta}$$

$$= \frac{G_1G_2G_3G_4G_5(1+G_8H_4) + G_6G_7G_8G_9G_{10}(1+G_3H_2+G_4H_3)}{\begin{aligned}1 + G_1H_1 + G_3H_2 + G_4H_3 + G_8H_4 + G_1G_3H_1H_2 + G_1G_4H_1H_3 + G_1G_8H_1H_4 \\ + G_3G_8H_2H_4 + G_4G_8H_3H_4 + G_1G_3G_8H_1H_2H_4 + G_1G_4G_8H_1H_3H_4\end{aligned}}$$

Example 7.12 Find the transfer function of the system shown in Fig. E7.41 using Mason's gain formula.

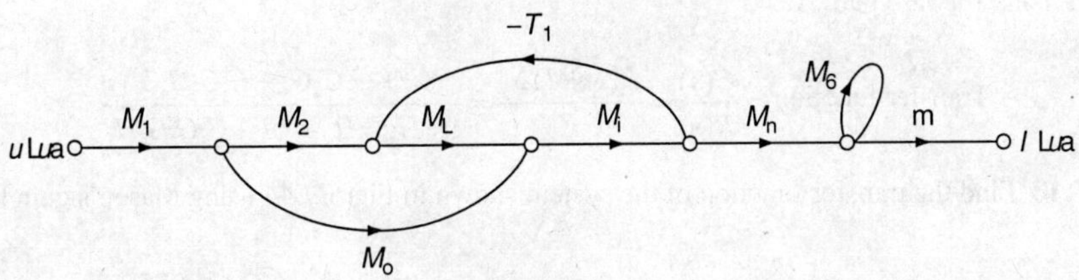

Fig. E7.41

Solution

Step 1: The forward paths of Fig. E7.41 are shown in Fig. E7.42.

The gain of the first forward path and that of the second forward path $P_1 = G_1G_2G_3G_4G_5$ and $P_2 = G_1G_4G_5G_6$ respectively.

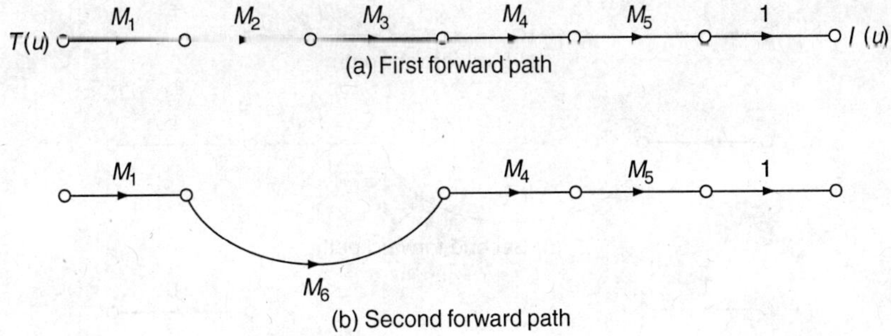

Fig. E7.42 Forward paths of Fig. E7.41

Step 2: The SPG shown in Fig. E7.41 has the following individual loops as shown in Fig. E7.43. The individual loop gains are

$L_1 = -G_3G_4H_1$ and $L_{21} = G_7$ respectively.

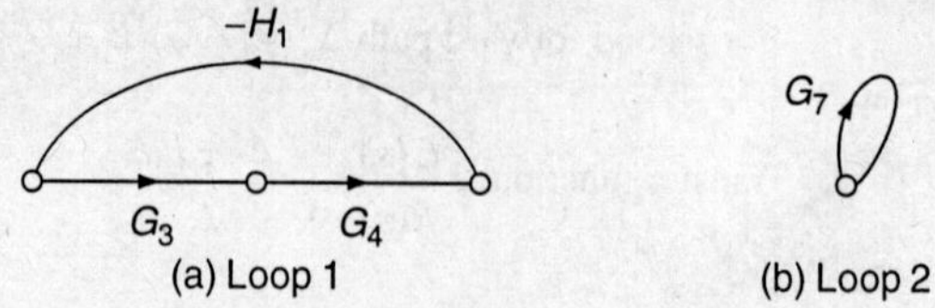

Fig. E7.43 Loops of Fig. E7.41

Step 3: Loop 1 and Loop 2 are non-touching.

$\therefore \quad L_{12}$ = Gain of Loop 1 × Gain of Loop

$$= -G_3G_4G_7H_1$$

Step 4: There are no higher-order non-touching loops.

Step 5:
$$\Delta = 1-(L_{11}+L_{21})+L_{12} = 1-(-G_3G_4H_1+G_7)-G_3G_4G_7H_1$$
$$= 1+G_3G_4H_1-G_7-G_3G_4G_7H_1$$

Step 6: For first forward path $\Delta_1 = 1$

For second forward path $\Delta_2 = 1$

Step 7:

$$\text{Transfer function} = \frac{C(s)}{R(s)} = \frac{P_1\Delta_1+P_2\Delta_2}{\Delta} = \frac{G_1G_2G_3G_4G_5+G_1G_4G_5G_6}{1+G_3G_4H_1-G_7-G_3G_4G_7H_1}$$

Example 7.13 Find the transfer function of the system shown in Fig. E7.44 using Mason's gain formula.

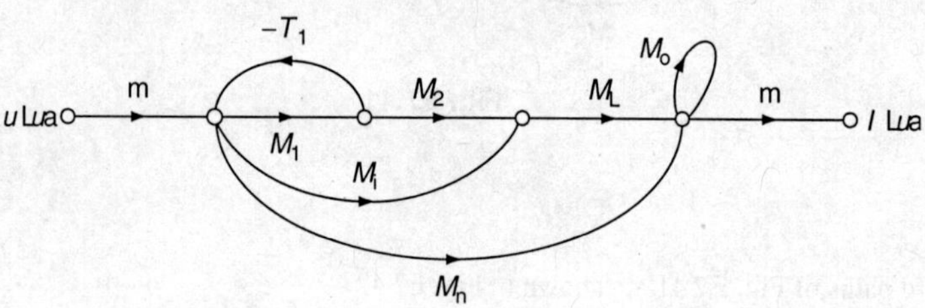

Fig. E7.44

Solution

Step 1: The SFG shown in Fig. E7.44 has the following three forward paths as shown in Fig. E7.45.

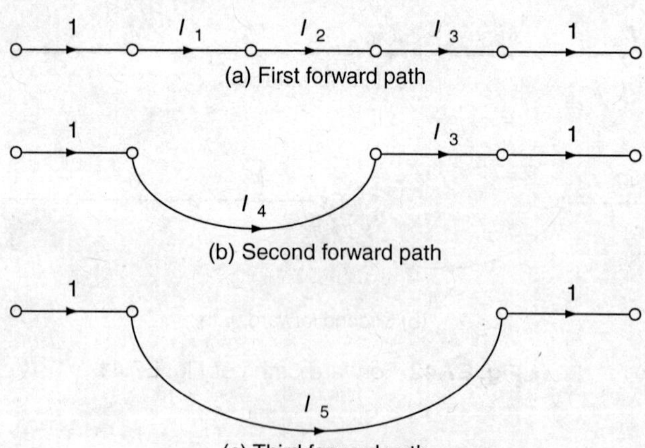

Fig. E7.45 Forward paths of Fig.E7.44

The gains of each forward path are

$P_1 = G_1G_2G_3$

$P_2 = G_3G_4$ and

$P_3 = G_5$

Step 2: The SPG shown in Fig. E7.44 has the following individual loops as shown in Fig. E7.46.
The gains of each individual loop are
$L_{11} = -G_1H_1$ and $L_{21} = G_6$ respectively.

Fig. E7.46 Loops of Fig. E7.44

Step 3: Loop 1 and Loop 2 are non-touching. Therefore,

$$L_{12} = -G_1H_1G_6 = -G_1G_6H_1$$

Step 4: There are no higher order non-touching loops.

Step 5: $\Delta = 1-(L_{11}+L_{21})+L_{12} = 1+G_1H_1-G_6-G_1G_6H_1$

Step 6: Since all loops touch each forward path,

$$\Delta_1 = \Delta_2 = \Delta_3 = 1.$$

Step 7:

$$\text{Transfer function } = \frac{C(s)}{R(s)} = \frac{P_1\Delta_1 + P_2\Delta_2 + P_3\Delta_3}{\Delta} = \frac{G_1G_2G_3 + G_3G_4 + G_5}{1+G_1H_1-G_6-G_1G_6H_1}$$

Example 7.14 Find the transfer function of the system shown in Fig. E7.47 using Mason's gain formula.

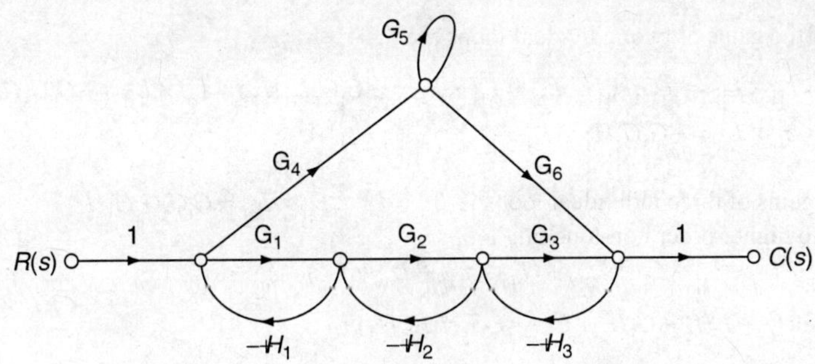

Fig. E7.47

Solution

Step 1: The forward paths of Fig. E7.47 are shown in Fig. E7.48.
The gains of each forward path are

$P_1 = G_1G_2G_3$ and $P_2 = G_4G_6$ respectively.

Step 2: The individual loops of Fig. E7.47 are shown in Fig. E7.49.
The gains of each individual loop are shown below.

$$L_{11} = -G_1H_1, L_{21} = -G_2H_2, L_{31} = -G_3H_3,\ L_{41} = G_5 \text{ and } L_{51} = -G_4G_6H_1H_2H_3$$

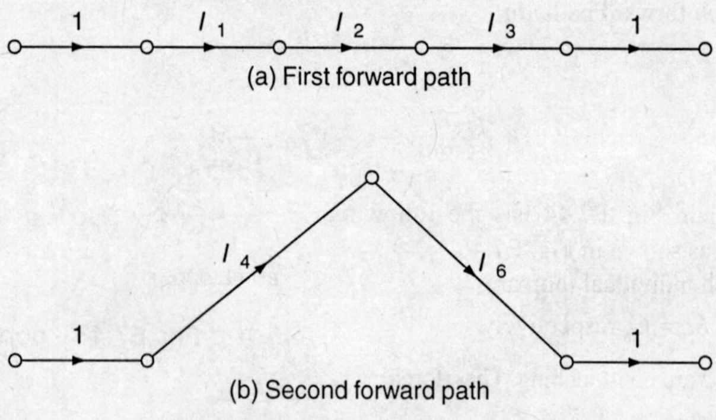

Fig. E7.48 Forward paths of Fig. E7.47

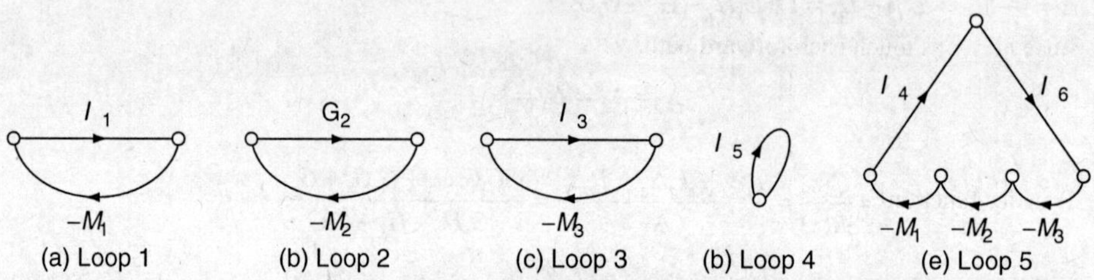

Fig. E7.49 Loops of Fig. E7.47

Step 3: Products of the gains of two individual loops are

$$L_{12} = L_{11} \times L_{31} = G_1G_3H_1H_3, L_{22} = L_{11} \times L_{41} = -G_1G_5H_1, L_{32} = L_{21} \times L_{41} = -G_2G_5H_2,$$
$$L_{43} = L_{41} \times L_{51} = -G_3G_5H_3$$

Step 4: Product of gains of three individual loops is $L_{13} = L_{11} \times L_{31} \times L_{41} = G_1G_3G_5H_1H_3$

Step 5: There are no higher-order non-touching loops.

Step 6: $\Delta = 1-(L_{11}+L_{21}+L_{31}+L_{41}+L_{51})+(L_{12}+L_{22}+L_{32}+L_{42})-L_{13}$

$$= 1-(-G_1H_1-G_2H_2-G_3H_3+G_5-G_4G_6H_1H_2H_3)$$
$$+(G_1G_3H_1H_3-G_1G_5H_1-G_2G_5H_2-G_3G_5H_3)-G_1G_3G_5H_1H_3$$

Step 7: For P_1, only loop 4 is non-touching. $\therefore \Delta_1 = 1-G_5$

For P_2, only loop 2 is non-touching. $\therefore \Delta_2 = 1-(-G_2H_2) = 1+G_2H_2$

Step 8:

$$\text{Transfer function} = \frac{C(s)}{R(s)} = \frac{P_1\Delta_1 + P_2\Delta_2}{\Delta}$$

$$= \frac{G_1G_2G_3(1-G_5)+G_4G_6(1+G_2H_2)}{1+G_1H_1+G_2H_2+G_3H_3-G_5+G_4G_6H_1H_2H_3+G_1G_3H_1H_3-G_1G_5H_1 -G_2G_5H_2-G_3G_5H_3-G_1G_3G_5H_1H_3}$$

Example 7.15 Construct the SFG for the following set of equations:

$$Y_2 = G_1Y_1 - G_2Y_4$$
$$Y_3 = G_3Y_2 + G_4Y_3$$
$$Y_4 = G_5Y_1 + G_6Y_3$$

where Y_1 and Y_4 are the input and output respectively.

Using Mason's gain formula, find the transfer function of the system.

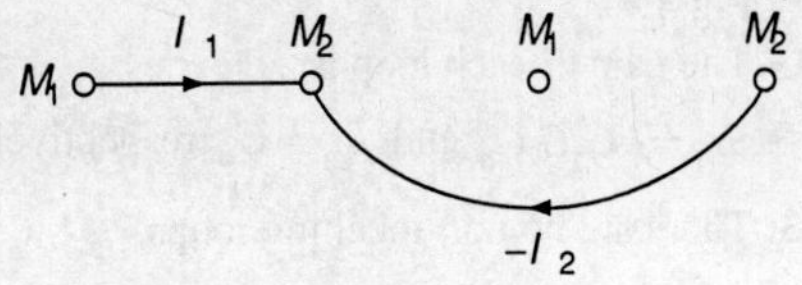

Fig. E7.50

Solution

From the equation $Y_2 = G_1Y_1 - G_2Y_4$, Fig. E7.50 is drawn.

Using the equation $Y_3 = G_3Y_2 + G_4Y_3$, Fig. E7.50 is modified and shown in Fig. E7.51.

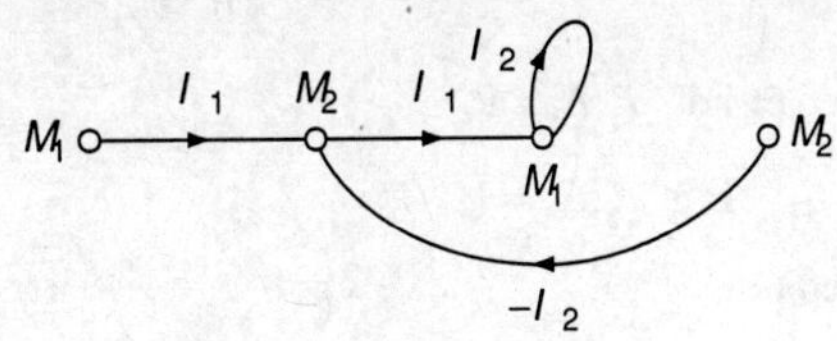

Fig. E7.51

Using the equation $Y_4 = G_5Y_1 + G_6Y_3$, Fig. E7.51 is modified and shown in Fig. E7.52.

Using dummy nodes for input and output, Fig. E7.52 is redrawn and shown in Fig. E7.53.

Figure E7.53 shows the SFG for the given equations of the system.

To get the transfer function of the system, the following steps are done.

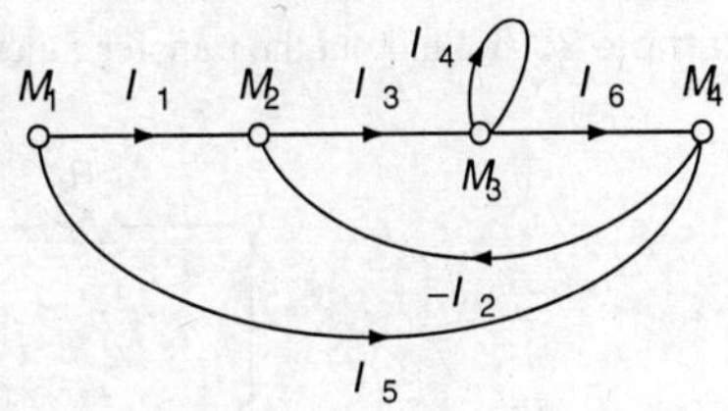

Fig. E7.52

Step 1: The forward paths of Fig.E7.53 are shown in Fig.E7.54. The gain of the first forward path and that of the second forward path are $P_1 = G_1G_3G_6$ and $P_2 = G_5$ respectively.

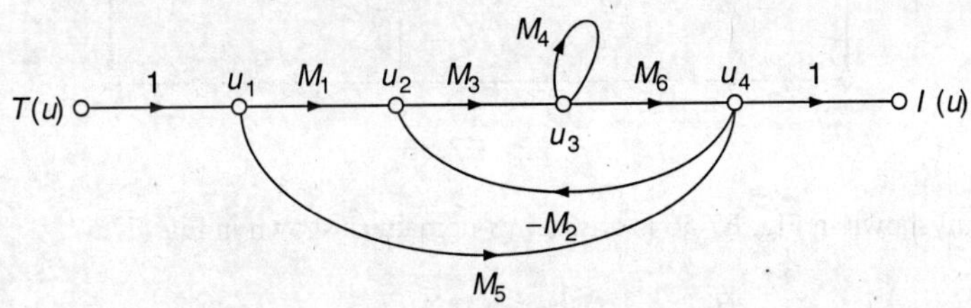

Fig. E7.53

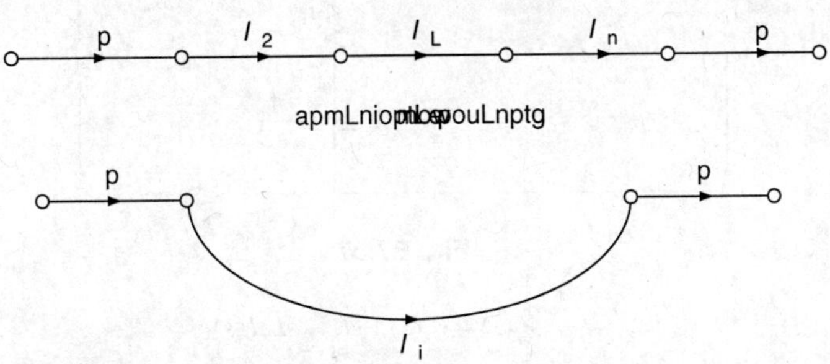

Fig. E7.54 Forward paths of Fig. E7.37

Step 2: The individual loops of Fig. E7.53 are shown in Fig. E7.55.

The gain of each loop is

$L_{11} = -G_2G_3G_6$ and $L_{21} = G_4$ respectively.

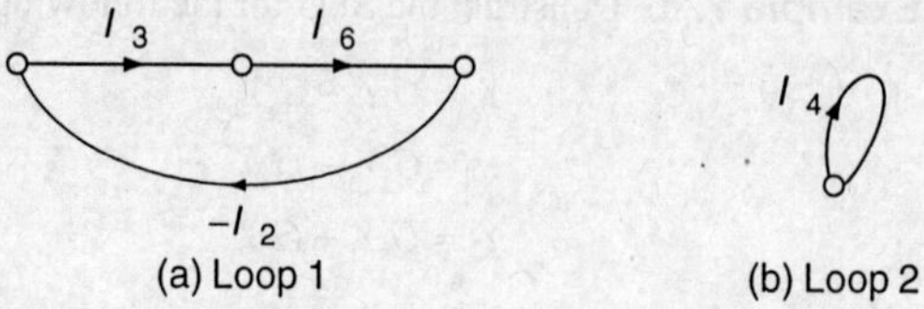

Fig. E7.55 Loops of Fig. E7.53

Step 3: There are no non-touching loops.

Step 4:

$$\Delta = 1-(L_{11}+L_{21}) = 1-(-G_2G_3G_6+G_4) = 1+G_2G_3G_6-G_4$$

Step 5: For $P_1, \Delta_1 = 1$

For $P_2, \Delta_2 = 1-G_4$

Step 6:

$$\text{Transfer function} = \frac{Y_1}{Y_4} = \frac{P_1\Delta_1+P_2\Delta_2}{\Delta} = \frac{G_1G_3G_6+G_5(1-G_4)}{1+G_2G_3G_6-G_4}$$

Example 7.16 Find out the transfer function of the electric network shown in Fig. E7.56.

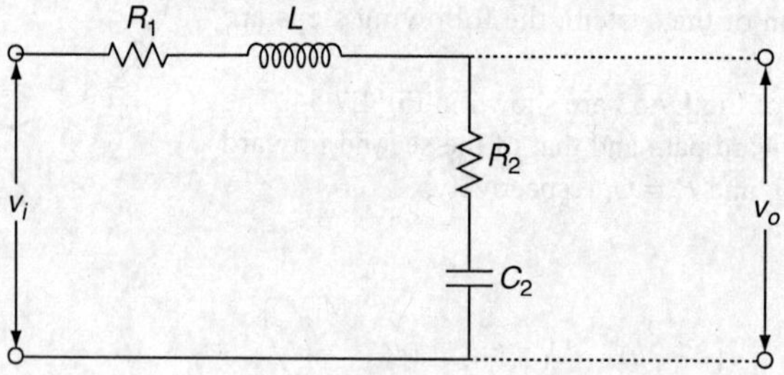

Fig. E7.56

Solution The circuit shown in Fig. E7.56 is drawn in *s*-domain as shown in Fig. E7.57.

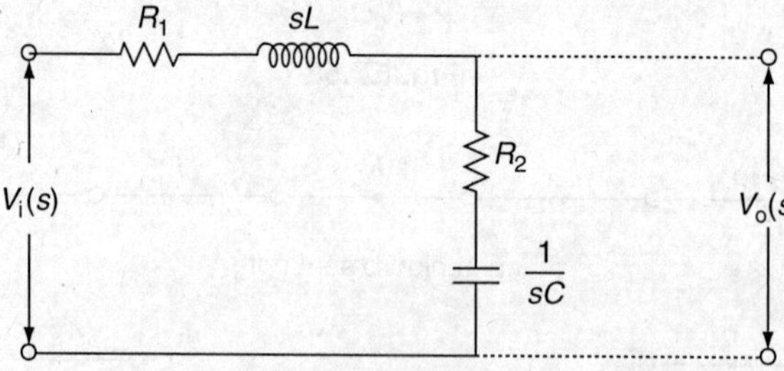

Fig. E7.57

$$\therefore \quad V_i(s) = V_o(s) + (R_1 + sL)I(s)$$

$$I(s) = \frac{V_i(s)-V_o(s)}{R_1+sL} \tag{1}$$

Using Eq. (1), the SFG can be drawn as shown in Fig. E7.58.

Again, $$V_o(s) = \left(R_2 + \frac{1}{sC}\right) I(s) \quad (2)$$

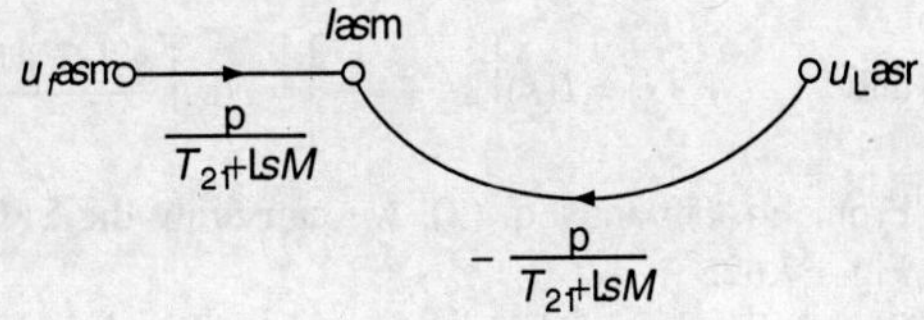

Fig. E7.58

Using Eq. (2), Fig. E7.58 is modified as shown in Fig. E7.59.
The SFG of Fig. E7.59 has only one forward path.

$$\therefore \text{ Path gain } \quad (P_1) = \left(R_2 + \frac{1}{sC}\right)\left(\frac{1}{(R_1 + sL)}\right) = \frac{1 + sR_2C}{sC(R_1 + sL)}$$

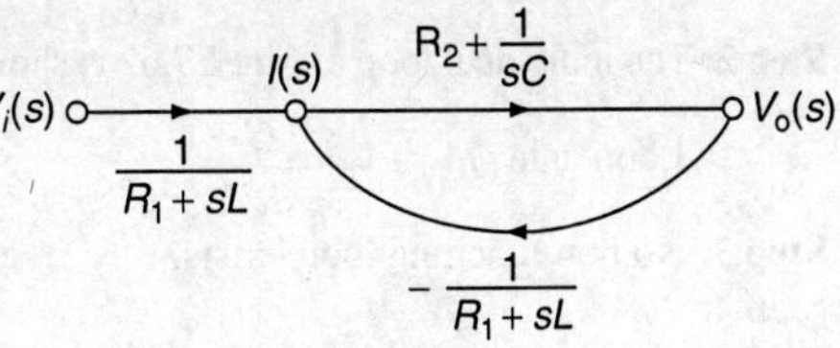

Fig. E7.59

Figure E7.59 has only one loop.

$$\therefore \text{ Loop gain } \quad (L_{11}) = -\frac{1 + sR_2C}{sC(R_1 + sL)}$$

$$\therefore \quad \Delta = 1 - L_{11} = 1 + \frac{1 + sR_2C}{sC(R_1 + sL)}$$

$\therefore$ $\Delta_1 = 1$ (since the loop is touching the forward path)

$$\therefore \text{ Transfer function} = \frac{P_1\Delta_1}{\Delta} = \frac{\dfrac{1 + sR_2C}{sC(R_1 + sL)} \times 1}{1 + \dfrac{1 + sR_2C}{sC(R_1 + sL)}} = \frac{1 + sR_2C}{s^2LC + sC(R_1 + R_2) + 1}$$

Example 7.17 Find the transfer function of the network shown in Fig. E7.60 using Mason's gain formula.

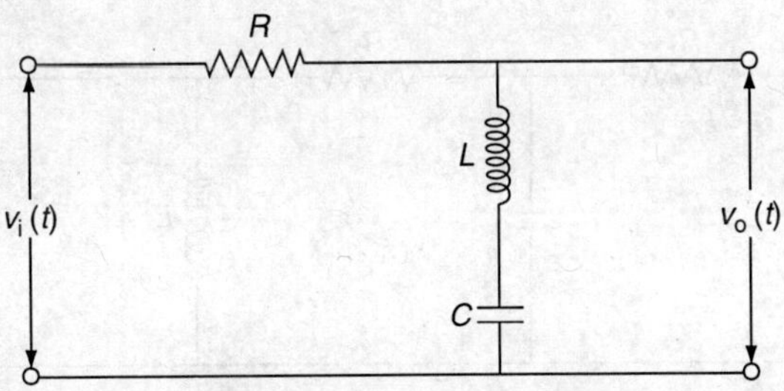

Fig. E7.60

Solution Figure 7.61 shows the Laplace transform of the network shown in Fig. E7.60.

From Fig. E7.61, we have $$I(s) = \frac{V_i(s) - V_o(s)}{R} = \frac{V_i(s)}{R} - \frac{V_o(s)}{R} \quad (1)$$

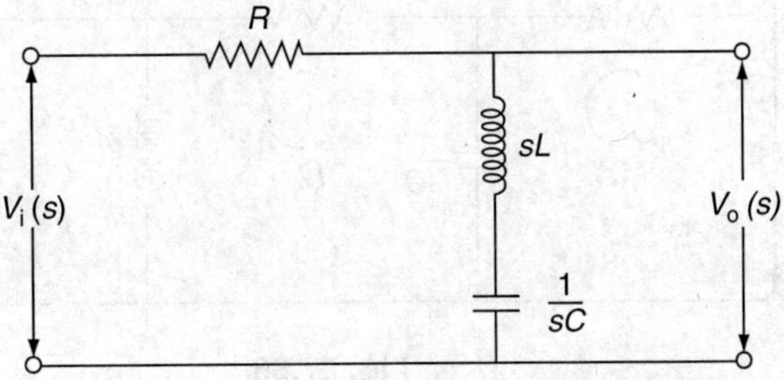

Fig. E7.61

and $$V_o(s) = I(s)\left[sL + \frac{1}{sC}\right] = I(s)\left[\frac{s^2LC+1}{sC}\right] \quad (2)$$

From Eq. (1) and Eq. (2), we can draw the SFG as shown in Fig. E7.62.

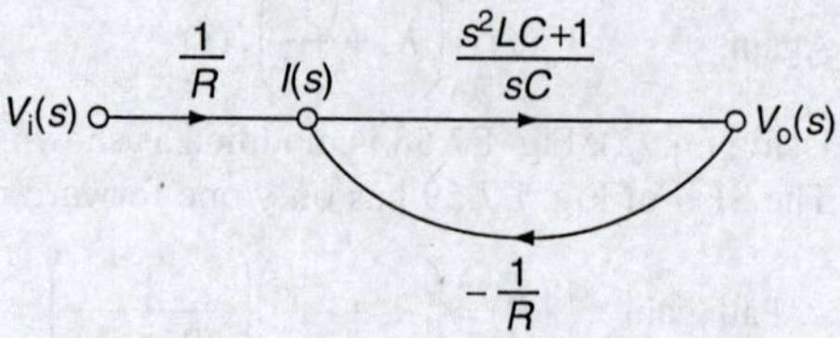

Fig. E7.62

Step 1: The forward path of Fig. E7.62 is shown in Fig. E7.63.

$$\text{Forward path gain } (P_1) = \frac{s^2LC+1}{sRC}$$

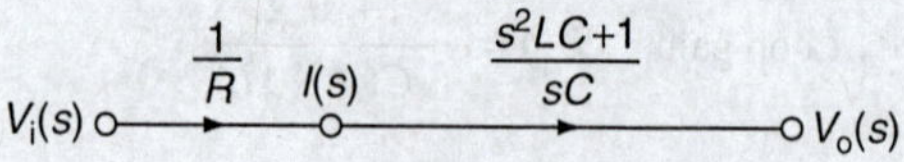

Fig. E7.63

Step 2: The individual loop of Fig. E7.62 is shown in Fig. E7.64.

$$\text{Loop gain } (L_{11}) = -\frac{s^2LC+1}{sRC}$$

Step 3: No non-touching loop exists.

Step 4:

$$\Delta = 1 - L_{11} = 1 + \frac{s^2LC+1}{sRC}$$

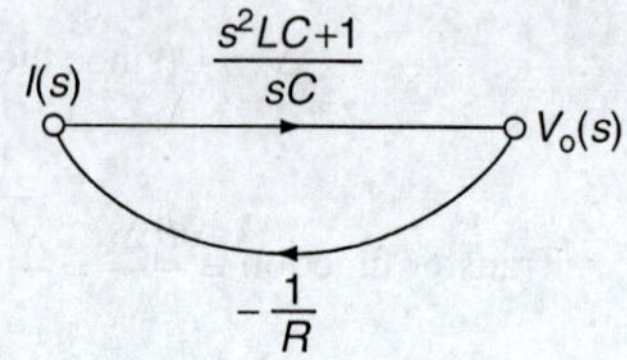

Fig. E7.64

Step 5: The loop touches the forward path, hence $\Delta_1 = 1$.

Step 6:

$$\text{Transfer function} = \frac{P_1\Delta_1}{\Delta} = \frac{s^2LC+1}{s^2LC+sRC+1}$$

Example 7.18 Determine the transfer function of the network shown in Fig. E7.65.

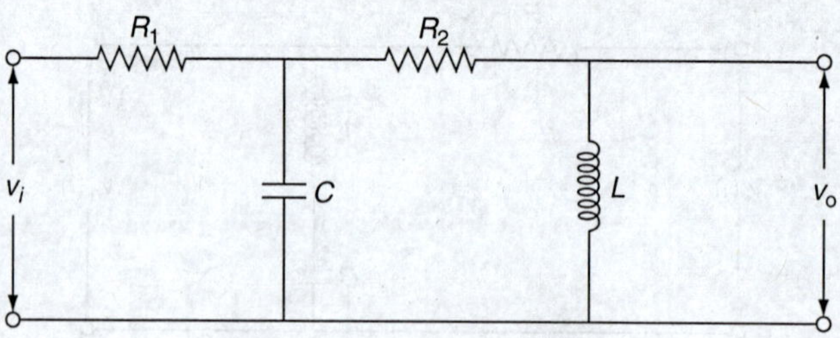

Fig. E7.65

Solution Figure 7.65 shows the Laplace transform of Fig. E7.66.

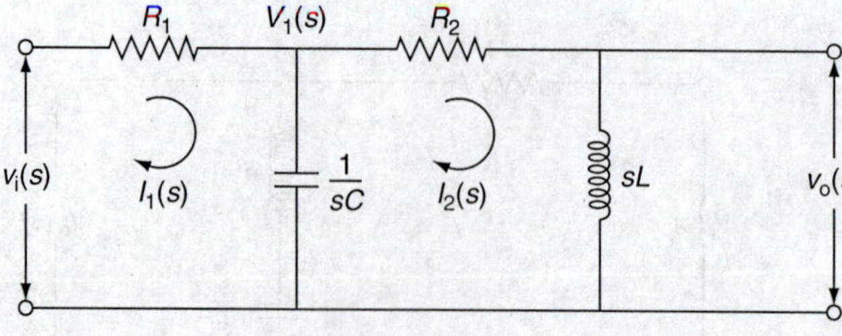

Fig. E7.66

From Fig. E7.66, we have,

$$I_1(s) = \frac{V_i(s) - V_1(s)}{R_1} \quad (1)$$

$$V_1(s) = \frac{I_1(s) - I_2(s)}{sC} \quad (2)$$

$$I_2(s) = \frac{V_1(s) - V_o(s)}{R_2} \quad (3)$$

$$V_o(s) = I_2(s)sL \quad (4)$$

From Eq. (1), Eq. (2), Eq. (3) and Eq. (4), we have the following SFG as shown in Fig. E7.67.

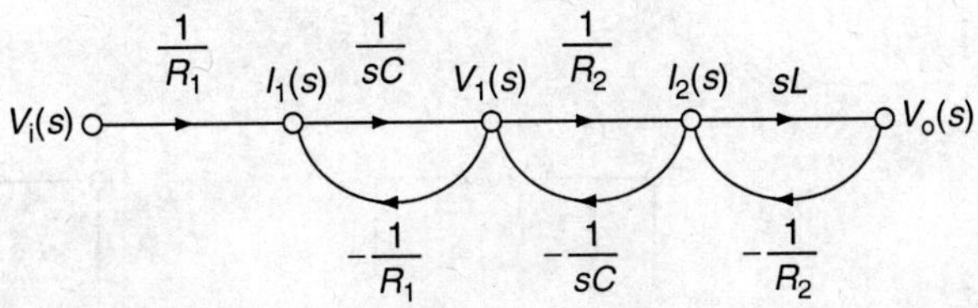

Fig. E7.67

Step 1: In Fig. E7.67, there is only one forward path as shown in Fig. E7.68.

The gain of the forward path is given by

$$P_1 = \frac{L}{R_1 R_2 C}$$

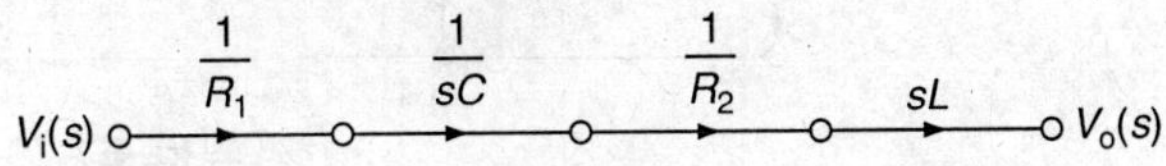

Fig. E7.68

Step2: Individual feedback loops are shown in Fig. E7.69.

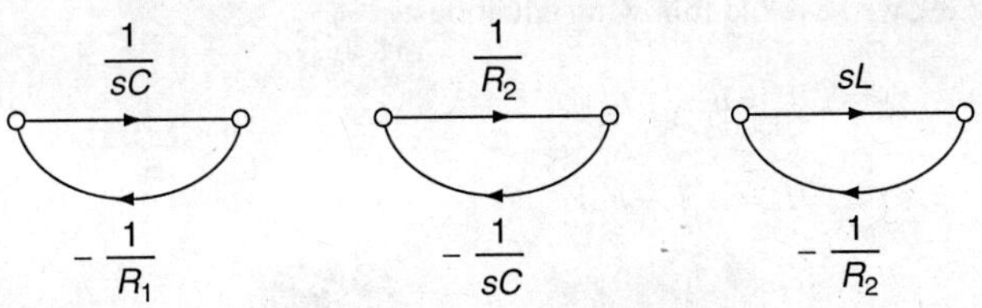

Fig. E7.69

$$L_{11} = \frac{-1}{sR_1C}, L_{21} = \frac{-1}{sR_2C}, \text{ and } L_{31} = -\frac{sL}{R_2}$$

Step 3: The loops L_{11} and L_{31} are non-touching.

Therefore,

$$L_{12} = L_{11}L_{31} = \frac{L}{R_1R_2C}$$

Step 4: There are no higher-order non-touching loops.

Step 5: Since all the loops touch the forward path, $\Delta_1 = 1$.

Step 6:

$$\Delta = 1 - \left(-\frac{1}{sR_1C} - \frac{1}{sR_2C} - \frac{sL}{R_2}\right) + \frac{L}{R_1R_2C} = 1 + \frac{1}{sR_1C} + \frac{1}{sR_2C} + \frac{sL}{R_2} + \frac{L}{R_1R_2C}$$

Step 7:

$$\text{Transfer function} = \frac{P_1\Delta_1}{\Delta} = \frac{\dfrac{L}{R_1R_2C}}{1+\dfrac{1}{sR_1C}+\dfrac{1}{sR_2C}+\dfrac{sL}{R_2}+\dfrac{L}{R_1R_2C}}$$

$$= \frac{sL}{s^2LR_1C + s(L+R_1R_2C)+(R_1+R_2)}$$

Example 7.19 Find the transfer function of the transistor shown in Fig. E7.70.

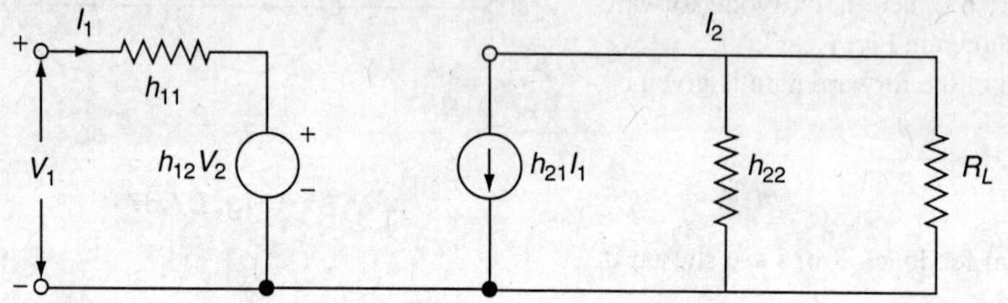

Fig. E7.70

Solution From Fig. E7.69, we have the following equations:

$$V_1 = h_{11}I_1 + h_{12}V_2 \tag{1}$$

$$I_2 = h_{21}I_1 - h_{22}V_2 \tag{2}$$

and

$$V_2 = -R_LI_2 \tag{3}$$

Taking V_2, I_1 and I_2 as dependent variables, we can draw the SGF as shown in Fig. E7.71.

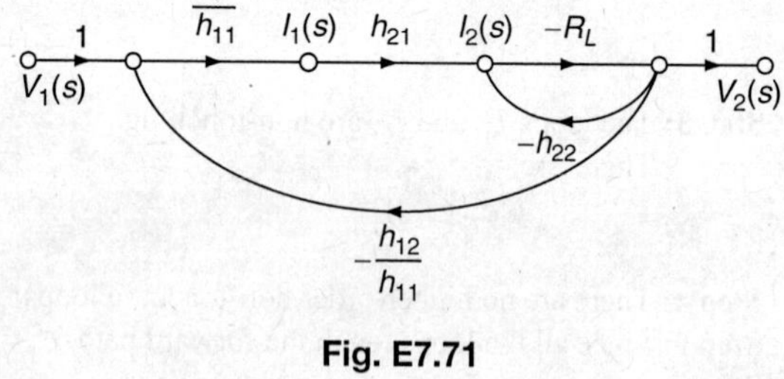

Fig. E7.71

Step 1: There is one forward path in Fig. E7.71 as shown in Fig. E7.72.

Gain of the forward path $(P_1) = -\dfrac{h_{21}R_L}{h_{11}}$

Step 2: There are two individual loops in Fig. E7.70 as shown in Fig. E7.73.

$$L_{11} = R_Lh_{22}$$

and

$$L_{21} = \frac{h_{12}h_{21}R_L}{h_{11}^2}$$

Fig. E7.72

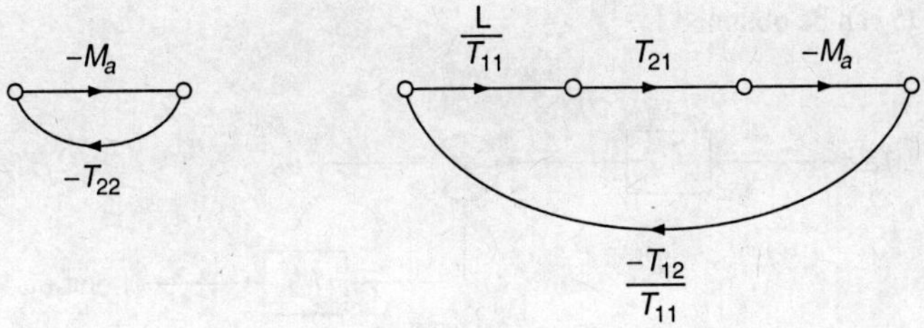

Fig. E7.73

Step 3: All the loops are touching each other.
Step 4:

$$\Delta = 1-(L_{11}+L_{21}) = 1 - R_L h_{22} - \frac{h_{12}h_{21}R_L}{h_{11}^2} = \frac{h_{11}^2 - R_L h_{11}^2 h_{22} - h_{12}h_{21}R_L}{h_{11}^2}$$

Step 5: Since all the loops touch the forward path, $\Delta_1 = 1$.
Step 6:

$$\text{Transfer function } = \frac{P_1\Delta_1}{\Delta} = \frac{-\dfrac{h_{21}R_L}{h_{11}}}{\dfrac{h_{11}^2 - R_L h_{11}^2 h_{22} - h_{12}h_{21}R_L}{h_{11}^2}}$$

$$= \frac{-h_{11}h_{21}R_L}{h_{11}^2 - R_L h_{11}^2 h_{22} - h_{12}h_{21}R_L}$$

Example 7.20 Obtain the block diagram for the SGF shown in Fig. E7.74.

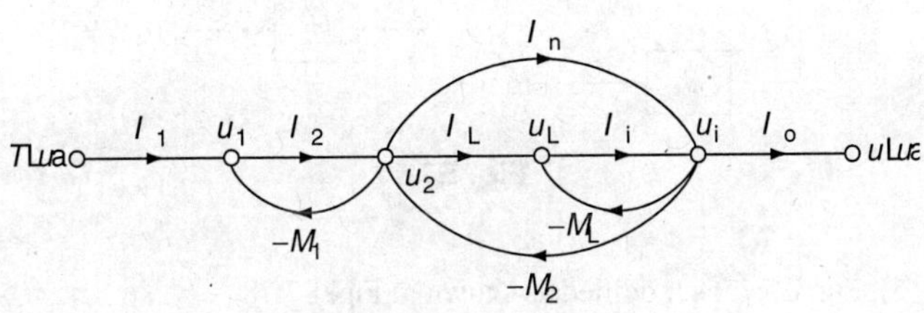

Fig. E7.74

Solution

$$Y_1 = G_1X - H_1Y_2 \tag{1}$$

$$Y_2 = G_2Y_1 - H_2Y_4 \tag{2}$$

$$Y_3 = G_3Y_2 - H_3Y_4 \tag{3}$$

$$Y_4 = G_4Y_3 + G_5Y_2 \tag{4}$$

$$Y(s) = G_6Y_4 \tag{5}$$

From Eq. (1), Fig. E7.75 can be obtained.

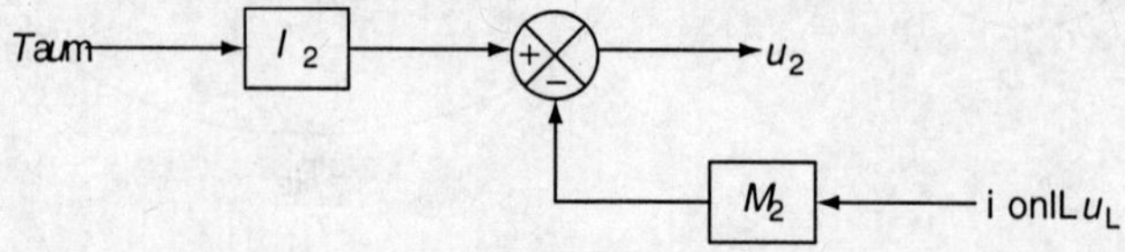

Fig. E7.75

Using Eq. (2), Fig. E7.75 is modified as shown in Fig. E7.76.

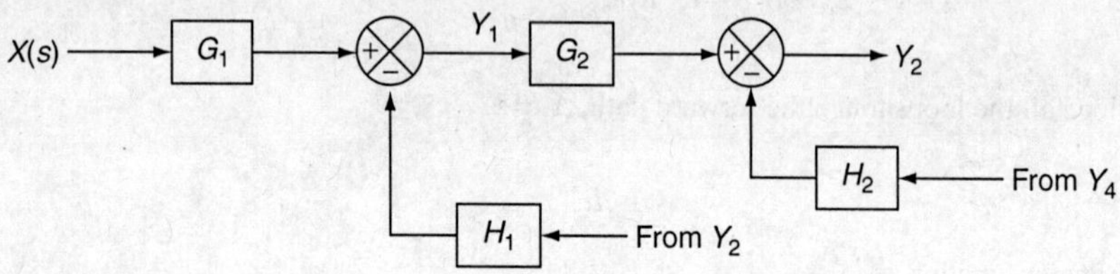

Fig. E7.76

Using Eq. (3), Fig. E7.76 is modified as shown in Fig. E7.77.

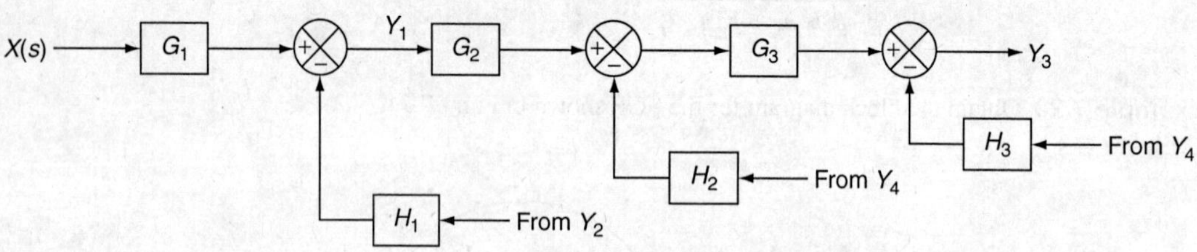

Fig. E7.77

Using Eq. (4) and Eq. (5), Fig. E7.77 is modified as shown in Fig. E7.78.

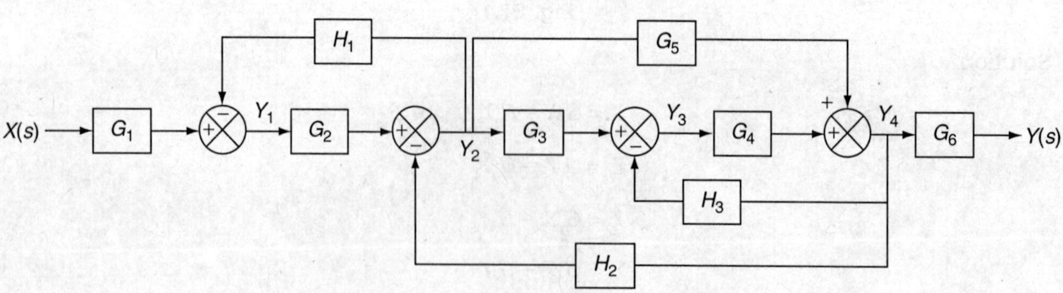

Fig. E7.78

Fig. E7.78 shows the block diagram representation of the SFG referred to in Fig. E4.74.

Example 7.21 Obtain the transfer function of the circuit shown in Fig. E7.79 using SFG.

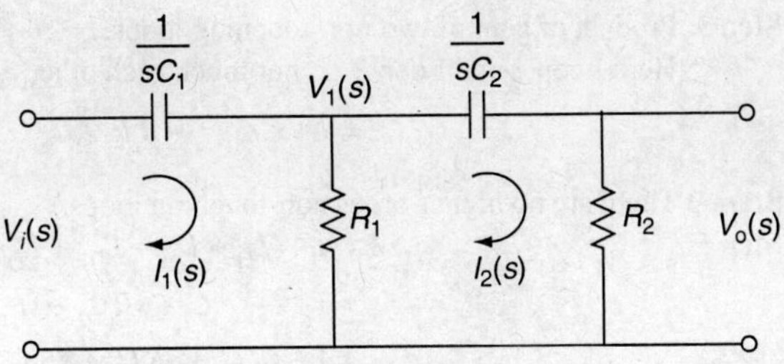

Fig. E7.79

Solutions From Fig. E7.79, the following equations can be obtained.

$$I_1(s) = \frac{V_i(s) - V_1(s)}{\dfrac{1}{sC_1}} = sC_1\left[V_i(s) - V_1(s)\right] \tag{1}$$

$$V_1(s) = \left[I_1(s) - I_2(s)\right]R_1 \tag{2}$$

$$I_2(s) = \left[V_1(s) - V_0(s)\right]sC_2 \tag{3}$$

$$V_0(s) = I_2(s)\,R_2 \tag{4}$$

Figure E7.80 shows the combined SFG.

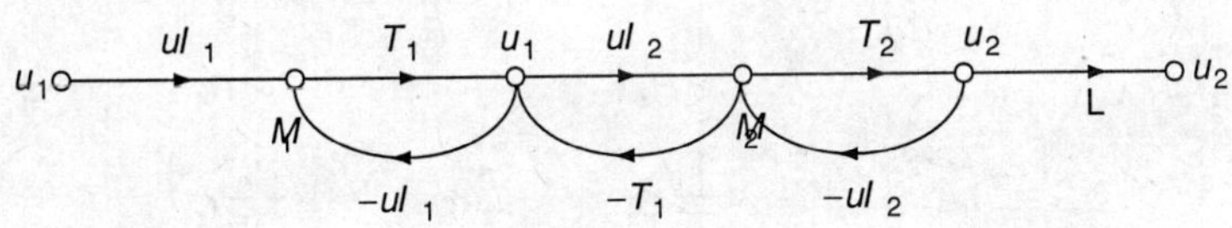

Fig. E7.80

Step 1: The forward path is shown in Fig. E7.81.

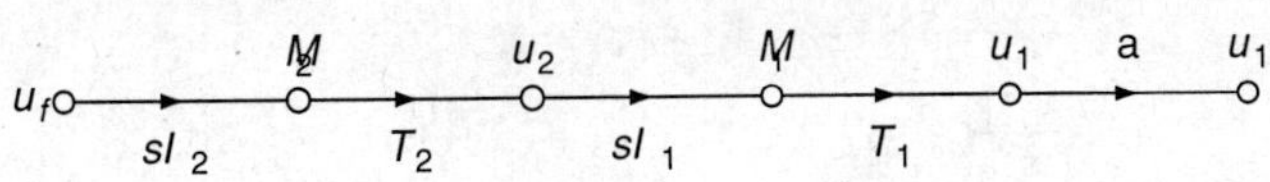

Fig. E7.81

Path gain $(P_1) = s^2 R_1 R_2 C_1 C_2$

Step 2: The individual loops are shown in Fig. E7.82. The individual loop gains are shown below.

$$L_{11} = -sR_1C_1$$

$$L_{21} = -sR_1C_2$$

and $$L_{31} = -sR_2C_2$$

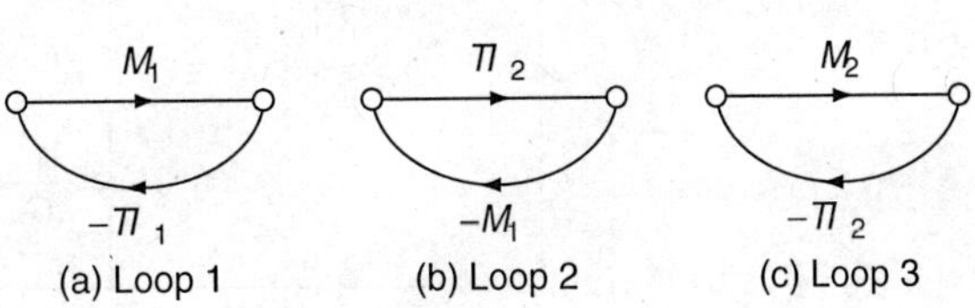

Fig. E7.82

Step 3: Product of gain of two non-touching loops.
Here Loop 1 and Loop 3 do not touch each other.

$$L_{12} = L_{11} \times L_{31} = s^2 R_1 R_2 C_1 C_2$$

Step 4: There are no higher-order non-touching loops.

Step 5:

$$\begin{aligned}\Delta &= 1-(L_{11}+L_{21}+L_{31})+L_{12}\\ &= 1-(-sR_1C_1 - sR_1C_2 - sR_2C_2)+s^2R_1R_2C_1C_2\\ &= 1+sR_1C_1+sR_1C_2+sR_2C_2+s^2R_1R_2C_1C_2\end{aligned}$$

Step 6: Since all the loops touch the forward path, $\Delta_1 = 1$.

Step 7: Transfer function $= \dfrac{P_1\Delta_1}{\Delta}$

$$= \frac{s^2R_1R_2C_1C_2}{1+sR_1C_1+sR_1C_2+sR_2C_2+s^2C_1C_2R_1R_2}$$

Example 7.22 Obtain the transfer function of the circuit shown in Fig. E7.83 using SFG.

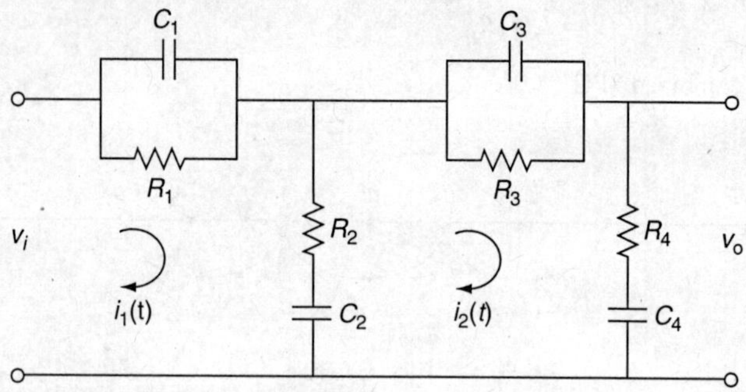

Fig. E7.83

Solution Figure E7.84 shows the Laplace domain of the circuit Fig. E7.83.

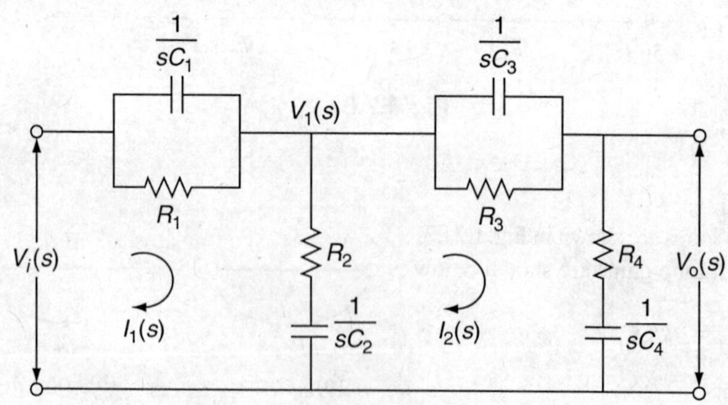

Fig. E7.84

From Fig. E7.83, the series RC gives equivalent

$$Z = R + \frac{1}{sC} = \frac{sCR+1}{sC}$$

and the parallel RC gives $$Y = sC + \frac{1}{R} = \frac{sCR+1}{R}$$

i.e., $$Z = \frac{1}{Y} = \frac{R}{sCR+1}$$

Using these Z-values, we obtain Fig. E7.85 from Fig. E7.84.

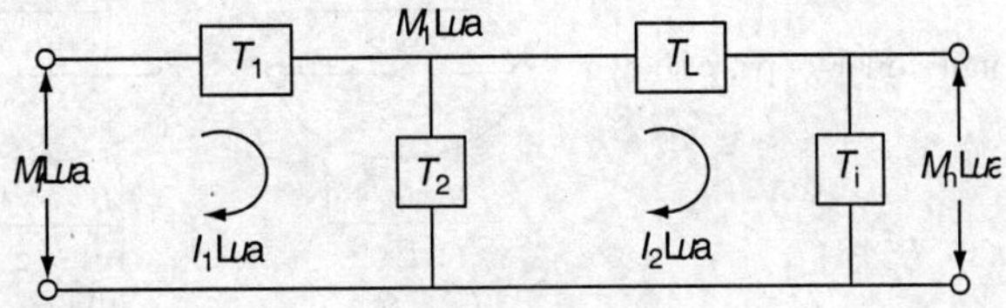

Fig. E7.85

From Fig. E7.85, we have

$$I_1 = \frac{V_i - V_1}{Z_1} = (V_i - V_1)\left(\frac{sC_1R_1+1}{R_1}\right) \quad (1)$$

$$V_1 = (I_1 - I_1)Z_2 = (I_1 - I_2)\left(\frac{sC_2R_2+1}{sC_2}\right) \quad (2)$$

$$I_2 = (V_1 - V_0)Z_3 = (V_1 - V_0)\left(\frac{sC_3R_3+1}{R_3}\right) \quad (3)$$

$$V_0 = I_2Z_4 = I_2\left(\frac{sC_4R_4+1}{sC_4}\right) \quad (4)$$

Figure E7.86 shows the combined SFG of Eq. (1), Eq. (2), Eq. (3) and Eq. (4).

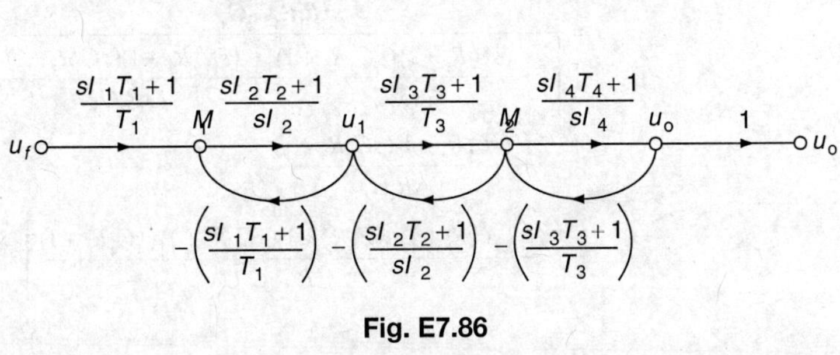

Fig. E7.86

Step 1: There is only one forward path in Fig. E7.87 as shown in Fig. E7.88.

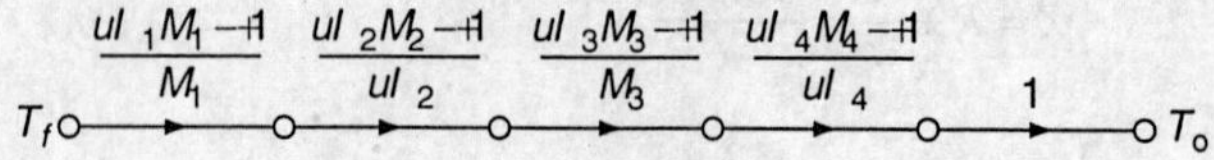

Fig. E7.87

The path gain is given by

$$P_1 = \frac{(sC_1R_1+1)(sC_2R_2+1)(sC_3R_3+1)(sC_4R_4+1)}{s^2R_1R_3C_2C_4}$$

Step 2: The individual loops are shown in Fig. E7.88.
The gains of each individual loop are shown below.

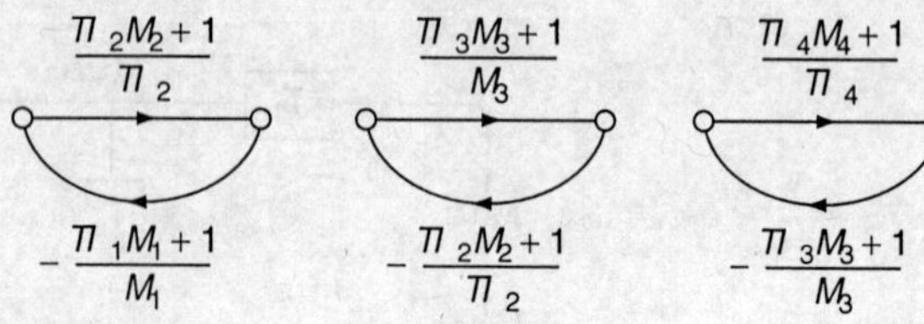

(a) Loop 1 (b) Loop 2 (c) Loop 3

Fig. E7.88

$$L_{11} = -\frac{(sC_1R_1+1)(sC_2R_2+1)}{sR_1C_2}$$

$$L_{21} = -\frac{(sC_2R_2+1)(sC_3R_3+1)}{sR_3C_2}$$

$$L_{31} = -\frac{(sC_3R_3+1)(sC_4R_4+1)}{sR_3C_4}$$

Step 3: Product of gains of two individual loops are shown below. Here Loop 1 and Loop 3 are non-touching.

$$\therefore \qquad L_{12} = L_{11} \times L_{31} = \frac{(sC_1R_1+1)(sC_2R_2+1)(sC_3R_3+1)(sC_4R_4+1)}{s^2R_1R_3C_2C_4}$$

Step 4: There are no higher-order non-touching loops.
Step 5: $\Delta = 1-(L_{11}+L_{21}+L_{31})+L_{12}$
Step 6: All loops touch the forward path P_1

$$\therefore \quad \Delta_1 = 1.$$

Step 7:

$$\text{Transfer function} = \frac{P_1\Delta_1}{\Delta}$$

$$= \frac{\dfrac{(sC_1R_1+1)(sC_2R_2+1)(sC_3R_3+1)(sC_4R_4+1)}{s^2R_1R_3C_2C_4}}{\left[1+\dfrac{(sC_1R_1+1)(sC_2R_2+1)}{sR_1C_2}+\dfrac{(sC_2R_2+1)(sC_3R_3+1)}{sR_3C_2} + \dfrac{(sC_3R_3+1)(sC_4R_4+1)}{sR_3C_4} + \dfrac{(sC_1R_1+1)(sC_2R_2+1)(sC_3R_3+1)(sC_4R_4+1)}{s^2R_1R_3C_2C_4}\right]}$$

$$= \frac{(sC_1R_1+1)(sC_2R_2+1)(sC_3R_3+1)(sC_4R_4+1)}{\begin{aligned}&[s^2R_1R_3C_2C_4 + sR_3C_4(sC_1R_1+1)(sC_2R_2+1)\\&+ sR_1C_4(sC_2R_2+1)(sC_3R_3+1)\\&+ sR_1C_2(sC_3R_3+1)(sC_4R_4+1)\\&+ (sC_1R_1+1)(sC_2R_2+1)(sC_3R_3+1)(sC_4R_4+1)]\end{aligned}}$$

Example 7.23 Find the transfer function of the SFG shown in Fig. E7.89.

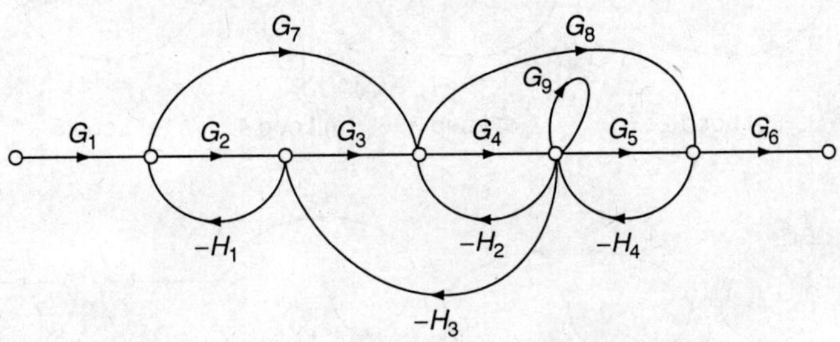

Fig. E7.89

Solution

Step 1: The SFG shown in Fig. E7.90 has the following four forward paths.

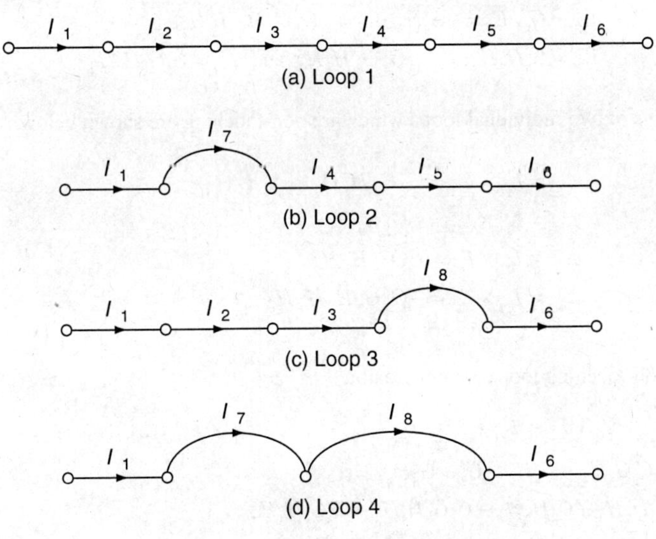

Fig. E7.90

The path gain of each forward path is

$$P_1 = G_1\, G_2\, G_3\, G_4\, G_5\, G_6$$
$$P_2 = G_1\, G_4\, G_5\, G_6\, G_7$$
$$P_3 = G_1\, G_2\, G_3\, G_6\, G_8$$

and
$$P_4 = G_1\, G_6\, G_7\, G_8$$

Step 2: The individual loops of the SFG shown in Fig. E7.89 are shown in Fig. E7.91.

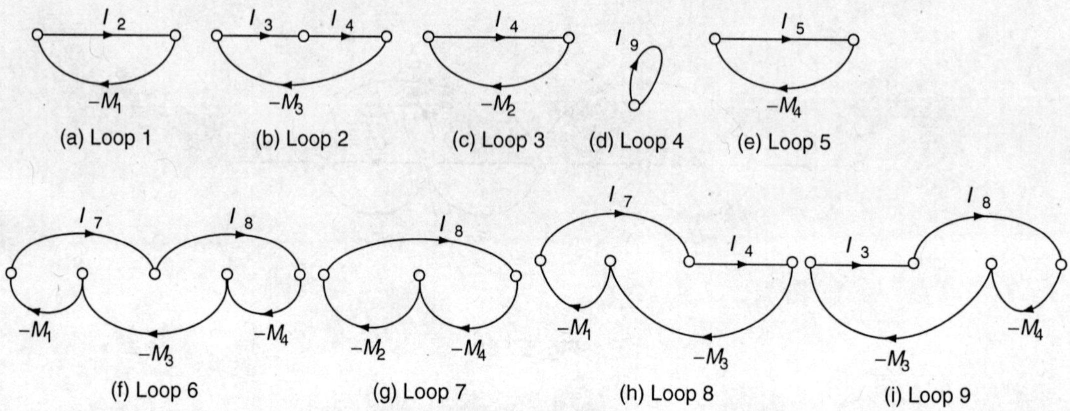

Fig. E7.91

The gain of each individual loop is shown below.

$$L_{11} = -G_2\, H_1,\ L_{21} = -G_3\, G_4\, H_3,\ L_{31} = -G_4\, H_2,$$
$$L_{41} = G_9,\ L_{51} = -G_5 H_4,\ L_{61} = -G_7\, G_8\, H_1\, H_3\, H_4,$$
$$L_{71} = G_8 H_2 H_4,\ L_{81} = G_4\, G_7\, H_1 H_3 \text{ and } L_{91} = G_3 G_8 H_3 H_4$$

Step 3: Products of gains of two individual loops which are non-touching are shown below.

$$L_{12} = L_{11} \times L_{13} = G_2\, G_4\, H_1\, H_2$$
$$L_{22} = L_{11} \times L_{14} = -G_2\, G_9\, H_1$$
$$L_{32} = L_{11} \times L_{15} = G_2\, G_5\, H_1\, H_2$$
$$L_{42} = L_{11} \times L_{17} = -G_2\, G_8\, H_1\, H_2\, H_4$$

Step 4: Higher-order non-touching loops are not feasible.

Step 5: $\Delta = 1 - (L_{11} + L_{21} + L_{31} + L_{41} + L_{51} + L_{61} + L_{71} + L_{81}) + (L_{12} + L_{22} + L_{32} + L_{42})$

$$= 1 + G_2 H_1 + G_3 G_4 H_3 + G_4 H_2 - G_9 + G_5 H_4$$
$$+ G_7 G_8 H_1 H_3 H_4 - G_8 H_2 H_4 - G_4 G_7 H_1 H_3 - G_3 G_8 H_3 H_4$$
$$+ G_2 G_4 H_1 H_2 - G_2 G_9 H_1 + G_2 G_5 H_1 H_2$$
$$- G_2 G_8 H_1 H_2 H_4$$

Step 6: All individual loops touch the forward paths P_1 and P_2. Hence $\Delta_1 = \Delta_2 = 1$.
Loop 4 does not touch the forward paths 3 and 4.

$\therefore\ \Delta_3 = \Delta_4 = 1 - L_{41} = 1 - G_9.$

Step 7: $TF = \dfrac{P_1\Delta_1 + P_2\Delta_2 + P_3\Delta_3 + P_4\Delta_4}{\Delta}$

$$= \frac{G_1G_2G_3G_4G_5G_6 + G_1G_4G_5G_6G_7 + G_1G_2G_3G_6G_8(1-G_9) + G_1G_6G_7G_8(1-G_9)}{\begin{array}{l}1 + G_2H_1 + G_3G_4H_3 + G_4H_2 - G_9 + G_5H_4 + G_7G_8H_1H_3H_4 - G_8H_2H_4 - G_4G_7H_1H_3 - G_3G_8H_3H_4 \\ + G_2G_4H_1H_2 - G_2G_9H_1 + G_2G_5H_1H_2 - G_2G_8H_1H_2H_4\end{array}}$$

Example 7.24 Find the transfer function of the SFG shown in Fig. E7.92

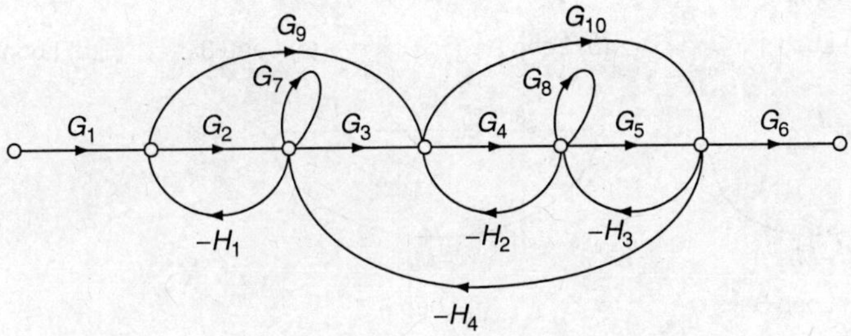

Fig. E7.92

Solution

Step 1: The SFG shown in Fig. E7.90 has four forward paths as shown in Fig. E7.93.

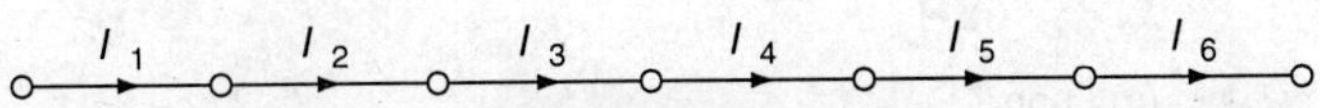

(a) Forward path P_1

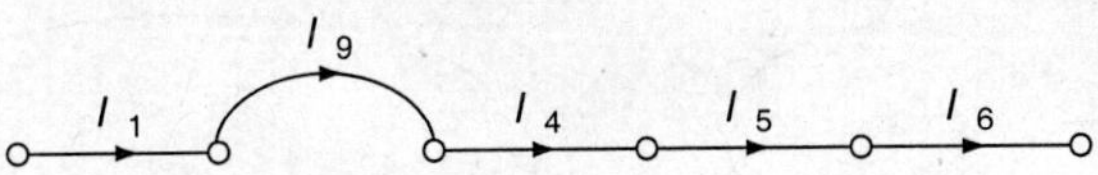

(b) Forward path P_2

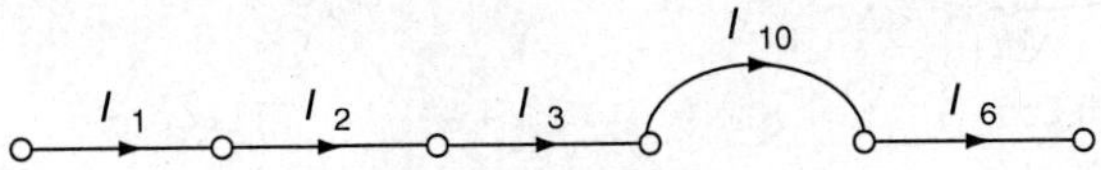

(c) Forward path P_3

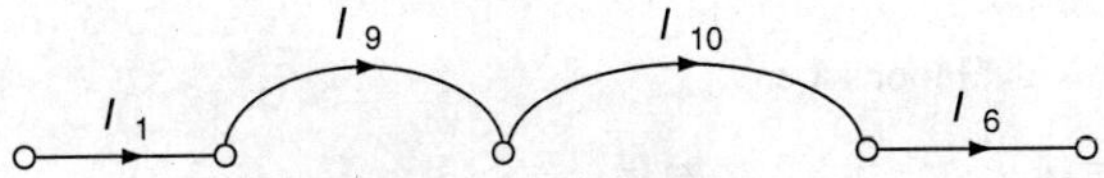

(d) Forward path P_4

Fig. E7.93

Here $\quad P_1 = G_1 G_2 G_3 G_4 G_5 G_6$

$\quad P_2 = G_1 G_4 G_5 G_6 G_9$

$\quad P_3 = G_1 G_2 G_3 G_6 G_{10}$

and $\quad P_4 = G_1 G_6 G_9 G_{10}$

Step 2: The individual loops of Fig. E7.92 are shown in Fig. E7.94.

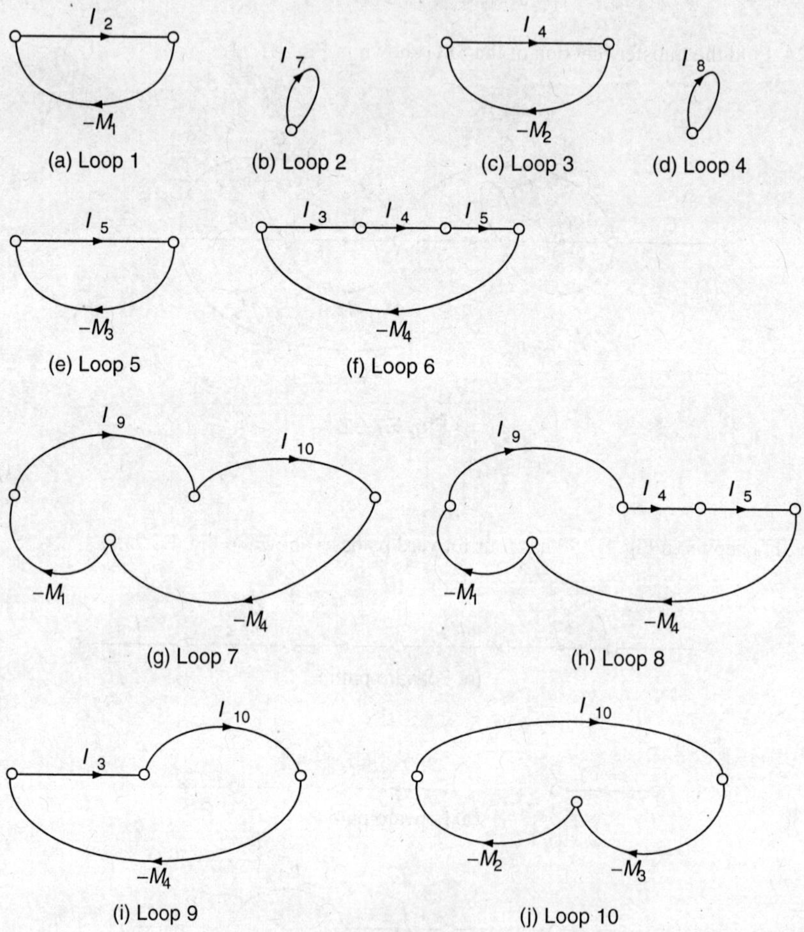

Fig. E7.94

The gain of the individual loops are

$L_{11} = -G_2 H_1,\ L_{21} = G_7, L_{31} = -G_4 H_2,\ L_{41} = G_8$

$L_{51} = -G_5 H_3, L_{61} = -G_3 G_4 G_5 H_4,\ L_{71} = G_9\, G_{10}\, H_1\, H_4,$

$L_{81} = G_4 G_5 G_9 H_1 H_4,\ L_{91} = -G_3\, G_{10}\, H_4$ and $L_{10,1} = G_{10} H_2 H_3$

Step 3: Products of gain of two individual loops which are non-touching are shown below.

$$L_{12} = L_{11}L_{31} = G_2 G_4 H_1 H_2$$
$$L_{22} = L_{11}L_{41} = -G_2 G_8 H_1$$
$$L_{32} = L_{11}L_{51} = G_2 G_5 H_1 H_3$$
$$L_{42} = L_{21}L_{31} = -G_4 G_7 H_2$$
$$L_{52} = L_{21}L_{41} = G_7 G_8$$
$$L_{62} = L_{21}L_{51} = -G_5 G_7 H_3$$
$$L_{72} = L_{41}L_{71} = G_8 G_9 G_{10} H_1 H_4$$
$$L_{82} = L_{41}L_{91} = -G_3 G_8 G_{10} H_4$$
$$L_{92} = L_{11}L_{10,1} = -G_2 G_{10} H_1 H_2 H_3$$
$$L_{10,2} = L_{21}L_{10,1} = +G_7 G_{10} H_2 H_3$$

Step 4: There are no further higher-order non-touching loops.

Step 5:
$$\begin{aligned}\Delta &= 1-(L_{11}+L_{21}+L_{31}+L_{41}+L_{51}+L_{61}+L_{71}+L_{81}+L_{91}+L_{10,1})\\ &\quad +(L_{12}+L_{22}+L_{32}+L_{42}+L_{52}+L_{62}+L_{72}+L_{82}+L_{92}+L_{10,2})\\ &= 1+G_2H_1-G_7+G_4H_2-G_8+G_5H_3+G_3G_4G_5H_4\\ &\quad -G_9G_{10}H_1H_4-G_4G_5G_9H_1H_4+G_3G_{10}H_4\\ &\quad -G_{10}H_2H_3+G_2G_4H_1H_2-G_2G_8H_1+G_2G_5H_1H_3-G_4G_7H_2+G_7G_8-G_5G_7H_3\\ &\quad +G_8G_9G_{10}H_1H_4-G_3G_8G_{10}H_4-G_2G_{10}H_1H_2H_3+G_7G_{10}H_2H_3\end{aligned}$$

Step 6: All the loops touch P_1. Therefore, $\Delta_1 = 1$. Loop 2 does not touch P_2. $\Delta_2 = 1 - G_7$ Loop 4 does not touch P_3. $\Delta_3 = 1 - G_8$. Loop 2 and Loop 4 do not touch P_4.

$$\therefore \Delta_3 = 1-(G_7+G_8)+G_7G_8$$

Step 7: Transfer function $= \dfrac{P_1\Delta_1 + P_2\Delta_2 + P_3\Delta_3 + P_4\Delta_4}{\Delta}$

Example 7.25 Find the transfer function of the SFG shown in Fig. E7.95 using Mason's gain formula.

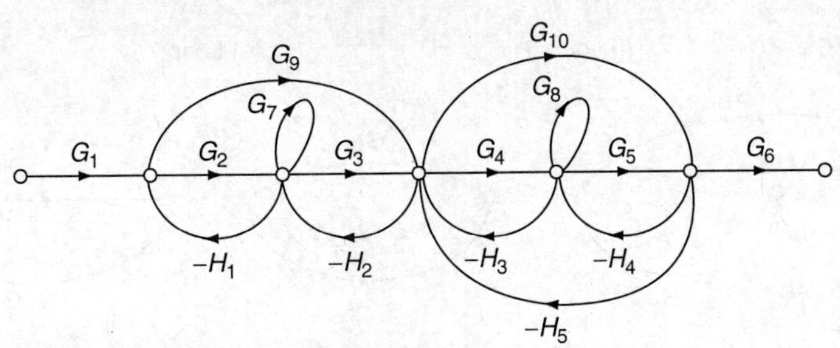

Fig. E7.95

Solution

Step 1: The SFG shown in Fig. E7.95 has the four following forward paths shown in Fig. E.7.96.

Here $P_1 = G_1 G_2 G_3 G_4 G_5 G_6$, $P_2 = G_1 G_4 G_5 G_6 G_9$, $P_3 = G_1 G_2 G_3 G_6 G_{10}$ and $P_4 = G_1 G_6 G_9 G_{10}$.

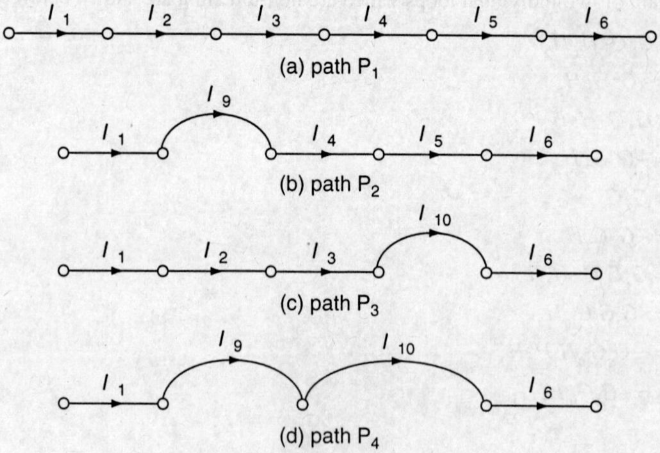

Fig. E7.96

Step 2: The individual loops of the SFG are shown in Fig. E.7.97.

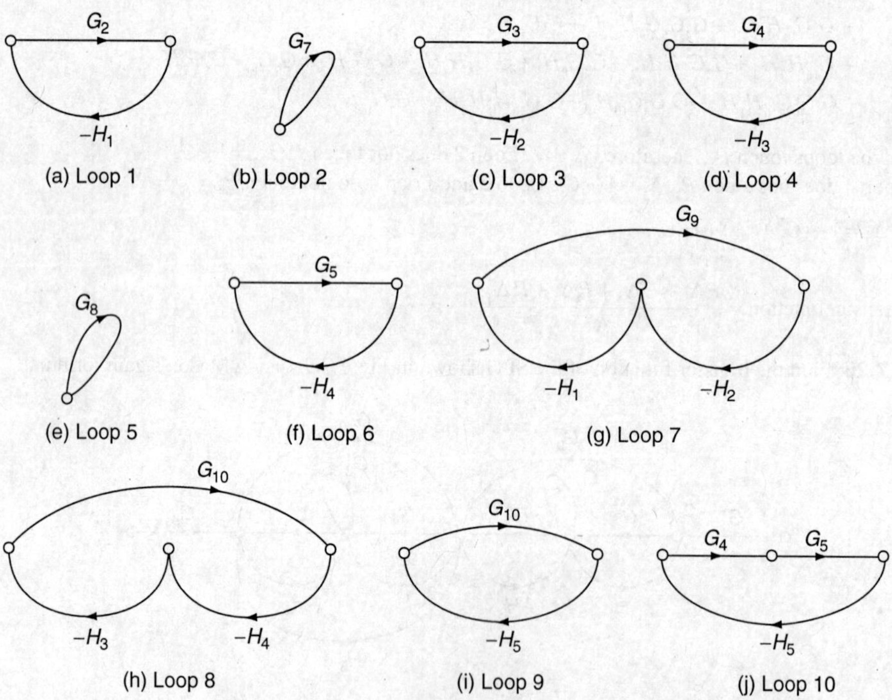

Fig. E7.97

The gain of each individual loop is

$L_{11} = -G_2 H_1, L_{21} = G_7, L_{31} = -G_3 H_2, L_{41} = -G_4 H_3,$

$L_{51} = G_8, L_{61} = -G_5 H_4,$

$L_{71} = G_9 H_1 H_2$, $L_{81} = G_{10} H_3 H_4$, $L_{91} = -G_{10} H_5$
and $L_{10,1} = -G_4 G_5 H_5$

Step 3: Product of gains of two non-touching loops are shown below.

$$L_{12} = L_{11} \times L_{41} = G_2 G_4 H_1 H_3$$
$$L_{22} = L_{11} \times L_{51} = -G_2 G_8 H_1$$
$$L_{32} = L_{11} \times L_{61} = G_2 G_5 H_1 H_4$$
$$L_{42} = L_{21} \times L_{41} = -G_4 G_7 H_3$$
$$L_{52} = L_{21} \times L_{51} = G_7 G_8$$
$$L_{62} = L_{21} \times L_{61} = -G_5 G_7 H_4$$
$$L_{72} = L_{31} \times L_{51} = -G_3 G_8 H_2$$
$$L_{82} = L_{31} \times L_{61} = G_3 G_5 H_2 H_4$$
$$L_{92} = L_{71} \times L_{51} = G_8 G_9 H_1 H_2$$
$$L_{10,2} = L_{71} \times L_{61} = -G_5 G_9 H_1 H_2 H_4$$
$$L_{11,2} = L_{81} \times L_{21} = G_7 G_{10} H_3 H_4$$
$$L_{12,2} = L_{81} \times L_{11} = -G_2 G_{10} H_1 H_3 H_4$$
$$L_{13,2} = L_{91} \times L_{11} = G_2 G_{10} H_1 H_5$$
$$L_{14,2} = L_{91} \times L_{21} = -G_7 G_{10} H_5$$
$$L_{15,2} = L_{91} \times L_{51} = -G_8 G_{10} H_5$$
$$L_{16,2} = L_{10,1} \times L_{11} = G_2 G_4 G_5 H_1 H_5$$
$$L_{17,2} = L_{10,1} \times L_{21} = -G_2 G_4 G_5 H_5$$

Step 4: The product of gains of three non-touching loops are

$$L_{13} = L_{11} L_{51} L_{91} = G_2 G_8 G_{10} H_1 H_5$$
$$L_{23} = L_{21} L_{51} L_{91} = -G_7 G_8 G_{10} H_5$$

Step 5: No higher-order non-touching loops exist.

Step 6:
$$\Delta = 1 - (L_{11} + L_{21} + L_{31} + L_{41} + L_{51} + L_{61} + L_{71} + L_{81} + L_{91} + L_{10,1})$$
$$+ (L_{12} + L_{22} + L_{32} + L_{42} + L_{52} + L_{62} + L_{72} + L_{82} + L_{92}$$
$$+ L_{10,2} + L_{11,2} + L_{12,2} + L_{13,2} + L_{14,2} + L_{15,2} + L_{16,2} + L_{17,2}) - L_{13}$$
$$= 1 + G_2 H_1 - G_7 + G_3 H_2 + G_4 H_3 - G_8 + G_5 H_4 - G_9 H_1 H_2$$
$$- G_{10} H_3 H_4 + G_{10} H_5 + G_4 G_5 H_5 + G_2 G_4 H_1 H_3$$
$$- G_2 G_8 H_1 + G_2 G_5 H_1 H_4 - G_4 G_7 H_3 + G_7 G_8$$
$$- G_5 H_7 H_4 - G_3 G_8 H_2 + G_3 G_5 H_2 H_4$$
$$+ G_8 G_9 H_1 H_2 - G_5 G_9 H_1 H_2 H_4 + G_7 G_{10} H_3 H_4$$
$$- G_2 G_{10} H_1 H_3 H_4 + G_2 G_{10} H_1 H_5 - G_7 G_{10} H_5 - G_8 G_{10} H_5$$
$$+ G_2 G_4 G_5 H_1 H_5 - G_4 G_5 G_7 H_5 - G_2 G_8 G_{10} H_1 H_5 + G_7 G_8 G_{10} H_5$$

Step 7: All loops touch P_1. $\quad \therefore \Delta_1 = 1$
Loop 2 does not touch P_2 $\quad \therefore \Delta_2 = 1 - G_7$

Loop 5 does not touch P_3 $\therefore \Delta_3 = 1 - G_8$
Loop 2 and Loop 5 do not touch P_4 $\therefore \Delta_4 = 1 - G_7 - G_8 + G_7 G_8$

Step 8: $TF = \dfrac{P_1\Delta_1 + P_2\Delta_2 + P_3\Delta_3 + P_4\Delta_4}{\Delta}$

SHORT QUESTIONS WITH ANSWERS

1. Define the following: (i) Signal-flow graph, (ii) source node, (iii) sink node, (iv) chain node, (v) path gain and (vi) dummy node.

Ans. (i) Signal-flow graph is a pictorial representation of a system, which graphically displays the transmission of signals in it.
(ii) The node having only outgoing branches is called input or source node.
(iii) The node having only incoming branches is called sink node or output node.
(iv) The node having both incoming and outgoing branches is known as chain node.
(v) The product of gains through a forward path is called path gain.
(vi) If the incoming as well as outgoing branches exist at the first and last nodes representing input and output variables, these nodes cannot be taken as input and output nodes.

2. State Mason's gain formula.

Ans. To find the transfer function of a complicated system of block diagram, reduction technique is a cumbersome process because step-by-step reduction of block diagram is required. On the other hand, it is possible to get the transfer function easily by using Mason's gain formula shown below:

$$\text{Transfer function} = \frac{C(s)}{R(s)} = \frac{\sum_{i=1}^{N} P_i\Delta_i}{\Delta}$$

where N = total number of forward paths
P_i = gain of the ith forward path
$\Delta = 1 - (\Sigma$ all individual feedback loop gain including self-loops)
+ (gain product of all possible combinations of two non-touching loops)
– (Σ gain product of all possible combinations of three non-touching loops)
+ …
Δ_i = value of Δ after eliminating all loops that touch its forward path.

EXERCISES

Questions For Practice

1. Find the transfer function of the following signal-flow graphs:

(a)

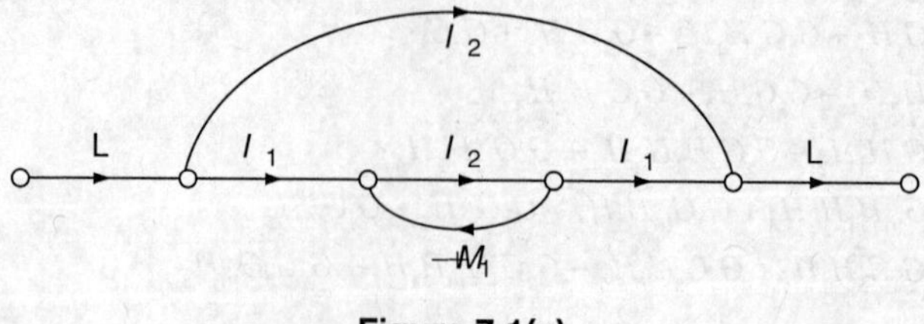

Figure 7.1(a)

(b)

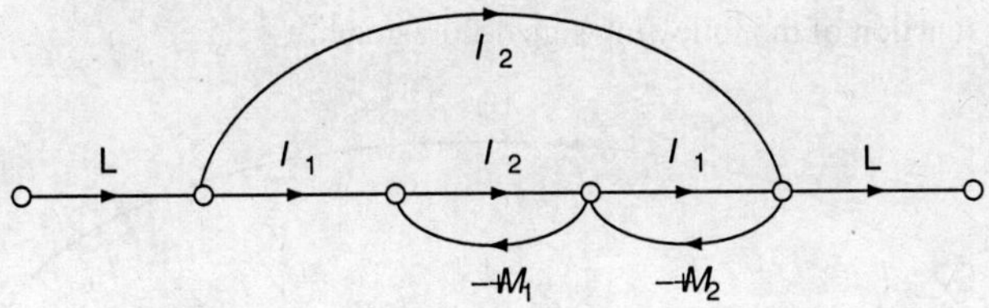

Figure 7.1(b)

(c)

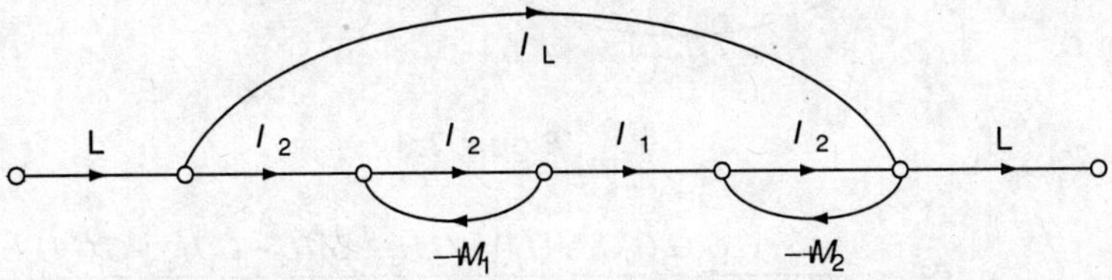

Figure 7.1(c)

Ans. (a) $\dfrac{G_1G_2G_3 + G_4(1+G_2H_1)}{1+G_2H_1}$ (b) $\dfrac{G_1G_2G_3 + G_4(1+G_2H_1)}{1+G_2H_1+G_2H_2}$

(c) $\dfrac{G_1G_2G_3G_4 + G_5(1+G_2H_1)}{1+G_2H_1+G_4H_2+G_2G_4H_1H_2}$

2. Find the transfer function of the following signal-flow graph.

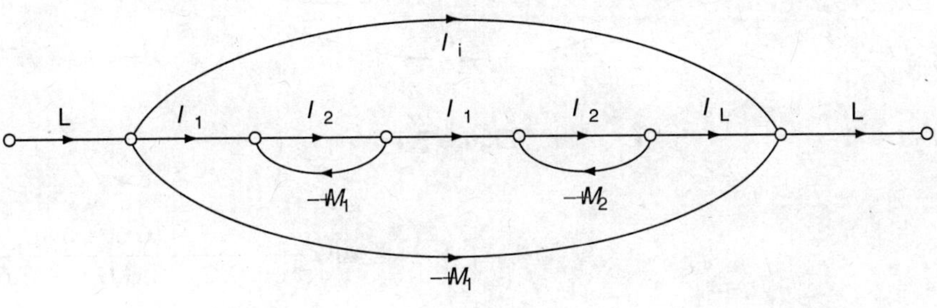

Figure 7.2

Ans. $\dfrac{G_1G_2G_3G_4G_5 + G_6(1+G_1H_1+G_4H_2+G_2G_4H_1H_2)}{\begin{array}{l}1+G_2H_1+G_4H_2+G_1G_2G_3G_4G_5H_3+G_6H_3+G_2G_4H_1H_2+G_2G_6H_1H_3\\ +G_4G_6H_2H_3+G_2G_4G_6H_1H_2H_3\end{array}}$

3. Find the transfer function of the following signal-flow graph.

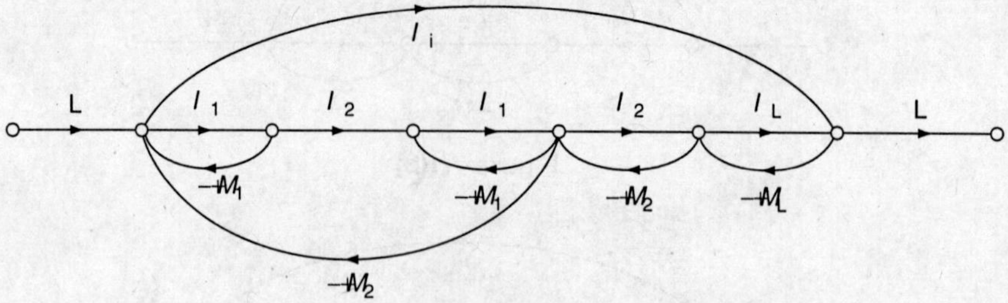

Figure 7.3

Ans. $$\frac{G_1G_2G_3G_4G_5 + G_6(1 + G_1H_1 + G_3H_3 + G_4H_4 + G_1G_3H_1H_3 + G_1G_4H_1H_4)}{\begin{array}{l}1 + G_1H_1 + G_3H_3 + G_1G_2G_3H_2 + G_4H_4 + G_5H_5 + G_6H_2H_4H_5 + G_1G_3H_1H_3 + G_1G_4H_1H_4 \\ + G_1G_5H_1H_5 + G_3G_5H_3H_5 + G_1G_2G_3G_5H_2H_5 + G_1G_3G_5H_1H_3H_5\end{array}}$$

4. Determine the transfer function of the system having the following block diagram.

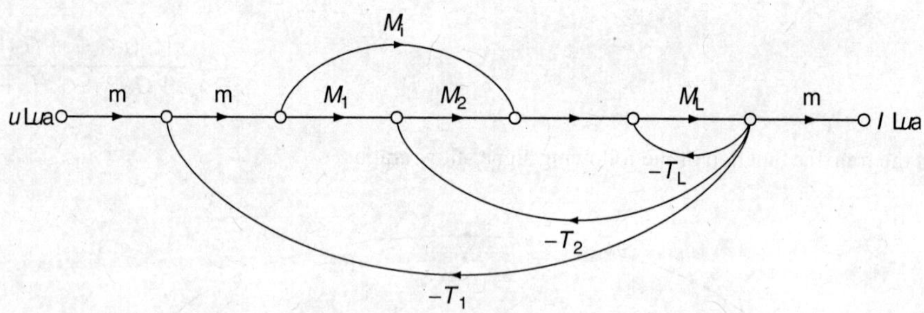

Figure 7.4

Ans. $$\frac{(G_1G_2 + G_4)G_3}{1 + G_1G_2G_3H_1 + G_4G_3H_1 + G_2G_3H_2 + G_3H_3}$$

5. Determine the transfer function of the system having the following block diagram.

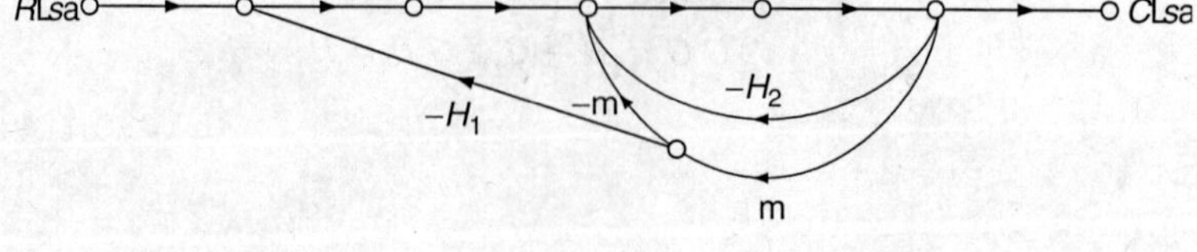

Figure 7.5

Ans. $\dfrac{G_1G_2G_3G_4 + G_1G_4H_3}{1 + G_3G_4 + G_3G_4H_2 + G_1G_4H_1H_2 + G_1G_2G_3G_4H_1}$

6. Determine the transfer function of the system having the following block diagram.

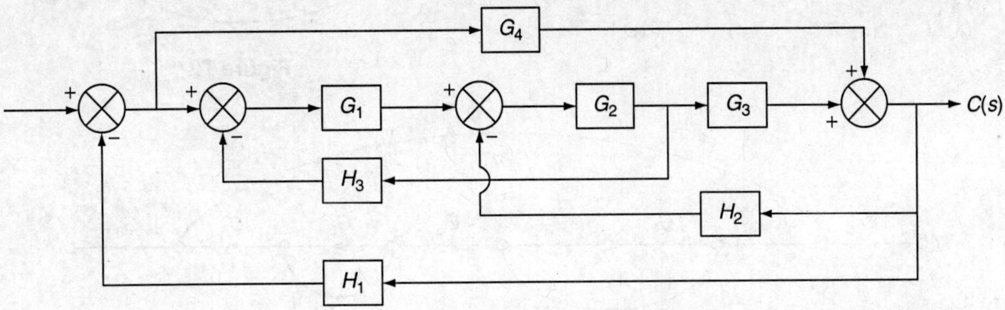

Figure 7.6

Ans. $\dfrac{G_2G_3G_4 + G_4}{1 + G_1G_2G_3H_1 + G_1G_2H_3 + G_2G_3H_3 + G_4H_1 - G_2G_4H_2H_3}$

7. Determine the transfer function of the system having the following block diagram.

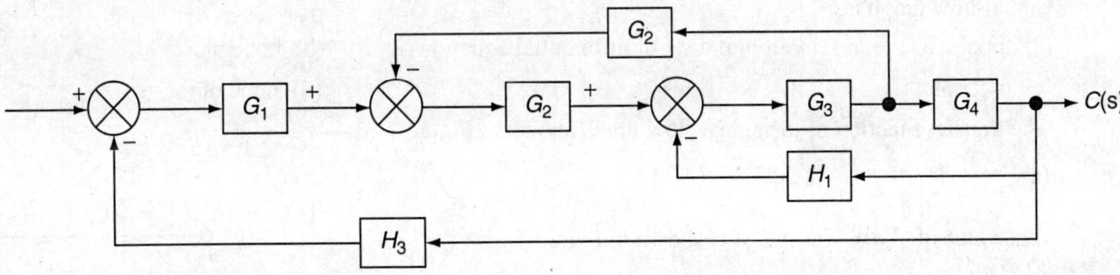

Figure 7.7

Ans. $\dfrac{G_1G_2G_3G_4}{1 + H_1G_2G_3 - H_1G_3G_4 + G_1G_2G_3G_4H_3}$

TEST YOURSELF

1.

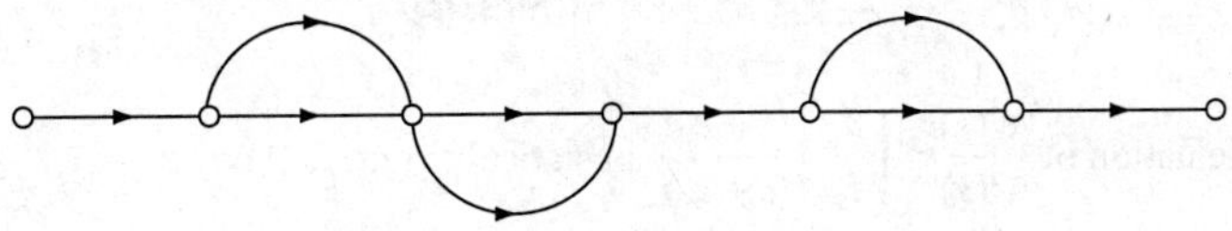

Figure T1

The number of forward paths of the SFG shown Figure T1 is

(a) 6 (b) 2 (c) 7 (d) 5

2. The number of individual loops in the signal-flow graph (Figure T2) is

(a) 5 (b) 6 (c) 3 (d) 7

3. The number of forward path(s) (Fig. 3) is given by

(a) 1 (b) 3 (c) 2 (d) 4

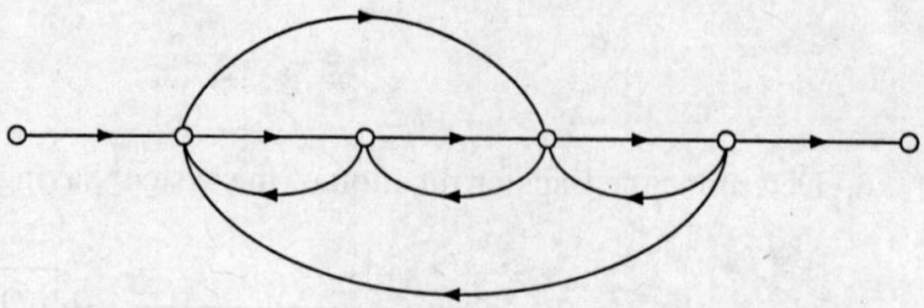

Figure T2

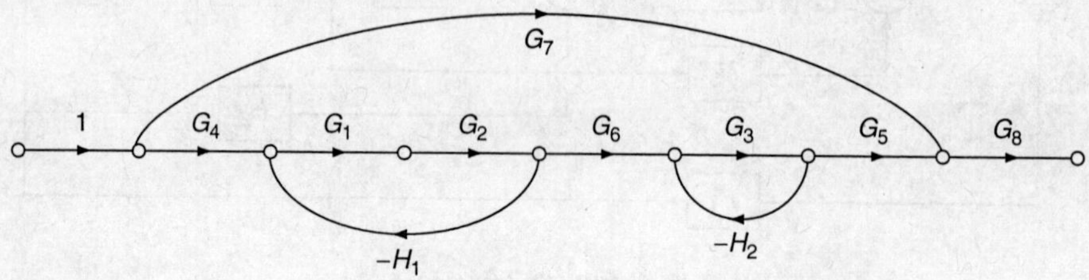

Figure T3

4. The loop gains of the SFG shown in Figure T3

(a) $-G_1G_2H_1, -G_3H_1$ (b) $-G_1G_2H_1, -G_3H_2$ (c) $G_4G_1, H_1H_2G_1$ (d) $-G_1G_2H_1, -G_3H_2$

5. Signal-flow graph is a

(a) topological representation of a set of differential equations (b) bode plot

(c) polar plot (d) none of these

6. The transfer function of the signal-flow graph shown in Figure T4 is

(a) 1 (b) 9 (c) 8 (d) 12

7. The value of X_3/X_1 for the SFG shown in Figure T5 is

(a) 8 (b) 2 (c) 12 (d) 5

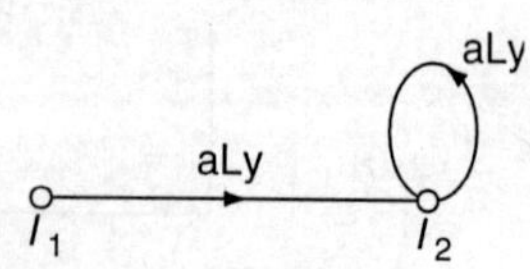

Figure T4

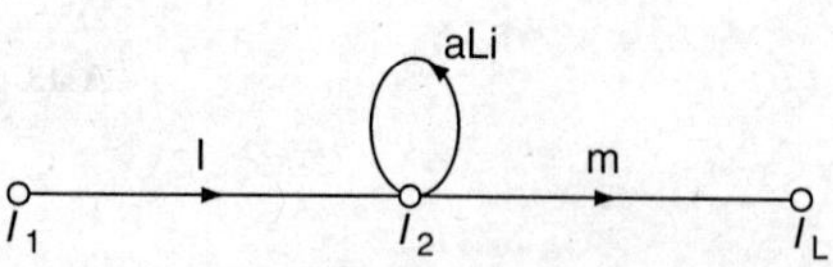

Figure T5

8. The representation of $\dfrac{C(s)}{R(s)} = \left(\dfrac{a_0}{b_0}\right)\dfrac{(s+c_1)}{(s+d_1)}$ in static diagram will be

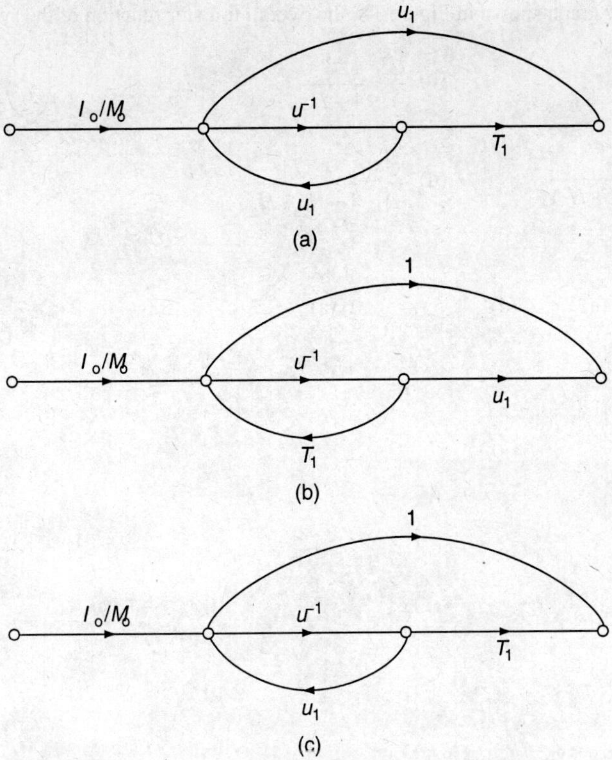

Figure T6

9. The signal-flow graph shown in Figure T7 has

   (a) three forward paths and two non-touching loops
   (b) three forward paths and three loops
   (c) two forward paths and two non-touching loops
   (d) two forward paths and three loops

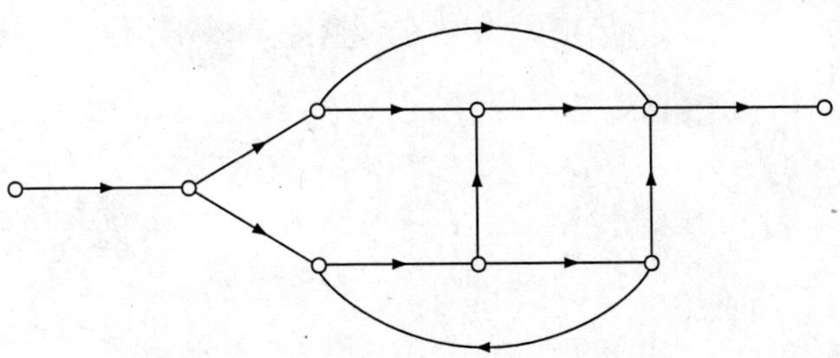

Figure T7

10. For the signal-flow graph shown in Figure T8, the overall transfer function of the system will be

(a) $\frac{C}{R} = G$

(b) $\frac{C}{R} = \frac{G}{1+H_2}$

(c) $\frac{C}{R} = \frac{G}{(1+H_1)(1+H_2)}$

(d) $\frac{C}{R} = \frac{G}{1+H_1+H_2}$

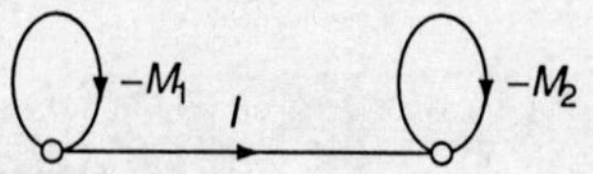

Figure T8

ANSWERS

1 (c) 2 (b) 3 (c) 4 (d) 5 (a) 6 (a) 7 (c) 8 (c) 9 (a) 10 (c)

Chapter 8

Time Domain Analysis of Control Systems

The time response has utmost importance for the design and analysis of control systems because these are inherently time domain systems where time is the independent variable. During the analysis of response, the variation of output with respect to time can be studied and it is known as time response. To obtain satisfactory performance of the system, the output behaviour of the system with respect to time must be within the specified limits. From time response analysis and corresponding results, the stability of system, accuracy of system and complete evaluation can be studied very easily. The standard test input have already been introduced in Chap. 1 and their Laplace transforms have been derived in Chap. 2. The aim of this chapter is to explore time response analysis of control systems, steady-state response, error, and transient response of control systems.

8.1 Classification of Time Responses

Due to the application of an excitation to a system, the response of the system is known as time response and it is a function of time. There are two parts of response of any system: (i) transient response and (ii) steady-state response.

8.1.1 Transient Response

The part of the time response, which goes to zero after large interval of time is known as transient response. In this case $\operatorname*{Lt}_{t\to\infty} c(t) = 0$. From transient response, we have the following information:

(a) The time interval after which the system responds taking the instant of application of excitation as reference.

(b) The total time that it takes to achieve the output for the first time.

(c) Whether or not the output shoots beyond the desired value and how much.

(d) Whether or not the output oscillates about its final value.

(e) The time that it takes to settle to the final value.

8.1.2 Steady-state Response

The part of response that remains even after the transients have died out is said to be steady-state response. From steady-state response, we have the following informations:

(a) The time that output takes to reach the steady-state value.
(b) Whether or not any error exists between the desired and the actual value.
(c) Whether this error is constant, zero or infinite.

The total response of a system is the sum of transient response and steady-state response:

$$c(t) = c_{tr}(t) + c_{ss}(t) \tag{8.1}$$

Figure 8.1 shows the transient and steady-state responses along with steady-state error.

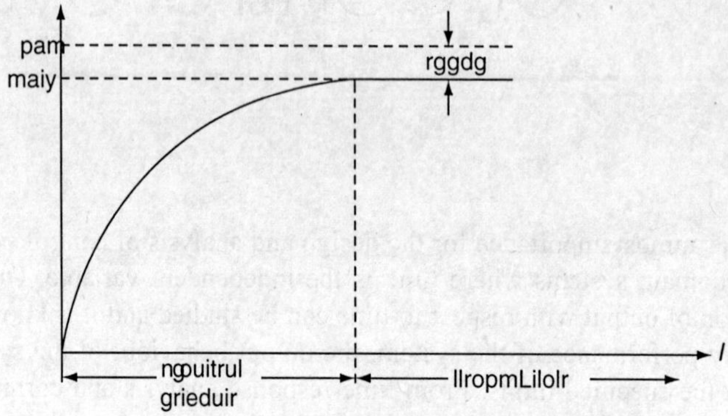

Fig. 8.1 Response of a system and steady-state error

8.2 System Time Response

After applying the excitation at the input terminals of a system, an output response $c(t)$ is produced at the output terminals, which varies with time as shown in Fig. 8.2.

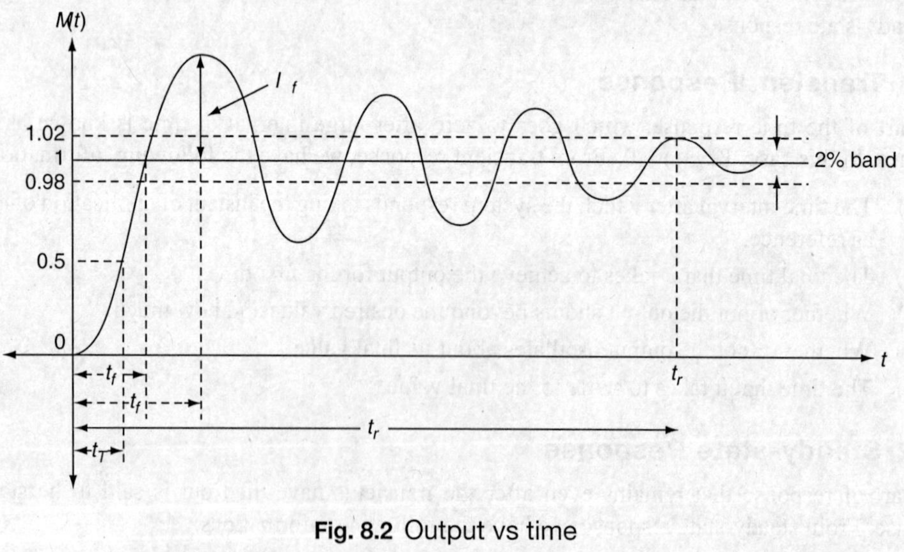

Fig. 8.2 Output vs time

Definitions of various terms related to Fig. 8.2 are shown below:

***Delay time*:** The time that the system output takes to reach 50% of its final value is known as delay time (T_d).

***Rise Time*:** The time that the system response takes to reach from 10% to 90% of the final value of output for overdamped system is called rise time (T_r). It is 0% to 100% for underdamped system. It is 5% to 95% for critically damped system.

***Settling time*:** The time taken by the system response to settle down and stay within ±2% or ± 5% of its final value is known as settling time (T_s).

***Peak time*:** The time taken by the system response to reach the first maximum value is known as peak time (T_p).

***Overshoot (M_p)*:** The ratio of the maximum value of the step-excited output measured from the final value to the final output is called overshoot of the system. The percentage peak overshoot (M_e) of a system is defined as follows:

$$\%\,\text{overshoot} = \left.\frac{\text{maximum output overshoot} - \text{final output}}{\text{Final value of output}}\right|_{\text{step input}} \times 100 \tag{8.2}$$

***Steady-state error*:** For a step excitation, the difference between the desired output and the final value of the output of the system is termed as steady-state error of the system.

8.3 Analysis of Steady-State Error

A simple closed-loop system using negative feedback is shown in Fig. 8.3.

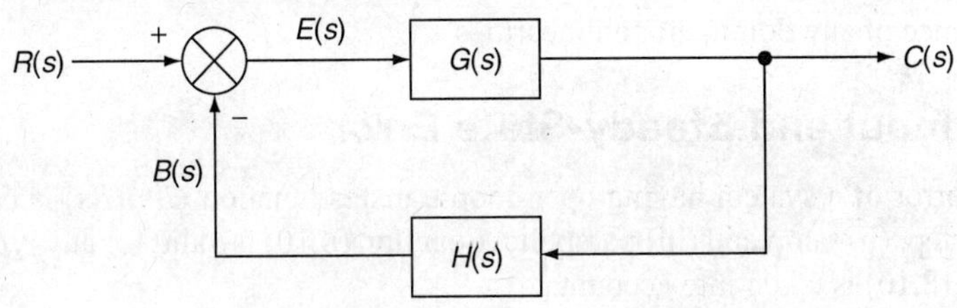

Fig. 8.3 Closed-loop system

In Fig. 8.3, $R(s)$ is the Laplace transform of the applied input $r(t)$, $C(s)$ is the Laplace transform of the output $c(t)$, $E(s)$ is the Laplace transform of the error signal and $B(s)$ is the Laplace transform of the feedback signal. From Fig. 8.3, we have

$$E(s) = R(s) - B(s) \tag{8.3}$$

$\therefore$

$$B(s) = C(s)\,H(s) \tag{8.4}$$

Using Eq. (8.4), from Eq. (8.3) we have

$$E(s) = R(s) - C(s)H(s) \tag{8.5}$$

Again,

$$C(s) = E(s)G(s) \tag{8.6}$$

Using Eq. (8.6), we can write Eq. (8.5) as follows:

$$E(s) = R(s) - E(s)G(s)H(s)$$

i.e.,
$$E(s)[1 + G(s)H(s)] = R(s)$$

i.e.,
$$E(s) = \frac{R(s)}{1 + G(s)H(s)} \tag{8.7}$$

For unity feedback,

$$E(s) = \frac{R(s)}{1 + G(s)} \tag{8.8}$$

For positive feedback, the denominator term in Eq. (8.7) and Eq. (8.8) will be $1 - G(s)\,H(s)$ and $1 - G(s)$ respectively. This error $E(s)$ is in Laplace domain. The corresponding error in time domain will be $e(t)$. During steady state, $t \to \infty$. The steady-state error is expressed by

Steady-state error $(e_{ss}) = \underset{t\to\infty}{\text{Lt}}\ e(t)$

$$\therefore \quad e_{ss} = \underset{t\to\infty}{\text{Lt}}\ e(t) = \underset{s\to 0}{\text{Lt}}\ sE(s) \tag{8.9}$$

Using Eq. (8.8), we have from Eq. (8.9)

$$e_{ss} = \underset{t\to\infty}{\text{Lt}}\ e(t) = \underset{s\to 0}{\text{Lt}}\ sE(s) = \underset{s\to 0}{\text{Lt}}\ \frac{sR(s)}{1 + G(s)H(s)} \tag{8.10}$$

From Eq. (8.10), we can say that the steady-state error depends on the following factors:

(i) type and magnitude of $R(s)$

(ii) open-loop transfer function $G(s)\,H(s)$ and

(iii) the presence of any dominant nonlinearities

8.4 Type of Input and Steady-State Error

The steady-state error of a system having open-loop transfer function $G(s)H(s)$ is calculated for three types of input (i) step, (ii) ramp and (iii) parabolic. Equation (8.10) is valid for any type of input. For the above inputs, Eq. (8.10) is taken into account.

8.4.1 Step Input

Let the step input of magnitude A be applied. The Laplace transform of step input $[r(t)]$ having magnitude A is given by

$$R(s) = \frac{A}{s} \tag{8.11}$$

Using Eq. (8.11), we have from Eq. (8.10),

$$e_{ss} = \underset{s\to 0}{\text{Lt}}\ \frac{sR(s)}{1 + G(s)H(s)} = \underset{s\to 0}{\text{Lt}}\ \frac{s\left(\frac{A}{s}\right)}{1 + G(s)\,H(s)}$$

i.e., $$e_{ss} = \frac{A}{1 + \operatorname*{Lt}_{s \to 0} G(s)H(s)} \tag{8.12}$$

Figure 8.4 shows the variation of $c(t)$ with time.

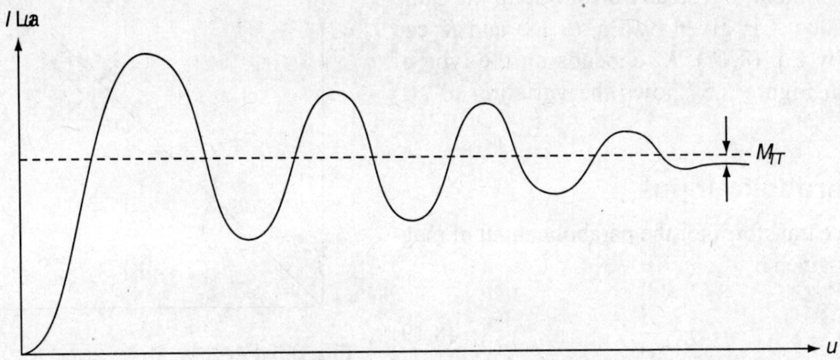

Fig. 8.4 Steady-state error for step input

For a system $\operatorname*{Lt}_{s \to 0} G(s)H(s)$ is constant and it is called positional error constant (K_p)

$$K_p = \operatorname*{Lt}_{s \to 0} G(s)H(s) \tag{8.13}$$

Using Eq. (8.13), Eq. (8.12) can be written as

$$e_{ss} = \frac{A}{1 + K_p} \tag{8.14}$$

For any given system, the steady-state error due to step input of value A is given by Eq. (8.14) and K_p is found by Eq. (8.13). K_p depends on the type of the system.

8.4.2 Ramp Input

The Laplace transform of a ramp input having magnitude A is given by

$$R(s) = \frac{A}{s^2} \tag{8.15}$$

Using Eq. (8.15), we have from Eq. (8.10),

$$e_{ss} = \operatorname*{Lt}_{s \to 0} \frac{s\left(\frac{A}{s^2}\right)}{1 + G(s)H(s)} = \operatorname*{Lt}_{s \to 0} \frac{A}{s + s\,G(s)H(s)} = \frac{A}{\operatorname*{Lt}_{s \to 0} sG(s)H(s)} \tag{8.16}$$

For a system $\operatorname*{Lt}_{s \to 0} sG(s)H(s)$ is constant and it is termed as velocity error coefficient. It is denoted by K_v.

$$\therefore \quad K_v = \operatorname*{Lt}_{s \to 0} sG(s)H(s) \tag{8.17}$$

Using Eq. (8.17), Eq. (8.16) can be expressed as

$$e_{ss} = \frac{A}{K_v} \tag{8.18}$$

For any system, the steady error due to the ramp input of value A is given by Eq. (8.18) and K_v can be found by Eq. (8.17). K_v depends on the type of the system. Figure 8.5 shows the variation of $c(t)$ with time.

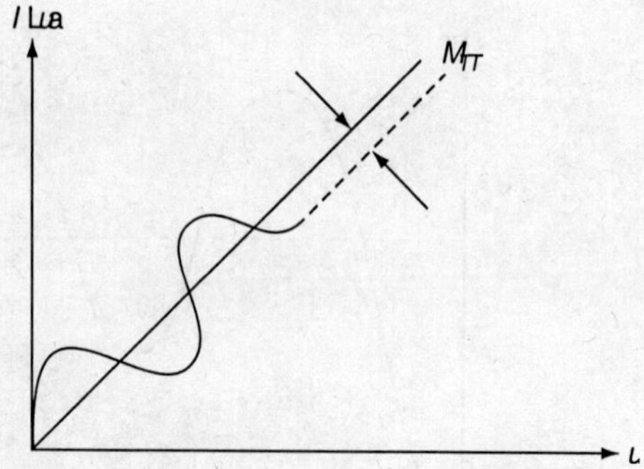

Fig. 8.5 Steady-state error for ramp input

8.4.3 Parabolic Input

The Laplace transform for the parabolic input of magnitude A is given by

$$R(s) = \frac{A}{s^3} \tag{8.19}$$

Using Eq. (8.19), we have from Eq. (8.10),

$$e_{ss} = \underset{s\to 0}{\text{Lt}} \frac{sR(s)}{1+G(s)H(s)} = \underset{s\to 0}{\text{Lt}} \frac{s\left(\frac{A}{s^3}\right)}{1+G(s)H(s)}$$

$$= \underset{s\to 0}{\text{Lt}} \frac{A}{s^2 + s^2G(s)H(s)}$$

i.e.,

$$e_{ss} = \frac{A}{\underset{s\to 0}{\text{Lt}}\, s^2G(s)H(s)} \tag{8.20}$$

For a system, $\underset{s\to 0}{\text{Lt}}\, s^2G(s)H(s)$ is constant and it is termed as acceleration error coefficient. It is denoted by K_a.

$$\therefore \quad K_a = \underset{s\to 0}{\text{Lt}}\, s^2G(s)H(s) \tag{8.21}$$

Using Eq. (8.21), Eq. (8.20) can be expressed as

$$e_{ss} = \frac{A}{K_a} \tag{8.22}$$

For any system, the steady-state error due to the parabolic input of value A is given by Eq. (8.22) and K_a can be determined by Eq. (8.21). K_a depends on the type of the system. Figure 8.6 shows the variation of $c(t)$ with time.

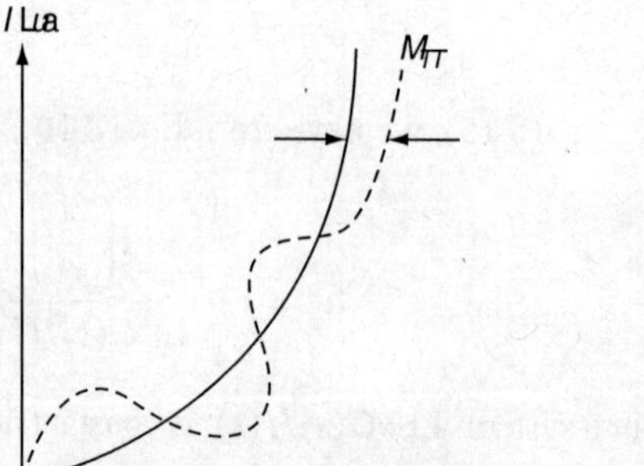

Fig. 8.6 Steady-state error for parabolic input

Table 8.1 summarizes the above results.

Table 8.1 Type of input and steady-state error

Constant	Equation	Steady-state error (e_{ss})
Position error constant (K_p)	$K_p = \underset{s\to 0}{Lt}\, G(s)H(s)$	$e_{ss} = \dfrac{A}{1+K_p}$
Velocity error constant (K_v)	$K_v = \underset{s\to 0}{Lt}\, sG(s)H(s)$	$e_{ss} = \dfrac{A}{K_v}$
Acceleration error constant (K_a)	$K_a = \underset{s\to 0}{Lt}\, s^2G(s)H(s)$	$e_{ss} = \dfrac{A}{K_a}$

8.5 Steady-State Error for Type 0, 1 and 2 Systems

The number of poles of a system having open-loop transfer function $G(s)H(s)$ at $s = 0$ specifies the type of the system. The general form of $G(s)H(s)$ is

$$G(s)H(s) = \frac{K(1+T_1 s)(1+T_2 s)\ldots(1+T_n s)}{s^j(1+T_a s)(1+T_b s)\ldots(1+T_m s)} \tag{8.23}$$

From Eq. (8.23), there are j poles at $s = 0$. Therefore, the type of the system is j.

Example 8.1 Identify the type number:

(i) $G(s) = \dfrac{K(1+3s)}{s^2}$ and $H(s) = \dfrac{1+5s}{(s^2+6s+7)}$

(ii) $G(s) = \dfrac{K}{s}$ and $H(s) = \dfrac{1}{s(s+2)}$

(iii) $G(s) = \dfrac{4}{s^2+6s+7}$ and $H(s) = s+3$

(iv) $G(s) = \dfrac{5}{s}$ and $H(s) = \dfrac{1}{s}$

Solution

(i) $G(s)H(s) = \dfrac{K(1+3s)}{s^2} \times \dfrac{(1+5s)}{(s^2+6s+7)} = \dfrac{K(1+3s)(1+5s)}{s^2(s^2+6s+7)}$ This is a type 2 system.

(ii) $G(s)H(s) = \dfrac{L}{s^2(s+2)}$. This is a type 2 system.

(iii) $G(s)H(s) = \dfrac{4(s+3)}{s^2+6s+7}$. This is a type 0 system.

(iv) $G(s)H(s) = \dfrac{5}{s^2}$. This is a type 2 system.

8.5.1 Error for Step Input

The error for a step input of value A is given by Eq. (8.14) as follows:

$$e_{ss} = \frac{A}{1+K_p} \tag{8.24}$$

Here $$K_p = \operatorname*{Lt}_{s\to 0} G(s)H(s) \tag{8.25}$$

(i) Type 0

In this case $j = 0$. We have from Eq. (8.23)

$$G(s)H(s) = \frac{K(1+T_1 s)(1+T_2 s)\ldots(1+T_n s)}{(1+T_a s)(1+T_b s)\ldots(1+T_m s)} \tag{8.26}$$

Since $K_p = \operatorname*{Lt}_{s\to 0} G(s)H(s)$, we have

∴ $$K_p = \frac{K.1\ldots 1}{1.1\ldots 1} \text{ [using Eq. (8.26)]}$$

i.e., $$K_p = K \tag{8.27}$$

Using Eq. (8.27), Eq. (8.24) can be expressed as

$$e_{ss} = \frac{A}{1+K} \tag{8.28}$$

Figure 8.7 shows the plot of $c(t)$ versus t for type 0 system and also shows the error e_{ss}.

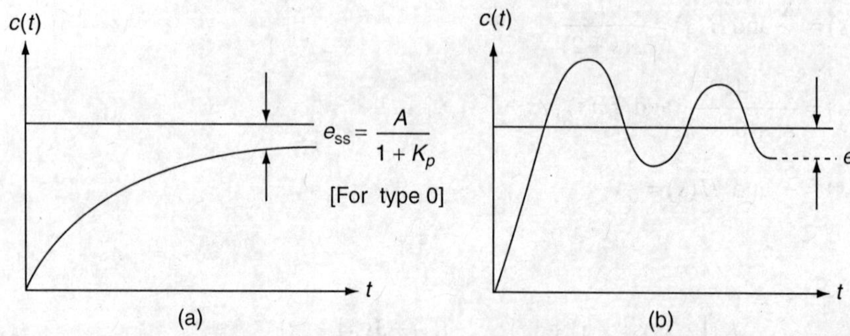

Fig. 8.7 Error for step input and type 0 system

(ii) Type 1

In this case $j = 1$. We have from Eq. (8.23)

∴ $$G(s)H(s) = \frac{K(1+T_1 s)(1+T_2 s)\ldots(1+T_n s)}{s(1+T_a s)(1+T_b s)\ldots(1+T_m s)} \tag{8.29}$$

$$\therefore \qquad K_p = \underset{s\to 0}{\text{Lt}}\, G(s)H(s) = \infty \tag{8.30}$$

$$\therefore \qquad e_{ss} = \frac{A}{1+K_p} = 0 \tag{8.31}$$

Figure 8.8 shows e_{ss} for type 1 system for step input and it shows that the type 1 system. tracks a unit step input without any steady-state error.

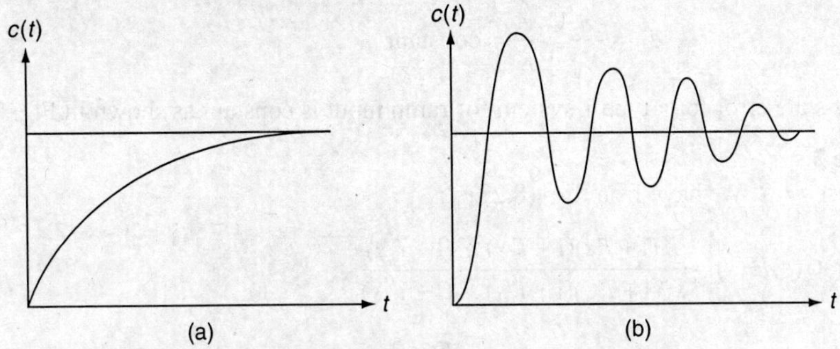

Fig. 8.8 Error for step input and type 1 system

Similarly, $K_p = \infty$ and $e_{ss} = 0$ for step input and for higher-order system. There are no steady-state error for type 1 and higher-order systems, i.e., $j \geq 1$.

8.5.2 Error for Ramp Input

The steady-state error for ramp input of value A is given by the equation

$$e_{ss} = \underset{s\to 0}{\text{Lt}} = \frac{A}{K_v} \tag{8.32}$$

where

$$K_v = \underset{s\to 0}{\text{Lt}}\, sG(s)H(s) \tag{8.33}$$

(i) For Type 0

In this case $j = 0$. We have from Eq. (8.23)

$$G(s)H(s) = \frac{K(1+T_1s)(1+T_2s)\ldots(1+T_ns)}{(1+T_es)(1+T_bs)\ldots(1+T_ms)} \tag{8.34}$$

$$\therefore \qquad K_v = \underset{s\to 0}{\text{Lt}}\, sG(s)H(s) = 0 \tag{8.35}$$

and

$$e_{ss} = \frac{A}{K_v} = \frac{A}{0} = \infty \tag{8.36}$$

Figure 8.9 shows e_{ss} for a type 0 system for ramp input.

Type 0 system fails to keep track of a ramp input due to increment of errors continuously.

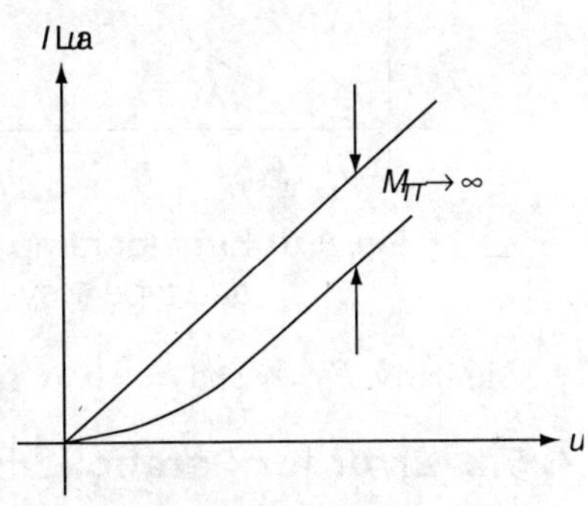

Fig. 8.9 Error for ramp input and type 0 system

(ii) For Type 1

In this case $j = 1$. We have from Eq. (8.23)

$$\therefore \quad G(s)H(s) = \frac{K(1+T_1 s)(1+T_2 s)\ldots(1+T_n s)}{s(1+T_a s)(1+T_b s)\ldots(1+T_m s)} \tag{8.37}$$

$$\therefore \quad K_v = \underset{s\to 0}{\text{Lt}}\, sG(s)H(s) = \underset{s\to 0}{\text{Lt}} \frac{K(1+T_1 s)(1+T_2 s)\ldots(1+T_n s)}{(1+T_a s)(1+T_b s)\ldots(1+T_m s)} = K \tag{8.38}$$

and
$$e_{ss} = \frac{A}{K_v} = \frac{A}{K} = \text{constant} \tag{8.39}$$

The steady-state error for a type 1 system for ramp input is constant as shown in Fig. 8.10.

(iii) For Type 2

In this case $j = 2$. We have from Eq. (8.23)

$$\therefore \quad G(s)H(s) = \frac{K(1+T_1 s)(1+T_2 s)\ldots(1+T_n s)}{s^2(1+T_a s)(1+T_b s)\ldots(1+T_m s)} \tag{8.40}$$

$$\therefore \quad K_v = \underset{s\to 0}{\text{Lt}}\, sG(s)H(s) = \underset{s\to 0}{\text{Lt}}\, s\left[\frac{K(1+T_1 s)(1+T_2 s)\ldots(1+T_n s)}{s^2(1+T_a s)(1+T_b s)\ldots(1+T_m s)}\right] = \infty \tag{8.41}$$

and
$$e_{ss} = \frac{A}{K_v} = \frac{A}{\infty} = 0 \tag{8.42}$$

The steady-error for type 2 for ramp input is zero as shown in Fig. 8.11.

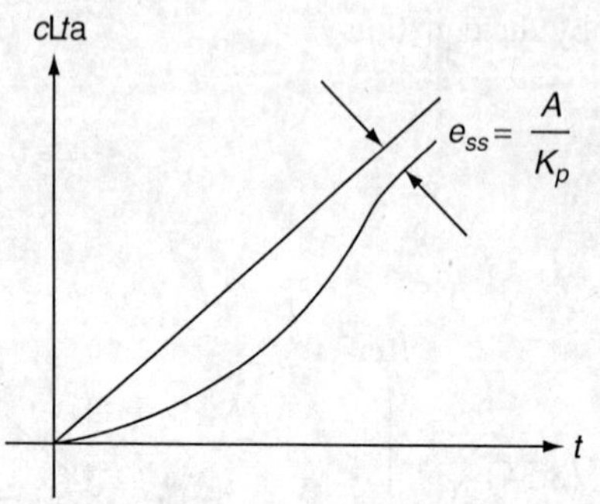

Fig. 8.10 Error for ramp input and type 1 system

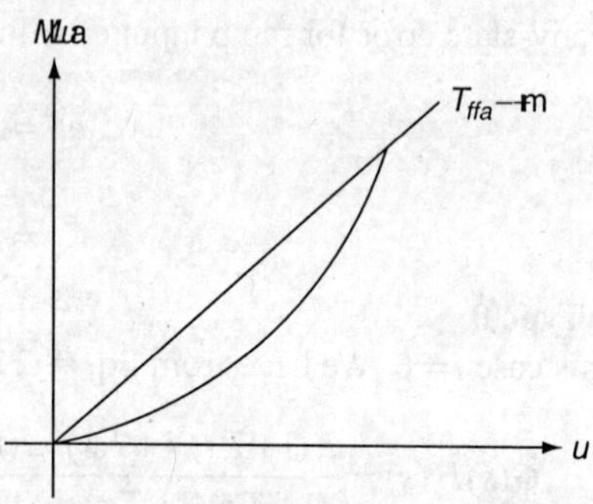

Fig. 8.11 Error for ramp input and type 2 system

Similarly, $K_v = \infty$ and $e_{ss} = 0$ for ramp input and higher-order systems, i.e., $j \geq 2$.

8.5.3 Error for Parabolic Input

The steady-state error for a parabolic input of value A is given by

$$e_{ss} = \frac{A}{K_a} \tag{8.43}$$

and $$K_a = \underset{s\to 0}{\text{Lt}}\, s^2G(s)H(s) \tag{8.44}$$

(i) For Type 0

In this case $j = 2$. We have from Eq. (8.23)

$$G(s)H(s) = \frac{K(1+T_1s)(1+T_2s)\ldots(1+T_ns)}{s(1+T_as)(1+T_bs)\ldots(1+T_ns)} \tag{8.45}$$

$$K_a = \underset{s\to 0}{\text{Lt}}\, s^2G(s)H(s) = \underset{s\to 0}{\text{Lt}}\, s^2\left[\frac{K(1+T_1s)(1+T_2s)\ldots(1+T_ns)}{(1+T_as)(1+T_ns)\ldots(1+T_ns)}\right] = 0 \tag{8.46}$$

$$\therefore \quad e_{ss} = \frac{a}{K_a} = \frac{A}{0} = \infty \tag{8.47}$$

The type 0 system fails to keep track of a parabolic input due to increment of error continuously.

(ii) For Type 1

In this case $j = 2$. We have from Eq. (8.23)

$$G(s)H(s) = \frac{K(1+T_1s)(1+T_2s)\ldots(1+T_ns)}{s(1+T_as)(1+T_bs)\ldots(1+T_ns)} \tag{8.48}$$

$$K_a = \underset{s\to 0}{\text{Lt}}\, sG(s)H(s) = \underset{s\to 0}{\text{Lt}}\, s^2\cdot\left[\frac{K(1+T_1s)(1+T_2s)\ldots(1+T_ns)}{s(1+T_as)(1+T_bs)\ldots(1+T_ms)}\right] = 0 \tag{8.49}$$

$$\therefore \quad e_{ss} = \frac{a}{K_a} = \frac{A}{0} = \infty \tag{8.50}$$

Figure 8.12 shows the error for parabolic input and for type 1 system fails to keep track of a parabolic input due to increment of error continuously.

(iii) For Type 2

In this case $j = 2$. We have from Eq. (8.23),

$$G(s)H(s) = \frac{K(1+T_1s)(1+T_2s)\ldots(1+T_ns)}{s^2(1+T_as)(1+T_bs)\ldots(1+T_ms)} \tag{8.51}$$

$$\underset{s\to 0}{\text{Lt}}\, s^2G(s)H(s) = \underset{s\to 0}{\text{Lt}}\, s^2\left[\frac{K(1+T_1s)(1+T_2s)\ldots(1+T_ns)}{s^2(1+T_as)(1+T_bs)\ldots(1+T_ms)}\right] = K \tag{8.52}$$

$$\therefore \quad e_{ss} = \frac{A}{K_a} = \frac{A}{K}\text{ constant} \tag{8.53}$$

Figure 8.13 shows error for parabolic input and type 2 system where finite error is present for a type 2 system.

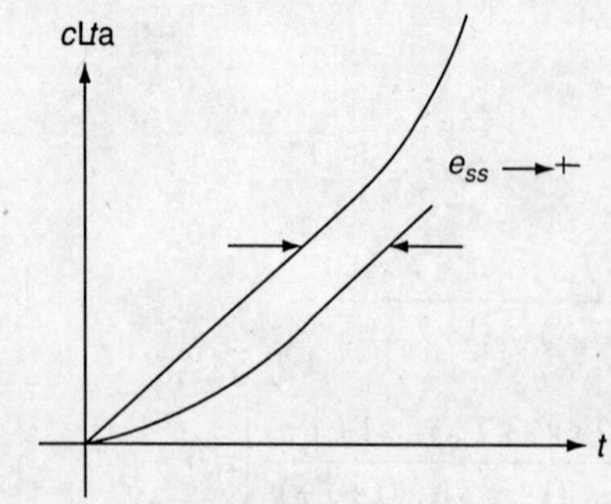

Fig. 8.12 Error for parabolic input and type 1 system

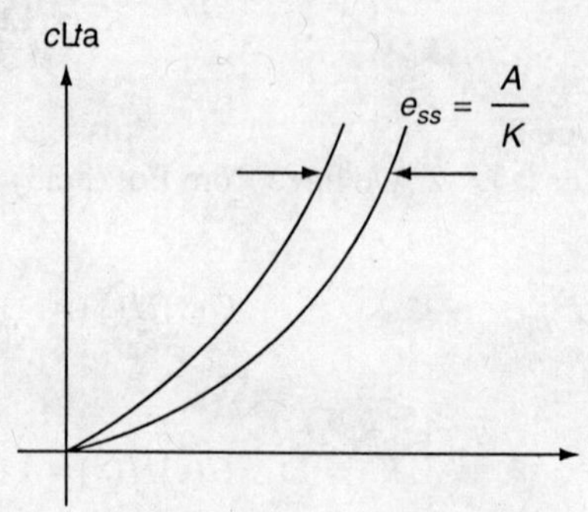

Fig. 8.13 Error for parabolic input and type 2 system

Table 8.2 summarises the above conclusions.

Table 8.2 Error constant and error for different types of input

Type	Step input		Ramp input		Parabolic input	
	K_p	e_{ss}	K_v	e_{ss}	K_a	e_{ss}
Type 0	K	$\frac{A}{1+K}$	0	∞	0	∞
Type 1	∞	0	K	$\frac{A}{K}$	0	∞
Type 2	∞	0	∞	0	K	$\frac{A}{K}$

The steady-state errors can be reduced by increasing the value of K, but there are limitations from stability point of view.

Example 8.2 Find the error coefficients for a system having

$$G(s)H(s) = \frac{(s+3)}{s(1+0.60s)(1+0.35s)}$$

Solution

$$K_p = \underset{s\to 0}{\text{Lt}}\, G(s)H(s) = \underset{s\to 0}{\text{Lt}} \frac{(s+3)}{s(1+0.60s)(1+0.35s)} = \infty$$

$$\therefore \quad K_v = \underset{s\to 0}{\text{Lt}}\, sG(s)H(s) = \underset{s\to 0}{\text{Lt}} \frac{(s+3)}{(1+0.60s)(1+0.35s)} = 3$$

and $$K_a = \underset{s\to 0}{\text{Lt}}\, s^2G(s)H(s) = \underset{s\to 0}{\text{Lt}} \frac{s(s+3)}{(1+0.60s)(1+0.35s)} = 0$$

Example 8.3 Find K_p, K_v, K_a for the system having

(i) $G(s) = \frac{10}{s^2}$ and $H(s) = 0.7$

(ii) $G(s)=\dfrac{5}{s^2+3s+5}$ and $H(s)=0.6$

(iii) $G(s)=\dfrac{10(1+0.5s)(1+0.8s)}{s^2(s^2+3s+5)}$ and $H(s)=0.8$

Solution

(i) $K_p=\underset{s\to 0}{\text{Lt}}\, G(s)H(s)=\underset{s\to 0}{\text{Lt}}\dfrac{7}{s^2}=\infty$

$K_v=\underset{s\to 0}{\text{Lt}}\, sG(s)H(s)=\underset{s\to 0}{\text{Lt}}\dfrac{7}{s}=\infty$

$K_a=\underset{s\to 0}{\text{Lt}}\, s^2G(s)H(s)=\underset{s\to 0}{\text{Lt}}\dfrac{7s^2}{s^2}=7$

(ii) $K_p=\underset{s\to 0}{\text{Lt}}\, G(s)H(s)=\underset{s\to 0}{\text{Lt}}\dfrac{3}{s^2+3s+5}=\dfrac{3}{5}$

$K_p=\underset{s\to 0}{\text{Lt}}\, sG(s)H(s)=\underset{s\to 0}{\text{Lt}}\dfrac{3s}{s^2+3s+5}=0$

$K_a=\underset{s\to 0}{\text{Lt}}\, s^2G(s)H(s)=\underset{s\to 0}{\text{Lt}}\dfrac{3s^2}{s^2+3s+5}=0$

(iii) $K_p=\underset{s\to 0}{\text{Lt}}\, G(s)H(s)=\underset{s\to 0}{\text{Lt}}\dfrac{8(1+0.5s)(1+0.8s)}{s^2(s^2+3s+5)}=\infty$

$K_v=\underset{s\to 0}{\text{Lt}}\, sG(s)H(s)=\underset{s\to 0}{\text{Lt}}\dfrac{8(1+0.5s)(1+0.8s)}{s(s^2+3s+5)}=\infty$

$K_a=\underset{s\to 0}{\text{Lt}}\, s^2G(s)H(s)=\underset{s\to 0}{\text{Lt}}\dfrac{8(1+0.5s)(1+0.8s)}{(s^2+3s+5)}=\dfrac{8}{5}$

Example 8.4 For a system having $G(s)H(s)=\dfrac{K(s+4)}{s(s^3+5s^2+6s)}$, find (i) the type of the system, (ii) error coefficients and (iii) error due to input $\dfrac{A}{2}t^2$.

Solution

(i) $G(s)H(s)=\dfrac{K(s+4)}{s(s^3+5s^2+6s)}=\dfrac{K(s+4)}{s^2(s^2+5s+6)}$

$$=\frac{K(s+4)}{s^2(s+2)(s+3)}=\frac{2K(1+0.25s)}{3s^2(1+0.5s)\left(1+\frac{1}{3}s\right)}$$

This is a type 2 system

(ii) Since it is a type 2 system, K_p and K_v both are ∞.

$$K_a = \underset{s\to 0}{\text{Lt}}\ s^2G(s)H(s) = \underset{s\to 0}{\text{Lt}}\ \frac{2K(1+0.25s)}{3(1+0.5s)\left(1+\frac{1}{3}s\right)} = \frac{2K}{3}$$

(iii) Since input is parabolic,

$$e_{ss} = \frac{A}{\frac{2K}{3}} = \frac{3A}{2K}$$

The steady-state error $= \dfrac{3A}{2K}$.

8.6 Advantages and Disadvantages of Static Error Coefficient Method

(i) This method is applicable only to the three standard input signals and cannot give error coefficients for other input signals.

(ii) Mathematical response of error, i.e., 0 or ∞, cannot give exact value of error.

(iii) This method can provide the important time variation of error.

(iv) This method can be used for a stable system.

8.7 Generalized Error Coefficient Method

The disadvantages of the static error coefficient method can be overcome by the generalized error coefficient method. This is also known as the dynamic error coefficient method.

From Eq. (8.7), we have

$$E(s) = \frac{R(s)}{1+G(s)H(s)} = \left[\frac{1}{1+G(s)H(s)}\right]R(s)$$

$$E(s) = F_1(s)F_2(s) \tag{8.54}$$

where $F_1(s) = \dfrac{1}{1+G(s)H(s)}$ and $F_2(s) = R(s)$

Using convolution integral, we have from Eq. (8.54)

$$e(t) = \int_0^t f_1(T)f_2(t-T)\,dT$$

i.e.,

$$e(t) = \int_0^t f_1(T)f_2(t-T)dT = \int_0^t f_1(T)r(t-T)dT \tag{8.55}$$

Using Taylor's series, we have

$$r(t-T) = r(t) - Tr'(t) + \frac{T^2}{2!}r''(t) - \frac{T^3}{3!}r'''(t) + \ldots. \tag{8.56}$$

Using Eq. (8.56), Eq. (8.55) can be expressed as

$$e(t) = \int_0^t f_1(T)r(t)\,dT - \int_0^t Tr'(t)f_1(T)dT + \frac{1}{2!}\int_0^t T^2r''(t)f_1(T)dT + \ldots.$$

Again $e_{ss} = \underset{t \to \infty}{\text{Lt}}\, e(t)$

$\therefore$ $$e_{ss} = \underset{t \to \infty}{\text{Lt}} \left[\int_0^t r(t) f_1(T)dT - \int_0^t r'(t) f_1(T)dT + \int_0^t \frac{T^2}{2!} r''(t) f_1(T)dT + \ldots \right]$$

$$= r(t) \int_0^\infty f_1(T)\, dT - r'(t) \int_0^\infty Tf_1(T)\, dT + r''(t) \int_0^\infty \frac{T^2}{2!} f_1\, (T)dT + \ldots \tag{8.57}$$

Putting $K_0 = \int_0^\infty f_1(T)\, dT, K_1 = -\int_0^\infty Tf_1(T)\, dT, K_2 = \int_0^\infty T^2 f_1(T)dT + \ldots$

Equation (8.57) can be expressed as

$$e_{ss} = K_0\, r(t) + K_1 r'(t) + \frac{K_2}{2!} r''(t) + \ldots \tag{8.58}$$

where K_0, K_1, K_2..., are known as dynamic error coefficients.

Now $$K_0 = \int_0^\infty f_1(T)\, dT$$

$\therefore$ $$\underset{s \to 0}{\text{Lt}}\, K_0 e^{-sT} = \underset{s \to 0}{\text{Lt}} \int_0^\infty f_1(T) e^{-sT} dT = \underset{s \to 0}{\text{Lt}}\, F_1(s) \tag{8.59}$$

Now $$\underset{s \to 0}{\text{Lt}}\, K_0 e^{-sT} = \underset{s \to 0}{\text{Lt}}\, F_1(s)$$

i.e., $$K_0 = \underset{s \to 0}{\text{Lt}}\, F_1(s) \tag{8.60}$$

Taking derivative of Eq. (8.59), we have

$$\underset{s \to 0}{\text{Lt}} \int_0^\infty -Tf_1(T) e^{-sT} dT = \underset{s \to 0}{\text{Lt}} \frac{dF_1(s)}{ds}$$

i.e., $$-\int_0^\infty T f_1(T) dT = \underset{s \to 0}{\text{Lt}} \frac{dF_1(s)}{ds}$$

i.e., $$K_1 = \underset{s \to 0}{\text{Lt}} \frac{dF_1(s)}{ds}$$

we have,

$$\left.\begin{aligned} K_0 &= \underset{s \to 0}{\text{Lt}}\, F_1(s) \\ K_1 &= \underset{s \to 0}{\text{Lt}} \frac{dF_1(s)}{ds} \\ K_2 &= \underset{s \to 0}{\text{Lt}} \frac{d^2 F_1(s)}{ds^2} \\ &\vdots \\ K_n &= \underset{s \to 0}{\text{Lt}} \frac{d^n F_1(s)}{ds^n} \end{aligned}\right\} \tag{8.61}$$

The disadvantages of the static error coefficient method are eliminated in this method. Equation (8.61) represents the dynamic error co-efficients for the error series shown by Eq. (8.58).

Example 8.5 For the system shown in Fig. E8.1, find the error using the dynamic error coefficient method for input of $4+3t+\frac{8}{2}t^2$.

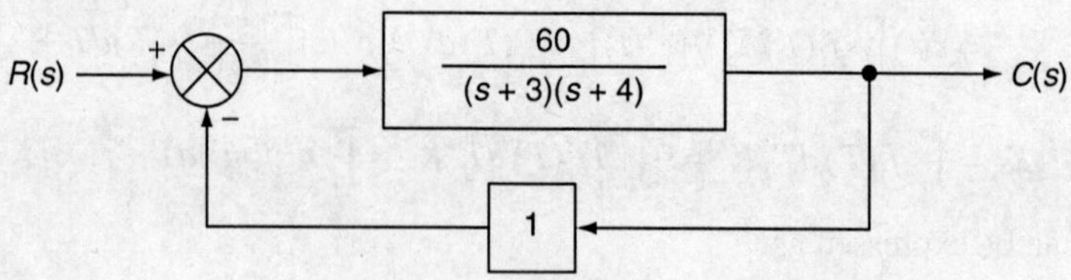

Fig. E8.1

Solution

Given

$$G(s)=\frac{60}{(s+3)(s+4)}, H(s)=1, \text{ and } r(t)=4+3t+\frac{8}{2}t^2$$

For the dynamic coefficient method,

$$e_{ss}(t)=K_0 r(t)+K_1 r'(t)+\frac{K_2}{2!}r''(t)+\ldots$$

where
$$K_0=\underset{s\to 0}{\text{Lt}}\,F_1(s), K_1=\underset{s\to 0}{\text{Lt}}\,\frac{dF_1(s)}{ds}, K_2=\underset{s\to 0}{\text{Lt}}\,\frac{d^2F_1(s)}{ds^2}$$

and
$$F_1(s)=\frac{1}{1+G(s)H(s)}=\frac{1}{1+\frac{60}{(s+3)(s+4)}}$$

$$=\frac{(s+3)(s+4)}{(s+3)(s+4)+60}=\frac{s^2+7s+12}{s^2+7s+72}$$

$$K_0=\underset{s\to 0}{\text{Lt}}\,F_1(s)=\underset{s\to 0}{\text{Lt}}\left[\frac{s^2+7s+12}{s^2+7s+72}\right]=\frac{12}{72}=\frac{1}{6}=0.1667$$

$$K_1=\underset{s\to 0}{\text{Lt}}\,\frac{dF_1(s)}{ds}=\underset{s\to 0}{\text{Lt}}\,\frac{d}{ds}\left[\frac{s^2+7s+12}{s^2+7s+72}\right]$$

$$=\underset{s\to 0}{\text{Lt}}\,\frac{(s^2+7s+72)(2s+7)-(s^2+7s+12)(2s+7)}{(s^2+7s+72)^2}$$

$$=\underset{s\to 0}{\text{Lt}}\,\frac{(s^2+7s+72-s^2-7s-12)(2s+7)}{(s^2+7s+72)^2}$$

$$=\underset{s\to 0}{\text{Lt}}\,\frac{60(2s+7)}{(s^2+7s+72)^2}=\frac{60\times 7}{(72)^2}=0.0810$$

$$K_2 = \underset{s\to 0}{\text{Lt}} \frac{d^2F_2(s)}{ds^2} = \underset{s\to 0}{\text{Lt}}\, 60 \frac{d}{ds}\left[\frac{2s+7}{(s^2+7s+72)^2}\right]$$

$$= 60 \underset{s\to 0}{\text{Lt}} \left\{\frac{(s^2+7s+72)^2(2) - 2(2s+7)(s^2+7s+72)(2s+7)}{(s^2+7s+72)^4}\right\}$$

$$= \underset{s\to 0}{\text{Lt}} \frac{(2\times 72^2 - 2\times 7\times 72\times 7)\times 60}{(72)^4} = 7.3945\times 10^{-3}$$

$\therefore$ $$e_{ss}(t) = K_0 r(t) + K_1 r'(t) + \frac{K_2}{2!} r'(t)$$

Since $$r(t) = 4 + 3t + \frac{8}{2}t^2$$

$\therefore$ $$r'(t) = 3 + 8t \text{ and } r''(t) = 8$$

$\therefore$ $$e_{ss} = 0.1667(4+3t+4t^2) + 0.0810(3+8t) + \frac{7.3945\times 10^{-3}\times 8}{2}$$

$$= 0.9394 + 1.1481t + 0.6668t^2$$

8.8 Analysis of a First-Order System

Figure 8.14 shows a first-order system.

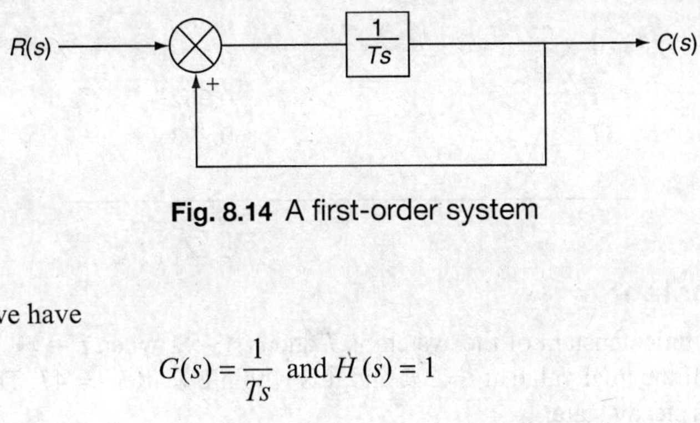

Fig. 8.14 A first-order system

From Fig. 8.14, we have

$$G(s) = \frac{1}{Ts} \text{ and } H(s) = 1$$

$\therefore$ $$\frac{C(s)}{R(s)} = \frac{G(s)}{1+G(s)H(s)} = \frac{\frac{1}{Ts}}{1+\frac{1}{Ts}} = \frac{1}{1+Ts} \tag{8.62}$$

Now $$G(s)H(s) = \frac{1}{sT}\times 1 = \frac{1}{sT} \tag{8.63}$$

The pole of $G(s)H(s)$ is at $s = 0$ and it is type 1. It should not have steady-state error for step input.

8.8.1 Unit Step Response

Here
$$r(t) = u(t)$$

$$\therefore \quad R(s) = \frac{1}{s} \tag{8.64}$$

Using Eq. (8.64), Eq. (8.62) can be expressed as

$$C(s) = \left(\frac{1}{1+Ts}\right)R(s) = \frac{1}{s(1+Ts)} = \frac{\frac{1}{T}}{s\left(s+\frac{1}{T}\right)} = \frac{1}{s} - \frac{1}{s+\frac{1}{T}}$$

$$\therefore \quad c(t) = 1 - e^{-\frac{t}{T}} \tag{8.65}$$

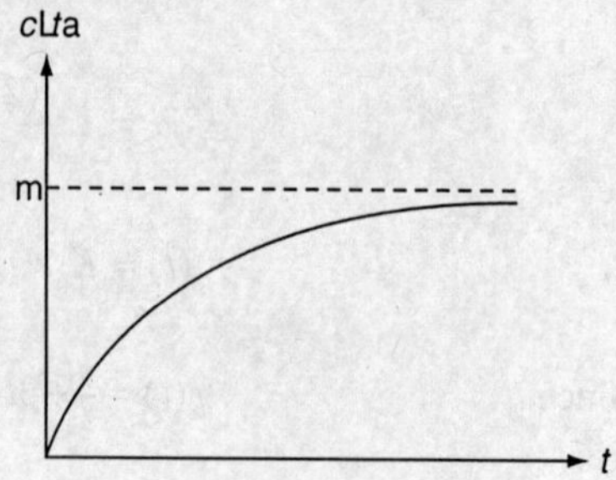

Fig. 8.15 Plot of $c(t)$ vs t for a first-order system

The plot $c(t)$ versus t is shown in Fig. 8.15.

From Fig. 8.15, it is obvious that the response is purely exponential. Table 8.3 shows the response values.

Table 8.3 Response values

t	c(t)
T	0.632
$2T$	0.860
$3T$	0.950
$4T$	0.982
$5T$	0.993
∞	1

8.8.2 Time Constant

From Eq. (8.65), the time constant of the system is T and $c(t) \to 1$ when $t \to \infty$. In practice, the value of $c(t)$ is within 5% of the final value at $t = 3T$, and it is within 2% after $t = 4T$. Therefore, $t = 3T$ or $4T$ may be considered as steady state.

8.8.3 Poles of Closed-Loop Transfer Function

From Eq. (8.59), the pole of this first-order system can be obtained by solving

$$1 + sT = 0$$

i.e.,
$$s = -\frac{1}{T} \tag{8.66}$$

Two cases are considered here for which $T_1 > T_2$, where T_1 and T_2 are two time constants for comparison. Figure 8.16 shows this plot on s-plane.

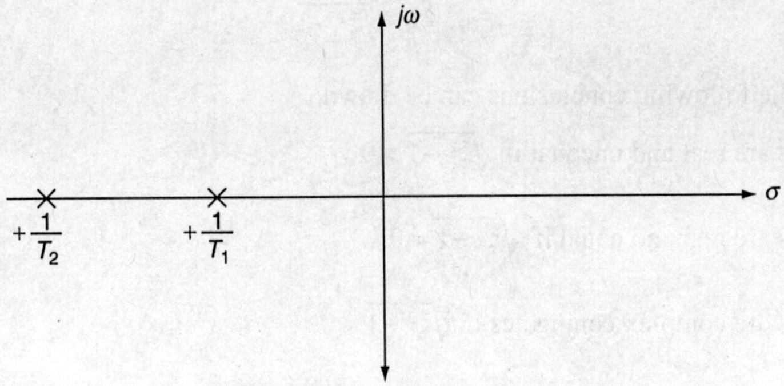

Fig. 8.16 Poles of closed-loop transfer function

8.9 Analysis of a Second-Order System

Figure 8.17 shows a general second-order system.

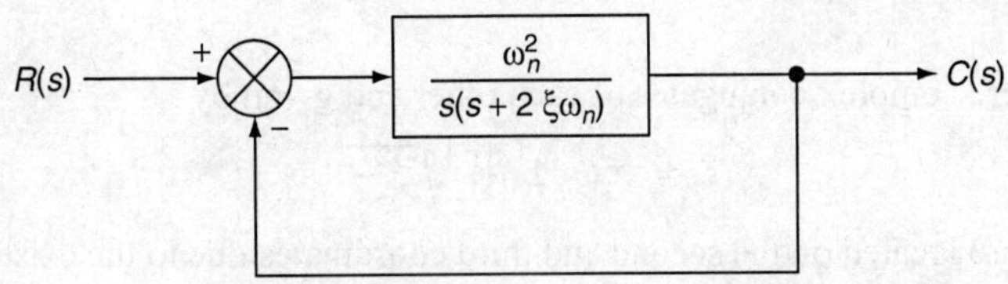

Fig. 8.17 A second-order system

From Fig. 8.17,

$$G(s) = \frac{\omega_n^2}{s(s+2\xi\omega_n)} \text{ and } H(s) = 1.$$

$$\therefore \qquad G(s)H(s) = \frac{\omega_n^2}{s^2 + 2s\xi\omega_n} \tag{8.67}$$

and

$$\frac{C(s)}{R(s)} = \frac{G(s)}{1+G(s)H(s)} = \frac{\dfrac{\omega_n^2}{s^2+2s\xi\omega_n}}{1+\dfrac{\omega_n^2}{s^2+2s\xi\omega_n}} \frac{\omega_n^2}{s^2+2s\xi\omega_n+\omega_n^2} \tag{8.68}$$

Let us study the effect of pole location before finding the unit step response. To find the poles of closed-loop transfer function of Fig. 8.17, let us put

$$s^2 + 2s\,\xi\omega_n + \omega_n^2 = 0$$

i.e.,

$$s = \frac{-2\xi\omega_n \pm \sqrt{(2\xi\omega_n)^2 - 4\omega_n^2}}{2} = -\xi\omega_n \pm \sqrt{\xi^2\omega_n^2 - \omega_n^2}$$

i.e., $$s = -\xi\omega_n \pm \omega_n\sqrt{\xi^2 - 1} \tag{8.69}$$

From Eq. (8.69), the following conclusions can be drawn:

(i) The poles are real and unequal if $\sqrt{\xi^2 - 1} > 0$
i.e., $\xi > 1$

(ii) The poles are real and equal if $\sqrt{\xi^2 - 1} = 0$
i.e., $\xi = 1$

(iii) The poles are complex conjugates if $\sqrt{\xi^2 - 1} < 0$
i.e., $\xi < 1$

where ξ is known as damping factor.

8.10 Effect of ξ on a Second-Order System

Let us analyse the effect of ξ on pole location as well as response of second-order system.

8.10.1 Effect of ξ on Pole location

Case 1: $0 < \xi < 1$

(a) The poles are complex conjugates of each other and given by

$$s_1, s_2 = -\xi\omega_n \pm j\omega_n\sqrt{\xi^2 - 1}$$

(b) The poles are located on the second and third coordinates due to the existence of both real and imaginary parts.

(c) The response is underdamped.

Case 2: $\xi = 1$

(a) The poles are real and equal. The poles are $s_1 = s_2 = -\xi\omega_n$.

(b) They lie on the negative σ-axis.

(c) The response is critically damped.

Case 3: $\xi > 1$

(a) The poles are real and unequal. They are given by

$$s_1, s_2 = -\xi\omega_n \pm \omega_n\sqrt{\xi^2 - 1}$$

(b) Since there are no imaginary terms ($\xi > 1$), the poles lie on the negative σ-axis and at unequal places.

(c) The response is overdamped.

Case 4: $\xi = 0$

(a) The poles are complex with only imaginary part and lie on $j\omega$ axis. The poles are conjugates of each other.

(b) The poles are given by

$$s_1, s_2 = \pm j\omega_n$$

(c) The response is undamped.

Case 5: $0 > \xi > -1$

(a) Since ξ is negative, the poles are given by

$$s_1, s_2 = -\xi\omega_n \pm j\omega_n\sqrt{\xi^2 - 1}$$

(b) If real parts of poles are positive, the poles lie on the right half plane (RHP) of *s*-plane.

(c) This is an unstable condition.

Case 6: $\xi = -1$

(a) The poles are given by $s_1, s_2 = -\xi\omega_n, -\xi\omega_n$.

(b) The poles are located in the RHP of *s*-plane.

(c) The system is unstable.

Case 7: $\xi < -1$

(a) The poles are given by $s_1, s_2 = -\xi\omega_n \pm \omega_n\sqrt{\xi^2 - 1}$

(b) The poles are located in the RHP of *s*-plane.

(c) The system is unstable.

Figure 8.18 shows the location of poles for a second-order system for the seven cases.

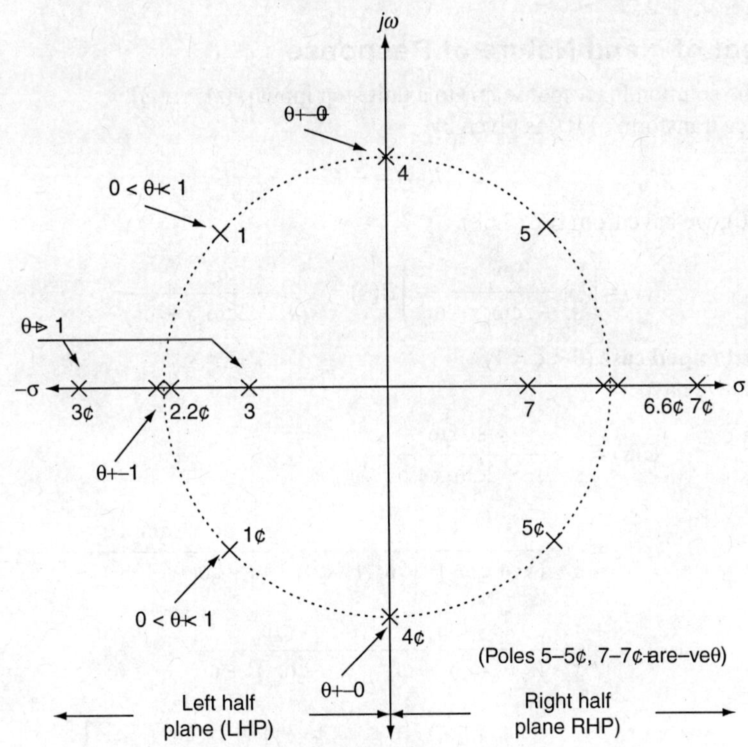

Fig. 8.18 Effect of ξ on pole location

Figure 8.19 shows ξ and nature of response.

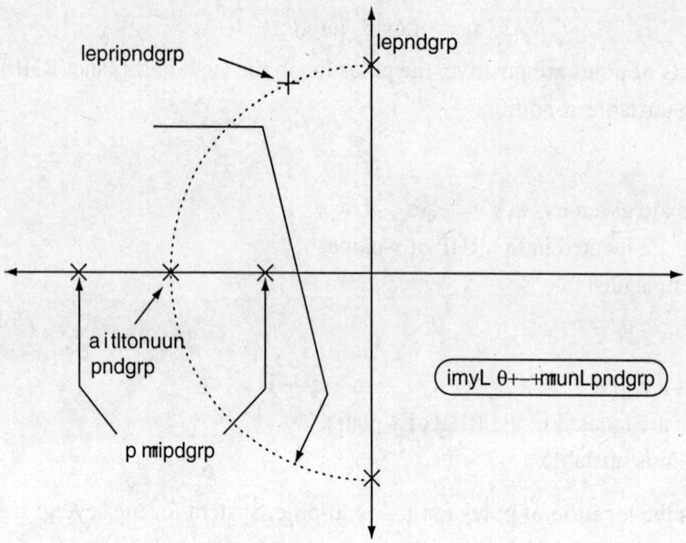

Fig. 8.19 ξ and nature of response

8.10.2 Effect of ξ and Nature of Response

Let us obtain the solution for response $c(t)$ to a unit step input $[r(t) = u(t)]$.

The Laplace transform of $r(t)$ is given by

$$R(s) = \frac{1}{s} \tag{8.70}$$

Using Eq. (8.70), we have from Eq. (8.68),

$$C(s) = \left(\frac{\omega_n^2}{s^2 + 2\xi\omega_n s + \omega_n^2}\right)R(s) = \frac{\omega_n^2}{s(s^2 + 2\xi\omega_n s + \omega_n^2)} \tag{8.71}$$

Case 1: Underdamped case (0 < ξ < 1)
From Eq. (8.71) we have

$$C(s) = \frac{1}{s} - \frac{s + 2\xi\omega_n}{s^2 + 2\xi\omega_n s + \omega_n^2}$$

$$= \frac{1}{s} - \frac{s + \xi\omega_n}{(s^2 + \xi\omega_n)^2 + \omega_n^2(1-\xi^2)} - \frac{\xi\omega_n}{(s + \xi\omega_n)^2 + \omega_n^2(1-\xi^2)}$$

$$= \frac{1}{s} - \frac{s + \xi\omega_n}{(s + \xi\omega_n)^2 + \omega_d^2} - \frac{\xi\omega_n}{(s + \xi\omega_n)^2 + \omega_d^2}$$

$$= \frac{1}{s} - \frac{s + \xi\omega_n}{(s + \xi\omega_n)^2 + \omega_d^2} - \frac{\xi}{\sqrt{1-\xi^2}}\left\{\frac{\omega_d}{(s + \xi\omega_n)^2 + \omega_d^2}\right\} \tag{8.72}$$

where $\omega_d^2 = \omega_n^2(1-\xi^2)$.

Taking inverse Laplace Transform of Eq. (8.72), we have

$$c(t) = 1 - e^{-\xi\omega_n t}\cos\omega_d t - \frac{\xi}{\sqrt{1-\xi^2}}e^{-\xi\omega_n t}\sin\omega_d t = 1 - e^{-\xi\omega_n t}\left\{\cos\omega_d t + \frac{\xi}{\sqrt{1-\xi^2}}\sin\omega_d t\right\}$$

$$= 1 - \frac{e^{-\xi\omega_n t}}{\sqrt{1-\xi^2}}\left\{\sqrt{1-\xi^2}\cos\omega_d t + \xi\sin\omega_d t\right\} = 1 - \frac{e^{-\xi\omega_n t}}{\sqrt{1-\xi^2}}\sin(\omega_d t + \alpha) \tag{8.73}$$

where

$$\left.\begin{aligned} \cos\alpha &= \xi \\ \sin\alpha &= \sqrt{1-\xi^2} \\ \alpha &= \tan^{-1}\frac{\sqrt{1-\xi^2}}{\xi} \end{aligned}\right\} \tag{8.74}$$

Equation (8.73) represents the solution for $0 < \xi < 1$. Figure 8.20(a) and 8.20(b) show the sine wave and an exponential decaying function, respectively. Figure 8.20(c) shows a sinusoidal decaying on an exponential envelope.

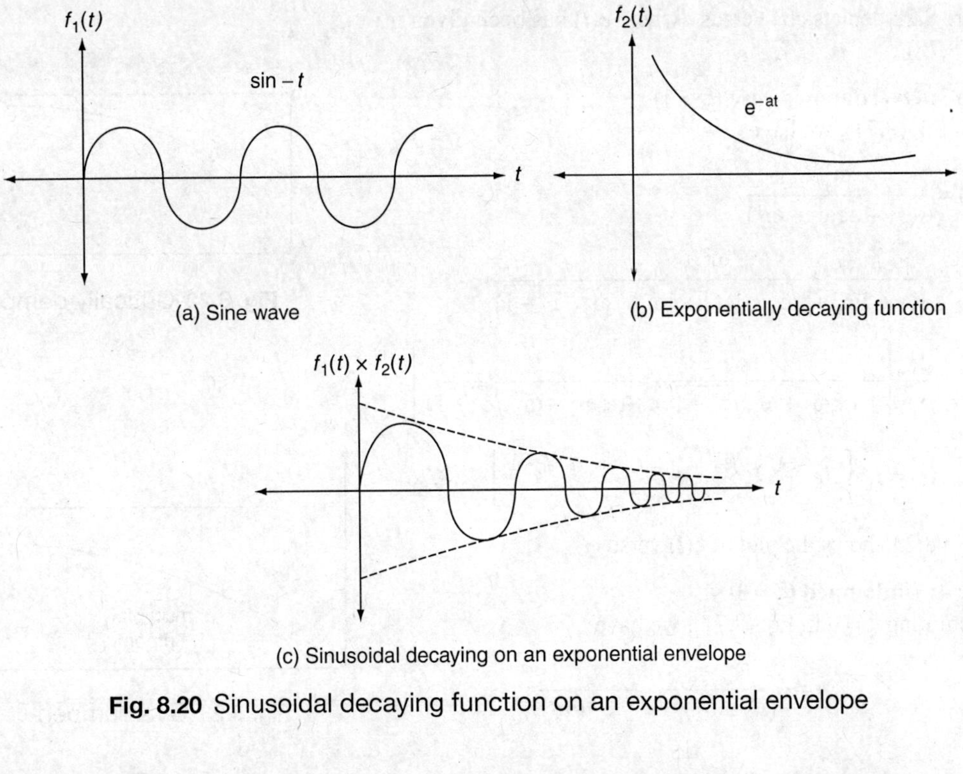

(a) Sine wave

(b) Exponentially decaying function

(c) Sinusoidal decaying on an exponential envelope

Fig. 8.20 Sinusoidal decaying function on an exponential envelope

Figure 8.21 shows the response $c(t)$ of Eq. (8.73) for $0 < \xi < 1$. The response overshoots the reference and oscillates before settling to the final value. The pole locations are in 2nd and 3rd quadrants for this response.

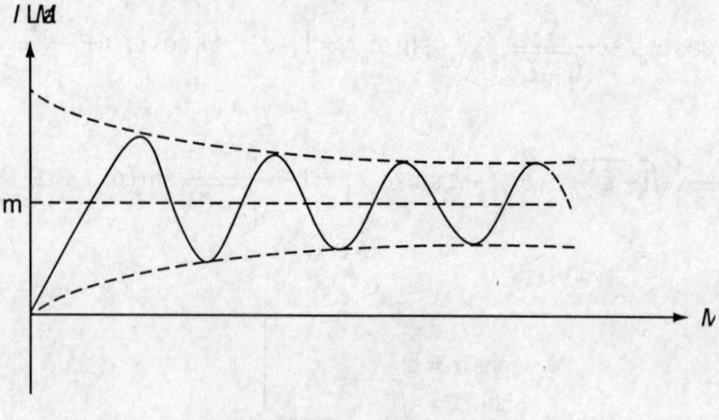

Fig. 8.21 Underdamped

Case 2: Critically damped ($\xi = 1$)

For critically damped systems, the roots are real and equal. Substituting $\xi = 1$ in Eq. (8.71), we have

$$C(s) = \frac{\omega_n^2}{s(s^2 + 2\omega_n s + \omega_n^2)} = \frac{\omega_n^2}{s(s+\omega_n)^2} = \frac{1}{s} - \frac{\omega_n}{(s+\omega_n)^2} - \frac{1}{s+\omega_n}$$

$$\therefore \qquad c(t) = 1 - \omega_n e^{-\omega_n t} t - e^{-\omega_n t} = 1 - (1 + \omega_n t)e^{-j\omega_n t} \qquad (8.75)$$

Figure 8.22 depicts $c(t)$ versus t where $c(t)$ has been given in Eq. (8.75).

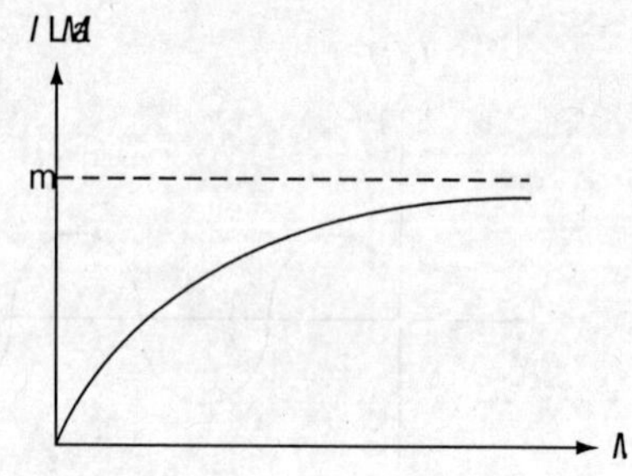

Fig. 8.22 Critically damped

Case 3: Overdamped case ($\xi > 1$)

From Eq. (8.71), we have

$$C(s) = \frac{\omega_n^2}{s(s^2 + 2\xi\omega_n + \omega_n^2)}$$

$$= \frac{\omega_n^2}{s(s + \xi\omega_n + \sqrt{\xi^2 - 1})(s + \xi\omega_n - \omega_n\sqrt{\xi^2 - 1})}$$

$$= \frac{1}{s} - \left[\frac{A}{s + \xi\omega_n + \omega_n\sqrt{\xi^2 - 1}} + \frac{B}{s + \xi\omega_n - \omega_n\sqrt{\xi^2 - 1})}\right]$$

$$\text{i.e.,} \quad c(t) = 1 - \left[Ae^{\left(-\xi\omega_n - \omega_n\sqrt{\xi^2-1}\right)t} + Be^{-\left(-\xi\omega_n + \omega_n\sqrt{\xi^2-1}\right)t}\right] \qquad (8.76)$$

Figure 8.23 shows the plot of $c(t)$ versus t.

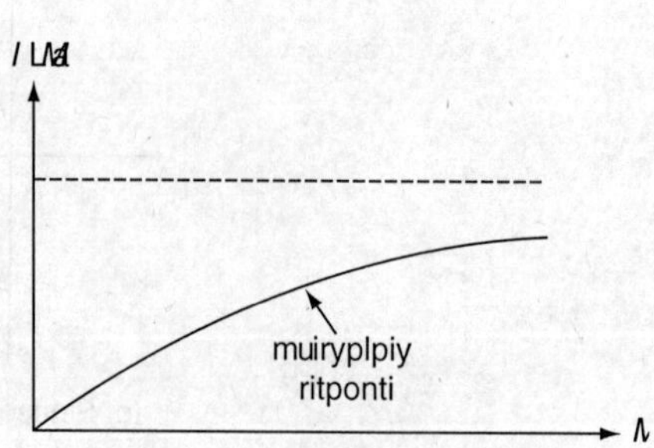

Fig. 8.23 Overdamped

Case 4: Undamped ($\xi = 0$)

Substituting $\xi = 0$ in Eq. (8.71), we have

$$C(s) = \frac{\omega_n^2}{s(s^2 + \omega_n^2)} = \frac{1}{s} - \frac{s}{s^2 + \omega_n^2}$$

i.e.,
$$c(t) = 1 - \cos \omega_n t \tag{8.77}$$

Since there is no time damping, this response does not die out with time. This response is known as undamped response. Figure 8.24 shows the plot of $c(t)$ versus t.

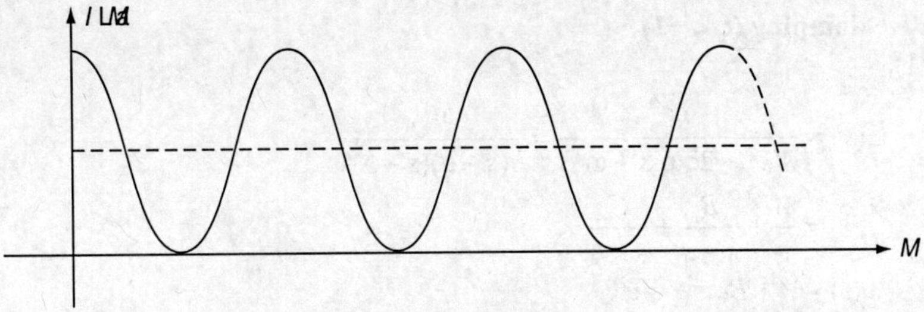

Fig. 8.24 Undamped

Case 5: Negative damping ($0 > \xi > -1$)
From Eq. (8.71), we have

$$C(s) = \frac{\omega_n^2}{s(s^2 + 2\xi\omega_n s + \omega_n^2)} = \frac{\omega_n^2}{s[(s-a)^2 + b^2]} = \frac{A}{s} + \frac{Bs + C}{(s-a)^2 + b^2}$$

where the poles are 0, $a \pm jb$ $(a, b > 0)$.

$$\therefore \quad c(t) = A + Be^{at}\cos bt + De^{at}\sin bt. \tag{8.78}$$

where $D = \dfrac{C}{b}$

Figure 8.25 shows the plot of $c(t)$ versus t. The response is unbounded.

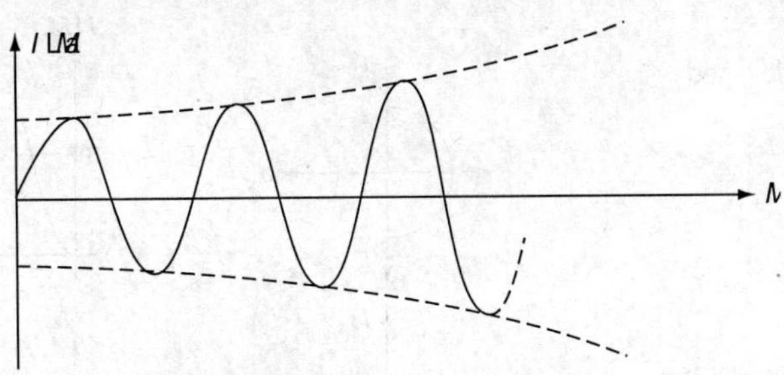

Fig. 8.25 Negative damping ($0 > \xi > -1$)

Case 6: Negative damping ($\xi = -1$)
From Eq. (8.71), we have

$$C(s) = \frac{\omega_n^2}{s(s^2 + 2\omega_n s + \omega_n^2)} = \frac{\omega_n^2}{s(s-\omega_n)^2} = \frac{1}{s} + \frac{\omega_n}{(s-\omega_n)^2} - \frac{1}{s-\omega_n}$$

i.e.,
$$c(t) = 1 + \omega_n t\, e^{\omega_n t} - e^{\omega_n t} = 1 + (\omega_n t - 1)\, e^{\omega_n t} \tag{8.79}$$

Figure 8.26 shows the plot of $c(t)$ versus t.

Case 7: Negative damping ($\xi < -1$)
From Eq. (8.71),

$$C(s) = \frac{\omega_n^2}{s(s^2 + 2\xi\omega_n s + \omega_n^2)} = \frac{\omega_n^2}{s(s-a)(s-b)}$$
$$= \frac{A}{s} + \frac{B}{s-a} + \frac{C}{s-b}$$

i.e.,
$$c(t) = A + Be^{at} + Ce^{bt} \tag{8.80}$$

Figure 8.27 shows the plot of $c(t)$ versus t.

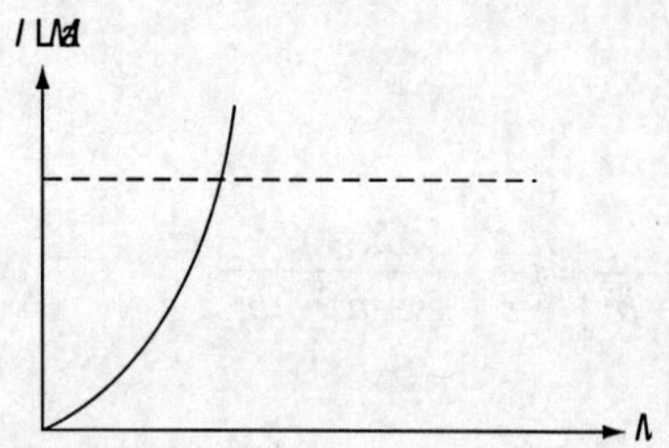

Fig. 8.26 Negative damping ($\xi = -1$)

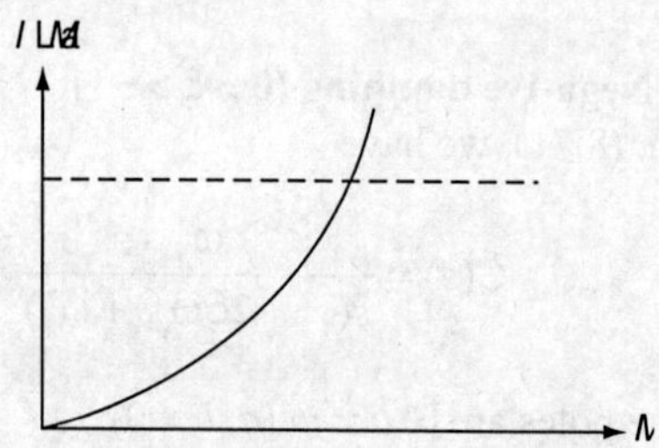

Fig. 8.27 Negative damping ($\xi < -1$)

Table 8.4 and Table 8.5 summarize the above result.

Table 8.4 Expression of ξ, pole location and step response

No.	ξ	Pole location	Step response $C_{(t)}$
1.	$0 < \xi < 1$ underdamped	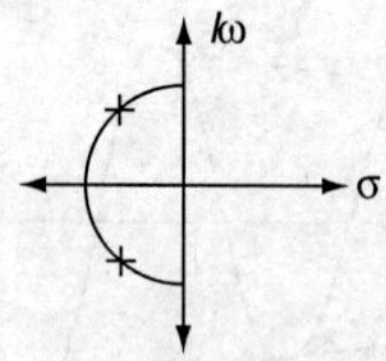	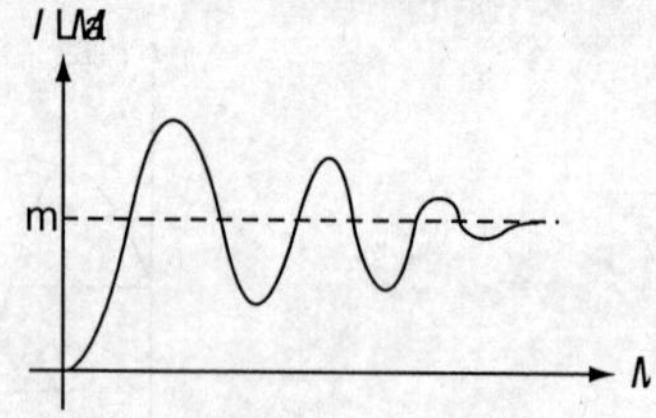
2.	$\xi = 1$ critically damped	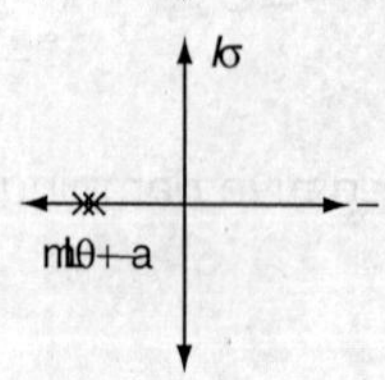	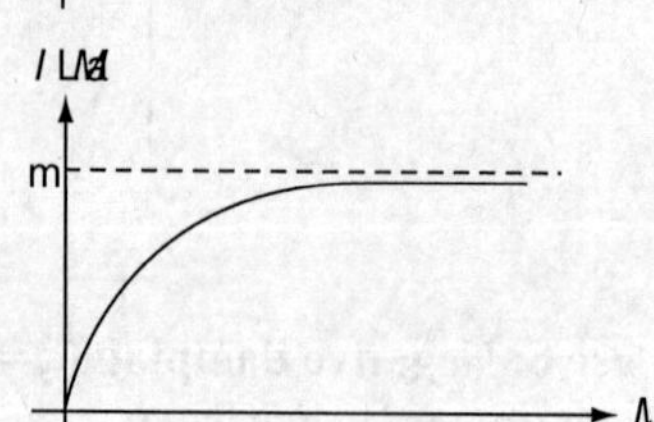

3.	$\xi > 1$ overdamped	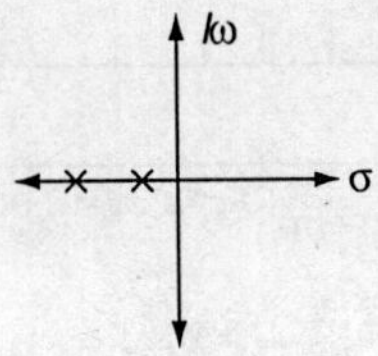	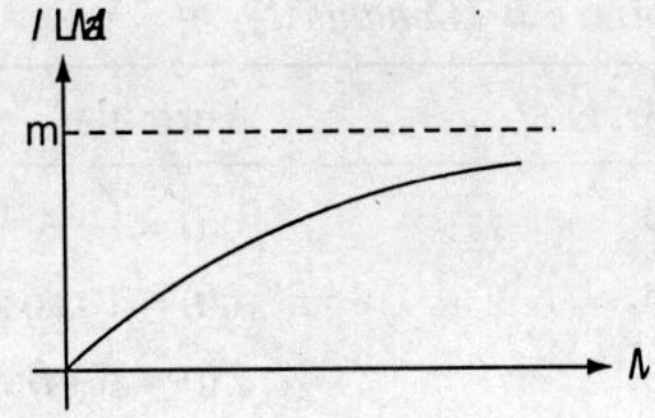
4.	$\xi = 0$	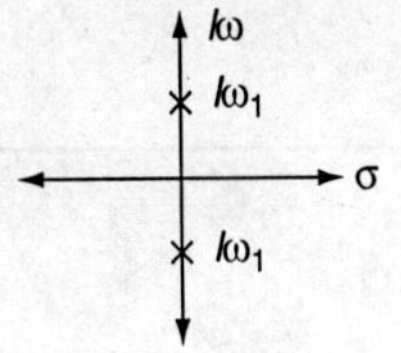	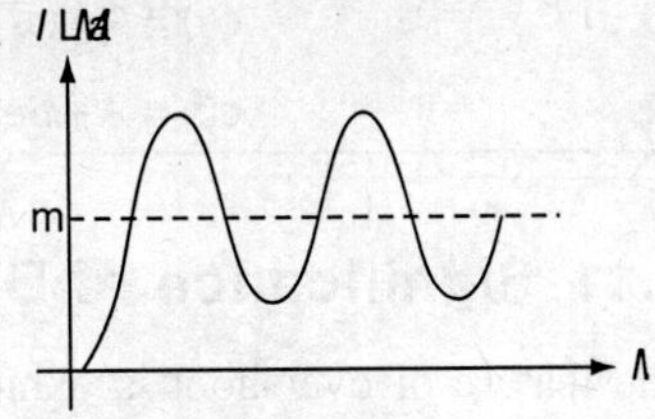
5.	$0 > \xi > -1$	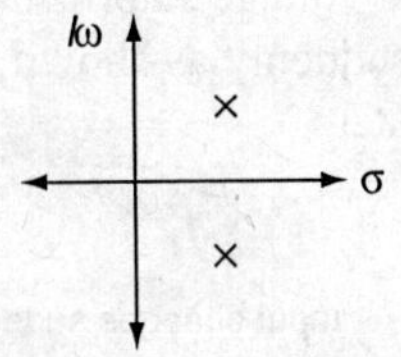	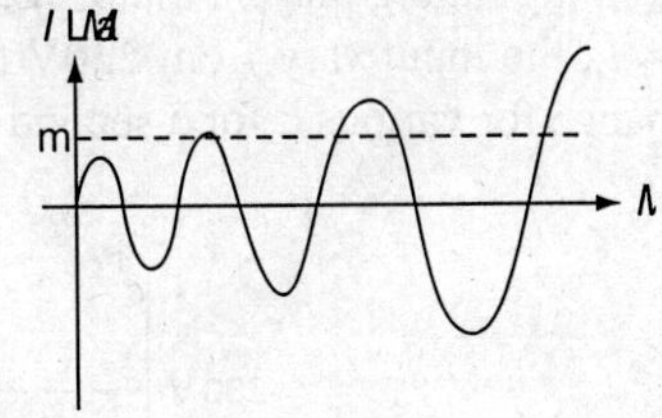
6.	$\xi = -1$	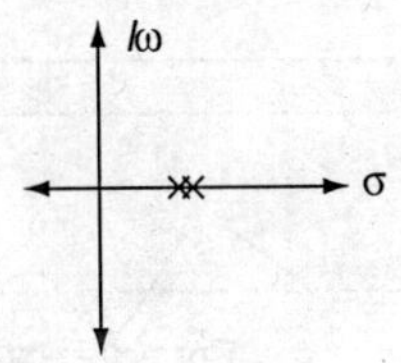	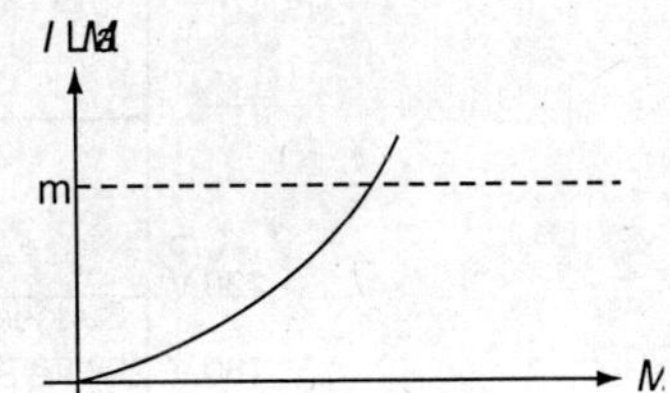
7.	$\xi < -1$	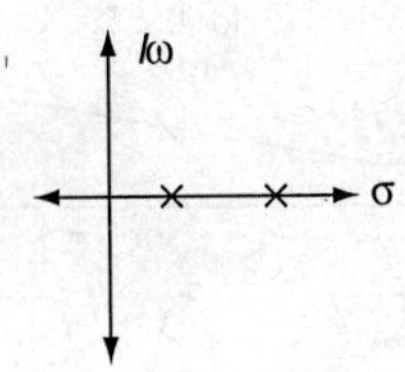	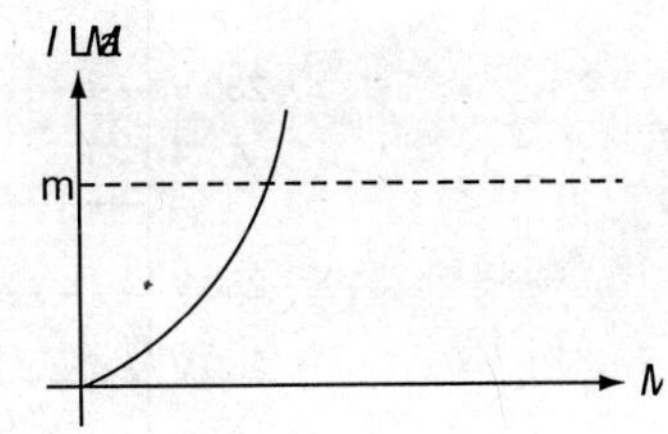

Table 8.5 Expression of *c(t)* for step response

Sr. No.	Formula	Valid For
1.	$c(t) = 1 - \frac{e^{-\xi\omega_n t}}{\sqrt{1-\xi^2}} \sin\left(\omega_d t + \tan^{-1}\frac{\sqrt{1-\xi^2}}{\xi}\right)$	$0 < \xi < 1$
2.	$c(t) = 1 - e^{-\xi\omega_n t}(1 + \omega_n t)$	$\xi = 1$

(Continued)

Table 8.5 (*Continued*)

Sr. No.	Formula	Valid For
3.	$c(t) = 1 - Ae^{\left(-\xi\omega_n - \omega_n\sqrt{\xi^2-1}\right)t} - Be^{\left(-\xi\omega_n + \omega_n\sqrt{\xi^2-1}\right)t}$	$\xi > 1$
4.	$c(t) = 1 - \cos\omega_n t$	$\xi = 0$
5.	$c(t) = A + Be^{at}\cos bt + De^{-at}\sin bt$	$0 > \xi > -1$
6.	$c(t) = 1 + (\omega_n t - 1)e^{\omega_n t}$	$\xi = -1$
7.	$c(t) = A + Be^{at} + Ce^{bt}$	$\xi = <-1$

8.11 Significance of Damping

The feature of overshoot is exhibited by the second-order and higher-order systems. Therefore, the damping ratio ξ plays a major role. A servo voltage stabilizer is considered here for illustration. Let at $t = t_1$, the input vary from 230V to 180V suddenly as shown in Fig. 8.28, which illustrates the output change for various ξ for a second-order system.

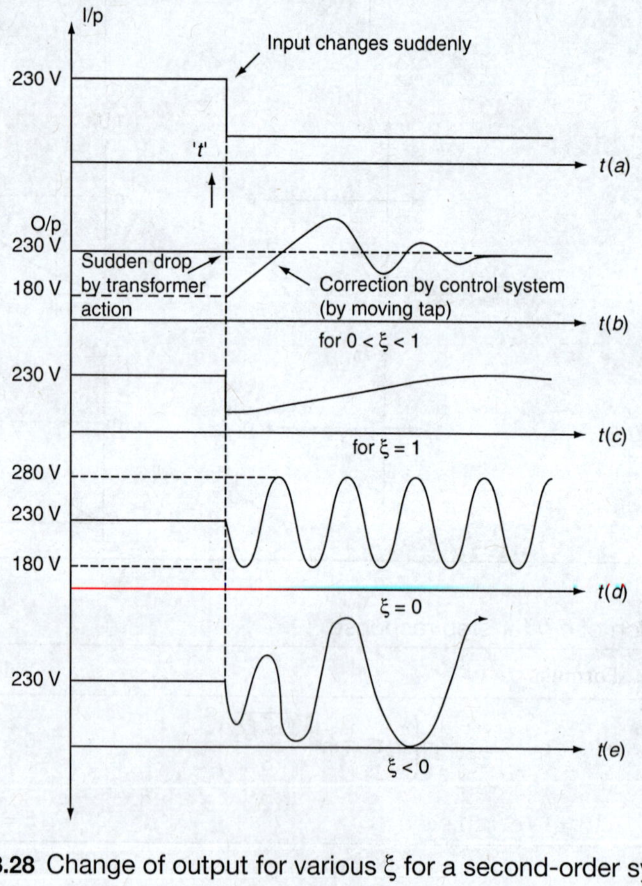

Fig. 8.28 Change of output for various ξ for a second-order system

The performance of the system is satisfactory for positive damping shown in Fig. 8.28(b) and Fig. 8.28(c). Figure 8.28(d) shows the undamped response ($\xi = 0$), which is not acceptable because the output rms values will be continuously changing. Figure 8.28(e) shows the response for $\xi < 0$, which is unstable. The cases for $\xi = 0$ and $\xi < 0$ give the unacceptable responses. The output settles only after quick correction. Hence, ξ should not have smaller values. Also for $\xi >> 1$, the response becomes slow and hence output will also suffer in this case.

8.12 Derivation of Time Response Specifications

Figure 8.2 shows the plot of $c(t)$ versus t for unit step input and different transient response specifications have also been pointed.

8.12.1 Delay Time (T_d)

The time required to reach 50% of output is known as delay time.

$$\therefore \qquad c(t) = \frac{1}{2} \text{ at } t = T_d$$

From Eq. (8.73),

$$\frac{1}{2} = 1 - \frac{e^{-\xi\omega_n T_d}}{\sqrt{1-\xi^2}} \sin\left[\omega_d T_d + \left(\tan^{-1}\frac{\sqrt{1-\xi^2}}{\xi}\right)\right]$$

The result of this equation is

$$T_d = \frac{1+0.7\xi}{\omega_n} \tag{8.81}$$

8.12.2 Rise Time (T_r)

Since the system is underdamped,

$$c(t) = 1 \text{ at } t = T_r$$

From Eq. (8.73),

$$1 = 1 - \frac{e^{-\xi\omega_n T_r}}{\sqrt{1-\xi^2}} \sin\left(\omega_d T_r + \tan^{-1}\frac{\sqrt{1-\xi^2}}{\xi}\right)$$

i.e.,

$$e^{-\xi\omega_n T_r} \sin\left(\omega_d T_r + \tan^{-1}\frac{\sqrt{1-\xi^2}}{\xi}\right) = 0 \tag{8.82}$$

since $e^{-\xi\omega_n T_r} \neq 0$, we have from Eq. (8.82)

$$\sin\left(\omega_d T_r + \tan^{-1}\frac{\sqrt{1-\xi^2}}{\xi}\right) = 0$$

i.e.,

$$\omega_d T_r + \tan^{-1}\left(\frac{\sqrt{1-\xi^2}}{\xi}\right) = 0$$

i.e.,

$$T_r = \frac{-1}{\omega_d}\tan^{-1}\left(\frac{\sqrt{1-\xi^2}}{\xi}\right) \tag{8.83}$$

Since $$\alpha = \tan^{-1}\left(\frac{\sqrt{1-\xi^2}}{\xi}\right) \tag{8.84}$$

and α is in radians, we have

$$\tan(\pi - \alpha) = -\tan\alpha$$

From Eq. (8.83), $$T_r = \frac{-\tan^{-1}\left(\frac{\sqrt{1-\xi^2}}{\xi}\right)}{\omega_d}$$

i.e., $$T_r = \frac{\pi - \alpha}{\omega_d} \tag{8.85}$$

8.12.3 Peak Time (T_p)

Differentiating Eq. (8.73) with respect to time, we have

$$\frac{dc(t)}{dt} = 0 - \left[\frac{e^{-\xi\omega_n t}(-\xi\omega_n)\sin(\omega_d t + \alpha)}{\sqrt{1-\xi^2}} + \frac{e^{-\xi\omega_n t}\omega_d \cos(\omega_d t + \alpha)}{\sqrt{1-\xi^2}}\right]$$

Putting $\frac{dc(t)}{dt} = 0$ and $t = T_p$, we have

$$\frac{\xi\omega_n e^{-\xi\omega_n T_p}\sin(\omega_d t_p + \alpha)}{\sqrt{1-\xi^2}} - \frac{e^{-\xi\omega_n T_p}\omega_d \cos(\omega_d T_p + \alpha)}{\sqrt{1-\xi^2}} = 0$$

i.e., $$\xi\omega_n \sin(\omega_d T_p + \alpha) - \omega_d \cos(\omega_d T_p + \alpha) = 0 \; [\because e^{-\xi\omega_n T_p} \neq 0]$$

i.e., $$\xi\omega_n \sin(\omega_d T_p + \alpha) - \omega_n\sqrt{1-\xi^2}\cos(\omega_d T_p + \alpha) = 0 \left[\text{since } \omega_d = \omega_n\sqrt{1-\xi^2}\right]$$

i.e., $$\tan(\omega_d T_p + \alpha) = \frac{\sqrt{1-\xi^2}}{\xi} = \tan\alpha \left[\because \alpha = \tan^{-1}\frac{\sqrt{1-\xi^2}}{\xi}\right]$$

i.e., $$\omega_d T_p + \alpha = n\pi + \alpha \; [\therefore \tan\alpha = \tan(n\pi + \alpha), \quad \text{where } n = 1, 2, 3\ldots]$$

i.e., $$T_p = \frac{n\pi}{\omega_d} \tag{8.86}$$

The first overshoot is obtained for $n = 1$ and the second overshoot is obtained for $n = 2$.

8.12.4 Peak Overshoot (M_p)

M_p is given by

$$M_p = c(t)|_{t=t_p} - 1$$

$\therefore$ $$c(t)\Big|_{t=T_p} = 1 - \frac{e^{-\xi\omega_n T_p}}{\sqrt{1-\xi^2}}\sin(\omega_d T_p + \alpha)$$

Putting $\omega_d T_p = \pi$

$$M_p = 1 - \frac{e^{-\xi\omega_n T_p}}{\sqrt{1-\xi^2}} \sin(\pi + \alpha) - 1$$

i.e.,

$$M_p = \frac{e^{-\xi\omega_n T_p}}{\sqrt{1-\xi^2}} \sin\alpha \tag{8.87}$$

Again

$$\sin\alpha = \sqrt{1-\xi^2}. \tag{8.88}$$

Using Eq. (8.88), we have from Eq. (8.87),

$$M_p = e^{-\xi\omega_n T_p} = e^{-\xi\omega_n\left(\frac{\pi}{\omega_d}\right)} = e^{-\xi\omega_n\left(\frac{\pi}{\omega_n\sqrt{1-\xi^2}}\right)} = e^{-\frac{\xi\pi}{\sqrt{1-\xi^2}}}$$

The value of M_p in percentages is given by

$$\%M_p = e^{-\frac{\xi\pi}{\sqrt{1-\xi^2}}} \times 100 \tag{8.89}$$

8.12.5 Settling Time (T_s)

For a 2% criterion,

$$T_s = 4T$$

∴

$$T_s = \frac{4}{\xi\omega_n} \tag{8.90}$$

where

$$\frac{1}{\xi\omega_n} = T = \text{time constant} \tag{8.91}$$

For 5% criterion

$$T_s = \frac{3}{\xi\omega_n}$$

Table 8.6 summarizes the time response specifications for a unit step response.

Table 8.6 Specification of time response

Sl. No.	Time response specification
1.	$T_d = \frac{1+0.7\xi}{\omega_n}$
2.	$T_r = \frac{\pi-\alpha}{\omega_d}\left[\text{where } \tan\alpha = \frac{\sqrt{1-\xi^2}}{\xi}\right], \alpha$ is in radians
3.	$T_p = \frac{\pi}{\omega_d}$
4.	$\%M_p = e^{-\frac{\xi x}{\sqrt{1-\xi^2}}} \times 100$
5.	$T_s = \frac{4}{\xi\omega_n}$ for 2% criterion and $T_s = \frac{3}{\xi\omega_n}$ for 5% criterion

Example 8.6 For a system having $G(s) = \dfrac{25}{s(s+10)}$ and unity feedback (negative), find (i) ω_n, (ii) ξ, (iii) ω_d, (iv) T_p, and (v) M_p.

Solution

$$G(s) = \frac{25}{s(s+10)} \text{ and } H(s) = 1$$

$\therefore$
$$G(s)H(s) = \frac{25}{s(s+10)}$$

and
$$\frac{C(s)}{R(s)} = \frac{G(s)}{1+G(s)H(s)} = \frac{\dfrac{25}{s(s+10)}}{1+\dfrac{25}{s(s+10)}}$$

i.e.,
$$\frac{C(s)}{R(s)} = \frac{25}{s^2+10s+25}$$

Comparing Eq. (1) with standard form $\dfrac{C(s)}{R(s)} = \dfrac{\omega_n^2}{s^2+2\xi\omega_n s+\omega_n^2}$,

we have $\omega_n^2 = 25$ i.e., $\omega_n = 5$ rad/sec and $2\xi\omega_n = 10$.

$\therefore$
$$2 \times \xi \times 5 = 10$$

i.e.,
$$\xi = 1$$

Hence the system is critically damped.

Since
$$\omega_d = \omega_n\sqrt{1-\xi^2}$$
$$\omega_d = 5\sqrt{1-1^2} = 0$$

Again,
$$T_p = \frac{\pi}{\omega_d}$$

i.e.,
$$T_p = \frac{\pi}{0} = \infty$$

Therefore, there is no peak.

Again,
$$M_p = 100e^{\frac{-\xi\pi}{\sqrt{1-\xi^2}}} = 100e^{\frac{\xi\pi}{\sqrt{1-1}}} = 100e^{-\infty} = 0$$

Thus,
$$\omega_n = 5\,\text{rad/sec},\ \xi = 1, \omega_d = 0, T_p = \infty, M_p = 0$$

Example 8.7 A system has $G(s) = 20/(s^2+5s+5)$ and unity feedback (negative). Find (i) ω_n, (ii) ξ, (iii) ω_d, (iv) T_d, (v) T_r, (vi) T_p, (vii) M_p, and (viii) T_s.

Solution Here $G(s) = \dfrac{20}{s^2+5s+5}$ and $H(s) = 1$(Feedback is negative).

$$\frac{C(s)}{R(s)} = \frac{G(s)}{1+G(s)H(s)} = \frac{\frac{20}{s^2+5s+5}}{1+\frac{20}{s^2+5s+5}} = \frac{20}{s^2+5s+25}$$

(i) Since the numerator may not be ω_n^2 due to the existence of a constant term in the denominator of $G(s)$, ω_n^2 must be compared with the denominator of $C(s)/R(s)$ only.

$\therefore \quad \omega_n^2 = 25$ i.e., $\quad \omega_n = 5 \text{ rad/sec}$

(ii) $2\xi\omega_n = 5$ or $\xi = \dfrac{5}{2\omega_n} = \dfrac{5}{2\times 5} = 0.5$

(iii) $\omega_d = \omega_n\sqrt{1-\xi^2} = 5\sqrt{1-0.5^2} = 4.33$ rad/sec

(iv) $T_d = \dfrac{1+0.7\xi}{\omega_n} = \dfrac{1+0.7\times 0.5}{5} = 0.27$ sec

(v) $T_r = \dfrac{\pi-\alpha}{\omega_d}$

$$\alpha = \tan^{-1}\frac{\sqrt{1-\xi^2}}{\xi} = \tan^{-1}\frac{\sqrt{1-0.5^2}}{0.5} = 1.0467 \text{ rad}$$

$\therefore \quad T_r = \dfrac{\pi - 1.0467}{4.33} = 0.4834$ sec

(vi) $T_p = \dfrac{\pi}{\omega_d} = \dfrac{3.14}{4.33} = 0.725$ sec

(vii) $M_p = 100\times e^{-\frac{\xi\pi}{\sqrt{1-\xi^2}}} = 100\times e^{-\frac{0.5\times 3.14}{\sqrt{1-0.5^2}}} = 16.32\%$

(viii) $T_s = \dfrac{4}{\xi\omega_n} = \dfrac{4}{0.5\times 5} = 1.6$ sec

8.13 Time Response of Second Order Control System Subjected to Unit Ramp Input Function

Laplace transform of unit ramp function $r(t) = t$ is given by

$$R(s) = \frac{1}{s^2}$$

The output of second order system is expressed by

$$C(s) = \frac{\omega_n^2 R(s)}{s^2+2\xi\omega_n s+\omega_n^2} = \frac{\omega_n^2}{s^2+2\xi\omega_n s+\omega_n^2}\times\frac{1}{s^2}$$

$$= \frac{1}{s^2} - \frac{s+2\xi\omega_n}{s[(s+2\xi\omega_n)^2+\omega_n^2(1-\xi^2)]}$$

$$= \frac{1}{s^2} - \frac{2\xi\omega_n}{s[(s+2\xi\omega_n)^2 + \omega_n^2(1-\xi^2)]} - \frac{1}{(s+2\xi\omega_n)^2 + \omega_n^2(1-\xi^2)}$$

$$= \frac{1}{s^2} - \frac{2\xi\omega_n}{\omega_n^2}\left[\frac{\omega_n^2}{s[(s+2\xi\omega_n)^2 + \omega_n^2(1-\xi^2)]}\right] - \frac{1}{(s+2\xi\omega_n)^2 + \omega_n^2(1-\xi^2)}$$

$$= \frac{1}{s^2} - \frac{2\xi}{\omega_n}\left[\frac{1}{s} - \frac{s+2\xi\omega_n}{[(s+2\xi\omega_n)^2 + \omega_n^2(1-\xi^2)]}\right] - \frac{1}{(s+2\xi\omega_n)^2 + \omega_n^2(1-\xi^2)}$$

$$= \frac{1}{s^2} - \frac{2\xi}{\omega_n s} + \frac{2\xi}{\omega_n} \times \frac{s+2\xi\omega_n}{[(s+2\xi\omega_n)^2 + \omega_n^2(1-\xi^2)]} + \frac{2\xi^2 - 1}{(s+2\xi\omega_n)^2 + \omega_n^2(1-\xi^2)}$$

$$\text{i.e., } C(s) = \frac{1}{s^2} - \frac{2\xi}{\omega_n s} + \frac{2\xi}{\omega_n} \times \frac{s+2\xi\omega_n}{[(s+2\xi\omega_n)^2 + \omega_n^2(1-\xi^2)]} + \frac{2\xi^2 - 1}{\omega_n\sqrt{1-\xi^2}} \times \frac{\omega_n\sqrt{1-\xi^2}}{(s+2\xi\omega_n)^2 + \omega_n^2(1-\xi^2)} \tag{8.92}$$

Taking Inverse Laplace transform of Eq.(8.92), we have

$$c(t) = t - \frac{2\xi}{\omega_n} + \frac{2\xi}{\omega_n} e^{-\xi\omega_n t} \cos\left\{\omega_n\sqrt{(1-\xi^2)}\right\}t + \left(\frac{2\xi^2 - 1}{\omega_n\sqrt{(1-\xi^2)}}\right) e^{-\xi\omega_n t} \sin\left\{\omega_n\sqrt{(1-\xi^2)}\right\}t$$

$$\text{i.e., } c(t) = t - \frac{2\xi}{\omega_n} + \frac{2\xi}{\omega_n} e^{-\xi\omega_n t} \cos\left\{\omega_n\sqrt{(1-\xi^2)}\right\}t + \left(\frac{2\xi^2 - 1}{\omega_n\sqrt{(1-\xi^2)}}\right) e^{-\xi\omega_n t} \sin\left\{\omega_n\sqrt{(1-\xi^2)}\right\}t$$

$$= t - \frac{2\xi}{\omega_n} + \frac{e^{-\xi\omega_n t}}{\omega_n\sqrt{(1-\xi^2)}}\left[2\xi\sqrt{(1-\xi^2)}\cos\left\{\omega_n\sqrt{(1-\xi^2)}\right\}t + (2\xi^2-1)\sin\left\{\omega_n\sqrt{(1-\xi^2)}\right\}t\right]$$

$$= t - \frac{2\xi}{\omega_n} + \frac{e^{-\xi\omega_n t}\sin\left(\omega_n\sqrt{(1-\xi^2)}t + \phi\right)}{\omega_n\sqrt{(1-\xi^2)}}$$

where $\phi = \tan^{-1}\left[\frac{2\xi\sqrt{1-\xi^2}}{2\xi^2-1}\right] = 2\tan^{-1}\left(\frac{\sqrt{1-\xi^2}}{\xi}\right)$

The error signal is given by

$$e(t) = r(t) - c(t) \tag{8.93}$$

Here $r(t) = t$, we have from Eq. (8.93),

$$e(t) = r(t) - c(t) = t - \left[t - \frac{2\xi}{\omega_n} + \frac{e^{-\xi\omega_n t}\sin\left(\omega_n\sqrt{(1-\xi^2)}t + \phi\right)}{\omega_n\sqrt{(1-\xi^2)}}\right]$$

$$= \frac{2\xi}{\omega_n} - \frac{e^{-\xi\omega_n t}\sin\left(\omega_n\sqrt{(1-\xi^2)}t + \phi\right)}{\omega_n\sqrt{(1-\xi^2)}} \tag{8.94}$$

Figure 8.29 shows the time response of a second order system for unit ramp input.

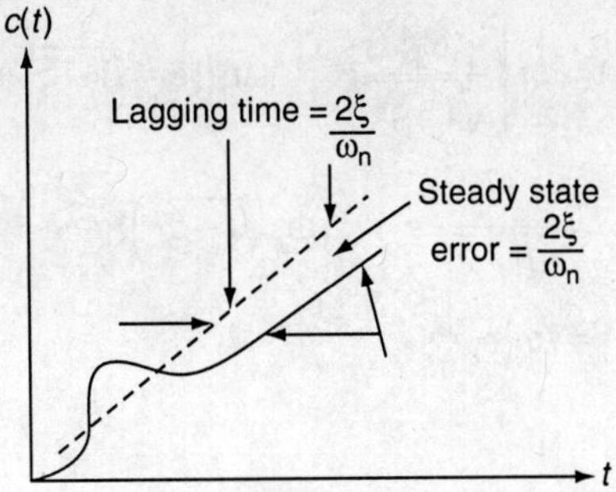

Fig. 8.29 Time response of a second order system for unit ramp input

For t → ∞, we have from Eq. (8.94)

$$e_{ss} = \frac{2\xi}{\omega_n} \tag{8.95}$$

8.14 Time Response of a Second Order Control System Subjected to Impulse Input Function

Laplace transform of a unit impulse function is

$$R(s) = 1$$

The output of a second order control system is given by

$$C(s) = \frac{\omega_n^2 R(s)}{s^2 + 2\xi\omega_n s + \omega_n^2} = \frac{\omega_n^2}{s^2 + 2\xi\omega_n s + \omega_n^2}$$

$$= \frac{\omega_n^2}{(s + 2\xi\omega_n)^2 + \omega_n^2(1-\xi^2)}$$

i.e.,

$$C(s) = \frac{\omega_n}{\sqrt{1-\xi^2}} \times \frac{\omega_n\sqrt{1-\xi^2}}{(s + 2\xi\omega_n)^2 + \omega_n^2(1-\xi^2)} \tag{8.96}$$

On taking inverse Laplace transform of Eq.(8.96), we have

$$c(t) = \frac{\omega_n}{\sqrt{1-\xi^2}} e^{-\xi\omega_n t} \sin\left(\omega_n\sqrt{1-\xi^2}\,t\right) \tag{8.97}$$

The following cases are discussed:

Case 1: $\xi < 1$

The output becomes

$$c(t) = \frac{\omega_n}{\sqrt{1-\xi^2}} e^{-\xi\omega_n t} \sin\left([\omega_n\sqrt{1-\xi^2}\,]t\right) \tag{8.98}$$

Case 2: $\xi = 1$
The output becomes

$$c(t) = \underset{\xi \to 1}{Lt}\left[\frac{\omega_n}{\sqrt{1-\xi^2}} e^{-\xi\omega_n t} \sin\left([\omega_n \sqrt{1-\xi^2}]t\right)\right]$$

$$= \frac{\omega_n}{\sqrt{1-\xi^2}} e^{-\xi\omega_n t}\left(\omega_n \sqrt{1-\xi^2}\right)t$$

$$= t\omega_n^2 e^{-\xi\omega_n t} \tag{8.99}$$

Case 3: $\xi > 1$
The output becomes

$$c(t) = \frac{\omega_n}{j\sqrt{\xi^2-1}} e^{-\xi\omega_n t} \sin\left([j\omega_n \sqrt{\xi^2-1}]t\right) = \frac{\omega_n}{j\sqrt{\xi^2-1}} e^{-\xi\omega_n t}\, j \sinh\left([\omega_n \sqrt{\xi^2-1}]t\right)$$

$$= \frac{\omega_n}{\sqrt{\xi^2-1}} e^{-\xi\omega_n t} \sinh\left([\omega_n \sqrt{\xi^2-1}]t\right)$$

$$= \frac{\omega_n}{\sqrt{\xi^2-1}} e^{-\xi\omega_n t} \frac{e^{[\omega_n \sqrt{\xi^2-1}]t} - e^{-[\omega_n \sqrt{\xi^2-1}]t}}{2}$$

$$c(t) = \left(\frac{\omega_n}{\sqrt{\xi^2-1}}\right)\frac{e^{[-\xi+\sqrt{\xi^2-1}]\omega_n t} - e^{[-\xi-\sqrt{\xi^2-1}]\omega_n t}}{2} \tag{8.100}$$

Figure 8.30 shows the time response of a second order system for impulse input.

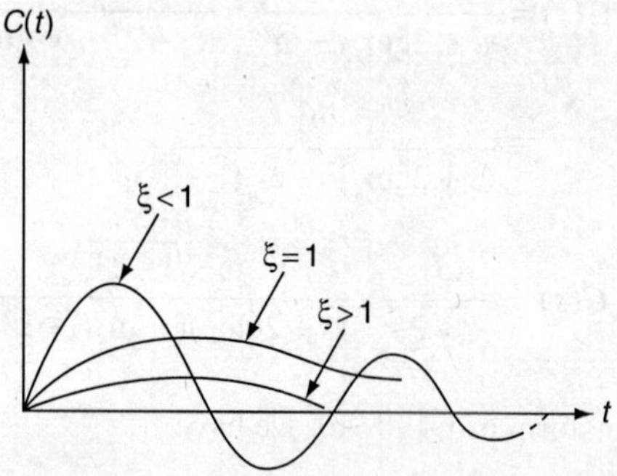

Fig. 8.30 Time response of a second order system for impulse input

8.15 Performance Indices

To achieve the best performance of control system, one or more parameters are selected, which are known as performance indices. The adjusted parameters are used to obtain the optimal control system. The following performance indices are discussed below in terms of error function:

(i) Integral square error criterion:

$$ISE = \int_0^\infty e^2(t)dt$$

(ii) Integral of time multiplied square error criterion:

$$ITSE = \int_0^\infty te^2(t)dt$$

(iii) Integral absolute error criterion:

$$IAE = \int_0^\infty |e(t)|\,dt$$

(iv) Integral of time multiplied absolute error criterion:

$$IATE = \int_0^\infty t|e(t)|\,dt$$

If $r(t)$ and $c(t)$ be the input and output respectively, the error signal in control system is given by

$$e(t) = r(t) - c(t)$$

For any stable system, the performance index will have finite value.
A second order system having $\xi < 1$ with the following transfer function is considered to evaluate the analytical evaluation of ISE and ITSE.

$$\frac{C(s)}{R(s)} = \frac{\omega_n^2}{s^2 + 2\xi\omega_n s + \omega_n^2} \tag{8.101}$$

Let the unit step input is taken.

$$\therefore \quad R(s) = \frac{1}{s} \tag{8.102}$$

$$C(s) = \frac{\omega_n^2 R(s)}{s^2 + 2\xi\omega_n s + \omega_n^2} = \frac{\omega_n^2}{s(s^2 + 2\xi\omega_n s + \omega_n^2)} \tag{8.103}$$

$$\therefore \quad E(s) = R(s) - C(s)$$

i.e.,

$$E(s) = \frac{1}{s} - \frac{\omega_n^2}{s(s^2 + 2\xi\omega_n s + \omega_n^2)} = \frac{s + 2\xi\omega_n}{s^2 + 2\xi\omega_n s + \omega_n^2} \tag{8.104}$$

$$= \frac{s + \xi\omega_n}{(s + \xi\omega_n)^2 + \omega_n^2(1 - \xi^2)} + \frac{\xi\omega_n}{(s + \xi\omega_n)^2 + \omega_n^2(1 - \xi^2)}$$

$$= \frac{s + \xi\omega_n}{(s + \xi\omega_n)^2 + \omega_n^2(1 - \xi^2)} + \frac{\xi}{\sqrt{1 - \xi^2}} \times \frac{\omega_n\sqrt{1 - \xi^2}}{(s + \xi\omega_n)^2 + \omega_n^2(1 - \xi^2)} \tag{8.105}$$

Taking inverse Laplace transform of the above Eq. (8.105), we have

$$e(t) = e^{-\xi\omega_n t}\left[\cos(\omega_n\sqrt{1-\xi^2}t) + \frac{\xi}{\sqrt{1-\xi^2}}\sin(\omega_n\sqrt{1-\xi^2}t)\right] \tag{8.106}$$

$$\therefore\ e^2(t) = e^{-2\xi\omega_n t}\left[\cos^2(\omega_n\sqrt{1-\xi^2}t) + \frac{\xi^2}{1-\xi^2}\sin^2(\omega_n\sqrt{1-\xi^2}t) + \frac{2\xi}{\sqrt{1-\xi^2}}\cos(\omega_n\sqrt{1-\xi^2}t)\sin(\omega_n\sqrt{1-\xi^2}t)\right]$$

$$= e^{-2\xi\omega_n t}\left[\frac{1+\cos(2\omega_n\sqrt{1-\xi^2}t)}{2} + \frac{\xi^2}{1-\xi^2}\times\frac{1-\cos(2\omega_n\sqrt{1-\xi^2}t)}{2} + \frac{\xi}{\sqrt{1-\xi^2}}\sin(2\omega_n\sqrt{1-\xi^2}t)\right]$$

$$= e^{-2\xi\omega_n t}\left[\frac{1}{2(1-\xi^2)} + \frac{1-2\xi^2}{2(1-\xi^2)}\cos(2\omega_n\sqrt{1-\xi^2}t) + \frac{\xi}{\sqrt{1-\xi^2}}\sin(2\omega_n\sqrt{1-\xi^2}t)\right] \tag{8.107}$$

Taking Laplace transform of both sides, we have

$$E^2(s) = \frac{1}{2(1-\xi^2)(s+2\xi\omega_n)} + \frac{1-2\xi^2}{2(1-\xi^2)}\times\frac{(s+2\xi\omega_n)}{(s+2\xi\omega_n)^2+4\omega_n^2(1-\xi^2)} + \frac{\xi}{\sqrt{1-\xi^2}}\times\frac{2\omega_n\sqrt{1-\xi^2}}{(s+2\xi\omega_n)^2+4\omega_n^2(1-\xi^2)} \tag{8.108}$$

We know that

$$\int_0^\infty f(t)dt = \underset{t\to\infty}{Lt}\int_0^t f(t)dt$$

Using final value theorem, we have

$$\underset{t\to\infty}{Lt}\int_0^t f(t)dt = \underset{s\to 0}{Lt}\ s\left\{LT\left[\int_0^t f(t)dt\right]\right\} = \underset{s\to 0}{Lt}\left[s\times\frac{F(s)}{s}\right] = \underset{s\to 0}{Lt}\ F(s)$$

Since $ISE = \int_0^\infty e^2(t)dt$, we have

$$ISE = \underset{s\to 0}{Lt}\ E^2(s)$$

$$\therefore\quad ISE = \underset{s\to 0}{Lt}\left[\frac{1}{2(1-\xi^2)(s+2\xi\omega_n)} + \frac{1-2\xi^2}{2(1-\xi^2)}\times\frac{(s+2\xi\omega_n)}{(s+2\xi\omega_n)^2+4\omega_n^2(1-\xi^2)} + \frac{\xi}{\sqrt{1-\xi^2}}\times\frac{2\omega_n\sqrt{1-\xi^2}}{(s+2\xi\omega_n)^2+4\omega_n^2(1-\xi^2)}\right]$$

$$\text{i.e.,}\quad ISE = \left[\frac{1}{2(1-\xi^2)(2\xi\omega_n)} + \frac{1-2\xi^2}{2(1-\xi^2)}\times\frac{(2\xi\omega_n)}{(2\xi\omega_n)^2+4\omega_n^2(1-\xi^2)} + \frac{\xi}{\sqrt{1-\xi^2}}\times\frac{2\omega_n\sqrt{1-\xi^2}}{(2\xi\omega_n)^2+4\omega_n^2(1-\xi^2)}\right]$$

i.e., $$ISE = \frac{1}{\omega_n}\left[\frac{1}{4\xi(1-\xi^2)} + \frac{\xi(1-2\xi^2)}{4(1-\xi^2)} + \frac{\xi}{2}\right]$$

i.e., $$ISE = \frac{1}{\omega_n}\left[\frac{1+3\xi^2-2\xi^4}{4\xi(1-\xi^2)}\right] = \frac{1}{\omega_n}\left[\frac{(4\xi^2+1)(1-\xi^2)}{4\xi(1-\xi^2)}\right] = \frac{4\xi^2+1}{4\xi\omega_n} = \frac{1}{\omega_n}\left(\xi + \frac{1}{4\xi}\right) \quad (8.109)$$

i.e., $$\omega_n(ISE) = \xi + \frac{1}{4\xi} \quad (8.110)$$

Figure 8.31 shows the plot of ISE with having ξ for $\omega_n = 1$. For minimization of ISE,

$$\frac{d(ISE)}{d\xi} = 0 \text{ with } \frac{d^2(ISE)}{d\xi^2} > 0$$

$$\frac{d(ISE)}{d\xi} = \frac{d}{d\xi}\left[\frac{1}{\omega_n}\left(\xi + \frac{1}{4\xi}\right)\right] = \frac{1}{\omega_n}\left(1 - \frac{1}{4\xi^2}\right)$$

$$\therefore \quad \frac{1}{\omega_n}\left(1 - \frac{1}{4\xi^2}\right) = 0$$

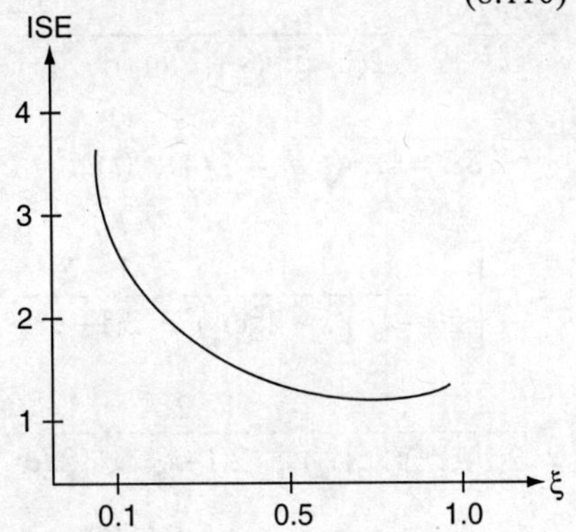

Fig.8.31 Plot of ISE as function of ξ

i.e., $\xi = 0.5$ for minimum ISE index.

Therefore, the minimum value of $\omega_n(ISE) = 0.5 + \frac{1}{4 \times 0.5} = 1.0$

From Eq. (8.106), we have

$$e(t) = e^{-\xi\omega_n t}\left[\cos(\omega_n\sqrt{1-\xi^2}t) + \frac{\xi}{\sqrt{1-\xi^2}}\sin(\omega_n\sqrt{1-\xi^2}t)\right]$$

Here $ITSE = \int_0^\infty te^2(t)dt$

We know that $LT[f(t)] = -\frac{dF(s)}{ds}$

Again from Eq. (8.107), we have

$$e^2(t) = e^{-2\xi\omega_n t}\left[\frac{1}{2(1-\xi^2)} + \frac{1-2\xi^2}{2(1-\xi^2)}\cos(2\omega_n\sqrt{1-\xi^2}t) + \frac{\xi}{\sqrt{1-\xi^2}}\sin(2\omega_n\sqrt{1-\xi^2}t)\right]$$

Let $f(t) = e^2(t)$

$$\therefore \quad f(t) = e^{-2\xi\omega_n t}\left[\frac{1}{2(1-\xi^2)} + \frac{1-2\xi^2}{2(1-\xi^2)}\cos(2\omega_n\sqrt{1-\xi^2}t) + \frac{\xi}{\sqrt{1-\xi^2}}\sin(2\omega_n\sqrt{1-\xi^2}t)\right]$$

$$\therefore \quad F(s) = \frac{1}{2(1-\xi^2)(s+2\xi\omega_n)} + \frac{1-2\xi^2}{2(1-\xi^2)} \times \frac{(s+2\xi\omega_n)}{(s+2\xi\omega_n)^2 + 4\omega_n^2(1-\xi^2)}$$
$$+ \frac{\xi}{\sqrt{1-\xi^2}} \times \frac{2\omega_n\sqrt{1-\xi^2}}{(s+2\xi\omega_n)^2 + 4\omega_n^2(1-\xi^2)}$$

i.e.,

$$\frac{dF(s)}{ds} = \frac{-1}{2(1-\xi^2)(s+2\xi\omega_n)^2} - \frac{1-2\xi^2}{2(1-\xi^2)} \times \frac{(s+2\xi\omega_n)^2 - 4\omega_n^2(1-\xi^2)}{[(s+2\xi\omega_n)^2 + 4\omega_n^2(1-\xi^2)]^2} - \frac{4\xi(s+2\xi\omega_n)}{[(s+2\xi\omega_n)^2 + 4\omega_n^2(1-\xi^2)]^2} \tag{8.111}$$

i.e.,

$$-\frac{dF(s)}{ds} = \frac{1}{2(1-\xi^2)(s+2\xi\omega_n)^2} + \frac{1-2\xi^2}{2(1-\xi^2)} \times \frac{(s+2\xi\omega_n)^2 - 4\omega_n^2(1-\xi^2)}{[(s+2\xi\omega_n)^2 + 4\omega_n^2(1-\xi^2)]^2} + \frac{4\xi(s+2\xi\omega_n)}{[(s+2\xi\omega_n)^2 + 4\omega_n^2(1-\xi^2)]^2}$$

$$ITSE = \int_0^\infty te^2(t)dt = \underset{t\to\infty}{Lt}\int_0^t te^2(t)dt = \underset{s\to 0}{Lt}\left[-\frac{dF(s)}{ds}\right]$$

$$= \underset{s\to 0}{Lt}\left[\frac{1}{2(1-\xi^2)(s+2\xi\omega_n)^2} + \frac{1-2\xi^2}{2(1-\xi^2)} \times \frac{(s+2\xi\omega_n)^2 - 4\omega_n^2(1-\xi^2)}{[(s+2\xi\omega_n)^2 + 4\omega_n^2(1-\xi^2)]^2} + \frac{4\xi(s+2\xi\omega_n)}{[(s+2\xi\omega_n)^2 + 4\omega_n^2(1-\xi^2)]^2}\right]$$

$$= \frac{1}{2(1-\xi^2)(2\xi\omega_n)^2} + \frac{1-2\xi^2}{2(1-\xi^2)} \times \frac{(2\xi\omega_n)^2 - 4\omega_n^2(1-\xi^2)}{[2\xi\omega_n)^2 + 4\omega_n^2(1-\xi^2)]^2} + \frac{4\xi(2\xi\omega_n)}{[(2\xi\omega_n)^2 + 4\omega_n^2(1-\xi^2)]^2}$$

$$= \frac{1}{2(1-\xi^2)(2\xi\omega_n)^2} + \frac{(1-2\xi^2)(2\xi^2-1)}{2(1-\xi^2)\times 4\omega_n^2} + \frac{\xi^2}{2\omega_n^2}$$

$$= \frac{1}{\omega_n^2}\left[\frac{1}{4\xi^2(1-\xi^2)} + \frac{(1-2\xi^2)(2\xi^2-1)}{8(1-\xi^2)} + \frac{\xi^2}{2}\right]$$

i.e.,

$$ITSE = \frac{1}{\omega_n^2}\left[\xi^2 + \frac{1}{8\xi^2}\right] \tag{8.112}$$

Figure 8.32 shows the plot of *ISTE* vs ξ for $\omega_n = 1$.

For minimization of *ISTE*,

$$\frac{d(ISTE)}{d\xi} = 0 \text{ with } \frac{d^2(ISTE)}{d\xi^2} > 0$$

$$\therefore \quad \frac{d(ISTE)}{d\xi} = \frac{d}{d\xi}\left[\frac{1}{\omega_n^2}\left(\xi^2 + \frac{1}{8\xi^2}\right)\right] = \frac{1}{\omega_n^2}\left(2\xi - \frac{1}{4\xi^3}\right)$$

Since $\frac{d(ISTE)}{d\xi} = 0$, we have

$$\frac{1}{\omega_n^2}\left(2\xi - \frac{1}{4\xi^3}\right) = 0$$

i.e.,

$$\xi = 0.595$$

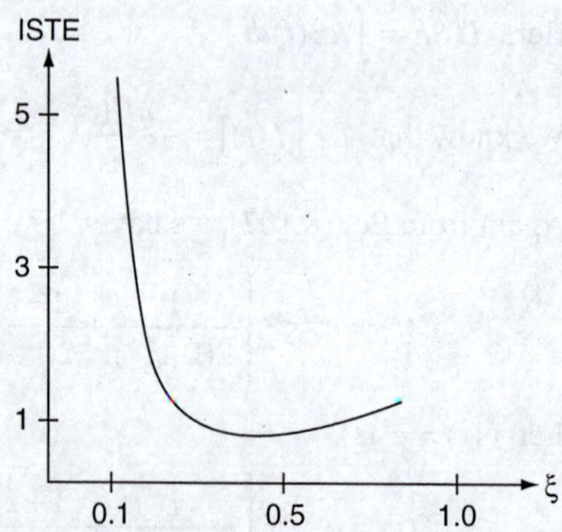

Fig. 8.32 Plot of ISTE as function of ξ

The minimum value of

$$\omega_n^2(ISTE) = \xi^2 + \frac{1}{8\xi^2} = 0.595^2 + \frac{1}{8\times 0.595^2} = 0.707$$

Figure 8.33 shows the error performance index graphs for a second order system.

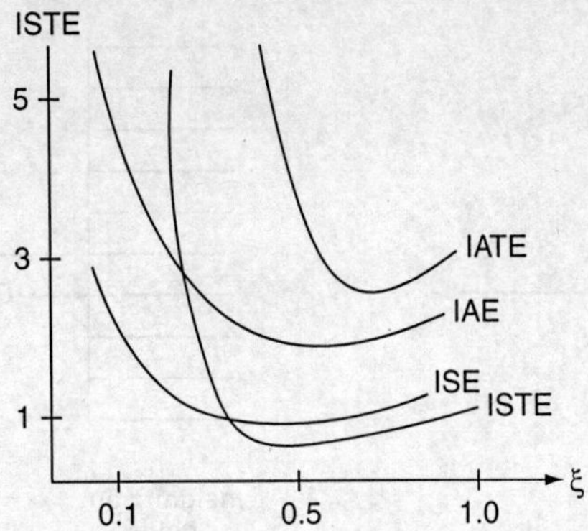

Fig. 8.33 Error performance index graph for a second order system

8.16 Discussion on Time Domain

The underdamped natural frequency (ω_n) corresponds to the system frequency for undamped system, i.e., $\xi = 0$. The magnitude of the imaginary part of the root for $0 < \xi < 1$ is expressed by

$$\omega_d = \omega_n\sqrt{1-\xi^2}$$

where ω_n is the radial distance from the roots to the origin.

The real part of the root is $-\xi\omega_n$ and settling time is inversely proportional to $\xi\omega_n$. We get idea about stability from the real part of the root. The cosine of the angle between the negative x-axis and a radial line drawn from origin to the root gives the damping factor (ξ) as shown in Fig. 8.34.

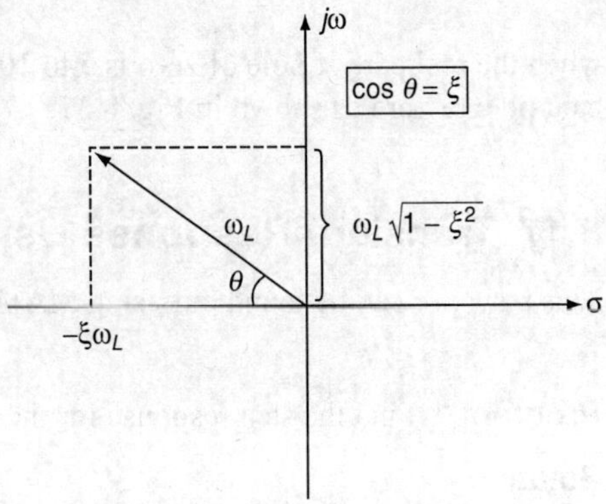

Fig. 8.34 Pictorial representation of ξ

The rise time and delay time are proportional to ξ and inversely proportional to ω_n. T_s, T_r and T_d increases when ω_n is decreased for constant value of ξ. T_r and T_d decreases but T_s increases when ω_n is kept fixed and ξ is decreased. This shows that the roots move towards the $j\omega$-axis. Therefore, settling time increases. Figure 8.35 and Fig. 8.36 show the variation of T_s, T_r and T_d for the above cases.

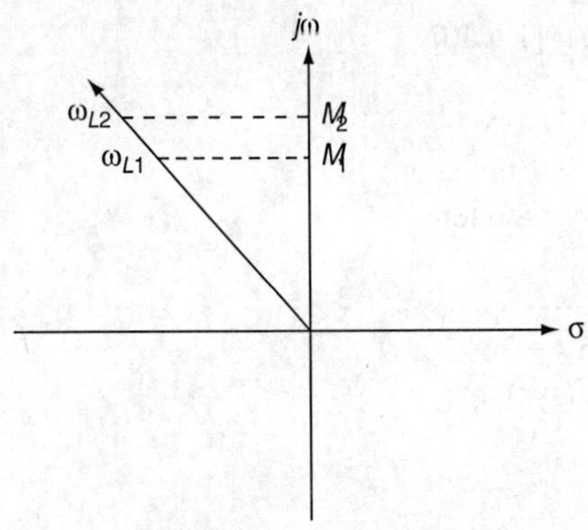

Fig. 8.35 Variation of T_s, T_r and T_d for ξ = constant and increasing value of ω_n

Since the poles and zeros near origin contribute significantly to transient response, these poles and zeros become dominant. The transient response due to the dominant poles and zeros decay slowly than that of poles or zeros located away from origin. Generally

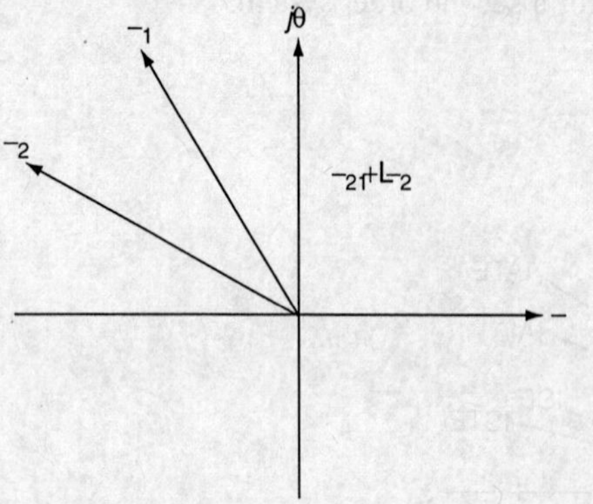

Fig. 8.36 Variation of T_s, T_r and T_d for ω_n = constant and ξ is decreased

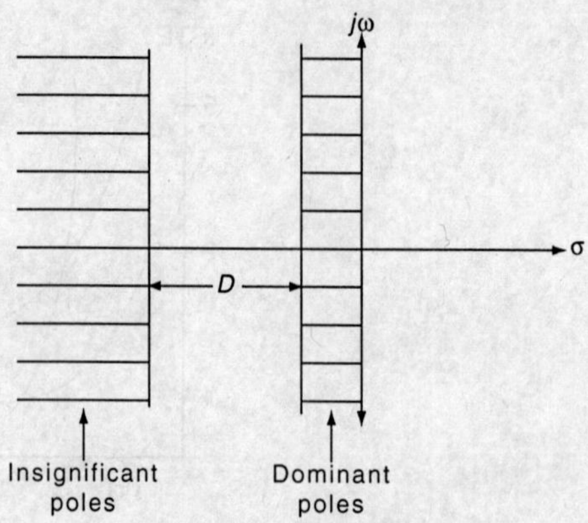

Fig. 8.37 Dominant poles

when the real part of pole of zero is 5 to 10 times more than a dominant pole, it is treated as insignificant pole or zero as shown in Fig. 8.37.

8.17 Transient Response Using MATLAB

This response can be found out using MATLAB program very easily as shown below.

Program 1 Find the step response for the following system $\dfrac{C(s)}{R(s)} = \dfrac{3s+20}{s^2+5s+36}$

Solution

```
>> num = [3 20]
num =
   3   20
>> den = [1 5 36]
den =
   1   5   36
>> sys = tf(num,den)
Transfer function:
  3 s + 20
--------------
s^2 + 5 s + 36
>> step(sys)
```

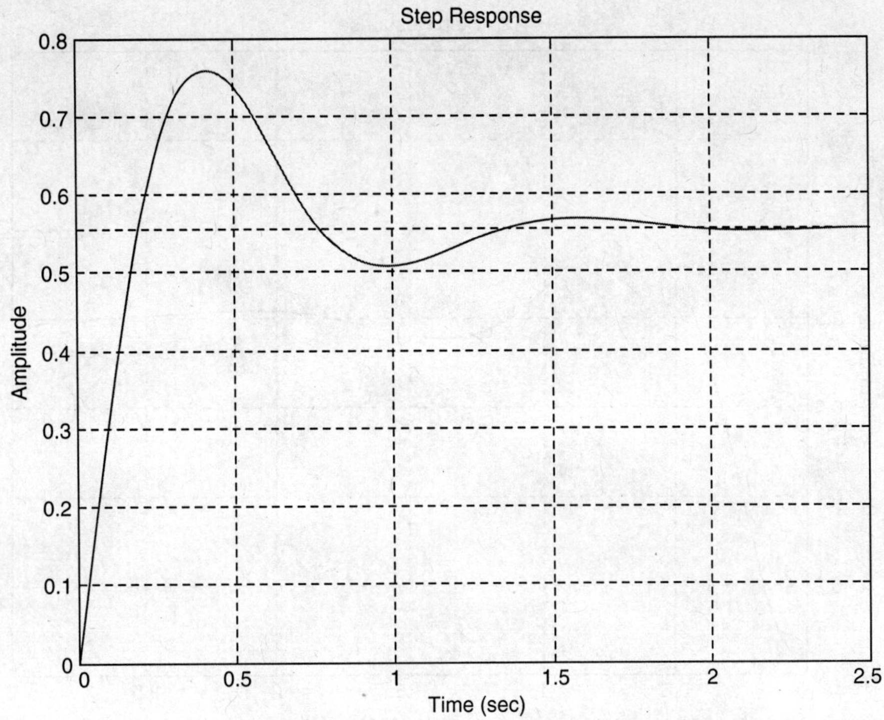

Fig. P1

Program 2 Find the step response for the following transfer function $\frac{C(s)}{R(s)} = \frac{20}{s^2 + 4s + 25}$

Solution

```
>> num = [20]
num =
   20
>> den = [1 4 25]
den =
   1    4   25
>> sys = tf(num,den)
Transfer function:
    20
--------------
s^2 + 4 s + 25
>> step(sys)
```

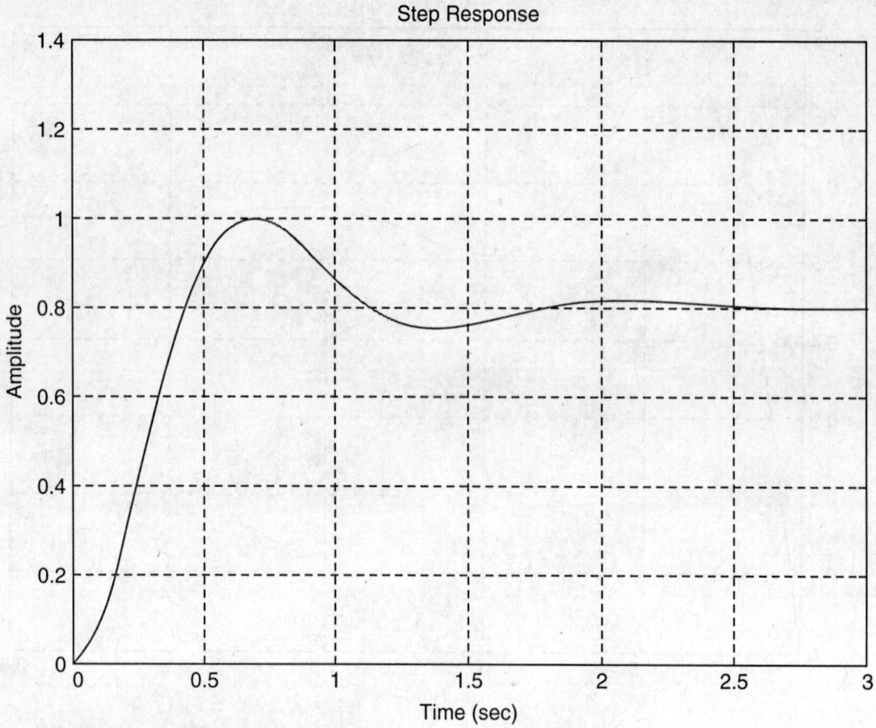

Fig. P2

SIGNIFICANT POINTS

Time Response: The variation of output with respect to time is known as time response.

Transient response: The part of the time response, which goes to zero after large interval of time, is known as transient response.

Steady-state response: The part of response that remains even after the transients have died out is said to be steady-state response.

Delay time: The time that the system output takes to reach its final value is known as delay time (T_d).

Rise time: The time that the system response takes to reach from 10% to 90% of the final value of output is called rise time (T_r).

Settling time: The time taken by the system response to settle down and stay within ± 2% of its final value is known as settling time (T_s).

Peak time: The time taken by the system response to reach the first maximum value is known as peak t ime (T_p).

Overshoot: The ratio of the maximum value of the step-excited output to the final output is called overshoot of the system. The percentage overshoot of a stem is defined as follows:

$$\%\,\text{overshoot} = \left.\frac{\text{Maximum output overshoot}}{\text{Final value of output}}\right|_{\text{step input}} \times 100$$

Sl. No.	Time response specification
1.	$T_d = \dfrac{1+0.7\xi}{\omega_n}$
2.	$T_r = \dfrac{\pi-\alpha}{\omega_d}\left[\text{where } \tan\alpha = \dfrac{\sqrt{1-\xi^2}}{\xi}\right]$, α is in radians
3.	$T_p = \dfrac{\pi}{\omega_d}$
4.	$\%M_p = e^{-\frac{\xi\pi}{\sqrt{1-\xi^2}}} \times 100$
5.	$T_s = \dfrac{4}{\xi\omega_n}$ for 2% criterion and $T_s = \dfrac{3}{\xi\omega_n}$ for 5% criterion

Steady-state error: For a step excitation, the difference between the desired output and the final value of any system is termed as steady-state error of the system:

$$E(s) = \frac{R(s)}{1+G(s)H(s)}$$

Position, velocity and acceleration error constant

Sl. No.	Constant	Equation	Steady-state error (e_{ss})
1.	Position error constant (K_p)	$K_p = \underset{s\to 0}{\text{Lt}}\, G(s)H(s)$	$e_{ss} = \dfrac{A}{1+K_p}$
2.	Velocity error constant (K_v)	$K_v = \underset{s\to 0}{\text{Lt}}\, sG(s)H(s)$	$e_{ss} = \dfrac{A}{K_v}$
3.	Acceleration error constant (K_a)	$K_a = \underset{s\to 0}{\text{Lt}}\, s^2G(s)H(s)$	$e_{ss} = \dfrac{A}{K_a}$

ADDITIONAL SOLVED EXAMPLES

Example 8.8 For a unity feedback system having $G(s) = \dfrac{35(s+4)}{s(s+2)(s+5)}$, find (i) the type of system, (ii) all error coefficients and (iii) error for ramp input with magnitude 5.

Solution

(i) $G(s) = \dfrac{35(s+4)}{s(s+2)(s+5)} \times 1 = \dfrac{35\times 4(1+0.25s)}{s(1+0.5s)(1+0.2s)\times 10} = \dfrac{14(1+0.25s)}{s(1+0.5s)(1+0.2s)}$

It is a type 1 system.

(ii) $K_p = \underset{s\to 0}{\text{Lt}}\, G(s)H(s) = \underset{s\to 0}{\text{Lt}}\dfrac{14(1+0.25s)}{s(1+0.5s)(1+0.2s)} = \infty$

$$K_v = \underset{s\to 0}{\text{Lt}}\, sG(s)H(s) = \underset{s\to 0}{\text{Lt}} \frac{14(1+0.25s)}{(1+0.5s)(1+0.2s)} = 14$$

$$K_a = \underset{s\to 0}{\text{Lt}}\, s^2G(s)H(s) = \underset{s\to 0}{\text{Lt}} \frac{14s(1+0.25s)}{(1+0.5s)(1+0.2s)} = 0$$

(iii) Since ramp input is used, K_v is to be taken into account. Since $e_{ss} = A/K_v$ and ramp magnitude $A = 5$, and $K_v = 14$, we get

$$e_{ss} = \frac{5}{14} = 0.357$$

Example 8.9 From a unity feedback system $G(s) = 20(s+2)/[s^2(s+1)(s+5)]$, determine (i) the type of system, (ii) error coefficients and (iii) steady-state error for the input $1 + 3t + t^2/2$.

Solution

(i) $$G(s)H(s) = \frac{20(s+2)}{s^2(s+1)(s+5)} = \frac{8(1+0.5s)}{s^2(1+s)(1+0.2s)}$$

This is a type 2 system.

(ii) $$K_p = \underset{s\to 0}{\text{Lt}}\, G(s)H(s) = \underset{s\to 0}{\text{Lt}} \frac{8(1+0.5s)}{s^2(1+s)(1+0.2s)} = \infty$$

$$K_v = \underset{s\to 0}{\text{Lt}}\, sG(s)H(s) = \underset{s\to 0}{\text{Lt}} \frac{8(1+0.5s)}{s(1+s)(1+0.2s)} = \infty$$

$$K_a = \underset{s\to 0}{\text{Lt}}\, s^2G(s)H(s) = \underset{s\to 0}{\text{Lt}} \frac{8(1+0.5s)}{(1+s)(1+0.2s)} = 8$$

(iii) Since the input is non-standard, let us use the formula

$$e_{ss} = \underset{s\to 0}{\text{Lt}} \left[\frac{sR(s)}{1+G(s)H(s)}\right]$$

Given $$r(t) = 1 + 3t + \frac{t^2}{2}$$

$\therefore$ $$R(s) = \frac{1}{s} + \frac{3}{s^2} + \frac{1}{s^3} = \frac{s^2+3s+1}{s^3}$$

$$e_{ss} = \underset{s\to 0}{\text{Lt}} \frac{sR(s)}{1+G(s)H(s)} = \underset{s\to 0}{\text{Lt}} \frac{(s^2+3s+1)}{s^2 + \dfrac{8(1+0.5s)}{(1+s)(1+0.2s)}}$$

$$= \frac{0+0+1}{0 + \dfrac{8(1+0)}{(1+0)(1+0)}} = \frac{1}{8} = 0.125$$

$\therefore$ $$e_{ss} = 0.125$$

Example 8.10 For the system shown in Fig. E8.2, determine the type of system, error coefficient and the error for the following inputs:

(i) $r(t) = 6$, (ii) $r(t) = 8t$ and (iii) $r(t) = 10 + 4t + \frac{3}{2}t^2$

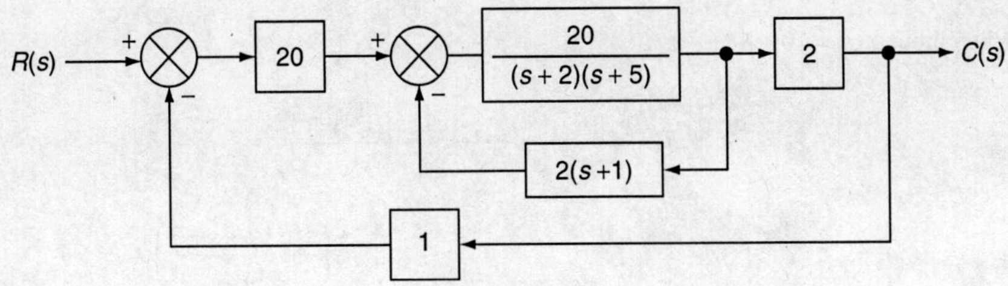

Fig. E8.2

Solution

Type of the system:

Simplifying internal feedback loop of Fig. E8.1, Fig. E 8.2(a) is obtained.

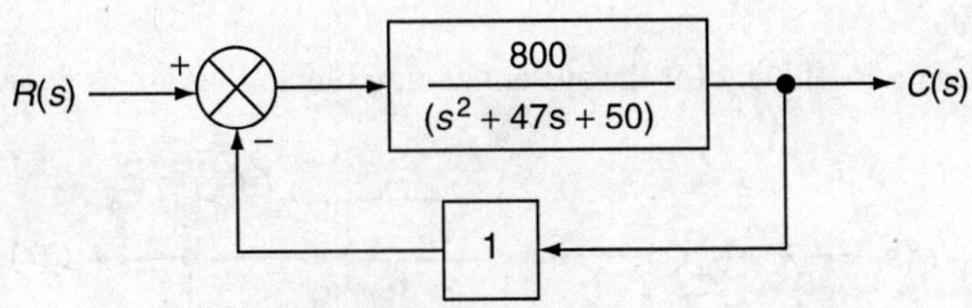

Fig. E8.2(a)

$$G(s) = \frac{800}{s^2 + 47s + 50} \text{ and } H(s) = 1$$

$$G(s)H(s) = \frac{800}{s^2 + 47s + 50}$$

The system is type 0.

Error coefficient

$$K_p = \underset{s\to 0}{\text{Lt}}\, G(s)H(s) = \underset{s\to 0}{\text{Lt}} \frac{800}{s^2 + 47s + 50} = \frac{800}{50} = 16$$

$$K_v = \underset{s\to 0}{\text{Lt}}\, G(s)H(s) = \underset{s\to 0}{\text{Lt}} \frac{800s}{s^2 + 47s + 50} = 0$$

$$K_a = \underset{s\to 0}{\text{Lt}}\, s^2 G(s)H(s) = \underset{s\to 0}{\text{Lt}} \frac{800s^2}{s^2 + 47s + 50} = 0$$

Error due to the following inputs

(i) When input $r(t) = 6$, $A = 6$

$$e_{ss} = \frac{A}{1 + K_p} = \frac{6}{1 + \frac{800}{50}} = \frac{300}{850} = \frac{6}{17}$$

(ii) When input $r(t) = 8t$, $A = 8$

$$e_{ss} = \frac{A}{K_v} = \frac{8}{0} = \infty$$

(iii) When input $r(t) = 10 + 4t + \frac{3}{2}t^2$

$$\therefore \quad R(s) = \frac{10}{s} + \frac{4}{s^2} + \frac{3}{s^3} = \frac{10s^2 + 4s + 3}{s^3}$$

$$\therefore \quad e_{ss} = \underset{s\to 0}{\text{Lt}} \frac{sR(s)}{1 + G(s)H(s)} = \underset{s\to 0}{\text{Lt}} \frac{s\left(\frac{10s^2+4s+3}{s^3}\right)}{1 + \frac{800}{s^2+47s+50}}$$

$$= \underset{s\to 0}{\text{Lt}} \frac{10s^2+4s+3}{s^2 + \frac{800s^2}{s^2+47s+50}} = \frac{3}{0} = \infty$$

Example 8.11 Find the error coefficient for the system having block diagram as shown in Fig. E8.3.

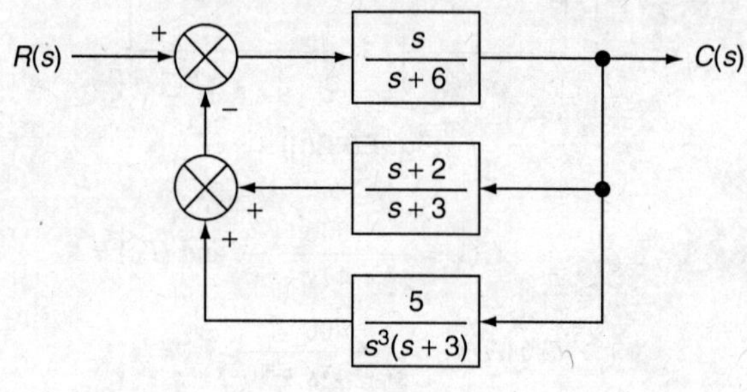

Fig. E8.3

Solution Simplifying Fig. E8.3, we have

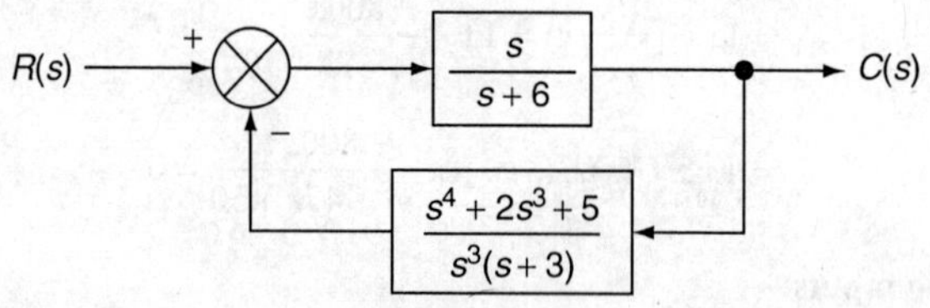

Fig. E8.3(a)

$$\therefore \quad G(s)H(s) = \frac{s(s^4 + 2s^3 + 5)}{s^3(s+3)(s+6)} = \frac{s^4 + 2s^3 + 5}{s^2(s+3)(s+6)}$$

It is a type 2 system.

$$K_p = \underset{s\to 0}{\text{Lt}}\, G(s)H(s) = \underset{s\to 0}{\text{Lt}} \frac{s^4 + 2s^3 + 5}{s^2(s+3)(s+6)} = \infty$$

$$K_v = \underset{s\to 0}{\text{Lt}}\, sG(s)H(s) = \underset{s\to 0}{\text{Lt}} \frac{s^4+2s^3+5}{s(s+3)(s+6)} = \infty$$

$$K_a = \underset{s\to 0}{\text{Lt}}\, s^2G(s)H(s) = \underset{s\to 0}{\text{Lt}} \frac{s^4+2s^3+5}{(s+3)(s+6)} = \frac{5}{18}$$

Example 8.12 Find K to limit the error of a system for input $1+8t+\frac{18}{2}t^2$ is 0.8 having $G(s)H(s) = \frac{K}{s^2(s+1)(s+4)}$

Solution Input $r(t) = 1+8t+\frac{18}{2}t^2$

$\therefore$
$$R(s) = \frac{1}{s}+\frac{8}{s^2}+\frac{18}{s^3} = \frac{s^2+8s+18}{s^3}$$

$$e_{ss} = \underset{s\to 0}{\text{Lt}} \frac{sR(s)}{1+G(s)H(s)} = \underset{s\to 0}{\text{Lt}} \frac{s\left(\frac{s^2+8s+18}{s^3}\right)}{1+\frac{K}{s^2(s+1)(s+4)}}$$

$$= \underset{s\to 0}{\text{Lt}} \frac{s^2+8s+10}{s^2+\frac{K}{(s+1)(s+4)}} = \frac{10}{\frac{K}{4}} = \frac{40}{K}$$

$\therefore$
$$\frac{40}{K} = 0.8$$

i.e.,
$$K = \frac{40}{0.8} = 50$$

Example 8.13 For $r(t) = 0.2t$, it is required that $e_{ss} \le 0.02$. Determine the value of K for the system shown in Fig. E8.4.

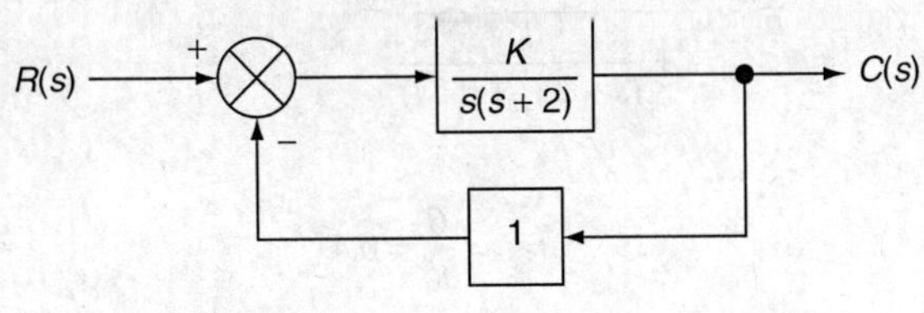

Fig. E8.4

Solution Here $G(s) = \frac{K}{s(s+2)}$ and $H(s) = 1$

$\therefore$
$$G(s)H(s) = \frac{K}{s(s+2)}$$

Again,
$$r(t) = 0.2t$$

$\therefore$ $$R(s) = \frac{0.2}{s^2}$$

Now, $$e_{ss} = \underset{s\to 0}{\text{Lt}} \frac{sR(s)}{1+G(s)H(s)} = \underset{s\to 0}{\text{Lt}} \frac{s\left(\frac{0.2}{s^2}\right)}{1+\frac{K}{s(s+2)}} = \underset{s\to 0}{\text{Lt}} \left(\frac{0.2}{s+\frac{K}{s+2}}\right) = \frac{0.4}{K}$$

For $e_{ss} \le 0.02$,

$\therefore$ $$\frac{0.4}{K} \le 0.02$$

i.e., $$K \ge \frac{0.4}{0.02}$$

i.e., $$K \ge 20$$

Example 8.14 A system has $G = K/[s(s+2)(1+0.5s)]$ and unity feedback negative. Find the steady-state error for $r(t) = 3t$ and $K = 4$. Also calculate K for $e_{ss} = 0.4$.

Solution

$$G(s) = \frac{K}{s(s+2)(1+0.5s)} \text{ and } H(s) = 1$$

$\therefore$ $$G(s)H(s) = \frac{K}{s(s+2)(1+0.5s)}$$

(i) $r(t) = 3t$

$\therefore$ $$R(s) = \frac{3}{s^2}$$

$\therefore$ $$e_{ss} = \underset{s\to 0}{\text{Lt}} \frac{sR(s)}{1+G(s)H(s)} = \underset{s\to 0}{\text{Lt}} \frac{s\left(\frac{3}{s^2}\right)}{1+\frac{K}{s(s+2)(1+0.5s)}}$$

$$= \underset{s\to 0}{\text{Lt}} \frac{3}{s+\frac{K}{(s+2)(1+0.5s)}} = \frac{6}{K}$$

For $e_{ss} = 0.4$, we have

$$e_{ss} = \frac{6}{K} = 0.4$$

i.e., $$K = \frac{6}{0.4} = \frac{60}{4} = 15$$

Example 8.15 A unity feedback system has

$$G(s) = \frac{180}{s(s+6)} \text{ and } r(t) = 4t$$

Determine (i) the steady-state error and (ii) the value of K to reduce the error by 6%.

Solution Given
$$G(s)H(s) = \frac{180}{s(s+6)}$$
$$r(t) = 4t$$
∴
$$R(s) = \frac{4}{s^2}$$
∴
$$e_{ss} = \underset{s\to 0}{\text{Lt}} \frac{sR(s)}{1+G(s)H(s)} = \underset{s\to 0}{\text{Lt}} \frac{s\left(\frac{4}{s^2}\right)}{1+\frac{180}{s(s+6)}} = \underset{s\to 0}{\text{Lt}} \frac{4}{s+\frac{180}{s+6}} = 0.134$$

Since new error is 6% less,

∴
$$(e_{ss})_{new} = 0.94e_{ss} = 0.12596$$

Again,
$$e_{ss} = \underset{s\to 0}{\text{Lt}} \frac{4}{s+\frac{K}{s+6}} = \frac{24}{K}$$
∴
$$\frac{24}{K} = 0.12596$$
i.e.,
$$K = \frac{24}{0.12596} = 190.537$$

Example 8.16 If the transfer function of a system is $\frac{C(s)}{R(s)} = \frac{81}{s^2+6s+81}$, find (i) ω_n, (ii) ξ, (iii) ω_d, (iv) T_p, and (v) M_p.

Solution

Comparing
$$\frac{C(s)}{R(s)} = \frac{81}{s^2+6s+81} \quad \text{with the standard form}$$
$$\frac{C(s)}{R(s)} = \frac{\omega_n}{s^2+2\xi\omega_n s+\omega_n^2},$$
we have

(i)
$$\omega_n^2 = 81$$
∴
$$\omega_n = 9\,\text{rad/sec}$$
(ii)
$$2\xi\omega_n = 6$$
∴
$$\xi = \frac{6}{2\omega_n} = \frac{1}{3} = 0.334$$

(iii) Again, $$\omega_d = \omega_n\sqrt{1-\xi^2} = 9\sqrt{1-(0.334)^2} = 8.483\,\text{rad/sec}.$$

(iv) Again, $$T_p = \frac{\pi}{\omega_d} = \frac{\pi}{8.483} = \frac{3.14}{8.483} = 0.3700\,\text{sec}.$$

(v) Again, $$M_p = 100\times e^{\frac{-\xi\pi}{\sqrt{1-\xi^2}}} = 100\times e^{-\frac{0.334\times 3.14}{\sqrt{1-0.334^2}}} = 32.86\%$$

Example 8.17 For a system having $\frac{C(s)}{R(s)} = \frac{20}{s^2+7s+25}$, find its time response specifications and expression for output.

Solution Here $\omega_n^2 = 25$. Therefore, $\omega_n = 5$ rad/sec and $2\xi\omega_n = 7$.

$$\xi = \frac{7}{2\omega_n} = \frac{7}{2\times 5} = 0.7$$

$$\omega_d = \omega_n\sqrt{1-\xi^2} = 5\sqrt{1-(0.7)^2} = 3.57\,\text{rad/sec}$$

Specifications:

(i) $$T_d = \frac{1+0.7\xi}{\omega_n} = \frac{1+0.7\times 0.7}{5} = 0.298\ \text{sec}$$

(ii) $$T_r = \frac{\pi-\alpha}{\omega_d}$$

Again, $$\alpha = \tan^{-1}\frac{\sqrt{1-\xi^2}}{\xi} = \tan^{-1}\frac{\sqrt{1-0.7^2}}{0.7} = 0.795\ \text{rad}$$

$$T_r = \frac{\pi-\alpha}{\omega_d} = \frac{3.14-0.795}{3.57} = 0.6568\ \text{sec}$$

(iii) $$T_p = \frac{\pi}{\omega_d} = \frac{3.14}{3.57} = 0.879\,\text{sec}$$

(iv) $$M_p = 100\times e^{\frac{-\xi\pi}{\sqrt{1-\xi^2}}} = 100\times e^{-\frac{0.7\times 3.14}{\sqrt{1-0.7^2}}} = 4.606\%$$

(v) $$T_s = \frac{4}{\xi\omega_n} = \frac{4}{0.7\times 5} = 1.1428\ \text{sec}$$

Expression for output,

$$\frac{Y(s)}{X(s)} = \frac{20}{s^2+7s+25} = \frac{20}{25}\left[\frac{25}{s^2+7s+25}\right]$$

For unit step input

$$y(t) = \frac{20}{25}\left[1 - \frac{e^{-\xi\omega_n t}}{\sqrt{1-\xi^2}}\sin\left(\omega_d t + \tan^{-1}\frac{\sqrt{1-\xi^2}}{\xi}\right)\right]$$

$$y(t) = 0.8\left[1 - \frac{e^{-3.5t}}{0.714}\sin(3.57t + 0.795\right]$$

Example 8.18 Determine the (a) time response specifications and expression for output for unit step input ($y \rightarrow$ output, $x \rightarrow$ input) to a system having system equation as follows:

$$\frac{d^2y}{dt^2}+5\frac{dy}{dt}+16y=9x$$

Solution The given system equation is

$$\frac{d^2y}{dt^2}+5\frac{dy}{dt}+16y=9x$$

Taking Laplace transform with zero initial conditions,

$$s^2Y(s)+5sY(s)+16Y(s)=9X(s)$$

i.e.,
$$(s^2+5s+16)Y(s)=9X(s)$$

i.e.,
$$\frac{Y(s)}{X(s)}=\frac{9}{s^2+5s+16}$$

Here
$$\omega_n^2=16$$

$$\omega_n=4 \text{ rad/sec}$$

and
$$2\xi\omega_n=5$$

i.e.,
$$\xi=\frac{5}{2\omega_n}=\frac{5}{2\times 4}=0.625$$

$\therefore$
$$\omega_d=\omega_n\sqrt{1-\xi^2}=4\sqrt{1-(0.625)^2}=3.1225.$$

(a) Time response specifications:

(i)
$$T_d=\frac{1+0.7\xi}{\omega_n}=\frac{1+0.7\times 0.625}{4}=0.3593\text{sec}$$

(ii)
$$T_r=\frac{\pi-\alpha}{\omega_d}$$

where
$$\alpha=\tan^{-1}\frac{\sqrt{1-\xi^2}}{\xi}=\tan^{-1}\frac{\sqrt{1-(0.625)^2}}{0.625}=0.8952\text{rad}$$

$\therefore$
$$T_r=\frac{3.14-0.8952}{3.1225}=0.7189\text{sec}$$

(iii)
$$T_p=\frac{\pi}{\omega_d}=\frac{3.14}{3.1225}=1.005 \text{ sec}$$

(iv)
$$\%M_p=100\times e^{\frac{-\xi\pi}{\sqrt{1-\xi^2}}}=100\times e^{\frac{-0.625\times 3.14}{\sqrt{1-0.625^2}}}=8.09\%$$

(v)
$$T_s=\frac{4}{\xi\omega_n}=\frac{4}{0.625\times 4}=1.6\text{sec}$$

(b) Expression for output:

$$\frac{Y(s)}{X(s)} = \frac{9}{s^2+5s+16} = \frac{9}{16}\left[\frac{16}{s^2+5s+16}\right]$$

$$c(t) = \frac{9}{16}\left[1 - \frac{e^{-\xi\omega_n t}}{\sqrt{1-\xi^2}}\sin(\omega_d t + \alpha)\right]$$

$$= \frac{9}{16}[1 - 1.2810\, e^{-2.5t}\sin(3.1225t + 0.8952)]$$

Example 8.19 Determine time response specifications for a unit step input to a unit negative feedback system having $G(s) = 144/[s(s+12)]$.

Solution

$$\frac{C(s)}{R(s)} = \frac{G(s)}{1+G(s)H(s)} = \frac{\dfrac{144}{s(s+12)}}{1+\dfrac{144}{s(s+12)}} = \frac{144}{s^2+12s+144}$$

Now,

$$\omega_n^2 = 144$$

$\therefore$

$$\omega_n = 12\,\text{rad/sec}$$

$$2\xi\omega_n = 12$$

$\therefore$

$$\xi = \frac{12}{2\omega_n} = \frac{12}{2\times 12} = 0.5$$

$$\omega_d = \omega_n\sqrt{1-\xi^2} = 12\sqrt{1-0.5^2} = 10.39\,\text{rad/sec}$$

Time response specifications:

(i)

$$T_d = \frac{1+0.7\xi}{\omega_n} = \frac{1+0.7\times 0.5}{12} = 0.1125\,\text{sec}$$

(ii)

$$T_r = \frac{\pi - \alpha}{\omega_d}$$

where

$$\alpha = \tan^{-1}\frac{\sqrt{1-\xi^2}}{\xi} = \tan^{-1}\frac{\sqrt{1-0.5^2}}{0.5} = 1.046\,\text{rad}$$

$\therefore$

$$T_r = \frac{3.14-1.046}{10.39} = 0.2015\,\text{sec}$$

(iii)

$$T_p = \frac{\pi}{\omega_d} = \frac{3.14}{10.39} = 0.3022\,\text{sec}$$

(iv)

$$\%M_p = 100\times e^{-\frac{\xi\pi}{\sqrt{1-\xi^2}}} = 100\times e^{-\frac{0.5\times 3.14}{\sqrt{1-0.5^2}}} = 16.32\%$$

(v)

$$T_s = \frac{4}{\xi\omega_n} = \frac{4}{0.5\times 12} = 0.667$$

Example 8.20 A system has $G(s) = K/[s(1+sT)]$ with unity feedback (negative) when K and T are constant. Determine the factor by which K should be multiplied to reduce the overshoot from 85% to 35%.

Solution

$$G(s) = \frac{K}{s(1+sT)} \text{ and } H(s) = 1$$

$$\therefore \quad G(s)H(s) = \frac{K}{s(1+sT)}$$

$$\frac{C(s)}{R(s)} = \frac{G(s)}{1+G(s)H(s)} = \frac{\frac{K}{s(1+sT)}}{1+\frac{K}{s(1+sT)}} = \frac{K}{s^2T+s+K} = \frac{\frac{K}{T}}{s^2+\frac{1}{T}s+\frac{K}{T}}$$

Now $$\omega_n^2 = \frac{K}{T}$$

i.e., $$\omega_n = \sqrt{\frac{K}{T}} \text{ rad/sec}$$

$$2\xi\omega_n = \frac{1}{T}$$

i.e., $$\xi = \frac{1}{2\omega_n T} = \frac{1}{2\left(\sqrt{\frac{K}{T}}\right)T} = \frac{1}{2\sqrt{KT}}$$

Let $\xi = \xi_1$ for $M_p = 85\%$ or 0.85.

$$\therefore \quad 0.85 = e^{-\frac{\xi\pi}{\sqrt{1-\xi_1^2}}}$$

i.e., $$ln(0.85) = ln\, e^{-\frac{\xi_1\pi}{\sqrt{1-\xi_1^2}}}$$

i.e., $$\sqrt{1-\xi_1^2} = 19.32\,\xi_1$$

i.e., $$1-\xi_1^{\ 2} = 373.2624\,\xi_1^2$$

i.e., $$374.2624\,\xi_1^{\ 2} = 1$$

$$\therefore \quad \xi_1 = 0.0517 \qquad (1)$$

Let $\xi = \xi_2$ for $M_p = 35\%$ or 0.35.

$$\therefore \quad 0.35 = e^{-\frac{\xi_1\pi}{\sqrt{1-\xi_2^2}}}$$

i.e., $$\sqrt{1-\xi_2^2} = 2.99\,\xi_2$$

i.c., $$1-\xi_2^{\ 2} = 8.94\,\xi_2^{\ 2}$$

i.e., $$9.94\xi_2^{\ 2} = 1$$

$$\therefore \quad \xi_2 = 0.317 \qquad (2)$$

From Eq. (1) and Eq. (2), we have

$$\frac{\xi_1}{\xi_2}=\frac{0.0517}{0.317}$$

i.e.,
$$\frac{\dfrac{1}{2\sqrt{K_1T}}}{\dfrac{1}{2\sqrt{K_2T}}}=\frac{0.0517}{0.317}\left[\because \xi=\frac{1}{2\sqrt{KT}}\right]$$

i.e.,
$$K_2=0.02659K_1$$

The required factor is 0.02659.

Example 8.21 For the feedback system shown in Fig. E8.5, find K_1 and K_2 so that peak overshoot of the system is 1.3% and the corresponding time is 1.06 sec.

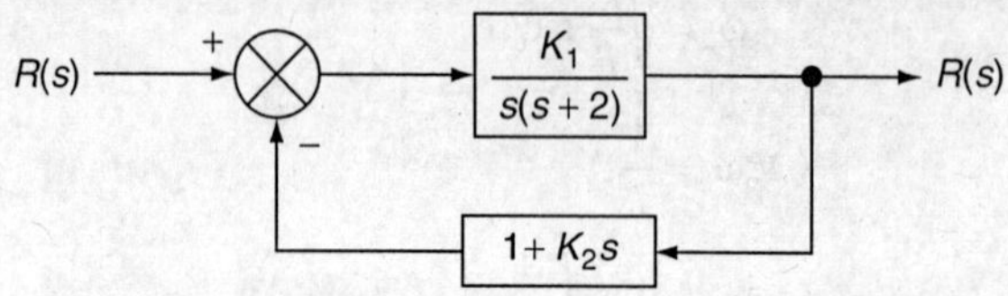

Fig. E8.5

Solution

$$\frac{G(s)}{R(s)}=\frac{G(s)}{1+G(s)H(s)}=\frac{\dfrac{K_1}{s(s+2)}}{1+\dfrac{K_1(1+K_2s)}{s(s+2)}}$$

$$=\frac{K_1}{s^2+2s+K_1K_2s+K_1}=\frac{K_1}{s^2+(K_1K_2+2)s+K_1}$$

$$\omega_n^2=K_1$$

i.e.,
$$\omega_n=\sqrt{K_1}$$

$$2\xi\omega_n=K_1K_2+2$$

i.e.,
$$\xi=\frac{K_1K_2+2}{2\omega_n}=\frac{K_1K_2+2}{2\sqrt{K_1}}$$

$$\omega_d=\omega_n\sqrt{1-\xi^2}=\omega_n\sqrt{1-\frac{(K_1K_2+2)^2}{4K_1}}=\sqrt{K_1}\sqrt{1-\frac{(K_1K_2+2)^2}{4K_1}}$$

$$\%M_p=100\times e^{-\frac{\xi\pi}{\sqrt{1-\xi^2}}}$$

i.e., $$0.013 = e^{-\frac{\xi\pi}{\sqrt{1-\xi^2}}}$$

i.e., $$ln(0.013) = -\frac{\xi\pi}{\sqrt{1-\xi^2}}$$

i.e., $$\sqrt{1-\xi^2} = \left(\frac{3.14}{4.3428}\right)\xi = 0.723\xi$$

i.e., $$1-\xi^2 = 0.5227\xi^2$$

∴ $$\xi = 0.8103$$

Again, $$T_p = \frac{\pi}{\omega_d} = 1.06$$

$$\omega_d = 2.96 \text{ rad/sec}$$

Now, $$\xi = \frac{K_1K_2+2}{2\sqrt{K_1}} = 0.8103 \qquad (1)$$

and $$\omega_d = \sqrt{K_1}\sqrt{1-\frac{(2+K_1K_2)^2}{4K_1}} = 2.96$$

i.e., $$\sqrt{K_1}\sqrt{1-\frac{(1.6206)^2K_1}{4K_1}} = 2.96 \qquad \text{[From Eq. (1)]}$$

i.e., $$\sqrt{K_1}\sqrt{1-0.6566} = 2.96$$

i.e., $$K_1 = 25.514$$

From Eq. (1), we have

$$K_1K_2 + 2 = 1.6206\sqrt{K_1}$$

i.e., $$K_2 = \frac{(1.6206\sqrt{K_1})-2}{K_1} = \frac{(1.6206\times\sqrt{25.514})-2}{25.514} = 0.2424$$

∴ $$K_1 = 25.514 \text{ and } K_2 = 0.2424.$$

Example 8.22 Determine the values of K, M and J if $M_p = 28\%$, $T_p = 2.2$ and output at steady-state is 6 units for the feedback system shown in Fig. E8.6, having unit step input.

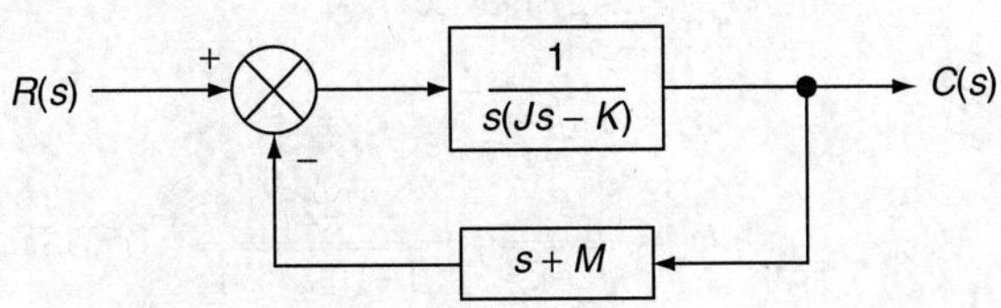

Fig. E8.6

Solution

Here, $$G(s)=\frac{1}{s(Js-K)} \text{ and } H(s)=s+M$$

$$\therefore \quad \frac{C(s)}{R(s)}=\frac{G(s)}{1+G(s)H(s)}=\frac{\frac{1}{s(Js-K)}}{1+\frac{(s+M)}{s(Js-K)}}=\frac{1}{s(Js-K)+(s+M)}=\frac{\frac{1}{J}}{s^2+s\left(\frac{1-K}{J}\right)+\frac{M}{J}}$$

$$\omega_n^2=\frac{M}{J} \text{ i.e., } \omega_n=\sqrt{\frac{M}{J}}$$

$$2\xi\omega_n=\frac{1-K}{J} \text{ i.e., } \xi=\frac{1}{2\omega_n}\left(\frac{1-K}{J}\right)=\frac{1-K}{2\sqrt{MJ}}$$

Since $$M_p=28\%,$$

$$\therefore \quad 28=100\times e^{-\frac{\xi\pi}{\sqrt{1-\xi^2}}}$$

i.e., $$0.28=e^{-\frac{\xi\pi}{\sqrt{1-\xi^2}}}$$

i.e., $$ln(0.28)=\frac{-\xi\pi}{\sqrt{1-\xi^2}}$$

i.e., $$-1.2729=-\frac{\xi\pi}{\sqrt{1-\xi^2}}$$

i.e., $$\sqrt{1-\xi^2}=2.4668\xi$$

i.e., $$1-\xi^2=6.085\xi^2$$

i.e., $$7.085\,\xi^2=1$$

$$\therefore \quad \xi=0.3757$$

Again, $$T_p=\pi/\omega_d$$

and here $T_p=2.2\,\text{sec}$. Therefore

$$2.2=\frac{\pi}{\omega_d}$$

Again, $$\omega_d=\frac{\pi}{2.2}=1.427 \text{ rad/sec}$$

i.e., $$\omega_d=\omega_n\sqrt{1-\xi^2}$$

$$\therefore \quad \omega_n=\frac{\omega_d}{\sqrt{1-\xi^2}}=\frac{1.427}{\sqrt{1-(0.3757)^2}}=1.5398 \text{ rad/sec}$$

Again, final output $(C_{ss})=\underset{t\to\infty}{\text{Lt}}\,c(t)=\underset{s\to\infty}{\text{Lt}}\,sC(s)$

i.e.,

$$C_{ss} = \underset{s\to 0}{\text{Lt}}\, s\left[\frac{\frac{1}{J}}{s^2+\left(\frac{1-K}{J}\right)s+\frac{M}{J}}\right]R(s) \tag{1}$$

Since input is unit step,

$$r(t) = u(t)$$

∴

$$R(s) = \frac{1}{s} \tag{2}$$

Using Eq. (2), Eq. (1) be expressed as

$$C_{ss} = \underset{s\to 0}{\text{Lt}}\, s\left[\frac{\frac{1}{J}}{s^2+\frac{1-K}{J}s+\frac{M}{J}}\right]\frac{1}{s} = \underset{s\to 0}{\text{Lt}}\frac{\frac{1}{J}}{s^2+\frac{1-K}{J}s+\frac{M}{J}} = \frac{1}{M} = 6$$

∴ $$M = \frac{1}{6} = 0.1667$$

Again, $$\omega_n = \sqrt{\frac{M}{J}}$$

∴ $$J = \frac{M}{\omega_n^2} = \frac{0.1667}{1.5398^2} = 0.0703$$

Example 8.23 Using generalized error series, calculate the steady-state error of a unity feedback system having $G(s) = 10/(s+2)$ for the following excitations:

(i) $r(t) = 3$, (ii) $r(t) = 2t$, (iii) $r(t) = t^2/2$ and (iv) $r(t) = 1 + 2t + \frac{t^2}{2}$.

Solution

(i) $$r(t) = 3$$

∴ $$e_{ss} = C_0 r(t) = 3C_0$$

Here, $$G(s) = \frac{10}{s+2}$$

and $$H(s) = 1$$

∴ $$C_0 = \underset{s\to 0}{\text{Lt}}\left(\frac{1}{1+G(s)H(s)}\right) = \underset{s\to 0}{\text{Lt}}\left(\frac{s+2}{s+12}\right) = \frac{1}{6}$$

∴ $$e_{ss} = 3\times\frac{1}{6} = \frac{1}{2} = 0.5$$

(ii) For $$r(t) = 2t,\ r'(t) = 2$$

$$e_{ss} = C_0 r(t) + C_1 r'(t)$$

Again, $$C_1 = \underset{s\to 0}{\text{Lt}}\frac{d}{ds}\left(\frac{s+2}{s+12}\right) = \underset{s\to 0}{\text{Lt}}\left[\frac{1}{s+12} - \frac{s+2}{(s+12)^2}\right] = \frac{1}{12} - \frac{2}{144} = \frac{10}{144}$$

$\therefore$ $$e_{ss} = \frac{1}{6} \times 2t + \frac{10}{144} \times 2 = \frac{24t + 10}{72}$$

(iii) Here, $$r(t) = t^2/2$$

$\therefore$ $$r'(t) = t \text{ and } r''(t) = 1$$

$\therefore$ $$e_{ss} = C_0 r(t) + C_1 r'(t) + \frac{C_2}{2!} r''(t) = \frac{t^2}{2} C_0 + tC_1 + \frac{C_2}{2!}$$

$$C_2 = \underset{s \to 0}{\text{Lt}} \frac{d^2}{ds^2}\left(\frac{s+2}{s+12}\right) = \underset{s \to 0}{\text{Lt}}\left[-\frac{1}{(s+12)^2} - \frac{1}{(s+12)^2} + \frac{2(s+2)}{(s+12)^3}\right]$$

$$= -\frac{1}{144} - \frac{1}{144} + \frac{4}{144 \times 12} = -\frac{20}{1728}$$

$\therefore$ $$e_{ss} = \frac{t^2}{12} + \frac{10}{144} t - \frac{20}{1728}$$

(iv) $$e(t) = 1 + 2t + \frac{t^2}{2}$$

$\therefore$ $$r'(t) = 2 + t \quad \text{and} \quad r''(t) = 1$$

$\therefore$ $$e_{ss} = C_0\left(1 + 2t + \frac{t^2}{2}\right) + C_1(2+t) + C_2 = \frac{1}{6}\left(1 + 2t + \frac{t^2}{2}\right) + \frac{10}{144}(2+t) - \frac{20}{1728} \times \frac{1}{2}$$

$$= \left(\frac{1}{6} + \frac{10}{72} - \frac{10}{1728}\right) + \left(\frac{1}{3} + \frac{10}{144}\right)t + \frac{t^2}{12}$$

$$= 0.2997 + 0.403t + 0.0834t^2$$

Example 8.24 Using the generalized error series, determine the steady-state error of the system for the input $r(t) = 1 + 2t + 5(t^2/2)$.

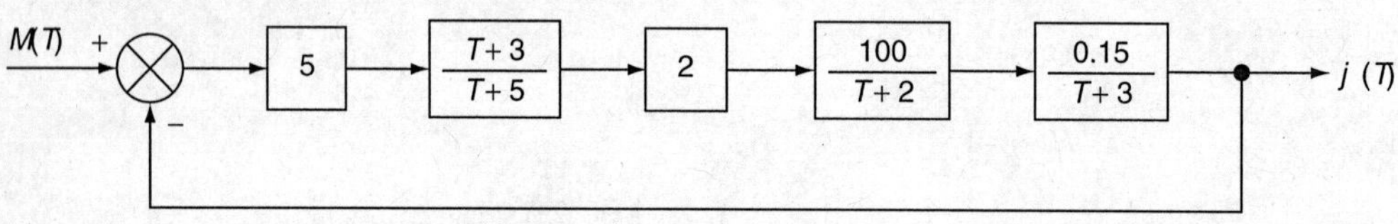

Fig. E8.7

Solution The open-loop transfer function of the system is given by

$$G(s) = 5\left(\frac{s+3}{s+5}\right)\left(\frac{2 \times 100}{s+2}\right)\left(\frac{0.15}{s+3}\right) = \frac{150}{(s+2)(s+5)} = \frac{150}{s^2 + 7s + 10}$$

The feedback is negative and $H(s) = 1$. Therefore,

$$E(s) = \frac{1}{1 + G(s)H(s)} = \frac{s^2 + 7s + 10}{s^2 + 7s + 160}$$

$$\therefore \quad C_0 = \underset{s\to 0}{\text{Lt}} \frac{s^2+7s+10}{s^2+7s+160} = \frac{1}{16}$$

$$\therefore \quad C_1 = \underset{s\to 0}{\text{Lt}} \frac{d}{ds}\left(\frac{s^2+7s+10}{s^2+7s+160}\right) = \underset{s\to 0}{\text{Lt}}\left[\frac{2s+7}{s^2+7s+160} - \frac{(2s+7)(s^2+7s+10)}{(s^2+7s+160)^2}\right]$$

$$= \frac{7}{160} - \frac{70}{25600} = 0.0410$$

$$\therefore \quad C_2 = \underset{s\to 0}{\text{Lt}} \frac{d^2}{ds^2}\left(\frac{s^2+7s+10}{s^2+7s+160}\right)$$

$$= \underset{s\to 0}{\text{Lt}}\left[\frac{2}{s^2+7s+160} - \frac{(2s+7)^2}{(s^2+7s+160)^2} - \frac{(2s+7)^2+2(s^2+7s+10)}{(s^2+7s+160)^2} + \frac{2(2s+7)^2(s^2+7s+10)}{(s^2+7s+160)^3}\right]$$

$$= \frac{2}{160} - \frac{49}{25600} - \frac{69}{25600} + \frac{980}{160^3} = 8.129\times 10^{-3}$$

$$\therefore \quad e_{ss} = \frac{1}{16}\left(1+2t+\frac{5t^2}{2}\right) + 0.0410(2+t) + 8.129\times 10^{-3}\times 5\times\frac{1}{2}$$

$$= \left(\frac{1}{16}+0.0820+0.0203225\right) + \left(\frac{1}{8}+0.0410\right)t + \frac{5t^2}{32} = 0.1648 + 0.166t + 0.15625t^2$$

Example 8.25 For the unity feedback system having $G(s) = K/[s(sT+2)]$, find the following:

(i) The factor by which the gain K should be multiplied to increase the damping ratio from 0.15 to 0.6.
(ii) The factor by which the time constant T should be multiplied to reduce the damping ratio from 0.8 to 0.4.

Solution:

Here, $$G(s) = \frac{K}{s(sT+2)} \text{ and } H(s) = 1$$

$$\therefore \quad \frac{C(s)}{R(s)} = \frac{G(s)}{1+G(s)H(s)} = \frac{K}{s^2T+2s+K} = \frac{\frac{K}{T}}{s^2+\frac{2}{T}s+\frac{K}{T}}$$

Now, $$\omega_n^2 = \frac{K}{T}$$

i.e., $$\omega_n = \sqrt{\frac{K}{T}}$$

and $$\xi = \frac{2}{T}\times\frac{1}{2\omega_n} = \frac{1}{T\omega_n} = \frac{1}{\sqrt{KT}}$$

(i) Let $$\xi_1 = \frac{1}{\sqrt{K_1T}}$$

and $$\xi_2 = \frac{1}{\sqrt{K_2T}} \text{ for } \xi = 0.15 \text{ and } \xi = 0.6 \text{ respectively.}$$

$$\frac{\xi_1}{\xi_2} = \sqrt{\frac{K_2 T}{K_1 T}} = \frac{0.15}{0.60} = \frac{1}{4}$$

i.e.,
$$\frac{K_2}{K_1} = \frac{1}{16}$$

The gain must be multiplied by factor 1/16 to increase the damping ratio from 0.15 to 0.6.

(ii) Let
$$\xi_1 = \frac{1}{\sqrt{T_1 K}}$$

and
$$\xi_2 = \frac{1}{\sqrt{T_2 K}} \quad \text{for } \xi = 0.8 \text{ and } \xi = 0.4 \text{ respectively.}$$

$$\frac{\xi_1}{\xi_2} = \sqrt{\frac{T_2 K}{T_1 K}} = \frac{0.8}{0.4} = 2$$

i.e.,
$$\frac{T_2}{T_1} = 4$$

i.e.,
$$T_2 = 4T_1$$

The time constant T must be multiplied by factor 4 to reduce the damping ratio from 0.8 to 0.4.

Example 8.26 For Fig. E8.8 shown below, determine the following:

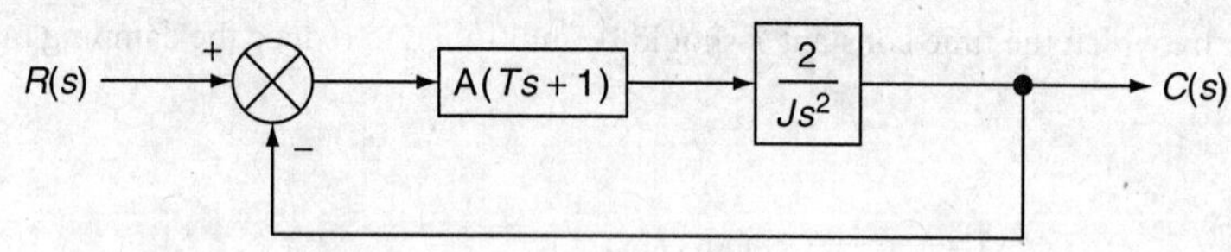

Fig. E8.8

(i) The value of A so that overshoot will be 1.9%.
(ii) The steady-state error of the system for a unit step input.
(iii) K_v and steady-state error for input $v(t) = t$.

Assume $T = 5$ sec and $J = 1500$ kg-m^2.

Solution

$$\frac{C(s)}{R(s)} = \frac{\dfrac{2A(Ts+1)}{Js^2}}{1+\dfrac{2A(Ts+1)}{Js^2}} = \frac{2A(Ts+1)}{Js^2+2A(Ts+1)} = \frac{2A(5s+1)}{1500s^2+10As+2A} = \frac{\dfrac{2A}{300}s+\dfrac{2A}{1500}}{s^2+\dfrac{10A}{1500}s+\dfrac{2A}{1500}}$$

$\therefore$
$$\omega_n^2 = \frac{2A}{1500}$$

i.e.,
$$\omega_n = \sqrt{\frac{2A}{1500}}$$

and $$2\xi\omega_n = \frac{10A}{1500}$$

∴ $$\xi = \left(\frac{10A}{1500}\right)\left(\frac{1}{2\omega_n}\right) = \frac{10A}{1500} \times \frac{1}{2} \times \sqrt{\frac{1500}{2A}} = \frac{10\sqrt{A}}{2\sqrt{3000}}$$

$$M_p = 100e^{-\frac{\pi\xi}{\sqrt{1-\xi^2}}}$$

i.e., $$1.9 = 100e^{-\frac{\pi\xi}{\sqrt{1-\xi^2}}}$$

i.e., $$-3.9633 = -\frac{\pi\xi}{\sqrt{1-\xi^2}}$$

i.e., $$\frac{1}{\xi^2} = 1 + \frac{\pi^2}{15.70} = 1.628$$

i.e., $$\frac{100A}{4 \times 3000} = \frac{1}{1.628}$$

i.e., $$A = \frac{120}{1.628} = 73.71$$

The required value of A is 73.71.

(ii) $$G(s)H(s) = \frac{2A(Ts+1)}{Js^2}$$

$$e_{ss} = \underset{s\to 0}{\text{Lt}} \frac{sR(s)}{1+G(s)H(s)} = \underset{s\to 0}{\text{Lt}} \frac{s\left(\frac{1}{s}\right)}{1+\frac{2A(Ts+1)}{Js^2}} = \underset{s\to 0}{\text{Lt}} \frac{Js^2}{Js^2 + 2A(Ts+1)} = 0$$

(iii) $$K_v = \underset{s\to 0}{\text{Lt}}\, sG(s)H(s) = \underset{s\to 0}{\text{Lt}} \frac{2A(Ts+1)}{Js} = \infty$$

∴ $$e_{ss} = \frac{1}{K_v} = 0$$

Example 8.27 For the system shown in Fig. E8.9, determine the steady-state error of the system when $K_a = 1/2$ and $K_t = 2$ when a step input of magnitude 15 V is applied, Find K_m if steady-state error for the above input is 0.02616. Assume $K_r = 1$ in both cases.

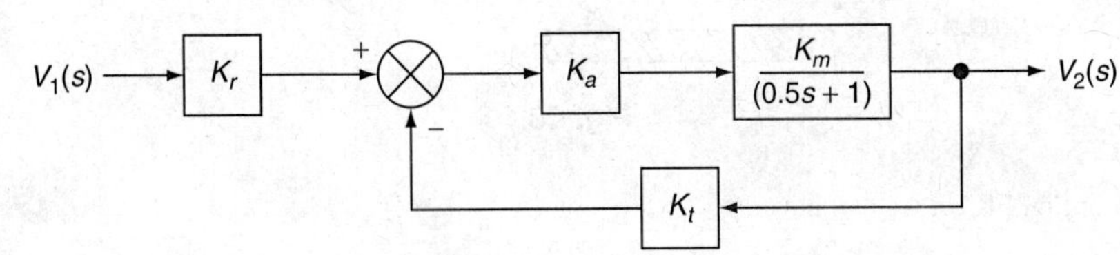

Fig. E8.9

Solution

(i) Here,
$$G(s) = \frac{K_a K_m}{0.5s + 1}$$

and
$$H(s) = K_t$$

Applied step input $r(t) = 15$V.

$\therefore$
$$R(s) = \frac{15}{s}$$

Now,
$$e_{ss} = \operatorname*{Lt}_{s\to 0} sR(s)\left[\frac{1}{1+G(s)H(s)}\right] = \operatorname*{Lt}_{s\to 0} s\left(\frac{15}{s}\right)\left[\frac{1}{1+\dfrac{K_a K_m K_t}{0.5s+1}}\right]$$

$$= \operatorname*{Lt}_{s\to 0}\left[\frac{15(0.5s+1)}{0.5s+1+K_a K_m K_t}\right] = \frac{15}{1+K_a K_m K_t} = \frac{15}{1+\frac{1}{2}\times K_m \times 2} = \frac{15}{1+K_m}$$

(ii) Given
$$e_{ss} = 0.02616$$

$\therefore$
$$\frac{15}{K_m + 1} = 0.02616$$

i.e.,
$$K_m = \frac{15}{0.02616} - 1 = 572.4$$

Example 8.28 In Fig. E8.10, $D_1 = 8$ N-m/(rad/sec), $D = 8$ N-m/(rad/sec), $F = 12$ N, $K = 169$ N-m, and $M = 1$ Kg Calculate the following:

(i) The final displacement of the car.
(ii) The maximum displacement of the car.
(iii) The time required to reach the maximum displacement.

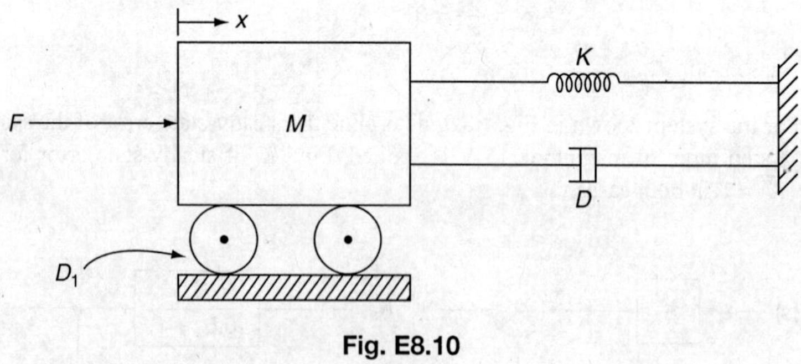

Fig. E8.10

Solution From Fig. E8.10, we can have,

$$F = M\frac{d^2x}{dt^2} + (D_1 + D)\frac{dx}{dt} + Kx$$

i.e.,
$$12 = \frac{d^2x}{dt^2} + 16\frac{dx}{dt} + 169x$$

Taking Laplace transform, we have
$$\frac{12}{s} = (s^2 + 16s + 169)X(s)$$

$\therefore$
$$X(s) = \frac{12}{s(s^2 + 16s + 169)}$$

(i) The final displacement of the car is
$$\underset{t\to\infty}{\text{Lt}}\, x(t) = \underset{s\to 0}{\text{Lt}}\, sX(s) = \underset{s\to 0}{\text{Lt}}\left(\frac{12}{s^2 + 16s + 169}\right) = \frac{12}{169} = 0.071$$

(ii) Here,
$$\omega_n = \sqrt{169} = 13 \text{ and } \xi = \frac{16}{2\times 13} = 0.6154.$$

The maximum displacement of the car is
$$= 1 + e^{-\frac{\pi\xi}{\sqrt{1-\xi^2}}} = 1 + e^{-\frac{3.14\times 0.6154}{\sqrt{1-0.6154^2}}} = 1.086 \text{ m}$$

(iii)
$$T_p = \frac{\pi}{\omega_n\sqrt{1-\xi^2}} = \frac{3.14}{13\times\sqrt{1-0.6154^2}} = 0.3064 \text{ sec}$$

Example 8.29 Assume the following:

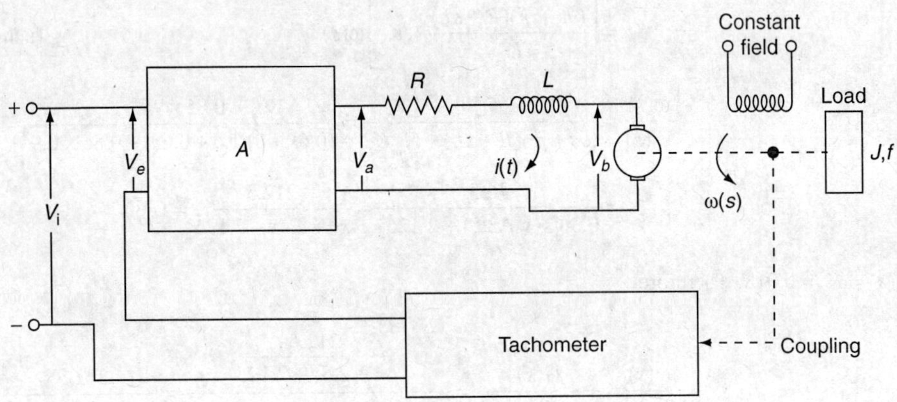

Fig. E8.11

V_i = applied voltage = 15V
A = voltage gain of the amplifier = 90
R = Armature resistance of the motor = 1.5 ohm
L = armature inductance of the motor = 0.6 H
K_b = electrical constant of the motor = 0.9 V/(rad/sec)
K_t = tachometer constant = 0.018 V/(rad/sec)
K_m = Motor torque constant = 0.85 N/A
f = 0 and J = 0.6 kg-m^2

Determine the following:

(i) The characteristic equation of the system.
(ii) The order of the system.
(iii) The magnitude of the steady-state value of the output.
(iv) The magnitude of the peak value of the output.
(v) The time required to reach the peak value of the output.

Solution From Fig. E8.11, we have

$$\frac{V_a(s)}{V_e(s)} = A \tag{1}$$

and

$$\frac{V_t(s)}{\omega(s)} = K_t \tag{2}$$

Again,

$$V_a(s) = (R + sL)\, I(s) + V_b(s) \tag{3}$$

$$V_b(s) = K_b \omega(s) \tag{4}$$

and

$$T(s) = K_m I(s) = (Js + f)\omega(s) \tag{5}$$

From Eq. (1) and Eq. (3), we have

$$AV_e(s) = V_a(s) = (R + sL)I(s) + V_b(s) = (R + sL)I(s) + K_b\omega(s) \qquad \text{[using Eq. (4)]}$$

$$= \left[\frac{(Js + f)(R + sL)}{K_m} + K_b\right]\omega(s) \qquad \text{[using Eq. (3)]}$$

$$\therefore \quad \frac{\omega(s)}{V_e(s)} = \frac{K_m A}{(Js + f)(R + sL) + K_m K_b} = \frac{0.85 \times 90}{(0.6s + 0)(1.5 + 0.6s) + 0.85 \times 0.9}$$

$$= \frac{76.5}{0.36[s(s + 2.5) + 2.125]} = \frac{212.5}{s^2 + 2.5s + 2.125} = G(s)$$

Here $H(s) = K_t = 0.018$. Therefore,

$$\frac{\omega(s)}{V_i(s)} = \frac{G(s)}{1 + G(s)H(s)} = \frac{\dfrac{212.5}{s^2 + 2.5s + 2.125}}{1 + \dfrac{212.5}{s^2 + 2.5s + 2.125} \times 0.018} = \frac{212.5}{s^2 + 2.5s + 5.95}.$$

$$\omega(s) = \frac{212.5\, V_i(s)}{s^2 + 2.5s + 5.95} = \frac{3187.5}{s(s^2 + 2.5s + 5.95)}$$

$$\omega_n = 2.448 \text{ rad/sec} \quad \text{and} \quad \xi = 0.5123.$$

(i) The characteristic equation of the system is

$$s^2 + 2.5s + 5.95 = 0$$

(ii) Since the highest power of the characteristic equation is 2, the order of the system is 2.

(iii) Using the final value theorem, we have

$$\omega_{ss} = \underset{s\to 0}{\text{Lt}}\, s\omega(s) = \underset{s\to 0}{\text{Lt}}\left(\frac{3187.5}{s^2+2.5s+5.95}\right) = 535.71$$

(iv) The magnitude of peak value

$$= 535.71\left(1+e^{-\frac{\pi\xi}{\sqrt{1-\xi^2}}}\right) = 535.71\left(1+e^{-\frac{3.14\times 0.5123}{\sqrt{1-0.5123^2}}}\right) = 618.02$$

(v)
$$t_p = \frac{\pi}{\omega_n\sqrt{1-\xi^2}} = \frac{3.14}{2.258\times\sqrt{1-0.5123^2}} = 1.62\,\text{sec}$$

Example 8.30 Calculate M, D and K for the system shown in Fig. E8.12(a) having output response shown in Fig. E8.12(b). The system was initially relaxed.

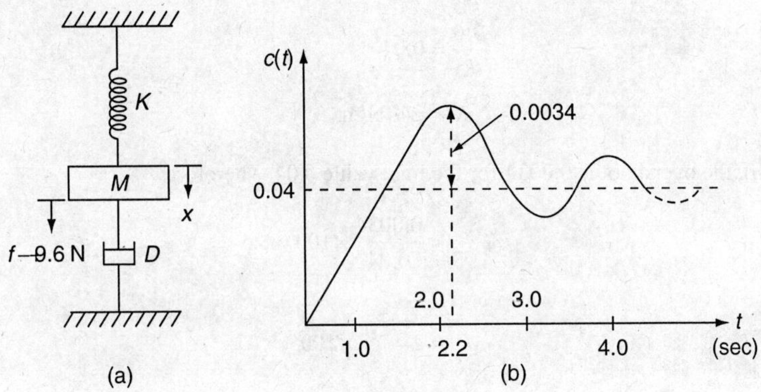

Fig. E8.12

Solution From Fig. E8.12 (a), we have

$$f(t) = M\frac{d^2x}{dt^2} + D\frac{dx}{dt} + Kx$$

Taking Laplace transform, we have

$$F(s) = s^2MX(s) + sDX(s) + KX(s) = (s^2M + sD + K)X(s)$$

i.e.,
$$X(s) = \frac{F(s)}{(s^2M + sD + K)}$$

Here $f(t) = 9.6$. Therefore,

$$F(s) = \frac{9.6}{s}$$

Therefore,
$$X(s) = \frac{9.6}{s(s^2M + sD + K)} = \frac{\frac{9.6}{M}}{s\left(s^2 + s\frac{D}{M} + \frac{K}{M}\right)}$$

Here, $$\omega_n^2 = \frac{K}{M}.$$

i.e., $$\omega_n = \sqrt{\frac{K}{M}}$$

and $$2\xi\omega_n = \frac{D}{M}$$

i.e., $$\xi = \frac{D}{M} \times \frac{1}{2\omega_n} = \frac{D}{2\sqrt{MK}}$$

From Fig. E8.12(b), the final value is 0.04. Therefore,

$$\underset{s\to 0}{\text{Lt}}\, sX(s) = \underset{s\to 0}{\text{Lt}} \frac{9.6}{s^2M + sD + K} = \frac{9.6}{K}$$

i.e., $$\frac{9.6}{K} = 0.04$$

i.e., $$K = 240\text{N/m}.$$

From Fig. E8.12(b), the overshoot is 0.0034 for the final value 0.04. Therefore,

$$\%M_p = \frac{0.0034}{0.04} \times 100 = 8.5$$

Since, $$\%M_p = e^{-\frac{\pi\xi}{\sqrt{1-\xi^2}}} \times 100$$

we can write

$$8.5 = e^{-\frac{\pi\xi}{\sqrt{1-\xi^2}}} \times 100$$

i.e., $$0.085 = e^{-\frac{\pi\xi}{\sqrt{1-\xi^2}}}$$

i.e., $$-2.465 = -\frac{\pi\xi}{\sqrt{1-\xi^2}}$$

i.e., $$\xi = 0.6175$$

Since $T_p = 2.2\,\text{sec}$, we have

$$\frac{\pi}{\omega_n\sqrt{1-\xi^2}} = 2.2$$

i.e., $$\omega_n = \frac{3.14}{2.2 \times \sqrt{1 - 0.6157^2}} = 1.8145 \text{ rad/sec}$$

Again, $$\omega_n = \sqrt{\frac{K}{M}} = 1.8145$$

i.e., $$M = \frac{K}{1.8145^2} = \frac{240}{1.8145^2} = 72.89 \text{ kg}$$

Since $$\xi = \frac{D}{2\sqrt{MK}}, \text{ we have}$$

i.e., $$D = 2\xi\sqrt{MK} = 2 \times 0.6145 \times \sqrt{72.89 \times 240} = 162.55 \text{ N/m/sec}$$

Example 8.31 Figure E8.13 shows step response of a second-order system having unity feedback (negative) when input is $3u(t)$. Determine the open-loop and closed-loop transfer functions of the system.

Solution Here $T_p = 2.5 \text{sec}$ and overshoot $= 3.6 - 3.0 = 0.6$ for final output 3.

$$\%M_p = \frac{0.6}{3} \times 100 = 20$$

Again, $\%M_p = e^{\frac{-\pi\sqrt{1-\xi^2}}{\xi}} \times 100$, we have

$$20 = e^{-\frac{\pi\xi}{\sqrt{1-\xi^2}}} \times 100$$

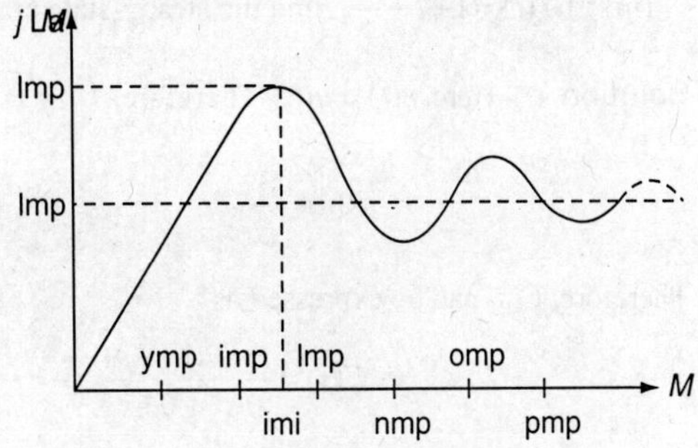

Fig. E8.13

i.e., $$0.2 = e^{-\frac{\pi\xi}{\sqrt{1-\xi^2}}}$$

i.e., $$-1.61 = -\frac{\pi\xi}{\sqrt{1-\xi^2}}$$

i.e., $$\xi = 0.456$$

Again, $$T_p = \frac{\pi}{\omega_n\sqrt{1-\xi^2}}$$

i.e., $$2.5 = \frac{3.14}{\omega_n\sqrt{1-0.456^2}}$$

i.e., $$\omega_n = 1.4114 \text{ rad/sec}$$

The closed-loop transfer function of the system is given by

$$\frac{C(s)}{R(s)} = \frac{\omega_n^2}{s^2 + 2\xi\omega_n s + \omega_n^2} = \frac{1.4112}{s^2 + 2 \times 0.456 \times 1.411s + 1.4112^2} = \frac{1.99}{s^2 + 1.28684s + 1.99}$$

Again, we have

$$\frac{C(s)}{R(s)} = \frac{\dfrac{1.99}{s^2 + 1.2868s}}{1 + \dfrac{1.99}{s^2 + 1.2868s} \times 1} = \frac{1.99}{s^2 + 1.2868s + 1.99}$$

The open-loop transfer function of the system is

$$G(s) = \frac{1.99}{s^2 + 1.2868s} \qquad [\because H(s) = 1]$$

Example 8.32 (i) A feedback control system is represented by the closed-loop transfer function given by

$$\frac{C(s)}{R(s)} = \frac{9(1+2s)}{s^2+0.6s+9}$$

Obtain the unit step response of the system.

(ii) Find K_p, K_v and K_a for the system having the following closed-loop transfer function:

$$\frac{C(s)}{R(s)} = \frac{9}{s^3+0.6s^2+9} \quad \text{with unity feedback.}$$

(iii) If $r(t) = 1 + t + \frac{t^2}{2}$, find the steady-state error in the system response.

Solution (i) Here $r(t) = u(t)$. Therefore,

$$R(s) = \frac{1}{s}$$

Therefore, $C(s)$ can be expressed as

$$C(s) = \frac{9(1+2s)}{s(s^2+0.6s+9)} = \frac{9}{s(s^2+0.6s+9)} + \frac{18}{s^2+0.6s+9} = C_1(s) + C_2(s)$$

Here $\xi = 0.1$ and $\omega_n = 3$ rad/sec. Therefore,

$$c_1(t) = 1 - e^{-\frac{\pi\xi}{\sqrt{1-\xi^2}}}\left(\sin \omega_n \sqrt{1-\xi^2}t + \tan^{-1}\frac{\sqrt{1-\xi^2}}{\xi}\right)$$

$$= 1 - \frac{e^{-\frac{3.14\times0.1}{\sqrt{1-0.1^2}}}}{\sqrt{1-\xi^2}} \sin \omega_n \sqrt{1-0.1^2}t + \tan^{-1}\frac{\sqrt{1-0.1^2}}{0.1}$$

$$= 1 - \frac{e^{-\frac{3.14\times0.1}{\sqrt{1-0.1^2}}}}{\sqrt{1-0.1^2}} \sin\left(3\sqrt{1-0.1^2}t + \tan^{-1}\frac{\sqrt{1-0.1^2}}{0.1}\right)$$

$$= 1 - 0.733\sin(2.984t + 84.26°)$$

$$\therefore \quad c_2(t) = \text{LT}^{-1}\left(\frac{18}{s^2+0.6s+9}\right) = \text{LT}^{-1}\left[\frac{6.03\sqrt{8.91}}{(s+0.3)^2+8.91}\right] = 6.03e^{-0.3t}\sin\sqrt{8.91}t$$

The required selecting is $c(t) = 1 - 0.733\sin(2.984t + 84.26°) + 6.03e^{-0.3t}\sin\sqrt{8.91}t$

(ii)

$$\frac{C(s)}{R(s)} = \frac{9}{s^3+0.6s^2+9} = \frac{\dfrac{9}{s^2+0.6s^2}}{1+\dfrac{9}{s^3+0.6s^2}\times 1}$$

$$G(s) = \frac{9}{s^3+0.6s^2}$$

(a) Therefore, $K_p = \underset{s\to0}{\text{Lt}}\, G(s)H(s) = \infty$

(b) Therefore, $K_v = \underset{s\to0}{\text{Lt}}\, sG(s)H(s) = \infty$

(c) $$K_a = \underset{s\to 0}{\text{Lt}}\, s^2G(s)H(s) = \underset{s\to 0}{\text{Lt}}\,\frac{9}{s+0.6} = 15$$

(iii) Since the input is $r(t) = 1 + t + t^2/2$, the steady-state error is given by

$$e_{ss} = \frac{1}{1+K_p} + \frac{1}{K_v} + \frac{1}{K_a} = 0 + 0 + \frac{1}{15} = \frac{1}{15}$$

MATLAB Programs

Program 3 Find the step response for the following transfer function $\frac{C(s)}{R(s)} = \frac{25}{s^2 + 10\xi + 25}$ for the following $\xi = 0.1, 0.2, 0.4, 0.6, 0.8, 1.0.$

Solution

```
>> % zeta = 0.1,0.2,0.4,0.6,0.8,1
>> zeta = [0.1 0.2 0.4 0.6 0.8 1]
zeta =
  0.1000  0.2000  0.4000  0.6000  0.8000  1.0000
>> t = 0:0.2:10;
>> for n = 1 : 6;
num = [0 0 25];
den = [1 10*zeta(n) 25];
[y(1:51,n),x,t] = step(num,den,t);
end
>> plot(t,y)
>> grid
> xlabel('t(sec)')
>> ylabel('Response')
>>
```

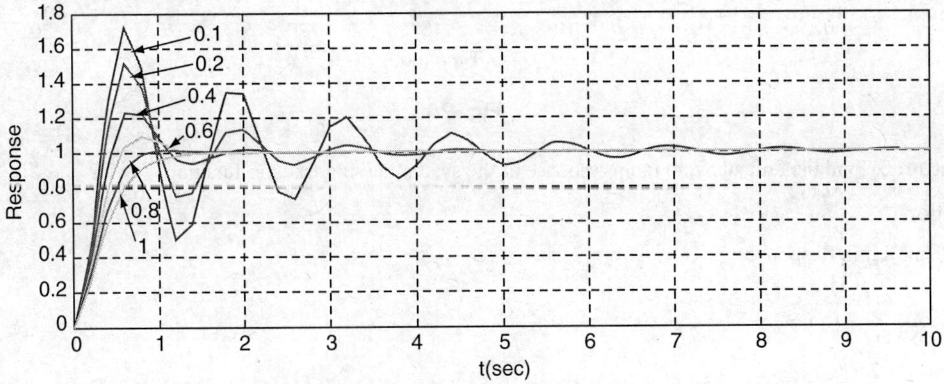

Fig. P3

Program 4 Find impulse response of the following system having transfer function $\frac{C(s)}{R(s)} = \frac{2}{s^2 + 0.4s + 1}$

Solution

```
>> %Unit Impulse Response
>> num = [0 0 2]
num =
```

```
   0   0   2
>> den = [1 0.4 1]
den =
   1.0000   0.4000   1.0000
>> impulse(num,den);
>> grid
```

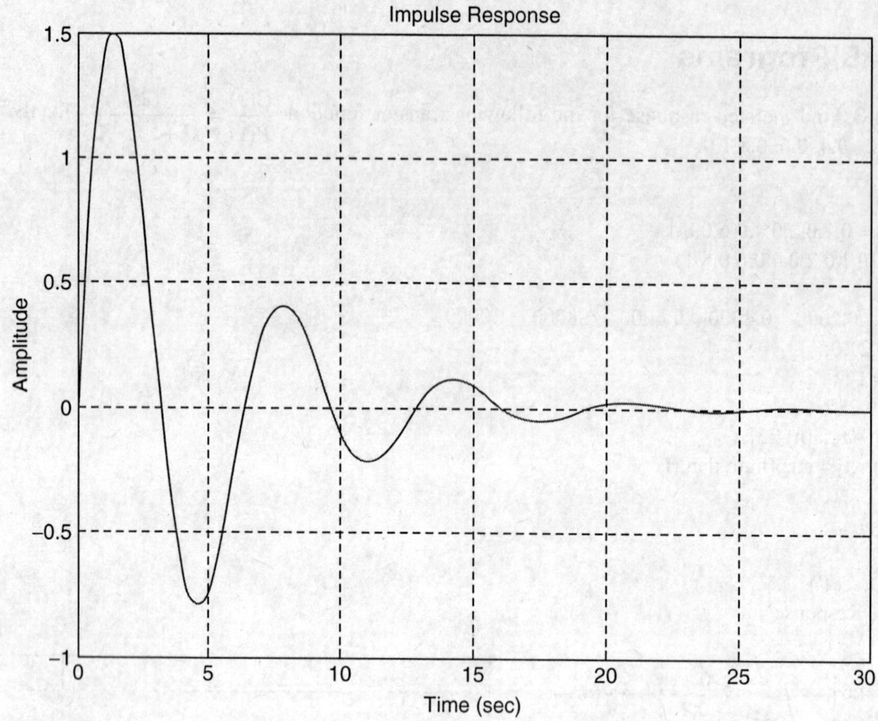

Fig. P4

Program 5 Find the unit step and ramp response of the system having transfer function $\frac{C(s)}{R(s)} = \frac{2}{s^2 + 2s + 8}$

Solution

```
>> %Unit Ramp Response
>> num = [ 0 0 1]
num =
    0   0   1
>> den = [1 2 1]
den =
    1   2   1
>> r = t;
>> y = lsim(num,den,r,t);
>> plot(t,r,'-',t,y,'o');grid
>> xlabel(' sec')
>> ylabel('Unit ramp input and System output')
>> title('lsim command has been used to get unit ramp response')
>>
```

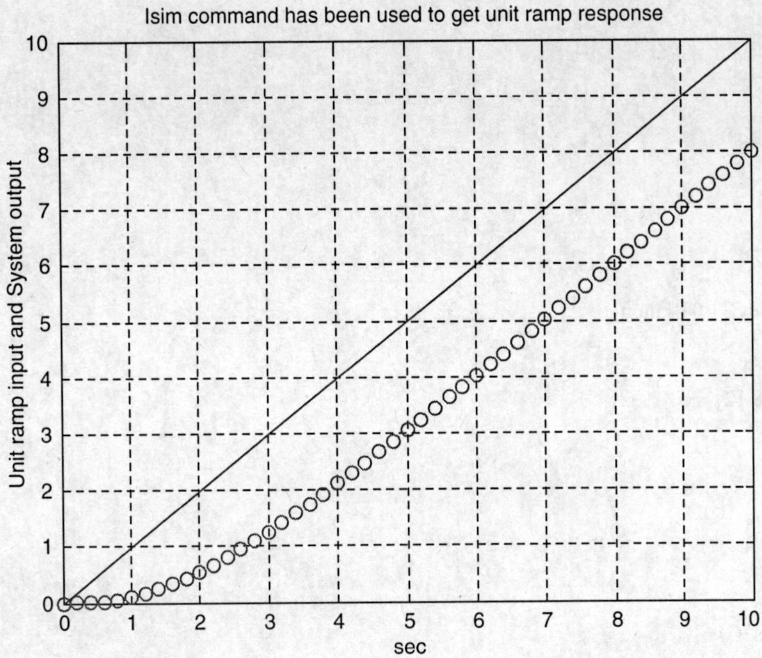

Fig. P5

Program 6 The open loop transfer function of a unity feedback control system is expressed by $G(s) = \dfrac{10}{s^2 + 8s + 12}$. Calculate (a) K_p, (ii) K_v and (iii) K_a.

Solution

```
>> num = [ 0 10];
>> den = [ 1 8 12];
>> G = tf(num,den)
Transfer function:
   10
--------------
s^2 + 8 s + 12
>> kp = dcgain(G)
kp =
   0.8333
>> ess = 1/(1+kp)
ess =
   0.5455
>> %Ramp input
>> numsg = conv([1 0],num)
numsg =
   0   10   0
>> densg = [1 8 12]
densg =
   1   8   12
>> sG = tf(numsg,densg)
```

```
Transfer function:
   10 s
--------------
s^2 + 8 s + 12
>> kv = dcgain(sG)
kv =
   0
>> ess = 1/kv
ess =
  Inf
>> num2sg = conv([1 0 0],num)
num2sg =
   0  10   0   0
>> den2sg = [1 8 12]
den2sg =
   1   8   12
>> s2G = tf(num2sg,den2sg)
Transfer function:
  10 s^2
--------------
s^2 + 8 s + 12
>> ka = dcgain(s2G)
ka =
   0
>> ess = 1/ka
ess =
  Inf
>>
```

Program 7 A unity feedback control system has open loop transfer function $G(s) = \dfrac{20}{(s+1)(s+5)}$. Determine the characteristic equation of the system, ω_n, ξ, ω_d, t_p, $\%M_p$, the time at which the first undershoot occurs, the time period of oscillations and the number of cycles completed before reaching the steady state.

Solution

```
>> num = [0 20];
>> q1 = [1 1]
q1 =
   1   1
>> q2 = [1 5]
q2 =
   1   5
>> den = conv(q1,q2)
den =
   1   6   5
>> [p,q] = cloop(num,den,–1)
p =
   0   0   20
```

```
q =
   1   6   25
>> tf(p,q)
Transfer function:
    20
--------------
s^2 + 6 s + 25
>> wn = sqrt(25)
wn =
   5
>> zeta = 5/(2*wn)
zeta =
   0.5000
>> wd = wn*sqrt(1-zeta^2)
wd =
   4.3301
>> tp = pi/wd
tp =
   0.7255
>> mp = exp(-zeta*wn*tp)*100
mp =
   16.3034
>> % time for first undershoot
>> tfirst = 2*pi/wd
tfirst =
   1.4510
>>  %period of oscillations
>> period = 2*pi/wd
period =
   1.4510
>> % number of oscillations before steady state
>> oscillations = (wd/(2*pi)*(4/(zeta*wn)))
oscillations =
   1.1027
>>
```

SHORT QUESTIONS WITH ANSWERS

1. Define the following: (i) time response, (ii) transient response, (iii) steady-state response, (iv) delay time, (v) rise time, (vi) settling time, (vii) peak time and (viii) overshoot.

Ans. (i) The variation of output with respect to time is known as time response.

(ii) The part of the time response which goes to zero after large interval of time is known as transient response.

(iii) The part of response that remains even after the transients have died out is said to be steady-state response.

(iv) The time that the system output takes to reach 50% of its final value is known as delay time (T_d).

(v) The time that the system response takes to reach from 10% to 90% of the final value of output is called rise time (T_r).

(vi) The time taken by the system response to settle down and stay within ± 2% of its final value is known as settling time (T_s).

(vii) The time taken by the system response to reach the first maximum value is known as peak time (T_p).

(viii) The ratio of the maximum value of the step-excited output to the final output is called overshoot of the system. The percentage overshoot of a system is defined as follows:

$$\%\text{overshoot} = \left.\frac{\text{Maximum output overshoot}}{\text{Final value of output}}\right|_{\text{step input}} \times 100$$

2. Write the expressions for $T_d, T_r, T_p, \%M_p$ and T_s.

Ans. $T_d = \dfrac{1+0.7\xi}{\omega_n}$, $T_r = \dfrac{\pi-\alpha}{\omega_d}\left[\text{where } \tan\alpha = \dfrac{\sqrt{1-\xi^2}}{\xi}\right]$ where α in radians, $T_p = \dfrac{\pi}{\omega_d}$, $\%M_p = e^{-\frac{\xi\pi}{\sqrt{1-\xi^2}}} \times 100$

and $T_s = \dfrac{4}{\xi\omega_n}$ for 2% criterion while it is $\dfrac{3}{\xi\omega_n}$ for 5% criterion.

3. Define steady-state error and position, velocity and acceleration error constants and corresponding steady-state error.

Ans. **Steady-state error:** For a step excitation, the difference between the desired output and the final value of any system is termed as steady-state error of the system.

Position, velocity and acceleration error constant

Sl. No.	Constant	Equation	Steady-state error (e_{ss})
1.	Position error constant (K_p)	$K_p = \underset{s\to 0}{\text{Lt}}\ G(s)H(s)$	$e_{ss} = \dfrac{A}{1+K_p}$
2.	Velocity error constant (K_v)	$K_v = \underset{s\to 0}{\text{Lt}}\ sG(s)H(s)$	$e_{ss} = \dfrac{A}{K_v}$
3.	Acceleration error constant (K_a)	$K_a = \underset{s\to 0}{\text{Lt}}\ s^2G(s)H(s)$	$e_{ss} = \dfrac{A}{K_a}$

EXERCISES

Questions for Practice

1. Determine the step, ramp and parabolic error constants of the following feedback control systems. The open-loop transfer functions are shown below:

(i) $G(s) = \dfrac{1200}{(1+0.2s)(1+11s)}$ (ii) $G(s) = \dfrac{120}{s(s^2+10s+120)}$

(iii) $G(s) = \dfrac{110}{s(1+0.2s)(1+12s)}$ (iv) $G(s) = \dfrac{120}{s^2(s^2+10s+120)}$

(v) $G(s) = \dfrac{1100}{s(s+10)(s+110)}$ (vi) $G(s) = \dfrac{10(1+4s)(1+6s)}{s^2(s^2+2s+1)}$

Ans. (i) $K_p = 1200, K_v = 0, K_a = 0$

(ii) $K_p = \infty, K_v = 1, K_a = 0$

(iii) $K_p = \infty, K_v = 110, K_a = 0$

(iv) $K_p = \infty, K_v = \infty, K_a = 1$

(v) $K_p = \infty, K_v = 1, K_a = 0$

(vi) $K_p = \infty, K_v = \infty, K_a = 10$

2. Find type, error coefficient and steady-state error of the system having unity feedback (negative) and open-loop transfer function $G(s) = \dfrac{K(s+4)}{s^2(s^3+8s^2+4)}$ when input is $\dfrac{A}{2}t^2$.

Ans. $2, \infty, \infty, K, \dfrac{A}{K}$

3. Determine the type, error coefficients and steady-state error of the system shown in Fig. 1 for the following inputs: (i) $r(t) = 12$, (ii) $r(t) = 6t$, (iii) $r(t) = \dfrac{8}{2}t^2$, (iv) $r(t) = 12 + 6t + \dfrac{8}{2}t^2$. In Fig. 1, $G_1(s) = \dfrac{25}{(s+5)(s+10)}$ and $H_1(s) = 5s$

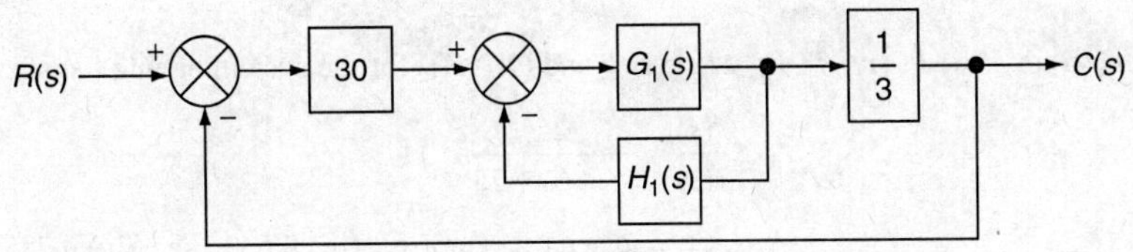

Figure 1

Ans. Type is $1, K_p = 5, K_v = 0, K_a = 0$, (i) 2 (ii) ∞ (iv) ∞

4. Determine the range of values of K for the system shown in Fig. 2 so that $e_{ss} < 0.004$ when $r(t) = 0.2t$.

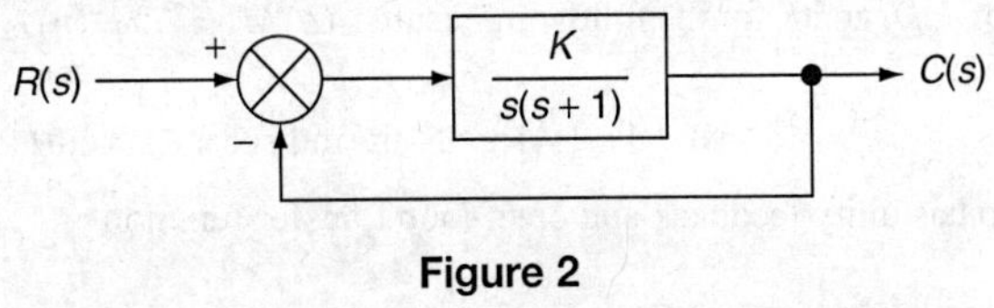

Figure 2

Ans. $50 < K < \infty$

5. A unity feedback control system has the following feedback path transfer function:

$$G(s) = \frac{K(2s+1)}{s(4s+1)(s^2+2s+3)}$$

If the input $r(t) = 1 + 5t$ is applied to the system, determine the minimum value of K to get the steady-state error equal to or less than 0.1. **Ans.** 150.

6. For the system shown in Fig. 3, determine error co-constants. Also calculate the steady-state error for the following inputs:

(i) A unit step input

(ii) A unit velocity step velocity input

(iii) A unit velocity acceleration input

Ans. $K_p = 12, K_v = 0, K_a = 0$. (i) 0.076923, (ii) ∞, (iii) ∞

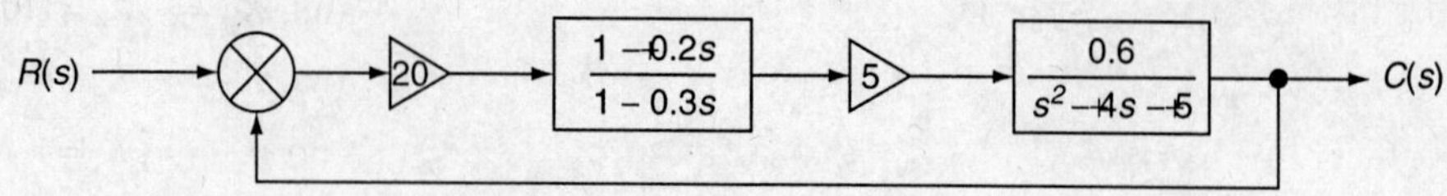

Figure 3

7. For a unity feedback control system having $G(s) = \dfrac{K}{s(sT+2)}$, find the following:
   (i) The factor by which the gain (K) must be multiplied so that the damping ratio increases from 0.3 to 0.9.
   (ii) The factor by which the time constant (T) must be multiplied so that the damping ratio decreases from 0.9 to 0.3.
   (iii) Show that $(TK_1 - 1)(TK_2 - 1) = 11.39$ when the system overshoot reduces from 70%, to 30% where K_1 and K_2 are values of K for 70% and 30% overshoot.

Ans. $\frac{1}{9}$, 9

8. Find (i) ω_n, (ii) ω_d, (iii) T_r, (iv) T_p, (v) M_p, (vi) T_s and (vii) $c(t)$ for the system having transfer function

$$\frac{C(s)}{R(s)} = \frac{25}{s^2 + 6s + 25}$$

Ans. (i) 5rad/sec, (ii) 4rad/sec, (iii) 0.5335 sec, (iv) 0.785 sec, (v) 9.84 %, (vi) 1.33 sec, (vii) $c(t) = 1 - 1.5625\,e^{-3t}\sin(4t + 0.9272)$

9. A second-order control system is represented by a transfer function given below:

$$\frac{\theta_o(s)}{T(s)} = \frac{1}{Js^2 + Ds + K}$$

where θ is the proportional output and T is the input signal. If a step input 12 N-m is applied to the system, calculate the values of J, D, and K for the following results: $(a) M_p = 7\%$, (b) $t_p = 1.2$ sec and (c) steady-state value of output is 0.6 rad.

Ans. J = 1.4185 N-m/(rad/sec²), D = 6.011 N-m (rad/sec.), K = 16.67 N

10. A second-order servo has unity feedback and open-loop transfer function

$$G(s) = \frac{500}{s(s+15)}$$

Find the following:
   (i) The numerical values of natural frequency and damping ratio.
   (ii) The value of % maximum overshoot and the time from start of the transient to maximum overshoot.
   (iii) The settling time of the system.
   (iv) The steady-state error if the system is subjected to a ramp input of 035 rad/sec.

Ans. (i) $10\sqrt{5}$ rad/sec, $0.15\sqrt{5}$ (ii) 32.67%, 0.149 sec. (iii) $\frac{8}{15}$ sec, and (iv) 0.03 rad/sec

11. A unity feedback servo-driven instrument has open-loop transfer function $G(s) = \dfrac{10}{s(s+2)}$. Find the following:
   (i) The time domain response for a unit step input,
   (ii) The natural frequency of oscillation,
   (iii) Maximum overshoot and peak time,
   (iv) Steady-state error to an input $1 + 4t$.

Ans. (i) $c(t) = 1 - 1.054928e^{-t}\sin(3t + 71.565°)$,
(ii) 3.16227 rad/sec, (iii) 35.09%, 11.0472 sec, (iv) 0.8

12. For a negative feedback control system having forward path transfer function $G(s) = \dfrac{K}{s(s+6)}$ and $H(s) = 1$, determine the value of gain K for the system to have damping ratio of 0.832. For this value of gain K, determine the completed time response to an input $r(t) = 2u(t)$, where $u(t)$ is unit step function.

Ans. $K = 13.00, c(t) = 2 - 3.605e^{-3t}\sin(2t + 33.7°)$

13. For a control system having $G(s) = \dfrac{K}{s(1+0.1s)}$ and a feedback gain $H(s) = 0.1$. Determine the following:
   (i) Value of gain K for the system to the damping ratio of 0.5.
   (ii) The unit step response of the system for this value of gain K.

Ans. (i) $K = 100$, (ii) $c(s) = t - \dfrac{1}{10}(1 - e^{-10t})$

14. Find the values of M, D and K for the mechanical vibratory system shown in Fig. 4(a) when force applied is 93 N. The output response of the system is shown in Fig. 4(b).

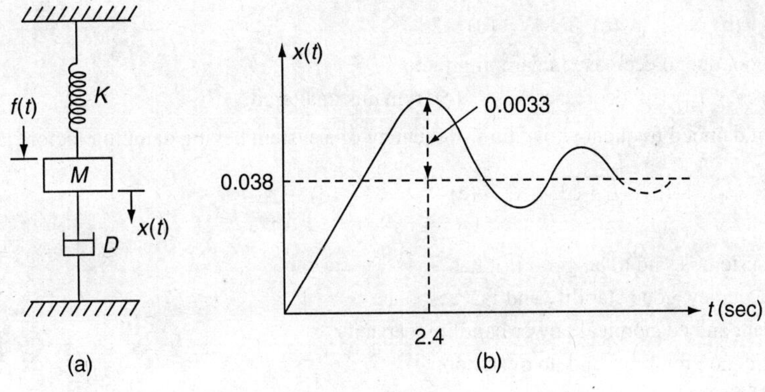

Figure 4

Ans. 89.04kg, 181.33 N/m/sec, 244.74 N/m

15. Figure 5 shows the step response of a second-order system for an input $3u(t)$. Assuming unity feedback, determine ξ, ω_n and open loop transfer function of the system.

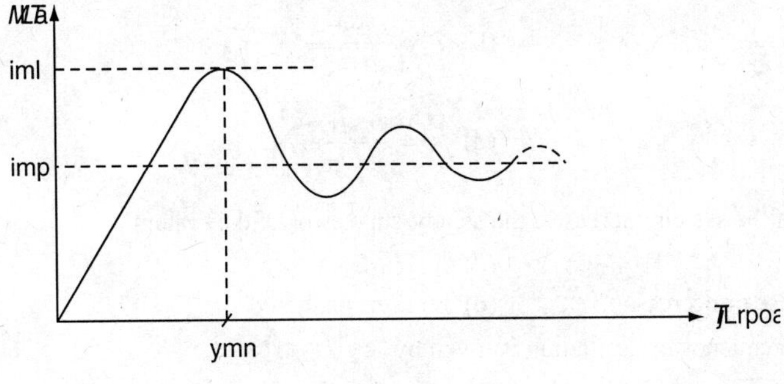

Figure 5

Ans. 0.495.2 rad/sec, $\dfrac{4}{s(s+1.98)}$

TEST YOURSELF

1. The transient response of a system with feedback compared to that of a system without feedback
   (a) decays more quickly (b) decays slowly
   (c) rises at a slower rate (d) rises at a faster rate
2. Match the following:
   (a) Rise time (i) 0 to 50%
   (b) Delay time (ii) 10 to 90%
   (c) Settling time (iii) ± 12 to 5%
   (d) Time constant (iv) 63.2%
3. The error coefficients for a unity feedback system having $G(s) = \dfrac{5(s+2)}{s(s+1)(s+5)}$ are
   (a) Position error constant (i) 5
   (b) Velocity error constant (ii) 0
   (c) Acceleration error constant (iii) ∞
4. The ratio of settling time to time constant for a tolerance of 2% of a underdamped system is approximately
   (a) 6 (b) 8 (c) 0 (d) 12
5. The overshoot due to decrease in damping factor
   (a) increases (b) decreases (c) remains unaltered.
6. The ratio of damped frequency to natural frequency of a system having damping factor ξ is
   (a) $\sqrt{1-\xi^2}$ (b) $\sqrt{1+\xi^2}$ (c) $\dfrac{1}{\xi}$ (d) $\dfrac{\xi}{1+\xi}$
7. A control system is said to be good if it has
   (a) good accuracy, good stability and fast response
   (b) sufficient and economical power handling capacity
   (c) no noise, no vibrations and no overshoot
   (d) none of these
8. For a type1, second-order system, the value of peak overshoot M_p is
   (a) $e^{-\frac{\pi\xi}{\sqrt{1-\xi^2}}}$ (b) $e^{\frac{-\pi\sqrt{1-\xi^2}}{\xi}}$ (c) $e^{\frac{\xi}{\sqrt{1-\xi^2}}}$ (d) none of these
9. In a type 1 second-order system, the first peak overshoot occurs at a time starting from zero to
   (a) $t_p = \dfrac{\omega}{\omega\sqrt{1-\xi^2}}$ (b) $t_p = \dfrac{\pi\omega_n}{\sqrt{1-\xi^2}}$
   (c) $t_p = \dfrac{\omega_n}{\pi\sqrt{1-\xi^2}}$ (d) $t_p = \dfrac{\omega_n\sqrt{1-\xi^2}}{\pi}$
10. If the gain K of the system increases, the steady-state error of the system
    (a) decreases (b) increases
    (c) may increase or decrease (d) remains unaltered
11. The damping frequency of oscillation is given by
    (a) $\omega_d = \omega_n\sqrt{1-\xi^2}$ (b) $\omega_d = \omega_n\sqrt{1-2\xi^2}$

(c) $\omega_d = \omega_r\sqrt{1-\xi^2}$ (d) $\omega_d = \dfrac{\omega_n}{\sqrt{1-\xi^2}}$

12. The first derivative control can be used to/for
    (a) decrease settling time (b) decrease damping
    (c) decrease velocity error (d) eigenvalue analysis
13. The system response can be tested better with
    (a) sinusoidal input signal (b) ramp input signal
    (c) unit impulse damping signal (d) exponentially decaying signal
14. For a second-order type 1 system, the resonance peak will occur when the system gain is at the
    (a) underdamping value (b) critical damping value (c) overdamping value
15. The open-loop gain for a unity feedback system is $G(s) = \dfrac{12}{s(s+12)}$. The steady-state velocity error of the system is
    (a) 0 (b) 1 (c) 12 (d) 10
16. For the system shown in Figure T1, the steady-state value of the output $c(t)$ is
    (a) 0 (b) 1
    (c) ∞ (d) dependent on the values of K and K_t

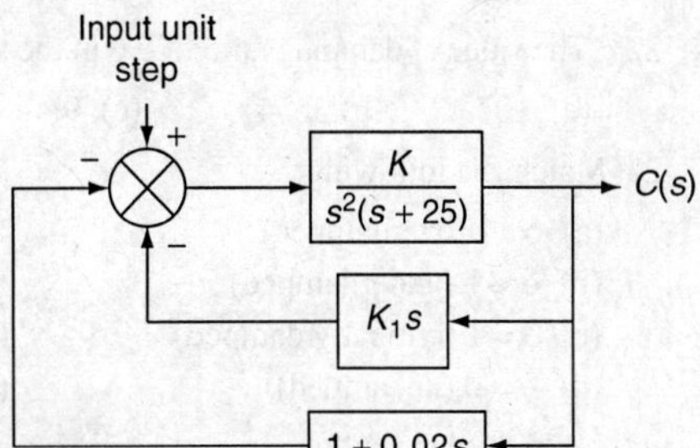

Figure T1

17. The unit impulse response of a linear time-invariant second-order system is $g(t) = 100e^{-8t}\sin 6t (t \geq 0)$.
    The natural frequency and the damping constant of the system are respectively
    (a) 10 rad/sec and 0.6 (b) 10 rad/sec and 0.8
    (c) 6 rad/sec and 0.6 (d) 6 rad/sec and 0.8
18. Match the following:
    (a) Single root at the origin (b) Single root not on the negative real axis
    (c) Two imaginary roots (d) The complex roots in the right half of an s-plane

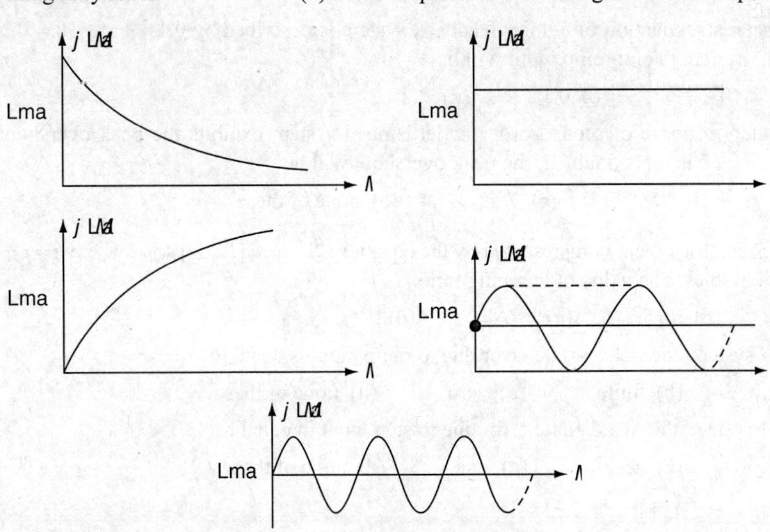

Figure T2

19. A linear second-order system with the transfer function $G(s) = \dfrac{49}{s^2 + 16s + 49}$ is initially at rest and is subjected to a step input signal. The response of the system will exhibit a peak overshoot of

(a) 16% (b) 9% (c) 2% (d) zero

20. $(-a \pm jb)$ are the complex conjugate roots of the characteristic equation of a second-order system. Its damping coefficient and natural frequency will be, respectively,

(a) $\dfrac{b}{\sqrt{a^2+b^2}}$ and $\sqrt{a^2+b^2}$ (b) $\dfrac{b}{\sqrt{a^2+b^2}}$ and a^2+b^2

(c) $\dfrac{a}{\sqrt{a^2+b^2}}$ and a^2+b^2 (d) $\dfrac{a}{\sqrt{a^2+b^2}}$ and a^2+b^2

21. A unity feedback control system has a forward path transfer function $G(s) = \dfrac{10(1+4s)}{s^2(1+s)}$. If the system is subjected to an input $r(t) = 1 + t + \dfrac{t^2}{2} (t \geq 0)$, the steady-state error of the system will be

(a) zero (b) 0.1 (c) 10 (d) infinity

22. The value of damping ratio of 0.6 in the step response of a second-order system results in maximum overshoot is

(a) 10 (b) 8.54 (c) 9.44 (d) 7.55

23. Match the following:

(a) $\delta > 1$ (overdamped)	(i) complex conjugate
(b) $\delta < 1$ (underdamped)	(ii) real and different
(c) $\delta = 1$ (critically damped)	(iii) purely imaginary
(d) $\delta = 0$ (undamped)	(iv) real and equal

24. Which of the following systems is generally preferred?

(a) critically damped (b) underdamped (c) overdamped (d) oscillatory

25. The unit impulse response of a second-order underdamped system having zero initial conditions is given by $c(t) = 12\exp(-3t)\sin 4t$. The natural frequency and damping ratio of the system are respectively,

(a) 10 and 0.6 (b) 4 and 0.3 (c) 5 and 0.6 (d) 8 and 0.5

26. The characteristic equation of a unity feedback system is described by $0.1s^2 + s + 10 = 0$. The steady-state error of the system to unit ramp input will be

(a) 1 (b) 10 (c) 0.1 (d) 0.2

27. The unit step response of second-order underdamped system exhibits the peak overshoot of 15%. If the magnitude of the input is doubled, the peak overshoot will be

(a) 30% (b) 15% (c) 7.5% (d) none of these

28. A feedback control system is represented by the equation $\dfrac{d^2e}{dt^2} + 6.4\dfrac{de}{dt} = 160e$, where $e = r - 0.4c$ and c and r are input variables. The value of damping ratio is

(a) 0.4 (b) 0.15 (c) 1/6.4 (d) 1

29. For type 2 system, the steady-state error due to ramp input is equal to

(a) infinity (b) finite (c) zero (d) none of these

30. For type 1 system, the steady-state error due to step input in equal to

(a) infinity (b) zero (c) one (d) none of these

ANSWERS

1(a)	2(a)→(ii) (c) →(iii)	(b)→(i) (d) →(iv)		3(a) →(iii) (c) →(ii) 3(b)	(b) →(i)		4(a)	5(a)	6(a), (b)
7(b)	8(a)	9(a)	10(a)	11(a)	12(d)	13(c)	14(b)	15(b)	16(b)
17(b)	18(a)→2 (c)→4	(b)→1 (d)→5	19(d)	20(c)	21(c)	22(c)	23(a)→(ii) (c) → (iv)	(b)→(i) (d)→(iii)	
24(b)	25(c)	26(c)	27(b)	28(a)	29(c)	30(b)			

Chapter **9**

Feedback Characteristics of Control Systems

The difference between the reference input $r(t)$ and controlled output $c(t)$ is known as error in control systems. To reduce this error, feedback is used. Using feedback, not only this error is reduced but also the sensitivity of the system due to parameter variations and unwanted internal and external disturbances are also reduced. Since the parameters of a system have a tendency to vary under a variety of changing conditions, these parameter variations have adverse affect on system performance. The aim of this chapter is

- to study the effect of parameter variation in open-loop and closed-loop control systems,
- to define sensitivity of a control systems,
- to study the effect of feedback on time constant and overall gain stability and disturbance of a control system and
- to introduce different controllers.

9.1 Effect of Parameter Variation in an Open-Loop Control System

Figure 9.1 shows an open-loop control system. The overall transfer function of the system can be written as follows:

Fig. 9.1 Open loop

$$\frac{C(s)}{R(s)} = G(s) \tag{9.1}$$

Let $\Delta G(s)$ be the change in $G(s)$due to parameter variation, and let the change in output be $\Delta C(s)$. We have

$$C(s)+\Delta C(s)=[G(s)+\Delta G(s)]R(s)=G(s)R(s)+\Delta G(s)R(s)$$

i.e.,
$$C(s)+\Delta C(s)=C(s)+\Delta G(s)R(s) \qquad \text{[from Eq. (9.1)]}$$

i.e.,
$$\Delta C(s)=\Delta G(s)R(s) \tag{9.2}$$

Equation (9.2) shows the effect of change of transfer function on output due to parameter variation.

9.2 Effect of Parameter Variation in a Closed-Loop System

Figure 9.2 shows a closed-loop control system. The overall transfer function of the system is

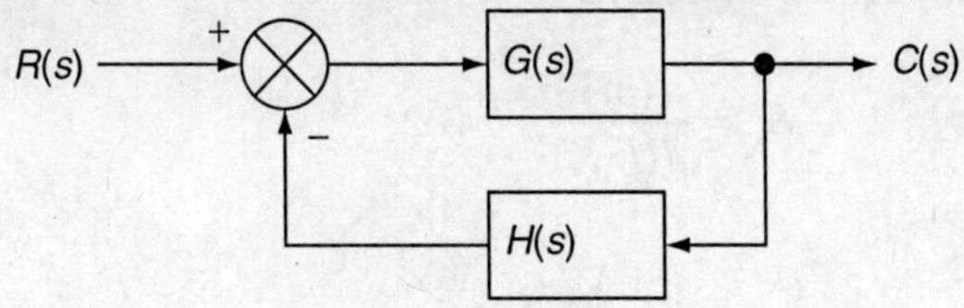

Fig. 9.2 Closed loop

$$\frac{C(s)}{R(s)} = \frac{G(s)}{1+G(s)H(s)} \tag{9.3}$$

Let $\Delta G(s)$ be the change in $G(s)$ due to parameter variation in the system and let the corresponding change in the output be $\Delta G(s)$:

$$C(s)+\Delta C(s) = \left[\frac{G(s)+\Delta G(s)}{1+\{G(s)+\Delta G(s)\}H(s)}\right]R(s)$$

$$C(s)+\Delta C(s) = \left[\frac{G(s)+\Delta G(s)}{1+G(s)H(s)+\Delta G(s)H(s)}\right]R(s) \tag{9.4}$$

The term $\Delta G(s)H(s)$ is very small compared to $G(s)H(s)$. Neglecting the term $\Delta G(s)H(s)$ in the denominator of Eq. (9.4), we have

$$C(s)+\Delta C(s) = \left[\frac{G(s)+\Delta G(s)}{1+G(s)H(s)}\right]R(s) = \frac{G(s)R(s)}{1+G(s)H(s)} + \frac{\Delta G(s)R(s)}{1+G(s)H(s)}$$

i.e., $$C(s)+\Delta C(s) = C(s) + \frac{\Delta G(s)R(s)}{1+G(s)H(s)}$$

i.e., $$\Delta C(s) = \left[\frac{\Delta G(s)}{1+G(s)H(s)}\right]R(s) \tag{9.5}$$

From Eq. (9.5) the change in output can be obtained due to parameter variation in $G(s)$ in a closed-loop system.

9.3 Sensitivity of a Control System

The parameters of a control system change due to environmental changes and other disturbances. This change has adverse affect on the system performance.

If the variable in a system be T due to the variation in parameter K of the system, the sensitivity of the system parameter T to the parameter K is expressed by

$$S = \frac{\%\text{ change in } T}{\%\text{ change in } K} \tag{9.6}$$

Mathematically, Eq. (9.6) can be expressed as

$$S_K^T = \frac{d(\ln T)}{d(\ln K)} \tag{9.7}$$

$$= \frac{\left(\frac{1}{T}\right)}{\left(\frac{1}{K}\right)} \frac{\partial T}{\partial K} \tag{9.7a}$$

$$= \frac{(\partial T/T)}{(\partial K/K)} \tag{9.7b}$$

S_K^T in Eq. (9.7) denotes the sensitivity of variable T with respect to the variations in parameter K.

Let $T(s)$ be the overall transfer function of a control system and its forward path gain $G(s)$ be varying.

The sensitivity of overall transfer function $T(s)$ with respect to the variations in $G(s)$ is given by

$$S_G^T = \frac{\partial T(s)/T(s)}{\partial G(s)/G(s)} \tag{9.8}$$

i.e.,

$$S_G^T = \frac{G(s)}{T(s)} \cdot \frac{\partial T(s)}{\partial G(s)} \tag{9.9}$$

For the open-loop system, $T(s) = G(s)$.

$$\therefore \quad \frac{\partial G(s)}{\partial T(s)} = 1$$

$$\therefore \quad S_G^T = \frac{G(s)}{T(s)} \cdot \frac{\partial G(s)}{\partial T(s)} = 1.1 = 1 \tag{9.10}$$

9.4 Sensitivity of Closed-Loop System

9.4.1 Sensitivity of *T*(*s*) with Respect to *G*(*s*)

The overall transfer function of a closed-loop system shown in Fig. 9.2 is represented by Eq. (9.3.):

$$T(s) = \frac{C(s)}{R(s)} = \frac{G(s)}{1 + G(s)H(s)}$$

$$\frac{\partial T(s)}{\partial G(s)} = \frac{[1+G(s)H(s)].1-[G(s)][H(s)]}{[1+G(s)H(s)]^2} = \frac{1}{[1+G(s)H(s)]^2} \tag{9.11}$$

The sensitivity of $T(s)$ with respect to the variation of $G(s)$ is given by

$$S_G^T = \frac{G(s)}{T(s)} \cdot \frac{\partial T(s)}{\partial G(s)} = \frac{G(s)}{\left[\dfrac{G(s)}{1+G(s)H(s)}\right]} \cdot \frac{1}{[1+G(s)H(s)]^2}$$

$$\therefore \qquad S_G^T = \frac{1}{1+G(s)H(s)} \tag{9.12}$$

The sensitivity function gets reduced by a factor $\dfrac{1}{1+G(s)H(s)}$ compared to open-loop sensitivity function due to the presence of feedback. In closed-loop systems, S_G^T is less sensitive to the variation of $G(s)$.

9.4.2 Sensitivity of *T*(*s*) with Respect to *H*(*s*)

$$\frac{\partial T(s)}{\partial H(s)} = \frac{[1+G(s)H(s)]\times 0 - G(s)G(s)}{[1+G(s)H(s)]^2}$$

$$= \frac{-[G(s)]^2}{[1+G(s)H(s)]^2} \tag{9.13}$$

The sensitivity of the transfer function $T(s)$ with respect the variation of $H(s)$ is expressed by

$$S_H^T = \frac{H(s)}{T(s)} \cdot \frac{\partial T(s)}{\partial H(s)} = \frac{H(s)}{\left[\dfrac{G(s)}{1+G(s)H(s)}\right]} \times \frac{-[G(s)]^2}{[1+G(s)H(s)]^2}$$

i.e.,

$$S_H^T = -\frac{G(s)H(s)}{1+G(s)H(s)} \tag{9.14}$$

Comparing Eq. (9.12) and Eq. (9.14), we can conclude that a closed–loop system is more sensitive to the variation in the feedback path parameters compared to the variations in the forward path parameters.

Example 9.1 Find the sensitivity of the overall transfer function of the system shown in Fig. E9.1 with respect to (i) forward path transfer function and (b) feedback path transfer function. The value of ω is 1.2 rad/sec.

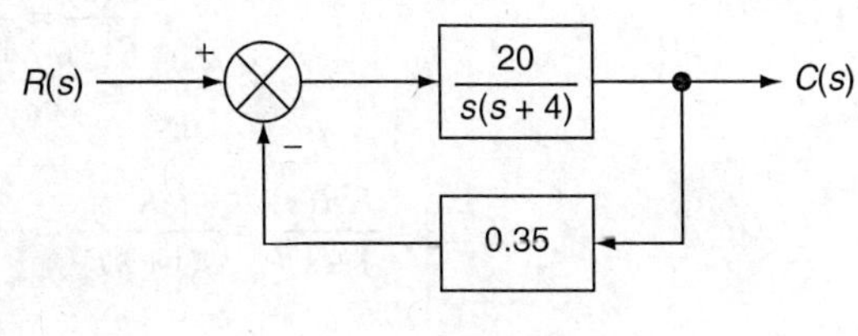

Fig. E9.1

Solution

(i) $$S_G^T = \frac{1}{1+G(s)H(s)} = \frac{1}{1+\dfrac{20}{s(s+4)}\times 0.35} = \frac{s^2+4s}{s^2+4s+7}$$

$$S_G^T\Big|_{s=j\omega} = \frac{-\omega^2+4j\omega}{-\omega^2+4j\omega+7}$$

Taking $\omega = 1.2$, we have

$$S_G^T\Big|_{s=j1.2} = \frac{-(1.2)^2+j4\times 1.2}{-(1.2)^2+j4\times 1.2+7} = \frac{-1.44+j4.8}{5.56+j4.8}$$

$\therefore$ $$\left|S_G^T\right| = 0.682$$

(ii) $$S_H^T = \frac{-G(s)H(s)}{1+G(s)H(s)} = \frac{-\dfrac{20}{s(s+4)}\times 0.35}{1+\dfrac{20}{s(s+4)}\times 0.35} = \frac{-7}{s^2+4s+7}$$

Taking $s = j\omega = j1.2$, we have

$$S_G^T\Big|_{s=j1.2} = \frac{-7}{-(1.2)^2+j4.8+7} = \frac{-7}{5.56+j4.8}$$

$\therefore$ $$\left|S_H^M\right| = 0.953$$

9.5 Effect of Feedback on Time Constant of a Control System

Let the transfer function of an open-loop control system be $G(s) = \dfrac{K}{1+sT}$

If the applied input to the system be unit step, its response in open loop can be obtained as follows:

$$\frac{C(s)}{R(s)} = \frac{K}{1+sT}$$

i.e., $$C(s) = \frac{KR(s)}{1+sT} = \frac{K}{s(1+sT)} = \frac{K/T}{s\left(s+\dfrac{1}{T}\right)} = K\left[\frac{1}{s} - \frac{1}{s+\dfrac{1}{T}}\right]$$

$$\therefore \quad c(t) = K(1 - e^{\frac{-t}{T}}) \qquad (9.15)$$

In Eq. (9.15) T is the time constant of the open-loop system. Introducing feedback in the system as shown in Fig. 9.3, the overall transfer function of the system can be expressed as

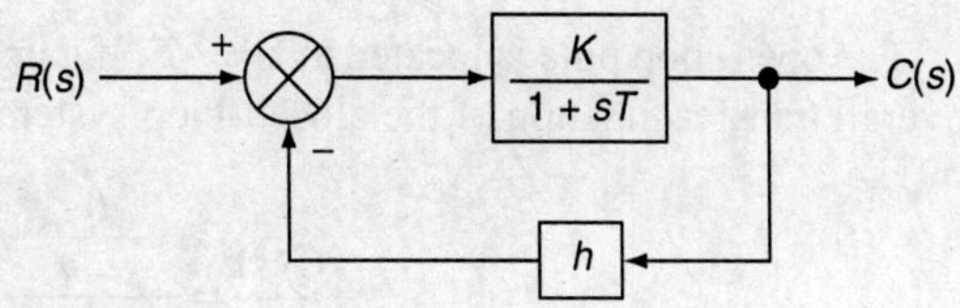

Fig. 9.3 Effect of feedback on time constant

$$\frac{C(s)}{R(s)} = \frac{\frac{K}{1+sT}}{1+\frac{Kh}{1+sT}} = \frac{K}{1+Kh+sT} = \frac{\frac{K}{T}}{s+\frac{1+Kh}{T}} \qquad (9.16)$$

Let the input of the closed-loop system be unit step.

$$\therefore \quad r(t) = u(t)$$

$$\therefore \quad R(s) = \frac{1}{s} \qquad (9.17)$$

Using Eq. (9.17), we can write Eq. (9.16) as follows:

$$C(s) = \frac{\left(\frac{K}{T}\right)R(s)}{s+\frac{1+Kh}{T}} = \frac{\frac{K}{T}}{s\left(s+\frac{1+Kh}{T}\right)} = \frac{K}{1+Kh}\left[\frac{1}{s} - \frac{1}{s+\frac{1+Kh}{T}}\right]$$

$$\therefore \quad c(t) = \frac{K}{1+Kh}\left[1 - e^{\frac{t(1+Kh)}{T}}\right] \qquad (9.18)$$

The new time constant of the feedback control system is $\frac{T}{(1+Kh)}$. If the value of h is positive and $K > 1$, the new time constant is less than 1 and response is faster. Therefore, feedback improves the time response of a system.

9.6 Effect of Feedback on Overall Gain

For an open-loop system with overall transfer function $G(s)$, the gain of such system is $G(s)$.

Due to introduction of feedback in such a system, the overall gain of the system becomes $\frac{G(s)}{1+G(s)H(s)}$.

If feedback is negative, gain $G(s)$ reduces by a factor $\frac{1}{1+G(s)H(s)}$. Therefore, due to negative feedback overall gain reduces.

9.7 Effect of Feedback on Stability

Due to introduction of feedback, the time constant of the circuit decreases and the transient response decays more rapidly. Let us consider an open-loop system with overall transfer function

$$G(s) = \frac{K}{s+T} \qquad (9.19)$$

The open-loop pole is located at $s = -T$. If a unity negative feedback is introduced in the system, the overall transfer function of the closed-loop system becomes

$$\frac{C(s)}{R(s)} = \frac{\dfrac{K}{s+T}}{1+\dfrac{K}{s+T}} = \frac{K}{s+(K+T)} \tag{9.20}$$

The closed-loop pole is located at $s = -(K+T)$.

Since the stability of a system depends upon the location of poles in s-plane, feedback may improve the stability of the system and may also be harmful to the system from stability point of view. Open-loop system is stable; the closed-loop system may be unstable. To control the stability of the system, a proper design and application of the feedback is required.

9.8 Effect of Feedback on Disturbance

Due to the presence of dominant non-linearities such as friction, dead zone, saturation, etc., the output of the system is adversely affected. In spite of these non-linearities, the output also becomes inaccurate due to some external disturbances such as high-frequency noise in electronic application, thermal noise in amplifier tubes, etc. The disturbance may be in forward path, feedback path or output of a system.

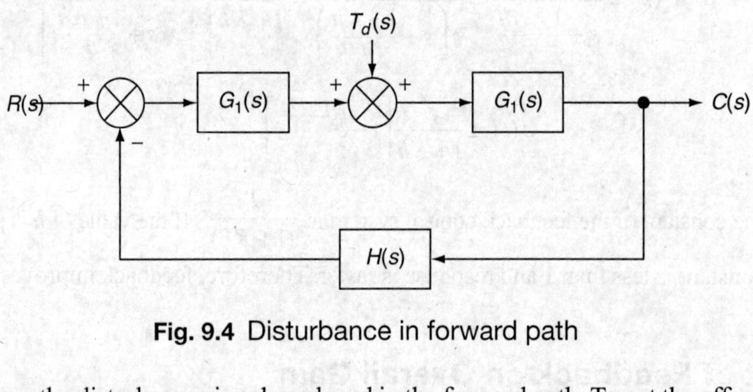

Fig. 9.4 Disturbance in forward path

Figure 9.4 shows the disturbance signal produced in the forward path. To get the effect of disturbance on output, let us put $R(s) = 0$ as shown in Fig. 9.5.

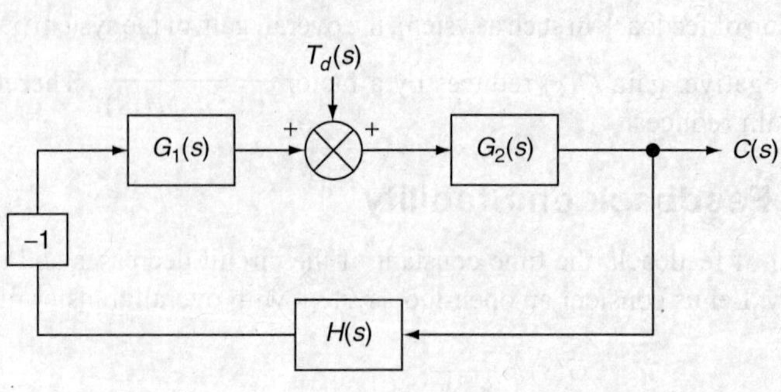

Fig. 9.5 Effect of disturbance on output

From Fig. 9.5,

$$\frac{C(s)}{T_d(s)} = \frac{G_2(s)}{1 - G_2(s)[-G_1(s)H(s)]}$$

$$\therefore \quad C(s) = \frac{T_d(s)G_2(s)}{1 + G_1(s)G_2(s)H(s)} \tag{9.21}$$

Assuming $G_1(s)G_2(s)H(s) >> 1$, Eq. (9.21) becomes

$$C(s) = \frac{T_d(s)}{G_1(s)H(s)} \tag{9.22}$$

Equation (9.22) suggests that $G_1(s)$ must be taken as large as possible to make the effect of disturbance on output as small as possible.

9.8.1 Disturbance in the Feedback Path

Figure 9.6 shows the disturbance $T_d(s)$ produced in the feedback loop. To study the effect of $T_d(s)$ on $C(s)$, let input be $R(s) = 0$. Figure 9.6 is redrawn and shown in Fig. 9.7.

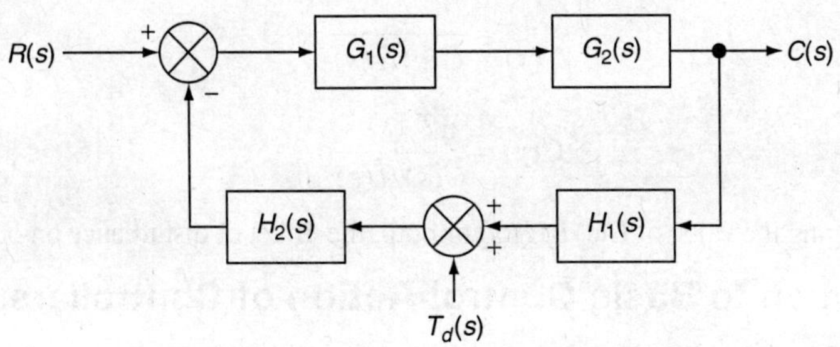

Fig. 9.6 Disturbance in feeback loop

$$\frac{C(s)}{T_d(s)} = \frac{-G_1(s)G_2(s)H_2(s)}{1 + G_1(s)G_2(s)H_1(s)H_2(s)} \tag{9.23}$$

Assuming $G_1(s)G_2(s)H_1(s)H_2(s) >> 1$, Eq. (9.23) becomes

$$\frac{C(s)}{R(s)} = -\frac{1}{H_1(s)} \tag{9.24}$$

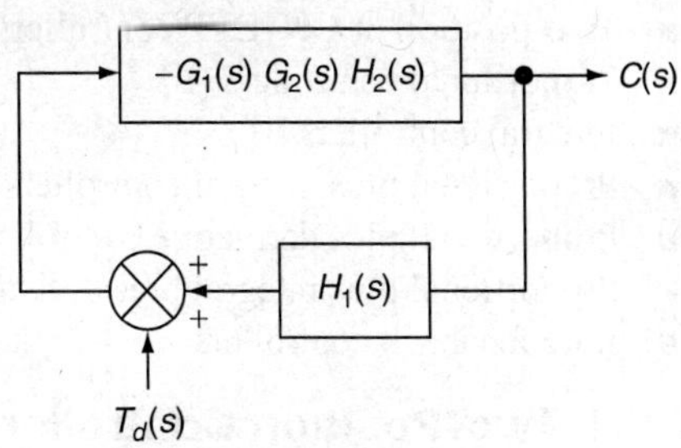

Fig. 9.7 Effect of disturbance on output

9.8.2 Disturbance at the Output

Figure 9.8 shows the disturbance produced at the output. To study the effect of disturbance on output only, let us put $R(s) = 0$. Figure 9.8 is redrawn in Figure 9.9.

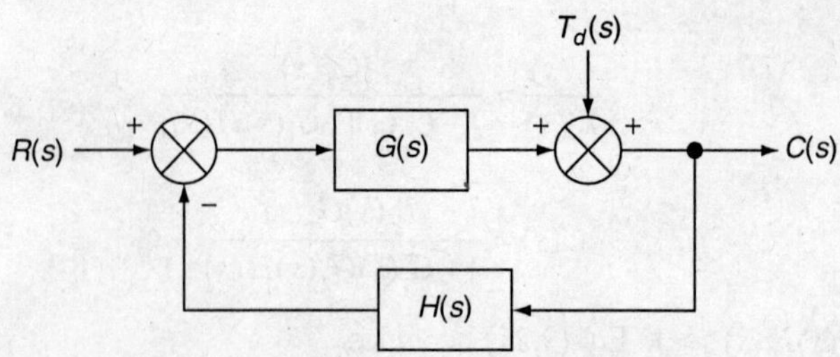

Fig. 9.8 Disturbance at the output

From Fig. 9.9,

$$\frac{C(s)}{T_d(s)} = \frac{1}{1-[-G(s)H(s)]} = \frac{1}{1+G(s)H(s)} \quad (9.25)$$

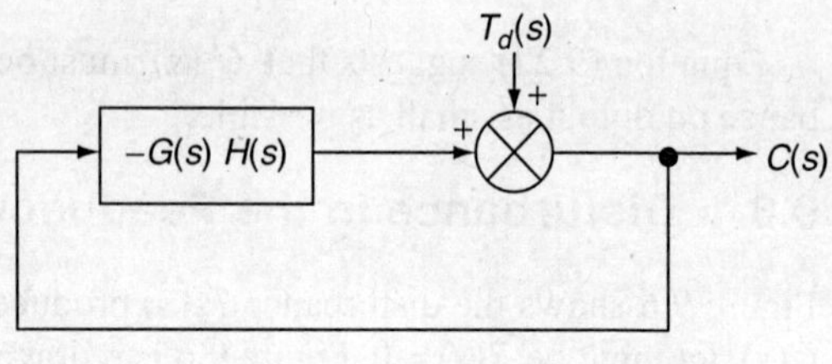

Fig. 9.9 Effect of disturbance on output

Assuming $G(s)H(s) >> 1$,

$$\frac{C(s)}{T_d(s)} = \frac{1}{G(s)H(s)}$$

i.e.,

$$C(s) = \frac{T_d(s)}{G(s)H(s)} \quad (9.26)$$

Therefore, changing the value of $G(s)$ or $H(s)$ or both, the effect of disturbance on output can be reduced.

9.9 Introduction to Basic Control Action of Controllers

Comparison of the actual value of plant output with the desired value is carried out by an automatic controller. It determines the deviation and generates a control signal to reduce the deviation to zero or a small value. Industrial controllers are classified on the basis of control action as follows:

- Two-position or ON–OFF controllers
- Proportional controllers [P]
- Integral controllers [I]
- Proportional plus integral controllers [PI]
- Proportional plus derivative controllers [PD]
- Proportional plus integral plus derivative controllers [PID]
- Rate feedback controllers

9.9.1 Two-Position Controller

Let the output of the controller be $c(t)$ and input to the controller be $e(t)$, which is the error signal. A two-position controller will produce an output given by

$$\left.\begin{aligned} c(t) &= M_1 e(t) > 0 \\ &= M_1 e(t) < 0 \end{aligned}\right\} \quad (9.27)$$

where M_1 and M_2 are constants.

Two position controllers are usually electric devices such as solenoid operated valves, relays, etc. Generally, a deadband is introduced to prevent chattering operation of the controllers shown in Fig. 9.10.

$$\therefore \quad \left.\begin{aligned} c(t) &= M_1 \quad e(t) > t_1 \\ &= M_2 \quad e(t) < t_2 \end{aligned}\right\} \tag{9.28}$$

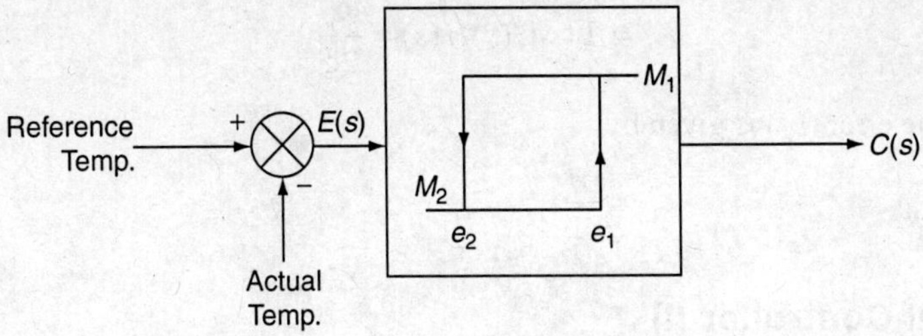

Fig. 9.10 Two position controllers

From Fig. 9.10, the following conclusions can be drawn:

(i) If the reference value falls below the actual value, the error is positive and the output of the controller is M_1.

(ii) If the reference value rises above the actual value, the error is negative and the output of the controller is M_2.

Advantages

- These are simple, economical controllers.

Disadvantages

- These are not suitable for complex system.

9.9.2 Proportional Controller (P)

In proportional controllers, the output $c(t)$ and the error $e(t)$ are related by

$$c(t) = K_p e(t) \tag{9.29}$$

Taking Laplace transform of Eq. (9.29), we have

$$C(s) = K_p E(s)$$

i.e.,

$$\frac{C(s)}{E(s)} = K_p \tag{9.30}$$

Figure 9.11 shows a second-order system.

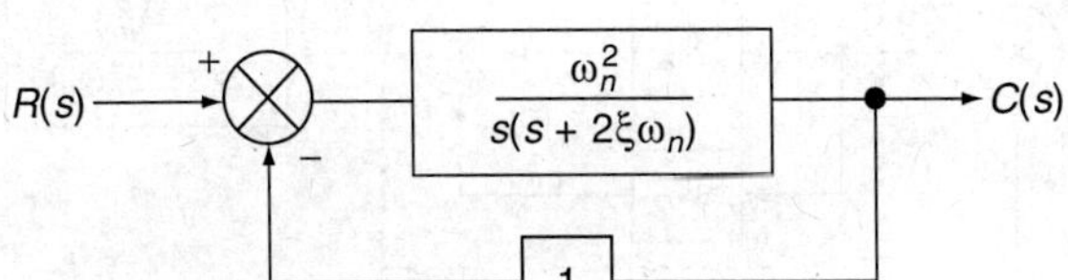

Fig. 9.11 Proportional controller

Here $$G(s)H(s) = \frac{\omega_n^2}{s(s+2\xi\omega_n)}$$

$\therefore$ $$K_p = \underset{s\to 0}{\text{Lt}}\, G(s)H(s) = \infty$$

and $$K_v = \underset{x\to 0}{\text{Lt}}\, sG(s)H(s) = \frac{\omega_n}{2\xi}$$

The steady-state error (e_{ss}) is given by

$$e_{ss} = \frac{A}{K_v} \tag{9.31}$$

9.9.3 Integral Controller (I)

Figure 9.12 shows an integral controller. From Fig. 9.12, we have

$$sC(s) = k_i E(s)$$

i.e., $$\frac{C(s)}{E(s)} = \frac{k_i}{s} \tag{9.32}$$

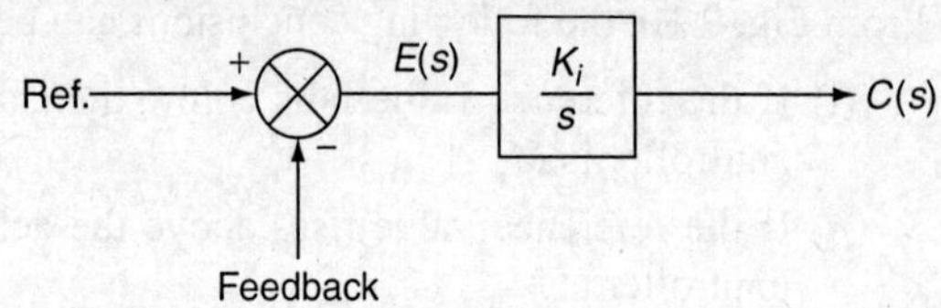

Fig. 9.12 Integral controller

i.e., $$c(t) = k_i \int_0^t e(t)dt \tag{9.33}$$

9.9.4 Proportional Plus Derivative Controller (PD)

Figure 9.13 shows the schematic diagram of a proportional plus derivative controller (PD). In PD controllers, output in proportional to (i) error and (ii) derivative of error. From Fig. 9.13

$$M(s) = (k_p + sk_D)E(s) \tag{9.34}$$

Taking inverse Laplace transform of Eq. (9.34), we have

$$m(t) = k_p e(t) + k_D \frac{de(t)}{dt} \tag{9.35}$$

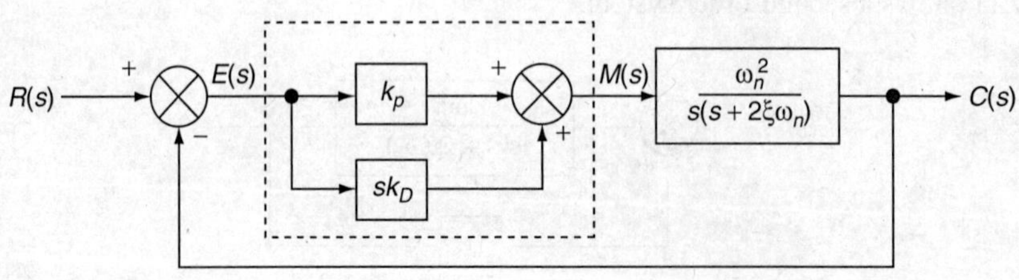

Fig. 9.13 PD controller

$$\therefore \qquad G(s) = \frac{(k_p + sk_D)\omega_n^2}{s(s+2\xi\omega_n)}$$

and $\qquad H(s) = 1.$

$$\therefore \qquad \frac{C(s)}{R(s)} = \frac{G(s)}{1+G(s)H(s)} = \frac{(k_p + k_D s)\omega_n^2}{s^2 + s(2\xi\omega_n + k_D\omega_n^2) + k_p\omega_n^2}$$

Now

$$2\xi\omega_n + k_D\omega_n^2 = 2\omega_n\left(\xi + \frac{k_D\omega_n}{2}\right) = 2\xi'\omega_n \tag{9.36}$$

where

$$\xi' = \left(\xi + \frac{k_D\omega_n}{2}\right) \tag{9.37}$$

Therefore, the damping factor increases.

Here

$$K_v = \underset{s\to 0}{\text{Lt}}\ sG(s)H(s) = \frac{k_p\omega_n}{2\xi} \tag{9.38}$$

In PD controllers, the following effects have been observed:

(i) Damping ratio improves and maximum overshoot reduces.
(ii) Rise time and settling time are reduced.
(iii) Bandwidth increases.
(iv) Gain margin, phase margin and resonant peak (**see Chapter 12**) are improved.
(v) Noise at higher frequency is accentuated.

Since PD controllers are sensitive to rate of change of error, they correct immediately any error by anticipating an error on the slope.

Example 9.2 Figure E9.2 shows a PD controller. Find T_D so that the system is critically damped. Also calculate its settling time.

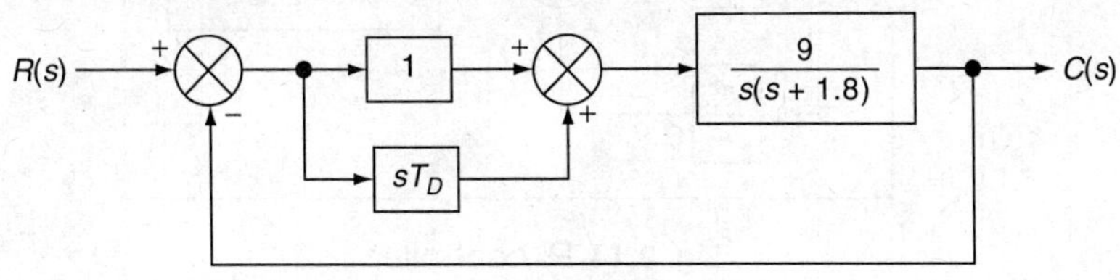

Fig. E9.2

Solution

$$G(s) = \frac{9(1+T_D s)}{s(s+1.8)}$$

and $$H(s) = 1$$

$$\therefore \quad \frac{C(s)}{R(s)} = \frac{G(s)}{1+G(s)H(s)} = \frac{\dfrac{9(1+T_D s)}{s(s+1.8)}}{1+\dfrac{9(1+T_D s)}{s(s+1.8)}\times 1} = \frac{9(1+T_D s)}{s^2+(9T_D+1.8)s+9}$$

Here $\omega_n^2 = 9$ and $\omega_n = 3$ rad/sec. For critically damped system, $\xi = 1$. Here

$$2\xi\omega_n = 9T_D + 1.8$$

i.e., $$\xi = \frac{9T_D+1.8}{2\omega_n}$$

i.e., $$\frac{9T_D+1.8}{2\omega_n} = 1$$

i.e., $$9T_D + 1.8 = 2\omega_n = 2\times 3 = 6$$

i.e., $$9T_D = 4.2$$

i.e., $$T_D = 0.467$$

$$\therefore \quad \text{Settling time} = \frac{4}{\xi\omega_n} = \frac{4}{1\times 3} = 1.34\,\text{sec}.$$

9.9.5 Proportional Integral Controllers (PI)

Figure 9.14 shows the schematic diagram of a proportional integral controller. The output is proportional to (i) error and (ii) integral of error

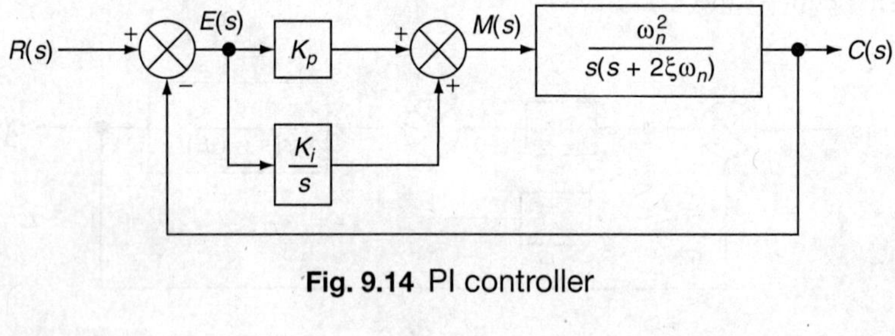

Fig. 9.14 PI controller

From Fig. 9.14, we have

$$M(s) = \left(k_p + \frac{k_i}{s}\right)E(s)$$

i.e., $$m(t) = k_p e(t) + k_i \int_0^t e(t)\,dt$$

Again

$$G(s) = \frac{\left(k_p + \frac{k_i}{s}\right)\omega_n^2}{s(s+2\xi\omega_n)} = \frac{(k_p s + k_i)\omega_n^2}{s^2(s+2\xi\omega_n)}$$

and

$$H(s) = 1$$

$$\therefore \quad G(s)H(s) = \frac{(k_p s + k_i)\omega_n^2}{s^2\left(s+2\xi\omega_n\right)}$$

Type of the system is 2.

$$\therefore \quad \frac{C(s)}{R(s)} = \frac{G(s)}{1+G(s)H(s)} = \frac{\dfrac{(k_p s + k_i)\omega_n^2}{s^2(s+2\xi\omega_n)}}{1+\dfrac{(k_p s + k_i)\omega_n^2}{s^2(s+2\xi\omega_n)}} = \frac{(k_p s + k_1)\omega_n^2}{s^3 + 2\xi\omega_n s^2 + k_p\omega_n^2 s + k_i\omega_n^2} \tag{9.39}$$

Since the order of the system increases, the system is susceptible from stability point of view. Therefore, K should be designed properly to make the system stable.

$$e_{ss} = \frac{A}{K_v} \tag{9.40}$$

$$K_v = \underset{s\to 0}{\text{Lt}}\, sG(s)H(s) = \infty \tag{9.41}$$

$$\therefore \quad e_{ss} = 0 \tag{9.42}$$

The main positive point is that system behaviour is accurate due to reduction of steady-state error. The main features are as follows:

- Type and order of the system increases by 1.
- Improvement of damping and reduction of overshoot.
- Bandwidth also increases.
- Noise is filtered out.
- K_i must be designed properly.

9.9.6 Proportional Plus Integral Plus Derivative Controller (PID)

Figure 9.15 shows the schematic diagram of a system having PID controller.

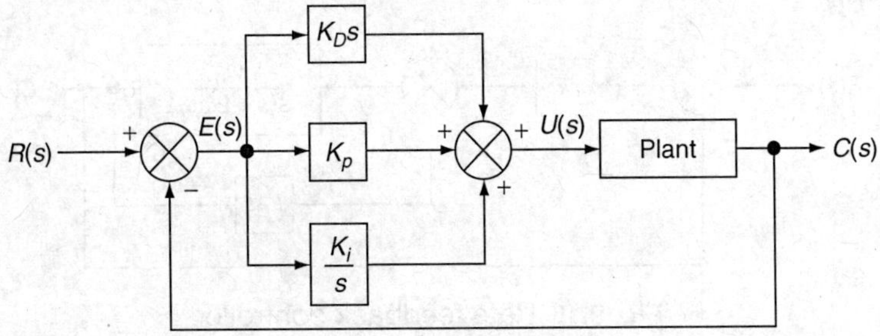

Fig. 9.15 PID controller

The output of a PID controller consists of the following three terms:

(i) One proportional to error signal,
(ii) Another one proportional to integral error signal and
(iii) The other proportional to derivative of error signal.

In a PID controller,

$$u(t) \propto \left[e(t) + \int e(t)dt + \frac{d}{dt}\{e(t)\} \right]$$

$$\therefore \quad u(t) = K_p e(t) + \frac{K_p}{T_i}\int e(t)dt + K_p T_d \frac{de(t)}{dt}, \tag{9.43}$$

where K_p is the proportional gain, T_i is the integral time and T_d is the derivative time.

Taking Laplace transform of Eq. (9.43) with zero initial condition, we have

$$U(s) = K_p E(s) + \frac{K_p}{T_i}\frac{E(s)}{s} + K_p T_d sE(s) \tag{9.44}$$

i.e.,

$$\frac{U(s)}{E(s)} = K_p \left(1 + \frac{1}{sT_i} + sT_d\right) \tag{9.45}$$

Equation (9.44) show the output of PID controller for the input $E(s)$. Eq. (9.45) shows the transfer function of the PID controller.

Any PID controller has the combination of proportional control, integral control action and derivative control action and hence the combined action has the advantages of each of the three individual control actions.

The gain is stabilized by the proportional controller and a steady-state error is produced. The steady-state error is reduced or eliminated by the integral controller. The rate of change of error is reduced by the derivative controller.

9.9.7 Rate Feedback Controller

If a PD controller is placed in the feedback path, a rate feedback controller will be obtained as shown in Fig. 9.16

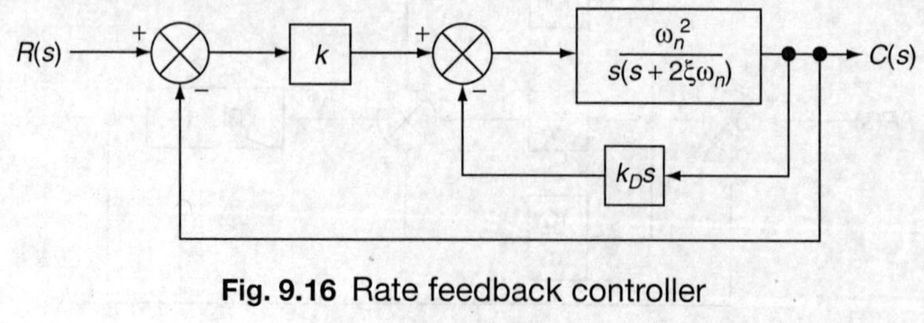

Fig. 9.16 Rate feedback controller

Rate feedback controller is also called output derivative controller. From Fig. 9.16

$$G_1(s) = \frac{\dfrac{\omega_n^2}{s(s+2\xi\omega_n)}}{1+\dfrac{\omega_n^2}{s(s+2\xi\omega_n)}\times k_D s} = \frac{\omega_n^2}{s^2+s(2\xi\omega_n+k_D\omega_n^2)}$$

$$\therefore \quad G(s) = kG_1(s) = \frac{k\omega_n^2}{s^2+s(2\xi\omega_n+k_D\omega_n^2)} \text{ and } H(s)=1.$$

Since $\quad \dfrac{C(s)}{R(s)} = \dfrac{G(s)}{1+G(s)H(s)}$

$$\therefore \quad \frac{C(s)}{R(s)} = \frac{\dfrac{k\omega_n^2}{s^2+s(2\xi\omega_n+k_D\omega_n^2)}}{1+\dfrac{k\omega_n^2}{s^2+s(2\xi\omega_n+k_D\omega_n^2)}}$$

$$= \frac{k\omega_n^2}{s^2+s(2\xi\omega_n+k_D\omega_n^2)+k\omega_n^2} \tag{9.46}$$

Equation (9.46) shows that there is change in the damping factor and improvement in ξ indicates improvement of transient response.

Again

$$K_V = \operatorname*{Lt}_{s\to 0} sG(s)H(s) = \operatorname*{Lt}_{s\to 0} \frac{k\omega_n^2}{s+2\xi\omega_n+k_D\omega_n^2} = \frac{k\omega_n}{2\xi+k_D\omega_n} \tag{9.47}$$

and

$$e_{ss} = \frac{A}{K_v} \tag{9.48}$$

Therefore, e_{ss} changes and hence ξ and K_v should be matched properly.

9.10 Realization of Controllers with Op-Amp

The realizations of different controllers with Op-Amp are discussed below.

9.10.1 Proportional Controller (P)

Figure 9.17 shows a P controller having a magnitude of

$$\left|\frac{e_o}{e_i}\right| = \frac{R_2}{R_1} \tag{9.49}$$

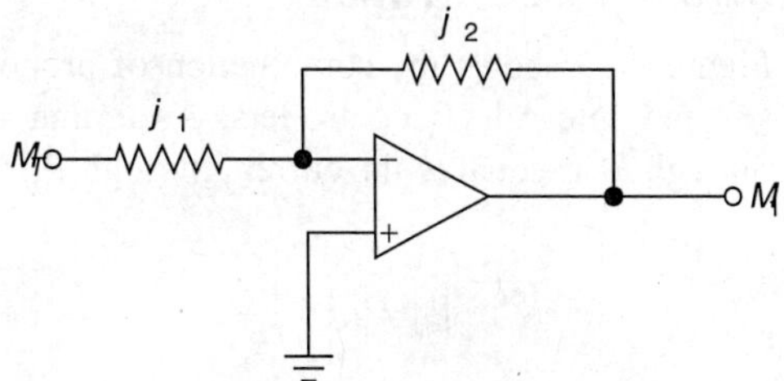

Fig. 9.17 P Controller

Here the output voltage is proportional to the input voltage. The main disadvantage of the *P* controller is that there exists a permanent error between the desired and the actual output voltage.

9.10.2 Integral Controller (I)

Figure 9.18 shows an integral controller. In ideal condition the current in the capacitor is equal to the current through the resistor. Using KCI at node *x*, we have

$$|i_1| = |i_2|$$

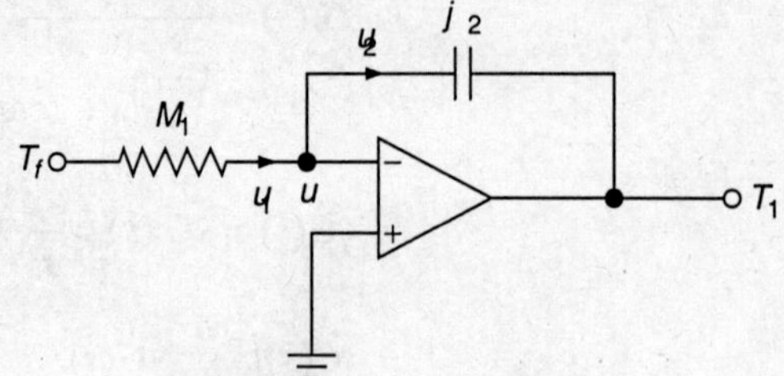

Fig. 9.18 Integral controller

where i_1 and i_2 are the currents in the resistor, and capacitor, respectively.

Here

$$|i_1| = \frac{e_i}{R_1}$$

Now

$$|e_o| = \frac{\int |i_2| dt}{C_1} = \frac{1}{C_1} \int |i_1| dt = \frac{1}{R_1 C_1} \int |e_i| dt \tag{9.50}$$

Equation (9.50) shows that the output voltage is $\frac{1}{R_1 C_1}$ times the integration of the input voltage. Taking Laplace transform of Eq. (9.50), we have

$$|E_o(s)| = \frac{|E_i(s)|}{sR_1 C_1}$$

or

Transfer function $$= \left|\frac{E_o(s)}{E_i(s)}\right| = \frac{1}{sR_1 C_1} \tag{9.51}$$

9.10.3 PI Controller

Figure 9.19 shows the combination of proportional (P) and integral (I) controllers. Assuming current through R_1 is equal to the current through R_2, we get

$$|i_1| = |i_2| = \frac{|e_i|}{R_1}$$

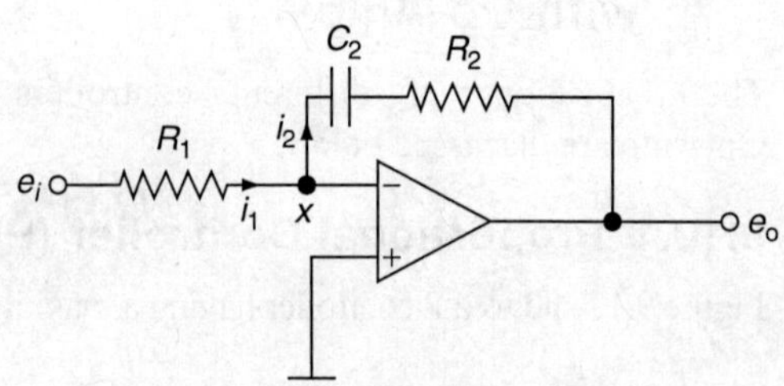

Fig. 9.19 PI controller

The output voltage $|e_o|$ can be written as

$$|e_o| = |i_2|R_2 + \frac{\int |i_2| dt}{C_2} = \frac{|e_i|}{R_1}R_2 + \frac{1}{R_1C_2}\int |e_i| dt \tag{9.52}$$

Equation (9.52) shows that the output voltage has the following two parts: (i) one proportional to actuating signal and (ii) other proportional to its integral.
Taking Laplace transform of Eq. (9.52), we have

$$|E_o(s)| = |E_i(s)|\left(\frac{R_2}{R_1}\right) + \frac{|E_i(s)|}{sR_1C_2}$$

i.e., Transfer function

$$= \frac{|E_o(s)|}{|E_i(s)|} = \frac{R_2}{R_1} + \frac{1}{sR_1C_2} = \left(\frac{R_2}{R_1}\right)\left[\frac{sR_2C_2 + 1}{sR_2C_2}\right] \tag{9.53}$$

9.10.4 PD Controller

Figure 9.20 shows the combination of proportional (P) and differential (D) controllers, which is represented in Laplace domain.

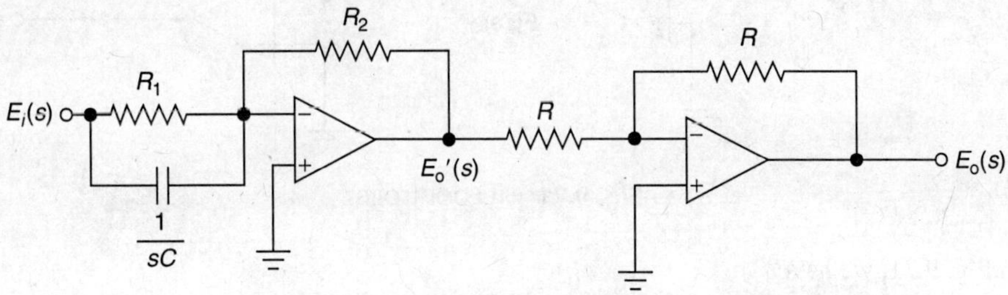

Fig. 9.20 PD controller

From Fig. 9.20, we have

$$-\frac{E_o'(s)}{R_2} = \frac{E_i(s)}{\dfrac{R_1 \times \dfrac{1}{sC}}{R_1 + \dfrac{1}{sC}}}$$

i.e.,

$$E_o'(s) = -R_2 \times \frac{E_i(s)}{\dfrac{R_1 \times \dfrac{1}{sC}}{R_1 + \dfrac{1}{sC}}} = -R_2 \times \frac{E_i(s)}{\dfrac{R_1}{R_1sC + 1}} = -R_2 \times \frac{(R_1sC + 1)E_i(s)}{R_1} \tag{9.54}$$

Since $E_o = -E_o'$, from Eq. (9.54) we have

$$E_o(s) = \frac{R_2}{R_1} \times (R_1 sC + 1) E_i(s) = \frac{R_2}{R_1} E_i(s) + sCR_2 E_i(s) \tag{9.55}$$

i.e.,

$$\frac{E_o(s)}{E_i(s)} = \frac{R_2}{R_1} + sCR_2 \tag{9.56}$$

Equation (9.55) shows that the output voltage consists of the following two parts: (i) first part is proportional to input voltage and (ii) the second part is the differential of input voltage. Hence its name is PD controller. Eq. (9.56) gives the transfer function of a PD controller.

9.10.5 PID Controller

Figure 9.21 shows the combination of a proportional (P) and integral (I) and differential (D) controllers, which is represented in Laplace domain.

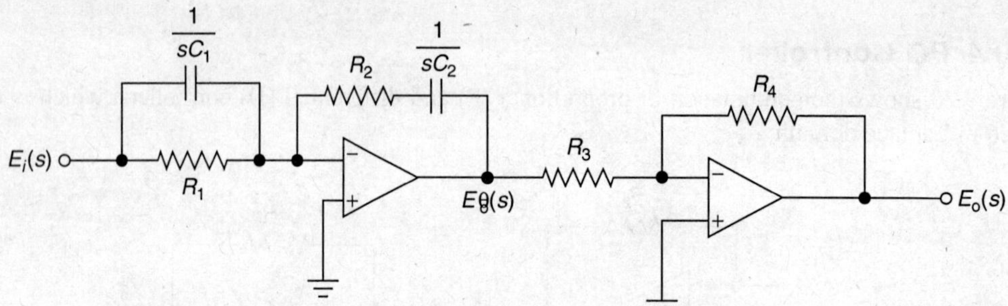

Fig. 9.21 PID controller

From Fig. 9.21, we have

$$\frac{E_i(s)}{\dfrac{R_1 \times \dfrac{1}{sC_1}}{R_1 + \dfrac{1}{sC_1}}} = -\frac{E_o'(s)}{R_2 + \dfrac{1}{sC_2}}$$

i.e.,

$$\frac{E_i(s)}{E_o'(s)} = -\frac{\dfrac{R_1 \times \dfrac{1}{sC_1}}{R_1 + \dfrac{1}{sC_1}}}{R_2 + \dfrac{1}{sC_2}} = -\frac{sR_1C_2}{(1 + sR_1C_1)(1 + sR_2C_2)} \tag{9.57}$$

Again,

$$\frac{E_o'(s)}{R_3} = -\frac{E_o(s)}{R_4}$$

i.e.,
$$E_o'(s) = -\left(\frac{R_3}{R_4}\right)E_o(s) \tag{9.58}$$

Using Eq. (9.58), we have from Eq. (9.57)

$$\frac{E_i(s)}{-\left(\frac{R_3}{R_4}\right)E_o(s)} = -\frac{sR_1C_2}{(1+sR_1C_1)(1+sR_1C_2)}$$

i.e.,
$$\frac{E_i(s)}{E_o(s)} = \left(\frac{R_3}{R_4}\right)\left[\frac{sR_1C_2}{(1+sR_1C_1)(1+sR_2C_2)}\right]$$

i.e.,
$$\frac{E_o(s)}{E_i(s)} = \frac{R_4(1+sR_1C_1)(1+sR_2C_2)}{sR_1R_3C_2} = \frac{R_2R_4(1+sR_1C_1)(1+sR_2C_2)}{sR_1R_2R_3C_2}$$

i.e.,
$$\frac{E_o(s)}{E_i(s)} = \left(\frac{R_2R_4}{R_1R_3}\right)\left[\frac{1+sR_1C_1+sR_2C_2+s^2R_1R_2C_1C_2}{sR_2C_2}\right]$$

i.e.,
$$\frac{E_o(s)}{E_i(s)} = \left(\frac{R_2R_4}{R_1R_3}\right)\left[\frac{sR_1C_1+sR_2C_2}{sR_2C_2}+\frac{1}{sR_2C_2}+sR_1C_1\right]$$

i.e.,
$$\frac{E_o(s)}{E_i(s)} = \frac{R_4(R_1C_1+R_2C_2)}{R_1R_3C_2}\left[1+\frac{1}{(R_1C_1+R_2C_2)s}+\frac{sR_1C_1R_2C_2}{R_1C_1+R_2C_2}\right] \tag{9.59}$$

Equation (9.59) gives the transfer function of the system shown in Fig. 9.21 and the transfer function has proportional, integral, and derivative parts. Hence Fig. 9.21 shows the PID controller.

SIGNIFICANT POINTS

Sensitivity: Let the variable in a system be T due to the variations in the parameter K of the system. The sensitivity of the system parameter T to parameter K is given by

$$S = \frac{\%\text{ change in } T}{\%\text{ change in } K}.$$

The sensitivity of overall transfer function $T(s)$ with respect to the variations in $G(s)$ is given by

$$S_G^T = \frac{G(s)}{T(s)}\cdot\frac{\partial T(s)}{\partial G(s)}$$

The sensitivity of the transfer function $T(s)$ with respect the variation of $H(s)$ is given by

$$S_H^T = \frac{H(s)\,\partial T(s)}{T(s)\,\partial H(s)}.$$

Basic Control Action of Controllers: Comparison of the actual value of plant output with the desired value is carried out by an automatic controller. It determines the deviation and generates a control signal to reduce the deviation to zero or a small value. Industrial controllers are classified on the basis of control action as follows:

- Two-position or ON–OFF controllers
- Proportional controllers [P]
- Integral controllers [I]
- Proportional plus integral controllers [PI]
- Proportional plus derivative controllers [PD]
- Proportional plus integral plus derivative controllers [PID]
- Rate feedback controllers

ADDITIONAL SOLVED EXAMPLES

Example 9.3 Calculate the sensitivity of the closed-loop system shown in Fig. E9.3 with respect to the forward path transfer function and feedback path transfer function at $\omega = 1.3$ rad/sec.

Solution

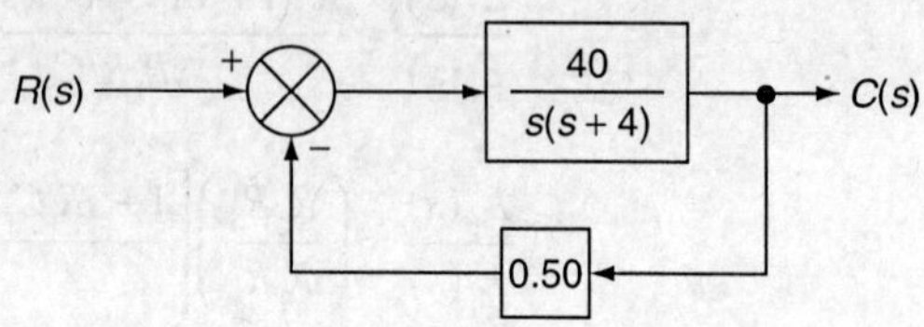

Fig. E9.3

(i) Here

$$G(s) = \frac{40}{s(s+4)}$$

and $H(s) = 0.50$

$$\therefore \quad S_G^T = \frac{1}{1+G(s)H(s)} = \frac{s^2+4s}{s^2+4s+20}$$

$$\therefore \quad S_G^T\Big|_{s=j\omega} = \frac{-\omega^2+j4\omega}{-\omega^2+j4\omega+20} = \frac{-1.69+j5.2}{-1.69+j5.2+20}[\because \omega = 1.3] = \frac{-1.69+j5.2}{18.31+j5.2}$$

Therefore, $|S_G^T| = 0.287$

(ii) $\therefore$

$$S_H^T = -\frac{G(s)H(s)}{1+G(s)H(s)} = -\frac{20}{s^2+4s+20}$$

$$\therefore \quad S_H^T\Big|_{s=j\omega} = \frac{-20}{-\omega^2+4j\omega+20} = \frac{-20}{18.31+j5.2}[\text{Taking } \omega = 1.3 \text{ rad/sec}]$$

Again,

$$|S_H^T| = \left|\frac{20}{18.31+j5.2}\right| = 1.05$$

Example 9.4 In Fig. E9.4,

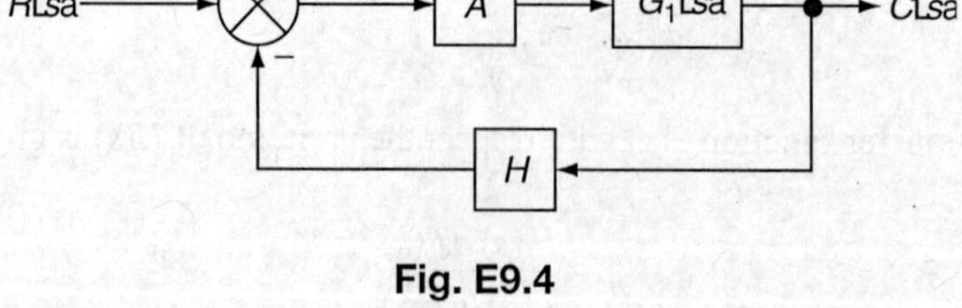

Fig. E9.4

$$G_1(s) = \frac{\omega_n^2}{s^2 + 2\xi\omega_n s + \omega_n^2}$$

where $\omega_n = \dfrac{8\pi}{T}$, $T = 6.28$ s and $\xi = 0.3$. Calculate the open-loop and closed-loop sensitivities for changes in A and H.

Solution

Here

$$G(s) = AG_1(s) = \frac{A\omega_n^2}{s^2 + 2\xi\omega_n s + \omega_n^2}$$

$$\therefore \quad S_A^G = \frac{\partial G(s)}{\partial A}\cdot\frac{A}{G(s)} = \left(\frac{\omega_n^2}{s^2 + 2\xi\omega_n s + \omega_n^2}\right)\left(\frac{s^2 + 2\xi\omega_n s + \omega_n^2}{\omega_n^2}\right) = 1 \qquad (1)$$

Here

$$T(s) = \frac{G(s)}{1 + G(s)H(s)} = \frac{\dfrac{A\omega_n^2}{s^2 + 2\xi\omega_n s + \omega_n^2}}{1 + \dfrac{A\omega_n^2}{s^2 + 2\xi\omega_n s + \omega_n^2}\times H} = \frac{A\omega_n^2}{s^2 + 2\xi\omega_n s + \omega_n^2 + A\omega_n^2 H}$$

Now

$$= \frac{16A}{s^2 + 2.4s + 16(1 + AH)}$$

$$\omega_n = \frac{8\pi}{T} = \frac{8\times 3.14}{6.28} = 4 \text{ rad/sec}$$

Again,

$$S_A^T = \frac{\partial T}{\partial A}\frac{A}{T}$$

$$\frac{\partial T}{\partial A} = \frac{16}{s^2 + 2.4s + 16(1 + AH)} - \frac{16^2 AH}{\{s^2 + 2.4s + 16(1 + AH)\}^2}$$

$$= \left[\frac{16}{s^2 + 2.4s + 16(1 + AH)}\right]\left[1 - \frac{16AH}{s^2 + 2.4s + 16(1 + AH)}\right]$$

$$S_A^T = \frac{\partial T}{\partial A}\frac{A}{T}$$

$$= \left[\frac{16}{s^2 + 2.4s + 16(1 + AH)}\right]\left[1 - \frac{16AH}{s^2 + 2.4s + 16(1 + AH)}\right]\left[\frac{s^2 + 2.4s + 16(1 + AH)}{16}\right]$$

$$= \left[1 - \frac{16AH}{s^2 + 2.4s + 16(1 + AH)}\right]$$

$$= \frac{s^2 + 2.4s + 16}{s^2 + 2.4s + 16(1 + AH)} \qquad (2)$$

From Eq. (1) and Eq. (2), we can conclude that the system sensitivity to variations in A is reduced in the feedback system.

Again,

$$S_H^T = \frac{\partial T}{\partial H}\cdot\frac{H}{T}$$

$$\frac{\partial T}{\partial H} = -\frac{16^2 A^2}{\{s^2+2.4s+16(1+AH)\}^2}$$

$$\therefore \quad S_H^T = \frac{\partial T}{\partial H}\cdot\frac{H}{T} = -\frac{16^2 A^2 H}{\{s^2+2.4s+16(1+AH)\}^2}\times\frac{s^2+2.4s+16(1+AH)}{16A}$$

$$= -\frac{16AH}{s^2+2.4s+16(1+AH)}$$

Example 9.5 In Fig. E9.5, $R_L = 10\ \text{k}\Omega$ and $r_p = 4\ \text{k}\Omega$. Find the following:

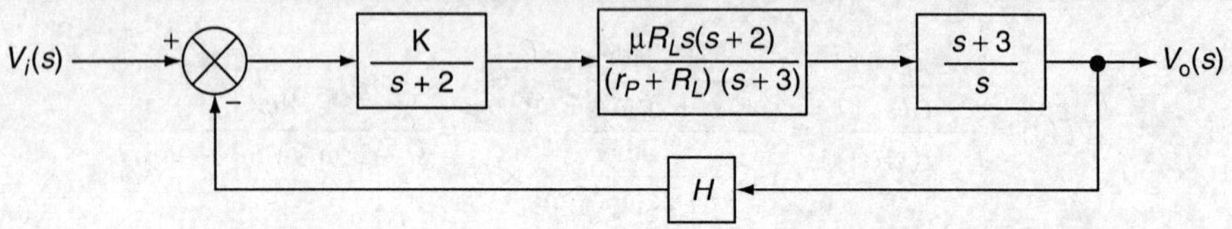

Fig. E9.5

(i) The value of K for 4% system sensitivity due to variation of μ, $H = 0.3$ and $\mu = 12$.
(ii) The value of K for 3% system sensitivity due to variation of H, $\mu = 18$ and $\mu = 0.25$.

Solution

Here

$$G(s) = \left(\frac{K}{s+2}\right)\left[\frac{\mu R_L s(s+2)}{(r_p+R_L)(s+3)}\right]\left(\frac{s+3}{s}\right) = \frac{\mu R_L K}{r_p+R_L} = \frac{\mu\times 10\times 10^3\times K}{(4+10)\times 10^3} = 0.7143\ \mu K$$

and

$$H(s) = H$$

$$T(s) = \frac{G(s)}{1+G(s)H(s)} = \frac{0.7143\mu K}{1+0.7143\mu KH}$$

(i) Again, $$S_\mu^T = \frac{\partial T}{\partial \mu}\left[\frac{\mu}{T(s)}\right]$$

$$\because \quad \frac{\partial T}{\partial \mu} = \frac{0.7143K}{1+0.7143\mu KH} - \frac{(0.7143K)^2\mu H}{(1+0.7143\mu KH)^2} = \frac{0.7143K}{(1+0.7143\mu KH)^2}$$

$$\therefore \quad S_\mu^T = \frac{\partial T}{\partial \mu}\frac{\mu}{T(s)} = \frac{0.7143K}{(1+0.7143\mu KH)^2}\times\left(\frac{1+0.7143\mu KH}{0.7143K}\right) = \frac{1}{1+0.7143\mu KH}$$

Here $\quad H = 0.3, \mu = 12$, and $S_\mu^T = 4\% = 0.04$.

$$\therefore \qquad \frac{1}{1+0.7143\times 12\times K\times 0.3}=0.04$$

i.e.,
$$\frac{1}{1+2.57148K}=0.04$$

$$K=\frac{25-1}{2.57148}=9.334$$

(ii) Again,
$$S_H^T=\frac{\partial T}{\partial H}\frac{H}{T(s)}$$

$$\therefore \qquad \frac{\partial T}{\partial H}=-\frac{(0.7143\mu K)^2}{(1+0.7143\mu KH)^2}$$

$$\therefore \qquad S_H^T=\frac{\partial T}{\partial H}\frac{H}{T(s)}=-\frac{(0.7143\mu K)^2}{(1+0.7143\mu KH)^2}\times\left(\frac{1+0.7143\mu KH}{0.7143\mu K}\right)\times H=-\frac{0.7143\mu KH}{1+0.7143\mu KH}$$

Here

$$H=0.25, \mu=18 \text{ and } \left|S_\mu^T\right|=3\%=0.03$$

$$\left|-\frac{0.7143\times 18\times K\times 0.25}{1+0.7143\times 18\times K\times 0.25}\right|=0.03$$

i.e.,
$$\frac{1}{3.21435K}+1=33.34$$

i.e.,
$$\frac{1}{3.21435K}=34.34$$

$$\therefore \qquad K=9.0595\times 10^{-3}$$

Example 9.6 In Fig. E9.6, $G=210$ and feedback factor $H=0.12$. Find the following:

(i) Reference voltage E_r when $E_o=240$V.

(ii) % change in the generator terminal voltage when the forward path gain has been reduced by 12%.

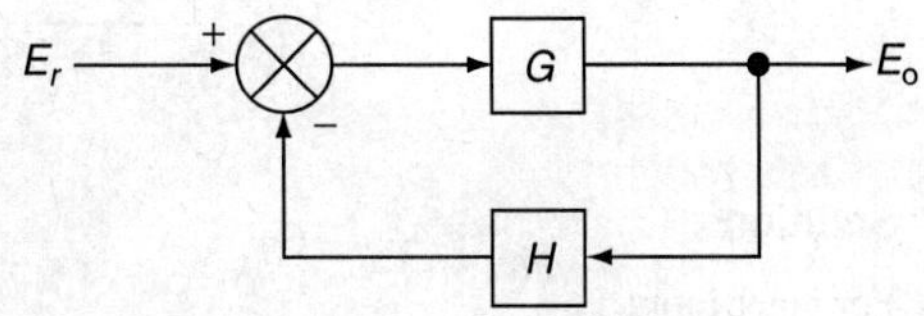

Fig. E9.6

Solution

(i) For a closed-loop system,

$$\frac{E_o}{E_r}=\frac{G}{1+GH}=\frac{210}{1+210\times 0.12}=\frac{210}{26.2}$$

$$\therefore \qquad E_r=\left(\frac{26.2}{210}\right)E_o=\frac{26.20}{210}\times 240=29.94\text{V}$$

For an open-loop system,

$$\frac{E_o}{E_r}=G$$

i.e.,
$$E_r = \frac{E_o}{G} = \frac{240}{210} = 1.143\text{V}$$

(ii) Since the forward path gain is reduced by 12%, the new value of gain is 184.8 V.

Now
$$S_G^T = \frac{1}{1+GH} = \frac{1}{1+184.8\times 0.12} = \frac{1}{23.176}$$

Again
$$S_G^T = \frac{\%\text{ change in }T}{\%\text{ change in }G}$$

i.e.,
$$\frac{1}{23.176} = \frac{\%\text{ change in }T}{12}$$

$\therefore$
$$\%\text{ change in }T = \frac{12}{23.176} = 0.5178$$

For an open-loop system:
$$\Delta E_o = (\Delta G)E_r = \frac{\Delta E_o}{E_o}\times 100 = \frac{28.8}{240}\times 100 = 12\%$$

Example 9.7 Determine K_t so that $\xi = 0.6$. Find the corresponding time domain specifications.

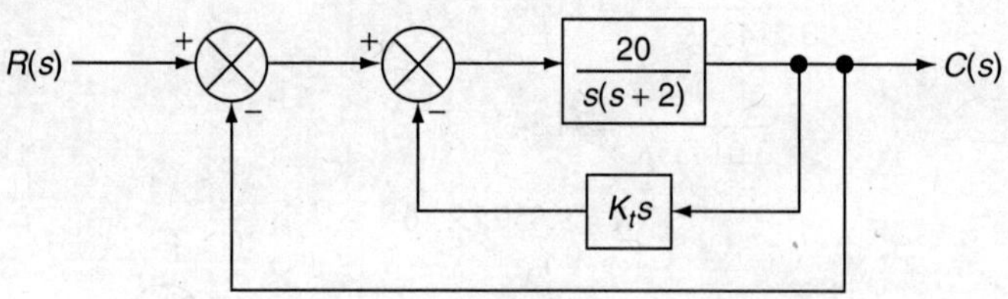

Fig. E9.7

Solution

For inner loop:
The closed loop is replaced by
$$G'(s) = \frac{\dfrac{20}{s(s+2)}}{1+\dfrac{20}{(s+2)s}\times K_t s} = \frac{20}{s^2+(2+20K_t)s}$$

Now for overall:
$$\frac{C(s)}{R(s)} = \frac{G'(s)}{1+G'(s)H(s)} = \frac{\dfrac{20}{s^2+(2+20K_t)s}}{1+\dfrac{20}{s^2+(2+20K_t)s}\times 1} \quad [\because H(s)=1]$$

$\therefore$
$$\frac{C(s)}{R(s)} = \frac{20}{s^2+(2+20K_t)s+20}$$

Here $\omega_n^2 = 20$

$\therefore$ $\omega_n = 4.472$ rad/sec

and

$$2\xi\omega_n = 2 + 20K_t$$

$$2 \times 0.6 \times 4.472 = 2 + 2K_t$$

i.e.,

i.e., $$K_t = \frac{2 \times 0.6 \times 4.472 - 2}{20} = 0.16832$$

Now,

$$\omega_d = \omega_n\sqrt{1-\xi^2} = 4.472 \times \sqrt{1-0.6^2} = 3.5776 \text{ rad/sec.}$$

The domain specifications:

$$T_p = \frac{\pi}{\omega_d} = 0.8777 \text{ sec.}$$

$$M_p = 100 \times e^{-\frac{\xi\pi}{\sqrt{1-\xi^2}}} = 9.49\%$$

$$T_s = \frac{4}{0.6 \times 4.472} = 1.49 \text{ sec}$$

Example 9.8 Refer to Fig. E9.8, calculate the following:

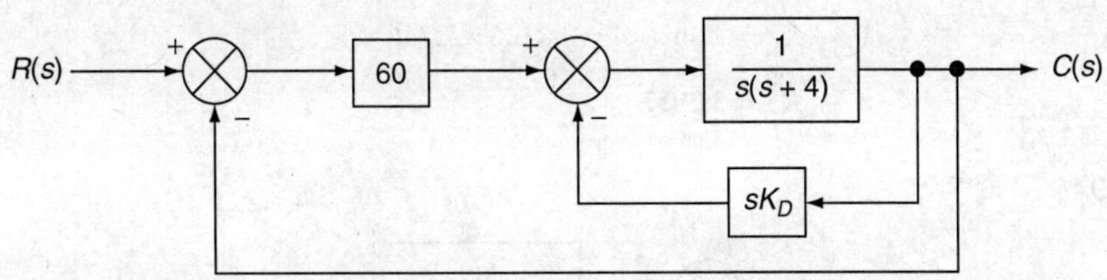

Fig. E9.8

(i) ξ and ω_n without controller K_D and (ii) K_D for $\xi = 0.60$ with controller.

Solution

(i) Without controller:

Here $$G(s) = \frac{60}{s(s+4)} \text{ and } H(s) = 1$$

$\therefore$ $$\frac{C(s)}{R(s)} = \frac{G(s)}{1+G(s)H(s)} = \frac{\frac{60}{s(s+4)}}{1+\frac{60}{s(s+4)}} = \frac{60}{s^2+4s+60} = \frac{\omega_n^2}{s^2+2\xi\omega_n+\omega_n^2}$$

Now $\omega_n^2 = 60$

i.e., $\omega_n = 7.746$ rad/sec

and $$2\xi\omega_n = 4$$

i.e., $$\xi = \frac{4}{2\omega_n} = \frac{4}{2 \times 7.746} = 0.258$$

(ii) with controller:

$$G'(s) = \frac{\dfrac{1}{s(s+4)}}{1 + \dfrac{sK_D}{s(s+4)}}$$

$$= \frac{1}{s^2 + (4 + K_D)s}$$

$\therefore$ $$G''(s) = 60\ G'(s)$$

$$= \frac{60}{s^2 + (4 + K_D)s}$$

Now $$\frac{C(s)}{R(s)} = \frac{G''(s)}{1 + G'(s)H(s)} = \frac{60}{s^2 + (4 + K_D)s + 60}$$

$\therefore$ $$\omega_n^2 = 60$$

i.e., $$\omega_n = 7.746 \text{ rad/sec}$$

and $$2\xi\omega_n = 4 + K_D$$

i.e., $$K_D = 2\xi\omega_n - 4 = 2 \times 0.6 \times 7.746 - 4$$

i.e., $$K_D = 5.2952$$

Example 9.9

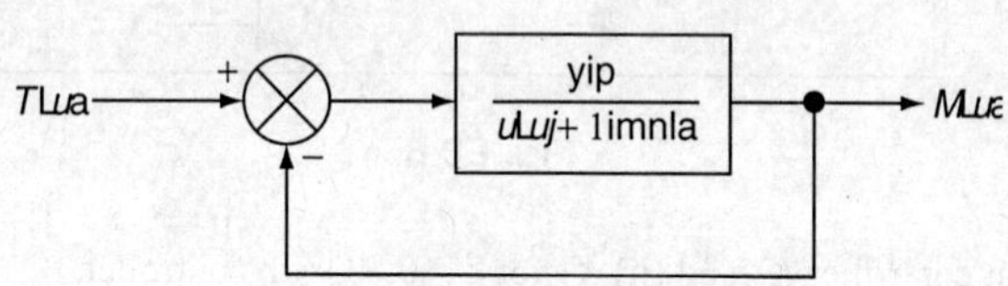

Fig. E9.9

(i) Determine ξ when a unit step input is applied in Fig. E9.9.

(ii) Determine ξ, % M_p, and T_s if a PD controller $\left(K_D = \dfrac{1}{40}\right)$ is introduced in Fig. E9.9 which has been shown in Fig. E9.9(a).

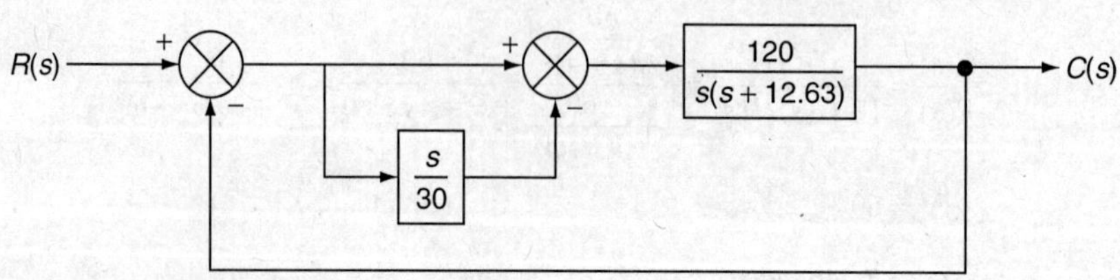

Fig. E9.9(a)

Solution

(i) Without controller:

$$\frac{C(s)}{R(s)} = \frac{120}{s^2 + 12.63s + 120}$$

$\therefore$ $$\omega_n^2 = 120$$

i.e., $$\omega_n = \sqrt{120} = 10.954 \text{ rad/sec}$$

and $$2\xi\omega_n = 12.63$$

i.e., $$\xi = \frac{12.63}{2\omega_n} = \frac{12.63}{2 \times 10.954} = 0.5765$$

(ii) With controller:

From Fig. E9.9(a), we have

$$G(s) = \frac{120}{s(s+12.63)} \times \left[1 + \frac{s}{30}\right] = \frac{4(s+30)}{s^2 + 12.63s}$$

$$\frac{C(s)}{R(s)} = \frac{\dfrac{4(s+30)}{s(s+12.63)}}{1 + \dfrac{4(s+30)}{s(s+12.63)}} = \frac{4(s+30)}{s^2 + 16.63s + 120}$$

$\therefore$ $$\omega_n^2 = 120,$$

i.e., $$\omega_n = 10.954 \text{ rad/sec}$$

and $$2\xi\omega_n = 16.63$$

i.e., $$\xi = \frac{16.63}{2\omega_n} = \frac{16.63}{2 \times 10.954} = 0.759$$

$\therefore$ $$\%M_p = e^{-\frac{\pi\xi}{\sqrt{1-\xi^2}}} \times 100 - 2.57\%$$

and $$T_s = \frac{4}{\xi\omega_n} = \frac{4}{0.759 \times 10.954} = 0.481 \text{sec}$$

Example 9.10 Figure E9.10 shows a unity feedback system. Calculate ξ and ω_n when $K = 0$. Also determine K when $\xi = 0.6$.

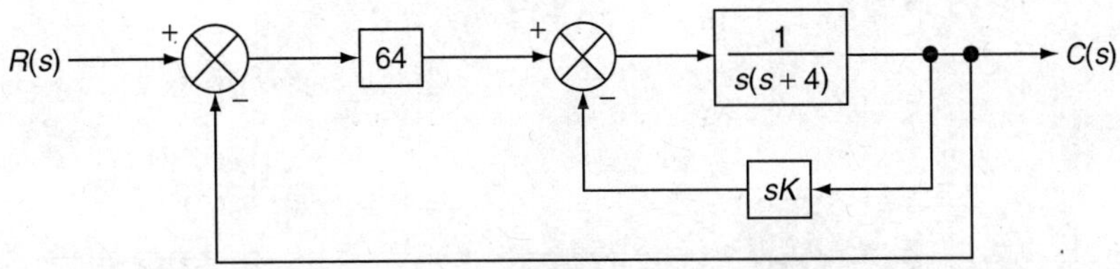

Fig. E9.10

Solution

(i) When $K = 0$:

$$G(s) = \frac{64}{s(s+4)} \text{ and}$$

$$H(s) = 1$$

$$\therefore \quad \frac{C(s)}{R(s)} = \frac{G(s)}{1+G(s)H(s)} = \frac{\frac{64}{s(s+4)}}{1+\frac{64}{s(s+4)}\times 1} = \frac{64}{s^2+4s+64}$$

i.e., $\omega_n^2 = 64$

i.e., $\omega_n = 8$ rad/sec

Again, $2\xi\omega_n = 4$ i.e., $\xi = \frac{4}{2\omega_n} = \frac{4}{2\times 8} = 0.25$

(ii) With controller

$$G(s) = 64 \times \frac{\frac{1}{s(s+4)}}{1+\frac{sK}{s(s+4)}} = \frac{64}{s^2+4s+sK} = \frac{64}{s^2+(4+K)s}$$

and

$$H(s) = 1$$

$$\therefore \quad \frac{C(s)}{R(s)} = \frac{G(s)}{1+G(s)H(s)} = \frac{\frac{64}{s^2+(4+K)s}}{1+\frac{64}{s^2+(4+K)s}\times 1} = \frac{64}{s^2+(4+K)s+64}$$

$\therefore \quad \omega_n^2 = 64$

i.e., $\omega_n = 8$ rad/sec

and $2\xi\omega_n = (4+\text{K})$

i.e., $\xi = \frac{4+K}{2\omega_n} = \frac{4+K}{16}$

Since $\xi = 0.6,$ we have

$$\frac{4+K}{16} = 0.6$$

i.e., $K = 9.6 - 4 = 5.6$

Example 9.11 A unity feedback system is shown in Fig. E9.11. Calculate K when $\xi = 0.56$. Determine (i) T_p, (ii) M_p, (iii) ω_d and T_s.

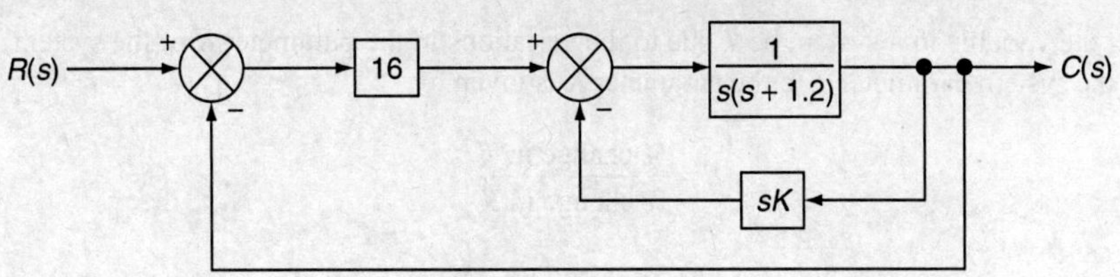

Fig. E9.11

Solution

$$\therefore \quad G(s) = 16 \times \frac{\dfrac{1}{s(s+1.2)}}{1 + \dfrac{1}{s(s+1.2)} \times sK} = \frac{16}{s(s+1.2+K)}$$

and $\quad H(s) = 1$

$$\therefore \quad \frac{C(s)}{R(s)} = \frac{G(s)}{1+G(s)H(s)} = \frac{\dfrac{16}{s(s+1.2+K)}}{1+\dfrac{16}{s(s+1.2+K)} \times 1} = \frac{16}{s^2 + (1.2+K)s + 16}$$

i.e., $\quad \omega_n^2 = 16$

i.e., $\quad \omega_n = 4$ rad/sec

$$2\xi\omega_n = (1.2 + K)$$

i.e., $\quad \xi = \dfrac{1.2+K}{2\omega_n} = \dfrac{1.2+K}{8}$

Since $\quad \xi = 0.56$, we can write

$$\frac{1.2+K}{8} = 0.56$$

i.e., $\quad K = 3.28$

(i) $\quad \omega_d = \omega_n\sqrt{1-\xi^2} = 4 \times \sqrt{1-0.56^2} = 3.314$ rad/sec

(ii) $\quad T_p = \dfrac{\pi}{\omega_d} = \dfrac{3.14}{3.314} = 0.9475$ rad/sec

(iii) $\quad \%M_p = e^{-\frac{\pi\xi}{\sqrt{1-\xi^2}}} \times 100 = e^{-\frac{3.14\times 0.56}{\sqrt{1-0.56^2}}} \times 100 = 11.97$

(iv) $\quad T_s = \dfrac{4}{\xi\omega_n} = \dfrac{4}{0.56.\times 4} = 1.7857\,\text{sec}$

SHORT QUESTIONS WITH ANSWERS

1. Define Sensitivity.

Ans. Let the variable in a system be T due to the variations in the parameter K of the system, the sensitivity of the system parameter T to the parameter K is given by

$$S = \frac{\%\text{ change in } T}{\%\text{ change in } K}$$

The sensitivity of overall transfer function $T(s)$ with respect to the variations in $G(s)$ is given by

$$S_G^T = \frac{G(s)\,\partial T(s)}{T(s)\,\partial G(s)}$$

The sensitivity of the transfer function $T(s)$ with respect the variation of $H(s)$ is given by

$$S_G^T = \frac{H(s)\,\partial T(s)}{T(s)\,\partial H(s)}$$

2. What do you mean by the basic control action of controllers? Also classify them.

Ans. Comparison of the actual value of plant output with the desired value is carried out by an automatic controller. It determines the deviation and generates a control signal to reduce the deviation to zero or a small value. Industrial controllers are classified on the basis of control action as follows:

- Two-position or ON–OFF controllers
- Proportional controllers [P]
- Integral controllers [I]
- Proportional plus integral controllers [PI]
- Proportional plus derivative controllers [PD]
- Proportional plus integral plus derivative controllers [PID]
- Rate feedback controllers.

EXERCISES

Questions for Practice

1. Figure 1 shows a closed-loop control system which regulates the generator terminal voltage E_o. The values of G and H are 220 and 0.12. Determine the following
   (i) The value of E_r to keep $E_o = 260\text{V}$.
   (ii) The percentage change in the generator terminal voltage if the forward path gain reduces by 12%. Also find the results (i) and (ii) when the system is open-loop and closed-loop.

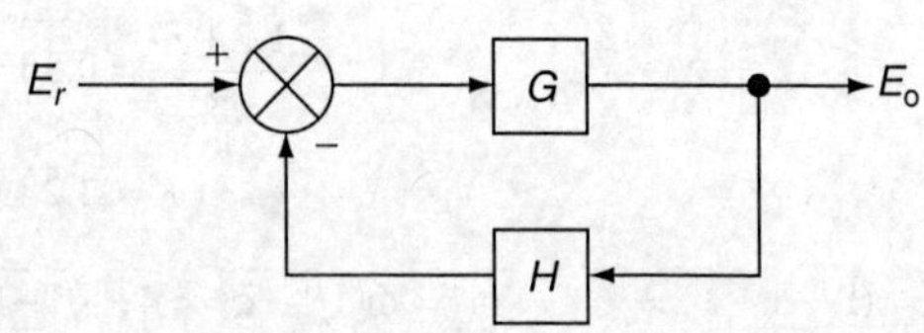

Figure 1

Ans. 1.18 V, 32.38 V and (ii) 12%, 0.495%

2. Figure 2 shows a PD controller used for a system. Determine the value of T_d so that $\xi = 0.85$. Find its settling time.

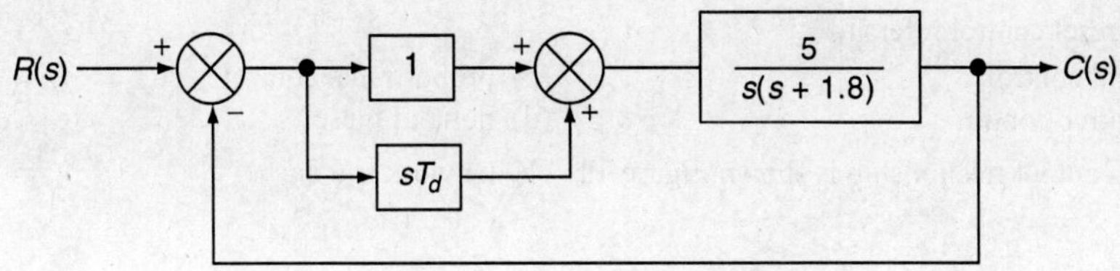

Figure 2

Ans. 0.4.2.10 sec

3. Figure 3 shows a unity feedback controller.
   (i) In the absence of a derivative controller $(K_o = 0)$, find ξ and ω_n.
   (ii) Also find K_o if $\xi = 0.56$ when controller is used.

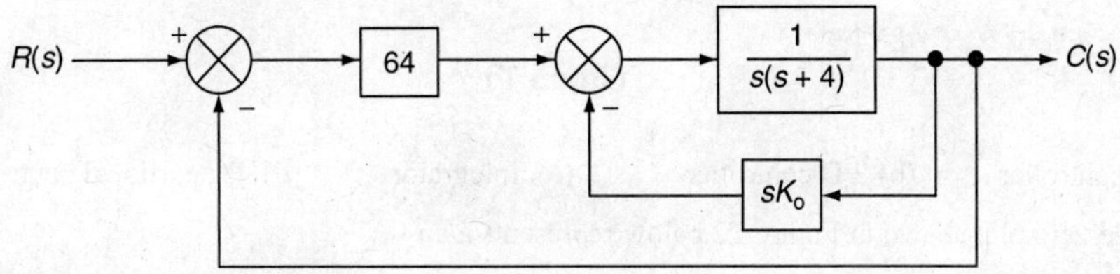

Figure 3

Ans. (i) 0.25, 8 rad/sec (ii) 4.96

4. Figure 4 shows a rate feedback controller. Find K_t so that $\xi = 0.45$. Also determine (i) $T_{p'}$, $M_{p'}$, ω_d and $T_{s'}$,

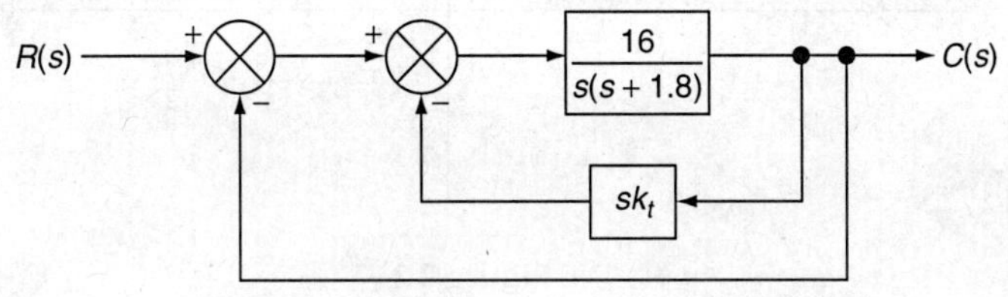

Figure 4

Ans. $K_t = 0.1125, T_p = 0.879\,\text{sec}, M_p = 20.55\%, \omega_d = 3.572\,\text{rad/sec}$ and $T_s = 2.22\,\text{sec}$

TEST YOURSELF

1. The system sensitivity to parameter variations
   (a) increases (b) becomes zero (c) decreases (d) none of these
2. To obtain high accuracy requirement, we use
   (a) derivative error compensation (b) integral error compensation
   (c) both (a) and (b) (d) none of these

3. The term reset control refers to
   (a) integral control (b) proportional control
   (c) derivative control (d) none of these
4. The input–output relationship is shown Figure T1. The network is a/an

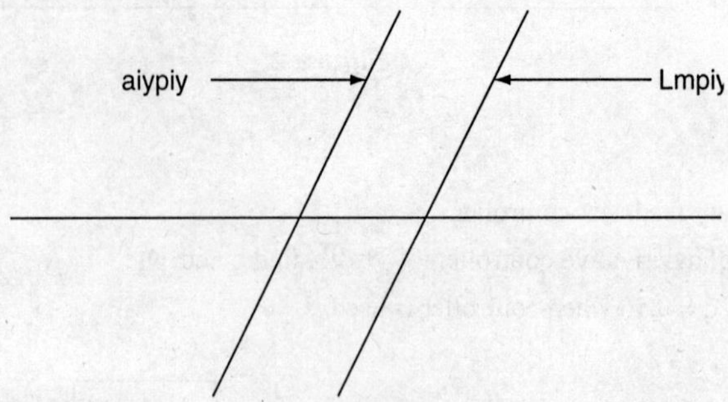

Figure T1

(a) PI controller (b) PD controller (c) Integrator (d) Proportional controller

5. The pole-zero plot shown in Figure T2 below represents a/an

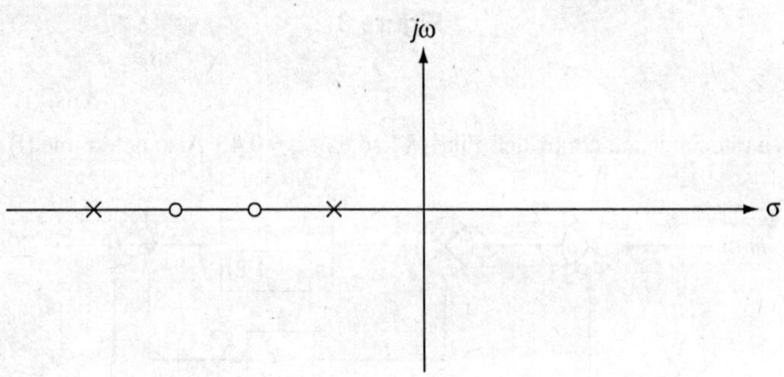

Figure T2

(a) PID controller (b) Lag-lead compensating network
(d) PD controller (d) none of these

6. Differentiators are not used in a system due to
   (a) large noise and saturation in the amplifier (b) large resistance and inductance
   (b) huge size and cost (d) none of these
7. The proportional error device has output as function of
   (a) derivative of error (b) integral of error (c) error (d) none of these
8. Find the correct sequence
   (a) IDP controller (b) DIP controller
   (c) IDP controller (d) PID controller

9. Find the following system which provides excellent transient and steady-state response
   (a) proportional action
   (b) proportional + integral action
   (c) proportional + differential action
   (d) proportional + integral + derivative action
10. Consider the systems shown in Figure T3(a) and Figure T3(b) below. If the forward path gain is reduced by 10% in each system, the variation in C_1 and C_2 will be, respectively,
    (a) 10% and 10%
    (b) 2% and 10%
    (c) 5% and 1%
    (d) 10% and 1%
11. To eliminate the effect of disturbance $D(s)$ on $C(s)$ in Figure T4, the transfer function $G_d(s)$ should be

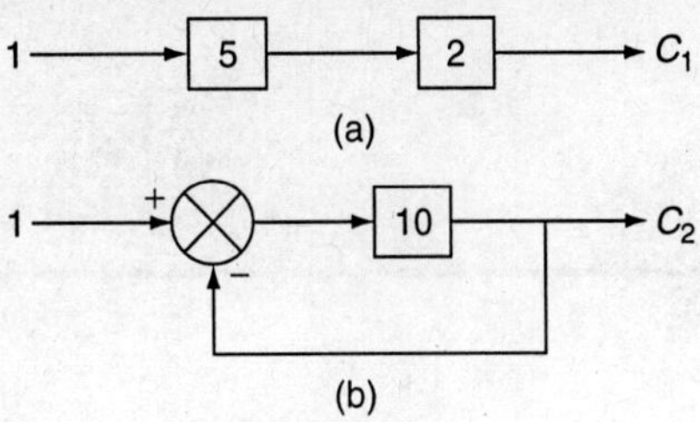

Figure T3

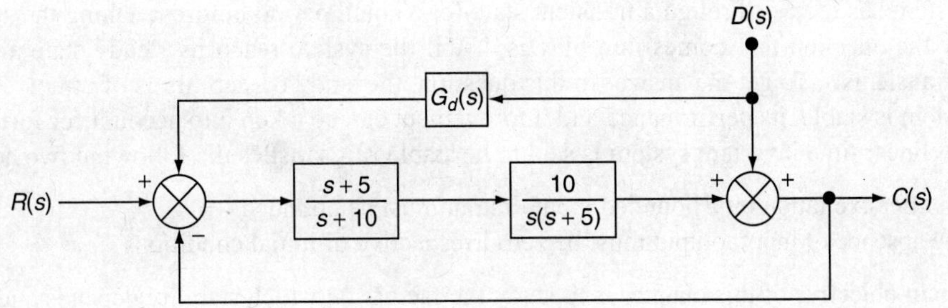

Figure T4

(a) $\dfrac{s+10}{10}$ (b) $\dfrac{s(s+10)}{10}$

(c) $\dfrac{10}{s+10}$ (d) $\dfrac{10}{s(s+10)}$

ANSWERS

1(c) 2(b) 3(a) 4(b) 5(b) 6(a) 7(c) 8(d) 9(d) 10(d) 11(b)

Chapter 10

Stability

Every system has to pass through a transient state for a small period before reaching the steady state. Naturally the question that comes into play is, "Will the system reach its steady state after passing through transients?" To get the answer to this question, the study of stability is of utmost importance. If the system is stable, its performance and improvement can be taken into account for further consideration. A linear time invariant system is said to be stable if it satisfies the following two conditions:

- Even after excitation by a bounded input, output must be bounded.
- In the absence of input, output must be zero irrespective of initial conditions.

The main objective of this chapter is to study the factors on which stability depends and to analyse the stability of systems using Routh-Hurwitz's criteria.

10.1 Effect of Location of Poles on Stability

Location of poles has a direct effect on stability. The entire s-plane is taken into account and it is divided as follows:

(a) Left half plane (LHP)
(b) $j\omega$-axis
(c) Right half plane (RHP)

Let us consider the following cases:

(a) LHP poles

(i) On real axis and simple
(ii) On real axis and multiple
(iii) Complex conjugates and simple

The system is stable for all of the above conditions.

(b) $j\omega$ axis

(i) On $j\omega$-axis and simple
(ii) On $j\omega$-axis and multiple
(iii) At origin
(iv) At origin and multiple

In this case, if the pole has zero real part and is not repeated, the system is called stable.

(c) RHP poles

(i) On real axis and simple
(ii) On real axis and multiple
(iii) Complex conjugates and simple

The system is unstable for all of the above conditions.
Figure 10.1 depicts the roots in s-plane as well as its impulse response.

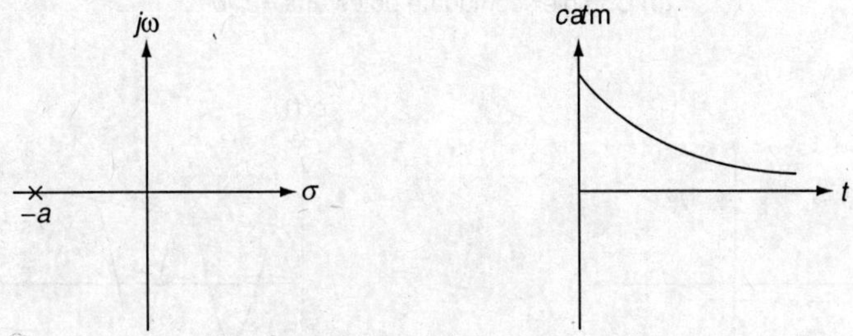

aymLpnploiLtroiLyslL= −a

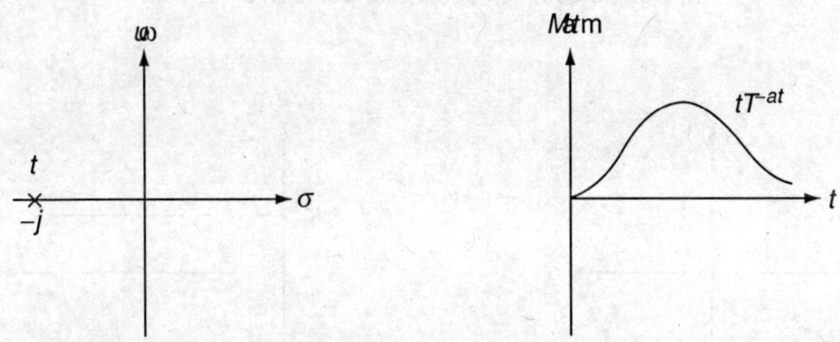

aimLpotinlLponlLyulL= −j

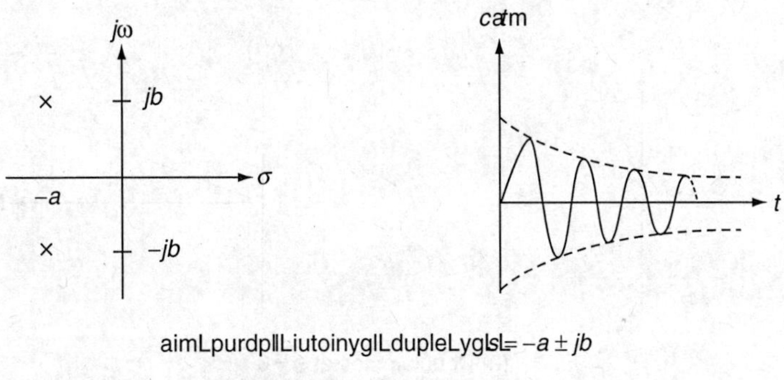

aimLpurdplLiutoinyglLdupleLyglsL= −a ± jb

Fig. 10.1 Continued

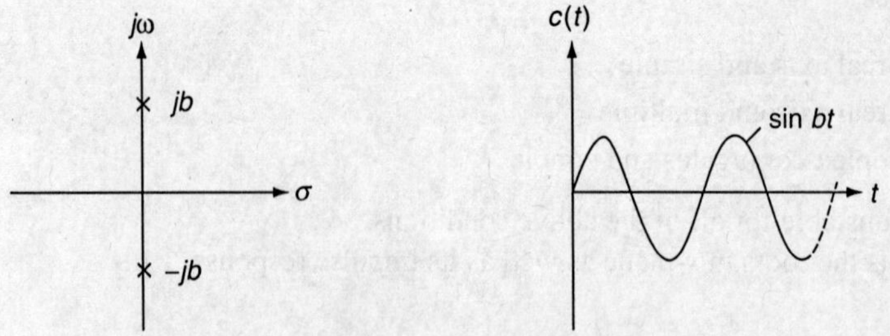

(d) Complex conjugate poles at $s = \pm jb$

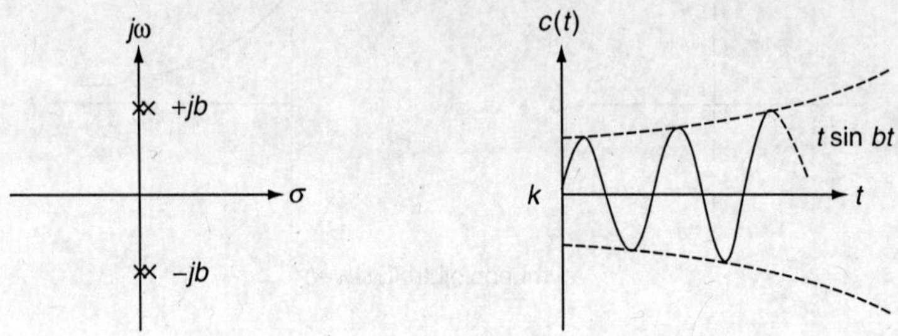

(e) Double complex conjugate poles at $s = \pm jb$

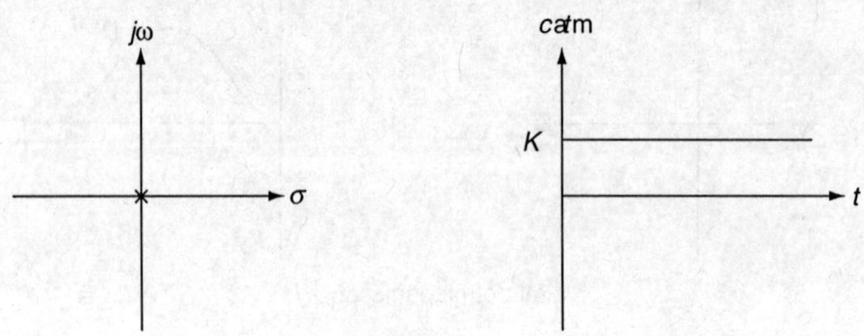

anmpbylrLitLsL= p

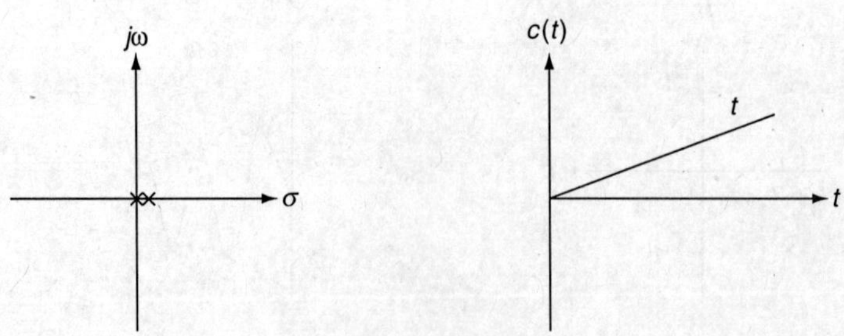

(g) Double poles at $s = 0$

Fig. 10.1 Continued

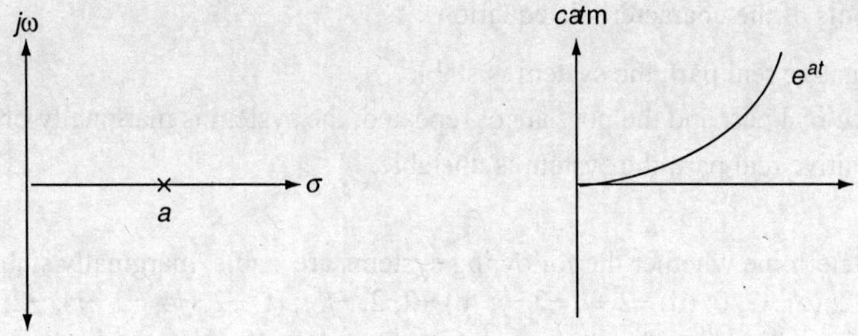

almLpnipLyrbL= a

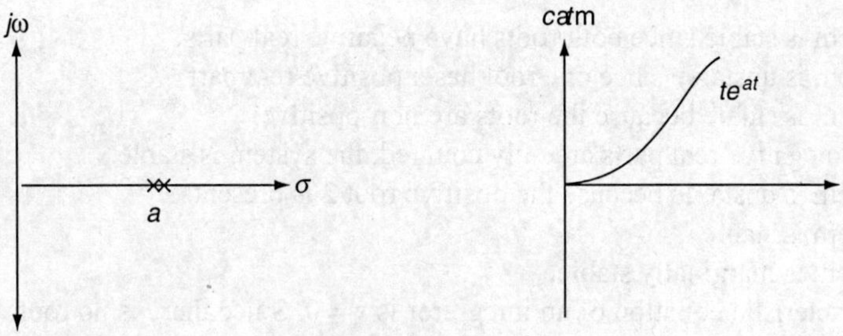

anmLppdiLrpoltLyubL= a

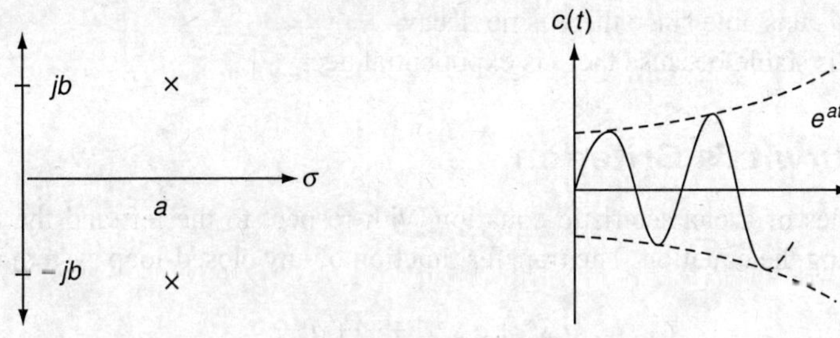

(j) Complex conjugate poles at $s = a \pm jb$

Fig. 10.1 Roots in *s*-plane and its impulse response

From the definition of stability, the impulse response must tend to zero as time approaches infinity. Therefore, from Fig. 10.1, the following conclusions can be drawn:

(i) For all the poles located in the left half of *s*-plane, the impulse response tends to zero and hence the system is stable.

(ii) For poles on the *j*ω axis and simple, the system is marginally stable or limitedly stable.

(iii) For all poles located in the right half of *s*-plane, the impulse response does not tend to zero and it increases with time. Therefore, the output becomes unbounded. Hence the system is unstable.

Therefore, if the roots of the characteristic equation

(i) have a negative real part, the system is stable.
(ii) have a zero real part and the pole is not repeated, the system is marginally or limitedly stable.
(iii) have a positive real part, the system is unstable.

Example 10.1 Determine whether the following systems are stable, marginally stable or unstable. (a) −2, −5; (b) −3, 2; (c) −2, 0; (d) $-2+j$, $-2-j$; (e) −2, 2, $-j$, j; (f) $-2+j4$, $-2-j4$, −2; (g) $-2+j2$, $-2-j2$, $-2j$, $2j$; (h) an integrator; (i) a step input; (j) a bounded input that produces an unbounded output; (k) $x(t) = \cos\omega t$; (l) $x(t) = e^{-t}\sin 4t$

Solution

(a) The system is stable since both roots have negative real parts.
(b) The system is unstable since one root has a positive real part.
(c) The system is stable because the roots are non-positive.
(d) Since the negative real parts are only counted, the system is stable.
(e) The system is unstable because the positive root 2 is present.
(f) The system is stable.
(g) The system is marginally stable.
(h) The characteristic equation of an integrator is $s = 0$. Since there is no root with a positive real part, an integrator is marginally stable.
(i) Since a step function gives $\frac{F(s)}{s}$, the system is marginally stable.
(j) The system is unstable.
(k) The system is unstable since there is no decay.
(l) The system is stable because there is exponential decay.

10.2 Routh-Hurwitz's Criterion

It determines the poles of a characteristic equation with respect to the left and the right half of the s-plane without solving the equation. The transfer function of any closed-loop system is given by

$$\frac{C(s)}{R(s)} = \frac{a_0 s^m + a_1 s^{m-1} + \cdots + a_m}{b_0 s^n + b_1 s^{n-1} + \cdots + b_n} \tag{10.1}$$

In Eq. (10.1), a and b are constants. To find the poles of this closed-loop system, let us take

$$b_0 s^n + b_1 s^{n-1} + \cdots + b_n = 0 \tag{10.2}$$

The roots of this characteristic equation represent the closed-loop poles. The stability of the system depends on these poles. The necessary, but no sufficient, conditions for the system having no roots in the right half of s-plane are shown below:

(i) All the coefficients of the polynomial must have the same sign.
(ii) All powers of s must be in descending order.
(iii) The above conditions are not sufficient.

Figure 10.2 depicts all the above conditions.

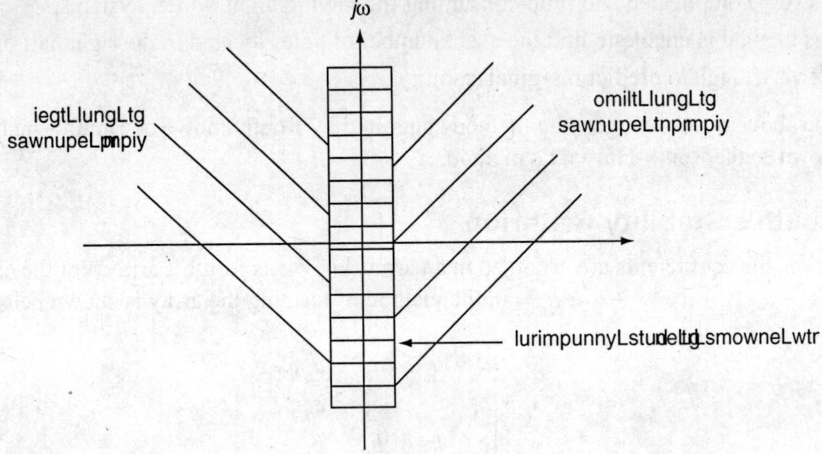

Fig. 10.2

10.2.1 Hurwitz's Criterion

Hurwitz gave the sufficient conditions for having all roots of the characteristic equation in the left half of s-plane. The polynomial $a_0 s^n + a_1 s^{n-1} + \cdots + a_n = 0$ has all roots in the left half of s-plane if and only if all the n sub-determinants D_n, $n = 1,2, \ldots, n$, of the Hurwitz determinant (H) are positive. H is given by

$$H = \begin{vmatrix} a_1 & a_3 & a_5 \ldots\ldots\ldots\ldots \\ a_0 & a_2 & a_4 \ldots\ldots\ldots\ldots \\ 0 & a_0 & a_2 \ldots\ldots\ldots\ldots \\ \ldots\ldots\ldots\ldots\ldots\ldots & & \\ \ldots\ldots\ldots\ldots\ldots\ldots & & \\ 0 & 0 & 0 \ldots\ldots\ldots\ldots a_n \end{vmatrix}$$

$$D_1 = |a_1|$$

$$D_2 = \begin{vmatrix} a_1 & a_3 \\ a_0 & a_2 \end{vmatrix}$$

$$D_3 = \begin{vmatrix} a_1 & a_3 & a_5 \\ a_0 & a_2 & a_4 \\ 0 & a_0 & a_2 \end{vmatrix}$$

$$\vdots$$

$$D_n = H$$

All the sub-determinants must be positive for a stable system.

The disadvantages of Hurwitz's criterion are as follows:

(i) It is very complicated and time-consuming for solving higher-order systems.
(ii) This method is unable to find the exact number of poles located in the right half of s-plane.
(iii) It is very tough to predict marginal stability.

Due to the above disadvantages, the method suggested by Routh known as Routh's method is used. It is also known as the Routh-Hurwitz's method.

10.2.2 Routh's Stability Criterion

In this criterion, the coefficients are arranged in an array, known as Routh's array. For the characteristic equation $a_0 s^n + a_1 s^{n-1} + a_2 s^{n-2} + \cdots + a_n = 0$, the method of forming the array is shown below.

$$\begin{array}{c|llll} s^n & a_0 & a_2 & a_4 & a_6 \cdots \\ s^{n-1} & a_1 & a_3 & a_5 & a_7 \cdots \\ s^{n-2} & b_1 & b_2 & b_3 & b_4 \cdots \\ s^{n-3} & c_1 & c_2 & c_3 \cdots & \\ \vdots & \vdots & & & \\ s^0 & a_n & & & \end{array}$$

For s^{n-2}:

$$b_1 = \frac{a_1 a_2 - a_0 a_3}{a_1}, \quad b_2 = \frac{a_1 a_4 - a_0 a_5}{a_1}, \; .....$$

This process is continued until s^0 is obtained, which is equal to a_n. The stability of the system can be predicted from this array as discussed below.

All the terms in the first column of Routh's array should have same sign. There should not be any change of sign. This is a necessary and sufficient condition for the system to be stable. Any change of sign in the first column of Routh's array indicates

(i) the system is unstable and
(ii) the number of changes in sign gives the number of roots lying in the right half of s-plane.

10.3 Applications of Routh's Criterion

10.3.1 Relative Stability Analysis

The Routh's method can be used effectively to find the relative stability of the system about a line $s = -\sigma$ i.e., to find the number of roots lie in the right half of this line $s = -\sigma$. In order to find this from Routh's array, the axis of s-plane is shifted and s is substituted by $s' = -\sigma$ where σ is a constant in the characteristic equation, Routh's array is applied. The number of sign changes in the first column gives the number of roots, which are located to the right of the vertical line $s = -\sigma$.

10.3.2 Determination Range of Values of *K*

Figure 10.3 shows a practical system in which an amplifier of variable gain K has been introduced.

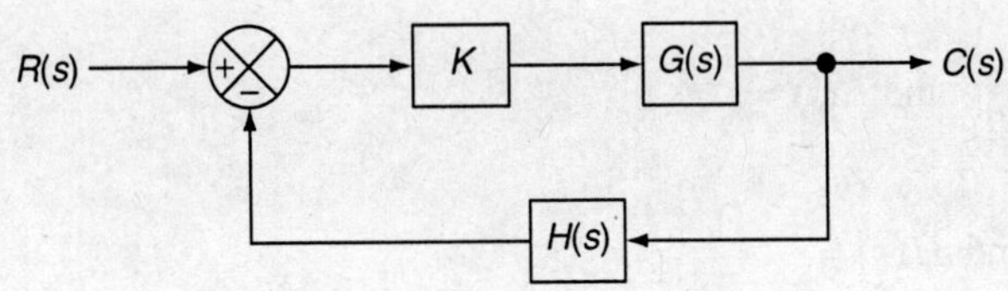

Fig. 10.3 Practical system

The closed-loop transfer function of the system shown in Fig.10.3 is expressed by

$$\frac{C(s)}{R(s)} = \frac{KG(s)}{1+KG(s)H(s)} \tag{10.3}$$

The characteristic equation is given by

$$1+KG(s)H(s)=0 \tag{10.4}$$

The roots of Eq. (10.4) depend on the proper selection of value of K. This unknown K is in the characteristic equation. Routh's array has to be constructed in terms of K. The range of values of K is to be determined in such a fashion so that any change of sign in the first column of the Routh's array does not occur. The range of values of K can be obtained for absolute stability of the system using Routh's criterion. A system is said to be a conditionally stable system if the stability depends on the condition of parameter K.

10.4 Advantages of Routh's Criterion

The following are the advantages of Routh's array method:

(i) It is possible to judge the stability of the system without solving the characteristic equation.
(ii) Since no evolution of determinants is required, calculation time is also reduced.
(iii) This method gives the number of roots of characteristic equation having positive real parts for unstable system.
(iv) It is possible to judge relative stability of the system very easily.
(v) It is possible to obtain critical values of system gain.
(vi) It is possible to find the range of values of K for system stability.
(vii) It is possible to find out intersection points of root locus with imaginary axis.

10.5 Limitations of Routh's Criterion

(i) This method is valid only for real coefficients of the characteristic equation.
(ii) This method is unable to give exact locations of the closed-loop poles in left or right half of s-plane.
(iii) This method is unable to suggest methods for stabilizing an unstable system.
(iv) This method is applicable to linear systems.

Example 10.2 Determine the characteristic equation of the following systems:

(i) $G(s) = \dfrac{12}{s(s^2+4s+2)}$ and $H(s) = 0.5$

(ii) $G(s) = \dfrac{7}{(s+1)(s+2)}$ and $H(s) = s$

(iii) $G(s) = \dfrac{2}{s(s+2)}$ and $H(s) = \dfrac{1}{s}$

Solution Characteristic equation is $1 + G(s)H(s) = 0$

(i) $$1 + \frac{12}{s(s^2+4s+2)} \times 0.5 = 0$$

i.e., $$s^3 + 4s^2 + 2s + 6 = 0$$

(ii) $$1 + \frac{7}{(s+1)(s+2)} \times s = 0$$

i.e., $$s^2 + 10s + 2 = 0$$

(iii) $$1 + \frac{2}{s(s+2)} \times \frac{1}{s} = 0$$

i.e., $$s^3 + 2s^2 + 2 = 0$$

Example 10.3 Examine the stability of $s^3 + 5s^2 + 10s + 3 = 0$ using Routh's method.

Solution

$$\begin{array}{c|cc} s^3 & 1 & 10 \\ s^2 & 5 & 3 \\ s & 9.4 & 0 \\ s^0 & 3 & \end{array}$$

In the first column, there is no change of sign. Therefore, the system is stable.

Example 10.4 Examine the stability of $s^3 + 2s^2 + 3s + 10 = 0$.

Solution

$$\begin{array}{c|cc} s^3 & +1 & 3 \\ s^2 & +2 & 10 \\ s & -2 & 0 \\ s^0 & +10 & \end{array}$$

There are two changes of sign in the first column. Hence the system is unstable.

Example 10.5 The characteristic equations of two systems are shown below:

(i) $s^4 + 6s^3 + 21s^2 + 36s + 20 = 0$ and (ii) $s^5 + 6s^4 + 3s^3 + 2s^2 + s + 1 = 0$

Find whether the systems are stable or not using Routh-Hurwitz's criteria.

Solution

(i) All the coefficients of $F(s) = s^4 + 6s^3 + 21s^2 + 36s + 20 = 0$ are positive and no coefficient missing. The Routh array is shown below:

$$\begin{array}{c|ccc} s^4 & 1 & 21 & 20 \\ s^3 & 6 & 36 & 0 \\ s^2 & 15 & 20 & 0 \\ s^1 & 28 & 0 & 0 \\ s^0 & 20 & 0 & 0 \end{array}$$

There is no change of sign in first column. Hence the system is stable.

(ii) All the coefficients of $F(s) = s^5 + 6s^4 + 3s^3 + 2s^2 + s + 1 = 0$ are positive and no coefficient is missing. The Routh array is shown below:

$$\begin{array}{c|ccc} s^5 & 1 & 3 & 1 \\ s^4 & 6 & 2 & 1 \\ s^3 & 2.67 & 0.83 & 0 \\ s^2 & 0.135 & 1 & 0 \\ s^1 & -18.95 & 0 & 0 \\ s^0 & 1 & 0 & 0 \end{array}$$

There are two changes of sign the first column. Hence the system is unstable.

Example 10.6 Examine the stability of $s^5 + 2s^4 + 4s^3 + 8s^2 + 3s + 1 = 0$ using Routh's criteria.

Solution

$$\begin{array}{c|ccc} s^5 & 1 & 4 & 3 \\ s^4 & 2 & 8 & 1 \\ s^3 & 0 & 2.5 & 0 \\ s^2 & \infty & \cdots & \cdots \leftarrow \text{Routh's array failed.} \end{array}$$

To remove the above difficulty, the following two methods can be used.

Method 1:

(i) Replace 0 by ε (a small number) and complete the array with ε.
(ii) Examine the sign change by taking $\varepsilon \to 0$.

The above example is solved below.

$$\begin{array}{c|ccc} s^5 & 1 & 4 & 3 \\ s^4 & 2 & 8 & 1 \\ s^3 & \varepsilon & 2.5 & 0 \\ s^2 & \dfrac{8\varepsilon - 5}{\varepsilon} & 1 & 0 \\ s^1 & \dfrac{2.5\left(\dfrac{8\varepsilon - 5}{\varepsilon}\right) - \varepsilon}{\dfrac{8\varepsilon - 5}{\varepsilon}} & 0 & \\ s^0 & 1 & & \end{array}$$

Now,

$$\underset{\varepsilon\to 0}{\text{Lt}}\frac{8\varepsilon-5}{\varepsilon}=8-\underset{\varepsilon\to 0}{\text{Lt}}\frac{5}{\varepsilon}=8-\infty=-\infty$$

Therefore, the sign is negative.

Again,

$$\underset{\varepsilon\to 0}{\text{Lt}}\frac{2.5\times\frac{8\varepsilon-5}{\varepsilon}-\varepsilon}{\frac{8\varepsilon-5}{\varepsilon}}=\underset{\varepsilon\to 0}{\text{Lt}}\frac{2.5(8\varepsilon-5)-\varepsilon^2}{8\varepsilon-5}=\underset{\varepsilon\to 0}{\text{Lt}}\frac{-12.5}{-5}=2.5$$

Routh's array is shown below.

$$\begin{array}{c|ccc} s^5 & 1 & 4 & 3 \\ s^4 & 2 & 8 & 1 \\ s^3 & \varepsilon & 2.5 & 0 \\ s^2 & -\infty & 1 & 0 \\ s^1 & 2.5 & 0 & 0 \\ s^0 & 1 & 0 & 0 \end{array}$$

There are two changes of sign in the first column, and hence the system is unstable.

Method 2:

Put $s=\frac{1}{Z}$. The polynomial $s^5+2s^4+4s^3+8s^2+3s+1=0$ becomes

$$\frac{1}{Z^5}+\frac{2}{Z^4}+\frac{4}{Z^3}+\frac{8}{Z^2}+\frac{3}{Z}+1=0$$

i.e.,

$$Z^5+3Z^4+8Z^3+4Z^2+2Z+1=0$$

∴ Routh's array becomes

$$\begin{array}{c|lll} Z^5 & 1 & 8 & 2 \\ Z^4 & 3 & 4 & 1 \\ Z^3 & 6.67 & 1.67 & 0 \\ Z^2 & 3.25 & 1 & 0 \\ Z^1 & -0.382 & 0 & 0 \\ Z^0 & 1 & 0 & 0 \end{array}$$

There are two changes of sign. Hence the system in unstable.

Example 10.7 Examine the stability of $s^5+2s^4+2s^3+4s^2+4s+8=0$ using Routh's method.

Solution

$$\begin{array}{c|ccc} s^5 & 1 & 2 & 4 \\ s^4 & 2 & 4 & 8 \\ s^3 & 0 & 0 & 0 \end{array}\ \text{Routh's array failed.}$$

To eliminate the above difficulty, form an equation by using the coefficients of the array which is just above the row of zeros. The equation is known as auxiliary equation.

$$\therefore \qquad A(s) = 2s^4 + 4s^2 + 8$$

$$\therefore \qquad \frac{dA(s)}{ds} = 8s^3 + 8s$$

Routh's array becomes

$$\begin{array}{c|ccc} s^5 & 1 & 2 & 3 \\ s^4 & 2 & 4 & 8 \\ s^3 & 8 & 8 & 0 \\ s^2 & 2 & 8 & 0 \\ s^1 & -24 & 0 & \\ s^0 & 8 & & \end{array}$$

There are two changes of sign. The system is unstable.

Example 10.8 Examine the stability of $s^4 + 5s^3 + 2s^2 + 3s + 2 = 0$ using Routh's method. Find the number of roots lying in the right half of the s-plane.

Solution

$$\begin{array}{c|ccc} s^4 & 1 & 2 & 2 \\ s^3 & 5 & 3 & 0 \\ s^2 & 1.4 & 2 & \\ s^1 & -4.14 & 0 & \\ s^0 & 2 & & \end{array}$$

There are two changes of sign. Hence the system is unstable. There are two roots in the right half of s-plane.

Example 10.9 Examine the stability of the system having characteristic equation $s^5 + s^4 + 3s^3 + 3s^2 + 4s + 8 = 0$.

Solution

$$\begin{array}{c|ccc} s^5 & 1 & 3 & 4 \\ s^4 & 1 & 3 & 8 \\ s^3 & 0 & -4 & 0 \end{array}$$

Hence Routh's array fails.

Method 1:

$$\begin{array}{c|ccc} s^5 & 1 & 3 & 4 \\ s^4 & 1 & 3 & 8 \\ s^3 & \varepsilon & -4 & 0 \\ s^2 & \dfrac{3\varepsilon+8}{\varepsilon} & 8 & 0 \\ s^1 & \dfrac{-4\left(\dfrac{3\varepsilon+8}{\varepsilon}\right)-8t}{\dfrac{3\varepsilon+8}{\varepsilon}} & 0 & \\ s^0 & 8 & & \end{array}$$

$$\underset{\varepsilon\to 0}{\mathrm{Lt}}\ \frac{3\varepsilon+8}{\varepsilon} = 3 + \underset{\varepsilon\to 0}{\mathrm{Lt}}\ \frac{8}{\varepsilon} = \infty$$

$$\underset{\varepsilon\to 0}{\mathrm{Lt}}\ \frac{-4\left(\dfrac{3\varepsilon+8}{\varepsilon}\right)-8\varepsilon}{\dfrac{3\varepsilon+8}{\varepsilon}} = \underset{\varepsilon\to 0}{\mathrm{Lt}}\ \frac{-4(3\varepsilon+8)-8\varepsilon^2}{3\varepsilon+8} = -4$$

Routh's array becomes

$$\begin{array}{c|ccc} s^5 & 1 & 2 & 4 \\ s^4 & 1 & 3 & 8 \\ s^3 & \varepsilon & -4 & 0 \\ s^2 & +\infty & 8 & 0 \\ s^1 & -4 & 0 & \\ s^0 & 8 & & \end{array}$$

There are two changes of sign. Hence the system is unstable. There are two roots lying in the right half of s-plane.

Method 2:

Put $s = \dfrac{1}{Z}$ in $(s^5 + s^4 + 3s^3 + 3s^2 + 4s + 8 = 0)$, we have

$$\frac{1}{Z^5} + \frac{1}{Z^4} + \frac{3}{Z^3} + \frac{3}{Z^2} + \frac{3}{Z} + 8 = 0$$

i.e., $$8Z^5 + 3Z^4 + 3Z^3 + 3Z^2 + Z + 1 = 0$$

Routh's array becomes

$$\begin{array}{l|lll} Z^5 & 8 & 3 & 1 \\ Z^4 & 4 & 3 & 1 \\ Z^3 & -3 & -1 & 0 \\ Z^2 & 1.67 & 1 & \\ Z^1 & 0.796 & 0 & \\ Z^0 & 1 & & \end{array}$$

There are two changes of sign. Hence the system is unstable. There are two roots lying in the right half of s-plane.

Example 10.10 Determine the range of values of K so that the system having the following characteristic equation will be stable:

$$s(s^2+2s+3)(s+2)+K=0$$

Solution

The given characteristic equation is

$$s(s^2+2s+3)(s+2)+K=0$$

i.e., $$s(s^3+4s^2+7s+6)+K=0$$

i.e., $$s^4+4s^3+7s^2+6s+K=0$$

Routh's array becomes

$$\begin{array}{l|lll} s^4 & 1 & 7 & K \\ s^3 & 4 & 6 & 0 \\ s^2 & 5.5 & K & 0 \\ s^1 & \dfrac{33-4K}{5.5} & 0 & \\ s^0 & K & & \end{array}$$

The system will be stable if no change of sign occurs in the first column. Therefore, $K > 0$

and $$\frac{33-4K}{5.5}>0$$

i.e., $$33-4K>0$$

i.e., $$4K<33$$

i.e., $$K<\frac{33}{4}$$

i.e., $$K<8.25$$

∴ Range of K is $0<K<8.25$.

Example 10.11 For a unity feedback system having forward transfer function $G(s) = \dfrac{K}{s(1+0.6s)(1+0.4s)}$, determine the range of values of K, marginal value of K and the frequency of sustained oscillation.

Solution The characteristic equation of the system is given by

$$1+\frac{K}{s(1+0.6s)(1+0.4s)}=0$$

i.e., $$s(1+0.6s)(1+0.4s)+K=0$$

i.e., $$s(1+s+0.24s^2)+K=0$$

i.e., $$0.24s^3+s^2+s+K=0$$

Routh's array becomes

$$\begin{array}{c|cc} s^3 & 0.24 & 1 \\ s^2 & 1 & K \\ s^1 & 1-0.24K & 0 \\ s^0 & K & \end{array}$$

The system will be stable if no change of sign occurs in the first column of Routh's array: i.e., $K > 0$ and $1 - 0.24K > 0$. Therefore,

$$K<\frac{100}{24}$$

i.e., $$K<4.167$$

∴ Range of values of K, $0 < K < 4.167$
For a marginally stable system,

$$1-0.24\,K=0$$

i.e., $$K=\frac{1}{0.24}=4.167$$

Routh's array becomes

$$\begin{array}{c|ccc} s^5 & 0.24 & 3 & 4 \\ s^4 & 1 & & K \\ s^3 & 0 & & 0 \end{array}$$

The auxiliary equation is

$$A(s)=s^2+K=0$$

i.e., $$s^2=-K$$

i.e., $$s^2=-4.167$$

i.e., $$s=\pm j2.04$$

∴ Frequency of oscillations is 2.04 rad/sec.

Example 10.12 The system shown in Fig. E10.1 has the oscillation of 2.5 rad/sec. Determine the values of K_{mar} and p. There are no poles in RHP.

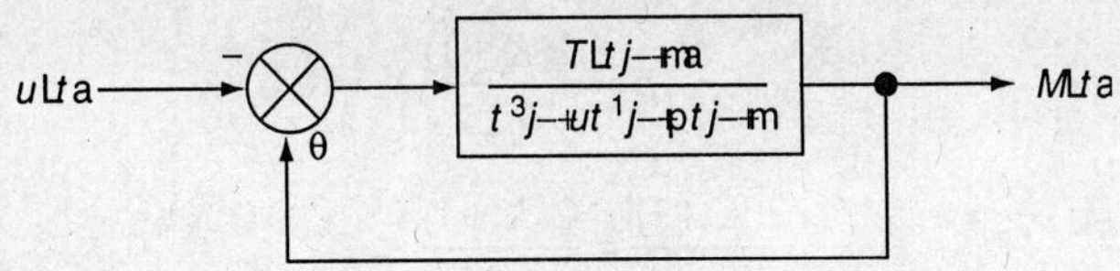

Fig. E10.1

Solution Since the system oscillates, it is marginally stable. The characteristic equation of the system becomes

$$1+\frac{K(s+2)}{s^3+ps^2+3s+2}=0$$

i.e.,
$$s^3+ps^2+3s+2+Ks+2K=0$$

i.e.,
$$s^3+ps^2+(K+3)s+2(K+1)=0$$

$$\begin{array}{c|cc} s^3 & 1 & K+3 \\ s^2 & p & 2(K+1) \\ s^1 & \dfrac{p(K+3)-2(K+1)}{p} & 0 \\ s^0 & 2(K+1) & \end{array}$$

At marginal value of K,

$$\frac{p(K+3)-2(K+1)}{p}=0$$

i.e.,
$$p(K+3)-2K-2=0$$

i.e.,
$$p=\frac{2(K+1)}{K+3} \tag{1}$$

Again, at this value of p,

$$A(s)=ps^2+2(K+1)=0$$

i.e.,
$$s^2=-\frac{2(K+1)}{p}$$

i.e.,
$$s=\pm j\sqrt{\frac{2(K+1)}{p}}=\pm j\omega$$

Here $\omega = 2.5$ rad/sec. Therefore,

$$\sqrt{\frac{2(K+1)}{p}}=2.5$$

i.e., $$\frac{2(K+1)}{p} = 6.25$$

i.e., $$p = \frac{2(K+1)}{6.25}$$

i.e. $$\frac{2(K+1)}{K+3} = \frac{2(K+1)}{6.25}$$ [using Eq. (1)]

i.e., $$K+3 = 6.25$$

i.e., $$K = 3.25$$

i.e., $$p = \frac{2(K+1)}{K+3} = \frac{2(3.25+1)}{6.25} = \frac{2\times 4.25}{6.25} = 1.36$$

Example 10.13 For a system having characteristic equation $2s^4 + 4s^2 + 1 = 0$, find the following:

(i) the number of roots in the left half of s-plane
(ii) the number of roots is the right half of s-plane
(iii) the number of roots on the imaginary axis

Use the Routh-Hurwitz's criterion.

Solution The system characteristic equation is given by

$$2s^4 + 4s^2 + 1 = 0$$

i.e., $$2s^4 + 0.s^3 + 4s^2 + 0.s + 1 = 0$$

Routh's array is

$$\begin{array}{c|ccc} s^5 & 2 & 4 & 1 \\ s^4 & 0 & 0 & \end{array} \quad \leftarrow \text{Row of zero}$$

∴ Auxiliary equation,

$$A(s) = 2s^4 + 4s^2 + 1$$

∴ $$\frac{dA(s)}{ds} = 8s^3 + 8s$$

$$\begin{array}{c|ccc} s^4 & 2 & 4 & 1 \\ s^3 & 8 & 8 & 0 \\ s^2 & 2 & 1 & \\ s^1 & 4 & 0 & \\ s^0 & 1 & & \end{array}$$

Since there is no change of sign, there are no roots lying on the right half of s-plane.

Again, solving $A(s) = 0$, we have

$$2s^4 + 4s^2 + 1 = 0$$

i.e., $$s^2 = \frac{-4 \pm \sqrt{4^2 - 4\times 2\times 1}}{2\times 2} = \frac{-4 \pm \sqrt{16-8}}{4} = \frac{-4 \pm \sqrt{8}}{4} = \frac{-4 \pm 2\sqrt{2}}{4}$$

i.e., $$s^2 = \frac{-4 + 2\sqrt{2}}{4} \text{ or } \frac{-4 - 2\sqrt{2}}{4} = -0.2929 \text{ or } -1.7071$$

i.e., $$s = \pm j0.5412, \pm 1.3065$$

Therefore,

(i) the number of roots lying in the left half of s-plane is 0
(ii) the number of roots lying in the right half of s-plane is 0
(iii) the number of roots lying on the imaginary axis is 4

Example 10.14 The open-loop transfer function of a feedback control system is given by

$$G(s)H(s) = \frac{K}{s(s+4)(s^2+2s+2)}$$

Determine the stability of the system when $K = 12$ and find the range of the value of K for stability.

Solution (a) The characteristic equation is

$$1 + G(s)H(s) = 0 \text{ i.e., } s^4 + 6s^3 + 10s^2 + 8s + K = 0$$

The Routh array is formed below:

$$\begin{array}{c|ccc} s^4 & 1 & 10 & K \\ s^3 & 6 & 8 & \\ s^2 & \frac{52}{6} & K & \\ s^1 & 8 - \frac{36K}{52} & 0 & \\ s^0 & K & & \end{array}$$

For a stable system, $8 - \frac{36K}{52} > 0$,

i.e., $$K < \frac{8\times 52}{36}$$

i.e., $$K < 11.56$$

For stability, $0 < K < 11.56$. The closed-loop system is unstable for $K = 12$ because one of the elements of the first column becomes negative.

Example 10.15 Find the stability of a unit feedback system having open-loop transfer function $G(s) = \dfrac{e^{-sT}}{s(s+2)}$. Use the Routh's criterion.

Solution The characteristic equation of the system is

$$1 + G(s)H(s) = 0$$

i.e.,
$$1 + \frac{e^{-sT}}{s(s+2)} = 0$$

i.e.,
$$s(s+2) + e^{-sT} = 0$$

i.e.,
$$s^2 + 2s + (1 - sT) = 0 \qquad [\because e^{-sT} \cong 1 - sT]$$

i.e.,
$$s^2 + (2 - T)s + 1 = 0$$

Routh's array becomes

$$\begin{array}{c|cc} s^2 & 1 & 1 \\ s^1 & 2-T & 0 \\ s^0 & 1 & \end{array}$$

The system will be stable if

$$2 - \mathrm{T} > 0$$

i.e.,
$$\mathrm{T} < 2.$$

Example 10.16 Find the stability of a system having characteristic equation $s^6 + 2s^5 + 7s^4 + 10s^3 + 14s^2 + 8s + 8 = 0$.

$$\begin{array}{c|cccc} s^6 & 1 & 7 & 14 & 8 \\ s^5 & 2 & 10 & 8 & 0 \\ s^4 & 2 & 10 & 8 & 0 \\ s^3 & 0 & 0 & 0 & 0 \leftarrow \text{Row of zero} \end{array}$$

$\therefore$
$$A(s) = 2s^4 + 10s^2 + 8 = 0$$

$\therefore$
$$\frac{dA(s)}{ds} = 8s^3 + 20s$$

Routh's array becomes

$$\begin{array}{c|cccc} s^6 & 1 & 7 & 14 & 8 \\ s^5 & 2 & 10 & 8 & 0 \\ s^4 & 2 & 10 & 8 & 0 \\ s^3 & 8 & 20 & 0 & 0 \\ s^2 & 5 & 8 & & \\ s^1 & 7.2 & 0 & & \\ s^0 & 8 & & & \end{array}$$

There are no changes of sign and hence the system is stable. Since there is a row of zero, the system may be marginally stable or unstable. To verify this, let us solve $A(s) = 0$

$\therefore$ $$2s^4 + 10s^2 + 8 = 0$$

i.e., $$s^2 = \frac{-10 \pm \sqrt{10^2 - 64}}{4} = \frac{-10 \pm \sqrt{36}}{4} = \frac{-10 \pm 6}{4}$$

i.e., $$s^2 = -1, -4$$

i.e., $$s = \pm j, \pm j2$$

Since the roots on the imaginary axis are not repeating, the system is stable.

Example 10.17 Determine (i) the range of K for stability and (ii) the frequency of oscillations when system is marginally stable for the unity feedback (negative) system having

$$G(s) = \frac{K}{(s+1)^3(s+3)}$$

Solution The characteristic equation of the system is expressed by

$$1 + G(s)H(s) = 0$$

i.e., $$1 + \frac{K}{(s+1)^3(s+3)} = 0$$

i.e., $$(s+1)^3(s+3) + K = 0$$

i.e., $$(s^3 + 3s^2 + 3s + 1)(s+3) + K = 0$$

i.e., $$s^4 + 3s^3 + 3s^2 + s + 3s^3 + 9s^2 + 9s + 3 + K = 0$$

i.e., $$s^4 + 6s^3 + 12s^2 + 10s + (3+K) = 0$$

Routh's array becomes

$$\begin{array}{c|ccc} s^4 & 1 & 12 & 3+K \\ s^3 & 6 & 10 & 0 \\ s^2 & 10.34 & 3+K & \\ s^1 & \dfrac{85.4-6K}{10.35} & 0 & \\ s^0 & 3+K & & \end{array}$$

From row of s^0, $3 + K > 0$ i.e., $K > -3$

From row of s^1, $85.4 - 6K > 0$ i.e., $K < 14.24$

(i) Range of K for stability is $-3 < K < 14.24$

(ii) For marginal stability, $K = K_{mar} = 14.24$

$$\therefore \qquad A(s) = 10.34s^2 + 3 + K = 0$$

i.e., $$10.34s^2 + 3 + 14.24 = 0 \qquad [\because K = K_{mar} = 14.24]$$

i.e., $$10.34s^2 + 17.24 = 0$$

i.e., $$s^2 = -\frac{17.24}{10.34} = -1.667$$

i.e., $$s = \pm j1.291$$

Therefore, the frequency of oscillation is 1.291 rad/sec.

Example 10.18 A unity feedback system has

$$G(s) = \frac{K}{s(s+1)(s+2)(s+4)}$$

Find the following using Routh-Hurwitz's criterion:

(i) the range of K for stability,
(ii) the value of K for marginal stability and
(iii) the actual location of closed-loop poles when the system is marginally stable.

Solution The characteristic equation is given by

$$1 + G(s)H(s) = 0$$

i.e., $$1 + \frac{K}{s(s+1)(s+2)(s+4)} = 0$$

i.e., $$s(s+1)(s+2)(s+4) + K = 0$$

i.e., $$s^4 + 7s^3 + 14s^2 + 8s + K = 0$$

Routh's array becomes,

$$\begin{array}{c|ccc} s^4 & 1 & 14 & K \\ s^3 & 7 & 8 & 0 \\ s^2 & 12.86 & K & \\ s^1 & \dfrac{102.88 - 7K}{12.86} & 0 & \\ s^0 & K & & \end{array}$$

From row of s^0, $K > 0$
From row of s^1, $102.88 - 7K > 0$ i.e., $K < 14.70$

(i) The range of K for stability is $0 < K < 14.70$
(ii) The value of K for marginal stability is $K = K_{mar} = 14.70$
(iii) At $K = K_{mar} = 14.70$, the auxiliary equation becomes

$$A(s) = 12.86s^2 + K = 0$$

i.e., $$s^2 = -\frac{K}{12.86} = -\frac{14.70}{12.86} = -1.143$$

i.e., $$s = \pm j1.07$$

The characteristic equation for K_{mar} is

$$s^4 + 7s^3 + 14s^2 + 8s + 14.70 = 0$$

The two roots are $s = \pm j1.07$

The remaining roots can be obtained as follows:

$$
\begin{array}{l}
s^2 + 1.143)\ s^4 + 7s^3 + 14s^2 + 8s + 14.70\ (s^2 + 7s + 12.857 \\
\quad s^4 \pm 1.143s^2 \\
\hline
\quad 7s^3 + 12.857s^2 + 8s + 14.70 \\
\quad 7s^3 \pm 8s \\
\hline
\quad 12.857s^2 + 14.70 \\
\quad 12.857s^2 + 14.70 \\
\hline
\quad 0
\end{array}
$$

The remaining two roots are the roots of the quadratic equation $s^2 + 7s + 12.857 = 0$

$$\therefore s = \frac{-7 \pm \sqrt{7^2 - 4 \times 12.857}}{2}$$

$$= \frac{-7 \pm j1.558}{2}$$

$$= -3.5 \pm j0.779$$

When the system is marginally stable, the actual locations of closed-loop poles are $-3.5 \pm j0.779$, $\pm j1.07$.

SIGNIFICANT POINTS

Stability: A linear time invariant system is said to be stable if it satisfies the following two conditions:

- Even after excitation by a bounded input, output must be bounded.
- In the absence of input, output must be zero irrespective of initial conditions.

Absolute stability: The stability of the system with respect to variation of parameters is known as absolute stability. It is absolutely stable when it is stable for all values of the parameters.

Relative stability: It measures the degree of stability. It is a quantitative measure of how fast the system transients die out in a system.

Conditional stability: The stability of the system with respect to its parameter variation is known as conditional stability.

Hurwitz's criterion: Hurwitz gave the sufficient conditions for having all roots of the characteristic equation in the left half of *s*-plane. The disadvantages of Hurwitz's criterion are as follows:

(i) It is very complicated and time-consuming for solving higher-order systems.

(ii) This method is unable to give the exact number of poles located in the right half of *s*-plane.

(iii) It is very tough to predict marginal stability.

Routh-Hurwitz's Criterion: It determines the poles of a characteristic equation with respect to the left and the right half of the s-plane without solving the equation. Any change of sign in the first column of Routh's array indicates

(i) the system is unstable and

(ii) the number of changes of sign gives the number of roots lying in the right half of s-plane.

SHORT QUESTIONS WITH ANSWERS

1. What do you mean by stability, absolute stability, and conditional stability?

   Ans. *Stability*: A linear time invariant system is said to be stable if it satisfies the following two conditions:

   - Even after excitation by a bounded input, output must be bounded.
   - In the absence of input, output must be zero irrespective of initial conditions.

   Absolute stability: The stability of the system with respect to variation of parameters is known as absolute stability. It said to be absolutely stable when it is stable for all values of the parameters.

   Relative stability: It measures the degree of stability and it is a quantitative measure of how fast the system transients die out in a system.

   Conditional stability: The stability of the system with respect to its parameter variation is known as conditional stability.

2. State Hurwitz's criterion.

   Ans. Hurwitz gave the sufficient conditions for having all roots of the characteristic equation in the left half of s-plane. The disadvantages of Hurwitz's criterion are as follows:

   (i) It is very complicated and time-consuming for solving higher-order systems.

   (ii) This method is unable to give the exact number of poles located in the right half of s-plane.

   (iii) It is very tough to predict marginal stability.

3. State Routh-Hurwitz's Criterion.

   Ans. It determines the poles of a characteristic equation with respect to the left and the right half of the s-plane without solving the equation. Any change of sign in the first column of Routh's array indicates

   (i) the system is unstable and

   (ii) the number of change of sign gives the number of roots lying in the right half of s-plane.

EXERCISES

Questions for Practice

1. A unity feedback servo-driven instrument has an open-loop transfer function

$$G(s) = \frac{10}{s(s+2)(s^2+3)}$$

   Find the absolute stability of the closed-loop system.

Ans. Unstable

2. A feedback control system has the following loop transfer function:

$$G(s) = \frac{K}{s(s^2+5s+9)}.$$

(i) Using Routh's criterion, find the ranges of values of K for which the system is stable. Also find the frequencies of self-sustained oscillations.

(ii) The element having a transportation lag of 1 sec is introduced in the forward path of the system described above. Will the system now be stable over the range of values of K as obtained in part (i) above?

Ans. For stability $0 < K < 45$, frequency of sustained oscillation is $\omega_0 = \pm 3$ rad/ sec, (ii) for stability $0 < K < 7.5$

3. A third-order control system has the characteristic equation

$$s^3 + 3408.3s^2 + 120400s + 1.5 \times 107K = 0.$$

Using Rough-Hurwitz's criterion, find the condition on K for the system to be stable. How many roots of the characteristic equation will lie in the right half of s-plane when $K = 280$.

Ans. $0 < K$ 273.6, two

4. A unity feedback control system has an open-loop transfer function

$$G(s) = \frac{Ke^{-s}}{s(s^2 + 5s + 8)}$$

Determine the maximum value of K for the closed-loop system to be stable.

Ans. 6.67

5. Investigate the stability of the system with characteristic equation

$$s^5 + 2s^4 + 24s^3 + 48s^2 - 25s - 50 = 0.$$

Also find all the roots of this equation.

Ans. Unstable, $\pm 1, \pm j\,5$, and -2

6. A unity feedback control system is characterized by the loop transfer function

$$G(s) = \frac{K(s+13)}{s(s+3)(s+7)}$$

Find the range of values of K for which the system is stable.

Ans. $0 < K < 70$

7. Find the characteristic equation of the system shown in Figure 1. Determine the range of T for which the system is unconditionally stable. $K > 0$.

Ans. $s^3 + 2s^2 + (10 + KT)s + K = 0, T > \left(\frac{1}{2} - \frac{10}{K}\right)$

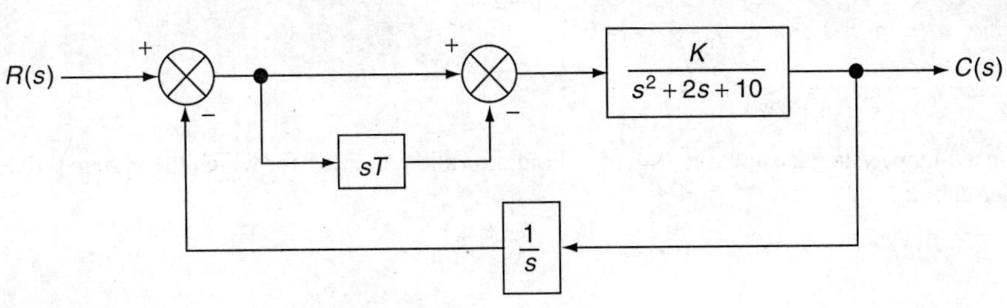

Figure 1

8. If a unity feedback system having $G(s) = \dfrac{5}{s(s^2 + ps + K)}$ is marginally stable and oscillates with a frequency 5 rad/sec, calculate K_{mar} and p.

Ans. $K_{mar} = 25, p = 0.2$

9. For the unity feedback system with $G(s) = \dfrac{K}{(s+1)^3(s+4)}$, determine the following:

   (i) The range of K for stability

   (ii) The frequency of oscillations when the system is marginally stable.

Ans. $-4 < K < 20.41$, (ii) 1.3628 rad/sec

TEST YOURSELF

1. A positive feedback signal improves the performance of an automatic control system.

   (a) False (b) True

2. Routh's array for a system is shown below:

$$\begin{array}{c|ccc} s^4 & 1 & 3 & 7 \\ s^3 & 1 & 2 & 0 \\ s^2 & 1 & 7 & \\ s^1 & -5 & & \\ s^0 & 7 & & \end{array}$$

   The system is

   (a) marginally stable (b) unstable (c) stable (d) conditionally stable

3. The number of sign changes in the entries in the first column of Routh's array denotes

   (a) the number of roots of the characteristic polynomial in RHP

   (b) the number of open-loop poles in RHP

   (c) the number of zeros of the system in RHP

   (d) the number of open-loop zeros in RHP

4. The transfer function of a system is $\dfrac{K}{a_3s^3 + a_2s^2 + a_1s + a_0}$. For the system to be absolutely stable, the valid conditions are

   (a) $a_3, a_2, a_1, a_0 > 0$ and $a_2\,a_1 - a_3\,a_0 < 0$

   (b) $a_3, a_2, a_1, a_0 > 0$ and $a_2\,a_1 - a_3\,a_0 > 0$

   (c) $a_3, a_2, a_1, a_0 > 0$ and $a_2\,a_1 - a_3\,a_0 = 0$

   (d) $a_2, a_0 > 0$ and $a_3, a_1 < 0$

5. A closed-loop system is shown in Figure 1. Find the valid condition for which the system will be very much under-damped.

   (a) $T_1 = T_2$ (b) $T_1 < T_2$ (c) $T_1 > T_2$ (d) $T_1 << T_2$

Figure 1

6. The Routh array of a characteristic equation is given below. The number of roots lying on the right hand side of s-plane is

$$\begin{array}{c|c} s^4 & 9 \\ s^3 & 8 \\ s^2 & -4 \\ s^1 & 2 \\ s^0 & 1 \end{array}$$

(a) 0 (b) 1 (c) 3 (d) 2

7. Routh's array is shown below:

$$\begin{array}{c|cccc} s^6 & 1 & 9 & 23 & 15 \\ s^5 & 3 & 18 & 15 & \\ s^4 & 3 & 18 & 15 & \\ s^3 & 0 & 0 & & \end{array}$$

The auxiliary equation of this array is

(a) $3s^4+15s^2+9=0$ (b) $2s^4+18s^2+15=0$

(c) $3s^4+18s^2+15=0$ (d) $s^4+9s^2+15=0$

8. If some pole of a system lies on the imaginary axis, the system is

(a) absolutely stable (b) conditionally stable

(c) marginally stable (d) unstable

9. The main cause of absolute instability is

(a) error detector where the two signals are compared

(b) parameters of the controlled system

(c) parameters of the uncontrolled system

(d) parameters of the controlling system

10. The characteristic equation of a feedback control system is $s^3+Ks^2+5s+10=0$.

For the system to be critically stable, the value of K should be

(a) 1 (b) 2 (c) 3 (d) 4

11. The open-loop transfer function of a unity feedback control system is

$$G(s)=\frac{K(s+10)(s+20)}{s^2(s+2)}$$

The closed-loop system will be stable if the value of K is

(a) 2 (b) 3 (c) 4 (d) 5

12. A feedback control system shown in Figure 2 is stable for all values of K, if

(a) $T=0$ (b) $T<0$ (c) $T>1$ (d) $0<T<1$

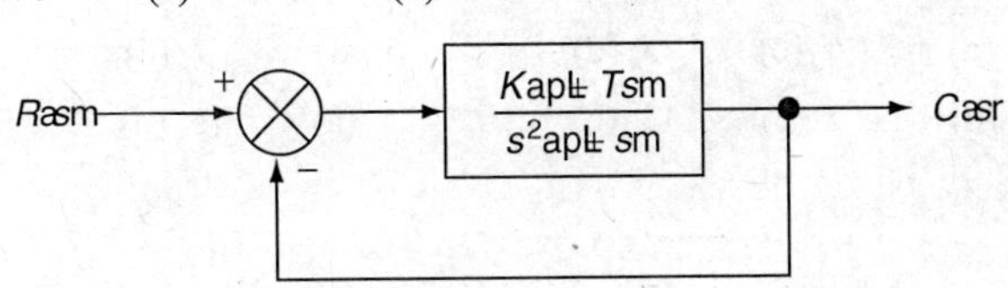

Figure 2

13. The characteristic equation of a unity feedback system is given by $s^3 + s^2 + 4s + 4 = 0$. The system
    (a) has one pole in the RH s-plane
    (b) has no poles in the RH s-plane.
    (c) exhibits oscillatory nature.
    (d) is asymptotically stable.

14. The control system shown in Figure 3 has an internal rate feedback loop. The closed-loop system for open and conditions of the switch will be, respectively,

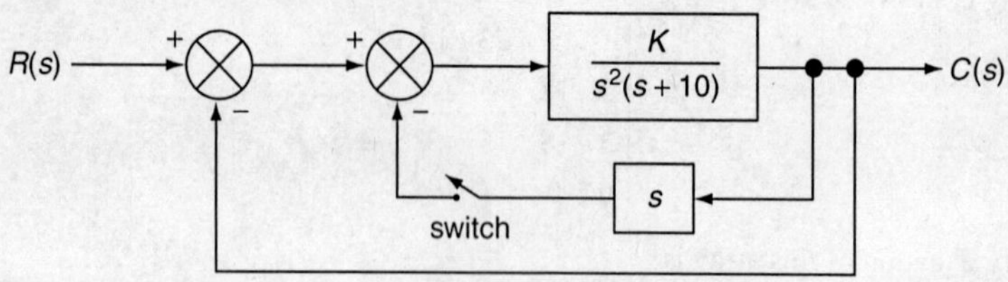

Figure 3

    (a) unstable and stable
    (b) unstable and unstable
    (c) stable and unstable
    (d) stable and stable

15. For the block diagram shown in Figure 4, the limiting value of K for stability of the inner loop is found to be $X < K < Y$. The overly system will be stable if and only if

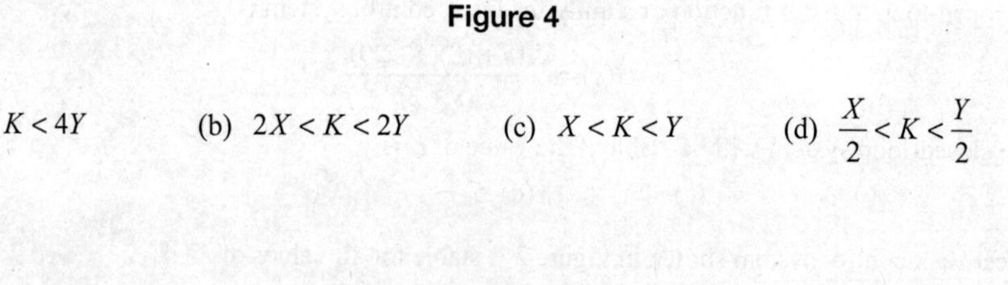

Figure 4

    (a) $4X < K < 4Y$
    (b) $2X < K < 2Y$
    (c) $X < K < Y$
    (d) $\frac{X}{2} < K < \frac{Y}{2}$

ANSWERS

1 (b)	2 (b)	3 (a)	4 (b)	5 (d)	6 (d)	7 (c)	8 (c)
9 (a)	10 (b)	11 (d)	12 (c)	13 (b) (c)	14 (a)	15 (d)	

Chapter 11

Root Locus Method

The stability as well as the transient response of any closed-loop system depends on the location of the closed-loop poles in the plane. If some parameters of the closed-loop system are varied, there will be corresponding movement of its poles. The cognizance of such movement of the poles with small variations of the parameters is remarkably useful in the design of closed-loop system. In 1948, W.R. Evans introduced the root locus technique, which provides knowledge of such movement of poles by a graphical method in the s-plane. In *root locus method*, the parameter (K) of the system is varied from zero to infinity. The aim of this chapter is to introduce the root locus method.

11.1 Fundamentals of Root Locus

Figure 11.1 shows a closed-loop system. The characteristic equation of this system is expressed by

$$1+G(s)H(s)=0 \qquad (11.1)$$

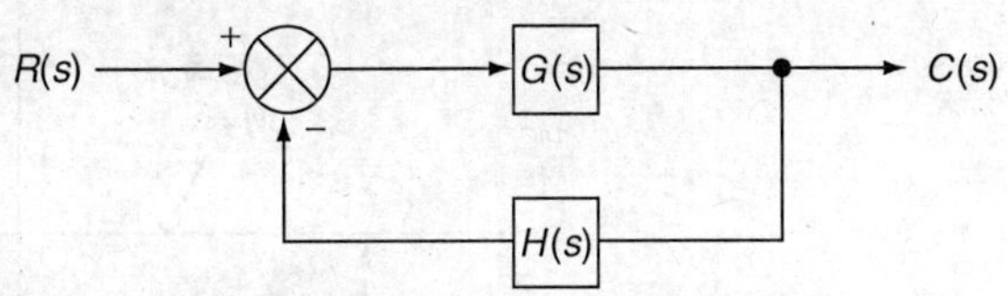

Fig. 11.1 Feedback system

Let the transfer function of the forward path be $KG'(s)$, where K is the gain of the system and is a variable parameter. Equation (11.1) becomes

$$1+KG'(s)H(s)=0 \qquad (11.2)$$

Therefore, the closed-loop poles of the system depend on the values of K. Varying K from $-\infty$ to $+\infty$, a separate set of locations of the roots of the characteristic equation will be obtained. The root locus can be achieved by joining all such locations. Root locus is defined as the plot of the locus of the closed-loop poles as a function of open-loop gain (K), when K is varied from $-\infty$ to $+\infty$. If K is varied from 0 to $+\infty$, it is known as direct root locus. On the other hand, if K is varied from $-\infty$ to 0, it is termed as inverse root locus. Usually, the root locus is used for direct root locus. **If nothing is specifically mentioned, the variation of K will be taken from 0 to +° and the plot will be termed as root locus.**

Example 11.1 Find the root locus of the unity feedback system having $G(s)=\dfrac{K}{s+1}$ as shown in Fig. E11.1.

Solution The characteristic equation of the system is given by

$$1+G(s)H(s)=0 \qquad (1)$$

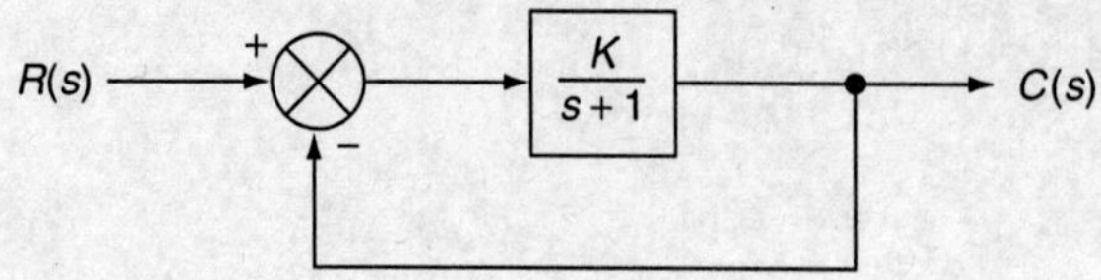

Fig. E11.1

We have $G(s)=\dfrac{K}{s+1}$ and $H(s)=1$

Equation (1) becomes

$$1+G(s)=0$$

i.e., $$1+\frac{K}{s+1}=0$$

i.e., $$s+K+1=0$$

$\therefore$ $$s=-(K+1)$$

For $K=0,\ s=-1$

For $K=\infty,\ s=-\infty$

If K is varied from 0 to ∞, root locus varies from -1 to $-\infty$ shown in Fig. E11.1(a)

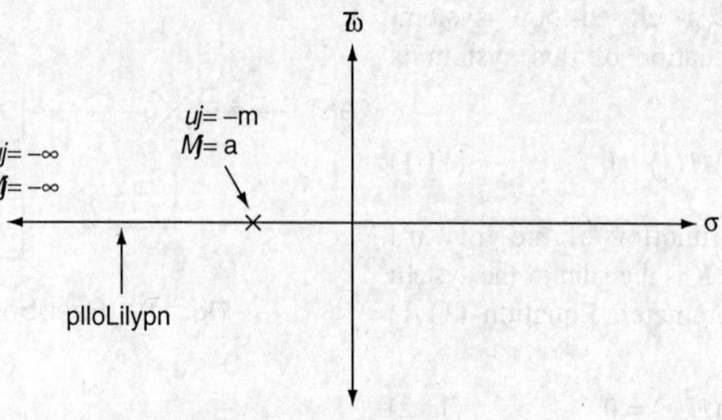

Fig. E11.1(a)

11.2 Relation between Poles and Zeros of OLTF with Poles of CLTF

To discuss the relation between open-loop transfer function (OLTF) poles and zeros with closed-loop transfer function (CLTF) poles, let us consider Fig. 11.1 and the OLTF be

$$G(s)H(s)=\frac{KN(s)}{D(s)} \tag{11.3}$$

where K = gain of OLTF, $N(s)$ is the numerator polynomial in s and $D(s)$ is the denominator polynomial in s.

(i) OLTF zeros can be obtained from

$$G(s)H(s) = 0$$

i.e.,
$$\frac{KN(s)}{D(s)} = 0$$

∴
$$N(s) = 0 \; [\because K \neq 0] \tag{11.4}$$

Therefore, OLTF zeros can be obtained from

$$N(s) = 0$$

(ii) OLTF poles can be obtained from

$$G(s)H(s) = \infty$$

i.e.,
$$K\frac{N(s)}{D(s)} = \infty$$

i.e.,
$$D(s) = 0 \tag{11.5}$$

Therefore, open-loop poles can be obtained from $D(s) = 0$.

(iii) CLTF poles can be obtained by using Eq. (11.1) i.e.,

$$1 + G(s)H(s) = 0$$

i.e.,
$$1 + K\frac{N(s)}{D(s)} = 0$$

i.e.,
$$D(s) + KN(s) = 0 \tag{11.6}$$

(a) If $K = 0$, $D(s) = 0$

It is same as Eq. (11.5) and gives the OLTF poles. Therefore, closed-loop poles are same as open-loop poles for $K = 0$.

(b) If $K = \infty$, $KN(s)$ should be larger than $D(s)$. Therefore, $KN(s) – 0$.

Since $K = \infty$, $N(s) = 0$.

This is same as Eq. (11.4) and gives open-loop zeros.

Therefore, closed-loop poles are same as open-loop zeros when $K = \infty$.

The root locus starts from open-loop pole at $K = 0$ and ends at open-loop zero at $K = \infty$.

Example 11.2 Determine where the root locus begins and ends for $G(s)H(s) = K(s+1)/(s+4)$.

Solution Since $G(s)H(s) = K(s+1)/(s+4)$, the open-loop transfer function pole is at $s = -4$ and zero is at $s = -1$.

Therefore, root locus begins at $s = -4$ at $K = 0$ and ends at $s = -1$ at $K = \infty$.

Example 11.3 Determine where the root locus begins and ends for $G(s)H(s)=\dfrac{K(s+3-j)\,(s+3+j)}{(s+2-j)\,(s+2+j)}$.

Solution Since $G(s)H(s)=[K(s+3-j)\,(s+3+j)]/[(s+2-j)\,(s+2+j)]$, the open-loop transfer function poles are at $s=-2\pm j$ and zeros are at $-3\pm j$. The root locus starts from $s=-2\pm j$ and ends at $s=-3\pm j$.

11.3 Angle and Magnitude Conditions

The CLTF poles can be obtained from Eq. (11.1), i.e.,

$$1+G(s)H(s)=0$$

i.e.,

$$G(s)H(s)=-1$$

i.e.,

$$G(s)H(s)=1\angle\pm180^\circ \tag{11.7}$$

From Eq. (11.7),

$$|G(s)H(s)|=1 \tag{11.8}$$

and

$$\angle G(s)H(s)=\pm(2q+1)180^\circ \tag{11.9}$$

Actually, $-1+j0=1\angle\pm180^\circ$, the point $-1+j0$ is a point on the negative real axis and it can be traced as magnitude 1 at an angle ± 180°, ± 540°, ± 900°, … ± (2*q* + 1) 180°.

Any point on the root locus must satisfy either the magnitude condition or the angle condition. Since *K* is not known in the function $G(s)H(s)$, it is very difficult to get $|G(s)H(s)|=1$ at any point in the *s*-plane and this condition is unsuitable to check the existence of any point on the root locus. If a point on the root locus is the function of $G(s)H(s)$, it helps to calculate the value of *K*.

Example 11.4 Check whether the points $-1+j$ and $-3+j$ lie on the root locus of a system, given by $G(s)H(s)=K/[(s+1)(s+2)]$. Use the angle condition.

Solution

$$G(s)H(s)\Big|_{s=-1+j}=\frac{K}{(s+1)\,(s+5)}\Bigg|_{s=-1+j}=\frac{K}{(-1+j+1)(-1+j+2)}$$

$$=\frac{K}{(1\angle90^\circ)\sqrt{2}\left(\dfrac{1}{\sqrt{2}}+j\dfrac{1}{\sqrt{2}}\right)}=\frac{K}{\sqrt{2}}\angle-135^\circ$$

Since the angle condition is not satisfied, the point $(-1+j)$ does no lie on the root locus.

$$G(s)H(s)\Big|_{s=-3+j}=\frac{K}{(s+1)\,(s+5)}\Bigg|_{s=-3+j}=\frac{K}{(-3+j+1)(-3+j2+2)}$$

$$= \frac{K}{(-2+j)(-1+j)} = \frac{K}{(2-3j-1)} = \frac{K}{(1-3j)} = \frac{K}{3.162\angle -71.56^\circ}$$

$$= \frac{K}{3.162}\angle 71.56^\circ$$

Since the angle condition is not satisfied, the point $(-3+j)$ does not lie on the root locus of the system.

Example 11.5 If $s = -0.85$ lies on the root locus of a system having $G(s)H(s) = \frac{K}{s(s+1)(s+3)}$, find the value of K using magnitude condition. Comment of the result.

Solution

$$|G(s)H(s)|_{s=-0.85} = \frac{K}{|-0.85||-0.85+1||-0.85+3|} = \frac{K}{0.274}$$

Since $$|G(s)H(s)|_{s=-0.85} = 1$$

i.e., $$K = 0.274$$

Comment

$$1 + G(s)\,H(s) = 0$$

i.e., $$1 + \frac{K}{s(s+1)(s+3)} = 0$$

i.e., $$s(s+1)(s+3) + K = 0$$

i.e., $$s^3 + 4s^2 + 3s + K = 0$$

This is a cubic equation. Using magnitude condition, the value of K has already been obtained for a particular point lying on the root locus of the system. Therefore, $s = -0.85$ is a root of the polynomial $s^3 + 4s^2 + 3s + K = 0$ and it is very easy to get the remaining two roots. Therefore, root locus also helps us to solve higher-order polynomials.

Example 11.6 Check whether the point $s = -3 + 5j$ lies on the root locus of a system having $G(s)H(s) = K/[(s+1)(s+5)]$. Determine the corresponding value of K.

Solution

$$G(s)H(s)\Big|_{s=-3+j5} = \frac{K}{(s+1)(s+5)}\Bigg|_{s=-3+j5} = \frac{K}{(-3+j5+1)(-3+j5+5)}$$

$$= \frac{K}{(-2+j5)(2+j5)} = \frac{K}{(5.3851\angle 111.8^\circ)(5.3851\angle 68.20^\circ)}$$

$$= \frac{K}{28.99\angle 180^\circ} \simeq \frac{K}{29}\angle -180^\circ$$

$$\angle G(s)H(s)\Big|_{s=-3+j5} = -180^\circ$$

Since the angle condition is satisfied, point $-3 + j5$ lies on the root locus of the given system. To find the value K, let us take

$$|G(s)H(s)|_{s=-3+j5} = 1$$

i.e.,
$$\frac{K}{29} = 1$$

i.e.,
$$K = 29$$

Example 11.7 Repeat Example 11.6 using graphical method.

Solution From Fig. E11.2,

$$\phi_{P1} = 90° + \tan^{-1}\frac{PE}{AE} = 90° + \tan^{-1}\left(\frac{2}{5}\right) = 90° + 21.8° = 111.8°$$

and
$$\phi_{P2} = \tan^{-1}\frac{BP}{CB} = \tan^{-1}\frac{5}{2} = 68.2° = \phi_P$$

$\therefore$
$$\phi_{P1} + \phi_{P2} = 111.8° + 68.2° = 180°$$

and
$$\phi_Z = 0°$$

$\therefore$
$$\phi_Z - \phi_p = \phi = 0° - 180° = -180°$$

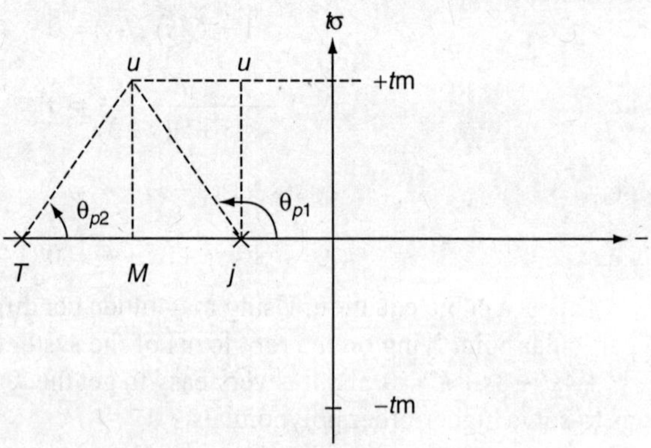

Fig. E11.2

Since the angle condition is satisfied, point $-3 + j5$ lies on the root locus of the given system. K can be determined using

$$K = \frac{\text{Product of phasor length of poles to the point}}{\text{Product of phasor lenghts of zeros to the point}}$$

From Fig. E11.2, we have

$$K = \frac{AP \times CP}{1} \quad \text{[since no zero is present, the denominator = 1]}$$

$$= \frac{\sqrt{2^2+5^2}\sqrt{2^2+5^2}}{1} = \frac{\sqrt{29}\sqrt{29}}{1} = 29$$

Note: If a point is at the pole, $K = 0$. If it is at zero, $K = \infty$.

11.4 General Methods for Construction of Root Locus

It is very difficult to obtain the value of K using an analytical method for higher-systems. The following steps are very useful and are usually applied to obtain the root locus.

Rule 1. Symmetry: Root locus is always symmetrical about the real axis (σ-axis). The roots are either real or complex conjugates or a combination of both. Therefore, the locus is symmetrical about the real axis of the s-plane.

Rule 2. The number of loci: Let the number of OLTF poles and OLTF zeros be n and m, respectively. If $n > m$, i.e., the number of OLTF poles is greater than number of OLTF zeros, the number of loci is n. Each locus will start from an OLTF pole and end on an OLTF zero. Here, $(n-m)$ loci will end at infinity and total loci are n.

Generally, $m > n$ is hardly found practically and more academic.

Example 11.8 Find the number of loci for the system having $G(s)H(s) = K(s+2)/[s(s+1)(s+4)]$.

Solution Here, $m = 1$ and $n = 3$. The number of loci ending at ∞ is $n - m = 3 - 1 = 2$. Three loci will start from $s = 0, -1$ and -4, one locus will end at -2 and the remaining two loci will end at ∞.

Rule 3. Real axis loci: Some of the loci will lie on the real axis. A point on the real axis will lie on the root locus if and only if the sum of the OLTF poles and zeros to the right of the point is odd.

Example 11.9 For a system having $G(s)H(s) = K(s+2)/[s(s+1)(s+4)]$, find on which sections of the real axis the root locus exists.

Solution The pole-zero plot of the system $G(s)H(s) = K(s+2)/[s(s+1)(s+4)]$ is shown in Fig. E11.3. The points P and Q lie on the root locus.

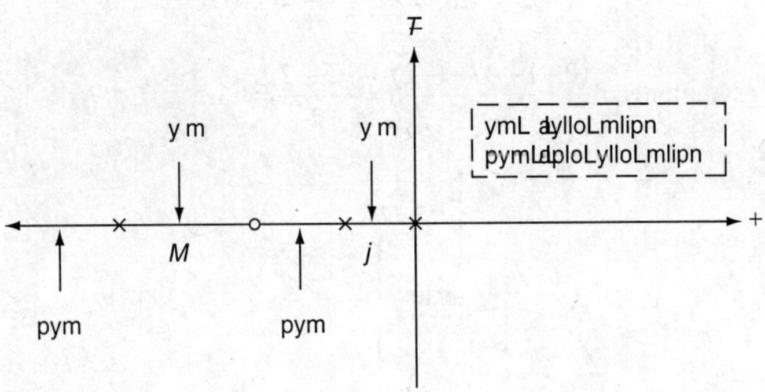

Fig. E11.3

Rule 4: Angle of asymptotes: Since the number of poles is greater than the number of zeros, i.e., $n > m$, $n - m$ branches will move to infinity and these branches move along an asymptote. An asymptote is defined as a line on which the root locus touches infinity.

The number of asymptotes = $n - m$.
The angle of asymptote is given by

$$\theta = \frac{2q+1}{n-m} \times 180° \text{ for } K > 0 \quad (11.10)$$

where $q = 0, 1, 2, ..., n - m - 1.$

and

$$\theta = \frac{2q}{n-m} \times 180° \text{ for } K < 0$$

where $q = 0, 1, 2, ..., n - m - 1$ (11.11)

Rule 5. Centre of asymptotes: Since only the angles of each asymptote are not sufficient, the location of asymptotes in s-plane has equal importance. The point where the asymptotes touch the real axis is called centroid (σ_c) and it is expressed by

$$\sigma_c = \frac{\sum \text{real parts of poles of } G(s)H(s) - \sum \text{real parts of zeros of } G(s)H(s)}{n-m}$$

Example 11.10 Calculate the angles of asymptotes and the centroid for the system having

$$G(s)H(s) = \frac{K(s+2)}{s(s+1)(s+4)} \text{ where } K > 0.$$

Solution Here, the number of poles (n) = 3 and number of zeros (m) = 1

The number of asymptotes = the number of branches approaching to infinity = $n - m = 2$.
Using Eq. (11.9),

For $q = 0, \theta_0 = \frac{1 \times 180°}{2} = 90°$

For $q = 1, \theta_1 = \frac{3 \times 180°}{2} = 270°$

$\therefore$ Centroid (σ_c) of asymptotes $= \frac{(0-1-4)-(-2)}{3-1} = \frac{-5+2}{2} = -1.5$

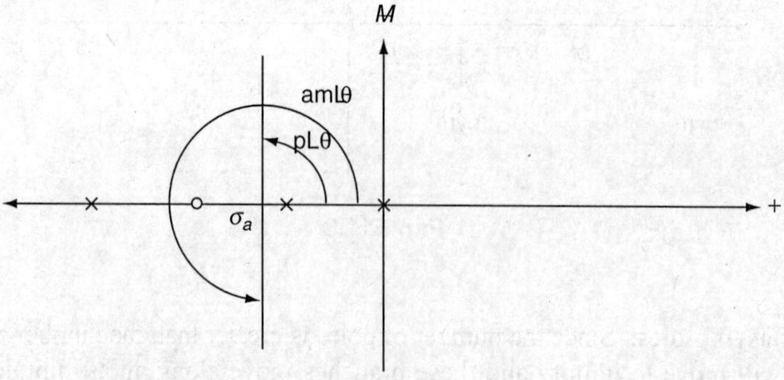

Fig. E11.4

Example 11.11 Calculate angles of asymptotes and the centroid for the system having

$$G(s)H(s) = \frac{K}{s(s+1-j)(s+1-j)} \quad \text{where } K > 0.$$

Solution Here,
$$n - m = 3 - 0 = 3,\ q = 0, 1, 2$$

$$\therefore \quad \theta_0 = \frac{(2\times 0+1)\times 180^\circ}{3} = 60^\circ, \theta_1 = \frac{(2\times 1+1)\times 180^\circ}{3} = 180^\circ, \theta_2 = \frac{(2\times 2+1)\times 180^\circ}{3} = 300^\circ,$$

and
$$\sigma_c = \frac{(0-1-1)-0}{3} = -\frac{2}{3} = -0.667$$

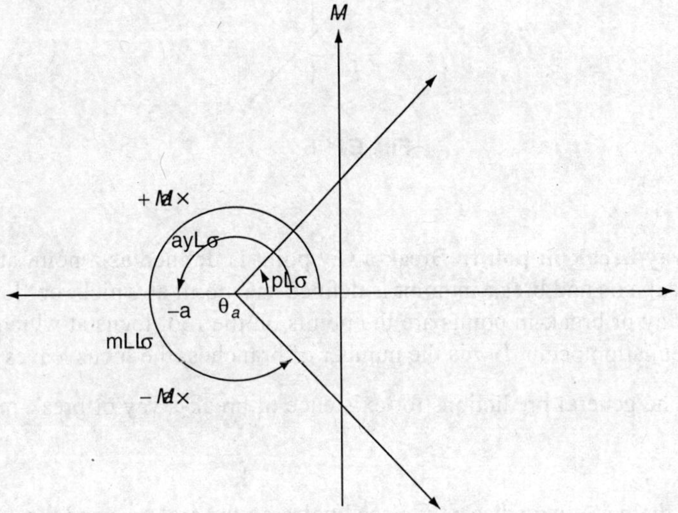

Fig. E11.5

Example 11.12 Calculate the angle of asymptotes and the centroid for the system having

$$G(s)H(s) = \frac{K(s+3)}{s(s+2)(s+4)(s+5)}.$$

Solution

$$n = 4 \text{ and } m = 1$$

$$\therefore \quad n - m = 4 - 1 = 3 \text{ Hence } q = 0,\ 1,\ 2$$

$$\therefore \quad \theta_0 = \frac{(2\times 0+1)\times 180^\circ}{3} = 60^\circ, \theta_1 = \frac{(2\times 1+1)\times 180^\circ}{3} = 180^\circ, \theta_2 = \frac{(2\times 2+1)\times 180^\circ}{3} = 300^\circ,$$

and
$$\sigma_c = \frac{(0-2-4-5)-(-3)}{3} = -\frac{11-3}{3} = -\frac{8}{3} = -2.667$$

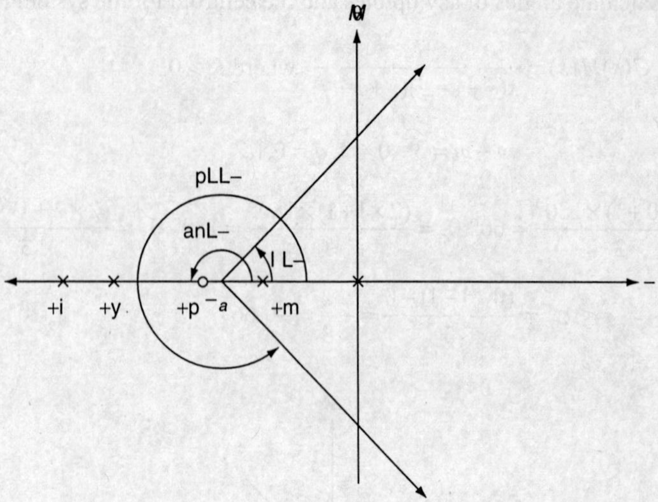

Fig. E11.6

Rule 6. Break-away/break-in point: Break-away point is defined as a point at which root locus comes out of the real axis and break-in point is defined as a point at which root locus enters the real axis. The break-away or break-in points are the points on the root locus at which multiple roots of the characteristic equation occur. If a is the number of branches, the locus leaves the real axis at an angle of $\pm\frac{180^\circ}{a}$. The general predictions for existence of break-away or break-in points are shown below.

Prediction 1 If there are two adjacently placed poles on the real axis and the real axis is a part of the root locus, one minimum break-away point exists in between adjacently placed poles as shown in Fig. 11.2.

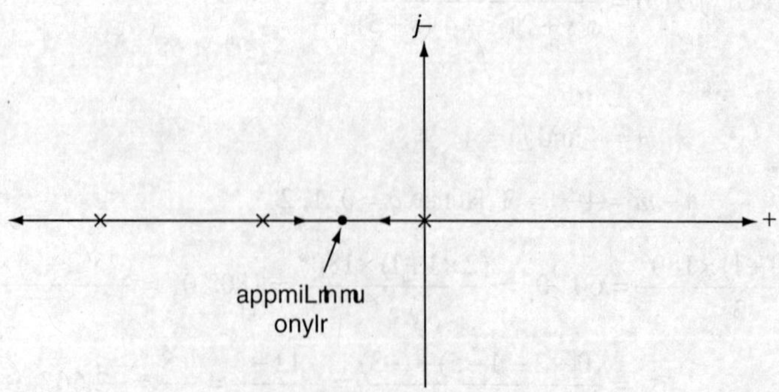

Fig. 11.2 Break-away point

Example 11.13 For a system having $G(s)H(s) = K/[s(s+1)(s+3)]$, determine the number of break-away points?

Solution The pole-zero plot of the system is shown is Fig. E11.7.

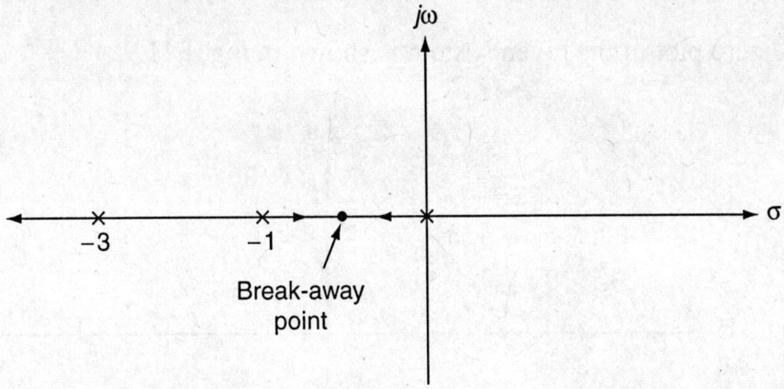

Fig. E11.7

In Fig. E11.7, $s = 0$ and $s = -1$ are the two adjacently placed poles and the real axis between $s = 0$ and $s = -1$ is a part of the root locus. Hence, one break-away point exists between $s = 0$ and $s = -1$.

Prediction 2 If there are two adjacently placed zeros on the real axis and the real axis is a part of the root locus, one minimum break-in point exists in between adjacently placed zeros.

Example 11.14 For the system having $G(s)H(s) = K(s+2)(s+4)/[s^2(s+5)]$, determine the number of break-in points.

Solution The pole-zero plot of the system is shown in Fig. E11.8.

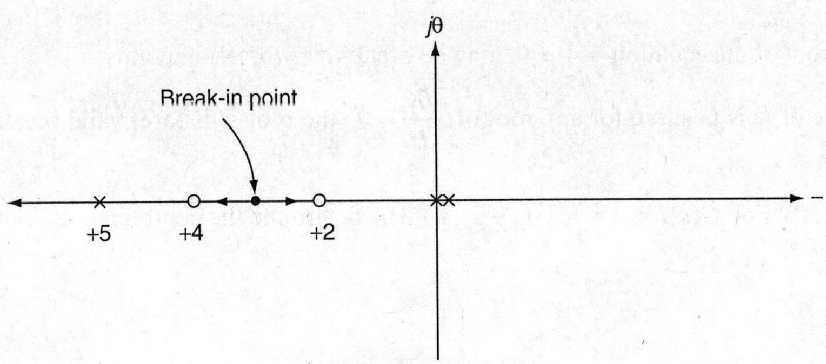

Fig. E11.8

In Fig. E11.8, the real axis between two adjacently placed zeros at $s = -2$ and $s = -4$ is a part of the root locus. Hence, one minimum break-in point exists between the above adjacently placed zeros.

Prediction 3 If there are no poles and zeros to the left of a zero on the real axis and the portion of the real axis to the left of zero is a part of the root locus, there exists minimum one break-in point to the left of that zero.

Example 11.15 For the system having $G(s)H(s) = \dfrac{K(s+6)}{(s+1)(s+3)}$, find the number of break-in points.

Solution The pole-zero plot of the given system is shown in Fig. E11.9.

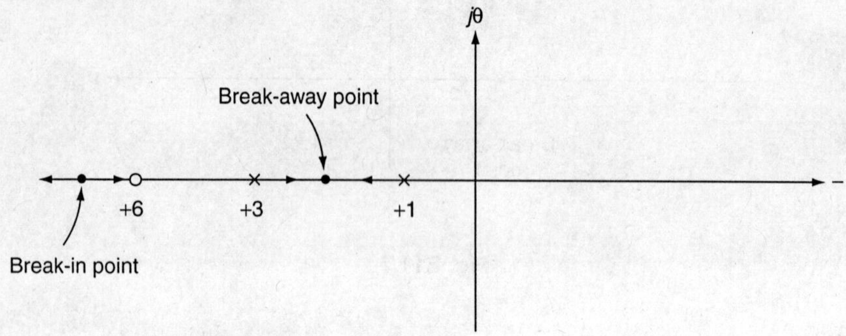

Fig. E11.9

To the left of the zero at $s=-6$, there are no poles and zeros and the portion of the real axis to the left of this zero is a part of the root locus. Hence, minimum one break-in point exists to the left of the zero.

Another break-away point exists between $s = 1$ and $s = -3$.

Steps to determine the break-away/break-in points

Step 1: Frame the characteristic equation $1+G(s)H(s)=0$ of the system.

Step 2: Write K in terms of s, i.e., $K=f(s)$.

Step 3: Derive $\dfrac{dK}{ds}$ and put $\dfrac{dK}{ds}=0$

Step 4: The root of the equation $\dfrac{dK}{ds}=0$ may be break-away/break-in points.

If the value of K is positive for any root of $\dfrac{dK}{ds}=0$, the root(s) is (are) valid break-away/break-in point(s).

Example 11.16 For $G(s)H(s) = K/[s(s+2)(s+3)]$, determine the coordinates of valid break-away/break-in point(s).

Solution

Step 1:
$$1+G(s)H(s)=0$$

i.e.,
$$1+\frac{K}{s(s+2)(s+3)}=0$$

i.e.,
$$s^3+5s^2+6s+K=0$$

Step 2: $K = -s^3 - 5s^2 - 6s$

Step 3: $\dfrac{dK}{ds} = -3s^2 - 10s - 6 = 0$

Step 4: $3s^2 + 10s + 6 = 0$

$\therefore$ roots are $= \dfrac{-10 \pm \sqrt{10^2 - 4\times3\times6}}{2\times3} = \dfrac{-10 \pm \sqrt{100-72}}{6} = \dfrac{-10\pm\sqrt{28}}{6} = \dfrac{-10\pm5.29}{6}$

$= -0.785, -2.55$

For $s = -0.785$, we have

$$K = -(-0.785)^3 - 5(-0.785)^2 - 6(-0.785) = 2.113$$

For $s = -2.55$, we have

$$K = -(-2.55)^3 - 5(-2.55)^2 - 6(-2.55) = -0.631$$

Thus for $s = -0.785$, K is positive and for $s = -2.55$, K is negative.
Therefore, $s = -0.785$ is a valid break-away point for the root locus.

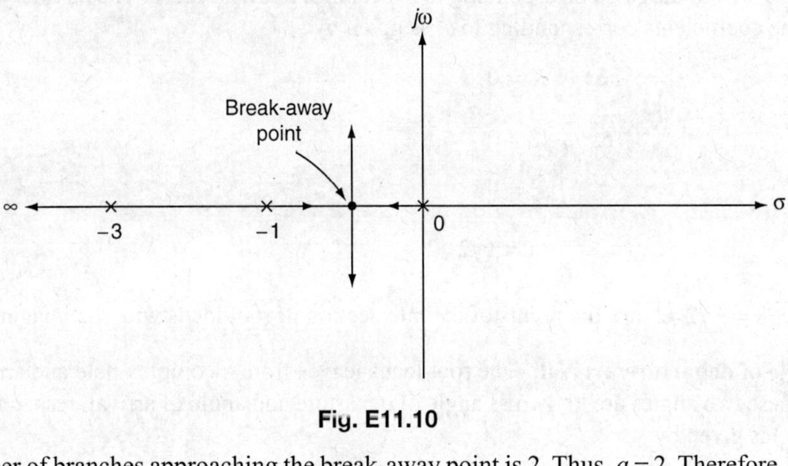

Fig. E11.10

The number of branches approaching the break-away point is 2. Thus, $a = 2$. Therefore, the loci leave the root locus at an angle of $\pm180°/2$, i.e., $\pm 90°$.

Rule 7. Intersection of root locus with $j\omega$-axis: To find the intersection of root locus with the imaginary axis, the following procedures are followed.

Step 1: Construct the characteristic equation $1 + G(s)H(s) = 0$.

Step 2: Develop Routh array in terms of K.

Step 3: Find K_{mar} that creates one of the roots of Routh array as a row of zeros.

Step 4: Frame auxiliary equation $A(s) = 0$ with the help of the coefficient of a row just above the row of zeros.

Step 5: The roots of the auxiliary equation $A(s) = 0$ for $K = K_{mar}$ give the intersection points of the root locus with the imaginary axis.

Example 11.17 For $G(s)H(s) = K/[s(s+2)(s+4)]$, find the point of the root locus with the $j\omega$-axis.

Solution The characteristic equation is given by

$$1 + G(s)H(s) = 0$$

i.e., $$1 + \frac{K}{s(s+2)(s+4)} = 0$$

i.e., $$s^3 + 6s^2 + 8s + K = 0$$

Routh array

$$\begin{array}{c|cc} s^3 & 1 & 8 \\ s^2 & 6 & K \\ s^1 & \dfrac{48-K}{6} & 0 \\ s^0 & K & \end{array}$$

Now, $K_{mar} = 48$ and the row corresponding to s^1 becomes a row of zeros. The auxiliary equation with the help of the coefficients corresponding to s^2 is given by

$$6s^2 + K = 0$$

i.e., $$6s^2 + 48 = 0$$

i.e., $$s^2 = -8$$

i.e., $$s = \pm j2\sqrt{2}$$

Therefore, $s = \pm j2\sqrt{2}$ are the points of the intersection of root locus with the imaginary axis.

Rule 8. Angle of departure/arrival: The root locus leaves from a complex pole and arrives at a complex zero. These two angles are known as angle of departure and angle of arrival, respectively. Angle of departure (θ_d) is given by

$$\theta_d = 180° + \arg[G(s)H(s)] \quad (11.11)$$

where arg $[G(s)H(s)]$ is the angle of $G(s)H(s)$ excluding the pole for whom the angle is to be calculated.

It is also possible to calculate arg $[G(s)H(s)]$ graphically. Similarly, the angle of arrival is given by

$$\theta_a = 180° - \arg[G(s)H(s)] \quad (11.12)$$

where arg $[G(s)H(s)]$ is the angle of $G(s)H(s)$ excluding the zero for whom the angle is to be calculated.

Example 11.18 Find the angle of departure for $G(s)H(s) = K(s+3)/[(s+2+j)(s+2-j)]$ where $K > 0$.

Solution

(i) Analytically:

To calculate θ_d analytically for $s = -2 + j$, the term $s + 2 - j$ should be excluded from $G(s)H(s)$.

$$\therefore\ G(s)H(s)\Big|_{s=-2+j} = \frac{K(-2+j+3)}{(-2+j+2+j)} = \frac{K(1+j)}{2j} = \frac{K\left(\frac{1}{\sqrt{2}}+j\frac{1}{\sqrt{2}}\right)}{\sqrt{2}\angle 90^\circ}$$

$$= \frac{K\angle 45^\circ}{\sqrt{2}\angle 90^\circ} = \frac{K}{\sqrt{2}}\angle -45^\circ$$

$$\therefore\ \theta_d = 180^\circ + \arg[G(s)H(s)] = 180^\circ - 45^\circ = 135^\circ$$

Since the poles are complex conjugates of each other, the angle of departure for the pole at $s = -2 - j$ will be -135°. The angle of departure is shown in Fig. E11.11.

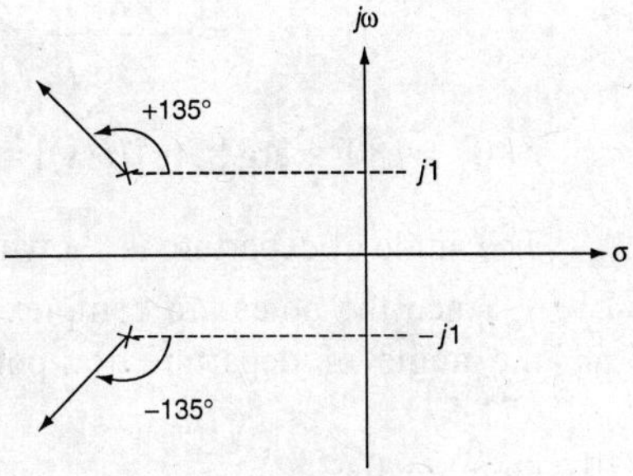

Fig. E11.11

(ii) Graphically:

The pole-zero plot of $G(s)H(s) = K(s+3)/$
$= [(s+2+j)(s+2-j)]$ is shown in Fig. E11.12.

$$\phi_Z = \tan^{-1}(1) = 45^\circ \text{and}\ \phi_P = 90^\circ$$

$$\therefore\quad \arg[G(s)H(s)] = \phi_Z - \phi_P = 45^\circ - 90^\circ = -45^\circ$$

$$\therefore\quad \phi_d = 180^\circ + \arg[G(s)H(s)] = 180^\circ - 45^\circ = 135^\circ$$

Therefore, the angle of departure at the pole $s = -2 + j$ is 135°. Similarly, the angle of departure at the pole $s = -2 - j$ is -135°.

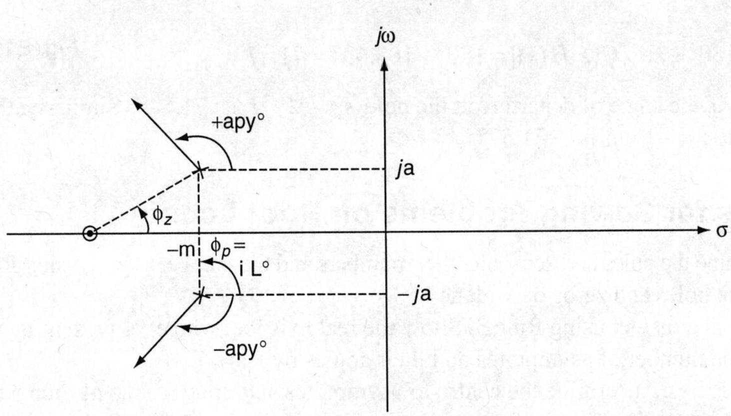

Fig. E11.12

Example 11.19 Find the angle of departure for

$$G(s)H(s) = \frac{K(s+1)}{s(s+2+j)(s+2-j)} \quad \text{where } K > 0.$$

Solution

(i) Analytically:

To calculate θ_d analytically for $s = -2 + j$, the term $s + 2 - j$ should be excluded from $G(s)H(s)$.

$$\therefore\ G(s)H(s)\big|_{s=-2+j} = \frac{K(-2+j+1)}{(-2+j)(-2+j+2+j)} = \frac{K(-1+j)}{(-2+j)(2j)}$$

$$= \frac{K\sqrt{2}\angle 135^\circ}{(2.236\angle 153.43^\circ)(2\angle 90^\circ)}$$

$$= \frac{K}{2.236\sqrt{2}}\angle -108.43^\circ$$

$$\therefore\ \theta_d = 180^\circ + \arg[G(s)H(s)] = 180^\circ - 108.43^\circ = 71.57^\circ$$

The angle of departure is shown in Fig. E11.13.

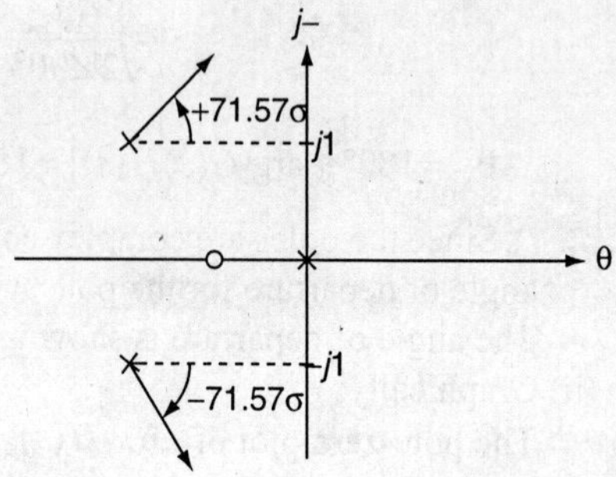

Fig. E11.13

Since the poles are complex conjugates of each other, the angle of departure for poles at $s = -2 - j$ will be -71.57°.

(ii) Graphically:

The pole-zero plot of $G(s)H(s) = \dfrac{K(s+1)}{s(s+2+j)(s+2-j)}$ is shown in Fig. E11.14.

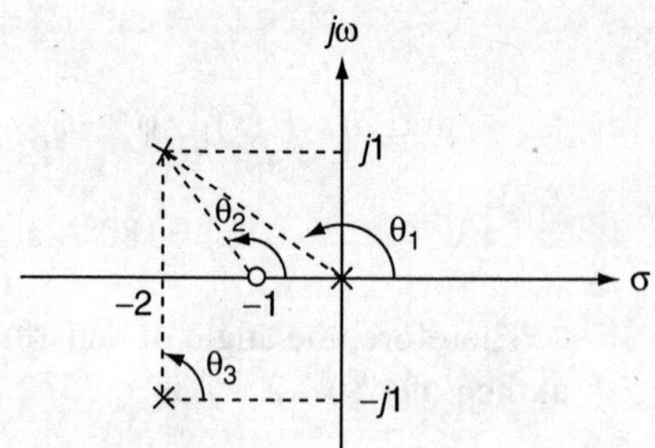

Fig. E11.14

$\therefore\quad \theta_1 = 90^\circ + \tan^{-1}(2) = 90^\circ + 63.43^\circ = 153.43^\circ,$

$\therefore\quad \theta_2 = 90^\circ + \tan^{-1}(1) = 90^\circ + 45^\circ = 135^\circ$ and $\theta_3 = 90^\circ$

$$\angle G(s)H(s)\big|_{s=-2+j} = \phi_z - \phi_p = 135^\circ - 153.43^\circ - 90^\circ = -108.43^\circ$$

$$\therefore\quad \theta_d = 180^\circ + \arg[G(s)H(s)] = 180^\circ - 108.43^\circ = 71.57^\circ.$$

Therefore, the angle of departure at the pole $s = -2 + j$ is 71.57°. Similarly, the angle of departure at pole $s = -2 - j$ is -71.57°.

11.5 Steps for Solving Problems on Root Locus

Step 1: Determine the poles and zeros and their numbers and the number of loci, using Rule 1 and Rule 2.
Step 2: Plot the poles and zeros on *s*-plane.
Step 3: Find real axis loci using Rule 3. Show the real axis loci wherever present by dark lines.
Step 4: Find the number of asymptotes and their angles by Rule 4.
Step 5: Using Rule 5, determine the centre of asymptotes and draw results of Step 4 and Step 5.
Step 6: Determine the break-away/break-in point if present using Rule 6 and mark the point only.
Step 7: Determine *jω* crossover using Rule 7 if the locus crosses the *jω* axis.

Step 8: Calculate the angle of departure or the angle of arrival due to complex poles or zeros, respectively, using Rule 8.

Step 9: Combine all the above steps to draw the root locus.

Step 10: Predict the stability and performance of the given system from its root locus.

Example 11.20 Sketch the root locus

$$G(s)H(s) = \frac{K}{s(s+1+j)(s+1-j)} (K > 0)$$

Comment on stability of the system.

Solution

Step 1: The number of poles = $n = 3$, the number of zeros = $m = 0$ and the number of loci = $n = 3$.

∴ The number of loci meeting at infinity $= n - m = 3 - 0 = 3$

Step 2: The pole-zero plot of $G(s)H(s)$ is shown in Fig. E11.15.

Step 3: Real axis loci are

(i) Present for $-1 < \sigma < 0$

(ii) Present for $-\infty < \sigma < -1$

These have been shown in Fig. E11.15(a)

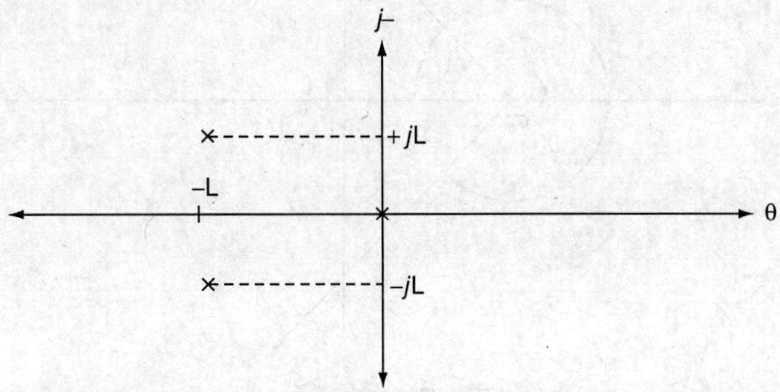

Fig. E11.15

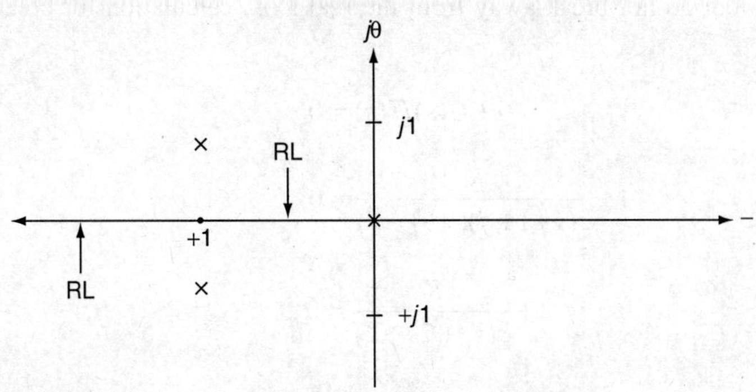

Fig. E11.15(a)

Step 4: Since $n-m=3$, the number of asymptotes is 3. Here, $q = 0, 1, 2$.

$$\therefore \qquad \theta_0 = \frac{(2\times 0+1)\times 180^\circ}{3} = 60^\circ,\ \theta_1 = \frac{(2\times 1+1)\times 180^\circ}{3} = 180^\circ \text{ and}$$

$$\theta_2 = \frac{(2\times 2\times 1)+180^\circ}{3} = 300^\circ$$

Step 5: Centroid (σ_c) $= \dfrac{(0-1-1)-0}{3} = \dfrac{-2}{3} = -0.667.$

Figure E11.15(b) shows the asymptotes.

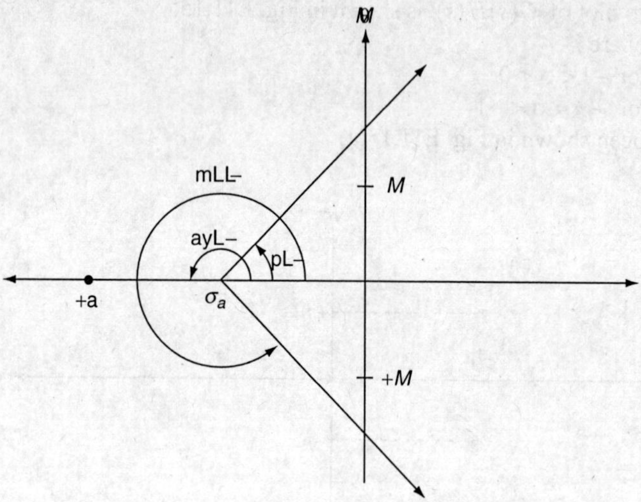

Fig. E11.15(b)

Step 6: Since the loci do not break-away from the real axis, calculation for break-away point is not required.

Step 7:
$$1+G(s)H(s)=0$$

i.e.,
$$1+\frac{K}{s(s+1+j)(s+1-j)}=0$$

i.e.,
$$1+\frac{K}{s(s^2+2s+2)}=0$$

i.e.,
$$s(s^2+2s+2)+K=0$$

i.e.,
$$s^3+2s^2+2s+K=0$$

Routh array

$$\begin{array}{c|cc} s^3 & 1 & 2 \\ s^2 & 2 & K \\ s^1 & \dfrac{4-K}{2} & 0 \\ s^0 & K & \end{array}$$

To get K_{mar}

$$\frac{4-K_{mar}}{2} = 0 \text{ i.e., } K_{mar} = 4$$

Here, $$A(s) = 2s^2 + K = 0$$

i.e., $$2s^2 = -K = -4$$

i.e., $$s^2 = -2$$

i.e., $$s = \pm j\sqrt{2}$$

$$G(s)H(s)\big|_{s=-1+j} = \frac{K}{(-1+j)(-1+j+1+j)} = \frac{K}{(-1+j)(2j)} = \frac{K}{2\sqrt{2}\left(\frac{-1}{\sqrt{2}} + j\frac{1}{\sqrt{2}}\right)(j)}$$

$$= \frac{K}{2\sqrt{2}\angle 135°\angle 90°} = \frac{K}{2\sqrt{2}\angle + 225°} = \frac{K}{\sqrt{2}}\angle - 225°$$

$\therefore$ $$\angle G(s)H(s) = -225°$$

$\therefore$ $$\theta_d = 180° + \arg[G(s)H(s)] = 180° - 225° = -45°.$$

By symmetry, the angle of departure from $s = -1 - j$ is $+45°$

Step 8: Figure E11.15(c) shows the complete root locus for

$$G(s)H(s) = \frac{K}{s(s+1+j)(s+1-j)} (K > 0)$$

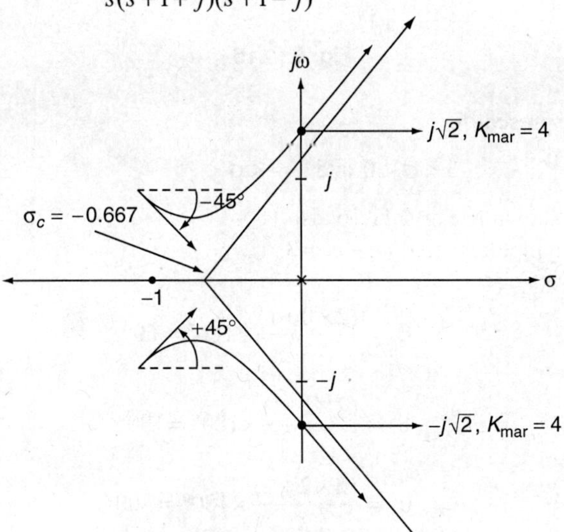

Fig. E11.15(c)

Step 9: Comment on stability: for $0 < K < 4$, all roots are in the left half of s-plane. Hence, the system is stable.

At $K = 4$, dominant roots are on the imaginary axis. Hence, the system is marginally stable, oscillating with 1.41 rad/sec. For $K > 4$, dominant roots lie on the right hand side of s-plane and hence system is unstable.

Example 11.21 Sketch the root locus for

$$G(s)H(s) = \frac{K}{s(s+3)(s+5)} (K > 0)$$

Comment on stability of the system.

Solution

Step 1: The number of poles $= n = 3$ and the number of zeros $= m = 0$.

$\therefore$ The number of loci meeting at infinity $= n - m = 3 - 0 = 3$.

Step 2: The pole-zero plot of $G(s)H(s) = \dfrac{K}{s(s+3)(s+5)}$ in shown in Fig. E11.16.

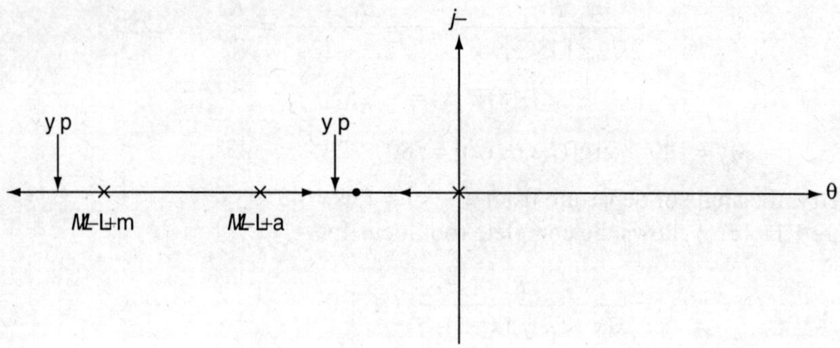

Fig. E11.16

Step 3: Real axis loci are present for

$$-3 < \sigma < 0 \text{ and } -\infty < \sigma < -5$$

These have been shown in Fig. E11.16 also.

Step 4: The number of asymptotes, $(a) = n - m = 3$.

$$\therefore \qquad \theta_0 = \frac{(2\times 0+1)}{3}\times 180° = 60°,$$

$$\theta_1 = \frac{(2\times 1+1)}{3}\times 180° = 180°$$

and

$$\theta_2 = \frac{(2\times 2+2)}{3}\times 180° = 300°$$

Step 5:

$$\text{Centroid } (\sigma_c) = \frac{(0-3-5)-0}{3} = -\frac{8}{3} = -2.667$$

Figure E11.16(a) depicts the asymptotes.

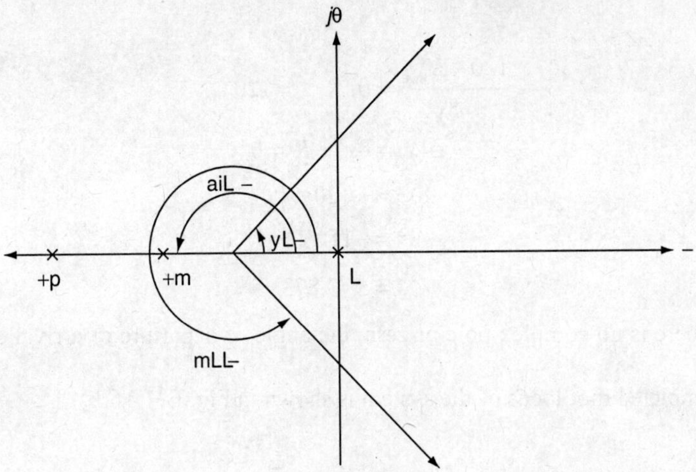

Fig. E11.16(a)

Step 6: $1+G(s)H(s)=0$

i.e., $$1+\frac{K}{s(s+3)(s+5)}=0$$

i.e., $$s(s+3)(s+5)+K=0$$

i.e., $$K = s^3-8s^2-15s$$

i.e., $$\frac{dK}{ds} = (-3s^2-16s-15)=0$$

i.e., $$3s^2+16s+15=0$$

i.e., $$s=\frac{-16\pm\sqrt{16^2-4\times3\times15}}{2\times3}=\frac{-16\pm 8.717}{6}$$

$\therefore$ s = either − 1.2138 or −4.1195. Now for $s=-1.2138$, $K=8.2088$ and for $s=-4.1195$, $K=-279.64$. Therefore, K is positive for $s=-1.2138$. At $s=-1.2138$, break-away point exits.

Step 7: $1+G(s)H(s)=0$

i.e., $$1+\frac{K}{s(s+3)(s+5)}=0$$

i.e., $$s^3+8s^2+15s+K=0$$

Routh array is shown below.

s^3	1	15
s^2	8	K
s^1	$\frac{120-K}{8}$	0
s^0	K	

Putting $$\frac{120-K_{mar}}{8}=0,\ K_{mar}=120$$

$\therefore$ $$A(s)=8s^2+K=0$$

i.e., $$8s^2=-120$$

i.e., $$s^2=-15$$

i.e., $$s=\pm j3.873$$

Step 8: Since there is no complex pole or zero, the angle of departure or arrival does not come into picture.

Step 9: The completed root locus of the system is shown in Fig. E11.16(b)

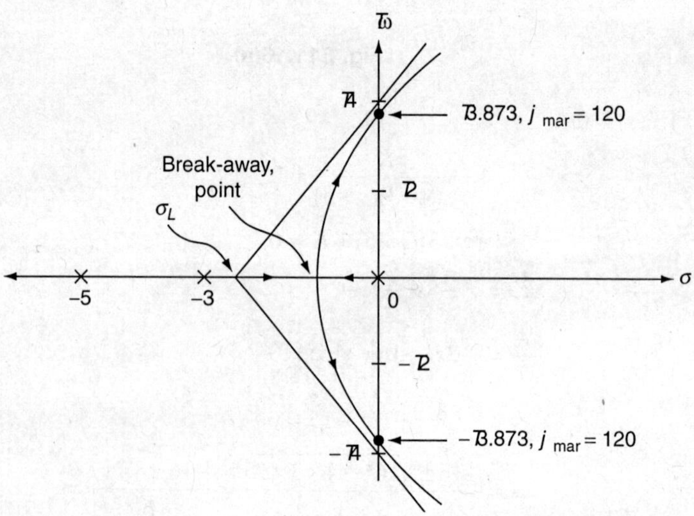

Fig. E11.16(b)

Step 10: Comment on stability: For $0<K<120$, all roots are in the left half of s-plane. System is absolutely stable.

At $K = 120$, dominant roots are on the imaginary axis. Hence, the system is marginally stable, oscillating with 3.873 rad/sec.

For $K>120$, dominant roots lie on the right hand side of s-plane and hence the system is unstable.

11.6 Graphical Determination of "K" for Specified Damping Ratio

The dominant pair of roots control the damping ratio "ξ" in higher-order systems. Figure 11.3 shows the dominant roots $-\xi\omega_n \pm j\omega_n\sqrt{1-\xi^2}$, which are complex conjugate to each other. From Fig. 11.3, the locations of the two roots are in the left half of s-plane. For the root $-\xi\omega_n + j\omega_n\sqrt{1-\xi^2}$ the phasor is drawn from such point to origin and it subtends the angle θ with respect to negative real axis.

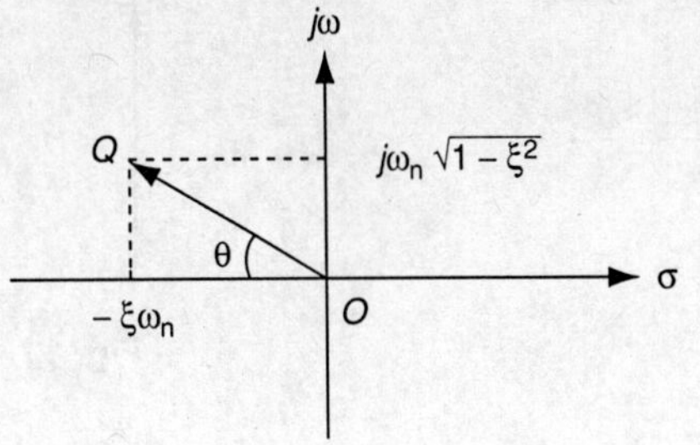

Fig. 11.3 Location of $-\xi\omega_n + j\omega_n\sqrt{1-\xi^2}$

From Fig. 11.3,

$$\cos\theta = \frac{\xi\omega_n}{OQ}$$

where $$OQ = \sqrt{(-\xi\omega_n)^2 + (\omega_n\sqrt{1-\xi^2})^2} = \omega_n$$

$\therefore$ $$\cos\theta = \frac{\xi\omega_n}{\omega_n} = \xi$$

The following steps are used to determine the value of system gain "K" by using root locus for designing a system with certain damping ratio "ξ" where "K" is the system gain for corresponding "ξ".

Step 1: The root locus is drawn to the scale on graph using the same scale for X and Y-axes.

Step 2: The value of $\theta = \cos^{-1}\xi$ is calculated.

Step 3: A line of an angle θ with negative real axis is drawn where "θ" is measured in the clockwise direction.

Step 4: The intersection point of this line with root locus is determined.

Step 5: The gain "K" is obtained using magnitude condition for this point.

11.7 Root Locus of Systems with Dead Time

The controller in some systems is put at a distance from the actual system and hence some time is consumed for changing the parameters of the system as per the controller action. Nothing is happened during the period, which is taken for controller action. This time is known as dead time. Measuring elements may be placed at some distance in some systems and some time is elapsed till measuring element measures the changes taken place in the system parameters. The time delay, which is present in measuring, in controller action or in actuation operation is known as transportation lag or dead time, which is present in most of the process control systems. The transfer function of the transportation lag or dead time is expressed by e^{-sT} where T indicates the amount of time delay. Figure 11.4 shows a system with a dead time. The effect of dead time on root locus will be discussed.

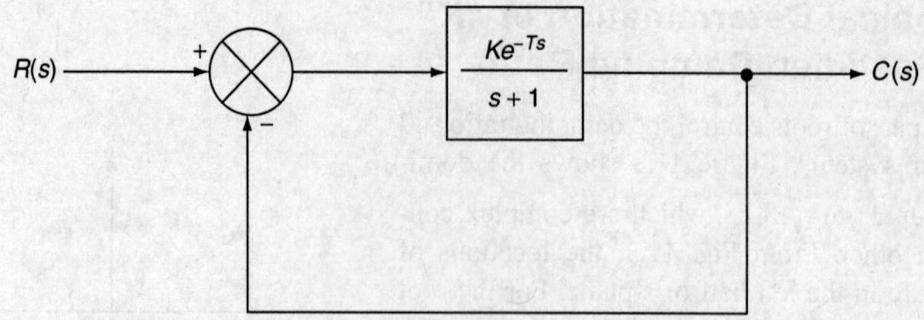

Fig. 11.4 System with dead time

The characteristic equation of the system is expressed by

$$1+\frac{Ke^{-sT}}{s+1}=0$$

i.e.,

$$\frac{Ke^{-Ts}}{s+1}=-1$$

From angle condition,

$$\text{Arg.}\left[\frac{Ke^{-Ts}}{s+1}\right]=\pm(2q+1)\times 180^\circ \quad \text{where } q=0, 1, 2,\ldots$$

i.e.,

$$\angle e^{-Ts}-\angle s+1=\pm(2q+1)\times 180^\circ. \tag{11.12}$$

Here,

$$\angle e^{-Ts}=\angle e^{-T(\sigma+j\omega)}=\angle(e^{-\sigma T}e^{-j\omega T})$$

Since $e^{-\sigma T}$ is real constant and its angle is zero.

$$\therefore \quad e^{-j\omega T}=\cos\omega T-j\sin\omega T$$

$$\therefore \quad \angle e^{-j\omega T}=\tan^{-1}\left[\frac{-\sin\omega T}{\cos\omega T}\right]=\tan^{-1}(-\tan^{-1}\omega T)$$

$$=-\omega T \text{ radians}$$

$$=-57.3\,\omega T \text{ degrees.} \tag{11.13}$$

Using Eq. (11.13), we have from Eq. (11.12)

$$-57.3^\circ\times\omega T-\angle(s+1)=\pm(2q+1)\times 180^\circ$$

$$\therefore \quad \angle(s+1)=\pm(2q+1)\times 180^\circ-57.3^\circ\times\omega T \tag{11.14}$$

The points on the s-plane, which satisfy Eq. (11.14), will lie on the root locus of the system. There is separate set of points in s-plane for each value of "q". Hence, there is infinite number of root locus branches. For $q = 0$, Eq. (11.14) becomes

$$\angle(s+1) = \pm 180° - 57.3° \times \omega T \qquad (11.15)$$

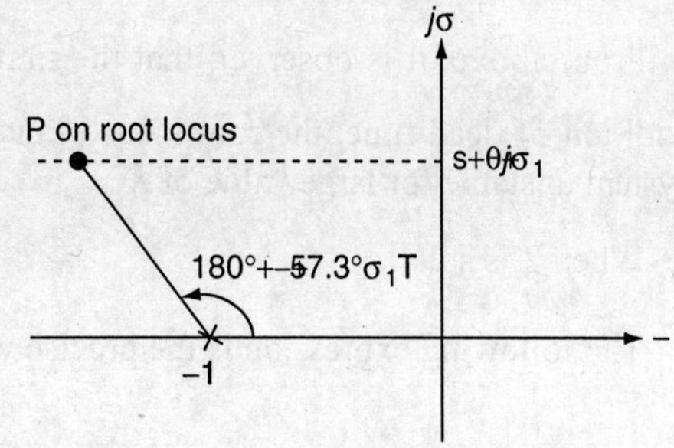

Fig. 11.5 Another point on root locus

The starting point of the root locus is $s = -1$, which is open-loop pole. For all values of $s < -1$, i.e., $s = -2, -3, -4, \cdots, -\infty$, $\angle(s+1)$ is 180°. For real value of "s", $\omega = 0$ and hence contribution by dead-time element is 0°. Eq. (11.15) is satisfied. The real axis starts from $s = -1$ to $s = -\infty$ is the part of the root locus.

Let us assume $\omega = \omega_1$ to another point on root locus. The angle due to dead element becomes $57.3° \times \omega_1 T$. A line is drawn at an angle $(180° - 57.3°\omega_1 T)$ at point -1. The point at which this line intersects the horizontal line of $s = j\omega_1$ is the point on root locus as shown in Fig. 11.5.

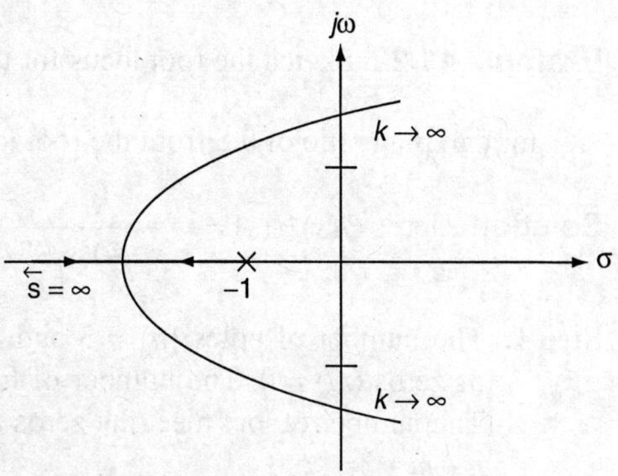

Fig. 11.6 Root locus for $q = 0$

Here, $s = -\infty$ also acts as a pole of the system. At $K = 0$, two branches starts to from $s = -1$ and $s = -\infty$. For $K \to \infty$, branches breakaway give complex conjugate roots, which travel towards right half of s-plane as shown in Fig. 11.6.

The magnitude condition for such system is given by

$$\left|\frac{Ke^{-Ts}}{s+1}\right| = 1$$

i.e.,

$$\frac{|K||e^{-Ts}|}{|s+1|} = 1$$

i.e.,

$$\frac{|K||e^{-\sigma T}|}{|s+1|} = 1 \quad \left[\because |e^{-j\omega T}| = 1\right]$$

i.e.,

$$\frac{|K|e^{-\sigma T}}{|s+1|} = 1 \quad \left[\because |e^{-\sigma T}| = e^{-\sigma T}\right]$$

i.e.,

$$|s+1| = |K|e^{-\sigma T} \qquad (11.16)$$

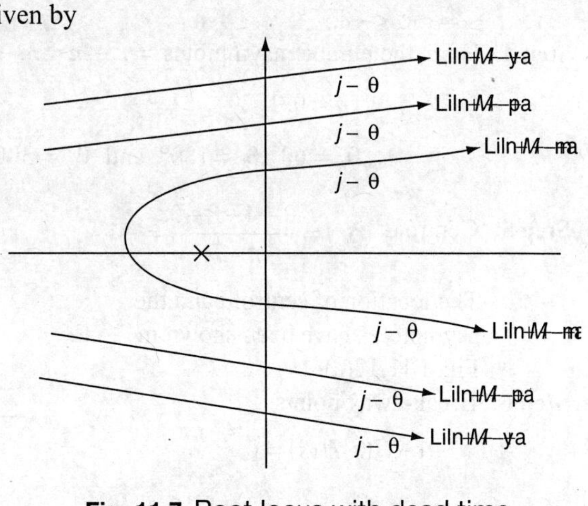

Fig. 11.7 Root locus with dead time

Figure 11.7 shows the root locus of such a system for $T = 1$ sec. and for various values of K and q. The root locus corresponding to $q = 0$ is most dominating in practice and hence taken into account for design viewpoint. Other branches are neglected.

From above, it is observed that the first-order system such as $\frac{K}{s+1}$ is always stable. Due to the addition of dead time, the root locus branches move to the right half of s-plane and hence make the system unstable for large value of K.

When T is small, $$e^{-Ts} \approx 1 - Ts \approx \frac{1}{1+Ts} \tag{11.17}$$

The following expression is the precise way to approximate the dead time.

$$e^{-Ts} \approx \frac{2-Ts}{2+Ts} \tag{11.18}$$

Example 11.22 Sketch the root locus for the system having $G(s)H(s) = \frac{K}{s(s+4)(s+8)}$. Determine "$K$" for damping ratio of 0.5 from the root locus. In addition, comment on the stability of the system.

Solution Here, $G(s)H(s) = \frac{K}{s(s+4)(s+8)}$.

Step 1: The number of poles (n) = 3 and the number of zeros (m) = 0. The number of loci = n = 3. The number of loci meet the zeros at infinity = $n - m = 3$.

Step 2: Figure E11.17 shows the pole-zero plot of the system.

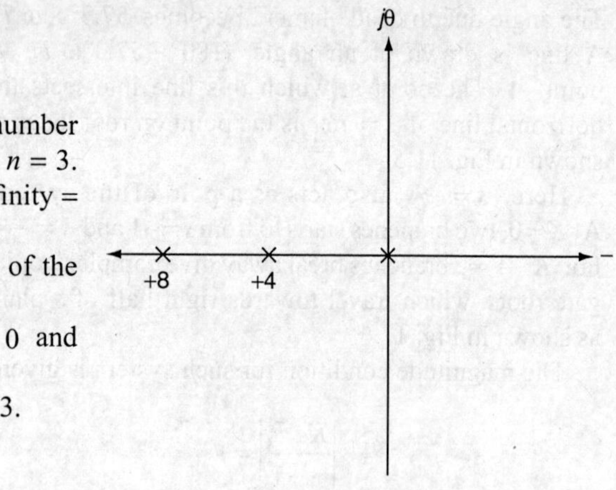

Fig. E11.17

Step 3: Real axis loci are present for $-4 < \sigma < 0$ and $-\infty < \sigma < -8$.

Step 4: Here, the number asymptotes $= n - m = 3$.

$$\therefore q = 0, 1, 2$$

$$\therefore \theta_0 = 60°, \theta_1 = 180° \text{ and } \theta_2 = 300°.$$

Step 5: Centroid $(\sigma_c) = \frac{0-4-8-0}{n-m} = -4$

The location of centroid and the asymptotes have been shown in Fig. E11.17(a).

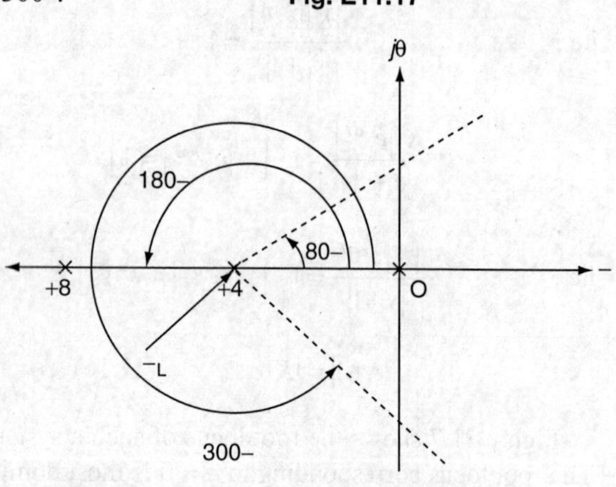

Fig. E11.7(a)

Step 6: Break-away points:

$$1 + G(s)H(s) = 0$$

i.e., $$1 + \frac{K}{s(s+4)(s+8)} = 0$$

i.e., $$K = -s(s+4)(s+8)$$

$$= -(s^3 + 12s^2 + 32s)$$

ie., $$\frac{dK}{ds} = -(3s^2 + 24s + 32) = 0$$

ie., $$3s^2 + 24s + 32 = 0$$

ie., $$s = \frac{-24 \pm \sqrt{24^2 - 4 \times 3 \times 32}}{2 \times 3}$$

$$= \frac{-24 \pm 13.86}{6}$$

$$= -6.31 \text{ or } -1.69.$$

For $s = -1.69,\ K = -(-1.69)^3 - 12(-1.69)^2 - 32(-1.69) = 24.63$

For $s = -6.31,\ K = -(-6.31)^3 - 12(-6.31)^2 - 32(-6.31) = -24.63$

Therefore, $s = -1.69$ is the valid break-away point.

Step 7: $j\omega$-axis cross-over:

$$1 + G(s)H(s) = 0$$

i.e., $$s^3 + 12s^2 + 32s + K = 0.$$

Routh array is shown below.

$$\begin{array}{c|cc} s^3 & 1 & 32 \\ s^2 & 12 & K \\ s^1 & \frac{384 - K}{12} & 0 \\ s^0 & K & \end{array}$$

To get K_{mar},

$$\frac{384 - K_{mar}}{12} = 0.$$

i.e., $$K_{mar} = 384.$$

Here, auxiliary equation

$$12s^2 + K = 0$$

i.e., $$s^2 = -\frac{K}{12} = -\frac{384}{12} = -32.$$

$\therefore$ $$s = \pm j5.65$$

Step 8: Since the roots are not complex conjugate, the angle of departure calculation is not required. Figure E11.17(b) shows the complete root locus of the system.

To determine K for $\xi = 0.5$, draw a line from origin at $\theta = \cos^{-1}\xi = 60°$ as shown in Fig. E11.17(b). The point of intersection (P) of the root locus by the line is found out. The co-ordinate (approximate) of the point P is $-1.31 + j2.34$. The magnitude condition is used to calculate the value of "K".

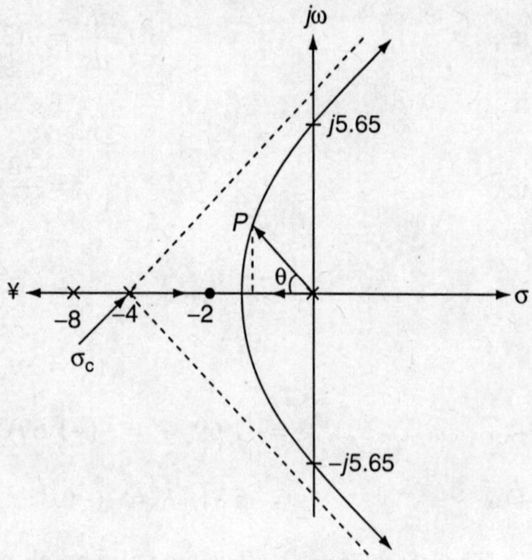

Fig. E11.17(b)

$$\therefore \quad |G(s)H(s)| = 1$$

i.e.,
$$\frac{K}{|s||s+4||s+8|} = 1.$$

i.e.,
$$\frac{K}{|-1.31 + j2.34||-1.31 + j2.34 + 4||-1.31 + j2.34 + 8|} = 1$$

i.e.,
$$\frac{K}{|-1.31 + j2.34||2.69 + j2.34||6.69 + j2.34|} = 1$$

i.e.,
$$\frac{K}{2.68 \times 3.56 \times 7.08} = 1$$

$$\therefore \quad K = 16.62.$$

The required value of K is 16.62.

For $0 < K < 384$, all roots are the left halves of s-plane. System is absolutely stable. At $K = 384$, dominant roots are on the imaginary axis. Hence, system is marginally stable, oscillating with 5.65 rad/sec. For $K > 384$, dominant roots lie on the right-hand side of s-plane and hence the system is unstable.

11.8 Cancellation of Poles of G(S) and Zeros of H(S)

Let us take
$$G(s) = \frac{K}{s(s+2)(s+4)}$$

and
$$H(s) = s + 2$$

$$\therefore \quad \frac{C(s)}{R(s)} = \frac{G(s)}{1 + G(s)H(s)} = \frac{\dfrac{K}{s(s+2)(s+4)}}{1 + \dfrac{K}{s(s+2)(s+4)} \times (s+2)} = \frac{K}{(s+2)[s(s+4) + K]}$$

In root locus, $G(s)H(s)$ is considered.

Here,
$$G(s)H(s) = \frac{K(s+2)}{s(s+2)(s+4)} = \frac{K}{s(s+4)}$$

The closed-loop pole is given by

$$1+G(s)H(s)=0$$

i.e.,

$$s(s+4)+K=0$$

The two terms (s + 2) and (s + 2) are cancelled. Again (s + 2) is a closed-loop pole and hence the equation is wrong. In the above case, the cancelled pole of $G(s)$ must be added as a closed-loop pole. If zero of $G(s)$ and pole of $H(s)$ are cancelled, no correction is required.

Example 11.23 Sketch the root locus for a system having

$$G(s)=\frac{K}{(s+1)} \text{ and } H(s)=\frac{(s+1)}{(s^2+4s+5)}$$

Solution

Here,

$$G(s)=\frac{K}{(s+1)}$$

And,

$$H(s)=\frac{(s+1)}{s^2+4s+5}$$

In this problem cancellation of pole of $G(s)$ at $s=-1$ and zero of $H(s)$ at $s=-1$ takes place. Therefore, as a closed loop at $s=-1$, pole must be added.

$$G(s)H(s)=\frac{K}{(s+1)(s^2+4s+5)}=\frac{K}{(s+1)(s+2+j)(s+2-j)}$$

Step 1: Here, the number of poles (n) = 3 and the number of zero (m) = 0. The three loci will end at ∞.
Step 2: The pole-zero plot is shown in Fig. E11.18.
Step 3: Real axis loci are present for $-\infty<\sigma<-1$ due to the presence of an odd number of poles to the right of any point in the zone $-\infty<s<-1$. This is shown in Fig. E11.18.
Step 4: The number of asymptotes $=n-m=3$. Here, $q=0, 1, 2$.

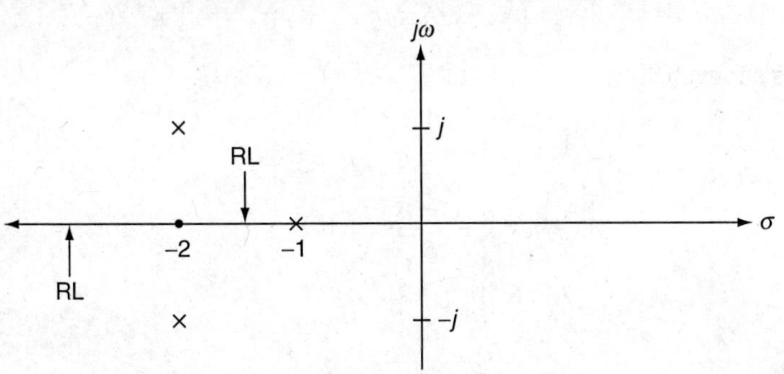

Fig. E11.18

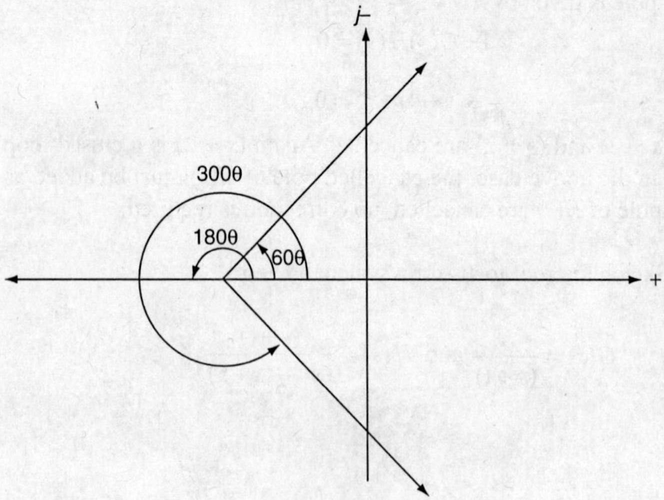

Fig. E11.18(a)

$$\therefore \qquad \theta_0 = \frac{(2\times 0+1)}{3}\times 180° = 60°, \theta_1 = \frac{(2\times 1+1)}{3}\times 180° = 180°$$

and
$$\theta_2 = \frac{(2\times 2+1)}{3}\times 180° = 300°$$

Step 5:

The asymptotes are shown in Fig. E11.18(a)

Step 6: No break-away point is needed because real axis loci present at ∞.

Step 7:
$$1+G(s)H(s) = 0$$

i.e.,
$$1+\frac{K}{(s+1)(s^2+4s+5)} = 0$$

i.e.,
$$(s+1)(s^2+4s+5)+K = 0$$

i.e.,
$$s^3+5s^2+9s+(5+K) = 0$$

Routh array is shown below.

$$\begin{array}{c|cc} s^3 & 1 & 9 \\ s^2 & 5 & K+5 \\ s^1 & \dfrac{45-K}{5} & 0 \\ s^0 & K+5 & \end{array}$$

To get K_{mar},

$$\frac{45-K_{mar}}{5} = 0$$

i.e., $$K_{mar} = 45$$

Here, the auxiliary equation

$$5s^2 + (K_{mar} + 5) = 0$$

i.e., $$5s^2 = -(K_{mar} + 5) = -50$$

i.e., $$s^2 = -10$$

i.e., $$s = \pm j3.16$$

Step 8: Angle departure from pole $s = -2 + j$.

$$G(s)H(s)|_{s=-2+j} = \left.\frac{K}{(s+1)(s+2+j)}\right|_{s=-2+j} = \frac{K}{(-2+j+1)(-2+j+2+j)}$$

$$= \frac{K}{2j(-1+j)} = \frac{K}{2\sqrt{2}j\left(-\frac{1}{\sqrt{2}} + \frac{1}{\sqrt{2}}\right)} = \frac{K}{2\sqrt{2}j\angle 90^\circ \angle 135^\circ}$$

$$= \frac{K}{2\sqrt{2}}\angle -225^\circ$$

$$\therefore \quad \phi_d = 180^\circ + \arg[G(s)H(s)] = 180^\circ - 225^\circ = -45^\circ$$

Similarly the angle of departure from the pole $s = -2 - j$ is +45°.
The completed root locus is shown in Fig. E11.18(b)

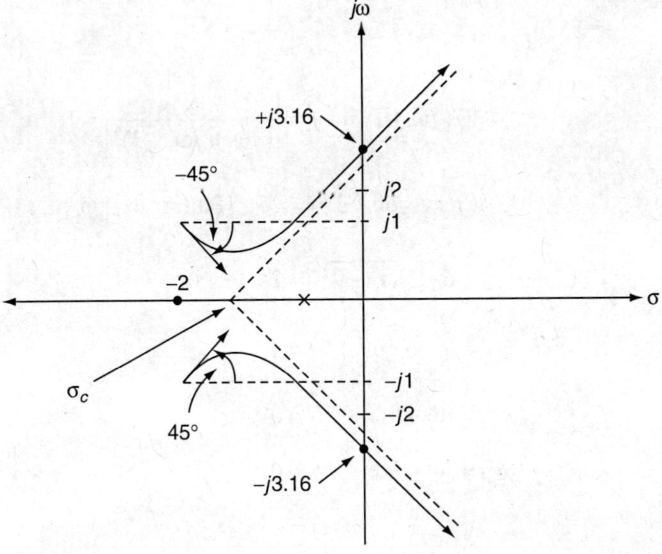

Fig. E11.18(b)

11.9 Value of Gain Margin

The ratio of K at imaginary crossover to the design value of K of the system is called the value of gain margin (GM) for the root locus of the system.

$$\text{Gain margin (GM)} = \frac{\text{Value of } K \text{ at imaginary corssover}}{\text{Design value of } K}$$

If the root locus does not cross the $j\omega$ axis, the gain is ∞. GM indicates the maximum gain can be multiplied for the system to be just on the instability.

Example 11.22 If $K_{design} = 1.5$ in Example 11.21, find gain margin (GM).

Solution

Here, $$GM = 3/1.5 = 2.$$

11.10 Phase Margin

Phase margin (PM) can be determined for a given value of K as follows:

- Calculate ω, for which $|G(j\omega)H(j\omega)| = 1$ for the given value of K.
- Calculate arg $[G(j\omega)H(j\omega)]$.
- Phase margin = 180° + arg $[G(j\omega)H(j\omega)]$.

Example 11.23 Find phase margin for the system having $G(s)H(s) = K/[s(s+3)^2]$ having $K_{design} = 25$.

Solution

$$|G(j\omega_1)H(j\omega_1)| = \left|\frac{K}{(j\omega_1)(j\omega_1+3)^2}\right| = 1$$

i.e., $$|(j\omega_1)(j\omega_1+3)^2| = 25 \quad [\text{Putting } \omega = \omega_1]$$

i.e., $$\omega_1\left(\sqrt{\omega_1^2+9}\right)^2 = 25$$

i.e., $$\omega_1\left(\omega_1^2+9\right) = 25$$

i.e., $$\omega_1^3 + 9\omega_1 - 25 = 0$$

Let $$f(\omega_1) = \omega_1^3 + 9\omega_1 - 25 = 0$$

∴ $$f(1) = -15$$

∴ $$f(2) = 8 + 18 - 25 = 1$$

Since there is change of sign, root lies between $\omega_1 = 1$ and $\omega_1 = 2$

Let $$\omega_1 = \frac{1+2}{2} = 1.5$$

$\therefore$ $$f(1.5) = 3.375 + 13.5 - 25 = -8.125$$

Let $$\omega_1 = \frac{1.5+2}{2} = 1.875$$

$\therefore$ $$f(1.875) = 6.592 + 16.875 = -1.532$$

Let $$\omega_1 = \frac{2+1.875}{2} = 1.9375$$

$\therefore$ $$f(1.9375) = 7.273 + 17.4375 - 25 = -0.2895$$

$\therefore$ $$\omega_1 = 1.9375$$

$\therefore$ $$G(j\omega_1)H(j\omega_1) = \frac{25}{(j1.9375)(j1.9375+3)} = \frac{25\angle 0^\circ}{(1.9375\angle 90^\circ)(3.571\angle 32.85^\circ)}$$

$$= 3.613\angle -122.85^\circ$$

$\therefore$ $$\arg G(j\omega_1)H(j\omega_1) = -122.85^\circ$$

$\therefore$ $$\text{Phase margin} = 180^\circ + \arg\ G(j\omega_1)H(j\omega_1) = 180^\circ - 122.85^\circ = 57.15^\circ$$

Therefore, for given $K_{\text{design}} = 25$, a phase of 57.15° can be added till the system reaches instability.

11.11 Root Locus with Positive Feedback

To sketch the root locus of a system with positive feedback, the following changes must be taken into account.

(i) The characteristic equation is

$$1 - G(j\omega)H(j\omega) = 0$$

i.e., $$1 - \frac{KN(s)}{D(s)} = 0 \left[\text{where } G(s)H(s) = K\frac{N(s)}{D(s)} \right]$$

i.e., $$D(s) - KN(s) = 0$$

(ii) The magnitude criterion remains unaltered.

(iii) The angle criterion is given by

$$\arg[G(s)H(s)] = 0^\circ$$

It is multiples of 360° instead of 180°.

(iv) Real axis loci will be present if the number of poles/zeros on right is even.

(v) The angle of asymptote $= \dfrac{q}{n-m} \times 360^\circ, q = 0,1,2,\ldots,n-m-1$

(vi) The angle of departure $(\phi_d) = 0 + \arg[G(j\omega_1)H(j\omega_1)]$ and angle of arrival $(\phi_a) = 0 - \arg[G(j\omega_1)H(j\omega_1)]$. Other rules are similar to the case of negative feedback.

Example 11.24 A positive feedback system has open-loop transfer function $G(s)H(s) = \dfrac{K(s+2)}{(s+3)(s^2+2s+2)}$ Sketch the root locus of the system.

Solution Here, $G(s)H(s) = \dfrac{K(s+2)}{(s+3)(s^2+2s+2)}$ and positive feedback has been used.

Step 1: The number of poles $(n) = 3$.
The number of zeros $(m) = 1$.
The number of loci $= n = 3$.
The number of loci at infinity $= n - m = 2$.

Step 2: The pole-zero plot has been shown in Fig. E11.19.

Step 3: Real axis loci are present for

(i) $-2 < \sigma < 0$ (the number of poles and zeros to the right of -2 is even)

(ii) $-\infty < \sigma < -3$ (the number of poles and zeros to the right of -3 is even).

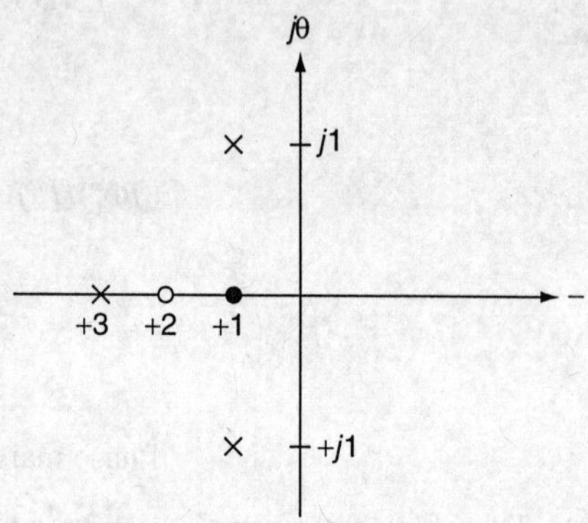

Fig. E11.19 Pole-zero plot

Step 4: The number of asymptotes $= n - m = 2$.

Here, $q = 0, 1$.

$$\therefore \qquad \theta_0 = \frac{0 \times 360°}{n-m} = 0°$$

and

$$\theta_1 = \frac{1 \times 360°}{n-m} = \frac{1 \times 360°}{2} = 180°.$$

Step 5: Centroid (σ_c)

$$\sigma_c = \frac{(-3-1-1)-(-2)}{3-1} = \frac{-3}{2} = -1.5$$

Step 6: Here, $1 - G(s)H(s) = 0$

$$\therefore \qquad 1 - \frac{K(s+2)}{(s+3)(s^2+2s+2)} = 0$$

i.e.,

$$K = \frac{(s+3)(s^2+2s+2)}{s+2}$$

$$\therefore \qquad \frac{dK}{ds} = 0$$

This gives $s = -2.35 \pm j0.77$ and $s = -0.8$.

$s = -0.8$ is accepted.

Step 7: Angle of departure:

$$G(s)H(s)\big|_{s=-1+j} = \frac{K(-1+j+2)}{(-1+j+3)(-1+j+1+j)} \text{[pole at } s=-1+j \text{ is exculded]}$$

$$= \frac{K(1+j)}{(2+j)(2j)}$$

$$\therefore \text{Arg}\left[G(s)H(s)\big|_{s=-1+j}\right] = \frac{\angle 0^\circ \angle 45^\circ}{\angle 26.56^\circ \angle 90^\circ}$$

$$= \angle -71.56^\circ$$

$$\therefore \qquad \phi_d = 0^\circ + (-71.56^\circ) = -71.56^\circ$$

Since $-1 + j$ and $-1 - j$ are complex conjugate poles, the angle of departure from the pole $-1 - j$ is 71.56°.

Figure E11.19(a) shows the root locus plot of the given system.

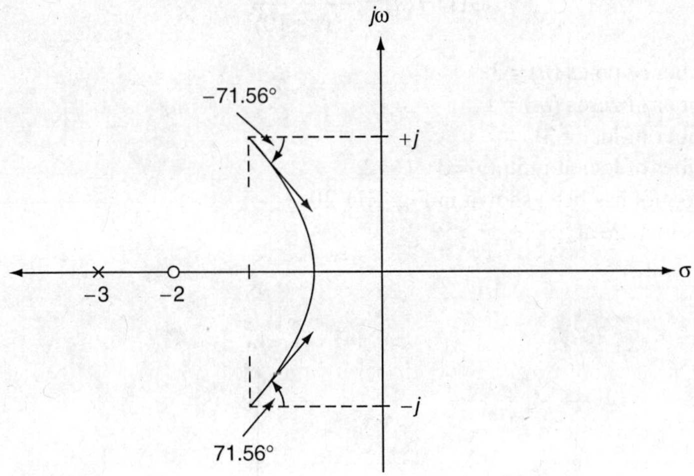

Fig. E11.19(a)

11.12 Root Contour

Figure 11.8 shown a closed-loop system.

For the above system, the study of stability via the root locus technique can be extended by varying any variable other than K from 0 to ∞.

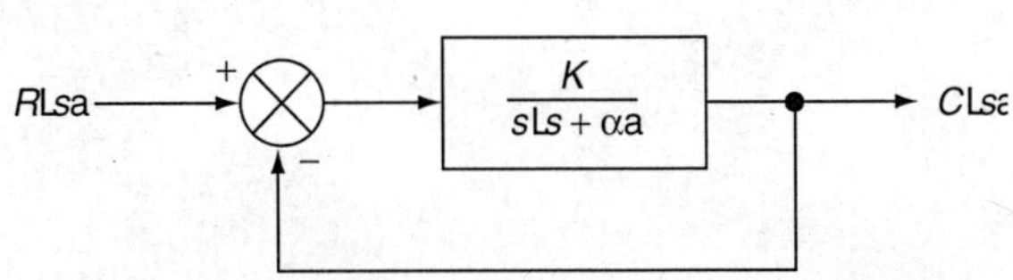

Fig. 11.8 Closed-loop system

Here,

$$G(s_1)H(s_1) = \frac{K}{s(s+\alpha)} \tag{11.19}$$

In Eq. (11.19) the parameter other than K (e.g., α) can be varied from 0 to ∞ to get the locus of the characteristic equation which is called root contour of the system. During sketching the root contours, the parameters, e.g., α, K, are to be varied simultaneously.

Example 11.25 The characteristic equation of a certain root locus is $s^3 + 10s^2 + Ks + K = 0$. Sketch the rough nature of the root locus of this system. In addition, comment on the stability.

Solution The characteristic equation of the given system is

$$s^3 + 10s^2 + Ks + K = 0$$

i.e.,

$$s^3 + 10s^2 + K(s+1) = 0$$

i.e.,

$$1 + \frac{K(s+1)}{s^2(s+10)} = 0 \tag{1}$$

Comparing Eq. (1) with $1 + G(s)H(s) = 0$, we have

$$G(s)H(s) = \frac{K(s+1)}{s^2(s+10)}$$

Step 1: The number of poles $(n) = 3$
The number of zeros $(m) = 1$.
The number of loci = 3.
The number of loci at infinity = 3 – 1 = 2.

Step 2: Pole-zero plot has been shown in Fig. E11.20.

Step 3: Real axis loci are at
$-10 < \sigma < -1$

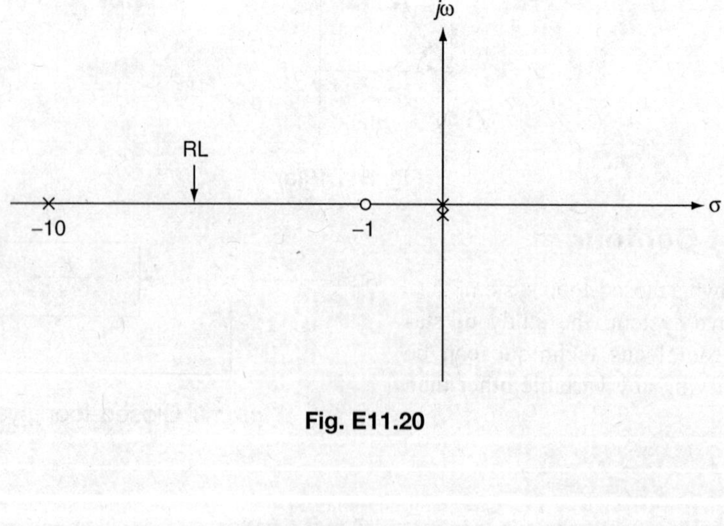

Fig. E11.20

Step 4: Here, $n-m=3-1=2$. Therefore, $q=0, 1$:
There are two asymptotes only.

$$\therefore \qquad \theta_1 = \frac{(2q+1)\times 180°}{n-m} = \frac{180°}{2} = 90°$$

and

$$\theta_1 = \frac{(2\times 1+1)\times 180°}{n-m} = \frac{3\times 180°}{2} = 270°$$

Step 5: Centroid $(\sigma_c) = \dfrac{0+0-10-(-1)}{2} = -4.5$

Figure E11.20(a) shows the centroid and asymptotes.

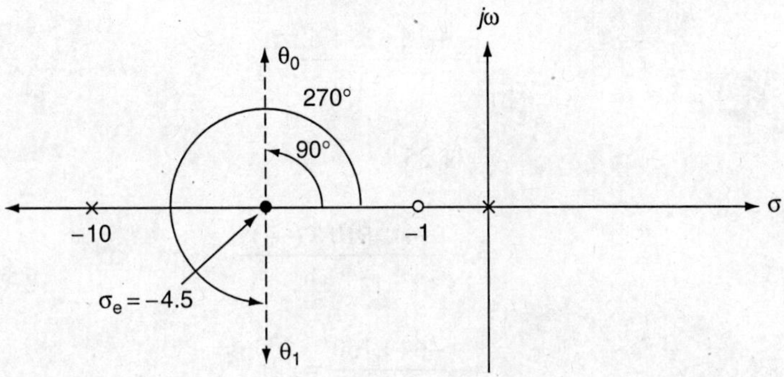

Fig. E11.20(a)

Step 6: From Fig. E11.20, we can predict that there are no break-away points because real axis root locus is at $-10<\sigma<-1$. In spite of the above prediction, let us find out the root of $\dfrac{dK}{ds}=0$. The characteristic equation of the system is

$$s^3+10s^2+Ks+K=0$$

i.e.,
$$s^2(s+10)+K(s+1)=0$$

i.e.,
$$K = -\frac{s^2(s+10)}{s+1} = -\frac{s^3+10s^2}{s+1}$$

i.e.,
$$\frac{dK}{ds} = -\frac{(s+1)(3s^2+20s)-(s^3+10s^2)}{(s+1)^2} = 0.$$

i.e.,
$$(s+1)(3s^2+20s)-(s^3+10s^2)=0$$

i.e.,
$$s[(s+1)(3s+20)-(s^2+10s)]=0$$

i.e.,
$$s[(3s^2+20s+3s+20-s^2-10s]=0$$

i.e.,
$$s(2s^2+13s+20)=0$$

i.e.,
$$s=0 \text{ and } 2s^2+13s+20=0$$

i.e., $$s = 0 \text{ and } s = \frac{-13 \pm \sqrt{13^2 - 4 \times 2 \times 20}}{2 \times 2}$$

$$= \frac{-13 \pm 3}{4} = -2.5, -4.$$

i.e., $$s = 0, -2.5, -4.$$

For $s = -2.5$, $$K = -\frac{(-2.5)^3 + 10 \times (-2.5)^2}{-2.5 + 1}$$

$$= -\frac{-15.625 + 62.5}{-1.5}$$

$$= 31.25$$

For $s = -4$, $$K = -\frac{(-4)^3 + 10 \times (-4)^2}{-4 + 1}$$

$$= -\frac{-64 + 160}{-3} = 32$$

Here, K is positive for $s = -2.5$ and -4. Both the points are on the real axis of root locus. This is a rare case where according to predictions, there are no break-away points.

Step 7: Intersection with imaginary axis:

The characteristic equation is $s^3 + 10s^2 + Ks + K = 0$

Routh array is

$$\begin{array}{c|cc} s^3 & 1 & K \\ s^2 & 10 & K \\ s^1 & \dfrac{9K}{10} & 0 \\ s^0 & K & \end{array}$$

The marginal value of K which will make row of s^1 as the row of zeros is $K = 0$. However, this is invalid. Since system cannot work with $K = 0$, this indicates that there is no intersection of root locus with the imaginary axis.

Step 8: Since there are no complex conjugate poles, there is no need to calculate angle of departure. Root locus will depart at angle $\pm 90°$ after meeting the breakaway $s = -4$ (last meet).

Step 9: Fig E11.20(b) shows the root locus of the system.

The system is absolutely stable for all values of "K" from 0 to ∞. Root locus lies in the left half of s-plane.

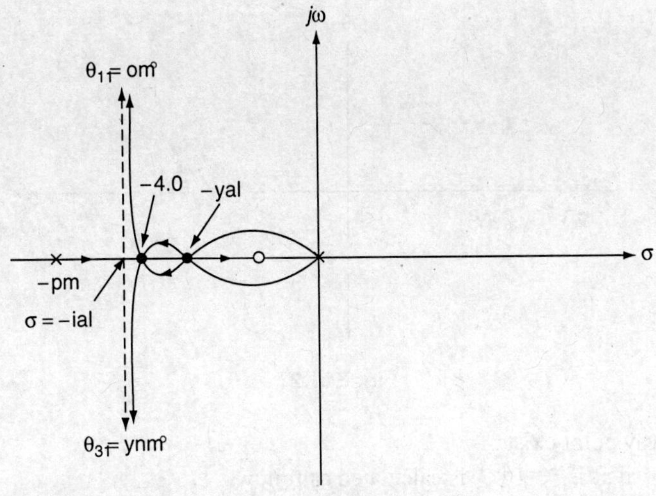

Fig. E11.20(b)

11.13 More Zeros and Less Poles

Here number of zeros (m) > number of poles (n). Number of loci = m. Out of these, n loci will start from pole and $m - n$ loci will start from infinity and all will terminate to finite zeros.

Example 11.26 Draw the root locus of the system having open-loop transfer function

$$G(s)H(s) = \frac{K(s+2)(s^2+2s+2)}{(s+1)}.$$

Solution Here, the open-loop transfer function of the system is

$$G(s)H(s) = \frac{K(s+2)(s^2+2s+2)}{(s+1)}$$

$$= \frac{K(s+2)(s+1+j)(s+1-j)}{(s+1)}$$

Step 1: Here, the number of zeros (m) = 3 and the number of poles (n) = 1.
The number of loci = m = 3.
The number of loci starting from infinity = $m - n$ = 2.

Step 2: The pole-zero plot is shown in Fig. E11.21.

Step 3: Real axis loci
(i) absent for $-1 < \sigma < 0$
(ii) present for $-2 < \sigma < -1$
(iii) absent for $-\infty < \sigma < -2$

Step 4: The number of asymptotes are = $m - n$ = 2.

$\therefore$ $q = 0, 1.$

The angles are 90° and 270°.

Step 5: Centroid $(\sigma_c) = \dfrac{(-1)-(-1-1-2)}{1-3} = -1.5$

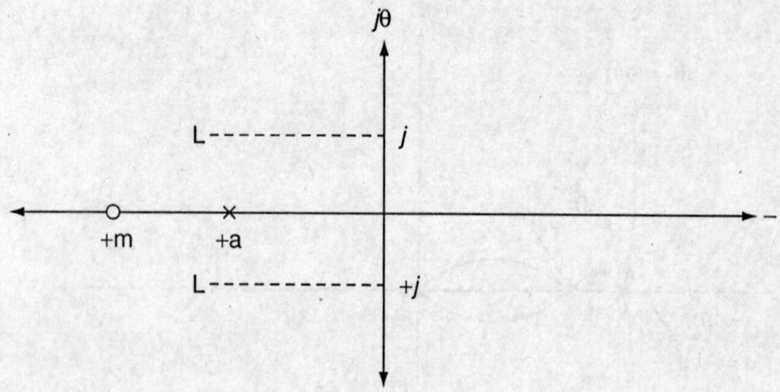

Fig. E11.21

Step 6: No break-away point exist.
Step 7: Angle of arrival at $s=-1+j$ is calculated as follows

$$G(s)H(s)\big|_{s=-1+j} = \left.\frac{K(s+2)(s+1+j)}{(s+1)}\right|_{s=-1+j}$$

$$= \frac{K(-1+j+2)(-1+j+1+j)}{(-1+j+1)}$$

$$= \frac{K(1+j)(2j)}{j}$$

$$= 2K(1+j)$$

$$= 2\sqrt{2}K\angle 45^\circ$$

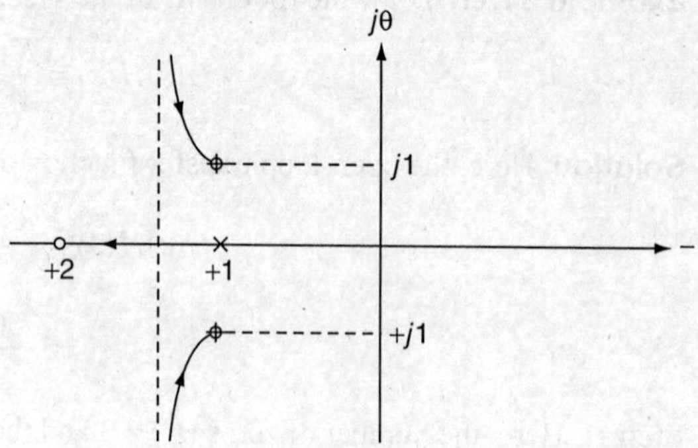

Fig. E11.21(a)

$$\therefore \quad \angle G(s)H(s)\big|_{s=-1+j} = 45^\circ$$

$$\therefore \quad \phi_A = 180^\circ - 45^\circ = 135^\circ.$$

$$\therefore \quad \phi_A \text{ at } s=-1-j \text{ is } -135^\circ.$$

Figure E11.21(a) shows the root locus.

11.14 Effects of Addition of Poles

If the pole is added on the root locus, the general effect of this addition has a tendency to shift the root locus towards the right of *s*-plane. This makes the stability of the system lower.

Let us consider a system with $G(s)H(s)=\dfrac{K}{s+3}$. The root locus plot of this system is shown in Fig. 11.9.

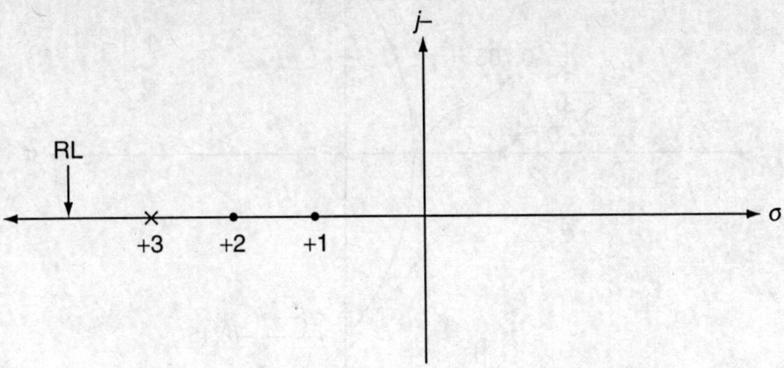

Fig. 11.9 Root locus plot of $G(s)H(s)=\dfrac{K}{s+3}$

If $G(s)H(s)$ of this system becomes $G(s)H(s)=\dfrac{K}{s(s+3)}$, the root locus plot of this system is shown in Fig. 11.10.

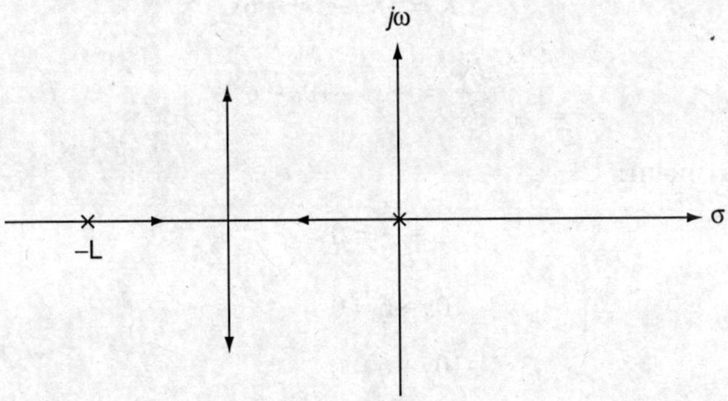

Fig. 11.10 Root locus plot of $G(s)H(s)=\dfrac{K}{s(s+3)}$

Let us add a plot at $s=-2$. The $G(s)H(s)$ of this system becomes $G(s)H(s)=\dfrac{K}{s(s+2)(s+3)}$. The root locus plot of this system is shown in Fig. 11.11.

Break-away point:

Characteristic equation is given by

$$1+G(s)H(s)=0$$

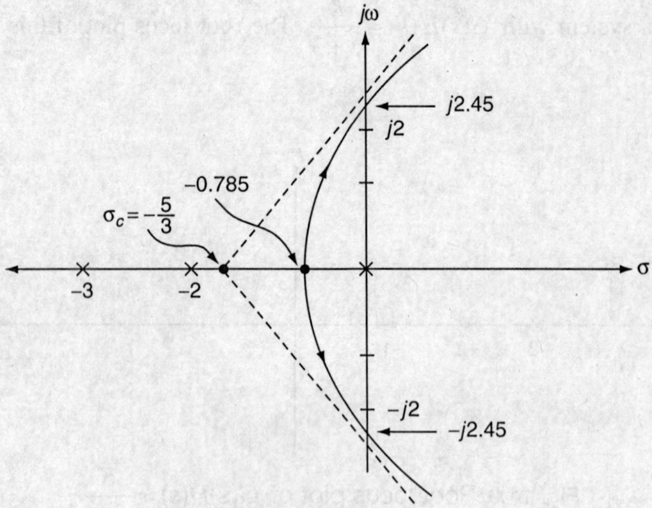

Fig. 11.11 Root locus plot of $G(s)\,H(s) = \dfrac{K}{s(s+2)\,(s+3)}$

i.e., $$s(s+2)(s+3)+K=0$$

i.e., $$K=-s^3-5s^2-6s$$

i.e., $$\frac{dK}{ds}=-3s^2-10s-6$$

To obtain break-away point:

$$\frac{dK}{ds}=0$$

i.e., $$-3s^2-10s-6=0$$

i.e., $$3s^2+10s+6=0$$

i.e., $$s=\frac{-10\pm\sqrt{10^2-72}}{6}$$

∴ $$s = \text{Either } -0.785 \text{ or } -2.55.$$

For $$s=-0.785,\ K=1.145$$

For $$s=-2.55,\ K=-66.32$$

The valid break-away point is $s=-0.785$.

***jω* - axis cross-over:**
The characteristic equation is given by

$$s^3+5s^2+6s+K=0$$

Routh array is given by

$$\begin{array}{c|cc} s^3 & 1 & 6 \\ s^2 & 5 & K \\ s^1 & \dfrac{30-K}{5} & 0 \\ s^0 & K & \end{array}$$

Here,
$$\frac{30-K}{5}=0$$

i.e.,
$$K = 30$$

The auxiliary equation becomes

$$5s^2 + K = 0$$

i.e.,
$$s^2 = -\frac{K}{5} = -\frac{30}{5} = -6$$

$\therefore$
$$s = \pm j\sqrt{6}$$

If a pole is added to the system having $G(s)H(s) = \dfrac{K}{s(s+3)}$ at $s = -5$, the OLTF of the system becomes

$$G(s)H(s) = \frac{K}{s(s+3)(s+5)}$$

The root locus plot of the system having $G(s)H(s) = \dfrac{K}{s(s+3)(s+5)}$ is shown in Fig. 11.12.

The valid break-away point is $s = -1.21$ on the σ axis and $j\omega$ axis cross-over is at $\pm j3.873$.

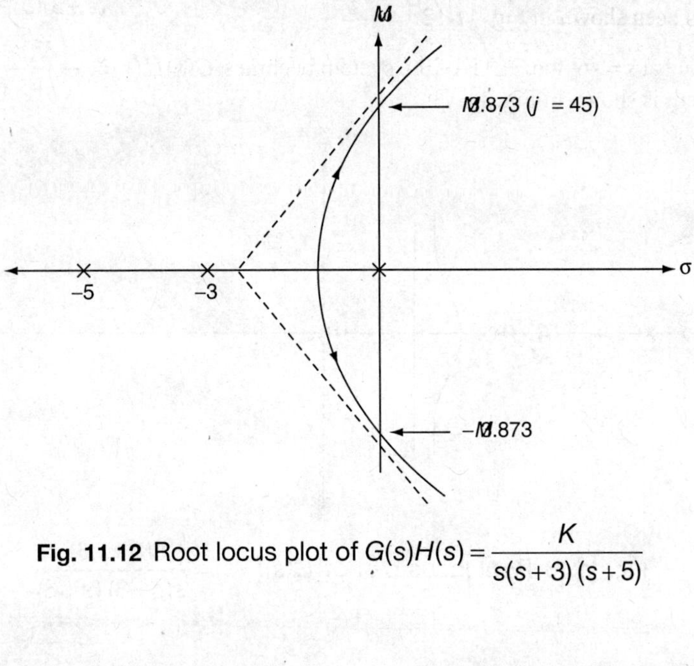

Fig. 11.12 Root locus plot of $G(s)H(s) = \dfrac{K}{s(s+3)(s+5)}$

From above, we can conclude that if a pole is added to the system having $G(s)H(s)=\dfrac{K}{s(s+3)}$ at $s=-2$, it crosses the $j\omega$-axis at $s=\pm j2.45$ having K = 30. If a pole is added to the system having $G(s)H(s)=\dfrac{K}{s(s+3)}$ at s = –5 instead of s = –2, the root locus crosses the $j\omega$-axis at $s=\pm j3.873$ having $K=45$.

The system remains stable and the cross-over will take place at a much higher value of K when the added pole lies to the right of the pole. Since the value of K is limited in general for an addition of a pole, the gain margin comes down.

From above, it can be concluded that there is a possibility of a stable system becoming unstable. The addition of a pole brings the closed-loop pole towards the $j\omega$-axis and makes the system more oscillatory. The pole at $s=-2$ for the above cases affects the change more on account of its nearness to $j\omega$-axis and origin.

The effects of addition of poles are shown below.

- There is change in shape of the root locus and it shifts towards the imaginary axis.
- The intercept on the jω-axis occurs for a lower value K because asymptotes angle being lower down.
- System becomes oscillatory.
- Gain margin and relative stability decrease.
- There is reduction in the range of K.
- A sluggish response can be changed to a quicker response for artful introduction of a pole.
- Settling time increases.

11.15 Effects of Addition of Zeros

If a zero is added to a system OLTF, the general effect of this addition is to pull the root locus to the left of s-plane and hence improves the stability. Let us consider $G(s)H(s)=\dfrac{K}{s(s+3)(s+5)}$. The root locus of this system has been shown in Fig. 11.12.

If a zero is added at $s=-6$, the OLTF of the system becomes $G(s)H(s)=\dfrac{K(s+6)}{s(s+3)(s+5)}$ and the root locus of this system is shown in Fig. 11.13.

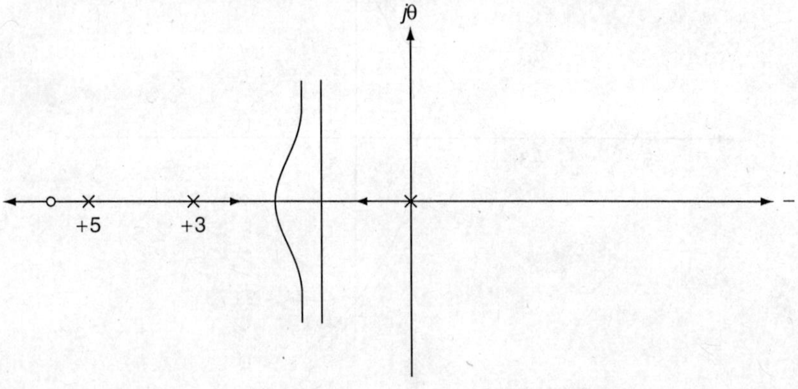

Fig. 11.13 Root locus plot of $G(s)H(s)=\dfrac{K(s+6)}{s(s+3)\,(s+6)}$

If a zero is added at $s = -4$, the OLTF of the system becomes $G(s)H(s) = \frac{K(s+4)}{s(s+3)(s+5)}$ and its root and its root locus plot has been shown in Fig. 11.14.

If a zero is added at $s = -2$, OLTF of the system becomes $G(s)H(s) = \frac{K(s+2)}{s(s+3)(s+5)}$ and its root locus plot has been shown in Fig. 11.15.

The break-away point shifts to the left. The system is made stable.

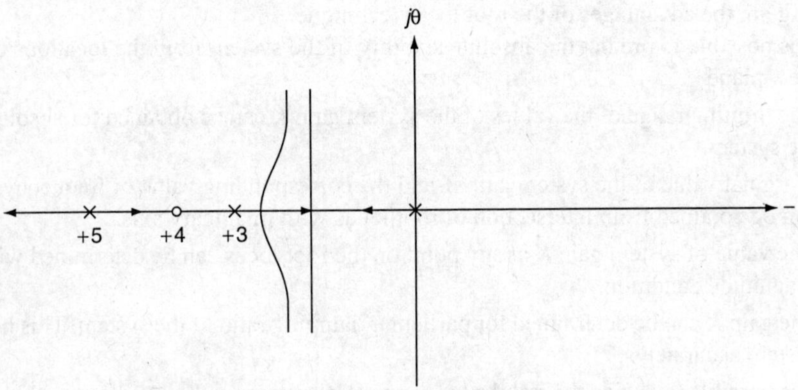

Fig. 11.14 Root locus plot of $G(s)\,H(s) = \frac{K(s+4)}{s(s+3)\,(s+5)}$

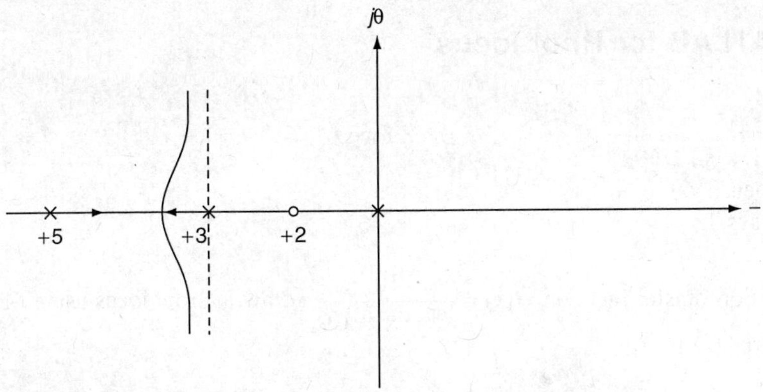

Fig. 11.15 Root locus plot of $G(s)\,H(s) = \frac{K(s+2)}{s(s+3)\,(s+5)}$

Therefore, the addition of zero speeds up settling time because it pulls locus to the left and hence stabilizes the system. Since the addition of zero means derivative action, this speeds up transient response.

The following are the effects of addition of zero to the system's OLTF:

- There is change in shape of the root locus and it shifts towards the left of s-plane.
- Stability of the system is enhanced.
- Range of K increases.
- Settling time speeds up.

11.16 Advantages of Root Locus Method

The following are the advantages of the root locus technique:

(i) It is possible to predict the absolute stability of the system from the locations of the roots in the *s*-plane.

(ii) The limiting range of the values of the system gain K can be obtained for absolute stability of the system.

(iii) Marginal value of the system gain K and the corresponding value of frequency of oscillation can be obtained from intersection of root locus with imaginary axis.

(iv) The value of system gain K at any point on the root locus can be determined with the help of magnitude condition.

(v) The gain K can be determined for particular damping ratio of the system. This helps to design system accurately.

(vi) It is possible to decide the stability of the control system with time delay with the help of root locus.

(vii) Gain margin and phase margin of the system can be determined from root locus.

(viii) Relative stability about a particular value of $s = -\sigma$ can also be focussed.

(ix) The information regarding settling time of the system can also be acquired from the root locus.

11.17 MATLAB for Root locus

```
rlocus(sys)
rlocus(sys, k)
rlocus(sys1, sys2,...)
[r, k] = rlocus(sys)
r = rlocus(sys, k)
```

Program 1 For transfer function $G(s) = \dfrac{10}{s^3 + 9s^2 + 10s}$, draw the root locus using MATLAB.

```
g = tf([100], [1 9 10 0])
```

Transfer function:

$$\frac{100}{s\text{^}3+9s\text{^}2+10s}$$

rlocus(g)

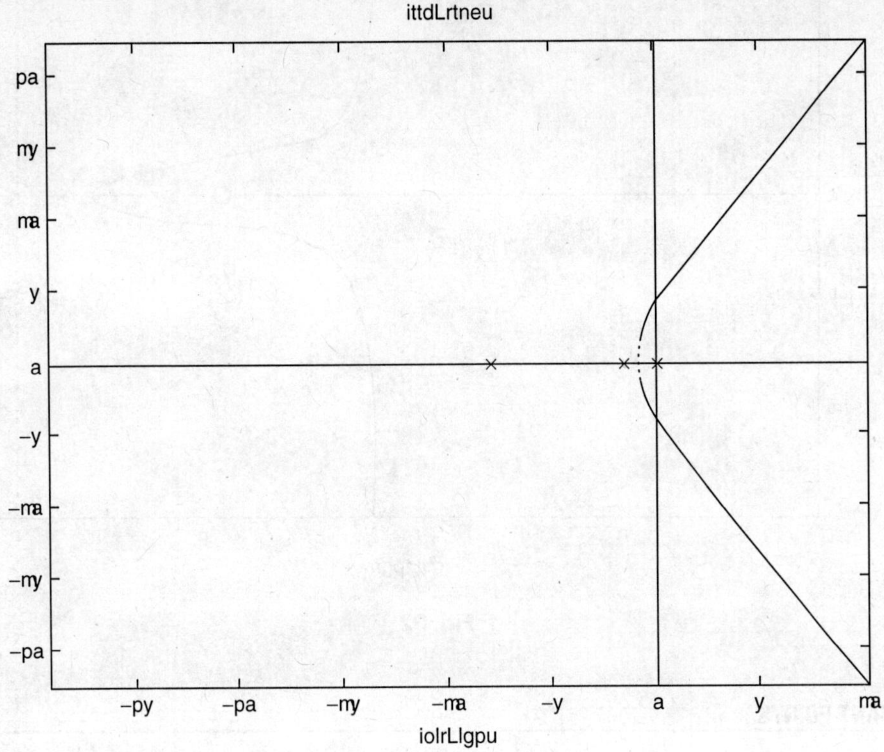

Fig. P1

Program 2 For transfer function $G(s)=\dfrac{50s+100}{s^3+9s^2+10s}$, draw the root locus using MATLAB.

g = tf([50 100], [1 9 10 0])

Transfer function:

$$\frac{50s+100}{s\text{^}3+9s\text{^}2+10s}$$

rlocus(g)

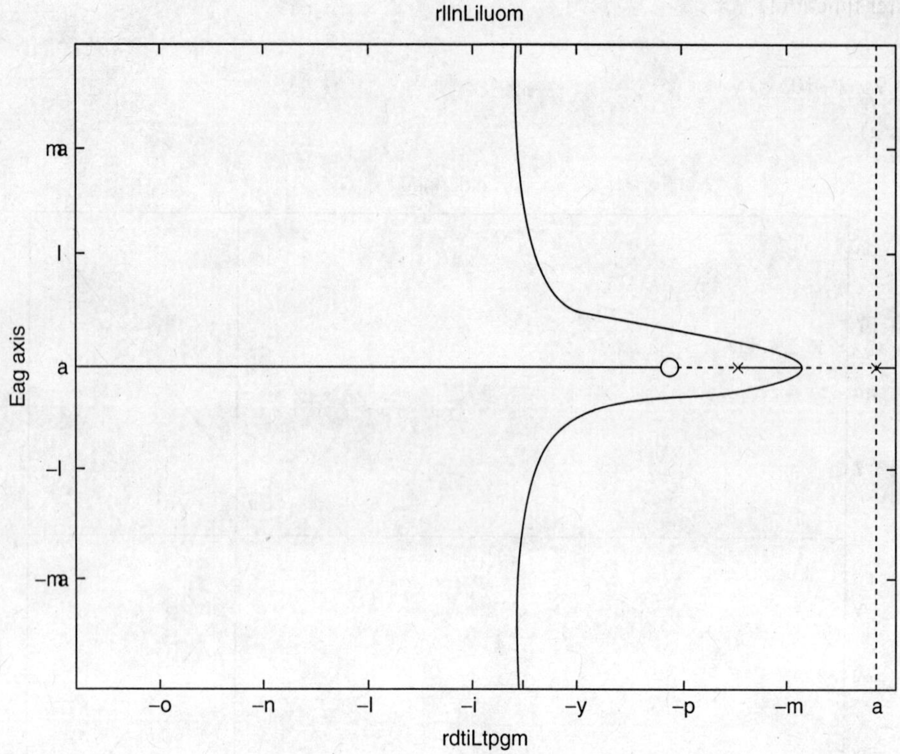

Fig. P2

SIGNIFICANT POINTS

Root Locus: It is defined as the locus of the plot of the closed-loop poles as a function of open-loop gain (K), when K is varied from $-\infty$ to $+\infty$. If K is varied from 0 to $+\infty$, it is known as direct root locus. On the other hand, if K is varied from $-\infty$ to 0, it is termed as inverse root locus.

Angle and Magnitude Condition

$$|G(s)H(s)| = 1 \text{ and } \angle G(s)H(s) = \pm(2q+1)180°.$$

General Methods for Construction of Roots Locus

Rule 1. Symmetry: Root locus is always symmetrical about the real axis (σ-axis). The roots are either real or complex conjugates or combination of both. Therefore, the locus is symmetrical about the real axis of the s-plane.

Rule 2. The number of loci: Let the number of OLTF poles and OLTF zeros be n and m, respectively, If $n > m$, i.e., the number of OLTF poles is greater than the number of OLTF zeros, the number of loci is n. Each locus will start from an OLTF pole and end on an OLTF zero. Here, $(n - m)$ loci will end at infinity and total loci $= n$.

Rule 3. Real axis loci: A part of the loci will lie on the real axis. A point on the real axis will lie on the root locus if and only if the sum of the OLTF poles and zeros to the right of the right of the point is odd.

Rule 4. Angle of asymptotes: The angle of asymptotes is given by

$$\theta = \frac{2q+1}{n-m} \times 180° \text{ for } K > 0$$

where $$q = 0, 1, 2, \ldots, n - m - 1.$$

and

$$\theta = \frac{2q}{n - m} \times 180^\circ \text{ for } K < 0$$

where $$q = 0, 1, 2, \ldots n - m - 1.$$

Rule 5. Centre of asymptotes: The point where the asymptotes touch the real axis is called centroid (σ_c) and it is given by

$$\sigma_c = \frac{\sum \text{real parts of poles of } G(s)H(s) - \sum \text{real parts of zeros of } G(s)H(s)}{n - m}$$

Rule 6. Break-away/Break-in point: Break-away point is defined as a point at which the root locus comes out of the real axis and break-in point is defined as a point at which the root locus enters the real axis. The break-away and break-in points are the points on the root locus at which multiple roots of the characteristic equation occur.

Rules 7. Intersection of the root locus with the jω-axis: To find the intersection of root locus with the imaginary axis, the following procedures are followed.

Step 1: Construct the characteristic equation $1 + G(s)H(s) = 0$
Step 2: Develop Routh array in terms of K.
Step 3: Find K_{mar} that creates one of the roots of Routh array as a row of zeros
Step 4: Frame auxiliary equation $A(s) = 0$ with the help of the coefficient of a row just above the row of zeros.
Step 5: The roots of the auxiliary equation $A(s) = 0$ for $K = K_{mar}$ give the intersection points of the root locus with the imaginary axis.

Rule 8. Angle of departure arrival: The root locus leaves from a complex pole and arrives at a complex zero. These two angles are known as angle of departure and angle of arrival, respectively. Angle of departure (θ_d) is given by $\theta_d = 180^\circ + \arg[G(s)H(s)]$, where $\arg[G(s)H(s)]$ is the angle of $G(s)H(s)$ excluding the pole where the angle is to be calculated. It is also possible to calculate $\arg[G(s)H(s)]$ graphically. Similarly, the angle of arrival is given by $\theta_a = 180^\circ - \arg[G(s)H(s)]$, where $\arg[G(s)H(s)]$ is the angle of $G(s)H(s)$ excluding the zero where the angle is to be calculated.

Steps for Solving Problems on Root Locus

Step 1: Determine the poles and zeros and their numbers, and the number of loci, using Rule 1 and Rule 2.
Step 2: Plot the poles and zeros on *s*-plane.
Step 3: Find real axis loci using Rule 3. Show the real axis loci wherever present by dark lines.
Step 4: Find the number of asymptotes and their angles by Rule 4.
Step 5: Using Rule 5, determine the centre of asymptotes and draw results of Step 4 and Step 5.
Step 6: Determine the break-away/break-in point if present using Rule 6 and mark the point only.
Step 7: Calculate the angle of departure or the angle of arrival due to complex poles or zeros, respectively, using Rule 6.
Step 8: Determine *j*ω crossover using Rule 7 if the locus crosses the *j*ω axis.

Value Gain Margin

$$\text{Gain margin (GM)} = \frac{\text{Value of } K \text{ at imaginary crossover}}{\text{Design value of } K}$$

If the root locus does not cross the *j*ω axis, the gain is ∞. GM represents the maximum gain that can be multiplied for the system to be just on the verge of instability.

Phase Margin: Phase margin (PM) can be determined for a given value of K as follows:

- Calculate ω, for which $|G(j\omega)H(j\omega)| = 1$ for the given value of K.
- Calculate $\arg[G(j\omega)H(j\omega)]$.
- Phase margin = $180° + \arg\ [G(j\omega)H(j\omega)]$.

Effects of Addition of Poles: The effects of addition of poles are as follows:

- There is change in shape of the root locus and it shifts towards the imaginary axis. The intercept on the jω axis occurs for a lower value of K because of asymptote's angle being lower down.
- System becomes oscillatory.
- Gain margin and relative stability decrease.
- There is reduction in the range of K.
- A sluggish response can be changed to a quicker response for artful introduction of a pole.
- Settling time increases.

Effects of Addition of Zeros: The effects of addition of zeros are as follows:

- There is change in shape of the root locus and it shifts towards the left of the s-plane.
- Stability of the system is enhanced.
- Range of K increases.
- Settling time speeds up.

ADDITIONAL SOLVED EXAMPLES

Example 11.27 Sketch the root locus for a system having

$$G(s)H(s) = \frac{K(s+3)}{(s+2+j\sqrt{5})(s+2-j\sqrt{5})} \quad \text{where } K > 0$$

Solution

Step 1: The number of poles $(n) = 2$

The number of zeros $(m) = 1$

Therefore, the number of loci terminating at infinity $= n - m = 2 - 1 = 1$

Step 2: The pole-zero plot is shown in Fig. E11.22

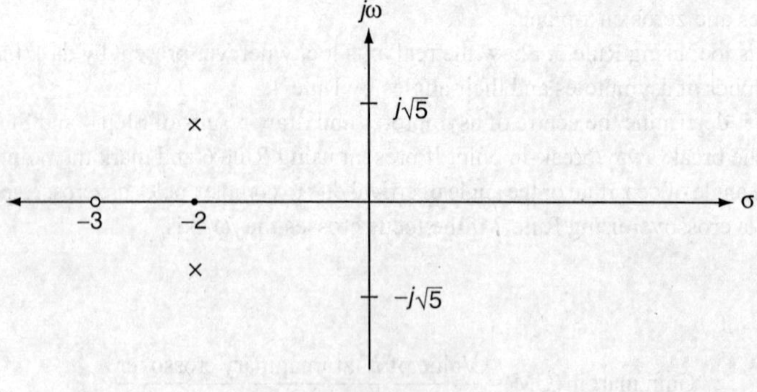

Fig. E11.22

Step 3: The real axis locus is

(i) absent for $(-2<\sigma<0)$
(ii) absent for $(-3<\sigma<-2)$
(iii) present for $(-\infty<\sigma<-3)$

Step 4: Here, the number of asymptotes $= n - m = 2 - 1 = 1$. Therefore, $q = 0$

$$\therefore \quad \theta_0 = \frac{2\times 0+1}{1}\times 180° = 180°$$

Step 5:

$$\sigma_C = \frac{(-2-2)-(-3)}{1} = \frac{-4+3}{1} = -1$$

Step 6:

$$1+G(s)H(s)=0$$

i.e.,

$$1+\frac{K(s+3)}{(s+2+j\sqrt{5})(s+2-j\sqrt{5})}=0$$

i.e.,

$$K = -\frac{(s+2+\sqrt{5})(s+2-j\sqrt{5})}{s+3} = -\frac{(s+2)^2-j^2 5}{s+3}$$

$$= \frac{-(s^2+4s+9)}{s+3}$$

$$\frac{dK}{ds} = -\left[\frac{(s+3)(2s+4)-(s^2+4s+9)\times 1}{(s+3)^2}\right]$$

i.e.,

$$2(s+3)(s+2)-(s^2+4s+9)=0$$

i.e.,

$$s^2+6s+3=0$$

i.e.,

$$s = \frac{-6\pm\sqrt{6^2-12}}{2} = \frac{-6\pm\sqrt{24}}{2} = \frac{-6\pm 4.89}{2}$$

i.e.,

$$s = \text{either} -0.555 \text{ or } -5.445$$

For $s=-0.555, K=-2.899$; for $s=-5.445$, $K=6.899$

$\therefore$ $s=-5.445$ is a valid break-away point.

Step 7:

$$G(s)H(s)\big|_{s=-2+j\sqrt{5}} = \left.\frac{K(s+3)}{(s+2+j\sqrt{5})}\right|_{s=-2+j\sqrt{5}}$$

$$= \frac{K(-2+j\sqrt{5}+3)}{(-2+j\sqrt{5}+2+j\sqrt{5})} = \frac{K(1+j\sqrt{5})}{j2\sqrt{5}} = \frac{K(2.949\angle 65.90°)}{2\sqrt{5}\angle 90°} = \frac{2.949K}{2\sqrt{5}}\angle -24.1°$$

$$\therefore \qquad \angle G(s)H(s) = -24.1°$$

$$\therefore \qquad \phi_d = 180° + \text{org}[G(s)H(s)] = 180° - 24.1° = 155.9°$$

Step 8: Figure E11.22(a) shows the completed root locus of the system.

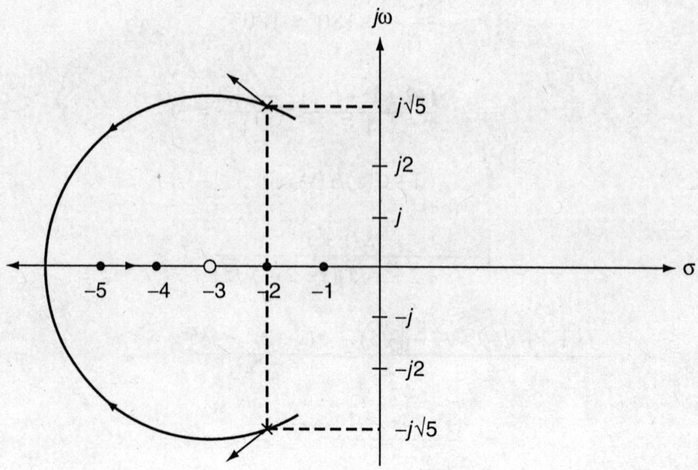

Fig. E11.22(a)

Example 11.28 Sketch the root locus for a system having

$$G(s)H(s) = \frac{K}{s(s+1)(s+2)(s+4)} \quad \text{where } K > 0$$

Solution

Step 1: Here, the number of poles (n) = 4 and the number of zeros (m) = 0.

Step 2: The pole-zero plot of the system is shown in Fig. E11.23.

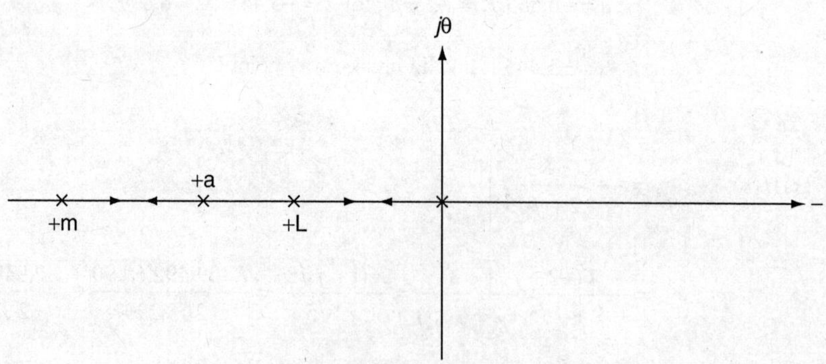

Fig. E11.23

Step 3: The real axis locus is

(i) Present for $-1 < \sigma < 0$
(ii) Absent for $-2 < \sigma < -1$
(iii) Present for $-4 < \sigma < -2$
(iv) Absent for $-\infty < \sigma < -4$

Step 4: The number of asymptotes $= n - m = 4 - 0 = 4$, Therefore, $q = 0$, 1, 2 and 3.

$$\therefore \qquad \theta = 45°, 135°, 225°, 315°$$

Step 5:
$$\sigma_c = \frac{(0-1-2-4)-0}{4} = \frac{-7}{4} = -1.75$$

Step 6:
$$1 + G(s)H(s) = 0$$

i.e.,
$$1 + \frac{K}{s(s+1)(s+2)(s+4)} = 0$$

i.e.,
$$K = -s^4 - 7s^3 - 14s^2 - 8s$$

i.e.,
$$\frac{dk}{ds} = -4s^3 - 21s^2 - 28s - 8 = 0$$

i.e.,
$$-4s^3 - 21s^2 - 28s - 8 = 0$$

The first break-away point can be expected between 0 and −1.

Let
$$f(s) = 4s^3 + 21s^2 + 28s + 8 = 0$$

$$\therefore \qquad f(0) = 8 \text{ and } f(-1) = -4 + 21 - 28 + 8 = -32 + 29 = -3$$

$$\therefore \qquad f(-0.5) = -1.25$$

Let
$$s = \frac{0 - 0.5}{2} = -0.25$$

$$\therefore \qquad f(-0.25) = 2.25$$

Since $f(-0.5) =$ is negative and $f(-0.25)$ is positive, the root lies between $s = -0.5$ and $s = -0.25$.

Let
$$s = \frac{-0.5 - 0.25}{2} = -0.375$$

Now,
$$f(-0.375) = 0.24218.$$

Again $f(-0.5)$ is negative and $f(-0.375)$ is positive. Let $(-0.5 - 0.375)/2 = -0.4375$.

Now,
$$f(-0.4375) = -0.56543.$$

Again $f(-04375)$ is negative and $f(-0.375)$ is positive.

Let
$$s = \frac{-0.4375 - 0.375}{2} = -0.40625$$

Now, $$f(-0.40625) = -0.177368.$$

Again $f(-0.40625)$ is negative and $f(-0.375)$ is positive.

Let $$s = \frac{-0.40625 - 0.375}{2} = -0.390625$$

Now, $$f(-0.390625) = 0.02842.$$

Therefore, $s = -0.39$ is **acceptable**

$$\begin{array}{r|rrrr} -0.39 & 4 & 21 & 28 & 8 \\ & & -1.56 & -7.5816 & -7.963176 \\ \hline & 4 & 19.44 & 20.4184 & (0.36824) \end{array}$$

The remainder is represented by

$$4 \ \ 19.44 \ \ 20.4184$$

which is $4s^2 + 19.44s + 20.4189 = 0$

$\therefore$ $$s = \frac{-19.44 \pm \sqrt{19.44^2 - 4 \times 4 \times 20.4184}}{2 \times 4} = \frac{-19.44 \pm 7.157}{8}$$

$\therefore$ $$s = -1.4345 \text{ and } -3.4355$$

$\therefore$ $s = -0.39$ and $s = -3.4355$ are the valid break-away points.

Step 8: $$1 + G(s)H(s) = 0$$

i.e., $$s^4 + 7s^3 + 14s^2 + 8s + K = 0$$

Routh array is

$$\begin{array}{c|ccc} s^4 & 1 & 14 & K \\ s^3 & 7 & 8 & 0 \\ s^2 & 12.857 & K & \\ s^1 & \dfrac{102.856 - 7K}{12.857} & 0 & \\ s^0 & K & & \end{array}$$

For K_{mar}, we have $$\frac{102.856 - 7K_{mar}}{12.857} = 0$$

i.e., $$7K_{mar} = 102.856$$

i.e., $$K_{mar} = 14.69$$

Therefore, $$s^2 = -1.1426$$

i.e., $$s = \pm j1.069$$

Step 9: Since there are no complex poles, the angle of departure does not exist.

Step 10: The complete root locus is shown in Fig. E11.23(a).

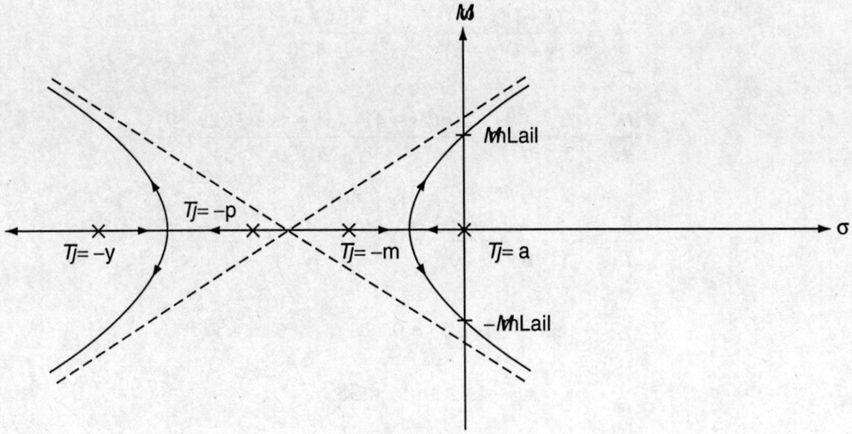

Fig. E11.23(a)

Example 11.29 Sketch the root locus for the system if possible and check poles to zeros for the system having

$$G(s)H(s) = \frac{K(s+4)(s+5)}{(s+3)(s+1)}(K > 0)$$

Solution

Step 1: Here, $m = 2$, $n = 2$.
Step 2: The pole-zero plot is shown in Fig. E11.24.

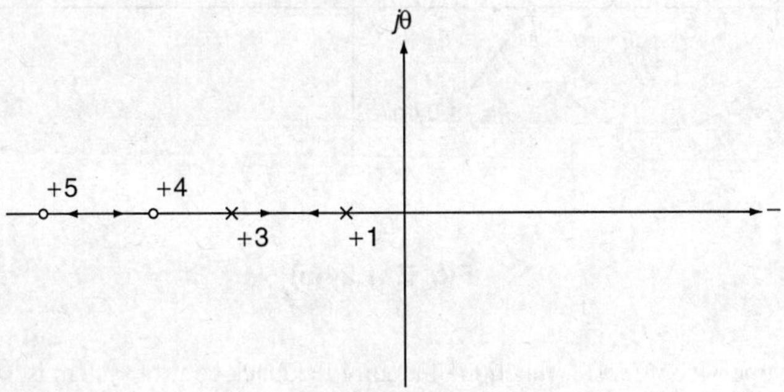

Fig. E11.24

Step 3: The real axis locus is present for $-1 < \sigma < -3$, and $-5 < s < -4$.
Step 4 and 5: Since $q = n - m = 0$, there steps are not required.

Step 6:
$$1 + G(s)H(s) = 0$$

i.e.,
$$1 + K\frac{(s+4)(s+5)}{(s+1)(s+3)} = 0$$

i.e., $$K = -\frac{(s+1)(s+3)}{(s+4)(s+5)} = -\frac{s^2+4s+3}{s^2+9s+20}$$

i.e., $$\frac{dK}{ds} = -\frac{(s^2+9s+20)(2s+4)-(s^2+4s+3)(2s+9)}{(s^2+9s+20)^2}$$

i.e., $$\frac{dK}{ds} = -\frac{5s^2+34s+53}{(s^2+9s+20)^2} = 0$$

i.e., $$5s^2+34s+53 = 0$$

$\therefore$ $$s = -2.42 \text{ and } -4.38$$

Step 7: Since there is no jω-axis cross over, this step is not required.
Step 8: Since there is no complex pole or zero, calculation of angle of departure or angle of arrival is not required.
Step 9: Figure E11.24(a) shows the completed root locus of the system.

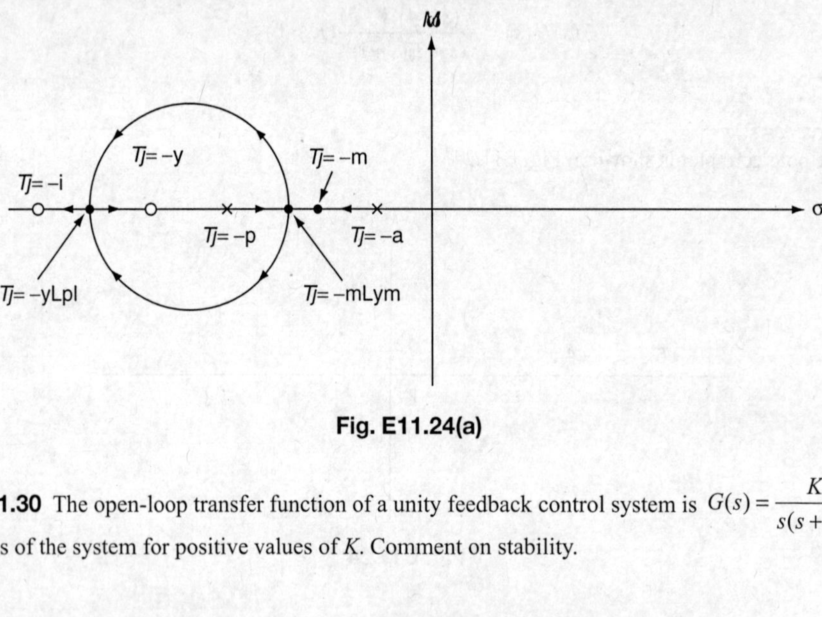

Fig. E11.24(a)

Example 11.30 The open-loop transfer function of a unity feedback control system is $G(s) = \frac{K}{s(s+3)^2}$. Sketch the root locus of the system for positive values of K. Comment on stability.

Solution

Step 1: Here, $m = 0$, $n = 3$. The number of root loci terminating at infinity $= n - m = 3$.
Step 2: The pole-zero plot is shown in Fig. E11.25.
Step 3: The real axis locus is present for $-3 < \sigma < 0$ and $\sigma < -3$.
Step 4: The number of asymptotes $(q) = n - m = 3$. Therefore, $q = 0$, 1 and 2.

Since $\theta_0 = \frac{(2\times q+1)}{3}\times 180° = 60°$, $\theta_1 = \frac{(2\times 0+1)}{3}\times 180° = 60°$ and $\theta_2 = \frac{(2\times 1+1)}{3}\times 180° = 180°$,

Step 5: $$\sigma_c = \frac{(0-3-3)}{3} = \frac{-6}{3} = -2$$

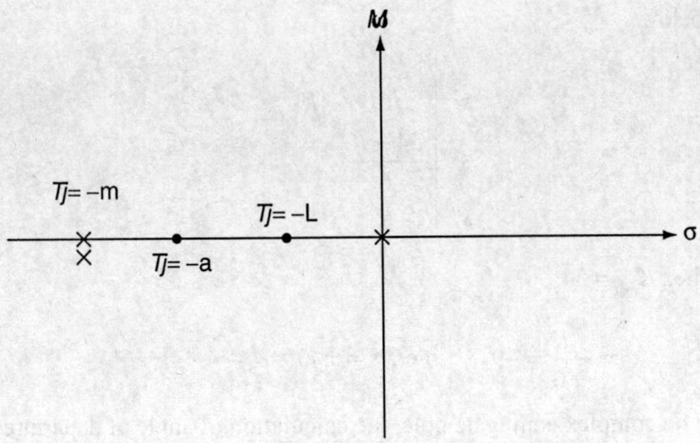

Fig. E11.25

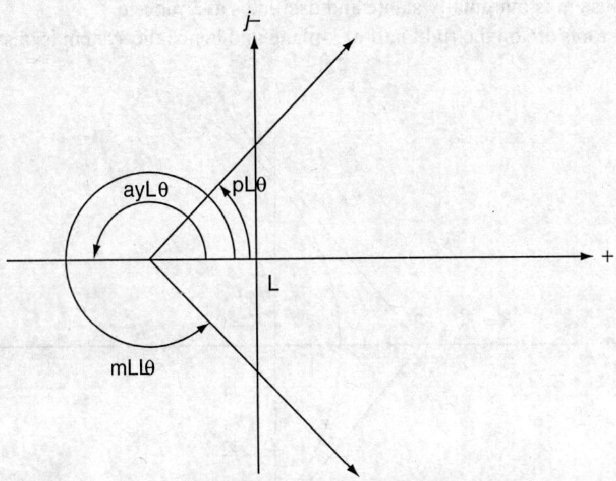

Fig. E11.25(a)

The asymptotes are shown in Fig. E11.25(a).

Step 6: $$1+G(s)H(s)=0$$

i.e., $$1+K\frac{1}{s(s+3)^2}=0$$

i.e., $$K=-s(s+3)^2=-(s^3+6s^2+9s)$$

i.e., $$\frac{dK}{ds}=-(3s^2+12s+9)=-3(s+1)(s+3)=0$$

$\therefore$ $$s=-1 \text{ and } -3.$$

Step 7: The characteristic equations $1+G(s)H(s)=0$ i.e., $s^3+6s^2+9s+K=0$

Routh array is shown below.

$$\begin{array}{c|cc} s^3 & 1 & 9 \\ s^2 & 6 & K \\ s^1 & \dfrac{54-K}{6} & \\ s^0 & K & \end{array}$$

For K_{mar}, $\dfrac{54-K_{mar}}{6}=0$ i.e., $K_{mar}=54$.

$A(s)=6s^2+K=0$ i.e., $s^2=-\dfrac{K}{6}=-9$ i.e., $s=\pm j3$.

Step 8: Since there are no complex conjugate pole, the calculation of angle of departure is not required.

Step 9: Figure E11.25(b) shows two complete root locus of this system.

Step 10: Comment on stability:

For $0<K<54$, all roots are in the left half of s-plane and hence system is stable. At $K=54$, dominant roots are on the imaginary axis. System is marginally stable and oscillates at 3 rad/sec.

For $K>54$, dominant roots are on the right half of s-plane and hence the system is unstable.

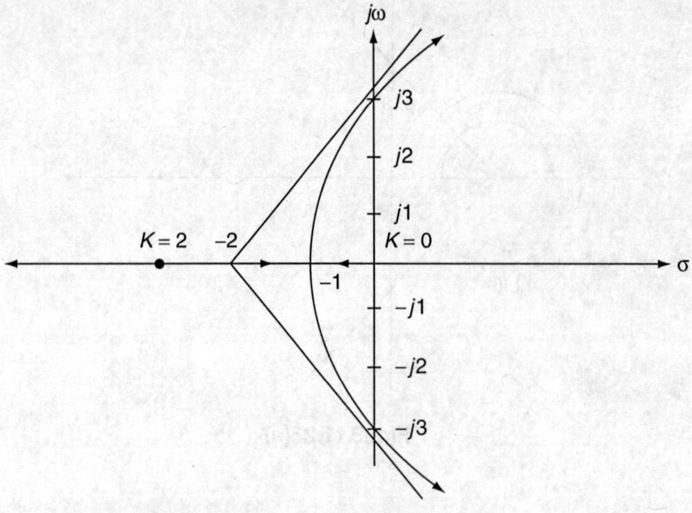

Fig. E11.25(b)

Example 11.31 The open-loop transfer function of a unity feedback control system is $G(s)=K/[s(s+4)(s+6)]$. Draw the root locus of the system.

Solution

Step 1: Here, $m=0, n=3$. The number of root loci terminating at infinity = 3.

Step 2: The pole-zero plot is shown in Fig. E11.26.

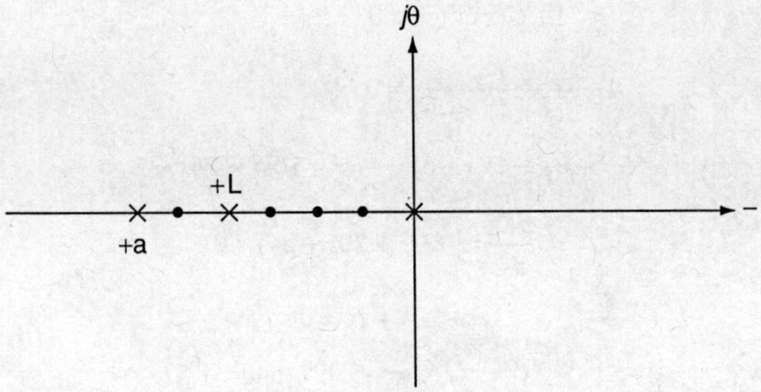

Fig. E11.26

Step 3: Real axis loci are present for $-4 < \sigma < 0$ and $-\infty < \sigma < -6$.

Step 4: $q = n - m = 3$.

The number of asymptotes $= n - m = 3$. Therefore, $q = 0, 1, 2$.

$$\therefore \qquad \theta_0 = \frac{(2\times 0+1)}{3}\times 180° = 60°,\ \theta_1 = \frac{(2\times 1+1)}{3}\times 180° = 180°,$$

and

$$\theta_2 = \frac{(2\times 2+1)}{3}\times 180° = 300°$$

Step 5:

$$\sigma_c = \frac{(0-4-6)-0}{3} = -\frac{10}{3} = -3.34$$

The asymptotes are shown in Fig. E11.26(a).

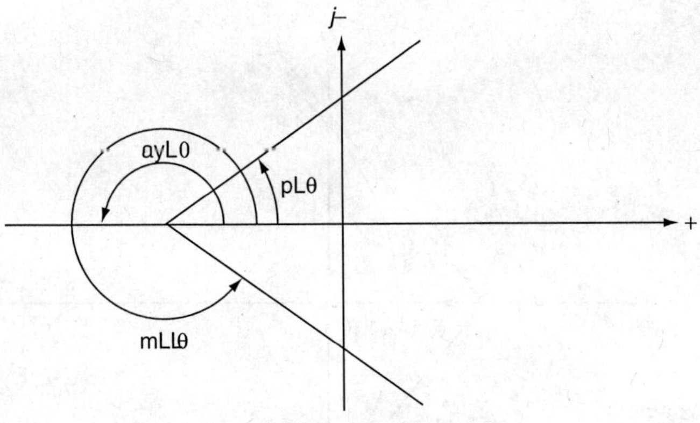

Fig. E11.26(a)

Step 6: $$1+G(s)H(s)=0$$

i.e., $$1+\frac{K}{s(s+4)(s+6)}=0$$

i.e., $$K=-s(s+4)(s+6)=-(s^3+10s^2+24s)$$

i.e., $$\frac{dK}{ds}=-(3s^2+20s+24)=0$$

i.e., $$3s^2+20s+24=0$$

$\therefore$ $$s=-5.097 \text{ and } -1.57.$$

Here s = −1.57 is the valid break-away point.

Step 7: The characteristic equation is $1+G(s)H(s)=0$ i.e., $s^3+10s^2+24s+K=0$

Routh array is shown below.

s^3	1	24
s^2	10	K
s^1	$\frac{240-K}{10}$	
s^0	K	

For K_{mar}, $\frac{240-K_{mar}}{10}=0$ i.e., $$K_{mar}=240.$$

$A(s)=10s^2+K=0$ i.e., $$s^2=-\frac{K}{10}=-24$$

i.e., $$s=\pm j4.89.$$

Step 8: Since there are no complex conjugat poles or zeros, calculation for angle of departure or angle of arrival is not required.

Step 9: Figure E11.26(b) shows the completed root locus of the system.

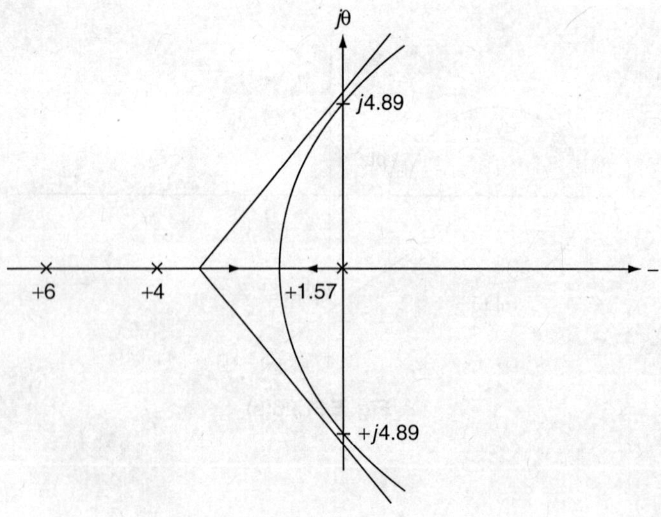

Fig. E11.26(b)

Example 11.32 Sketch root locus for $G(s)H(s) = \dfrac{K}{s(s+2)(s^2+2s+5)}$ In addition, find K for $\xi = 0.5$.

Solution Here,

$$G(s)H(s) = \frac{K}{s(s+2)(s^2+2s+5)}$$

$$= \frac{K}{s(s+2)\{(s+1)^2+4\}}$$

$$= \frac{K}{s(s+2)(s+1+j2)(s+1-j2)}$$

Step 1: Here, the number of poles $(m) = 0$ and the number of zeros $(n) = 4$.

$\therefore$ The number of loci $= n - m = 4$.

$\therefore$ The number of loci terminating at infinity $= n - m = 4 - 0 = 4$.

Step 2: The pole-zero plot has been shown in Fig. E11.27.

Step 3: Real axis loci are

(i) present for $-2 < \sigma < 0$

(ii) absent for $-\infty < \sigma < -2$

Step 4: The number of asymptotes $= n - m = 4 - 0 = 4$

$$\therefore \quad q = 0, 1, 2, 3.$$

$$\therefore \quad \theta_0 = \frac{2\times 0+1}{n-m}\times 180° = \frac{180°}{4} = 45°$$

$$\therefore \quad \theta_1 = \frac{2\times 1+1}{n-m}\times 180° = \frac{3\times 180°}{4} = 135°$$

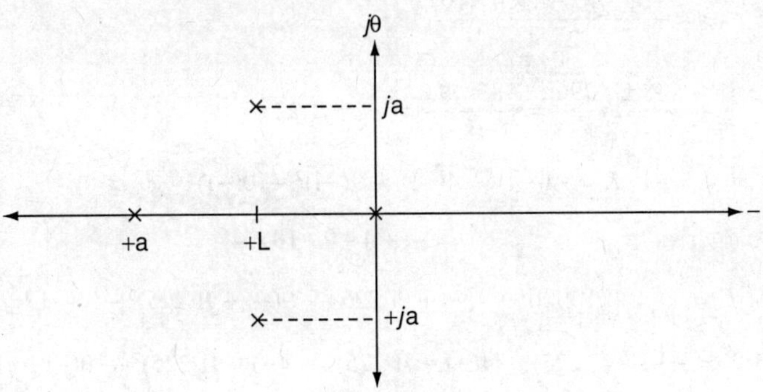

Fig. E11.27

$$\therefore \quad \theta_2 = \frac{2\times 2+1}{n-m}\times 180° = \frac{5\times 180°}{4} = 225°$$

$$\therefore \quad \theta_3 = \frac{2\times 3+1}{n-m}\times 180° = \frac{7\times 180°}{4} = 315°$$

Step 5:

$$\text{Centroid } (\sigma_c) = \frac{0-2-2-2-0}{n-m} = \frac{-6}{4} = -1.5$$

Step 6: Break-away point

Here, $1+G(s)H(s)=0$

$$\therefore \qquad 1+\frac{K}{s(s+2)(s^2+2s+5)}=0$$

i.e.,
$$K=-\frac{s(s+2)(s^2+2s+5)}{1}$$
$$=-s(s^3+2s^2+5s+2s^2+4s+10)$$
$$=-s(s^3+4s^2+9s+10)$$
$$=-s^4-4s^3-9s^2-10s$$

$$\therefore \qquad \frac{dK}{ds}=-(4s^3+12s^2+18s+10)=0$$

$$\therefore \qquad 4s^3+12s^2+18s+10=0$$

The break-away point between 0 and –2. Let us try $s=-1$ as starting point.

–1	4	12	18	10
		–4	–8	–10
	4	8	10	0

Other break-away points are obtained by solving

$$4s^2+8s+10=0$$

i.e.,
$$s=\frac{-8\pm\sqrt{8^2-4\times4\times10}}{2\times4}$$
$$=\frac{-8\pm j\sqrt{96}}{8}=\frac{-8\pm j9.8}{8}=-1\pm j1.225$$

For $\quad s=-1,\ K=-1(-1)^4-4(-1)^3-9(-1)^2-10(-1)$
$$=-1+4-9+10=4$$

For $\quad s=-1+j1.225,\ K=-(-1+j1.225)^4-4(-1+j1.225)^3-9(-1+j1.225)^2-10(-1+j1.225)=6.25$

For $\quad s=-1-j1.225,\ K=-(-1-j1.225)^4-4(-1-j1.225)^3-9(-1-j1.225)^2-10(-1-j1.225)=6.25$

For $\quad s=-1\pm j1.225,\ K$ is positive. Hence, the points are valid break-away points.

Step 7: ϕ_D at $s=-1+j2$ is calculated graphically. Figure E11.27(a) is considered.

$$\sum\phi_{\text{zero}}=\sum\phi_z=0$$

$$\sum\phi_{\text{poles}}=\sum\phi_p=\theta_1+\theta_2+\theta_3$$

$$\therefore \quad \theta_1=180^\circ-\tan^{-1}\frac{2}{1}=180^\circ-63.44^\circ=116.56^\circ$$

$$\therefore \qquad \theta_2=90^\circ$$

$$\therefore \quad \theta_3 = \tan^{-1}\frac{2}{1} = 63.44°$$

$$\therefore \quad \theta_1 + \theta_2 + \theta_3 = 116.56° + 90° + 63.44° = 270°$$

$$\therefore \quad \angle GH = \Sigma\phi_z - \Sigma\phi_p = 0 - 270° = -270°$$

$$\therefore \quad \phi_D = 180° + \text{Arg(GH)} = 180° - 270° = -90°$$

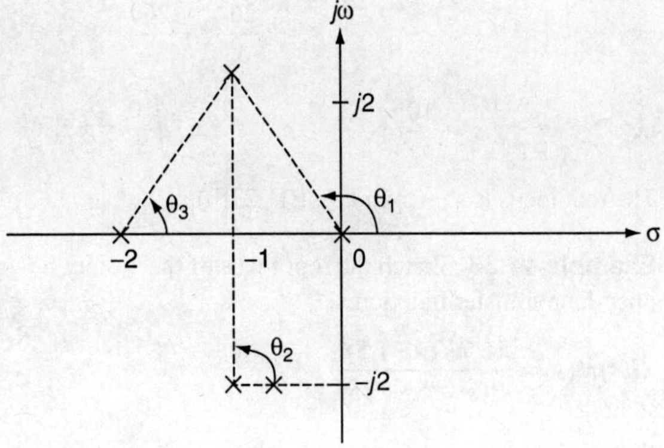

Fig. E11.27(a)

Since $-1 - j2$ is the complex conjugate of $-1 + j2$.

$$\therefore \quad \phi'_D \text{ for } -1 - j2 \text{ is } 90.$$

Step 8: Cross-over at $j\omega$-axis

The characteristic equation of the system is given by

$$1 + G(s)H(s) = 0$$

i.e., $$1 + \frac{K}{s(s+2)(s^2+2s+5)} = 0$$

i.e., $$s(s+2)(s^2+2s+5) + K = 0$$

i.e., $$s(s^3 + 2s^2 + 5s + 2s^2 + 4s + 10) + K = 0$$

i.e., $$s(s^3 + 4s^2 + 9s + 10) + K = 0$$

i.e., $$s^4 + 4s^2 + 9s^2 + 10s + K = 0$$

Routh array is given by

$$\begin{array}{c|ccc} s^4 & 1 & 9 & K \\ s^3 & 4 & 10 & \\ s^2 & \frac{26}{4} & K & \\ s^1 & \dfrac{\frac{260}{4} - 4K}{\frac{26}{4}} & & \\ s^0 & K & & \end{array}$$

Cross-over at $j\omega$-axis is at

$$\frac{\frac{260}{4} - 4K}{\frac{26}{4}} = 0$$

$$\therefore \quad K = \frac{260}{16}$$

Auxiliary equation at s^2 row is

$$\frac{26}{4}s^2 + K = 0$$

i.e., $$\frac{26}{4}s^2 + \frac{260}{16} = 0$$

i.e., $$s^2 + \frac{10}{4} = 0 \quad \text{i.e.,} \quad s = \pm j\sqrt{\frac{10}{4}} = \pm j1.58$$

The root locus is shown in Fig. E11.27(b).

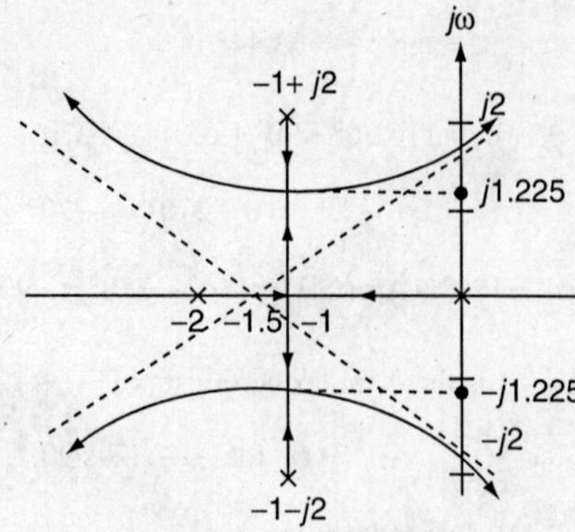

Fig. E11.27(b)

Example 11.33 Sketch the root locus of the system having open-loop transfer function

$$G(s)H(s) = \frac{16.7K(s+1.5)}{s^2(s^2+5s+25)}$$

Solution

Here, the open-loop transfer function of the system is

$$G(s)H(s) = \frac{16.7K(s+1.5)}{s^2(s^2+5s+25)}$$

$$= \frac{16.7K(s+1.5)}{s^2(s+2.5+j4.33)(s+2.5-j4.33)}$$

Step 1: Here, the number of poles $(n) = 4$ and the number of zeros $(m) = 1$. The total number of loci = 4. The total number of loci terminating at infinity $= n - m = 4 - 1 = 3$ because one locus will terminate at the zero at $s = -1.5$.

Step 2: Figure E11.28 shows the pole-zero plot of the open-loop transfer function of the system.

Step 3: Real axis loci are only present for $-\infty < \sigma < -1.5$.

Step 4: The number of asymptotes are $n - m = 4 - 1 = 3$.

$$\therefore \quad q = 0, 1, 2.$$

$$\therefore \quad \theta_0 = \frac{2\times 0+1}{3}\times 180° = 60°$$

$$\therefore \quad \theta_1 = \frac{2\times 1+1}{3}\times 180° = 180°$$

$$\therefore \quad \theta_2 = \frac{2\times 2+1}{3}\times 180° = 300°$$

Step 5: Centroid $(\sigma_c) = \dfrac{(-2.5-2.5)-(-1.5)}{3} = -1.167.$

Step 6: Break-away point:

$$1 + G(s)H(s) = 0$$

i.e., $$1 + \frac{16.5K(s+1.5)}{s^2(s^2+5s+25)} = 0$$

i.e., $$\frac{16.5K(s+1.5)}{s^2(s^2+5s+25)} = -1$$

i.e., $$K = \frac{s^2(s^2+5s+25)}{16.5(s+1.5)}$$

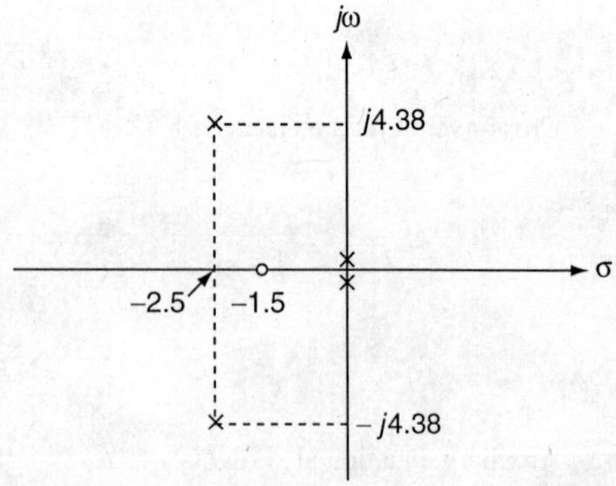

Fig. E11.28

$$\therefore \qquad \frac{dK}{ds} = 0$$

i.e.,
$$\frac{(s+1.5)(4s^3+15s^2+50s)-(s^4+5s^3+25s^2)(1)}{16.5(s+1.5)} = 0$$

i.e.,
$$3s^4 + 16s^3 + 47.5s^2 + 75s = 0$$

i.e.,
$$3s^3 + 16s^2 + 47.5s + 75 = 0.$$

Since the break-away point will be beyond the zero on the left, let us take $s = -3$ as first trial.

−3	3	16	47.5	75
		−9	− 21	−79.5
	3	7	26.5	(−4.5)

Let us take a trial $s = -4$ as second trial.

−4	3	16	47.5	75
		−12	− 16	−126
	3	4	31.5	(−51)

Here, there are no changes of sign. Therefore, break-away point is before −3. Let us take $s = -2$ as third trial.

−2	3	16	47.5	75
		−6	− 20	− 55
	3	10	27.5	(+20)

There are changes sign here. The root is between −3 and 2. Let us take $s = -2.5$ as first guess the fourth trial.

−2.5	3	16	47.5	75
		−8.25	−21.31	−72
	3	7.75	26.19	(+3)

At $s = -3, f(-3) = -4.5$ while at $s = -2.5, f(-2.5)$ is +3. The root will be near $s = -2.85$ or so. This much error is acceptable. The approximate break-away point is at $\sigma_b = -2.8$.

Step 8: Intersection of jω-axis

Let us put A = 16.5 *K*.

The characteristic equation of the system becomes $1 + G(s)H(s) = 0$

i.e.,
$$s^2(s^2 + 5s + 25) + A(s + 1.5) = 0$$

i.e.,
$$s^4 + 5s^3 + 25s^2 + As + 1.5A = 0$$

Routh array is given by

$$\begin{array}{l|lll}
s^4 & 1 & 25 & 1.5A \\
s^3 & 5 & A & 0 \\
s^2 & \dfrac{125-A}{5} & 1.5A & \\
s^1 & \dfrac{\dfrac{125-A}{5}\times A - 7.5A}{\dfrac{125-A}{5}} & 0 & \\
s^0 & 1.5A & &
\end{array}$$

Equating 1st term of s^1 to zero, we have

$$\frac{\frac{(125-A)A}{5}-7.5A}{\frac{125-A}{5}}=0$$

i.e., $$125A-A^2-37.5A=0$$

i.e., $$A^2-87.5A=0$$

i.e., $$A=\text{either } 0 \text{ or } 87.5$$

Taking $A=87.5$,

$$16.7K=87.5 \quad [\because A=16.5K]$$

i.e., $$K=5.23$$

If A is taken zero, s^1 row will become full of zeros.
Auxiliary equation for s^2 row becomes

$$\left(\frac{125-87.5}{5}\right)s^2+1.5\times 87.5=0$$

i.e., $$7.5s^2+131.25=0$$

i.e., $$s^2=-17.5$$

i.e., $$s=\pm j\,4.18$$

Step 9: To calculate the angle of departure at $s=-2.5+j\,4.33$ Figure E11.28(a) is considered.

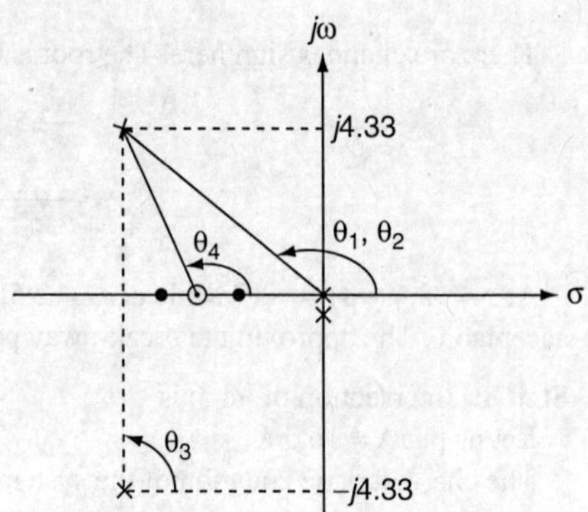

Fig. E11.28(a)

Here, $$\Sigma\phi_p=\theta_1+\theta_2+\theta_3 \quad and \quad \Sigma\phi_z=\theta_4$$

$$\theta_1=\theta_2=180°-\tan^{-1}\frac{4.33}{2.5}=180°-60°=120°.$$

$$\theta_3=90°.$$

$$\therefore \quad \sum\theta_p=\theta_1+\theta_2+\theta_3=120°+120°+90°=330°.$$

$$\theta_4=180°-\tan^{-1}\frac{4.33}{1}=180°-77°=103°.$$

$$\therefore \quad \sum\phi_z=\theta_4=103°.$$

$$\text{Arg } GH=\sum\phi_z-\sum\phi_p=103°-330°=-227°.$$

$$\therefore \quad \phi_D=180°+\text{Arg } GH=180°-227°=-47°$$

$$\therefore \quad \phi_D \text{ at } s=-2+j\,4.33 \text{ is } -47°.$$

Since $-2-j\,4.33$ is complex conjugate of $-2+j\,4.33$, ϕ_D *at* $s=-2-j\,4.33$ will be 47°. The root locus is shown in Fig. E11.28(b).

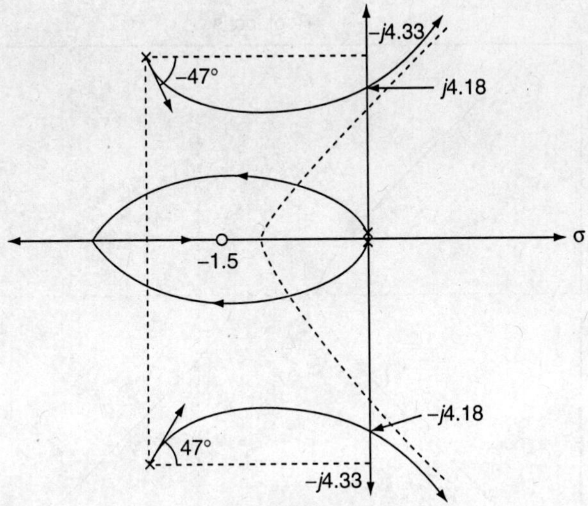

Fig. E11.28(b)

Program 3 Sketch the root locus for the open loop transfer function $G(s)H(s) = \dfrac{K}{(s+1)\,(s+2)(s+4)(s+5)}$.

Solution

```
>> num = [0 1]
num =
   0   1
>> q1 = [1 1];
>> q2 = [1 2];
>> q3 = [1 4];
>> q4 = [ 1 5];
>> den = conv(q1,q2);
>> den = conv(den,q3);
>> den = conv(den,q4)
den =
   1   12   49   78   40
>> sys = tf(num,den)

Transfer function:

          1
---------------------------------
s^4 + 12 s^3 + 49 s^2 + 78 s + 40
>> rlocus(sys)
>>
```

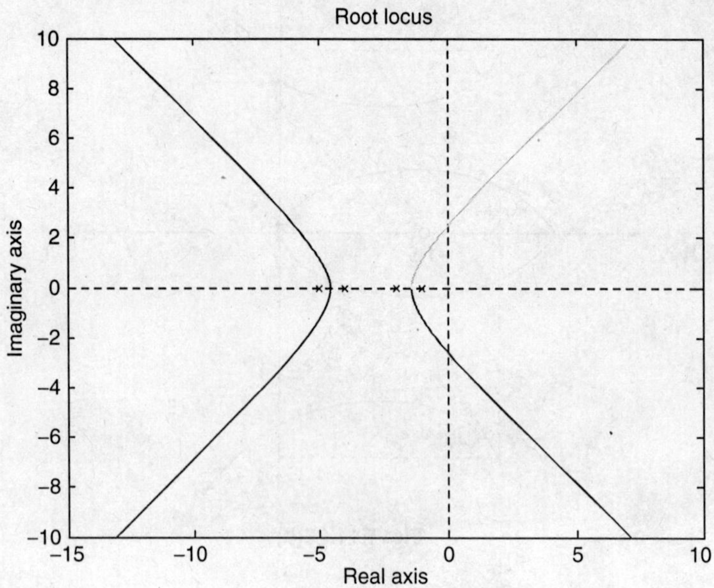

Fig. P3

SHORT QUESTIONS WITH ANSWERS

1. What do you mean by root locus?

   Ans. Root locus is defined as the plot of the locus of the closed-loop poles as a function of open-loop gain (K), when K is varied from $-\infty$ to ∞. If K is varied from 0 to $+\infty$, it is known as direct root locus. On the other hand, it K is varied from $-\infty$ to 0, it is termed as inverse root locus.

2. State the angle and magnitude condition of root locus?

   Ans. Magnitude condition: $|G(s)H(s)| = 1$.

   Angle condition: $\angle G(s)H(s) = \pm(2q+1)180°$.

3. Define the number of loci.

   Ans. Let the number of OLTF poles and OLTF zeros be n and m, respectively. If $n > m$, the number of loci in n. Each locus will start from OLTF pole and end on OLTF zero. Here, $(n - m)$ loci will end at infinity and total loci $= n$.

4. What do you mean by the real axis loci?

   Ans. Part of the loci will lie on the real axis. A point on the real axis will lie on the root locus if and only if the sum of the OLTF poles and zeros to the right of the point is odd.

5. Write the expression for angle of asymptotes.

   Ans. The angle of asymptotes is given by

$$\theta = \frac{2q+1}{n-m} \times 180° \text{ for } K > 0$$

   where $q = 0, 1, 2, \ldots, n-m-1.$

$$\theta = \frac{2q}{n-m} \times 180° \text{ for } K < 0$$

where $q = 0, 1, 2, \ldots, n-m-1.$

6. Define centre of asymptotes.

Ans. The point where the asymptotes touch the real axis is called centroid (σ_c) and it is given by

$$\sigma_c = \frac{\sum \text{real parts of poles of } G(s)H(s) - \sum \text{real parts of zeros of } G(s)H(s)}{n-m}$$

7. What do you mean by break-away/break-in point?

Ans. Break-away point is defined as a point at which the root locus comes out of the real axis and break-in point is defined as a point at which the root locus enters the real axis. The break-away or break-in points are also as the points on the root locus at which multiple roots of the characteristic equation occur.

8. What do mean by angle of departure and arrival?

Ans. The root locus leaves from a complex pole and arrives at a complex zero. These two angles are known as angle of departure or angle of arrival, respectively. The angle of departure (θ_d) is given by $\theta_d = 180° + \arg[G(s)H(s)]$, where $\arg G(s)H(s)$ is the angle of $G(s)H(s)$ excluding the pole for whom the angle is to be calculated. It is also possible to calculate $\arg G(s)H(s)$, graphically. Similarly, the angle of arrival is given by $\theta_a = 180° - \arg[G(s)H(s)]$, where $\arg G(s)H(s)$ is the angle of $G(s)H(s)$ excluding the zero for whom the angle is to be calculated.

9. What do you mean by gain margin?

Ans. $\text{Gain margin (GM)} = \dfrac{\text{Value of } K \text{ at imaginary crossover}}{\text{Design value of } K}$

If the root locus does not cross the $j\omega$ axis, gain is ∞, GM represents the maximum gain that can be multiplied to the system to be just on the verge of instability.

10. What do you mean by phase margin?

Ans. Phase margin (PM) can be determined for a given value of K as follows:

- Calculate ω, for which $|G(j\omega)H(j\omega)| = 1$ for the given value of K.
- Calculate arg $[G(j\omega)H(j\omega)]$.
- Phase margin = $180° + \arg\ [G(j\omega)H(j\omega)]$.

11. State the effects of addition of poles.

Ans. The effects of addition of the poles are as follows:

- There is change in shape of the root locus and it shifts towards the imaginary axis.
- The intercept on the jω axis occurs for lower values of K because of asymptote's angle being lower down.
- System becomes oscillatory.
- Gain margin and relative stability decreases.
- There is reduction in the range of K.
- A sluggish response can be changed to a quicker response for artful introduction of a pole.
- Setting time increases.

12. State the effect of addition of zeros.

Ans. The effects of addition of zeros are as follows:

- There is change in shape of the root locus and it shifts towards the left of s-plane.
- Stability of the system is enhanced.
- Range of K increases.
- Settling time speeds up.

EXERCISES

Questions for Practice

1. Calculate the angle and magnitude of $G(s)H(s) = 18(s+2)/[s(s+1)(s+4)]$ at the following points:

   (i) $s = j3$, (ii) $s = j8$, (iii) $s = -2 + j3$, (iv) $s = -5$,

   (v) $s = -7$ and (vi) $s = -5 + j3$.

   Ans. (i) 1.368, −142.12°; (ii) 0.2573, −160.35°; (iii) 1.3136, 161.56°; (iv) 2.7,0°; (v) 0.714, 0° and (vi) 0.828, 94.4°

2. Determine the break-away point on each root locus for the following transfer functions:

   (i) $G(s)H(s) = \dfrac{K}{s(s+3)(s+5)}$ and (ii) $G(s)H(s) = \dfrac{K}{s^2(s+10)}$

   Ans. (i) −1.2137, (ii) − 5

3. Find the angle of departure of the root locus from the pole $s = -8 + j8$ for

   $$G(s)H(s) = \frac{K(s+6)}{(s+4)(s+8+j8)(s+8-j8)} \quad (K > 0)$$

   Ans. 77.48°

4. Find the angle of arrival at the complex zero $s = -2 + j3$ for

   $$G(s)H(s) = \frac{K(s+2+j3)(s+2-j3)}{(s+4)(s+1+j2)(s+1-j2)} \quad (K > 0)$$

   Ans. 22.62°

5. Sketch the root locus of a unity feedback system with open-loop transfer function $G(s) = K/[s(s+2)(s+4)]$. Find the range of values of K for which the system has damped oscillatory response.

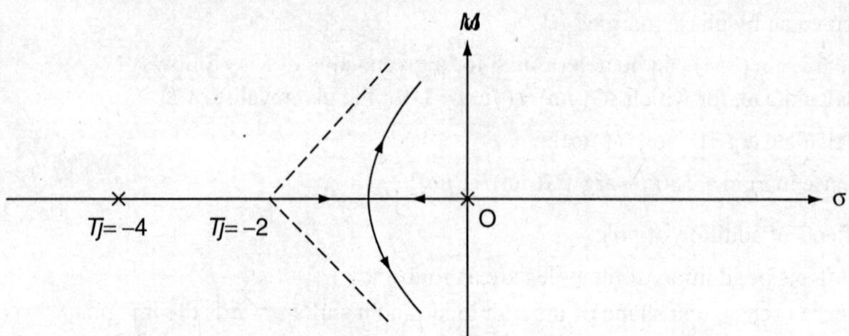

Figure 1

   Ans. $3.0792 < K < 48$

6. For the control system shown in Fig. 2, draw the root locus diagram.

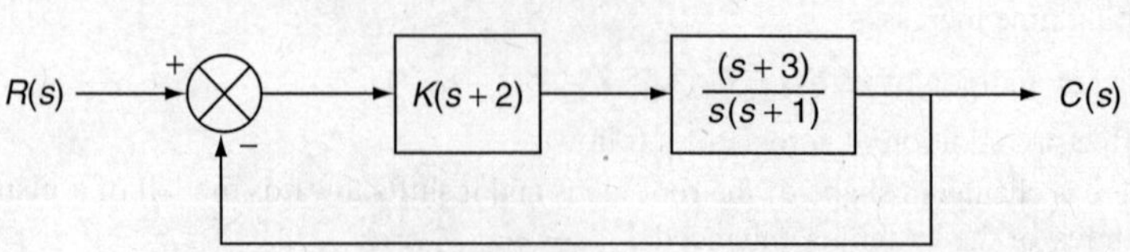

Figure 2

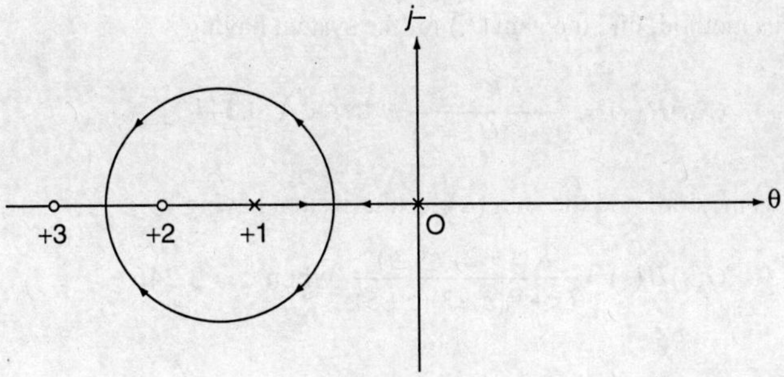

Figure 3

Ans.

7. The open-loop control system transfer is $G(s) = K/[s(s^2 + 4s + 8)]$. Sketch the root loci of the system.

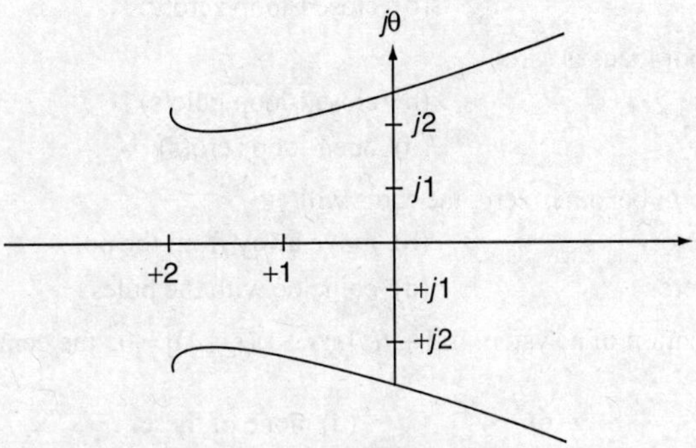

Figure 4

Ans.

8. Sketch the root locus diagram for the feedback control system having the following open-loop transfer function. Assume that K will take all positive values from 0 to ∞.

$$G(s) = \frac{K}{s(s+1)(s^2+4s+8)}$$

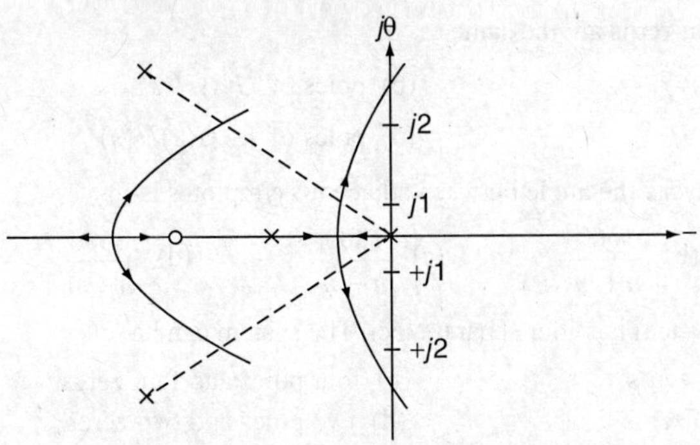

Figure 5

Ans.

9. Using the root locus method, find the gain (K) for the system having

$$G(s)H(s)=\frac{K}{s(s+2)(s+3)} \text{ when } \xi = 0.341$$

Ans. $K = 8.28$

10. Using the root locus method, find the gain (K) for the system having

$$G(s)H(s)=\frac{K(s+2)(s+4)}{(s+1)(s+3)(s+5)} \text{ when } \xi = 0.246$$

Ans. $K = 33.2$

TEST YOURSELF

1. The starting point(s) of a root locus is (are)
   (a) open-loop pole(s) (b) closed-loop pole(s)
   (c) open-loop zero(s) (d) closed-loop zero(s)
2. The end point(s) of a root locus is (are)
   (a) open-loop pole(s) (b) closed-loop pole(s)
   (c) closed-loop zero(s) (d) open-loop zero(s)
3. If the gain (K) of a system becomes zero, the roots will
   (a) move away from zeros (b) move away from the poles
   (c) coincide with the zeros (d) coincide with the poles
4. If the characteristic equation of a system is $1+K/[s(s+1)(s+4)]=0$, the centroid of the asymptotes in the root locus is given by
   (a) −1 (b) −2 (c) −4 (d) none of these
5. The root locus plot is symmetrical about the real axis because
   (a) roots occur simultaneously in LH and RH planes
   (b) complex roots occur in conjugate pairs
   (c) all roots occur in pairs
   (d) none of these
6. The break-away points of the root locus occur at
   (a) imaginary axis (b) real axis
   (c) multiple roots of characteristic equation (d) either (a) or (b)
7. Loop transfer function zeros are the same as
   (a) zeros of $G(s)H(s)$ (b) poles of $G(s)H(s)$
   (c) zeros of $1+G(s)H(s)$ (d) poles of $1+G(s)H(s)$
8. In the root locus analysis, the angle between adjacent asymptotes is
   (a) $\frac{180°}{n+m}$ (b) $\frac{360°}{n+m}$ (c) $\frac{180°}{n-m}$ (d) $\frac{360°}{n-m}$
9. The root locus of a system has four separate loci. The system can have
   (a) four poles if four zeros (b) four poles and four zeros
   (c) six poles and two zeros (d) two poles and two zeros
10. The algebraic sum of the angles of the vectors from all poles and zeros to the point on any root locus segment is
    (a) 180° (b) 150°
    (c) 180° or its odd multiple (d) 80° or its odd multiple

11. Intersection of root locus branches with the imaginary axis may be determined by the use of the
(a) Nyquist criterion (b) Routh criterion (c) polar plot (d) Bode plot

12. The root locus of a unity feedback system is shown is Figure T1. The open-loop transfer function of the system is
(a) $\dfrac{K}{s(s+1)(s+3)}$ (b) $\dfrac{K(s+1)}{s(s+3)}$
(c) $\dfrac{K(s+3)}{s(s+1)}$ (d) $\dfrac{Ks}{(s+1)(s+3)}$

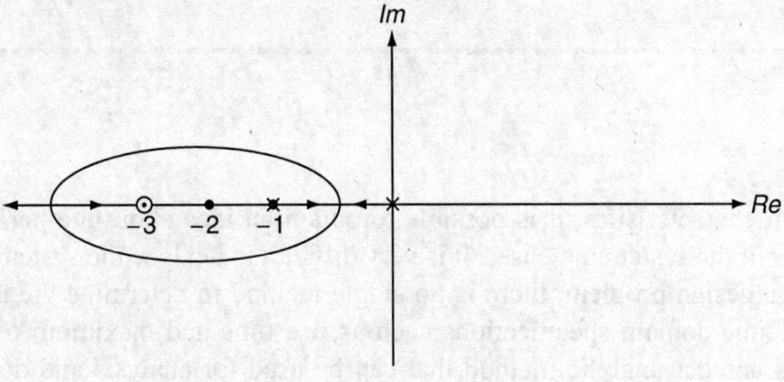

Figure T1

13. Match List I with List II in respect of the open-loop transfer function

$$G(s)H(s) = \frac{K(s+10)(s^2+20s+500)}{s(s+20)(s+50)(s^2+4s+5)}$$

List I (types of loci)	List II (numbers)
A. Separate	1. One
B. Loci on the real axis	2. Two
C. Asymptotes	3. Three
D. Break-away points	4. Five

14. If the characteristic equation of a closed-loop system is $1+K/[s(s+1)(s+2)]+1=0$, the centroid the asymptotes of the root locus will be at
(a) 0 (b) 2 (c) −1 (d) −2

15. The open-loop transfer function of a feedback control system $K/[(s^2+3s+6)]$. The break-away points(s) of its root locus plot
(a) exist at $(-1 \pm j)$ (b) exist at $\left[-\frac{3}{2} \pm j\sqrt{\frac{15}{16}}\right]$ (c) exist at origin (d) do not exist

ANSWERS

1 (a) 2 (d) 3 (d) 4 (d) 5 (b) 6 (c) 7 (a) 8 (d) 9 (a) 10 (c) 11 (b)
12 (c) 13 A→4 B→3 C→2 D→1 14 (c) 15 (b)

Chapter 12

Frequency Domain Analysis

Using time domain characteristics, it is possible to obtain an idea about the performance of a given system. If the order of the system increases, it is very difficult to analyze the system by the time domain approach. Also in a design problem, there is no single method to determine the design solutions that obeys the various time domain specifications such as rise time and maximum overshoot. Frequency domain analysis is another analytic method that can be used for analysis and design. During 1930s, Bode and Nyquist proposed methods to analyze the stability of a system. The root locus method (i.e., time domain method) was developed during 1940s. The frequency response method was developed earlier than the time domain method. The aim of this chapter is to discuss the basics of the frequency domain analysis of systems.

12.1 Frequency Response

It is defined as the steady-state response of a system due to a sinusoidal input. Fig. 12.1 shows a linear system with a sinusoidal input. Let us study its output and derive the relation between its input and output.

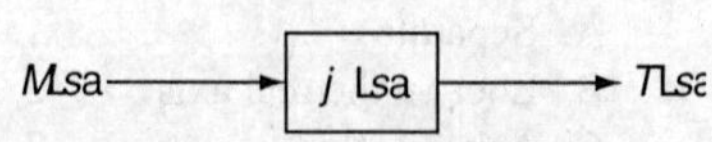

Fig. 12.1 A linear system with sinusoidal input

In Fig. 12.1, $x(t)$ is the sinusoidal input function, $y(t)$ is the output function and $g(t)$ is the transfer function. Let $X(s)$, $Y(s)$ and $G(s)$ be the Laplace transforms of $x(t)$, $y(t)$ and $g(t)$, respectively. $G(s)$ is the transfer function of the system and it is defined for the zero initial condition. Let $G(s)$ be given by

$$G(s) = \frac{N(s)}{D(s)} = \frac{N(s)}{(s+a)(s+b)(s+c)\cdots} \tag{12.1}$$

In Eq. (12.1), $N(s)$ is a polynomial of s and the poles of $G(s)$ are at $-a$, $-b$, $-c$, … etc.

Again $$G(s) = \frac{Y(s)}{X(s)} \tag{12.2}$$

Using Eq. (12.1) and Eq. (12.2), we have

$$Y(s) = \frac{N(s)X(s)}{(s+a)(s+b)(s+c)\cdots} \tag{12.3}$$

Let $x(t)$ = A sin ωt be the input to the system. The Laplace transform of $x(t)$ is given by

$$X(s) = \frac{A\omega}{s^2 + \omega^2} \tag{12.4}$$

Using Eq. (12.4), Eq. (12.3) can be expressed as follows:

$$Y(s) = \left[\frac{N(s)}{(s+a)(s+b)(s+c)\cdots}\right]\left(\frac{A\omega}{s^2+\omega^2}\right)$$

$$= \frac{A_1}{s+a} + \frac{A_2}{s+b} + \frac{A_3}{s+c} + \ldots + \frac{B_1}{s+j\omega} + \frac{B_2}{s-j\omega}$$

i.e.,
$$y(t) = A_1 e^{-at} + A_2 e^{-bt} + A_3 e^{-ct} + \ldots + B_1 e^{-j\omega t} + B_2 e^{+j\omega t} \tag{12.5}$$

The terms with coefficients A_1 to A_n will decay to zero as time tends to infinity, because these terms contribute to decaying exponents. Therefore, these terms constitute transient response. The terms with coefficients B_1 and B_2 will contribute steady-state response. Therefore, the steady-state response of Eq. (12.5) is given by

$$y_{ss}(t) = B_1 e^{-j\omega t} + B_2 e^{j\omega t} \tag{12.6}$$

$\therefore$
$$B_1 = \left[G(s)\frac{A\omega}{(s^2+\omega^2)}(s+j\omega)\right]_{s=-j\omega} = -\frac{A}{2j}G(-j\omega) \tag{12.7}$$

and
$$B_2 = \left[G(s)\frac{A\omega}{(s^2+\omega^2)}(s-j\omega)\right]_{s=j\omega} = \frac{A}{2j}G(j\omega) \tag{12.8}$$

$G(j\omega)$ is a complex quantity and we can write $G(j\omega)$ as follows:

$$G(j\omega) = |G(j\omega)|e^{j\phi}, \tag{12.9}$$

where $|G(j\omega)|$ is the magnitude of $G(j\omega)$ and ϕ its angle.
Using Eq. (12.7) and Eq. (12.8), we have from Eq. (12.6):

$$y_{ss}(t) = -\frac{A}{2j}|G(j\omega)|e^{-j\phi}e^{-j\omega t} + \frac{A}{2j}|G(j\omega)|e^{j\phi}e^{j\omega t}$$

$$= A|G(j\omega)|\left\{\frac{e^{j(\omega t+\phi)} - e^{-j(\omega t+\phi)}}{2j}\right\}$$

$$= A|G(j\omega)|\sin(\omega t+\phi)$$

i.e.,
$$y_{ss}(t) = |Y(j\omega)|\phi \tag{12.10}$$

where
$$|Y(j\omega)| = A|G(j\omega)| \tag{12.11}$$

From Eq. (12.11),

$$|Y(j\omega)| = |G(j\omega)||X(j\omega)| \tag{12.12}$$

i.e.,
$$|G(j\omega)| = \frac{|Y(j\omega)|}{|X(j\omega)|} \tag{12.13}$$

and $$\angle|Y(j\omega)| = \phi = \phi + 0° \tag{12.14}$$

i.e., $$\angle|Y(j\omega)| = \angle|G(j\omega)| + |X(j\omega)| \tag{12.14a}$$

i.e., $$\angle|G(j\omega)| = \angle|Y(j\omega)| - \angle|X(j\omega)| \tag{12.14b}$$

Therefore, the frequency of sinusoidal output is the same as the input and there is merely any change in the amplitude and phase of the output with respect to input.

12.2 Methods in Frequency Response

The magnitude function and phase function are the two plots in the frequency response characteristics. The angular frequency (ω) is the common variable for both these plots. The following curves are commonly used in frequency response analysis:

(i) *Polar plot:* The plot of the magnitude versus phase angle in polar coordinate where ω is varied from 0 to ∞ is called polar plots.

(ii) *Bode plot:* The plots of both (i) magnitude and (ii) phase angle versus frequency ω in logarithmic values are called Bode plots.

(iii) *Magnitude versus phase angle plot:* The magnitude versus phase angle plot on rectangular coordinate as ω varied is known as magnitude versus phase angle plot.

12.3 Definitions of Frequency Domain Specifications

Response peak (M_r): From Fig. 12.2, the closed-loop transfer function is expressed by

$$\frac{C(s)}{R(s)} = \frac{G(s)}{1+G(s)H(s)} = M(s)$$

i.e., $$M(j\omega) = \frac{G(j\omega)}{1+G(j\omega)H(j\omega)} \tag{12.15}$$

The maximum value of $M(j\omega)$ as ω varied is termed as resonance peak (M_r). The relative stability of the closed-loop system can be indicated by M_r. If M_r is large, there will be large maximum overshoot in a step response. Generally, M_r is between 1.1 and 1.5.

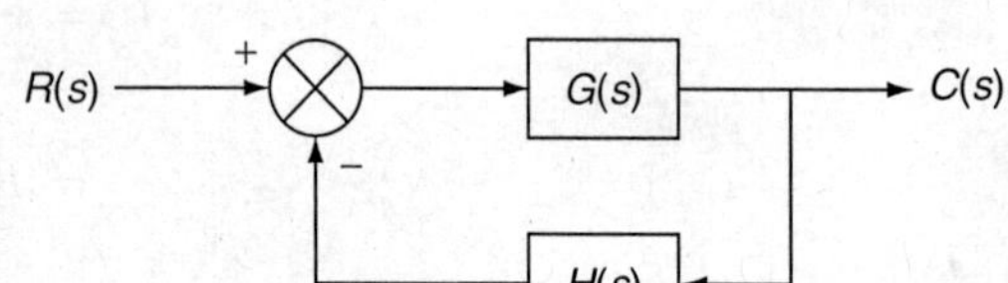

Fig. 12.2 Closed-loop system

Resonance frequency: The frequency at which peak resonance M_r occurs is known as resonance frequency (ω_r).

Bandwidth (BW): The frequency at which the magnitude of $M(j\omega)$ drops by 30 dB below a certain specified level, generally 0 dB level is known as bandwidth. If bandwidth is large, higher frequencies will pass through the system. This indicates a faster rise in time. Therefore, bandwidth gives an idea about transient response. Again, small bandwidth means lower frequencies are passed and hence response is sluggish. From bandwidth, noise filtering can be known because large bandwidth indicates susceptibility to noise. Bandwidth also implies the sensitivity to parameter variations.

***Cut-off rate*:** The slope of the log magnitude curve near the cut-off frequency is known as cut-off rate. The ability of the system to distinguish a signal from noise can be determined by cut-off rate.

***Phase Margin*:** If ϕ is the phase angle of a system at unity gain, the phase margin is given by $180° + \phi$. The frequency (ω_1) at which gain is unity is known as gain crossover frequency. The increment in the system angle that caused the system to be unstable is indicated by phase margin. It is a measure of relative stability.

***Gain margin*:** The frequency (ω_1) at which the phase angle of the transfer function is $-180°$ is known as phase crossover frequency. The gain at this frequency ω_1 is $|G(j\omega_1)H(j\omega_1)|$ and gain margin is given by $1/|G(j\omega_1)H(j\omega_1)|$. The increment in gain required to cause the system to become unstable is indicated by gain margin.

Nyquist analysis, Bode plot, Nichols chart and root locus method are the graphical methods. Out of these, root locus method is a time domain method whereas the others are frequency domain methods.

12.4 Correlation between Time and Frequency Domain Specifications for a Second-Order System

The relationship between frequency domain specifications and time domain specifications is important to get these ideas before designing the system. For a standard second order system with open loop,

$$G(s) = \frac{\omega_n^2}{s(s+2\xi\omega_n)} \quad \text{and} \quad H(s) = 1$$

Transfer function
$$T(s) = \frac{C(s)}{R(s)} = \frac{\omega_n^2}{s^2 + 2\xi\omega_n s + \omega_n^2}$$

$\therefore$
$$T(j\omega) = \frac{C(j\omega)}{R(j\omega)} = \frac{\omega_n^2}{(j\omega)^2 + 2\xi\omega_n(j\omega) + \omega_n^2}$$

i.e.,
$$T(j\omega) = \frac{1}{1 + 2j\xi\left(\dfrac{\omega}{\omega_n}\right) - \left(\dfrac{\omega}{\omega_n}\right)^2} \tag{12.16}$$

Let $u = \dfrac{\omega}{\omega_n}$ from Eq. (12.16), we have

$$T(j\omega) = \frac{1}{1 + 2ju\xi - u^2} \tag{12.17}$$

i.e.,
$$|T(j\omega)| = M = \frac{1}{\sqrt{(1-u^2)^2 + (2u\xi)^2}} \tag{12.18}$$

and
$$\angle T(j\omega) = -\tan^{-1}\frac{2u\xi}{1-u^2} \tag{12.19}$$

Putting $\left.\dfrac{dM}{du}\right|_{u=u_r} = 0$, the resonant frequency ω_r can be obtained.

Now $\qquad \left.\dfrac{dM}{du}\right|_{u=u_r} = -\dfrac{1}{2}\dfrac{[2(1-u_r^2)(-2u_r)+8\xi^2 u_r]}{[(1-u_r^2)^2+(2u_r\xi)^2]^{3/2}} = 0$

i.e., $\qquad 4u_r - 8u_r\xi^2 - 4u_r^3 = 0$

i.e., $\qquad 4u_r(1-2\xi^2-u_r^2) = 0$ i.e., $u_r(1-2\xi^2-u_r^2) = 0$

Either $\qquad u_r = 0$ or $1-2\xi^2-4u_r^2 = 0$

Now, $\qquad u_r = 0$ gives $\omega_r/\omega_n = 0$ i.e., $\omega_r = 0$.

Again $\qquad 1-2\xi^2-u_r^2 = 0$

i.e., $\qquad u_r^2 = 1-2\xi^2$ i.e., $u_r = \sqrt{1-2\xi^2}$ (12.20)

i.e., $\qquad \dfrac{\omega_r}{\omega_n} = \sqrt{1-2\xi^2}$

i.e., $\qquad \omega_r = \omega_n\sqrt{1-2\xi^2}$ (12.21)

$$\therefore \qquad M_r = \left.G(j\omega)\right|_{u=u_r} = \frac{1}{\sqrt{(1-u_r^2)^2+(2u_r\xi)^2}} = \frac{1}{\sqrt{(2\xi^2)^2+4\xi^2(1-2\xi^2)}}$$

$$= \frac{1}{\sqrt{4\xi^4+4\xi^2-8\xi^4}} = \frac{1}{\sqrt{4\xi^2-4\xi^4}} = \frac{1}{\sqrt{4\xi^2(1-\xi^2)}}$$

$$\therefore \qquad M_r = \frac{1}{2\xi\sqrt{1-\xi^2}} \qquad (12.22)$$

From Eq. (12.20), u_r is valid as long as $1-2\xi^2 \geq 0$, i.e., $\xi \leq 0.707$ and Eq. (12.22) shows that M_r is a function of ξ. From Eq. (12.19), we have

$$\phi_r = \left.G(ju)\right|_{u=u_r} = -\tan^{-1}\frac{\sqrt{1-2\xi}}{\xi} \qquad (12.23)$$

Equation (12.21) shows that ω_r is a function of ξ and ω_n. From Eq. (12.21) and Eq. (12.22), the following conclusions can be drawn:

(i) M_r approaches infinity and ω_r approaches ω_n when ξ approaches zero.
(ii) M_r is greater than unity and ω_r is less than ω_n for $0 < \xi < 0.707$.
(iii) There is no resonant peak and the maximum value of M_r is unity and $\omega_r = 0$ when $\xi > 0.707$.

Figure 12.3(a) and Fig. 12.3(b) show the plot of magnitude (M) versus normalized frequency (u) and phase angle (ϕ) versus normalized frequency (u) respectively.

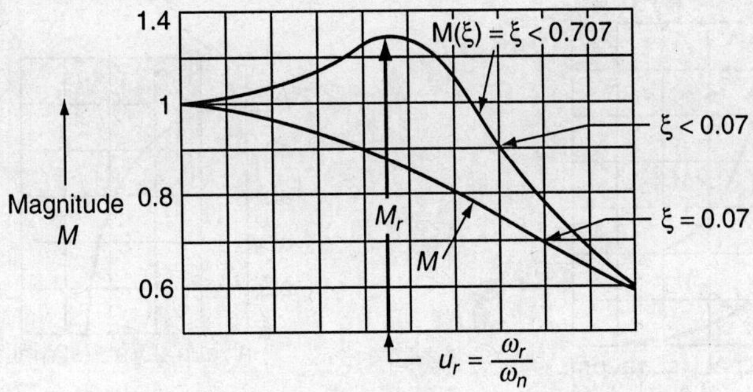

(a) Frequency response magnitude characteristics

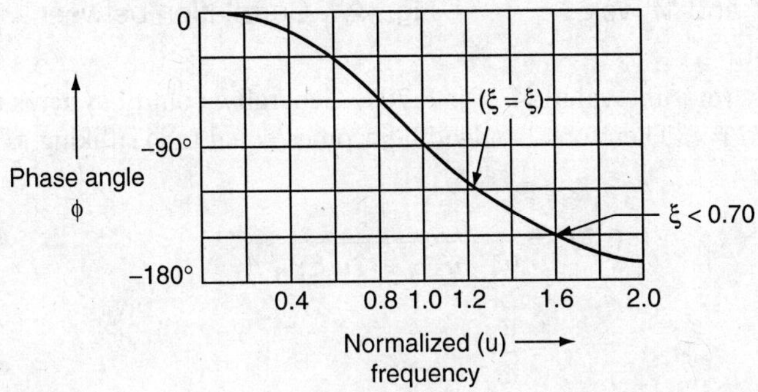

(b) Frequency response characteristics

Fig. 12.3 Frequency response

To determine cut-off rate and BW, let us consider the following figures. Figure 12.4 shows the plot of M_r versus ω known as the typical magnification curve of a feedback control system. Figure 12.5 shows the plot of bandwidth versus damping factor. Figure 12.6 shows the plot of M_r and M_p versus ξ. Figure 12.7 represents the correlation between ω_r and ω_d.

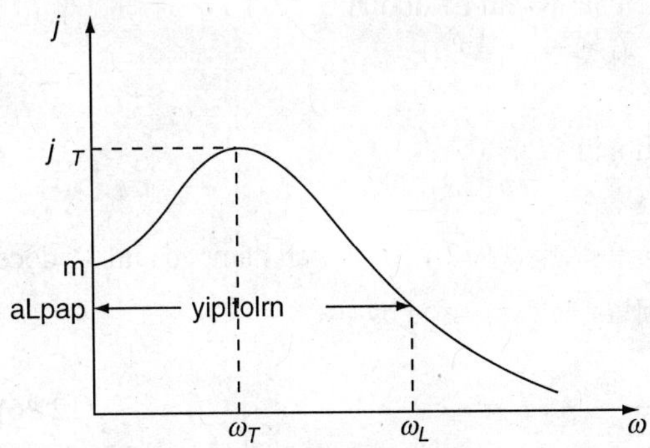

Fig. 12.4 Plot of M_r vs ω

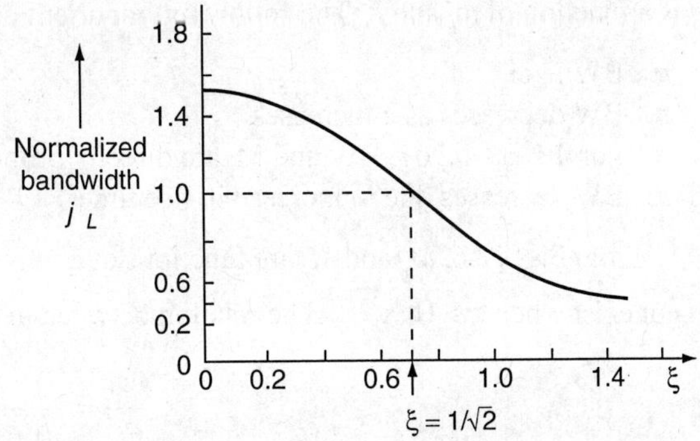

Fig. 12.5 Bandwidth vs ξ

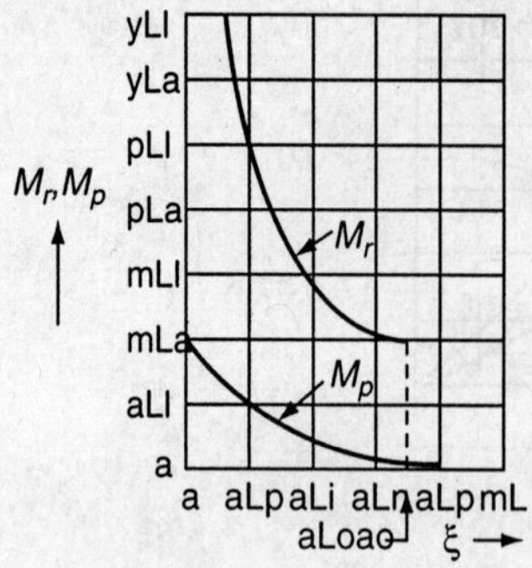

Fig. 12.6 M_r and M_p vs ξ

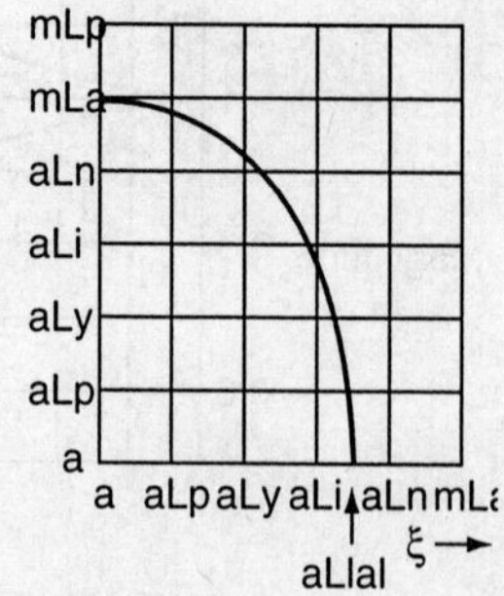

Fig. 12.7 Correlation between ω_r and ω_d

At cut-off frequency (ω_c), the value of M is 0.707. Generally, control systems are low pass-filters and at zero frequency $M = 1$. Therefore, bandwidth becomes equal to ω_c. Taking $M = 0.707$ and $u = u_c$ in Eq. (12.18), we have

$$0.707 = \frac{1}{[(1-u_c^2)^2 + (2u_c\xi)^2]^{1/2}}$$

where $$u_c = \frac{\omega_c}{\omega_n} \quad (12.24)$$

i.e., $$(1-u_c^2)^2 + (2u_c\xi)^2 = 2$$

Solving Eq. (12.24), we have

$$u_c^2 = (1-2\xi^2) \pm \sqrt{4\xi^4 - 4\xi^2 + 2}$$

i.e., $$\omega_c = \omega_n\left\{(1-2\xi^2) + \sqrt{4\xi^4 - 4\xi^2 + 2}\right\}^{1/2}$$

i.e., $$\text{BW} = \omega_c = \omega_n\left\{(1-2\xi^2) + \sqrt{4\xi^4 - 4\xi^2 + 2}\right\}^{1/2} \quad (12.25)$$

Positive sign is considered because BW is a positive quantity. Therefore, Eq. (12.25) shows that BW is a function of ω_n and ξ. The following important conclusions can be drawn:

- BW □ ω_n
- BW decreases as ξ increases
- For $0 < \xi < 0.707$, BW and M_r are directly proportional
- BW decreases due to increase in rise time

From Fig. 12.6, M_p and M_r are functions of ξ only for $0 < \xi < (1/\sqrt{2})$. M_p is well damped and M_r does not exist when $\xi > (1/\sqrt{2})$. The relation between ω_r and ω_d is expressed by

$$\frac{\omega_d}{\omega_r} = \frac{1-\xi^2}{\sqrt{1-2\xi^2}} \quad (12.26)$$

If a zero is added in the forward path of a second-order system, the following effects occur:

- Rise time decreases due to increase of bandwidth.
- Settling time will increase.
- System will be more stable.

On the other hand, if a pole is added in the forward path of a second-order system, the following effects occur:

- Rise time will increase and bandwidth will decrease.
- M_r will increase.
- System will be less stable.

Example 12.1 Determine the frequency domain specifications for a second-order system with unity feedback and $G(s) = 225/[s(s+6)]$.

Solution The transfer function $M(s)$ is given by

$$M(s) = \frac{C(s)}{R(s)} = \frac{G(s)}{1+G(s)H(s)} = \frac{\dfrac{225}{s(s+6)}}{1+\dfrac{225}{s(s+6)}} = \frac{225}{s^2+6s+225} \tag{1}$$

Comparing Eq. (1) with $\dfrac{\omega_n^2}{s^2+2\xi\omega_n s+\omega_n^2}$, we have

$$\omega_n^2 = 225 \text{ or } \omega_n = 15 \text{ rad/sec}$$

and

$$2\xi\omega_n = 6$$

i.e.,

$$\xi = \frac{6}{2\omega_n} = \frac{6}{2\times 15} = 0.2$$

$$M_r = \frac{1}{2\xi\sqrt{1-\xi^2}} = \frac{1}{2\times 0.2\times\sqrt{1-(0.2)^2}} = 2.60$$

Again

$$\omega_r = \omega_n\sqrt{1-2\xi^2} = 15\sqrt{1-2\times(0.2)^2} = 14.39 \text{ rad/sec}$$

and

$$\omega_c = \omega_n\left\{(1-2\xi^2)+\sqrt{4\xi^4-4\xi^2+2}\right\}^{1/2} = 15\times(0.92+1.359)^{1/2}$$

$$= 22.64 \text{ rad/sec}$$

Therefore,

$$\text{BW} = \omega_c = 22.64 \text{ rad/sec}$$

12.5 Determination of ω_{gc} and PM for Standard Second-Order System

Let us consider a system with $G(s)=\dfrac{\omega_n^2}{s(s+2\xi\omega_n)}$ and $H(s)=1$. The closed-loop transfer function of the system becomes

$$\frac{C(s)}{R(s)}=\frac{G(s)}{1+G(s)H(s)}=\frac{\omega_n^2}{s^2+2\xi\omega_n s+\omega_n^2}. \tag{12.27}$$

The open-loop transfer function of the system i.e., $G(s)H(s)$ is to be taken into account to find ω_{gc} and PM.

$$\therefore \quad G(s)H(s)=\frac{\omega_n^2}{s(s+2\xi\omega_n)}=\frac{\omega_n^2}{s^2+2\xi\omega_n s}$$

$$\therefore \quad G(j\omega)H(j\omega)=\frac{\omega_n^2}{(j\omega)^2+2\xi\omega_n(j\omega)}=\frac{\omega_n^2}{-\omega^2+2j\xi\omega\omega_n} \tag{12.28}$$

The magnitude of $G(j\omega)H(j\omega)$ at the gain cross-over frequency (ω_{gc}) is 0 dB

i.e., $20\log_{10}|G(j\omega)H(j\omega)|=0 \therefore |G(j\omega)H(j\omega)|=1$ for $\omega=\omega_{gc}$

$$\therefore \quad G(j\omega)H(j\omega)=\frac{\omega_n^2}{\left\{(-\omega_{gc}^2)^2+(2\xi\omega_{gc}\omega_n)^2\right\}^{1/2}}=1$$

i.e.,
$$\frac{\omega_n^4}{\omega_{gc}^4+4\xi^2\omega_{gc}^2\omega_n^2}=1$$

i.e.,
$$\omega_{gc}^4+4\xi^2\omega_{gc}^2\omega_n^2-\omega_n^4=0$$

i.e.,
$$\omega_{gc}^2=\frac{-4\xi^2\omega_n^2\pm\sqrt{(4\xi^2\omega_n^2)^2-4\times1\times(-\omega_n^4)}}{2\times1}$$

i.e.,
$$\omega_{gc}^2=\frac{-4\xi^2\omega_n^2\pm\sqrt{16\xi^4\omega_n^4+4\omega_n^4}}{2}$$

i.e.,
$$\omega_{gc}^2=-2\xi^2\omega_n^2\pm\frac{\omega_n^2\sqrt{16\xi^4+4}}{2}$$

i.e.,
$$\omega_{gc}^2=-2\xi^2\omega_n^2\pm\omega_n^2\sqrt{4\xi^4+1} \tag{12.29}$$

Since ω_{gc} cannot be negative, we have from Eq. (12.29)

$$\omega_{gc}=\sqrt{-2\xi^2\omega_n^2+\omega_n^2\sqrt{4\xi^4+1}}$$

$$=\omega_n\sqrt{-2\xi^2+\sqrt{4\xi^4+1}}$$

PM is determined by

$$\text{PM} = 180° + \angle G(j\omega)H(j\omega)\big|_{\omega=\omega_{gc}} \tag{12.30}$$

From Eq. (12.28), we have

$$\angle G(j\omega)H(j\omega) = -\tan^{-1}\left\{\frac{2\xi\omega_{gc}\omega_n}{-\omega_{gc}^2}\right\} = -\tan^{-1}\left\{\frac{2\xi\omega_n}{-\omega_{gc}}\right\} = \tan^{-1}\left\{\frac{2\xi\omega_n}{\omega_{gc}}\right\} - 180°$$

[180° has been subtracted from it to absorb the negative sign]

$$= \tan^{-1}\left\{\frac{2\xi\omega_n}{\omega_n\sqrt{-2\xi^2+\sqrt{4\xi^4+1}}}\right\} - 180°$$

$$= \tan^{-1}\left\{\frac{2\xi}{\sqrt{-2\xi^2+\sqrt{4\xi^4+1}}}\right\} - 180°$$

$$\therefore\quad PM = 180° + \angle G(j\omega)H(j\omega)\big|_{\omega=\omega_{gc}}$$

$$= 180° + \tan^{-1}\left\{\frac{2\xi}{\sqrt{-2\xi^2+\sqrt{4\xi^4+1}}}\right\} - 180°$$

$$\therefore\quad PM = \tan^{-1}\left\{\frac{2\xi}{\sqrt{-2\xi^2+\sqrt{4\xi^4+1}}}\right\} \tag{12.31}$$

Example 12.2 A unity feedback system having $G(s) = \dfrac{K}{s(1+Ts)}$ is subjected to unit step input. It is given 23% overshoot and shows the resonant frequency 9 rad/sec in frequency domain. Calculate K and T. Also determine (i) gain crossover frequency, (ii) phase margin and (iii) resonant peak.

Solution

Here $$G(s) = \frac{K}{s(1+Ts)} \text{ and } H(s) = 1.$$

$$\therefore\quad \frac{C(s)}{R(s)} = \frac{G(s)}{1+G(s)H(s)} = \frac{\dfrac{K}{s(1+Ts)}}{1+\dfrac{K}{s(1+Ts)}\times 1}$$

i.e., $$\frac{C(s)}{R(s)} = \frac{K}{Ts^2+s+K} = \frac{\dfrac{K}{T}}{s^2+\dfrac{1}{T}s+\dfrac{K}{T}} \tag{1}$$

Comparing Eq. (1) with $\frac{C(s)}{R(s)} = \frac{\omega_n^2}{s^2 + 2\xi\omega_n + \omega_n^2}$, we have

$$\omega_n^2 = \frac{K}{T} \quad \text{i.e.,} \quad \omega_n = \sqrt{\frac{K}{T}} \tag{2}$$

and

$$2\xi\omega_n = \frac{1}{T} \quad \text{i.e.,} \quad \xi = \frac{1}{2\sqrt{KT}} \tag{3}$$

For unit step input, $M_P = 23\%$.

∴ $$23 = e^{-\frac{\pi\xi}{\sqrt{1-\xi^2}}} \times 100$$

i.e., $$0.23 = e^{-\frac{\pi\xi}{\sqrt{1-\xi^2}}}$$

i.e., $$ln\,(0.23) = -\frac{\pi\xi}{\sqrt{1-\xi^2}}$$

i.e., $$-1.4696 = -\frac{\pi\xi}{\sqrt{1-\xi^2}}$$

i.e., $$0.4680 = \frac{\xi}{\sqrt{1-\xi^2}}$$

i.e., $$0.219(1-\xi^2) = \xi^2$$

i.e., $$0.219 = 1.219\xi^2$$

i.e., $$\xi^2 = \frac{0.219}{1.219} = 0.1796$$

i.e., $$\xi = 0.4238$$

In frequency domain,

$$\omega_r = 9 = \omega_n\sqrt{1-2\xi^2}$$

∴ $$9 = \omega_n\sqrt{1-2\times 0.4238^2}$$

i.e., $$\omega_n = \frac{9}{0.8005} = 11.243 \text{ rad/sec.}$$

From Eq. (2), we have after squaring both sides.

$$K = T\omega_n^2 \tag{4}$$

From Eq. (3), we have

$$\xi^2 = \frac{1}{4KT} = \frac{1}{4T^2\omega_n^2} \qquad \text{[using Eq. (4)]}$$

i.e.,
$$\xi^2 = \frac{1}{4\times T^2 \times 11.243^2}$$

i.e.,
$$T^2 = \frac{1}{4\times\xi^2\times 11.243^2} = \frac{1}{4\times 0.4238^2 \times 11.243^2}$$

i.e.,
$$T^2 = 0.0110$$

i.e.,
$$T = 0.1048 \qquad (5)$$

Using Eq. (5), we have from Eq. (4)

$$K = 0.1048\times 11.243^2 = 13.247$$

Now,
$$\omega_{gc} = \omega_n\sqrt{\sqrt{4\xi^4+1}-2\xi^2}$$
$$= 11.243\sqrt{\sqrt{4\times 0.4238^4+1}-2\times 0.4238^2}$$
$$= 9.429 \text{ rad/sec.}$$

$$\text{P.M.} = \tan^{-1}\left\{\frac{2\xi}{\sqrt{-2\xi^2+\sqrt{4\xi^4+1}}}\right\}$$
$$= \tan^{-1}\left\{\frac{2\times 0.4238}{\sqrt{-2\times 0.4238^2+\sqrt{4\times 0.4238^4+1}}}\right\}$$
$$= \tan^{-1}\left\{\frac{2\times 0.4238}{\sqrt{-2\times 0.4238^2+1.0625}}\right\}$$
$$= \tan^{-1}\left\{\frac{2\times 0.4238}{\sqrt{-0.3592+1.0625}}\right\}$$
$$= \tan^{-1}(1.0107)$$
$$= 45.30^\circ$$

$$M_r = \frac{1}{2\xi\sqrt{1-\xi^2}} = \frac{1}{2\times 0.4238\times\sqrt{1-0.4238^2}}$$
$$= 1.3025$$

12.6 Advantages of Frequency Domain Analysis

- In laboratory, it is possible to obtain a frequency response test with good accuracy and it is very useful when it is very difficult to obtain transfer function by an analytical technique.

- It is very easy to design open-loop transfer function for specified closed-loop performance in frequency domain compared to time domain.
- In frequency domain it is very easy to visualize the effects of noise disturbance and parameter variations.

12.7 Disadvantages of Frequency Response Methods

- These methods are basically applicable only to linear systems.
- It is possible to obtain frequency response for an existing system if the time constant is up to few minutes.
- These are time-consuming processes.
- These methods are backdated compared to the extensive methods developed for digital computer simulation and modelling.

SIGNIFICANT POINTS

$$\omega_r = \omega_n\sqrt{1-2\xi^2} \qquad \text{for } 0<\xi<\frac{1}{\sqrt{2}}$$

$$= 0 \qquad \text{for } \xi>\frac{1}{\sqrt{2}}$$

$$= \omega_n \qquad \text{for } \xi = 0$$

$$M_r = \frac{1}{2\xi\sqrt{1-\xi^2}} \qquad \text{for } 0<\xi<\frac{1}{\sqrt{2}}$$

$$= 1 \qquad \text{for } \xi>\frac{1}{\sqrt{2}}$$

$$= \infty \qquad \text{for } \xi = 0$$

$$= \text{system is unstable} \qquad \text{for } \xi < 0$$

$$\omega_c = \text{BW} = \omega_n\left\{(1-2\xi^2)+\sqrt{4\xi^4-4\xi^2+2}\right\}^{1/2}$$

ADDITIONAL SOLVED EXAMPLES

Example 12.3 Find frequency domain specifications with a unity feedback system having $G(s) = 36/[s(s+8)]$.

Solution

$$\frac{C(s)}{R(s)} = \frac{G(s)}{1+G(s)H(s)} = \frac{\frac{36}{s(s+8)}}{1+\frac{36}{s(s+8)}} = \frac{36}{s^2+8s+36}$$

$\therefore$
$$\omega_n^2 = 36$$

i.e.,
$$\omega_n = 6 \text{ rad/sec}$$

and
$$2\xi\omega_n = 8$$

i.e.,
$$\xi = \frac{8}{2\omega_n} = \frac{8}{2\times 6} = 0.667$$

Now, $$\omega_r = \omega_n\sqrt{1-2\xi^2} = 6\sqrt{1-2\times 0.667^2} = 1.99 \text{ rad/sec}$$

And, $$M_r = \frac{1}{2\xi\sqrt{1-\xi^2}} = \frac{1}{2\times 0.667\times\sqrt{1-0.667^2}} = 1.006$$

Again, $$BW = \omega_n\left\{(1-2\xi^2)+\sqrt{4\xi^4-4\xi^2+2}\right\}^{1/2}$$

$$= 6\left\{1-2(0.667)^2+\sqrt{4\times(0.667)^4-4\times(0.667)^2+2}\right\}^{1/2} = 6.34 \text{ rad/sec}$$

Example 12.4 Find frequency domain specifications for a system having

$$\frac{C(s)}{R(s)} = \frac{84}{s^2+7s+81}$$

Solution

Here, $$\omega_n^2 = 81,\ \omega_n = 9 \text{ rad/sec, and } 2\xi\omega_n = 7$$

i.e., $$\xi = \frac{7}{2\omega_n} = 0.389$$

Now, $$M_r = \frac{1}{2\xi\sqrt{1-\xi^2}} = \frac{1}{2\times 0.398\times\sqrt{1-(0.389)^2}} = 1.395$$

and $$\omega_r = \omega_n\sqrt{1-2\xi^2} = 9\sqrt{1-2\times(0.389)^2} = 7.515 \text{ rad/sec}$$

Again, $$\omega_r = \omega_n\left\{(1-2\xi^2)+\sqrt{4\xi^4-4\xi^2+2}\right\}^{1/2}$$

$$= 9\left[\{1-2\times(0.389)^2\}+\sqrt{4\times(0.389)^2-4\times(0.389)^2+2}\right]^{1/2} = 12.46 \text{ rad/sec}$$

∴ $$BW = \omega_c = 12.46 \text{ rad/sec}$$

Example 12.5 Find frequency domain specifications for a system having $G(s) = 81/[s(s+18)]$ with unity feedback.

Solution

$$G(s) = \frac{81}{s(s+18)} = \frac{81}{s^2+18s}$$

∴ $$\frac{C(s)}{R(s)} = \frac{C(s)}{1+C(s)H(s)} = \frac{81}{s^2+18s+81}$$

Now, $$\omega_n^2 = 81$$

i.e., $$\omega_n = 9 \text{ rad/sec}$$

and $$2\xi\omega_n = 18$$

i.e., $$\xi = \frac{18}{2\times 9} = 1$$

Now,

$u_r = \sqrt{1-2\xi^2}$ becomes complex.

The other solution is $u_r = 0$. Hence $\omega_r = 0$ and $M_r = 1$ (this is true for $\xi > 0.707$). Again,

$\omega_c = \omega_n\left\{1-2\xi^2+\sqrt{4\xi^4-4\xi^2+2}\right\}^{\frac{1}{2}} = 5.979$ rad/sec. BW $= \omega_c = 5.792$ rad/sec.

BW $< \omega_n$ for $\xi > 0.707$.

Example 12.6 Find K and a for the feedback system shown in Fig. E12.1 so that $M_r = 1.25$ and $\omega_r = 12.65$ rad/sec will be satisfied. Also determine the settling time and bandwidth.

Solution

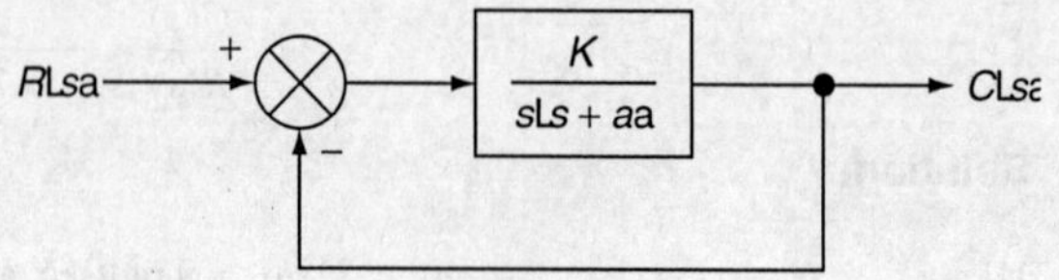

Fig. E12.1

$$\frac{C(s)}{R(s)} = \frac{C(s)}{1+G(s)H(s)} = \frac{\dfrac{K}{s(s+a)}}{1+\dfrac{K}{s(s+a)}}$$

i.e., $$\frac{C(s)}{R(s)} = \frac{K}{s^2+as+K}$$

Comparing it with $$\frac{C(s)}{R(s)} = \frac{\omega_n^2}{s^2+2\xi\omega_n s+\omega_n^2}$$

we have $$\omega_n^2 = K$$

i.e., $$\omega_n = \sqrt{K} \text{ rad/sec}$$

and $$2\xi\omega_n = a$$

i.e., $$\xi = \frac{a}{2\omega_n} = \frac{a}{2\sqrt{K}}$$

Now, $$M_r = \frac{1}{2\xi\sqrt{1-\xi^2}}$$

i.e., $$1.25 = \frac{1}{\dfrac{a}{\sqrt{K}}\sqrt{1-\dfrac{a^2}{4K}}}$$

i.e., $$1.25 = \frac{2K}{a\sqrt{4K-a^2}}$$

i.e., $$1.5625a^2(4K-a^2) = 4K^2 \qquad (1)$$

i.e., $$4K^2 - 6.25a^2K + 1.5625a^4 = 0$$

Solving $$K = \frac{6.25a^2 \pm 3.75a^2}{8} = \text{either } 1.25a^2 \text{ or } 0.3125a^2$$

Again, $$\omega_r = \omega_n\sqrt{1-2\xi^2}$$

i.e., $$12.65 = \sqrt{K}\sqrt{1-2\left(\frac{a}{2\sqrt{K}}\right)^2} = \frac{1}{\sqrt{2}}\sqrt{2K-a^2}$$

i.e., $$2K - a^2 = 320 \qquad (2)$$

Taking $K = 1.25a^2$ in Eq. (2)

$$2.5a^2 - a^2 = 320$$

i.e., $$a^2 = 213.34$$

$\therefore$ $$a = 14.60 \text{ and } K = 1.25a^2 = 1.25 \times 213.34 = 266.675$$

Taking $K = 0.3125a^2$ in Eq. (2),

$$0.3125a^2 - a^2 = 320$$

It is not possible to get a real value of a.

$\therefore$ $$K = 266.67 \text{ and } a = 14.60$$

Now,

$$\text{Settling time} = \frac{4}{\xi\omega_n} = \frac{4}{a/2} = \frac{8}{a} = \frac{8}{14.60} = 0.5479 \text{ sec}$$

Now, $$\omega_n = \sqrt{K} = \sqrt{266.67} = 16.33 \text{ rad/sec}$$

and $$\xi = \frac{a}{2\sqrt{K}} = \frac{14.60}{2\times 16.33} = 0.4470$$

Again, $$\text{BW} = \omega_c = \omega_n\left\{1 - 2\xi^2 + \sqrt{4\xi^4 - 4\xi^2 + 2}\right\}$$

$$= 16.33\left\{1 - 2(0.4470)^2 + \sqrt{4\times(0.4470)^4 - 4\times(0.4470)^2 + 2}\right\}^{1/2}$$

$$= 21.70 \text{ rad/sec.}$$

Example 12.7 A second-order system has overshoot of 40% and period of oscillations 0.25 sec. when subjected to unit step response. Determine the (i) resonant peak, (ii) resonant frequency and (iii) bandwidth.

Solution For step response,

$$\%M_P = e^{-\frac{\pi\xi}{\sqrt{1-\xi^2}}} \times 100$$

i.e., $$40 = e^{-\frac{\pi\xi}{\sqrt{1-\xi^2}}} \times 100$$

i.e., $$0.4 = e^{-\frac{\pi\xi}{\sqrt{1-\xi^2}}}$$

i.e., $$-0.9163 = -\frac{\pi\xi}{\sqrt{1-\xi^2}}$$

i.e., $$-0.2918 = \frac{\xi}{\sqrt{1-\xi^2}}$$

i.e., $$0.085(1-\xi^2)=\xi^2$$

i.e., $$\xi^2=\frac{0.085}{1.085}=0.07834$$

i.e., $$\xi=0.2799$$

Time period (T) = 0.25 sec.

$\therefore$ $$f_d=\frac{1}{T}=4Hz$$

$\therefore$ $$\omega_d=2\pi f_d=2\times 3.14\times 4=25.12 \text{ rad/sec.}$$

Since $\omega_d=\omega_n\sqrt{1-\xi^2}$, we have

$$\omega_n=\frac{\omega_d}{\sqrt{1-\xi^2}}=\frac{25.12}{\sqrt{1-0.2799^2}}=27.25 \text{ rad/sec.}$$

$$\text{Resonant peak } (M_r)=\frac{1}{2\xi\sqrt{1-\xi^2}}=\frac{1}{2\times 0.2799\times\sqrt{1-0.2799^2}}=1.860$$

$$\text{Resonant frequency } (\omega_r)=\omega_n\sqrt{1-2\xi^2}=27.25\sqrt{1-2\times 0.2799^2}=25.02 \text{ rad/sec.}$$

For bandwidth,

$$-20\log_{10}\frac{1}{\sqrt{\left(1-\frac{\omega^2}{\omega_n^2}\right)^2+4\xi^2\left(\frac{\omega}{\omega_n}\right)^2}}=-3$$

i.e., $$\sqrt{\left(1-\frac{\omega^2}{\omega_n^2}\right)^2+4\xi^2\left(\frac{\omega}{\omega_n}\right)^2}=1.4125 \qquad (1)$$

Substituting $\omega_n = 27.25$ rad/sec, we have from Eq. (1),

$$\left(1-\frac{\omega^2}{742.56}\right)^2+4\xi^2\times 1.3466\times 10^{-3}\omega^2=1.9952$$

i.e., $$1-\frac{2\omega^2}{742.56}+\frac{\omega^4}{742.56^2}+4\times(0.2799)^2\times 1.3466\times 10^{-3}\omega^2=1.9952$$

i.e., $$\omega^4-1485.12\omega^2+551395.35+232.68\omega^2=1.9952\times 551395.35$$

i.e., $$\omega^4-1252.44\omega^2-548748.65=0$$

$\therefore$ $$\omega^2=\frac{1252.44\pm\sqrt{1252.44^2+4\times 548748.65}}{2}$$

$$=\frac{1252.44\pm 1940.00}{2}$$

$$=1596.22 \text{ or } -343.78$$

Here, $\omega^2 = 1596.22$ is accepted.

$\therefore$ $$\omega = 39.95 \text{ rad/sec.}$$

$\therefore$ $$\omega_c = \omega = 39.95 \text{ rad/sec}$$

$\therefore$ $$BW = 39.95 \text{ rad/sec.}$$

Example 12.8 Figure E 12.2 shows a feedback control system. The peak overshoot to a step input is 12%. Determine the following:

(i) Resonant peak M_r and value of K
(ii) Normalized resonant frequency W_r
(iii) Bandwidth

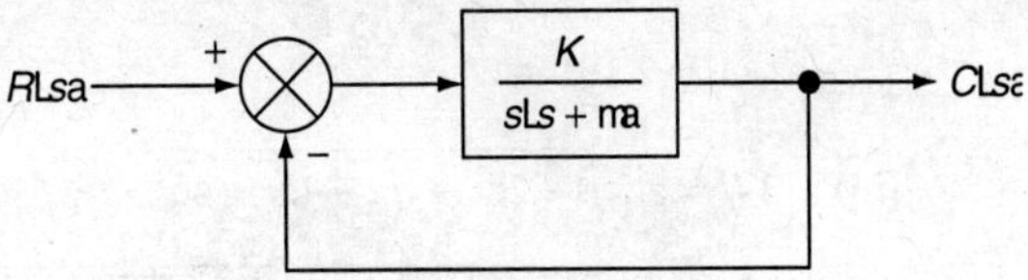

Fig. E 12.2

Solution

$$\frac{C(s)}{R(s)} = \frac{\frac{K}{s(s+4)}}{1+\frac{K}{s(s+4)}} = \frac{K}{s^2+4s+K} \tag{1}$$

Comparing Eq. (1) with $\dfrac{C(s)}{R(s)} = \dfrac{\omega_n^2}{s^2+2\xi\omega_n s+\omega_n^2}$, we have

$$\omega_n^2 = K \quad \text{i.e.,} \quad \omega_n = \sqrt{K} \tag{2}$$

and $$2\xi\omega_n = 4 \quad \text{i.e.,} \quad \xi = \frac{4}{2\sqrt{K}} \tag{3}$$

$$M_P = 12\% = e^{-\frac{\pi\xi}{\sqrt{1-\xi^2}}} \times 100$$

$\therefore$ $$ln\,(0.12) = -\frac{\pi\xi}{\sqrt{1-\xi^2}}$$

i.e., $$-2.120 = -\frac{\pi\xi}{\sqrt{1-\xi^2}}$$

i.e., $$0.675 = \frac{\xi}{\sqrt{1-\xi^2}}$$

i.e., $$0.456(1-\xi^2) = \xi^2$$

i.e., $$\xi^2 = \frac{0.456}{1.456} = 0.313$$

i.e., $$\xi = 0.559$$

From Eq. (3), we have

$$K = \frac{16}{4\xi^2} = \frac{4}{0.559^2} \simeq 12.80$$

From Eq. (2), we have

$$\omega_n = \sqrt{K} = 3.578 \text{ rad/sec.}$$

(i) $M_r = \dfrac{1}{2\xi\sqrt{1-\xi^2}} = \dfrac{1}{2\times 0.559\times\sqrt{1-0.559^2}} = 1.078$

(ii) $\omega_r = \omega_n\sqrt{1-2\xi^2} = 3.578\times\sqrt{1-2\times 0.559^2}$

$\therefore \omega_r = 2.191$ rad/sec.

(iii) $\text{BW} = \omega_n\sqrt{1-2\xi^2+\sqrt{2-4\xi^2+4\xi^4}}$

$= 3.578\sqrt{1-2\times(0.559)^2+\sqrt{2-4\times(0.559)^2+4\times(0.559)^4}}$

$= 4.298$

$\therefore \text{BW} = 20\log_{10} 4.298 \text{ dB} = 12.66 \text{ dB}.$

SHORT QUESTIONS WITH ANSWERS

1. What do you mean by frequency response?

   Ans. It is defined as the steady-state response of a system due to a sinusoidal input.

2. What do you mean by resonance peak? How is it related to the relative stability of a closed-loop system?

   Ans. The maximum value of $M(j\omega) = G(j\omega)/[1+G(j\omega)H(j\omega)]$ as ω varied termed as resonance peak (M_r). The relative stability of the closed-loop system is indicated by M_r. If M_r is large, there will be large maximum overshoot in a step response. Generally, M_r is between 1.1 and 1.5.

3. Define the following: (i) Resonance frequency, (ii) bandwidth, (iii) cut-off rate, (iv) phase margin and (v) gain margin.

   Ans. (i) The frequency at which peak resonance M_r occurs is known as resonance frequency (ω_r).

   (ii) The frequency at which the magnitude of $M(j\omega)$ drops by 30 dB below a certain specified level, generally the 0 dB level. If bandwidth is large, higher frequencies will pass through the system. This indicates a faster rise in time. Therefore, bandwidth gives an idea about transient response. Again, small bandwidth means lower frequencies are passed and hence response is sluggish. From bandwidth, noise filtering can be known because large bandwidth indicates susceptibility to noise. Bandwidth also implies the sensitivity to parameter variations.

   (iii) The slope of the log magnitude curve near the cut-off frequency is known as cut-off rate. The ability of the system to distinguish a signal from noise can be determined by cut-off rate.

   (iv) If ϕ is the phase angle of a system at unity gain, the phase margin is given by 180° + ϕ. The frequency (ω_1) at which gain is unity is known as gain crossover frequency. The increment in the system angle that causes the system to be unstable is indicated by phase margin. It is a measure of relative stability.

   (v) The frequency (ω_1) at which the phase angle of the transfer function is –180° is known as phase crossover frequency. The gain at this frequency (ω_1) is $|G(j\omega_1)H(j\omega_1)|$ and gain margin is given by $1/|G(j\omega_1)H(j\omega_1)|$. The increment in gain required to cause the system to become unstable is indicated by gain margin.

4. State the advantages of frequency domain analysis.

   Ans. Advantages of frequency domain analysis are:

   - In laboratory, it is possible to obtain a frequency response test with good accuracy and it is very useful when it is very difficult to obtain transfer function by an analytical technique.
   - It is very easy to design open-loop transfer function for specified closed-loop performance in frequency domain compared to time domain.
   - In frequency domain it is very easy to visualize the effects of noise disturbance and parameter variations.

5. State the disadvantages of frequency response methods.

   Ans. Disadvantages of frequency domain methods are:

   - These methods are basically applicable only to linear system.
   - It is possible to obtain frequency response for an existing system if the time constants are up to few minutes.
   - These are time-consuming processes.
   - These methods are backdated compared to the extensive methods developed for digital computer simulation and modelling.

6. What will happen if a zero is added in the forward path of a second-order system?

   Ans.

   - Rise time decreases due to increase in bandwidth.
   - Settling time will increase
   - The system will be more stable.

7. What will happen if a pole is added in the forward path of a second-order system?

   Ans.

   - Rise time will increase and bandwidth will decrease.
   - M_r will increase
   - The system will be less stable.

EXERCISES

Questions for Practice

1. Find frequency domain specification for a second-order system with unity feedback and $G(s) = 169/[s(s+7)]$.

   Ans. $M_r = 1.930, \omega_r = 12.02$ rad/sec, $\omega_c = 19.15$ rad/sec, BW $= 19.15$ rad/sec.

2. Find frequency domain specification for a second-order system with unity feedback and $G(s) = 121/[s(s+11)]$.

   Ans. $M_r = 1.155, \omega_r = 7.778$ rad/sec, $\omega_c = 14$ rad/sec, BW $= 14$ rad/sec

3. Find K and a for the feedback system shown in Figure 1 so that $M_r = 1.04$ and $\omega_r = 11.55$ rad/sec will be satisfied. Also determine the bandwidth.

   Ans. $K = 491, a = 26.77$, BW $= 38.62$ rad/sec

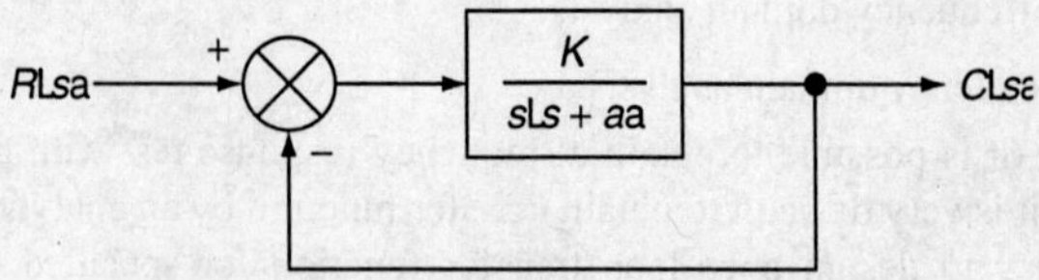

Figure 1

4. A unit feedback system has $G(s) = \dfrac{K}{s(1+Ts)}$. For a unit step input its peak overshoot and resonant frequency are 25% and 8 rad/sec respectively in frequency domain. Calculate K, T, gain crossover frequency, phase Margin and resonant beak.

Ans. 12.071, 0.1271, 8.3021 rad/sec, + 43.45°, 1.354

TEST YOURSELF

1. For $0 < \xi < 1/2$, ω_r is equal to

   (a) 0 (b) $\dfrac{\omega_n}{\sqrt{1-\xi^2}}$ (c) $\omega_n\sqrt{1-\xi^2}$ (d) $\omega_n\sqrt{1-2\xi^2}$

2. For $\xi = 0$, ω_r is equal to

   (a) 0 (b) ω_n (c) $\dfrac{\omega_n}{\sqrt{1-2\xi^2}}$ (d) $\omega_n\sqrt{1-3\xi^2}$

3. $\xi = 0$, M_r is given by

   (a) 0 (b) 1 (c) 4 (d) ∞

4. For $0 < \xi < 1/\sqrt{2}$, M_r is equal to

   (a) $\dfrac{1}{2\xi\sqrt{1-\xi^2}}$ (b) $\dfrac{1}{2\xi\sqrt{1-\xi^2}}$ (c) $\dfrac{\xi}{\sqrt{1-2\xi^2}}$ (d) $\dfrac{\xi}{3}$

5. For $\xi > 1/\sqrt{2}$, M_r is equal to

   (a) $\dfrac{1}{2}$ (b) 1 (c) $\dfrac{1}{3}$ (d) 2

ANSWERS

1 (d) 2 (b) 3 (d) 4 (b) 5 (b)

Chapter 13

Bode Plot

The plot of magnitude as well as phase angle versus frequency may represent a sinusoidal transfer function. H.W. Bode used the logarithmic scale extensively for the study of the magnitude of transfer function and the frequency variable. That's why a logarithmic plot is also called a Bode plot. A Bode plot consists of (i) a plot of the logarithm (base 10) of the magnitude in decibels of a sinusoidal transfer function versus frequency in logarithmic scale and (ii) a plot of phase angle versus frequency in logarithmic scale. The aim of this chapter is to discuss the Bode plot.

13.1 Logarithmic Plot

Any sinusoidal transfer function $G(j\omega)$ can be expressed as

$$G(j\omega) = |G(j\omega)| \angle G(j\omega) \tag{13.1}$$

Now,

$$ln|G(j\omega)| = ln[M(\omega)] \angle G(j\omega) \tag{13.2}$$

where

$$M(\omega) = |G(j\omega)| \tag{13.3}$$

In Eq. (13.2), $ln\,[M(\omega)]$ is the gain or logarithmic gain in nepers and ϕ is the angle in radians. Generally, the unit for gain is decibel (dB) and the angle ϕ is in degrees. The gain in decibels is given by

$$\text{Gain in dB} = 20\, log_{10}\, [M(\omega)] \tag{13.4}$$

The following relations are well valid.

$$\text{Number of decibels} = 8.68 \times \text{number of nepers} \tag{13.5}$$

$$\text{Number of degrees} = 57.3 \times \text{number of radians} \tag{13.6}$$

Let

$$u = log_{10}\omega \tag{13.7}$$

i.e.,

$$\omega = 10^{u} \tag{13.8}$$

Now,

$$u_2 - u_1 = log_{10}\omega_2 \quad log_{10}\omega_1$$

i.e., $$u_2 - u_1 = log_{10}\left(\frac{\omega_2}{\omega_1}\right) \tag{13.9}$$

Equation (13.9) shows the two frequency intervals. These intervals will be in octave and decade if $\omega_2 = 2\omega_1$ and $\omega_2 = 10\omega_1$, respectively. The number of decades between two frequencies ω_1 and ω_2 is given by

$$\text{Number of decades} = \frac{log_{10}\frac{\omega_2}{\omega_1}}{log_{10}10} = log_{10}\frac{\omega_2}{\omega_1} \tag{13.10}$$

The number of octaves between two frequencies is given by

$$\text{Number of octaves} = \frac{log_{10}\frac{\omega_2}{\omega_1}}{log_{10}2} = \frac{log_{10}\frac{\omega_2}{\omega_1}}{0.301} \tag{13.11}$$

To convert dB/octave to dB/decade, let us consider Fig. 13.1.

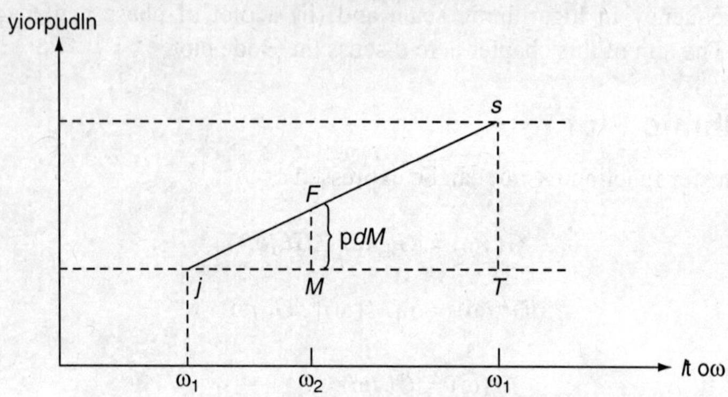

Fig. 13.1 Relation between octave and decade

From Fig. 13.1,

$$\text{Slope} = \frac{BF}{AB} = \frac{CD}{AC} \tag{13.12}$$

$$AB = 1 \text{ octave} = log_{10}\omega_2 - log_{10}\omega_1 = log_{10}\left(\frac{\omega_2}{\omega_1}\right)$$

$$AC = 1 \text{ decade} = log_{10}\omega_3 - log_{10}\omega_1 = log_{10}\left(\frac{\omega_3}{\omega_1}\right)$$

$$CD = 20 \text{ dB}$$

$$\therefore \quad BF = \frac{CD}{AC} \times AB = 20\left(\frac{log_{10}\frac{\omega_2}{\omega_1}}{log_{10}\frac{\omega_3}{\omega_1}}\right) = 20\left(\frac{log_{10}2}{log_{10}10}\right) \approx 6 \text{ dB} \tag{13.13}$$

$$\therefore \quad \frac{6\,\text{dB}}{\text{octave}} = \frac{20\,\text{dB}}{\text{decade}}$$

Let the generalized expression for the open-loop transfer function of a system be given by

$$G(s)H(s) = \frac{K(1+T_1 s)(1+T_2 s)\omega_n^2}{s(1+T_a s)(s^2+2\xi\omega_n s+\omega_n^2)} \tag{13.14}$$

where K, T_1, T_2, T_a, ξ and ω_n are all real coefficients.

Putting $s = j\omega$ in Eq. (13.14), we have

$$G(s)H(s) = \frac{K(1+j\omega T_1)(1+j\omega T_2)}{j\omega(1+j\omega T_a)(1+j2\xi\,\mu-\mu^2)} \tag{13.15}$$

where

$$\mu = \omega/\omega_n \tag{13.16}$$

In Eq. (13.15), ω_n is the natural frequency and ξ is the damping coefficient (<1) of the polynomial $1+j2\xi\mu-\mu^2$ having two complex conjugate zeros. From Eq. (13.15), the magnitude of $G(j\omega)H(j\omega)$ in decibels is given by

$$\begin{aligned} 20\log_{10}|G(j\omega)H(j\omega)| &= 20\log_{10} K + 20\log_{10}|1+j\omega T_1| + 20\log_{10}|1+j\omega T_2| \\ &\quad -20\log_{10}|j\omega| - 20\log_{10}|1+j\omega T_a| - 20\log_{10}|1+j2\xi\mu-\mu^2| \end{aligned} \tag{13.17}$$

The phase angle of $G(j\omega)H(j\omega)$ is given by

$$\angle G(j\omega)H(j\omega) = \angle K + \angle(1+j\omega T_1) + \angle(1+j\omega T_2) - \angle j\omega - \angle(1+j\omega T_a) - \angle(1+2j\xi\mu-\mu^2) \tag{13.18}$$

Generally, there one or more than one of the following four simple types of factors in any $G(j\omega)H(j\omega)$:

(i) Constant	K
(ii) Zeros or poles at origin	$(j\omega)^{\pm n}$
(iii) Simple zeros or poles	$(1+j\omega T)^{\pm 1}$
(iv) Quadratic of zeros or poles	$(1+j2\xi\mu-\mu^2)^{\pm 1}$

13.1.1 The Constant Term *K*

$$K\text{ (in dB)} = 20\log_{10}|K| = \text{constant} \tag{13.19}$$

and

$$\phi = \arg K = 0^\circ \tag{13.20}$$

Figure 13.2 shows the Bode plot of constant (K).

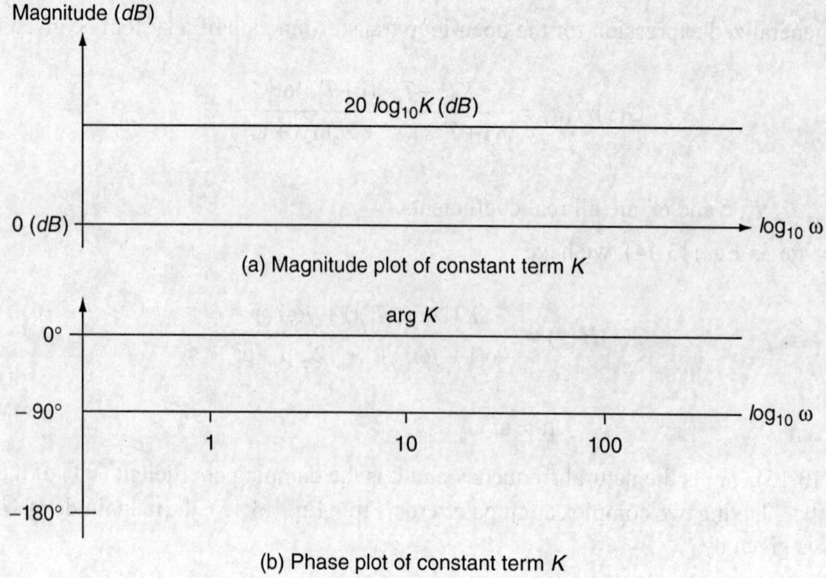

Fig. 13.2 Bode plot of Constant (K)

Example 13.1 A unity feedback control system has $G(s) = 40$. Draw the Bode plot.

Solution Here, $G(s) = 20$ and $H(s) = 1$.

$$\therefore \qquad G(s)\,H(s) = 40$$

$$\therefore \qquad 20\,log_{10}|G(s)\,H(s)| = 20\,log_{10}\,40 = 32\text{ dB}$$

and

$$\phi = \text{Arg}\,(40) = 0°.$$

Figure E13.1(a) shows the magnitude plot and Fig. E13.1(b) shows the phase angle plot.

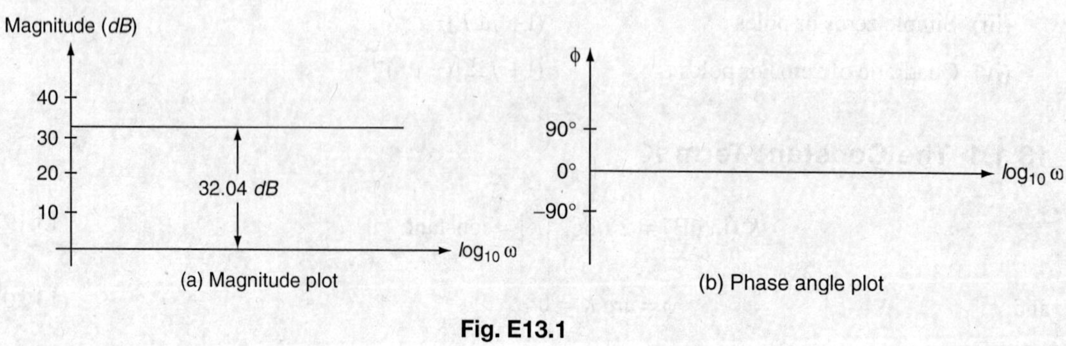

Fig. E13.1

13.1.2 Zeros at the Origin, i.e., $(j\omega)^{+n}$

The magnitude in decibels is given by

$$20\,log_{10}\left|(j\omega)^{+n}\right| = +20n\ log_{10}\ \omega\ dB \tag{13.21}$$

Equation (13.26) represents the equation of a straight line having slope of 20*n* dB/decade or 6*n* dB/octave, which passes through the 0 dB point at $\omega = 1$ in the logarithmic frequency scale on semi-log paper.

The phase shift of $(j\omega)^{+n}$ is given by

$$\phi = (+n)\times 90° \tag{13.22}$$

Figure 13.3(a) shows the magnitude plot and Fig. 13.3(b) shows in phase plot of the term $(j\omega)^{+n}$.

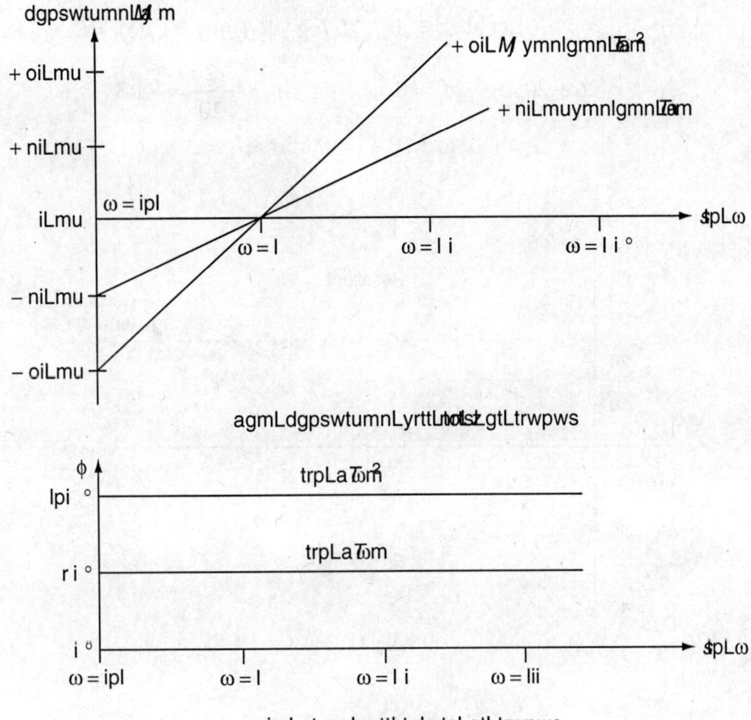

Fig. 13.3 Magnitude and phase curves of $(j\omega)^{+n}$

Example 13.2 A unity feedback control system has $G(s) = 20$ s. Draw the Bode plot of the system.

Solution Here, $G(s) = 20s$ and $H(s) = 1$.

$$\therefore \qquad G(s)\,H(s) = 20s$$

$$\therefore \qquad G(j\omega)\ H(j\omega) = 20(j\omega)$$

$$\therefore \quad 20\, log_{10} |G(j\omega)\, H(j\omega)| = 20\, log_{10} |20(j\omega)|$$
$$= 20\, log_{10} 20 + 20\, log_{10} |j\omega|$$
$$= 20\, log_{10} 20 + 20\, log_{10} \omega$$
$$= 26 + 20\, log_{10} \omega \qquad (1)$$

The Bode plot of the above system consists of the plot of a constant term and a zero at the origin. The magnitude plot is shown in Fig. E13.2(a) whereas the phase plot has been shown in Fig. E13.2(b).

From Eq. (1),

$$\text{at } \omega = 0.1,\ 20\, log_{10} |G(j\omega)\, H(j\omega)| = 6$$
$$\text{at } \omega = 1,\ 20\, log_{10} |G(j\omega)\, H(j\omega)| = 26$$
$$\text{at } \omega = 10,\ 20\, log_{10} |G(j\omega)\, H(j\omega)| = 46$$
$$\text{Let at } \omega = \omega_1,\ 20\, log_{10} |G(j\omega)\, H(j\omega)| = 0.$$

$$\therefore \quad 26 + 20\, log_{10}\, \omega_1 = 0 \text{ i.e., } log_{10}\, \omega_1 = \frac{-26}{20} = -1.3$$
$$\text{i.e., } \omega = 10^{-1.3} = 0.05 \text{ rad/sec.}$$

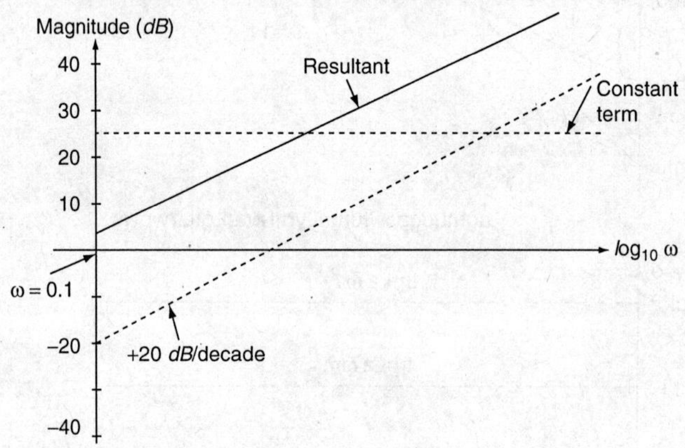

(a) Magnitude plot

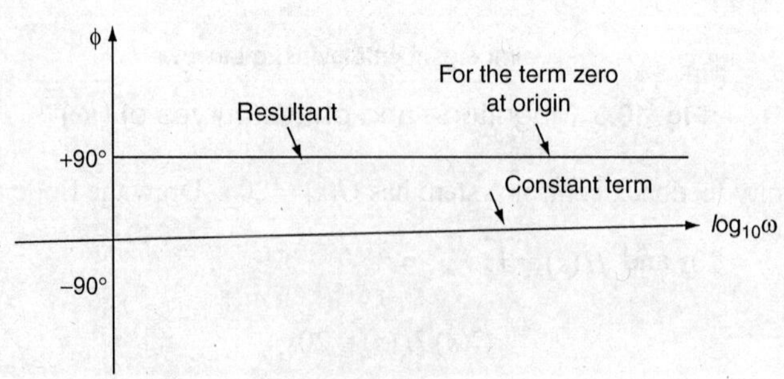

(b) Phase plot

Fig. E13.2

For the constant term, $\phi = 0°$ throughout whereas $\phi = +90°$ for the term zero at origin with $n = 1$ throughout. Therefore, ϕ (resultant) is 90° throughout.

13.1.3 Poles at the Origin, i.e., $(j\omega)^{-n}$

The magnitude in decibels in given by

$$20\log_{10}|(j\omega)^{-n}| = -20n\log_{10}\omega \tag{13.23}$$

Equation (13.23) represents the equation of a straight line having slope of $-20n$ dB/decade or $-6n$ dB/octave and it passes through the 0 dB point at $\omega = 1$ in the logarithmic frequency scale on semilog paper. The phase shift of $(j\omega)^{-n}$ is given by

$$\phi = (-n)\times 90° \tag{13.24}$$

Figure 13.4(a) shows the magnitude plot and Fig. 13.4(b) shows is phase curves of the term $(j\omega)^{-n}$.

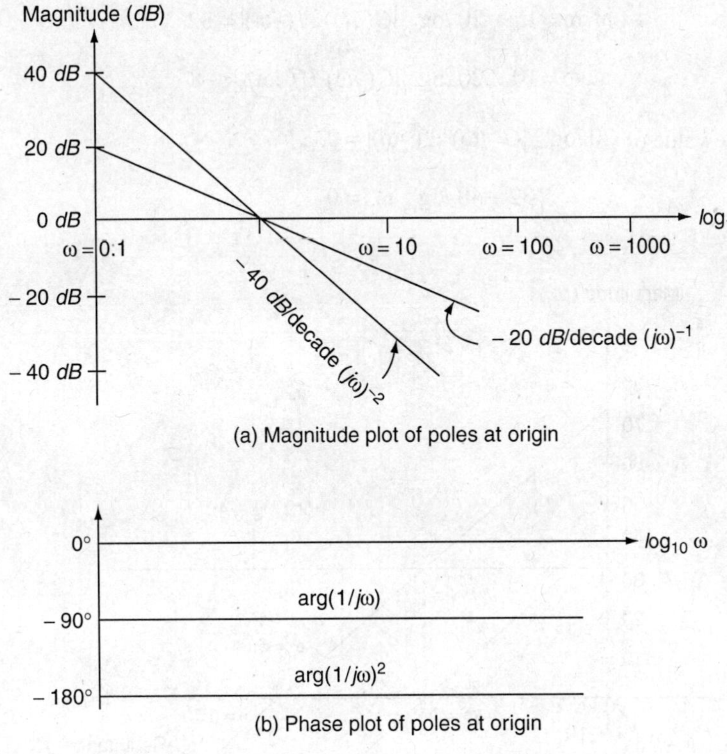

Fig. 13.4 Magnitude and phase curves of $(j\omega)^{-n}$

Example 13.3 A unity feedback control system has $G(s) = \dfrac{40}{s^2}$. Draw the Bode plot.

Solution Here, $G(s) = \dfrac{40}{s^2}$ and $H(s) = 1$.

$$\therefore \quad G(s)\,H(s) = \frac{40}{s^2} \qquad \therefore G(j\omega)\,H(j\omega) = \frac{40}{j^2\omega^2} = \frac{-40}{\omega^2}$$

$$\therefore \quad 20\,log_{10}\left|G(j\omega)\,H(j\omega)\right| = 20\,log_{10}\left|\frac{-40}{\omega^2}\right|$$

$$= 20\,log_{10}\,40 - 20\,log_{10}\,\omega^2$$

$$= 20\,log_{10}\,40 - 20\,log_{10}\,\omega$$

$$\therefore \quad 20\,log_{10}\left|G(j\omega)\,H(j\omega)\right| = 32 - 40\,log_{10}\,\omega \qquad (1)$$

The above plot consists of the Bode plot of a constant term and poles at origin. The magnitude plot is shown in Fig. E13.3(a) whereas the phase magnitude plot has been shown in Fig. E13.3(b).

From Eq. (1),

$$\text{at } \omega = 0.1, \quad 20\,log_{10}\left|G(j\omega)\,H(j\omega)\right| = 72$$

$$\text{at } \omega = 1, \quad 20\,log_{10}\left|G(j\omega)\,H(j\omega)\right| = 32$$

$$\text{at } \omega = 10, \quad 20\,log_{10}\left|G(j\omega)\,H(j\omega)\right| = -8$$

Let at $\omega = \omega_1$, the value of $20\,log_{10}\left|G(j\omega)\,H(j\omega)\right| = 0$.

$$\therefore \quad 32 - 40\,log_{10}\,\omega_1 = 0$$

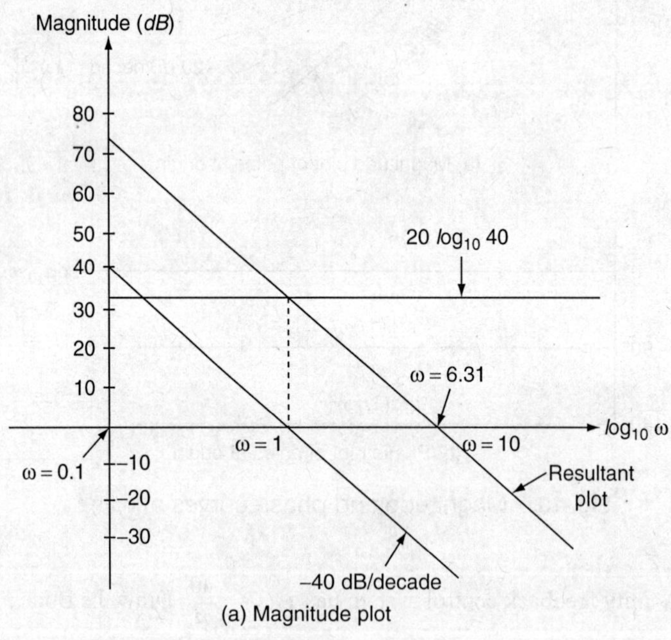

(a) Magnitude plot

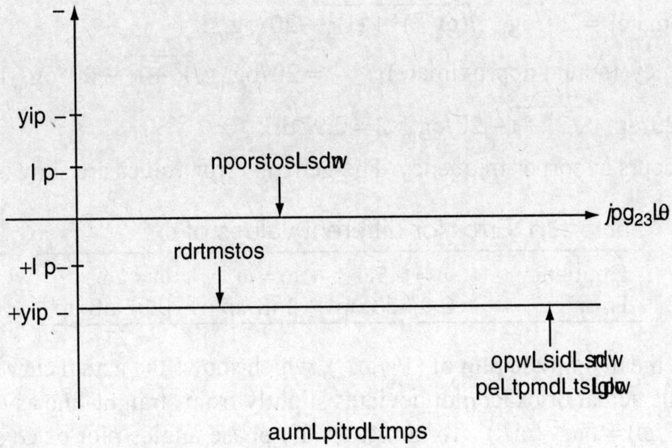

Fig. E13.3

i.e., $\log_{10} \omega_1 = 0.8$ i.e., $\omega_1 = 10^{0.8} = 6.31$ rad/sec.

For constant term, $\phi = 0°$ whereas $\phi = -180°$ for the pole at origin, i.e., $\frac{1}{s^2}$. Here, $n = 2$. Therefore, the resultant phase angle in 180°.

13.1.4 Simple Zero (1 + *jωT*)

$$G(j\omega)H(j\omega) = 1 + j\omega T$$

Now,
$$20\log_{10}|G(j\omega)H(j\omega)| = 20\log_{10}\sqrt{1+\omega^2T^2} \tag{13.25}$$

To plot the magnitude curve, usually a linear asymptotic relation is used. At very low frequencies, $\omega T << 1$ and Eq. (13.25) can be expressed as

$$20\log_{10}|G(j\omega)H(j\omega)| = 20\log_{10}1 = 0\,\text{dB} \tag{13.26}$$

Therefore, for low frequencies the magnitude plot will be a straight line of 0 dB. The contribution by such factor for low frequency range can be neglected completely.

At very high frequency, i.e., $\omega T >> 1$, Eq. (13.25) can be written as

$$20\log_{10}|G(j\omega)H(j\omega)| = 20\log_{10}\sqrt{\omega^2T^2} = 20\log_{10}(\omega T)$$

$$20\log_{10}|G(j\omega)H(j\omega)| = 20\log_{10}\omega + 20\log_{10}T \tag{13.27}$$

Equation (13.27) represents a straight line having slope of 20 dB/decade (or 6dB/octave). For every decade, magnitude decrease by 20 dB because slope is −20 dB/decade. Corner frequency (ω_c) can be obtained from the intersection of the low-frequency and high-frequency asymptotes as follows:

$$20\log_{10}(\omega_c T) = 20\log_{10}1$$

i.e.,
$$\omega_c = \frac{1}{T} \tag{13.28}$$

At $\omega = 0.5\,\omega_c$, error $= 20\log_{10}\sqrt{0.5^2+1} - 20\log_{10}1 = 0.97\,\text{dB}$.

Now, (actual-approximate) $= 20\,log_{10}\sqrt{(\omega_c T)^2 + (1)^2} - 20\,log_{10}\,1$ (13.29)

Now at corner frequency, (actual-approximate)$\Big|_{\omega=\omega_c} = 20\,log_{10}\sqrt{1^2+1^2} - 20\,log_{10}\,1 = 3.03$ dB.

At $\omega = 2\omega_c$, error $= 20\,log_{10}\sqrt{2^2+1} - 20\,log_{10}\,2 = 0.97$ dB.

Maximum error occurs at corner frequency. The general error values are shown in Table 13.1.

Table 13.1 Error for different values of ω

Frequency	$\omega = 0.5\omega_c$	$\omega = \omega_c$	$\omega = 2\omega_c$
Error	0.97 dB up	3dB up	0.97 dB up

Figure 13.5 shows the magnitude plot of $(1+j\omega T)$, which shows the actual curve as well as the linear asymptotic curves. The actual or exact plot deviates slightly from straight-line asymptotes.

Now, $\angle G(j\omega)H(j\omega) = \tan^{-1}(\omega T)$. To construct the phase angles plot of $G(j\omega)H(j\omega)$ the phase angles $G(j\omega)H(j\omega)$ for different values of ω are calculated and these are shown in Table 13.2.

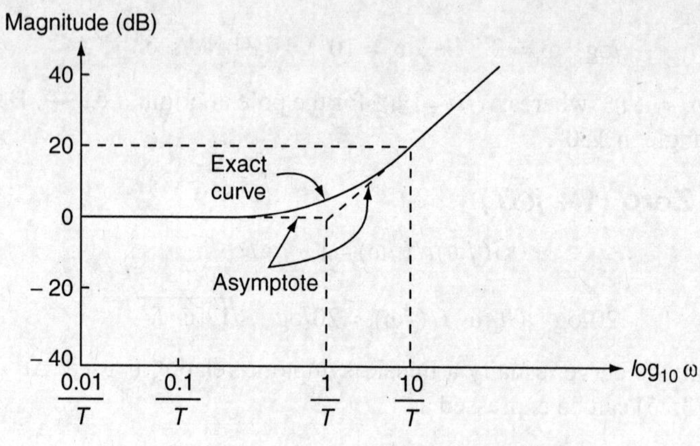

Fig. 13.5 Magnitude plot of simple zero

Table 13.2 Phase angle for different values of ω

ω	$0.1\omega_c$	$0.5\omega_c$	$1\omega_c$	$2\omega_c$	$10\omega_c$	∞
∠G(jω)H(jω)	+5.71°	+26.6°	+45°	+63.4°	+84.3°	+90°

Figure 13.6 shows the phase angle plot of the simple zero $1+j\omega T$.

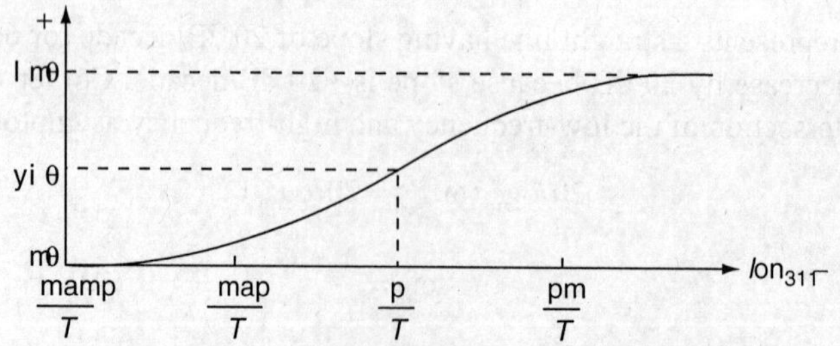

Fig. 13.6 Phase plot of simple zero

Example 13.4 A unity feedback control system has $G(s) = \left(1 + \frac{s}{3}\right)$. Draw the Bode plot.

Solution Here, $G(s) = \left(1 + \frac{s}{3}\right)$ and $H(s) = 1$.

$$\therefore \quad G(s)\,H(s) = 1 + \frac{s}{3}$$

$$\therefore \quad G(j\omega)\,H(j\omega) = 1 + \frac{j\omega}{3}$$

$$\therefore \quad M = 20\ log_{10}\left|G(j\omega)\,H(j\omega)\right| = 20\ log_{10}\left|1 + \frac{j\omega}{3}\right|$$

$$= 20\ log_{10}\sqrt{1 + \frac{\omega^2}{9}} \tag{1}$$

At very low frequency, i.e., $\frac{\omega}{3} << 1$, the above expression, i.e., Eq. (1) can be expressed as

$$\therefore \quad M = 20\ log_{10}\left|G(j\omega)\,H(j\omega)\right| = 20\ log_{10}(1) = 0 \tag{2}$$

At very high frequency, i.e., $\frac{\omega}{3} >> 1$, the above expression, i.e., Eq. (1) can be expressed as

$$\therefore \quad M = 20\ log_{10}\left|G(j\omega)\,H(j\omega)\right| = 20\ log_{10}\frac{\omega}{3} \tag{3}$$

At $\omega = \omega_c = 3$ rad/sec., the magnitude M of Eq. (3) becomes zero.
Here, phase angle becomes

$$\phi = \tan^{-1}\frac{\omega}{3}$$

At $\omega = 0.01$, $\quad \phi = 0.19° \cong 0°$

At $\omega = 3$, $\quad \phi = \tan^{-1}(1) = 45°$

At $\omega = 30$, $\quad \phi = \tan^{-1}(10) = 84.29°$.

Figure E13.4(a) shows the magnitude plot whereas Fig. E13.4(b) shows the phase angle plot.

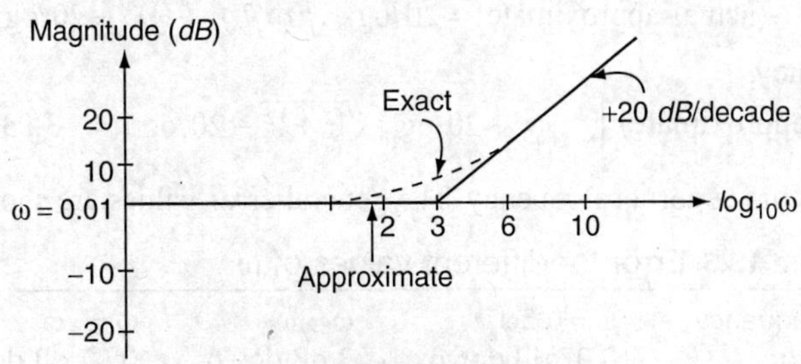

(a) Magnitude plot

Fig. E13.4(a)

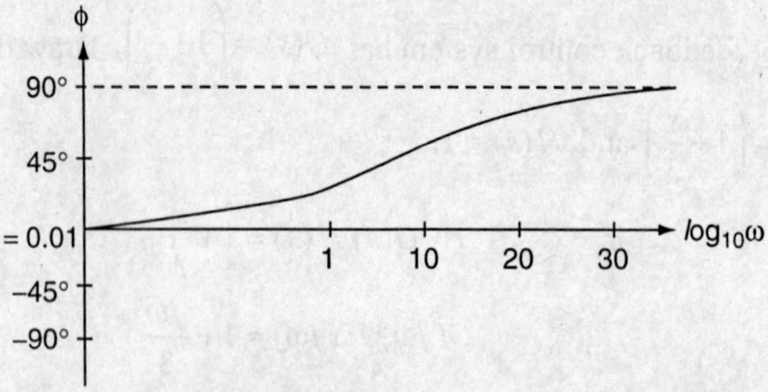

(b) Phase plot

Fig. E13.4(b)

13.1.5 Simple Pole $(1 + j\omega T)^{-1}$

Here, $$G(j\omega)H(j\omega) = (1 + j\omega T)^{-1}$$

Now, $$20\,log_{10}|G(j\omega)H(j\omega)| = -20\,log_{10}\sqrt{1+\omega^2T^2} \qquad (13.30)$$

At very low frequencies, $\omega T << 1$, Eq. (13.30) can be expressed as

$$20\,log_{10}|G(j\omega)H(j\omega)| = -20\,log_{10}1 = 0\,\text{dB} \qquad (13.31)$$

At very high frequencies, $\omega T >> 1$, Eq. (13.30) can be expressed as

$$20\,log_{10}|G(j\omega)H(j\omega)| = -20\,log_{10}\sqrt{\omega^2T^2} = -20\,log_{10}(\omega T)$$

Now, $$20\,log_{10}|G(j\omega)H(j\omega)| = -20\,log_{10}\,\omega - log_{10}T \qquad (13.32)$$

Equation (13.32) represents a straight line having slope of −20 dB/decade (or −6dB/octave). Corner frequency (ω_c) can be obtained from the intersection of the low-frequency and high-frequency asymptotes as follows:

Corner frequency (ω_c) can be obtained from

$$-20\,log_{10}(\omega_c T) = -20\,log_{10}1$$

i.e., $$\omega_c T = 1$$

i.e., $$\omega_c = \frac{1}{T} \qquad (13.33)$$

Now,

$$\text{(actual-approximate)} = 20\,log_{10}\sqrt{(\omega_c T)^2 + (1)^2} + 20\,log_{10}1$$

Now, at corner frequency,

$$\text{(actual-approximate)}\Big|_{\omega=\omega_c} = -20\,log_{10}\sqrt{1^2+1^2} - 20\,log_{10}1 = -3.03\,\text{dB}$$

Maximum error occurs at corner frequency. The general error values are shown in Table 13.3.

Table 13.3 Error for different values of ω

Frequency	$\omega = 0.5\omega_c$	$\omega = \omega_c$	$\omega = 2\omega_c$
Error	0.97 dB down	3 dB down	0.97 dB down

Figure 13.7 shows the magnitude plot of $1/(1+j\omega T)$, which shows the actual curve as well as the linear asymptotic curves. The actual or exact plot deviates slightly from straight-line asymptotes.

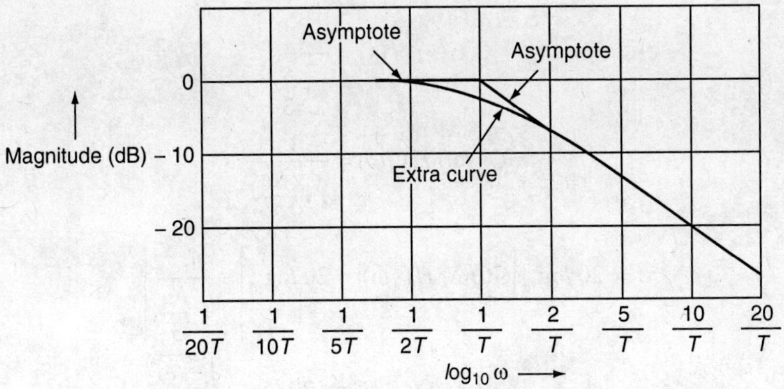

Fig. 13.7 Magnitude plot of simple pole

The phase shift of $G(j\omega)H(j\omega)$ is given by

$$\arg[G(j\omega)H(j\omega)] = \phi = -\tan^{-1}\omega T \tag{13.34}$$

To construct the phase angle plot of $G(j\omega)H(j\omega)$, the phase angle of $G(j\omega)H(j\omega)$ for different values of ω are calculated and these are shown in Table 13.4.

Table 13.4 Phase angle for different values of ω

ω	$0.1\omega_c$	$0.5\omega_c$	$1\omega_c$	$2\omega_c$	$10\omega_c$	∞
$\angle G(j\omega)H(j\omega)$	$-5.71°$	$-26.6°$	$-45°$	$-63.4°$	$-84.3°$	$-90°$

Figure 13.8 shows the phase angle plot of the simple pole $1/(1+j\omega T)$.

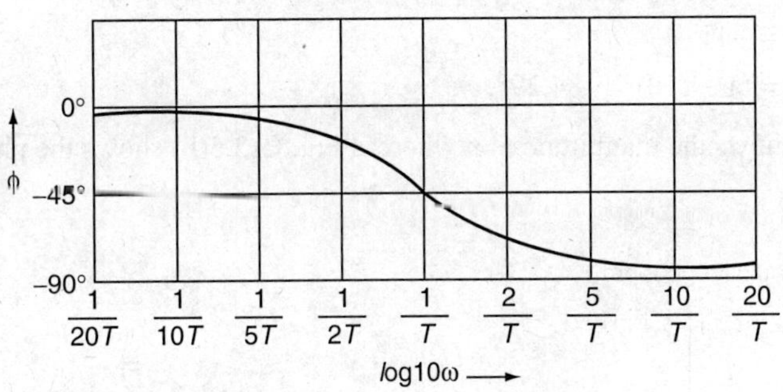

Fig. 13.8 Phase plot of simple pole

Example 13.5 A unity feedback system has $G(s) = \dfrac{1}{1+\dfrac{s}{3}}$. Draw its Bode plot.

Solution

Here, $\qquad G(s) = \dfrac{1}{1+\dfrac{s}{3}}$ and $H(s) = 1$.

$$\therefore \qquad G(s)\,H(s) = \frac{1}{1+\dfrac{s}{3}}$$

$$\therefore \qquad G(j\omega)\,H(j\omega) = \frac{1}{1+\dfrac{j\omega}{3}}$$

$$\therefore \qquad M = 20\,log_{10}\left|G(j\omega)\,H(j\omega)\right| = 20\,log_{10}\left|\frac{1}{1+\dfrac{j\omega}{3}}\right|$$

$$= -20\,log_{10}\sqrt{1+\frac{\omega^2}{9}} \qquad (1)$$

At very low frequency, i.e., $\frac{\omega}{3} << 1,$ the above expression, i.e., Eq. (1) can be expressed as

$$M = -20\,log_{10}\left|G(j\omega)\,H(j\omega)\right| = -20\,log_{10}(1) = 0. \qquad (2)$$

At very high frequency, i.e., $\frac{\omega}{3} >> 1,$ the above expression, i.e., Eq. (1) can be expressed as

$$M = 20\log_{10}\left|G(j\omega)\,H(j\omega)\right| = -20\log_{10}\frac{\omega}{3} \qquad (3)$$

At $\omega = \omega_c = 3$ rad/sec., the magnitude M of Eq. (3) becomes zero.

Here, phase angle $\phi = -\tan^{-1}\frac{\omega}{3}$

At $\omega = 0.01,$ $\qquad \phi = -0.19° \simeq 0$

At $\omega = 3,$ $\qquad \phi = -45°$

At $\omega = 30,$ $\qquad \phi = \tan^{-1}(10) = -84.29°.$

Figure E13.5(a) shows the magnitude plot whereas Fig. E13.5(b) shows the phase angle plot.

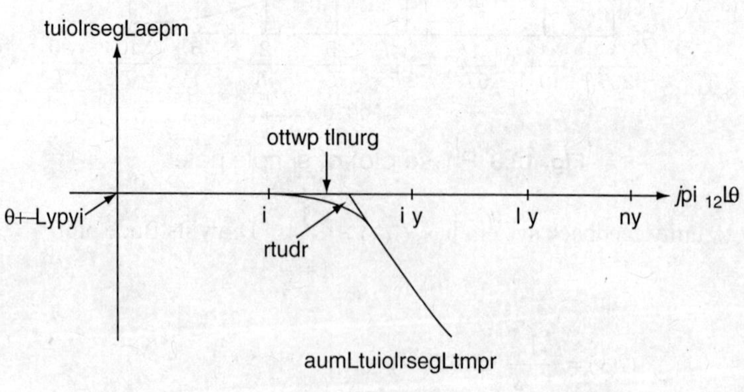

Fig. E13.5a

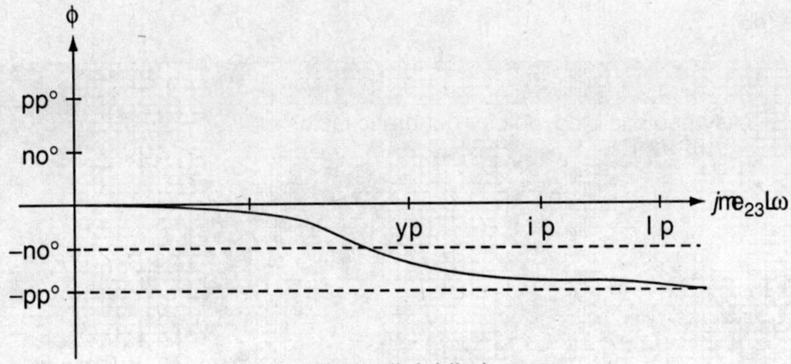

aumLrgtodLtleidLnimp

Fig. E13.5

13.1.6 Quadratic Poles i.e., $\omega_n^2/(s^2+2\xi\omega_n s+\omega_n^2)$

Here, $$G(j\omega)H(j\omega)=\frac{\omega_n^2}{(\omega_n^2-\omega^2)+2j\xi\omega_n\omega}$$

i.e., $$G(j\omega)H(j\omega)=\frac{1}{\left(1-\dfrac{\omega^2}{\omega_n^2}\right)+2j\xi\left(\dfrac{\omega}{\omega_n}\right)}$$

$$\therefore \quad 20\log_{10}|G(j\omega)H(j\omega)|=20\log_{10}\left|\frac{1}{\left(1-\dfrac{\omega^2}{\omega_n^2}\right)+2j\xi\left(\dfrac{\omega}{\omega_n}\right)}\right| \tag{13.35}$$

At lower frequency, i.e., $\omega/\omega_n \ll 1$, we have from Eq. (13.35)

$$20\,log_{10}|G(j\omega)H(j\omega)|=-20\,log_{10}1=0\,\text{dB} \tag{13.36}$$

At higher frequency, i.e., $\omega/\omega_n \gg 1$, we have from Eq. (13.35)

$$20\,log_{10}|G(j\omega)H(j\omega)|=-20\log_{10}\sqrt{\left(\frac{\omega^2}{\omega_n^2}\right)}=-20\,log_{10}\left(\frac{\omega^2}{\omega_n^2}\right)$$

i.e., $$20\,log_{10}|G(j\omega)H(j\omega)|=-40\,log_{10}\,\omega+40\,log_{10}\,\omega_n \tag{13.37}$$

Figure 13.9 shows the Bode plot for the above quadratic pole. Equating Eq. (13.37) to zero, we have

$$-40\,log_{10}\,\omega_c+40\,log_{10}\,\omega_n=0$$

i.e., $$\omega_c=\omega_n$$

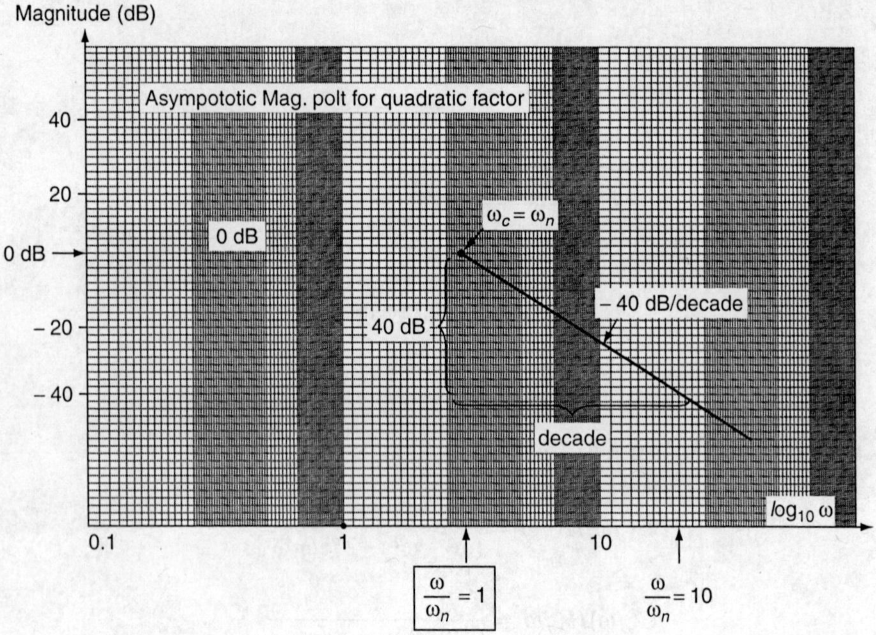

Fig. 13.9 Bode plot for quadratic poles

Therefore, ω_n is the corner frequency for the given quadratic term.

The above asymptote is not accurate because the error of the quadratic factor not only depends on the value ω but also on the value of ξ.

$$\text{Actual magnitude in dB} = -20\,log_{10}\sqrt{\left[1-\left(\frac{\omega}{\omega_n}\right)^2\right]^2 + 4\xi^2\left(\frac{\omega}{\omega_n}\right)^2} \tag{13.38}$$

Now, at $\omega = \omega_n$,

actual magnitude in $\quad \text{dB} = -20\,log_{10}\sqrt{4\xi^2} = -20\,log_{10}(2\xi)$

Table 13.5 shows the error values for various values of ξ.

Table 13.5 Error for various values of ξ

ξ	Accurate magnitude in dB	Approximate magnitude in dB	Error
0.1	+13.97	0	13.97 up
0.2	+7.95	0	7.95 up
0.3	+4.43	0	4.43 up
0.4	+1.93	0	1.93 up
0.7	−2.92	0	2.92 down
0.9	−5.10	0	5.10 down
1	−6.02	0	−6.02 down

Figure 13.10 shows the exact plot of the quadratic poles for different values of ξ.

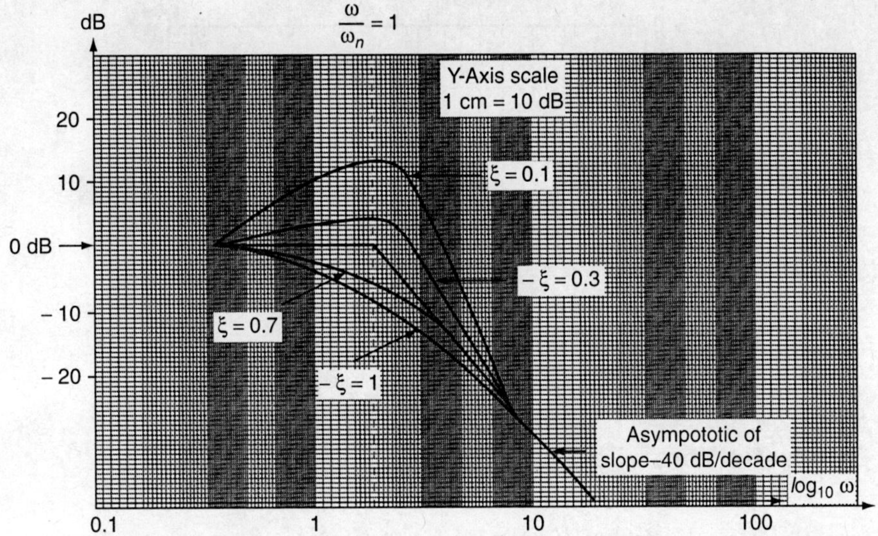

Fig. 13.10 Exact plot of quadratic pole for different values of ξ

The phase angle of the quadratic pole is expressed by

$$\angle G(j\omega)H(j\omega) = -\tan^{-1}\left[\frac{2\xi\left(\frac{\omega}{\omega_n}\right)}{1-\left(\frac{\omega}{\omega_n}\right)^2}\right] \tag{13.39}$$

At $\omega = \omega_n$, the phase angle of $G(j\omega)H(j\omega)$ is $-90°$ and at $\omega = \infty$, the phase angle of $G(j\omega)H(j\omega)$ becomes $-180°$. When $\omega/\omega_n > 1$, the phase angle of $G(j\omega)H(j\omega)$ becomes positive. In this case, the correct value of the phase angle is obtained after subtracting $-180°$ from the obtained phase angle. Table 13.6 Phase angle for different values of ω/ω_n.

Table 13.6 Phase angle different values of ω/ω_n

$\frac{\omega}{\omega_n}$	$\angle G(j\omega)H(j\omega)$
0.1	−3.46°
0.5	−21.8°
1	−90°
2	+21.8° − 180° = −158.19°
4	+10.09° − 180° = −170.9°
10	+3.46° − 180° = −176.53°
∞	−180°

Figure 13.11 shows the plot of phase angle versus $log\omega$.

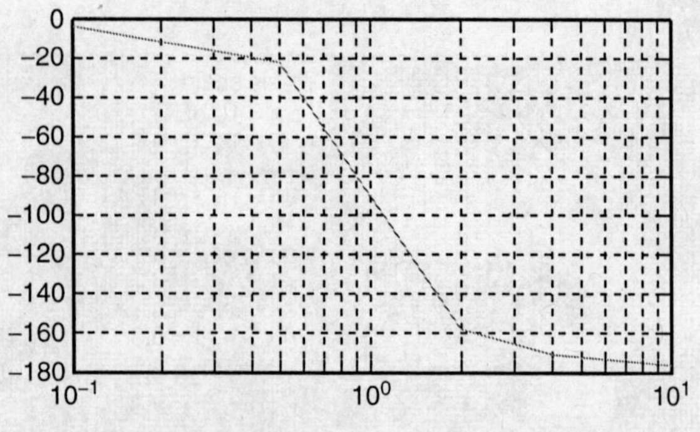

Fig. 13.11 Plot of phase angle versus logω

13.1.7 Quadratic Zeros i.e., $(s^2+2\xi\omega_n s+\omega_n^2)/\omega_n^2$

Here,
$$G(j\omega)H(j\omega)=(s^2+2\xi\omega_n s+\omega_n^2)/\omega_n^2$$

$$\therefore \quad G(j\omega)H(j\omega)=\left(1-\frac{\omega^2}{\omega_n^2}\right)+2j\xi\frac{\omega}{\omega_n} \tag{13.40}$$

$$20\,log_{10}\left|G(j\omega)H(j\omega)\right|=20\,log_{10}\left|\left(1-\frac{\omega^2}{\omega_n^2}\right)+2j\xi\frac{\omega}{\omega_n}\right| \tag{13.41}$$

At lower frequency, i.e., $\omega/\omega_n << 1$, we have from Eq. (13.41)

$$20\,log_{10}\left|G(j\omega)H(j\omega)\right|=20\,log_{10}1=0\ dB \tag{13.42}$$

At higher frequency, i.e., $\omega/\omega_n >> 1$, we have from Eq. (13.41)

$$20\ log_{10}\left|G(j\omega)H(j\omega)\right|=20\ log_{10}\sqrt{\left(\frac{\omega^2}{\omega_n^2}\right)^2}=20\,log_{10}\left(\frac{\omega^2}{\omega_n^2}\right)$$

i.e.,
$$20\,log_{10}\left|G(j\omega)H(j\omega)\right|=40\,log_{10}\omega\ -40\,log_{10}\,\omega_n \tag{13.43}$$

Figure 13.12 shows the Bode plot for the above quadratic zero. Equating Eq. (13.43) to zero, we have

$$40\ log_{10}\,\omega_c-40\ log_{10}\,\omega_n=0$$

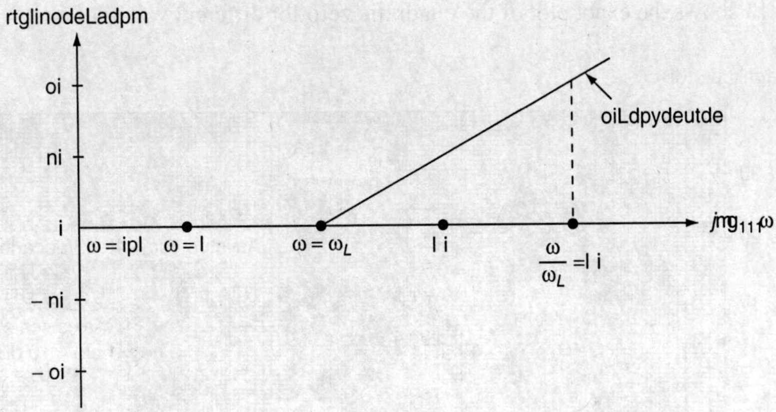

Fig. 13.12 Bode plot for quadratic zero

i.e., $$\omega_c = \omega_n$$

Therefore, ω_n is corner frequency for the given quadratic term.

The above asymptotic is not accurate because the error of the quadratic factor not only depends upon on the value ω but also on the value of ξ.

$$\text{Actual magnitude in dB} = 20\, log_{10} \sqrt{\left[1-\left(\frac{\omega}{\omega_n}\right)^2\right]^2 + 4\xi^2\left(\frac{\omega}{\omega_n}\right)^2} \tag{13.44}$$

Now, at $\omega = \omega_n$,

$$\text{actual magnitude in dB} = 20\, log_{10}\sqrt{4\xi^2} = 20\, log_{10}(2\xi)$$

Table 13.7 shows the error values for various values of ξ.

Table 13.7 Error values for various values of ξ

ξ	Accurate magnitude in dB	Approximate magnitude in dB	Error
0.1	−13.97	0	13.97 down
0.2	−7.95	0	7.95 down
0.3	−4.43	0	4.43 down
0.4	−1.93	0	1.93 down
0.7	+2.92	0	2.92 up
0.9	+5.10	0	5.10 up
1	+6.02	0	6.02 up

Figure 13.13 shows the exact plot of the quadratic zero for different values of ξ.

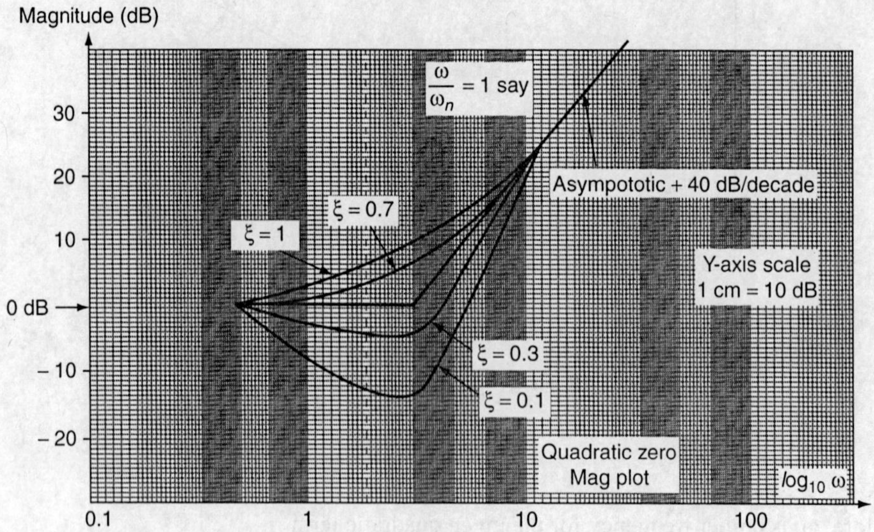

Fig.13.13 Exact plot of quadratic zeros for different values of ξ

The phase angle of the quadratic zero is expressed by

$$\angle G(j\omega)H(j\omega) = \tan^{-1}\left[\frac{2\xi\left(\dfrac{\omega}{\omega_n}\right)}{1-\left(\dfrac{\omega}{\omega_n}\right)^2}\right] \tag{13.45}$$

At $\omega = \omega_n$, the phase angle of $G(j\omega)H(j\omega)$ is 90° and at $\omega = \infty$, the phase angle of $G(j\omega)H(j\omega)$ becomes 180°. When $\omega/\omega_n > 1$, the phase angle of $G(j\omega)H(j\omega)$ becomes negative. In this case the phase angle is obtained after adding 180° from the obtained phase angle.

Table 13.8 shows the phase angle for different values of ω/ω_n and $\xi = 0.3$.

Table 13.8 Phase angle for different value of ω/ω_n and $\xi = 0.3$

$\dfrac{\omega}{\omega_n}$	$\angle G(j\omega)H(j\omega)$
0.1	3.46°
0.5	21.8°
1	90°
2	− 21.8° + 180° = 158.19°
4	− 10.09° + 180° = 170.9°
10	− 3.46° + 180° = 176.53°
∞	+ 180°

Figure 13.14 shows the plot of phase angle vs $log_{10}\omega$.

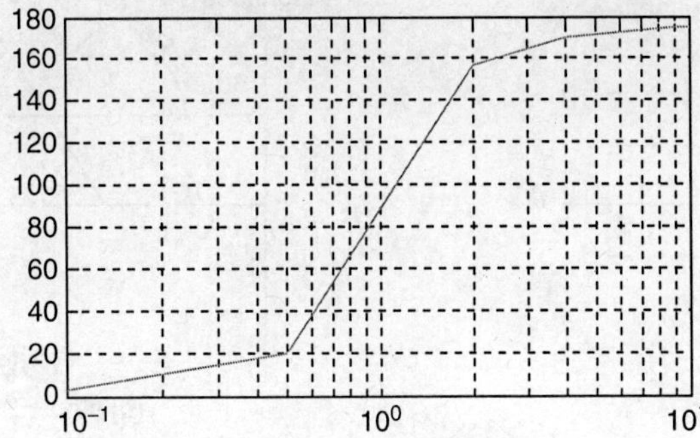

Fig. 13.14 Plot of phase angle vs $log_{10}\omega$

13.2 Steps to Plot Bode Plot

Step 1: Express $G(s)H(s)$ in standard form.
Step 2: Each factor must be identified.
Step 3: Magnitude plot for each factor is to be drawn.
Step 4: All points are added to obtain the resultant plot.
Step 5: Phase plot for each factor is drawn.
Step 6: All points are added clearly to obtain the resultant curve.

Asymptotic Bode plots are easy to draw compared to a precise Bode plot and there will not be large errors from these approximated Bode plots.

13.3 Resultant Curve

Generally, there are the following three cases:

Step ± Step: When two step functions, say S_1 and S_2, are added, the resultant is a step value of $S_1 \pm S_2$ at the point of addition as shown in Fig. 13.15(a).

Step ± Ramp: When a step function is added to a ramp function, the resultant is a ramp function at the point of addition as shown in Fig. 13.15(b).

Ramp ± Ramp: When a ramp function is added to a ramp function, the resultant is a ramp function at the point of addition as shown in Fig. 13.15(c).

Example 13.6 A unity feedback control system has $G(s) = 20/(s+2)$. Draw the Bode plot.

Solution

Step 1: Here, $G(s) = 20/(s+2)$ and $H(s) = 1$.

$$\therefore \quad G(s)H(s) = \frac{20}{s+2} = \frac{10}{1+\frac{s}{2}}$$

Step 2: Here $K = 10$ and there is one simple pole having corner frequency at $\omega = 2$ rad/sec.

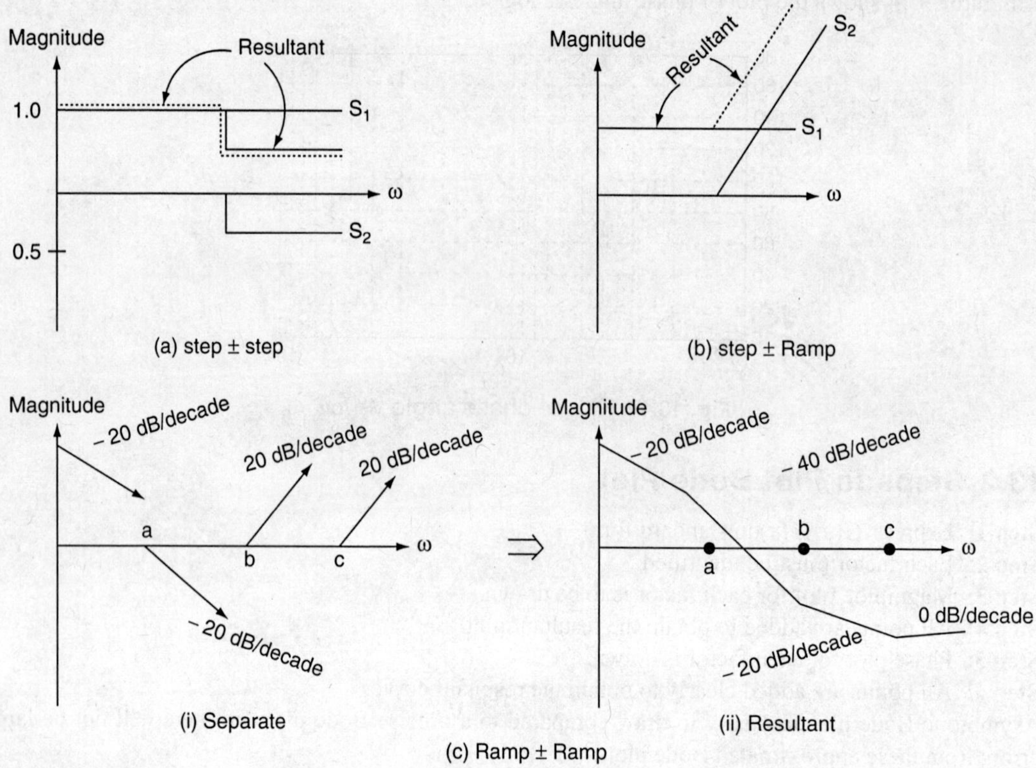

Fig. 13.15 Resultant curve

Step 3: (i) $20 \, log_{10} K = 20 \, log_{10} 10 = 20$ dB and its Bode plot is a straight line having slope 0 dB/decade.

(ii) There is one simple pole having corner frequency at $\omega = 2$ rad/sec. Its Bode plot is a straight line of slope 0 dB/decade up to $\omega = 2$ rad/sec and from the point $\omega = 2$ rad/sec and 0 dB, its slope is −20 dB/decade.

(iii) Now, let us shift the pint $\omega = 2$ at 0 dB on $20 \, log_{10} K$ line and draw parallel to −20 dB/decade line, which has been drawn already and this is the resultant line of K and $1/(s+2)$. This resultant line passes through the point $\omega = 20$ at 0 dB.

Step 4: Phase angle plot.

$$G(j\omega)H(j\omega) = \frac{10}{1 + \dfrac{j\omega}{2}}$$

∴

$$\angle G(j\omega)H(j\omega) = \phi = -\tan^{-1}\left(\frac{\omega}{2}\right)$$

Phase angle table:

Table E13.1

ω	0.2	2	8	10	20	40	80	∞
φ	−5.7°	−45°	−75.9°	−78.7°	−84.3°	−87.1°	−88.6°	−90°

Figure E13.6 shows the Bode plot.

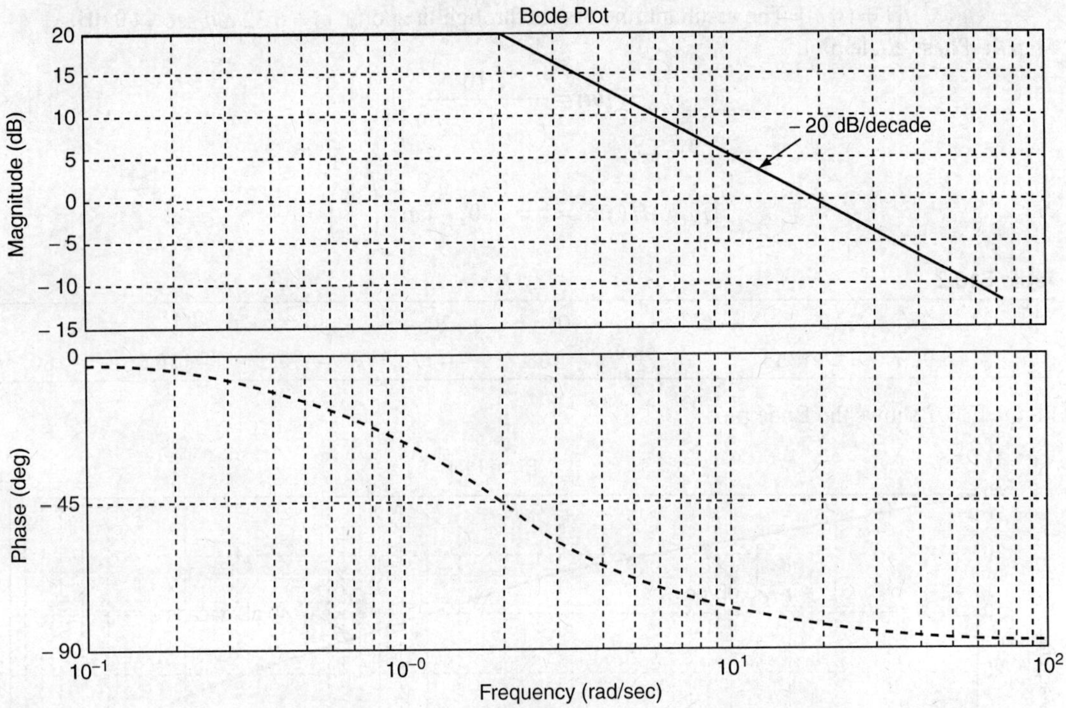

Fig. E13.6

Example 13.7 A unity feedback control system has $G(s) = 20/[s(s+2)]$. Draw the Bode plot.

Solution

Step 1: Here, $G(s) = 20/[s(s+2)]$ and $H(s) = 1$. Therefore

$$G(s)H(s) = \frac{20}{s(s+2)} = \frac{10}{s\left(1+\dfrac{s}{2}\right)}$$

Step 2: For $K = 10$, there is one pole at origin and one simple pole.

Step 3: (i) $20\,log_{10}K = 20\,log_{10}10 = 20$ dB and its Bode plot is a straight line of slope 0 dB/decade.

(ii) There is one pole at origin. Its Bode plot is a straight line of slope −20 dB/decade, which passes through the intersection point of $\omega = 1$ at 0 dB.

(iii) Now, let us shift intersection point $\omega = 1$ and 0 dB on the $20\,log_{10}\,K$ line and draw parallel to −20 dB/decade line. This is the resultant line of *K* and 1/*s* and it continues till first corner frequency occurs at $\omega_{c1} = 2$.

(iv) Now,at $\omega_{c1} = 2$,thereisasimplepoleanditcontributesattherateof−20dB/decade,Therefore, theresultantslopeafter $\omega_{c1} = 2$ becomes−40dB/decade.Thisistheresultantcurveof*K*,1/*s*and 1/[1 + (*s*/2)]. The resultant line passes through the point $\omega = 6.32$ rad/sec at 0 dB.

Step 4: Phase angle plot:

$$G(j\omega)H(j\omega) = \frac{10}{(j\omega)\left(1+\frac{j\omega}{2}\right)}$$

$$\therefore \qquad G(j\omega)H(j\omega) = \phi = -90° - \tan^{-1}\left(\frac{\omega}{2}\right)$$

Table E13.2

w	0.2	2	8	10	20	40	80	∞
f	−95.7°	−135°	−165.9°	−168.7°	−174.3°	−177.1°	−178.6°	−180°

Figure E13.7 shows the Bode plot

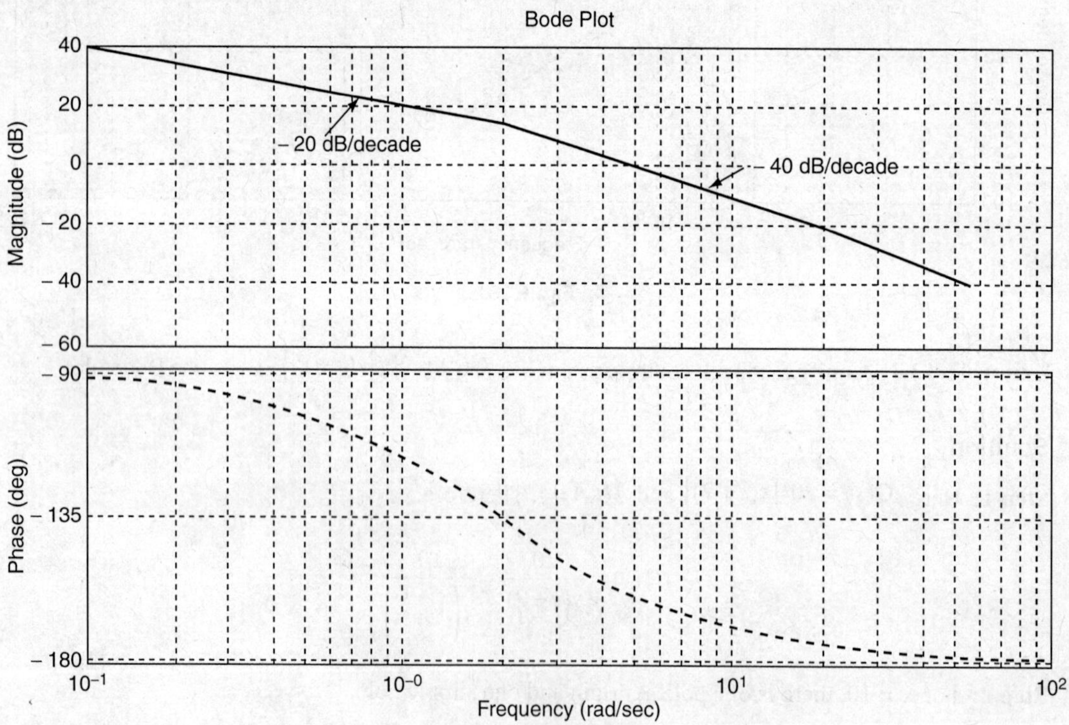

Fig. E13.7

13.4 Stability Consideration

Gain margin and phase margin give us idea as to how far the system is away from the condition of instability. If a system having $G(s) = K/[s(1+p_1s)(1+p_2s)]$ is unstable for $K > 90$, the system is marginally stable for $K = 90$ and stable for $K < 90$ and

$$|G(s)H(s)|_{\omega=\omega_{pc}=\omega_{gc}} = 1 \tag{13.46}$$

where ω_{pc} is the phase crossover frequency and ω_{gc} is the gain crossover frequency. On the other hand, if the actual gain is less than 90, say $K = 40$, the system will be stable. Gain margin (GM) is defined as the ratio of the critical value of gain to the actual value of gain.

$$\therefore \qquad \text{GM} = \frac{\text{Critical value of } K}{\text{Actual value of } K} \tag{13.47}$$

The phase margin (PM) is defined as

$$\text{PM} = 180° + \angle[G(j\omega)H(j\omega)]_{\omega=\omega_{gc}}$$

From the Bode plot, we can find out the gain crossover frequency, phase crossover frequency, gain margin as well as phase margin.

13.4.1 Phase Margin

To find out phase margin, the frequency at which the resultant magnitude is 0 dB is found out at first. This frequency is called gain crossover frequency (ω_{gc}). Now, from the phase plot, the phase angle (ϕ) is calculated the phase crossover frequency. The phase margin is given by

$$\begin{aligned} \text{PM} &= [\angle G(j\omega)H(j\omega)]_{\omega=\omega_{gc}} - (-180°) \\ &= \phi + 180° \\ &= 180° + \phi \end{aligned} \tag{13.48}$$

PM has been shown in Fig. 13.16.

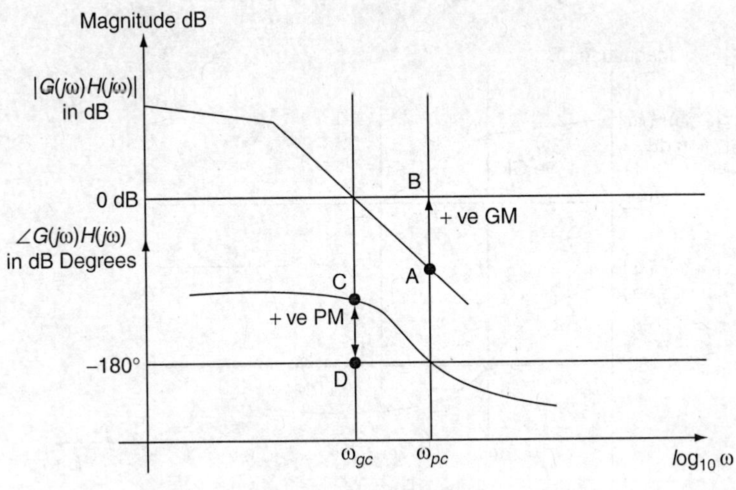

Fig.13.16 $\omega_{gc} < \omega_{pc}$ GM and PM positive, stable system

13.4.2 Gain Margin

To find out gain margin, the frequency at which the resultant phase curve has a phase of −180° is found out at first. This frequency is known as phase crossover frequency (ω_{pc}). Next, the value of the resultant curve is calculated at phase crossover frequency. If the value of the resultant curve is below 0 dB, it is considered positive. Otherwise, it is negative.
GM is expressed by

$$GM = [0 - 20\ log_{10}|G(j\omega)H(j\omega)|_{\omega=\omega_{pc}}]\text{dB} \tag{13.49}$$

For $\omega_{gc} < \omega_{pc}$, both gain margin and phase margin are positive. Therefore, the system is stable as shown in Fig. 13.16. For $\omega_{gc} = \omega_{pc}$, the system is marginally stable as shown in Fig. 13.17. For $\omega_{gc} > \omega_{pc}$, both gain margin and phase margin are negative and hence the system is unstable as shown in Fig. 13.18.

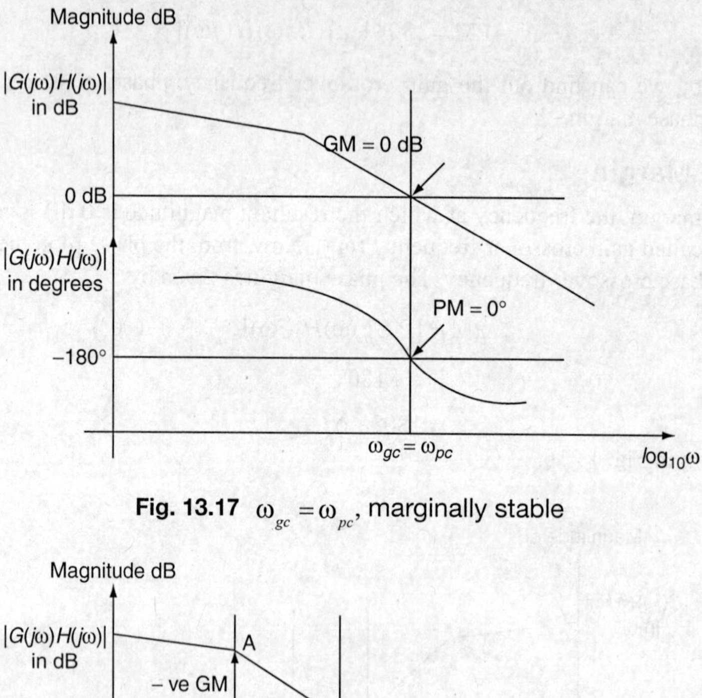

Fig. 13.17 $\omega_{gc} = \omega_{pc}$, marginally stable

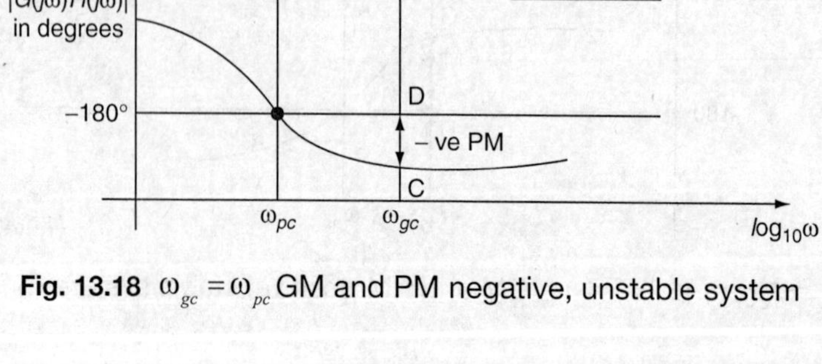

Fig. 13.18 $\omega_{gc} = \omega_{pc}$ GM and PM negative, unstable system

Generally, we take gain as high as possible so that steady-state error is reduced and acceptable response speed is obtained. Besides these, adequate GM and PM will be maintained when gain is high. If low gain is considered, GM and PM will be high and this will cause higher steady-state error, higher values of rise time and settling time. Therefore, the system will be sluggish. Designing a controller is a tremendous task for any engineer. Generally, a gain margin of 2 dB and phase margin of 30° is taken, which is good enough as a thumb rule.

Example 13.8 A unity feedback control system has $G(s) = 40/[s(s+2)(s+5)]$. Draw the Bode plot. Find GM and PM.

Solution

Step 1: Here, $$G(s) = \frac{40}{s(s+2)(s+5)} \text{ and } H(s) = 1$$

$$\therefore \quad G(s)H(s) = \frac{40}{s(s+1)(s+5)} = \frac{4}{s\left(1+\frac{s}{2}\right)\left(1+\frac{s}{5}\right)}$$

Step 2: For $K = 4$, there is one pole at origin and two simple poles.

Step 3: (i) $20\ log_{10}\ K = 20\ log_{10}\ 4 = 12\ \text{dB}$.

(ii) There is one pole at origin. The Bode plot of it is a straight line of slope −20 dB/decade, which passes through the intersection point of $\omega = 1$ at 0 dB.

(iii) Now, let us shift intersection point $\omega = 1$ at 0 dB on the $20\ log_{10}\ K$ line and draw a line parallel to thc −20 dB/decade line. This is the resultant line of K and $1/s$ and it continues till first corner frequency occurs at $\omega_{c1} = 2$.

(iv) Now, at $\omega_{c1} = 2$, there is a simple pole and it contributes at the rate of −20dB/decade. Therefore, the resultant slope after $\omega_{c1} = 2$ becomes −40 dB/decade. This is the resultant curve of K, $1/s$ and $1/(1 + s/2)$; it continues till second corner frequency occurs at $\omega_{c2} = 5$.

(v) Once again, at $\omega_{c2} = 5$, there is a simple pole contributing at the rate of −20 dB/decade and hence the resultant slope after $\omega_{c2} = 5$ become −60 dB/decade. Since there are no factors present, the final slope becomes −60 dB/decade. The resultant line passes it though the point $\omega = 3.12$ rad/sec at 0 dB.

Step 4: Phase angle plot:

Now, $$G(j\omega)H(j\omega) = \frac{4}{(j\omega)\left(1+\frac{j\omega}{2}\right)\left(1+\frac{j\omega}{5}\right)}$$

$$\therefore \quad \angle G(j\omega)H(j\omega) = \phi = -90° - \tan^{-1}\left(\frac{\omega}{2}\right) - \tan^{-1}\left(\frac{\omega}{5}\right)$$

Phase angle table:

Table E13.3

ω	0.2	2	8	10	20	40	80	∞
ϕ	−98°	−156.8°	−223.9°	−232.1°	−250.2°	−260°	−265°	−270°

Figure 13.8 shows the Bode plot.

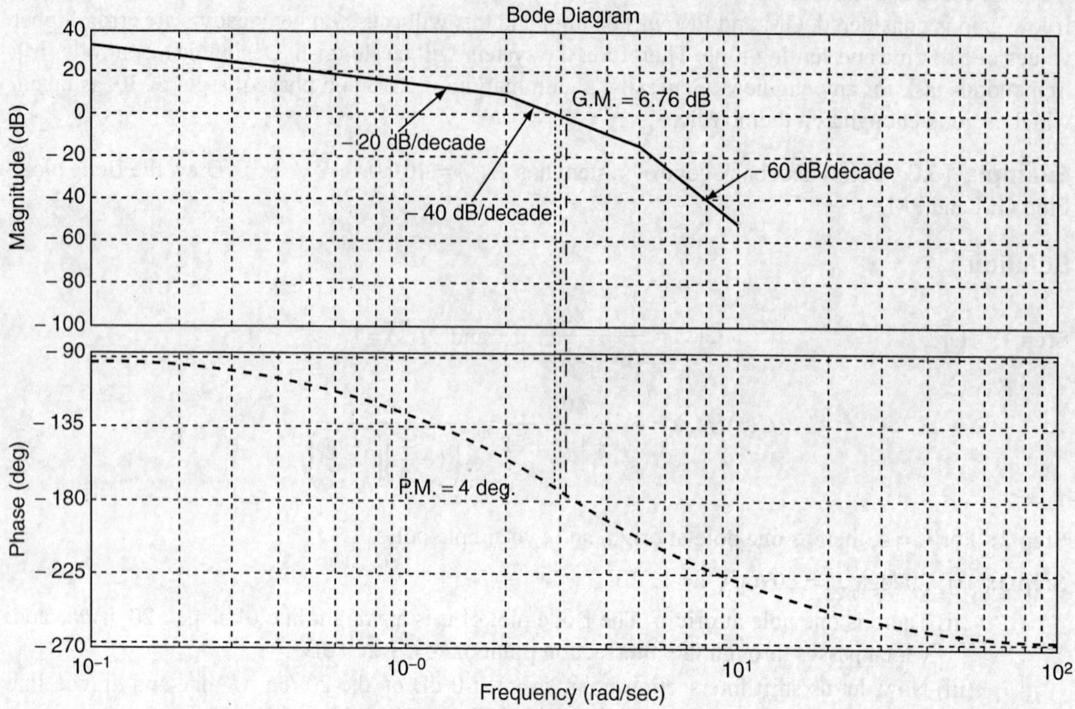

Fig. E13.8

13.4.3 All-Pass and Minimum Phase Functions

If all poles and zeros of any transfer function lie on the left half of the s-plane, this transfer function is known as minimum phase transfer function. The functions that we have already discussed earlier for Bode plot is minimum phase functions.

If all the zeros of a transfer function lie on the right half of the s-plane, all poles lie on the left half of the s-plane and the location of every pole zero pair is symmetric about the $j\omega$ axis, the transfer function is termed as all-pass function. Equation (13.48) shows the transfer function of an all-pass function where pole-zero plot has been shown in Fig. 13.19(a).

$$T(s) = \left(\frac{s-a}{s+a}\right)\left(\frac{s-b}{s+b}\right) \tag{13.48}$$

Figure 13.19(a) shows the pole zero plot for all-pass function whereas Fig. 13.19(b) shows the magnitude and phase characteristic of an all-pass transfer function. If the transfer function has one or more zeros on the right half of the s-plane, it is known as a non-minimum transfer function. A non-minimum transfer function is equivalent to the product of one minimum phase function and one all-pass function.

$$G(j\omega) = \frac{1 - j\omega T}{(1 + j\omega T_1)(1 + j\omega T_2)}$$

$$= \left[\frac{1 + j\omega T}{(1 + j\omega T_1)(1 + j\omega T_2)}\right]\left[\frac{1 - j\omega T}{(1 + j\omega T)}\right] = G_1(j\omega)G_2(j\omega)$$

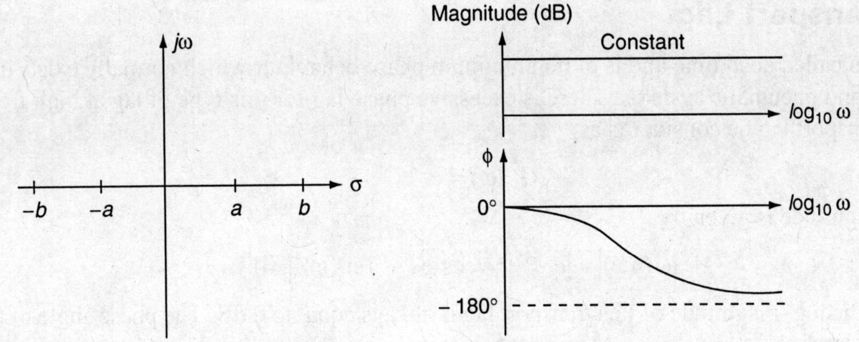

(a) Pole and zero's for all pass network

(b) Magnitude and phase characteristic for an all pass network

Fig. 13.19 All-pass network

where $G_1(j\omega)$ and $G_2(j\omega)$ are minimum and all-pass function respectively. Figure 13.20(a), Fig. 13.20(b) and Fig. 13.20(c) show the pole zero plots of $G(j\omega)$, $G_1(j\omega)$, and $G_2(j\omega)$ respect ively. Figure 13.21 shows the Bode plot of $G(j\omega)$.

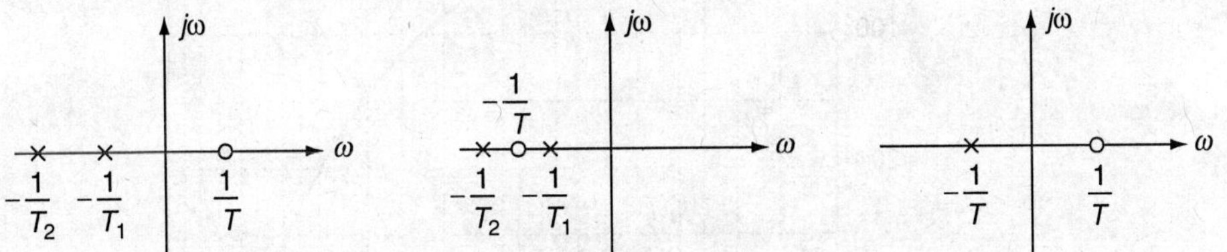

(a) $G(j\omega)$ (b) $G_1(j\omega)$ (c) $G_2(j\omega)$

Fig.13.20 Pole zero plot of $G(j\omega)$, $G_1(j\omega)$ and $G_2(j\omega)$

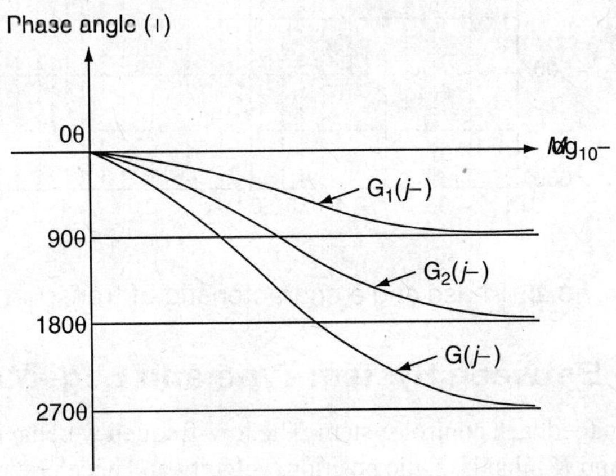

Fig. 13.21 Bode plots of $G(j\omega)$, $G_1(j\omega)$ and $G_2(j\omega)$

13.5 Transport Lag

This is also called dead time and is of nonminimum phase behaviour which normally exists in thermal, hydraulic and pneumatic systems. There is excessive phase lag for this type of lag at high frequencies. Let the transport lag be considered as

$$G(j\omega) = e^{-j\omega T} \tag{13.49}$$

The magnitude is given by

$$|G(j\omega)| = |e^{-j\omega T}| = |\cos \omega T - j \sin \omega T| = 1$$

Hence the log–magnitude of this transport lag is always equal to 0 dB. The phase angle of this transport lag becomes

$$\angle G(j\omega) = -\omega T \qquad \text{(radians)} \tag{13.50}$$

$$= -57.3\omega T \qquad \text{(degrees)} \tag{13.51}$$

Equations (13.50) and (13.51) show that the phase angle varies with frequency ω. Figure 13.22 shows the phase angle characteristic of the transport lag.

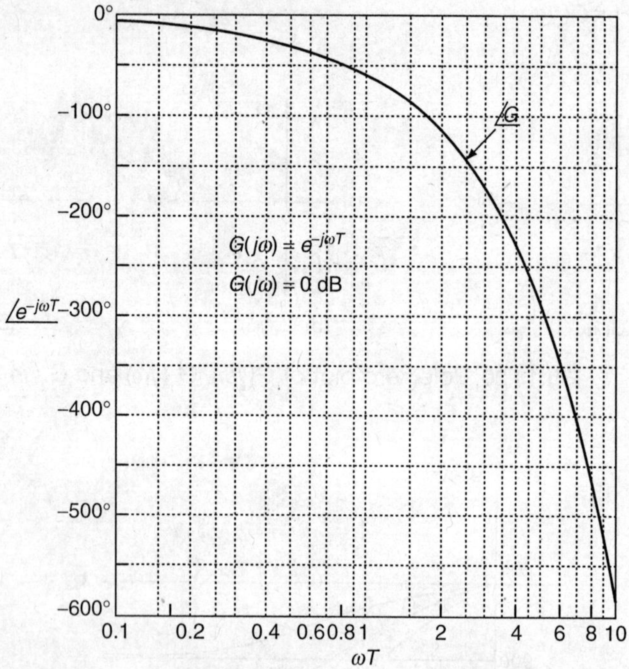

Fig. 13.22 Phase angle characteristic of transport lag

13.6 Relationship Between System Type and Log–Magnitude Curve

Figure 13.23 shows a unity feedback control system. The low-frequency behaviour of type 0, type 1 and type 2 systems are K_p, K_v and K_a, that is, static position, velocity and acceleration error coefficients. It is to be noted that only one static error constant is finite and significant for a given system. The slope of log–magnitude curve at the low frequency is determined by the type of the system.

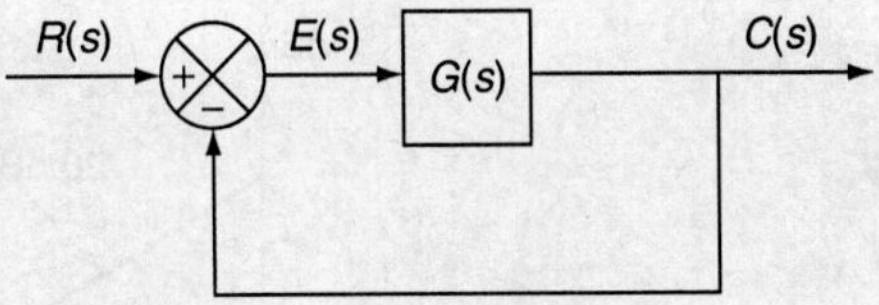

Fig. 13.23 Unity feedback system

13.6.1 Determination of Static Position Error Constants

A unity feedback system shown in Fig. 13.23 is considered here. The open loop transfer function is taken as

$$G(s) = \frac{K(1+T_a s)(1+T_b s)\cdots\cdots(1+T_q s)}{s^N(1+T_1 s)(1+T_2 s)\cdots\cdots(1+T_p s)} \tag{13.52}$$

i.e.,

$$G(j\omega) = \frac{K(1+j\omega T_a)(1+j\omega T_b)\cdots\cdots(1+j\omega T_q)}{(j\omega)^N(1+j\omega T_1)(1+j\omega T_2)\cdots\cdots(1+j\omega T_p)} \tag{13.53}$$

For '0' type system,

$$\underset{\omega\to 0}{Lt}\; G(j\omega) = K = K_p \tag{13.54}$$

Figure 13.24 shows the log–magnitude plot for such system.

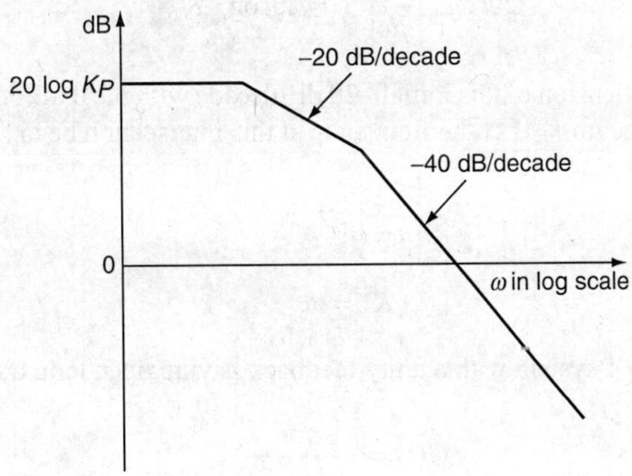

Fig. 13.24 Log–magnitude curve for type 0 system

13.6.2 Determination of Static Velocity Error Constants

A unity feedback system shown in Fig. 13.23 is considered here. Figure 13.25 shows the log–magnitude plot for type 1 system.

Since the magnitude of the intersection of initial −20 dB/decade with the line $\omega = 1$ has the magnitude $20\log_{10}K_v$, we have

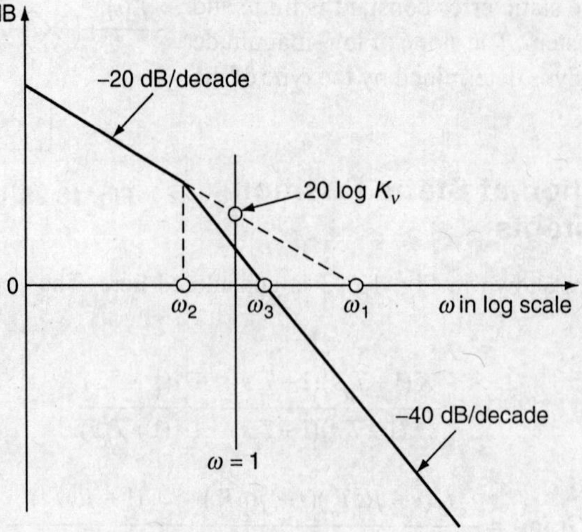

Fig. 13.25 Log–magnitude curve for type 1 system

$$G(j\omega) = \frac{K_v}{j\omega} \quad \text{for } \omega \le 1 \tag{13.55}$$

Therefore,

$$20\log_{10}\left|\frac{K_v}{j\omega}\right|_{\omega=1} = 20\log_{10} K_v \tag{13.56}$$

The intersection of the extension of this initial −20 dB/decade with the 0 dB/decade line has a frequency which is numerically equal to K_v. Let the frequency at this intersection be ω_1. Therefore, we have

$$\left|\frac{K_v}{j\omega_1}\right| = 1$$

i.e.,

$$K_v = \omega_1 \tag{13.57}$$

Let us consider a type 1 system with a unity feedback having open loop transfer function:

$$G(s) = \frac{K}{s(Js+F)} \tag{13.58}$$

Let the corner frequency be ω_2 and the frequency of intersection of the −40 dB/decade line with 0 dB line be ω_3. We have

$$\omega_2 = \frac{F}{J} \tag{13.59}$$

and

$$\omega_3^2 = \frac{K}{J} \tag{13.60}$$

Again,

$$\omega_1 = \frac{K}{F} \tag{13.61}$$

From Eqs. (13.59) to Eq. (13.61), we have

$$\omega_1\omega_2 = \omega_3^2$$

i.e.,

$$\frac{\omega_1}{\omega_3} = \frac{\omega_3}{\omega_2} \tag{13.62}$$

On Bode plot, we have

$$l\,og_{10}\,\omega_1 - l\,og_{10}\,\omega_3 = l\,og_{10}\,\omega_3 - l\,og_{10}\,\omega_2 \tag{13.63}$$

Therefore, ω_3 is the mid way between ω_1 and ω_2 points. The damping ratio ξ of the system is given by

$$\xi = \frac{F}{2\sqrt{KJ}} = \frac{\omega_2}{2\omega_3} \tag{13.64}$$

13.6.3 Determination of Static Acceleration Constants

A unity feedback system shown in Fig. 13.23 is considered here. Figure 13.26 shows the log–magnitude plot for type 2 system.

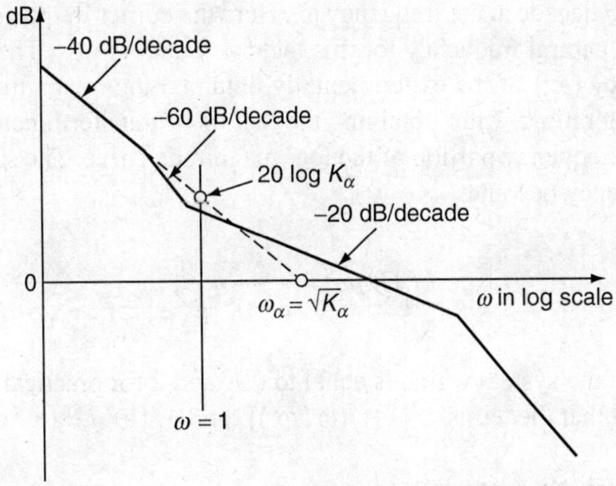

Fig. 13.26 Log–magnitude curve for type 2 system

Since the magnitude of the intersection of initial –40 dB/decade with the line $\omega = 1$ has the magnitude $20l\text{og}_{10}K_a$, we have

$$G(j\omega) = \frac{K_a}{(j\omega)^2} \quad \text{for } \omega \le 1 \tag{13.65}$$

Therefore,

$$20l\,og_{10}\left|\frac{K_v}{(j\omega)^2}\right|_{\omega=1} = 20l\,og_{10}\,K_a \tag{13.66}$$

The frequency of intersection of the –40 dB/decade line with 0 dB line be ω_a which is square root of K_a numerically.

13.7 Experimental Determination of Transfer Function

The derivation of mathematical modelling of the plant is very important for its analysis and design, which is quite difficult. Using experimental analysis it is possible to get it. The simple frequency response measurements can be done to determine the transfer function of the plant to be considered.

The Bode diagram can be done after measurement of amplitude ratio and phase shift at different frequencies in the frequency range and using asymptotic approximations, the transfer function can be determined. The asymptotic log–magnitude curves for several segments can be built up. The very close fit to the curve can be made after trial and error juggling of corner frequencies. The suitable sinusoidal signal generators which must be free from harmonics are required for performing the frequency response test where the signal may be in mechanical, electrical and pneumatic form. The large and small time constant systems, the frequency range may be 0.001–10 Hz or 0.1–1000 Hz.

A mechanical signal generator is used for low frequency range (< 0.01 Hz) whereas an electrical signal generator may be used for the frequency range of 0.01–1000 Hz.

The asymptotes corresponding to the experimentally obtained log–magnitude curve is drawn at first where the asymptotes should have the slope multiple of ±20 dB/decade. A factor $1/[1+j(\omega/\omega_1)]$ exist in the transfer function when the slope changes from –20 dB/decade to –40 dB/decade at the frequency ω_1. A factor $1/[1+j2\xi(\omega/\omega_2)+(j\omega/\omega_2)^2]$ exist in the transfer function when the slope changes from –20 dB/decade to –40 db/decade at the frequency ω_2. Here the corner frequency for this quadratic factor is ω_2 and the undamped natural frequency for this factor is equal to ω_2. The amount of resonant peak near the corner frequency (ω_2) of the experimentally obtained log–magnitude curve is measured for determining the damping ratio ξ. After obtaining the factors of transfer function $G(j\omega)$, the gain can be found out from the low frequency portion of the log–magnitude curve. The sinusoidal transfer function $G(j\omega)$ at very low frequency becomes

$$\underset{\omega\to 0}{Lt}\ G(j\omega)=\frac{K}{(j\omega)^n} \tag{13.67}$$

where n is the type of the system which is equal to 0,1, and 2 for practical systems.

This is due to the fact that the terms $1/[1+j(\omega/\omega_1)]$ and $1/[1+j2\xi(\omega/\omega_2)+(j\omega/\omega_2)^2]$ are unity as $\omega\to 0$.

The following cases are considered:

(i) n = 0, that is type 0 systems.

$$G(j\omega)=K \text{ for } \omega\le 0.$$

Therefore, $20\ log_{10}|G(j\omega)| = 20\ log_{10}K$ for $\omega \le 0$.

(ii) $n = 1$, that is, type 1 systems.

$$G(j\omega)=\frac{K}{j\omega} \text{ for } \omega\le 1.$$

Therefore, $20\ l\log_{10}|G(j\omega)| = 20\ l\log_{10} K - 20\ l\log_{10}\omega$ for $\omega \le 1$.

This shows that the slope of low frequency is –20dB/.decade. The cross over frequency at the 0 dB line by this asymptote is numerically equal to K.

(iii) $n = 2$, that is, type 2 systems.

$$G(j\omega) = \frac{K}{(j\omega)^2} \text{ for } \omega \leq 1.$$

Therefore, $20\,log_{10}|G(j\omega)| = 20log_{10}K - 40log_{10}\omega$ for $\omega \leq 1$.

This shows that the slope of low frequency is –40dB/decade. The cross over frequency at the 0 dB line by this asymptote is numerically equal to square root of K.

The phase angle curve obtained experimentally helps to check the transfer function obtained from the log–magnitude curve. The phase angle curve obtained experimentally must match reasonably with the theoretically obtained phase angle curve obtained from the derived transfer function for non-minimum phase transfer function. There should be exact agreement between the phase angle curve obtained theoretically and experimentally in the low as well as in the high frequency range. If the value of experimentally obtained phase angle is not equal to $-90^0\,(p-q)$ where p and q are the degrees of the numerator and denominator, the transfer function becomes a nonminimum transfer function.

The zero at the right half of s–plane is needed to add the transfer function when the computed phase lag at the high frequency end is 180^0 less than the experimentally obtained phase lag. If there is difference between the computed phase lag and experimentally obtained phase lag by a constant rate of change of phase, the transport lag or dead time is present. Let the transfer function be of the form $G(s)e^{-Ts}$ where G(s) is the ratio of polynomials of s.

Therefore, $$\underset{\omega\to\infty}{Lt}\frac{d}{d\omega}\angle\left[G(j\omega)e^{-j\omega T}\right] = \underset{\omega\to\infty}{Lt}\frac{d}{d\omega}\angle\left[G(j\omega)\right] + \underset{\omega\to\infty}{Lt}\frac{d}{d\omega}\angle\left[e^{-j\omega T}\right]$$

$$= \underset{\omega\to\infty}{Lt}\frac{d}{d\omega}\angle\left[G(j\omega)\right] - \omega T = 0 - T = -T$$

Since $\angle G(j\omega)$ is constant, therefore its derivative with respect to ω is taken as 0.

The following points may be noted:

(i) Since phase shift measurements may contain errors caused by instruments or by misinterpretations, therefore accurate amplitude measurements is possible compared to phase shift measurements.

(ii) The instruments used to measure the system output should have flat magnitude vs frequency curve and the phase angle must be proportional to the frequency.

(iii) Since there may be different types of nonlinearities in physical system, the amplitude of input sinusoidal signal should be considered carefully.

(iv) The normal operation of the instruments should not be stopped for the frequency–response tests for the systems operating for a long time.

13.8 Advantages of a Bode Plot

(i) It is possible to show both the high-frequency and the low-frequency characteristic of a transfer function in a single diagram.

(ii) Using valid assumptions, the Bode plots can be constructed very easily.

(iii) From a Bode plot, it is possible to calculate the GM and PM very easily; hence, the relative stability of the system can be studied.

(iv) It is possible to determine the various other frequency specifications such as the cut-off frequency and bandwidth.

(v) From a Bode plot, we get the transfer function of the system.

(vi) Using a Bode plot, it is possible to compensate the system to obtain the desired response.

(vii) From a Bode plot, it is also possible to design the value of K from the knowledge of GM or PM.

(viii) Experimentally, it is possible to draw a Bode plot without the knowledge of the transfer function.

13.9 MATLAB Commands for Bode Plot

BODE

Compute the Bode frequency response of LTI models

Syntax

bode(sys)

bode(sys, w)

bode(sys1, sys2,…, sysN)

bode(sys1, sys2,…, sysN, w)

bode(sys1, 'PlotStyle1',…, sysN, 'PlotStyleN')

[mag, Phase, w] = bode (sys)

or to plot the response on a wider frequency range, for example, from 0.1 to 100 0rad/s.

type

bode(g,{0.1, 1000})

where

g = tf([100],[19 10 0])

Transfer function:

$$\frac{100}{s\hat{}3+9s\hat{}2+10s}$$

bode(g)

bode(g,(0.1,1000))

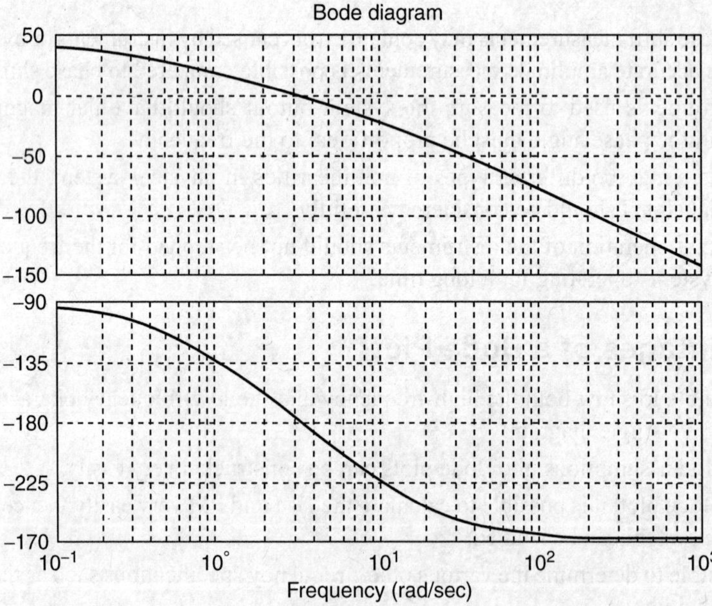

Fig. 13.27

MARGIN

Compute gain and phase margins and associated crossover frequencies

Syntax

[Gm, Pm, Wcg, Wcp] = margin(sys)

[Gm, Pm, Wcg, Wcp] = margin(mag, phase, w)

margin(sys)

g = tf([100],[1 9 10 0])

Transfer function:

$$\frac{100}{s\hat{}3+9s\hat{} +10s}$$

margin(g)

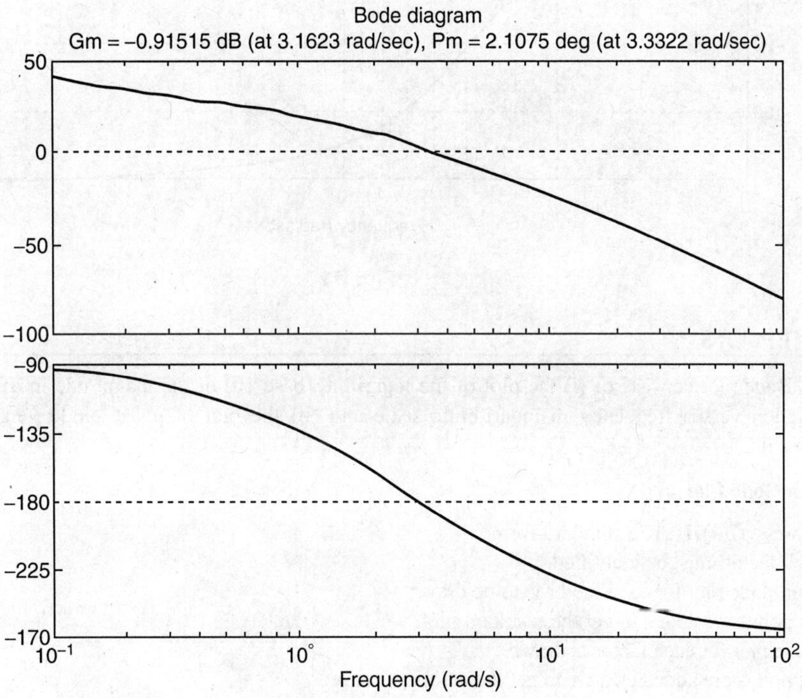

Fig. 13.28

Program 1 Draw the bode plot of $\dfrac{25}{s^2+4s+25}$

Solution

```
>>num = [0 0 25];
>>den = [1 4 25];
>>bode(num, den);
```

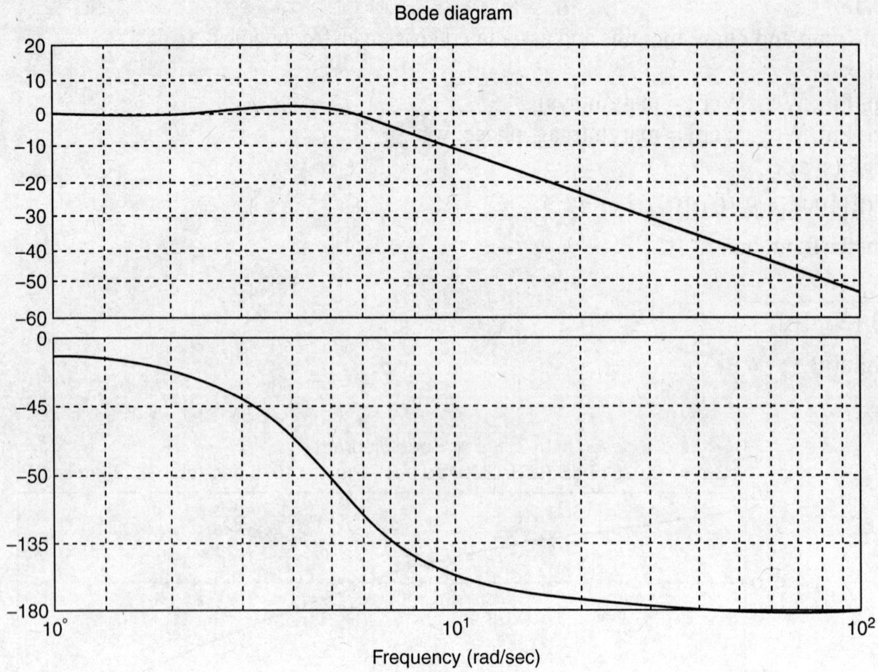

Fig. P1

SIGNIFICANT POINTS

Bode Plot: Bode plot consists of (i) the plot of the logarithm (base 10) of the magnitude in dB of a sinusoidal transfer function versus frequency in logarithmic scale and (ii) the plot of phase angle versus frequency in logarithmic scale.

Steps to Plot Bode Plot

Step 1: Express $G(s)H(s)$ in standard form.
Step 2: Each factor must be identified.
Step 3: Magnitude plot for each factor is to be drawn.
Step 4: All points are added to get the resultant plot.
Step 5: Phase plot for each factor is drawn.
Step 6: All points are added clea rly to get the resultant curve.

Asymptotic Bode plots are easy to draw compared to a precise Bode plot and there will not be large errors from these approximated Bode plots.

Gain Margin (GM): Gain margin is defined as the ratio of the critical gain to the actual gain:

$$\text{GM} = \frac{\text{Critical value of } K}{\text{Actual value of } K}$$

Phase Margin (PM): Phase margin (PM) is given by

$$\text{PM} = 180° + \angle |[G(j\omega)H(j\omega)]|_{\omega=\omega_{gc}}$$

From a Bode plot, we can obtain the gain crossover frequency, phase crossover frequency, gain margin as well as phase margin.

Minimum Phase Functions: If all poles and zeros of any transfer function lie on the left half of the s-plane, this transfer function is known as the minimum phase transfer function. The functions that we have already discussed earlier for a Bode are minimum phase transfer functions.

All-Pass Phase Functions: If all the zeros of a transfer function lie on the right half of the s-plane, all poles lie on the left half of the s-plane and the location of every pole zero pair is symmetric about the $j\omega$ axis, the transfer function is termed as all-pass function.

ADDITIONAL SOLVED EXAMPLES

Example 13.9 A unity feedback control system has $G(s) = 40/[s^2(s+2)(s+5)]$ and $H(s) = 1$. Draw the Bode plot.

Solution

Step 1: Here $G(s) = 40/[s^2(s+2)(s+5)]$ and $H(s) = 1$.

$$\therefore \qquad G(s)H(s) = \frac{40}{s^2(s+1)(s+5)} = \frac{4}{s^2\left(1+\frac{s}{2}\right)\left(1+\frac{s}{5}\right)}$$

Step 2: For $K = 4$, there is two pole at origin and two simple poles.

Step 3: (i) $20\,log_{10}K = 20\,log_{10}4 = 12$ dB and it Bode plot is a straight line with slope 0 dB/decade.

(ii) There are two poles at the origin. The Bode plot of it is a straight line of slope -40dB/decade, which passes through the intersection point of $\omega = 1$ and 0 dB.

(iii) Now, let us shift the intersection point $\omega = 1$ at 0 dB on the $20\,log_{10}K$ line and draw a line parallel to the -40 dB/decade line. This is the result ant line of K and $1/s^2$ and it continues till first corner frequency occurs at $\omega_{c1} = 2$.

(iv) Now, at $\omega_{c1} = 2$, there is a simple pole and it contributes at the rate of -20 dB/decade. Therefore, the resultant slope after $\omega_{c1} = 2$ becomes -60 dB/decade. This is the resultant curve of K, $1/s^2$ and $1/(1 + s/2)$; it continues till second corner frequency occurs at $\omega_{c2} = 5$.

(v) Once again, at $\omega_{c2} = 5$, there is a simple pole contributing at the rate of -20 dB/decade and hence the resultant slope after $\omega_{c2} = 5$ becomes -80 dB/decade. Since there are no factors present, the final slope becomes -80 dB/decade.

Step 4: Phase angle plot:

$$\text{Now,} \qquad G(j\omega)H(j\omega) = \frac{4}{(j\omega)^2\left(1+\frac{j\omega}{2}\right)\left(1+\frac{j\omega}{5}\right)}$$

$$\therefore \qquad \angle G(j\omega)H(j\omega) = \phi = -180^\circ - \tan^{-1}\left(\frac{\omega}{2}\right) - \tan^{-1}\left(\frac{\omega}{5}\right)$$

Phase angle table:

Table E13.4

ω	0.2	2	8	10	20	40	80	∞
ϕ	-188°	-246.8°	-313.9°	-322.1°	-340.2°	-350°	-355°	-360°

Figure E13.9 shows the Bode plot.

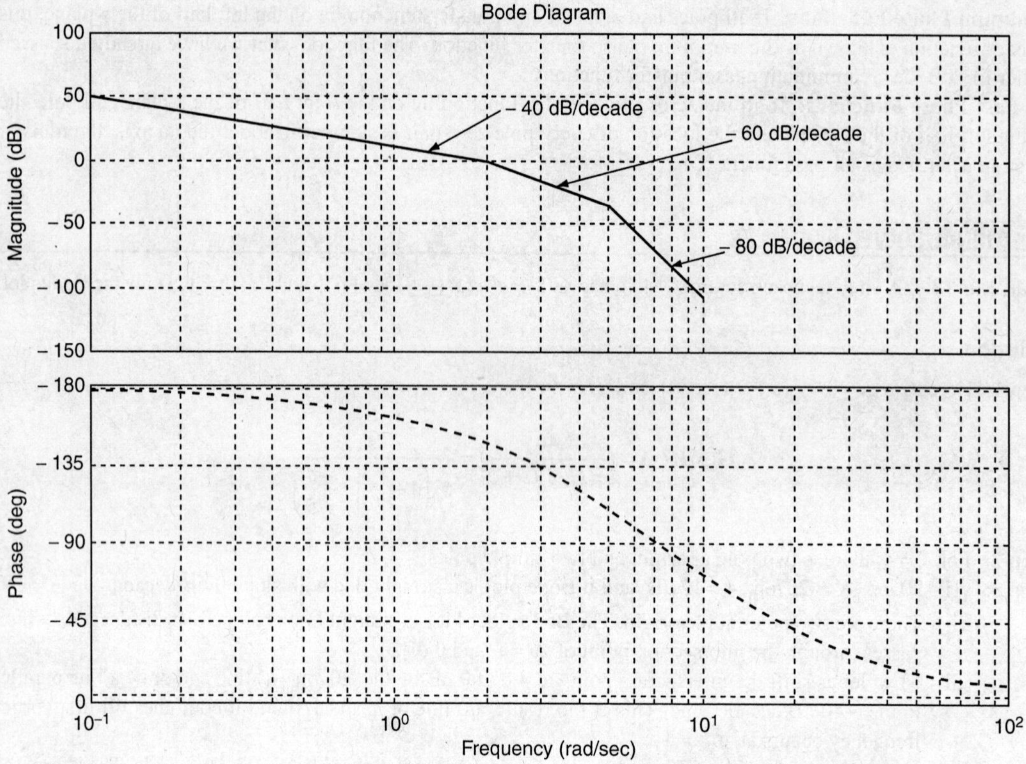

Fig. E13.9

Example 13.10 A unity feedback control system has $G(s) = \dfrac{400(s+2)}{s^2(s+5)(s+10)}$. Draw the Bode plot.

Solution

Step 1: Here, $$G(s) = \frac{40(s+2)}{s^2(s+5)(s+10)} \text{ and } H(s) = 1$$

$\therefore$ $$G(s)H(s) = \frac{400(s+2)}{s^2(s+5)(s+10)} = \frac{16\left(1+\dfrac{s}{2}\right)}{s^2\left(1+\dfrac{s}{5}\right)\left(1+\dfrac{s}{10}\right)}$$

Step 2: For $K = 16$, there are two poles at the origin, one simple zero and two simple poles.

For simple zero $\left(1+\dfrac{s}{2}\right), \omega_{c1} = 2$ rad/sec

For simple pole $\left(1+\dfrac{s}{5}\right), \omega_{c2} = 5$ rad/sec

For simple pole $\left(1+\dfrac{s}{10}\right), \omega_{c3} = 10$ rad/sec

Step 3: (i) $20\, log_{10} K = 20\, log_{10} 16 = 24$ dB.

(ii) There are two poles at the origin. Its Bode plot is a straight line of slope −40 dB/decade, which passes through the intersection point of $\omega = 1$ and 0 dB.

(iii) Now, let us shift the intersection point $\omega = 1$ and 0 dB on the 20 log_{10} K line and draw a line parallel to the −40 dB/decade line. This is the resultant line of K and $1/s^2$ and it continues till the first corner frequency occurs at $\omega_{c1} = 2$.

(iv) Now, at $\omega_{c1} = 2$, there is a simple zero and it contributes at the rate of 20 dB/decade. Therefore, the resultant slope after $\omega_{c1} = 2$ becomes −20 dB/decade. This is the resultant curve of K, $1/s^2$ and $\left(1+\frac{s}{2}\right)$; it continues till second corner frequency occurs at $\omega_{c2} = 5$.

(v) Once again, at $\omega_{c2} = 5$, there is a simple pole contributing −20 dB/decade; hence, the resultant slope after $\omega_{c2} = 5$ becomes −40 dB/decade. This is the resultant curve of K, $1/s^2$ and $\left(1+\frac{s}{2}\right)$ and $\frac{1}{(1+s/5)}$; it continues till the third corner frequency occurs at $\omega_{c3} = 10$.

(vi) Once again, at $\omega_{c3} = 10$, there is a simple pole contributing at the rate of −20 dB/decade; hence, the resultant slope after $\omega_{c3} = 5$ becomes −60 dB/decade. Since there are no factors present, the final slope becomes −60 dB/decade.

Step 4: Phase angle plot:

$$G(j\omega)H(j\omega) = \frac{4\left(1+\frac{j\omega}{2}\right)}{(j\omega)^2\left(1+\frac{j\omega}{5}\right)\left(1+\frac{j\omega}{10}\right)}$$

$$\therefore \quad G(j\omega)H(j\omega) = \phi = -180^\circ + \tan^{-1}\left(\frac{\omega}{2}\right) - \tan^{-1}\left(\frac{\omega}{5}\right) - \tan^{-1}\left(\frac{\omega}{10}\right)$$

Phase angle table:

Table E13.5

ω	W	0.2	2	8	10	20	40	80	∞
ϕ	F	−184.5°	−168.1°	−200.7°	−209.7°	−235.1°	−251.7°	−260.7°	−270°

Figure E13.10 shows the Bode plot

Example 13.11 A unity feedback control system has $G(s) = 288(s+4)/s(s+1)(s^2+4.8s+144)]$. Draw the Bode plot.

Solution

Step 1: Here,

$$G(s) = \frac{288(s+4)}{s(s+2)(s^2+4.8s+144)} \text{ and } H(s) = 1$$

$$G(s)H(s) = \frac{288(s+4)}{s(s+2)(s^2+4.8s+144)}$$

$$= \frac{4\left(1+\frac{s}{4}\right)}{s\left(1+\frac{s}{2}\right)\left(1+0.0334s+\frac{s^2}{144}\right)}$$

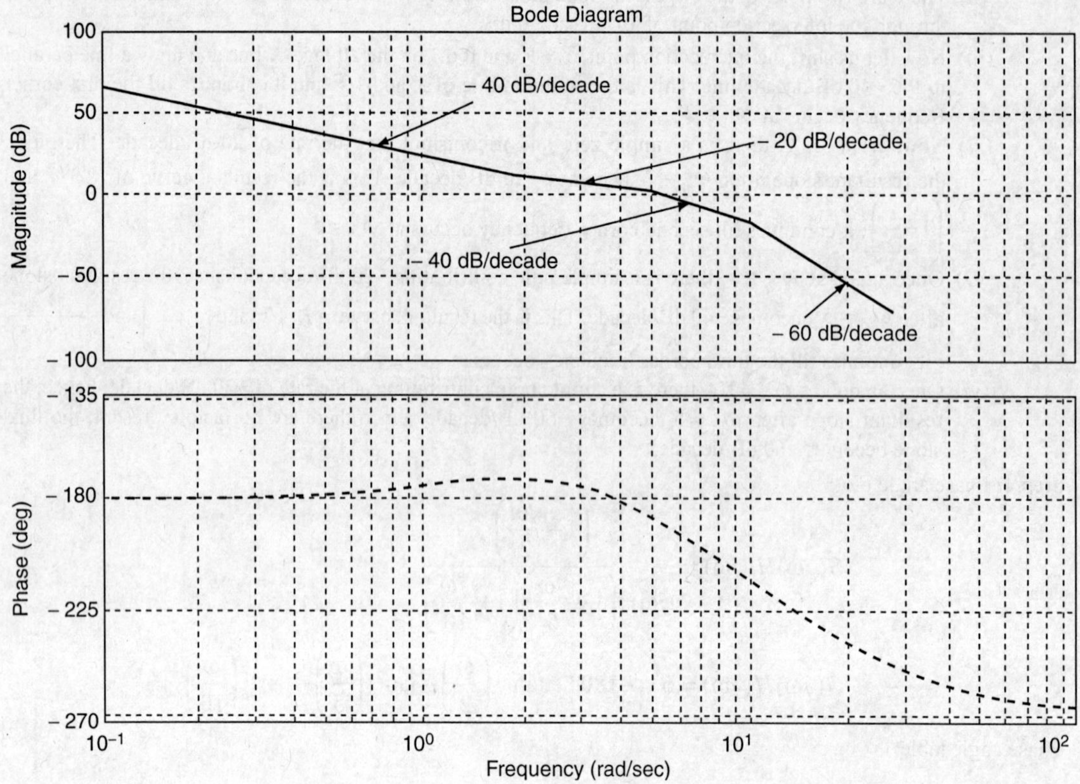

Fig. E13.10

$\omega_n^2 = 144$ or $\omega_n = 12$

and $2\xi\omega_n = 4.8$ or $\xi = 0.2$

Therefore, $\omega_n = 12\,\text{rad/sec}$. The required correction is + 7.95 dB upwards.

Step 2: For $K = 1\ 6$, there is one pole at the origin, one simple pole and one quadratic pole.

$$\text{For simple zero } \left(1+\frac{s}{4}\right), \omega_{c1} = 2 \text{ rad/sec}$$

$$\text{For simple zero } \left(1+\frac{s}{2}\right), \omega_{c2} = 5 \text{ rad/sec}$$

$$\text{Quadratic pole } \frac{1}{\left(1+0.0334s+\frac{s^2}{144}\right)}, \omega_{c3} = 12 \text{ rad/sec}$$

Step 3: (i) $20\,log_{10} K = 20\,log_{10} 4 = 12\,\text{dB}$.

(ii) There is one pole at origin. Its Bode plot is a straight line of slope −20 dB/decade, which passes through the intersection point of $\omega = 1$ and 0 dB.

(iii) Now, let us shift the intersection point $\omega = 1$ and 0 dB on the $20\,log_{10} K$ line and draw a line parallel to the −20 dB/decade line which has been drawn already. This is the resultant line of K and $1/s^2$ and it continues till the first corner frequency occurs at $\omega_{c1} = 2$.

(iv) Now, at $\omega_{c1} = 2$, there is a simple pole and it contributes at the rate of −20 dB/decade. Therefore, the resultant slope after $\omega_{c1} = 2$ becomes −40 dB/decade. This is the resultant curve of K, $1/s$ and $\frac{1}{(1+s/2)}$; it continues till be second corner frequency occurs at $\omega_{c2} = 4$.

(v) Once again, $\omega_{c2} = 4$, there is a simple zero contributing at the rate of 20 dB/decade and hence the resultant slope after $\omega_{c2} = 4$ becomes −20 dB/decade. This is the resultant curve of K, $1/s$, $\frac{1}{(1+s/2)}$ and $1+(s/4)$. It continues till the first corner frequency occurs at $\omega_{c3} = 12$.

(vi) Once again, at $\omega_{c3} = 12$, a quadratic pole with $\xi = 0.2$ occurs so it contributes −40 dB/decade individually and hence the resultant slope after $\omega_{c2} = 12$ becomes −60 dB/decade. Since there are no factors present, the final slope becomes −60 dB/decade.

Step 4: Phase angle plot:

$$G(j\omega)H(j\omega) = \frac{4\left(1+\frac{j\omega}{4}\right)}{(j\omega)\left(1+\frac{j\omega}{2}\right)\left(1+j0.0334\omega-\frac{\omega^2}{144}\right)}$$

$$\therefore \quad \angle G(j\omega)H(j\omega) = \phi = -90° + \tan^{-1}\left(\frac{\omega}{2}\right) - \tan^{-1}\left(\frac{\omega}{5}\right) - \tan^{-1}\left[\frac{0.0334\omega}{1-\frac{\omega^2}{144}}\right]$$

Phase angle table:

Table E13.6

ω	0.2	2	8	10	20	40	∞
ϕ	−94.7°	−122.9°	−129.5°	−135.1°	−153.6°	−257.8°	−270°

Figure E13.11 shows the Bode plot

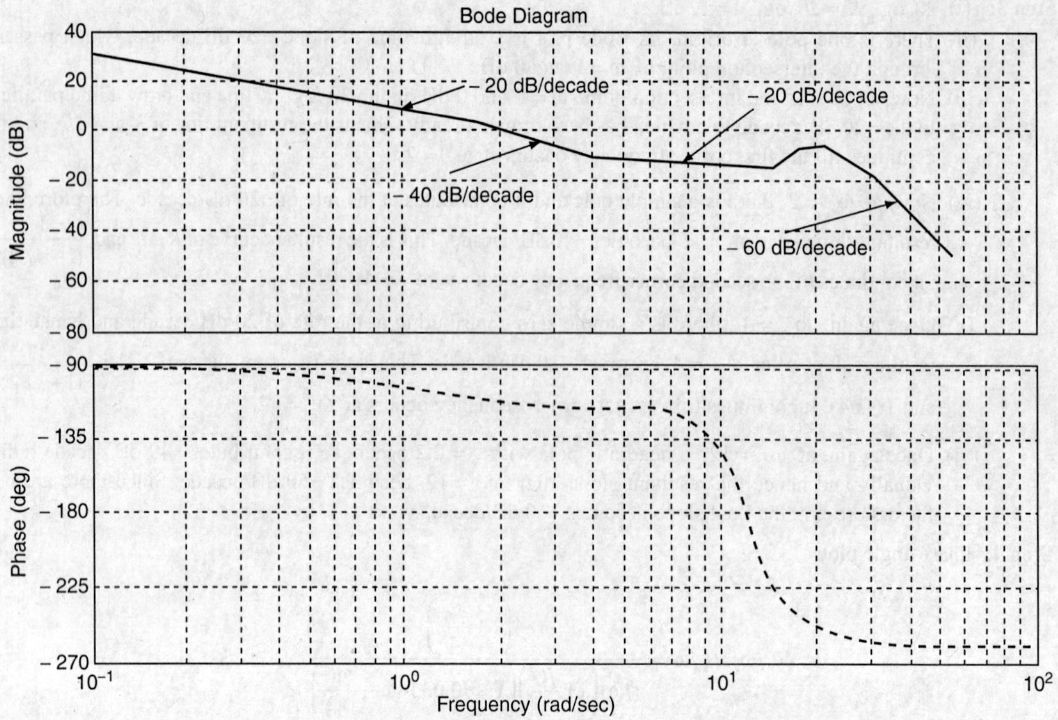

Fig. E13.11

Example 13.12 A unity feedback control system has $G(s) = K[s(s+4)(s+10)]$. Draw the Bode plot. Find K when the system is marginally stable.

Solution

Step 1: Here,

$$G(s) = \frac{K}{s(s+4)(s+10)} \text{ and } H(s) = 1$$

$$\therefore \qquad G(s)H(s) = \frac{K}{s(s+4)(s+10)} = \frac{\frac{K}{40}}{s\left(1+\frac{s}{4}\right)\left(1+\frac{s}{10}\right)} = \frac{K'}{s+\left(1+\frac{s}{4}\right)\left(1+\frac{s}{10}\right)}$$

Step 2: For unknown K', there is one pole at the origin and two simple poles.

Step 3: (i) 20 log_{10} K' is unknown.

(ii) There is one pole at the origin. Its Bode plot is a straight line of slope −20 dB/decade, which passes through the intersection point of $\omega = 1$ and 0 dB.

(iii) Now, the shift intersection point $\omega = 1$ and 0 dB on the 20 log_{10} K line is not required because K' is unknown. This line 1/s continues till the first corner frequency occurs at $\omega_{c1} = 4$.

(iv) Now, at $\omega_{c1} = 4$, there is a simple pole and it contributes at the rate of −20 dB/decade. Therefore, the resultant slope after $\omega_{c1} = 10$ becomes −40 dB/decade. This is the resultant curve of 1/s and 1/[1+(s/4). It continues till the second corner frequency occurs at $\omega_{c2} = 10$.

(v) Once again, at $\omega_{c2} = 10$, there is a simple pole contributing at the rate of −20 dB/decade; hence, the resultant slope after $\omega_{c2} = 10$ becomes −60 dB/decade. Since there are no factors present, the final slope becomes −60 dB/decade.

Step 4: Phase angle plot:

$$\angle G(j\omega)H(j\omega) = \frac{K'}{(j\omega)\left(1+\dfrac{j\omega}{4}\right)\left(1+\dfrac{j\omega}{10}\right)}$$

$$\therefore \qquad \angle G(j\omega)H(j\omega) = \phi = -90° - \tan^{-1}\left(\frac{\omega}{4}\right) - \tan^{-1}\left(\frac{\omega}{10}\right)$$

Phase angle table:

Table E13.7

ω	0.2	1	2	4	8	10	20	40	∞
ϕ	−92°	−109.7°	−127.9°	−156.8°	−192.1°	−203.2°	−232.12°	−250.2°	−270°

Step 5: The Bode plot is sketched, In this problem, the system is marginally stable. Therefore, $\omega_{gc} = \omega_{pc}$. But ω_{pc} is independent of K'. So we will shift the magnitude plot upwards and downwards till $\omega_{gc} = \omega_{pc}$. To obtain K' we will equate this shift to 20 $\log_{10} K'$. Therefore, K' will be obtained. Hence we will get K also.

$$\therefore \qquad 20\log_{10} K' = 20 \text{ dB i.e., } K' = 10$$

$$\therefore \qquad K = 400.$$

Figure E13.12 shows the Bode plot

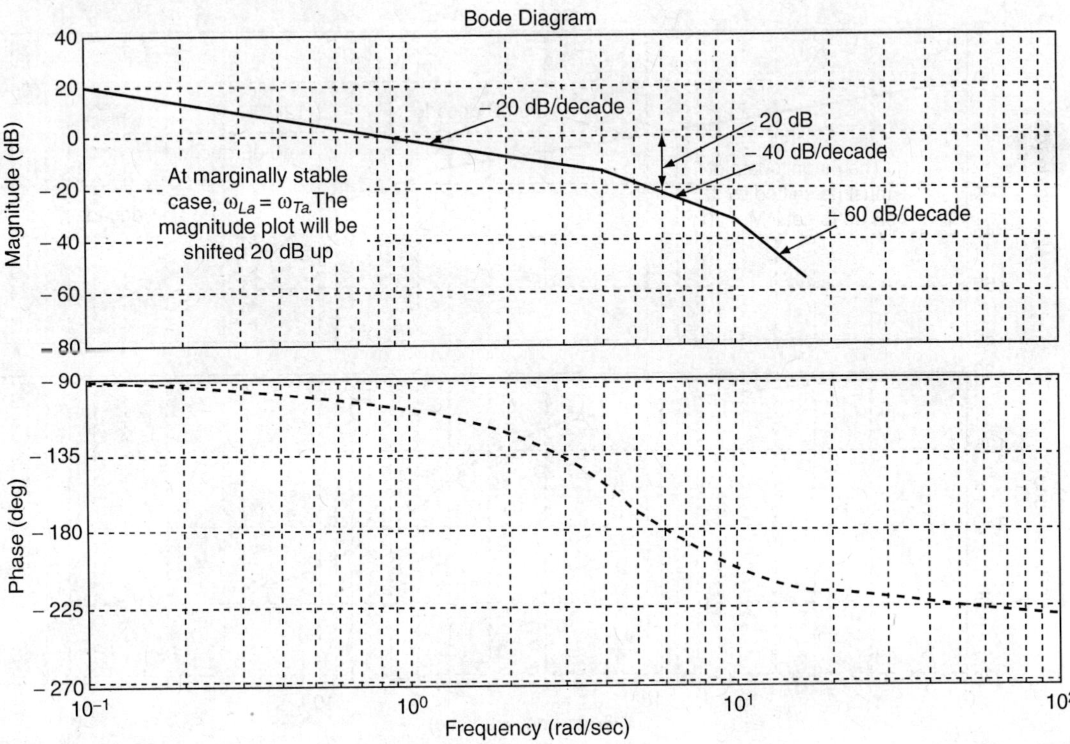

Fig. E13.12

Example 13.13 A unity feedback control system has $G(s) = \dfrac{K}{s(s+4)(s+10)}$. Draw the Bode plot. Find K when PM = 30°.

Solution From Example 13.12 we get the Bode plot. Here PM = 30°. Therefore, the point of intersection of ω_{gc} with the resultant phase plot must be 30° above the −180° line. From Fig. E13.13,

$$\text{PM} = 180° + \angle G(j\omega)H(j\omega)\Big|_{\omega=\omega_{gc}}$$

$$30° = 180° + \angle G(j\omega)H(j\omega)\Big|_{\omega=\omega_{gc}}$$

$$\angle G(j\omega)H(j\omega)\Big|_{\omega=\omega_{gc}} = -150°$$

The magnitude plot will be shifted such that, $\angle G(j\omega)H(j\omega)\Big|_{\omega=\omega_{gc}} = -150°$ will be satisfied.

$$20\log_{10} K' = 12$$

$$\therefore \quad K' = 398$$

$$K = 159.2$$

Equating this required dB to 20 log_{10} K', the value of K' will be obtained. Hence the value of K will also be obtained.

Figure E13.13 shows the Bode plot

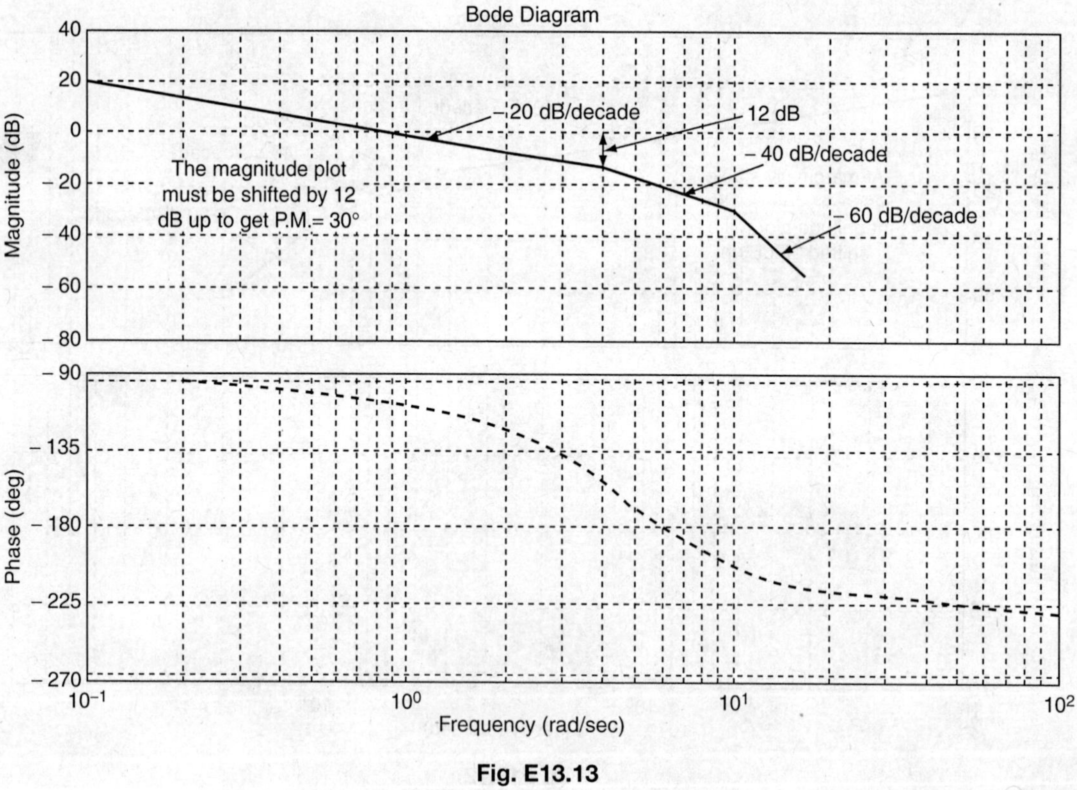

Fig. E13.13

Example 13.14 A unity feedback control system has $G(s) = k/[s(s+4)(s+10)]$. Draw the Bode plot. Find K when GM = 10 dB.

Solution In this problem, GM = 10 dB. From the phase angle plot, the phase crossover frequency (ω_{pc}) is obtained. The magnitude plot is shifted so that at ω_{pc} the GM becomes 10 dB.

$$20 \log_{10} K' = 7$$

$$\therefore \qquad K' = 2.238$$

$$K = 89.52$$

This shifted dB is equated to $20 \log_{10} K'$ and the value of K' will be obtained. Hence, the value of K will also be obtained.

Figure E13.14 shows the Bode plot

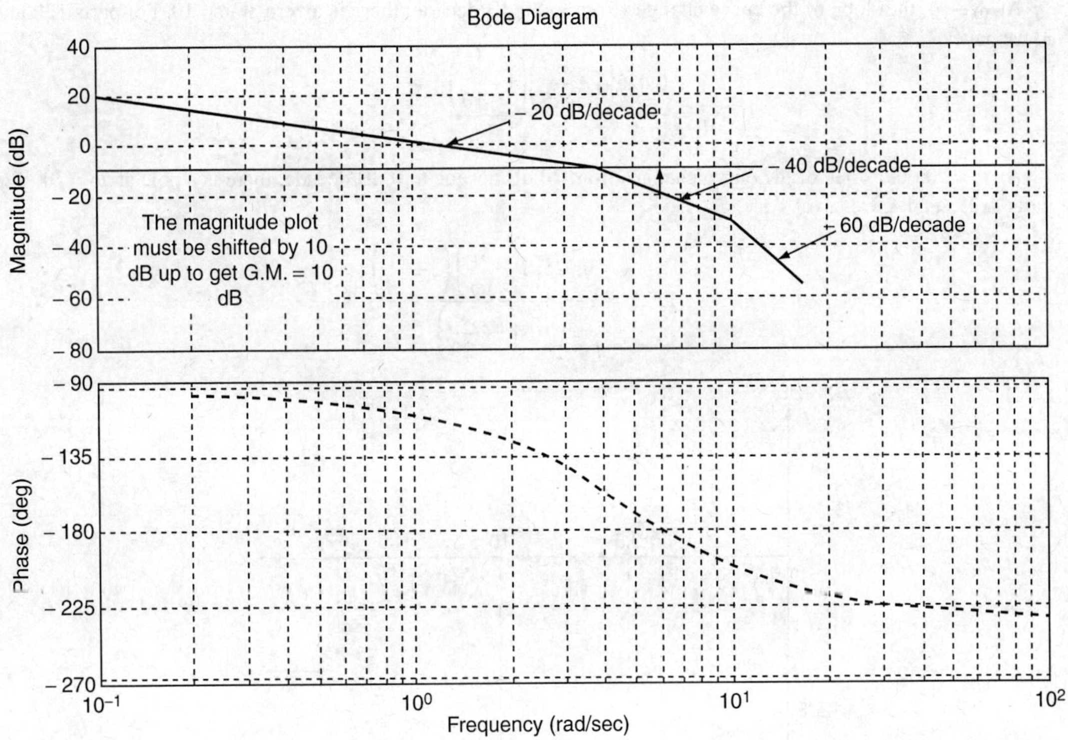

Fig. E13.14

Example 13.15 Find the transfer function of a system having Bode plot shown in Fig. E13.15

Solution For Part 1, at $\omega = 0.1$. dB = –12 and the slope of the curve is –20 dB/decade.

Therefore, $$TF_1 = \frac{K}{s}.$$

$\therefore$ $$\text{dB} = 20\,log_{10}\,K - 20\,log_{10}\,\omega$$

i.e., $$-12 = 20\,log_{10}\,K - 20\,log_{10}(0.1) = 20\,log_{10}\,K + 20$$

i.e., $$20\,log_{10}\,K = -32$$

i.e., $$K = 0.025$$

$\therefore$ $$TF_1 = \frac{0.025}{s}$$

At $\omega = 3.5$, the slope of the curve changes from −20 dB/decade to 0 dB/decade; there is a zero at $\omega = 3.5$. For parts 1 and 2, we have

$$TF_{12} = \frac{0.025\left(1+\frac{s}{3.5}\right)}{s}$$

At $\omega = 10$, the slope of the curve changes from to 20 dB/decade; there is a zero at $\omega = 10$. For parts 1, 2 and 3, we have

$$TF_{123} = \frac{0.025\left(1+\frac{s}{3.5}\right)\left(1+\frac{s}{10}\right)}{s}$$

At $\omega = 20$, the slope of the curve changes from 20 dB/decade to 0 dB/decade; there is a pole at $\omega = 20$. For parts 1, 2, 3 and 4, we have

$$TF_{1234} = \frac{0.025\left(1+\frac{s}{3.5}\right)\left(1+\frac{s}{10}\right)}{s\left(1+\frac{s}{20}\right)}$$

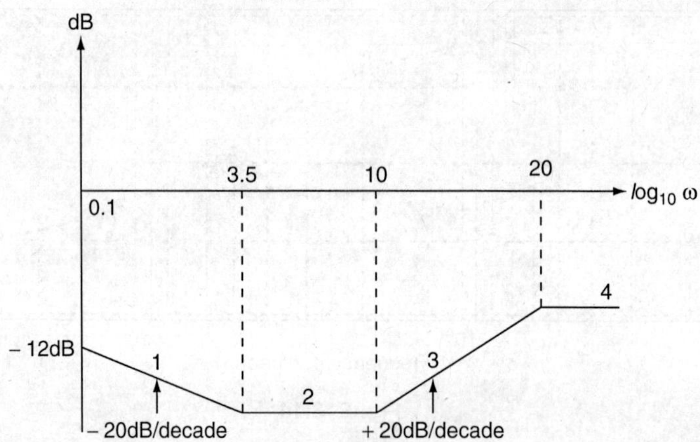

Fig. E13.15 shows the Bode plot

Example 13.16 Find the transfer function of a system having Bode plot shown in Fig. E13.16

Solution For part 1, at $\omega = 2$, dB =34 and the slope of the curve is −20 dB/decade.

Therefore, $$TF_1 = \frac{K}{s}$$

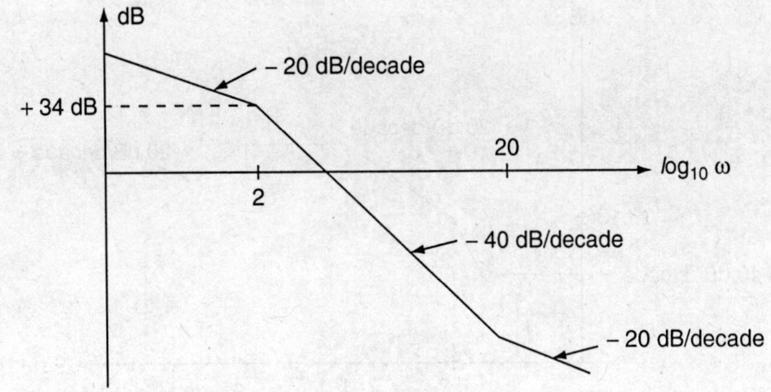

Fig. E13.16

$$\therefore \quad \text{dB} = 20\,log_{10} K - 20\,log_{10}\,\omega$$

i.e., $$34 = 20\,log_{10} K - 20\,log_{10} 2 = 20\,log_{10} K - 6$$

i.e., $$20\,log_{10} K = 40$$

i.e., $$K = 100$$

$$\therefore \quad TF_1 = \frac{100}{s}$$

At $\omega = 2$, the slope of the curve changes from -20 dB/decade to -40 dB/decade; there is a pole at $\omega = 2$. For parts 1 and 2, we have

$$\text{TF}_{12} = \frac{100}{s\left(1+\dfrac{s}{2}\right)}$$

At $\omega = 20$, the slope of the curve changes from -40 dB/decade to -20 dB/decade; Hence, there is a zero at $\omega = 20$. For parts 1, 2 and 3, we have

$$\text{TF}_{123} = \frac{100\left(1+\dfrac{s}{20}\right)}{s\left(1+\dfrac{s}{2}\right)}$$

Example 13.17 Find the transfer function of a system having Bode plot shown in Fig. E13.17

Solution

For part 1, $$\text{TF}_1 = Ks^2$$

$$\therefore \quad \text{dB} = 20\,log_{10} K + 40\,log_{10}\,\omega$$

At $$\omega = 0.8, \text{dB} = 36.13$$

i.e., $$36.13 = 20\,log_{10} K + 40\,log_{10}(0.8) = 40\,log_{10} K - 3.87$$

i.e., $$20\,log_{10} K = 40$$

i.e., $$K = 100$$

$$\therefore \quad \text{TF}_1 = 100s^2$$

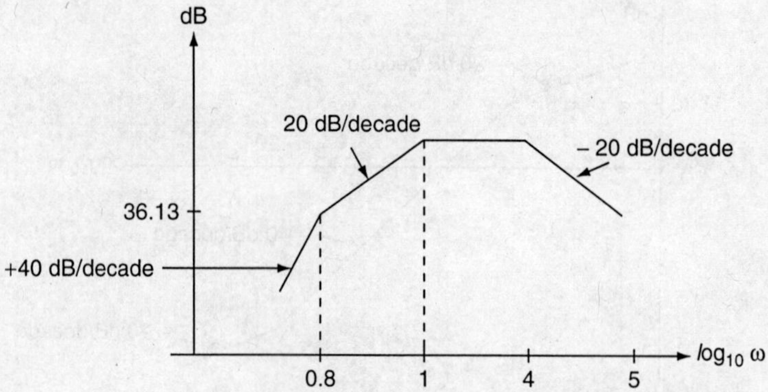

Fig. 13.17

At $\omega = 0.8$, the slope of the curve changes from 40 dB/decade to 20 dB/decade, there is a pole at $\omega = 0.8$. For parts 1 and 2, we have

$$TF_{12} = \frac{100s^2}{\left(1+\frac{s}{0.8}\right)}$$

At $\omega = 1$, the slope of the curve changes from 20 dB/decade to 0 dB/decade, there is a pole at $\omega = 1$. For parts 1, 2 and 3, we have

$$TF_{123} = \frac{100s^2}{\left(1+\frac{s}{0.8}\right)\left(1+\frac{s}{1}\right)}$$

At $\omega = 4$, the slope of the curve changes from 0 dB/decade to −20 dB/decade, there is a pole at $\omega = 4$. For parts 1, 2, 3 and 4, we have

$$TF_{1234} = \frac{100s^2}{\left(1+\frac{s}{0.8}\right)\left(1+\frac{s}{1}\right)\left(1+\frac{s}{4}\right)}$$

Program 2 Sketch the bode plot for the open loop transfer function $G(s)H(s) = \frac{40}{s(s+1)(s+4)}$. Also determine the gain margin, phase margin, gain cross over frequency and phase cross over frequency.

Soultion

```
>> num = [0 40]
num =
   0   40
>> q1 = [1 0];
>> q2 = [1 1];
>> q3 = [1 4];
>> den = conv(q1,q2);
>> den = conv(den,q3);
den =
```

```
1 9 24 16 0
>> sys=tf(num,den)
Transfer function:
          40
---------------------------------
s^4 + 9 s^3 + 24 s^2 + 16 s
>> bode(num,den)
>> margin(sys)
>>
```

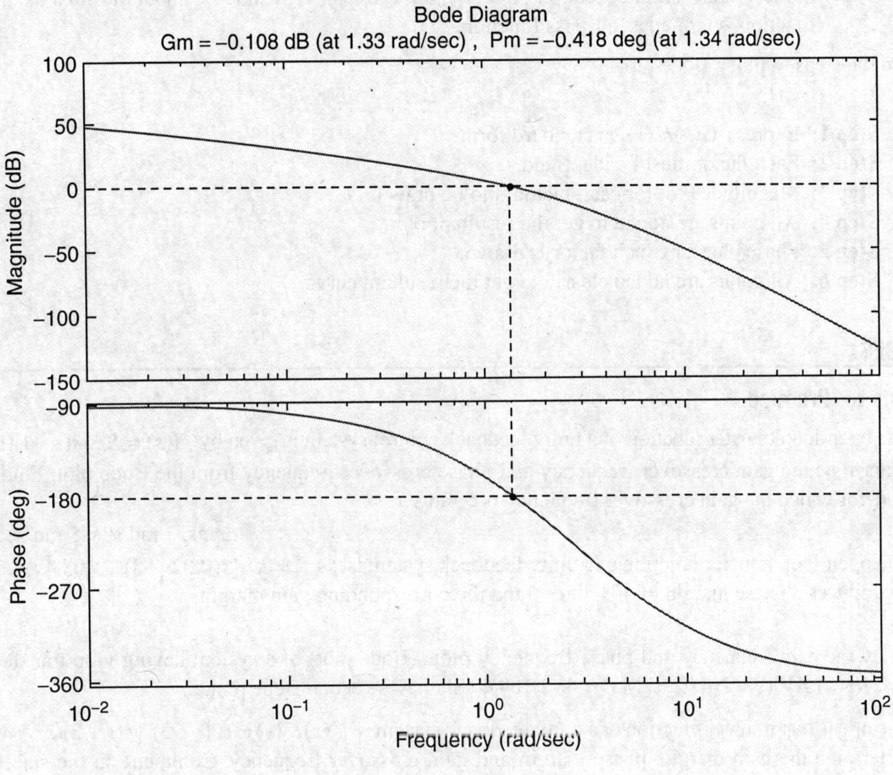

Fig. P2

SHORT QUESTIONS WITH ANSWERS

1. Define the following.
   (i) Bode plot, (ii) gain margin, (iii) phase margin, (iv) minimum phase function and (v) all-pass function.

   Ans. (i) A Bode plot consists of (i) a plot of the logarithm (base 10) of the magnitude in dB of a sinusoidal transfer function versus frequency in logarithmic scale and (ii) a plot of phase angle versus frequency in logarithmic scale.

(ii) Gain margin (GM) is defined as the ratio of the critical gain to the actual gain.

$$\therefore \qquad \text{GM} = \frac{\text{Critical value of } K}{\text{Actual value of } K}$$

(iii) Phase margin (PM) is given by

$$\text{PM} = 180° + \angle[G(j\omega)H(j\omega)]_{\omega=\omega_{gc}}$$

(iv) If all the poles and zeros of any transfer function lie on the left half of the s-plane, this transfer function is known as minimum phase transfer function.

(v) If all the zeros of a transfer function lie on the right half of the s-plane, all poles lie on the left half of the s-plane and the location of every pole zero pair is symmetric about the $j\omega$ axis, the transfer function is termed as all-pass function.

2. State the steps to plot a Bode plot.

Ans.

Step 1: Express $G(s)H(s)$ in standard form.
Step 2: Each factor must be identified.
Step 3: Magnitude plot for each factor is to be drawn.
Step 4: All points are added to get the resultant plot.
Step 5: Phase plot for e ach factor is drawn.
Step 6: All points are added clearly to get the resultant curve.

EXERCISES

Questions for Practice

1. The open-loop transfer function of a unity feedback control system is given by $G(s) = K/s[s(s+1)(1+0.1s)]$. Determine the gain crossover frequency and phase crossover frequency from the Bode plot. Find the value of K for which the gain crossover frequency is 5 rad/sec.

Ans. 1 rad/sec, 5 rad/sec. 0.03981

2. The open-loop transfer function of a unity feedback system is $G(s) = K/s[s(0.02s+1)(1+0.04s)]$. If the gain K produces a phase margin of 45°, find K and the corresponding gain margin.

Ans. $K = 12.125$, 12 dB

3. Draw the gain frequency and phase frequency plots (Bode plot) of a system having loop transfer function $G(s)H(s) = 10(s+3)/[s(s+2)(s^2+4s+100)]$. Sketch the actual Bode plot.

4. The open-loop transfer function of a unity feedback system is $G(s)H(s) = 10(s+3)/[s(s+2)(s^2+4s+100)]$. Determine the gain margin, phase margin and gain crossover frequency. Comment on the stability of the closed-loop system.

Ans. Gain crossover frequency = 4.5 rad/sec,
Phase crossover frequency = 9.4 rad/sec,
Gain margin = 14 dB, phase margin = 45°

5. Find the transfer function of a system having the Bode plot (magnitude) shown in Figure. 1.

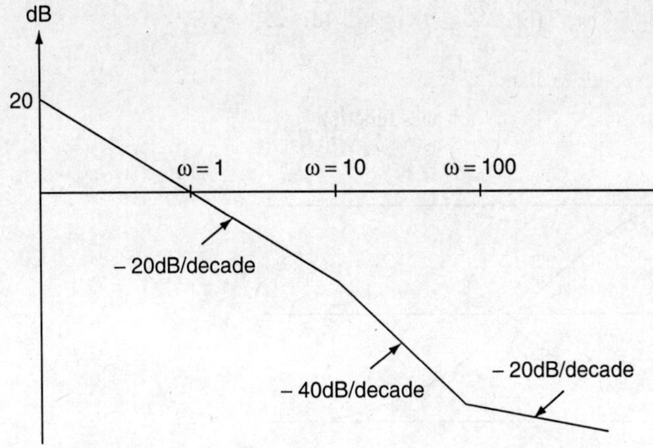

Ans. $\dfrac{s+100}{s(s+10)}$

TEST YOURSELF

1. A system has a transfer function $(1-s)/(1+s)$. It is called
   (a) low-pass system (b) all-pass system (c) high-pass system (d) none of these
2. The following log-magnitude plot of a system shown in Figure T1 represents a
   (a) PID controller (b) PD controller
   (c) PI controller (d) Proportional controller

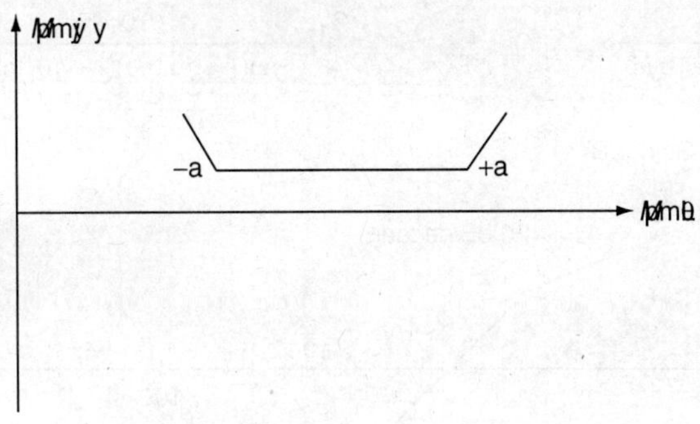

Figure T1

3. The Bode amplitude plot of a constant is a
   (a) line with slope –40 dB/decade (b) line with slope –20 dB/decade
   (c) line with slope +20 dB/decade (d) straight line parallel to the frequency axis
4. The pole factor $1/(1+j\omega T)$ has a slope of
   (a) 20 dB/decade (b) –20 dB/decade
   (c) 40 dB/decade (d) –40 dB/decade
5. Gain crossover frequency is the one at which $|G(j\omega)H(j\omega)|$ is
   (a) equal to 1 (b) equal to –1 (c) >1 (d) < –1

6. An octave frequency range is specified by

(a) $\frac{\omega^2}{\omega_1} = 8$ (b) $\frac{\omega_2}{\omega_1} = 6$ (c) $\frac{\omega_2}{\omega_1} = 7$ (d) $\frac{\omega_2}{\omega_1} = 2$

7. The graph shown in Figure T2 corresponds to a/an

(a) minimum phase function (b) all-pass function

(c) non-minimum phase function

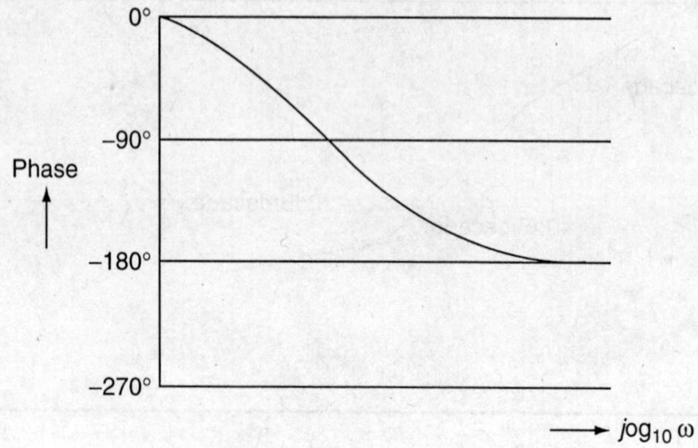

Figure T2

8. The transfer function of the system having the Bode plot as shown in Figure T3 is given by

(a) $\frac{100}{j\omega(1+j0.5\omega)(1+j0.1\omega)}$ (b) $\frac{100}{j\omega(2+j\omega)(10+j\omega)}$

(c) $\frac{10}{j\omega(1+2j\omega)(1+10j\omega)}$ (d) $\frac{10}{j\omega(1+j0.5\omega)(1+j0.1\omega)}$

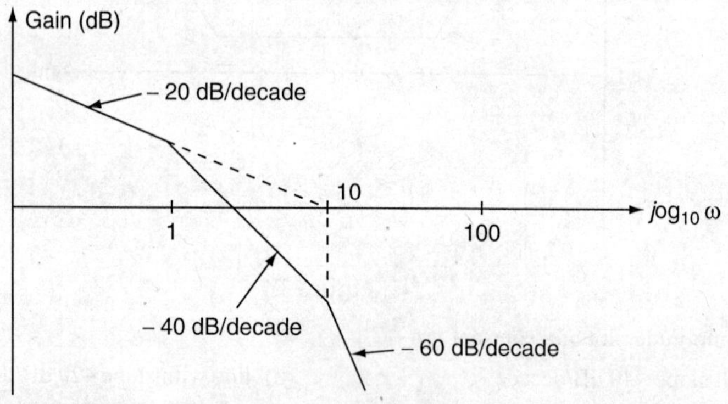

Figure T3

9. A system with transfer function $G(s) = s/(s+1)$ is subjected to a sinusoidal input $r(t) = \sin \omega t$. In steady state, the phase angles of the output relative to the input at $\omega = 0$ and $\omega = \infty$ will be, respectively,

(a) 0° and −90° (b) 0° and 0° (c) 90° and 0° (d) 90° and 0°

10. A system has 14 poles and 2 zeros. The slope of its highest frequency asymptote in its magnitude plot is
    (a) –40 dB/decade (b) –240 dB/decade
    (c) –280 dB/decade (d) –320 dB/decade
11. The function corresponding to the Bode plot shown in Figure T4 is

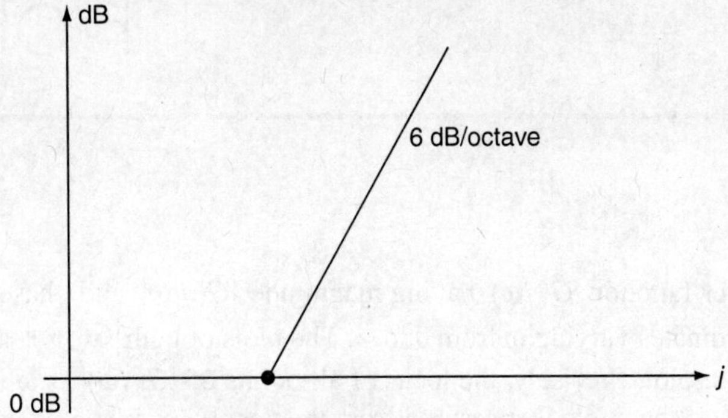

Figure T4

    (a) $A = jf_1/f_1$ (b) $A = 1/(-jf_1/f_1)$
    (c) $A = 1/(1 + jf_1/f_1)$ (d) $A = 1 + j\,f_1/f_1$
12. For a type-3 system, the asymptote at a lower frequency will have a slope of
    (a) –6 dB/decade (b) –12 dB/decade
    (c) –18 dB/decade (d) –40 dB/decade
13. The open-loop transfer function of a unity feedback control system is $10/(s+5)^3$. The gain margin of the system will be
    (a) 20 dB (b) 40 dB (c) 60 dB (d) 80 dB
14. If a transfer function has one or more zeros in the S plane, it is called
    (a) all-pass transfer function (b) non-minimum phase transfer function
    (c) minimum phase transfer function (d) none of these
15. A decade frequency range is specified by
    (a) $\frac{\omega^2}{\omega_1} = 2$ (b) $\frac{\omega_2}{\omega_1} = 10$ (c) $\frac{\omega_2}{\omega_1} = 3$ (d) $\frac{\omega_2}{\omega_1} = 6$
16. The slope a Bode plot within the region $5 < \omega < 20$ for the open-loop system function $G(s) = [s(s+5)(s+20)]$ is
    (a) –6 dB/octave (b) –12 dB/octave (c) 6 dB/octave (d) 12 dB/octave
17. For a type-3 system, the asymptote at a lower frequency will have a slope of
    (a) –6 dB/octave (b) –12 dB/octave
    (c) –24 dB/octave (d) –40 dB/octave

ANSWERS

1 (b)	2 (a)	3 (d)	4 (b)	5 (a)	6 (d)	7 (b)	8 (d)	9 (b)	10 (b)
11 (d)	12 (c)	13 (b)	14 (b)	15 (b)	16 (b)	17 (c)			

Chapter 14

Polar Plot

Any sinusoidal transfer function $G(j\omega)$ having magnitude $|G(j\omega)|$ and phase angle $\angle G(j\omega)$ can be plotted on polar co-ordinates varying ω from 0 to ∞. The plots of both $|G(j\omega)|$ and $\angle G(j\omega)$ as a function of ω show a polar plot. Precisely, the locus of all points of $|G(j\omega)|$ as ω is varied from 0 to ∞ is known as a polar plot. If the angle is measured from the positive *x*-axis, it is positive in anticlockwise direction, otherwise it is negative. The polar plot is very useful in determining the stability of a closed-loop system from its open-loop frequency response. For closed-loop system stability, the open-loop transfer function is $G(s)H(s)$ instead of $G(s)$ unless $H(s)=1$. Unless stated otherwise, the open-loop transfer function is taken as $G(s)$. The aim of this chapter is to introduce the polar plot.

14.1 Polar Plot of Standard Functions

14.1.1 G(s) = 1/s

Putting $s = j\omega$ in $G(s)=\frac{1}{s}$, we have

$$G(j\omega)=\frac{1}{j\omega}=\frac{1}{\omega}\angle -90°$$

Here

$$|G(j\omega)|=\frac{1}{\omega} \tag{14.1}$$

and

$$\angle G(j\omega)=-90° \tag{14.2}$$

Table 14.1 shows a few values of ω and the corresponding values of $|G(j\omega)|$.

Table 14.1 ω and $|G(j\omega)|$

ω	$\|G(j\omega)\|$
0	∞
1	1
10	0.1
100	0.01
⋮	⋮
⋮	⋮
∞	0

The polar plot of the given function $G(s) = 1/s$ is shown in Fig. 14.1

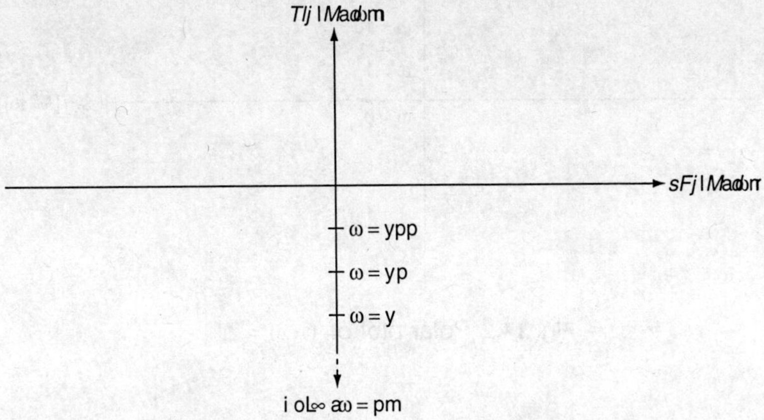

Fig. 14.1 Polar plot of $G(s) = 1/s$

14.1.2 G(s) = s

Putting $s = j\omega$ in $G(s) = s$, bb, we have

$$G(j\omega) = j\omega, |G(j\omega)| = \omega \text{ and } \angle G(j\omega) = 90° \tag{14.3}$$

Table 14.2 shows a few values of ω and the corresponding values of $|G(j\omega)|$.

Table 14.2 ω and $|G(j\omega)|$

ω	$\|G(j\omega)\|$
0	0
1	1
10	10
100	100
:	:
:	:
∞	∞

Figure 14.2 shows the polar plot of $G(s) = s$.

14.1.3 G(s) = 1/(1 + sa)

Putting $s = j\omega$ in $G(s) = \dfrac{1}{1+sa}$, we have

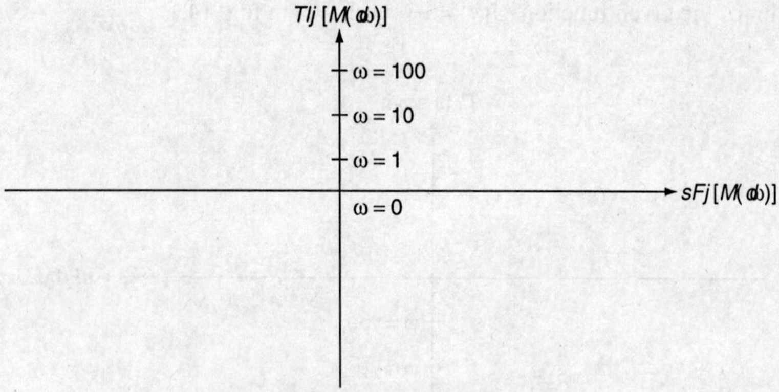

Fig. 14.2 Polar plot of $G(s) = 2$

$$G(j\omega) = \frac{1}{1 + j\omega a} = \frac{1}{\sqrt{1+\omega^2 a^2}} \angle - \tan^{-1} \omega a \tag{14.4}$$

Here,

$$|G(j\omega)| = \frac{1}{\sqrt{1+\omega^2 a^2}} \tag{14.5}$$

$$\angle G(j\omega) = -\tan^{-1} \omega a \tag{14.6}$$

Table 14.3 shows a few values of ω and the corresponding values of $|G(j\omega)|$

Table 14.3 ω and $|G(j\omega)|$ and $\angle G(j\omega)$

ω	$\|G(j\omega)\|$	$\angle G(j\omega)$
0	1	0°
1/a	0.707	−45°
10/a	0.1	−84.28°
100/a	0.01	−89.42°
:	:	:
:	:	:
∞	0	−90°

Figure 14.3 shows the polar plot of the function $G(s) = 1/(1 + sa)$. From Table 14.3, we have

At $\omega = 0$, $\quad G(j\omega) = 1\angle 0°$

At $\omega = \infty$, $\quad G(j\omega) = 1\angle -90°$

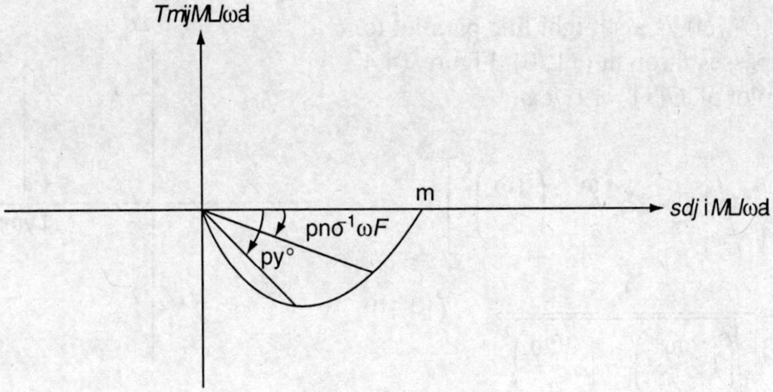

Fig. 14.3 Polar plot of 1/(1 + *sa*)

Therefore, for $\omega = \infty$, the polar plot approaches the origin by −90°, i.e., the *Y* axis. Again, from Table 14.3 and Fig. 14.3, it is clear that from $\omega = 1/a$ to $\omega = \infty$, 45° is covered. Therefore, the locus is a semicircle. The upper semicircle is obtained for $\omega = 0$ to $\omega = -\infty$.

14.1.4 *G*(*s*) = 1 + *sa*

Putting $s = j\omega$ in $G(s) = 1 + sa$, we have

$$G(j\omega) = 1 + j\omega a = \sqrt{1+\omega^2 a^2} \;\angle \tan^{-1} \omega a$$

$$\therefore \quad |G(j\omega)| = \sqrt{1+\omega^2 a^2} \tag{14.7}$$

$$\angle G(j\omega) = \tan^{-1} \omega a \tag{14.8}$$

Table 14.4 shows ω and the corresponding values of $\sqrt{1+\omega^2 a^2}$ and $\angle G(j\omega)$

Table 14.4 ω and the corresponding $|G(j\omega)|$ and $\angle |G(j\omega)|$

ω	|G(jω)|	∠G(jω)
0	1	0°
1	$\sqrt{1+a^2}$	$\tan^{-1} \omega a$
:	:	:
:	:	:
∞	∞	90°

Phase angle $$\phi = \tan^{-1} \frac{\omega a}{1} \tag{14.9}$$

It is obvious that $G(j\omega)$ is a straight line parallel to the Y axis and it passes through (+1, 0). Figure 14.4 depicts the polar plot of $G(s) = 1 + j\omega a$.

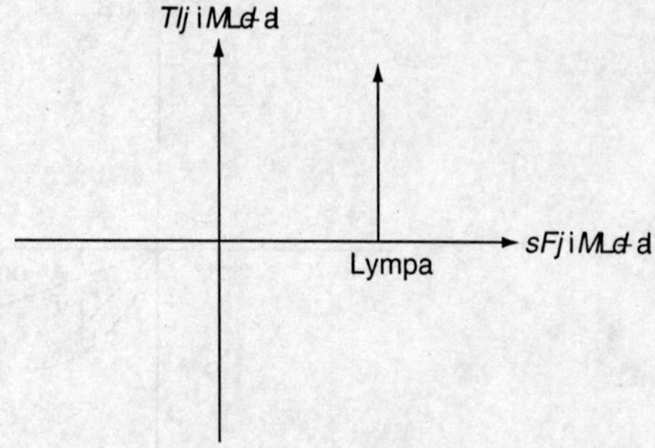

Fig. 14.4 Polar plot of $G(j\omega) = 1 + j\omega a$

14.1.5 $\mathbf{G(j\omega) = 1\Big/\left[1 + 2\xi\frac{j\omega}{a} + \left(\frac{j\omega}{a}\right)^2\right]}$

Here,

$$|G(j\omega)| = \frac{1}{\sqrt{\left(1 - \frac{\omega^2}{a^2}\right)^2 + \left(\frac{2\xi\omega}{a}\right)^2}} \qquad (14.10)$$

$$\angle G(j\omega) = -\tan^{-1}\frac{\frac{2\xi\omega}{a}}{1 - \frac{\omega^2}{a^2}} \qquad (14.11)$$

At $\omega = 0, |G(j\omega)| = 1$ and $\angle G(j\omega) = 0°$

At $\omega = \infty, |G(j\omega)| = \infty$ and $\angle G(j\omega) = -180°$

For this function, the exact shape of the polar plot totally depends on the value of ξ. Fig. 14.5 shows the polar plot of the given function.

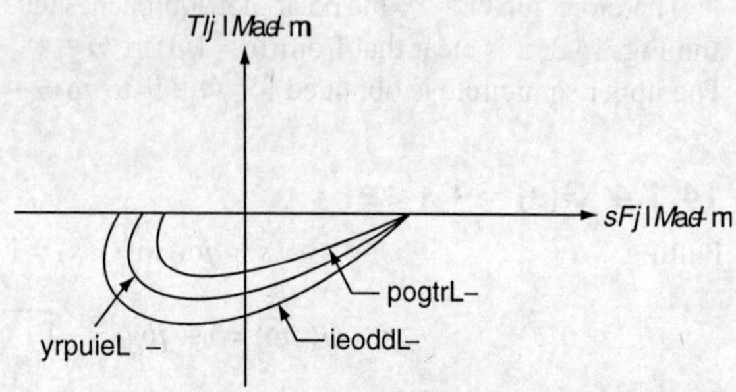

Fig. 14.5 Polar plot of $\left[1 + 2\xi\left(\frac{j\omega}{a}\right) + \left(j\frac{\omega}{a}\right)^2\right]^{-1}$

14.1.6 $\mathbf{G(j\omega) = 1 + 2\xi\left(\frac{j\omega}{a}\right) + \left(\frac{j\omega}{a}\right)^2}$

Here,

$$|G(j\omega)| = \sqrt{\left(1 - \frac{\omega^2}{a^2}\right)^2 + \left(\frac{2\ \omega}{a}\right)^2} \qquad (14.12)$$

$$\angle G(j\omega) = \tan^{-1}\left[\frac{\frac{2\xi\omega}{a}}{1 - \frac{\omega^2}{a^2}}\right] \qquad (14.13)$$

At $\omega = 0, |G(j\omega)| = 1$ and $\angle G(j\omega) = 0°$

At $\omega = \infty, |G(j\omega)| = \infty$ and $\angle G(j\omega) = 180°$.

Figure 14.6 shows the polar plot of the given function.

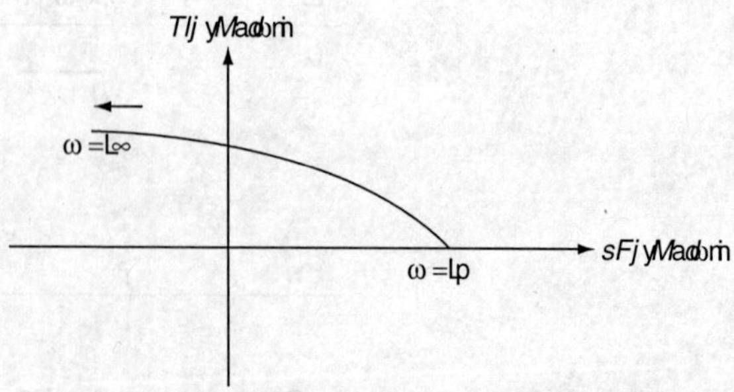

Fig. 14.6 Polar plot of $G(j\omega) = 1 + 2\xi\left(\frac{j\omega}{a}\right) + \left(\frac{j\omega}{a}\right)^2$

14.1.7 $G(j\omega) = e^{-j\omega a}$ (transport lag)

$$G(j\omega) = \cos\omega a - j\sin\omega a = 1\angle -\omega a$$

$$\therefore \qquad |G(j\omega)| = 1 \qquad (14.14)$$

$$\angle G(j\omega) = -\omega a \qquad (14.15)$$

From Eq. (14.12) and Eq. (14.13), it is clear that the magnitude is unity and the angle varies with ω. Therefore, the plot is a circle in $G(j\omega)$ (Figure. 14.7).

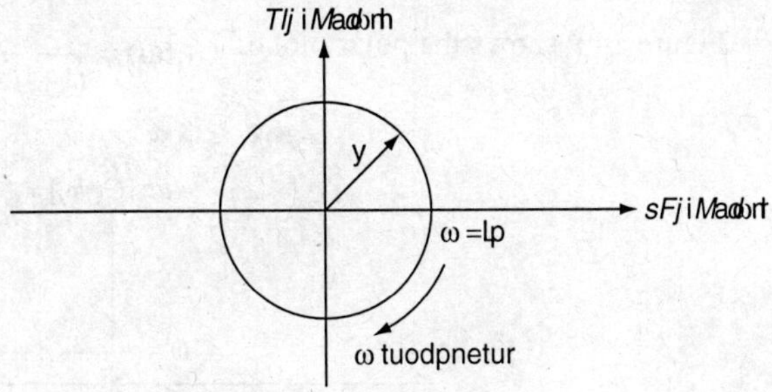

Fig. 14.7 Polar plot of $e^{-j\omega a}$

14.1.8 $G(j\omega) = \dfrac{e^{-j\omega a}}{1+j\omega b}$

Here, $$G(j\omega) = \frac{e^{-j\omega a}}{1+j\omega b}$$

$$\therefore \qquad |G(j\omega)| = \frac{1}{\sqrt{1+\omega^2 b^2}} \qquad (14.16)$$

$$\angle G(j\omega) = -\omega a - \tan^{-1}\omega b \qquad (14.17)$$

Figure 14.8 shows the polar plot of $G(j\omega) = e^{-j\omega a}/(1+j\omega b)$. The plot is a spiral and the curve tends to origin as $\omega \to \infty$.

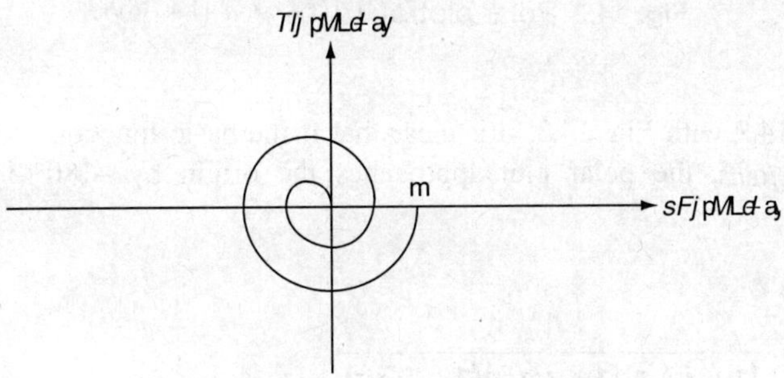

Fig. 14.8 Plot of $G(j\omega) = \dfrac{e^{-j\omega a}}{1+j\omega b}$

14.1.9 $G(j\omega) = \dfrac{1}{(1+j\omega a)(1+j\omega b)}$

Here, $$G(j\omega) = \frac{1}{(1+j\omega a)(1+j\omega b)}$$

$$|G(j\omega)| = \frac{1}{\sqrt{1+\omega^2 a^2}\sqrt{1+\omega^2 b^2}} \qquad (14.18)$$

$$\angle G(j\omega) = -\tan^{-1}\omega a - \tan^{-1}\omega b \tag{14.19}$$

At $\omega = 0, |G(j\omega)| = 1$ and $\angle G(j\omega) = 0°$

At $\omega = \infty, |G(j\omega)| = 0$ and $\angle G(j\omega) = -180°$

At $\omega = \dfrac{1}{\sqrt{ab}}, |G(j\omega)| = \dfrac{\sqrt{ab}}{a+b}$, and $\angle G(j\omega) = -90°$

Figure 14.9 shows the polar plot of $G(j\omega) = \dfrac{1}{(1+j\omega a)(1+j\omega b)}$

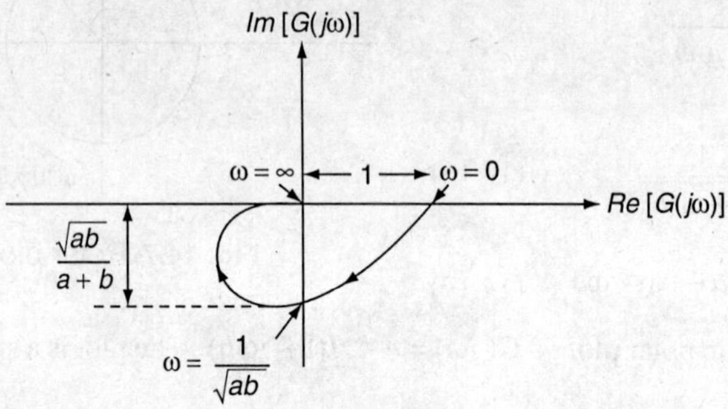

Fig. 14.9 Polar plot of $1/[(1+j\omega a)(1+j\omega b)]$

Comparing Fig. 14.9 with Fig. 14.3, it is clear that if the basic function $1/(1+j\omega a)$ is multiplied by the term $1/(1+j\omega b)$, the polar plot approaches the origin by $-180°$ instead of $-90°$ as in Sec. 14.1.3.

14.1.10 $G(j\omega) = \dfrac{1}{(1+j\omega a)(1+j\omega b)(1+j\omega c)}$

Here, $G(j\omega) = \dfrac{1}{(1+j\omega a)(1+j\omega b)(1+j\omega c)}$

i.e.,
$$|G(j\omega)| = \frac{1}{\sqrt{1+\omega^2 a^2}\sqrt{1+\omega^2 b^2}\sqrt{1+\omega^2 c^2}} \tag{14.20}$$

$\therefore$
$$|G(j\omega)| = \frac{1}{\{1-\omega^2(ab+bc+ca)\}^2 + \{\omega(a+b+c-\omega^2 abc)\}^2}$$

$$\text{and} \qquad \angle G(j\omega) = -\tan^{-1}\frac{\omega(a+b+c-\omega^2 abc)}{1-\omega^2(ab+bc+ca)} = -\tan^{-1}\omega a - \tan^{-1}\omega b - \tan^{-1}\omega c \tag{14.21}$$

At $\omega = 0,\ |G(j\omega)| = 1$ and $\angle G(j\omega) = 0°$

At $\omega = \infty,\ |G(j\omega)| = \infty$ and $\angle G(j\omega) = -270°$

At $\omega = \dfrac{1}{\sqrt{ab+bc+ca}},\ |G(j\omega)| = \sqrt{\dfrac{a+b+c}{abc}}$, and $\angle G(j\omega) = -90°$

The polar plot of the given $G(j\omega)$ is shown in Fig. 14.10.

Comparing Fig. 14.10 with Fig. 14.9, it is clear that if $1/[(1+j\omega a)(1+j\omega b)]$ is multiplied by $1/(1+j\omega c)$, the polar plot approaches the origin by –270° instead of –180° as in Sec. 14.1.9.

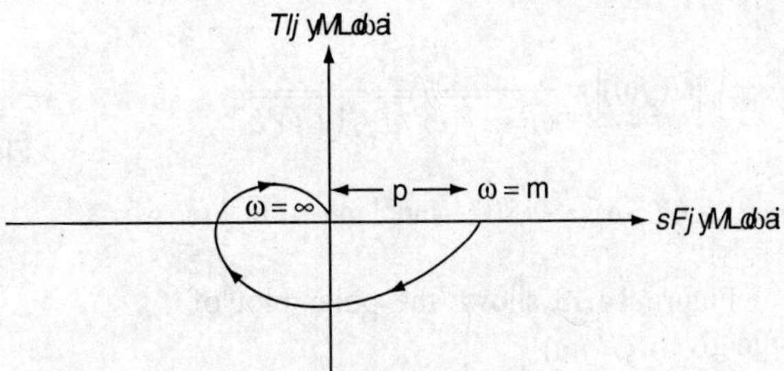

Fig. 14.10 Polar plot of $1/[(1+j\omega a)(1+j\omega b)(1+j\omega c)]$

14.1.11 $G(j\omega) = \dfrac{1}{j\omega(1+j\omega a)}$

$$G(j\omega) = \frac{1}{j\omega(1+j\omega a)}$$

$$|G(j\omega)| = \frac{1}{\omega\sqrt{1+\omega^2 a^2}} \tag{14.22}$$

$$\angle G(j\omega) = -90° - \tan^{-1}\omega a \tag{14.23}$$

At $\omega = 0, |G(j\omega)| = \infty$ and $\angle G(j\omega) = -90°$

At $\omega = \infty, |G(j\omega)| = 0$ and $\angle G(j\omega) = -180°$

Figure 14.11 shows the polar plot of the given $G(j\omega)$.

Comparing Fig. 14.11 with Fig. 14.3, it is clear that if the basic function $1/(1+j\omega a)$ is multiplied by the term $1/j\omega$, the polar plot shift shifts by an angle –90°.

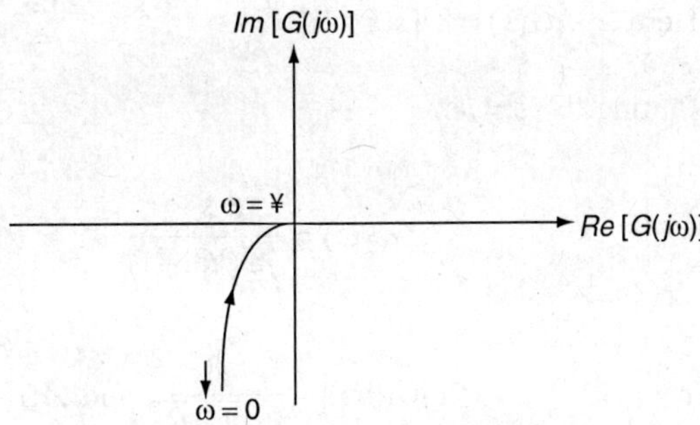

Fig. 14.11 Polar plot of $\dfrac{1}{j\omega(1+j\omega a)}$

14.1.12 $G(j\omega) = \dfrac{1}{j\omega(1+j\omega a)(1+j\omega b)}$

$$\text{Here} \qquad |G(j\omega)| = \frac{1}{\omega\sqrt{1+\omega^2 a^2}\sqrt{1+\omega^2 b^2}} \tag{14.24}$$

and $$\angle G(j\omega) = -90° - \tan^{-1} \omega a - \tan^{-1} \omega b \tag{14.25}$$

Figure 14.12 shows the polar plot of the given $G(j\omega)$.

Comparing Fig. 14.12 with Fig. 14.9, it is clear that if $1/[(1+j\omega a)(1+j\omega b)]$ is multiplied by $1/j\omega$, the polar plot shifts by −90°.

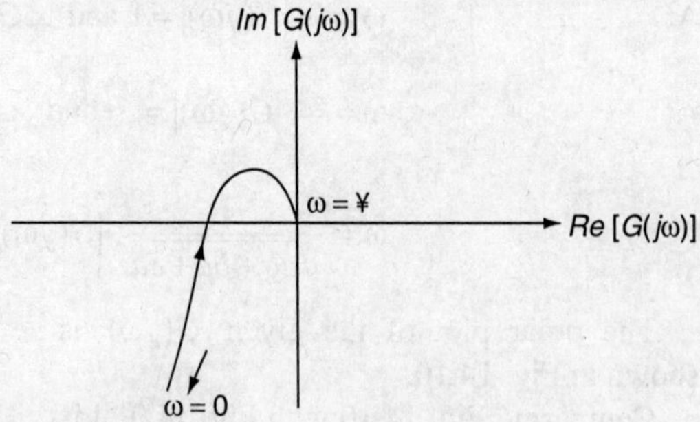

Fig. 14.12 Polar plot of $\frac{1}{j\omega(1+j\omega a)(1+j\omega b)}$

14.1.13 $G(j\omega) = \frac{1}{(j\omega)^2(1+j\omega a)(1+j\omega b)}$

$$|G(j\omega)| = \frac{1}{\omega^2\sqrt{1+\omega^2 a^2}\sqrt{1+\omega^2 b^2}}$$

$$\angle G(j\omega) = -180° - \tan^{-1} \omega a - \tan^{-1} \omega b$$

Figure 14.13 shows the polar plot of the given $G(j\omega)$.

Comparing Fig.14.13 with Fig. 14.12, it is clear that there is another shift of −90° due to another $1/j\omega$.

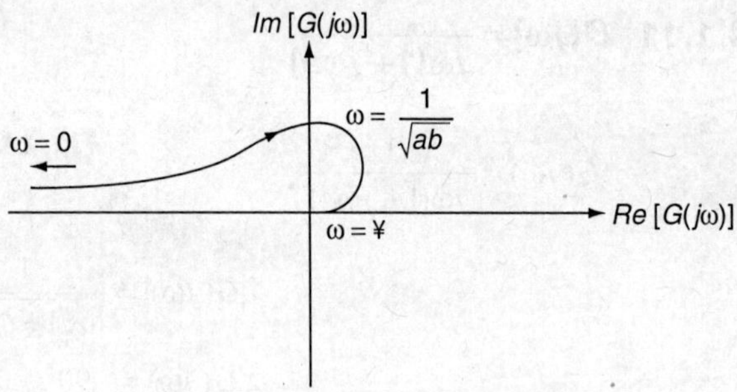

Fig. 14.13 Polar plot of $G(j\omega) = \frac{1}{(j\omega)^2(1+j\omega a)(1+j\omega b)}$

Example 14.1 Find the polar plot of $8/[s(s+1)]$.

Solution

Here $G(s) = 8/[s(s+1)]$.

Putting $s = j\omega$,

$$G(j\omega) = \frac{8}{j\omega(j\omega+1)}$$

∴ $$|G(j\omega)| = \frac{8}{\omega\sqrt{1+\omega^2}} \text{ and } \angle G(j\omega) = -90° - \tan^{-1}\omega$$

At $\omega = 0$, $|G(j\omega)| = \infty$ and $\angle G(j\omega) = -90°$

At $\omega = \infty$, $|G(j\omega)| = 0$, and $\angle G(j\omega) = -180°$

Figure E14.1 shows the polar plot of the given function

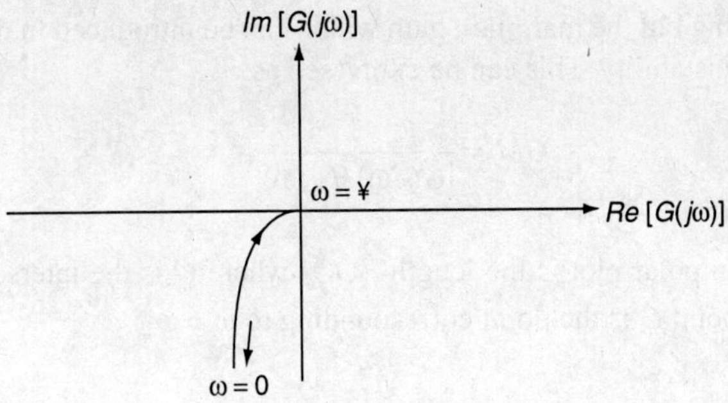

Fig. E14.1

14.2 ω_{gc} and ω_{pc} from Polar Plot

The gain crossover frequency in Bode plot has been defined as the frequency at which $|G(j\omega)H(j\omega)| = 0$ dB. In polar plot, there is no value of magnitude in dB. Therefore, $|G(j\omega)H(j\omega)| = 1$ corresponding to 0dB at $\omega = \omega_{gc}$ from polar plot viewpoint. To locate the point on polar plot having $M = |G(j\omega)H(j\omega)| = 1$, the polar plot is shown in Fig. 14.14.

A circle with unit radius is drawn having centre at origin to obtain a point with $M = |G(j\omega)H(j\omega)| = 1$. The point of intersection of the circle with polar plot gives the point at which $|G(j\omega)H(j\omega)| = 1$ and the corresponding frequency in $\omega = \omega_{gc}$. Such a point P has been shown in Fig. 14.14.

The phase crossover frequency is the frequency at which $\angle G(j\omega)H(j\omega) = -180°$. The angle of the point on the polar plot is $-180°$, i.e., the point lies on the negative real axis. Therefore, at $\omega = \omega_{pc}$, the polar plot intersects negative real axis. Such a point Q has been shown in Fig. 14.14. To determine the stability, $|G(j\omega)H(j\omega)| = 1$ and $\angle G(j\omega)H(j\omega) = -180°$ play a very important role. The point $1\angle -180°$ is nothing but the point $-1 + j0$ on the negative real axis. This point is known as critical point.

Fig. 14.14 Determination of ω_{gc} and ω_{pc} from polar plot

14.3 Determination of GM and PM from Polar Plot

The gain margin is nothing but the margin in gain which can be introduced in the system till the system reaches on the verge of instability. This can be expressed as

$$GM = \frac{1}{|G(j\omega)H(j\omega)|_{\omega = \omega_{pc}}}$$

$|G(j\omega)H(j\omega)|_{\omega = \omega_{pc}}$ in polar plot is the length $l(OQ)$ where Q is the intersection of polar plot with negative real axis. The point Q is the point corresponding to $\omega = \omega_{pc}$.

$$\therefore \qquad GM = \frac{1}{l(OQ)}$$

The point Q is the intersection of polar plot with negative real axis.
GM in dB is expressed as

$$\therefore \qquad \text{GM} = 20\log_{10}\frac{1}{|OQ|}$$

From the definition of phase margin,

$$\text{PM} = 180° + \angle G(j\omega)H(j\omega)|_{\omega=\omega_{gc}}$$

The point P is the point on the polar plot corresponding to $\omega = \omega_{gc}$ which is obtained by drawing unit radius circle. Therefore, if ϕ is the angle of the point P (at $\omega = \omega_{gc}$), the PM can be calculated as shown in Fig. 14.15.

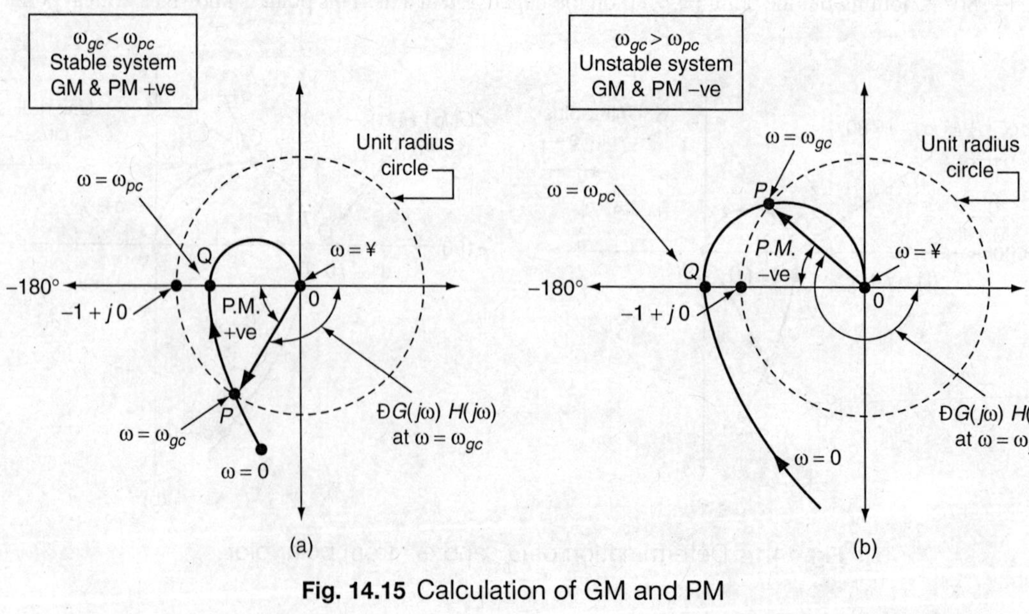

Fig. 14.15 Calculation of GM and PM

To get accurate value of PM, precise polar plot is required and it must be in scale. Otherwise, accurate value of PM cannot be obtained. Since the point Q is the intersection of polar plot with the negative real axis, this point can be obtained mathematically.

14.4 Determination of Stability from Polar Plot

The comparison of magnitudes of ω_{gc} and ω_{pc} decides the stability of the system and it has been shown in Bode plot that

(i) $\omega_{gc} < \omega_{pc}$, GM and PM are positive; hence, the system is stable.

(ii) $\omega_{gc} > \omega_{pc}$, GM and PM are negative; hence, the system is unstable.

(iii) $\omega_{gc} = \omega_{pc}$, GM = PM = 0; hence, the system is marginally stable.

From Fig. 14.14(a), as ω is varied from 0 to ∞, ω_{gc} at point P occurs first and then ω_{pc} at point Q occurs. Therefore, $\omega_{gc} < \omega_{pc}$. The polar plot shows the stable behaviour of system; hence, PM is positive for such case shown in Fig. 14.15(a).

From Fig. 14.14(b), as ω is varied from 0 to ∞, the point Q corresponding to $\omega = \omega_{pc}$ occurs first and then point P corresponding to $\omega = \omega_{gc}$. Hence, $\omega_{pc} < \omega_{gc}$. The polar plot shows an unstable behaviour of system. PM for such a system is negative as shown in Fig. 14.15(b).

The points P and Q coincide at the critical point $-1 + j\,0$ for zero values of GM and PM. Therefore, the polar plot intersects the negative real axis at a critical point $-1 + j\,0$. The system is marginally stable in nature.

The critical point $-1 + j\,0$ is outside the polar plot for stable system as shown in Fig. 14.15(a) whereas it is inside the plot for unstable system as shown in Fig. 14.15(b).

This is known as enclosement. If the critical point is enclosed by the polar plot, the system is unstable. If it is not enclosed by the polar plot, the system is stable. A portion will be enclosed within polar plot when it lies to the right-hand side during travelling, along the polar plot from the point corresponding to $\omega = 0$ to point corresponding to $\omega = \infty$.

GM and PM are very simple concepts, which can be applied to find stability. The polar plot for a system is shown in Fig. 14.16. It is very difficult to determine GM, PM and stability for such a system.

These systems are conditionally stable. The polar plot of these systems cannot give the clear picture of stability. The Nyquist plot of these systems is able to make it clear about the stability. Therefore, Nyquist creations are used for stability analysis of the system instead of polar plot.

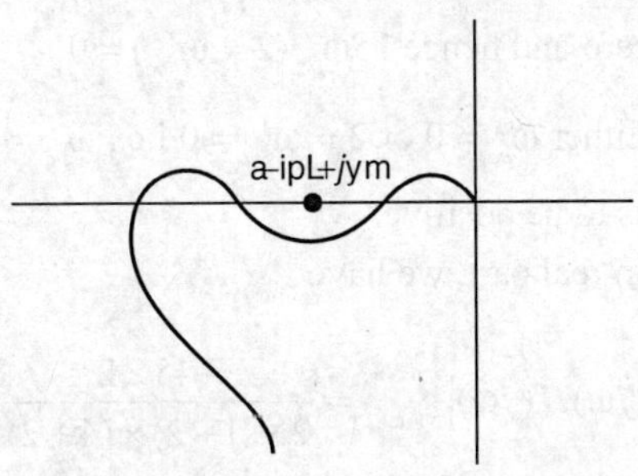

Fig. 14.16 Polar plot of system

Example 14.2 Find the polar plot for system with $G(s)H(s)=\dfrac{15}{s(s+1)(s+2)}$. Calculate its gain margin in dB and comment on its stability.

Solution

Here $$G(s)H(s)=\frac{15}{s(s+1)(s+2)}$$

$\therefore$ $$G(j\omega)H(j\omega)=\frac{15}{(j\omega)(j\omega+1)(j\omega+2)}$$

$\therefore$ $$|G(j\omega)H(j\omega)|=\frac{15}{\omega\sqrt{1+\omega^2}\sqrt{4+\omega^2}}$$

$\therefore$ $$\angle G(j\omega)H(j\omega)=\phi=-90^\circ-\tan^{-1}\omega-\tan^{-1}\frac{\omega}{2}$$

For $\omega\to 0,\quad G(j\omega)H(j\omega)=\infty\angle-90^\circ$

For $\omega\to\infty,\quad G(j\omega)H(j\omega)=0\angle-270^\circ$

$\therefore$ Rotation of plot $=-270^\circ-(-90^\circ)=-180^\circ$ clockwise.

Let us find out the intersection of the polar plot with negative real axis.

$\therefore$ $$G(j\omega)H(j\omega)=\frac{15}{(j\omega)(j\omega+1)(j\omega+2)}$$

$$=\frac{15(-j\omega)(-j\omega+1)(-j\omega+2)}{(j\omega)(j\omega+1)(j\omega+2)(-j\omega)(-j\omega+1)(-j\omega+2)}$$

$$=\frac{-15j\omega(2-3j\omega-\omega^2)}{\omega^2(1+\omega^2)(4+\omega^2)}$$

$$=\frac{-45\omega^2}{D}-\frac{15j\omega(2-\omega^2)}{D}$$

where $$D=\omega^2(1+\omega^2)(4+\omega^2).$$

At $\omega=\omega_{pc}$, imaginary part is zero and hence $15\omega_{pc}(2-\omega_{pc}^2)=0$

$\therefore$ Either $\omega_{pc}=0$ or $2-\omega_{pc}^2=0$ i.e., $\omega_{pc}=\sqrt{2}$.

Therefore, $\omega_{pc}=\sqrt{2}$, as it is to be positive.

Substituting $\omega=\omega_{pc}=\sqrt{2}$ in real part, we have

$$G(j\omega)H(j\omega)\big|_{\omega=\omega_{pc}}=\frac{-45\times 2}{2\times(1+2)\times(4+2)}+j0$$

$$=-2.5+j0$$

Figure E14.2 shows the polar plot of the given system.

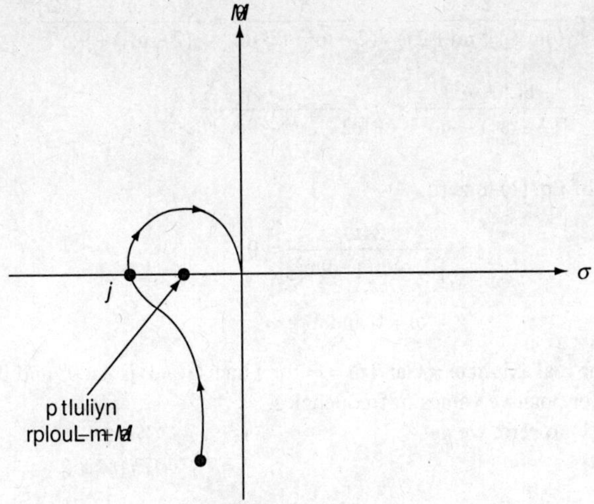

Fig. E14.2

$$\therefore \qquad GM = \frac{1}{|OQ|} = \frac{1}{2.5} = 0.4$$

$$\therefore \qquad GM\,(\text{in dB}) = 20\,log_{10}\,0.4 = -7.96$$

Since GM is negative, critical point is enclosed; hence, the system is unstable.

ADDITIONAL SOLVED EXAMPLES

Example 14.3 Find the polar plot of $G(s) = 8/[(s+1)(s+2)]$.

Solution

This is a type 0 system.

$$s = j\omega, \text{ therefore, } G(s) = \frac{8}{(s+1)(s+2)}$$

It can be written as

$$G(j\omega) = \frac{8}{(j\omega+1)(j\omega+2)}$$

$$\therefore \qquad |G(j\omega)| = \frac{8}{\sqrt{1+\omega^2}\,\sqrt{4+\omega^2}}$$

and

$$\angle G(j\omega) = -\tan^{-1}\omega - \tan^{-1}\frac{\omega}{2}$$

At

$$\omega = 0, \quad |G(j\omega)| = 4 \text{ and } \angle G(j\omega) = 0^\circ$$

At $$\omega=\infty,\quad |G(j\omega)|=0 \text{ and } \angle G(j\omega)=-180°$$

Again $$G(j\omega)=\frac{8}{(j\omega+1)(j\omega+2)}=\frac{8}{(2-\omega^2)+3j\omega}=\frac{8[(2-\omega^2)-3j\omega]}{(2-\omega^2)+9\omega^2}$$

$$=\frac{8(2-\omega^2)}{[(2-\omega^2)-9\omega^2]}-j\frac{24\omega}{[(2-\omega^2)-9\omega^2]} \quad (1)$$

Equating imaginary parts of Eq. (1) to zero,

$$\frac{24\omega}{[(2-\omega^2)+9\omega^2]}=0$$

∴ $$\omega=0 \text{ and } \omega\pm\infty$$

The polar plot intersects the real axis at $\omega=0$ and $\omega=\infty$ for which $G(j\omega)$ is $4\angle 0°$ and $0\angle -180°$. The polar plot intersects the real axis only for positive values of frequencies.

Equating real part of Eq. (1) to zero, we get

$$\frac{8(2-\omega^2)}{[(2-\omega^2)-9\omega^2]}=0$$

∴ $$\omega=\pm\sqrt{2} \text{ and } \omega=\infty.$$

At $\omega=\sqrt{2}, |G(j\omega)|=\frac{8}{3\sqrt{2}}$ and $\angle G(j\omega)=-90°$

At $\omega=\infty, |G(j\omega)|=0$ and $\angle G(j\omega)=-180°$

Figure E14.3 shows the polar plot of the given $G(s)$.

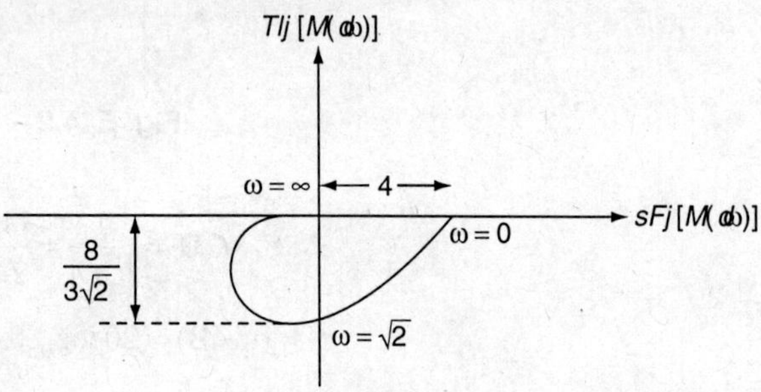

Fig. E14.3

Example 14.4 Find the polar plot of $14/[s(s+1)(s+2)]$

Solution

This is of type 1 system.

Putting $$G(s)=\frac{14}{s(s+1)(s+2)}$$

$$G(j\omega)=\frac{14}{(j\omega)(j\omega+1)(j\omega+2)}$$

∴ $$|G(j\omega)|=\frac{14}{\omega\sqrt{1+\omega^2}\sqrt{4+\omega^2}}$$

and $$\angle G(j\omega)=-90°-\tan^{-1}\omega-\tan^{-1}\frac{\omega}{2}$$

At $$\omega=0,\quad |G(j\omega)|=\infty \text{ and } \angle G(j\omega)=-90°$$

At $$\omega=\infty,\quad |G(j\omega)|=0 \text{ and } \angle G(j\omega)=-270°$$

Again

$$G(j\omega) = \frac{14}{(j\omega)[(2-\omega^2)+3j\omega]} = \frac{-42}{(2-\omega^2)^2+9\omega^2} - \frac{14j(2-\omega^2)}{[(2-\omega^2)^2+9\omega^2]\omega} \quad (1)$$

Equating the imaginary part of Eq. (1) to zero, ω can be obtained as follows:

$$\omega = \pm\sqrt{2} \quad \text{and} \quad \omega = \pm\infty$$

Since the polar plot intersects the real axis at $\omega = \sqrt{2}$ and $\omega = \infty$ for only positive frequencies. Again equating the real part of Eq. (1) to zero, we get $\omega = \pm\infty$. Considering only positive values, $\omega = \sqrt{2}$ and ∞. At $\omega = \sqrt{2}$, $|G(j\omega)| = 14/6$ and $\angle G(j\omega) = 180°$. Figure E14.4 shows the polar plot of the given function.

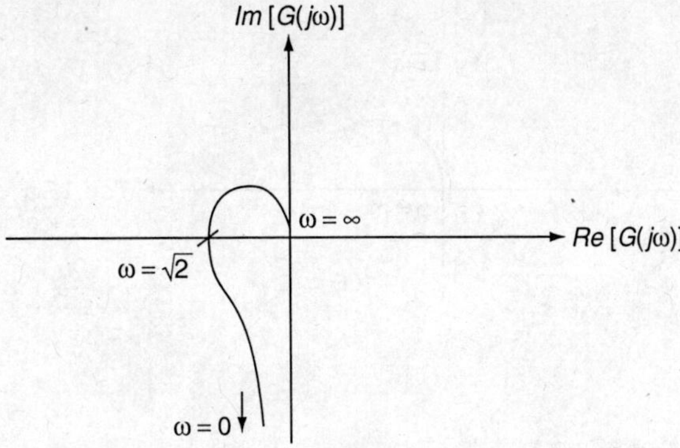

Fig. E14.4

Example 14.5 Find the polar plot of $G(s) = (1+6s)/(1+2s)$.

Solution

Putting $s = j\omega$ in $G(s) = (1+6s)/(1+2s)$

$$G(j\omega) = \frac{1+6j\omega}{1+2j\omega} = \frac{\sqrt{1+36\omega^2}}{\sqrt{1+4\omega^2}} \angle \tan^{-1} 6\omega - \tan^{-1} 2\omega$$

$$\therefore \quad |G(j\omega)| = \frac{\sqrt{1+36\omega^2}}{\sqrt{1+4\omega^2}}$$

and

$$\angle G(j\omega) = \tan^{-1} 6\omega - \tan^{-1} 2\omega$$

At $\omega = 0$, $|G(j\omega)| = 1$, $\angle G(j\omega) = 0°$

At $\omega = 1$, $|G(j\omega)| = 2.72$, $\angle G(j\omega) = 17.10°$

At $\omega = \infty$, $|G(j\omega)| = 3.00$, $\angle G(j\omega) = 0°$

Figure E14.5 shows the polar plot of the given function.

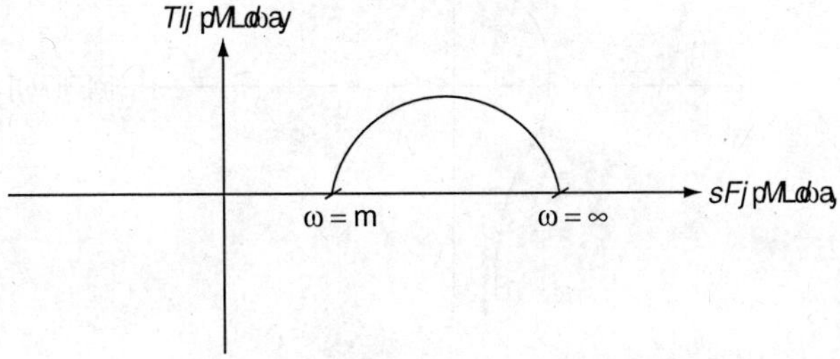

Fig. E14.5

EXERCISES

Questions for Practice

1. Sketch the polar plot for the following functions:

(i) $G(s) = \dfrac{sa}{1+sa}$ (ii) $G(s) = \dfrac{1+sa}{1+sb}$

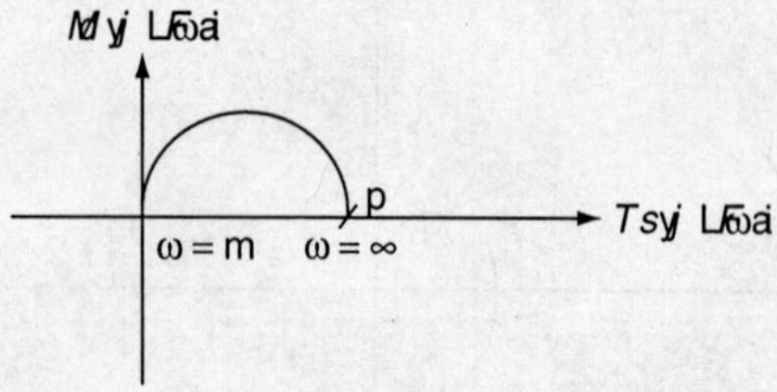

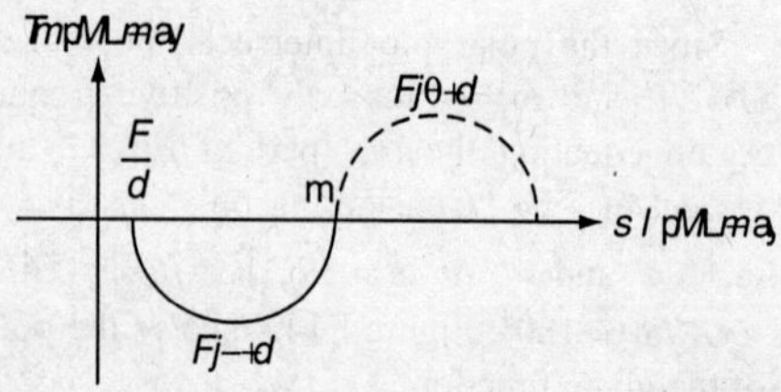

Ans. (i)
Ans. (ii)

2. Sketch the polar plot of $G(s) = \dfrac{12}{s(s+1)}$

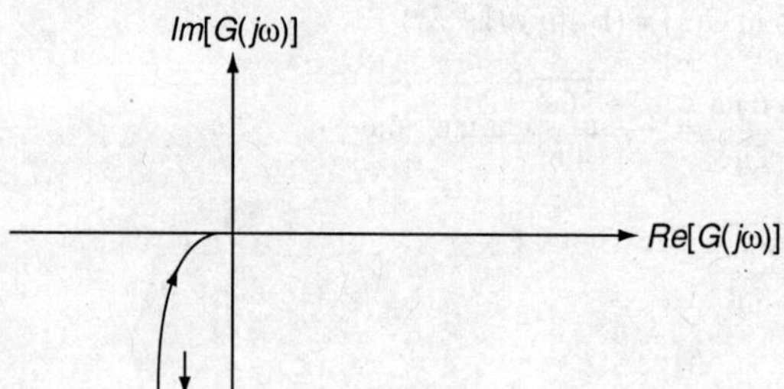

Ans.

3. Sketch the polar plot for $G(s) = \dfrac{1}{s(1+sb)(1+sb)(1+sc)}$

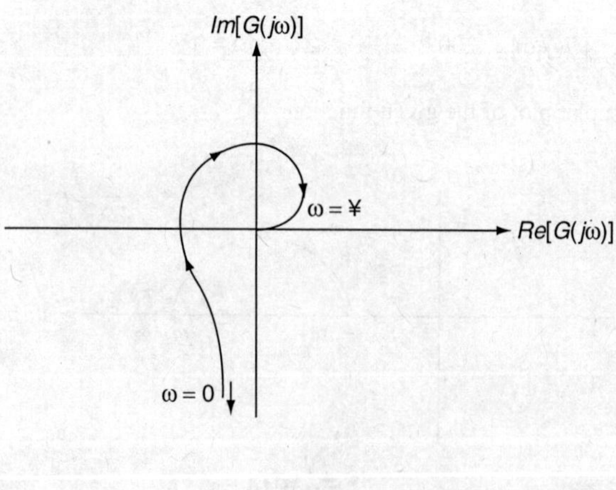

Ans.

4. Find the plot for $G(s) = \dfrac{1}{s^2(1+sb)(1+sb)(1+sc)}$

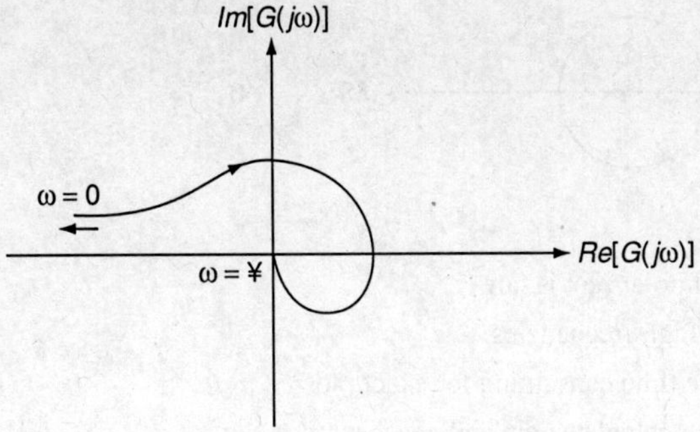

Ans.

TEST YOURSELF

1. "Polar plot is plot of magnitude versus phase angle varying frequency from 0 to ∞."
   This statement is
   (a) True (b) False
2. Match the following:

(a) $\dfrac{s^2+1}{s^3}$

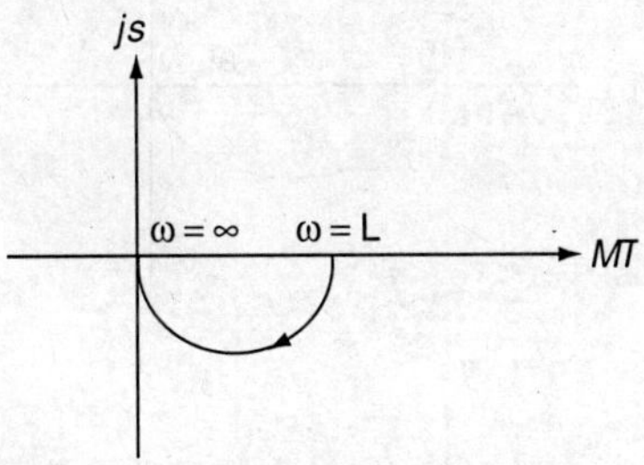

(i)

(b) $\dfrac{1}{s+6}$

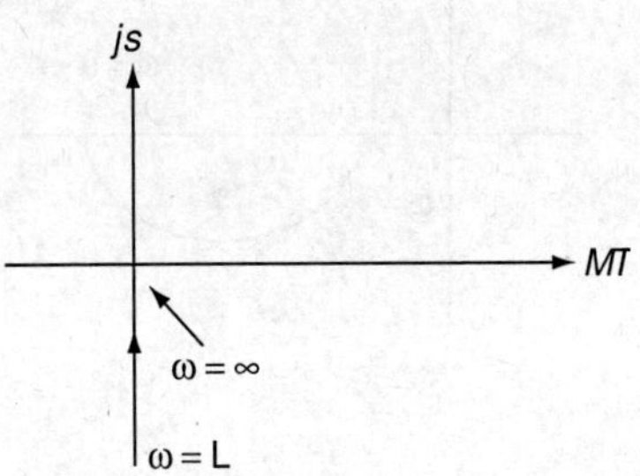

(ii)

(c) $\dfrac{s}{(s+1)(s+2)}$

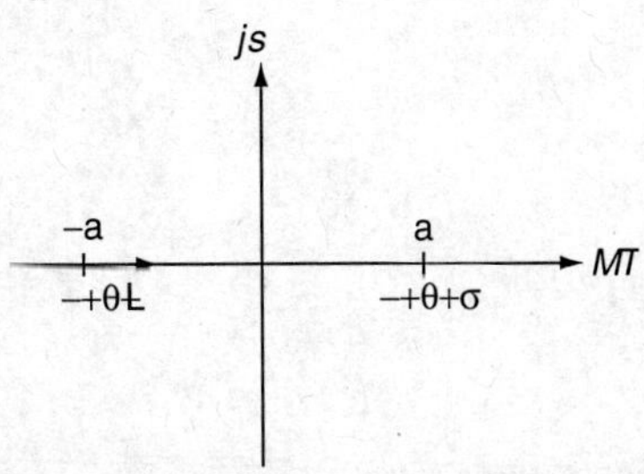

(iii)

(d) $\dfrac{s^2-1}{s^2+1}$ 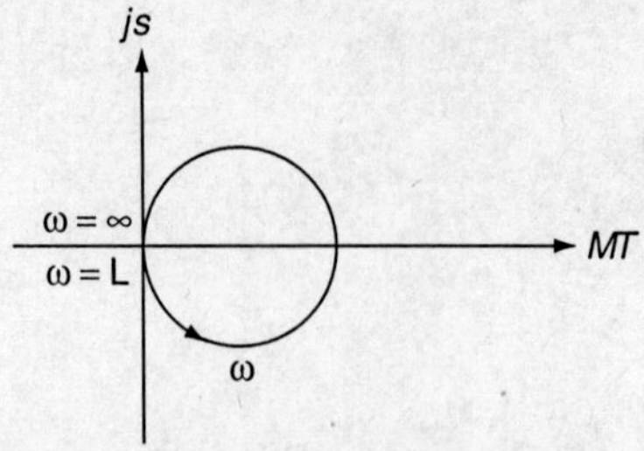

(iv)

3. The disadvantage(s) of polar plot is (are)
   (a) plot is cramped at high frequencies
   (b) the calculations are time consuming for exact plot
   (c) it is very difficult to calculate gain and phase margin
   (d) both (a) and (b)
   (e) both (b) and (c)
   (f) all of these
4. The polar plot of $G(s) = (1+s)/1+4s$ for $0 \leq \omega \leq \infty$ in G-plane is

(a) 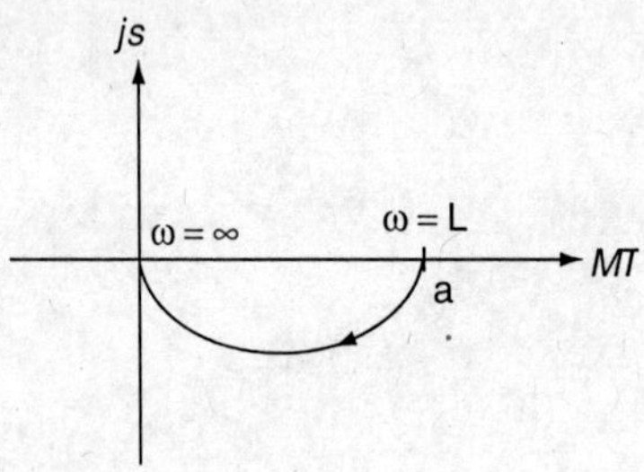

(b) 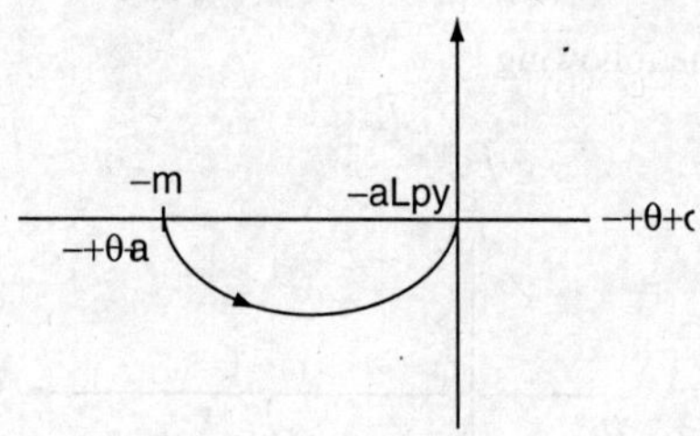

(c) 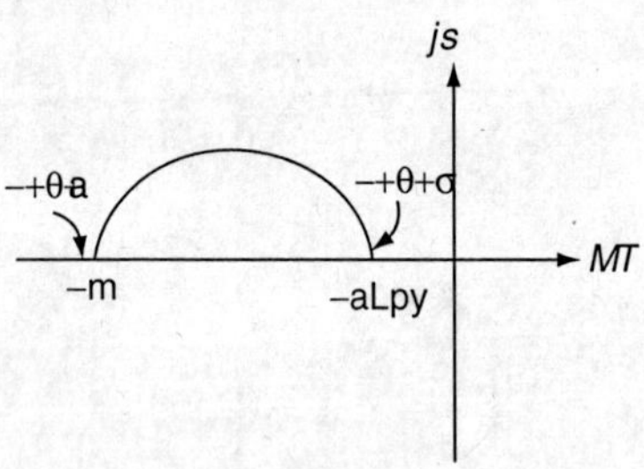

(d) 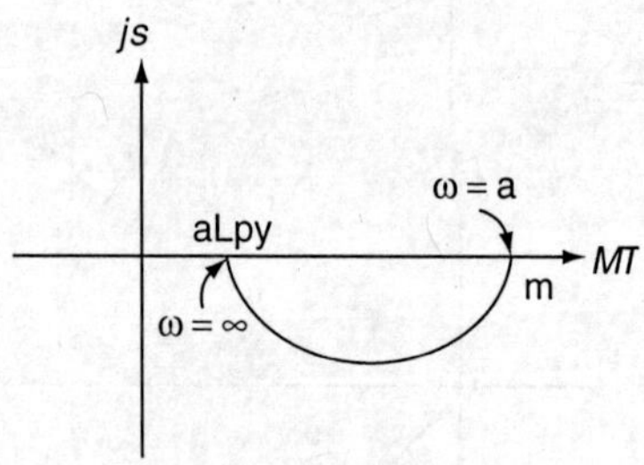

ANSWERS

1 (a) 2 a→2, b→1, c→4, d→3 3(f) 4 (d)

Chapter 15

Nyquist Plot

Using Routh–Hurwitz criterion, only the absolute stability of a system can be found out. On the other hand, the Nyquist criterion uses a different approach for finding the stability of a system. It focusses also on the relative stability of the system. It is possible to determine the stability of closed-loop pole from open-loop pole without knowing the roots of the closed-loop system. Nyquist plot is based on polar plot. The aim of this chapter is to introduce the Nyquist plot as well as the stability analysis of the system.

15.1 Basic Definitions

Encircled: If a point is found to be inside the path, the point is said to be encircled by the closed path. Figure 15.1 shows that point X is encircled by the closed path whereas the point Y is not encircled by the closed path.

Enclosed: If a path is traversed in the clockwise direction and the point is found to lie to the right of the path, the point is said to be enclosed by the path. Figure 15.2(a) shows that point X is enclosed by the path, whereas point Y is enclosed by the path shown in Figure 15.2(b).

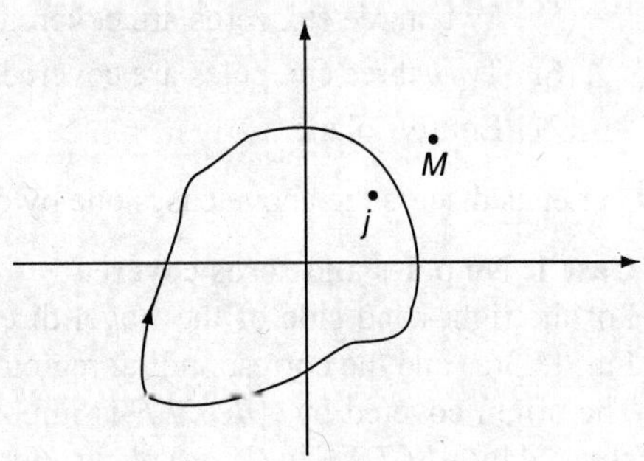

Fig. 15.1 Point X is encircled

Analytic Function and Singularities: A function is said to be analytic at a point in a plane if its value and its derivative have finite existence at that point. If $F(s) = K[s(s+a)]$, then function $F(s)$ is analytic at all points in s-plane except at $s = 0$ and $s = -a$. On the other hand, at $s = 0$ and $s = -a$, the value of $F(s)$ is ∞. Therefore, the pole of the function is its singularities.

Single-Valued Function: A function is said to be single valued if it has only one value for each separate value of s.

If $F(s) = \sqrt{s}$ and $s = 36$, the values of $F(s)$ are -6 and $+6$. The function $F(s) = \sqrt{s}$ is not a single-valued function. On the other hand, the function $F(s) = s$ is a single-valued function.

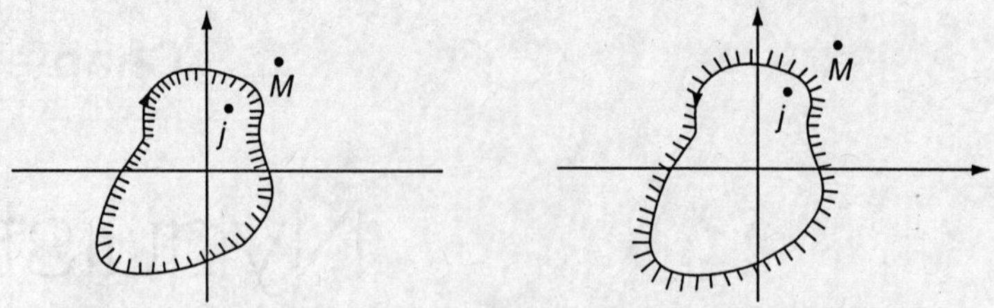

aymLpteJLpdLotlrudon aimLppteMLpdLotlrudon

Fig. 15.2 Definition of enclosed

15.2 Nyquist Analysis

15.2.1 Mapping Theorem and the Principle of Argument

Let us consider a function as follows:

$$G(s) = \frac{(s-a)(s-b)\cdots}{(s-a_1)(s-b_1)\cdots}$$

Let us take a path of s in the s-plane for the following cases:

(1) No poles or zeros of $G(s)$ are covered.
(2) Only zero is covered.
(3) Only pole is covered.
(4) One pole and one zero are covered.
(5) Two, three etc. zeros are covered.
(6) Two, three etc. poles are covered.
(7) Entire s-plane.

Let us discuss the above cases one by one.

Case 1: No poles and zeros covered
For the right-hand side of the travel direction, the region covered by *ABCDEF* in s-plane shown in Fig. 15.3(a) and the corresponding region may be *ABCDEF* in the $G(j\omega)$ plane shown in Fig. 15.3(b). The origin covered by *ABCDEF* is important rather than its shape. In the present case, origin is not covered by *ABCDEF* in $G(j\omega)$ plane. Here $Z = P = 0$. $\therefore N = Z - P = 0$.

Case 2: One zero only is covered
Figure 15.4(a) shows that one zero is covered by *ABCDEF* in the clockwise direction in s-plane and the origin is covered by the *ABCDEF* in the clockwise direction (CW) as shown Fig. 15.4(b) because $N = Z - P = 1 - 0 = 1$.

Case 3: One pole only is covered
Figure 15.5(a) shows that one pole is covered by *ABCDEF* in the clockwise direction in s-plane and the origin in covered by the *ABCDEF* once in the $G(j\omega)$ plane in the counter clockwise (CCW) direction, as shown in Fig. 15.5(b) because $N = Z - P = 0 - 1 = -1$.

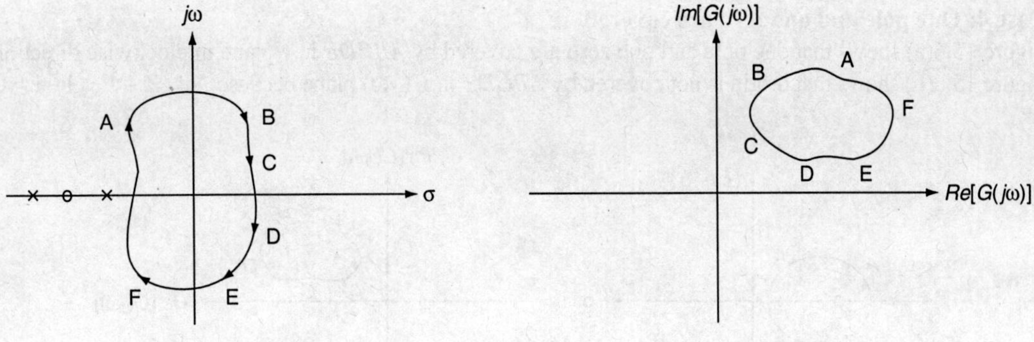

(a) s-plane (b) G(jω) plane

Fig. 15.3 Case 1

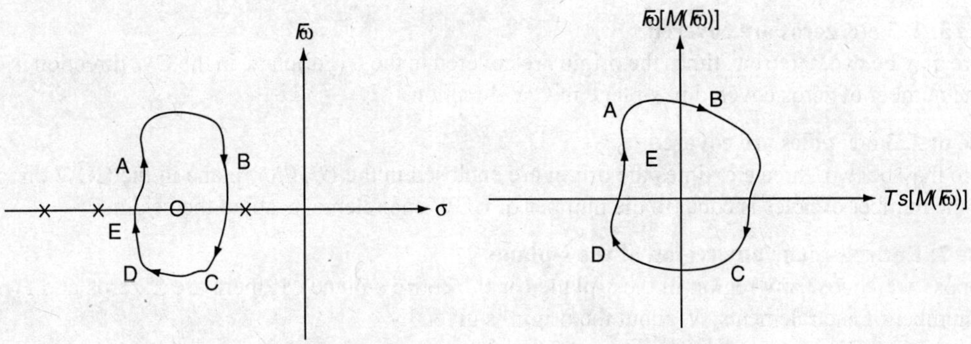

(a) jσ-plane (b) jM plane

Fig. 15.4 Case 2

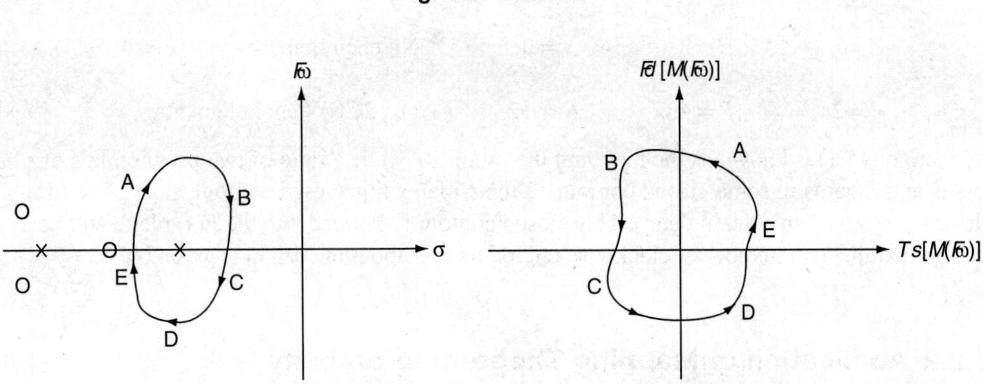

(a) jl-plane (b) jM plane

Fig. 15.5 Case 3

Case 4: One pole and one zero are covered

Figure 15.6(a) shows that one pole and one zero are covered by $ABCDE$ in s-plane in clockwise direction. Figure 15.6(b) shows that origin is not covered by $ABCDE$ in $G(j\omega)$ plane because $N = Z - P = 1 - 1 = 0$.

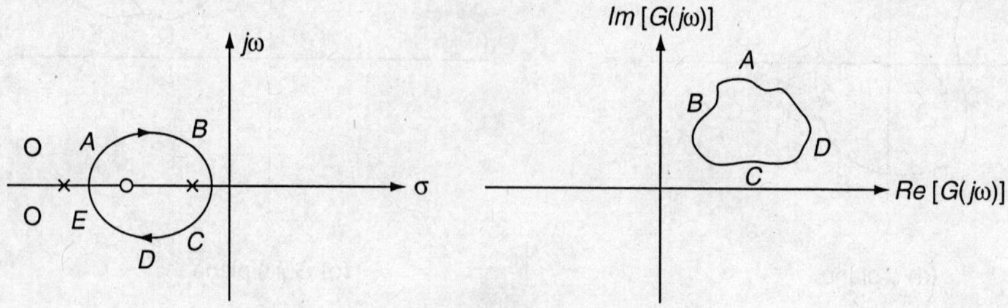

Fig. 15.6 Case 4

Case 5: 2, 3 etc. zeros are covered

There may be two, three etc. times the origin are covered in the $G(j\omega)$plane in the CW direction is equal to the number of zeros covered in s-plane in CW direction.

Case 6: 2, 3 etc. poles are covered

There may be two, three etc. times the origin are enclosed in the $GH(j\omega)$ plane in the CCW direction, i.e., the number of poles is equal to the number of CCW encirclements about the origin.

Case 7: Entire s-plane/any region of the s-plane

Suppose we choose any region in the s-plane (or the entire s-plane). If there are Z zeros and P poles, the number of encirclements, N, about the origin is given by

$$N = Z - P \tag{15.1}$$

For	$Z = 4,$	$P = 2,$	$N = 2,$	2 CW encirclements
	$Z = 2,$	$P = 2,$	$N = 0,$	No encirclements
	$Z = 2,$	$P = 4,$	$N = -2$	i.e., 2CCW encirclements

Equation (15.1) is known as the mapping theorem. If $F(s)$ be a ratio of two polynomials of s having P poles and Z zeros in some closed contour in the s-plane, this closed contour must pass through any pole or zero (but it can contain them). This closed contour if mapped into the $F(s)$ plane will be a closed curve so that the total number of clockwise encirclements about the origin is given by

$$N = Z - P$$

15.2.2 Application of Mapping Theorem to Stability

In polar plot, ω is varied from 0 to ∞ and in the mapping theorem discussion, we made s to vary in some closed contour. We can apply the above concepts to determine the stability.

$$G(s)H(s) = \frac{N(s)}{D(s)} = \frac{(s-a)(s-b)}{(s-a_1)(s-b_1)}$$

The poles and zeros of the open-loop transfer function are easy to locate. Now the closed-loop transfer function (CLTF) is given by

$$\frac{C(s)}{R(s)} = \frac{G(s)}{1+G(s)H(s)}$$

The poles of the CLTF must not lie in the RHP from stability considerations. The closed poles are obtained from

$$1+G(s)H(s)=0$$

Let $$q(s)=1+G(s)H(s)=\frac{N_1(s)}{D_1(s)} \tag{15.2}$$

The poles of the closed-loop transfer function are given by $q(s)=0$

i.e., $$\frac{N_1(s)}{D_1(s)}=0 \tag{15.3}$$

i.e., $$N_1(s)=0$$

i.e., the zeros of $q(s)$ are poles of a CLTF.

Therefore, the zeros of $q(s)$ [i.e., the zeros of $1+G(s)H(s)$] should not lie in the RHP from stability point of view.

For a given $q(s)$, let us choose a closed contour such that it covers the entire RHP shown in Fig. 15.7. Plotting the Nyquist plot of $q(s)$ for such a closed contour, we can conclude whether there is zero in the RHP by examining the number of encirclements about the origin using the mapping theorem.

For a given $q(s)$, let us choose the RHP as path $ABCDA$ with limit $R \to \infty$, so that we can map the RHP.

We can decide the stability after getting the encirclement(s) of the polar plot about the origin.

If the contour of $q(s)$ encloses the origin once in the clockwise direction, $Z-P=1$. If $q(s)$ has three poles in the RHP, from $Z-P=1$ we get $Z=4$ or four zeros of $q(s)$ are in the RHP. Here, we can conclude the following:

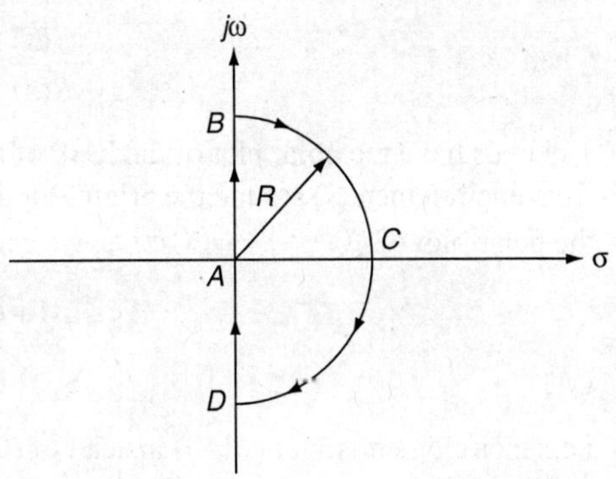

Fig. 15.7 Path covering the RHP of the s-plane ($R \to \infty$)

- Take $G(s)H(s)$ as an open-loop transfer function.
- Define poles of CLTF.
- Define a function $q(s)$ whose zeros are the poles of the CLTF.
- Map the RHP of the s-plane in $q(s)$.
- Examine for the number of zeros-poles by noting encirclements about the origin.
- If there are any zeros in RHP, we have concluded that the system is unstable.

Now, for $N = Z - P$ (Z must be zero for stability), there are the following values of N which are possible.

$$N = 0 \quad \text{(no encirclement)}$$
$$> 0 \quad \text{(clockwise encirclement, i.e., } Z > P\text{)}$$
$$< 0 \quad \text{(CCW encirclement } Z < P\text{)}$$

Case 1: If $N = 0$, either $Z = 0$ and $P = 0$ or $Z = P$.
If $N = 0$, P must be 0 for stability. This is the first condition.

Case 2: If N > 0, either $P = 0$ and $N = Z$ or $Z > P$.
In both cases, the number of zeros is more than poles. It indicates that there are zeros in the RHP given by $Z = N + P$. This is an unstable condition.

Case 3: If $N < 0$, we can conclude that either $Z = 0$ so $N = -P$ or $N = Z - P$ with $P > Z$. If $N < 0$, for stability, the number of encirclements must be equal to the number of poles. This is another condition.

A given system will be stable if

- There are no clockwise encirclements about the origin and the number of poles in the RHP is zero.
- There is anticlockwise encirclement about the origin with the number of poles equal to the number of encirclements.

15.3 Polar Plots of *G*(*s*)*H*(*s*) and Stability

From Eq. (15.2), we have

$$q(s) = 1 + G(s)H(s)$$

when

$$\frac{C(s)}{R(s)} = \frac{G(s)}{1 + G(s)H(s)}$$

Let us have the polar plot of the RHP of the *s*-plane for $G(s)H(s)$. For any zero(s) in the RHP we get from encirclement(s) around the origin. Modifying the function to $1 + G(s)H(s)$ and instead of redrawing the polar plot of $q(s) = 1 + G(s)H(s)$, we can examine the encirclement from the polar plot of

$$1 + G(s)H(s) = 0$$

i.e.,

$$G(s)H(s) = -1 \tag{15.4}$$

i.e., encirclement about (−1, 0) instead of (0, 0).

For a given $G(s)H(s)$, the stability of an open-loop transfer function system can be carried out by (i) choosing a Nyquist path that maps RHP and (ii) plotting corresponding polar plot of $G(s)H(s)$. After getting the encirclements about (−1, 0) say N', in order to distinguish it from N which is the encirclement about (0, 0) where $N' = Z - P$ where Z is zeros of $1 + G(s)H(s)$ and P is poles $G(s)H(s)$ as well as of $1 + G(s)H(s)$.

15.4 Nyquist Path

This path should not pass through a pole/zero, but it may contain poles and zeros. The path must map the RHP. Figure 15.8 shows a general Nyquist path based on these constraints.

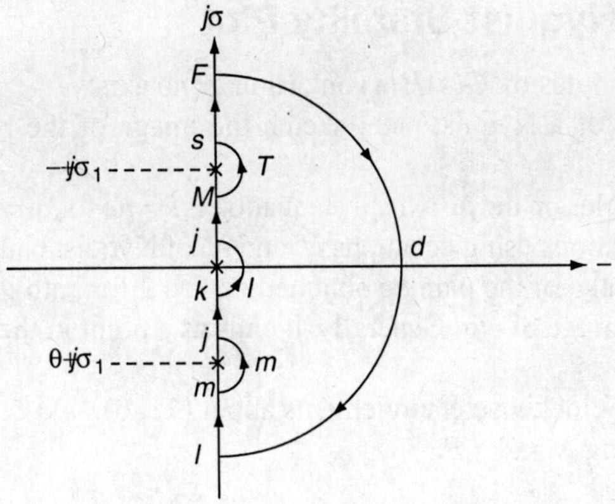

Fig. 15.8 Nyquist path

Table 15.1 shows the description of Fig. 15.8.

Table 15.1 Description of Fig. 15.8

Path	Equation	Equation No.	Limits
ab	$s = j\omega$	N_1	$0 < \omega < \omega_o$
bcd	$s = \underset{\rho \to 0}{\text{Lt}}\,(\rho e^{j\theta} + j\omega_o)$	N_2	$-90° < \theta < 90°$
de	$s = j\omega$	N_3	$\omega_0 < \omega < \infty$
efg	$s = \underset{R \to \infty}{\text{Lt}}\,(Re^{j\theta})$	N_4	$+\,90° > \theta > 90°$
gh	$s = j\omega$	N_5	$-\infty < \omega < -\omega_o$
hij	$\underset{\rho \to 0}{\text{Lt}}\,(\rho e^{j\theta} - j\omega_o)$	N_6	$-90° < \theta < 90°$
jk	$s = j\omega$	N_7	$-\omega_o < \omega < 0$
kla	$\underset{\rho \to 0}{\text{Lt}}\,(\rho e^{j\theta})$	N_8	$-90° < \theta < 90°$

15.4.1 Nyquist Stability Criterion

A closed-loop control system having an open-loop transfer function $G(s)H(s)$ will be stable if and only if

$$N = -P_0$$

where

N = the number of counterclockwise encirclements about (−1, 0) point in the $G(s)H(s)$ plane
P_0 = the number of poles $G(s)H(s)$ in the RHP.

If $N > 0$, the system is unstable with the number of zeros of $1 + G(s)H(s)$ in RHP given by $Z = N + P_0$

If $N \le 0$, i.e., (−1, 0) is not enclosed, the system is stable only for $N = 0$ and $P = 0$ or $N = -P_0$

15.5 Drawing of a Nyquist Stability Plot

Step 1: Find the number of poles of $G(s)H(s)$ on origin or $j\omega$ axis.

Step 2: Using the equation of a Nyquist path, sketch the image of the path *ae* depending on step 1 above.
For the number of poles on the $j\omega$ axis [the equation is $s = j\omega$ for $\omega = 0$ to ∞], sketch the polar plot.
For poles, take deviations using equations N_1 and N_8 of Nyquist path, as shown in Table 15.1.

Step 3: Draw the mirror image of the path *ae* obtained in step 2 for path *gh*.

Step 4: Using N_4, find the image of *efg*. Generally, it ends as a point in the origin.

Step 5: Connect all the curves.

Step 6: Note the number of clockwise encirclements about (−1, 0).

Step 7: Decide stability using $N = Z - P$.

Example 15.1 For $G(s)H(s) = 1/(s+2)$, draw the Nyquist plot and decide stability.

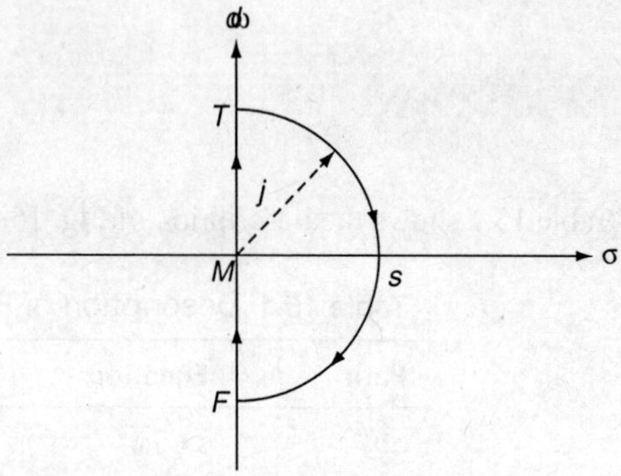

Fig. E15.1

Solution

Step 1: Pole of $G(s)H(s)$ is at $s = -2$ and there are no poles either on the $j\omega$ axis or on the origin

Step 2: The Nyquist path is shown in Fig. E15.1.
For path *ae* $s = j\omega$ where $0 < \omega < \infty$.

$$G(j\omega)H(j\omega) = \frac{1}{j\omega + 2} = \frac{1}{\sqrt{4+\omega^2}} \angle -\tan^{-1}\left(\frac{\omega}{2}\right)$$

$$\therefore \quad |G(j\omega)H(j\omega)| = \frac{1}{\sqrt{4+\omega^2}} \text{ and } \angle G(j\omega)H(j\omega) = -\tan^{-1}\left(\frac{\omega}{2}\right)$$

At $\omega = 0$, $|G(j\omega)H(j\omega)| = 0.5$ and $\angle G(j\omega)H(j\omega) = 0°$

At $\omega = \infty$, $|G(j\omega)H(j\omega)| = 0$ and $\angle G(j\omega)H(j\omega) = -90°$

The polar plot of path *ae* is shown in Fig. E15.1(a).

Step 3: The dotted mirror path of *ae* is the polar plot of *ga*.

Step 4: For path *efg*, $s = Re^{j\theta}$ where $R \to \infty$ and $+90° < \theta < -90°$.

$$G(s)H(s)\Big|_{efg} = \frac{1}{\underset{R\to\infty}{\text{Lt}}\,(Re^{j\theta} + 1)} = 0$$

Therefore, the infinite semicircle *efg* maps onto a point.

Step 5: Connected

Step 6: There are no encirclements about (−1, 0).

Step 7: Since $P = 0$ (number of poles in the RHP) = poles of $G(s)H(s)$, here, the number of zeros of $1 + G(s)H(s)$ in the RHP is zero.

$\therefore$ $Z = 0$ and hence system is stable.

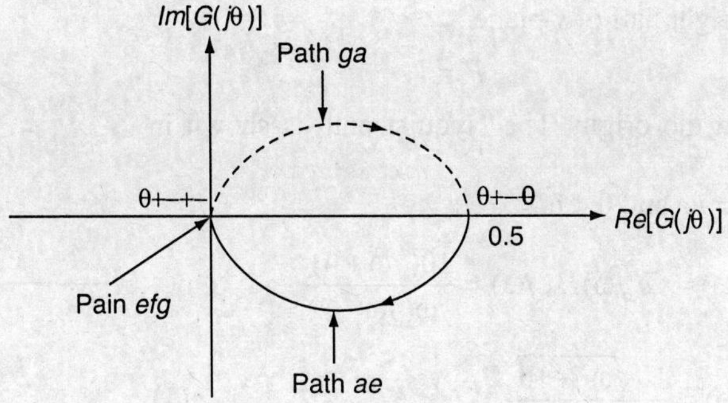

Fig. E15.1(a)

15.6 Behaviour of Right-Half Pole

Let us have $\frac{1}{s-4}$ i.e., a pole in the right half of s-plane which is a factor of $G(s)\,H(s)$.

$$\therefore \qquad G(j\omega)\,H(j\omega) = \frac{1}{j\omega - 4} = \frac{1}{-4 + j\omega}$$

For $\omega \to \infty$, $-4 + j\omega$ will approach to the point P which has been shown in Fig. 15.9. $\theta \approx 90°$ in this limiting case and hence this factor will contribute +90° in the denominator. Therefore, we have

$$\angle G(j\omega)\,H(j\omega)\Big|_{\omega + \infty} = \frac{0°}{+90°} = -90°$$

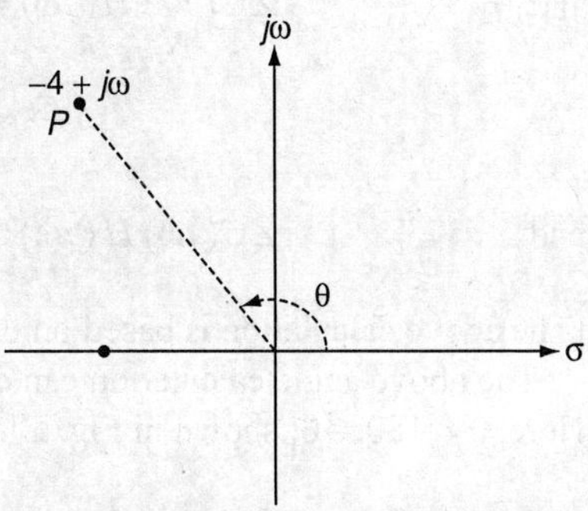

Fig. 15.9 Limiting cases of $-4 + j\omega$

When $\omega \to +0$, $-4 + j\omega$ will approach to a point $-4 + j0$ on the negative real axis having angle +180°. Since we are travelling in the anticlockwise direction, this is considered as +180°. For $\omega \to -0$, $-4 + j\omega$ will be at position $-4 - j0$, i.e., on the negative real axis which contributes angle +180° again.

For $\omega \to +0$ or, -0, $\angle G(j\omega)\,H(j\omega) = \frac{0°}{180°} = -180°$

If this factor is pole, the angles will be in the denominator and the effective contribution will be –90° and –180°. If this factor is zero, effective contribution of this angle will be +90° and +180°.

Example 15.2 Sketch the Nyquist plot for a system having $G(s)H(s) = \frac{10(s+4)}{s(s-2)}$. In addition, comment on the closed-loop stability.

Solution Here, $G(s)H(s) = \frac{10(s+4)}{s(s-2)}$.

Step 1: One pole is in right half of s-plane.

$$\therefore \qquad P = 1.$$

Step 2: Only pole is at the origin. The Nyquist path is shown in Fig. E15.2.

Step 3: For path ae: $s = j\omega$ and $0 < \omega < \infty$.

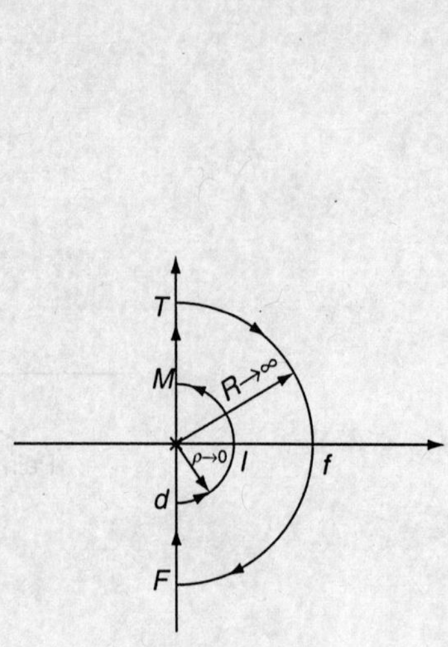

Fig. E15.2

Here, $$M(j\omega) = G(j\omega)H(j\omega) = \frac{10(j\omega+4)}{j\omega(j\omega-2)}$$

$$\therefore \qquad |G(j\omega)H(j\omega)| = \frac{10\sqrt{\omega^2+16}}{\omega\sqrt{\omega^2+4}} \tag{1}$$

and $$\angle G(j\omega)\,H(j\omega) = \tan^{-1}\frac{\omega}{4} - 90° - \theta \tag{2}$$

where θ is the angle contributed effectively by the pole $\dfrac{1}{j\omega-2}$.

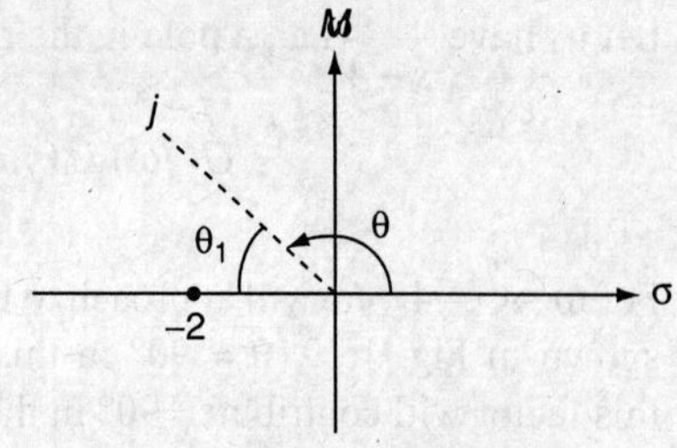

Fig. E15.2(a)

For $$\omega \to +0, \quad |G(j\omega)H(j\omega)| = \infty$$

and $$\angle G(j\omega)\,H(j\omega) = 0° + 0° - 90° - 180° = -270°$$

For $$\omega \to \infty, \quad |G(j\omega)H(j\omega)| = 0.$$

and $$\angle G(j\omega)\,H(j\omega) = 0° + 90° - 90° - 90° = -90°.$$

[The above calculation is based on the discussion in Art. 15.6.]

The above angle calculation can done using Eq. (3) as shown below. Here, $\theta = 180° - \theta_1$ shown in Fig. E15.2(a).

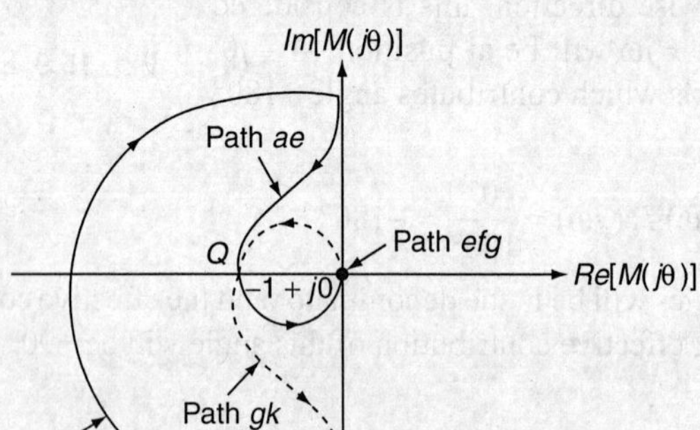

Fig. E15.2(b)

Here, $\theta_1 = \tan^{-1}\frac{\omega}{2}$.

$$\angle G(j\omega)H(j\omega) = \tan^{-1}\frac{\omega}{4} - 90° - (180° - \theta_1)$$

$$\therefore \quad = \tan^{-1}\frac{\omega}{4} - 270° + \theta_1 \qquad (3)$$

$$= \tan^{-1}\frac{\omega}{4} - 270° + \tan^{-1}\frac{\omega}{2}$$

For $\quad \omega \to +0, \quad |G(j\omega)H(j\omega)| = \infty.$

and $\quad \angle G(j\omega)H(j\omega) = 0° - 270° + 0° = -270°$

For $\quad \omega \to \infty, \quad |G(j\omega)H(j\omega)| = 0$

and $\quad \angle G(j\omega)H(j\omega) = 90° - 270° + 90° = -90°$

The polar plot of path *ae* is shown in Fig. E15.2(b)

Step 4: The polar plot of path *gk* is shown in Fig. 15.2(b) by dotted line, which is the mirror image of the path *ae*.

Step 5: For path *efg*.

Let us put $s = \text{R}e^{j\theta}$ in the expression of $G(s)\,H(s)$ where $90° > \theta > -90°$ and $\text{R} \to \infty$.

$$\therefore \quad G(s)\,H(s)\Big|_{efg} = \underset{R\to\infty}{\text{Lt}} \frac{10(\text{R}e^{j\theta}+4)}{\text{R}e^{j\theta}(\text{R}e^{j\theta}-2)}$$

$$= \underset{R\to\infty}{\text{Lt}} \frac{10}{\text{R}e^{j\theta}}$$

$$= \frac{10}{\infty e^{j\theta}} = 0e^{-j\theta}$$

Therefore, the infinite semicircle maps onto a point.

Step 6: For path *kla*:

Let us put $s = \rho e^{j\theta}$ in the expression of $G(s)\,H(s)$ where $-90° < \theta < 90°$ and $\rho \to 0$.

$$\therefore \quad G(s)H(s)\Big|_{kla} = \underset{\rho\to 0}{\text{Lt}} \frac{10(\rho e^{j\theta}+4)}{\rho e^{j\theta}(\rho e^{j\theta}-2)}$$

$$= -\infty e^{-j\theta}$$

$$= \infty e^{-j(180°+\theta)} \qquad [\because e^{-j(180°+\theta)} = e^{-j\theta}]$$

Since θ varies from $-90°$ to $+90°$ through $0°$, the angle of $G(j\omega)\,H(j\omega)$ for this path varies from $-90°$ to $-270°$ through $-180°$.

The above calculation for angle can also be obtained as follows:

For $\omega \to +0, \quad \angle G(j\omega)H(j\omega) = 0° + 0° - 90° - 180° = -270°$

For $\omega \to -0, \quad \angle G(j\omega)H(j\omega) = 0° + 0° - (-90°) - 180° = -90°.$

$\omega \to +0$, is in the upward direction of second quadrant and $\omega \to -0$ is in the downward direction of third quadrant.

Step 7: Connected.

Step 8: Q point

$$G(j\omega)H(j\omega) = \frac{10(j\omega+4)}{(j\omega)(j\omega-2)}$$

$$= \frac{10(j\omega+4)(-j\omega)(j\omega+2)}{(j\omega)(j\omega-2)(-j\omega)(j\omega+2)}$$

$$= \frac{10(j\omega+4)(-j\omega)(j\omega+2)}{\omega^2(-\omega^2-4)}$$

$$= \frac{10(j\omega+4)(j\omega)(j\omega+2)}{\omega^2(4+\omega^2)}$$

$$= \frac{10(-\omega^2+6j\omega+8)(j\omega)}{\omega^2(4+\omega^2)}$$

$$= \frac{10j\omega(-\omega^2+6j\omega+8)}{\omega^2(4+\omega^2)}$$

$$= \frac{-60\omega^2}{\omega^2(4+\omega^2)} + j\frac{10\omega(8-\omega^2)}{\omega^2(4+\omega^2)} \tag{4}$$

Equating imaginary part of Eq. (4) to zero, we have

$$\frac{10\omega(8-\omega^2)}{\omega^2(4+\omega^2)} = 0$$

i.e., $$8 - \omega^2 = 0$$

i.e., $$\omega = 2\sqrt{2}$$

∴ $$\omega_{pc} = \omega = 2\sqrt{2}$$

Substituting $\omega = \omega_{pc} = 2\sqrt{2}$ in real part of Eq. (4), we have

$$Q = \frac{-60\times 8}{8\times(4+8)} = -5 + j0.$$

Step 9: From Fig. E15.2(a), the net circlement of $-1 + j0$ is one in CCW direction. Therefore, $N = -1$.

Step 10: Since, $N = Z - P$

$\therefore$ $$-1 = Z - 1$$

i.e., $$Z = 0.$$

Since $Z = 0$, this matches with the stability criterion. Therefore, the given system is stable in nature, which is the rare example of open-loop unstable system that behaves as stable closed-loop system.

15.7 Effects of Addition of Poles and Zeros to *G*(*s*)*H*(*s*) on the Shape of Nyquist Plot

The shape of Nyquist plot due to addition of poles and zero to $G(s)H(s)$ is discussed below.

15.7.1 Addition of Poles at s = 0

The addition of a pole at $s = 0$ is equivalent to divide $G(s)H(s)$ by $j\omega$ and hence the phase angle of $G(s)\, H(s)$ is reduced by 90° at both zero and infinite frequencies. This suggests that the addition of a pole at the origin to a loop transfer function gives the rotation of the Nyquist plot at zero and infinite frequencies by a further angle of –90°. The magnitude of $G(j\omega)\, H(j\omega)$ at $\omega = 0$ becomes infinite. Therefore, the addition of a pole at $s = 0$ will affect the stability of the closed system adversely. Therefore, if a system has more than one pole at $s = 0$ in loop transfer function, it becomes likely to be unstable and difficult to stabilize.

15.7.2 Addition of Finite Non-Zero Poles

If a non-zero pole is added to the loop transfer function, it shifts the phase of the Nyquist plot by –90° at $\omega = \infty$. This results in further rotation of the polar plot through an angle of –90° as $\omega \to \infty$ affecting the stability adversely.

15.7.3 Addition of a Zero

If a zero is added to the loop transfer function, it rotates the high-frequency portion of the Nyquist plot by an angle of 90° in CCW direction without effecting the value at $\omega = 0$. This reduces the effect of overshoot and hence the general effect of stabilization.

15.8 Advantages of Nyquist Plot

(i) The information regarding absolute stability as provided by Routh's criterion is also informed by Nyquist plot.

(ii) This is useful to obtain the stability from open-loop transfer function for a closed-loop system.

(iii) This plot also gives the relative stability of the system.

(iv) The values of GM and PM are also possible to calculate from this plot.

(v) This plot also gives the reality and the manner in which system should be compensated to get desired response.

(vi) It is also possible to obtain information regarding frequency response.

(vii) This plot is very useful to analyze conditionally stable systems.

15.9 MATLAB for Nyquist Plot

The following programs illustrates the required MATLAB command for drawing nyquist plot.

Program 1 Develop the Nyquist diagram of the following unity feedback system $G(s) = \dfrac{1}{s^2 + 0.8s + 1}$

Solution

```
>>num = [0 0 1]
>>den = [1 0.8 1];
>> nyquist (num, den);
```

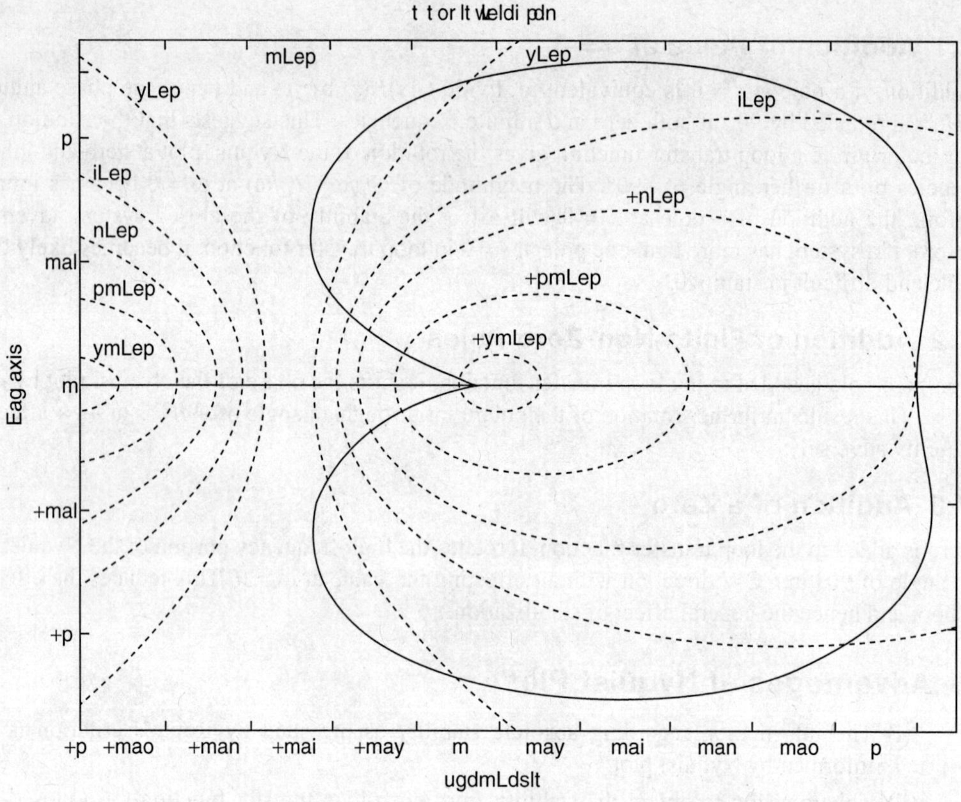

Fig. P1

Program 2 Develop the Nyquist diagram of the unity feedback system $G(s) = \dfrac{(s+2)}{(s+1)(s-1)}$

Solution

```
>>num = [1 2];
>>den = [1 0 −1];
>> nyquist (num, den);
```

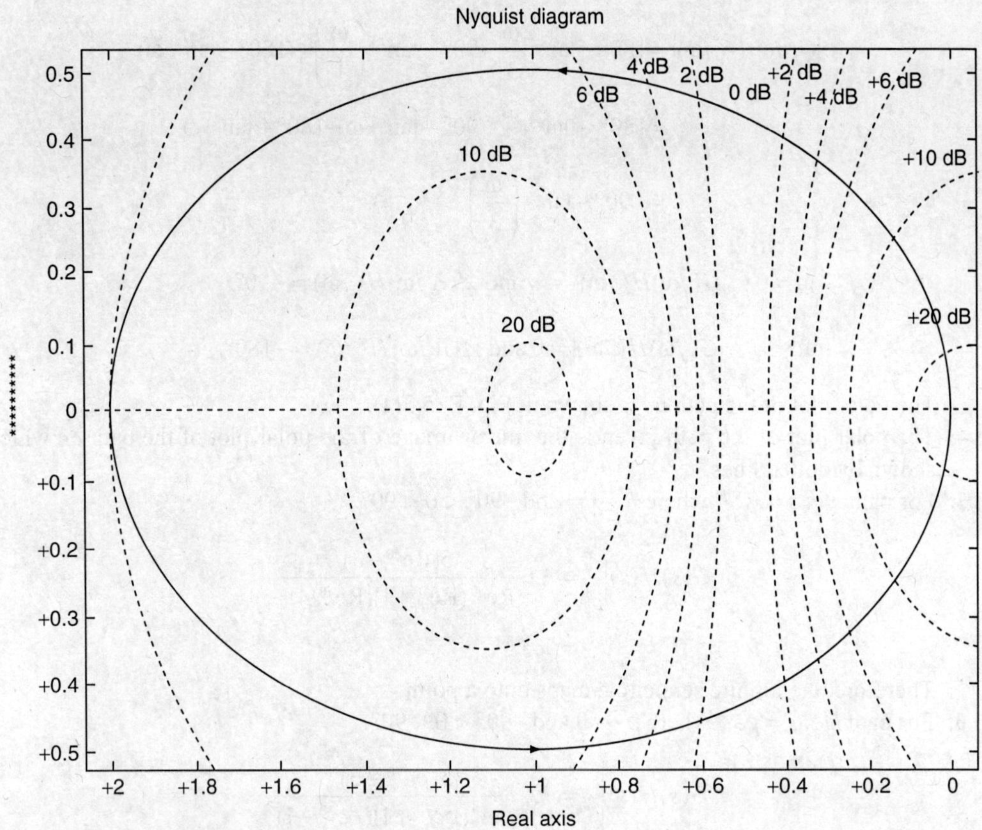

Fig. P2

Example 15.3 For $G(s)\,H(s) = \dfrac{5(s-3)}{s(s+1)(s-1)}$, draw the Nyquist plot and decide the stability.

Solution

Step 1: Poles of $G(s)\,H(s)$ are at $s = 0, -1$ and 1. There is a pole at origin and no pole on the $j\omega$-axis.

Step 2: The Nyquist path is shown in Fig. E15.3.

Step 3: For path ae, $s = j\omega$ and $0 < \omega < \infty$.

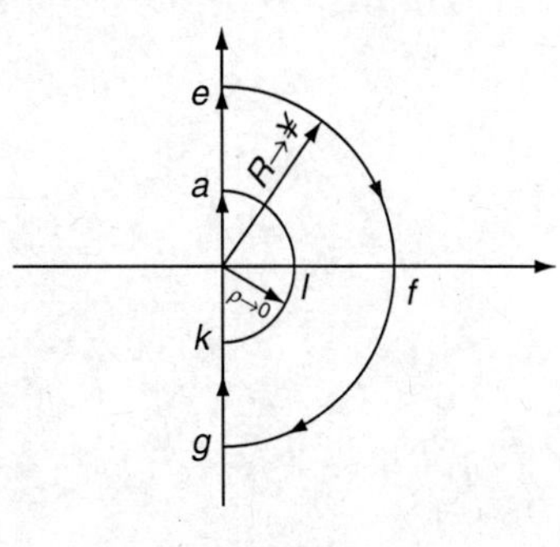

Fig. E15.3

Here, $$M(j\omega) = G(j\omega)\,H(j\omega) = \frac{5(j\omega - 3)}{j\omega(j\omega+1)(j\omega-1)}.$$

$$\therefore \qquad |G(j\omega)H(j\omega)| = \frac{5\sqrt{\omega^2+9}}{\omega\sqrt{\omega^2+1}\sqrt{\omega^2+1}}$$

$$\therefore \quad |G(j\omega)H(j\omega)| = 180° - \tan^{-1}\frac{\omega}{3} - 90° - \tan^{-1}\left(\frac{\omega}{1}\right) - (180° - \tan^{-1}\omega)$$

$$= 180° - \tan^{-1}\frac{\omega}{3} - 90° - \tan^{-1}\omega - 180° + \tan^{-1}\omega$$

$$= -90° - \tan^{-1}\left(\frac{\omega}{3}\right)$$

For $\omega = 0, \quad |G(j\omega)H(j\omega)| = \infty$ and $\angle G(j\omega)H(j\omega) = -90°$

For $\omega \pm \infty, \quad |G(j\omega)H(j\omega)| = 0$ and $\angle G(j\omega)H(j\omega) = -180°$.

The polar plot of the path *ae* is shown in Fig. E15.3(a).

Step 4: The polar plot of the path given is the mirror image of the polar plot of the path *ae* which is shown by dotted lines.

Step 5: For path *efg*: $s = \text{Re}^{j\theta}$ where $R \to \infty$ and $-90° < \theta < 90°$.

$$\therefore \quad G(s)H(s)\Big|_{efg} = \underset{R\to\infty}{\text{Lt}} \frac{5(\text{Re}^{j\theta} - 3)}{\text{Re}^{j\theta}(\text{Re}^{j\theta}+1)(\text{Re}^{j\theta}-1)}$$

$$= 0e^{-j2\theta}$$

Therefore, the infinite semicircle maps onto a point.

Step 6: For path *kla*: $s = \rho e^{j\theta}$ where $\rho \to 0$ and $-90° < \theta < 90°$.

$$\therefore \quad G(s)H(s)\Big|_{kla} = \underset{\rho\to 0}{\text{Lt}} \frac{5(\rho e^{j\theta} - 3)}{\rho e^{j\theta}(\rho e^{j\theta}+1)(\rho e^{j\theta}-1)}$$

$$= +\infty e^{-j\theta}$$

since θ varies from $-90°$ to $+90°$ through $0°$, the angle of $G(j\omega)\,H(j\omega)$ for this path varies from $+90°$ to $-90°$ through $0°$.

Step 7: Connected. Figure E15.3(a) shows the Nyquist plot.

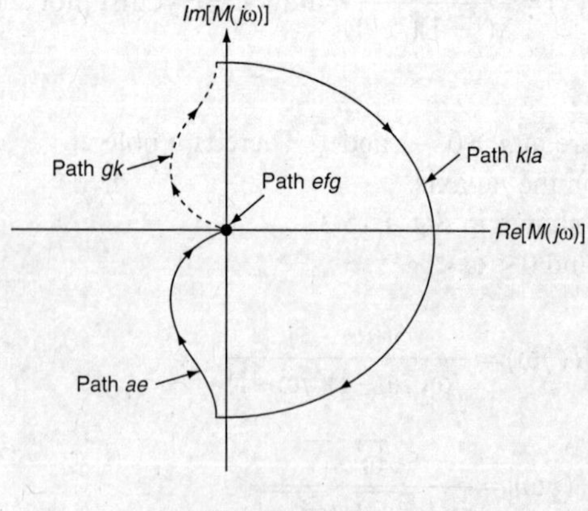

Fig. E15.3(a)

Since the Nyquist plot does not encircle $-1 + j0$, $N = 0$.
Again, $N = Z - P$. Here, $N = 0$ and $P = 1$.

$\therefore \quad Z = N + P = 0 + 1 = 1.$

The system is stable.

Example 15.4 For $G(s)\,H(s) = \dfrac{50}{s(s+4)(s-1)}$ draw Nyquist plot. Comment on the stability.

Solution

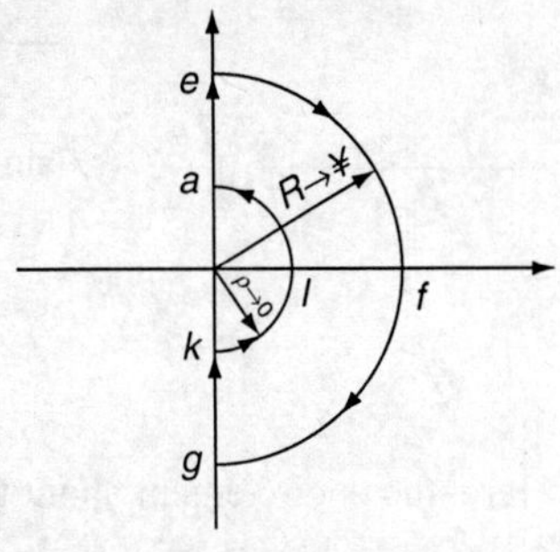

Fig. E15.4

Step 1: Poles are at $s = 0$, -4 and $+1$. There is one pole in the RHP.
$\therefore \quad P = 1$. There is no other pole on the $j\omega$-axis.
Step 2: Figure E15.4 shows the Nyquist path.
Step 3: For the path ae: $s = j\omega$ and $0 < \omega < \infty$.

$$\therefore \quad G(j\omega)H(j\omega) = \frac{50}{j\omega(j\omega+4)(j\omega-1)}$$

$$\therefore \quad |G(j\omega)H(j\omega)| = \frac{50}{\omega\sqrt{\omega^2+16}\sqrt{\omega^2+1}}$$

$$\therefore \quad \angle G(j\omega)H(j\omega) = 0^\circ - 90^\circ - \tan^{-1}\frac{\omega}{4} - \left[180^\circ - \tan^{-1}\left(\frac{\omega}{1}\right)\right]$$

$$= -270^\circ - \tan^{-1}\frac{\omega}{4} + \tan^{-1}\omega.$$

For $\quad \omega \to 0, \quad |G(j\omega)H(j\omega)| = \infty$ and $\angle G(j\omega)H(j\omega) = -270^\circ$

For $\quad \omega \to \infty, \quad |G(j\omega)H(j\omega)| = 0$ and $\angle G(j\omega)H(j\omega) = -270^\circ$.

The polar plot of the path ae is shown in Fig. E15.4(a)

Step 4: The polar plot of the path gk is the mirror image of the polar plot of the path ae, which has been shown by the dotted lines in Fig. E15.4(a).

Step 5: For path efg: $s = Re^{j\theta}$ where $R \to \infty$ and $-90^\circ < \theta < 90^\circ$.

$$\therefore \quad G(s)H(s)\Big|_{efg} = \underset{R\to\infty}{\text{Lt}} \frac{50}{\text{Re}^{j\theta}(\text{Re}^{j\theta}+4)(\text{Re}^{j\theta}-1)}$$

$$= 0e^{j3\theta} = 0\,\angle -3\theta$$

Therefore, the infinite semicircle maps onto a point as shown in Fig. E15.4(a).

Step 6: For path kla: $s = \rho e^{j\theta}$ where $\rho \to 0$ and $-90^\circ < \theta < 90^\circ$.

$$\therefore \quad G(s)H(s)\Big|_{kla} = \underset{\rho\to 0}{\text{Lt}} \frac{5}{\rho e^{j\theta}(\rho e^{j\theta}+4)(\rho e^{j\theta}-1)}$$

$$= -\infty e^{-j\theta} = \infty e^{-j(180^\circ+\theta)}$$

Since the angle of θ varies from -90° to $+90^\circ$ through 0°, the angle of $G(j\omega)\,H(j\omega)$ varies from -90° to -270° through -180°.

Step 7: Connected. Figure E15.4(a) shows the Nyquist plot.

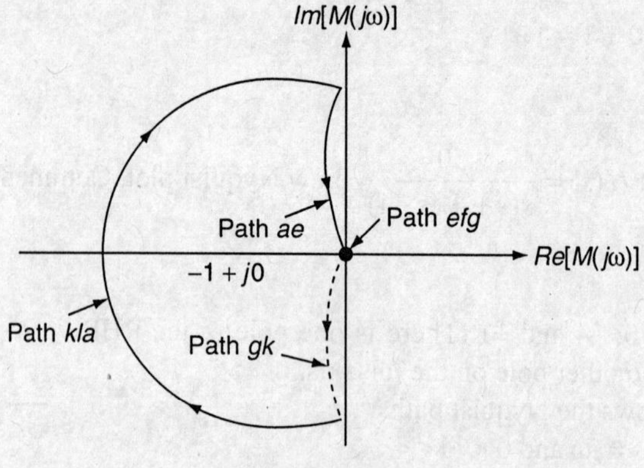

Fig. E15.4(a)

Here, the encirclement about the origin is 1 in CW direction. $\therefore N = 1$.

Since, $$N = Z - P,$$

$$Z = N + P = 1 + 1 = 2.$$

The system is unstable.

Example 15.5 For the system $G(s)H(s) = \dfrac{10K(s+0.6)}{s^2(s+2)(s+10)}$, draw the Nyquist plot and using Nyquist stability criterion determine stability with $K = 1$, $K = 10$ and $K = 100$.

Solution

Step 1: The poles are at $s = 0, 0, -2$ and -10. There is no pole in the RHP.

Step 2: The Nyquist path is shown in Fig. E15.5.

Step 3: For path *ae*: $s = j\omega$ where $0 < \omega < \infty$.

Fig. E15.5

$$\therefore \quad M(j\omega) = G(j\omega)H(j\omega) = \frac{10K(j\omega+0.6)}{(j\omega)^2(j\omega+2)(j\omega+10)}$$

$$\therefore \quad |G(j\omega)H(j\omega)| = \frac{10K\sqrt{\omega^2+0.36}}{\omega^2\sqrt{\omega^2+4}\sqrt{\omega^2+100}}.$$

and $$\angle G(j\omega)H(j\omega) = \tan^{-1}\frac{\omega}{0.6} - 180° - \tan^{-1}\frac{\omega}{2} - \tan^{-1}\frac{\omega}{10}$$

For $\omega = 0$, $|G(j\omega)H(j\omega)| = \infty$ and $\angle G(j\omega)H(j\omega) = -180°$

For $\omega = \infty$, $|G(j\omega)H(j\omega)| = 0$ and $\angle G(j\omega)H(j\omega) = -270°$

The polar plots of the path *ae* are shown in Fig. E15.5(a), Fig. E15.5(b) and Fig. E15.5(c)

Step 4: The polar plot of the path *gk* is the mirror image of the polar plot of the path *ae* shown by dotted lines in Fig. 15.5(a), Fig. E15.5(b) and Fig. E15.5(c)

Step 5: For path *efg*: $s = \text{Re}^{j\theta}$ where $R \to \infty$ and $-90° < \theta < 90°$.

$$\therefore \quad G(s)H(s)\Big|_{efg} = \underset{R\to\infty}{\text{Lt}} \frac{10K(\text{Re}^{j\theta}+0.6)}{(Re^{j\theta})^2(Re^{j\theta}+2)(Re^{j\theta}+10)}$$

$$= 0e^{-3j\theta} = 0\angle -3\theta$$

The infinite semicircle maps onto a point. This has been shown in Fig. E15.5(a); Fig. E15.5(b) and Fig. E15.5(c).

Step 6: For path *kla*: $s = \rho e^{j\theta}$ where $\rho \to 0$ and θ varies from $-90°$ to $+90°$ through $0°$.

$$\therefore \quad G(s)H(s)\Big|_{kla} = \underset{\rho\to 0}{\text{Lt}} \frac{10K(\rho e^{j\theta}+0.6)}{(\rho e^{j\theta})^2(\rho e^{j\theta}+2)(\rho e^{j\theta}+10)}$$

$$= \infty\, e^{-2j\theta}$$

$\therefore$ The angle of $G(j\omega)H(j\omega)$ varies from $180°$ to $-180°$ through $0°$.

Step 7: To find the intersection with negative real axis.

$$G(j\omega)H(j\omega) = \frac{10K(j\omega+0.6)}{(-\omega^2)(j\omega+2)(j\omega+10)}$$

$$= \frac{10K(j\omega+0.6)(-j\omega+2)(-j\omega+10)}{(-\omega^2)(\omega^2+4)(\omega^2+100)}$$

$$= \frac{10K(j\omega+0.6)(j\omega-2)(j\omega-10)}{(-\omega^2)(\omega^2+4)(\omega^2+100)}$$

$$= \frac{10K(j\omega+0.6)(-\omega^2-12j\omega+20)}{(-\omega^2)(\omega^2+4)(\omega^2+100)}$$

$$= \frac{10K(-j\omega^3+12\omega^2+20j\omega-0.6\omega^2-7.2j\omega+12)}{(-\omega^2)(\omega^2+4)(\omega^2+100)}$$

$$= \frac{10K(11.4\omega^2+12)+10jk\,\omega(12.8-\omega^2)}{(-\omega^2)(\omega^2+4)(\omega^2+100)}$$

Equating imaginary part to zero, we have

$$12.8-\omega^2 = 0 \quad \text{i.e.,} \quad \omega = \sqrt{12.8}$$

Substituting this value in real part of $G(j\omega)H(j\omega)$, we have the point Q.

$$Q = \frac{10K\,(11.4\omega^2+12)}{(-\omega^2)(\omega^2+4)(\omega^2+100)}$$

$$= \frac{10K(11.4\times 12.8+12)}{(-12.8)(12.8+4)(12.8+100)}$$

$$= -0.065K$$

Step 8: Let us obtain Nynquist plot.

Case (i): $K = 1$.

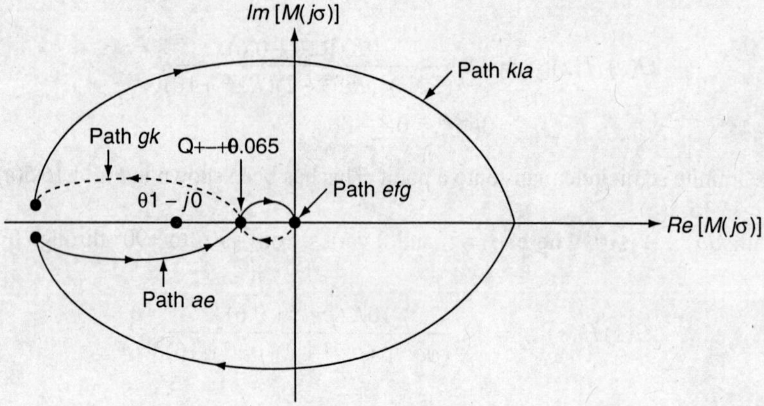

Fig. E15.5(a)

Here, $N = 0$. Again, $P = 0$. $\therefore\ Z = N + P = 0$. This satisfies the stability criterion. Hence, the system is stable for $K = 1$.

Case (ii): $K = 10$
$Q = -0.65$

Figure E15.5(b) shows the Nyquist plot.

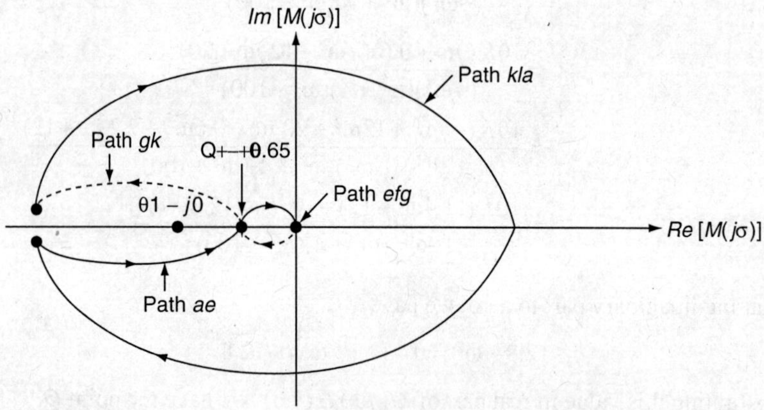

Fig. E15.5(b)

From this plot, $N = 0$.
Hence, the system in stable for $K = 10$.

Case (iii): $Q = -6.5$
Figure E15.5(c) shows the polar plot.

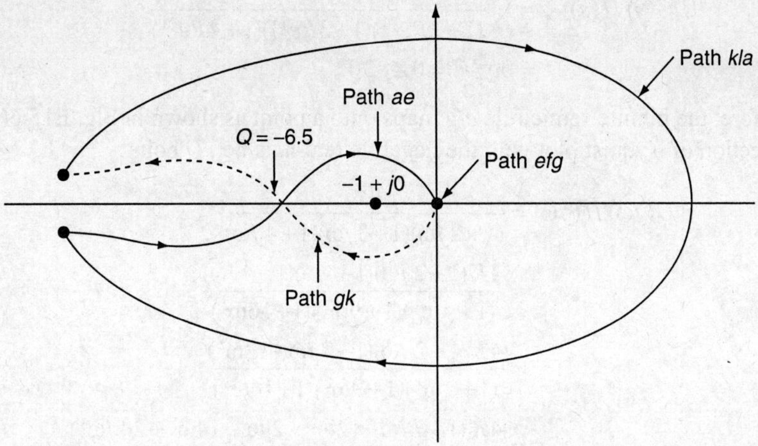

Fig. E15.5(c)

Here, $N = 2$. $P = 0$. Therefore, $Z = 2$.
The system is unstable.

Example 15.6 For a system $G(s)H(s) = \dfrac{4K}{(1+2s)(1+3s)(1+4s)}$, draw the Nyquist plot of the system. Comment on the stability of the system for $K = 1$, $K = 2$ and $K = 10$. In addition, calculate gain margin in each case and find the range of values of K for stability.

Solution

Step 1: The system has three poles in LHP. There are no poles in RHP and also on the $j\omega$-axis.
Step 2: The Nyquist path is shown in Fig. E15.6.
Step 3: For path ae, $s = j\omega$ and $0 < \omega < \infty$.

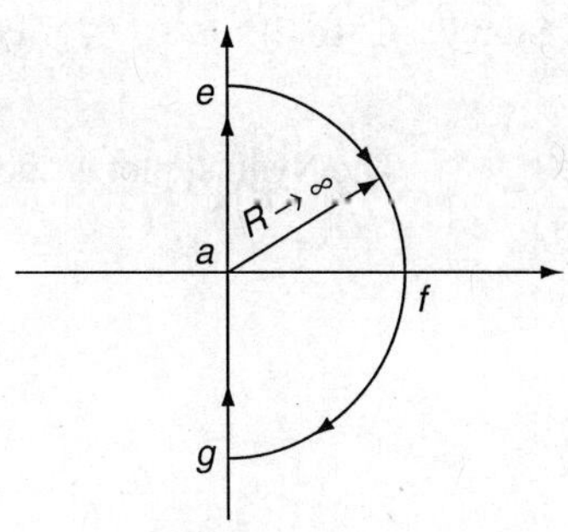

Fig. E15.6

$$\therefore \quad M(j\omega) = G(j\omega)H(j\omega) = \frac{4K}{(1+2j\omega)(1+3j\omega)(1+4j\omega)}$$

At $\quad \omega = 0, \quad |M(j\omega)| = 1 \quad \text{and} \quad \angle M(j\omega) = 0°$

At $\quad \omega = \infty, \quad |M(j\omega)| = 0 \quad \text{and} \quad \angle M(j\omega) = -270°$

The polar plot of the path ae is shown in Figs. E15.6(a), E15.6(b) and E15.6(c).

Step 4: The polar plot of the path ga is the mirror image of the polar plot of the path ae which has been shown by dotted line in Figs. E15.6(a), E15.6(b) and E15.6(c).

Step 5: For path *efg*: $s = Re^{j\theta}$ where $R \to \infty$ and $-90° < \theta < 90°$

$$\therefore \quad G(s)H(s)\Big|_{efg} = \underset{R\to\infty}{\text{Lt}} \frac{4K}{(1+2Re^{j\theta})(1+3Re^{j\theta})(1+4Re^{j\theta})}$$

$$= 0e^{-3j\theta} = 0\angle -3\theta$$

Therefore, the infinite semicircle *efg* maps onto a point as shown in Fig. E15.6(a).

Step 6: Intersection of Nyquist plot with the negative real-axis i.e., Q point:

$$G(j\omega)H(j\omega) = \frac{4K}{(1+2j\omega)(1+3j\omega)(1+4j\omega)}$$

$$= \frac{4K(1-2j\omega)(1-j\omega)(1-4j\omega)}{(1+4\omega^2)(1+9\omega^2)(1+16\omega^2)}$$

$$= \frac{4K(1-2j\omega)(1-7j\omega-12\omega^2)}{(1+4\omega^2)(1+9\omega^2)(1+16\omega^2)}$$

$$= \frac{4K(1-7j\omega-12\omega^2-2j\omega-14\omega^2+24j\omega^3)}{(1+4\omega^2)(1+9\omega^2)(1+16\omega^2)}$$

$$= \frac{4K(1-26\omega^2)+4Kj\omega(24\omega^2-9)}{(1+4\omega^2)(1+9\omega^2)(1+16\omega^2)}$$

Equating imaginary part of zero, we have $24\omega^2 - 9 = 0$

i.e., $\omega = 0.6123$ rad/rec.

Substituting ω in real part of $G(j\omega)H(j\omega)$, we have

$$Q = \frac{4K(1-26\times 0.375)}{(1+4\times 0.375)(1+9\times 0.375)(1+16\times 0.375)}$$

$$= -0.1142 \times 4K$$

$$= -0.4568\,K$$

Step 7: Case (i): For $K = 1$

$$Q = -0.4568$$

The Nyquist plot is shown in Fig. E15.6(a).

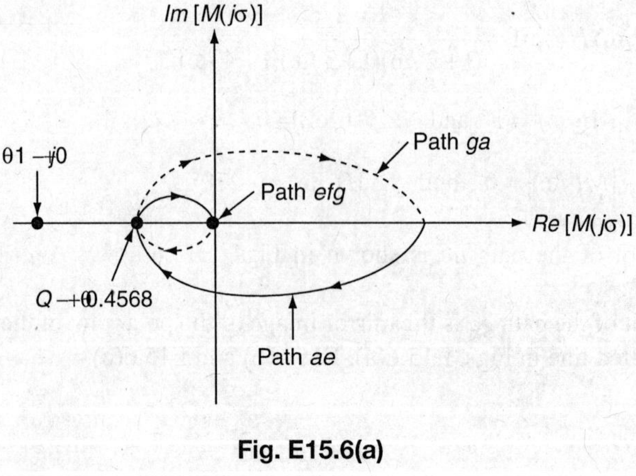

Fig. E15.6(a)

Here, the polar plot does not encircle the $-1+j0$ point. Hence, $N = 0$. Since $P = 0$, the system is stable.

Case (ii): For $K = 2$

$$Q = -0.9136$$

The Nyquist plot is shown in Fig. E15.6(b).

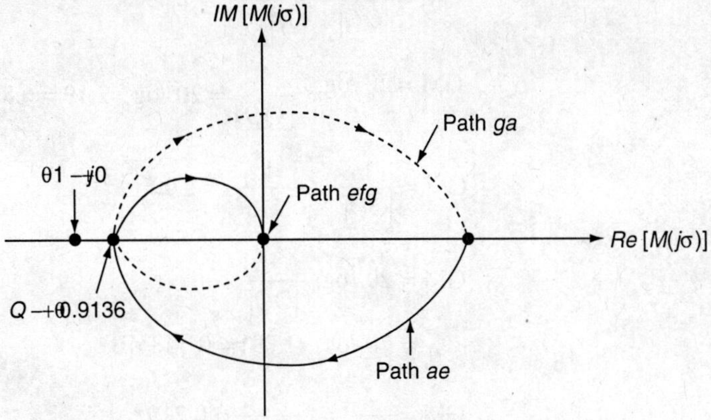

Fig. E15.6(b)

The polar plot does not encircle the $-1+j0$ point. The system is stable

Case (iii): For $K = 10$.

$$Q = -4.568$$

The Nyquist plot is shown in Fig. E15.6(c).

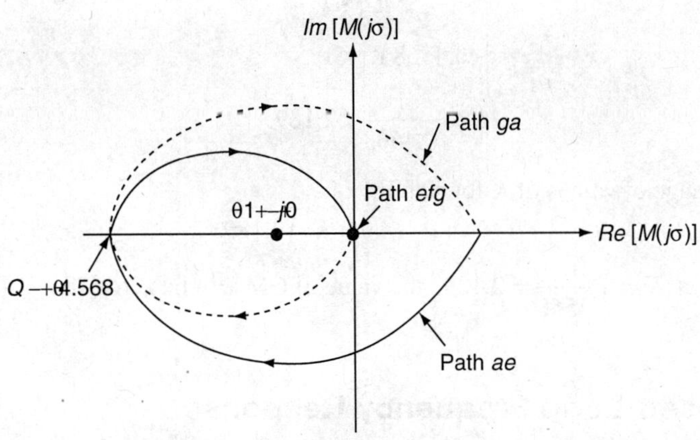

Fig. E15.6(c)

The Nyquist plot encircles the $-1+j0$ point twice in the clockwise direction and hence $N = 2$.

$\therefore$ $$N = Z - P$$

i.e., $$2 = Z - 0$$

$\therefore$ $$Z = 2$$

The system is stable.

Gain margin:

Case (i): $$\text{GM} = \frac{1}{|OQ|} = \frac{1}{|0.4568|} = 2.19$$

or, $$\text{GM} = 20\ log_{10}\ \frac{1}{|OQ|} = 20\ log_{10}\ 2.19 = 6.80\,\text{dB}$$

Case (ii): $$\text{GM} = \frac{1}{|OQ|} = \frac{1}{|0.9136|} = 1.09$$

or, $$\text{GM} = 20\ log_{10}\ \frac{1}{|OQ|}$$
$$= 20\ log_{10}(1.09) = 0.748\ \text{dB}.$$

Case (iii): $$\text{GM} = \frac{1}{|OQ|} = \frac{1}{|4.568|} = 0.219$$

or, $$\text{GM} = 20\ log_{10}\ \frac{1}{|OQ|}.$$
$$= 20\ log_{10}(0.219) = -13.19\ \text{dB}$$

The values of gain margin (in dB) are positive in Case (i) and Case (ii) whereas its value is negative in Case (iii).

Step 8: For absolute stability,

$$|OQ| < 1$$

i.e., $$|-0.4568\ \text{K}| < 1$$

i.e., $$K < \frac{1}{0.4568} < 2.189.$$

The range of values of K for stability is

$$0 < K < 2.189.$$

Here, at $K = \dfrac{1}{0.4568} = 2.189$, the value of GM will be zero (dB).

15.10 Closed-Loop Frequency Response

We have already discussed the closed-loop stability. Plotting the open-loop frequency response of the system, we have obtained frequency response specifications. The open-loop frequency response is the graph of the magnitude of $G(j\omega)H(j\omega)$ and the phase angle of $G(j\omega)H(j\omega)$ against frequency ω. The methods such as the bode plot, polar plot and nyquist plot have already been discussed to obtain open-loop frequency response. For a unity feedback closed-loop sable system, the closed-loop transfer function of a unity feedback $[H(s)=1]$ system is given by

$$M(s)=\frac{C(s)}{R(s)}=\frac{G(s)}{1+G(s)}$$

$$M(j\omega)=\frac{C(j\omega)}{R(j\omega)}=\frac{G(j\omega)}{1+G(j\omega)}$$

The plot of magnitude of $C(j\omega)/R(j\omega)$ and the phase angle of $C(j\omega)/R(j\omega)$ against frequency ω. The data required to obtain closed-loop frequency response can be obtained from the polar plotted from an open-loop transfer function $G(j\omega)$. Figure 15.10 shows the polar plot of $G(j\omega)$.

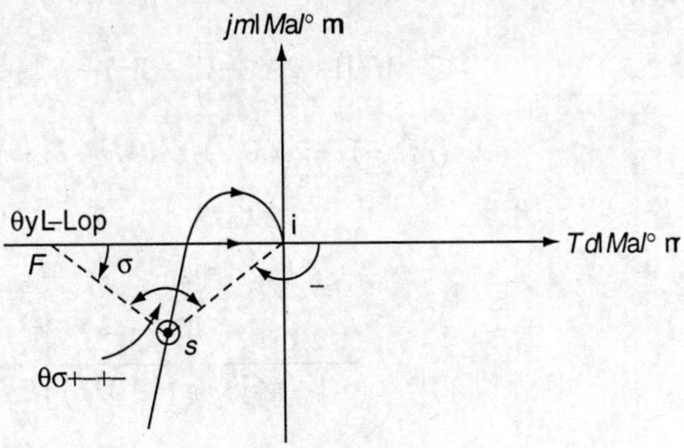

Fig. 15.10

At $\omega=\omega_1$, the point X is on the polar plot of $G(j\omega)$.

$$\therefore \qquad \vec{OX}=|G(j\omega)|$$

The point Y is the critical point $-1+j0$. Let us join point Y to point X to get vector YX. Therefore, $\vec{YX}$ represent $[1+G(j\omega)]_{\omega=\omega_1}$. The magnitude of $C(j\omega)/R(j\omega)$ is given by

$$|M(j\omega)|_{\omega=\omega_1}=\left.\frac{C(j\omega)}{R(j\omega)}\right|_{\omega=\omega_1}=\alpha=\frac{\vec{OX}}{\vec{YX}}=\left.\frac{G(j\omega)}{1+G(j\omega)}\right|_{\omega=\omega_1}$$

The phase angle of $C(j\omega)/R(j\omega)$ is given by

$$M(j\omega)|_{\omega=\omega_1}=\angle\left.\frac{C(j\omega)}{R(j\omega)}\right|_{\omega=\omega_1}=\frac{\angle\vec{OX}}{\angle\vec{YX}}=\frac{\phi}{\theta}$$

$$\therefore \qquad \alpha=(\phi-\theta)$$

Figure 15.10 shows the angle $\alpha=\phi-\theta$.

The above method is very complicated to obtain closed-loop frequency response. Using M and N circles, we can very easily obtain constant magnitude loci and constant phase loci respectively.

15.10.1 M Circles [Constant Magnitude Loci]

The closed loop transfer function of a unity feedback system is

$$M(s)=\frac{C(s)}{R(s)}=\frac{G(s)}{1+G(s)}$$

Let $\qquad G(j\omega)=X+jY$. Therefore,

$$M(j\omega)=\frac{G(j\omega)}{1+G(j\omega)}=\frac{X+jY}{1+X+jY}$$

Let M be the magnitude of $C(j\omega)/R(j\omega)$, we have

$$M = \left|\frac{X + jY}{1 + X + jY}\right| = \frac{\sqrt{X^2 + Y^2}}{\sqrt{(1+X)^2 + Y^2}}$$

i.e.,
$$M^2 = \frac{X^2 + Y^2}{(1+X)^2 + Y^2}$$

i.e.,
$$M^2[(1+X^2) + Y^2] = X^2 + Y^2$$

i.e.,
$$M^2[1 + 2X + X^2 + Y^2] - X^2 - Y^2 = 0$$

i.e.,
$$X^2(M^2 - 1) + 2XM^2 + Y^2(M^2 - 1) + M^2 = 0 \quad (15.5)$$

i.e.,
$$X^2 + \frac{2XM^2}{(M^2-1)} + Y^2 + \frac{M^2}{(M^2-1)} = 0$$

i.e.,
$$X^2 - \frac{2XM^2}{(1-M^2)} + \left[\frac{M^2}{(1-M^2)}\right]^2 + Y^2 = \frac{M^2}{(1-M^2)} + \left[\frac{M^2}{(1-M^2)}\right]^2$$

i.e.,
$$\left[X - \frac{M^2}{(1-M^2)}\right]^2 + Y^2 = \frac{M^2}{(1-M^2)}\left[1 + \frac{M^2}{1-M^2}\right]$$

i.e.,
$$\left[X - \frac{M^2}{1-M^2}\right] + Y^2 = \left[\frac{M}{1-M^2}\right]^2 \quad (15.6)$$

Equation (15.6) represents the equation of a circle with the centre at $X = M^2/(1-M^2)$ and $Y = 0$ having radius $= M/(1-M^2)$.

For a particular value of M, a circle will be obtained. The magnitude of a closed-loop transfer function is constant for that particular value of M. Therefore, with various values of M, we get a family of circles. These circles are known as constant magnitude loci or M circles. Table 15.2 shows the centre and radius of such circles for various values of M.
From Table 15.2, we can conclude the following:

Table 15.2 Centre and radius for various values of M

M	Centre $X = \frac{M^2}{1-M^2}, Y = 0$	Radius $R = \left\|\frac{M}{1-M^2}\right\|$
0.2	(0.041,0)	0.208
0.5	(0.33,0)	0.667
1	∞	∞
2	(−1.33, 0)	0.667
5	(−1.041, 0)	0.208

(1) If M becomes larger and larger compared to 1, the radius becomes smaller and smaller and, ultimately, circles converge to the critical point $-1+j0$.

(2) If $M > 1$, the centres are to the left of point $-1+j0$.

(3) If M becomes smaller and, smaller compared to 1, the radius becomes smaller and smaller and ultimately, circles converge to the origin.

(4) If $M < 1$, the centres are to the right of the origin.

(5) If $M = 1$, the locus is of the points which are equidistant from the origin as well as from the $-1+j0$ point.

Figure 15.11 shows the family of constant M circles.

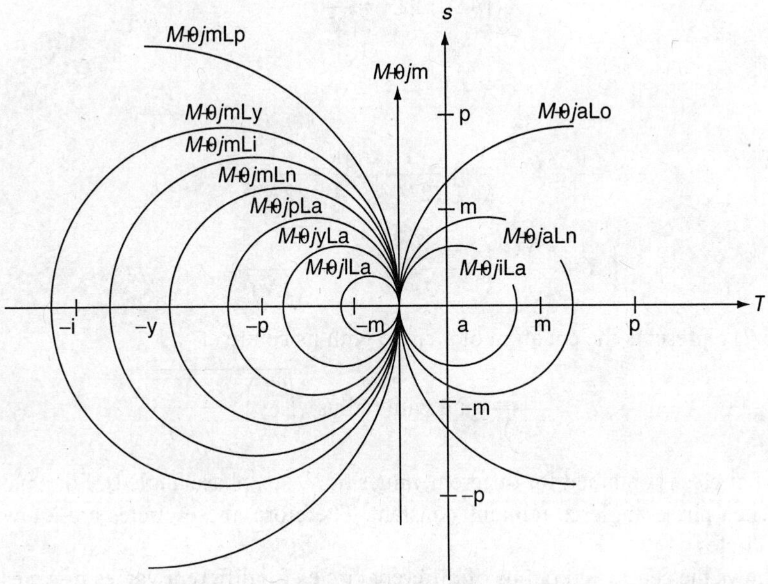

Fig. 15.11 A family of constant M circles

15.10.2 N Circles [Constant Phase Loci]

Let us obtain the phase angle α of $C(s)/R(s)$ in terms of X and Y. We have

$$\angle\frac{C(j\omega)}{R(j\omega)} = \angle\frac{G(j\omega)}{1+G(j\omega)} = \angle\frac{X+jY}{(1+X)+jY}$$

i.e.,

$$\alpha = \tan^{-1}\left(\frac{Y}{X}\right) - \tan^{-1}\left(\frac{Y}{1+X}\right)$$

i.e.,

$$\tan\alpha = N = \tan\left[\tan^{-1}\left(\frac{Y}{X}\right) - \tan^{-1}\left(\frac{Y}{1+X}\right)\right]$$

i.e.,
$$N=\frac{\tan\left[\tan^{-1}\left(\frac{Y}{X}\right)\right]-\tan\left[\tan^{-1}\left(\frac{Y}{1+Y}\right)\right]}{1+\left[\tan^{-1}\left(\frac{Y}{X}\right)\right]\left[\tan^{-1}\left(\frac{Y}{1+X}\right)\right]}$$

i.e.,
$$N=\frac{\tan\left[\tan^{-1}\left(\frac{Y}{X}\right)\right]-\tan\left[\tan^{-1}\left(\frac{Y}{1+X}\right)\right]}{1+\left[\tan\left\{\tan^{-1}\left(\frac{Y}{X}\right)\right\}\right]\left[\tan\left\{\tan^{-1}\left(\frac{Y}{1+X}\right)\right\}\right]}$$

i.e.,
$$X(1+X)+Y^2=\frac{Y}{N}$$

i.e.,
$$X^2+X+Y^2-\frac{Y}{N}=0$$

i.e.,
$$X^2+X+\frac{1}{4}+Y^2-\frac{Y}{N}+\frac{1}{(2N)^2}=\frac{1}{4}+\frac{1}{(2N)^2}$$

i.e.,
$$\left(X+\frac{1}{2}\right)^2+\left(Y-\frac{1}{2N}\right)^2=\frac{1}{4}+\frac{1}{(2N)^2} \tag{15.7}$$

Equation (15.7) represents the equation of a circle with its centre at

$$X=-\frac{1}{2},\ Y=\frac{1}{(2N)^2}\ \text{and with radius}\ \sqrt{\left(\frac{1}{4}\right)+\frac{1}{(2N)^2}}$$

The family of circles is obtained for different values of N, i.e., phase angle α. For a particular circle, the value of N, i.e., phase angle α, remains constant. Therefore, these circles are known as constant phase loci or N circles.

Table 15.3 shows the centre and radius of different circles for different values of α i.e., N.

Table 15.3 Centre and radius of different circles for different values of N

α	N = tan α	Centre $X=-\frac{1}{2}, Y=\frac{1}{2N}$	Radius $R=\sqrt{\frac{1}{4}+\frac{1}{(2N)^2}}$
– 90°	∞	$-\frac{1}{2}, 0$	$\frac{1}{2}=0.5$
– 60°	– 1.732	$\left(-\frac{1}{2}, -0.288\right)$	0.577
– 30°	– 0.5773	$\left(-\frac{1}{2}, -0.866\right)$	1

0°	0	$\left(-\frac{1}{2},\infty\right)$	∞
+30°	0.5773	$\left(-\frac{1}{2},0.866\right)$	1
+60°	1.732	$\left(-\frac{1}{2},-0.288\right)$	0.577
+90°	∞	$\left(-\frac{1}{2},0\right)$	0.5

Equation (15.7) is satisfied for $X = 0$, $Y = 0$ as well as $X = 1$, $Y = 0$ irrespective of the value of N. We conclude that each circle represented by Eq. (15.7) passes through the points (0, 0) and $(-1 + j0)$. Figure 15.12 shows the family of constant N circles.

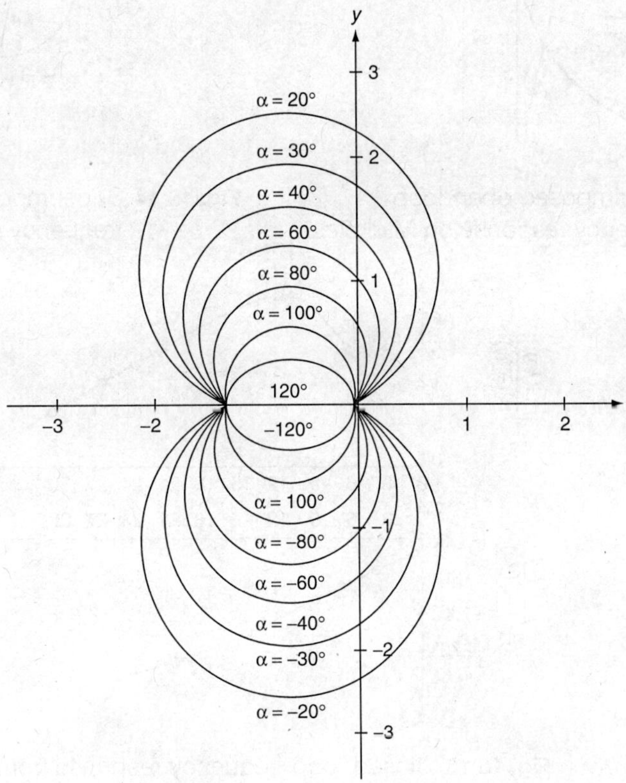

Fig. 15.12 A family of constant N circles

15.11 Closed-loop Frequency Response of Control System from M and N Circles

It is possible to obtain closed-loop frequency response from open-loop response. There is no requirement of calculation of the magnitude and phase angle. Superimposing the open-loop frequency response on M and N circles, the magnitude and phase angle are read from the point of intersection. On the other hand, the frequency is obtained from the open-loop frequency response plot.

Figure 15.13 shows the superimposed open-loop frequency response on M circles whereas Fig.15.14 shows the superimposed open-loop frequency response on N circles.

Figure 15.15 shows the corresponding frequency response.

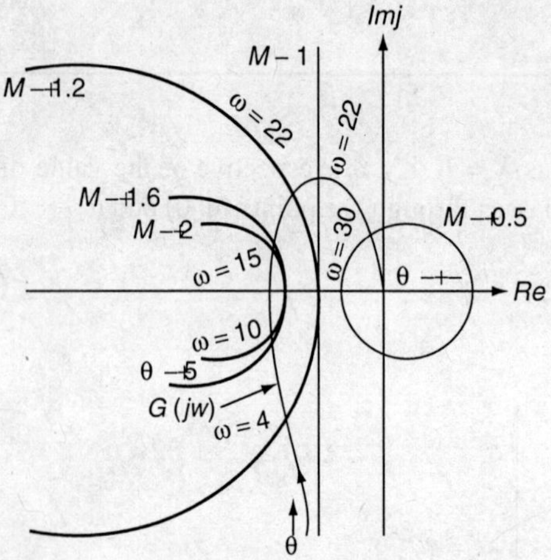

Fig. 15.13 Superimposed open-loop frequency response on M circles

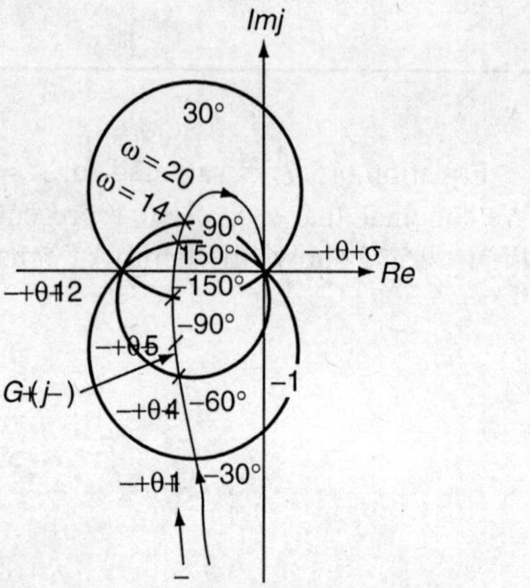

Fig. 15.14 Superimposed open-loop frequency response on N circles

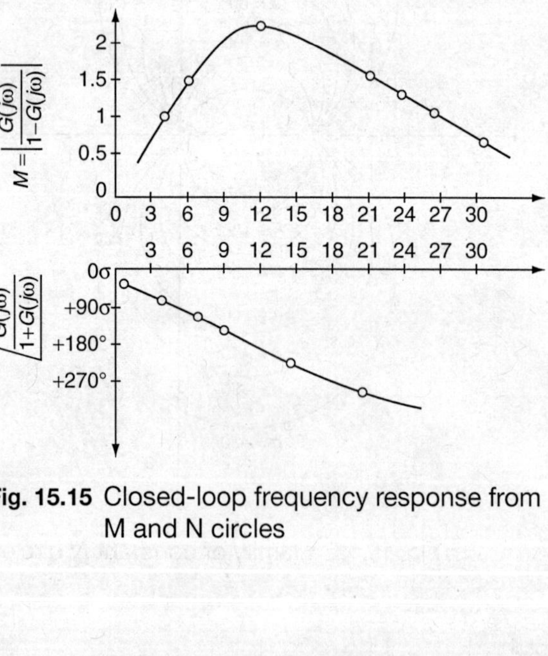

Fig. 15.15 Closed-loop frequency response from M and N circles

15.12 Gain Adjustment Using M Circle

Figure 15.16 shows a M circle where M > 1. A tangential line from origin is drawn to M circle which meets the M circle at the point X. The line OX makes angle ψ with the negative real axis.

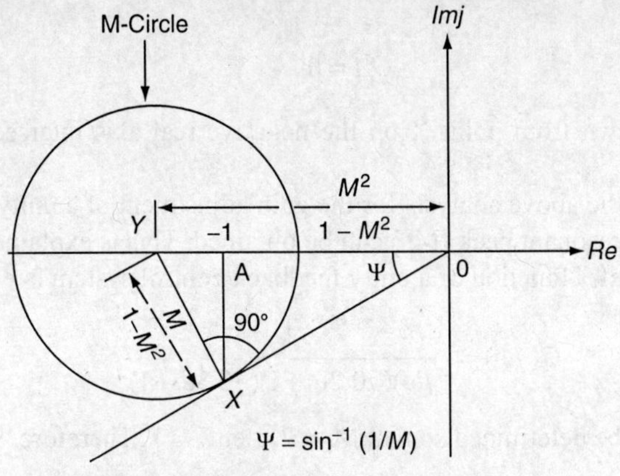

Fig. 15.16 M circle and tangent from origin

From Fig. 15.16, we have

$$\sin\psi = \left|\frac{\dfrac{M}{1-M^2}}{\dfrac{M^2}{1-M^2}}\right| = \frac{1}{M} \tag{15.8}$$

Again,
$$OX^2 = OY^2 - XY^2 \tag{15.9}$$

and
$$AX = OX\sin\psi \tag{15.10}$$

Also
$$OA^2 = OX^2 - AX^2 \tag{15.11}$$

Using Eq. (15.10) we have from Eq. (15.11)

$$OA^2 = OX^2 - OX^2\sin^2\psi$$

i.e.,
$$OA^2 = OX^2(1-\sin^2\psi)$$

i.e.,
$$OA^2 = (OY^2 - XY^2)\,(1-\sin^2\psi) \tag{15.12}$$

Again
$$OY = \frac{M^2}{1-M^2} \tag{15.13}$$

and
$$XY = \frac{M}{1-M^2} \tag{15.14}$$

Using Eq. (15.8), Eq. (15.13) and Eq. (15.14), we have from Eq. (15.12)

$$OA^2 = \left\{\left(\frac{M^2}{1-M^2}\right)^2 - \left(\frac{M}{1-M^2}\right)^2\right\}\left(1-\frac{1}{M^2}\right)$$

i.e.,
$$OA^2 = 1$$

i.e.,
$$OA = |1|$$

The perpendicular drawn from point X on the negative real axis intersects the negative real axis at $(-1 + j0)$.

It is possible to apply the above analysis for the gain adjustment of a unity feedback control system and the desired value for resonant peak (M_r) can be obtained. This is explained below.

Let the open-loop transfer function of a unity feedback control system is expressed by

$$G(j\omega) = \frac{1}{(j\omega)(j0.2\omega+1)(j0.3\omega+1)} \tag{15.15}$$

The gain value K should be determined so that $M_r = 2$. Let K = 1. Therefore, we have

$$G(j\omega) = \frac{1}{(j\omega)(j0.2\omega+1)(j0.3\omega+1)}$$

$$= \frac{(-j\omega)(-j0.2\omega+1)(-j0.3\omega+1)}{(\omega^2)(0.04\omega^2+1)(0.09\omega^2+1)}$$

$$= \frac{(-j\omega)(1-j0.5\omega-0.06\omega^2)}{(\omega^2)(0.04\omega^2+1)(0.09\omega^2+1)}$$

$$= -\frac{0.5}{(0.04\omega^2+1)(0.09\omega^2+1)} - \frac{j(1-0.06\omega^2)}{\omega(0.04\omega^2+1)(0.09\omega^2+1)}$$

ω (rad/sec)	2	3	4.08	10	∞
Re[G(jω)]	−0.3169	−0.2031	−0.1201	−0.0100	−0
Im[G(jω)]	−j0.2409	−j0.0623	−j0.0	+j0.010	+j0

The resonant peak $M_r = 2.0$. Therefore,

$$\Psi = \sin^{-1}\left(\frac{1}{M_r}\right) = \sin^{-1}\left(\frac{1}{2}\right) = 30^0$$

Figure 15.17 shows the required polar plot and a line at angle 30^0 to the negative real axis and the circle which is tangential to the polar plot as well as the line. The line touches the circle at a point X and XA is perpendicular to the negative real axis where OA = 0.24.

The circle drawn in Fig. 15.17 will be the desired M circle corresponding to $M_r = 2$, then $OA = 1 \angle -180°$ i.e., the point A will be $-1+j0$.

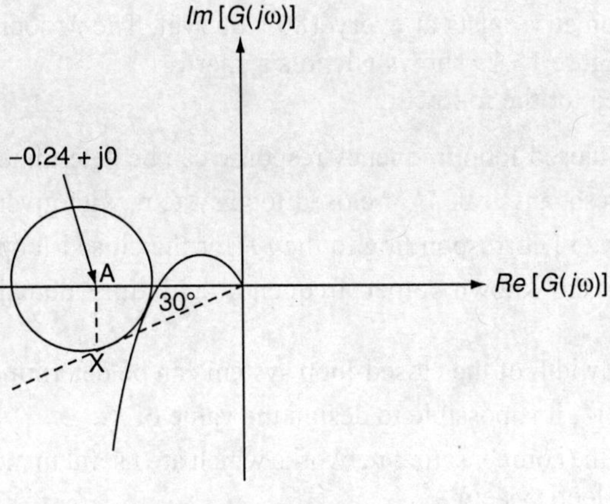

Fig. 15.17

The required value of K to obtain the desired $M_r = 2$ will be

$$K = \frac{1}{|OA|} = \frac{1}{0.24} = 4.167.$$

15.13 Nichol's Chart

It is possible to obtain the frequency response by sketching the magnitude in dB against the phase angle for various frequencies. The plot obtained is called gain-phase plot or log magnitude versus phase plot. Figure 15.18 shows such a gain-phase plot for $G(j\omega) = 1/(1 + j\omega T)$, i.e., $G(s) = 1/(1 + sT)$.

After transferring the constant magnitude loci (M circles) and constant phase angle loci (N circles) to the gain-phase plot, the resultant chart is known as Nichol's chart.

If we superimpose the gain-phase plot of an open-loop transfer function on Nichol's chart, we get very easily the closed-loop frequency response. The magnitude is expressed in dB while the phase angle is in degrees. Nichol's chart gives the points of intersection of the gain-phase plot of an open-loop transfer function, which are symmetrical about the −180° axis. The M and N loci also repeat for

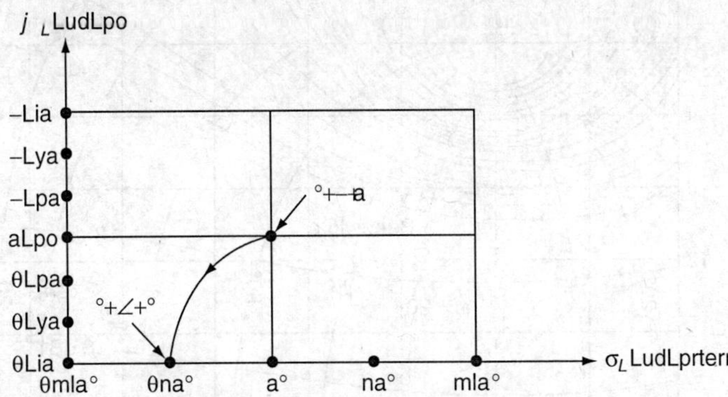

Fig. 15.18 Gain-phase plot for $G(j\omega)$

every 360°. Again symmetry exists at every 180° interval. The M loci are centred about the critical point (0 dB, –180°). Figure 15.19 shows Nichols's chart.

Nichol's chart is used for the following:

(1) The complete closed-loop frequency response can be determined.

(2) The value of resonant peak M or closed-loop system, with given $G(j\omega)$ can be obtained.

(3) The frequency ω_r, corresponding to the M_r for the closed-loop system can be also obtained.

(4) If M_r and ω_r are known, other frequency and time domain specifications can also be calculated.

(5) The 3 dB bandwidth of the closed-loop system can be determined.

(6) For the given M_r, it is possible to design the value of K.

(7) For designing the compensating networks which are useful to meet more than one specification for the closed-loop system.

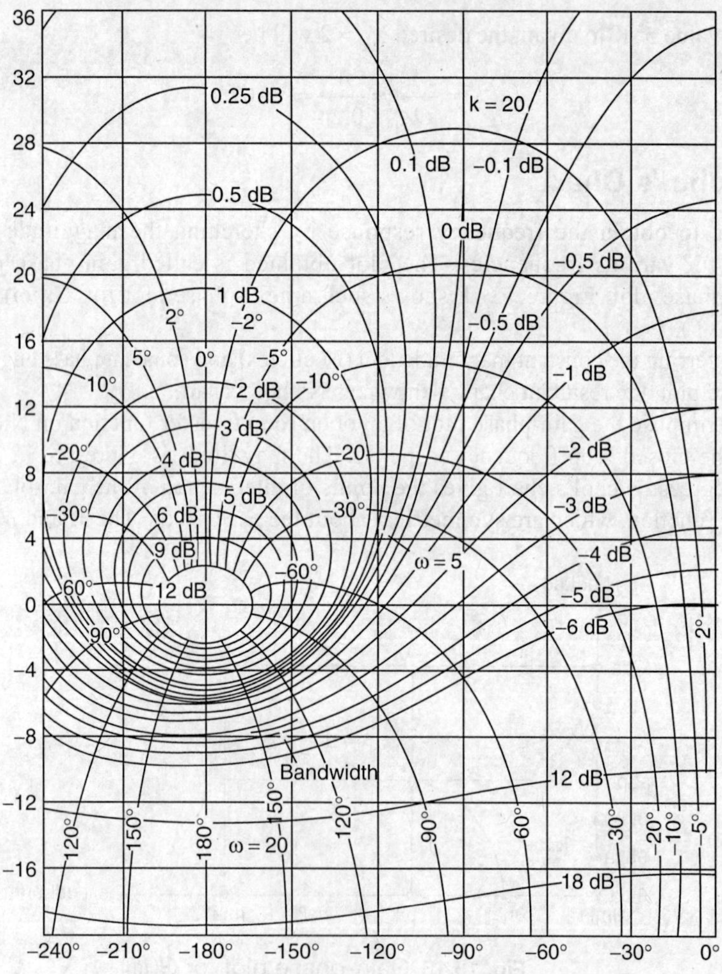

Fig. 15.19 Nichols chart

SIGNIFICANT POINTS

Enclosed: If point is found to be inside the path, the point is said to be encircled by the closed path.
Enclosed: If a path is traversed in the clockwise direction and the point is found to lie to the right of the path, the point is said to be enclosed by the path.
Analytic Function: A function is said to be analytic at a point in a plane if its value and derivative have finite existence at that point.
Single-Valued Function: A function is said to be single valued if it has only one value for each separate value of s.
If $F(s)=\sqrt{s}$ and $s=36$, the values of $F(s)$ are −6 and +6. The function $F(s)=s$ is not a single-valued function. On the other hand, the function $F(s)=s$ is a single-valued function.
Nyquist Criterion: It focusses also on the relative stability of the system. It is possible to determine the stability of the closed-loop pole from the open-loop pole without knowing the roots of the closed-loop system. A Nyquist plot is based on a polar plot.

ADDITIONAL SOLVED EXAMPLES

Example 15.7 For $G(s)H(s)=1/[s(s+2)]$, draw the Nyquist plot and decide stability.

Solution

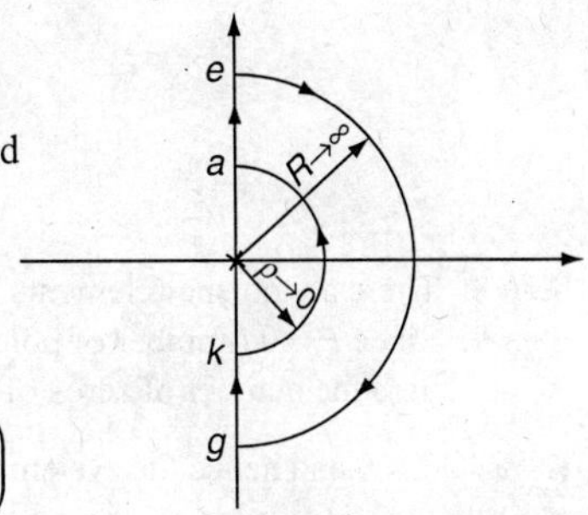

Fig. E15.7

Step 1: Poles of $G(s)H(s)$ are at $s=0, s=-2$. There is a pole at the origin and no pole on the $j\omega$ axis.
Step 2: The Nyquist path is shown in Fig. E15.7.
Step 3: For path *ae*

$$s=j\omega \quad \text{and} \quad 0<\omega<\infty$$

$$\text{Here, } M(j\omega)=G(j\omega)H(j\omega)=\frac{1}{j\omega(j\omega+2)}=\frac{1}{\omega\sqrt{\omega^2+4}}\angle\left(-90^\circ-\tan^{-1}\frac{\omega}{2}\right)$$

$$\therefore \qquad |G(j\omega)H(j\omega)|=\frac{1}{\omega\sqrt{\omega^2+4}}$$

$$\text{and} \qquad \angle G(j\omega)H(j\omega)=-90^\circ-\tan^{-1}\frac{\omega}{2}$$

$$\text{At} \qquad \omega=0,\ |G(j\omega)H(j\omega)|=\infty \quad \text{and} \quad \angle G(j\omega)H(j\omega)=-90^\circ$$

$$\text{At} \qquad \omega=\infty,\ |G(j\omega)H(j\omega)|=0 \quad \text{and} \quad \angle G(j\omega)H(j\omega)=-180^\circ$$

The polar plot of path *ae* is shown in Fig. E15.7(a).

Step 4: The polar plot path is of *gk* is shown in Fig. E15.7(a) by dotted line, which is the mirror image of the polar plot of path *ae*.
Step 5: For path *efg*:

$$s=Re^{j\theta} \text{ where } R\to\infty \text{ and } +90^\circ>\theta>-90^\circ.$$

$$G(s)H(s)\Big|_{efg}=\frac{1}{\underset{R\to\infty}{\text{Lt}}\ Re^{j\theta}(PRe^{j\theta}+1)}=0e^{-i2\theta}$$

Therefore, the infinite semicircle *efg* maps onto a point.

Step 6: For path *kla*

$$s=\rho e^{j\theta} \text{ where } \rho\to 0 \text{ and } -90^\circ<\theta<+90^\circ.$$

$$G(s)H(s)\Big|_{kla} = \frac{1}{\underset{\rho\to 0}{\text{Lt}}\ \rho e^{j\theta}(\rho e^{j\theta}+1)} = \infty e^{-j\theta}$$

Therefore, the phase angle of $G(s)H(s)\big|_{kla}$ varies from +90° to −90° since θ varies from −90° to +90°.

Step 7: Connected

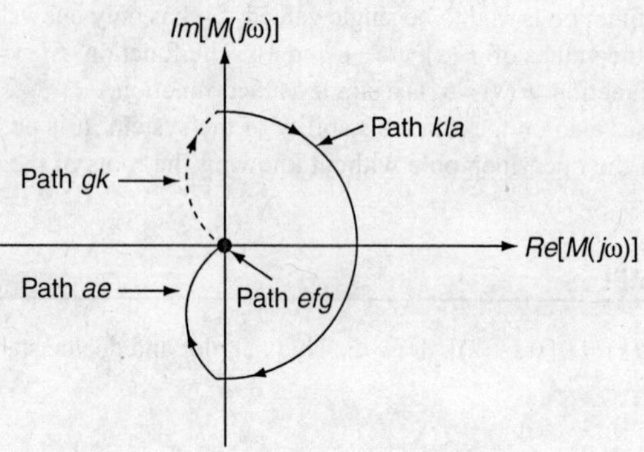

Fig. E15.7(a)

Step 8: There are no encirclements about (−1, 0).

Step 9: Since $P = 0$ (number of poles in the RHP) = poles of $G(s)H(s)$, here the number of zeros of $1+G(s)H(s)$ in the RHP is zero.

∴ $Z = 0$ and hence the system is stable.

Example 15.8 For $G(s)H(s) = 1/[s^2(s+2)]$, sketch the Nyquist plot and determine the stability of the system.

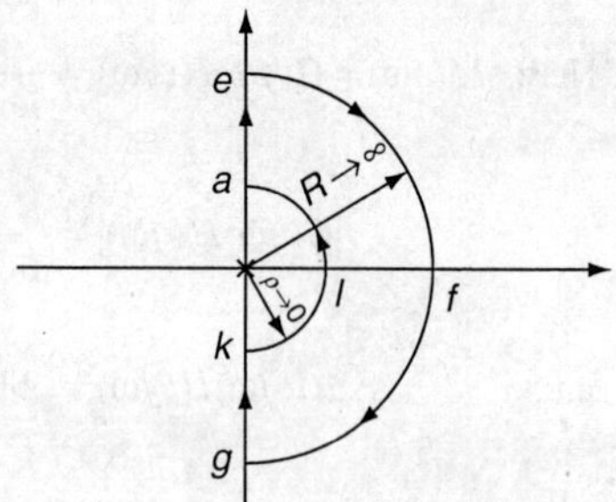

Fig. E15.8

Solution

Step 1: Poles are at $s = 0, 0, -2$. There are two poles at the origin.

Step 2: The Nyquist path is shown in Fig. E15.8.

For Path *ae*,

$$s = j\omega \text{ where } 0 < \omega < \infty$$

Here
$$M(j\omega) = G(j\omega)H(j\omega) = \frac{1}{(j\omega)^2(j\omega+2)} = \frac{1}{\omega^2\sqrt{\omega^2+4}}\angle\left(-180° - \tan^{-1}\frac{\omega}{2}\right)$$

∴
$$|G(j\omega)H(j\omega)| = \frac{1}{\omega^2\sqrt{\omega^2+4}}$$

and
$$\angle G(j\omega)H(j\omega) = -180° - \tan^{-1}\frac{\omega}{2}$$

At
$$\omega = 0, \quad |G(j\omega)H(j\omega)| = \infty \text{ and } \angle G(j\omega)H(j\omega) = -180°$$

At
$$\omega = \infty, \quad |G(j\omega)H(j\omega)| = 0 \text{ and } \angle G(j\omega)H(j\omega) = -270°$$

The polar plot of path *ae* is shown in Fig. E15.8(a)

Step 3: The polar plot of path of *gk* is shown in Fig. E15.8(a) by dotted line.

Step 4: For path *efg*

$$s = Re^{j\theta} \text{ where } R \to \infty \text{ and } +90° > \theta > -90°.$$

$$G(s)H(s)\Big|_{efg} = \frac{1}{\underset{R\to\infty}{\text{Lt}}\,(Re^{j\theta})^2(Re^{j\theta}+1)} = 0.e^{-3j\theta}$$

Therefore, the infinite semicircle *efg* maps onto a point.

Step 5: For path *kla*

$$s = \rho e^{j\theta} \text{ where } \rho \to 0 \text{ and } -90° < \theta < +90°$$

$$G(s)H(s)\Big|_{kla} = \frac{1}{\underset{\rho\to 0}{\text{Lt}}\,(\rho e^{j\theta})^2(\rho e^{j\theta}+1)} = \infty\rho e^{-2j\theta}$$

Therefore, the phase angle of $G(s)H(s)\big|_{kla}$ varies from +180° to −180° since θ varies from −90° to +90°. There are two infinite semicircles.

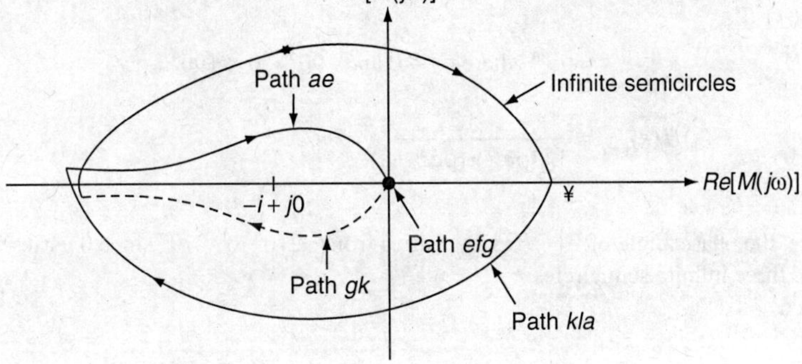

Fig. E15.8(a)

Step 6: Connected

Step 7: There are encirclements about (−1, 0) and $N > 0$.

Step 8: Since $P = 0$ (number of poles in the RHP) = poles of $G(s)H(s)$, here, the number of zeros of $1 + G(s)H(s)$ in the RHP is $Z = N$, and hence system is unstable.

Example 15.9 For $G(s)H(s) = 1/[s^3(s+2)]$, sketch the Nyquist plot and determine the stability of the system.

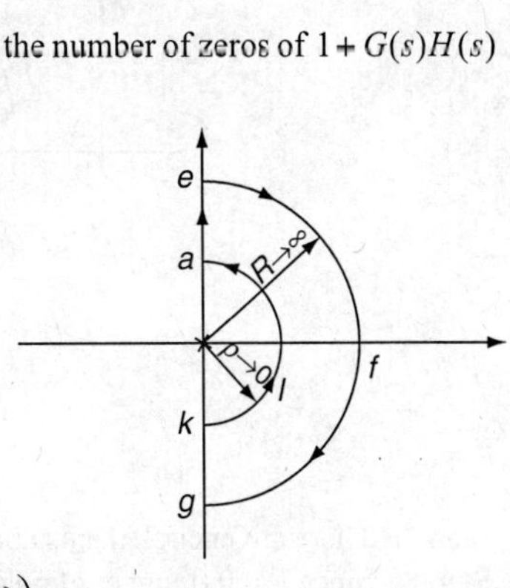

Fig. E15.9

Solution

Step 1: Poles are at $s = 0, 0, 0, -2$. There are three poles at the origin.

Step 2: The Nyquist path is shown in Fig. E15.9

For path *ae*,

$$s = j\omega \text{ where } 0 < \omega < \infty.$$

Here,

$$M(j\omega) = G(j\omega)H(j\omega) = \frac{1}{(j\omega)^3(j\omega+2)} = \frac{1}{\omega^3\sqrt{\omega^2+4}}\angle\left(-270° - \tan^{-1}\frac{\omega}{2}\right)$$

$\therefore$
$$|G(j\omega)H(j\omega)| = \frac{1}{\omega^3\sqrt{\omega^2+4}}$$

and
$$\angle G(j\omega)H(j\omega) = -270° - \tan^{-1}\frac{\omega}{2}$$

At $\omega = 0, \quad |G(j\omega)H(j\omega)| = \infty \quad \text{and} \quad \angle G(j\omega)H(j\omega) = -270°$

At $\omega = \infty, \quad |G(j\omega)H(j\omega)| = 0 \quad \text{and} \quad \angle G(j\omega)H(j\omega) = -360°$

The polar plot of path *ae* is shown in Fig. E15.9(a).

Step 3: The polar plot path *gk* has been shown in Fig. E15.9(a), long dotted line, which is the mirror image of *ae*.

Step 4: For path *efg*

$$s = Re^{j\theta} \text{ where } R \to \infty \text{ and } +90° > \theta > -90°.$$

$$G(s)H(s)\Big|_{efg} = \frac{1}{\underset{R\to\infty}{\text{Lt}}\,(Re^{j\theta})^3(Re^{j\theta}+1)} = 0$$

Therefore, the infinite semicircle *efg* maps onto a point.

Step 5: For path *kla*

$$s = \rho e^{j\theta} \text{ where } \rho \to 0 \text{ and } -90° < \theta < +90°.$$

$$G(s)H(s)\Big|_{kla} = \frac{1}{\underset{\rho\to 0}{\text{Lt}}\,(\rho e^{j\theta})^3(\rho e^{j\theta}+1)} = \infty\rho e^{-3j\theta}$$

Therefore, the phase angle of $G(s)H(s)\Big|_{kla}$ varies from +270° to –270° since θ varies from –90° to +90°. There are three infinite semicircles.

Step 6: Connected

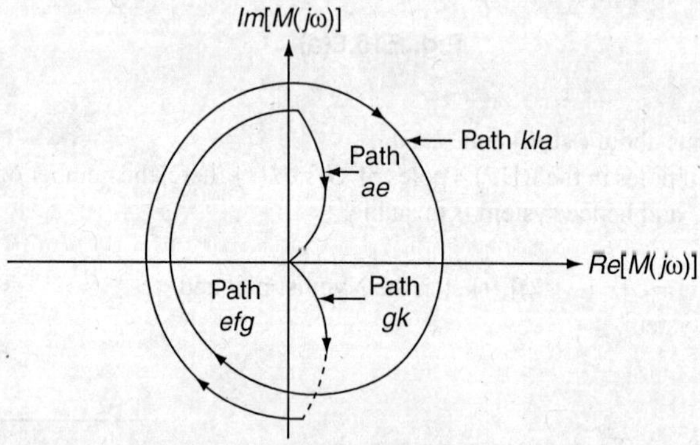

Fig. E15.9(a)

Step 7: There are encirclements about (–1, 0) and $N > 0$.

Step 8: Since $P = 0$ (number of poles in the RHP) = poles of $G(s)H(s)$, here, the number of zeros of $1+G(s)H(s)$ in the RHP is $N > 0$ and hence the system is unstable.

Example 15.10 For $G(s) = 1/[s(s-2)]$, sketch the Nyquist plot and determine the stability of the system.

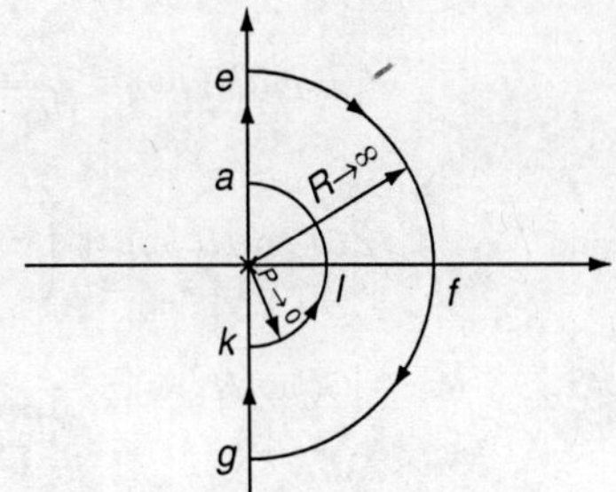

Fig. E15.10

Solution

Step 1: Poles of $G(s)H(s)$ are at $s = 0, s = -2$. There is a pole at the origin and no pole on the $j\omega$ axis.

Step 2: The Nyquist path is shown in Fig. E15.10

For path *ae*

$$s = j\omega \text{ and } 0 < \omega < \infty$$

Here, $M(j\omega) = G(j\omega)H(j\omega) = \dfrac{1}{j\omega(j\omega-2)} = \dfrac{1}{\omega\sqrt{\omega^2+4}}\angle\left[-90^\circ + \tan^{-1}\left(-\dfrac{\omega}{2}\right)\right]$

$\therefore$ $|G(j\omega)H(j\omega)| = \dfrac{1}{\omega\sqrt{\omega^2+4}}$ and

$$\angle G(j\omega)H(j\omega) = -90^\circ - \tan^{-1}\left(-\frac{\omega}{2}\right) = -90^\circ - 180^\circ + \tan^{-1}\frac{\omega}{2}$$

At $\omega = +0, |G(j\omega)H(j\omega)| = \infty$ and $\angle G(j\omega)H(j\omega) = -270^\circ$

At $\omega = \infty, |G(j\omega)H(j\omega)| = 0$ and $\angle G(j\omega)H(j\omega) = -180^\circ$

The polar plot of path *ae* is shown in Fig. E15.10(a).

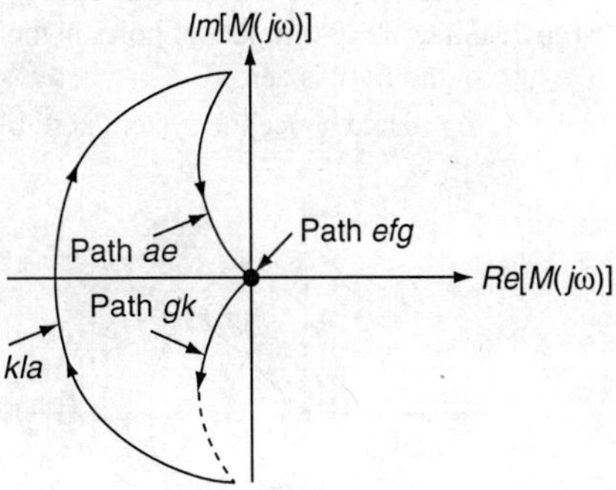

Fig. E15.10(a)

Step 3: The dotted mirror path is of polar plot of the path *ae* is the polar plot of the path *gk*.

Step 4: For path *efg*

$$s = Re^{j\theta} \text{ where } R \to \infty \text{ and } +90^\circ > \theta > -90^\circ.$$

$$G(s)H(s)\Big|_{efg} = \frac{1}{\underset{R\to\infty}{\text{Lt}}\,(Re^{j\theta})^3(Re^{j\theta}-1)} = 0$$

Therefore, the infinite semicircle *efg* maps onto a point.

Step 5: For path *kla* where $\rho \to 0$ and $-90^\circ < \theta < +90^\circ$.

$$G(s)H(s)\Big|_{kla} = \frac{1}{\underset{\rho\to 0}{\text{Lt}}\,(\rho e^{j\theta})^3(\rho e^{j\theta}-1)} = -\infty^{-3j\theta} = \infty e^{-j(180^\circ+3\theta)}$$

Therefore, the phase angle of $G(s)H(s)\big|_{kla}$ varies from -90° to -270°, since θ varies from -90° to $+90^\circ$.

Step 6: Connected

Step 7: There are encirclements about $(-1, 0)$ in CW direction and $N = 1$.

Step 8: Since $P = 1$ (number of poles in the RHP) = poles of $G(s)H(s)$, here, the number of zeros of $1 + G(s)H(s)$ in the RHP is $N = Z - P$ i.e., $+1 = Z - 1$ i.e., $Z = 2$. Here, the given system is unstable.

Example 15.11 For $G(s)H(s) = \dfrac{1}{(s+P_1)(s+P_2)}$ $P_1, P_2 > 0$, sketch the Nyquist plot and determine the stability of the system.

Solution

Step 1: Poles are at $P_1 - P_2$ are in the LHP and there is no pole at the origin.

Step 2: The Nyquist path is shown in Fig. E15.11.

For path *ae*

$$s = j\omega \text{ where } 0 < \omega < \infty.$$

Here, $M(j\omega) = G(j\omega)H(j\omega) = \dfrac{1}{(j\omega+P_1)(j\omega+P_2)} = \dfrac{1}{\sqrt{\omega^2+P_2^2}\sqrt{\omega^2+P_2^2}}\angle-\left(\tan^{-1}\dfrac{\omega}{P_1} + \tan^{-1}\dfrac{\omega}{P_1}\right)$

$$|G(j\omega)H(j\omega)| = \frac{1}{\sqrt{\omega^2 + P_2^2}\sqrt{\omega^2 + P_2^2}}$$

and $$\angle G(j\omega)H(j\omega) = -\left(\tan^{-1}\frac{\omega}{P_1} + \tan^{-1}\frac{\omega}{P_1}\right)$$

At $\omega = 0, |G(j\omega)H(j\omega)| = \frac{1}{P_1P_2}$ and $\angle G(j\omega)H(j\omega) = 0°$

At $\omega = \infty, |G(j\omega)H(j\omega)| = 0$ and $\angle G(j\omega)H(j\omega) = -180°$

The polar plot of path *ae* is shown in Fig. E15.11(a).

Step 3: The dotted mirror path is of polar plot of *ae* is the polar plot of the path *ga*.

Step 4: For path *efg*

$$s = Re^{j\theta} \text{ where } R \to \infty \text{ and } +90° < \theta < -90°.$$

$$G(s)H(s)\Big|_{efg} = \frac{1}{\underset{R\to\infty}{\text{Lt}} (Re^{j\theta} + P_1)^3 (Re^{j\theta} - P_2)} = 0$$

Therefore, the infinite semicircle *efg* maps onto a point.

Step 5: Connected

Step 6: There are no encirclements about (–1, 0).

Step 7: Since $P = 0$ (number of poles in the RHP) = poles of $G(s)H(s)$, here, the number of zeros of $1 + G(s)H(s)$ in the RHP is zero.

∴ $Z = 0$ and hence the system is stable.

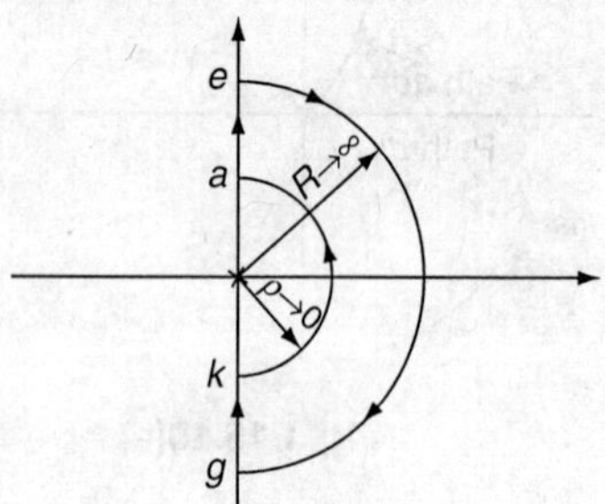

Fig. E15.11

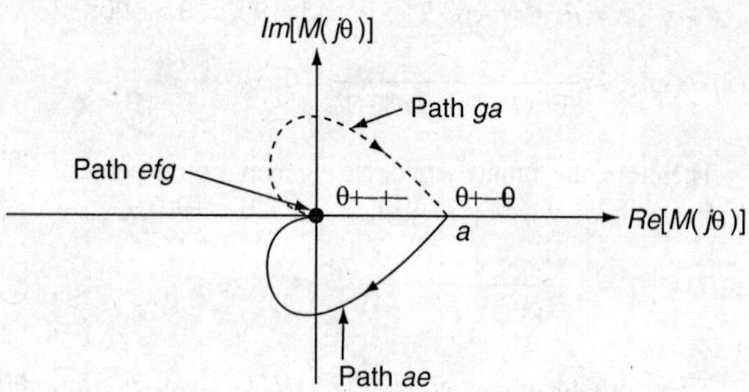

Fig. E15.11(a)

Example 15.12 For $G(s)H(s) = 12/[s(s+1)(s+2)]$, draw the Nyquist plot and decide stability. Calculate gain margin also.

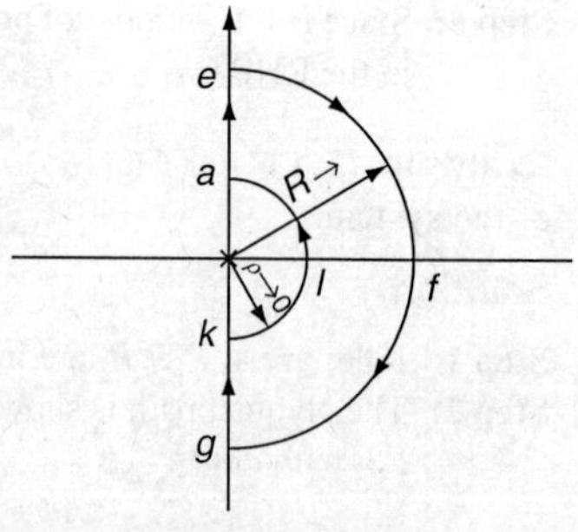

Fig. E15.12

Solution

Step 1: Poles of $G(s)H(s)$ are at $s = 0$, $s = -1$ and $s = -2$. There is a pole at the origin and no pole on the $j\omega$ axis.

Step 2: The Nyquist path is shown in Fig. E15.12.

For path *ae*

$$s = j\omega \text{ and } 0 < \omega < \infty$$

Here, $$M(j\omega)=G(j\omega)H(j\omega)=\frac{12}{j\omega(j\omega+1)(j\omega+2)}=\frac{12}{\omega\sqrt{\omega^2+1}\sqrt{\omega^2+4}}\angle\left(-90^\circ-\tan^{-1}\omega-\tan^{-1}\frac{\omega}{2}\right)$$

$\therefore$ $$|G(j\omega)H(j\omega)|=\frac{12}{\omega\sqrt{\omega^2+1}\sqrt{\omega^2+4}}$$

and $$\angle G(j\omega)H(j\omega)=-90^\circ-\tan^{-1}\omega-\tan^{-1}\frac{\omega}{2}$$

At $$\omega=0,\ \ |G(j\omega)H(j\omega)|=\infty \ \text{ and } \ \angle G(j\omega)H(j\omega)=-90^\circ$$

At $$\omega=\infty,\ \ |G(j\omega)H(j\omega)|=0 \ \text{ and } \ \angle G(j\omega)H(j\omega)=-270^\circ$$

The polar plot of path *ae* is shown in Fig. E15.12(a).

Step 3: The dotted mirror path is of the polar plot of the *ae* is the polar plot of the image *ga*.

Step 4: For path *efg*

$$s=Re^{j\theta} \text{ where } R\to\infty \ \text{ and } +90^\circ<\theta<-90^\circ.$$

$$G(s)H(s)\Big|_{efg}=\frac{12}{\underset{R\to\infty}{\text{Lt}}\,(Re^{j\theta}+1)^3(Re^{j\theta}-2)}=0$$

Therefore, the infinite semicircle *efg* maps onto a point.

Step 5: For path *kla*

$$s=\rho e^{j\theta} \text{ where } \rho\to 0 \text{ and } -90^\circ<\theta<+90^\circ.$$

$$G(s)H(s)\Big|_{kla}=\frac{12}{\underset{\rho\to 0}{\text{Lt}}\,\rho e^{j\theta}(\rho e^{j\theta}+1)^3(\rho e^{j\theta}+2)}=\infty e^{-j\theta}$$

Therefore, the phase angle of $G(s)H(s)\big|_{kla}$ varies from +90° to –90° since θ varies from −90° to +90°.

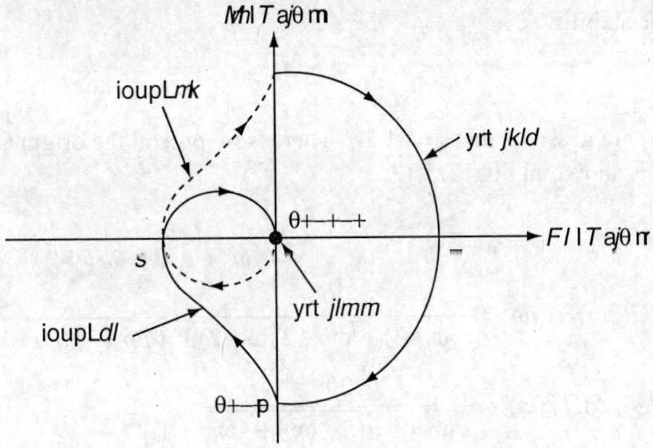

Fig. E15.12(a)

Step 6: Connected

Step 7: *Q* point

$$G(j\omega)H(j\omega)=\frac{12}{(j\omega)(j\omega+1)(j\omega+2)}=\frac{12(-j\omega)(-j\omega+1)(-j\omega+2)}{(-j^2\omega^2)(-j^2\omega^2+1)(-j^2\omega^2+2)}$$

$$= \frac{-36\omega^2 - j12\omega(-\omega^2+2)}{(\omega^2)(\omega^2+1)(\omega^2+2)}$$

$$= \frac{-36\omega^2}{(\omega^2)(\omega^2+1)(\omega^2+2)} - \frac{j12\omega(-\omega^2+2)}{(\omega^2)(\omega^2+1)(\omega^2+2)}$$

Equating the imaginary part to zero, we have

$$\frac{j12\omega(-\omega^2+2)}{(\omega^2)(\omega^2+1)(\omega^2+2)} = 0$$

i.e., $\omega = \text{either } 0 \text{ or } \sqrt{2}$

Therefore, $\omega_{pc} = \sqrt{2}$ because ω_{pc} is positive.

$$G(j\omega)H(j\omega)\Big|_{\omega_{pc}=\sqrt{2}} = -2 + j0$$

Therefore $-1 + j0$ is encircled. There are two clockwise encirclements about $(-1, 0)$. Here, $N = 2$.

Step 8: Since $P = 0$ (number of poles in the RHP) = poles of $G(s)H(s)$, here, the number of zeros of $1 + G(s)H(s)$ in the RHP is 2.

$Z = 2$ and hence system is unstable.

$$GM = \frac{1}{|OQ|} = \frac{1}{2} = 0.5$$

i.e., $GM = 20\log_{10}(0.5) = -6.20 \text{ dB}$

Since GM is negative, the system is unstable.

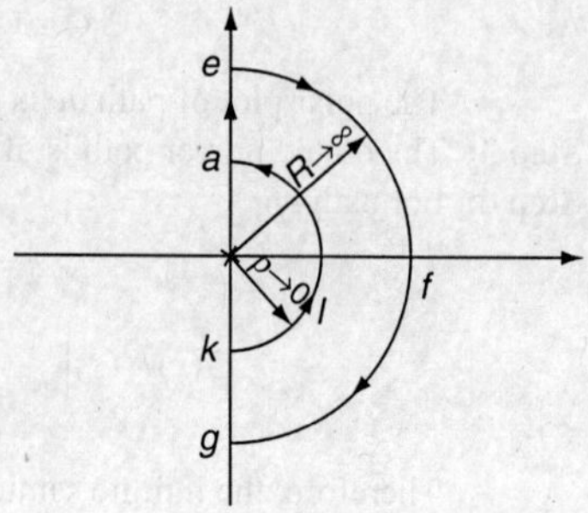

Fig. E15.13

Example 15.13 For $G(s)H(s) = 20/[(s+3)(s^2+2s+2)]$, draw the Nyquist plot and decide the stability.

Solution

Step 1: Poles of $G(s)H(s)$ are at $s = -3$ and $s = -1 \pm j$. There is no pole at the origin and on the $j\omega$ axis.

Step 2: The Nyquist path is shown in Fig. E15.13.

For path *ae*

$$s = j\omega \text{ and } 0 < \omega < \infty$$

Here, $$M(j\omega) = G(j\omega)H(j\omega) = \frac{20}{(j\omega+3)(j^2\omega^2+2j\omega+2)} = \frac{20}{(j\omega+3)(j\omega+1+j)(j\omega+1-j)}$$

$\therefore$ $$|G(j\omega)H(j\omega)| = \frac{20}{\sqrt{\omega^2+9}\sqrt{(2-\omega^2)+4\omega^2}}$$

and $$\angle G(j\omega)H(j\omega) = -\tan^{-1}\frac{\omega}{3} - \tan^{-1}\left(\frac{2\omega}{2-\omega^2}\right) = -\tan^{-1}\frac{\omega}{3} - \tan^{-1}(\omega+1) - \tan^{-1}(\omega-1)$$

At $\omega = 0$, $|G(j\omega)H(j\omega)| = \frac{10}{3}$ and $\angle G(j\omega)H(j\omega) = 0°$

At $\omega = \infty$, $|G(j\omega)H(j\omega)| = 0$ and $\angle G(j\omega)H(j\omega) = -270°$

The polar plot of path *ae* is shown in Fig. E15.13(a).

Step 3: The dotted mirror path is of the polar plot of the path *ae* is the polar plot of the path *ga*.

Step 4: For path *efg*

$$s = Re^{j\theta} \text{ where } R \to \infty \quad \text{and} + 90° < \theta < -90°.$$

$$G(s)H(s)\Big|_{efg} = \frac{20}{\underset{R\to\infty}{\text{Lt}}\ (Re^{j\theta}+3)[(Re^{j\theta})^2 + Re^{j\theta} + 2]} = 0$$

Therefore, the infinite semicircle *efg* maps onto a point.

Step 5: Connected.

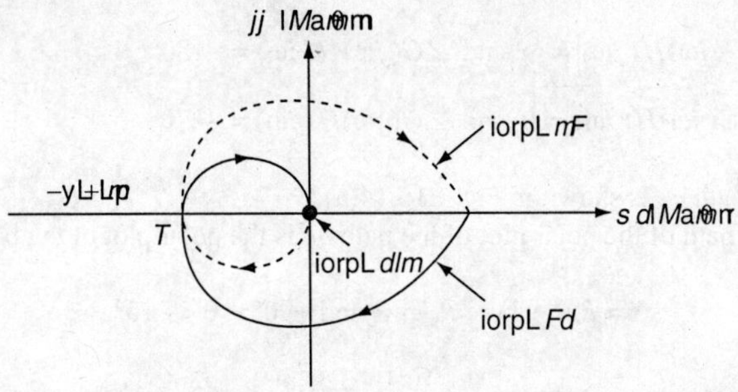

Fig. E15.13(a)

Step 6: To get Q of Fig. E15.13(a),

$$G(j\omega)H(j\omega) = \frac{20}{(j\omega+3)(2-\omega^2+2j\omega)} = \frac{20(3-j\omega)(2-\omega^2-2j\omega)}{(\omega^2+9)[(2-\omega^2)^2+4\omega^2]}$$

$$= \frac{20(6-7\omega^2)}{(\omega^2+9)[(2-\omega^2)^2+4\omega^2]} + j\frac{20(\omega^2-8)}{(\omega^2+9)[(2-\omega^2)^2+4\omega^2]}$$

Equating the imaginary part to zero, we get $\omega = 2\sqrt{2}$.

$\therefore$ Point $$Q = \frac{20(6-7\times 8)}{(8+9)[(2-8)^2+4\times 8]} = -0.865$$

The point (–1, 0) is outside the plot. Hence, there is no encirclement about the origin and $N = 0$.

Step 7: Since $P = 0$ (number of poles in the RHP) = poles of $G(s)H(s)$, here, the number of zeros of $1+G(s)H(s)$ in the RHP is 0,

$\therefore$ $Z = 0$ and hence the system is stable.

Gain margin

$$\text{GM} = \frac{1}{|OQ|} = \frac{1}{0.865} = 1.156$$

$$= 20\log_{10}\frac{1}{|OQ|} = 20\log_{10}1.156 = +1.259\,\text{dB}$$

Example 15.14 For $G(s)H(s) = 10(1+0.5s)/[s^2(1+0.1s)(1+0.03s)]$, draw the Nyquist plot and decide stability.

Solution

Step 1: Poles of $G(s)H(s)$ are at $s = 0, 0, -10$ and -33.33. There are two poles at the origin and no pole on the $j\omega$ axis.

Step 2: The Nyquist path is shown in Fig. E15.14.

For path ae, $s = j\omega$ and $0 < \omega < \infty$

Here, $$M(j\omega) = G(j\omega)H(j\omega) = \frac{10(1+j0.5\omega)}{-\omega^2(1+j0.1\omega)(1+j0.03\omega)}$$

$$|G(j\omega)H(j\omega)| = \frac{10\sqrt{1+0.25\omega^2}}{\omega^2\sqrt{1+0.01\omega^2}\sqrt{1+9\times10^{-4}\omega^2}}$$

and $\angle G(j\omega)H(j\omega) = +\tan^{-1}0.5\omega - 2\times 90° - \tan^{-1}0.1\omega - \tan^{-1}0.03\omega$

At $\omega = 0$, $|G(j\omega)H(j\omega)| = \infty$ and $\angle G(j\omega)H(j\omega) = -180°$

At $\omega = \infty$, $|G(j\omega)H(j\omega)| = 0$ and $\angle G(j\omega)H(j\omega) = -270°$

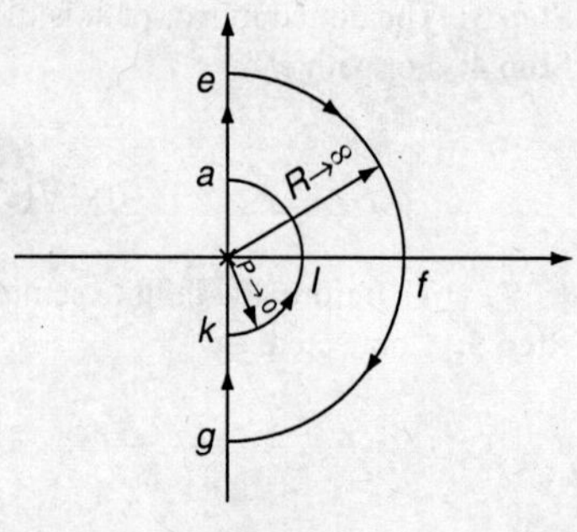

Fig. E15.14

The polar plot of path ae is shown in Fig. E15.14(a),

Step 3: The dotted mirror path of the polar plot of the path ae is the polar plot of the path gk.

Step 4: For path efg

$$s = Re^{j\theta} \text{ where } R \to \infty \text{ and } +90° < \theta < -90°.$$

$$G(s)H(s)\Big|_{efg} = \operatorname*{Lt}_{R\to\infty} \frac{10(1+0.5Re^{j\theta})}{(Re^{j\theta})^2(1+0.1Re^{j\theta})(1+0.03Re^{j\theta})} = 0$$

Therefore, the infinite semicircle efg maps onto a point.

Step 5: For path kla

$$s = \rho e^{j\theta} \text{ where } \rho \to 0 \text{ and } -90° < \theta < +90°.$$

$$G(s)H(s)\Big|_{kla} = \operatorname*{Lt}_{\rho\to 0} \frac{10(1+0.5\rho e^{j\theta})}{(\rho e^{j\theta})^2(1+0.1\rho e^{j\theta})(1+0.03\rho e^{j\theta})} = \infty e^{-2j\theta}$$

Therefore, the phase angle of $G(s)H(s)\big|_{kla}$ varies from +180° to −180° since θ varies from −90° to +90°.

Step 6: Connected

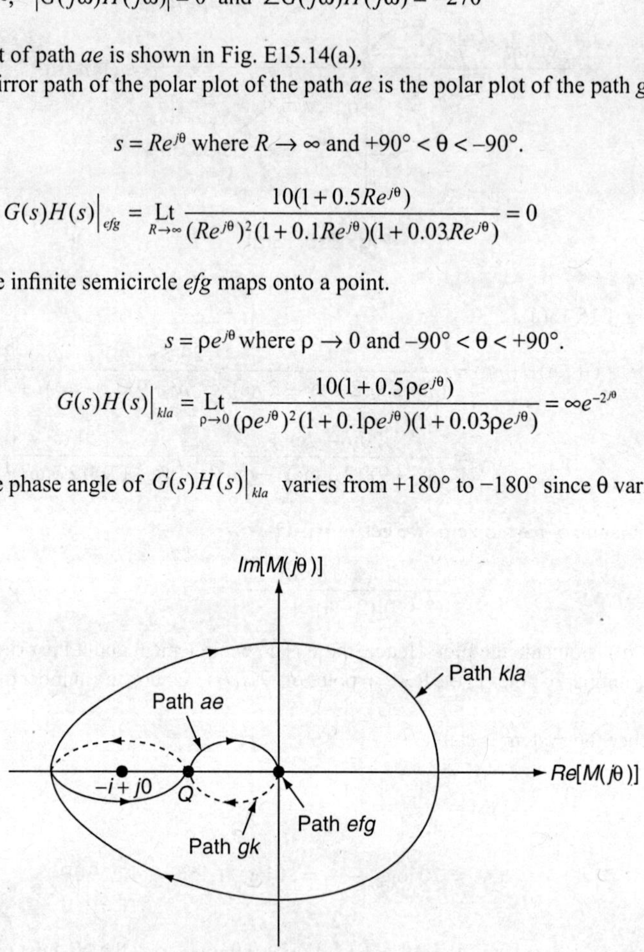

Fig. E15.14(a)

Step 7: To calculate Q.

$$G(j\omega)H(j\omega) = \frac{10(1+j\,0.5\omega)}{-\omega^2(1+j\,0.1\omega)(1+j\,0.03\omega)} = \frac{10(1+0.062\omega^2)}{D}\, j\frac{(0.37\omega - 0.0015\omega^3)}{D}$$

where $$D = -\omega^2(1+0.01\omega^2)(1+9\times10^{-4}\omega^2)$$

Equating the imaginary part to zero, we get

$$\omega = 15.70 \text{ rad/sec}.$$

Point $$Q = \left.\frac{10(1+0.062\omega^2)}{-\omega^2(1+0.01\omega^2)(1+9\times10^{-4}\omega^2)}\right|_{\omega=10.64} = -0.156$$

The point (−1, 0) is encircled once in the clockwise and once in the anticlockwise direction. Here, $N = 0$.

Step 8: Since $P = 0$ (number of poles in the RHP) = poles of $G(s)H(s)$, the number of zeros of $1+G(s)H(s)$ in the RHP is 0.

∴ $Z = 0$ and hence the system is stable.

Example 15.15 For $G(s)H(s) = K/[s(s+1)(s+3)]$, draw the Nyquist plot and hence calculate the range of values of K for stability.

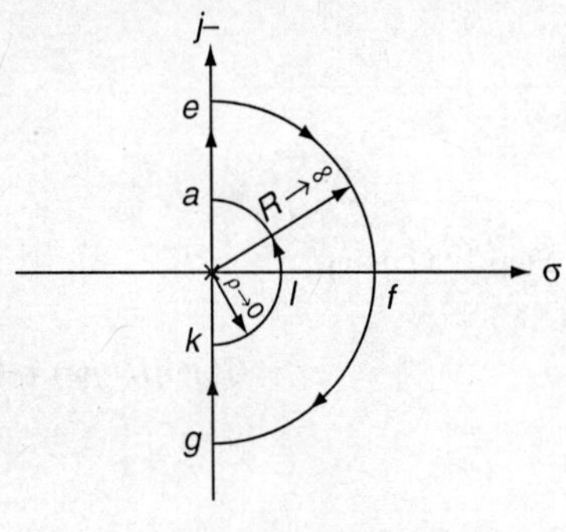

Fig. E15.15

Solution

Step 1: Poles of $G(s)H(s)$ are at $s = 0$, $s = -1$ and $s = -3$. There is a pole at the origin and no pole on the $j\omega$ axis.

Step 2: The Nyquist path is shown in Fig. E15.15.

For path *ae*

$$s = j\omega \text{ and } 0 < \omega < \infty$$

Here, $$M(j\omega) = G(j\omega)H(j\omega) = \frac{K}{j\omega(j\omega+1)(j\omega+3)} = \frac{K}{\omega\sqrt{\omega^2+1}\sqrt{\omega^2+9}}\angle\left(-90^\circ - \tan^{-1}\omega - \tan^{-1}\frac{\omega}{3}\right)$$

∴ $$|G(j\omega)H(j\omega)| = \frac{K}{\omega\sqrt{\omega^2+1}\sqrt{\omega^2+9}}$$

and $$\angle G(j\omega)H(j\omega) = -90^\circ - \tan^{-1}\omega - \tan^{-1}\frac{\omega}{2}$$

At $$\omega = 0,\quad |G(j\omega)H(j\omega)| = \infty \text{ and } \angle G(j\omega)H(j\omega) = -90^\circ$$

At $$\omega = \infty,\quad |G(j\omega)H(j\omega)| = 0 \text{ and } \angle G(j\omega)H(j\omega) = -270^\circ$$

The polar plot of path *ae* is shown in Fig. E15.15(a).

Step 3: The dotted mirror path of the polar plot of the path *ae* is the polar plot of the path *gk*.

Step 4: For path *efg*

$$s = Re^{j\theta} \text{ where } R \to \infty \text{ and } +90^\circ < \theta < -90^\circ.$$

$$G(s)H(s)\Big|_{efg} = \frac{K}{\operatorname*{Lt}_{R\to\infty} Re^{j\theta}(Re^{j\theta}+1)(Re^{j\theta}+3)} = 0$$

Therefore, the infinite semicircle *efg* maps onto a point.

Step 5: For path *kla*

$$s = \rho e^{j\theta} \text{ where } \rho \to 0 \text{ and } -90^\circ < \theta < +90^\circ.$$

$$G(s)H(s)\Big|_{kla} = \frac{K}{\operatorname*{Lt}_{\rho\to 0} (\rho e^{j\theta})^2(\rho e^{j\theta}+1)(\rho e^{j\theta}+3)} = \infty e^{-j\theta}$$

Therefore, the phase angel of $G(s)H(s)\Big|_{kla}$ varies from +90° to −90°, since θ varies from −90° to +90°.

Step 6: Connected

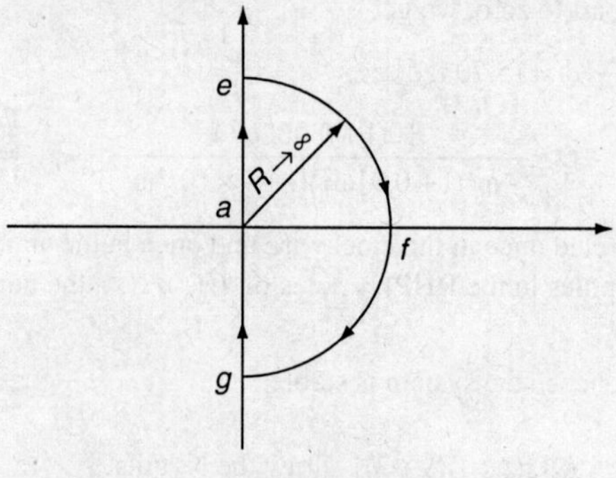

Fig. E15.15(a)

Step 7: Q point

$$G(j\omega)H(j\omega)=\frac{K}{(j\omega)(j\omega+1)(j\omega+3)}=\frac{K(-j\omega)(-j\omega+1)(-j\omega+3)}{(-j^2\omega^2)(-j^2\omega^2+1)(-j^2\omega^2+9)}$$

$$=\frac{-4K\omega^2-jK\omega(-\omega^2+3)}{(\omega^2)(\omega^2+1)(\omega^2+9)}$$

$$=\frac{-4K\omega^2}{(\omega^2)(\omega^2+1)(\omega^2+9)}+\frac{jK\omega(-\omega^2+3)}{(\omega^2)(\omega^2+1)(\omega^2+9)}$$

Equating the imaginary part to zero, we have

$$\frac{jK\omega(-\omega^2+3)}{(\omega^2)(\omega^2+1)(\omega^2+9)}=0$$

i.e., $$\omega=\text{either } 0 \text{ or } \sqrt{3}$$

Therefore, $\omega_{pc}=\sqrt{3}$ because ω_{pc} is positive.

$$G(j\omega)H(j\omega)\Big|_{\omega_{pc}=\sqrt{2}}=-\frac{4K\times 3}{3\times 4\times 12}+j0=-\frac{K}{12}+j0$$

Step 8: For absolute stability, $N=0$, i.e., the point $-1+j0$ should locate to the left side of point Q.

Therefore, $$|OQ|<1$$

i.e., $$\left|-\frac{K}{12}\right|<1$$

$$K<12.$$

For stability, the range of K is $0<K<12$.

MATLAB Program

Program 3 Sketch the nyquist plot for the open loop transfer function $G(s)H(s) = \dfrac{30}{s(s+1)(s+4)}$.

Solution

```
>> num = [ 0 30]
num =
   0   30
>> q1 = [1 0];
>> q2 = [1 1];
>> q3 = [1 4];
>> den = conv(q1,q2);
>> den = conv(den,q3)
den =
   1   5   4   0
>> sys=tf(num,den)
Transfer function:
   30
-----------------
s^3 + 5 s^2 + 4 s
>> nyquist(sys)
>>
```

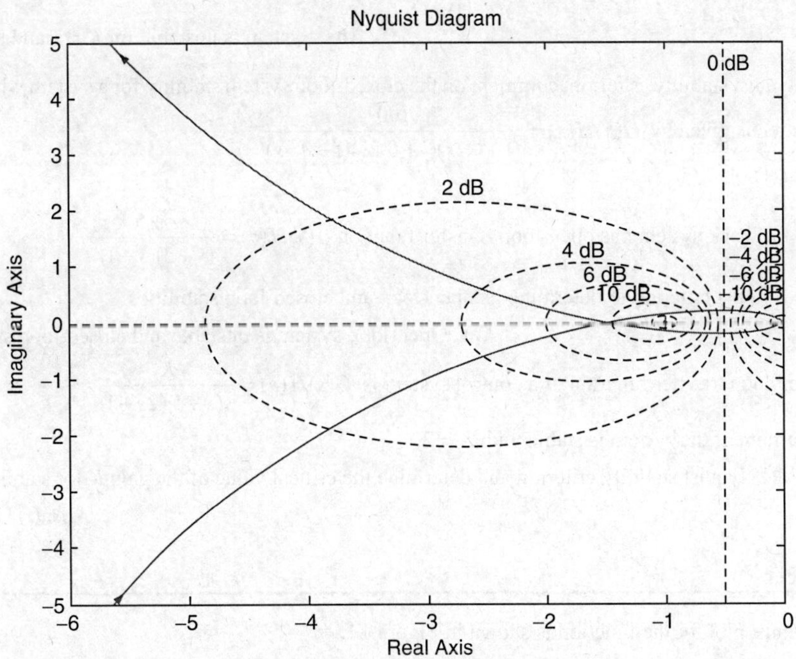

Fig. P3

SHORT QUESTIONS WITH ANSWERS

1. Define the following:
(i) Encircled, (ii) enclosed, (iii) analytic function and (iv) single-valued function.

Ans. (i) If point is found to be inside the path, the point is said to be encircled by the closed path.
(ii) If a path is traversed in the clockwise direction and a point is found to lie to the right of the path, the point is said to be enclosed by the path.
(iii) A function is said to be analytic at a point in a plane if its value and derivative have finite existence at that point.
(iv) A function is said to be single valued if it has only one value for each separate value of s. If $F(s)=\sqrt{s}$ and $s=36$, the values of $F(s)$ are -6 and $+6$. The function $F(s)=\sqrt{s}$ is a not a single-valued function. On the other hand, the function $F(s)=s$ is a single-valued function.

2. What do you mean by Nyquist criterion?

Ans. If focusses also on the relative stability of the system. It is possible to determine the stability of a closed-loop pole from an open-loop pole without knowing the roots of the closed-loop system. A Nyquist plot is based on a polar plot.

EXERCISES

Questions For Practice

1. Using Nyquist stability criterion, comment on the closed-loop system stability for a system whose loop transfer function is given by $G(s)H(s)=\dfrac{K(s+3)}{s(s-1)}$.

Ans. System is unstable for $K<1$ and stable for $K>1$

2. Using Nyquist stability criterion, comment on the closed-loop system stability for a system whose loop transfer function is given by $G(s)H(s)=\dfrac{100}{(1+0.1s)(1+0.2s)(1+0.3s)}$.

Ans. Stable

3. A unity feedback system has open-loop transfer function $G(s)H(s)=\dfrac{s+5}{(s+3)(s-1)}$.

Use the Nyquist criterion to determine its open-loop and closed-loop stabilities.

Ans. Open-loop system is unstable and closed-loop system is stable.

4. The open-loop transfer function of a control system is $G(s)H(s)=\dfrac{K}{s(s+1)(2s+1)}$.

(i) Determine if the system is stable with $K=2$.
(ii) Use the Nyquist stability criterion and determine the critical value of the gain K for stability.

Ans. (i) Unstable (ii) 1.5

TEST YOURSELF

1. The Nyquist plot for the function is shown in Figure T1.

$$G(s)H(s)=\frac{K}{s(1+T_1s)(1+T_2s)}$$

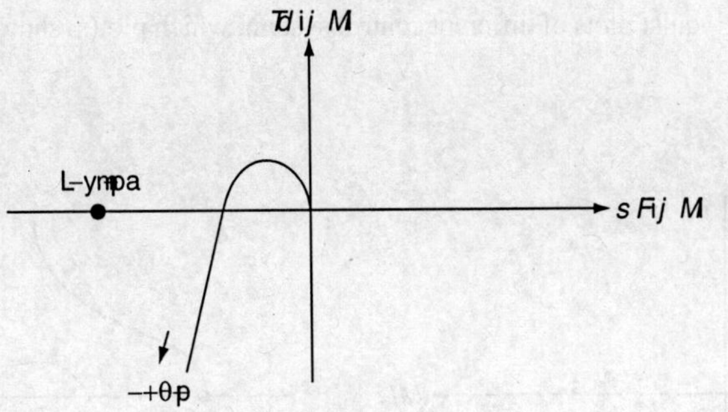

Figure T1

The condition for stability is given by

(a) $\frac{K(T_1+T_2)}{T_1T_2}<1$ (b) $\frac{K(T_1+T_2)}{T_1T_2(T_1-T_2)}<1$ (c) $\frac{T_1T_2}{K(T_1+T_2)}<1$ (d) $\frac{T_1-T_2}{K(T_1+T_2)}<1$

2. The number of points encircled by *X* and *Y* in Figure T2 is

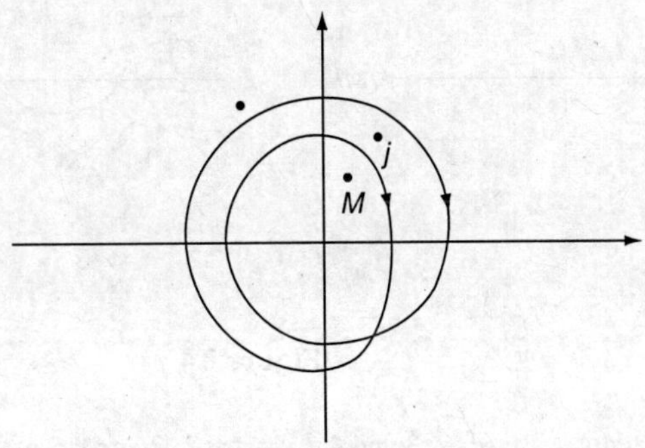

Figure T2

(a) 2, 1 (b) 1, 2 (b) 1, 1 (d) 2, 2

3. Gain margin is the reciprocal of the gain at the frequency at which the phase angle is

(a) 90° (b) 180° (b) –180° (d) 0°

4. Nichol's chart is useful for determining the
   (a) open-loop frequency response
   (b) open-loop and closed-loop frequency responses
   (c) closed-loop frequency response
5. For positive frequencies, the Nyquist plot of an OLTF is shown in Figure T3. The gain margin is

   (a) 6 (b) 8 (c) 5 (d) 10

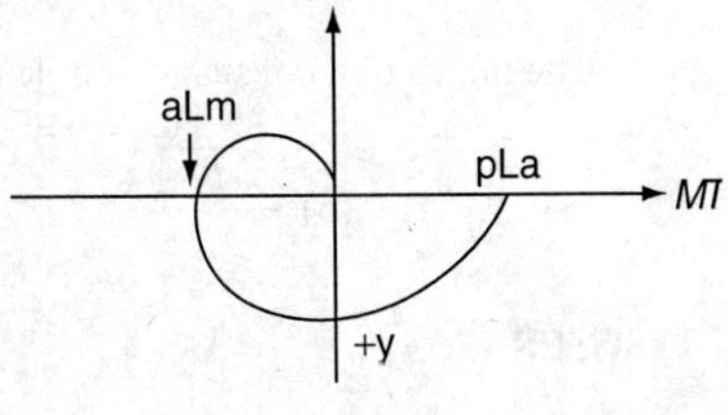

Figure T3

6. For the following Nyquist plots of different control system, which plot(s) shown in Figure T4 represents(s) a stable system.

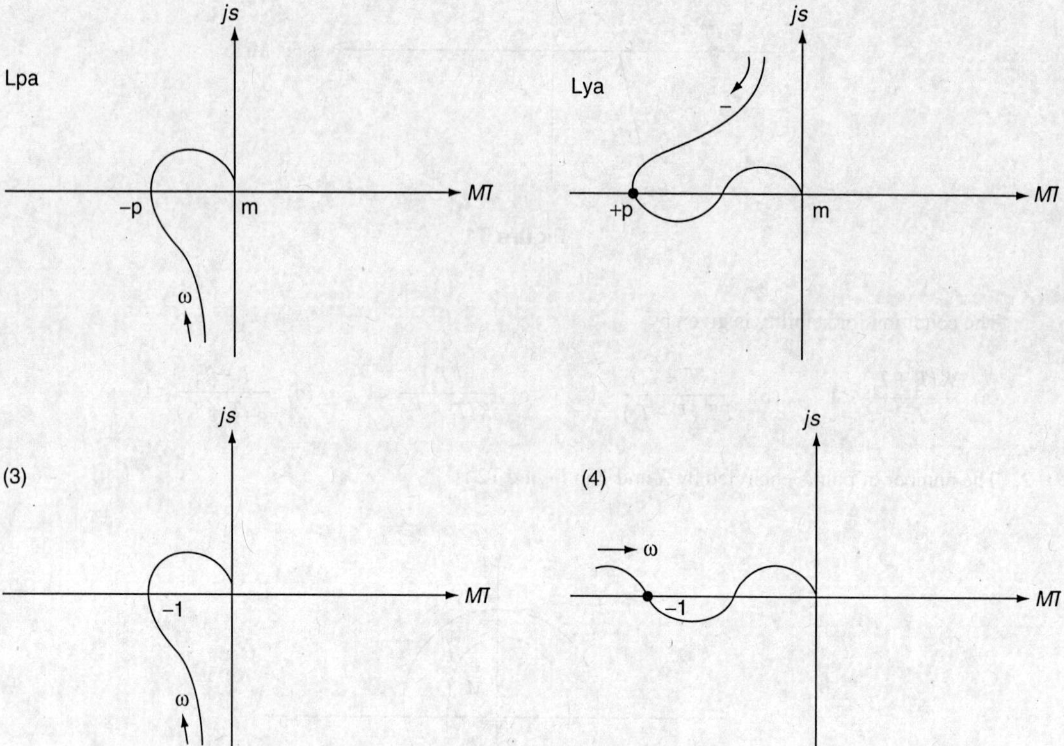

Figure T4

(a) 1 (b) 1, 3 and 4 (c) 2, 3 and 4 (d) 1, 2 and 4

7. For a constant M circle the centre will lie at $[M^2/(1-M^2),0]$ in G-plane and has a radius of

(a) M (b) $\frac{M}{1-M^2}$ (c) $1-M^2$ (d) $\frac{M}{1+M^2}$

8. For $M=1$ the constant M circle will be a straight line parallel to the imaginary axis at x equal to

(a) $-\frac{1}{2}$ (b) $\frac{1}{2}$ (c)1 (d) –2

9. The radius of a constant N circle at centre $\left(-\frac{1}{2},\frac{1}{2N}\right)$ for a given phase angle is

(a) $\frac{1}{N}$ (b) $\frac{\sqrt{N+1}}{2N}$ (c) $\frac{2N}{\sqrt{N+1}}$ (d) N

ANSWERS

1 (a) 2(b) 3(b) 4(c) 5(c) 6(c) 7(d) 8(a) 9(b)

Chapter **16**

Introduction to Compensators

A well-designed control system must have less error, good accuracy, good speed of response, good relative stability, good damping so that undue overshoots etc. can be avoided. Usually, first priority is given to gain adjustment. Practically, it has been found that adjusting the gain alone, it is very difficult to obtain a satisfactory result. A higher value of gain improves the steady-state behaviour, but transient response of the system becomes poor. Therefore, design of a control system is a challenging task for a control system engineer. To fulfil the required specification, it is required to redesign a control system. An external device is added to the system. This process of redesign by adding an external device is called compensation of control system, whereas the external device is known as a compensator. The aim of this chapter is to introduce compensators.

16.1 Types of Compensation

An external device termed compensator can be added in a system anywhere depending upon the requirement. Depending upon the introduction of a compensator, there are the following types of compensation:

1. Series compensation,
2. Parallel compensation and
3. Series-parallel compensation.

16.1.1 Series Cascade Compensation

When the compensator is placed in series with the forward path transfer function of the plant, the scheme is known as series compensation. It is also termed as cascade compensation. Figure 16.1 shows the arrangement of series compensation.

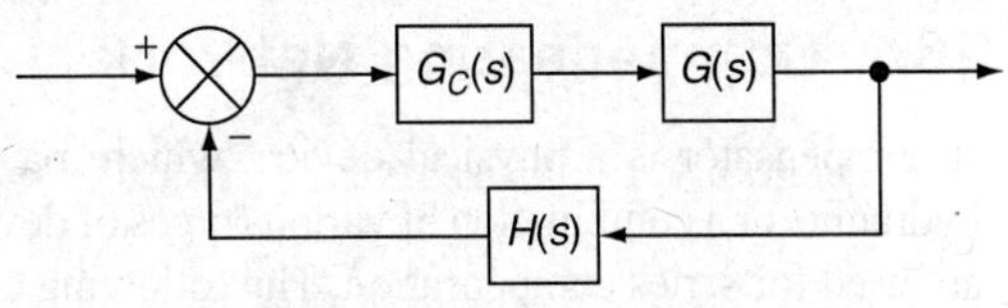

Fig. 16.1 Series compensation

In this scheme, the signal flows from lower energy level towards a higher energy level. Therefore, additional amplifiers are required to increase the gain as well as to provide necessary isolation. A series scheme has more number of components compared to a parallel scheme.

16.1.2 Parallel Compensation

If the feedback is taken from some internal element and a compensator is introduced in such a feedback path to provide an additional internal feedback loop, this scheme is known as feedback compensation or parallel compensation. Figure 16.2 shows the parallel compensation.

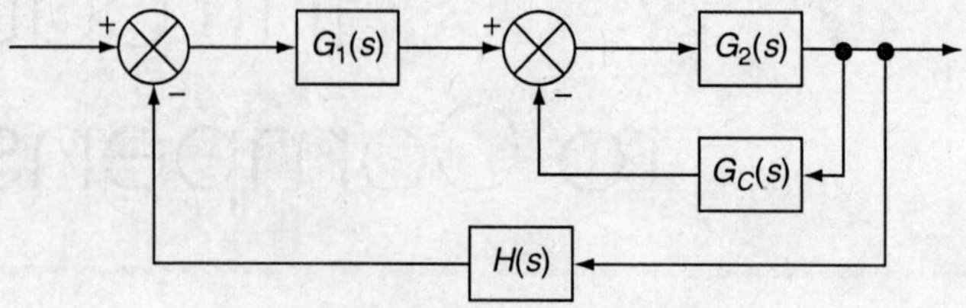

Fig.16.2 Parallel compensation

In this scheme the flow of signal is from a higher energy level to a lower energy level point. Therefore, the additional amplifiers are not required.

16.1.3 Series-Parallel Compensation

Sometimes a situation demands to provide both types of compensation, series as well as feedback. This scheme is known as series-parallel compensation. Figure 16.3 shows series-parallel compensation.

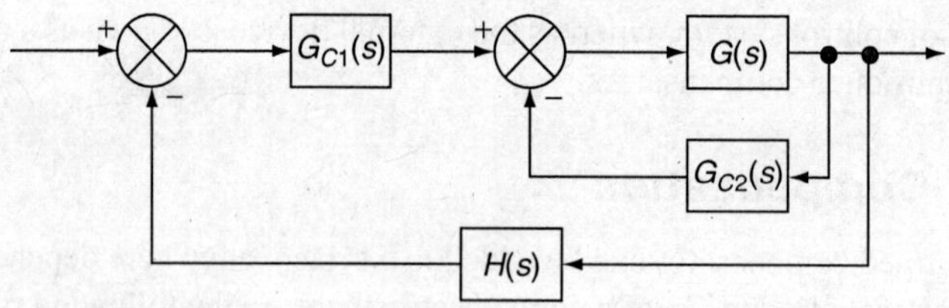

Fig.16.3 Series-parallel compensation

To select the proper compensation scheme, the nature of the signals available in the system, the power levels at the various points, available components, the economic considerations and the designer's experience are important.

16.2 Compensating Network

A compensator is a physical device, which may be an electrical network, mechanical unit, pneumatic, hydraulic, or a combination of various types of devices. Here, we discuss only the electrical networks, which are used for series compensation. The following electrical compensating networks are generally used:

1. Lead network or lead compensator
2. Lag network or lag compensator
3. Lag-lead network or lag-lead compensator

If a sinusoidal input is applied to a network, a sinusoidal steady-state output having a phase is obtained. If the steady-state output leads the input, the network is called lead network. If the steady-state output lags behind the input, the network is called lag network. Both phase lag and lead occur in the lag-lead network in different frequency regions. The phase lag occurs in the low frequency region, whereas the phase lead occurs in the high frequency region. The design of each compensator as a cascade using bode plot is discussed in respective article. Root locus technique for designing each as a cascade is discussed later on (Art. 16.3 onwards).

16.2.1 Lead Compensator

Figure 16.4 shows an electrical network, which is a lead compensating network.

To obtain the transfer function of such an electrical lead network, let us assume the circuit is unloaded. Applying, KCL for the output node, we have

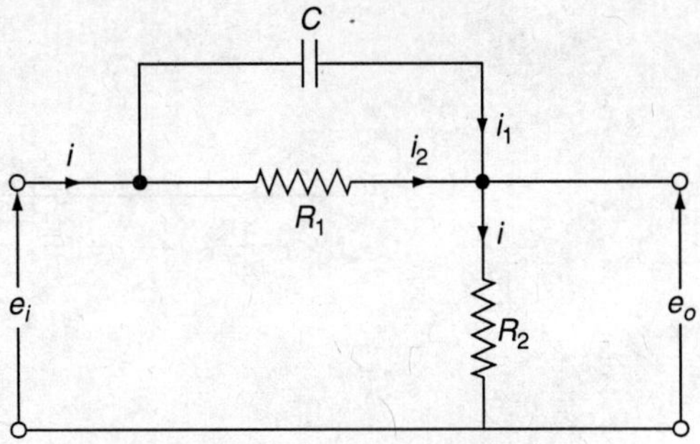

Fig. 16.4 Lead compensation

$$i_1 + i_2 = i$$

$$C\frac{d(e_i - e_o)}{dt} + \frac{1}{R_1}(e_i - e_o) = \frac{1}{R_2}e_o \qquad (16.1)$$

Taking Laplace transform of Eq. (16.1), we have

$$sCE_i(s) - sCE_o(s) + \frac{1}{R_1}E_i(s) - \frac{1}{R_1}E_o(s) = \frac{1}{R_2}E_o(s)$$

i.e.,
$$E_i(s)\left[sC + \frac{1}{R_1}\right] = E_o(s)\left[sC + \frac{1}{R_1} + \frac{1}{R_2}\right]$$

i.e.,
$$\frac{E_o(s)}{E_i(s)} = \frac{\left(s + \frac{1}{R_1C}\right)}{s + \frac{(R_1 + R_2)}{R_1R_2C}} = \frac{\left(s + \frac{1}{R_1C}\right)}{\left[s + \frac{1}{\left(\frac{R_2}{R_1 + R_2}\right)R_1C}\right]}$$

i.e.,
$$\frac{E_o(s)}{E_i(s)} = \frac{s + \frac{1}{T}}{s + \frac{1}{\alpha T}} \qquad (16.2)$$

where
$$T = R_1C \text{ and } \alpha = \frac{R_2}{R_1 + R_2} < 1.$$

The minimum value of α is generally 0.05.

The lead compensator has a zero at $s = -1/T$ and a pole at $s = -\dfrac{1}{\alpha T}$.

Since $0 < \alpha < 1$, the zero is always located to the right of the pole. Figure 16.5 shows the pole zero plot of a lead compensator.

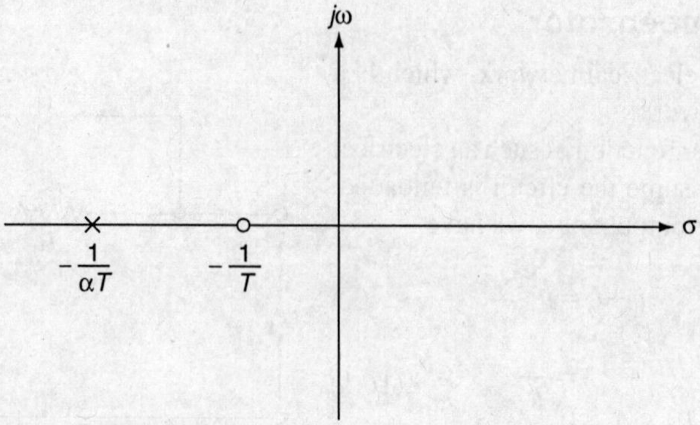

Fig. 16.5 Pole-zero plot of lead compensator

Maximum lead angle ϕ_m and α To obtain the maximum lead angle ϕ_m, substituting $s = j\omega$ in Eq. (16.2), we have

$$\frac{E_o(j\omega)}{E_i(j\omega)} = \frac{\alpha(1+j\omega T)}{(1+j\omega\alpha T)} \tag{16.3}$$

$\therefore$

$$\left|\frac{E_o(j\omega)}{E_i(j\omega)}\right| = M = \frac{\alpha\sqrt{1+\omega^2T^2}}{\sqrt{1+\omega^2\alpha^2T}} \tag{16.4}$$

The phase angle ϕ is given by

$$\phi = \tan^{-1}\omega T - \tan^{-1}\omega\alpha T \tag{16.5}$$

To get ω_m for which the phase angle ϕ is maximum, let us put

$$\left.\frac{d\phi}{d\omega}\right|_{\omega=\omega_m} = 0$$

i.e.,

$$\left.\frac{d\phi}{d\omega}\right|_{\omega=\omega_m} = \frac{d}{d\omega}\left[\tan^{-1}\omega T - \tan^{-1}\omega\alpha T\right]\Big|_{\omega=\omega_m} = 0$$

i.e.,

$$\frac{T}{1+\omega_m^2T^2} - \frac{\alpha T}{1+\alpha^2\omega_m^2T^2} = 0$$

i.e.,

$$T(1+\alpha^2\omega_m^2T^2) - \alpha T(1+\omega_m^2T^2) = 0$$

i.e.,

$$(1+\alpha^2\omega_m^2T^2 - \alpha - \alpha\omega_m^2T^2) = 0$$

$$\omega_m^2\alpha T^2(\alpha-1) = -(1-\alpha)$$

i.e., $$\omega_m^2 = \frac{1}{\alpha T^2}$$

i.e., $$\omega_m = \frac{1}{T\sqrt{\alpha}} \tag{16.6}$$

Equation (16.6) shows the frequency at which the phase lead is at its maximum and ω_m is the geometric mean of the two corner frequencies of the compensator, which are $\omega_{c1} = 1/T$ and $\omega_{c2} = \dfrac{1}{\alpha T}$.

From Eq. (16.5), we have

$$\tan\phi = \tan[\tan^{-1} - \omega T - \tan^{-1}\alpha\omega T]$$

$\therefore$ $$\tan\phi = \frac{\omega T - \alpha\omega T}{1+\omega T.\alpha\omega T} = \frac{\omega T(1-\alpha)}{1+\omega^2 T^2\alpha}$$

At $$\omega = \omega_m = \frac{1}{T\sqrt{\alpha}}$$

$$\tan\phi_m = \frac{(1-\alpha)}{\sqrt{\alpha}(1+1)} = \frac{1-\alpha}{2\sqrt{\alpha}} \tag{16.7}$$

$\therefore$ $$\sin\phi_m = \frac{1-\alpha}{1+\alpha} \tag{16.8}$$

i.e., $$\phi_m = \sin^{-1}\left(\frac{1-\alpha}{1+\alpha}\right) \tag{16.8a}$$

Equation (16.8) relates α and the maximum lead angle ϕ_m.

Polar plot of a lead compensator From Eq. (16.4) and Eq. (16.5), we get

At $\omega = 0$, $M = \alpha$ and $\phi = 0°$

At $\omega = \infty$, $M = 1$ and $\phi = 0°$

Here both the points, starting as well as terminating, are on the positive real axis. The magnitude is always positive for $0 < \omega < \infty$ and ϕ will be always positive for $\alpha < 1, \tan^{-1}\omega T > \tan^{-1}\alpha\omega T$. Therefore, all points of a polar plot are always in the first quadrant of the complex plane.

Figure 16.6 shows the polar plot of a lead compensator.

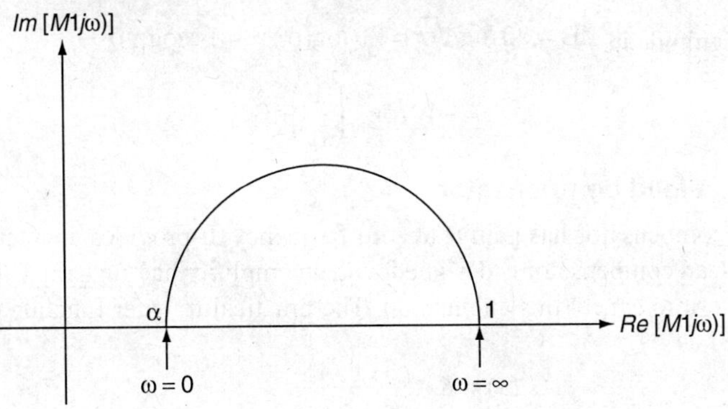

Fig. 16.6 Polar plot of lead compensator

Bode plot of a lead compensator The lead compensator has the following corner frequencies:

$$\omega_{c1} = \frac{1}{T} \text{ for a simple zero at } s = -\frac{1}{T},$$

$$\omega_{c2} = \frac{1}{\alpha T} \text{ for a simple zero at } s = -\frac{1}{\alpha T}$$

and $K = \alpha$.

Figure 16.7 shows the Bode plot of a lead compensator. The line with zero slope till $\omega_{c1} = \frac{1}{T}$. After that a line of slope +20 dB/decade till $\omega_{c2} = \frac{1}{\alpha T}$. Again there will be a line of zero slope from ω_{c2} to α. Since $\alpha < 1$, the straight line $20\,log_{10}\,\alpha$ is below 0dB.

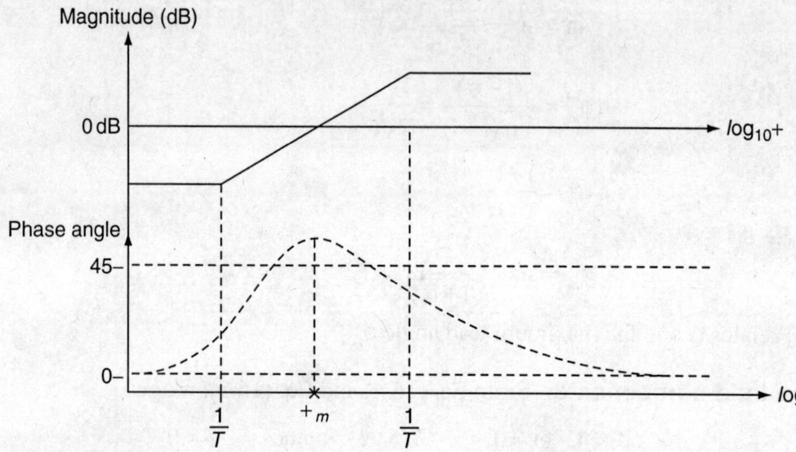

Fig. 16.7 Bode plot of a lead compensator

At $\omega = \omega_m$,

$$M = \frac{\alpha\sqrt{1+\omega_m^2 T^2}}{\sqrt{1+\omega_m^2 \alpha^2 T^2}} = \frac{\alpha\sqrt{1+\frac{1}{\alpha}}}{\sqrt{1+\alpha}}$$

At $\omega = \omega_m$, the magnitude in $\text{dB} = 20\,log\sqrt{\alpha} = 20\,log(\alpha)^{1/2} = 10\,log_{10}\,\alpha$

$$= -10\,log_{10}\left(\frac{1}{\alpha}\right)\text{dB}$$

Steps to design a lead compensator

Step 1: The lead compensator has gain α at zero frequency. It provides an attenuation for $\alpha < 1$. The practical lead compensator is designed with an amplifier having gain K_c in series with the basic lead network to cancel this attenuation. The practical transfer function of a lead compensator is assumed as

$$G_c(s) = K_c\alpha\frac{(1+Ts)}{(1+\alpha Ts)}$$

$$G_C(s) = \frac{K(1+Ts)}{(1+\alpha Ts)} \quad (16.9)$$

where $$K = \text{DC gain} = K_c\alpha \quad (16.10)$$

$$\therefore \quad G_C(s)G(s) = \frac{K(1+Ts)}{(1+\alpha Ts)}G(s) = \frac{(1+Ts)}{(1+\alpha Ts)}KG(s) = \frac{(1+Ts)}{(1+\alpha Ts)}G_1(s) \quad (16.11)$$

where $$G_1(s) = KG(s)$$

Equation (16.11) shows the open-loop transfer function of the compensated network. Usually, one of the error constants is given as specification and the value of K is determined from the given value of K.

Step 2: The Bode plot of $G_1(j\omega)$ is drawn using the value of K. The phase margin is computed from the Bode plot.

Step 3: Generally, PM is given for a design problem

Let $$\phi = \text{PM given}$$

$$\phi = \text{PM obtained in the step 2}$$

The necessary phase lead ϕ_m required to be added is determined using relation $\phi_m = \phi + \varepsilon$ where ε is the margin of safety as crossover frequency may shift due to compensation and $\varepsilon = 5° - 15°$.

Step 4: From the Eq. (16.8)

$$\sin\phi_m = \frac{1-\alpha}{1+\alpha},$$

the value of α can be determined.

Step 5: The frequency ω_m is calculated so that the magnitude of the uncompensated system is $-10\,log_{10}\left(\frac{1}{\alpha}\right)$ dB at $\omega = \omega_m$. This frequency is taken as new gain crossover frequency, where

$$\omega_m = \frac{1}{T\sqrt{\alpha}}$$

Here, α is known. Therefore, we can determine $1/T$.

Step 6: The two following corner frequencies of the lead compensator are determined.

$$\omega_{c1} = \frac{1}{T} \text{ and } \omega_{c2} = \frac{1}{\alpha T}$$

Step 7: Since $K = K_c\alpha$, we can obtain the value of K_c.

Step 8: If the gain margin of the compensated system is not satisfactory, the design is repeated by modifying the pole-zero location of the compensator till a satisfactory result is obtained.

Effects of a lead compensation The various effects of a lead compensation are as follows:

1. Since a lead compensator adds a dominant zero and a pole, the damping of the closed-loop system is increased.
2. The less overshoot, less rise time and less settling time are obtained due to increase of damping and hence there is improvement in the transient response.
3. It improves the phase margin of the closed-loop system.

4. Improvement of gain and phase margins and hence the relative stability occurs due to the reduction in the slope of the magnitude plot in a Bode diagram of the forward path transfer function at the gain crossover frequency.
5. Bandwidth of the closed-loop system is increased and hence the response is faster.
6. The steady-state error does not get affected.

Limitations of a lead compensation The various limitations of the lead compensation are as follows:

1. Since an additional increase in gain to offset the attenuation inherent in the lead network is required, the effect of a larger gain requirement in most of the cases implies larger space, more elements, greater weight and higher cost.
2. The noise entering the system is more susceptible to the noise signals due to increase in the high frequency gain and hence more bandwidth is desirable sometimes.
3. Since the compensated system may have a larger undershoot than overshoot, there is tendency to over compensate a system, which may lead to a conditionally stable system.
4. From a single lead network, the maximum lead angle available is about 60°. For lead of more than 70°–90°, a multistage lead compensator is required.

Example 16.1 A system is shown in Fig. E16.1. Design a lead compensation such that the closed-loop system satisfies the following specifications:

Static velocity error constant = 24 s^{-1}.
Phase margin = 55°.
Gain Margin ≥ 13 dB

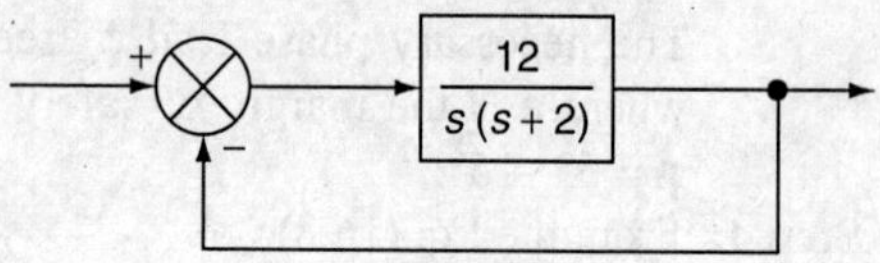

Fig. E16.1

Solution

Step 1: A lead compensator is assumed as follows:

$$G_c(s) = K_c\alpha\frac{(1+Ts)}{(1+\alpha Ts)} = \frac{K(1+Ts)}{(1+\alpha Ts)}$$

and
$$G_1(s) = KG(s) = \frac{12K}{s(s+2)}$$

$$K_v = 24 = \underset{s\to 0}{\text{LT}}\; sG(s)E_c(s)$$

i.e.,
$$24 = \underset{s\to 0}{\text{LT}}\; s\cdot\frac{12K(1+Ts)}{s(s+2)(1+\alpha Ts)}$$

i.e.,
$$24 = 6K$$

i.e.,
$$K = 4$$

∴
$$G_1(s) = \frac{48}{s(s+2)}$$

Step 2: The Bode plot of $G_1(s)$ is shown in Fig. E16.2,

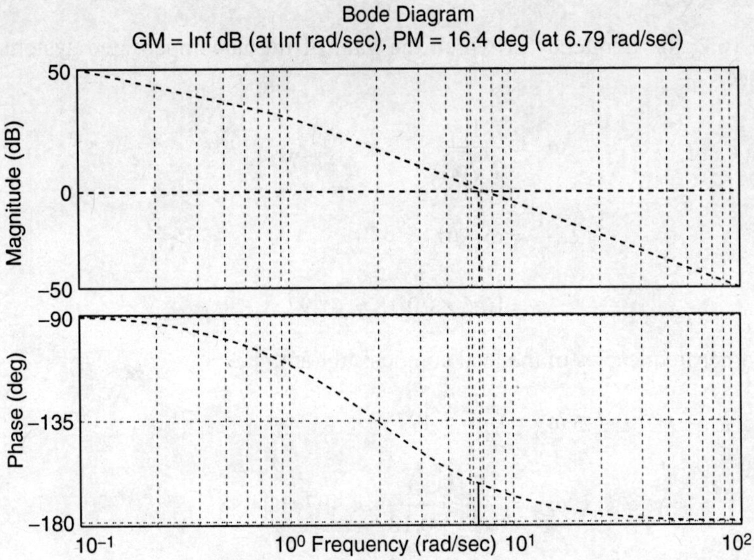

Fig. E16.2

From Fig. E16.2, $\phi_1 = \text{PM} = 16.4°,\ \omega_{gc} = 6.8$ rad/sec.,

$$\text{GM} = +\infty\ dB$$

Step 3: $\phi_s = 55°$

$\therefore$ $\phi_m = \phi_s - \phi_1 + E_r$

Let $E_r = 5°$

$\therefore$ $\phi_m = 55° - 16.4° + 5°$

$$= 43.6°$$

Step 4: $\sin\phi_m = \dfrac{1-\alpha}{1+\alpha}$

$\therefore$ $\sin 43.6° = \dfrac{1-\alpha}{1+\alpha} = 0.6896$

i.e., $\dfrac{1+\alpha-1+\alpha}{1+\alpha+1-\alpha} = \dfrac{1-0.6896}{1+0.6896}$

i.e., $\alpha = 0.1837$

Let $\alpha = 0.18$

Step 5: $-10\, \log_{10}\left(\dfrac{1}{\alpha}\right) = -7.44$ dB for $\alpha = 0.18$

From Fig. E16.2, the frequency at which the gain of the uncompensated system is –7.44 dB is $\omega_m = 10.6$ rad/sec.
We know that

$$\omega_m = \frac{1}{T\sqrt{\alpha}}$$

$$\text{i.e.,} \frac{1}{T} = \omega_m \sqrt{\alpha}$$

$$= 10.6 \times \sqrt{0.18} = 4.497$$

Step 6: Two corner frequencies of the lead compensator are,

$$\omega_{c1} = \frac{1}{T} = 4.497$$

and

$$\omega_{c2} = \frac{1}{\alpha T} = \frac{1}{0.18} \times 4.497 = 24.98$$

Step 7:

$$K = K_c \alpha$$

∴

$$K_c = \frac{K}{\alpha} = \frac{4}{0.18} = 22.22$$

Step 8:

$$G_c(s) = 22.22 \times 0.18 \times \frac{(1+0.2224s)}{(1+0.0400s)}$$

$$= \frac{3.99\,(1+0.2224s)}{(1+0.0400s)}$$

This is the designed lead compensator.

∴

$$G_c(s)\, G(s) = \frac{12}{s(s+2)} \times \frac{3.99\,(1+0.2224s)}{(1+0.0400s)}$$

$$= \frac{47.88\,(1+0.2224s)}{s(s+2)\,(1+0.0400s)}$$

The Bode plot for this transfer function is shown in Fig. E16.3.
From Fig. E16.3, for compensated system, we have

$$\omega_{gc} = 10.5 \text{ rad/sec.}$$

$$\text{PM} = 54.8° \simeq 55°$$

$$\text{GM} = \infty.$$

Therefore, the compensated system satisfies all the specifications.

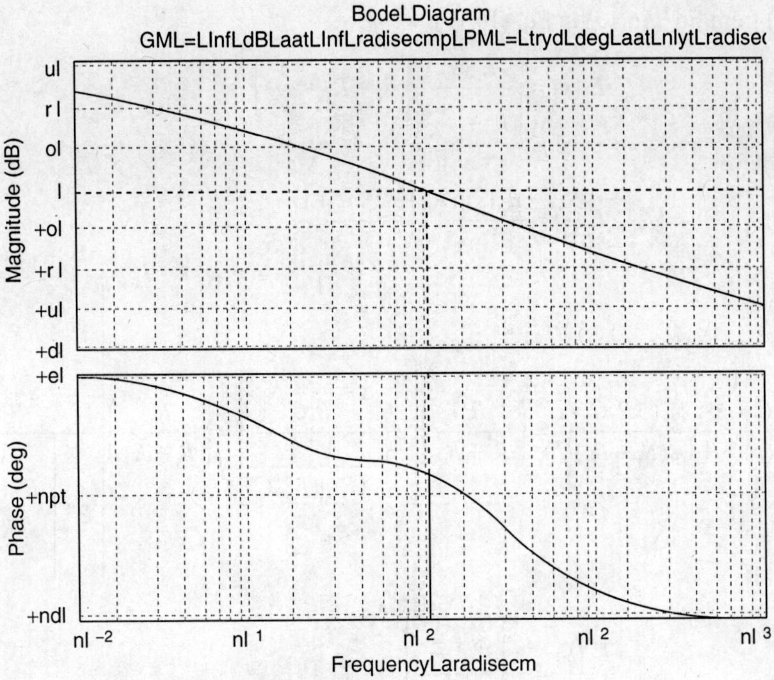

Fig. E16.3

16.2.2 Lag Compensator

Figure 16.8 shows an unloaded electrical network which is a lag compensating network.

To obtain the transfer function of such an electrical lag network, let us apply KVL to the loop.

$$e_i(t) = i(t)R_1 + i(t)R_2 + \frac{1}{C}\int i(t)dt \tag{16.12}$$

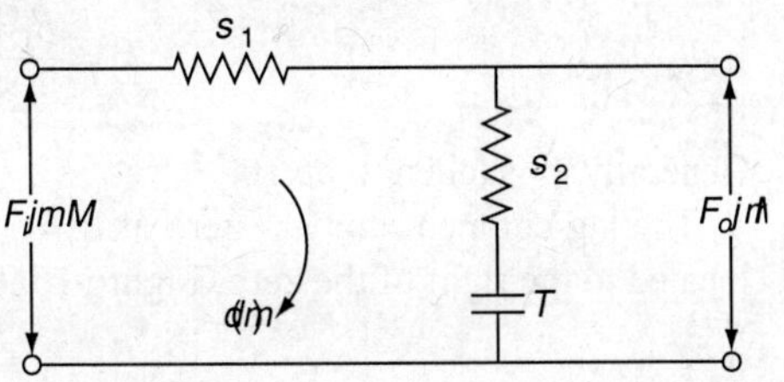

Fig. 16.8 Lag compensator

Taking Laplace transform of Eq. (16.12), we have

$$E_i(s) = I(s)\left[R_1 + R_2 + \frac{1}{sC}\right] \tag{16.13}$$

The output equation becomes

$$e_o(t) = i(t)R_2 + \frac{1}{C}\int i(t)dt \tag{16.14}$$

Taking Laplace transform of Eq. (16.14), we have

$$E_o(s) = I(s)\left[R_2 + \frac{1}{sC}\right] \tag{16.15}$$

Substituting $I(s)$ from Eq. (16.15) in Eq. (16.16), we have

$$E_i(s) = \frac{E_o(s)}{\left[R_2 + \frac{1}{sC}\right]}\left[R_1 + R_2 + \frac{1}{sC}\right]$$

i.e.,
$$E_i(s) = \frac{E_o(s)[(R_1 + R_2)sC + 1]}{(1 + R_2 sC)}$$

i.e.,
$$\frac{E_o(s)}{E_i(s)} = \frac{1 + sR_2C}{1 + s(R_1 + R_2)C} = \left(\frac{R_2}{R_1 + R_2}\right)\left[\frac{s + \frac{1}{R_2C}}{s + \frac{1}{(R_1 + R_2)C}}\right] = \left(\frac{R_2}{R_1 + R_2}\right)\left[\frac{s + \frac{1}{R_2C}}{s + \frac{1}{\left[\frac{R_1 + R_2}{R_2}\right]R_2C}}\right]$$

i.e.,
$$\frac{E_o(s)}{E_i(s)} = \frac{1}{\beta}\left(\frac{s + \frac{1}{T}}{s + \frac{1}{\beta T}}\right) \tag{16.16}$$

where
$$T = R_2C$$

and
$$\beta = \frac{R_1 + R_2}{R_2} > 1$$

Generally, β is taken to be 10.

The lag compensator has zero at $s = -1/T$ and a pole at $s = 1/\beta T$. Since $\beta > 1$, the pole is always located to the right of the zero. Figure 16.9 shows the pole-zero plot of a lag compensator.

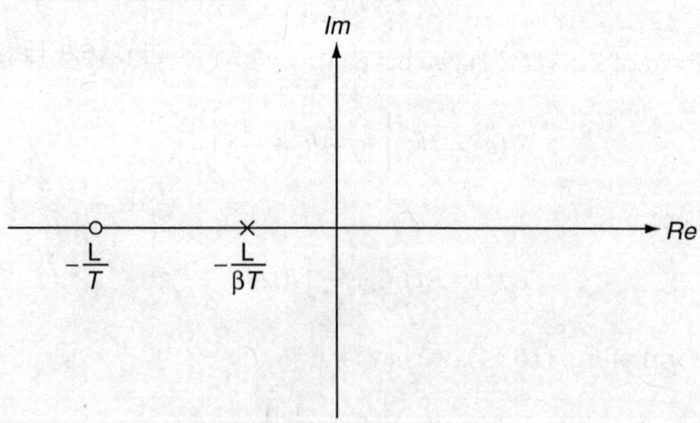

Fig. 16.9 Pole-zero plot of lag compensator

Maximum lag angle and β Substituting $s = j\omega$ in Eq. (16.16), we have

$$\frac{E_o(j\omega)}{E_i(j\omega)} = \frac{1+j\omega T}{1+j\omega\beta T} \quad (16.17)$$

i.e.,
$$M = \left|\frac{E_o(j\omega)}{E_i(j\omega)}\right| \frac{\sqrt{1+\omega^2T^2}}{\sqrt{1+j\omega^2\beta^2T^2}} \quad (16.18)$$

The phase angle is given by,

$$\phi = \tan^{-1}(\omega T) - \tan^{-1}(\omega\beta T) \quad (16.19)$$

The equation is exactly similar to the lead network. In a lag network, $\beta > 1$.

To obtain ω_m which maximizes ϕ, let us put

$$\left.\frac{d\phi}{d\omega}\right|_{\omega=\omega_m} = 0$$

i.e.,
$$\frac{d}{d\omega}\left[\tan^{-1}\omega T - \tan^{-1}\omega\beta T\right]_{\omega=\omega_m} = 0$$

After solving,

$$\omega_m = \frac{1}{T\sqrt{\beta}} = \sqrt{\frac{1}{T}\left(\frac{1}{\beta T}\right)} \quad (16.20)$$

Equation (16.20) shows the frequency at which the phase lag is at its maximum and ω_m is the geometric mean of the two frequencies.

Here,
$$\omega_{c1} = 1/T$$

and
$\omega_{c2} = 1/\beta T$ are the two corner frequencies of the lag compensator.

A lag compensator provides attenuation in the high frequency range to give a system a sufficient phase margin and there is no role of the phase angle in the lag compensation.

Polar plot of lag compensator From Eqs. (16.18) and (16.19), we have

At $\omega = 0, \quad M = 1$ and $\phi = 0°$

At $\omega = \infty, \quad M = \dfrac{1}{\beta}$ and $\phi = 0°$

Here both points, starting and terminating points, are on the positive real axis. The magnitude is always positive for $0 < \omega < \infty$, while the phase angle ϕ is negative for $\beta > 1, \tan^{-1}\omega T < \tan^{-1}\omega\beta T$. Thus, the resultant ϕ will be always negative for any value of ω between 0 and ∞. Therefore, all the points of a polar plot are always in the fourth quadrant of the complex plane.

Figure 16.10 shows the polar plot of a lag compensator.

Bode plot of a lag compensator The lag compensator has the following corner frequencies:

$$\omega_{c1} = \frac{1}{\beta T} \text{ for a simple pole at } s = -\frac{1}{\beta T}$$

$$\omega_{c2} = \frac{1}{T} \text{ for a simple zero at } s = -\frac{1}{T}$$

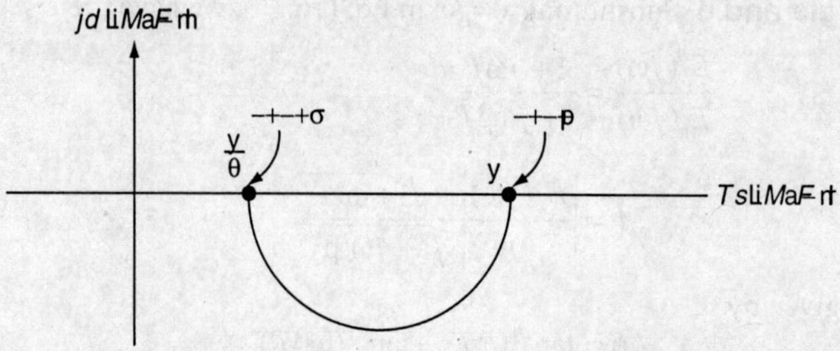

Fig. 16.10 Polar plot of lag compensator

The pole dominates over zero. The transfer function of a basic lag network is

$$T(s) = \frac{1+sT}{1+s\beta T}$$

Figure 16.11 shows the Bode plot of the lag compensator.

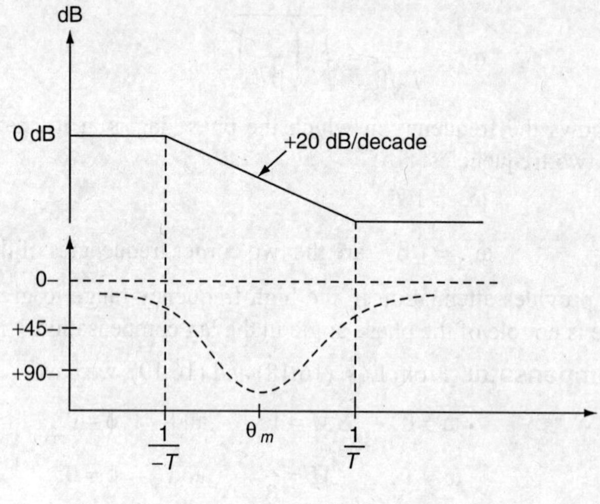

Fig. 16.11 Bode plot of lag compensator

Steps to design a lag compensator

Step 1: Assume the transfer function of a lag compensator as

$$G_c(s) = \frac{E_o(s)}{E_i(s)} = \frac{1}{\beta}\left(\frac{s+\frac{1}{T}}{s+\frac{1}{\beta T}}\right) = \frac{(1+Ts)}{(1+\beta Ts)}$$

It is taken as $G_1(s) = KG(s)$. The value of K is given from the error constant and it must satisfy the steady-state performance.

Step 2: The Bode plot of $G_1(j\omega)$ is drawn using the value of K determined in step 1. Obtain the phase margin (ϕ_1).

Step 3: Usually, phase margin is specified. Let the given phase margin be ϕ_s.
Therefore, $\phi_2 = \phi_s + \varepsilon$. Here ε compensates for the phase lag of the lag compensator and ε is the margin of safety = 5° to 15°.

Step 4: Corresponding to the phase margin of ϕ_2 degrees, the frequency ω_2 is determined, i.e., the frequency at which the phase angle of an open-loop transfer function is $180° + \phi_2$. This frequency is taken as the new gain crossover frequency.

Step 5: For ω_2 as the new gain crossover frequency, the attenuation necessary to shift the magnitude curve up of down to 0 dB is determined and this shift is due to the contribution of β which is $20\, log_{10}(1/\beta)$.

$$\therefore \text{ Shift to have } \omega_2 \text{ as new gain crossover} = 20\, log_{10}\frac{1}{\beta} = -20\, log_{10}\,\beta.$$

Down shift must be taken negative while up shift must be taken positive. The value of β is determined.

Step 6: The upper corner frequency $1/T$ is chosen which 1/2 or 1/10 below the ω_2 determined in step 4.

$$\therefore \qquad \omega_{c2} = \frac{1}{T} = \frac{\omega_2}{2} \text{ or } \frac{\omega_2}{10}$$

This determines the value of T.
For the lag compensator, the other corner frequency is

$$\omega_{c1} = \frac{1}{\beta T}$$

Step 7: The Bode plot of the lag compensated system is drawn after defining its transfer function. If specifications are not satisfied, then redesign by modifying the pole-zero locations of the compensator is carried out till a satisfactory result is obtained.

Effects and limitations of a lag compensator The lag compensator has the following effects and limitations:

1. Since the lag compensator allows high gain at low frequencies, it is basically a low pass filter. Therefore, it improves the steady-state performance.
2. In lag compensation, the attenuation characteristics is used for the compensation, whereas the phase lag characteristics are of no use in compensation.
3. Since the attenuation due to a lag compensator shifts the gain crossover frequency to a lower frequency point, the bandwidth of the system is reduced. Hence a slower response occurs. Therefore, rise time and settling time are usually longer and hence the transient response lasts for longer time.
4. The system is very sensitive to parameter variation.
5. Since a lag compensator approximately acts as proportional plus integral controller, it thus tends to make a system less stable.

Example 16.2 A system has $G(s) = \dfrac{0.035}{s(1+0.5s)(1+0.04s)}$

Design a suitable lag compensator to give velocity error constant 27.3 s^{-1} and phase margin = 45°.

Solution

Step 1: Assume

$$G_1(s) = KG(s) = \frac{0.035K}{s(1+0.5s)(1+0.04s)}$$

$\therefore$
$$K_v = 27.3 = \underset{s\to 0}{\mathrm{LT}}\; sG_1(s)G_c(s)$$

i.e.,
$$27.3 = \underset{s\to 0}{\mathrm{LT}}\; s \cdot \frac{0.035K}{s(1+0.5s)(1+0.04s)}$$

i.e.,
$$27.3 = 0.035K$$

i.e.,
$$K = 780.$$

$\therefore$
$$G_1(s) = \frac{0.035\times 780}{s(1+0.5s)(1+0.04s)} = \frac{27.3}{s(1+0.5s)(1+0.04s)}$$

Step 2: The Bode plot of $G_1(s)$ has been shown in Fig. E16.4.

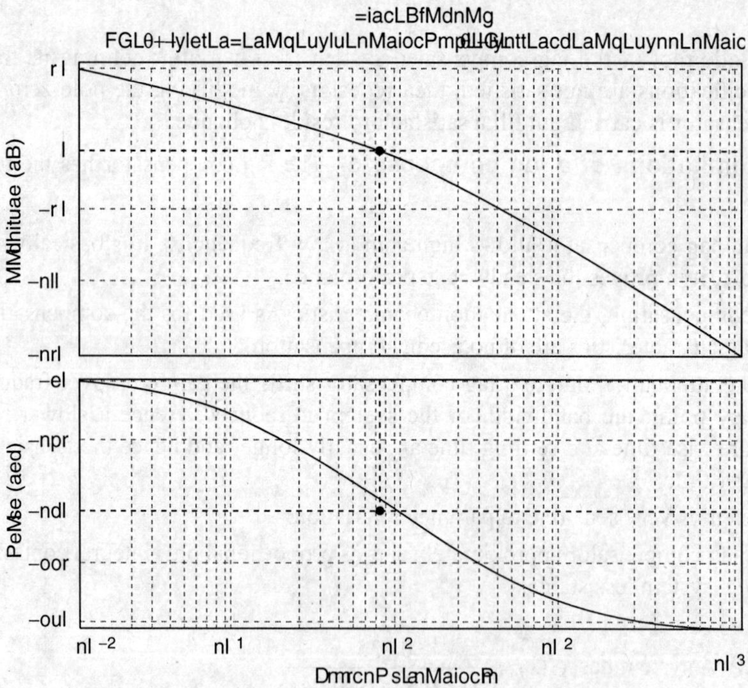

Fig. E16.4

From Fig. E16.4,

$$\omega_{gc} = 7.11 \text{ rad/sec.}$$

$$\omega_{pc} = 7.07 \text{ rad/sec.}$$

$$GM = -0.096 \text{ dB}$$

$$PM = -0.166°$$

Uncompensated system is unstable.

Step 3: $\phi_s = 45°, \in = 5°$

$\therefore$ $\phi_2 = \phi_s + \in = 45° + 5° = 50°.$

Step 4: The frequency ω_2 at which phase margin becomes 50° is given by $\omega_2 = 1.48$ rad/sec.

Step 5: Let $\omega_2 = 1.48$ rad/sec be the new gain crossover. It is necessary to bring the magnitude curve down by 23.3 dB.

$\therefore$ $$-20 \log \beta = -23.3$$

i.e., $$\beta = 14.62$$

Step 6: Choose $\omega_{c2} = \dfrac{\omega_2}{10} = \dfrac{1.48}{10} = 0.148$ rad/sec.

Now, $$\omega_{c2} = \frac{1}{T} = 0.148$$

i.e., $$T = \frac{1}{0.148} = 6.75$$

and $$\omega_{c1} = \frac{1}{\beta T} = \frac{1}{14.62 \times 6.75} = 0.0101$$

Step 7: The log compensator is

$$G_c(s) = \frac{1 + 6.75s}{1 + 98.68s}$$

The transfer function for this uncompensated system is

$$G_c(s)\, G(s) = \frac{27.3(1 + 6.75s)}{s(1 + 0.5s)(1 + 0.04s)(1 + 98.68s)}$$

The Bode plot of the compensated system is shown in Fig. E16.5.
From Fig. E16.5, we have

$$\omega_{gc} = 1.5 \text{ rad/sec.}$$

$$\omega_{pc} = 6.8 \text{ rad/sec.}$$

$$GM = 22.5 \text{ dB}$$

$$PM = 44.5°.$$

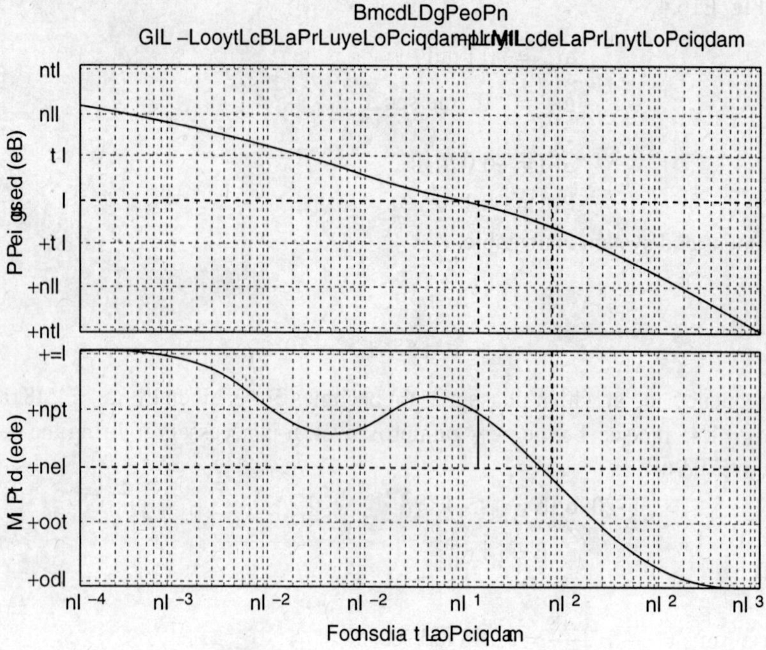

Fig. E16.5

16.2.3 Lag-Lead Compensator

Figure 16.12 shows an electrical network which is the combination of a lag and a lead compensator.

To obtain the transfer function of the electrical lag-lead network, let us apply KCL.

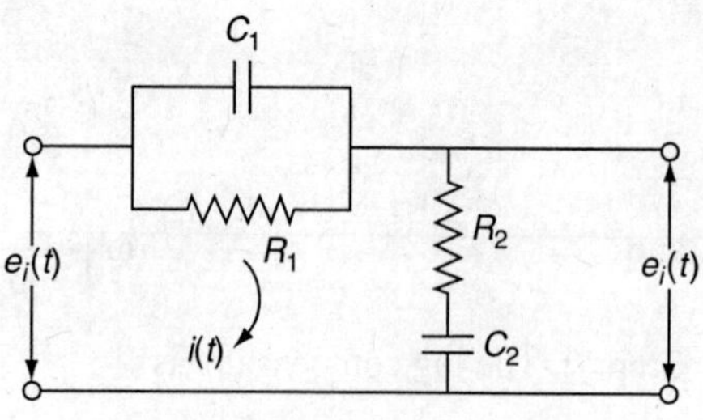

Fig. 16.12 Lag-lead network

$$\therefore \quad \frac{e_i - e_0}{R_1} + C_1 \frac{d(e_i - e_o)}{dt} = i(t) \tag{16.21}$$

Taking Laplace transform of Eq. (16.21), we have

$$\frac{1}{R_1}E_i(s) - \frac{1}{R_1}E_o(s) + sC_1E_i(s) - sC_1E_o(s) = I(s) \tag{16.22}$$

The output equation is given by

$$i(t)R_2 + \frac{1}{C_2}\int i(t)dt = e_o(t) \tag{16.23}$$

Taking Laplace transform of Eq. (16.23), we have

$$E(s)\left(R_2 + \frac{1}{sC_2}\right) = E_o(s) \tag{16.24}$$

Substituting $I(s)$ from Eq. (16.24) in Eq. (16.22), we have

$$\left\{E_i(s)\left[\frac{1}{R_1}+sC_1\right]-E_o(s)\left[\frac{1}{R_1}+sC_1\right]\right\}\left[R_2+\frac{1}{sC_2}\right]=E_o(s)$$

i.e.,
$$\left[\frac{E_i(s)(1+sR_1C_1)-E_o(s)(1+sR_1C_1)}{R_1}\right]\left[\frac{(1+sR_2C_2)}{sC_2}\right]=E_o(s)$$

i.e.,
$$E_i(s)\left[\frac{(1+sR_1C_1)(1+sR_2C_2)}{sR_1C_2}\right]=E_o(s)\left[1+\frac{(1+sR_1C_1)(1+sR_2C_2)}{sR_1C_2}\right]$$

i.e.,
$$\frac{E_o(s)}{E_i(s)}=\frac{(1+sR_1C_1)(1+sR_2C_2)}{sR_1C_2+(1+sR_1C_1)(1+sR_2C_2)}$$

i.e.,
$$\frac{E_o(s)}{E_i(s)}=\frac{(1+sR_1C_1)(1+sR_2C_2)}{s^2R_1R_2C_1C_2+s[R_1C_1+R_2C_2+R_1C_2])+1}$$

i.e.,
$$\frac{E_o(s)}{E_i(s)}=\frac{R_1R_2C_1C_2\left(s+\frac{1}{R_1C_1}\right)\left(s+\frac{1}{R_2C_2}\right)}{R_1R_2C_1C_2\left[s^2+s\left(\frac{1}{R_1C_1}+\frac{1}{R_2C_2}+\frac{1}{R_2C_1}\right)+\frac{1}{R_1R_2C_1C_2}\right]}$$

i.e.,
$$\frac{E_o(s)}{E_i(s)}=\frac{\left(s+\frac{1}{T_1}\right)\left(s+\frac{1}{T_2}\right)}{\left(s+\frac{\beta}{T_1}\right)\left(s+\frac{1}{\beta T_2}\right)} \tag{16.25}$$

where $T_1=R_1C_1, T_2=R_2C_2$

$$\frac{\beta}{T_1}+\frac{1}{\beta T_2}=\frac{1}{\alpha T_1}+\frac{1}{\beta T_2}=\frac{1}{R_1C_1}+\frac{1}{R_2C_2}+\frac{1}{R_2C}$$

$\therefore$ $\alpha\beta T_1T_2=R_1R_2C_1C_2$

i.e., $\alpha\beta=1$

Again
$$\frac{E_o(s)}{E_i(s)}=\frac{(1+T_1s)(1+T_2s)}{\left(1+\frac{T_1}{\beta}s\right)(1+T_2\beta s)} \tag{16.26}$$

where $\beta>1$.

The phase lead portion adds the phase lead angle and the phase lag portion provides attenuation near and above the gain crossover frequency.

Figure 16.13 shows the pole-zero plot for the lag-lead compensator.

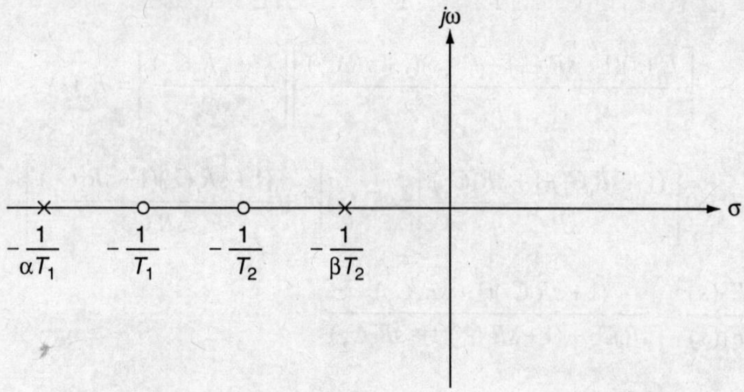

Fig. 16.13 Pole-zero plot of lag-lead compensator

Polar plot of a lag-lead compensator Substituting $s = j\omega$ in Eq. (16.26), we have

$$\frac{E_o(j\omega)}{E_i(j\omega)} = \frac{(1+T_1 j\omega)(1+T_2 j\omega)}{\left(1+\frac{T_1}{\beta} j\omega\right)(1+jT_2\beta\omega)} \tag{16.27}$$

The magnitude of the transfer functions is

$$M = \left|\frac{E_o(j\omega)}{E_i(j\omega)}\right| = \frac{\sqrt{1+\omega^2 T_1^2}\,\sqrt{1+\omega^2 T_2^2}}{\sqrt{1+\frac{T_1^2\omega^2}{\beta^2}}\,\sqrt{1+\beta^2 T_2^2\omega^2}} \tag{16.28}$$

The phase angle is given by

$$\phi = \tan^{-1}\omega T_1 + \tan^{-1}\omega T_2 - \tan^{-1}\frac{\omega T_1}{\beta} - \tan^{-1}\beta T_2\omega$$

Now for $\omega = 0,\quad M = 1\quad$ and $\quad\phi = 0°$

While for $\omega = \infty,\quad M = 1\quad$ and $\quad\phi = 0°$

Figure 16.14 shows the Polar plot of Lag-lead compensator.

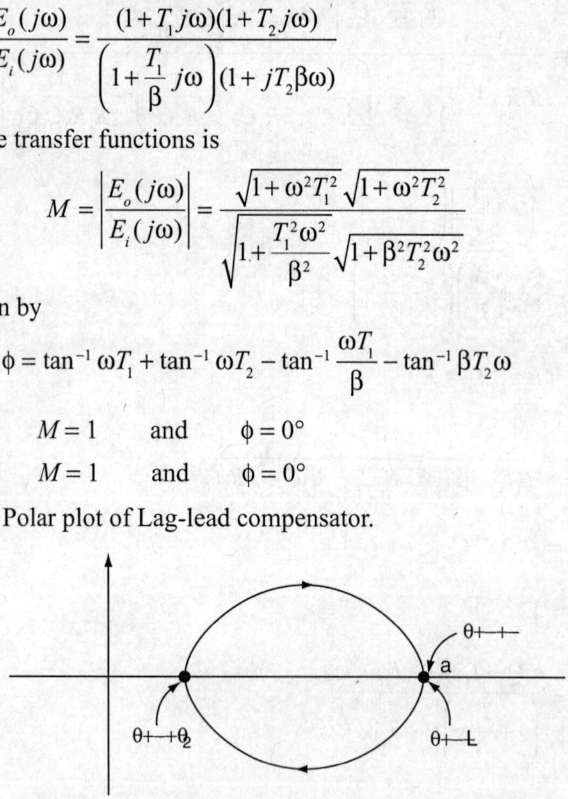

Fig. 16.14 Polar plot of lag-lead compensator

The frequency ω_1 is given by

$$\omega_1 = \frac{1}{\sqrt{T_1 T_2}}$$

Bode plot of lag-lead compensator The various corner frequencies of the lag-lead compensator are

$$\omega_{c1} = -\frac{\beta}{T_1} \text{ for simple pole, } \omega_{c2} = -\frac{1}{T_1} \text{ for a simple zero}$$

$$\omega_{c3} = - \text{ for simple zero, } \omega_{c4} = -\frac{1}{\beta T_2}\text{, for a simple pole}$$

Figure 16.14 shows the Bode plot of a lag-lead compensator.

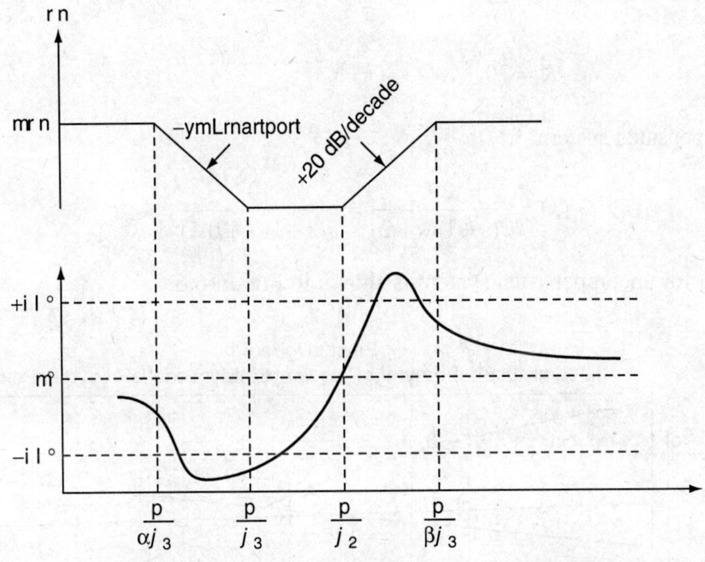

Fig. 16.15 Bode plot of lag-lead compensator

Effects of lag-lead compensator To get first response and good static accuracy, a lag-lead compensator is used. It also increases the low frequency gain which improves the steady state. Since it increases the bandwidth of the system, the system response becomes very fast. In general, the phase lead portion of this compensator provides large bandwidth and hence shorter rise time and settling time, while the phase lag portion provides the major damping of the system.

Example 16.3 The open-loop transfer function of a unity feedback system $G(s) = \frac{K}{s(s+1)(s+2)}$.

Design suitable lag-lead compensator to achieve the following:

Static velocity error constant = 10 s^{-1}.

Phase margin = 50° and

Gain margin ≥ 10 dB.

Solution

Let the transfer function of the compensated system be

$$G_c(s) = \frac{(1+T_1 s)(1+T_2 s)}{\left(1+\frac{T_1}{\beta}s\right)(1+\beta T_2 s)}$$

$$K_v = \underset{s\to 0}{\text{LT}}\ s\,G_c(s)G(s)$$

i.e.,

$$10 = \underset{s\to 0}{\text{LT}}\ s\cdot\frac{(1+T_1 s)(1+T_2 s)}{\left(1+\frac{T_1}{\beta}s\right)(1+\beta T_2 s)} \times \frac{K}{s(s+1)(s+2)}$$

i.e.,

$$10 = \frac{K}{2}$$

i.e.,

$$K = 20$$

$\therefore$ For uncompensated system

$$G_1(s) = \frac{20}{s(s+1)(s+2)} = \frac{10}{s(s+1)(s+0.5)}$$

The Bode plot for uncompensated system is shown in Fig. E16.6.

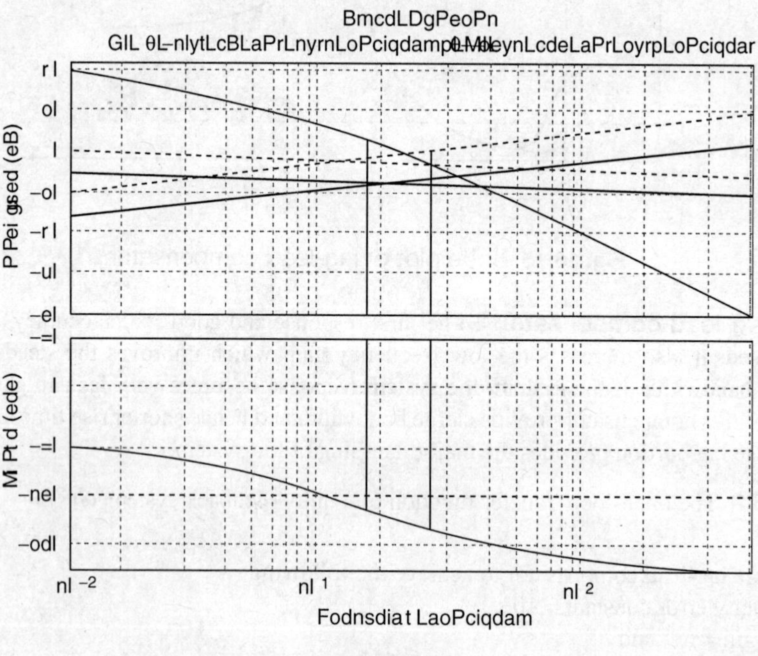

Fig. E16.6

From Fig. E16.6, we have

$$\omega_{gc} = 2.43 \text{ rad/sec.}$$

$$\omega_{pc} = 1.41 \text{ rad/sec.}$$

$$GM = -10.5 \text{ dB}$$

$$PM = -28.1°$$

It is necessary to add phase lead of 50º at ω = 1.41 rad/sec. to get PM of 50º. At the same, new gain crossover frequency becomes ω = 1.41 rad/sec.

$$\therefore \qquad \phi_m = 50°$$

Now,
$$\sin\phi_m = \frac{1-\alpha}{1+\alpha} \text{ and } \alpha\beta = 1$$

$$\therefore \qquad \alpha = \frac{1}{\beta}.$$

$$\therefore \qquad \sin\alpha = \frac{1-\frac{1}{\beta}}{1+\frac{1}{\beta}} = \frac{\beta-1}{\beta+1}$$

i.e.,
$$\sin 50° = \frac{\beta-1}{\beta+1}$$

i.e.,
$$0.766 = \frac{\beta-1}{\beta+1}$$

i.e.,
$$\beta = 7.584.$$

Practically maximum value of β is 10.

Let $\beta = 10$.

The new gain crossover frequency is 1.41 rad/sec.

The corner frequency of lag portion is usually $\frac{1}{10}$ th of the gain cross-over frequency.

$$\therefore \text{ Corner frequency } (\omega) = \frac{1}{T_2} = \frac{1.41}{10} = 0.14 \text{ rad/sec.}$$

Another corner frequency is

$$\omega = \frac{1}{\beta T_2} = 0.014 \text{ rad/sec.}$$

The transfer function of the phase lag portion becomes $\dfrac{1+T_2s}{1+\beta T_2 s} = \dfrac{1+7.14s}{1+71.43s}$

From Fig. E16.6, $|G(j\omega)|$ at $\omega = 1.41$ rad/sec. is + 10.5 dB.

Therefore, the lead compensator must contribute –10.5 dB at ω = 1.41 rad/sec. So that ω = 1.41 becomes new gain crossover frequency. Therefore, a line is drawn from the point having co-ordinate (1.41 rad/sec., – 10.5 dB) of slop = 20 dB /decade.

Its intersection with 0 dB line is 6 rad/sec and with –20 dB line is 0.50 rad/sec.

These are the corner frequency for the lead portion. The transfer function of the lead portion is

$$\frac{1+T_1 s}{1+\dfrac{T_1}{\beta}s} = \frac{1+2s}{1+0.2s}$$

where $$T_1 = \frac{1}{\omega_{c1}} = \frac{1}{0.5} = 2$$

and $$\frac{T_1}{\beta} = \frac{2}{10} = 0.2$$

The transfer function of lag-lead compensator is

$$G_c(s) = \frac{(1+2s)(1+7.14s)}{(1+0.2s)(1+71.43s)}$$

The transfer function of the compensated system is

$$G_c(s)G_1(s) = \frac{10(1+2s)(1+7.14s)}{s(1+0.2s)(1+71.43s)(1+s)(1+05s)}$$

Figure E16.7 shows the Bode plot of the compensated system.

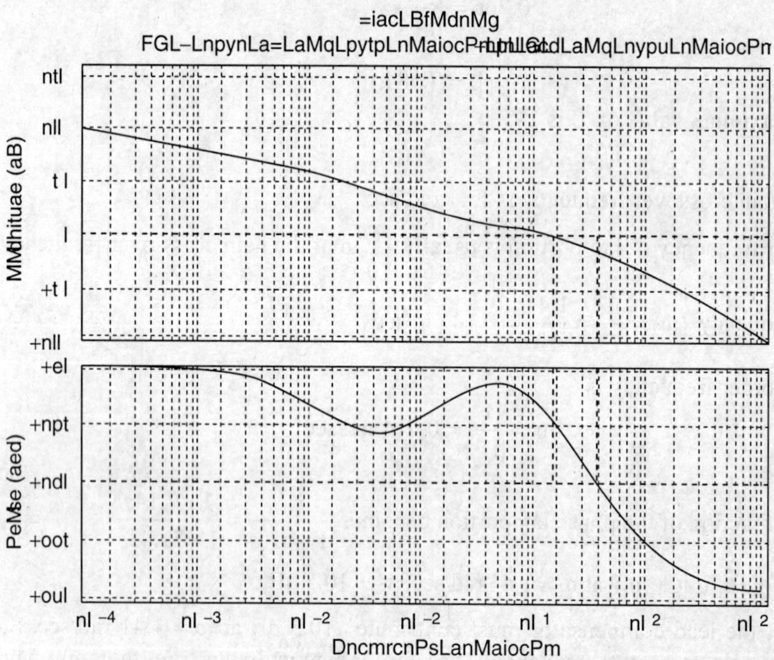

Fig. E16.7

The specification is as follows:

Phase crossover frequency = 3.63 rad/sec.
Gain crossover frequency = 1.37 rad/sec.

PM = 51°

GM = 13.1/dB.

16.3 Compensation Using Root Locus

The compensation using Bode plot had been discussed earlier. The specifications are given in frequency domain. The root locus approach is used for design purpose when the specification is given in time domain. This approach is also very powerful. The two main specifications which are used in the root locus compensation are damping factor and undamped natural frequency. The dominant closed-loop poles near $j\omega$-axis are decided by these two specifications. Therefore, the compensation using root locus reshapes the root locus near $j\omega$-axis and origin, so that the dominant closed-loop poles at the desired locations are placed. It is possible to achieve root locus compensation using lead, lag or lag-lead network.

If the dominant complex conjugate poles are expressed by $-\xi\omega_n \pm j\omega_d$, we have $\xi = \sin\theta$ in complex plan where θ is measured from $j\omega$-axis or $\xi = \omega_s\theta$ when θ is measured from negative real axis. This has shown in Fig. 16.16.

Figure 16.17 shows the lines of constant ξ, which are radial lines passing through the origin.

For the design purpose, ξ = sin θ will be used where θ is measured from $j\omega$–axis.

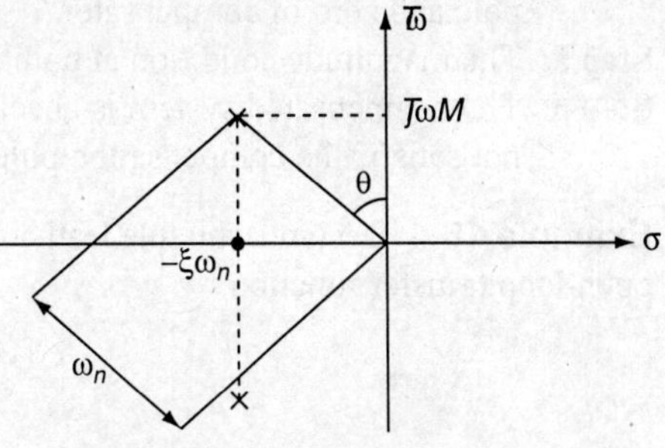

Fig. 16.16 Complex conjugate poles

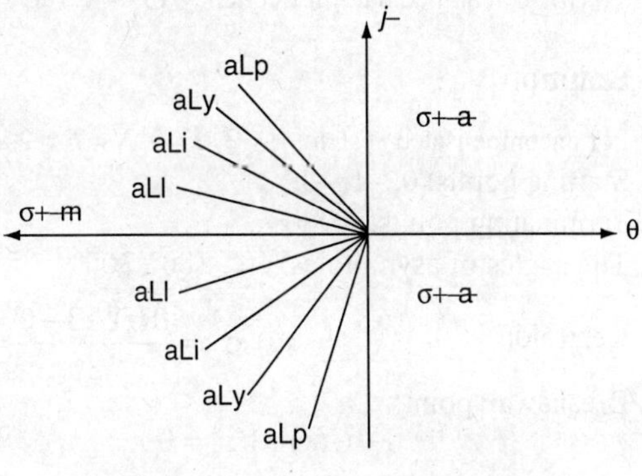

Fig. 16.17 Lines of constant ξ

16.4 Design of Lead Compensator Using Root Locus

The following steps are used to design lead compensator.

Step 1: The desired locations of the dominant closed-loop poles are found from the given specifications.

Step 2: The lead compensator is assumed as

$$G_c(s) = \frac{K_c\alpha(1+Ts)}{1+\alpha Ts}$$

where $\alpha < 1$ and K_c is determined from the requirement of open-loop gain.

Step 3: The sum of the angles at the desired location of one of the dominant closed-loop poles with the open-loop poles and zeros of the original system is to be calculated, which must be odd multiple of 180°. If it is not equal to odd multiple of 180°, the necessary angle ϕ is calculated which is to be added to get the sum as an odd multiple of 180°. This angle ϕ is to be contributed by the lead compensator. If this angle ϕ is more than 60°, two or more lead networks may be used. This angle ϕ also helps to determine values of α and T.

Step 4: To calculate α and T for the known angle ϕ, a horizontal line from one of the dominant closed-loop pole say P is drawn as shown in Fig. 16.18. The origin is joined to P. The angle between PA and PO is bisected. The two lines PC and PD are drawn, which makes an angle $\pm\frac{\phi}{2}$ with the bisector PB. The intersection of PC and PD with negative real axis will give the required pole and zero of compensator.

Step 5: The magnitude condition at point P is applied to determine the open-loop gain.

Step 6: The compensated system is checked whether it satisfies all the specifications or not. If it does not satisfy, the compensation pole and zero are adjusted till all the specifications are satisfied.

Example 16.4 Design a suitable lead compensation for a system with unity feedback and having open-loop transfer function

$$G(s) = \frac{K}{s(s+1)(s+3)}$$

This has to meet the following specifications

(i) Damping ratio $\xi = 0.6$ and

(ii) Undamped natural frequency $\omega_n = 2$ rad/sec.

Solution

For uncompensated system, $P = 3$, $z = 0$, $N = P = 3$ and $P - z = 3$.
Starting points: 0, –1, –3
Terminating points: ∞, ∞, ∞,
The angles of asymptotes: 60°, 180°, 300°.

Centroid: $$\sigma_c = \frac{0-1-3-0}{3} = -1.34$$

Breakaway point:

$$1 + G(s) + 1(s) = 0$$

i.e., $$1 + \frac{K}{s(s+1)(s+3)} = 0$$

i.e., $$s^3 + 4s^2 + 3s + K = 0$$

i.e., $$K = -s^3 - 4s^2 - 3s$$

i.e., $$\frac{dK}{ds} = -3s^2 - s - 3 = 0$$

i.e., $$s^2 + 2.67s + 1 = 0$$

∴ $$s = \frac{-2.67 \pm \sqrt{2.67^2 - 4}}{2} = \frac{-2.67 \pm 1.77}{2}$$

$$= -2.22, -0.45$$

For $s = -2.22, \; K = -23.99$

For $s = -0.45, \; K = 0.4488$

$\therefore s = 0.45$ is a valid break away point.

Intersection with imaginary axis:

The Routh array is given by

$$\begin{array}{c|cc} s^3 & 1 & 3 \\ s^2 & 4 & K \\ s^1 & \dfrac{12+K}{4} & 0 \\ s^0 & K & \end{array}$$

$$\therefore \quad \frac{12 - K_{mar}}{4} = 0$$

i.e.,

$$K_{mar} = 12$$

The auxiliary equation becomes

$$4s^2 + K_{mar} = 0$$

i.e.,

$$4s^2 + 12 = 0$$

i.e. $s = \pm\sqrt{3}$ are the intersection points.

The root locus is shown in Fig. E16.8.

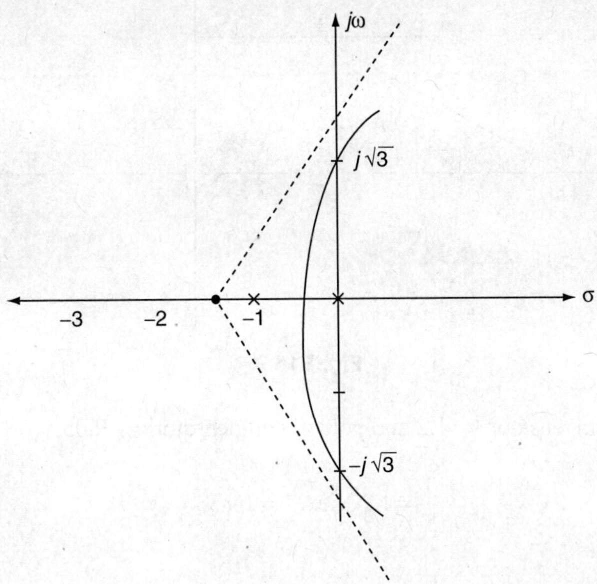

Fig. E16.8

Step 1: $\xi = 0.6$ and $\omega_n = 2$ rad/sec.

The desired dominant closed-loop poles are

$$= -\xi\omega_n \pm j\omega_n\sqrt{1-\xi^2}$$

$$= -1.2 \pm j1.6$$

The dominant poles are not on the root locus shown in Fig. E16.

Step 2: A lead compensation is assumed with

$$G_c(s) = K_c\alpha\left(\frac{1+Ts}{1+\alpha Ts}\right)$$

Step 3: $\angle G(s)$ at the dominant pole $s = -1.2 + j1.6$ becomes

$$\angle G(s)\Big|\ s = -1.2 + j1.6 = \frac{\angle K}{\angle(1.2 + j1.6)\angle(-1.2 + j1.6 + 1)\angle(-1.2 + j1.6 + 3)}$$

$$= 0° - 126.86° - 90° - 41.63° = -258.49°$$

$\therefore$ Angle contributed by the lead compensator is

$$= -180° - (-258.49°) = 78.49°$$

$\therefore \qquad \phi = 78.49°$

Step 4: Find the location of poles and zeros:

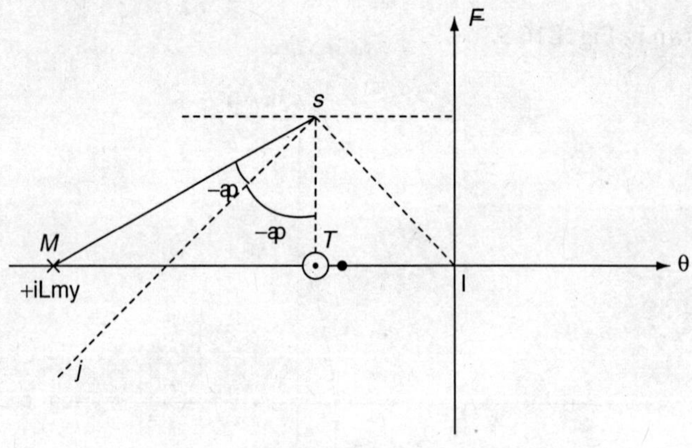

Fig. E16.9

$\therefore$ Zero of compensator = –1.2 and pole of compensator = – 9.05.

$\therefore \qquad \frac{1}{T} = 1.2 \quad \text{i.e.,} T = 0.833$

and $\qquad \frac{1}{\alpha T} = 9.05 \quad \text{i.e.,} \alpha T = 0.1105$

$$\therefore \qquad \alpha = \frac{0.1105}{T} = 0.133 < 1.$$

The transfer function of the compensated system is

$$G_c G_s = \frac{K(1+0.833s)}{s(1+s)(s+3)(s+0.1105)}$$

Step 5: Use magnitude condition to obtain value of K at $s = -1.2 + j\,1.6$.

$$\left|G_c(s)G(s)\right|_{s=-1.2+j1.6} = 1$$

i.e.,
$$\frac{K\left|1+0.833(-1.2+j1.6)\right|}{\left|-12+j1.6\right|\left|-1.2+j1.6+1\right|\left|-1.2+j1.6+3\right|\left|1+0.133(-1.2+j1.6)\right|}$$

i.e.,
$$\frac{K \times 1.3328}{2 \times 1.612 \times 2.408 \times 0.867} = 1$$

i.e.,
$$K = 5.05$$

$$\therefore \qquad \therefore G_c(s)G(s) = \frac{5.05(1+0.833s)}{s(s+1)(s+3)(1+0.1105s)}$$

The root locus of the compensated system is shown in Fig. E16.10

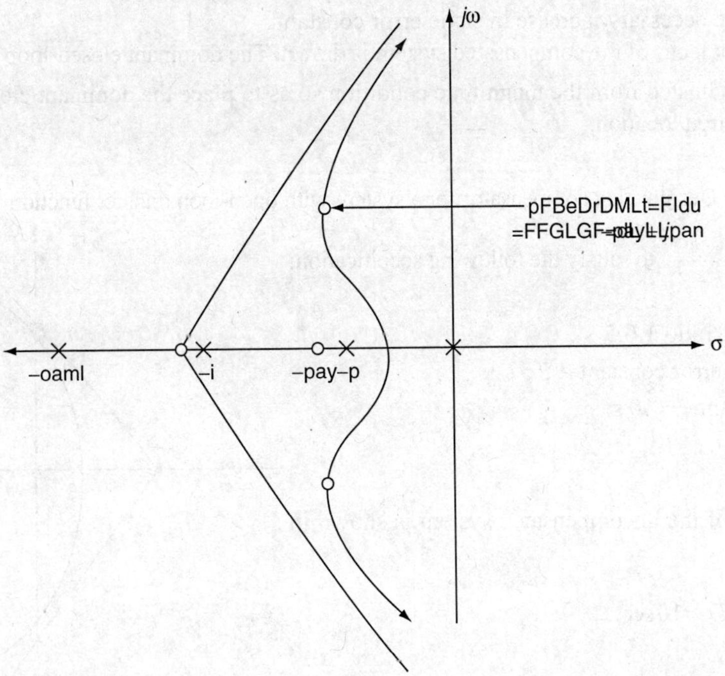

Fig. E16.10

16.5 Design of Lag Compensator Using Root Locus

If a system shows good transient response which is satisfactory but it is unsatisfactory for steady-state characteristics, the lag compensator is used in this case. This compensator increases the open-loop gain but it does not change the root locus appreciably near the dominant closed-loop poles. The angle contribution by the lag compensator is usually 5°, which is very small. The pole and zero of the lag compensator are placed close together which are near the origin of the s-plane in order to achieve this.

If gain is increased, the static error constant is also increased. Lag compensator is used for the case to satisfy static error constant specification using root locus.

The following steps are followed to design the lag compensator.

Step 1: The root locus of the uncompensated system is drawn and the dominant closed-loop poles on the root locus are located.

Step 2: A lag compensator having transfer function

$$G_c(s) = \widehat{K}_c \frac{s+\frac{1}{T}}{s+\frac{1}{\beta T}}, \beta > 1$$

is assumed.

Step 3: The static error constant specified in the problem is calculated.

Step 4: The amount of increase in static error constant which is necessary to satisfy the specification is calculated.

Step 5: The pole and zero of the compensator do not produce appreciable change in original root locus but give necessary increase in static error constant.

Step 6: The root locus of the compensated system is drawn. The dominant closed-loop poles are located.

Step 7: $\widehat{K}_c$ is adjusted from the magnitude condition so as to place the dominant closed-loop poles at the desired location.

Example 16.5 Design a lag compensator for a system with open-loop transfer function

$$G(s) = \frac{K}{s(s+1)(s+3)}$$ to satisfy the following specification:

1. Damping ratio + 0.5
2. Velocity error constant ≥ 5.5 s^{-1}.
3. Settling time = 10 s.

Solution

The root locus of the uncompensated system is shown in Fig. E16.11.

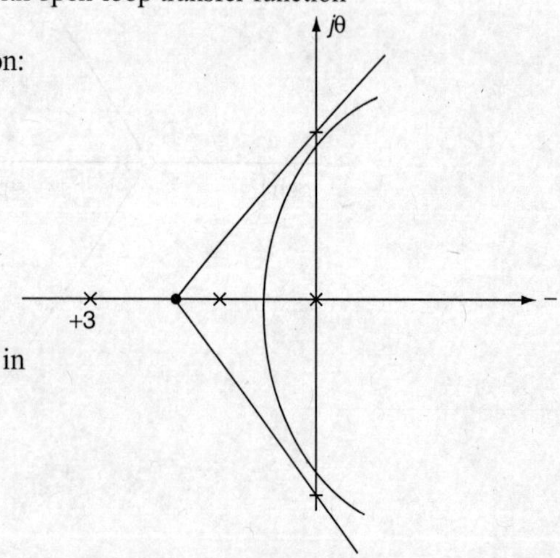

Fig. E16.11

Here, $\xi = 0.5,\ T_s = 10\,\text{sec.},$

$$\therefore T_s = \frac{4}{\xi\omega_n} = 10 \quad \text{i.e., } \omega_n = 0.8 \text{ rad/sec.}$$

Hence dominant closed-loop poles are

$$= -\xi\omega_n \pm j\omega_n\sqrt{1-\xi^2}$$

$$= -0.4 \pm j0.693$$

The gain K at dominant pole $-0.4 + j0.693$ is to be obtained by magnitude condition.

$$|G(s)H(s)|_{s=-0.4+j0.693} = 1$$

i.e.,

$$\frac{K}{|(0.4 + j0.693)||-0.4 + j0.693 + 1||-0.4 + j0.693 + 3|} = 1$$

i.e.,

$$\frac{K}{0.8 \times 0.9166 \times 2.69} = 1$$

i.e.,

$$K = 1.973$$

The transfer function of the uncompensated system becomes

$$G(s) = \frac{1.973}{s(s+1)(s+3)}$$

$\therefore$

$$K_v = \underset{s\to 0}{\text{Lt}}\, sG(s) = 0.657$$

The factor by which static error constant to be increased to keep desired K_v of 5 s^{-1} is given by the

factor $= \dfrac{5.5}{0.657} = 8.37$

Let us take this factor as 10.

$\therefore$

$$\beta = 10.$$

The pole and zero of the compensator are placed very close to the origin.
Let zero of the compensator at $s = -0.1$.
$\therefore$ The pole of the compensator at $s = -0.01$.
$\therefore$ The transfer function of the lag compensator becomes

$$G_c(s) = \widehat{K}_c\left(\frac{s+0.1}{s+0.01}\right)$$

The transfer function of the compensated system becomes

$$G(s)G_c(s) = \frac{1.973\widehat{K}_c(s+0.1)}{s(s+1)(s+3)(s+0.01)}$$

$$= \frac{K(s+0.1)}{s(s+1)(s+3)(s+0.01)}$$

where $K = 1.973\widehat{K}_c$.

The root locus of the compensated system is shown in Fig. E16.12.

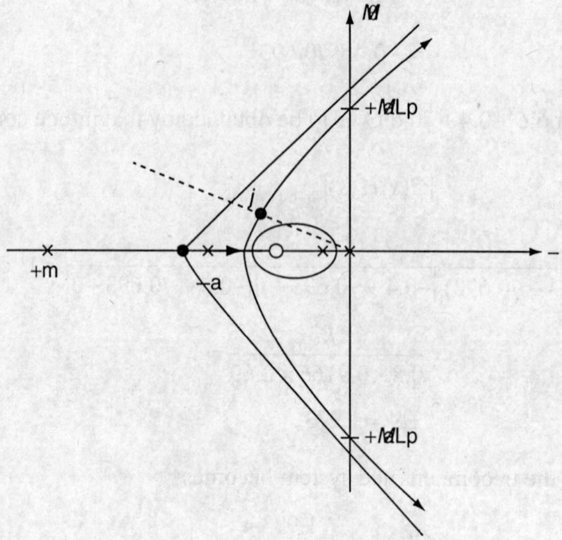

Fig. E16.12

The co-ordinate of the point P is $-0.326 + j0.669$.

The new dominant pole is at $-0.326 + j0.669$.

$$\left|G(s)G_c(s)\right|_{s=-0.326+j0.669} = 1.$$

i.e., i.e., $$\frac{K(-0.326+j0.669+0.1)}{\left|-0.326+j0.669\right|\left|-0.326+j0.669+1\right|\left|-0.326+j0.669+3\right|\left|-0.326+j0.669+0.01\right|} = 1$$

i.e., $$\frac{K \times 0.706}{0.744 \times 0.9496 \times 2.756 \times 0.7398} = 1.$$

i.e., $$K = 2.040$$

However, $$K = 1.973\widehat{K}_c$$

$\therefore$ $$\widehat{K}_c = \frac{K}{1.973} = \frac{2.040}{1.973} = 1.034$$

Hence, the compensated transfer function becomes

$$G_c(s)G(s) = \frac{2.040(s+0.1)}{s(s+1)(s+3)(s+0.01)}.$$

This gives a static error constant of 6.8, which is greater than 5.5.

16.6 Designing Lag-Lead Compensator Using Root Locus

The lead compensator increases the speed of the response and improves stability while lag compensator improves steady-state response. The lag-lead compensator is used when the requirement of the improvement both transient and steady-state responses is desired.

Let the transfer of the lag-lead compensator be assumed as

$$G_c(s) = K_c \frac{\left(s+\frac{1}{T_1}\right)}{\left(s+\frac{r}{T_1}\right)} \times \frac{\left(s+\frac{1}{T_2}\right)}{\left(s+\frac{1}{\beta T_2}\right)}$$

where $\beta > 1$ and $r > 1$.

The following two cases are discussed to design such compensator.

Case I: $r \neq \beta$

The locations of dominant closed-loop poles are determined from the given specification.

The angle of deficiency (ϕ) is to be calculated from the uncompensated system which must be contributed by the lead problem of the compensator.

T_2 is to be chosen sufficiently large, so that the magnitude of lag portion becomes unity.

$$\left| \frac{s_1 + \frac{1}{T_2}}{s_1 + \frac{1}{\beta T_2}} \right| = 1.$$

where $s = s_1$ is one of the dominant closed-loop poles. The values of T_1 and r are to be determined from the angle of deficiency (ϕ).

$$\angle \frac{s_1 + \frac{1}{T_1}}{s_1 + \frac{1}{\beta T_1}} = \phi$$

The value K_c is determined from the magnitude condition.

$$\left| K_c \frac{s_1 + \frac{1}{T_1}}{s_1 + \frac{r}{T_1}} G(s_1) \right| = 1$$

Static velocity error constant K_v is given by

$$K_v = \underset{s\to 0}{\text{Lt}}\, sG_c(s)G(s)$$

$$= \underset{s\to 0}{\text{Lt}}\, sK_c \left(\frac{s+\frac{1}{T_1}}{s+\frac{r}{T_1}} \right) \left(\frac{s+\frac{1}{T_2}}{s+\frac{1}{\beta T_2}} \right) \cdot G(s)$$

$$= \underset{s \to 0}{\text{Lt}}\ sK_c \frac{\beta}{r} G(s)$$

Since K_c and r are known, β can be determined.

The value of T_2 should be chosen using the above value, such that the magnitude of lag portion becomes unit.

Case II: $r = \beta$

The locations of dominant closed-loop poles are to be determined as per the given specifications.

Since $r = \beta$, the transfer function of the compensator becomes

$$G_c(s) = \left(\frac{s + \frac{1}{T_1}}{s + \frac{\beta}{T_1}} \right) \left(\frac{s + \frac{1}{T_2}}{s + \frac{1}{\beta T_2}} \right), \beta > 1$$

K_c can be determined from the requirement of K_v.

The angle of deficiency ϕ is to be determined, so that the dominant closed-loop poles will be in the desired locations.

T_2 is to be selected large at the end, so that

$$\left| \frac{s_1 + \frac{1}{T_2}}{s_1 + \frac{1}{\beta T_2}} \right| = 1$$

where $s = s_1$ is one of the dominant closed-loop poles.

Now,
$$\left| K_c \frac{s_1 + \frac{1}{T_1}}{s_1 + \frac{\beta}{T_1}} G(s_1) \right| = 1$$

and
$$\angle \frac{s_1 + \frac{1}{T_1}}{s_1 + \frac{\beta}{T_1}} = \phi$$

The above two equations give the value of T_1 and β.

After getting β, T_2 is selected large, so that the given magnitude of lag portion becomes approximately equal to unity and

$$-5° < \angle \frac{s_1 + \frac{1}{T_2}}{s_1 + \frac{1}{\beta T_2}} < 0°$$

16.7 Tuning Rules for PID Controllers

A PID control of plant is shown in Fig. 16.18. Before applications of different design techniques to determine parameters of the controller for meeting the transient and steady state specifications, the

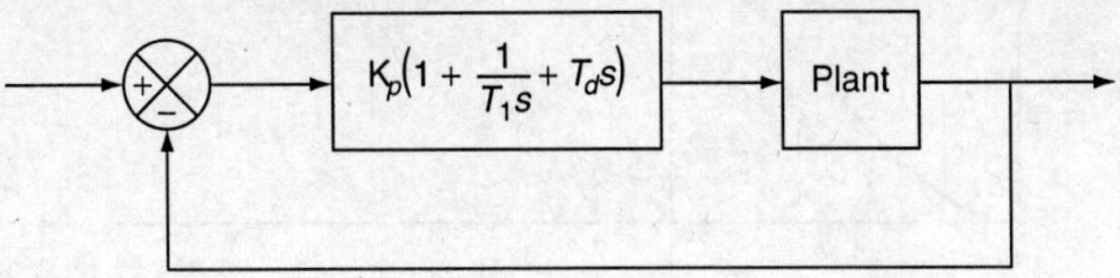

Fig. 16.18 PID controllers

mathematical model of the plant should be known to us. For a complicated plant, it is difficult to obtain its mathematical model and hence analytical approach to design the PID controller is not feasible. The experimental approach for tuning the PID controllers is to be followed. Hence the process for selection of the controller parameters to satisfy the given performance specifications is termed as controller tuning. The rules suggested by Ziegler and Nichols to tune PID controller parameters (K_p, T_i and T_d) is based on the experimental step responses or based on the values of K_p resulting in marginal stability when proportional control action is used. Ziegler and Nichols rules are very useful for the plants whose mathematical model is not known which suggest set a of values of K_p, T_i and T_d that result stable operation of the system. There may be a large number of overshoots for the system due to step response. These are not acceptable. A series of fine tunings are to be done until the acceptable result is obtained. Ziegler and Nichols rules only can give a prime guess for parameter values which provide the starting point of the tuning instead of providing the final settings of K_p, T_i and T_d in a single shot.

The rules proposed by Ziegler and Nichols determine the values of proportional gain (K_p), integral time (T_i) and derivative time (T_d) based on the transient response characteristic of the plant. The following two methods are discussed for Ziegler and Nichols rule.

First Rule:

Figure 16.19 shows unit step response of the plant which is obtained experimentally. The response may be in S shape when the plant does not have integrator or dominant complex conjugate poles. This

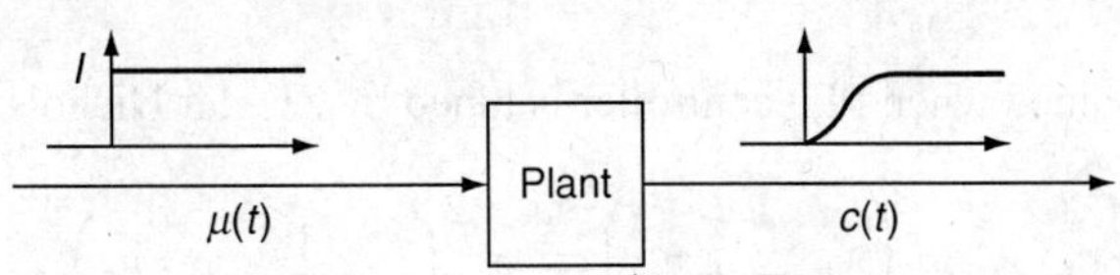

Fig. 16.19 Unit step response of a plant

has been shown in Fig. 16.20 which may be generated experimentally or by dynamic simulation of the plant. The delay time (L) and time constant (T) are used to characterize this S shaped curve which are determined by drawing the tangent line at the inflection point of the S shaped curve and the point of intersections of the tangent line with the time axis line $c(t) = \text{K}$ respectively as shown in Fig. 16.20.

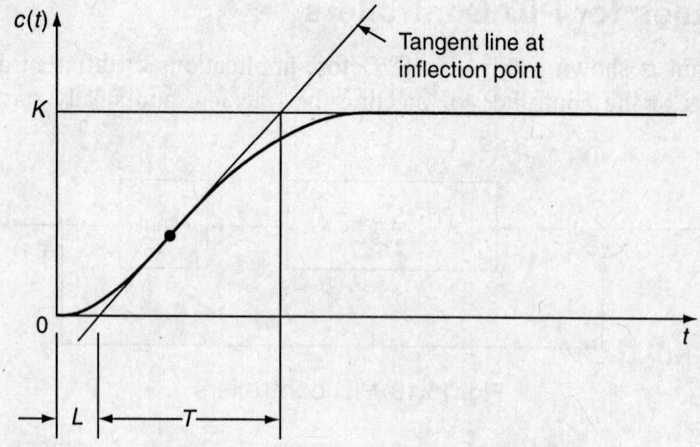

Fig. 16.20 S–shaped response curve

The transfer function $C(s)/U(s)$ can be approximated by a first order transfer function shown below.

$$\frac{C(s)}{U(s)} = \frac{Ke^{-Ls}}{1+Ts} \tag{16.29}$$

The suggested set of values of K_p, T_i and T_d is shown in Table 16.1.

Table 16.1 Values of K_p, T_i and T_d based on the tuning rules

Type of controller	K_p	T_i	T_d
P	$\frac{T}{L}$	∞	0
PI	$0.9\frac{T}{L}$	$\frac{L}{0.3}$	0
PID	$1.2\frac{T}{L}$	$2L$	$0.5L$

The following is obtained when PID controller is tuned by Ziegler Nichols rules:

$$G_c(s) = K_p\left(1+\frac{1}{T_i s}+T_d s\right) = 1.2\frac{T}{L}\left(1+\frac{1}{2Ls}+0.5Ls\right)$$

$$= 0.6T\frac{\left(s+\frac{1}{L}\right)^2}{s} \tag{16.30}$$

Equation (16.30) shows that the PID controller has a pole at origin and zeros are at $-\frac{1}{L}$.

Second Rule:

In this method $T_i = \infty$ and $T_d = 0$ is set at first. The proportional controller is used in the closed loop system shown in Fig. 16.21 and K_p is increased from 0 to a critical value K_{cr} so that the system exhibit sustained oscillations. The critical gain is K_{cr} and the corresponding period is T_{cr} as shown in Fig. 16.22 are determined experimentally.

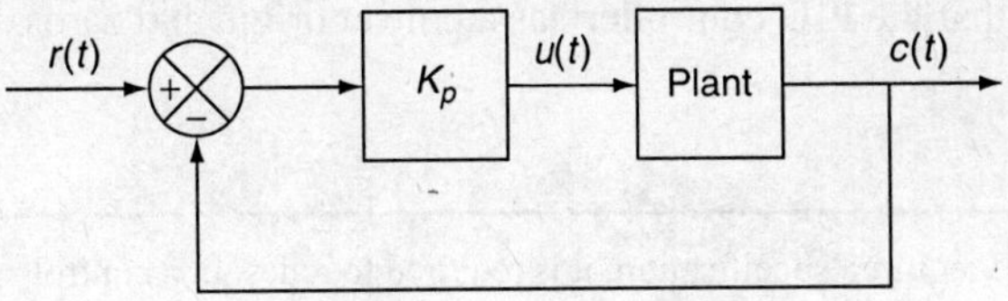

Fig. 16.21 PID controllers

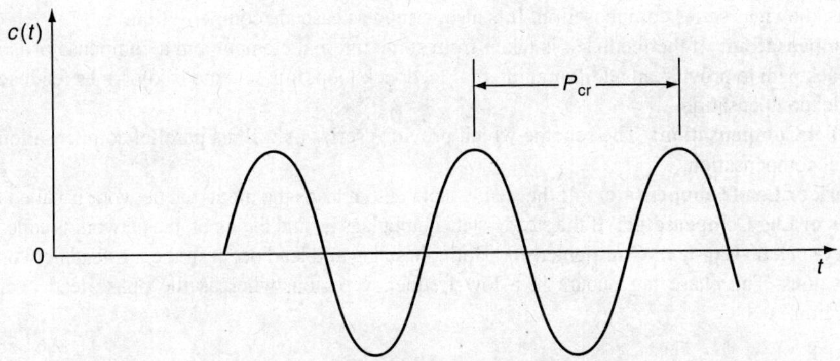

Fig. 16.22 Sustained oscillations

The suggested set of values of K_p, T_i and T_d is shown in Table 16.2.

Table 16.2 Values of K_p, T_i and T_d (second method)

Type of controller	K_p	T_i	T_d
P	$0.5K_{cr}$	∞	0
PI	$0.45K_{cr}$	$\frac{P_{cr}}{1.2}$	0
PID	$0.6K_{cr}$	$0.5P_{cr}$	$0.125P_{cr}$

The following is obtained when PID controller is tuned by second method of Ziegler Nichols rules:

$$G_c(s) = K_p\left(1 + \frac{1}{T_i s} + T_d s\right) = 0.6K_{cr}\left(1 + \frac{1}{0.5P_{cr}s} + 0.125P_{cr}s\right)$$

$$= 0.075 K_{cr} P_{cr} \frac{\left(s + \frac{4}{P_{cr}}\right)^2}{s} \tag{16.31}$$

Equation (16.31) shows that the PID controller has a pole at origin and zeros are at $-\frac{4}{P_{cr}}$.

SIGNIFICANT POINTS

Compensator: To fulfil the required specification, it is required to redesign a control system by adding an external device with the system. This process of redesigning by adding an external device is called compensation of a control system. The external device is known as a compensator.

Series Compensation: If the compensator is placed in series with the forward path transfer function of the plant, the scheme is known as series compensation. It is also termed as cascade compensation.

Parallel Compensation: If the feedback is taken from some internal element and a compensator is introduced in such a feedback path to provide an additional internal feedback loop, this scheme is known as feedback compensation or parallel compensation.

Series-Parallel Compensation: The scheme which provides series as well as parallel compensation is known as series-parallel compensation.

Lead Network or Lead Compensator: If the steady-state output leads the input, the network is called lead network.

Lag Network or Lag Compensator: If the steady-state output lags behind the input, the network is called lag network.

Lag-lead Network or Lag-lead Compensator: Both phase lag and lead occur in the lag-lead network in different frequency regions. The phase lag occurs in a low frequency region, whereas the phase lead occurs in a high frequency region.

SHORT QUESTIONS WITH ANSWERS

1. What do you mean by compensator?

   Ans. The process of redesigning a system by adding an external device is called compensation of a control system. The external device is known as a compensator.

2. Define the following:
   (i) Series compensation, (ii) parallel compensation, (iii) series-parallel compensation.

   Ans. (i) If the compensator is placed in series with the forward path transfer function of the plant, the scheme is known as series compensation. It is also termed as cascade compensation.

   (ii) If the feedback is taken from some internal element and a compensator is introduced in such a feedback path to provide an additional internal feedback loop, this scheme is known as feedback compensation or parallel compensation.

   (iii) Sometimes situation demands to provide series as well as feedback from some internal element. This scheme is known as series-parallel compensation.

   Define (i) Lead network, (ii) lag network and (iii) lag-lead network.

   1. Lead network or lead compensator: If the steady-state output leads the input, the network is called lead network.
   2. Lag network or lag compensator: If the steady-state output lags behind the input, the network is called lag network.

3. Lag-lead network or lag-lead compensator: Both phase lag and lead occur in the lag-lead network in different frequency regions. The phase lag occurs in a low frequency region, whereas the phase lead occurs in a high frequency region.

EXERCISES

Questions for Practice

1. What do you mean by compensator. Discuss series and parallel compensators.
2. Discuss lead compensator. Sketch the Bode plot of a lead compensator. Give the design steps of a lead compensator.
3. Discuss lag compensator. Sketch the Bode plot of a lag compensator. Give the design steps of a lag compensator.
4. Sketch the Bode plot and pole-zero plot of a lag-lead compensator.

TEST YOURSELF

1. The circuit shown in Figure T1 is a

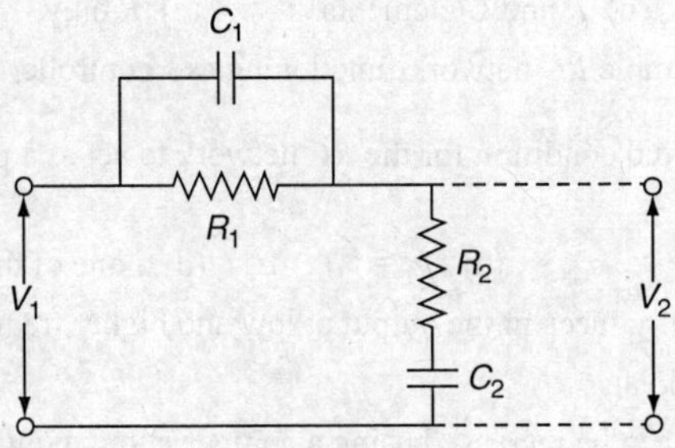

Figure T1

(a) lead network. (b) lag network. (c) lag-lead network. (d) none of these.

2. The circuit shown in Figure T2 is a

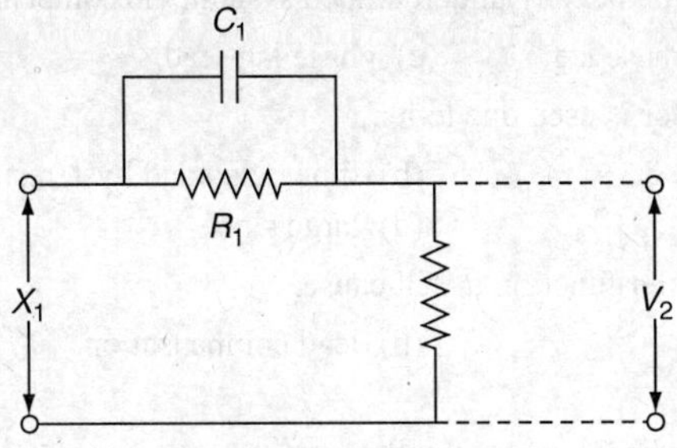

Figure T2

(a) lag-lead network. (b) lag network. (c) lead network. (d) none of these.

3. The circuit shown in Figure T3 is a

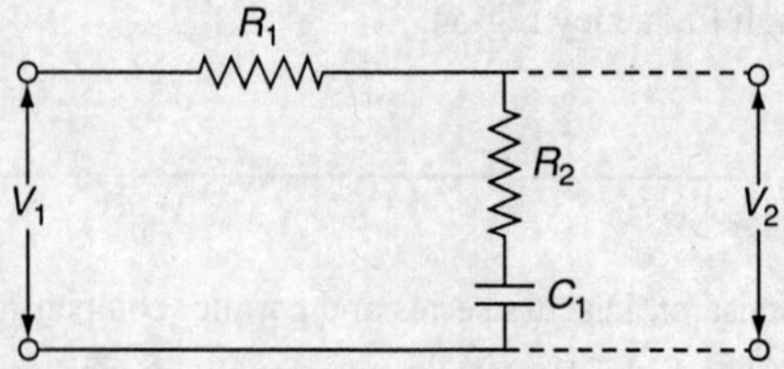

Figure T3

(a) lead network. (b) lag network. (c) lag-lead network. (d) none of these.

4. The transfer function $\dfrac{(1+0.5s)}{(1+s)}$ represents a

(a) lag network. (b) lead network.
(c) proportional controller. (d) lag-lead controller.

5. A lag network for compensation generally consists of

(a) R and L elements. (b) R and C elements. (c) R only. (d) none of these.

6. The transfer function of a simple RC network functioning as s controller is

$G_c(s) = \dfrac{(s+z_1)}{(s+p_1)}$. The required condition for the RC network to act as a phase lead controller is

(a) $p_1 < z_1$. (b) $p_1 > z_1$. (c) $p_1 = z_1$. (d) none of these.

7. A phase lag-lead network introduces in the output at low and higher frequencies, respectively,

(a) lag, lead (b) lead, lag

8. The damping of the system can be increased using a compensator having a pair of complex roots as

(a) phase, lead. (b) phase lag, lead. (c) phase lag. (d) none of these.

9. Phase lag network

(a) maintains constant velocity gain (b) decreases bandwidth.
(c) increases system stability. (d) all of these.

10. To increases the damping of a heavily underdamped system, the compensator used is

(a) phase lead. (b) phase lag (c) phase lag-lead.

11. In a lag network, the inductor is used due to its.

(a) high reactance. (b) time delay and hysteresis loss.
(c) high cost. (d) large size.

12. If poles are added in a transfer function it will cause

(a) lag compensation. (b) lead compensation.
(c) lag-lead compensation.

13. If zeros are added in a transfer function, it will cause

(a) lag compensation. (b) lead compensation.
(c) lag-lead compensation.

14. To increase the bandwidth of a control system, we may use

(a) phase lag compensator. (b) phase lead compensator.
(c) phase lag-lead compensator.

15. Lead-lag compensation is required for

    (a) steady-state response improvement. (b) transient response improvement.
    (c) both (a) and (b).

16. The transfer function of a lead compensator is

$$G_C(s) = \frac{1+0.12s}{1+0.04s}$$

The maximum phase shift that can be obtained from this compensator is

(a) 60° (b) 45° (c) 30° (d) 15°

17. Match the following

(a) Constant M loci	(i)	Constant gain and phase shift loci of the closed-loop system.
(b) Constant N loci	(ii)	Plot of loop gain with variation of ω.
(c) Nichol's chart	(iii)	Circles of constant gain for closed-loop transfer function.
(d) Nyquist plot	(iv)	Circle of constant phase shift of a closed transfer function.

ANSWERS

1 (c) 2 (c) 3 (b) 4 (a) 5 (b) 6 (b) 7 (a) 8 (a) 9 (d) 10 (a) 11 (b)
12 (b) 13 (a) 14 (b) 15 (a) 16 (c) 17 (a) →(iii) (b)→(iv) (c)→(i) (d)→(ii)

Chapter 17

State Variable Approach

The conventional control theory, which is applicable to linear time-invariant systems having single input and single output, has already been discussed. The conventional control theory cannot be applied for time-varying and/or non-linear systems. In the conventional control theory the main theme is to formulate the transfer function putting all initial conditions to zero. On the other hand, state variable analysis, i.e., the modern control theory automatically, takes care of initial conditions and it is also possible to analyze time-variant or time-invariant, linear or non-linear, single or multiple input–output systems. The main purpose of this chapter is to introduce state variable analysis.

17.1 Limitations of Conventional Approach

The following are the limitations of conventional approach:

(i) The precise solution of any system is naturally obtained after incorporating the initial conditions which is lost in conventional approach.

(ii) To get complete time domain solution of higher order system using the conventional approach is troublesome.

(iii) The conventional approach is not very much convenient for analysis of multiple input and multiple output systems.

(iv) This method gives analysis of system for specific inputs like step, ramp etc.

(v) This method is applicable to linear time invariant (*LTI*) systems.

(vi) It is not feasible to obtain the optimal solution using the classical methods like root locus, bode plot etc because these are the trial and error techniques.

From above discussion, it is obvious that a method which can overcome all the above limitations is required. Hence a method which uses the concept of total internal state of the system considering all initial conditions is to be discussed. This technique is known as state variable analysis or state space analysis because it uses the concept of state. This analysis is a time domain approach.

17.2 Advantages of State Variable Analysis

The following are the advantages of state variable analysis:

(i) This method has the capacity to take into account the effect of all initial conditions.

(ii) This method is applicable to nonlinear as well as time varying systems.

(iii) This method is applicable to multiple input multiple output (MIMO) systems.
(iv) It is possible to design the system for the optimal conditions precisely using this method.
(v) It is possible to consider any type of input for designing the system.
(vi) This method can be conveniently adopted for digital computers because it involves matrix algebra.
(vii) It is not mandatory that the selected state variables will be the physical quantities of the system.
(viii) The mathematical representation of the system is simplified greatly by the vector matrix notation.

17.3 Concept of State

Let us consider a network shown in Figure 17.1. The knowledge of initial voltage of the capacitor is required to calculate the output voltage (V_{out}). The knowledge of V_{in} only is not sufficient in this case for precise calculation of V_{out} at any time $t \geq 0$. Therefore, the output of this system is not only dependent on the input but also dependent on the initial conditions. Hence these type of systems are called systems with memory or dynamic systems.

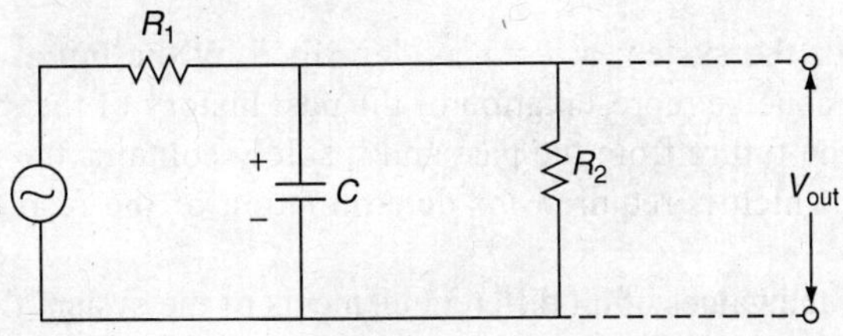

Fig. 17.1

If any resistance (R) replaces the capacitor of the circuit shown in Fig. 17.1, the output in this case solely depends on the input (V_{in}). Since, the output in these type of systems depend only on the input, this type of system is known as systems with zero memory or static system.

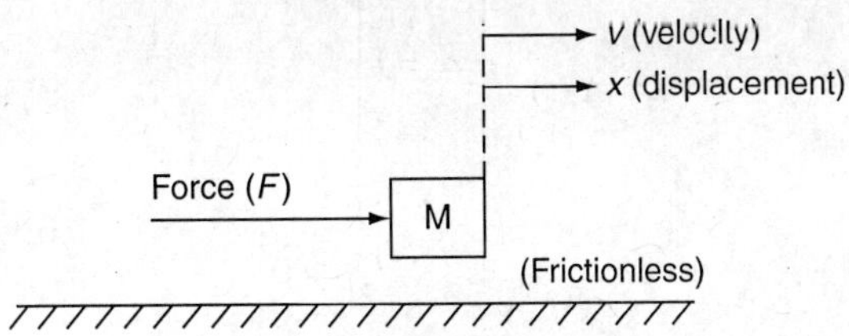

Fig. 17.2

Figure 17.2 shows another example which is a mechanical system. As per Newton's second law of motion, we can write

$$F = Ma \tag{17.1}$$

where 'a' is the acceleration of mass M.

Equation (17.1) can be expressed as

$$F(t) = M\frac{dv(t)}{dt}$$

i.e.,

$$v(t) = \frac{1}{M}\int_{-\infty}^{t} F(t)dt \tag{17.2}$$

Equation (17.2) shows that the velocity any time t is the response of the force applied to the mass in the past. Therefore, we have from Eq. (17.2)

$$v(t) = \frac{1}{M}\int_{-\infty}^{t_0} F(t)dt + \frac{1}{M}\int_{t_0}^{t} F(t)dt \tag{17.3}$$

where t_0 is the initial time in Eq. (17.3).

Equation (17.3) shows that there are different values of $v(t)$ for the same value of $F(t)$ and this is due to the choice of the parameter t_0 and hence the value of $v(t_0)$.

For known value of $v(t_0)$ and $F(t)$ from t_0 to t, it is possible to get the unique value of $v(t)$ at any time $t > t_0$.

From above we can conclude that the initial conditions affects the system characterization and subsequent behaviour.

The status or state of the system at $t = t_0$ is described by the initial conditions and hence the state is the compact or concise representation of the past history of the system. Therefore, the state of a system separates the future from the past and it solely contains the information about the past history of the system, which is required for determination of the response of the system for any mentioned input.

The overall effect of the values of all different elements of the system those are associated with the initial conditions of the system is the state of the system. Therefore, the complete state of the system is considered as a vector, which has components those are variables of the system and associated with the initial conditions. Therefore, a state is defined as a vector $X(t)$ known as state vector, which is a state at any time 't'. It is a column matrix (say $n \times 1$) as shown below.

$$X(t) = \begin{bmatrix} X_1(t) \\ X_2(t) \\ X_3(t) \\ \vdots \\ \vdots \\ X_n(t) \end{bmatrix}_{n\times 1} \tag{17.4}$$

Therefore, the variables $X_1(t)$, $X_2(t)$, ……, $X_n(t)$ constitute the state vector $X(t)$ and they are known as the state variables of the system.

The value of $X(t_0)$ and the knowledge of the input between t_0 and t_1 must be known before determining the state at $t = t_1$. The new state will be $X(t_1)$ which will be the initial state for the state to be found out at $t = t_2$ where $t_2 > t_1$. This is termed as updating of the state. For any system the number of state variables and the order of the system is equal. The number of independent state variables is usually decided by the number of energy storing elements (capacitor, inductor) contained in the system.

17.4 Basic Terms

State: The state of a dynamic system is the smallest set of variables and the knowledge of these variables at $t = t_0$ together with inputs for $t \geq t_0$ completely determines the behaviour of the system at $t \geq t_0$. A compact and concise representation of the past history of the system can be termed as the state of the system.

State variables: The smallest set of variables that determine the state of the system are known as state variables. The knowledge of capacitor, voltage at $t = 0$, i.e., the initial voltage of the capacitor is a history-dependent term and it forms a state variable. Similarly, initial current of an inductor is treated as a state variable.

State vector: The n state variables that completely describe the behaviour of a given system are said to be n components of a vector.

State space: The n-dimensional space whose coordinate axes consist of the x_1 axis, x_2 axis, ..., x_n axis is known as a state space.

State trajectory: The locus of the tips of the state vectors with time as implicit variables is known as the state trajectory.

A second order system is considered to explain the above definitions. Since the order of the system is two, the number of state variables will be two. Therefore, state vector will be a 2×1 matrix and it is given by

$$\boldsymbol{X}(t) = \begin{bmatrix} \boldsymbol{X}_1(t) \\ \boldsymbol{X}_2(t) \end{bmatrix}$$

i.e.,

$$\vec{X}(t) = \vec{X}_1(t) + \vec{X}_2(t) \tag{17.5}$$

In this case state space will be a plane as shown in Fig. 17.3 because the number of state variables is two.

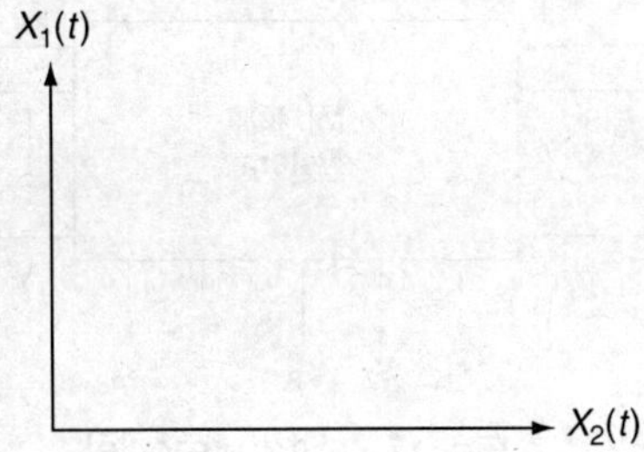

Fig. 17.3 State space

At time $t = t_1$, the state vector $X(t_1)$ is shown in Fig. 17.4 where as Fig. 17.5 shows the state vector $X(t_2)$ at time $t = t_2$.

Here

$$\vec{X}(t_1) = \vec{X}_1(t_1) + \vec{X}_2(t_1) \tag{17.6}$$

and

$$\vec{X}(t_2) = \vec{X}_1(t_2) + \vec{X}_2(t_2) \tag{17.7}$$

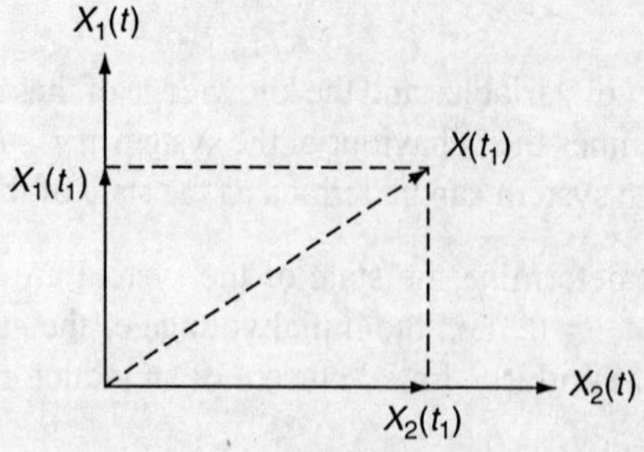

Fig. 17.4 State vector $X(t_1)$

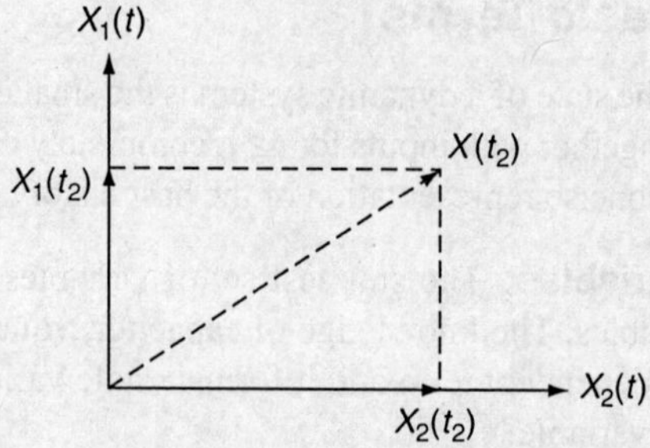

Fig. 17.5 State vector $X(t_2)$

If the tips of the two state vectors $X(t_1)$ and $X(t_2)$ are joined, the state trajectory will be obtained as shown in Fig. 17.6.

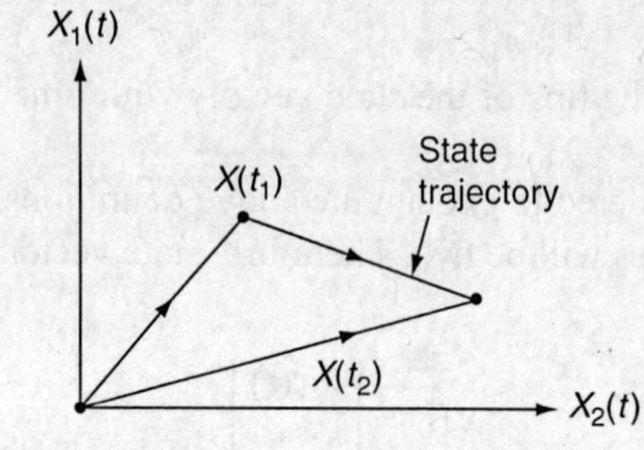

Fig. 17.6 State trajectory

17.5 State Model of MIMO Systems

Figure 17.7 shows an nth-order system having multiple input and output.

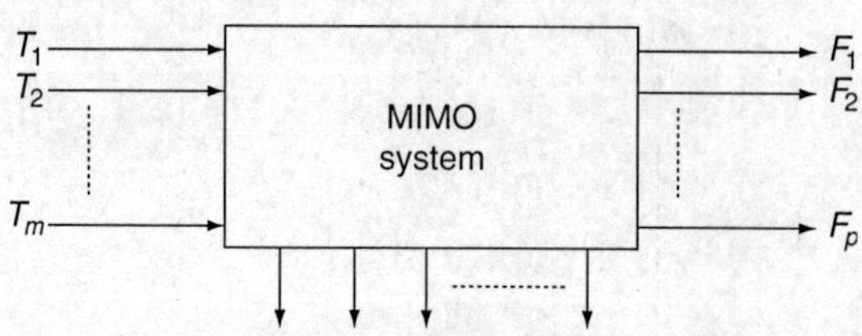

Fig. 17.7 State model

Here,

$$u(t) = \begin{bmatrix} u_1(t) \\ u_2(t) \\ \vdots \\ \vdots \\ u_m(t) \end{bmatrix}_{m\times 1}, x(t) = \begin{bmatrix} x_1(t) \\ x_2(t) \\ \vdots \\ \vdots \\ x_n(t) \end{bmatrix}_{n\times 1} \text{ and } y(t) = \begin{bmatrix} y_1(t) \\ y_2(t) \\ \vdots \\ \vdots \\ y_p(t) \end{bmatrix}_{p\times 1}$$

The state variable representation of such a system in the form of n first-order differential equations is shown below:

$$\dot{x}_1(t) = f_1(x_1, x_2, \cdots, x_n, u_1, u_2, \cdots, u_m)$$
$$\dot{x}_2(t) = f_2(x_1, x_2, \cdots, x_n, u_1, u_2, \cdots, u_m)$$
$$\vdots$$
$$\dot{x}_n(t) = f_n(x_1, x_2, \cdots, x_n, u_1, u_2, \cdots, u_m)$$

where

$$f(t) = \begin{bmatrix} f_1 \\ f_2 \\ \vdots \\ \vdots \\ f_n \end{bmatrix},$$

is a functional operator.

On integration, we have

$$x_i(t) = x_i(t_0) \times \int_{t_0}^{t} f_i(x_1, x_2, \cdots, x_n, u_1, u_2, \cdots, u_m)dt \tag{17.8}$$

where $i = 1, 2, \cdots, n$

Therefore, n state variables and hence state vectors at any time t can be determined uniquely. The functional form of any n-dimensional time-invariant system is given by

$$\dot{x}(t) = f(x, u) \tag{17.9}$$

The output of such a system is dependent on the state of the system and the instantaneous input. The functional output equation can be written as

$$y(t) = g(x, u) \tag{17.10}$$

where g is the functional operator. For a time-invariant system, we have

$$\dot{x}(t) = f(x, u, t) \tag{17.11}$$
$$y(t) = g(x, u, t) \tag{17.12}$$

Equation (17.11) and Eq. (17.12) show the state and output equations, respectively. For a linear time-invariant system, the functional equations can be written in the form as below:

$$\dot{x}(t) = Ax(t) + Bu(t) \tag{17.13}$$
$$y(t) = Cx(t) + Du(t) \tag{17.14}$$

where A, B, C and D are constant matrices.

For linear time-variant system, we have

$$\dot{x}(t) = A(t)x(t) + B(t)u(t) \tag{17.15}$$
$$y(t) = C(t)x(t) + D(t)u(t) \tag{17.16}$$

where A, B, C and D are time-dependent matrices.

Figure 17.8 shows the diagrammatic representation.

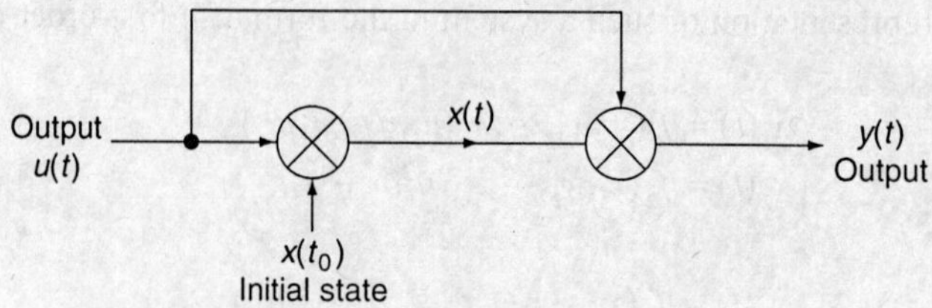

Fig. 17.8 Diagrammatic representation of Fig. 17.1

The order of the above matrices is shown below:

$A \rightarrow n \times n$ known as evolution matrix

$B \rightarrow n \times m$ known as control matrix

$C \rightarrow p \times n$ known as observation matrix

$D \rightarrow p \times m$ known as direct transmission matrix

This is known as the state model of the given system.

17.6 State Model of SISO System

A single input and single output system is considered here i.e., $m = 1$ and $p = 1$. The n state variables are required to define the state of the system because its order is n. The state model becomes

$$\dot{x}(t) = Ax(t) + Bu(t) \tag{17.17}$$

and

$$y(t) = Cx(t) + Du(t) \tag{17.18}$$

where A is a $n \times n$ matrix, B is a $n \times 1$ matrix, C is $1 \times n$ matrix, D is constant and $u(t)$ is a single scalar input variable.

17.7 State Diagram Representation

State diagram shows a close relationship amongst the state model, differential equations of the system and its solution and it is a pictorial representation which is basically a block diagram approach. State diagram is a proper interconnection of scalars, adders and integrators shown in Fig. 17.9 (a), Fig. 17.9(b) and Fig. 17.9(c) respectively.

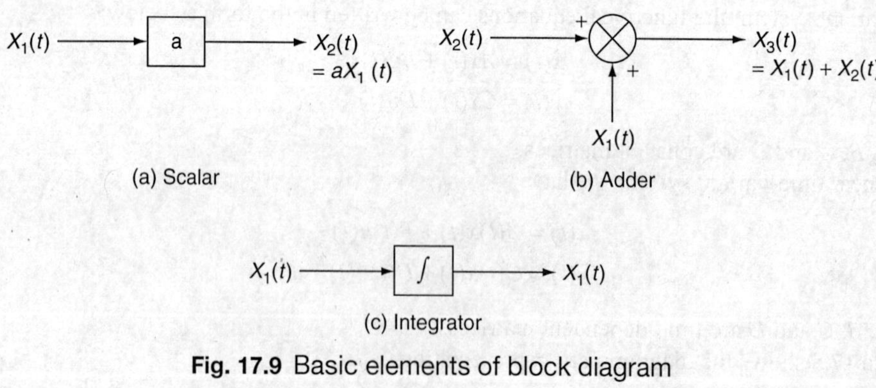

Fig. 17.9 Basic elements of block diagram

17.8 State Diagram of Standard State Model

The standard state model is considered here which is represented by the following equations.

$$\dot{x}(t) = Ax(t) + Bu(t) \tag{17.19}$$

and

$$y(t) = Cx(t) + Du(t) \tag{17.20}$$

Figure 17.10 shows its state diagram.

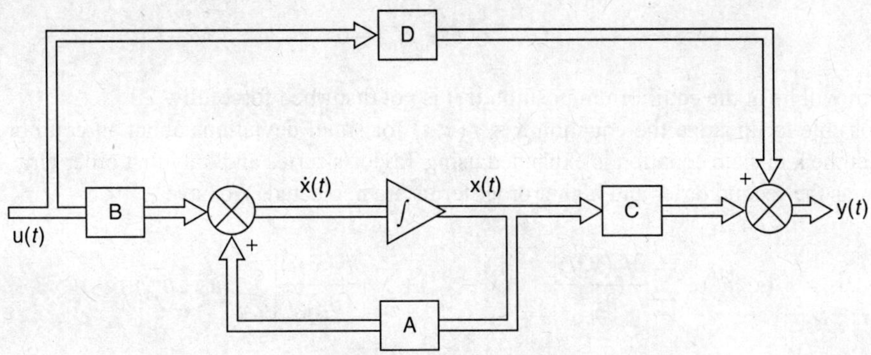

Fig. 17.10 State diagram of MIMO system

Since the system has multiple number of input, output and state variables, this has been represented by thick arrow as shown in Fig. 17.10. Therefore, n parallel integrators must be present as there are n state variables where the output of each integrator is a separate variable. After obtaining the state diagram, the state model can be easily determined from it.

17.9 Non-Uniqueness of State Model

The state model of any system is not unique. Alternate representations are possible for any given system. Let us consider a state model of a system as given below:

$$\dot{x} = Ax + Bu \tag{17.21}$$

$$y = Cx + Du \tag{17.22}$$

Let the transformation be

$$x = Mz \tag{17.23}$$

where M is a non-singular constant matrix. Therefore, using Eq. (17.23), Eq. (17.21) can be written as

$$M\dot{z} = AMz + Bu$$

$$\dot{z} = M^{-1}AMz + M^{-1}Bu = Pz + Qu \tag{17.24}$$

where $P = M^{-1}AM$ and $Q = M^{-1}B$

Using Eq. (17.23), Eq. (17.22) can be written as

$$y = CMz + Du = Rz + Du \tag{17.25}$$

where $$R = CM$$

Equation (17.24) and Eq. (17.25) show that any given state model is not unique.

17.10 Linearization of State Equation

During equilibrium position at the point (x_0,u_0) of any time invariant system, the derivatives of all state variables must be zero. Therefore, we have

$$\dot{x} = f(x_0, u_0) = 0 \tag{17.26}$$

The system will lie at the equilibrium position if it is not disturbed forcefully.

It is possible to linearize the equation $\dot{x} = f(x,u)$ for small deviations about an equilibrium point (x_0, u_0). Let the k th state equation is expanded using Taylor's series and only first order term is considered whereas the second order and higher order terms are neglected. We have

$$\dot{x}_k = f_k(x_0,u_0) + \sum_{l=0}^{n} \left.\frac{\partial f_k(x,u)}{\partial x_l}\right|_{\substack{x=x_0\\u=u_0}} (x_l - x_{l0}) + \sum_{l=0}^{n} \left.\frac{\partial f_k(x,u)}{\partial u_j}\right|_{\substack{x=x_0\\u=u_0}} (u_j - u_{j0}) \tag{17.27}$$

Since $f_k(x_0,u_0) = 0$ at the equilibrium point and let us take the variation about the operating point be

$$\tilde{x}_l = x_l - x_{l0} \text{ hence } \dot{\tilde{x}}_l = \dot{x}_l$$

and
$$\tilde{u}_j = u_j - u_{j0}$$

Therefore, kth equation can be linearized as

$$\dot{x}_k = f_k(x_0, u_0) + \sum_{l=0}^{n} \left.\frac{\partial f_k(x,u)}{\partial x_l}\right|_{\substack{x=x_0\\u=u_0}} \tilde{x}_l + \sum_{l=0}^{n} \left.\frac{\partial f_k(x,u)}{\partial u_j}\right|_{\substack{x=x_0\\u=u_0}} \tilde{u}_j \tag{17.28}$$

Here $\dot{x}_k, \tilde{x}_l$ and $\tilde{u}_j$ are vector matrices and the remaining terms in Eq. (17.28) are the matices of order of n × n and n × m respectively. Therefore, the linearized equation can be expressed as

$$\dot{\tilde{x}} = A\tilde{x} + B\bar{u} \tag{17.29}$$

where

$$A = \begin{bmatrix} \frac{\partial f_1}{\partial x_1} & \frac{\partial f_1}{\partial x_2} & \cdots\cdots & \frac{\partial f_1}{\partial x_n} \\ \frac{\partial f_2}{\partial x_1} & \frac{\partial f_2}{\partial x_2} & \cdots\cdots & \frac{\partial f_2}{\partial x_n} \\ \vdots & \vdots & \vdots & \vdots \\ \vdots & \vdots & \vdots & \vdots \\ \frac{\partial f_n}{\partial x_1} & \frac{\partial f_n}{\partial x_2} & \cdots\cdots & \frac{\partial f_n}{\partial x_n} \end{bmatrix}_{n \times n} \tag{17.30}$$

and
$$B = \begin{bmatrix} \frac{\partial f_1}{\partial u_1} & \frac{\partial f_1}{\partial u_2} & \cdots\cdots & \frac{\partial f_1}{\partial u_m} \\ \frac{\partial f_2}{\partial u_1} & \frac{\partial f_2}{\partial u_2} & \cdots\cdots & \frac{\partial f_2}{\partial u_m} \\ \vdots & \vdots & \vdots & \vdots \\ \vdots & \vdots & \vdots & \vdots \\ \frac{\partial f_n}{\partial u_1} & \frac{\partial f_n}{\partial u_2} & \cdots\cdots & \frac{\partial f_n}{\partial u_m} \end{bmatrix}_{n \times m} \tag{17.31}$$

The partial derivatives present in the matrices A and B are to be obtained at an equilibrium state (x_0, u_0) and these matrices are called the Jacobian matrices.

17.11 Representation of State Variable using Phase Variables

The minimum number of variables associated with the initial conditions of the system are known as the state variables. Since the state model is not unique as shown in Art. 17.8, their sequence is unimportant. The number of state variables is equal to the order of the system. The selection of state variable is required for obtaining the state model of the system and the various physical quantities of the system are selected as the state variables. The current through the inductors and voltage across the capacitors in electrical systems are taken as state variables. The equation of the network is to be written in terms of the selected state variables, their derivatives and the inputs. To obtain the required state model, the equations are arranged in proper form.

The displacements and velocities of energy storing elements such as spring and friction in mechanical systems are taken as the state variables.

Example 17.1 Obtain the state model of the parallel RLC network as shown in Fig. E17.1.

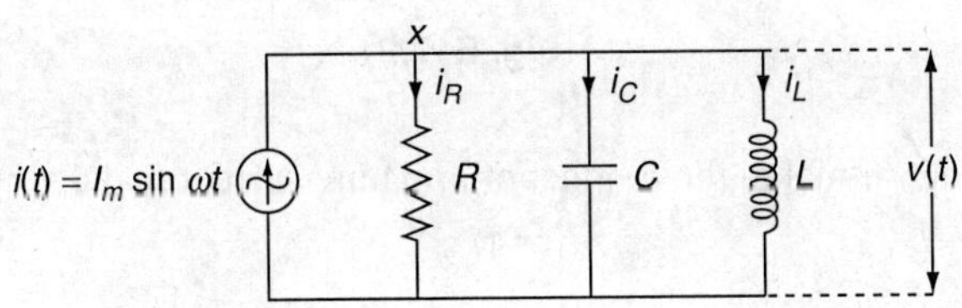

Fig. E17.1

Solution

Applying KCL at node x, we have

Let
$$i(t) = i_R(t) + i_L(t) + i_C(t)$$

$\therefore$
$$i(t) = \frac{v(t)}{R} + \frac{1}{L}\int v(t)dt + C\frac{dv(t)}{dt}$$

$\therefore$
$$C\frac{d^2v(t)}{dt^2} + \frac{1}{R}\frac{dv(t)}{dt} + \frac{1}{L}v(t) = \frac{di(t)}{dt} = I_m \omega \cos \omega t$$

Let $$v(t) = x_1(t) \tag{1}$$

and $$\dot{x}_1(t) = x_2(t) \tag{2}$$

∴ $$\dot{x}_2(t) = -\frac{1}{LC}x_1(t) - \frac{1}{RC}x_2(t) + \frac{\omega I_m}{C}\cos\omega t \tag{3}$$

From Eq. (2) and Eq. (3), the state equation can be expressed as

$$\begin{bmatrix}\dot{x}_1(t)\\ \dot{x}_2(t)\end{bmatrix} = \begin{bmatrix}0 & 1\\ -\frac{1}{LC} & -\frac{1}{RC}\end{bmatrix}\begin{bmatrix}x_1(t)\\ x_2(t)\end{bmatrix} + \begin{bmatrix}0\\ \frac{\omega I_m}{C}\cos\omega t\end{bmatrix}[u] \tag{4}$$

From Eq. (1), the output equation can be written as

$$v(t) = [1\ 0]\begin{bmatrix}x_1(t)\\ x_2(t)\end{bmatrix} \tag{5}$$

Example 17.2 Obtain the state model of the armature controlled DC motor shown in Fig. E17.2, where notations have their usual meaning.

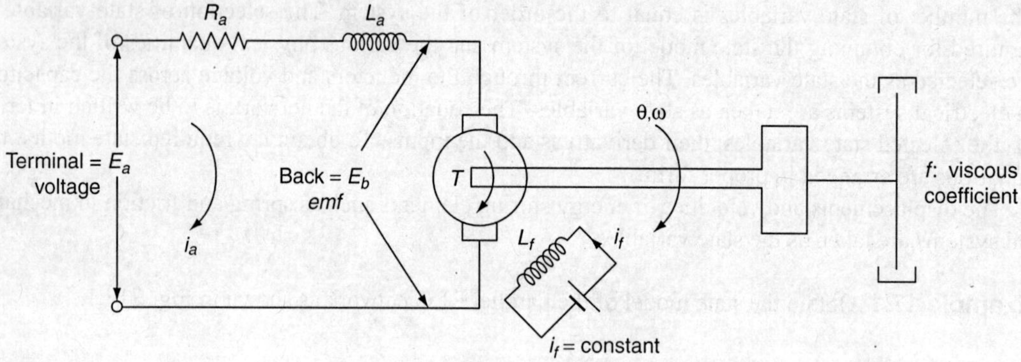

Fig. E17.2

Solution

Since the torque T is proportional to the product of armature current i_a and air gap flux ϕ, we can write

$$T \propto \phi i_a \tag{1}$$

Again

$$\phi \propto i_f \tag{2}$$

Using Eq. (2), we have from Eq. (1) as follows:

$$T \propto i_f i_a \tag{3}$$

∴ $$T = k_1 i_f i_a \tag{4}$$

In armature controlled DC motor, the field current is held constant. From Eq. (4), we have

$$T = k i_a \tag{5}$$

Again, back emf can be given by

$$E_b = k_b \omega \tag{6}$$

The differential equation for the armature circuit is

$$L_a \frac{di_a}{dt} + R_a i_a + E_b = E_a \tag{7}$$

The torque produced by the armature current overcomes the inertia and friction, hence we have

$$J \frac{d^2\theta}{dt^2} + f \frac{d\theta}{dt} = k i_a \tag{8}$$

From Eq. (7) and Eq. (8), we have

$$\frac{di_a}{dt} = -\frac{R_a}{L_a} i_a - \frac{k_b}{L_a} \omega + \frac{1}{L_a} E_a \tag{9}$$

and

$$\frac{d\omega}{dt} = \frac{k}{J} i_a - \frac{f}{J} \omega \tag{10}$$

Also

$$\frac{d\theta}{dt} = \omega \tag{11}$$

We have also

$$\theta = 0.i_a + 0.\omega + \theta \tag{12}$$

$$\therefore \quad \begin{bmatrix} \dfrac{di_a}{dt} \\ \dfrac{d\omega}{dt} \\ \dfrac{d\theta}{dt} \end{bmatrix} = \begin{bmatrix} -\dfrac{R_a}{L_a} & -\dfrac{k_a}{L_a} & 0 \\ \dfrac{k}{J} & -\dfrac{f}{J} & 0 \\ 0 & 1 & 0 \end{bmatrix} \begin{bmatrix} i_a \\ \omega \\ \theta \end{bmatrix} + \begin{bmatrix} \dfrac{1}{L_a} \\ 0 \\ 0 \end{bmatrix} E_a$$

and

$$\theta = [0 \quad 0 \quad 1] \begin{bmatrix} i_a \\ \omega \\ \theta \end{bmatrix}$$

17.12 Different Representations of a State Model

17.12.1 State Space Representation Using Phase Variables in Controllable Canonical Form (CCF)

Let us consider a transfer function of the following form:

$$\frac{Y(s)}{U(s)} = \frac{b}{s^n + a_1 s^{n-1} + a_2 s^{n-2} + \cdots + a_{n-1} s + a_n} \tag{17.32}$$

The transfer function given in Eq. (17.32) can also be represented by the differential equation as follows:

$$\frac{d^n y}{dt^n} + a_1 \frac{d^{n-1} y}{dt^{n-1}} + a_2 \frac{d^{n-2} y}{dt^{n-2}} + \cdots + a_{n-1} \frac{dy}{dt} + a_n y = bu \tag{17.33}$$

Let us put

$$x_1 = y$$
$$\dot{x}_1 = x_2$$
$$\dot{x}_2 = x_3$$
$$\vdots$$
$$\dot{x}_{n-1} = x_n$$

With the help of the above representations, Eq. (17.33) can be written as

$$\dot{x}_n = -a_n x_1 - a_{n-1} x_2 - a_{n-3} x_3 \cdots - a_1 x_n + bu$$

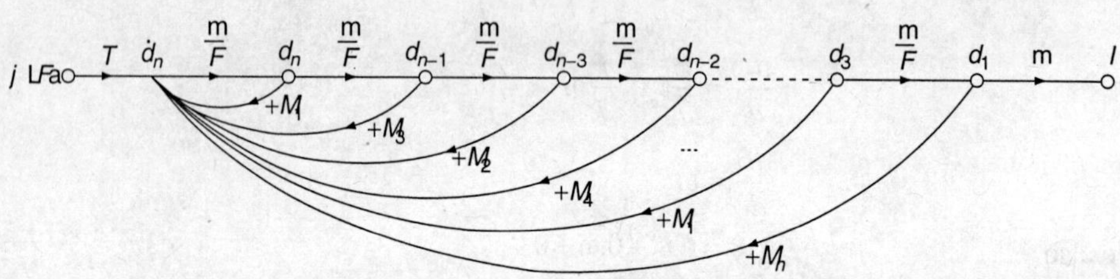

Fig. 17.11 SFG of CCF

Figure 17.11 shows the signal flow graph of CCF. The state equation is represented by

$$\begin{bmatrix} \dot{x}_1 \\ \dot{x}_2 \\ \vdots \\ \dot{x}_{n-1} \\ \dot{x}_n \end{bmatrix} = \begin{bmatrix} 0 & 1 & 0 & \cdots & 0 \\ 0 & 0 & 1 & \cdots & 0 \\ \vdots & \vdots & \vdots & & \\ & & & \cdots & 1 \\ -a_n & -a_{n-1} & -a_{n-2} & \cdots & -a_1 \end{bmatrix} \begin{bmatrix} x_1 \\ x_2 \\ \vdots \\ x_{n-1} \\ x_n \end{bmatrix} + \begin{bmatrix} 0 \\ 0 \\ \vdots \\ 0 \\ b \end{bmatrix} [u] \tag{17.34}$$

$$\therefore \quad \dot{x} = Ax + Bu \tag{17.35}$$

The matrix for the output equation of the system is given below:

$$y = [1 \ 0 \ 0 \ \cdots \ 0] \begin{bmatrix} x_1 \\ x_2 \\ \vdots \\ x_{n-1} \\ x_n \end{bmatrix} + [0][u] \tag{17.36}$$

$$\therefore \quad y = Cx + Du \tag{17.37}$$

The matrix A has all 1s in the upper off-diagonal and other elements except the last row. The coefficient of the last row has a negative coefficient. This form is also known as bush or comparison form.

Example 17.3 Express the following transfer function in a CCF form. Draw the signal flow graph (SFG).

$$\frac{Y(s)}{U(s)} = \frac{3}{s^4 + 2s^3 + 3s + 2}$$

Solution

From the given transfer function, we have

$$(s^4 + 2s^3 + 3s + 2)Y(s) = 3U(s)$$

i.e.,
$$\frac{d^4y}{dt^4} + 2\frac{d^3y}{dt^3} + 3\frac{dy}{dt} + 2y = 3u$$

Let
$$x_1 = y \tag{1}$$
$$\dot{x}_1 = \frac{dy}{dt} = x_2 \tag{2}$$
$$\dot{x}_2 = \frac{d^2y}{dt^2} = x_3 \tag{3}$$
$$\dot{x}_3 = \frac{d^3y}{dt^3} = x_4 \tag{4}$$
$$\dot{x}_4 = \frac{d^4y}{dt^4} = 3u - 2\frac{d^3y}{dt^3} - 3\frac{dy}{dt} - 2y = 3u - 2x_4 - 3x_2 - 2x_1 \tag{5}$$

From Eq. (2) to Eq. (5), we have

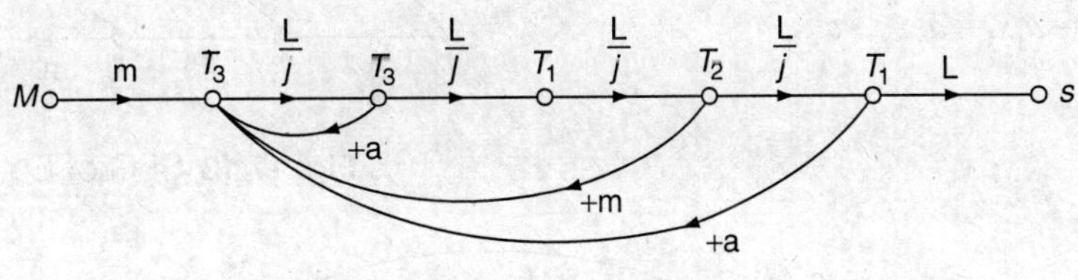

Fig. E17.3 SFG

$$\begin{bmatrix} \dot{x}_1 \\ \dot{x}_2 \\ \dot{x}_3 \\ \dot{x}_4 \end{bmatrix} = \begin{bmatrix} 0 & 1 & 0 & 0 \\ 0 & 0 & 1 & 0 \\ 0 & 0 & 0 & 1 \\ -2 & -3 & 0 & -2 \end{bmatrix} \begin{bmatrix} x_1 \\ x_2 \\ x_3 \\ x_4 \end{bmatrix} + \begin{bmatrix} 0 \\ 0 \\ 0 \\ 3 \end{bmatrix} [u] \tag{6}$$

and from Eq. (1),

$$y = [1\ 0\ 0\ 0] \begin{bmatrix} x_1 \\ x_2 \\ x_3 \\ x_4 \end{bmatrix} + [0][u] \tag{7}$$

Equation (6) and Eq. (7) represent the required CCF from. The signal flow graph is shown in Fig. E17.3.

17.12.2 Phase Variable CCF Form for Numerator Terms

Let the transform of the system be

$$\frac{Y(s)}{U(s)}=\frac{b_0s^3+b_1s^2+b_2s+b_3}{s^3+a_1s^2+a_2s+a_3}=\frac{(b_0+b_1s^{-1}+b_2s^{-2}+b_3s^{-3})X(s)}{(1+a_1s^{-1}+a_2s^{-2}+a_3s^{-3})X(s)} \quad (17.38)$$

Let $$Y(s)=(b_0+b_1s^{-1}+b_2s^{-2}+b_3s^{-3})X(s) \quad (17.39)$$

and $$U(s)=(1+a_1s^{-1}+a_2s^{-2}+a_3s^{-3})X(s) \quad (17.40)$$

From Eq. (17.40), we have

$$X(s)=U(s)-(a_1s^{-1}+a_2s^{-2}+a_3s^{-3})X(s) \quad (17.41)$$

The signal flow graph of Eq. (17.41) is shown in Fig. 17.12.

The SFG of Eq. (17.39) is shown in Fig. 17.13.

Combining Fig. 17.12 and Fig. 17.13, the state diagram is shown in Fig. 17.14.

The relations are given below:

$$\dot{x}_1=x_2$$

$$\dot{x}_2=x_3$$

$$\dot{x}_3=u-a_1x_3-a_2x_2-a_3x_1$$

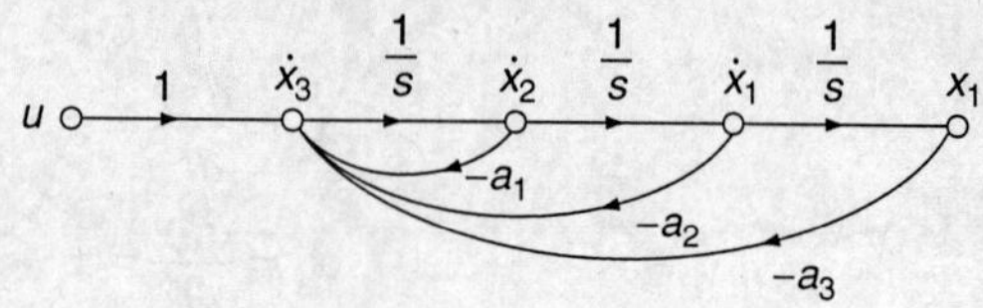

Fig. 17.12 SFG of Eq. (17.41)

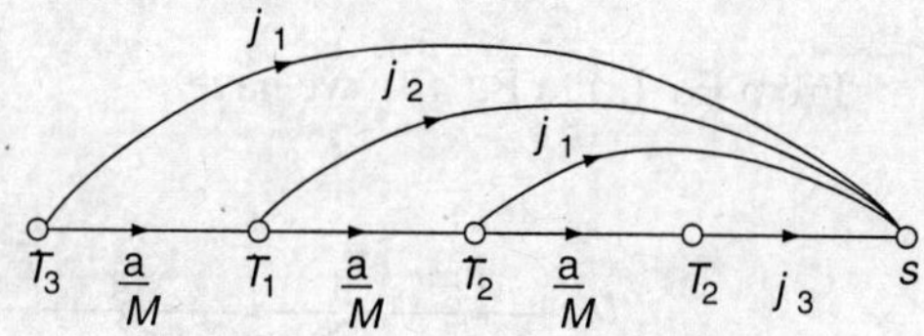

Fig. 17.13 SFG of Eq. (17.39)

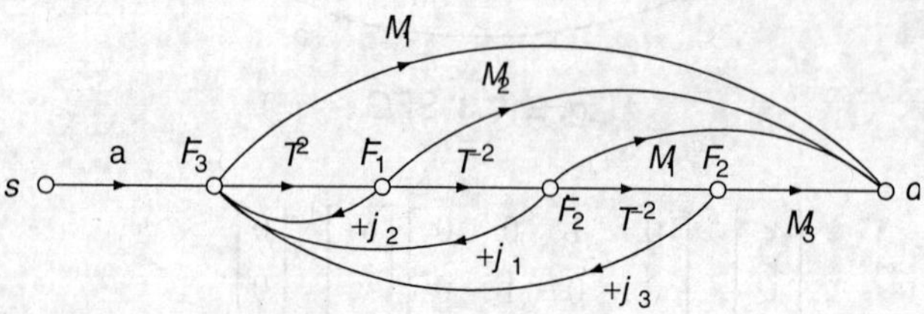

Fig. 17.14 SFG of CCF

The state equation is represented by

$$\therefore \quad \begin{bmatrix}\dot{x}_1\\ \dot{x}_2\\ \dot{x}_3\end{bmatrix}=\begin{bmatrix}0 & 1 & 0\\ 0 & 0 & 1\\ -a_3 & -a_2 & -a_1\end{bmatrix}\begin{bmatrix}x_1\\ x_2\\ x_3\end{bmatrix}+\begin{bmatrix}0\\ 0\\ 1\end{bmatrix}[u] \quad (17.42)$$

$$\therefore \quad \dot{x}=Ax+Bx \quad (17.43)$$

Now $$y=b_3x_1+b_2x_2+b_1x_3+b_0\dot{x}_3$$

$$= b_3x_1 + b_2x_2 + b_1x_3 + b_0(u - a_1x_3 - a_2x_2 - a_3x_1)$$
$$= (b_3 - b_0a_3)x_1 + (b_2 - b_0a_2)x_2 + (b_1 - b_0a_1)x_3 + b_0u$$

The output equation is represented by

$$\therefore \qquad y = [(b_3 - b_0a_3)(b_2 - b_0a_2)(b_1 - b_0a_1)] + \begin{bmatrix} x_1 \\ x_2 \\ x_3 \end{bmatrix} + [b_0][u] \qquad (17.44)$$

i.e.,
$$y = Cx + Du \qquad (17.45)$$

Equation (17.42) and Eq. (17.44) are state models in phase variable form (CCF). Here, matrices A and B in Eq. (17.43) are apparently of CCF. Generally, the degree of numerator is less than the degree of denominator.

Example 17.4 The transfer function of a system is given below:

$$\frac{Y(s)}{U(s)} = \frac{3s^2 + 2s + 7}{s^3 + 5s^2 + 12s + 5}$$

Express in CCF. Draw a state diagram and state matrix.

Solution $\dfrac{Y(s)}{U(s)} = \dfrac{3s^2 + 2s + 7}{s^3 + 5s^2 + 12s + 5} = \dfrac{(3s^{-1} + 2s^{-2} + 7s^{-3})X(s)}{(1 + 5s^{-1} + 12s^{-2} + 5s^{-3})X(s)}$

Let
$$Y(s) = (3s^{-1} + 2s^{-2} + 7s^{-3})X(s) \qquad (1)$$

and
$$U(s) = (1 + 5s^{-1} + 12s^{-2} + 5s^{-3})X(s) \qquad (2)$$

$$\therefore \qquad X(s) = U(s) - (5s^{-1} + 12s^{-2} + 5s^{-3})X(s) \qquad (3)$$

The SFG is shown in Fig. E17.4.

Let
$$\dot{x}_1 = x_2$$
$$\dot{x}_2 = x_3$$
$$\therefore \qquad \dot{x}_3 = u - 5x_3 - 12x_2 - 5x_1$$

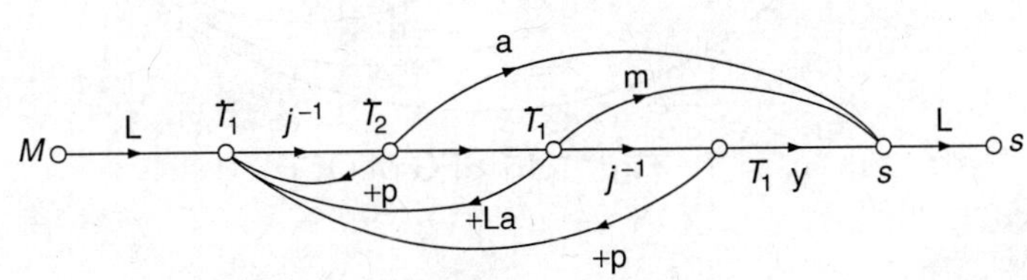

Fig. E17.4 SFG

$$\therefore \quad \begin{bmatrix} \dot{x}_1 \\ \dot{x}_2 \\ \dot{x}_3 \end{bmatrix} = \begin{bmatrix} 0 & 1 & 0 \\ 0 & 0 & 1 \\ -5 & -12 & -15 \end{bmatrix} \begin{bmatrix} x_1 \\ x_2 \\ x_3 \end{bmatrix} + \begin{bmatrix} 0 \\ 0 \\ 1 \end{bmatrix} [u]$$

and $$y = 7x_1 + 2x_2 + 3x_3$$

$$\therefore \quad [y] = [7 \; 2 \; 3] \begin{bmatrix} x_1 \\ x_2 \\ x_3 \end{bmatrix} + [0][u]$$

17.12.3 Phase Variable in Observable Controllable Form (OCF)

Let $$\frac{Y(s)}{U(s)} = \frac{b_{n-1}s^{-1} + b_{n-2}s^{-2} + \cdots + b_1 s^{-n+1} + b_0 s^{-n}}{1 + a_{n-1}s^{-1} + a_{n-2}s^{-2} + \cdots + a_1 s^{-n+1} + a_0 s^{-n}}$$

$$\therefore \quad (1 + a_{n-1}s^{-1} + a_{n-2}s^{-2} + \cdots + a_1 s^{-n+1} + a_0 s^{-n})Y(s)$$
$$= (b_{n-1}s^{-1} + b_{n-2}s^{-2} + \cdots + b_1 s^{-n+1} + b_0 s^{-n} +)U(s)$$

$$\therefore \quad Y(s) = -(a_{n-1}s^{-1} + a_{n-2}s^{-2} + \cdots + a_1 s^{-n+1} + a_0 s^{-n})Y(s)$$
$$+ (b_{n-1}s^{-1} + b_{n-2}s^{-2} + \cdots + b_1 s^{-n+1} + b_0 s^{-n})U(s) \qquad (17.46)$$

Figure 17.15 shows SFG of Eq. (17.46).

$$\therefore \quad \dot{x}_1 = -a_0 x_n + b_0 u$$
$$\dot{x}_2 = x_1 - a_1 x_n + b_1 u$$
$$\dot{x}_3 = x_2 - a_2 x_n + b_2 u$$
$$\vdots$$

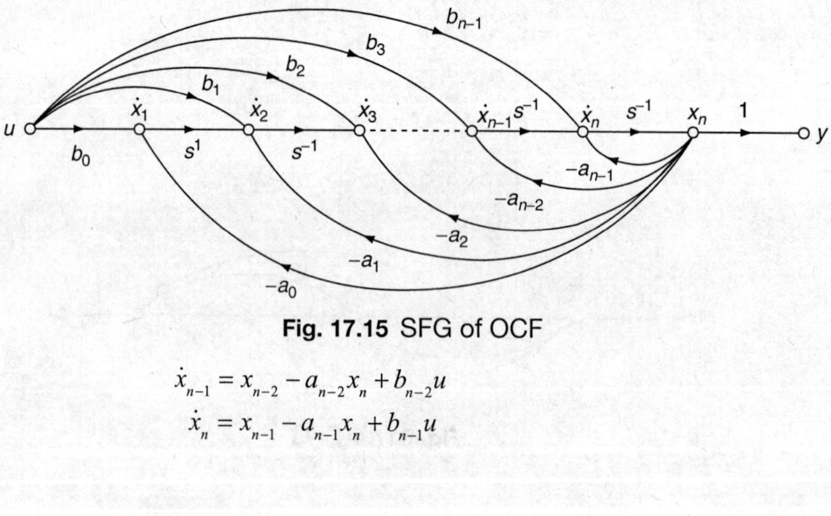

Fig. 17.15 SFG of OCF

$$\dot{x}_{n-1} = x_{n-2} - a_{n-2}x_n + b_{n-2}u$$
$$\dot{x}_n = x_{n-1} - a_{n-1}x_n + b_{n-1}u$$

$$\therefore \quad \begin{bmatrix} \dot{x}_1 \\ \dot{x}_2 \\ \dot{x}3 \\ \vdots \\ \dot{x}_n \end{bmatrix} = \begin{bmatrix} 0 & 0 & \cdots & \cdots & -a_0 \\ 1 & 0 & \cdots & \cdots & -a_1 \\ 0 & 1 & \cdots & \cdots & -a_2 \\ \vdots & \vdots & \vdots & \vdots & \\ 0 & 0 & \cdots & 1 & -a_{n-1} \end{bmatrix} \begin{bmatrix} x_1 \\ x_2 \\ x_3 \\ \vdots \\ x_n \end{bmatrix} + \begin{bmatrix} b_0 \\ b_1 \\ b_2 \\ \vdots \\ b_{n-1} \end{bmatrix} [u] \tag{17.47}$$

$$\therefore \quad \dot{x} = Ax + Bu \tag{17.48}$$

and

$$y = x_n$$

$$[y] = [0 \quad 0 \cdots 1] \begin{bmatrix} x_1 \\ x_2 \\ x_3 \\ \vdots \\ x_n \end{bmatrix} + [0][u] \tag{17.49}$$

i.e.,

$$y = Cx + Du \tag{17.50}$$

The matrices A and C are of OCF.

Example 17.5 Express the following transfer function in OCF form:

$$2\frac{d^3y}{dt^3} + 4\frac{d^2y}{dt^2} + 6\frac{dy}{dt} + 3y = 4u$$

Solution

The given differential equation

$$2\frac{d^3y}{dt^3} + 4\frac{d^2y}{dt^2} + 6\frac{dy}{dt} + 3y = 4u \tag{1}$$

Taking Laplace transform of both sides of Eq. (1), we have

$$2s^3Y(s) + 4s^2Y(s) + 6sY(s) + 3Y(s) = 4U(s)$$

or

$$\frac{Y(s)}{U(s)} = \frac{4}{2s^3 + 4s^2 + 6s + 3} = \frac{2s^{-3}}{1 + 2s^{-1} + 3s^{-2} + 1.5s^{-3}}$$

or

$$Y(s)[1 + 2s^{-1} + 3s^{-2} + 1.5s^{-3}] = 2s^{-3}U(s)$$

or

$$Y(s) = 2s^{-3}U(s) - Y(s)[2s^{-1} + 3s^{-2} + 1.5s^{-3}]$$

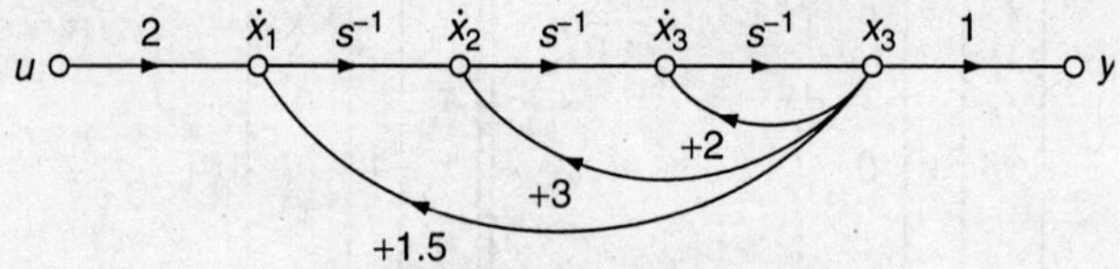

Fig. E17.5 SFG

From Fig. E17.5, we can write

$$\dot{x}_1 = 2u - 1.5x_3 \quad (2)$$

$$\dot{x}_2 = x_1 - 3x_3 \quad (3)$$

$$\dot{x}_3 = x_2 - 2x_3 \quad (4)$$

and

$$y = x_3 \quad (5)$$

From Eqs. (2) to (5), we can write

$$\begin{bmatrix}\dot{x}_1\\ \dot{x}_2\\ \dot{x}_3\end{bmatrix} = \begin{bmatrix}0 & 0 & -1.5\\ 1 & 0 & -3\\ 0 & 1 & -2\end{bmatrix}\begin{bmatrix}x_1\\ x_2\\ x_3\end{bmatrix} + \begin{bmatrix}2\\ 0\\ 0\end{bmatrix}[u]$$

and

$$[y] = [0 \quad 0 \quad 1] + \begin{bmatrix}x_1\\ x_2\\ x_3\end{bmatrix} + [0][u]$$

17.12.4 Cascade Decomposition

This form is also known as Pole-zero form on Gullemin's form, which can be used effectively when both numerator and denominator can be factorised.

The transfer function is written in terms of the product of first-order product in cascade decomposition.

$$\therefore \quad \frac{Y(s)}{U(s)} = k\frac{(s+b_1)(s+b_2)\ldots}{(s+a_1)(s+a_2)\ldots} \quad (17.51)$$

Let us consider a general term for discussing the basic principle in cascade decomposition, which is given below:

$$\frac{Y(s)}{X(s)} = \frac{s+b}{s+a} = \frac{1+bs^{-1}}{1+as^{-1}} = \frac{1+bs^{-1}}{1-(-as^{-1})} \quad (17.52)$$

The two forward paths between nodes $X(s)$ and $Y(s)$ are 1 and bs^{-1} and the loop gain is $-as^{-1}$. It touches both the forward paths. This is shown in Fig. 17.16.

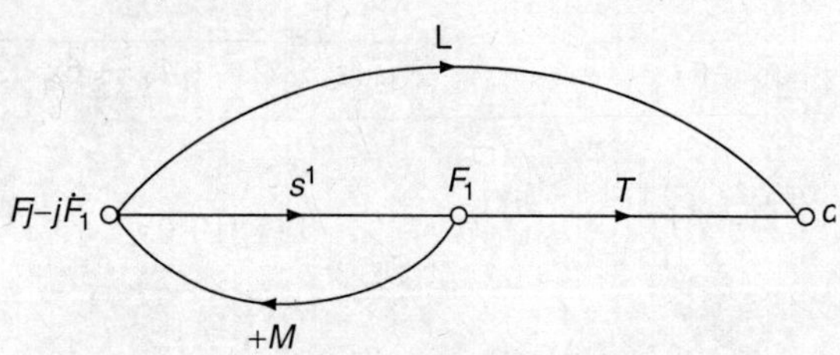

Fig. 17.16 SFG

Let

$$\frac{Y(s)}{U(s)} = \frac{Y(s)}{X(s)}\frac{X(s)}{X^1(s)}\frac{X^1(s)}{U(s)} = \left(\frac{s+b_1}{s+a_1}\right)\left(\frac{s+b_2}{s+a_2}\right)k$$

where
$$\frac{Y(s)}{X(s)} = \frac{1+b_1 s^{-1}}{1+a_1 s^{-1}}$$
$$\frac{X(s)}{X^1(s)} = \frac{1+b_2 s^{-1}}{1+a_2 s^{-1}}$$

and
$$\frac{X^1(s)}{U(s)} = k$$

The SFG is shown is Fig. 17.17.

$$\dot{x}_1 = b_2 x_2 + \dot{x}_2 - a_1 x_1$$

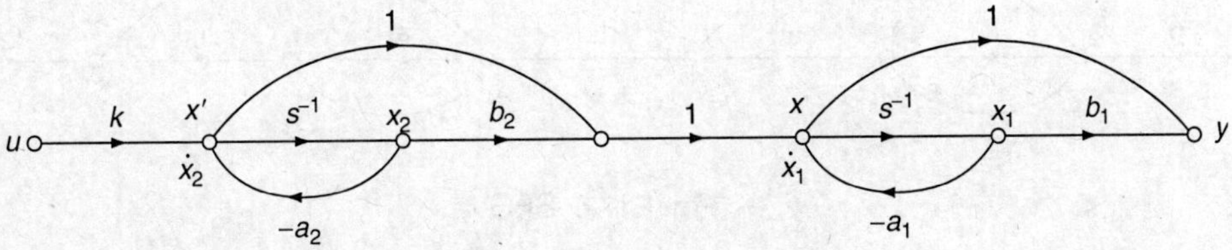

Fig. 17.17 SFG of cascade decomposition

and
$$\dot{x}_2 = -a_2 x_2 + ku$$

$\therefore$
$$\dot{x}_1 = b_2 x_2 - a_2 x_2 + ku - a_1 x_1 = (b_2 - a_2)x_2 - a_1 x_1 + ku$$

$\therefore$
$$\begin{bmatrix} \dot{x}_1 \\ \dot{x}_2 \end{bmatrix} = \begin{bmatrix} -a_1 & b_2 - a_2 \\ 0 & -a_2 \end{bmatrix} \begin{bmatrix} x_1 \\ x_2 \end{bmatrix} + [k][u] \tag{17.53}$$

$$y = b_1 x_1 + \dot{x}_1 = b_1 x_1 + (b_2 - a_2)x_2 - a_1 x_1 + ku$$
$$= (b_1 - a_1)x_1 + (b_2 - a_2)x_2 + ku$$

$\therefore$
$$[y] = [b_1 - a_1 \;\; b_2 - a_2] \begin{bmatrix} x_1 \\ x_2 \end{bmatrix} + [k][u] \tag{17.54}$$

Feature of matrix A is that its principle diagonal contains gains of all feedback paths which are associated with integrators. On the other hand, all terms below the principle diagonal are zero. This method is also known as iterative programming.

Example 17.6 Find cascade decomposition of the transfer function given below:
$$\frac{Y(s)}{U(s)} = \frac{5(s+1)(s+2)}{(s+4)(s+5)}$$

Solution

$$\frac{Y(s)}{U(s)} = \frac{5(s+1)(s+2)}{(s+4)(s+5)} = \frac{(s+1)}{(s+4)}\frac{(s+2)}{(s+5)} \times 5 = \frac{Y(s)}{X(s)}\frac{X(s)}{X^1(s)}\frac{X^1(s)}{U(s)}$$

Now,
$$\frac{Y(s)}{X(s)} = \frac{s+1}{s+4} = \frac{1+s^{-1}}{1+4s^{-1}} \tag{1}$$

$$\frac{X(s)}{X^1(s)} = \frac{s+2}{s+5} = \frac{1+2s^{-1}}{1+5s^{-1}} \tag{2}$$

∴
$$\frac{X^1(s)}{U(s)} = 5 \tag{3}$$

Figure E17.6 shows the SFG of Eqs. (3), (2) and (1).

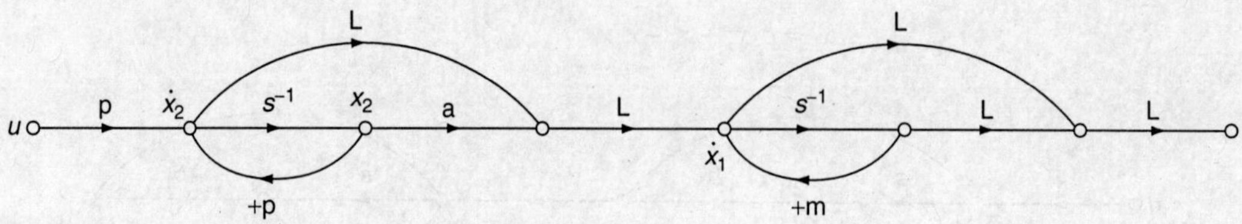

Fig. E17.6 SFG

Therefore
$$\dot{x}_2 = -5x_2 + u \tag{4}$$

∴
$$\dot{x}_1 = -4x_1 + 2x_2 + \dot{x}_2 = -4x_1 - 5x_2 + 2x_2 + u = -4x_1 - 3x_2 + u \tag{5}$$

and
$$y = x_1 + \dot{x}_1 = x_1 - 4x_1 - 3x_2 + u = -3x_1 - 3x_2 + u \tag{6}$$

From Eqs. (4) to (6), we have

$$\begin{bmatrix}\dot{x}_1\\ \dot{x}_2\end{bmatrix} = \begin{bmatrix}-4 & -3\\ 0 & -5\end{bmatrix}\begin{bmatrix}x_1\\ x_2\end{bmatrix} + \begin{bmatrix}1\\ 1\end{bmatrix}[u]$$

and

$$[y] = [-3 \quad -3]\begin{bmatrix}x_1\\ x_2\end{bmatrix} + [1][u]$$

17.12.5 Parallel Decomposition

This method uses canonical variables and is used to obtain the state model. The matrix A obtained in this method is said to have canonical form, normal form or Foster's form. This matrix A is a diagonal matrix and has an important role in the state space analysis. This method is based on partial fraction approach.

Let

$$\frac{Y(s)}{U(s)} = \frac{P(s)}{(s+a_1)(s+a_2)\cdots(s+a_n)} \tag{17.55}$$

Where $P(s) = \text{bo}s^m + b_1 s^{m-1} + b_2 s^{m-2} + \cdots + b_m$.

Case 1: Degree of numerator is less than the degree of denominator i.e., $m < n$. From Eq. (17.55), we have

$$\frac{Y(s)}{U(s)} = \frac{d_1}{s+a_1} + \frac{d_2}{s+a_2} + \cdots + \frac{d_n}{s+a_n} \tag{17.56}$$

where d_i s are constants.

$$\therefore \quad Y(s) = \frac{d_1}{s+a_1}U(s) + \frac{d_2}{s+a_2}U(s) + \cdots + \frac{d_n}{s+a_n}U(s) \tag{17.57}$$

$$= d_1Y_1(s) + d_2Y_2(s) + \cdots + d_nY_n(s) \tag{17.58}$$

where

$$\left.\begin{aligned} Y_1(s) &= \frac{1}{s+a_1}U(s) \quad \text{i.e., } sY_1(s) = -a_1Y_1(s) + U(s) \\ Y_2(s) &= \frac{1}{s+a_2}U(s) \quad \text{i.e., } sY_2(s) = -a_2Y_2(s) + U(s) \\ &\vdots \qquad\qquad\qquad\qquad \vdots \\ Y_n(s) &= \frac{1}{s+a_n}U(s) \quad \text{i.e., } sY_n(s) = -a_nY_n(s) + U(s) \end{aligned}\right\} \tag{17.59}$$

The inverse Laplace transformation of the above equation is

$$\left.\begin{aligned} \dot{y}_1 &= -a_1y_1 + u \\ \dot{y}_2 &= -a_2y_2 + u \\ &\vdots \\ \dot{y}_n &= -a_ny_n + u \end{aligned}\right\} \tag{17.60}$$

Figure 17.18 shows the SFG of Eq. (17.38).

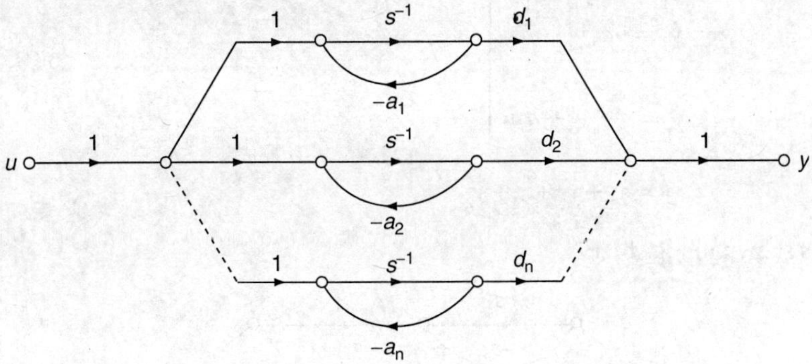

Fig. 17.18 SFG of parallel decomposition

$$\begin{bmatrix} \dot{y}_1 \\ \dot{y}_2 \\ \vdots \\ \dot{y}_n \end{bmatrix} = \begin{bmatrix} -a_1 & 0 & \cdots & 0 \\ 0 & -a_2 & \cdots & 0 \\ \vdots & \vdots & \cdots & 0 \\ 0 & 0 & \cdots & -a_n \end{bmatrix} \begin{bmatrix} y_1 \\ y_2 \\ \vdots \\ y_n \end{bmatrix} + \begin{bmatrix} 1 \\ 1 \\ \vdots \\ 1 \end{bmatrix} \dot{u} \tag{17.61}$$

$$\therefore \quad \dot{y} = Ay + Bu \tag{17.62}$$

and inverse Laplace transform of Eq. (17.62) is

$$y = d_1 y_1 + d_2 y_2 + \ldots + d_n y_n$$

$$\therefore \qquad [y] = [d_1\, d_2 \ldots d_n]\begin{bmatrix} y_1 \\ y_2 \\ \vdots \\ y_n \end{bmatrix} + [0][u] \tag{17.63}$$

$$\therefore \qquad \dot{y} = Cy + Du \tag{17.64}$$

Another form of parallel decomposition is discussed below.
Equation (17.57) can be written as follows:

$$Y(s) = Y_1(s) + Y_2(s) + \ldots + Y_n(s) \tag{17.65}$$

$$\left.\begin{aligned} Y_1(s) &= \frac{d_1}{s+a_1}U(s) \quad \text{i.e., } sY_1(s) = -a_1 Y_1(s) + d_1\, U(s) \\ Y_2(s) &= \frac{d_2}{s+a_2}U(s) \quad \text{i.e., } sY_2(s) = -a_2 Y_2(s) + d_2\, U(s) \\ &\vdots \qquad\qquad\qquad\qquad \vdots \\ Y_n(s) &= \frac{d_n}{s+a_n}U(s) \quad \text{i.e., } sY_n(s) = -a_n Y_n(s) + d_n\, U(s) \end{aligned}\right\} \tag{17.66}$$

The inverse Laplace transform of the above Y_i's are given by

$$\left.\begin{aligned} \dot{y}_1 &= -a_1 y_1 + d_1 u \\ \dot{y}_2 &= -a_2 y_2 + d_2 u \\ &\vdots \\ \dot{y}_n &= -a_n y_n + d_n u \end{aligned}\right\} \tag{17.67}$$

and

$$y = y_1 + y_2 + \ldots + y_n \tag{17.68}$$

The SFG is shown in Fig. 17.19.

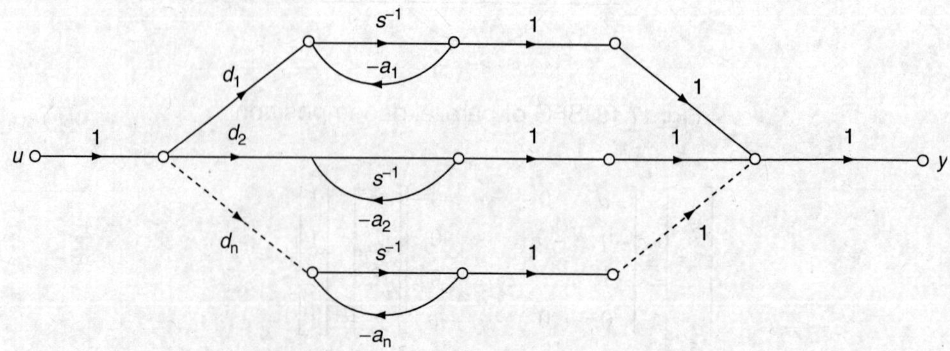

Fig. 17.19 SFG of parallel decomposition

$$\begin{bmatrix} \dot{y}_1 \\ \dot{y}_2 \\ \vdots \\ \dot{y}_n \end{bmatrix} = \begin{bmatrix} -a_1 & 0 & \cdots & 0 \\ 0 & -a_2 & \cdots & 0 \\ \vdots & \vdots & \vdots & 0 \\ 0 & 0 & \cdots & -a_n \end{bmatrix} \begin{bmatrix} y_1 \\ y_2 \\ \vdots \\ y_n \end{bmatrix} + \begin{bmatrix} d_1 \\ d_2 \\ \vdots \\ d_n \end{bmatrix} [u] \tag{17.69}$$

$\therefore$

$$\dot{y} = Ay + Bu \tag{17.70}$$

and

$$[y] = [1 \;\; 1 \; \cdots \; 1] \begin{bmatrix} y_1 \\ y_2 \\ \vdots \\ y_n \end{bmatrix} + [0][u] \tag{17.71}$$

$\therefore$

$$y = Cy + Du \tag{17.72}$$

Here D is does not exist.

Case 2: If the degree of numerator is equal to the degree of the denominator i.e., m = n, then we have from Eq. (17.55)

$$Y(s) = d_0 U(s) + \sum_{i=1}^{n} \frac{d_i}{s + a_i} U(s) \tag{17.73}$$

where d_0 is the constant obtained due to $m = n$.

The state diagram remains as before i.e., Fig.17.18 and Fig. 17.19 except an additional element is added to obtain the resultant output as shown in Fig. 17.20 (a) and Fig.17.20(b).

Example 17.7 Find parallel decomposition of the transfer function given below:

$$\frac{Y(s)}{U(s)} = \frac{(s+1)}{(s+2)(s+3)(s+5)}$$

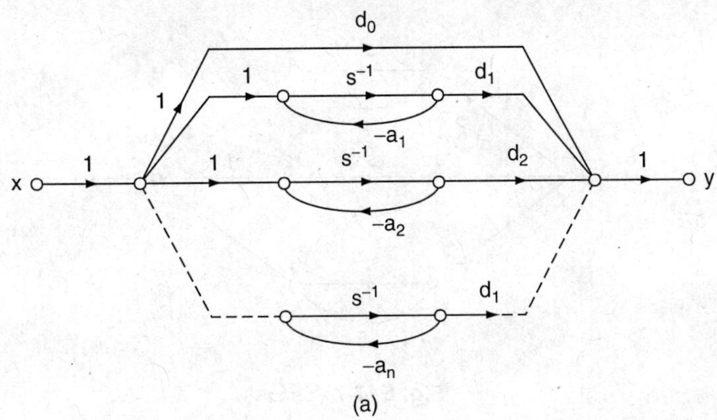

(a)

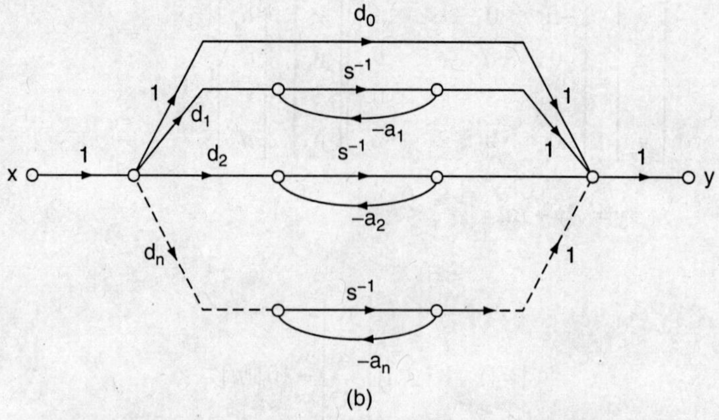

Fig. 17.20 SFG of parallel decomposition for $m = n$

Solution

Let

$$\frac{Y(s)}{U(s)} = \frac{A}{s+2} + \frac{B}{s+3} + \frac{C}{s+5}$$

$\therefore$

$$A = \left.\frac{(s+1)}{(s+3)(s+5)}\right|_{s=-2} = -\frac{1}{3}$$

$$B = \left.\frac{(s+1)}{(s+2)(s+5)}\right|_{s=-3} = 1$$

$$C = \left.\frac{(s+1)}{(s+2)(s+3)}\right|_{s=-5} = -\frac{2}{3}$$

$$Y(s) = \frac{-\frac{1}{3}}{s+2}U(s) + \frac{1}{s+3}U(s) + \frac{-\frac{2}{3}}{s+5}U(s)$$

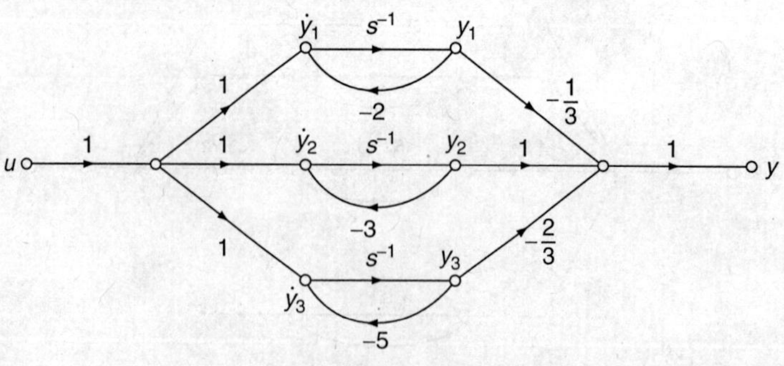

Fig. E17.7 SFG

From Fig. E17.7, we can write

$$\dot{y}_1 = -2y_1 - u$$

$$\dot{y}_2 = -3y_2 + u$$

$$\dot{y}_3 = -5y_3 - u$$

and

$$y = -\frac{1}{3}y_1 + y_2 - \frac{2}{3}y_3$$

$$\therefore \quad \begin{bmatrix} \dot{x}_1 \\ \dot{x}_2 \\ \dot{x}_3 \end{bmatrix} = \begin{bmatrix} -2 & 0 & 0 \\ 0 & -3 & 0 \\ 0 & 0 & -5 \end{bmatrix} \begin{bmatrix} y_1 \\ y_2 \\ y_3 \end{bmatrix} + \begin{bmatrix} 1 \\ 1 \\ 1 \end{bmatrix} [u]$$

and

$$[y] = \begin{bmatrix} -\frac{1}{3} & 1 & -\frac{2}{3} \end{bmatrix} \begin{bmatrix} y_1 \\ y_2 \\ y_3 \end{bmatrix} + [0][u]$$

Other way of solving:

$$Y(s) = \frac{-\frac{1}{3}}{s+2}U(s) + \frac{1}{s+3}U(s) + \frac{-\frac{2}{3}}{s+5}U(s)$$

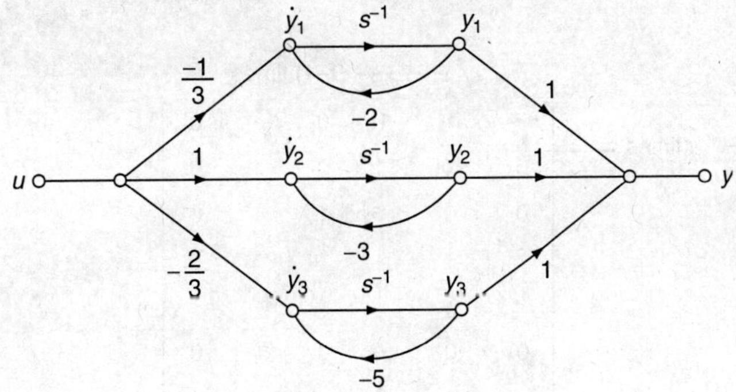

Fig. E17.7(a) SFG

From Fig. E17.7(a), we have

$$\dot{y}_1 = -2y_1 - \frac{1}{3}u$$

$$\dot{y}_2 = -3y_2 + u$$

$$\dot{y}_3 = -5y_3 - \frac{2}{3}u$$

and $$y = y_1 + y_2 + y_3$$

$$\therefore \quad \begin{bmatrix} \dot{y}_1 \\ \dot{y}_2 \\ \dot{y}_3 \end{bmatrix} = \begin{bmatrix} -2 & 0 & 0 \\ 0 & -3 & 0 \\ 0 & 0 & -5 \end{bmatrix} \begin{bmatrix} y_1 \\ y_2 \\ y_3 \end{bmatrix} + \begin{bmatrix} -\frac{1}{3} \\ 1 \\ -\frac{2}{3} \end{bmatrix} [u]$$

and $$[y] = [1 \ \ 1 \ \ 1] \begin{bmatrix} y_1 \\ y_2 \\ y_3 \end{bmatrix} + [0][u]$$

17.12.6 Jordan's Canonical Form

If there are repeated roots of the denominator of a transfer function for parallel decomposition, let the transfer function be

$$\frac{Y(s)}{U(s)} = \frac{P(s)}{(s+a_1)^r (s+a_2) \cdots (s+a_n)}$$

$$= \frac{d_{11}}{(s+a_1)^r} + \frac{d_{12}}{(s+a_1)^{r-1}} + \cdots + \frac{d_{1r}}{(s+a_1)} + \frac{d_2}{(s+a_2)}$$

$$+ \cdots + \frac{d_n}{(s+a_n)} + \frac{d_2}{s+a_2} + \cdots + \frac{d_n}{s+a_n} \qquad (17.74)$$

(r times along the diagonal; $(r-1)$ times along the superdiagonal)

$$A = \begin{bmatrix} -a_1 & 1 & 0 & 0 & \cdots & 0 \\ 0 & -a_1 & 1 & 0 & \cdots & 0 \\ 0 & 0 & -a_1 & 1 & \cdots & 0 \\ \vdots & \vdots & \vdots & \vdots & \cdots & \vdots \\ \vdots & \vdots & \vdots & \vdots & \cdots & \vdots \\ 0 & 0 & 0 & 0 & -a_r & 0 \\ \vdots & \vdots & \vdots & \vdots & \cdots & \vdots \\ \vdots & \vdots & \vdots & \vdots & \cdots & \vdots \\ 0 & 0 & 0 & 0 & \cdots & -a_n \end{bmatrix} \qquad (17.75)$$

This is called the Jordan canonical form.

Example 17.8 Derive Jordan's canonical form for the following transfer function:

$$\frac{Y(s)}{U(s)} = \frac{6}{(s+1)^2 (s+2)}$$

Solution

Let

$$\frac{Y(s)}{U(s)} = \frac{6}{(s+1)^2(s+2)} = \frac{A_{11}}{(s+1)^2} + \frac{A_{12}}{s+1} + \frac{A_2}{s+2}$$

Now,

$$A_{11} = \left[(s+1)^2\left\{\frac{6}{(s+1)^2(s+2)}\right\}\right]_{s=-1} = 6$$

$$A_{12} = \left[\frac{d}{ds}\left\{(s+1)^2 \times \frac{6}{(s+1)^2(s+2)}\right\}\right]_{s=-1} = \frac{d}{ds}\left[\frac{6}{(s+2)}\right]_{s=-1} = -6$$

$$A_2 = \left[(s+2) \times \frac{6}{(s+1)^2(s+2)}\right]_{s=-2} = \left[\frac{6}{(s+1)^2}\right]_{s=-2} = 6$$

$\therefore$
$$\frac{Y(s)}{U(s)} = \frac{6}{(s+1)^2} + \frac{-6}{s+1} + \frac{6}{s+2}$$

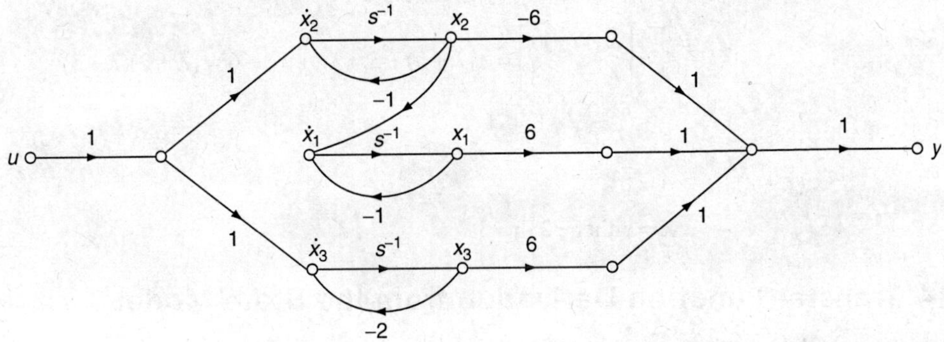

Fig. E17.8 SFG

From Fig. E17.8, we have

$$\dot{x}_1 = x_2 - x_1 = -x_1 + x_2 \tag{1}$$

$$\dot{x}_2 = -x_2 + u \tag{2}$$

$$\dot{x}_3 = -2x_3 + u \tag{3}$$

and

$$y = 6x_1 - 6x_2 + 6x_3 \tag{4}$$

From Eqs. (1) to (4), we get

$$\begin{bmatrix}\dot{x}_1\\ \dot{x}_2\\ \dot{x}_3\end{bmatrix} = \begin{bmatrix}-1 & 1 & 0\\ 0 & -1 & 0\\ 0 & 0 & -2\end{bmatrix}\begin{bmatrix}x_1\\ x_2\\ x_3\end{bmatrix} + \begin{bmatrix}0\\ 1\\ 1\end{bmatrix}[u]$$

and

$$[y] = [6 \quad -6 \quad 6]\begin{bmatrix} x_1 \\ x_2 \\ x_3 \end{bmatrix} + [0][u]$$

17.13 Eigen Value

The roots of the characteristic equation $|\lambda_i I - A| = 0$ are called the eigenvalues of the matrix $A(n \times n)$. The $n \times 1$ vector P_i that satisfies the matrix equation $|\lambda_i I - A| p_i = 0$ is called eigenvector of $A(n \times n)$ associated with the eigenvalue of λ_i where λ_i is the eigenvalue of A.

Example 17.7 Find the eigenvalues of $A = \begin{bmatrix} 0 & 1 \\ -6 & -5 \end{bmatrix}$

Solution

$$\lambda I - A = \lambda\begin{bmatrix} 1 & 0 \\ 0 & 1 \end{bmatrix} - \begin{bmatrix} 0 & 1 \\ -6 & -5 \end{bmatrix} = \begin{bmatrix} \lambda & -1 \\ 6 & \lambda+5 \end{bmatrix}$$

$$\therefore \qquad |\lambda I - A| = \begin{bmatrix} \lambda & -1 \\ 6 & \lambda+5 \end{bmatrix} = \lambda(\lambda+5) + 6 = \lambda^2 + 5\lambda + 6 = (\lambda+2)(\lambda+3)$$

Now, $|\lambda I - A| = 0$ gives

$$(\lambda+2)(\lambda+3) = 0$$

$$\therefore \qquad \lambda = \text{either } -2 \text{ or } -3$$

17.14 Transfer Function Derivation from the State Model

From the state model equations, we can derive the transfer function of the system.

From the definition of a transfer function, we can write

$$\text{Transfer function} = \left.\frac{\text{Laplace transform of output}}{\text{Laplace transform of input}}\right|_{\text{initial conditions are zero}} \tag{17.76}$$

For a linear time-invariant system, the dynamics can be written in a vector differential form as follows:

$$\dot{x} = Ax + Bu \tag{17.77}$$

$$y = Cx + Du \tag{17.78}$$

Taking Laplace transform of Eq. (17.77), we have

$$sX(s) - x(0) = AX(s) + BU(s)$$

$$\therefore \qquad [sI - A]X(s) = BU(s) \qquad [\text{Taking } x(0) = 0]$$

$$\therefore \qquad X(s) = [sI - A]^{-1}BU(s) \tag{17.79}$$

Taking Laplace transform of Eq. (17.78), we have

$$Y(s) = CX(s) + DU(s) = C[sI - A]^{-1}BU(s) + DU(s)$$

$$\therefore \quad \text{Transfer function} = \frac{Y(s)}{U(s)} = C[sI - A]^{-1}B + D \tag{17.80}$$

If there is no direct coupling between input and output, $D = 0$. Equation (17.80) becomes

$$\text{Transfer function} = \frac{Y(s)}{U(s)} = C[sI - A]^{-1}B \tag{17.81}$$

Therefore, the transfer function is the ratio of polynomials of s. The denominator polynomial of s of the transfer function is known as characteristic equation. The characteristic equation is given by

$$|sI - A| = 0 \tag{17.82}$$

Under similarity transformation, the invariance of a transfer function can be proved. It is possible to show that the transfer functions $C[sI - A]^{-1}B$ and $C^*[sI - A]^{-1}B^*$ are identical.

$$\begin{aligned}
\therefore \quad C^*[sI - A]^{-1}B^* &= CP[sP^{-1}P - P^{-1}AP]^{-1}P^{-1}B = CP[P^{-1}(sI - A)P]^{-1}P^{-1}B \\
&= CPP^{-1}(sI - A)^{-1}PP^{-1}B \, [\because (XY)^{-1} = Y^{-1}X^{-1}] \\
&= C[sI - A]^{-1}B
\end{aligned}$$

Therefore, under similarity transformation, the transfer function remains unchanged.

Example 17.10 Find the transfer function when

$$A = \begin{bmatrix} -2 & 1 \\ 0 & -3 \end{bmatrix}, B = \begin{bmatrix} 0 \\ 1 \end{bmatrix} \text{ and } C = [1 \quad 1]$$

Solution

$$sI - A = s\begin{bmatrix} 1 & 0 \\ 0 & 1 \end{bmatrix} - \begin{bmatrix} -2 & 1 \\ 0 & -3 \end{bmatrix} = \begin{bmatrix} s+2 & -1 \\ 0 & s+3 \end{bmatrix}$$

Now,

$$|sI - A| = \begin{vmatrix} s+2 & -1 \\ 0 & s+3 \end{vmatrix} = (s+2)(s+3) \neq 0$$

Therefore, $(sI - A)^{-1}$ exists.

Now,

$$(sI - A)^{-1} = \frac{Adj(sI - A)}{|sI - A|} = \frac{\begin{bmatrix} s+3 & +1 \\ 0 & s+2 \end{bmatrix}}{(s+2)(s+3)}$$

$$\therefore \quad C(sI - A)^{-1}B = [1 \quad 1]\frac{\begin{bmatrix} s+3 & 1 \\ 0 & s+2 \end{bmatrix}}{(s+2)(s+3)}\begin{bmatrix} 0 \\ 1 \end{bmatrix} = \frac{[1 \quad 1]\begin{bmatrix} 1 \\ s+2 \end{bmatrix}}{(s+2)(s+3)} = \frac{(s+3)}{(s+2)(s+3)} = \frac{1}{s+2}$$

17.15 Solution of the State Equation

For a time-invariant system, the state equation is given by $\dot{x}(t) = Ax(t) + Bu(t)$, which is divided into two types as follows:

(i) Homogeneous state equation
(ii) Non-homogeneous state equation

The equation is said to be homogeneous if A is a constant matrix and U is a zero vector. In this case no control forces are applied. The equation is said to be non-homogenous when A is a constant matrix and U is a non-zero vector. In the first case, discharging a charged capacitor is an example, while in the second case total response is an example.

17.15.1 Solution of Homogeneous State Equation

In the homogeneous case, the term $u(t)$ is not present.

Equation (17.77) becomes $$\dot{x} = Ax \tag{17.83}$$

Taking Laplace transform of Eq. (17.83) we have

$$sX(s) - x(0) = AX(s)$$

$$(sI - A)X(s) = x(0)$$

$$X(s) = (sI - A)^{-1}x(0) \tag{17.84}$$

$$X(s) = \phi(s)x(0) \tag{17.85}$$

where

$$\phi(s) = (sI - A)^{-1} \tag{17.86}$$

is called the resolvent matrix. Now $\phi(s) = (sI - A)^{-1}$

$$= \frac{1}{s}\left(I - \frac{A}{s}\right)^{-1} = \frac{1}{s}\left[I + \frac{A}{s} + \frac{A^2}{s^2} + \cdots\right]$$

$$\phi(s) = \left(\frac{I}{s} + \frac{A}{s^2} + \frac{A^2}{s^3} - \cdots\right) \tag{17.87}$$

Taking inverse Laplace transform of Eq. (17.87), we have

$$\phi(t) = I + At + \frac{A^2t^2}{2!} + \cdots\cdots \tag{17.88}$$

Taking inverse Laplace transform of Eq. (17.85), we have

$$x(t) = \phi(t)x(0) \tag{17.89}$$

where $\phi(t)$ is given in Eq. (17.88). If $\phi(t)$ is known, it is possible to transit from $x(0)$ to $x(t)$. So $\phi(t)$ is called state transition matrix (STM). To evaluate STM, the methods are classified into two groups as follows:

(i) The methods that need the knowledge of eigenvalues are
  (a) inverse Laplace transform method,
  (b) Caylay Hamilton method and
  (c) Sylvarster's method.

(ii) The methods that do not need the knowledge of eigenvalues are:
 (a) Infinite power series method
 (b) Taylor's series expansion method.

17.15.2 State Transition Matrix

$$x(t) = \phi(t)x(0) = e^{At}x(0) \tag{17.90}$$

is the solution of $\dot{x} = Ax$

Taking derivative of Eq. (17.90), we have

$$\dot{x}(t) = Ae^{At}x(0) = Ax(t) \tag{17.91}$$

Now, $\phi(t) = e^{At}$

$\therefore$ $\phi(0) = e^{A0} = I$

At $t = 0$,

$$\dot{x}(0) = \phi(0)x(0) = Ix(0) \tag{17.92}$$

and $$\phi^{-1}(t) = (e^{At})^{-1} = e^{-At} = \phi(-t) \tag{17.93}$$

17.15.3 Properties of STM, $\phi(t)$

(i) $\phi(0) = e^{A0} = I$

(ii) $\phi(t) = e^{At} = (e^{-At})^{-1} = [\phi(-t)]^{-1}$

(iii) $\phi^{-1}(t) = \phi(-t)$

(iv) $\phi(t_1 + t_2) = e^{A[t_1+t_2]} = e^{At_1}e^{At_2} = \phi(t_1)\phi(t_2) = e^{At_2}e^{At_1} = \phi(t_2)\phi(t_1)$

(v) $[\phi(t)]^n = (e^{At})^n = e^{Ant} = \phi(nt)$

(vi) $\phi(t_1 - t_2)\phi(t_2 - t_0) = e^{A(t_1-t_2)}e^{A(t_2-t_0)}$
$= e^{(t_1-t_2+t_2-t_0)} = e^{A(t_1-t_0)} = \phi(t_1 - t_0)$

Example 17.11 A linear time-invariant system is characterized by the homogeneous state equation

$$\begin{bmatrix} \dot{x}_1 \\ \dot{x}_2 \end{bmatrix} = \begin{bmatrix} 0 & 1 \\ -6 & -5 \end{bmatrix}\begin{bmatrix} x_1 \\ x_2 \end{bmatrix}$$

The initial state is $\begin{bmatrix} x_1 \\ x_2 \end{bmatrix} = \begin{bmatrix} 1 \\ 0 \end{bmatrix}$

Find the resolvent matrix $\phi(s)$, state transition matrix $\phi(t)$, $\phi^{-1}(t)$, and the solution of the given equation.

Solution

This is a homogeneous equation because $u = 0$.
Here

$$A = \begin{bmatrix} 0 & 1 \\ -6 & -5 \end{bmatrix}$$

$$\therefore \quad sI - A = s\begin{bmatrix} 1 & 0 \\ 0 & 1 \end{bmatrix} - \begin{bmatrix} 0 & 1 \\ -6 & -5 \end{bmatrix} = \begin{bmatrix} s & -1 \\ 6 & s+5 \end{bmatrix}$$

$$\therefore \quad |sI - A| = \begin{vmatrix} s & -1 \\ 6 & s+5 \end{vmatrix} = (s+2)(s+3)$$

$$\therefore \quad \phi(s) = (sI - A)^{-1} = \begin{bmatrix} s & -1 \\ 6 & s+5 \end{bmatrix}^{-1} = \frac{\begin{bmatrix} s+5 & 1 \\ -6 & s \end{bmatrix}}{(s+2)(s+3)}$$

$$= \begin{bmatrix} \dfrac{s+5}{(s+2)(s+3)} & \dfrac{1}{(s+2)(s+3)} \\ \dfrac{-6}{(s+2)(s+3)} & \dfrac{s}{(s+2)(s+3)} \end{bmatrix} \begin{bmatrix} \phi_{11}(s) & \phi_{12}(s) \\ \phi_{21}(s) & \phi_{22}(s) \end{bmatrix} \tag{1}$$

Taking inverse Laplace transform, we get

$$\phi(t) = \begin{bmatrix} \phi_{11}(t) & \phi_{12}(t) \\ \phi_{21}(t) & \phi_{22}(t) \end{bmatrix}$$

where

$$\phi_{11}(s) = \frac{s+5}{(s+3)(s+1)} = \frac{3}{s+2} - \frac{2}{s+3}$$

$$\therefore \quad \phi_{11}(t) = 3e^{-2t} - 2e^{-3t}$$

$$\phi_{12}(s) = \frac{1}{(s+2)(s+3)} = \frac{1}{s+2} - \frac{1}{s+3}$$

$$\therefore \quad \phi_{12}(t) = e^{-2t} - e^{-3t}$$

$$\phi_{21}(s) = \frac{-6}{(s+2)(s+3)} = \frac{-6}{s+2} + \frac{6}{s+3}$$

$$\therefore \quad \phi_{21}(s) = -6(e^{-2t} - e^{-3t})$$

and

$$\phi_{22}(s) = \frac{s}{(s+2)(s+3)} = \frac{-2}{s+2} + \frac{3}{s+3}$$

$$\phi_{22}(s) = -2e^{-2t} + 3e^{-3t}$$

$$\therefore \quad \phi(t) = \begin{bmatrix} 3e^{-2t} - 2e^{-3t} & e^{-2t} - e^{-3t} \\ 6(e^{-3t} - 2e^{-2t}) & 3e^{-3t} - 2e^{-2t} \end{bmatrix} \tag{2}$$

$$\because \quad \phi^{-1}(t) = \phi(-t)$$

$$\therefore \qquad \phi^{-1}(t) = \phi(-t) = \begin{bmatrix} 3e^{2t} - 2e^{3t} & e^{2t} - e^{3t} \\ 6(e^{3t} - 2e^{2t}) & 3e^{3t} - 2e^{2t} \end{bmatrix}$$

Now,

$$x(t) = \phi(t)x(t) = \phi(t) = \begin{bmatrix} 3e^{-2t} - 2e^{-3t} & e^{-2t} - e^{-3t} \\ 6(e^{-3t} - 2e^{-2t}) & 3e^{-3t} - 2e^{-2t} \end{bmatrix}\begin{bmatrix} 1 \\ 0 \end{bmatrix} = \begin{bmatrix} 3e^{-2t} - 2e^{-3t} \\ 6(e^{-3t} - 2e^{-2t}) \end{bmatrix}$$

17.15.4 Solution of Non-Homogeneous State Equation

In this case $u(t)$ is taken into account.

Taking Laplace transform of Eq. (17.52), we have

$$sX(s) - x(0) = AX(s) + BU(s)$$

$$\therefore \qquad (sI - A)X(s) = x(0) + BU(s) \tag{17.94}$$

$$\therefore \qquad X(s) = [sI - A]^{-1}x(0) + (sI - A)^{-1}BU(s) \tag{17.95}$$

Taking inverse Laplace transform of Eq. (17.95), we have

$$x(t) = LT^{-1}[sI - A]^{-1}x(0) + (sI - A)^{-1}BU(s)]$$
$$= LT^{-1}[(sI - A)^{-1}x(0) + LT^{-1}[sI - A)^{-1}BU(s)]$$

Now,

$$LT^{-1}[(sI - A)^{-1}x(0)] = \phi(t)x(0) = e^{At}x(0)$$

and

$$LT^{-1}[(sI - A)^{-1}BU(s)] = \int_0^t e^{A(t-x)}Bu(x)dx$$

$$\therefore \qquad x(t) = e^{At}x(0) + \int_0^t e^{A(t-x)}Bu(x)dx \tag{17.96}$$

Equation (17.96) represents the solution of the non-homogeneous equation. It consists of (i) the term $e^{At}x(0)$ called homogeneous solution or free response zero input response (ZIR) and (ii) the term, $\int_0^t e^{A(t-x)}Bu(x)dx$ called forced response or non-homogeneous response or zero state response (ZSR).

Example 17.12 A linear time–invariant system is characterized by the non-homogeneous state equation

$$\begin{bmatrix} \dot{x}_1 \\ \dot{x}_2 \end{bmatrix} = \begin{bmatrix} 0 & 1 \\ -6 & -5 \end{bmatrix}\begin{bmatrix} x_1 \\ x_2 \end{bmatrix} + \begin{bmatrix} 0 \\ 1 \end{bmatrix}u$$

Determine the non-homogenous solution if $\begin{bmatrix} x_1(0) \\ x_2(0) \end{bmatrix} = \begin{bmatrix} 1 \\ 0 \end{bmatrix}$ and u is a unit step function.

Solution For non-homogeneous equation, the solution is given by

$$x(t) = \phi(t)x(0) + \int_0^t \phi(t - \tau)Bu(\tau)d\tau$$

From Example 12.11, $\phi(\tau)$ can be written as follows:

$$\therefore \qquad \phi(\tau) = \begin{bmatrix} 3e^{-2t} - 2e^{-3t} & e^{-2t} - e^{-3t} \\ 6(e^{-3t} - e^{-2t}) & 3e^{-3t} - 2e^{-2t} \end{bmatrix}$$

$$\therefore \quad \phi(t-\tau)=\begin{bmatrix} 3e^{-2(t-\tau)}-2e^{-3(t-\tau)} & e^{-2(t-\tau)}-e^{-3(t-\tau)} \\ 6(e^{-3(t-\tau)}-e^{-2(t-\tau)}) & 3e^{-3(t-\tau)}-2e^{-2(t-\tau)} \end{bmatrix}$$

Now

$$\int_0^t \phi(t-\tau)Bu(\tau)d\tau=\int_0^t \begin{bmatrix} e^{-2(t-\tau)}-e^{-3(t-\tau)} \\ 3e^{-3(t-\tau)}-2e^{-2(t-\tau)} \end{bmatrix}[1]d\tau$$

$$=\int_0^t \begin{bmatrix} e^{-2(t-\tau)}-e^{-3(t-\tau)} \\ 3e^{-3(t-\tau)}-2e^{-2(t-\tau)} \end{bmatrix}d\tau=\begin{bmatrix} \frac{1}{6}-\frac{1}{2}e^{-2t}+\frac{1}{3}e^{-3t} \\ e^{-2t}-e^{-3t} \end{bmatrix}$$

$$\therefore \quad x(t)=\phi(t)x(0)+\int_0^t \phi(t-\tau)Bu(\tau)d\tau$$

$$=\begin{bmatrix} 3e^{-2t}-2e^{-3t} \\ 6(e^{-3t}-2e^{-2t}) \end{bmatrix}+\begin{bmatrix} \frac{1}{6}-\frac{1}{2}e^{-2t}+\frac{1}{3}e^{-3t} \\ e^{-2t}-e^{-3t} \end{bmatrix}=\begin{bmatrix} \frac{1}{6}+\frac{10}{3}e^{-3t}-\frac{5}{2}e^{-2t} \\ 5(e^{-2t}-e^{-3t}) \end{bmatrix}$$

17.16 Computation of State Transition Matrix: e^{At}

The computation of state transition matrix e^{At} has an important role for obtaining the solution of state equation. There are various methods to obtain state transition matrix which are

(i) Laplace transform method
(ii) Power series method
(iii) Cayley Hamilton method
(iv) Similarity transformation method

These methods are discussed below.

17.16.1 Laplace Method

From Eq. (17.52), we have

$$\dot{x}(t)=Ax(t)+Bu(t) \tag{17.97}$$

$u(t)=0$ for unforced system. From Eq. (17.97), we have for $u(t)=0$

$$\dot{x}(t)=Ax(t) \tag{17.98}$$

Taking Laplace transform of both sides of Eq. (17.98), we have

$$sX(s)-x(0)=AX(s)$$

i.e.,

$$[sI-A]X(s)=\lambda(0)$$

i.e.,

$$X(s)=[sI-A]^{-1}x(0 \tag{17.99}$$

In Eq. (17.99),

$[sI-A]^{-1}=\phi(s)=$ resolvent matrix

$$\therefore \quad \phi(t)=e^{AT}=LT^{-1}[\phi(s)]=LT^{-1}[sI-A]^{-1} \tag{17.100}$$

where $\phi(t)=e^{At}$ is called state transition matrix.

Example 17.13 Determine the STM when

$$A = \begin{bmatrix} -3 & 1 \\ 0 & -4 \end{bmatrix}$$

Solution

$$sI - A = s\begin{bmatrix} 1 & 0 \\ 0 & 1 \end{bmatrix} - \begin{bmatrix} -3 & 1 \\ 0 & -4 \end{bmatrix}$$

$$= \begin{bmatrix} s+3 & -1 \\ 0 & s+4 \end{bmatrix}$$

$\therefore$ $$|sI - A| = (s+3)(s+4)$$

$\therefore$ $$[sI - A]^{-1} = \begin{bmatrix} s+4 & 1 \\ 0 & s+3 \end{bmatrix} \frac{1}{(s+3)(s+4)}$$

$$= \begin{bmatrix} \dfrac{1}{s+3} & \dfrac{1}{(s+3)(s+4)} \\ 0 & \dfrac{1}{s+4} \end{bmatrix}$$

$$= \begin{bmatrix} \dfrac{1}{s+3} & \dfrac{1}{s+3} - \dfrac{1}{s+4} \\ 0 & \dfrac{1}{s+4} \end{bmatrix}$$

$\therefore$ $$\phi(t) = LT^{-1}\left[sI - A\right]^{-1} = \begin{bmatrix} e^{-3t} & e^{-3t} - e^{-4t} \\ 0 & e^{-4t} \end{bmatrix}$$

17.16.2 Power Series Method

Since the state transition matrix is matrix exponential, it can be expressed in power series as

$$e^{At} = I + At + \frac{1}{2!}A^2t^2 + \cdots\cdots + \frac{1}{k!}A^kt^k + \cdots\cdots$$

i.e., $$e^{At} = \sum_{i=0}^{\infty} \frac{1}{i!} A^i t^i \qquad (17.101)$$

After calculation of various power of A and few terms of power series, e^{At} can be obtained. Since the power series is infinite, it is better to stop after computation of first few terms of the series.

This method is not used practically because it consumes time for computation of high powers of A.

Example 17.14 Evaluate STM by series summation method, when

$$A = \begin{bmatrix} 3 & 2 & 3 \\ 1 & 4 & 5 \\ 6 & 3 & 2 \end{bmatrix}$$

Solution Given that
$$A = \begin{bmatrix} 3 & 2 & 3 \\ 1 & 4 & 5 \\ 6 & 3 & 2 \end{bmatrix}$$

$\therefore$
$$A^2 = \begin{bmatrix} 29 & 23 & 25 \\ 37 & 33 & 33 \\ 33 & 30 & 37 \end{bmatrix}, A^3 = \begin{bmatrix} 260 & 225 & 252 \\ 342 & 305 & 342 \\ 351 & 297 & 323 \end{bmatrix}$$

and so on.
$$e^{At} = \begin{bmatrix} 1 & 0 & 0 \\ 0 & 1 & 0 \\ 0 & 0 & 1 \end{bmatrix} + \begin{bmatrix} 3 & 2 & 3 \\ 1 & 4 & 5 \\ 6 & 3 & 2 \end{bmatrix} t + \begin{bmatrix} 29 & 33 & 25 \\ 37 & 33 & 33 \\ 33 & 30 & 37 \end{bmatrix} \frac{t^2}{2} + \begin{bmatrix} 260 & 225 & 252 \\ 342 & 305 & 342 \\ 351 & 297 & 323 \end{bmatrix} \frac{t^3}{6}$$

$$= \begin{bmatrix} 1+3t+\frac{29}{2}t^2+\frac{260}{6}t^3+\cdots & 2t+\frac{33}{2}t^2+\frac{225}{6}t^3+\cdots & 3t+\frac{25}{2}t^2+\frac{252}{6}t^3+\cdots \\ t+\frac{37}{2}t^2+\frac{342}{6}t^3+\cdots & 1+4t+\frac{33}{2}t^2+\frac{305}{6}t^3+\cdots & 5t+\frac{33}{2}t^2+\frac{342}{6}t^3+\cdots \\ 6t+\frac{33}{2}t^2+\frac{351}{6}t^3+\cdots & 3t+\frac{30}{2}t^2+\frac{297}{6}t^3+\cdots & 1+2t+\frac{37}{2}t^2+\frac{323}{6}t^3+\cdots \end{bmatrix}$$

17.16.3 Cayley Hamilton Method

Cayley Hamilton theorem states that every square matrix A satisfies its own characteristic equation which is the base of this method. Let $f(\lambda)$ be the characteristic equation which is represented by
$$f(\lambda) = |\lambda I - A| = 0$$

i.e.,
$$f(\lambda) = \lambda^n + a_1\lambda^{n-1} + a_2\lambda^{n-2} + \cdots\cdots + a_{n-1}\lambda + a_n = 0 \tag{17.102}$$

The matrix A has to satisfy the characteristic equation as per Cayley Hamilton theorem. Therefore, we have
$$f(A) = A^n + a_1A^{n-1} + a_2A^{n-2} + \cdots\cdots + a_{n-1}A + a_nI = 0 \tag{17.103}$$

where I is the identity matrix.

Let us consider the following matrix polynomial
$$p(A) = \beta_0 I + \beta_1 A + \beta_2 A^2 + \cdots\cdots + \beta_n A^n + \beta_{n+1}A^{n+1} + \cdots\cdots \tag{17.104}$$

The polynomial represented by Eq. (17.104) has higher degree than A. The scalar polynomial is considered to calculate the above polynomial.
$$p(\lambda) = \beta_0 + \beta_1\lambda + \beta_2\lambda^2 + \cdots\cdots + \beta_n\lambda^n + \beta_{n+1}\lambda^{n+1} + \cdots\cdots \tag{17.105}$$

From Eq. (17.103) and Eq. (17.105) we have
$$\frac{p(\lambda)}{f(\lambda)} = Q(\lambda) + \frac{R(\lambda)}{f(\lambda)} \tag{17.106}$$

where $R(\lambda)$ is the remainder polynomial.
$$\therefore p(\lambda) = f(\lambda)Q(\lambda) + R(\lambda) \tag{17.107}$$

The order of the remainder polynomial is less than that of the characteristic polynomial and it can be represented as

$$R(\lambda) = \alpha_0 + \alpha_1\lambda + \alpha_2\lambda^2 + \cdots\cdots + +\alpha_{n-1}\lambda^{n-1} \tag{17.108}$$

Let $\lambda_1, \lambda_2, \ldots, \lambda_n$ be the distinct eigen values of matrix A and $f(\lambda_i) = 0$ for $i = 1,2, .., n$ as $\lambda_1, \lambda_2, \ldots, \lambda_n$ are roots of $f(\lambda) = 0$. $p(\lambda)$ is to be evaluated for all eigen values i.e., $\lambda_1, \lambda_2, \ldots, \lambda_n$. Therefore we have

$$p(\lambda_i) = f(\lambda_i)Q(\lambda_i) + R(\lambda_i) \text{ for } i = 1,2, \ldots, n \tag{17.109}$$

Since $f(\lambda_i) = 0$, we have from Eq. (17.109)

$$p(\lambda_i) = R(\lambda_i) = \alpha_0 + \alpha_1\lambda_i + \alpha_2\lambda_i^2 + \cdots\cdots + \alpha_{n-1}\lambda_i^{n-1} \tag{17.110}$$

If $\lambda_1, \lambda_2, \ldots, \lambda_n$ are substituted in Eq. (17.110), the set of simultaneous equation will be obtained which is to be solved to obtain $\alpha_0, \alpha_1, \alpha_2, \ldots, \alpha_n$.

Replacing λ in scalar polynomial by matrix A in Eq. (17.107), we have

$$p(A) = f(A)Q(A) + R(A)$$

Since $f(A) = 0$ by Caley Hamilton theorem, we have

$$p(A) = R(A) = \alpha_0 + \alpha_1 A + \alpha_2 A^2 + \cdots\cdots + \alpha_{n-1}A^{n-1} \tag{17.111}$$

The matrix polynomial $p(A)$ can be evaluated.

17.16.3.1 Procedure of Calculation of e^{At}

The following steps are followed to compute e^{At} using Caley Hamilton theorem:

Step 1: The eigen values of matrix A is to found out from $|\lambda I - A| = 0$.

Step 2: The remainder polynomial $R(\lambda)$ of order $n - 1$ is formed where

$$R(\lambda_i) = \alpha_0 + \alpha_1\lambda_i + \alpha_2\lambda_i^2 + \cdots\cdots + \alpha_{n-1}\lambda_i^{n-1} = p(\lambda_i)$$

Step 3: For distinct eigen values i.e., $\lambda_1, \lambda_2, \ldots, \lambda_n$, the simultaneous equations for are found out to compute $\alpha_0, \alpha_1, \alpha_2, \ldots, \alpha_n$.

If eigen values λ_k are repeated for m times say, one independent equation is obtained for λ_k by substituting $\lambda = \lambda_k$ in $R(\lambda)$. The remaining $m - 1$ equations by differentiating equation

$$R(\lambda) = p(\lambda)$$

$$\therefore \left.\frac{d^i p(\lambda)}{d\lambda^i}\right|_{\lambda=\lambda_k} = \left.\frac{d^i R(\lambda)}{d\lambda^i}\right|_{\lambda=\lambda_k} \quad \text{for } i = 1,2,\ldots\ldots, m-1. \tag{17.112}$$

Step 4: The simultaneous equations are to be solved to obtain $\alpha_0, \alpha_1, \alpha_2, \ldots, \alpha_n$.

Step 5: Find $p(A) = R(A) = \alpha_0 + \alpha_1 A + \alpha_2 A^2 + \cdots\cdots + \alpha_{n-1}A^{n-1}$

Example 17.15 Find $f(A) = A^{10}$ for $A = \begin{bmatrix} 0 & 1 \\ -3 & -4 \end{bmatrix}$ using Cayley Hamilton theorem.

Solution The function is $f(A) = A^{10}$. $\therefore f(\lambda) = \lambda^{10}$

Step 1: Find eigen values.

$$|\lambda I - A| = \begin{vmatrix} \lambda & -1 \\ 3 & \lambda + 4 \end{vmatrix} = 0$$

i.e., $$\lambda(\lambda+4)+3=0$$

i.e., $$(\lambda+1)(\lambda+3)=0$$

$$\therefore \lambda_1=-1 \quad \text{and} \quad \lambda_2=-3$$

Step 2: $R(\lambda)=\alpha_0+\alpha_1\lambda=p(\lambda)$ as $n=2$.

Step 3: Substitute λ_1 and λ_2 in $R(\lambda)$.

$$\therefore \alpha_0-\alpha_1=(-1)^{10}=1 \tag{1}$$

and $$\alpha_0-3\alpha_1=(-3)^{10}=59049 \tag{2}$$

Step 4: Solving Eq. (1) and Eq. (2), we have

$$\alpha_0=-29523 \quad \textit{and} \quad \alpha_1=-29524$$

Step 5: $f(A)=A^{10}=R(A)=\alpha_0 I+\alpha_1 A$

$$-29523\begin{bmatrix}1 & 0\\ 0 & 1\end{bmatrix}-29524\begin{bmatrix}0 & 1\\ -3 & -4\end{bmatrix}=\begin{bmatrix}-29523 & -29524\\ 88572 & 88573\end{bmatrix}$$

$$\therefore A^{10}=\begin{bmatrix}-29523 & -29524\\ 88572 & 88573\end{bmatrix}$$

Example 17.16 Find the state transition matrix for $A=\begin{bmatrix}0 & 1\\ -2 & -3\end{bmatrix}$ using Cayley Hamilton theorem.

Solution The function to be evaluated is $f(A)=e^{At}$

$$\therefore f(\lambda)=e^{\lambda t}$$

Step 1: Find eigen values.

$$|\lambda I-A|=\begin{vmatrix}\lambda & -1\\ 2 & \lambda+3\end{vmatrix}=0$$

i.e., $$\lambda(\lambda+3)+2=0$$

i.e., $$(\lambda+1)(\lambda+2)=0$$

$$\therefore \lambda_1=-1 \text{ and } \lambda_2=-2$$

Step 2: $R(\lambda)=\alpha_0+\alpha_1\lambda=p(\lambda)$ as $n=2$.

Step 3: Substitute λ_1 and λ_2 in $R(\lambda)$.

$$\therefore \alpha_0-\alpha_1=e^{-t} \tag{1}$$

and $$\alpha_0-3\alpha_1=e^{-2t} \tag{2}$$

Step 4: Solving Eq. (1) and Eq. (2), we have

$$\alpha_0=2e^{-t}-e^{-2t} \quad \text{and} \quad \alpha_1=e^{-t}-e^{-2t}$$

Step 5: $f(A)=A^{10}=R(A)=\alpha_0 I+\alpha_1 A$

$$= (2e^{-t} - e^{-2t})\begin{bmatrix} 1 & 0 \\ 0 & 1 \end{bmatrix} + (e^{-t} - e^{-2t})\begin{bmatrix} 0 & 1 \\ -3 & -4 \end{bmatrix}$$

$$= \begin{bmatrix} 2e^{-t} - e^{-2t} & e^{-t} - e^{-2t} \\ -2e^{-t} + e^{-2t} & -e^{-t} + 2e^{-2t} \end{bmatrix}$$

Example 17.17 Find the state transition matrix of $A = \begin{bmatrix} 0 & 0 & 2 \\ 0 & 1 & 0 \\ -1 & 0 & 3 \end{bmatrix}$ by Cayley Hamilton theorem.

Solution The function to be evaluated is $f(A) = e^{At}$.

Hence $$f(\lambda) = e^{\lambda t}$$

Step 1: Find the eigen values.

$$|\lambda I - A| = \begin{vmatrix} \lambda & 0 & -2 \\ 0 & \lambda - 1 & 0 \\ -1 & 0 & \lambda - 3 \end{vmatrix} = 0$$

i.e., $$\lambda(\lambda - 1)(\lambda - 3) + 2(\lambda - 1) = 0$$

i.e., $$(\lambda - 1)(\lambda^2 - 3\lambda + 2) = 0$$

i.e., $$(\lambda - 1)(\lambda - 1)(\lambda - 2) = 0$$

$$\therefore \lambda_1 = 1, \lambda_2 = 1 \text{ and } \lambda_3 = 2$$

Here $\lambda_1 = 1$ is repeated twice.

Step 2: $R(\lambda) = \alpha_0 + \alpha_1\lambda + \alpha_2\lambda^2 = p(\alpha_1\lambda) = e^{\lambda t}$

Step 3: For $\lambda_1 = 1$,

$$\alpha_0 + \alpha_1 + \alpha_2 = e^t \tag{1}$$

For $\lambda_2 = 1$ which is repeated use.

$$\frac{d}{d\lambda}[p(\lambda)]_{\lambda=\lambda_2} = \frac{d}{d\lambda}[R(\lambda)]_{\lambda=\lambda_2}$$

i.e., $$\frac{d}{d\lambda}[e^{\lambda t}]_{\lambda=\lambda_2} = \frac{d}{d\lambda}[\alpha_0 + \alpha_1\lambda + \alpha_2\lambda^2]_{\lambda=\lambda_2}$$

i.e., $$[te^{\lambda t}]_{\lambda=\lambda_2} = [\alpha_1 + 2\alpha_2\lambda]_{\lambda=\lambda_2}$$

i.e., $$\alpha_1 + 2\alpha_2 = te^{\lambda} \tag{2}$$

For $$\lambda_3 = 2$$

$$\alpha_0 + 2\alpha_1 + 4\alpha_2 = e^{2t} \tag{3}$$

Step 4: Solving Eq. (1), Eq. (2) and Eq. (3) we have

$$\alpha_0 = -2te^t + e^{2t}$$
$$\alpha_1 = 3te^t + 2e^t - 2e^{2t}$$
$$\alpha_2 = e^{2t} - e^t - te^t$$

Step 5: $f(A) = R(A) = \alpha_0 I + \alpha_1 A + \alpha_2 A^2$

$$= (-2te^t + e^{2t})\begin{bmatrix} 1 & 0 & 0 \\ 0 & 1 & 0 \\ 0 & 0 & 1 \end{bmatrix} + (3te^t + 2e^t - 2e^{2t})\begin{bmatrix} 0 & 0 & 2 \\ 0 & 1 & 0 \\ -1 & 0 & 3 \end{bmatrix}$$

$$+ (e^{2t} - e^t - te^t)\begin{bmatrix} -2 & 0 & 6 \\ 0 & 1 & 0 \\ -3 & 0 & 7 \end{bmatrix}$$

$$= \begin{bmatrix} 2e^t - e^{2t} & 0 & 2e^{2t} - 2e^t \\ 0 & e^t & 0 \\ e^t - e^{2t} & 0 & 2e^{2t} - e^t \end{bmatrix}$$

17.16.4 Similarity Transformation Method

A matrix A of order $n \times n$ is considered having eigen values $\lambda_1, \lambda_2, \ldots, \lambda_n$. A diagonal matrix Λ will exist such that its diagonal elements are the eigen values of the matrix A.

$$\Lambda = \begin{bmatrix} \lambda_1 & 0 & \cdots\cdots & 0 \\ 0 & \lambda_2 & \cdots\cdots & 0 \\ \vdots & \vdots & \cdots\cdots & \vdots \\ \vdots & \vdots & \cdots\cdots & \vdots \\ 0 & 0 & \cdots\cdots & \lambda_n \end{bmatrix} = \textit{Diagonal Matrix} \qquad (17.113)$$

The matrices A and Λ have the same eigen values and are called similar matrices.
To obtain diagonal matrix Λ, a similarity transformation of state variables is defined as

$$x(t) = Mz(t) \qquad (17.114)$$

In Eq. (17.114) M is the modal of matrix A. The diagonal matrix is obtained as

$$\Lambda = M^{-1}AM \qquad (17.115)$$

Now

$$e^{\Lambda t} = I + \Lambda t + \frac{1}{2!}\Lambda^2 t^2 + \cdots\cdots$$

$$= \begin{bmatrix} 1 & 0 & \cdots\cdots & 0 \\ 0 & 1 & \cdots\cdots & 0 \\ \vdots & \vdots & \cdots\cdots & \vdots \\ \vdots & \vdots & \cdots\cdots & \vdots \\ 0 & 0 & \cdots\cdots & 1 \end{bmatrix} + \begin{bmatrix} \lambda_1 & 0 & \cdots\cdots & 0 \\ 0 & \lambda_2 & \cdots\cdots & 0 \\ \vdots & \vdots & \cdots\cdots & \vdots \\ \vdots & \vdots & \cdots\cdots & \vdots \\ 0 & 0 & \cdots\cdots & \lambda_n \end{bmatrix} t + \frac{1}{2!}\begin{bmatrix} \lambda_1^2 & 0 & \cdots\cdots & 0 \\ 0 & \lambda_2^2 & \cdots\cdots & 0 \\ \vdots & \vdots & \cdots\cdots & \vdots \\ \vdots & \vdots & \cdots\cdots & \vdots \\ 0 & 0 & \cdots\cdots & \lambda_n^2 \end{bmatrix} t^2 + \cdots$$

$$= \begin{bmatrix} 1+\lambda_1 t+\frac{1}{2!}\lambda_1^2 t^2+\cdots\cdots & 0 & 0 \cdots\cdots\cdots\cdots\cdots\cdots & 0 \\ 0 & 1+\lambda_2 t+\frac{1}{2!}\lambda_2^2 t^2+\cdots\cdots & 0 \cdots\cdots\cdots\cdots\cdots\cdots & 0 \\ \vdots & \vdots & \vdots \cdots\cdots\cdots\cdots\cdots\cdots & \vdots \\ \vdots & \vdots & \vdots \cdots\cdots\cdots\cdots\cdots\cdots & \vdots \\ 0 & \vdots & 0 \cdots\cdots\cdots\cdots\cdots\cdots & 1+\lambda_n t+\frac{1}{2!}\lambda_n^2 t^2+\cdots\cdots \end{bmatrix}$$

$$= \begin{bmatrix} e^{\lambda_1 t} & 0 & 0 & \cdots\cdots\cdots\cdots\cdots\cdots & 0 \\ 0 & e^{\lambda_2 t} & 0 & \cdots\cdots\cdots\cdots\cdots\cdots & 0 \\ \vdots & \vdots & \vdots & \cdots\cdots\cdots\cdots\cdots\cdots & \vdots \\ \vdots & \vdots & \vdots & \cdots\cdots\cdots\cdots\cdots\cdots & \vdots \\ 0 & \vdots & 0 & \cdots\cdots\cdots\cdots\cdots\cdots & e^{\lambda_n t} \end{bmatrix} \tag{17.116}$$

The transformation used is

$$x(t) = Mz(t)$$

$\therefore$

$$x(0) = Mz(0)$$

$\therefore$

$$M^{-1}x(0) = M^{-1}Mz(0) = z(0) \tag{17.117}$$

and

$$\dot{x}(t) = M\dot{z}(t)$$

i.e.,

$$\dot{x}(t) = Ax(t) = AMz(t)$$

i.e.,

$$M\dot{z}(t) = AMz(t)$$

i.e.,

$$\dot{z}(t) = M^{-1}AMz(t)$$

i.e.,

$$\dot{z}(t) = \Lambda z(t) \tag{17.118}$$

Equation (17.118) shows the required transformed model.
The solution of homogeneous state equation in $x(t)$ is given by

$$x(t) = e^{At}x(0) \tag{17.119}$$

The solution of Eq. (17.118) is

$$z(t) = e^{\Lambda t}z(0) \tag{17.120}$$

Using Eq. (17.117), we have from Eq. (17.119)

$$z(t) = e^{\Lambda t}M^{-1}x(0)$$

i.e.,

$$Mz(t) = Me^{\Lambda t}M^{-1}x(0)$$

i.e.,

$$x(t) = Me^{\Lambda t}M^{-1}x(0) \tag{17.121}$$

From Eq. (17.119) and Eq. (17.120) we have

$$e^{At} = Me^{\Lambda t}M^{-1} \tag{17.122}$$

From Eq. (17.121), e^{At} can be computed where M is the modal matrix. The matrix M is also known as Vandermonde's matrix.

Example 17.18 Determine e^{At} where $A = \begin{bmatrix} 0 & 1 \\ -2 & -3 \end{bmatrix}$ using similarity transformation method.

Solution

$$\lambda I - A = \lambda \begin{bmatrix} 1 & 0 \\ 0 & 1 \end{bmatrix} - \begin{bmatrix} 0 & 1 \\ -2 & -3 \end{bmatrix} = \begin{bmatrix} \lambda & -1 \\ 2 & \lambda+3 \end{bmatrix} \tag{1}$$

$\therefore$ $$|\lambda I - A| = \lambda(\lambda+3)+2 = \lambda^2+3\lambda+2$$

$$= (\lambda+1)(\lambda+2) = 0$$

$\therefore$ $$\lambda_1 = -1$$

and $$\lambda_2 = -2$$

Let the eigenvector $$P_1 = \begin{bmatrix} P_{11} \\ P_{21} \end{bmatrix} \text{ and } P_2 = \begin{bmatrix} P_{12} \\ P_{22} \end{bmatrix}$$

Putting $\lambda = -1$ in Eq. (1), we have $\lambda I - A = \begin{bmatrix} -1 & -1 \\ 2 & 2 \end{bmatrix}$

$\therefore$ $$\begin{bmatrix} -1 & -1 \\ 2 & 2 \end{bmatrix}\begin{bmatrix} P_{11} \\ P_{21} \end{bmatrix} = \begin{bmatrix} 0 \\ 0 \end{bmatrix}$$

$\therefore$ $$-P_{11} - P_{21} = 0$$

and $$2P_{11} + 2P_{21} = 0$$

Let $P_{11} = 1$, then $P_{21} = -1$

$\therefore$ $$P_1 = \begin{bmatrix} P_{11} \\ P_{21} \end{bmatrix} = \begin{bmatrix} 1 \\ -1 \end{bmatrix}$$

Let as put $\lambda = -2$, in Eq. (1), we have

$$\lambda I - A = \begin{bmatrix} -2 & -1 \\ 2 & 1 \end{bmatrix}$$

$\therefore$ $$\begin{bmatrix} -2 & -1 \\ 2 & 1 \end{bmatrix}\begin{bmatrix} P_{12} \\ P_{22} \end{bmatrix} = 0$$

$\therefore$ $$-2P_{12} - P_{22} = 0$$

and $$2P_{12} + P_{22} = 0$$

Let $P_{12} = 1$, then $P_{22} = -2$

$$P_2 = \begin{bmatrix} P_{12} \\ P_{22} \end{bmatrix} = \begin{bmatrix} 1 \\ -2 \end{bmatrix}$$

$\therefore$ $$M = \begin{bmatrix} P_1 & P_2 \end{bmatrix} = \begin{bmatrix} 1 & 1 \\ -1 & -2 \end{bmatrix}$$

$\therefore$ $$M^{-1} = \begin{bmatrix} 2 & 1 \\ -1 & -1 \end{bmatrix} \text{ and } A = \begin{bmatrix} -1 & 0 \\ 0 & -2 \end{bmatrix}$$

$\therefore$ $$e^{At} = Me^{At}M^{-1}$$

$$= \begin{bmatrix} 1 & 1 \\ -1 & -2 \end{bmatrix} \begin{bmatrix} e^{-t} & 0 \\ 0 & e^{-2t} \end{bmatrix} \begin{bmatrix} 2 & 1 \\ -1 & -1 \end{bmatrix}$$

$$= \begin{bmatrix} 1 & 1 \\ -1 & -2 \end{bmatrix} \begin{bmatrix} 2e^{-t} & e^{-t} \\ -e^{-2t} & e^{-2t} \end{bmatrix}$$

$$= \begin{bmatrix} 2e^{-t} - e^{-2t} & e^{-t} + e^{-2t} \\ -2e^{-t} + 2e^{-2t} & -e^{-t} - 2e^{-2t} \end{bmatrix}$$

17.17 Controllability and Observability

The main aim in control system is to obtain the optimal solution for a given control problem. The following points are important from the point of view of the existence of such solution.

The first one is that the possibility of transferring any initial condition to any other desired state in finite time under the effect of suitable control input force for any given system.

The second one is that possibility of determining initial state of the system when output is measured for finite time with the knowledge of input.

17.18 Controllability

The answer of the first point is the meaning of controllability which is discussed below.

Any linear time-invariant system can be represented by the dynamic equation

$$\dot{x}(t) = Ax(t) + Bu(t) \tag{17.123}$$

and

$$y(t) = Cx(t) + Du(t) \tag{17.124}$$

The state $x(t)$ will be controllable at $t = t_0$ when there exists a piecewise continuous input $u(t)$ which will drive the state to any final state $x(t)$ for finite time interval $(t–t_0) \geq 0$. The system is called controllable if every state of the system is controllable in a finite time interval.

If there is no connection between a certain state and input, the state is not controllable. Figure 17.21 shows that the state x_2 is not controllable.

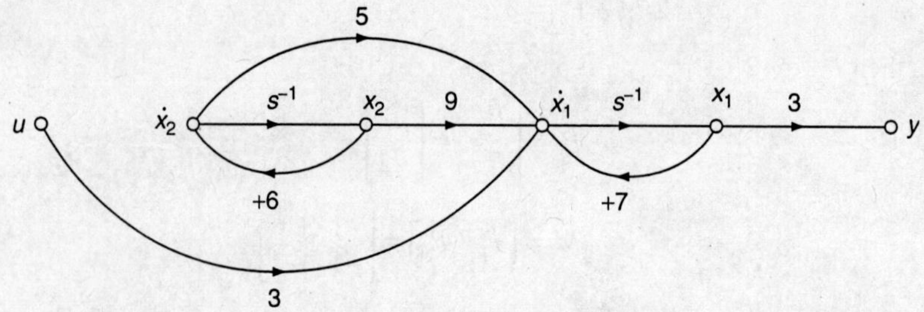

Fig. 17.21 x_2 is not controllable

17.18.1 Kalman's Test for Controllability

An the order multiple input and multiple time invariant system is considered which is represented by its state equation

$$\dot{x}(t) = Ax(t) + Bu(t) \tag{17.125}$$

where A is a $n \times n$ matrix, $u(t)$ is a $m \times 1$ matrix having m inputs and $x(t)$ is a $n \times 1$ state vector.

The necessary and sufficient condition for any system to be completely state controllable when the rank of the composite matrix (Q_c) is 'n'. The composite matrix is given by

$$Q_c = [B : AB : A^2B : \cdots\cdots : A^{n-1}B] \tag{17.126}$$

In this composite matrix B, AB, A^2B,.... Are the various columns.

Proof: Let the final state is the origin of the state space and the initial time be zero i.e., $t_0 = 0$. Since the final state $t = t_f$ is the origin of the state space, we have

$$x(t_f) = 0 \tag{17.127}$$

The solution of the state equation i.e., Eq. (17.124) is given by

$$x(t) = e^{At}x(0) + \int_0^t e^{A(t-\tau)}Bu(\tau)d\tau \tag{17.128}$$

For final state $t = t_f$, we have from Eq. (17.124)

$$x(t_f) = e^{At_f}x(0) + \int_0^{t_f} e^{A(t_f-\tau)}Bu(\tau)d\tau \tag{17.129}$$

Using Eq. (3), we have from Eq. (17.127)

$$0 = e^{At_f}x(0) + \int_0^{t_f} e^{A(t_f-\tau)}Bu(\tau)d\tau$$

i.e.,

$$e^{At_f}x(0) = -\int_0^{t_f} e^{A(t_f-\tau)}Bu(\tau)d\tau$$

i.e.,

$$x(0) = -\int_0^{t_f} e^{-A\tau}Bu(\tau)d\tau \tag{17.130}$$

Equation (17.128) must be satisfied by the initial state $x(0)$ for the complete state controllability. Therefore, the controllability depends only on the matrices A and B and is independent on matrix C.

We know that

$$e^{-A\tau} = \sum_{k=0}^{n-1} \alpha_k(\tau) A^k \tag{17.131}$$

Using Eq. (17.129) in Eq. (17.128), we have

$$x(0) = -\sum_{k=0}^{n-1} A^k \int_0^{t_f} \alpha_k(\tau) Bu(\tau) d\tau$$

$$= -\sum_{k=0}^{n-1} A^k B \beta_k$$

where
$$\beta_k = \int_0^{t_f} \alpha_k(\tau) u(\tau) d\tau \tag{17.132}$$

The vector matrix β_k is of order $n \times 1$.

$\therefore$
$$x(0) = -\left[B : AB : \cdots\cdots : A^{n-1}B\right] \begin{bmatrix} \beta_0 \\ \beta_1 \\ \beta_2 \\ \vdots \\ \vdots \\ \beta_{n-1} \end{bmatrix} \tag{17.133}$$

i.e.,
$$x(0) = -Q_c \begin{bmatrix} \beta_0 \\ \beta_1 \\ \beta_2 \\ \vdots \\ \vdots \\ \beta_{n-1} \end{bmatrix} \tag{17.134}$$

The system has to be completely controllable when $x(0)$ satisfies Eq. (17.131) or Eq. (17.132) and the tank of Q_c must be n which is $n \times n$ matrix. If the rank of Q_c is less than n, this indicates that some of the states are not controllable.

From above, it can be concluded that the rank of the composite matrix Q_c must be 'n' for complete state of controllability of the system. The main advantage of Kalman's test is that this test can be applied for any form of matrix A whether it is canonical or otherwise. This test does not show the physical feel of the problem which is the main limitation of this test. If the system is uncontrollable as per Kalman's test, this method is unable to give which mode is uncontrollable. This test is used for controllability analysis due to its mathematical simplicity.

Example 17.19 Find the controllability of the system described by the state equation

$$\begin{bmatrix} \dot{x}_1 \\ \dot{x}_2 \end{bmatrix} = \begin{bmatrix} 3 & 0 \\ 2 & 4 \end{bmatrix} \begin{bmatrix} x_1 \\ x_2 \end{bmatrix} + \begin{bmatrix} 0 \\ 1 \end{bmatrix} u$$

Solution

Here $$A = \begin{bmatrix} 3 & 0 \\ 2 & 4 \end{bmatrix} \text{ and } B = \begin{bmatrix} 0 \\ 1 \end{bmatrix}$$

$\therefore$ $$S = [B \quad AB]$$

Now $$AB = \begin{bmatrix} 3 & 0 \\ 2 & 4 \end{bmatrix} \begin{bmatrix} 0 \\ 1 \end{bmatrix} = \begin{bmatrix} 0 \\ 4 \end{bmatrix}$$

$\therefore$ $$S = [B \quad AB] = \begin{bmatrix} 0 & 0 \\ 1 & 4 \end{bmatrix}$$

Since the matrix is singular, the system is not controllable.

Example 17.20 Find the controllability of the system represented by

$$\dot{x} = \begin{bmatrix} 1 & 2 \\ 0 & -1 \end{bmatrix} x + \begin{bmatrix} 1 \\ 0 \end{bmatrix} u$$

Solution Here $A = \begin{bmatrix} 1 & 2 \\ 0 & -1 \end{bmatrix}$ and $B = \begin{bmatrix} 1 \\ 0 \end{bmatrix}$ and $n = 2$.

$\therefore$ $$Q_c = [B \quad AB]$$

$$AB = \begin{bmatrix} 1 & 2 \\ 0 & -1 \end{bmatrix} \begin{bmatrix} 1 \\ 0 \end{bmatrix} = \begin{bmatrix} 1 \\ 0 \end{bmatrix}$$

$\therefore$ $$Q_c = \begin{bmatrix} 1 & 1 \\ 0 & 0 \end{bmatrix}$$

$$|Q_c| = \begin{vmatrix} 1 & 1 \\ 0 & 0 \end{vmatrix} = 0$$

$\therefore$ Rank of $Q_c < 2$.

Hence system is uncontrollable.

17.18.2 Condition for State Controllability in *s*–Plane

The system is represented by the transfer function and it is s–plane. If there is no common factor in the numerator and denominator polynomials of the transfer function, the system is completely state controllable. On the other hand, we can say that there should not be any pole zero cancellation in the s–domain transfer function of the system for complete state controllability. Due to cancellation of pole zero in the transfer function, it is not possible to carry out the information about the dynamic behaviour of the system. The system cannot be controlled in the direction of the cancelled mode.

17.18.3 Gilbert's Test for Controllability in *s*–Plane

The matrix A must in canonical form for the Gilbert's test. The given state model is to be transformed to the canonical form first before applying Gilbert's test. Let a single input linear time invariant system be represented by

$$\dot{x}(t) = Ax(t) + Bu(t) \tag{17.135}$$

The matrix A in Eq. (17.133) is not in canonical form. The following transformation is used to transform it in canonical form.

$$x(t) = Mz(t) \tag{17.136}$$

where M is the modal matrix.

The derived transformed state model takes the form

$$\dot{z}(t) = \Lambda z(t) + \tilde{B}u(t) \tag{17.137}$$

where Λ is the diagonal matrix and

$$\tilde{B} = M^{-1}B \tag{17.138}$$

The eigen values of A are assumed to be distinct.

The necessary and sufficient condition for the complete state controllability becomes that the matrix $\tilde{B}$ must not have any zero elements. If the matrix $\tilde{B}$ has zero elements, the corresponding state variables are not controllable.

If the eigen values are $\lambda_1, \lambda_1, \lambda_2, \lambda_2, \lambda_2, \lambda_3, \ldots\ldots, \lambda_n$, the transformation results Jordan canonical form as

$$J = \begin{bmatrix} \lambda_1 & 1 & 0 & 0 & 0 & 0 \cdots\cdots & 0 \\ 0 & \lambda_1 & 0 & 0 & 0 & 0 \cdots\cdots & 0 \\ 0 & 0 & \lambda_2 & 1 & 0 & 0 \cdots\cdots & 0 \\ 0 & 0 & 0 & \lambda_2 & 1 & 0 \cdots\cdots & 0 \\ 0 & 0 & 0 & 0 & \lambda_2 & 0 \cdots\cdots & 0 \\ 0 & 0 & 0 & 0 & 0 & \lambda_3 \cdots\cdots & 0 \\ \vdots & \vdots & \vdots & \vdots & \vdots & \vdots \cdots\cdots & 0 \\ \vdots & \vdots & \vdots & \vdots & \vdots & \vdots \cdots\cdots & \lambda_n \end{bmatrix}$$

The condition for complete controllability in this case is that the elements of any row of $\tilde{B}$ which corresponds to the last row of each Jordan block are not all zero.

Gilbert's test requires to transform matrix A into canonical form which is time consuming. Therefore, Kalman's test is used to test controllability.

Example 17.21 Estimate the controllability of the system having state equation

$$\dot{x} = \begin{bmatrix} 0 & 1 & 0 \\ 0 & 0 & 1 \\ -8 & -14 & -7 \end{bmatrix} x + \begin{bmatrix} 0 \\ 0 \\ 1 \end{bmatrix} u$$

by (i) Kalman's test and (ii) Gilbert's test.

Solution

(i) Kalman's test:

Here $$A = \begin{bmatrix} 0 & 1 & 0 \\ 0 & 0 & 1 \\ -8 & -14 & -7 \end{bmatrix} \text{ and } B = \begin{bmatrix} 0 \\ 0 \\ 1 \end{bmatrix}$$

$$\therefore AB = \begin{bmatrix} 0 & 1 & 0 \\ 0 & 0 & 1 \\ -8 & -14 & -7 \end{bmatrix} \begin{bmatrix} 0 \\ 0 \\ 1 \end{bmatrix} = \begin{bmatrix} 0 \\ 1 \\ -7 \end{bmatrix}$$

$$\therefore A^2 = \begin{bmatrix} 0 & 1 & 0 \\ 0 & 0 & 1 \\ -8 & -14 & -7 \end{bmatrix} \begin{bmatrix} 0 & 1 & 0 \\ 0 & 0 & 1 \\ -8 & -14 & -7 \end{bmatrix} = \begin{bmatrix} 0 & 0 & 1 \\ -8 & -14 & -7 \\ 56 & 90 & 35 \end{bmatrix}$$

$$\therefore A^2B = \begin{bmatrix} 0 & 0 & 1 \\ -8 & -14 & -7 \\ 56 & 90 & 35 \end{bmatrix} \begin{bmatrix} 0 \\ 0 \\ 1 \end{bmatrix} = \begin{bmatrix} 1 \\ -7 \\ 35 \end{bmatrix}$$

$$\therefore Q_c = [B : AB : A^2B] = \begin{bmatrix} 0 & 0 & 1 \\ 0 & 1 & -7 \\ 1 & -7 & 35 \end{bmatrix}$$

$$\therefore |Q_c| = \begin{vmatrix} 0 & 0 & 1 \\ 0 & 1 & -7 \\ 1 & -7 & 35 \end{vmatrix} = -1$$

$\therefore$ Rank of $Q_c = 3$

(ii) Gilbert's Test:

The matrix A must e in canonical form. The eigen values of A is found out.

$$|\lambda I - A| = 0$$

i.e., $$|\lambda I - A| = \begin{vmatrix} \lambda & 0 & 0 \\ 0 & \lambda & -1 \\ 8 & 14 & \lambda + 7 \end{vmatrix} = 0$$

i.e., $$\lambda[\lambda(\lambda+8)+14]+7 = 0$$

i.e., $$(\lambda+1)(\lambda+2)(\lambda+4) = 0$$

i.e., $$\lambda^3 + 8\lambda^2 + 14\lambda + 7 = 0$$

$$\therefore \lambda_1 = -1, \lambda_2 = -2 \text{ and } \lambda_3 = -4$$

The modal matrix M is given by

$$M = \begin{bmatrix} 1 & 1 & 1 \\ \lambda_1 & \lambda_2 & \lambda_3 \\ \lambda_1^2 & \lambda_2^2 & \lambda_3^2 \end{bmatrix} = \begin{bmatrix} 1 & 1 & 1 \\ -1 & -2 & -4 \\ 1 & 4 & 16 \end{bmatrix}$$

$$\therefore M^{-1} = \begin{bmatrix} 2.6667 & 2.0000 & 0.3333 \\ -2.0000 & -2.5000 & -0.5000 \\ 0.3333 & 0.5000 & 0.1667 \end{bmatrix}$$

$$\tilde{B} = M^{-1}B = \begin{bmatrix} 2.6667 & 2.0000 & 0.3333 \\ -2.0000 & -2.5000 & -0.5000 \\ 0.3333 & 0.5000 & 0.1667 \end{bmatrix} \begin{bmatrix} 0 \\ 0 \\ 1 \end{bmatrix} = \begin{bmatrix} 0.3333 \\ -0.5000 \\ 0.1667 \end{bmatrix}$$

Since none of the elements of $\tilde{B}$ are zero, the system is completely controllable.

17.18.4 Output Controllability

During designing of some practical systems, it is required to control the output variables rather than the state variables. Let a linear time invariant system be represented by

$$\dot{x}(t) = Ax(t) + Bu(t) \tag{17.139}$$

$$y(t) = Cx(t) + Du(t) \tag{17.140}$$

If any unconstrained input vector $u(t)$ transfer any given initial output $y(t_0)$ to any final input $y(t_f)$ in a finite time interval $t_0 \le t \le t_f$. The test matrix Q_c is to be constructed as follows:

$$Q_c = \left[CB : CAB : CA^2B : \cdots\cdots CA^{n-1}B : D\right] \tag{17.141}$$

If the order of matrix C is $p \times n$ and there are p outputs, for complete controllability the rank of test matrix Q_c should be p.

17.19 Observability

Determination of the system state by measuring the output for finite length of time given the observability of the system. When every state $X(t_0)$ can be completely identified by measurements of the output $Y(t)$ over a finite interval of time, the system is said to be completely observable.

If any state of a system cannot be observed from measurements, it is called unobservable. From Fig. 17.22, it is clear that the state x_2 is not connected to output. Therefore, the state x_2 is called unobservable.

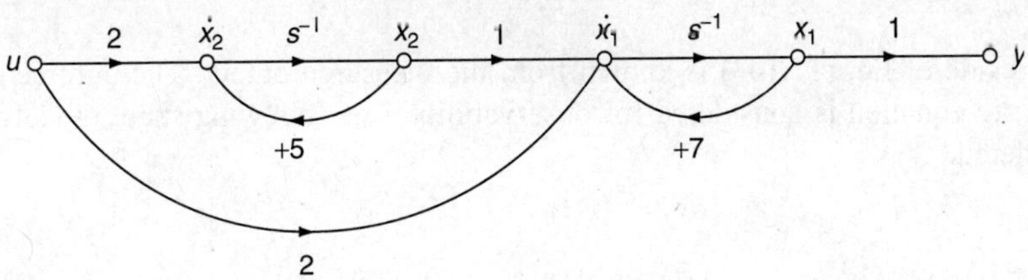

Fig. 17.22 x_2 is unobservable

Any linear time-invariant system can be represented by the dynamic equ ation

$$\dot{x}(t) = Ax(t) + Bu(t)$$

and

$$y(t) = Cx(t) + Du(t)$$

The state $x(t)$ is said to be observable if there exists a finite time $t_1 \geq 0$ for any given input $u(t)$ so that the knowledge of $u(t)$ for $t_1 \geq t \geq t_0$, the matrices A, B, C, D and output $y(t)$ for $t_1 \geq t \geq t_0$ are sufficient to determine $x(t_0)$.

The observability of the system can also be obtained by using Kalman's test.

17.19.1 Kalman's Test for Observability

Let us consider a n th order multiple input multiple output linear time invariant system which is represented by the following state equation:

$$\dot{x}(t) = Ax(t) + Bu(t) \tag{17.142}$$

$$y(t) = Cx(t) \tag{17.143}$$

where $y(t)$ is a $p \times 1$ output vector and C is a $1 \times n$ matrix.

The system is said to be completely observable when the rank of the composite matrix Q_c is of order n. The composite matrix Q_c is given by

$$Q_c = [C^T : A^T C^T : \cdots\cdots\cdots (A^T)^{n-1} C^T]$$

where C^T is the transpose of matrix C and A^T is the transpose of matrix A.

Proof: When the system is described by Eq. (17.140) and Eq. (17.141), the solution becomes

$$x(t) = e^{At}x(0) + \int_0^t e^{A(t-\tau)} Bu(\tau)d\tau \tag{17.144}$$

and

$$y(t) = Ce^{At}x(0) + C\int_0^t e^{A(t-\tau)} Bu(\tau)d\tau \tag{17.145}$$

The matrices A, B and C are known and the input vector $u(t)$ is known, the integral term in Eq. (17.146) is also known. This integral term can be subtracted from the observed value of $y(t)$ so that the left hand side is modified. The following equation is considered to investigate the observability of the system.

$$y(t) = Ce^{At}x(0) \tag{17.146}$$

The left hand side of Eq. (17.143) is known from the measured output. The unforced system i.e., homogeneous state equation is considered for observability. The following state model for unforced system is considered.

$$\dot{x}(t) = Ax(t) \tag{17.147}$$

and

$$y(t) = Cx(t) \tag{17.148}$$

The output is given by

$$y(t) = Ce^{At}x(0) \tag{17.149}$$

We know that

$$e^{-A\tau} = \sum_{k=0}^{n-1} \alpha_k(t) A^k \tag{17.150}$$

Using Eq. (17.147), we have from Eq. (17.146)

$$y(t) = \sum_{k=0}^{n-1} C\alpha_k(t)A^k x(0) \tag{17.151}$$

i.e.,

$$y(t) = \alpha_0(t)Cx(0) + \alpha_1(t)CAx(0) + \cdots\cdots\cdots + \alpha_{n-1}(t)CA^{n-1}x(0) \tag{17.152}$$

The system becomes completely observavabale, when $x(0)$ must be determined uniquely from the given output $y(t)$ over the interval $0 \le t \le t_f$. This will be possible if the rank of the matrix

$$\begin{bmatrix} C \\ CA \\ CA^2 \\ \vdots \\ \vdots \\ CA^{n-1} \end{bmatrix}$$

becomes 'n'.

The transpose of the above matrix becomes

$Q_c = [C^T : A^T C^T : \cdots\cdots\cdots\cdots (A^T)^{n-1} C^T]$ and the rank Q_c must be 'n'.

Example 17.22 Example 17.23 The dynamic equation of a system is represented by

$$\begin{bmatrix} \dot{x}_1 \\ \dot{x}_2 \end{bmatrix} = \begin{bmatrix} -2 & 1 \\ 0 & -3 \end{bmatrix} \begin{bmatrix} x_1 \\ x_2 \end{bmatrix} + \begin{bmatrix} 4 \\ 1 \end{bmatrix} u$$

$$y = [1 \quad 0] \begin{bmatrix} x_1 \\ x_2 \end{bmatrix}$$

Is the system observable?

Solution

$$A = \begin{bmatrix} -2 & 1 \\ 0 & -3 \end{bmatrix}, C = [1 \quad 0]$$

$$\therefore \quad CA = [1 \quad 0] \begin{bmatrix} -2 & 1 \\ 0 & -3 \end{bmatrix} = [-2 \quad 1]$$

$$\therefore \quad V = \begin{bmatrix} C \\ CA \end{bmatrix} = \begin{bmatrix} 1 & 0 \\ -2 & 1 \end{bmatrix}$$

Since V is not singular, the system is observable.

17.19.2 Condition for Complete Observability in *s*–Plane

The condition for complete observability of the system in s–plane can also be obtained from the transfer function. The numerator and denominator polynomial of the transfer function must be co–prime. Any cancellation of pole and zero of the transfer function should not be there. Otherwies, the system will not be observable and the cancelled mode cannot be observed.

17.19.3 Gilbert's Test for Observability

The state model must be in canonical form for Gilbert's test. Let the state model for linear time invariant system be described by

$$\dot{x}(t) = Ax(t) + Bu(t) \tag{17.153}$$

and

$$y(t) = Cx(t) \tag{17.154}$$

The following transformation is used.

$$x(t) = CMz(t) \tag{17.155}$$

In Eq. (17.152) M is the modal matrix.

$$\therefore \quad y(t) = CMz(t) = \tilde{C}z(t) \tag{17.156}$$

where

$$\tilde{C} = CM \tag{17.157}$$

For single input and single output system, we have

$$y(t) = \tilde{C}z(t) = [\tilde{C}_{11} \quad \tilde{C}_{12} \quad \cdots\cdots \quad \tilde{C}_{1n}] \begin{bmatrix} Z_1(t) \\ Z_2(t) \\ \vdots \\ \vdots \\ Z_3(t) \end{bmatrix}$$

$$= \tilde{C}_{11}Z_1(t) + \tilde{C}_{12}Z_2(t) + \cdots\cdots + \tilde{C}_{1n}Z_n(t) \tag{17.158}$$

All the states are decoupled since the system state model is in canonical form and hence not linked to each other. If each term corresponding to each state is observed from output, the system becomes observable. Therefore, all the coefficients of $\tilde{C}$ must be nonzero for complete observability of the system. If any element of $\tilde{C}$ becomes zero, the corresponding state will be unobserved and it is shielded from observation.

Example 17.23 Evaluate the observability of the system with

$$A = \begin{bmatrix} 0 & 1 & 0 \\ 0 & 0 & 1 \\ 0 & -3 & -4 \end{bmatrix}, \; B = \begin{bmatrix} 0 \\ 0 \\ 1 \end{bmatrix} \text{ and } C = \begin{bmatrix} 2 & 3 & 1 \end{bmatrix}$$

by (i) Kalman's test and by (ii) Gilbert's test.

Solution

(i) Kalman's test:

Here

$$C = \begin{bmatrix} 2 \\ 3 \\ 1 \end{bmatrix} \text{ and } A = \begin{bmatrix} 0 & 1 & 0 \\ 0 & 0 & 1 \\ 0 & -3 & -4 \end{bmatrix}$$

$$CA = [0\ \ -1\ \ -1]$$

$$A^2 = \begin{bmatrix} 0 & 0 & 1 \\ 0 & -3 & -4 \\ 0 & 12 & 13 \end{bmatrix}$$

$$CA^2 = [2\ \ 3\ \ 1]\begin{bmatrix} 0 & 0 & 1 \\ 0 & -3 & -4 \\ 0 & 12 & 13 \end{bmatrix} = [0\ \ 3\ \ 3]$$

$$\therefore Q_o = \begin{bmatrix} C \\ CA \\ CA^2 \end{bmatrix} = \begin{bmatrix} 2 & 3 & 1 \\ 0 & -1 & -1 \\ 0 & 3 & 3 \end{bmatrix}$$

$$\therefore |Q_o| = \begin{vmatrix} 2 & 3 & 1 \\ 0 & -1 & -1 \\ 0 & 3 & 3 \end{vmatrix} = 0$$

$\therefore$ Rank of $Q < 3$.

Hence the system is not completely observable.

(ii) Gilbert's Test:

The matrix A must e in canonical form. The eigen values of A is found out.

$$|\lambda I - A| = 0$$

i.e.,

$$|\lambda I - A| = \begin{vmatrix} \lambda & -1 & 0 \\ 0 & \lambda & -1 \\ 0 & 3 & \lambda + 4 \end{vmatrix} = 0$$

i.e.,

$$\lambda^3 + 4\lambda^2 + 3\lambda = 0$$

i.e.,

$$\lambda(\lambda + 1)(\lambda + 3) = 0$$

$$\therefore \lambda_1 = 0, \lambda_2 = -1 \text{ and } \lambda_3 = -3$$

The modal matrix M is given by

$$M = \begin{bmatrix} 1 & 1 & 1 \\ \lambda_1 & \lambda_2 & \lambda_3 \\ \lambda_1^2 & \lambda_2^2 & \lambda_3^2 \end{bmatrix} = \begin{bmatrix} 1 & 1 & 1 \\ 0 & -1 & -3 \\ 0 & 1 & 9 \end{bmatrix}$$

$$\tilde{C} = CM = [2 \quad 3 \quad 1]\begin{bmatrix} 1 & 1 & 1 \\ 0 & -1 & -3 \\ 0 & 1 & 9 \end{bmatrix} = [2 \quad 0 \quad 2]$$

Since there is one zero element in $\tilde{C}$, the system is not completely observable.

SIGNIFICANT POINTS

Modern Control Theory: Modern control theory takes care of initial conditions automatically and it is also possible to analyze time-va rying or time-invariant, linear or non-linear, single or multiple input–output systems.
State: The state of a dynamic system is the smallest set of variables and the knowledge of these variables at $t = t_0$ together with inputs for $t \geq t_0$ completely determines the behaviour of the system at $t \geq t_0$.
State Variables: The smallest set of variables that determine the state of the system are known as state variables.
State Vector: The n state variables that completely describe the behaviour of a given system are said to be n components of a vector.
State Space: The n-dimensional space whose coordinate axes consist of the x_1 axis, x_2 axis, …, x_n axis is known as a state space.
State Equation: The equations which describe the relationships between the derivations of the state variables as a function of the state variable, input and parameter time are called the state equations.
State Transition Matrix: The matrix operator which determines the transition of any initial state $x(0)$ at $t = 0$ to state $x(t)$ at time t is known as state transition matrix.
Controllability: The state x(t) will be controllable at $t = t_0$ when there exists a piecewise continuous input $u(t)$ which will drive the state to any final state $x(t)$ for a finite time interval $(t - t_0) \geq 0$. The system is called controllable if every state of the system is controllable in a finite time interval.
Observability: If any state of a system cannot be observed from measurements, it is called unobservable.
Eigenvalue: The roots of the characteristic equation $|\lambda_i I - A| = 0$ are called the eigenvalues of the matrix $A(n \times n)$.

Transfer function $= \dfrac{Y(s)}{U(s)} = C[sI - A]^{-1}B + D$

Resolvent matrix $\phi(s) = (sI - A)^{-1} = \dfrac{1}{s}\left(I - \dfrac{A}{s}\right)^{-1} = \dfrac{1}{s}\left[1 + \dfrac{A}{s} + \dfrac{A^2}{s^2} + \cdots\right]$

State transition matrix $\phi(t) = e^{At}$

Properties of STM

(i) $\phi(0) = e^{A0} = I$

(ii) $\phi(t) = e^{At} = (e^{-At})^{-1} = [\phi(-t)]^{-1}$

(iii) $\phi^{-1}(t) = \phi(-t)$

(iv) $\phi(t_1 + t_2) = e^{A(t_1+t_2)} = e^{At_1}e^{At_2} = \phi(t_1)\phi(t_2) = e^{At_2}e^{At_1} = \phi(t_2)\phi(t_1)$

(v) $[\phi(t)]^n = (e^{At})^n = e^{Ant} = \phi(nt)$

(vi) $\phi(t_1 - t_2)\phi(t_2 - t_0) = e^{A(t_1-t_2)}e^{A(t_2-t_0)}$
$= e^{[t_1-t_2+t_2-t_0]} = e^{A(t_1-t_0)} = \phi(t_1 - t_0)$

ADDITIONAL SOLVED EXAMPLES

Example 17.24 Obtain the state model of the network shown in Fig. E17.9.

Solution

Using KCL at note x, we can write

$$i = i_c + i_L = C\frac{dv_c}{dt} + i_L$$

i.e., $$\frac{dv_C}{dt} = \frac{1}{C}i - \frac{1}{C}i_L \quad (1)$$

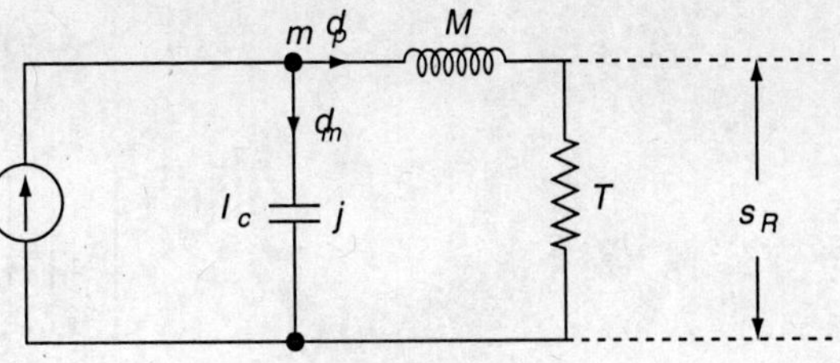

Fig. E17.9

Again by mesh equation, we get

$$v_C = L\frac{di_L}{dt} + Ri_L$$

i.e., $$\frac{di_L}{dt} = \frac{1}{L}v_C - \frac{R}{L}i_L \quad (2)$$

From Eqs. (1) and (2), we get

$$\begin{bmatrix} \dfrac{dv_C}{dt} \\ \dfrac{di_L}{dt} \end{bmatrix} = \begin{bmatrix} 0 & -\dfrac{1}{C} \\ \dfrac{1}{L} & -\dfrac{R}{L} \end{bmatrix} \begin{bmatrix} v_C \\ i_L \end{bmatrix} + \begin{bmatrix} \dfrac{1}{C} \\ 0 \end{bmatrix} i$$

and output is

$$v_R = i_L R = [0 \quad R]\begin{bmatrix} v_C \\ i_L \end{bmatrix}$$

Example 17.25 Express the following transfer function in (i) CCF and (ii) OCF forms.

$$\frac{Y(s)}{U(s)} = \frac{5s^2 + 2s + 6}{s^3 + 7s^2 + 11s + 8}$$

Solution

(i) CCF Form

$$\frac{Y(s)}{U(s)} = \frac{5s^2 + 2s + 6}{s^3 + 7s^2 + 11s + 8} = \frac{(5s^{-1} + 2s^{-2} + 6s^{-3})X(s)}{(1 + 7s^{-1} + 11s^{-2} + 8s^{-3})X(s)}$$

Let $$Y(s) = (5s^{-1} + 2s^{-2} + 6s^{-3})X(s) \quad (1)$$

and $$U(s) = (1 + 7s^{-1} + 11s^{-2} + 8s^{-3})X(s) \quad (2)$$

$\therefore$ $$X(s) = U(s) - (7s^{-1} + 11s^{-2} + 8s^{-3})X(s) \quad (3)$$

The SFG is shown in Fig. E17.10

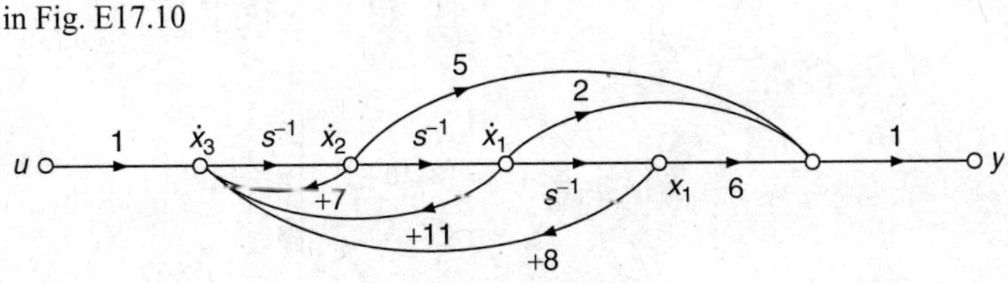

Fig. E17.10

Let $\dot{x}_1 = x_2$

$$\dot{x}_2 = x_3$$

$\therefore$ $\dot{x}_3 = u - 7x_3 - 11x_2 - 8x_1$

$\therefore$ $y = 6x_1 + 2x_2 + 5x_3$

and

$$\begin{bmatrix}\dot{x}_1\\ \dot{x}_2\\ \dot{x}_3\end{bmatrix} = \begin{bmatrix}0 & 1 & 0\\ 0 & 0 & 1\\ -7 & -11 & -8\end{bmatrix}\begin{bmatrix}x_1\\ x_2\\ x_3\end{bmatrix} + \begin{bmatrix}0\\ 0\\ 1\end{bmatrix}u \tag{4}$$

$$[y] = [6 \quad 2 \quad 5]\begin{bmatrix}x_1\\ x_2\\ x_3\end{bmatrix} + [0][u] \tag{5}$$

Equations (4) and (5) represent the model in the CCF form.

(ii) OCF form

$$\frac{Y(s)}{U(s)} = \frac{5s^2 + 2s + 6}{s^3 + 7s^2 + 11s + 8} = \frac{(5s^{-1} + 2s^{-2} + 6s^{-3})X(s)}{(1 + 7s^{-1} + 11s^{-2} + 8s^{-3})X(s)}$$

$\therefore$ $Y(s) = (5s^{-1} + 2s^{-2} + 6s^{-3})\,U(s) - (7s^{-1} + 11s^{-2} + 8s^{-3})Y(s)$

$\therefore$ $\dot{x}_1 = -8x_3 + 6u$ (7)

$$\dot{x}_2 = -11x_3 + 2u + x_1 \tag{8}$$

$$\dot{x}_3 = -7x_3 + 5u + x_2 \tag{9}$$

and $y = x_3$ (10)

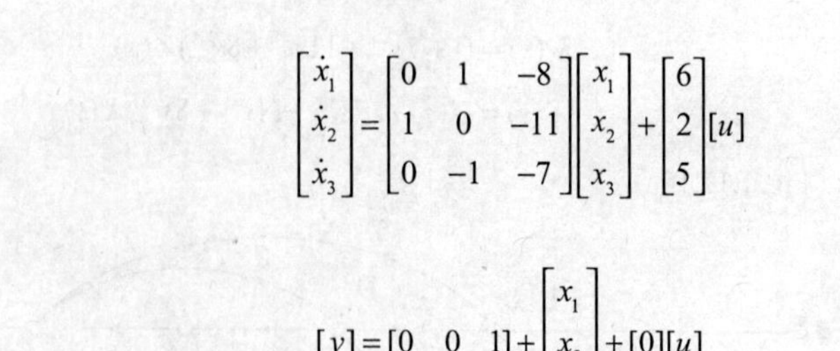

Fig. E17.11

$\therefore$

$$\begin{bmatrix}\dot{x}_1\\ \dot{x}_2\\ \dot{x}_3\end{bmatrix} = \begin{bmatrix}0 & 1 & -8\\ 1 & 0 & -11\\ 0 & -1 & -7\end{bmatrix}\begin{bmatrix}x_1\\ x_2\\ x_3\end{bmatrix} + \begin{bmatrix}6\\ 2\\ 5\end{bmatrix}[u] \tag{11}$$

and

$$[y] = [0 \quad 0 \quad 1] + \begin{bmatrix}x_1\\ x_2\\ x_3\end{bmatrix} + [0][u] \tag{12}$$

Equations (11) and (12) represent the OCF form.

Example 17.26 Derive the state equation of a system having transfer function as follows:

$$\frac{Y(s)}{U(s)} = \frac{8}{s(s+2)(s+3)}$$

Use (i) cascade and (ii) parallel decomposition.

Solution

(i) Cascade decomposition

$$\frac{Y(s)}{U(s)} = \frac{Y(s)}{X_1(s)}\frac{X_1(s)}{X_2(s)}\frac{X_2(s)}{X_3(s)}\frac{X_3(s)}{U(s)} = \frac{8}{s(s+2)(s+3)}$$

where

$$\frac{Y(s)}{X_1(s)} = \frac{1}{s}$$

$$\frac{X_1(s)}{X_2(s)} = \frac{1}{s+2}$$

$$\frac{X_2(s)}{X_3(s)} = \frac{1}{s+3}$$

$$\frac{X_3(s)}{U(s)} = 8$$

$\therefore$

$$\dot{x}_3 = -3x_3 + 8u$$

$$\dot{x}_2 = x_3 - 2x_2$$

$$\dot{x}_1 = x_2$$

$$y = x_1$$

and

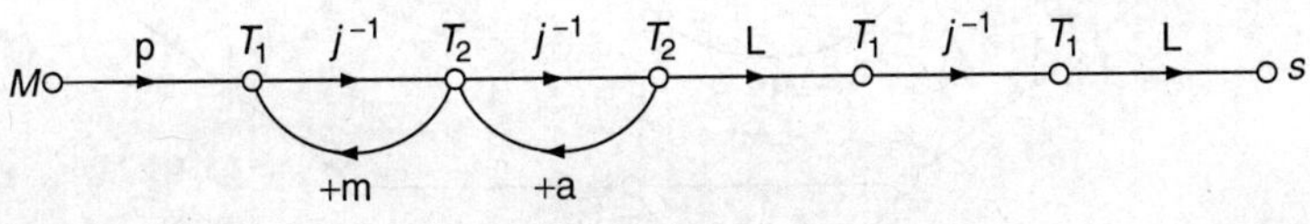

Fig. E17.12

$\therefore$

$$\begin{bmatrix} \dot{x}_1 \\ \dot{x}_2 \\ \dot{x}_3 \end{bmatrix} = \begin{bmatrix} 0 & 1 & 0 \\ 0 & -2 & 1 \\ 0 & 0 & -3 \end{bmatrix} \begin{bmatrix} x_1 \\ x_2 \\ x_3 \end{bmatrix} + \begin{bmatrix} 0 \\ 0 \\ 8 \end{bmatrix} [u]$$

and

$$y = [1 \quad 0 \quad 0] \begin{bmatrix} x_1 \\ x_2 \\ x_3 \end{bmatrix} + [0][u]$$

(ii) Parallel decomposition
Let

$$\frac{Y(s)}{U(s)} = \frac{8}{s(s+2)(s+3)} = \frac{d_1}{s} + \frac{d_2}{s} + \frac{d_3}{s}$$

where

$$d_1 = \left[s\left\{ \frac{8}{s(s+2)(s+3)} \right\} \right]_{s=0} = \frac{8}{6}$$

$$d_2 = \left[(s+2)\left\{ \frac{8}{s(s+2)(s+3)} \right\} \right]_{s=-2} = -\frac{8}{2} = -4$$

$$d_3 = \left[(s+3)\left\{ \frac{8}{s(s+2)(s+3)} \right\} \right]_{s=-3} = \frac{4}{3}$$

$\therefore$

$$\frac{Y(s)}{U(s)} = \frac{\frac{8}{6}}{s} + \frac{-4}{s+2} + \frac{\frac{4}{3}}{s+3}$$

where,

$$\therefore Y(s) = \frac{8}{6} \times \frac{U(s)}{s} - 4 \times \frac{U(s)}{s+2} + \frac{4}{3} \times \frac{U(s)}{s+3}$$

$$= \frac{8}{6} x_1(s) - 4x_2(s) + \frac{4}{3} x_3(s)$$

where

$$x_1(s) = \frac{U(S)}{s}, \; x_2(s) = \frac{U(s)}{s+2} \text{ and } x_3(s) = \frac{U(s)}{s+3}$$

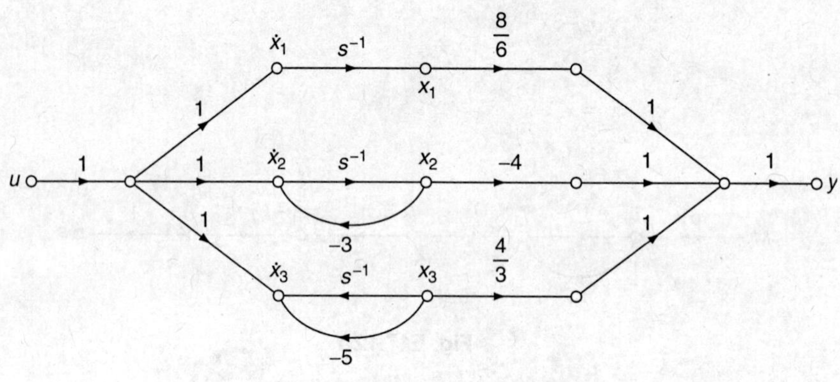

Fig. E17.13

$\therefore$

$$\dot{x}_1 = u$$
$$\dot{x}_2 = -3x_2 + u$$
$$\dot{x}_3 = -5x_3 + u$$

and

$$y = \frac{8}{6} x_1 - 4x_2 + \frac{4}{3} x_3$$

$\therefore$

$$\begin{bmatrix} \dot{x}_1 \\ \dot{x}_2 \\ \dot{x}_3 \end{bmatrix} = \begin{bmatrix} 0 & 1 & 0 \\ 0 & -3 & 0 \\ 0 & 0 & -5 \end{bmatrix} \begin{bmatrix} x_1 \\ x_2 \\ x_3 \end{bmatrix} + \begin{bmatrix} 1 \\ 1 \\ 1 \end{bmatrix} [u]$$

and

$$[y] = \begin{bmatrix} \frac{8}{6} & -4 & \frac{4}{3} \end{bmatrix} \begin{bmatrix} x_1 \\ x_2 \\ x_3 \end{bmatrix} + [0][u]$$

Example 17.27 Determine the resolvent matrix $\phi(s)$, state transition $[\phi(t)]$ and $\phi^{-1}(t)$ of the following system:

$$\begin{bmatrix} \dot{x}_1 \\ \dot{x}_2 \end{bmatrix} = \begin{bmatrix} 0 & 1 \\ -3 & -4 \end{bmatrix} \begin{bmatrix} x_1 \\ x_2 \end{bmatrix} + \begin{bmatrix} 0 \\ 1 \end{bmatrix} [u]$$

Solution

For the given system

$$A = \begin{bmatrix} 0 & 1 \\ -3 & -4 \end{bmatrix}$$

$$\therefore \quad [sI - A] = s\begin{bmatrix} 1 & 0 \\ 0 & 1 \end{bmatrix} - \begin{bmatrix} 0 & 1 \\ -3 & -4 \end{bmatrix} = \begin{bmatrix} s & -1 \\ 3 & s+4 \end{bmatrix}$$

$$\therefore \quad \phi(s) = (sI - A)^{-1} = \begin{bmatrix} s & -1 \\ 3 & s+4 \end{bmatrix}^{-1} = \frac{\begin{bmatrix} s+4 & 1 \\ -3 & s \end{bmatrix}}{s^2 + 4s + 3} = \frac{\begin{bmatrix} s+4 & 1 \\ -3 & s \end{bmatrix}}{(s+3)(s+1)}$$

$$\therefore \quad \phi(s) = \begin{bmatrix} \frac{s+4}{(s+3)(s+1)} & \frac{1}{(s+3)(s+1)} \\ \frac{-3}{(s+3)(s+1)} & \frac{s}{(s+3)(s+1)} \end{bmatrix} = \begin{bmatrix} \phi_{11}(s) & \phi_{12}(s) \\ \phi_{21}(s) & \phi_{22}(s) \end{bmatrix}$$

Taking inverse Laplace transform, we get

$$\phi(t) = \begin{bmatrix} \phi_{11}(t) & \phi_{12}(t) \\ \phi_{21}(t) & \phi_{22}(t) \end{bmatrix}$$

where

$$\phi_{11}(s) = \frac{s+4}{(s+3)(s+1)} = \frac{-\frac{1}{2}}{s+3} + \frac{\frac{3}{2}}{s+1}$$

$$\therefore \quad \phi_{11}(t) = -\frac{1}{2}e^{-3t} + \frac{3}{2}e^{-t}$$

$$\phi_{12}(s) = \frac{1}{(s+3)(s+1)} = \frac{1}{2}\left(\frac{1}{s+1} - \frac{1}{s+3}\right)$$

$$\therefore \quad \phi_{12}(t) = \frac{1}{2}(e^{-t} - e^{-3t})$$

$$\phi_{21}(s) = \frac{-3}{(s+3)(s+1)} = -\frac{3}{2}\left(\frac{1}{s+1} - \frac{1}{s+3}\right)$$

$$\therefore \quad \phi_{21}(s) = -\frac{3}{2}(e^{-t} - e^{-3t})$$

and

$$\phi_{22}(s) = \frac{s}{(s+3)(s+1)} = \frac{\frac{3}{2}}{s+3} + \frac{-\frac{1}{2}}{s+1}$$

$\therefore$
$$\phi_{22}(s) = \frac{3}{2}\left(e^{-3t} - \frac{1}{3}e^{-t}\right)$$

$\therefore$
$$\phi(t) = \begin{bmatrix} -\frac{1}{2}e^{-3t} + \frac{3}{2}e^{-t} & \frac{1}{2}(e^{-t} - e^{-3t}) \\ -\frac{3}{2}(e^{-t} - e^{-3t}) & \frac{3}{2}e^{-3t} - \frac{1}{2}e^{-t} \end{bmatrix}$$

$\therefore$
$$\phi^{-1}(t) = \phi(-t)$$

$\therefore$
$$\phi^{-1}(t) = \phi(-t) = \begin{bmatrix} -\frac{1}{2}e^{3t} + \frac{3}{2}e^{t} & \frac{1}{2}(e^{t} - e^{3t}) \\ -\frac{3}{2}(e^{t} - e^{3t}) & \frac{3}{2}e^{3t} - \frac{1}{2}e^{t} \end{bmatrix}$$

Example 17.28 Find the output response of the following system:

$$\begin{bmatrix} \dot{x}_1 \\ \dot{x}_2 \end{bmatrix} = \begin{bmatrix} 0 & 1 \\ -3 & -4 \end{bmatrix} \begin{bmatrix} x_1 \\ x_2 \end{bmatrix} + \begin{bmatrix} 0 \\ 1 \end{bmatrix} [u]$$

and output $y = [1 \;\; 0]\, x$, where $u(t)$ is the unit step input and $x_1(0) = 0$ and $x_2(0) = 0$.

Solution

In Example 17.11 it has already been obtained

$\therefore$
$$\phi(t) = \begin{bmatrix} -\frac{1}{2}e^{-3t} + \frac{3}{2}e^{-t} & \frac{1}{2}(e^{-t} - e^{-3t}) \\ -\frac{3}{2}(e^{-t} - e^{-3t}) & \frac{3}{2}e^{-3t} - \frac{1}{2}e^{-t} \end{bmatrix}$$

$\therefore$
$$x(t) = \phi(t)x(0) + \int_0^t e^{A(t-\tau)}Bu(\tau)d\tau$$

$\therefore$
$$\int_0^t e^{A(t-\tau)}Bu(\tau)dT = \int_0^t \phi(t-\tau)Bu(\tau)dT$$

Now,
$$\phi(t-\tau)Bu(\tau)dT = \begin{bmatrix} \frac{1}{2}e^{-(t-\tau)} & -\frac{1}{2}e^{-3(t-\tau)} \\ \frac{3}{2}e^{-3(t-\tau)} & -\frac{1}{2}e^{-(t-\tau)} \end{bmatrix}$$

$\therefore$
$$\int_0^t e^{A(t-\tau)}Bu(\tau)dT = \int_0^t \phi(t-\tau)Bu(\tau)dT$$

$$= \begin{bmatrix} \frac{1}{2}e^{-t}\int_0^t e^{\tau}d\tau - \frac{1}{2}e^{-3t}\int_0^t e^{3\tau}d\tau \\ \frac{3}{2}e^{-t}\int_0^t e^{3\tau}d\tau - \frac{1}{2}e^{-t}\int_0^t e^{\tau}d\tau \end{bmatrix} = \begin{bmatrix} \frac{1}{2}e^{-t}(e^{t}-1) - \frac{1}{6}e^{-3t}(e^{-3t}-1) \\ \frac{1}{2}e^{-3t}(e^{-3t}-1) - \frac{1}{2}e^{-t}(e^{-t}-1) \end{bmatrix}$$

$$= \begin{bmatrix} \frac{1}{3} - \frac{1}{2}e^{-t} + \frac{1}{6}e^{-3t} \\ \frac{1}{2}(e^{-t} - e^{-3t}) \end{bmatrix}$$

Therefore, the output response is given by

$$y(t) = [1 \quad 0]\begin{bmatrix} x_1(t) \\ x_2(t) \end{bmatrix} = [1 \quad 0]\begin{bmatrix} \frac{1}{3} - \frac{1}{2}e^{-t} + \frac{1}{6}e^{-3t} \\ \frac{1}{2}(e^{-t} - e^{-3t}) \end{bmatrix} = \frac{1}{3} - \frac{1}{2}e^{-t} + \frac{1}{6}e^{-3t}$$

Example 17.29 Find the controllability of the system described by the state equation

$$\begin{bmatrix} \dot{x}_1 \\ \dot{x}_2 \end{bmatrix} = \begin{bmatrix} 0 & 1 \\ -2 & 0 \end{bmatrix}\begin{bmatrix} x_1 \\ x_2 \end{bmatrix} + \begin{bmatrix} 0 \\ 3 \end{bmatrix}u$$

Solution

Here, $$A = \begin{bmatrix} 0 & 1 \\ -2 & 0 \end{bmatrix} \text{ and } B = \begin{bmatrix} 0 \\ 3 \end{bmatrix}$$

$\therefore$ $$S = [B \quad AB]$$

Now, $$AB = \begin{bmatrix} 0 & 1 \\ -2 & 0 \end{bmatrix}\begin{bmatrix} 0 \\ 3 \end{bmatrix} = \begin{bmatrix} 3 \\ 0 \end{bmatrix}$$

$\therefore$ $$S = [B \quad AB] = \begin{bmatrix} 0 & 3 \\ 3 & 0 \end{bmatrix}$$

Since the matrix is non-singular, the system is controllable.

Example 17.30 The dynamic equation of a system is represented by

$$\begin{bmatrix} \dot{x}_1 \\ \dot{x}_2 \end{bmatrix} = \begin{bmatrix} -3 & 0 \\ 0 & 2 \end{bmatrix}\begin{bmatrix} x_1 \\ x_2 \end{bmatrix} + \begin{bmatrix} 4 \\ 1 \end{bmatrix}u$$

$$y = [2 \quad 0]\begin{bmatrix} x_1 \\ x_2 \end{bmatrix}$$

Is the system observable?

Solution

$$A = \begin{bmatrix} -3 & 0 \\ 0 & -2 \end{bmatrix}, C = [2 \quad 0]$$

$\therefore$ $$CA = [2 \quad 0]\begin{bmatrix} -3 & 0 \\ 0 & -2 \end{bmatrix} = [-6 \quad 0]$$

$$\therefore \qquad V = \begin{bmatrix} C \\ CA \end{bmatrix} = \begin{bmatrix} 2 & 0 \\ -6 & 0 \end{bmatrix}$$

Since V is singular, the system is not observable.

Example 17.31 The state equation of LTIV system is given below:

$$\dot{x}(t) = Ax(t) + bu(t)$$

where

$$A = \begin{bmatrix} -3 & 0 & 0 \\ 0 & -1 & 1 \\ 0 & 0 & -1 \end{bmatrix} \text{ and } B = \begin{bmatrix} 0 \\ 1 \\ 0 \end{bmatrix}$$

Find the state transition matrix $\phi(t)$.

Solution

$$sI - A = s\begin{bmatrix} 1 & 0 & 0 \\ 0 & 1 & 0 \\ 0 & 0 & 1 \end{bmatrix} - \begin{bmatrix} -3 & 0 & 0 \\ 0 & -1 & 1 \\ 0 & 0 & -1 \end{bmatrix} = \begin{bmatrix} s+3 & 0 & 0 \\ 0 & s+1 & -1 \\ 0 & 0 & s+1 \end{bmatrix}$$

$$|sI - A| = (s+3)(s+1)^2$$

$$(sI - A)^{-1} = \frac{\text{Adjoint}(sI - A)}{|sI - A|}$$

$$= \begin{bmatrix} \frac{(s+1)^2}{(s+3)(s+1)^2} & 0 & 0 \\ 0 & \frac{(s+1)(s+3)}{(s+3)(s+1)^2} & \frac{(s+3)}{(s+3)(s+1)^2} \\ 0 & 0 & \frac{(s+1)(s+3)}{(s+3)(s+1)^2} \end{bmatrix}$$

$$\phi(s) = \begin{bmatrix} \frac{1}{(s+3)} & 0 & 0 \\ 0 & \frac{1}{(s+1)} & \frac{1}{(s+1)^2} \\ 0 & 0 & \frac{1}{(s+1)} \end{bmatrix}$$

Taking Laplace inverse transform, we get

$$\phi(t) = \begin{bmatrix} e^{-3t} & 0 & 0 \\ 0 & e^{-t} & te^{-t} \\ 0 & 0 & e^{-t} \end{bmatrix}$$

Example 17.32 The state equation of a linear time-invariant system is given below:

$$\begin{bmatrix} \dot{x}_1 \\ \dot{x}_2 \end{bmatrix} = \begin{bmatrix} -2 & 0 \\ 1 & -1 \end{bmatrix}\begin{bmatrix} x_1 \\ x_2 \end{bmatrix} + \begin{bmatrix} 0 \\ 1 \end{bmatrix} u$$

where $u > 0$.
Determine the following:

(i) The state transition matrix.
(ii) $x_1(1)$ under zero initial conditions and a step input.
(iii) Controllability and observability of the system.

Solution

(i)

$$sI - A = s\begin{bmatrix} 1 & 0 \\ 0 & 1 \end{bmatrix} - \begin{bmatrix} -2 & 0 \\ 1 & -1 \end{bmatrix} = \begin{bmatrix} s+2 & 0 \\ -1 & s+1 \end{bmatrix}$$

$$|sI - A| = \begin{vmatrix} s+2 & 0 \\ -1 & s+1 \end{vmatrix} = (s+1)(s+2)$$

$$\phi(s) = (sI - A)^{-1} = \frac{sI - A}{|sI - A|} = \begin{bmatrix} \dfrac{s+1}{(s+1)(s+2)} & 0 \\ \dfrac{1}{(s+1)(s+2)} & \dfrac{s+2}{(s+1)(s+2)} \end{bmatrix}$$

$$= \begin{bmatrix} \dfrac{1}{(s+2)} & 0 \\ \dfrac{1}{(s+1)(s+2)} & \dfrac{1}{(s+1)} \end{bmatrix} = \begin{bmatrix} \dfrac{1}{s+2} & 0 \\ \dfrac{1}{s+1} - \dfrac{1}{s+2} & \dfrac{1}{s+1} \end{bmatrix}$$

Taking Laplace inverse transform, we will get the state transition matrix.

$$\phi(t) = \begin{bmatrix} e^{-t} & 0 \\ e^{-t} - e^{-t} & e^{-t} \end{bmatrix}$$

(ii) The completed solution of the state equation at $t > 0$ due to application of unit step function at $t > 0$ is

$$x(t) = e^{At}x(0) + \int_0^t e^{A(t-\tau)}Bu(\tau)d\tau$$

$$= \begin{bmatrix} e^{-2t} & 0 \\ e^{-t} - e^{-2t} & e^{-t} \end{bmatrix}\begin{bmatrix} x_1(0) \\ x_2(0) \end{bmatrix} + \int_0^t \begin{bmatrix} e^{-2(t-\tau)} & 0 \\ e^{-(t-\tau)} & -e^{-2(t-\tau)} \; e^{-(t-\tau)} \end{bmatrix}\begin{bmatrix} 0 \\ 1 \end{bmatrix} d\tau$$

$$= 0 + \int_0^t e^{-(t-\tau)} d\tau = (1 - e^{-t})$$

(iii)

$$A = \begin{bmatrix} -2 & 0 \\ 1 & -1 \end{bmatrix} \text{ and } B = \begin{bmatrix} 0 \\ 1 \end{bmatrix}$$

$$\therefore \quad AB = \begin{bmatrix} -2 & 0 \\ 1 & -1 \end{bmatrix} \begin{bmatrix} 0 \\ 1 \end{bmatrix} = \begin{bmatrix} 0 \\ -1 \end{bmatrix}$$

$$\therefore \quad [B\ AB] = \begin{bmatrix} 0 & 0 \\ 1 & -1 \end{bmatrix}$$

$$|B\ AB| = \begin{vmatrix} 0 & 0 \\ 1 & -1 \end{vmatrix} = 0$$

Therefore, the system is uncontrollable.

MATLAB Programs

Program 1 Find the state space model of the following transfer function using MATLAB $\frac{Y(s)}{U(s)} = \frac{s}{(s+5)(s^2+5s+8)}$. Also obtain the phase variable form.

Solution

```
>> num = [1 0]
num =
   1   0
>> q1 = [ 1 5]
q1 =
   1   5
>> q2 = [ 1 5 8]
q2 =
   1   5   8
>> den = conv(q1,q2)
den =
   1  10  33  40
>> [A B C D] =  tf2ss (num,den)
A =
  -10  -33  -40
    1    0    0
    0    1    0
B =
   1
   0
   0
C =
   0   1   0
D =
   0
>> P = [1 0 0;0 1 0;0 0 1]
P =
   1   0   0
   0   1   0
   0   0   1
```

```
>> Ap=inv(P)*A*P
Ap =
  -10  -33  -40
    1    0    0
    0    1    0
>> Bp=inv(P)*B
Bp =
    1
    0
    0
>> Cp=C*P
Cp =
    0    1    0
>> Dp=D
Dp =
    0
>>
```

Program 2 Obtain the transfer function of the system defined by the following state space equation:

$$\begin{bmatrix} \dot{x}_1 \\ \dot{x}_2 \\ \dot{x}_3 \end{bmatrix} = \begin{bmatrix} 0 & 1 & 0 \\ 0 & 0 & 1 \\ -6 & -21 & -4 \end{bmatrix} \begin{bmatrix} x_1 \\ x_2 \\ x_3 \end{bmatrix} + \begin{bmatrix} 0 \\ 22 \\ -129 \end{bmatrix} u$$

$$y = [1 \quad 0 \quad 0] \begin{bmatrix} x_1 \\ x_2 \\ x_3 \end{bmatrix}$$

Solution

```
>> A = [0 1 0; 0 0 1; -6 -21 -4];
>> B = [ 0 ; 22 ; -129];
>> C = [ 1 0 0];
>> D = [0];
% This command produces the transfer function
>> [num,den] = ss2tf(A,B,C,D)
num =
      0   0.0000  22.0000  -41.0000
den =
   1.0000   4.0000  21.0000   6.0000
% The other command
>> [num,den] = ss2tf(A,B,C,D,1)
num =
      0   0.0000  22.0000  -41.0000
den =
   1.0000   4.0000  21.0000   6.0000
>>
```

Program 3 Obtain the unit step response for the following system:

$$\begin{bmatrix} \dot{x}_1 \\ \dot{x}_2 \end{bmatrix} = \begin{bmatrix} -1 & -2 \\ 2.6 & 0 \end{bmatrix} \begin{bmatrix} x_1 \\ x_2 \end{bmatrix} + \begin{bmatrix} 1 & 1 \\ 1 & 0 \end{bmatrix} \begin{bmatrix} u_1 \\ u_2 \end{bmatrix}$$

$$\begin{bmatrix} y_1 \\ y_2 \end{bmatrix} = \begin{bmatrix} 1.2 & 0 \\ 0 & 1 \end{bmatrix} \begin{bmatrix} x_1 \\ x_2 \end{bmatrix} + \begin{bmatrix} 0 & 0 \\ 0 & 0 \end{bmatrix} \begin{bmatrix} u_1 \\ u_2 \end{bmatrix}$$

Solution

```
>> A = [−1 −2;2.6 0];
>> B = [1  1; 1 0];
>> C = [1.2 0;0 1];
>> D = [0 0;0 0];
>> step(A,B,C,D)
>>
```

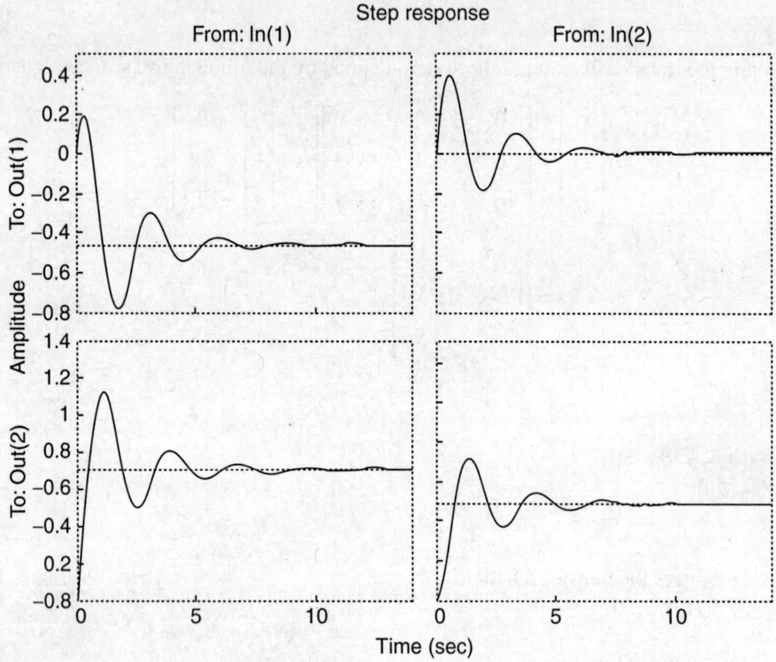

Fig. P3

This programme plots two step response curve:

```
>> A = [−1 −2;2.6 0];
>> B = [1  1; 1 0];
>> C = [1.2 0;0 1];
>> D = [0 0;0 0];
>> step(A,B,C,D,1)
>> step(A,B,C,D,2)
>>
```

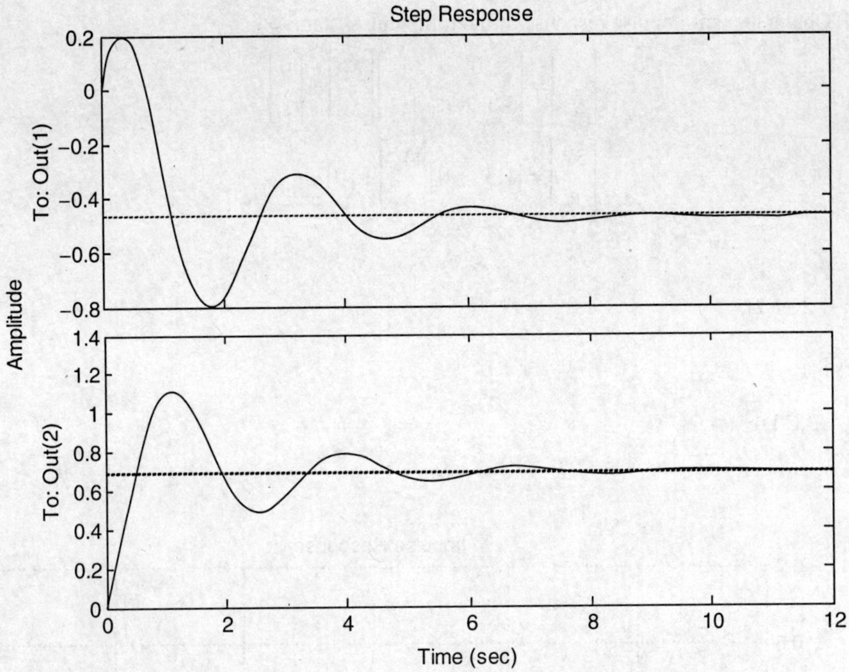
Step Response
Amplitude
To: Out(1)
To: Out(2)
0.2
0
−0.2
−0.4
−0.6
−0.8
1.4
1.2
1
0.8
0.6
0.4
0.2
0
0
2
4
6
8
10
12
Time (sec)

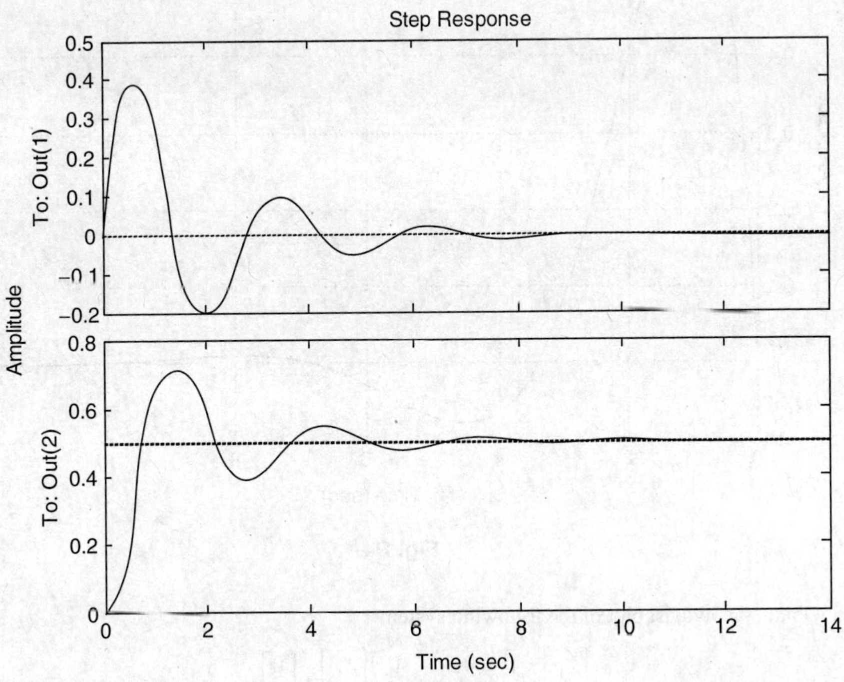
Step Response
Amplitude
To: Out(1)
To: Out(2)
0.5
0.4
0.3
0.2
0.1
0
−0.1
−0.2
0.8
0.6
0.4
0.2
0
0
2
4
6
8
10
12
14
Time (sec)

Fig. P3(a)

Program 4 Obtain the unit impulse response of the following system:

$$\begin{bmatrix} \dot{x}_1 \\ \dot{x}_2 \end{bmatrix} = \begin{bmatrix} 0 & 1 \\ -1.2 & -1.2 \end{bmatrix} \begin{bmatrix} x_1 \\ x_2 \end{bmatrix} + \begin{bmatrix} 0 \\ 1 \end{bmatrix} u$$

$$\begin{bmatrix} y_1 \\ y_2 \end{bmatrix} = \begin{bmatrix} 1.3 & 0 \end{bmatrix} \begin{bmatrix} x_1 \\ x_2 \end{bmatrix} + [0]u$$

Solution

```
>> A = [0 1;–1.2 –1.2];
>> B = [0 ; 1];
>> C = [1.3 0];
>> D=[0];
>> impulse(A,B,C,D);
>>grid
>>
```

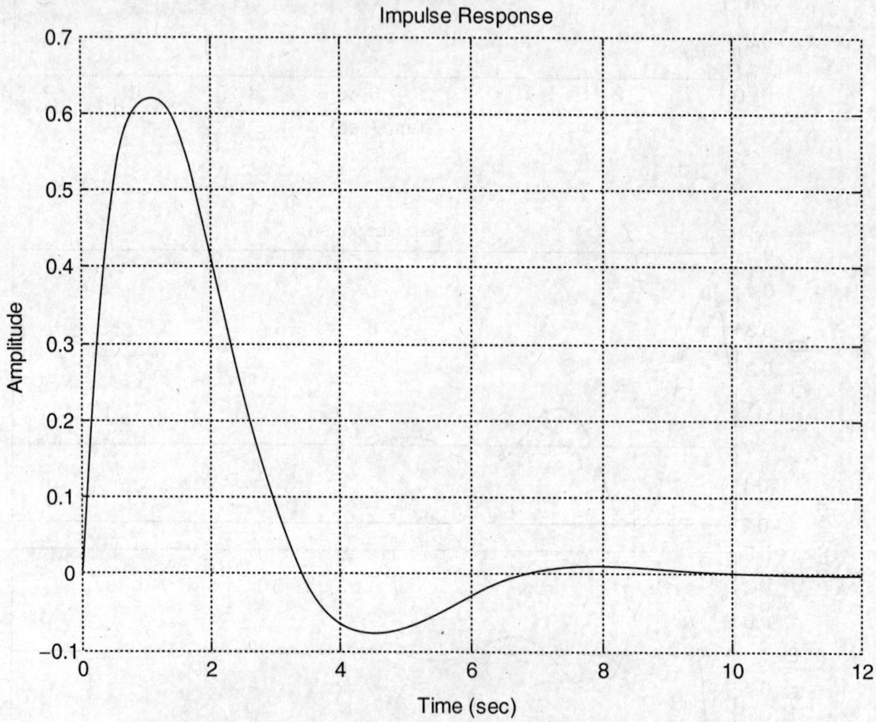

Fig. P4

Program 5 Obtain the nyquist plot of the following system:

$$\begin{bmatrix} \dot{x}_1 \\ \dot{x}_2 \end{bmatrix} = \begin{bmatrix} 0 & 1 \\ -1.2 & -1.2 \end{bmatrix} \begin{bmatrix} x_1 \\ x_2 \end{bmatrix} + \begin{bmatrix} 0 \\ 1 \end{bmatrix} u$$

$$\begin{bmatrix} y_1 \\ y_2 \end{bmatrix} = [1.3 \quad 0] \begin{bmatrix} x_1 \\ x_2 \end{bmatrix} + [0]u$$

Solution

```
>> A = [0 1;–1.2 –1.2];
>> B = [0 ; 1];
>> C = [1.3 0];
>> D=[0];
>> nyquist(A,B,C,D);
>> grid
>>
```

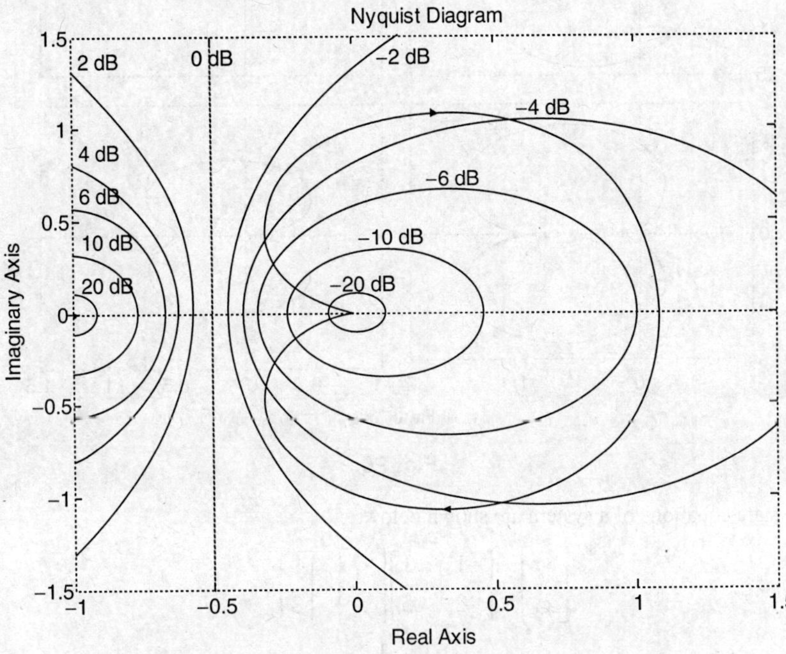

Fig. P5

Program 6 Obtain nyquist plot for the following system:

$$\begin{bmatrix} \dot{x}_1 \\ \dot{x}_2 \end{bmatrix} = \begin{bmatrix} -1 & -2 \\ 2.6 & 0 \end{bmatrix} \begin{bmatrix} x_1 \\ x_2 \end{bmatrix} + \begin{bmatrix} 1 & 1 \\ 1 & 0 \end{bmatrix} \begin{bmatrix} u_1 \\ u_2 \end{bmatrix}$$

$$\begin{bmatrix} y_1 \\ y_2 \end{bmatrix} = \begin{bmatrix} 1.2 & 0 \\ 0 & 1 \end{bmatrix} \begin{bmatrix} x_1 \\ x_2 \end{bmatrix} + \begin{bmatrix} 0 & 0 \\ 0 & 0 \end{bmatrix} \begin{bmatrix} u_1 \\ u_2 \end{bmatrix}$$

Solution

```
>> A = [–1 –2;2.6 0];
>> B = [1  1; 1 0];
>> C = [1.2 0;0 1];
```

```
>> D = [0 0;0 0];
>>nyquist(A,B,C,D);
>>
```

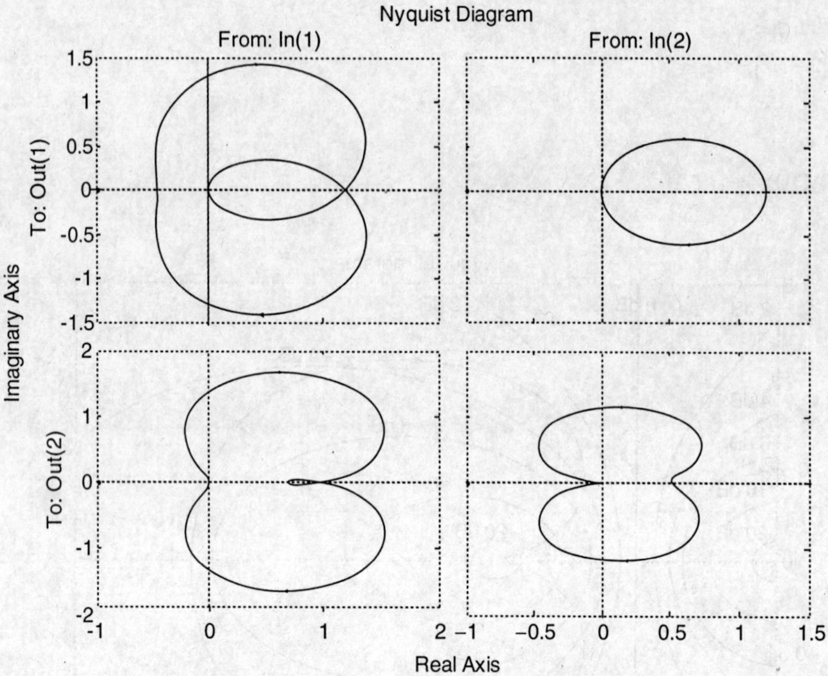

Fig. P6

Program 7 The state equations of a system are shown below:

$$\begin{bmatrix} \dot{x}_1 \\ \dot{x}_2 \end{bmatrix} = \begin{bmatrix} -1 & 3 \\ -2 & 5 \end{bmatrix} \begin{bmatrix} x_1 \\ x_2 \end{bmatrix} + \begin{bmatrix} 1 \\ 3 \end{bmatrix} u$$

$$y = \begin{bmatrix} 1 & 5 \end{bmatrix} \begin{bmatrix} x_1 \\ x_2 \end{bmatrix} + [0]\, u$$

Obtain the transfer function of the system. Check the controllability and observability of the system.

Solution

```
>> A = [–1 3;–2 5]
A =
   –1   3
   –2   5
>> B = [1 ; 3]
B =
    1
    3
>> C = [1 5]
```

```
C =
   1   5
>> D = [0]
D =
   0
>> [num,den]=ss2tf(A,B,C,D)
num =
       0  16.0000   9.0000
den =
   1.0000  -4.0000   1.0000
>> sys = tf(num,den)
Transfer function:
 16 s + 9
-------------
s^2 - 4 s + 1
>> P=ctrb(A,B)
P =
   1   8
   3  13
>> rank(P)
ans =
   2
>> %the system is controllable
>> Q = obsv(A,C)
Q =
    1   5
  -11  28
>> rank(Q)
ans =
   2
>> %the system is observable
>>
```

SHORT QUESTIONS WITH ANSWERS

1. What do you mean by modern control theory?

   Ans. Modern control theory takes care of initial conditions automatically and it is also possible to analyze time-variant or time-invariant, linear or non-linear, single or multiple input–output systems.

2. Define the following terms:
   (i) state, (ii) state variable, (iii) state vector, (iv) state space, (v) state equation, (vi) state transition matrix.

   Ans. (i) The state of a dynamic system is the smallest set of variables and the knowledge of these variables at $t = t_0$ together with inputs for $t \geq t_0$ completely determines the behaviour of the system at $t \geq t_0$.
   (ii) The smallest set of variables that determine the state of the system are known as state variables.
   (iii) The n state variables that completely describe the behaviour of a given system are said to be n components of a vector.
   (iv) The n-dimensional space whose coordinate axes consist of the x_1 axis, x_2 axis, …, x_n axis is known as a state space.

(v) The equation, which describe the relationships between the derivations of the state variables as a function of the state variable, input and parameter time, are called the state equations.

(vi) The matrix operator, which determines the transition of any initial state $x(0)$ at $t = 0$ to a state $x(t)$ at time t, is known as state transition matrix.

3. What do you mean by eigenvalue?

Ans. The roots of the characteristic equation $|\lambda_i I - A| = 0$ are called the eigenvalues of a matrix $A(n \times n)$.

4. Write the expression for transfer function, resolvent matrix and state transition matrix.

Ans. Transfer function $= \dfrac{Y(s)}{U(s)} = C[sI - A]^{-1}B + D$

Resolvent matrix $\phi(s) = (sI - A)^{-1} = \dfrac{1}{s}\left(I - \dfrac{A}{s}\right)^{-1} = \dfrac{1}{s}\left[I + \dfrac{A}{s} + \dfrac{A^2}{s^2} + \ldots\right]$

State Transition Matrix

$$A = e^{At}$$

5. Write the properties of state transition matrix.

Ans.

(i) $\phi(0) = e^{A0} = I$

(ii) $\phi(t) = e^{At} = (e^{-At})^{-1} = [\phi(-t)]^{-1}$

(iii) $\phi^{-1}(t) = \phi(-t)$

(iv) $\phi(t_1 + t_2) = e^{A(t_1+t_2)} = e^{At_1}e^{At_2} = \phi(t_1)\phi(t_2) = e^{At_2}e^{At_1} = \phi(t_2)\phi(t_1)$

(v) $[\phi(t)]^n = (e^{At})^n = e^{Ant} = \phi(nt)$

(vi) $\phi(t_1 - t_2)\phi(t_2 - t_0) = e^{A(t_1-t_2)}e^{A(t_2-t_0)}$

$$= e^{[t_1-t_2+t_2-t_0]} = e^{A(t_1-t_0)} = \phi(t_1 - t_0)$$

6. What do you mean by (i) controllability and (ii) observability.

Ans. (i) The state $x(t)$ will be controllable at $t = t_0$ when there exists a piecewise continuous input $u(t)$, which will drive the state to any final state $x(t)$ for a finite time interval $(t = t_0) \geq 0$. The system is called controllable if every state of the system is controllable in a finite time interval.

(ii) If any state of a system cannot be observed form measurements, it is called unobservable.

EXERCISES

Questions For Practice

1. For the circuit shown in Figure 1 choose a set of variables and derive the voltage/current equations necessary for solving the circuit in terms of state variables.

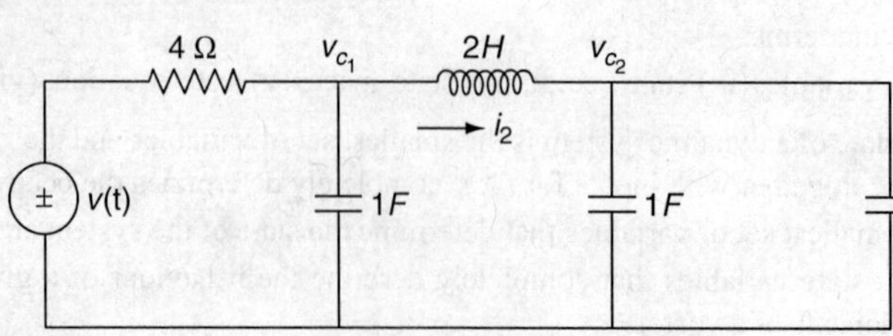

Figure 1

Ans.

$$\begin{bmatrix} \dot{v}_{c_1} \\ \dot{v}_{c_1} \\ \dot{v}_{c_1} \end{bmatrix} = \begin{bmatrix} -\frac{1}{4} & 0 & -1 \\ 0 & 0 & \frac{1}{2} \\ \frac{1}{5} & -\frac{1}{5} & 0 \end{bmatrix} \begin{bmatrix} v_{c_1} \\ v_{c_1} \\ v_{c_1} \end{bmatrix} + \begin{bmatrix} \frac{1}{4} \\ 0 \\ 0 \end{bmatrix} [u]$$

2. Obtain the matrices A, B, C and D for Figure 2.

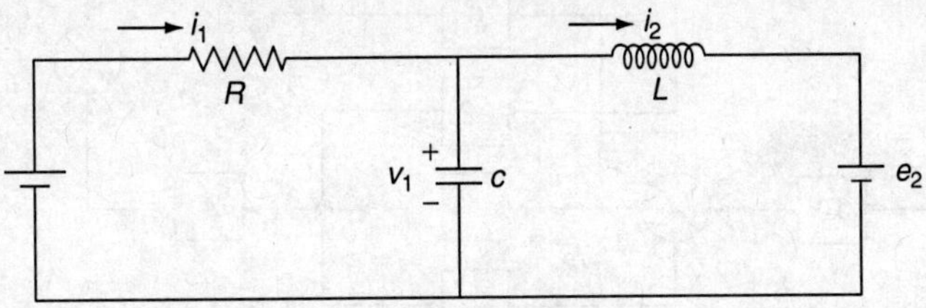

Figure 2

Ans.

$$A = \begin{bmatrix} -\frac{1}{RC} & -\frac{1}{C} \\ \frac{1}{L} & 0 \end{bmatrix} B = \begin{bmatrix} \frac{1}{RC} & -\frac{1}{C} \\ 0 & \frac{1}{L} \end{bmatrix} C = \begin{bmatrix} 1 & 0 \\ 1 & 0 \end{bmatrix} D = \begin{bmatrix} 0 & 0 \\ 0 & -1 \end{bmatrix}$$

3. The transfer function of a system is given by

$$\frac{C(s)}{R(s)} = \frac{s^2 + 3s + 3}{s^3 + 6s^2 + 11s + 6}.$$

Obtain the state model in phase variable form.

$$\dot{x} = \begin{bmatrix} 0 & 1 & 0 \\ 0 & 0 & 1 \\ -6 & 11 & -6 \end{bmatrix} x + \begin{bmatrix} 0 \\ 0 \\ 1 \end{bmatrix} u \text{ and } y = [3 \;\; 3 \;\; 1]x + [0]u$$ **Ans.**

4. A feedback system has a closed-loop transfer function $\frac{C(s)}{R(s)} = \frac{10(s+4)}{s(s+1)(s+3)}$. Construct the state model using parallel decomposition.

$$\dot{x} = \begin{bmatrix} 0 & 0 & 0 \\ 0 & -1 & 0 \\ 0 & 0 & -3 \end{bmatrix} x + \begin{bmatrix} \frac{40}{3} \\ -15 \\ \frac{-5}{3} \end{bmatrix} u \text{ and } y = [1 \;\; 1 \;\; 1]x + [0]u$$

Ans.

5. Find the state model and output equation of the system shown in Figure 3. Obtain its transfer function.

$$\dot{x}_1 = -10x_1 - 2x_2 + 4r \text{ and } \dot{x}_2 = 6x_1 - 2x_2 + 10r$$

$$\text{Transfer function} = \frac{44s^2 + 582s + 988}{(s+2)(s^2+12s+26)}$$

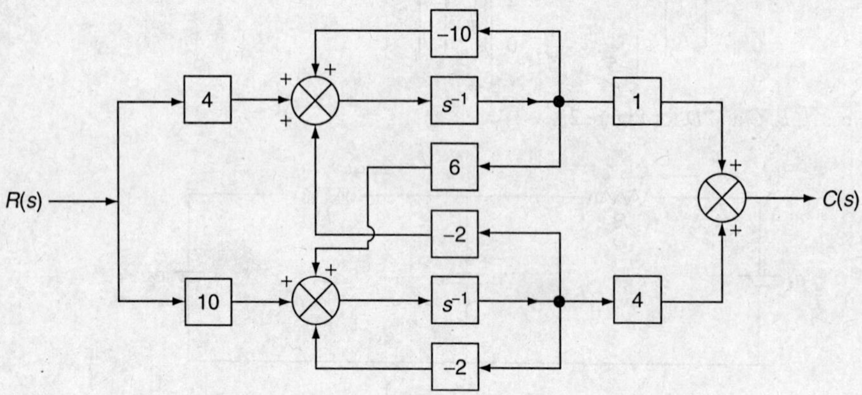

Figure 3

Ans.

$$\dot{x}_1 = -10x_1 - 2x_2 + 4r \text{ and } \dot{x}_2 = 6x_1 - 2x_2 + 10r$$

$$\text{Transfer function} = \frac{44s^2 + 582s + 988}{(s+2)(s^2+12s+26)}$$

6. The state equation of a linear time-invariant system is given by

$$\dot{x} = \begin{bmatrix} 0 & 0 \\ -1 & 1 \end{bmatrix} \begin{bmatrix} x_1 \\ x_2 \end{bmatrix} + \begin{bmatrix} 1 \\ 1 \end{bmatrix} u$$

Find the state vector $x(t)$ for $t \geq 0$ where $u(t)$ is a unit step input.

Ans.

$$x(t) = \begin{bmatrix} -t\{1 - t^2 e^t (2-t)\} \\ -te^{-1}\{1 - te^t (2-t)\} \end{bmatrix}$$

7. A system is described by

$$\frac{d^3y}{dt^3} + 6\frac{d^2y}{dt^2} + 11\frac{dy}{dt} + 2y = 6u$$

where y is the output and u is the input of the system. Obtain the state space representation of the system.

Ans.

$$\dot{x} = \begin{bmatrix} 0 & 1 & 0 \\ 0 & 0 & 1 \\ -6 & -11 & -22 \end{bmatrix} x + \begin{bmatrix} 0 \\ 0 \\ 6 \end{bmatrix} u \text{ and } y = [1 \quad 0 \quad 0]x + [0]u$$

8. Obtain the state transition matrix for the system given by $\dot{x} = Ax$, where

$$A = \begin{bmatrix} 0 & 1 & 0 \\ 3 & 0 & 2 \\ -12 & -7 & -6 \end{bmatrix}$$

Ans.

$$\begin{bmatrix} 4.5e^{-t} - 6e^{-2t} + 2.5e^{-3t} & 2.5e^{-t} - 4e^{-2t} + 1.5e^{-3t} & e^{-t} - 2e^{-2t} + e^{-3t} \\ -4.5e^{-t} + 12e^{-2t} - 7.5e^{-3t} & -2.5e^{-t} + 8e^{-2t} - 4.5e^{-3t} & -e^{-t} + 4e^{-2t} - 3e^{-3t} \\ -4.5e^{-t} - 3e^{-2t} + 7.5e^{-3t} & -2.5e^{-t} - 2e^{-2t} + 4.5e^{-3t} & -e^{-t} - e^{-2t} + 3e^{-3t} \end{bmatrix}$$

TEST YOURSELF

1. The state equation of a linear system is given by $\dot{x} = Ax + Bu$. The state transition matrix of the system is

(a) $\begin{bmatrix} e^{2t} & 0 \\ 0 & e^{2t} \end{bmatrix}$ (b) $\begin{bmatrix} e^{-2t} & 0 \\ 0 & e^{-2t} \end{bmatrix}$ (c) $\begin{bmatrix} \sin 2t & \cos 2t \\ -\cos 2t & \sin 2t \end{bmatrix}$ (d) $\begin{bmatrix} \cos 2t & \sin 2t \\ -\sin 2t & \cos 2t \end{bmatrix}$

2. The state and output equations of a system are as under:

State equation

$$\begin{bmatrix} \dot{x}_1(t) \\ \dot{x}_1(t) \end{bmatrix} = \begin{bmatrix} 0 & 1 \\ -1 & -2 \end{bmatrix} \begin{bmatrix} x_1(t) \\ x_1(t) \end{bmatrix} + \begin{bmatrix} 0 \\ 1 \end{bmatrix} u(t)$$

Output equation

$$c(t) = [1 \quad 1] \begin{bmatrix} x_1(t) \\ x_1(t) \end{bmatrix}$$

The system is

(a) neither state controllable nor output controllable.

(b) state controllable but not output controllable.

(c) output controllable but not state controllable.

(d) both sate controllable and output controllable.

3. The value of A matrix in $X = Ax$ for the system described by the differential equation $\ddot{y} + 2\dot{y} + 3y = 0$ is

(a) $\begin{bmatrix} 1 & 0 \\ -2 & -1 \end{bmatrix}$ (b) $\begin{bmatrix} 1 & 0 \\ -1 & -2 \end{bmatrix}$ (c) $\begin{bmatrix} 0 & 1 \\ -2 & -1 \end{bmatrix}$ (d) $\begin{bmatrix} 0 & 1 \\ -3 & -2 \end{bmatrix}$

4. The minimum number of states necessary to describe the network shown in Figure T1 in a state variable form is

(a) 2 (b) 3

(c) 4 (d) 6

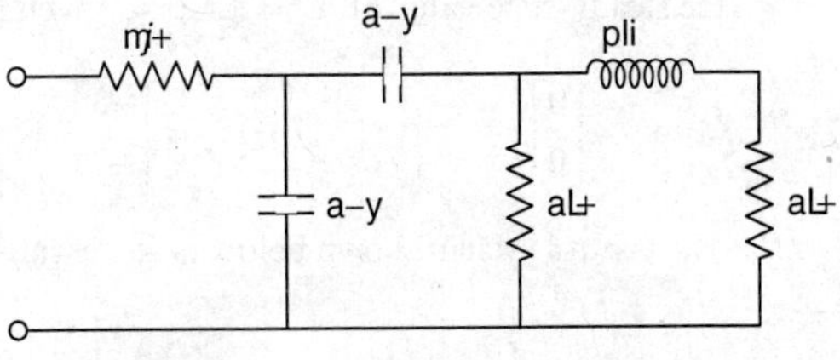

Figure T1

5. The State diagram of a system is shown in Figure T2. The system is

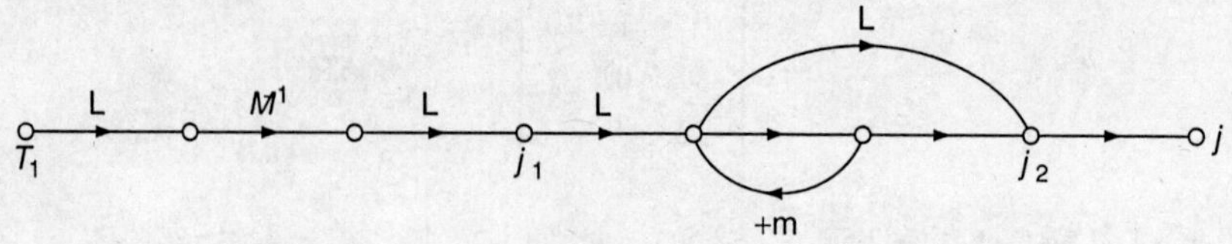

Figure T2

(a) controllable and observable.

(b) controllable but not observable.

(c) observable but not controllable.

(d) neither controllable nor observable.

6. For the state equations

$$\dot{x}(t) = Px(t) + Qu(t)$$

$$y(t) = Rx(t) + Su(t), \text{ match the following}$$

List I	List II
a. P	(1) $n \times p$
b. Q	(2) $q \times n$
c. R	(3) $n \times n$
d. S	(4) $V \times p$

7. Consider the state transition matrix

$$Q(s) = \begin{bmatrix} \dfrac{s+6}{s^2+6s+5} & \dfrac{-1}{s^2+6s+5} \\ \dfrac{-5}{s^2+6s+5} & \dfrac{s}{s^2+6s+5} \end{bmatrix}$$

The eigenvalues of the system are

(a) 0 and −6. (b) −1 and −5. (c) 1 and −5. (d) 0 and 6.

8. Given the homogenous state space equation.

$$x = \begin{bmatrix} -3 & 1 \\ -2 & -1 \end{bmatrix}$$

the steady-state value of $x_{ss} = \underset{t\to\infty}{\text{Lt}}\, x(t)$, given the initial state value of $x(0) = [10 - 10]^T$, is

(a) $x_{ss} = \begin{bmatrix} 0 \\ 0 \end{bmatrix}$ (b) $x_{ss} = \begin{bmatrix} -3 \\ -2 \end{bmatrix}$ (c) $x_{ss} = \begin{bmatrix} -10 \\ 10 \end{bmatrix}$ (d) $x_{ss} = \begin{bmatrix} \infty \\ \infty \end{bmatrix}$

9. The control system shown below is represented by the equation

$$\begin{bmatrix} y_1 \\ y_2 \end{bmatrix} = [G] \begin{bmatrix} u_1 \\ u_2 \end{bmatrix}$$

The matrix G of the system shown in Figure T3 is

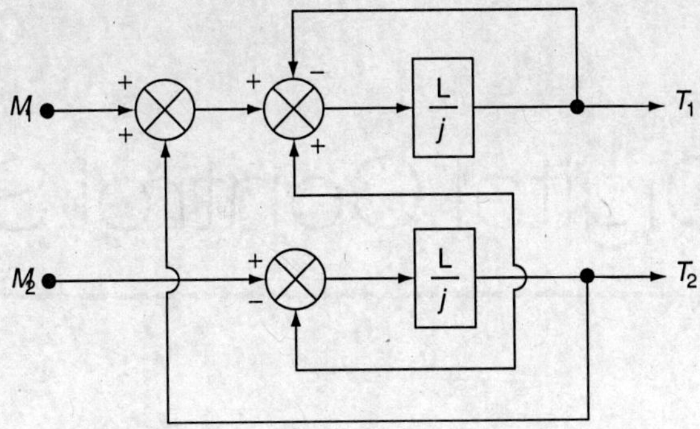

Figure T3

(a) $\begin{bmatrix} \frac{1}{2} & -\frac{1}{s^2} \\ 0 & \frac{1}{s} \end{bmatrix}$ (b) $\begin{bmatrix} \frac{1}{s} & \frac{1}{s^2} \\ 0 & -\frac{1}{s} \end{bmatrix}$ (c) $\begin{bmatrix} \frac{1}{s+1} & -\frac{1}{(s+1)^2} \\ 0 & \frac{1}{s+1} \end{bmatrix}$ (d) $\begin{bmatrix} \frac{1}{s+1} & \frac{1}{(s+2)^2} \\ 0. & -\frac{1}{s+1} \end{bmatrix}$

ANSWERS

1 (d) 2 (b) 3 (d) 4 (b) 5 (a) 6 (a) → (3) (b) → (1) (c) → (4) (d) → (2)

7 (b) 8 (c) 9 (d)

Chapter **18**

Digital Control Systems

We have already discussed control systems in which signals were analog in nature. Due to advancement of digital computers, their use as a compensator device has grown dramatically. In digital or discrete control systems the signals are discrete in nature. *Z* transform has a vital role in digital control systems. The aim of this chapter is to introduce *Z* transform and the digital control systems.

18.1 Z Transform

The *Z* transform of a discrete time signal $x(n)$ is generally defined as

$$\text{ZT}[x(n)] = \sum_{n=-\infty}^{+\infty} x(n)Z^{-n}$$

For a causal sequence, $x(n) = 0$, for $n < 0$. Hence

$$X(Z) = \sum_{n=0}^{\infty} x(n)Z^{-n}$$

18.2 Z Transform of Some Common Sequence

(a) The unit impulse sequence $\delta(n)$

$$\text{ZT}\,[\delta(n)] = \sum_{n=0}^{\infty} \delta(n)Z^{-n} = Z^{-0} = 1$$

(b) The unit step sequence $u(n)$

$$\text{ZT}\,[u(n)] = \sum_{n=0}^{\infty} 1 \times Z^{-n}$$

Using the fact that for a geometric series

$$\sum_{n=0}^{\infty} x^n = \frac{1}{1-x} |x| < 1 \text{ for convergence}$$

then

$$\text{ZT}\,[u(n)] = \sum_{n=0}^{\infty} Z^{-n} = \sum_{n=0}^{\infty}\left(\frac{1}{Z}\right)^n$$

$$= 1 - \frac{1}{Z} = \frac{Z}{Z-1} \quad \left|\frac{1}{Z}\right| < 1$$

(c) The sample exponential sequence $e^{-an}u(n)$

$$\text{ZT}[e^{-an}u(n)] = \sum_{n=0}^{\infty} e^{-an}Z^{-an} = \sum_{n=0}^{\infty}\left(\frac{1}{Ze^{an}}\right)^n = \frac{1}{1-\dfrac{1}{Ze^{an}}}$$

$$= \frac{Z}{Z-e^{-an}}$$

(d) The sampled sinusoidal sequence sin $(bn)u(n)$

$$\text{ZT}[\sin(bn)u(n)] = \sum_{n=0}^{\infty}\left(\frac{e^{jbn}-e^{-jbn}}{2j}\right)Z^{-n}$$

$$= \frac{1}{2j}\sum_{n=0}^{\infty} e^{jbn}Z^{-n} - \frac{1}{2j}\sum_{n=0}^{\infty} e^{jbn}Z^{-n}$$

$$= \frac{\dfrac{Z}{2j}}{Z-e^{jb}} - \frac{\dfrac{Z}{2j}}{Z-e^{-jb}}$$

$$= \frac{\left(\dfrac{Z}{2j}\right)(Z-e^{-jb}-Z+e^{jb})}{Z^2-(e^{jb}+e^{-jb})Z+1}$$

$$= \frac{Z\left(\dfrac{e^{jb}-e^{jb}}{2j}\right)}{Z^2-2\left(\dfrac{e^{jb}+e^{-jb}}{2}\right)Z+1} = \frac{Z\sin b}{Z^2-2Z\cos b+1}$$

18.3 Properties of *Z* Transform

18.3.1 Linearity

Consider two sequence $x(n)$ and $y(n)$ with Z transforms $X(Z)$ and $Y(Z)$, respectively, i.e.,

$$\text{ZT}[x(n)] = X(Z)$$

and

$$\text{ZT}[y(n)] = Y(Z)$$

Then

$$\text{ZT}[ax(n)+by(n)] = aX(Z)+bY(Z)$$

18.3.2 Shift of a Sequence

Let $x(n)$ be a casual sequence whose Z transform is $X(Z)$. Then

$$ZT[x(n-1)] = \sum_{n=0}^{\infty} x(n-1)^{-n} = \sum_{n=-1}^{\infty} x(n)Z^{-(n+1)}$$

$$= Z^{-1}X(Z)$$

and

$$\text{ZT}[x(n-2)] = Z^{-2}X(Z)$$

Similarly,

$$ZT[x(n-x)] = Z^{-k}X(Z)$$

18.3.3 Differentiation of *X*(*Z*)

Let $x(n)$ be a causal sequence whose Z transform is $X(Z)$. Then

$$X(Z) = \sum_{n=0}^{\infty} x(n)Z^{-n}$$

$$\frac{dX(Z)}{dZ} = \sum_{n=0}^{\infty} x(n)\frac{d}{dZ}(Z^{-n}) = -Z^{-1}\sum_{n=0}^{\infty} x(n)(-n)Z^{-n}$$

$$= -Z^{-1}\sum_{n=0}^{\infty}[n\,x(n)]Z^{-n}$$

i.e.,
$$\sum_{n=0}^{\infty}[n\,x(n)]Z^{-n} = -\frac{ZdX(z)}{dZ}$$

Thus,
$$ZT[n\,x(n)] = -\frac{ZdX(z)}{dZ}$$

18.3.4 Initial Value Theorem

If $x(Z)$ is the Z transform of $x(n)$, then

$$x(0) = \underset{z\to 0}{\text{LT}}\,X(Z)$$

18.3.5 Final Value Theorem

If $X(Z)$ is the Z transform of $x(n)$, then

$$\underset{n\to\infty}{\text{Lt}}\, x(n) = \underset{Z\to 1}{\text{Lt}}\left[\frac{(Z-1)}{Z}X(Z)\right]$$

provided that $\underset{n\to\infty}{\text{LT}}\, x(n)$ exists and is finite.

Example 18.1 Find Z transforms of the following sequences:

(a) $x(n) = [(-0.5)^n - 4(0.2)^n]u(n)$

(b) $x(n) = \begin{cases} (1)^n\, n = 3,4,5,\ldots \\ 0 \quad \text{otherwise} \end{cases}$

Solution

(a)
$$X(Z) = \sum_{n=0}^{\infty}[(-0.5)^n - 4(0.2^n)]Z^{-n}$$

$$= \sum_{n=0}^{\infty}\left(-\frac{0.5}{Z}\right) - \sum_{n=0}^{\infty}\left(\frac{0.2}{Z}\right)^n$$

$$= \frac{1}{1+\dfrac{0.5}{Z}} - 4\left(\frac{1}{1-\dfrac{0.2}{Z}}\right) = \frac{Z}{Z+0.5} - \frac{4Z}{Z-0.2}$$

$$X(Z) = \sum_{n=3}^{\infty} (-1)^n Z^{-n}$$

$$= \frac{\left(-\dfrac{1}{Z}\right)^3}{1+\dfrac{1}{Z}} = \frac{Z\left(\dfrac{-1}{Z^3}\right)}{Z^3+1} = \frac{-Z^{-2}}{(Z+1)}$$

18.4 Inverse *Z* Transform

The sequence $x(n)$ corresponding to a rational Z transform $X(Z)$ may be obtained by the following methods.

(a) Power series method If the Z transform is available in power series form, we can simply observe the sequence value $x(n)$ in the coefficient of the terms involving Z^{-n} in the power series.

Example 18.2 Find inverse Z-transform of $x(z) = \dfrac{1}{1-az^{-1}}$ where $|z| > |a|$.

Solution

Consider the Z transform $X(Z) = \log(1+aZ)$ convergent for $|a| < |z|$. Using the power series expression for log (1 + Z), we obtain

$$X(Z) = \sum_{n=1}^{\infty} \frac{(-1)^{n+1} a^n Z^{-n}}{n}$$

Therefore $$x(n) = \frac{(-1)^{n+1} a^n}{n} \quad n \geq 1$$

$$= 0 \text{ otherwise.}$$

For a rational z transform, a power series expansion can be obtained by long division.

Consider the transform $X(Z) = \dfrac{1}{1-aZ^{-1}}, |Z| > |a|$

Carrying out a long division, we have

$$\frac{1}{1-aZ^{-1}} = 1 + aZ^{-1} + a^2 z^{-2} \ldots$$

i.e., $$x(n) = a^n u(n)$$

(b) Partial fraction expansion Another technique which is often useful for rational Z transforms is to carry out a partial fraction expansion and identify the inverse Z transform of the simpler terms. If $X(Z)$ is a ratio of polynomials in Z with the order of the numerator less than the order of the denominator, then

$$X(Z) = \frac{P(Z)}{Q(Z)}$$

If $X(Z)$ has, for example, N first-order poles and 1 pole or order s, then $X(Z)$ can be expressed as

$$X(Z) = \sum_{k=1}^{N} \frac{A_k}{Z - Z_k} + \sum_{k=1}^{s} \frac{C_k}{(Z - Z_i)^k}$$

where
$$A_k = (Z - Z_k)X(Z)\Big|_{Z+Z_k}$$

and
$$C_k = \frac{1}{(s-k)!}\left\{\frac{d^{s-k}}{dZ^{s-k}}\left[Z - Z_i\right]^s X(Z)\right\}_{Z=Z_i}$$

Example 18.3 Find the inverse Z transform of $F(Z) = (Z^3 - 3)/[Z(Z - 0.25)(Z - 0.5)]$.

Solution

$$\frac{F(Z)}{Z} = \frac{(Z^3 - 3)}{Z^2(Z - 0.25)(Z - 0.5)}$$

$$\frac{F(Z)}{Z} = \frac{A}{Z^2} + \frac{B}{Z} + \frac{C}{(Z - 0.25)} + \frac{D}{(Z - 0.5)}$$

$$A = \left\{Z^2 \times \frac{(Z^3 - 3)}{Z^2(Z - 0.25)(Z - 0.5)}\right\}_{Z=0} \frac{-3}{(-0.25)(-0.5)} = -24$$

$$B = \frac{1}{1!}\left\{\frac{d}{dZ}\left[\frac{(Z^3 - 3)}{Z(Z - 0.25)(Z - 0.5)}\right]\right\}_{Z=0} = -144$$

$$C = \frac{(Z^3 - 3)}{Z^2(Z - 0.5)}\bigg|_{z=0.25} = 191$$

$$D = \frac{(Z^3 - 3)}{Z^2(Z - 0.25)}\bigg|_{z=0.25} = -46$$

$$\frac{F(Z)}{Z} = -\frac{144}{Z} - \frac{24}{Z^2} + \frac{191}{Z-0.25} - \frac{46}{Z-0.5}$$

$$F(Z) = -144 - \frac{24}{Z} + \frac{191Z}{Z-0.25} - \frac{46Z}{Z-0.5}$$

$$f(n) = -[-144\delta(n) - 24\delta(n-1) + 191(0.25)^n + 46(0.5)^n]u(n)$$

18.5 Digital/Discrete-Time Systems

A signal can be defined as a function that conveys information, generally about the state or behaviour of a physical system. Continuous-time signals or analog signals are defined as functions of a continuous time variable, and analog system are defined or described in terms of such signals. Discrete time or digital time have the same meaning; they refer to signals which are defined only for specific instants of time and such systems have the input and output as discrete-time signals.

Discrete-time signals may arise in the following two ways:

(a) By sampling a continuous time as shown in Fig. 18.1.

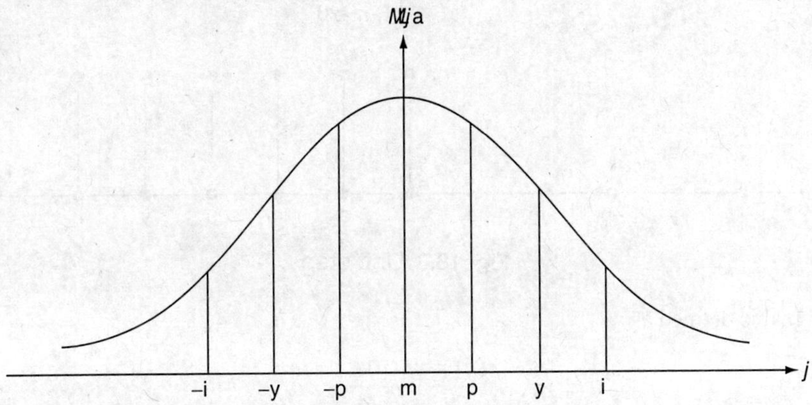

Fig. 18.1 Sampling of a continuous time signal

A discrete-time signal is represented by

$$x(n) = x(nT) \text{ for } -\infty < n < \infty$$

(b) Directly by some discrete-time process. An example is a light source flashing at intervals of T seconds. An observer would record signals which are present only for these intervals of time.

18.6 Discrete-Time Signals

Some examples of discrete-time signals are given below:

(a) Unit-sample signal is defined as

$$x(n) = \begin{cases} 0, & n \neq 0 \\ 1, & n = 0 \end{cases}$$

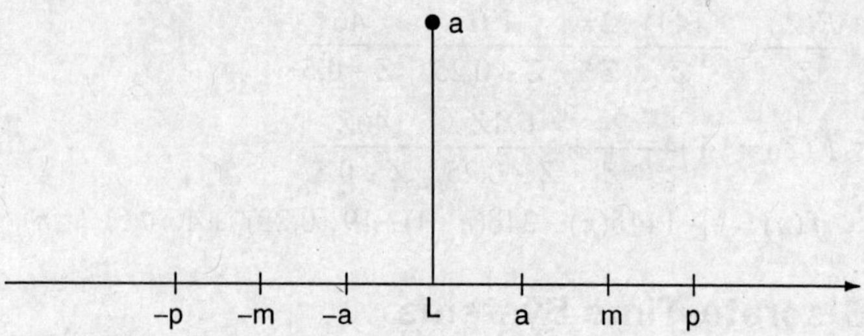

Fig. 18.2 Discrete-time signals

By definition,

$$\delta(n-1) = \begin{cases} 0, & n \neq 1 \\ 1, & n = 1 \end{cases}$$

(b) Unit step signal is defined as

$$x(n) = u(n) = \begin{cases} 0, & n < 0 \\ 1, & n \geq 0 \end{cases}$$

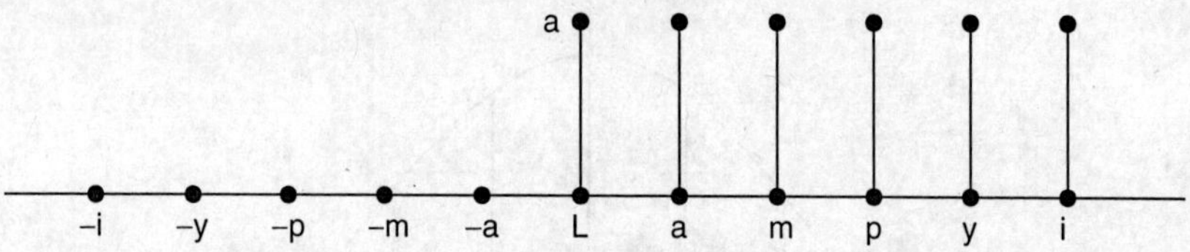

Fig. 18.3 Unit step

(c) Ramp signal is defined as

$$x(n) = nu(n)$$

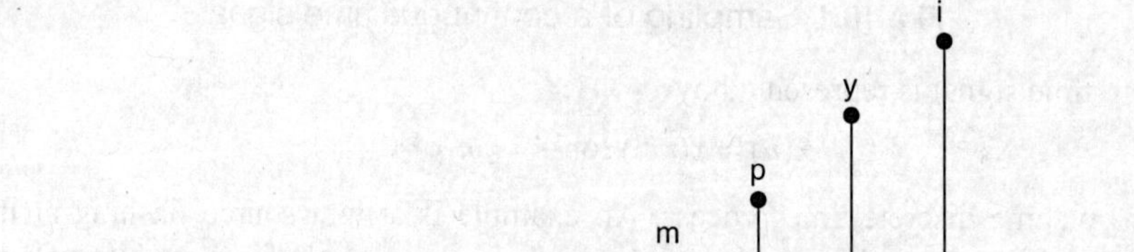

Fig. 18.4 Ramp

i.e.,

$$x(n) = \begin{cases} n, & n \geq 0 \\ 0, & n < 0 \end{cases}$$

(d) Real exponential has values of the form

$x(n) = a^n u(n)$, where a is a real number.

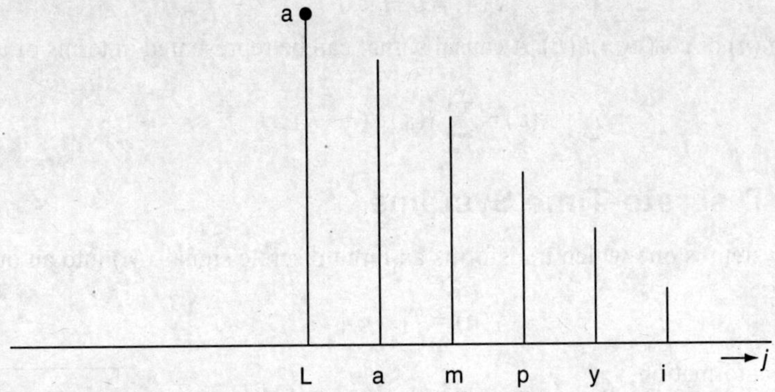

Fig. 18.5 Real exponential

(e) Sinusoidal signal

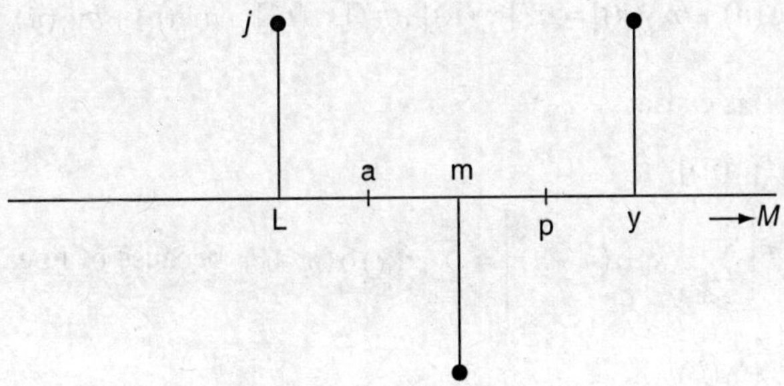

Fig. 18.6 Sinusoidal

$$x(n) = [Acos(\omega_0 n)]u(n)$$

Example 18.4 Sketch the function $f(t)$ and the samples $f(n)$ for a sampling period of $T = 0.5$ for the following cases:

(1) $f(t) = e^{0.5t}u(t)$

(2) $f(t) = [\sin(2\pi t)]u(t)$

Solution

(1) $$f(t) = e^{0.5t}u(t)$$
$$f(n) = e^{-(0.5)n(0.5)}u(n) = E^{-0.25n}u(n)$$

(2) $$f(t) = [\sin(2\pi t)]u(t)$$
$$f(n) = [\sin(2\pi \times n \times 0.5)]u(t) = [\sin(\pi n)]u(n) = 0$$

18.7 Causal Signals

A causal discrete-time signal $x(n)$ is one which is zero for $n < 0$, i.e.,

$$x\,(n) = 0,\ n < 0$$

Examples are $u(n)$ or $\cos(\omega_0 n)u(n)$. A causal signal can be represented in terms of a unit samples as

$$x(n) = \sum_{k=1}^{\infty} x(k)\,\delta\,(n-k)$$

18.8 Linear Discrete-Time Systems

A discrete-time system is one which transforms an input discrete signal $x(n)$ into an output signal $y(n)$. This is denoted as

$$y(n) = T[x(n)]$$

where T is the transformation.

The class of linear systems is one which obeys the law of superposition, i.e., if $y_1(n)$ and $y_2(n)$ are the outputs of the system to $x_1(n)$ and $x_2(n)$ respectively, then

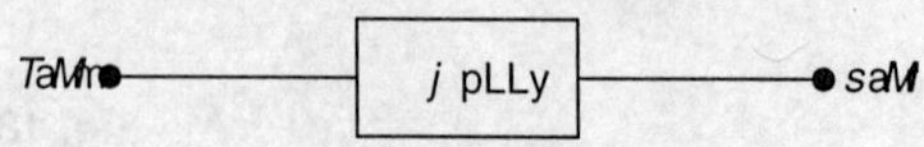

Fig. 18.7 Discrete-time system

$$T[ax_1(n) + bx_2(n)] = aT[x_1(n)] + bT[x_2(n)] = ay_1(n) + by_2(n)$$

For such a system if $x(n)$ is causal

$$\begin{aligned} y(n) &= T[x(n)] \\ &= T\left[\sum_{k=0}^{\infty} x(k)\,\delta\,(n-k)\right] = \sum_{k=0}^{\infty} x(k)\,\delta\,(n-k) \text{ because of linearity} \\ &= \sum_{k=0}^{\infty} x(k) h_n n \end{aligned}$$

Thus $h(n)$ is the output of the system when the input is $\delta(n)$.

18.9 Difference Equations

Discrete-time systems are described by difference equations of the form

$$\sum_{k=0}^{N} a_k y(n-k) = \sum_{k=0}^{N} b_r x(n-r) \tag{1}$$

If the system is causal, a linear difference equation provides an explicit relationship between the input and output. This can be seen by rewriting.

$$y(n) = \sum_{k=0}^{N} \frac{a_k}{a_0} y(n-k) = \sum_{k=0}^{N} \frac{b_r}{a_0} x(n-r) \tag{2}$$

Thus the nth value of the output can be computed from the nth input value and the N and M past values of the output and input, respectively.

Example 18.5 A first-order difference equation is of the form $y(n) = ay(n-1) + x(n)$. Obtain the unit sample response $h(n)$.

Solution

Let $x(n) = \delta(n)$ and assume initial rest conditions.

Then
$$h(n) = 0, n < 0$$
$$h(0) = ah(-1) + 1 = 1$$
$$h(1) = ah(0) + 0 = a$$
$$\vdots$$
$$h(n) = ah(n-1) = a^n$$

An important role in the analysis and representation of discrete-time systems is played by the Z transform. The Z transform plays the same role that the Laplace transform plays in the analysis of continuous-time systems.

18.10 Role of Z Transform in Linear Difference Equations

Equation (1) gives us the form of the linear difference equation that describes the system. Taking Z transform on either side and assuming zero initial conditions, we have

$$\sum_{k=0}^{N} a_k Z^{-k} Y(Z) = \sum_{k=0}^{N} b_r Z^{-r} X(Z)$$

or
$$H(Z) = \frac{Y(Z)}{X(Z)} = \frac{\sum_{r=0}^{M} b_r Z^{-r}}{\sum_{k=0}^{N} a_k Z^{-k}}$$

where $H(Z)$ is the Z transform of the unit sample response $h(n)$.

Example 18.6 A discrete-time system is given by the equation

$$y(k+3) = -0.3y(k+2) + y(k+1) - 0.5y(k) + 4x(k+3) - x(k+1).$$

Find its transfer function $H(Z)$. The initial conditions are zero.

Solution Z transforming on either side with zero initial conditions gives

$$(Z^3 + 0.3Z^2 - Z + 0.5)y(Z) = (4Z^3 - Z)X(Z)$$

Hence
$$H(Z) = \frac{Y(Z)}{X(Z)} = \frac{4Z^3 - Z}{Z^3 + 0.3Z^2 - Z + 0.5}$$

18.11 Stability of a Discrete-Time System

A discrete-time system is said to be stable if and only if its unit sample response $h(n)$ decays with n. This gives the condition for stability that all the poles of the Z-transfer function $H(Z)$ should lie within the unit circle on the complex plane.

Example 18.7 Determine whether or not each of the following discrete-time systems is stable:

(a) $H(Z)=\dfrac{-3Z^2+1}{(4Z^2+2Z-1)}$

(b) $H(Z)=\dfrac{Z^3}{(Z-15)(Z^2-3Z+2)}=\dfrac{Z^3}{(Z-15)(Z-2)(Z-1)}$

Solution

(a)
$$H(Z)=\frac{-3Z^2+1}{(4Z^2+2Z-1)}$$

The roots of the denominator polynomial are

$$Z=0.31 \text{ and } Z=-0.81$$

Since $|Z|<1$ for all the roots of the denominator polynomial, the system is stable.

(b)
$$H(Z)=\frac{Z^3}{(Z-15)(Z^2-3Z+2)}=\frac{Z^3}{(Z-15)(Z-2)(Z-1)}$$

The pole locations of $H(z)$ are $Z = 1.5, 2, 1$. Since the first two poles are greater than 1, the system is unstable.

SIGNIFICANT POINTS

The Z-transform of a discrete-time signal $x(n)$ is generally defined as

$$ZT[x(n)]=\sum_{n=-\infty}^{+\infty} x(n)Z^{-n}$$

For a causal sequence, $x(n) = 0$, for $n < 0$. Hence

$$X(Z)=\sum_{n=0}^{\infty} x(n)Z^{-n}$$

Linearity

$$ZT[ax(n)+by(n)]=aX(Z)+bY(Z)$$

Shift of a Sequence

$$ZT[x(n-k)]=Z^{-k}X(Z)$$

Differentiation of $X(Z)$

$$ZT[nx(n)]=-\frac{ZdX(Z)}{dZ}$$

Initial Value Theorem

$$x(0) = \underset{Z \to 0}{\text{Lt}}\ X(Z)$$

Final Value Theorem

$$\underset{n \to \infty}{\text{Lt}}\ x(n) = \lim_{Z \to 1}\left[\frac{(Z-1)}{Z} X(Z)\right]$$

provided that $\underset{n \to \infty}{\text{Lt}}\ x(n)$ exists and is finite.

Digital and Discrete-Time Systems: Continuous-time signals or analog signals are defined as functions of a continuous-time variable, and analog systems are defined or described in terms of such signals.

Discrete-time or digital systems have the same meaning; they refer to signals which are defined only for specific instants of time and such systems have the input and output as discrete-time signals.

Causal Signals: A causal discrete-time signal $x(n)$ is one which is zero for $n < 0$, i.e.,

i.e. $$x(n) = 0, \ n < 0$$

A causal signal can be represented in terms of a unit sample as

$$x(n) = \sum_{k=1}^{\infty} x(k)\delta(n-k)$$

Linear Discrete-Time System: A discrete-time system is one which transforms an input discrete signal $x(n)$ into an output signal $y(n)$. This is denoted as

$$y(n) = T[x(n)]$$

where T is the transformation.

Difference Equations: Discrete-time systems are described by difference equations of the form

$$\sum_{k=0}^{N} a_k y(n-k) = \sum_{k=0}^{N} b_r x(n-r)$$

Stability of a Discrete-Time System: A discrete-time system is said to be stable if and only if its unit sample response $h(n)$ decays with n. This gives the condition for stability that all the poles of the Z transfer function $H(Z)$ should lie within the unit circle on the complex plane.

ADDITIONAL SOLVED EXAMPLES

Example 18.8 Derive Z transform of the following:

(a) $x(1) = 2, x(4) = -3, x(7) = 8,$ and all other samples are zero

(b) $x(n) = u(n) \sin(3n) - 2u(n-4) \sin[3(n-4)]$

Solution

(a)

$$X(Z) = \sum_{n=1,4,7}^{\infty} x(n)Z^{-n}$$

$$= 2Z^{-1} + (-3)Z^{-4} + 8Z^{-7}$$

$$= 2Z^{-1} - 3Z^{-4} + 8Z^{-7}$$

(b)

$$X(Z) = \sum_{n=0}^{\infty} \sin(3n)Z^{-n} - 2\sum_{n=4}^{\infty} \sin[3(n-4)]Z^{-n}$$

$$= \frac{Z\sin 3}{Z^2 - 2Z\cos 3 + 1} - \frac{(2Z^{-4})(Z\sin 3)}{Z^2 - 2Z\cos 3 + 1}$$

$$= \frac{Z\sin 3(1 - 2Z^{-4})}{Z^2 - 2Z\cos 3 + 1}$$

Example 18.9 Determine the Z transform function for the following system:

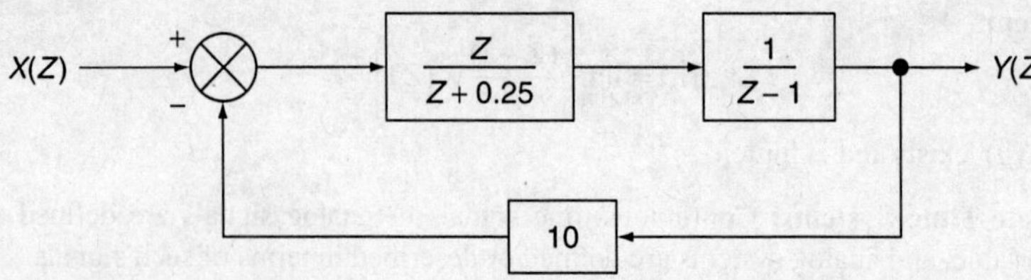

Fig. E18.1

Also determine whether or not the system is stable.

Solution The system can be simplifier to

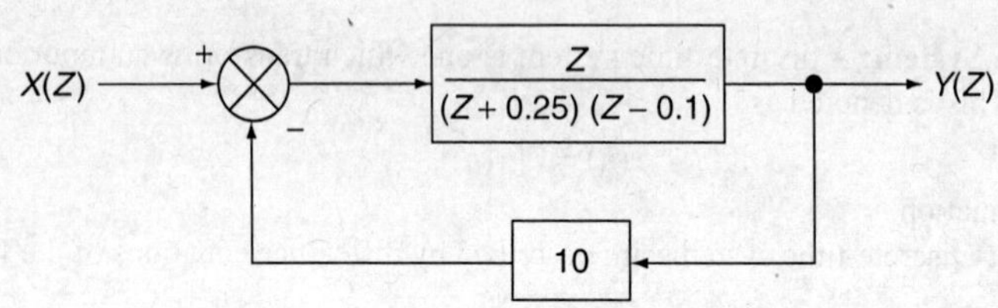

Fig. E18.1(a)

Hence
$$[-Y(Z)+X(Z)]G(Z)=Y(Z)$$
$$Y(Z)[1+G(Z)H(Z)]=X(Z)G(Z)$$

$$H(Z)=\frac{Y(Z)}{X(Z)}=\frac{G(Z)}{1+G(Z)H(Z)}=\frac{\dfrac{Z}{(Z+0.25)(Z-1)}}{\dfrac{10Z}{(Z+0.25)(Z-1)}}$$

$$=\frac{Z}{Z^2-0.75Z-0.25+10Z}=\frac{Z}{Z^2+9.25Z-0.25}$$

The pole position is

$$Z=\frac{-9.25\pm\sqrt{(9.25)^2+1}}{2}=-9.28,0.027$$

Since one of the poles lies in $|Z|>1$, *the system is unstable.*

SHORT QUESTIONS WITH ANSWERS

1. Define Z transformation.

   Ans. The Z transform of a discrete-time signal $x(n)$ is generally defined as

$$ZT[x(n)]=\sum_{n=-\infty}^{+\infty}x(n)Z^{-n}$$

For a causal sequence, $x(n) = 0$, for $n < 0$. Hence

$$X(Z) = \sum_{n=0}^{\infty} x(n)Z^{-n}$$

2. What do you mean by digital and discrete-time systems?

   Ans. Continuous-time signals or analog signals are defined as functions of a continuous-time variable, and analog systems are defined or described in terms of such signals. Discrete-time or digital systems have the same meaning; they refer to signals which are defined only for specific instants of time and such systems have the input and output as discrete-time signals.

3. What do you mean by causal signals?

   Ans. A causal discrete-time signal $x(n)$ is one which is zero for $n < 0$.

   i.e. $$x(n) = 0, n < 0$$

   A causal signal can be represented in terms of a unit sample as

   $$x(n) = \sum_{k=1}^{\infty} x(k)\delta(n-k)$$

4. Define linear Discrete-time systems.

   Ans. A discrete-time system is one which transforms an input discrete signal $x(n)$ into an output signal $y(n)$. This is denoted as

   $$y(n) = T[x(n)],$$

   where T is the transformation.

5. Define the stability of a discrete-time system.

   Ans. A discrete-time system is said to be stable if and only if its unit sample response $h(n)$ decays with n. This gives the condition for stability that all the poles of the Z transfer function $H(Z)$ should lie within the unit circle on the complex plane.

EXERCISES

Questions for Practice

1. Find the Z transform and ROC of the following finite duration sequences:

   (i) $x(n) = \{2 \;\; 1 \;\; 1 \;\; 2 \;\; 5\}$ (ii) $x(n) = \left(\frac{1}{4}\right)^n u(n)$

   Ans. (i) $x(z) = 2z^2 + z + 1 + 2z^{-1} + 5z^{-2}$. ROC is entire z plane except at $z = 0$ and $z = \infty$, and $z - \infty$,

   (ii) $x(z) = \dfrac{z}{z - \frac{1}{4}}$, ROC $|z| > \frac{1}{4}$

2. Find the characteristic equation of the Z domain for the sample data system shown in Figure 1. State whether the system is stable.

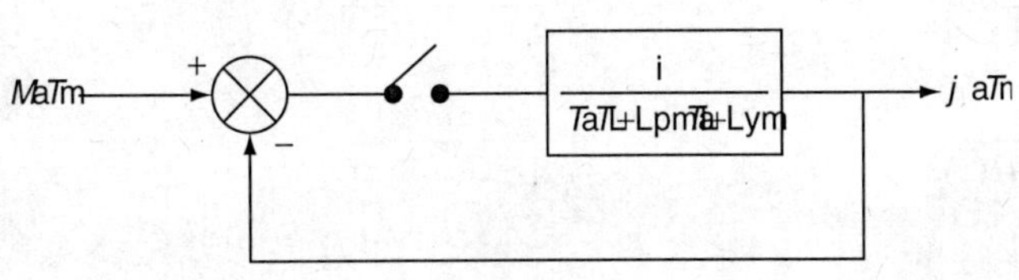

Figure 1

Ans. $z^3 - 0.5z^2 + 2.5z - 0.5 = 0$,

TEST YOURSELF

1. Consider the following statements regarding hold circuits for the reconstruction of sampled signals:
   (i) Hold circuits are essentially low pass filters.
   (ii) A first order hold circuit introduces less phase lag in comparison to a zero hold circuit.
   (iii) A zero-order hold has a flat gain frequency response over the frequency range of $0 \le \omega \le \frac{2\pi}{T}$, where T is the sampling period.

   Which of the following statement(s) is/are correct?

   (a) 3 alone (b) 1 and 2 (c) 2 and 3 (d) 1 alone

2. The overall pulse transfer function of the system shown in Figure T1 below is

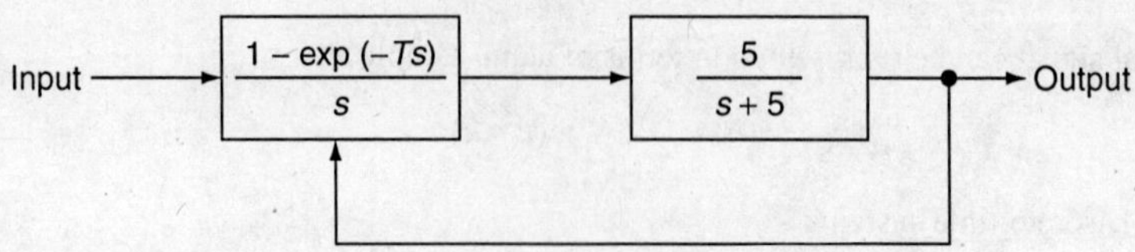

Figure T1

(a) $\dfrac{1-\exp(-1)}{Z-\exp(-1)}$ (b) $\dfrac{Z[1+\exp(-1)]}{(Z-1)[Z+\exp(-1)]}$

(c) $\dfrac{1+\exp(-1)}{Z+\exp(-1)}$ (d) $\dfrac{Z[1-\exp(-1)]}{(Z-1)[Z-\exp(-1)]}$

3. The block diagram of a sampled data system is shown in Figure T2 below.

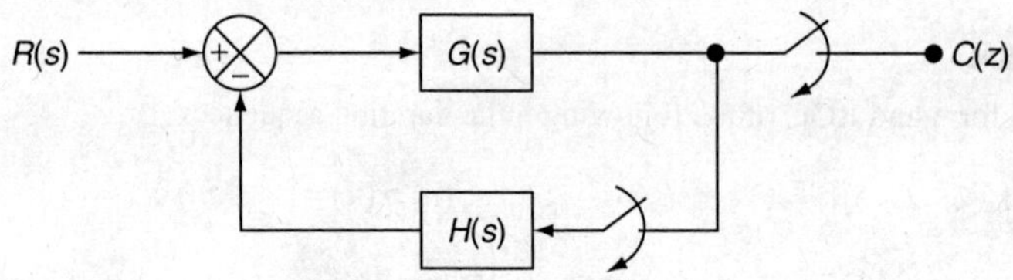

Figure T2

The output $C(Z)$ of the system is given by

(a) $\dfrac{GR(Z)}{1+GH(z)}$ (b) $\dfrac{G(z)R(z)}{1+G(z)H(z)}$

(c) $\dfrac{GR(z)}{1+G(z)H(z)}$ (d) $\dfrac{G(z)R(z)}{1+G(z)R(z)}$

ANSWERS

1 (d) 2 (a) 3 (a)

Chapter 19

Non-Linear Control Systems

We have already introduced linear control systems and studied them from their pole-zero configuration, which is an extremely powerful tool both in frequency and time-domain analysis and synthesis. Unfortunately, systems that exist in nature are non-linear. Hence, the superposition principle which is well valid in linear systems is totally invalid in non-linear systems. There are control systems which do not have any linearized version and it is very difficult to linearize these control system. In some cases, it has been found that the non-linearities such as dead zone and hysteresis and combination of these are undesirable and unavoidable. The aim of this chapter is to discuss the methods to study non-linear systems as well as their stability.

19.1 The State Space Approach

Figure 19.1 shows the block diagram representation of a class of non-linear control systems. The plant dynamics is represented by the state vector x. In Fig. 19.1, r is the reference, u is plant input and c is the controlled output vector.

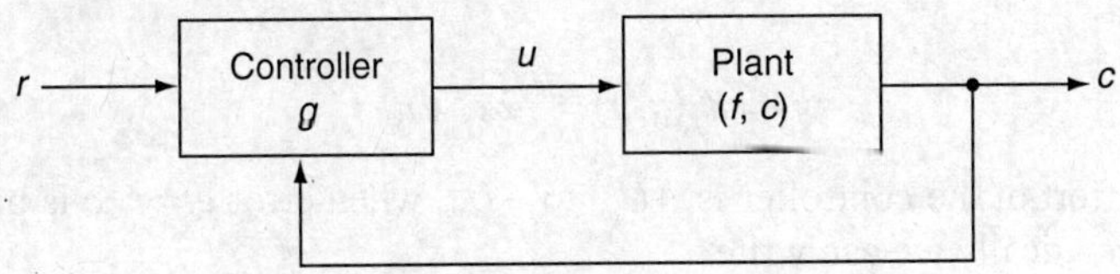

Fig. 19.1 Non-linear control system

The mathematical models to represent the plant and controller are given below.

Plant

$$\dot{x} = f(x,u) \tag{19.1}$$

and

$$c = Cx \tag{19.2}$$

Controller

$$u = g\,(c, r) \tag{19.3}$$

In Eqs. (19.1) and (19.2), the vectors are in general non-linear functions, and we have assumed that the controller does not contain any energy storing elements. Substituting Eq. (19.2) in Eq. (19.3), we get

$$u = g(Cx, r) \tag{19.4}$$

Substituting Eq. (19.4) in Eq. (19.1), we get

$$\dot{x} = f(x, g(Cx, r)) = F(x, r) \tag{19.5}$$

If the input vector r is assumed to be constant, Eq. (19.5) can be written as

$$\dot{x} = F(x, r) = F_0(x) \tag{19.6}$$

Let us take a single output system as shown in Fig. 19.2 to obtain the state space equations for the given system.

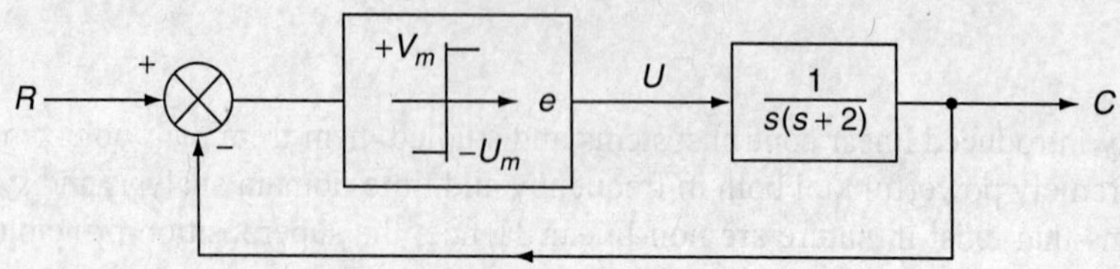

Fig. 19.2 Single output system

For plant we can write

$$C(s) = \frac{1}{s(s+2)} U(s)$$

This gives the differential equation $u = \ddot{c} + 2\dot{c}$. Putting $c = x_1$ and $x_2 = \dot{x}_1 = \dot{c}$. Therefore, $\dot{x}_2 = u - 2\dot{c} = u - 2x_2$.

In state space form we can write the plant differential equations as follows:

$$\dot{x}_1 = f_1(x, u) = x_2$$

and

$$\dot{x}_2 = f_2(x, u) = -2x_2 + u$$

The maximum effort of the controller is $+U_m$ to $-U_m$ when error $e = r - c$ is positive and negative, respectively. Mathematically we can write

$$u = \frac{r - c}{|r - c|} U_m = \frac{r - x_1}{|r - x_1|} U_m = U_m \, \text{sgn}(r - x_1)$$

where the function sgn$(r - x_1)$ has a value + 1 when $(r - x_1)$ is positive and it has a negative value when $(r - x_1)$ is negative.

In state space form we can write the controller differential equations as follows:

$$\dot{x}_1 = F_1(x, r) = x_2$$

and

$$\dot{x}_2 = F_2(x, r) = -2x_2 + U_m \, \text{sgn}(r - x_1)$$

19.2 Phase Plane Method

It has been observed that non-linear system may be simplified to a non-linear second-order system with reasonable approximations. We can apply very easily a graphical representation of the phase trajectory of the above second-order system. Isoclines in a phase plane are a very general and convenient graphical method among other existing methods. The main disadvantage in the phase plane technique is that the time does not appear explicitly in the trajectory. Let us discuss a graphical method for determining the phase trajectory for a second-order system represented by

$$\ddot{c} + f_1(c,\dot{c})\dot{c} + f_2(c,\dot{c})c = r \tag{19.7}$$

where $f_1(c,\dot{c})\dot{c}$ and $f_2(c,\dot{c})c$ are functions of c and $\dot{c}$ respectively.

Putting $x_1 = c$ and $x_2 = \dot{c}$, we get form Eq. (19.7)

$$\dot{x}_1 = x_2 \tag{19.8}$$

and

$$\dot{x}_2 = r - f_1(x_1,x_2)x_2 - f_2(x_1,x_2)x_1 \tag{19.9}$$

From Eqs. (19.8) and (19.9) we have

$$\frac{dx_2}{dx_1} = \frac{\dot{x}_2}{\dot{x}_1} = -f_1(x_1,x_2) - f_2(x_1,x_2)\frac{x_1}{x_2} + \frac{r}{x_2} \tag{19.10}$$

Equation (19.10) gives the slope of the trajectory at any point (x_1, x_2). Let us take

$$\frac{dx_2}{dx_1} = m \tag{19.11}$$

where m is constant. We can write

$$F(x_1,x_2) = m$$

$$F_1(x_1,x_2) = \frac{\dot{x}_2}{\dot{x}_1} = -f_1(x_1,x_2) - f_2(x_1,x_2)\frac{x_1}{x_2} + \frac{r}{x_2} \tag{19.12}$$

Any trajectory can cross the curve described by Eq. (19.12) and this curve is known as an isocline.

Let us consider a second-order linear system as shown in Fig. 19.3 with input $r(t) = 2u(t)$ and output $c(t)$ having the following initial conditions:

$$c(0) = -1 \text{ and } \dot{c}(0) = 0$$

From Fig. 19.2, we can write

$$(s^2 + 2s + 5)C(s) = 5R(s)$$

Taking inverse Laplace transform, we get

$$\ddot{c} + 2\dot{c} + 5c = 5r = 10 \tag{19.13}$$

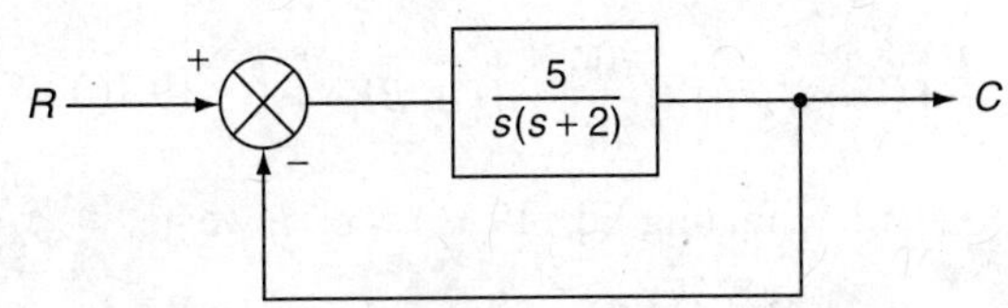

Fig. 19.3 Second-order linear system

Putting $x_1 = c$ and $x_2 = \dot{c}$, comparing Eq. (19.13) with Eq. (19.7) we get

$$f_1(x_1, x_2) = 2$$
$$f_2(x_1, x_2) = 5$$

As per Eq. (19.13), we get

$$m = -2 - 5\frac{x_1}{x_2} + \frac{10}{x_2}$$

The family of isoclines will have the following equation:

$$x_2 = -\frac{5x_1}{m+2} + \frac{10}{m+2} \quad (19.14)$$

which describes a family of straight lines. The slope of these lines is given by

$$m' = -\frac{5}{m+2} \quad (19.15)$$

These lines intersect the x axis at a point having the following value at the x coordinate:

$$0 = -\frac{5x_1}{m+2} = \frac{10}{m+2}$$

i.e., $x_1 = 2.$

The common point of all isoclines is (2, 0). If $m = 0$, the equation for the isoclines will be

$$x_2 = -\frac{5}{2}x_1 + 5$$

Figure 19.4 shows the isoclines for different values of m.

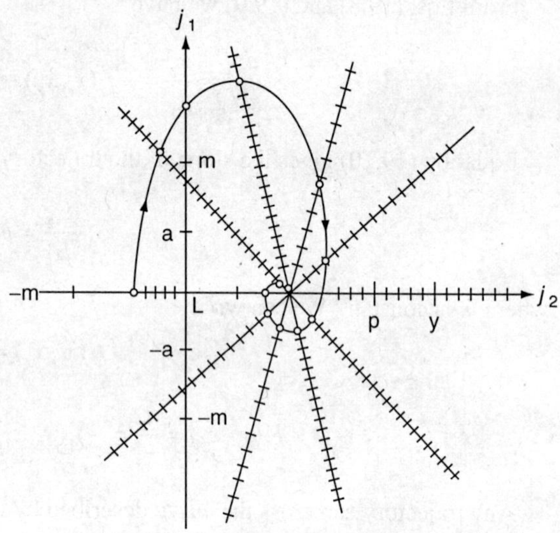

Fig. 19.4 Isoclines of Fig. 19.3

To get the phase plane trajectory for a non-linear system, let us consider Fig. 19.5 where inductance L and capacitance C are linear but resistance R is non-linear. Let the current characteristics of the non-linear resistance be given by

$$i = -AV + BV^3$$

From Fig. 19.5, we get

$$\frac{1}{L}\int (V - E)dt + C\frac{dV}{dt} - AV + BV^3 = 0 \quad (19.16)$$

Differentiating Eq. (19.16) w.r.t. t, we get

$$LC\frac{d^2V}{dt^2} + \frac{dV}{dt}[3LBV^2 - AL] + V = E$$

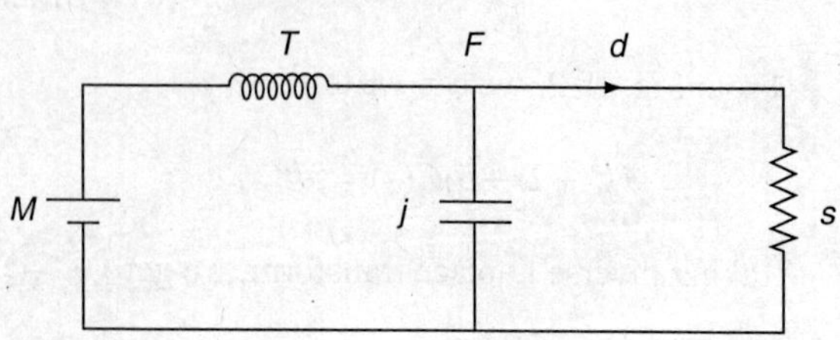

Fig. 19.5 A system

Let us take

$$\tau = \frac{t}{\sqrt{LC}}$$

$$\frac{dV}{dt} = \frac{dV}{d\tau}\frac{d\tau}{dt} = \frac{dV}{d\tau}\frac{1}{\sqrt{LC}} \tag{19.17}$$

and

$$\frac{d^2V}{dt^2} = \left(\frac{1}{LC}\right)\frac{d^2V}{d\tau^2} \tag{19.18}$$

Using Eqs. (19.17) and (19.18), we get from Eq. (19.16) as

$$E = \frac{d^2V}{d\tau^2} - \frac{dV}{d\tau}\left[A\sqrt{\frac{L}{C}} - 3B\sqrt{\frac{L}{C}}V^2\right] + V$$

i.e.,

$$E = \frac{d^2V}{d\tau^2} - \frac{dV}{d\tau}[a - abV^2] + V \tag{19.19}$$

where

$$a = A\sqrt{\frac{L}{C}}$$

and

$$b = \frac{3B}{A}$$

Form Eq. (19.19), we have

$$\frac{d^2V}{d\tau^2} - \frac{dV}{d\tau}[a(1 - bV^2)] + V = E$$

Setting $x_1(\tau) = v(\tau)$ and $x_2(\tau) = \dot{v}(\tau)$ and assuming the standard values $a = 0.2$ and $b = 1$, we can write the following:

$$\dot{x}_1 = x_2$$

$$\dot{x}_2 = 0.2(1 - x_1^2) - x_1$$

and

$$m = \frac{dx_2}{dx_1} = \frac{\dot{x}_2}{x_1} = 0.2(1 - x_1^2) - \frac{x_1}{x_2}$$

The equations of isoclines are given by

$$x_2 = \frac{5x_1}{(1 - 5m) - x_1^2} \tag{19.20}$$

For $m = 0$, the isoclines are given by

$$x_2 = \frac{5x_1}{1 - x_1^2}$$

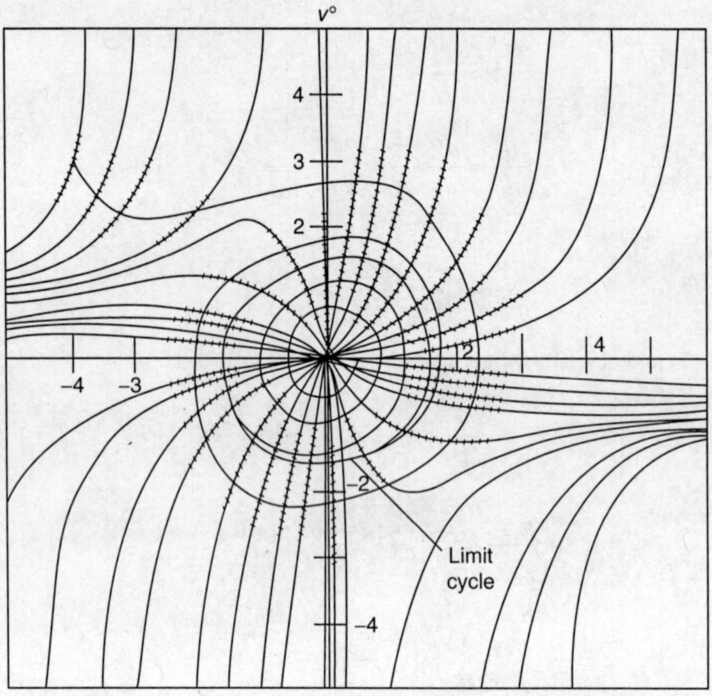

Fig. 19.6 Isoclines and trajectories for a non-linear system

Figure 19.6 shows the isoclines and trajectories for a non-linear system for different values of m. For a non-linear system, we get a family of trajectories. Therefore, this diagram is also called *phase portrait*.

19.3 Describing Function Analysis

The steady-state output of a non-linear device will have the same period of output, but there is no guarantee that the output will be sinusoidal in spite of sinusoidal input. We know that the Fourier analysis of the fundamental wave form will lead to a fundamental component as well as several harmonics. It is assumed that subharmonics do not exist. If the feedback system shown in Fig. 19.7 is of low pass nature, the harmonics will be attenuated and may be neglected to get a reasonable approximation to the performance of the system. There are possibilities of oscillations if the loop gain is one with a phase shift of 180°.

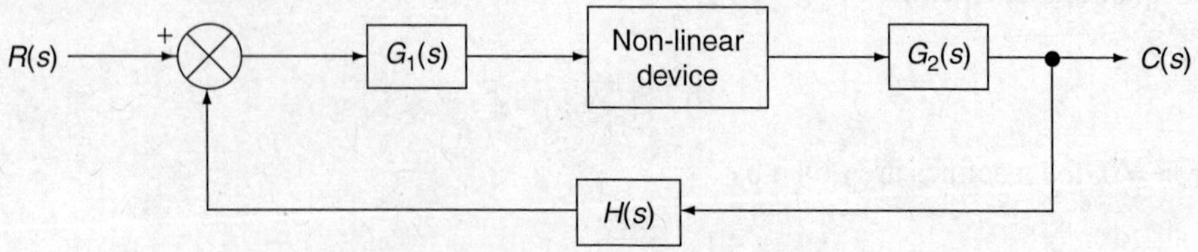

Fig. 19.7 Non-linear system

The complex ratio of the fundamental harmonic component of the output to the input defines the describing function of a non-linear element and it is given by

$$X = \frac{Y_1}{X} \angle \phi_1$$

where X is the amplitude of the input sinusoid, Y_1 is the amplitude of the fundamental harmonic component of the output, and ϕ_1 is the phase shift of the fundamental harmonic component of the output with respect to the input.

Let $x = X \sin \omega t$ be the input to a non-linear element, by Fourier series the output is given by

$$y(t) = A_0 + \sum_{n=1}^{\infty} (A_n \sin \omega t + B_n \cos \omega t) = A_0 + \sum_{n=1}^{\infty} Y_n \sin(n\omega t + \phi_n)$$

where

$$A_n = \frac{1}{\pi}\int_0^{2\pi} y(t) \cos n\,\omega t \, d(\omega t), B_n = \frac{1}{\pi}\int_0^{2\pi} y(t) \sin n\,\omega t \, d(\omega t)$$

$$Y_n = \sqrt{A_n^2 + B_n^2}$$

$$\phi_n = \tan^{-1}\left(\frac{A_n}{B_n}\right)$$

$A_0 = 0$ when non-linearity is symmetric. The fundamental component of the output is

$$y_1 = A_1 \cos \omega t + B_1 \sin \omega t = Y_1 \sin(\omega t + \phi_1),$$

where

$$A_1 = \frac{1}{\pi}\int_0^{2\pi} y(t) \cos \omega t \, d(\omega t) \quad \text{and} \quad B_1 = \frac{1}{\pi}\int_0^{2\pi} y(t) \sin \omega t \, d(\omega t)$$

The amplitude and phase angle of the fundamental component of the output are given by

$$Y_1 = \sqrt{A_1^2 + B_1^2} \tag{19.21}$$

and

$$\phi_1 = \tan^{-1}\left(\frac{A_1}{B_1}\right) \tag{19.22}$$

respectively.

19.3.1 Dead-Zone Non-Linearity

Figure 19.8 shows idealized characteristics of non-linearity having dead-zone and its response due to sinusoidal input. No output can be obtained within the dead-zone amplitude for inputs to a dead-zone element.

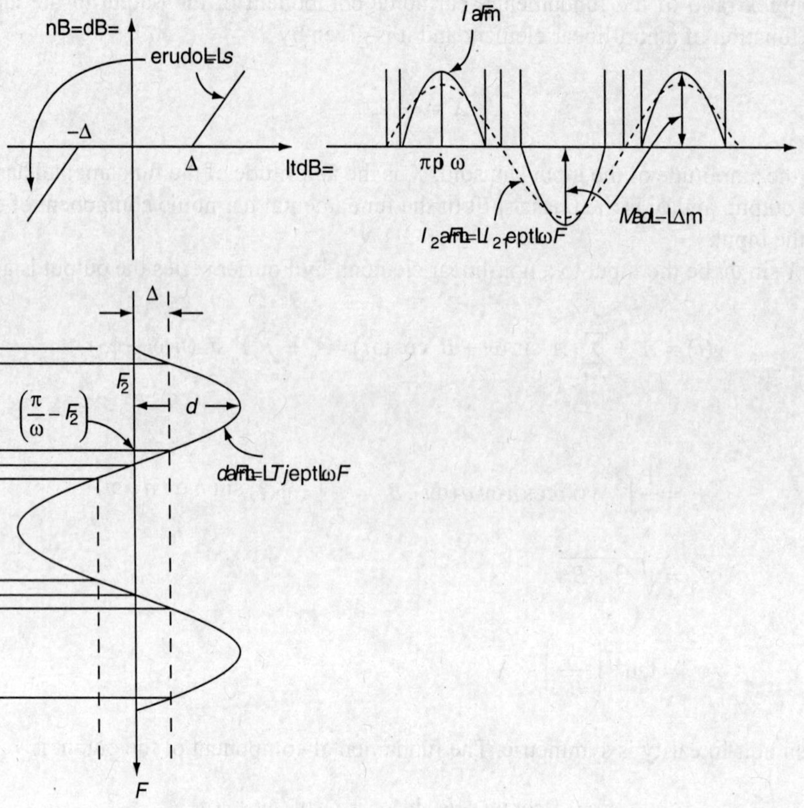

Fig. 19.8 Sinusoidal response of non-linearity with dead-zone linearity

From Fig. 19.8 the output $y(t)$ for $0 \le \omega t \le \pi$ is given by

$$y(t) = \begin{cases} 0 & \text{for } 0 < t < t_1 \\ K(X \sin \omega t - \Delta) & \text{for } t_1 < t < \dfrac{\pi}{\omega} - t_1 \\ 0 & \text{for } \dfrac{\pi}{\omega} - t_1 < t < \dfrac{\pi}{\omega} \end{cases} \tag{19.23}$$

The output function is an odd function and hence A_n terms will be absent. The Fourier expansion contains only sine terms. The fundamental harmonic component of the output is given by

$$y_1(t) = B_1 \sin \omega t$$

where

$$B_1 = \frac{1}{\pi} \int_0^{2\pi} y(t) \sin \omega t \, d(\omega t)$$

Due to half-wave symmetry, we can write

$$B_1 = \frac{4}{\pi} \int_0^{\frac{\pi}{2}} y(t) \sin \omega t \, d(\omega t) = \frac{4}{\pi} \int_{\omega t_1}^{\frac{\pi}{2}} K(X \sin \omega t - \Delta) \sin \omega t \, d(\omega t) \tag{19.24}$$

where $\Delta = X \sin \omega t_1$ i.e., $\omega t_1 = \sin^{-1}(\Delta/X)$

From Eq. (19.24), we get

$$B_1 = \frac{4XK}{\pi}\int_{\omega t_1}^{\frac{\pi}{2}} (\sin \omega t - \sin \omega t_1)\sin \omega t \, d(\omega t)$$

$$= \frac{4XK}{\pi}\left[\frac{1}{2}\left\{\frac{\pi}{2} - \sin^{-1}\left(\frac{\Delta}{X}\right)\right\} - \frac{\Delta}{X}\sqrt{1-\left(\frac{\Delta}{X}\right)^2}\right] \tag{19.25}$$

The describing function of an element with the dead zone is given by

$$N = \frac{B_1}{X}\angle 0° = \left[K - \frac{4K}{\pi}\left\{\frac{1}{2}\sin^{-1}\left(\frac{\Delta}{X}\right) - \frac{\Delta}{X}\sqrt{1-\left(\frac{\Delta}{X}\right)^2}\right\}\right]\angle 0° \tag{19.26}$$

Figure 19.9 shows the plot of N/K versus $\frac{\Delta}{X}$.

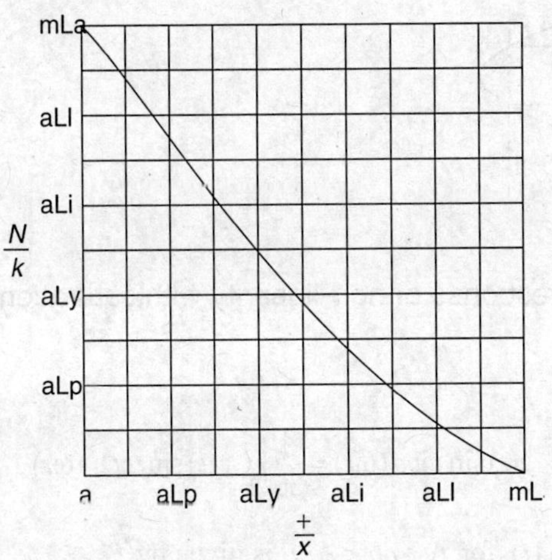

Fig. 19.9 Describing function for the dead-zone non-linearity

19.3.2 Saturation Non-Linearity

Figure 19.10 shows a typical input–output characteristic for the saturation non-linearity. In this case, the output of the saturation element is proportional to the input for small input signals, the output will not increase proportionally for larger input signals and at last the output is constant for very large signals.

Figure 19.10 shows that the output is odd and the fundamental component contains only sine terms. The fundamental harmonic component of output is given by

$$y(t) = B_1 \sin \omega t$$

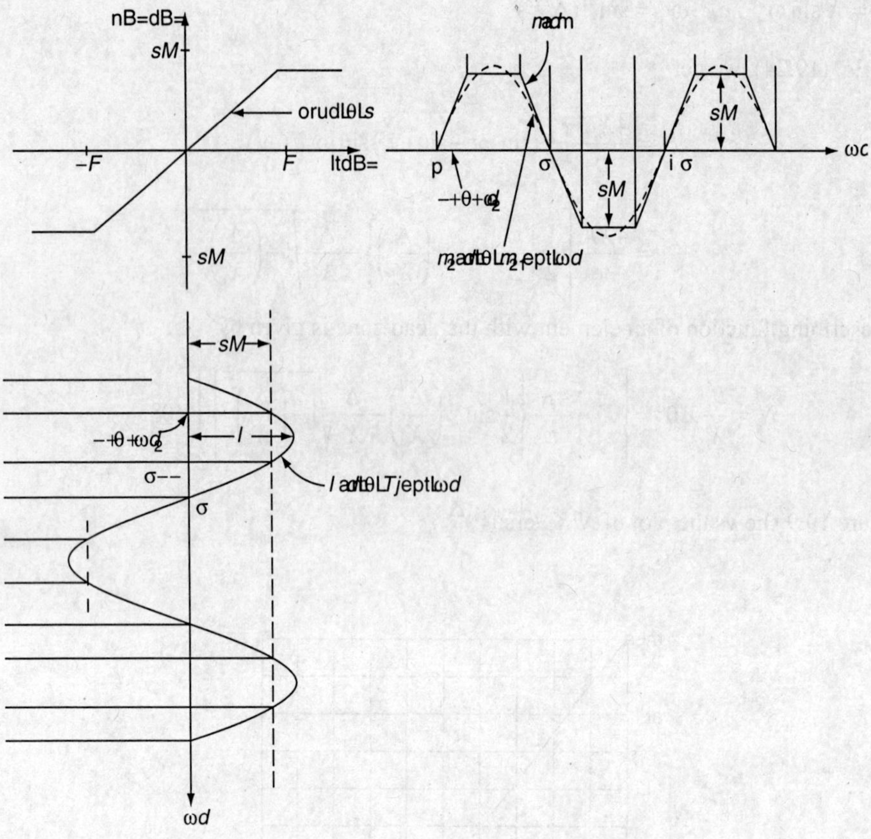

Fig. 19.10 Sinusoidal response of non-linearity with saturation non-linearity

where

$$B_1 = \frac{1}{\pi}\int_0^{2\pi} y(t)\sin\omega t\, d(\omega t) = \frac{4}{\pi}\int_0^{\frac{\pi}{2}} y(t)\sin\omega t\, d(\omega t) \tag{19.27}$$

From Fig. 19.10, the output $y(t)$ for $0 \le \omega t \le \pi/2$ is given by

$$y(t) = \begin{cases} KX\sin\omega t & \text{for } 0 < \omega t < \beta \\ KS & \text{for } \beta < \omega t < \dfrac{\pi}{2} \end{cases} \tag{19.28}$$

Using Eq. (19.28), we get form Eq. (19.27)

$$B_1 = \frac{4}{\pi}\int_0^{\frac{\pi}{2}} y(t)\sin\omega t\, d(\omega t)$$

$$= \frac{4}{\pi}\left[\int_0^{\beta} KX\sin^2\omega t\, d(\omega t) + \int_{\beta}^{\frac{\pi}{2}} KS\sin\omega t\, d(\omega t)\right]$$

$$B_1 = \frac{4K}{\pi}\left[\frac{X}{2} + \beta + S\cos\beta\right] \tag{19.29}$$

where $\beta = \sin^{-1}(S/X)$.

The describing function of such an element is given by

$$N = \frac{B_1}{X}\angle 0^\circ$$

$$= \frac{4K}{\pi}\left[\frac{1}{2}\sin^{-1}\left(\frac{S}{X}\right) + \frac{S}{X}\sqrt{1-\left(\frac{S}{X}\right)^{1}}\right]\angle 0^\circ \tag{19.30}$$

The plot of N/K as a function of S/X is shown in Fig. 19.11.

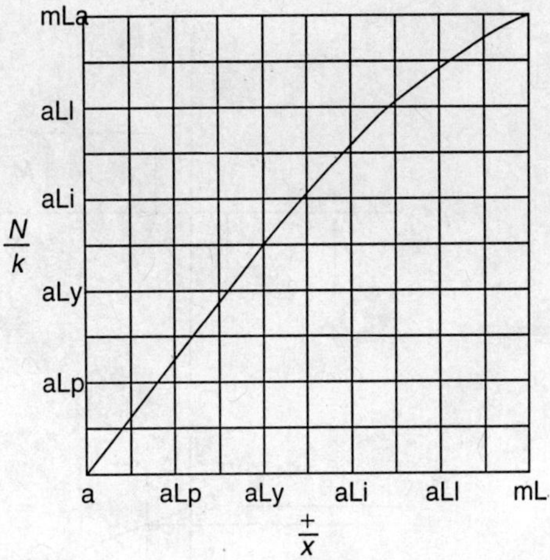

Fig. 19.11 Describing function for the saturation non-linearity

The relation between the describing function for the dead zone non-linearity and that of saturation non-linearity is given by

$$N_{\text{dead-zone}} = K - N_{\text{saturation}} \quad \text{for } \Delta = S \tag{19.31}$$

19.3.3 On–Off Non-Linearity with Dead Zone

Figure 19.12 shows the describing function for the on–off non-linearity with dead zone.

Figure 19.13 shows the input–output characteristics for the on–off non-linearity with dead zone. It shows that the output is odd and the fundamental contains only sine terms.

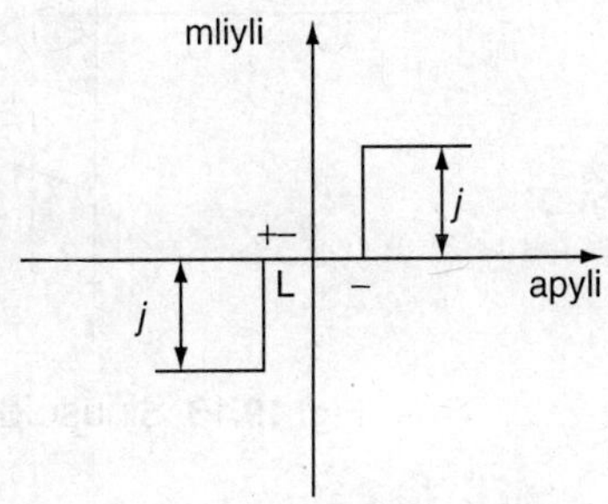

Fig. 19.12 On–off non-linearity with dead zone

Form Fig. 19.13, the output $y(t)$ for $0 \le \omega t \le \pi$ is given by

$$y(t) = \begin{cases} 0 & \text{for } 0 < t \le t_1 \\ M & \text{for } t_1 < t < \left(\dfrac{\pi}{\omega} - t_1\right) \\ 0 & \text{for } \left(\dfrac{\pi}{\omega} - t_1\right) < t < \dfrac{\pi}{\omega} \end{cases} \tag{19.32}$$

The fundamental harmonic component of output is given by

$$y(t) = B_1 \sin\omega t$$

where

$$B_1 = \frac{1}{\pi}\int_0^{2\pi} y(t)\sin\omega t\, d(\omega t) = \frac{4}{\pi}\int_0^{\frac{\pi}{2}} y(t)\sin\omega t\, d(\omega t)$$

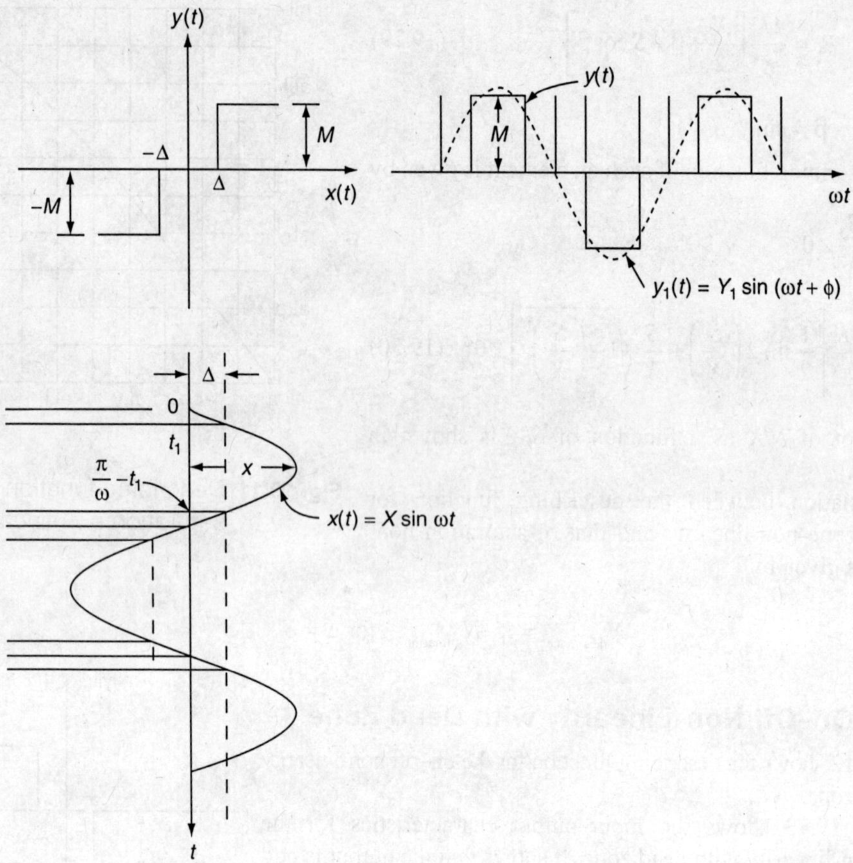

Fig. 19.13 Sinusoidal response of on–off with dead-zone non-linearity

or

$$B_1 = \frac{4}{\pi}\int_{\omega t_1}^{\frac{\pi}{2}} M \sin \omega t \; d(\omega t) = \frac{4M}{\pi} \cos \omega t_1$$

Now $\quad \sin \omega t = \dfrac{\Delta}{X} \quad$ and $\quad \cos \omega t_1 = \sqrt{1 - \left(\dfrac{\Delta}{X}\right)^2}.$

Therefore, the describing function is given by

$$N = \frac{4M}{\pi X}\left(\sqrt{1 - \left(\frac{\Delta}{X}\right)^2}\right) \angle 0° \qquad (19.33)$$

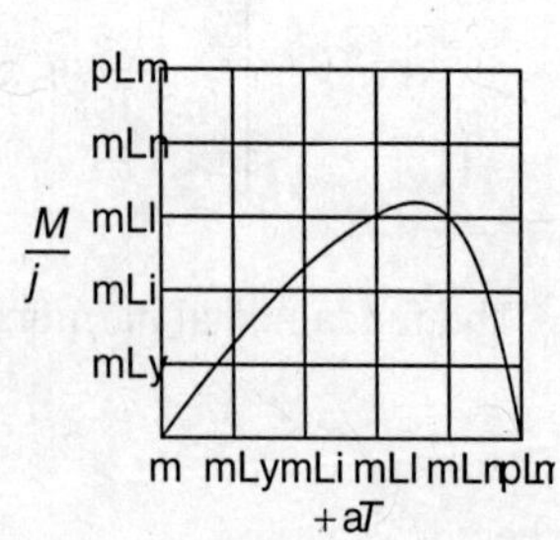

Fig. 19.14 Plot of describing function

The describing function is only function of the input amplitude X. Fig. 19.14 shows the plot of N/M versus Δ/X.

19.3.4 On–Off Non-Linearity

Figure 19.15 shows the input–output characteristic curve of an element with on–off non-linearity where the output of this element is either a positive constant or a negative constant. The on–off non-linearity is sometimes called two-position non-linearity.

Let $x = X \sin \omega t$ be the input to a non-linear element. By Fourier series the output is given by

$$y(t) = A_0 + \sum_{n=1}^{\infty} (A_n \sin \omega t + B_n \cos \omega t)$$

Since the output is an odd term, $y(t) = \sum_{n=1}^{\infty} B_n \cos \omega t$. The fundamental harmonic component is given by

$$y(t) = B_1 \sin \omega t$$

where

$$B_1 = \frac{1}{\pi}\int_0^{2\pi} y(t) \sin \omega t \, d(\omega t) = \frac{2}{\pi}\int_0^{\pi} y(t) \sin \omega t \, d(\omega t)$$

$$= \frac{2}{\pi}\int_0^{\pi} M \sin \omega t \, d(\omega t) = \frac{4M}{\pi}$$

Therefore, $y(t) = B_1 \sin \omega t = \dfrac{4M}{\pi} \sin \omega t$. The describing function N is given by

$$N = \frac{Y_1}{X} \angle 0° = \frac{4M}{\pi X} \angle 0° \tag{19.34}$$

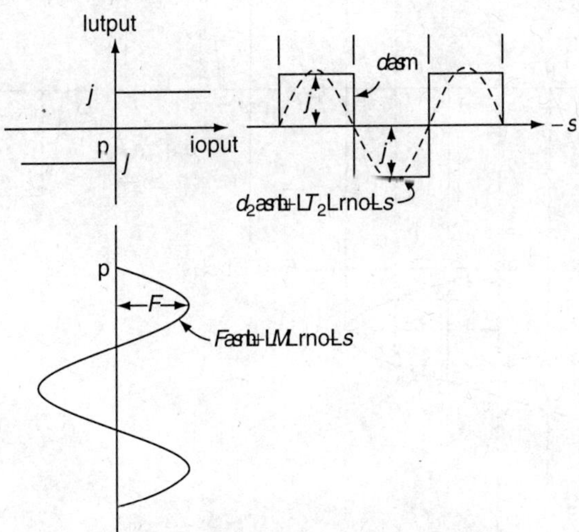

Fig.19.15 Sinusoidal response of non-linearity with on–off non-linearity

Equation (19.34) shows that the describing function for the on–off element is a real quantity and is a function of X only. Figure 19.16 shows the plot of N versus M/X.

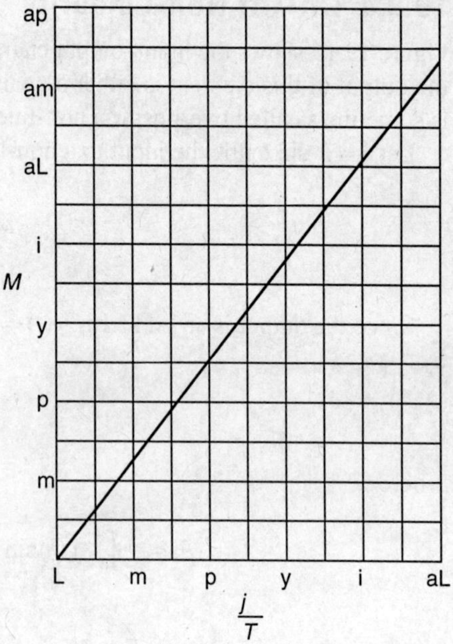

Fig. 19.16 Describing function for the on–off non-linearity

19.3.5 Dead Zone and Saturation

Figure 19.17 shows the idealized characteristic of a non-linearity having dead zone and its sinusoidal response. The output is given by

$$y(t) = \begin{cases} 0 & \text{for } 0 \geq \omega t \geq \alpha \\ K(X-\Delta) & \text{for } \alpha \geq \omega t \geq \beta \\ K(S-\Delta) & \text{for } \beta \geq \omega t \geq (\pi-\beta) \\ K(X-\Delta) & \text{for } (\pi-\beta) \geq \omega t \geq (\pi-\alpha) \\ 0 & \text{for } (\pi-\alpha) \geq \omega t \geq \pi \end{cases}$$

Since the output is an odd term, $y(t) = \sum_{n=1}^{\infty} B_n \cos \omega t.$

The output has also half-wave as well as quarter-wave symmetry also. The fundamental harmonic component is given by

$$y(t) = B_1 \sin \omega t$$

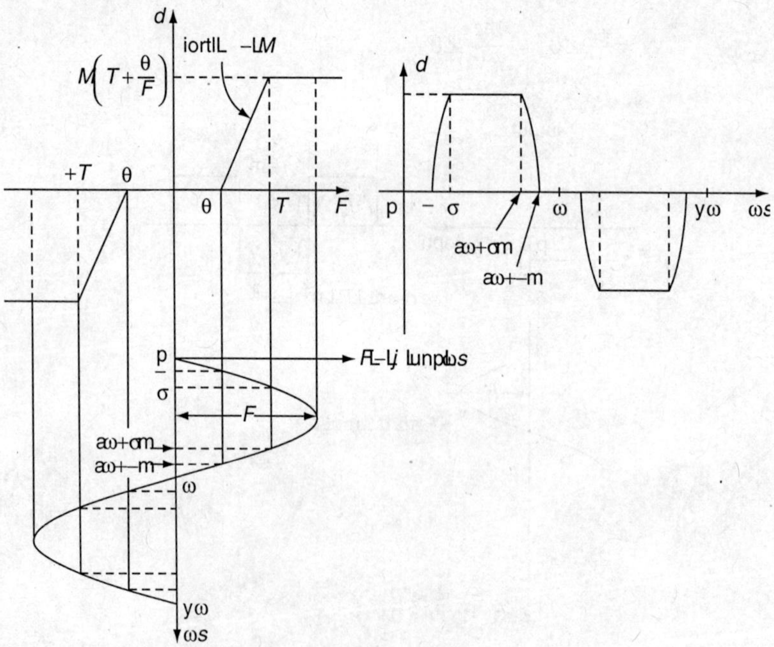

Fig. 19.17 Sinusoidal response of non-linearity with dead zone and saturation

where

$$B_1 = \frac{1}{\pi}\int_0^{2\pi} y(t)\sin\omega t\, d(\omega t) = \frac{4}{\pi}\int_0^{\frac{\pi}{2}} y(t)\sin\omega t\, d(\omega t)$$

or

$$B_1 = \frac{4}{\pi}\int_\alpha^\beta K(X\sin\omega t - \Delta)\sin\omega t\, d(\omega t) + \frac{4}{\pi}\int_\beta^{\frac{\pi}{2}} K(S-\Delta)\sin\omega t\, d(\omega t)$$

$$= \frac{KX}{\pi}[2(\beta-\alpha) + (\sin 2\beta - \sin 2\alpha)]$$

The describing function is given by

$$\frac{N}{K} = \begin{cases} 0 & \text{for } |X| < S, \alpha = \beta = \dfrac{\pi}{2} \\ 1 - \dfrac{2}{\pi}(\alpha + \sin\alpha\cos\alpha) & \text{for } \Delta < |X| < S, \beta = \dfrac{\pi}{2} \\ \dfrac{1}{\pi}[2(\beta-\alpha) + (\sin 2\beta - \sin 2\alpha)] & \text{for} |X| > S \end{cases} \tag{19.35}$$

For saturation non-linearity

$$\Delta = 0 \quad \text{and} \quad \alpha = 0$$

Therefore, the describing function in given by

$$\frac{N}{K} = \begin{cases} 1 & \text{for} |X| < S \\ \dfrac{2}{\pi}(\beta + \sin\beta\cos\beta) & \text{for} |X| \geq S \\ \dfrac{2}{\pi}\left[\sin^{-1}\left(\dfrac{S}{X}\right) + \left(\dfrac{S}{X}\right)\sqrt{1-\left(\dfrac{S}{X}\right)^2}\right] & \end{cases} \tag{19.36}$$

Figure 19.18 shows the describing function of saturation non-linearity.

Dead-zone non-linearity In this case $S \to \infty$, $\beta = \pi/2$. Therefore, the describing function is given by

$$\frac{N}{K} = \begin{cases} 0 & \text{for } |X| < \Delta \\ 1 - \dfrac{2}{\pi}(\alpha + \sin\alpha\cos\alpha) = 1 - \dfrac{2}{\pi}\sin^{-1}\left(\dfrac{\Delta}{X}\right) + \left(\dfrac{\Delta}{X}\right)\sqrt{1-\left(\dfrac{\Delta}{X}\right)^2} & \text{for } |X| > \Delta \end{cases} \tag{19.37}$$

Figure 19.19 shows the describing function with dead-zone non-linearity.

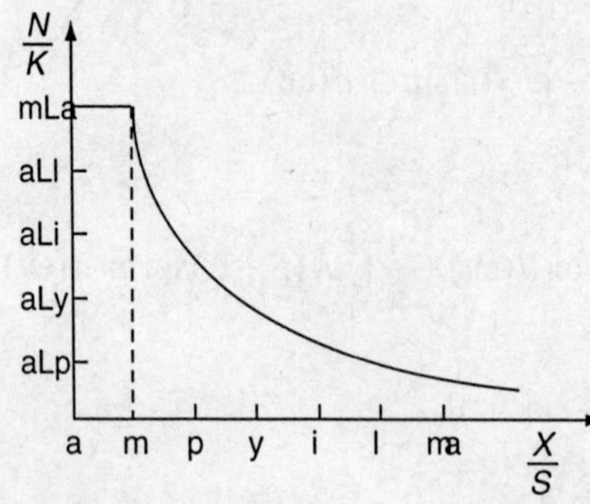

Fig. 19.18 Describing function of saturation non-linearity

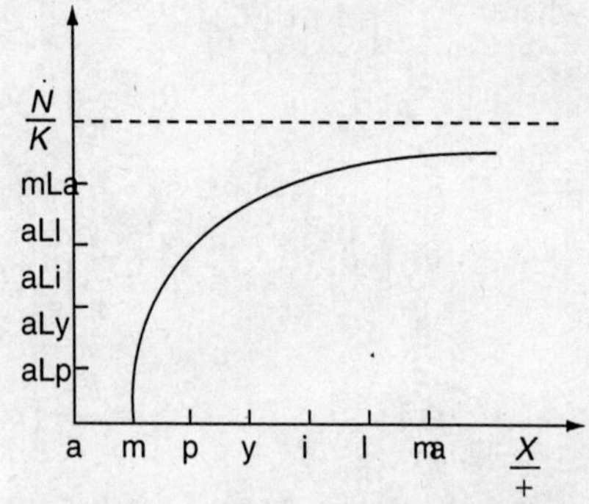

Fig. 19.19 Describing function of dead-zone non-linearity

For relay with dead zone In this case $S = \Delta$ and slope $K = \infty$ and the saturation value is $|M|$. Taking $a = \beta$, the expression for N/M is given by

$$\frac{N}{M} = \begin{cases} 0 & \text{for } |X| < \Delta \\ \dfrac{4}{\pi X}\sqrt{1-\left(\dfrac{\Delta}{X}\right)^2} & \text{for } |X| > \Delta \end{cases} \tag{19.38}$$

19.3.6 On–Off Non-Linearity with Hysteresis

Figure 19.20 shows the input–output characteristic curve of an on–off element. Let the input $x(t) = X\sin\omega t$, $X > h$ where h is the dead zone of the on–off non-linearity with hysteresis. The output is given by

$$y(t) = \begin{cases} M & \text{for } t_1 \le t \le \left(\dfrac{\pi}{\omega}+t_1\right) \\ -M & \text{for } \left(\dfrac{\pi}{\omega}+t_1\right) \le t \le \left(\dfrac{2\pi}{\omega}+t_1\right) \end{cases} \tag{19.39}$$

where $\omega t_1 = \sin^{-1}\left(\dfrac{h}{X}\right)$. Due to odd symmetry of the output, its fundamental contains only sine terms. The fundamental harmonic component is given by

$$y(t) = B_1 \sin\omega t$$

where

$$B_1 = \frac{1}{\pi}\int_0^{2\pi} y(t)\sin\omega t\, d(\omega t) = \frac{2}{\pi}\int_{\omega t_1}^{\pi+\omega t_1} y(t)\sin\omega t\, d(\omega t)$$

$$= \frac{2}{\pi}\int_{\omega t_1}^{\pi+\omega t_1} M\sin\omega t\, d(\omega t) = \frac{4M}{\pi}$$

$$\therefore \qquad \frac{B}{X} = \frac{4M}{\pi X}$$

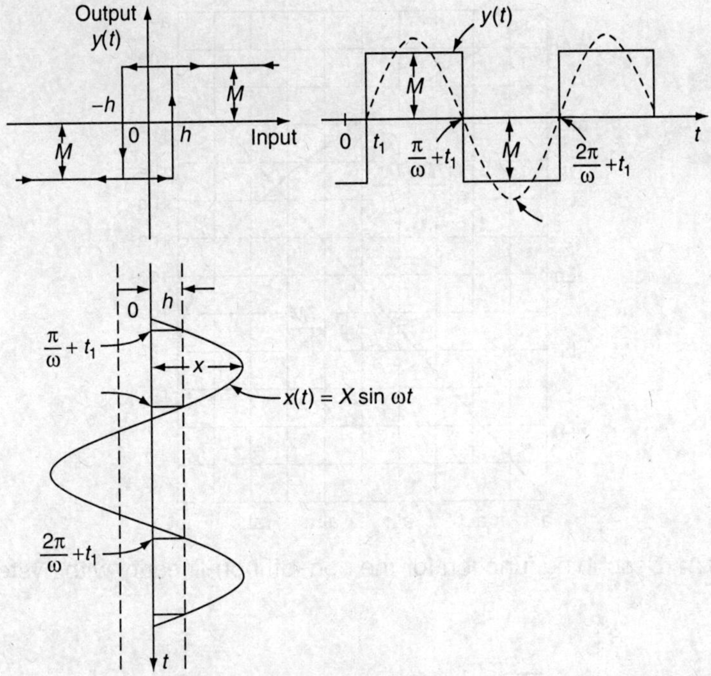

Fig. 19.20 Sinusoidal response of relay with hysteresis-type non-linearity

The output is a square wave and it lags behind the input by an angle $-\sin^{-1}(h/X)$ and the describing function for the non-linear element is given by

$$N = \frac{4M}{\pi X} \angle -\sin^{-1}\left(\frac{h}{X}\right)$$

Generally it is convenient to plot $(h/M)N = \left(\frac{4h}{\pi X}\right) \angle -\sin^{-1}(h/X)$ versus $\frac{h}{X}$ instead of N versus h/X. Figure 19.21 shows the plot of hN/M versus h/X.

19.3.7 Relay with Dead Zone and Hysteresis

Figure 19.22 shows the input–output characteristics of a relay with dead-zone hysteresis. The output $y(t)$ is given by

$$y(t) = \begin{cases} 0 & \text{for } 0 \le \omega t \le \pi \\ +M & \text{for } 0 \le \omega t \le \pi \\ 0 & \text{for } (\pi - \beta) \le \omega t \le (\pi + \alpha) \\ -M & \text{for } (\pi + \alpha) \le \omega t \le (2\pi - \beta) \\ 0 & \text{for } (2\pi - \beta) \le \omega t \le 2\pi \end{cases}$$

where $\alpha = \sin^{-1}(\Delta/X)$ and $\beta = \sin^{-1}[\Delta - 2h/X]$.

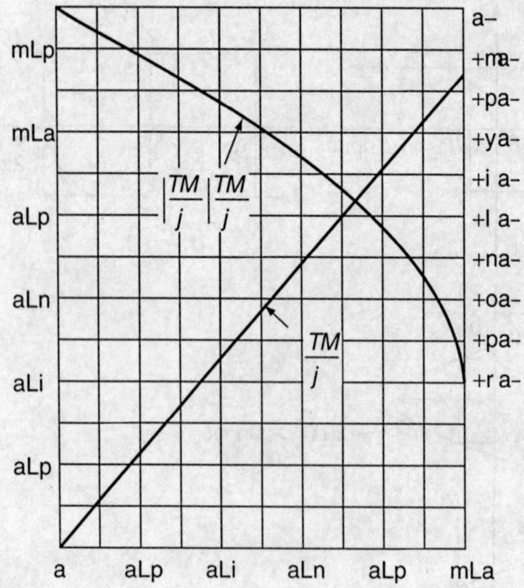

Fig. 19.21 Describing function for the non–off non-linearity with hysteresis

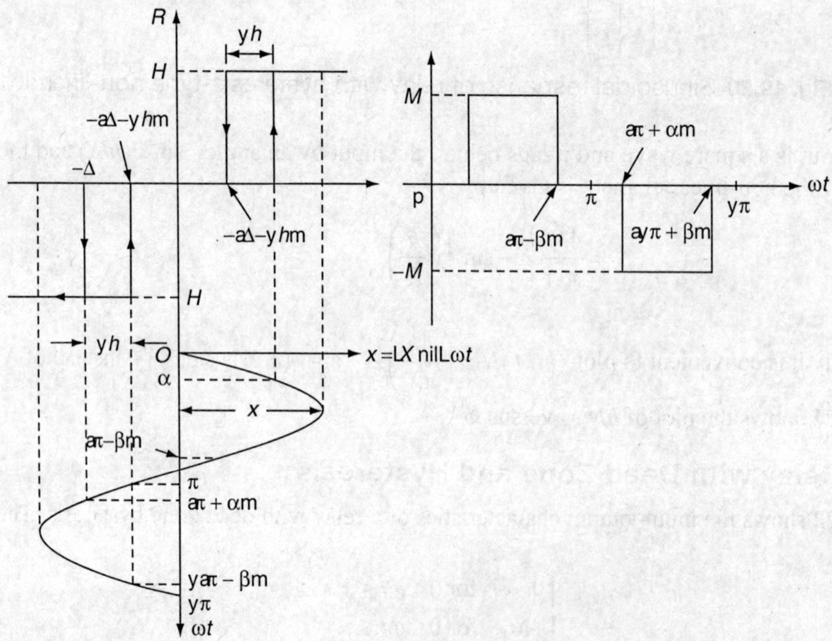

Fig. 19.22 Sinusoidal response of relay with dead-zone and hysteresis

$$A_1 = \frac{2}{\pi}\int_0^{\pi} y(t)\cos\omega t \, d(\omega t) = \frac{2}{\pi}\int_{\alpha}^{\pi-\beta} M\cos\omega t \; d(\omega t) = -\frac{4Mh}{\pi X}$$

and

$$B_1 = \frac{2}{\pi}\int_0^{\pi} y(t)\sin\omega t\, d(\omega t) = \frac{2}{\pi}\int_{\alpha}^{\pi-\beta} M \sin\omega t\ d(\omega t)$$

$$= \frac{2M}{\pi}(\cos\alpha + \cos\beta)$$

$$= \frac{2M}{\pi}\left[\sqrt{1-\left(\frac{\Delta}{X}\right)^2} + \sqrt{1-\left(\frac{\Delta-2h}{X}\right)^2}\right]$$

$$\therefore \quad \frac{A_1}{X} = \frac{2M}{\pi X}\left(-\frac{2h}{X}\right) \quad \text{and} \quad \frac{B_1}{X} = \frac{2M}{\pi}\left[\sqrt{1-\left(\frac{\Delta}{X}\right)^2} + \sqrt{1-\left(\frac{\Delta-2h}{X}\right)^2}\right]$$

Therefore,

$$N = \begin{cases} \sqrt{\left(\dfrac{A_1}{X}\right)^2 + \left(\dfrac{B_1}{X}\right)^2} \angle \tan^{-1}\left(\dfrac{B_1}{A_1}\right) & \text{for } |X| < \Delta \\ 0 & \text{for } |X| < \Delta \end{cases} \tag{19.40}$$

Equation (19.39) shows that output is independent of frequency and dependent on the history of input, i.e., it is a memory type non-linearity.

Figure 19.23 shows the plot of $|\Delta N/M|$ and $\angle N$ versus $X/2\Delta$ for various values of $\dfrac{h}{\Delta}$.

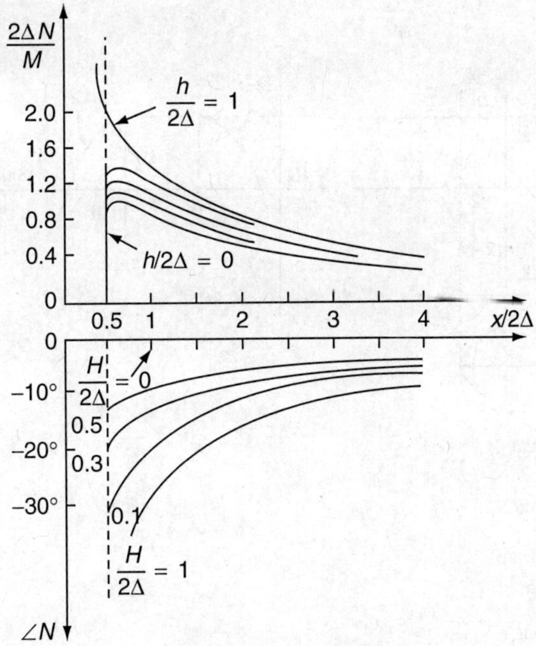

Fig. 19.23 Describing function of relay with dead-zone and hysteresis

19.3.8 Backlash

Figure 19.24 shows the input–output characteristic of a backlash characteristic. The output is given below.

$$y(t)=\begin{cases} x-\dfrac{b}{2} & \text{for } 0<\omega t<\dfrac{\pi}{2} \\ X-\dfrac{b}{2} & \text{for } \dfrac{\pi}{2}<\omega t<(\pi-\beta) \\ x+\dfrac{b}{2} & \text{for } (\pi-\beta)<\omega t<\dfrac{3\pi}{2} \\ -X+\dfrac{b}{2} & \text{for } \dfrac{3\pi}{2}<\omega t<(2\pi-\beta) \\ x-\dfrac{b}{2} & \text{for } (2\pi-\beta)<\omega t<2\pi \end{cases}$$

where $$\beta=\sin^{-1}(1-b/X)$$

Now,

$$A_1=\frac{2}{\pi}\int_0^x y(t)\cos\omega t\, d(\omega t)$$

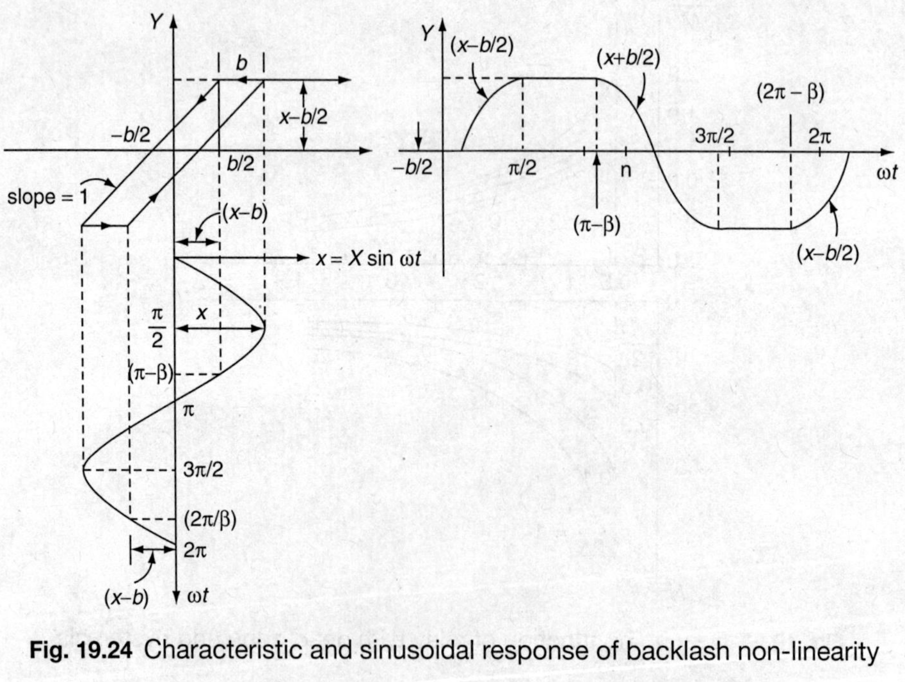

Fig. 19.24 Characteristic and sinusoidal response of backlash non-linearity

$$= \frac{2}{\pi}\int_0^{\frac{\pi}{2}}\left(X\sin\omega t - \frac{b}{2}\right)\cos\omega t\, d(\omega t) + \frac{2}{\pi}\int_{\frac{\pi}{2}}^{\pi-\beta}\left(X - \frac{b}{2}\right)\cos\omega t\, d(\omega t)$$

$$+\frac{2}{\pi}\int_{\pi-\beta}^{\pi}\left(X\sin\omega t + \frac{b}{2}\right)\cos\omega t\, d(\omega t) = -\frac{X}{\pi}\cos^2\beta \tag{19.41}$$

and

$$B_1 = \frac{2}{\pi}\int_0^{\pi} y(t)\sin\omega t\, d(\omega t)$$

$$= \frac{2}{\pi}\int_0^{\frac{\pi}{2}}\left(X\sin\omega t - \frac{b}{2}\right)\sin\omega t\, d(\omega t) + \frac{2}{\pi}\int_{\frac{\pi}{2}}^{\pi-\beta}\left(X - \frac{b}{2}\right)\sin\omega t\, d(\omega t)$$

$$+\frac{2}{\pi}\int_{\pi-\beta}^{\pi}\left(X\sin\omega t + \frac{b}{2}\right)\sin\omega t\, d(\omega t) = \frac{X}{\pi}\left[\left(\frac{\pi}{2} + \beta\right) + \frac{1}{2}\sin 2\beta\right] \tag{19.42}$$

Therefore,

$$N = \begin{cases} 0 & \text{for } |X| < \dfrac{b}{2} \\ \sqrt{\left(\dfrac{A_1}{X}\right)^2 + \left(\dfrac{B_1}{X}\right)^2}\ \tan^{-1}\dfrac{B_1}{A_1} & \text{for } |X| > \dfrac{b}{2} \end{cases}$$

Figure 19.25 shows the magnitude and phase characteristics of the describing function of a backlash non-linearity.

19.4 Stability

The stability characteristic of a non-linear system can be determined using Lyapunov method. The following definitions are important before studying the Lyapunov method.

Let us consider a scalar function F of state variable x, i.e., $F(x) = F_1(x_1, x_2, x_3, \ldots, x_n)$. If the function F has always a positive or negative sign in a region about the origin except at the origin, the function is said to be positive or negative definite. The function $F = x_1^2 + x_2^2 + x_3^2$ is positive definite in three-dimensional space except at the origin. The function $F = -2x_1^2 - 3x_2^2 - 5x_3^2$ is negative definite in three-dimensional space except at the origin. The function $F = (x_1 - 5) + (x_2 - 7)^2$ is not positive definite because it has zero values at $x_1 = 5$ and $x_2 = 7$ also. This function is called positive semi-definite. Similarly, the function $F = -2(x_1 - 5)^2 - 3(x_2 - 7)^2$ is not negative definite because it has zero values at $x_1 = 5$ and $x_2 = 7$ also. This function is called negative semi-definite.

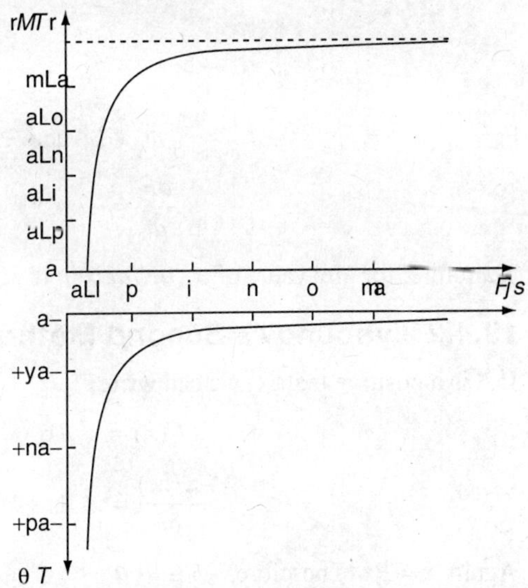

Fig. 19.25 Describing function of backlash non-linearity

19.4.1 Lyapunov's First Theorem

If there exists a scalar function $F(x)$ for an autonomous system $\dot{x} = F(x)$ such $F(x)$ is real, continuous with continuous first order partial derivatives and

$$F(0) = 0$$

$$F(x) > 0 \quad \text{for all} \quad x \neq 0, \quad \text{and}$$

$$\frac{dF(x)}{dt} < \quad \text{for all} \quad x \neq 0$$

Then, the system is asymptotically stable in the neighbourhood of the origin. On the other hand, if $\frac{dF(x)}{dt} < 0$, the system is stable but there is no assurance for asymptotic stability. The function $F(x)$ is known as Lyapunov's function. Let us consider and autonomous system expressed as

$$\dot{x}_1 = x_2$$

$$\dot{x}_2 = -m_1 x_1 - m_2 x_2$$

Let us take Lyapunov's function as

$$F = x_1^2 + x_2^2$$

Here $F = x_1^2 + x_2^2$ is positive definite. Therefore, $F(0) = 0$. Now,

$$\frac{dF}{dt} = \frac{\partial F}{\partial x_1} \cdot \frac{\partial x_1}{\partial t} + \frac{\partial F}{\partial x_2} \cdot \frac{\partial x_2}{\partial t} = 2x_1\dot{x}_1 + 2x_2\dot{x}_2$$

$$= 2x_1\dot{x}_1 + 2x_2(-m_1x_1 - m_2x_2) = 2x_1x_2 - 2x_2m_2x_2 - 2x_2m_1x_1$$

$$= 2(x_1 - m_2x_2 - m_1x_1)x_2 = [2(1 - m_1)x_1 - 2m_2x_2]x_2$$

For $\quad m_1 = 1 \quad \text{and} \quad m_2 > 0, \quad$ we get

$$\frac{dF}{dt} = -2m_2x_2^2$$

Therefore, for any value of $x_1, dF/dt = 0$ if $x_2 = 0$.

19.4.2 Lyapunov's Second Method

If A is a positive matrix, we can write

$$F(x) = x^T Ax$$

$$\therefore \quad \frac{\delta F(x)}{\delta t} = \dot{x}^T Ax + x^T A\dot{x}$$

Again $\dot{x} = Bx$. Therefore, $\dot{x}^T = x^T B^T$

$$\therefore \quad \frac{\delta F(x)}{\delta t} = \dot{x}^T Ax + x^T A\dot{x} = x^T B^T Ax + x^T ABx = x^T (B^T A + AB)x$$

Let $$C = -(B^T A + AB)$$

Therefore, $$\frac{\delta F(x)}{\delta t} = \dot{F}(x) = -x^T Cx.$$

Selecting arbitrarily a positive definite or semi-definite matrix for C, we can solve for C. The necessary and sufficient condition for asymptotic stability is that the matrix C must be positive definite.

Any matrix is said to be positive definite if its symmetric component is positive definite.

Now

$$C = \frac{C + C^T}{2} = -\left[\frac{1}{2}(B^T A + AB) + \frac{1}{2}(A^T B + BA)\right]$$

$$= -\left[B^T\left(\frac{A + A^T}{2}\right) + \left(\frac{A + A^T}{2}\right)B\right]$$

Let

$$D = \left(\frac{P + P^T}{2}\right),$$

where D is symmetric.

Therefore, we can write $$C = -[B^T D + DA]$$

Let us consider an autonomous system expressed as

$$\dot{x}_1 = x_2$$

$$\dot{x}_2 = -m_1 x_1 - m_2 x_2$$

For the system we can write

$$\begin{bmatrix}\dot{x}_1 \\ \dot{x}_2\end{bmatrix} = \begin{bmatrix}0 & 1 \\ -m_1 & -m_2\end{bmatrix}\begin{bmatrix}x_1 \\ x_2\end{bmatrix} = Bx$$

Let $$C = (B^T D + DB) = I$$

i.e., $$\begin{bmatrix}0 & -m_1 \\ 1 & -m_2\end{bmatrix}\begin{bmatrix}D_{11} & D_{12} \\ D_{21} & D_{22}\end{bmatrix} + \begin{bmatrix}D_{11} & D_{12} \\ D_{21} & D_{22}\end{bmatrix}\begin{bmatrix}0 & 1 \\ -m_1 & -m_2\end{bmatrix} = \begin{bmatrix}-1 & 0 \\ 0 & -1\end{bmatrix}$$

From the above equation, we can write

$$-2m_2 D_{12} = -1 \tag{19.43}$$

$$D_{11} - m_1 D_{12} - m_2 D_{22} = 0 \tag{19.44}$$

$$2D_{12} - 2m_1 D_{22} = -1 \tag{19.45}$$

From Eqs. (19.43)–(19.45) we get

$$D_{11} = \frac{m_1^2 + m_2^2 + m_1}{2m_1 m_2} = \frac{m_2}{2m_1} + \frac{m_1 + 1}{2m_2}$$

$$D_{11} = \frac{1}{2m_1} \quad \text{and} \quad D_{22} = \frac{m_2 + 1}{2m_1 m_2}$$

To make D positive definite

$$D_{11} > 0 \tag{19.46}$$

and

$$D_{11}D_{22} - D_{12}^2 > 0 \tag{19.47}$$

From Eqs. (19.46) and (19.47) we get

$$\frac{m_1^2 + m_2^2 + m_1}{2m_1 m_2} > 0 \tag{19.48}$$

and

$$\frac{1}{m_1 m_2^2}[m_2^2 + (m_1 + 1)^2] > 0 \tag{19.49}$$

To make Eq. (19.49) true, $m_1 > 0$ and $[\{m_2^2 + (m_1 + 1)^2\}/m_2^2]$ is greater than zero.

Here $m_1 > 0$ and to make D positive definite, m_2 must be greater than zero. For stability from Lyapunov's second method, $m_1 > 0$ and $m_2 > 0$. In Lyapunov's first method we have already considered $m_1 > 0$ and m_2 must be greater than zero. Therefore, Lyapunov's first method is highly conservative.

SIGNIFICANT POINTS

Non-Linear Control Systems: There are control systems which do not have any linearized version and it is very difficult to linearize these control systems. In some cases it has been found that non-linearities such as dead zone and hysteresis and combinations of these are undesirable and unavoidable.

Dead-Zone Non-Linearity

$$N = \frac{B_1}{X}\angle 0° = \left[K - \frac{4K}{\pi}\left\{\frac{1}{2}\sin^{-1}\left(\frac{\Delta}{X}\right) - \frac{\Delta}{X}\sqrt{1 - \left(\frac{\Delta}{X}\right)^2}\right\}\right]\angle 0°$$

Saturation-Non-Linearity

$$N = \frac{B_1}{X}\angle 0° = \frac{4K}{\pi}\left[\frac{1}{2}\sin^{-1}\left(\frac{S}{X}\right) + \frac{S}{X}\sqrt{1 - \left(\frac{S}{X}\right)^2}\right]\angle 0°$$

On–Off Non-Linearity with Dead Zone

$$N = \frac{4M}{\pi X}\left(\sqrt{1 - \left(\frac{\Delta}{X}\right)^2}\right)\angle 0°$$

On–Off Non-Linearity

$$N = \frac{Y_1}{x} < 0° = \frac{4M}{\pi x} < 0°$$

Dead Zone and Saturation

$$\frac{N}{K} = \begin{cases} 0 & \text{for } |X| < S, \quad \alpha = \beta = \frac{\pi}{2} \\ 1 - \frac{2}{\pi}(\alpha + \sin\alpha\cos\alpha) & \text{for } \Delta < |X| < S, \quad \beta = \frac{\pi}{2} \\ \frac{1}{\pi}[2(\beta - \alpha) + (\sin 2\beta - \sin 2\alpha)] & \text{for } |X| > S \end{cases}$$

For Saturation Non-Linearity

$$D = 0 \quad \text{and} \quad \alpha = 0$$

Therefore, the describing function is given by

$$\frac{N}{K} = \begin{cases} 1 & \text{for } |X| < S \\ \frac{2}{\pi}(\beta + \sin\beta\cos\beta) & \text{for } |X| \geq S \\ \frac{2}{\pi}\left[\sin^{-1}\left(\frac{S}{X}\right) + \left(\frac{S}{X}\right)\sqrt{1 - \left(\frac{S}{X}\right)^2}\right] & \end{cases}$$

Dead-Zone Non-linearity

In this case $S \to \infty, \beta = \pi/2$. Therefore, the describing function is given by

$$\frac{N}{K} = \begin{cases} 0 & \text{for } |X| < \Delta \\ 1 - \frac{2}{\pi}(\alpha + \sin\alpha\cos\alpha) = 1 - \frac{2}{\pi}\sin^{-1}\left(\frac{\Delta}{X}\right) + \left(\frac{\Delta}{X}\right)\sqrt{1 - \left(\frac{\Delta}{X}\right)^2} & \text{for } |X| > \Delta \end{cases}$$

For Relay with Dead Zone

In this case $S = \Delta$ and slope $K = \infty$ and the saturation value is $|M|$. Taking $\alpha = \beta$, the expression for N/M is given by

$$\frac{N}{M} = \begin{cases} 0 & \text{for } |X| < \Delta \\ \frac{4}{\pi X}\sqrt{1 - \left(\frac{\Delta}{X}\right)^2} & \text{for } |X| > \Delta \end{cases}$$

On–Off Non-Linearity with Hysteresis

$$N = \frac{4M}{\pi X} \angle -\sin^{-1}\left(\frac{h}{X}\right)$$

Relay with Dead-Zone Hysteresis

$$N = \begin{cases} \sqrt{\left(\frac{A_1}{X}\right)^2 + \left(\frac{B_1}{X}\right)^2} \angle \tan^{-1}\left(\frac{B_1}{A_1}\right) & \text{for } |X| > \Delta \\ 0 & \text{for } |X| < \Delta \end{cases}$$

Backlash

$$N = \begin{cases} 0 & \text{for } |X| < \frac{b}{2} \\ \sqrt{\left(\frac{A_1}{X}\right)^2 + \left(\frac{B_1}{X}\right)^2} \tan^{-1}\frac{B_1}{A_1} & \text{for } |X| > \frac{b}{2} \end{cases}$$

Lyapunov's First Theorem: If there exists a scalar function $F(x)$ for an autonomous system $\dot{x} = f(x)$ such that $F(x)$ is real, continuous with continuous first partial derivatives and

$$F(0) = 0$$

$$F(x) > 0 \quad \text{for all } x \neq 0$$

$$\frac{dF(x)}{dt} < 0 \quad \text{for all } x \neq 0$$

Then, the system is asymptotically stable in the neighbourhood of the origin. On the other hand, if $\frac{dF(x)}{dt} \leq 0$, the system is stable but there is no assurance for asymptotic stability. The function $F(x)$ is known as Lyapunov's function.

Lyapunov's Second Method: If A is a positive matrix, we can write

$$F(x) = x^T A x$$

$$\therefore \quad \frac{\delta F(x)}{\delta t} = \dot{x}^T A x + x^T A \dot{x}$$

Again $\dot{x} = Bx$. Therefore $\dot{x}^T = x^T B^T$,

$$\therefore \quad \frac{\delta F(x)}{\delta t} = \dot{x}^T A x + x^T A \dot{x} = x^T B^T A x + x^T A B x = x^T (B^T A + AB) x$$

Let $$C = -(B^T A + AB)$$

Therefore, $$\frac{\delta F(x)}{\delta t} = \dot{F}(x) = -x^T C x$$

Selecting arbitrarily a positive definite or semi-definite matrix for C, we can solve for C. The necessary and sufficient condition for asymptotic stability is that the matrix C must be positive definite.

SHORT QUESTIONS WITH ANSWERS

1. What do you mean by non-linear control systems?

   Ans. There are control systems which do not have any linearized version and it is very difficult to linearize these control systems. In some cases it has been found that non-linearities such as dead-zone and hysteresis and combinations of these are undesirable and unavoidable.

2. State Lyapunov's first theorem.

   Ans. If there exists a scalar function $F(x)$ for an autonomous system such that $F(x)$ is real, continuous with continuous first partial derivatives and

   (i) $F(0) = 0$

   (ii) $F(x) > 0$ for all $x \neq 0$

   (iii) $\dfrac{dF(x)}{dt} <$ for all $x \neq 0$

   Then, the system is asymptotically stable in the neighbourhood of the origin. On the other hand, if $\dfrac{dF(x)}{dt} \leq 0$, the system is stable but there is no assurance for asymptotic stability. The function $F(x)$ is known as Lyapunov's function.

3. State Lyapunov's second theorem.

   Ans. If A is a positive matrix, we can write

   $$F(x) = x^T Ax$$

   $$\therefore \quad \frac{\delta F(x)}{\delta t} = \dot{x}^T Ax + x^T A\dot{x}$$

   Again $\dot{x} = Bx$. Therefore, $\dot{x}^T = x^T B^T$

   $$\therefore \quad \frac{\delta F(x)}{\delta t} = \dot{x}^T Ax + x^T A\dot{x} = x^T B^T Ax + x^T ABx = x^T (B^T A + AB)x$$

   Let $$C = (B^T A + AB)$$

   Therefore, $$\frac{\delta F(x)}{\delta t} = \dot{F}(x) = -x^T Cx$$

   Selecting arbitrarily a positive definite or semi-definite matrix for C, we can solve for C. The necessary and sufficient condition for asymptotic stability is that the matrix C must be positive definite.

EXERCISES

Questions for Practice

1. What do you mean by describing function?
2. Define the phase plane technique.
3. What do you mean by relay? Given an example.
4. Discuss backlash with an example.
5. Discuss on–off non-linearity with an example.
6. Discuss the Lyapunov method for defining the stability of a non-linear system. Which one is a superior method?

TEST YOURSELF

1. A non-linear control system is described by the equation $\dot{\theta} + K.\sin\theta = 0$. The type of singular points at $A = (0,0)$ and $B = (\pi, 0)$ will be, respectively,
   (a) centre and saddle (b) centre and focus (c) focus and saddle (d) saddle and centre
2. A linear control system is shown below. If the intercept of $G(j\omega)$ on the negative real axis of the G plane is $\left(-\frac{\pi}{2}, j0\right)$, the amplitude of the limit cycle is

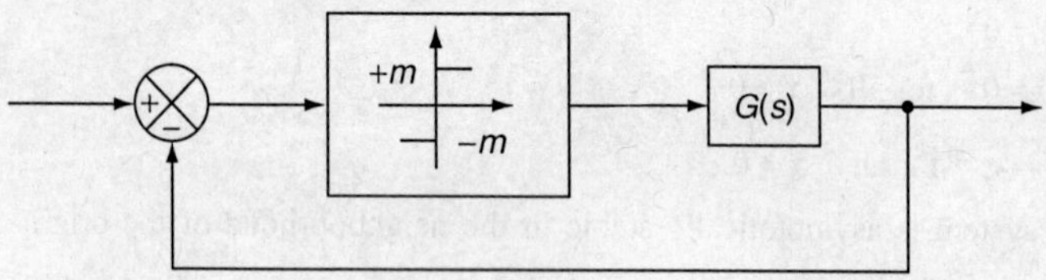

Figure 1

   (a) m (b) $2m$ (c) $3m$ (d) $4m$
3. For determination of stability of a non-linear control system by the describing function method, the plots of $G(j\omega)$ and $-1/N(s)$ drawn as shown below for an input $x = X \sin \omega t$. The types of limit cycles at the points P and Q will be respectively,
   (a) stable and stable (b) stable and unstable (c) unstable and stable (d) unstable and unstable
4. Non-linearities can be
   (a) incidental (b) intentional (c) linearized (d) either (a) or (b)
5. The intentional non-linear systems are
   (a) On–off transducers (b) contractors servos
   (c) non-linear pitch or dampers in aircraft control (d) all of these
6. Non-linearity occurs in servomotor due to
   (a) dead space (b) static friction (c) coulomb friction
   (d) backlash (e) saturation
7. Introduction of non-linearities
   (a) produces non-linear motion of the system (b) achieves special characteristics
   (c) simplifies the system (d) all of theses (e) both (b) and (c)

ANSWERS

1 (a) 2 (b) 3 (a) 4 (d) 5 (d) 6 (e) 7 (e)

Appendix A

Fuzzy Logic

Dr Latif Zadeh rightly said, "As the complexity of a system increases, our ability to make precise and yet significant statement about its behaviour diminishes until a threshold is reached beyond which precision and significance (or relevance) become almost mutually exclusive characteristics." The real world is complex. Due to uncertainty in the form of ambiguity, the complexity of the world increases. As the above quote of Dr Latif Zadeh's principle of incompatibility suggests, complexity and ambiguity (imprecision) are correlated: "The closer one looks at a real world problem, the fuzzier become its solution." [Zadeh, 1973]. Fuzzy set theory provides a means for representing uncertainties. Probability is the primary tool for representing uncertainty as a mathematical model because all uncertainties are assumed to follow the characteristics of random uncertainty. Since all uncertainties are not random, the probability theory does not suit non-random uncertainty. Fuzzy set theory is an excellent tool for these types of uncertainty. Fuzzy logic has rapidly become one of the most successful technologies of today's for developing sophisticated control systems because it addresses such applications perfectly as it resembles human decision making with an ability to generate precise solutions from certain or approximate information. It fills an important gap in engineering design methods left vacant by purely mathematical approaches.

The first applications of the fuzzy theory were primarily industrial, such as process control for cement kilns. In 1987, the first fuzzy logic-controlled subway was opened in Sendai in northern Japan. Fuzzy-logic controllers make subway journeys more comfortable with smooth braking and acceleration. In the *fuzzy washing machine*, load clothes in it and press start, and the machine begins to churn, automatically choosing the best cycle. While the US and certain parts of Europe ignored it, Japan, China, and most oriental countries welcomed expected fuzzy logic with open arms. The aim of this chapter is to introduce fuzzy logic and its application in control system engineering.

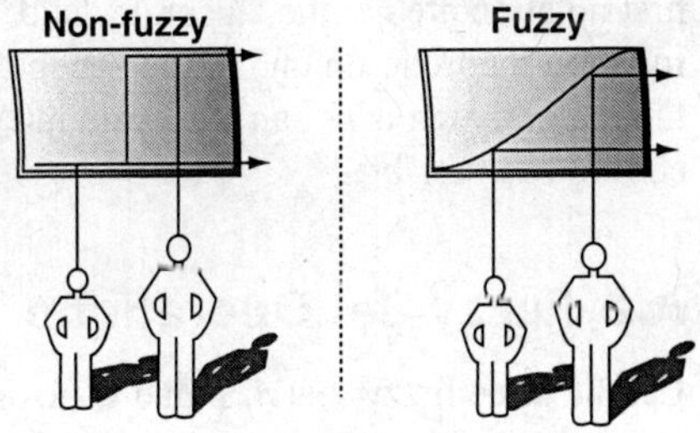

Fig. A.1

A.1 Fuzzy Sets

Before discussing fuzzy sets, let us glance over the following essential characteristics of fuzzy logic:

- In fuzzy logic, exact reasoning is viewed as a limiting case of approximate reasoning.
- In fuzzy logic everything is a matter of degree.

- Any logical system can be fuzzified.
- In fuzzy logic, knowledge is interpreted as a collection of elastic or, equivalently, fuzzy constraint on a collection of variables.
- Inference is viewed as a process of propagation of elastic constraints.

The third statement defines Boolean logic as a subset of fuzzy logic.

In classical, or crisp, sets the transition for an element in the universe between membership and non-membership in a given set is abrupt and well defined is said to be "Crisp". A fuzzy set, then, is a set containing elements that have varying degrees of membership in the set. A notation convention for fuzzy sets when the universe of discourse, X, is discrete and finite, is as follows for a fuzzy A:

$$A = \left\{ \frac{\mu_A(x_1)}{x_1} + \frac{\mu_A(x_2)}{x_2} + \cdots \right\} = \left\{ \sum_i \frac{\mu_A(x_i)}{x_i} \right\}$$

If the universe X is continuous and infinite, the fuzzy set A is denoted by

$$A = \left\{ \int \frac{\mu_A(x)}{x} \right\}$$

In both notations, the horizontal bar is not a quotient but rather a delimiter. The numerator in each term in the membership value is set A associated with the element of the universe indicated in the denominator. In the first notation, the summation symbol is no for algebraic summation but rather denotes the collection or aggregation of each element; hence the "+" signs in the first notation are not the algebraic "add" but are a function-theoretic union. In the second notation, the integral sign is not an algebraic integral but a continuous variable.

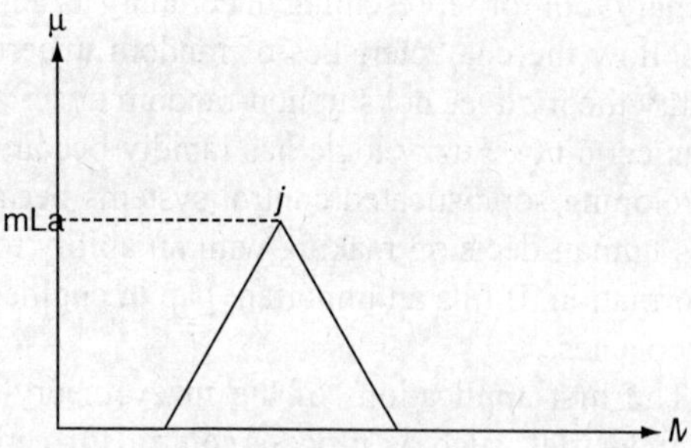

Fig. A.2 Membership function for fuzzy set

A.2 Fuzzy Set Operations

Let the three fuzzy sets A, B and C be on the universe X. For a given element x of the universe, we can write

Union $\quad \mu_{A\cup B}(x) = \mu_A(x) \vee \mu_B(x)$

Intersection $\quad \mu_{A\cap B}(x) = \mu_A(x) \wedge \mu_B(x)$

Complement $\quad \mu_{\bar{A}}(x) = 1 - \mu_A(x)$

Figure A.3 shows the Venn diagrams for these operations, extended to consider fuzzy sets. Any fuzzy set A defined on a universe X is a subset of that universe. De Morgan's law for classical sets also holds for fuzzy sets, as denoted by these expressions:

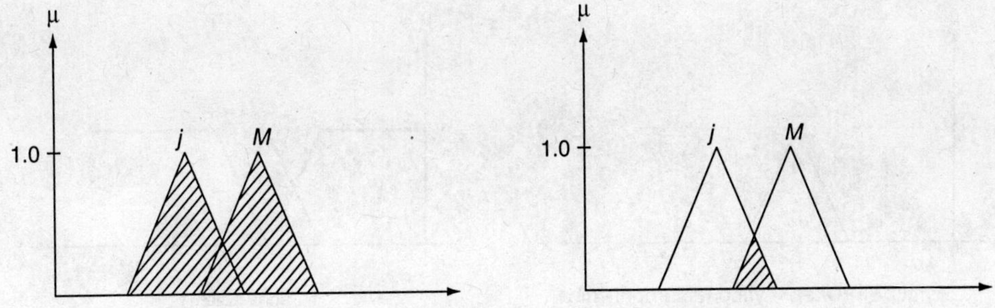

(a) Union of fuzzy sets *j* and *M* (b) Intersection of fuzzy sets *j* and *M*

Fig. A.3

$$\overline{A \cap B} = \bar{A} \cup \bar{B}$$
$$\overline{A \cup B} = \bar{A} \cap \bar{B}$$

A.3 Properties of Fuzzy Sets

Fuzzy sets follow the same properties as crisp sets. The following properties are frequently used on fuzzy sets:

Communicative $A \cap B = B \cap A$ $A \cup \bar{A}$

$A \cup B = B \cup A$

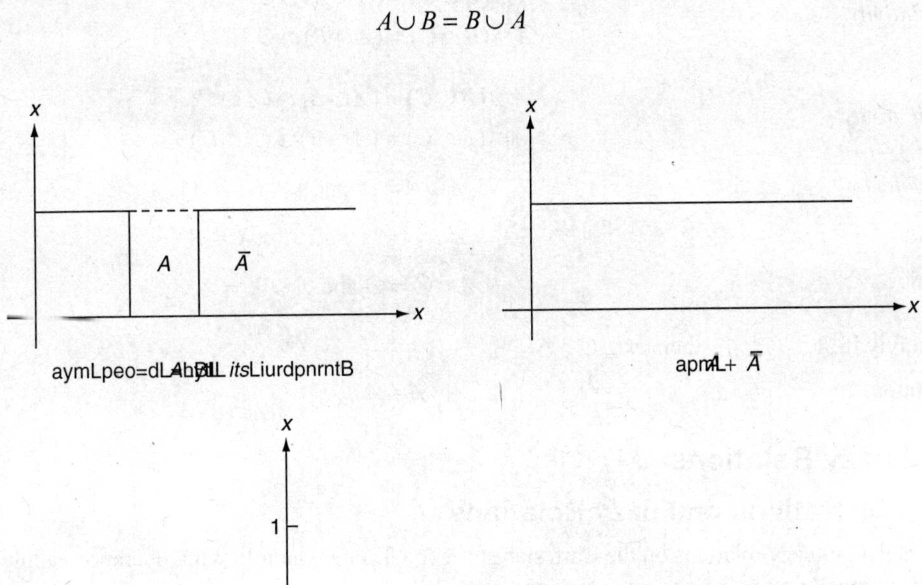

(c) $A + \bar{A}$

Fig. A.4 Crisp set

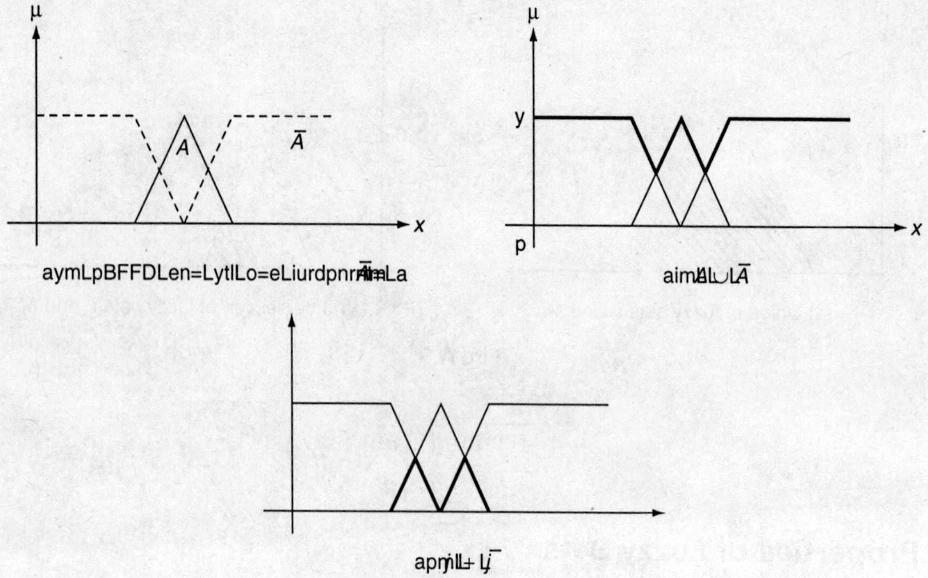

Fig. A.5 Fuzzy sets

Associativity
$$A \cup (B \cup C) = (A \cup B) \cup C$$
$$A \cap (B \cap C) = (A \cap B) \cap C$$

Distributivity
$$A \cup (B \cup C) = (A \cup B) \cap (A \cup C)$$
$$A \cap (B \cap C) = (A \cap B) \cup (A \cap C)$$

Idempotency $\quad A \cup A = A$ and $A \cap A = A$

Identity $\quad A \cup \varnothing = A$ and $A \cap X = A$

$A \cap \varnothing = \varnothing$ and $A \cup X = X$

Transitiviy if $A \subseteq B \subseteq C$, then $A \subseteq C$

Involution $\quad A = A$

A.4 Fuzzy Relations

A.4.1 Operations on Fuzzy Relations

Let R and S be fuzzy relations on the Cartesian space $X \times Y$. Then the following operations apply for the membership values for various set operations:

Union $\quad \mu_{R \cup S}(x, y) = \max \mu_R(x, y), \mu_S(x, y)$

Intersection $\quad \mu_{R \cap S}(x, y) = \min \mu_R(x, y), \mu_S(x, y)$

Complement $\mu_{\bar{R}}(x, y) = 1 - \mu_R(x, y)$

Containment $R \subset S \Rightarrow \mu_R(x, y) \leq \mu_S(x, y)$

A.5 Fuzzy Tolerance and Equivalence Relation

A fuzzy relation R, on a single universe X, is also a relation from X to X. It is a fuzzy equivalence relation if all three of the following properties for matrix relations define it; e.g.,

Reflexivity $\mu_R(x_i, x_i) = 1$

Symmetry $\mu_R(x_i, x_j) = \mu_R(x_j, x_i)$

Transitivity $\mu_R(x_i, x_j) = \lambda_1$ and $\mu_R(x_j, x_K) = \lambda_2 \rightarrow \mu_R(x_i, x_K) = \lambda$

where $\lambda \geq \min[\lambda_1, \lambda_2]$.

A.6 Features of the Membership Function

The core of the membership function for some fuzzy set A is the region of the universe that is characterized by complete and full membership in the set A. Therefore, the core comprises elements x of the universe such that $\mu_A(x) = 1$.

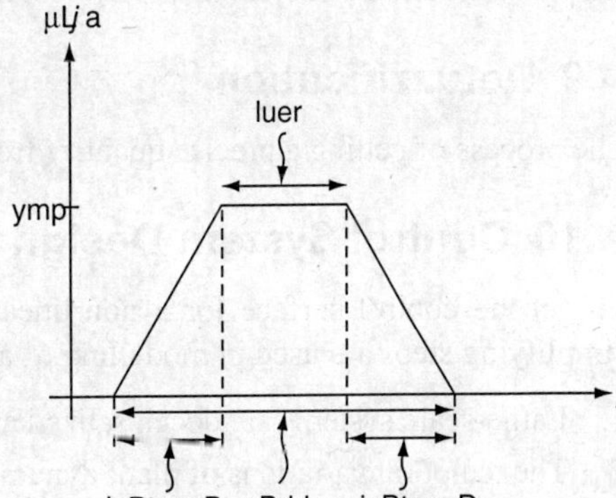

Fig. A.6 Core, support, and boundary of a fuzzy set

[The support of a membership function for some fuzzy set A is the region of the universe which is characterized by complete and full membership in the set A. Therefore, the support comprises elements x of the universe such that $\mu_A(x) > 0$].

The boundaries of a membership function for some fuzzy set A are the region of the universe containing elements that have a nonzero membership but not complete membership, that is, the boundaries comprise elements x of the universe such that $0 < \mu_A(x) < 1$.

If the membership function has at least one element x in the universe whose membership value is unity, it is called normal fuzzy set. Figure A.7 shows typical normal and subnormal fuzzy sets.

If the membership function whose membership function values are strictly monotonically decreasing, or whose membership values are strictly monotonically increasing then strictly monotonically decreasing with increasing values for elements in the universe, is known as convex fuzzy set.

$$\mu_A(v) \geq \min[\mu_A(x), \mu_A(z)]$$

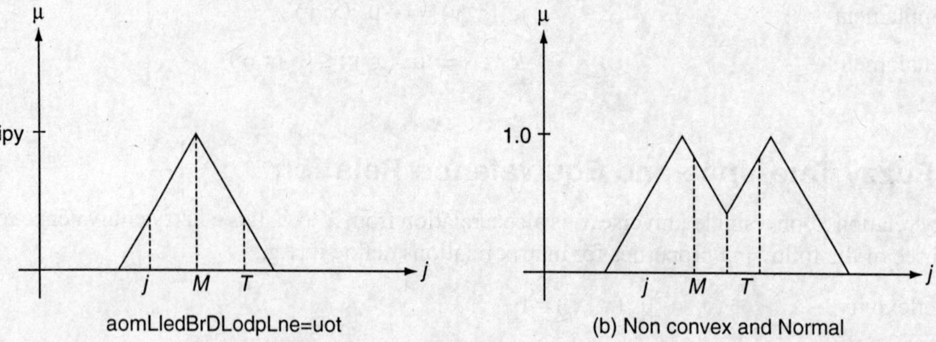

Fig. A.7

A.7 Standard Forms and Boundaries

The normal and convex are the most common forms of membership function. However, operations on a membership function result in fuzzy sets that are subnormal and nonconvex. Membership functions can be symmetrical or asymmetrical.

A.8 Fuzzification

The process of making a crisp quantity fuzzy is known as fuzzification. A digital voltmeter generates crisp data which are subject to experimental error.

A.9 Defuzzification

The process of getting a precise quantity from fuzzy quantity is known as defuzzification.

A.10 Control System Design Stages

To get the control surface for a non-linear time-varying real-world dynamic system, the following simplifying steps are used in modelling of a controller for the system:

1. Large-scale systems are decentralized and decomposed into a collection of decoupled subsystems.
2. The temporal variations of plant dynamics are assumed to be "slowly varying".
3. The non–linear plant dynamics are locally linearized about a set of operating points (see rotation assumptions in Example 1A.A).
4. A set of state variables, control variables, or output features are made available.
5. A simple P, PD, PID (output-feedback), or state-feedback controller is designed for each decoupled system. The controllers are of regulatory type and are fast enough to perform satisfactorily under tracking control situations. Optimal controllers might also prove useful.
6. In addition to the uncertainties introduced in the first five steps, there are uncertainties due to the external environment. The controller design should be made close as possible to the optimal one based on the control engineers knowledge, in the form of input-output numerical observations data and analysis, linguistic, intuitive and other kinds of information regarding the plant dynamics and the external environment.

7. A supervisory control system, either automatic or a human expert operator, forms an additional feedback control loop to tune and adjust the controller's parameters, in order to compensate for the effects of variations caused by unmodelled dynamics.

A.11 Assumptions in a Fuzzy Control System Design

The following assumptions are implicit in a fuzzy control system design:

1. The plant is observable and controllable: State, input and output variables are usually available for observation and measurement or computation.
2. There exists a body of knowledge comprising a set of expert production linguistic rules, engineering common sense, initiation, a set of input-output measurement data, or an analytic model that can be fuzzified and from which rules can be extracted.
3. A solution exits.
4. The control engineer is looking for a "good enough" solution, not necessarily the optimum one.
5. We will design a controller to the best of our available knowledge and within an acceptable range of precision.
6. The problems of stability and optimality are still open problems in fuzzy controller design.

A.12 Fuzzy Rules

Fuzzy rules operate using a series of If-Then statements. For instance, if *X* then *A*, if *y* then *b*, where *A* and *B* are all sets of *X* and *Y*. Fuzzy rules define fuzzy *patches*, which is the key idea in fuzzy logic. A machine is made smarter using a concept designed by Bart Kosko called the fuzzy approximation theorem (FAT). The FAT generally states a finite number of patches can cover a curve as seen in Fig. A.8. If the patches are large, then the rules are sloppy. If the patches are small, then the rules are fine.

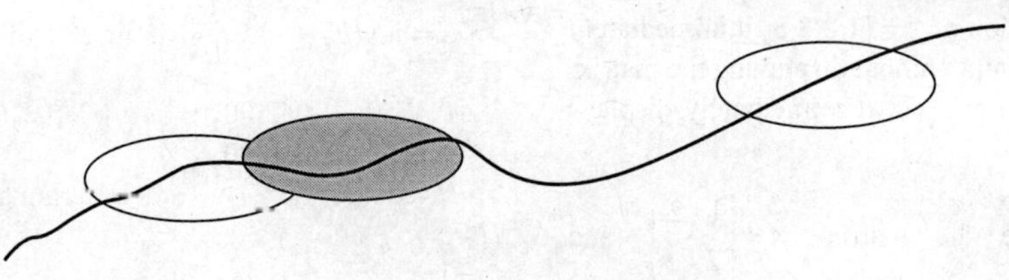

Fig. A.8 Fuzzy patches

In a fuzzy system this simply means that all our rules can be seen as patches and the input and output of the machine can be associated together using these patches.

A fuzzy production rule system consists of four structures [Weiss & Donnel, 1979].

1. A set of rules that represent the policies and heuristic strategies of the expert decision maker.
2. A set of input data assessed immediately prior to the actual decision.
3. A method for evaluating any proposed action in terms of its conformity to the expressed rules, given the available data.
4. A method for generating promising actions and for determining when to stop searching for better ones.

MATLAB Problems

Problem 1 Write the matrix $A = \begin{bmatrix} 1 & 2 \\ 2 & 3 \end{bmatrix}$ and display it.

Solution

```
>>% no display of array
>> A = [1 2;2 3];
>> % display of array
>> A = [1 2;2 3]
A =
   1  2
   2  3
>>
```

From above, we can conclude that using the semicolon after the expression of $A = [1\ 2;2\ 3]$ it immediately gives the matlab prompt without displaying the matrix. If the semicolon is not used, it immediately displays the matrix.

Problem 2 Write the matrix $A = \begin{bmatrix} 1 & 2 \\ 2 & 3 \end{bmatrix}$ and $B = \begin{bmatrix} 1 & 2 \\ 2 & 3 \end{bmatrix}$, Calculate $A + B$, $B - A$, $A*B$, inverse of A and B, Determinant of A and B, rank of A and B.

Solution

```
>> A = [1 2; 2 3]
A =
   1  2
   2  3
>> B = [4 2; 2 3]
B =
   4  2
   2  3
>> C = A + B                % Addition of A and B
C =
   5  4
   4  6
>> D = B − A                % Subtraction of A from B
D =
   3  0
   0  0
>> E = A*B                  % Multiplication of A and B
E =
   8  8
  14  13
>> F = inv(A)               % Inverse of A
F =
  −3  2
   2  −1
>> G = inv(B)               % Inverse of B
G =
   0.3750  −0.2500
  −0.2500   0.5000
>> H = det(A)               % Determinant of A
H =
  −1
>> H = det(B)               % Determinant of B
H =
   8
>> K = rank(A)              % Rank of A
K =
   2
>> K = rank(B)              % Rank of B
K =
   2
```

Problem 3 Get the determinant, rank, inverse and transpose of the matrix $A = \begin{bmatrix} 1 & 2 & 4 \\ 5 & 7 & 2 \\ 3 & 8 & 4 \end{bmatrix}$.

Solution

Get the matrix $A = \begin{bmatrix} 1 & 2 & 4 \\ 5 & 7 & 2 \\ 3 & 8 & 4 \end{bmatrix}$

```
>> A = [1 2 4; 5 7 2; 3 8 4]
A =
   1   2   4
   5   7   2
   3   8   4
>> det(A)
ans =
  60
>> rank(A)
ans =
  3
>> B = inv(A)
B =
    0.2000    0.4000   -0.4000
   -0.2333   -0.1333    0.3000
    0.3167   -0.0333   -0.0500
>> A'      % Transpose of the matrix A
ans =
   1   5   3
   2   7   8
   4   2   4
```

Problem 4 Compute the positive integer power (2) of the matrix $A = \begin{bmatrix} 1 & 2 & 4 \\ 5 & 7 & 2 \\ 3 & 8 & 4 \end{bmatrix}$.

Solution

Here $A = \begin{bmatrix} 1 & 2 & 4 \\ 5 & 7 & 2 \\ 3 & 8 & 4 \end{bmatrix}$.

```
% Multiplication of A by one time i.e., 2 – 1 = 1 time
>> X = A ^ 2
X =
   23   48   24
   46   75   42
   55   94   44
```

Problem 5 Compute the Positive integer power (5) of the matrix $A = \begin{bmatrix} 1 & 2 & 4 \\ 5 & 7 & 2 \\ 3 & 8 & 4 \end{bmatrix}$.

Solution

Here $A = \begin{bmatrix} 1 & 2 & 4 \\ 5 & 7 & 2 \\ 3 & 8 & 4 \end{bmatrix}$

```
% Multiplication of A by one time i.e., 2 – 1 = 1 time
>> X = A ^ 5
X =
   49729    85826   44644
   84029   144919   75242
   98751   170432   88504
```

Problem 6 Compute the negative integer power (2) and integer power (3) of the matrix $A = \begin{bmatrix} 1 & 2 & 4 \\ 5 & 7 & 2 \\ 3 & 8 & 4 \end{bmatrix}$

Solution

```
>> A = [1 2 4; 5 7 2; 3 8 4]
A =
   1   2   4
   5   7   2
   3   8   4
% A ^ (–2) effectively multiplies inv(A) by itself 2 – 1
times
>> A ^ (–2)
ans =
   – 0.1800    0.0400    0.0600
     0.0794   –0.0856    0.0383
     0.0553    0.1328  – 0.1342
% A ^ (–3) effectively multiplies inv(A) by itself 3 – 1
times
>> A ^ (–3)
ans =
   –0.0263    0.0793    0.0810
    0.0480    0.0419   –0.0594
   –0.0624    0.0089    0.0244
```

Problem 7 Multiply the polynomials $4s^4 + 3s^3 + 4s^2 + 2s + 1$ and $5s^3 + 3s^2 + 4s + 3$ and get the coefficients of the new polynomials.

Solution

```
>> p = [ 4 3 4 2 1]
p =
   4   3   4   2   1
>> q = [ 5 3 4 3]
q =
   5   3   4   3
```

```
>> m = conv(p, q)
m =
  20  27  45  46  36  23  10  3
```

Here m = **conv(p, q)** multiplies p and q and gives the coefficients of the new polynomials.

The new polynomial is $20s^7 + 27s^6 + 45s^5 + 46s^4 + 6s^3 + 23s^2 + 10s + 3$.

Problem 8 The impulse response of a system is expressed by $g(t) = e^{-2t}(1 - \sin 2t)$. Find the transfer function of the system.

Solution

```
>> syms t
>> a = exp(-2*t)*(1 - sin (2*t))
a =
exp(-2*t)*(1 - sin(2*t))
>> a1 = exp(-2*t)
a1 =
exp(-2*t)
>> a2= -exp(-2*t)*sin(2*t)
a2 =
-exp(-2*t)*sin(2*t)
>> a = laplace(a1) + laplace(a2)
a =
1/(s + 2) -2/(s ^ 2 + 4*s + 8)
```

Problem 9 The impulse response of a system is expressed by $g(t) = e^{-t}(1 - \cos 3t)$. Find the transfer function of the system.

Solution

```
>> syms t
>> a = exp(-t)
>> a1 = exp(-t)
a1 =
exp(-t)
>> a2 = -exp(-t)*cos(3*t)
a2 =
-exp(-t)*cos(3*t)
>> a = laplace(a1) + laplace(a2)
a =
1/(1 + s) - (1 + s)/(10 + 2*s + s ^ 2)
>>
```

Problem 10 The unit step response of a system is expressed by $c(t) = e^{-t}(1 - \cos 3t)$. Find the transfer function of the system.

Solution

```
>> syms t s
>> c = exp (-t)*(1 - cos(3*t))
c =
exp (-t)*(1 - cos(3*t))
>> laplace (c)
9/((1 + s)*(10 + 2*s + s ^ 2))
>> a = [1 1]
a =
  1  1
>> b = [1 2 10]
b =
  1  2  10
>> d = conv(a, b)
d =
  1  3  12  10
>> e = [0 1]
e =
  0  1
>> f= [1 0]
f=
  1  0
>> g = conv(d, f)
g =
  1  3  12  10  0
>> tf(e, g)
Transfer function:
          1
---------------------------
s ^ 4 + 3 s ^ 3 + 12 s ^ 2 + 10 s
>>
```

Problem 11 The transfer function of a system is shown below:

$$G(s) = \frac{6(s+2)(s+3)}{s(s+1)(s^2+3s+4)}$$

Solution

```
>> p1 = [0 6]
p1 =
  0  6
>> q1 = [1 2]
q1 =
  1  2
>> r1 = [1 3]
r1 =
  1  3
```

```
>> p2 = [1 0]
p2 =
   1   0
>> q2 = [1 1]
q2 =
   1   1
>> r2 = [1 3 4]
r2 =
   1   3   4
>> num = conv(p1, q1)
num =
   0   6   12
>> num = conv(num, r1)
num =
   0   6   30   36
>> den = conv(p2, q2)
den =
   1   1   0
>> den = conv(den, r2)
den =
   1   4   7   4   0
```

>> % command tf : It prints Transfer function: and then its expression

```
>> sys = tf(num, den)
Transfer function:
   6 s ^ 2 + 30 s + 36
-------------------------
s ^ 4 + 4 s ^ 3 + 7 s ^ 2 + 4 s
```

>> % command printsys : It prints num/den = and then its expression

```
>> printsys(num, den)
num/den =
     6 s ^ 2 + 30 s + 36
  -------------------------
  s ^ 4 + 4 s ^ 3 + 7 s ^ 2 + 4 s
>> pzmap(sys);
>>
```

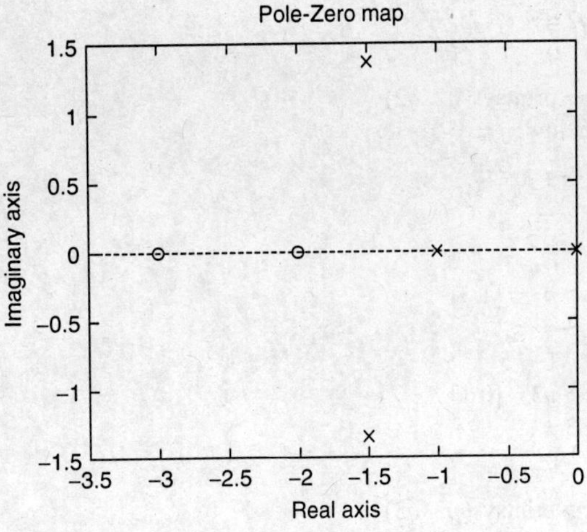

Fig. P11

Problem 12 Determine the transfer function of the control system shown in **Fig. 1**.

Solution

```
>> p1 = [0 3]
p1 = 0  3
>> q1 = [ 1 2 0]
q1 =
   1   2   0
>> printsys(p1, q1)
num/den =

      3
  ---------
  s ^ 2 + 2 s

>> p2 = [1 3]
p2 =
   1   3
>> q2 = [0 1]
```

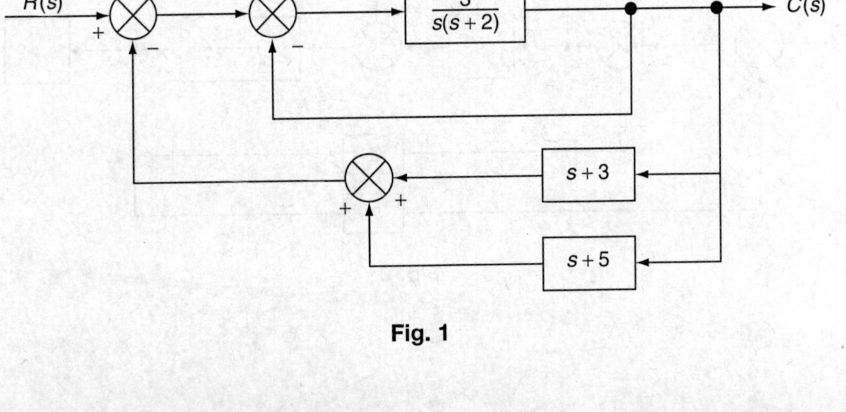

Fig. 1

```
q2 =
   0   1
>> printsys(p2, q2)
num/den =

 s + 3
 -----
   1
>> p3 = [1 5]
p3 =
   1   5
>> q3 = [0 1]
q3 =
   0   1
>> printsys(p3, q3)
num/den =

 s + 5
 -----
   1
>> [p4 q4] = cloop(p1, q1, -1)
p4 =
   0   0   3
q4 =
   1   2   3
>> [p5 q5] = parallel(p2, q2, p3, q3)
p5 =
   0   2   8
q5 =
   0   0   1
>> [p6 q6] = feedback(p4, q4, p5, q5, -1)
p6 =
   0   0   0   0   3
q6 =
   0   0   1   8   27
>> printsys(p6, q6)
num/den =

        3
  --------------
  s ^ 2 + 8 s + 27
>> tf(p6, q6)
Transfer function:

     3
--------------
s ^ 2 + 8 s + 27
>>
```

Problem 13 Determine the transfer function of the system shown in **Fig. 2**.

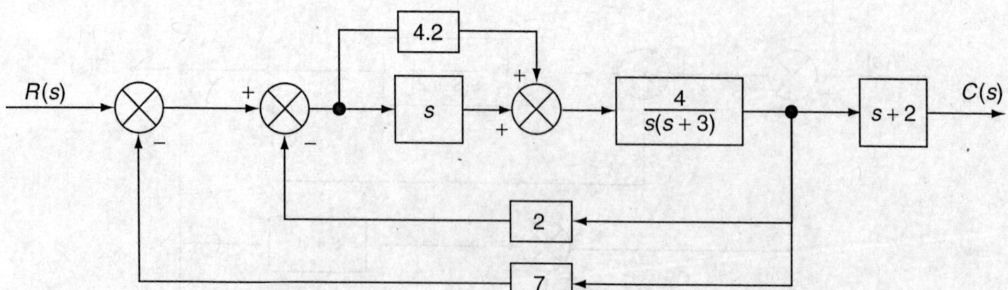

Fig. 2

Solution

```
>> p1 = [1 0]
p1 =
   1   0
>> q1 = [0 1]
q1 =
   0   1
>> printsys(p1, q1)
num/den =
 s
 -
 1
>> p2 = [ 0 4.2]
p2 =
      0   4.2000
>> q2 = [ 0 1]
q2 =
   0   1
>> printsys(p2, q2)
num/den =

 4.2
 ---
  1
```

```
>> [p3, q3] = parallel(p1, q1, p2, q2)
p3 =
     0   1.0000   4.2000
q3 =
   0   0   1
>> p4 = [0 4]
p4 =
   0   4
>> q4 = [1 3 0]
q4 =
   1   3   0
>> printsys (p4, q4)
num/den =

     4
  ---------
  s ^ 2 + 3 s

>> [p5, q5] = series(p3, q3, p4, q4)
p5 =
     0      0      0   4.0000   16.8000
q5 =
   0   0   1   3   0
>> p6 = [0 2]
p6 =
   0   2
>> q6 = [ 0 1]
q6 =
   0   1
>> printsys(p6, q6)
num/den =
  2
  -
  1
>> [p7 q7] = feedback (p5, q5, p6, q6, -1)
p7 =
     0      0      0      0   4.0000   16.8000
q7 =
     0      0      0   1.0000   11.0000   33.6000
>> p8 = [ 0 7]
p8 =
   0   7
>> q8 = [0 1]
q8 =
   0   1
>> printsys(p8, q8)
num/den =

  7
  -
  1
>> [p9, q9] = feedback (p7, q7, p8, q8, -1)
p9 =
   0   0   0   0       0     4.0000     16.8000
q9 =
   0   0   0   0   1.0000   39.0000   151.2000
>> p10 = [1 2]
p10 =
   1   2
>> q10 = [0 1]
q10 =
   0   1
>> printsys(p10, q10)
num/den =

  s + 2
  -----
    1

>> [p11, q11] = series(p9, q9, p10, q10)
p11 =
   0   0   0   0   0   4.0000   24.8000   33.6000
q11 =
   0   0   0   0   0   1.0000   39.0000   151.2000
>> printsys(p11, q11)
num/den =

  4 s ^ 2 + 24.8 s + 33.6
  ---------------------
    s ^ 2 + 39 s + 151.2
>> tf(p11, q11)
Transfer function:

4 s ^ 2 + 24.8 s + 33.6
---------------------
s ^ 2 + 39 s + 151.2
>>
```

Problem 14 Get the unit step response for a system having negative feedback and the transfer function of the system is given by

$$\frac{C(s)}{R(s)} = \frac{15}{s^3 + 9s^2 + 11s + 15}.$$

Solution

```
>> p = [0 15]
p =
   0   15
>> q = [1 9 11 15]
q =
   1   9   11   15
```

```
>> sys = tf(p, q)
Transfer function:
        15
----------------------
s^3 + 9 s^2 + 11 s + 15
>> ltiview(sys);
```

Problem 15 Get the unit step response for a system having negative feedback and the transfer function of the system is given by

$$\frac{C(s)}{R(s)} = \frac{10}{s^3 + 6s^2 + 8s + 10}$$

Solution

```
>> p = [0 10]
p =
   0  10
>> q = [1 6 8 10]
q =
   1   6   8  10
>> sys = tf(p, q)
Transfer function:
        10
--------------------
s^3 + 6 s^2 + 8 s + 10
>> ltiview(sys);
```

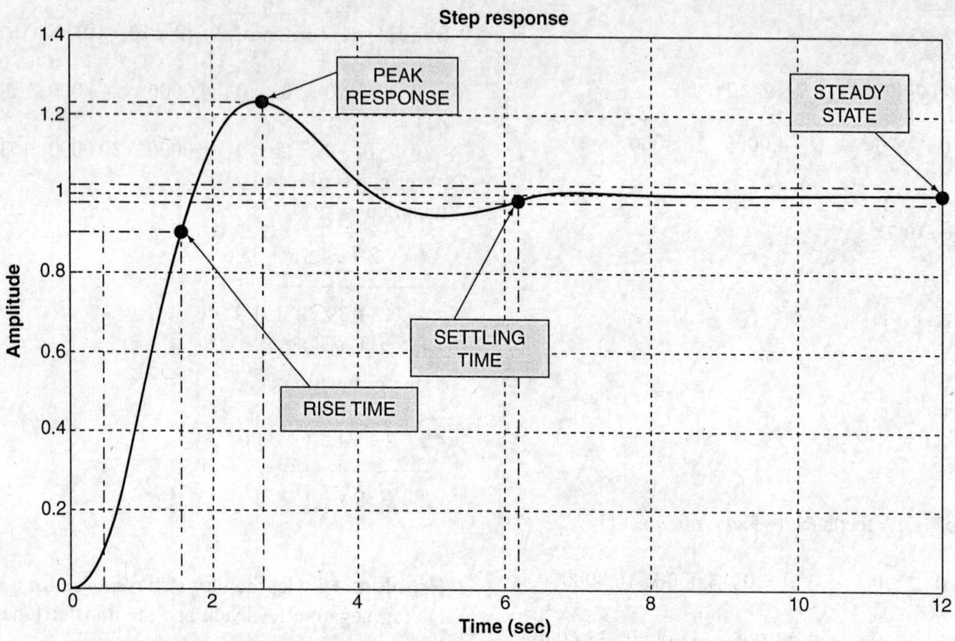

Fig. P14

Problem 16 The forward path transfer function of a unity feedback control system is given by $G(s) = \frac{3}{s(s+2)}$. Find the time response for unit step response, the rise time and the maximum overshoot.

Solution

```
>> p = [0 3]
p =
   0   3
>> q = [1 2 0]
q =
   1   2   0
>> printsys(p, q)
num/den =
     3
  --------
  s^2 + 2 s
```

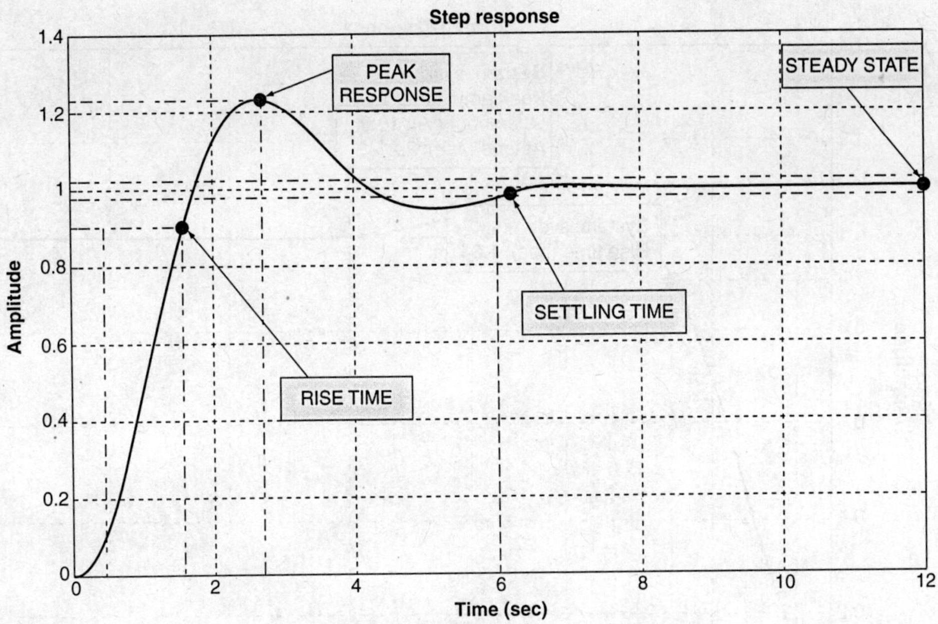

Fig. P15

```
>> [p1, q1] = cloop(p, q, -1)
p1 =
   0   0   3
q1 =
   1   2   3
>> printsys(p1, q1)
num/den =
      3
   ------------
   s ^ 2 + 2 s + 3
>> step(p1, q1);grid
>>
```

Problem 17 Get the unit step response for a system having negative feedback and the transfer function of the system is given by

$$\frac{C(s)}{R(s)} = \frac{12}{s^3 + 6s^2 + 8s + 12}.$$

Solution

```
>> p = [0 12]
p =
   0   12
>> q = [1  6  8  12]
q =
   1   6   8   12
>> printsys(p, q)
num/den =

         12
   ----------------------
   s ^ 3 + 6 s ^ 2 + 8 s + 12
>> step(p, q);grid
>>
```

Problem 18 Determine the time response of the following transfer function:

$$\frac{C(s)}{R(s)} = \frac{s+3}{s^2 + 3s + 3}$$

Solution

```
>> p = [1  3]
p =
   1   3
>> q = [1  3  3]
q =
   1   3   3
```

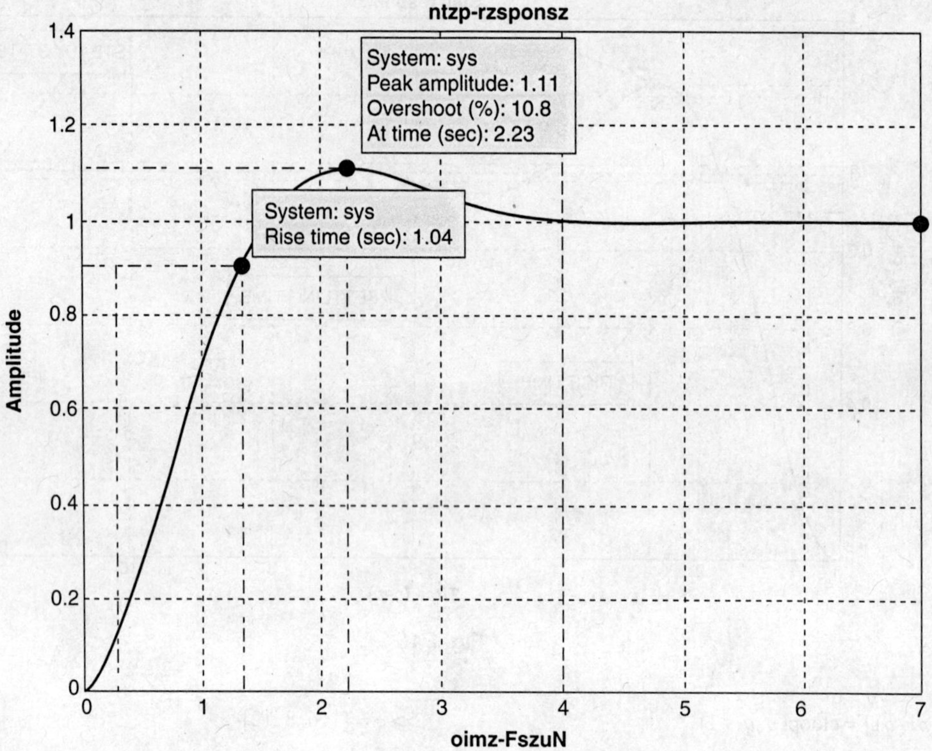

Fig. P16

```
>> printsys(p, q)
num/den =
    s + 3
  ------------
  s ^ 2 + 3 s + 3
>> step(p, q);grid
>>
```

Problem 19 The forward path transfer function of a unity feedback control system is given by $G(s) = \frac{22}{(s+1).(s+4)}$. Find the time response for unit step response, the rise time, the maximum overshoot, settling time and maximum overshoot.

Solution

```
>> p = [0 22]
p =
   0   22
>> q = [1 5 4]
q =
   1   5   4
>> printsys(p, q)
num/den =
        22
   ------------
   s ^ 2 + 5 s + 4
>> [p1, q1] = cloop(p, q, –1)
p1 =
   0   0   22
q1 =
   1   5   26
>> printsys(p1, q1)
num/den =
```

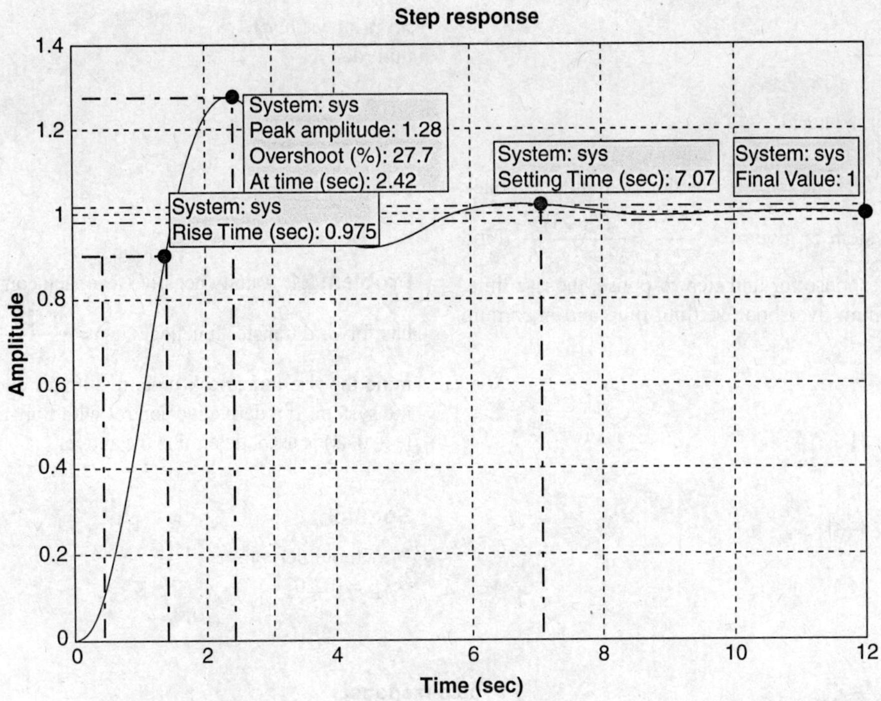

Fig. P17

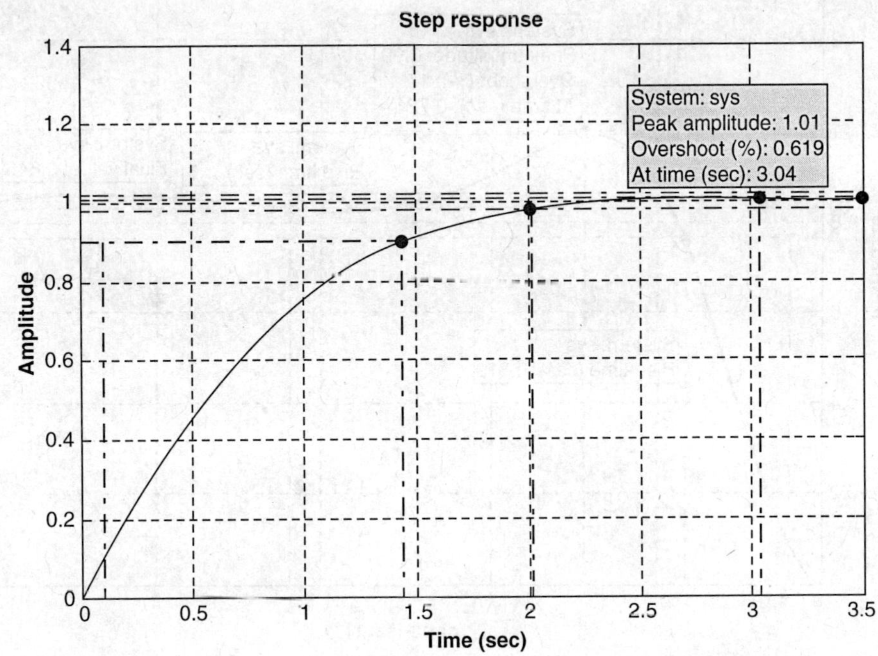

Fig. P18

```
        22
 --------------
 s^2 + 5 s + 26
>> step(p1, q1);grid
```

Problem 20 The transfer function of a unity feedback control system is given by $\frac{C(s)}{R(s)} = \frac{5(s+1)}{s^2+3s+6}$. Find the time response for unit step response, the rise time, the maximum overshoot, settling time and maximum overshoot.

Solution

```
>> p = [5 5]
p =
   5  5
>> q = [1 3 6]
q =
   1  3  6
>> printsys(p, q)
num/den =
      5 s + 5
   ------------
   s^2 + 3 s + 6
>> step(p, q);grid
>>
```

Problem 21 A unity negative feedback control system has forward transfer function $G(s) = \frac{20}{s(s+1.8)}$. Determine the rise time, peak time, maximum overshoot for the system. If a derivative control with transfer function (1 + 0.4s) is used, determine the above.

Solution

```
% without derivative
>> p = [0 20]
```

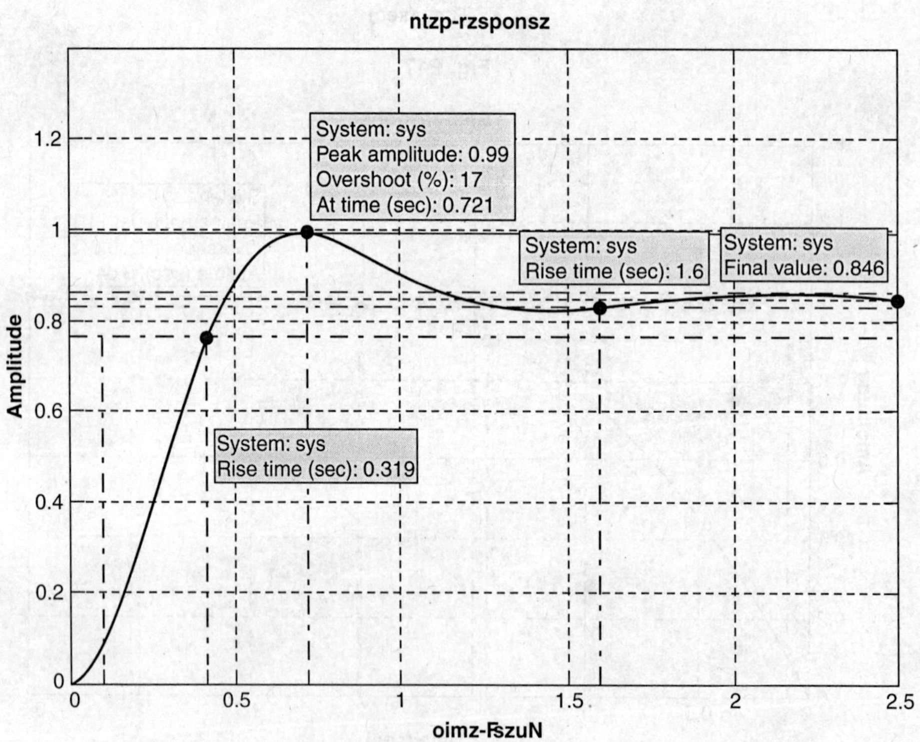

Fig. P19

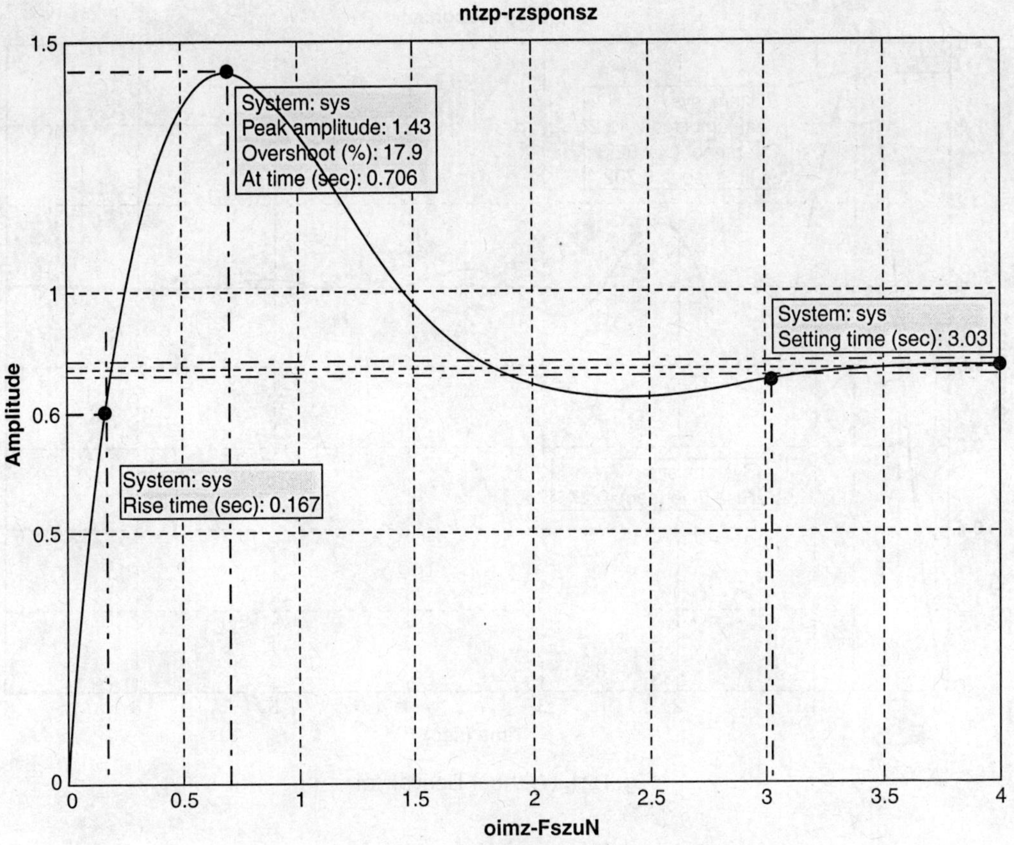

Fig. P20

```
p =
   0   20
>> q = [1 1.8 0]
q =
   1.0000   1.8000        0
>> printsys(p, q)

num/den =

       20
   ----------
   s ^ 2 + 1.8 s

>> [p1, q1] = cloop(p, q, -1)
p1 =
   0   0   20
q1 =
   1.0000   1.8000   20.0000
>> sys = tf(p1, q1)
Transfer function:

      20
---------------
s ^ 2 + 1.8 s + 20

>> ltiview(sys);
%With derivative
>> p1 = [0.4 1]
p1 =
   0.4000   1.0000
>> q1 = [0 1]
q1 =
   0   1
>> printsys(p1, q1)
num/den =
```

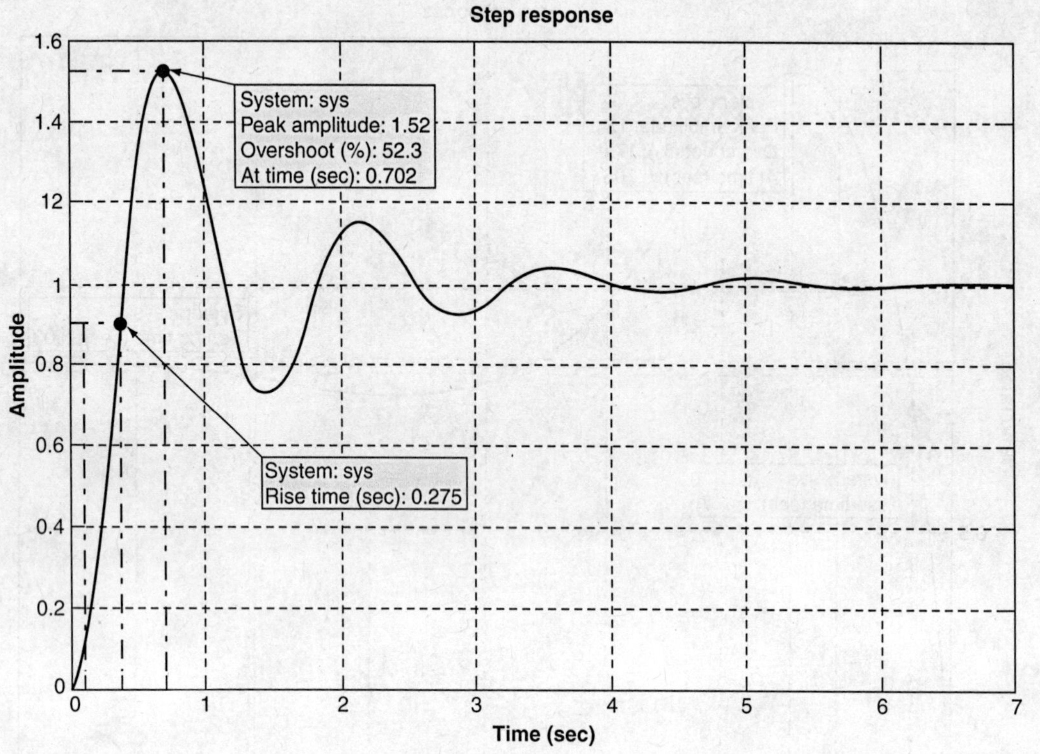

Fig. P21 (Without Derivative)

```
  0.4 s + 1
  ---------
      1

>> p2 = [0 20]
p2 =
   0   20
>> q2 = [1 1.8 0]
q2 =
  1.0000   1.8000        0
>> [p3, q3] = series(p1, q1, p2, q2)
p3 =
   0    0    8   20
q3 =
        0   1.0000   1.8000        0
>> [p4, q4] = cloop(p3, q3, -1)
p4 =
   0    0    8   20
q4 =
    0    1.0000    9.8000   20.0000
>> printsys(p4, q4)
num/den =

      8 s + 20
  ----------------
  s ^ 2 + 9.8 s + 20

>> sys = tf(p4, q4)
Transfer function:

   8 s + 20
----------------
s ^ 2 + 9.8 s + 20

>> ltiview(sys)
>>
```

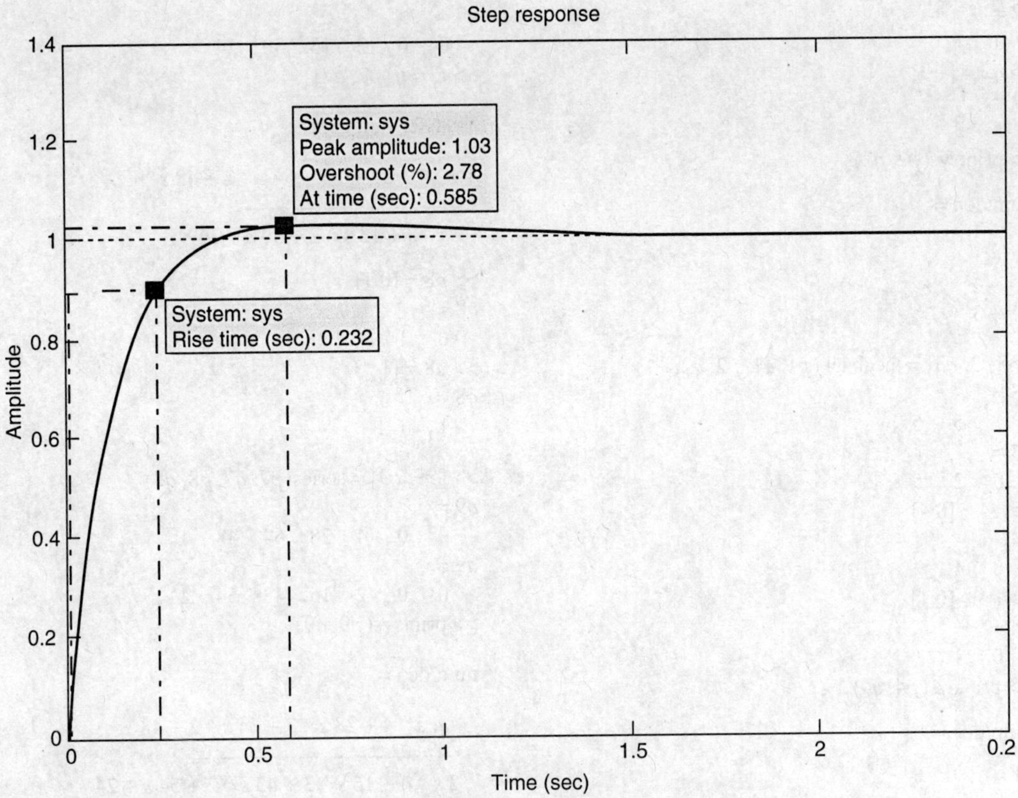

Fig. P21 (With Derivative)

Problem 22 Find the overall transfer function of the system shown in Fig. 3. Find the expression of output for unit step input.

Solution

```
>> p1 = [1 2]
p1 =
  1    2
>> q1 = [0 1]
q1 =
   0    1
>> printsys(p1, q1)
num/den =
   s + 2
   -----
     1
>> p2 = [0 1]
```

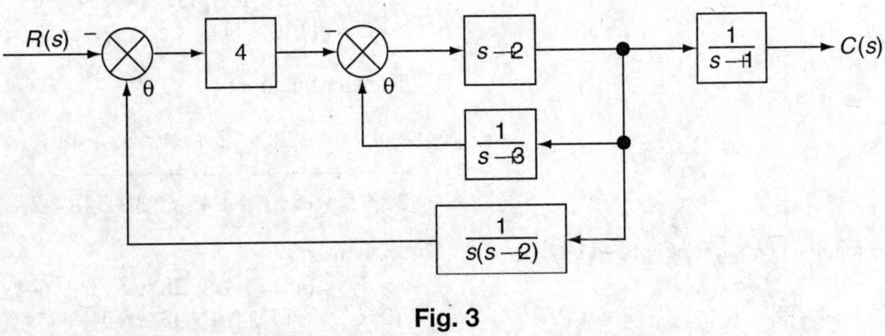

Fig. 3

```
p2 =
   0   1
>> q2 = [1 3]
q2 =
   1   3
>> printsys(p2, q2)

num/den =

   1
  -----
  s + 3

>> [p3, q3] = feedback(p1, q1, p2, q2, -1)
p3 =
   1   5   6
q3 =
   0   2   5
>> p4 = [0 4]
p4 =
   0   4
>> q4 = [0 1]
q4 =
   0   1
>> printsys(p4, q4)

num/den =

  4
  -
  1

>> [p5, q5] = series(p3, q3, p4, q4)
p5 =
   0   4   20   24
q5 =
   0   0   2   5
>> p6 = [0 1]
p6 =
   0   1
>> q6 = [1 2 0]
q6 =
   1   2   0
>> printsys(p6, q6)

num/den =

     1
  ---------
  s ^ 2 + 2 s

>> [p7, q7] = feedback(p5, q5, p6, q6, -1)
p7 =
   0   4   28   64   48   0
q7 =
   0   0   2   13   30   24
>> C = tf(p7, q7)

Transfer function:

4 s ^ 4 + 28 s ^ 3 + 64 s ^ 2 + 48 s
------------------------------
 2 s ^ 3 + 13 s ^ 2 + 30 s + 24

>> p8 = [0 1]
p8 =
   0   1
>> q8 = [1 1]
q8 =
   1   1

>> [p9, q9] = series(p7, q7, p8, q8)
p9 =
   0   0   4   28   64   48   0
q9 =
   0   0   2   15   43   54   24
>> printsys(p9, q9)

num/den =

   4 s ^ 4 + 28 s ^ 3 + 64 s ^ 2 + 48 s
  -----------------------------------
  2 s ^ 4 + 15 s ^ 3 + 43 s ^ 2 + 54 s + 24

>> p10 = [0 1]
p10 =
   0   1
>> q10 = [1 0]
q10 =
   1   0

>> [p11, q11] = series(p9, q9, p10, q10)
p11 =
   0   0   0   4   28   64   48   0
q11 =
   0   0   2   15   43   54   24   0
>> C = tf(p11, q11)

Transfer function:

   4 s ^ 4 + 28 s ^ 3 + 64 s ^ 2 + 48 s
-------------------------------------
2 s ^ 5 + 15 s ^ 4 + 43 s ^ 3 + 54 s ^ 2 + 24 s

>> syms s t
>> F = (4*s ^ 4 + 28*s ^ 3 + 64*s ^ 2 + 48*s)
/(2*s ^ 5 + 15*s ^ 4 + 43*s ^ 3 + 54*s ^ 2 + 24*s)
```

```
F =

(4*s^4 + 28*s^3 + 64*s^2 + 48*s)
/(2*s^5 + 15*s^4 + 43*s^3 + 54*s^2 + 24*s)

>> ilaplace(F)

ans =

8/5*exp(-t) + 2/5*exp(-9/4*t)*(cos(1/4*15^(1/2)*t) +
15^(1/2)*sin(1/4*15^(1/2)*t))

>>
```

Problem 23 The open loop transfer function of a unity feedback control system is $G(s)=\frac{5s+1}{5s^2}$. Find the expression for time response when the system is subjected to

(i) unit impulse response
(ii) unit step input function

Solution

```
>> syms s t
>> p1 = [ 5 1]

p1 =
   5   1
>> q1 = [5 0 0]

q1 =
   5   0   0

>> printsys(p1, q1)

num/den =

  5 s + 1
  -------
  5 s^2
>> [p, q] = cloop(p1, q1, -1)
p =
   0   5   1
q =
   5   5   1

>> sys = tf(p, q)

Transfer function:

  5 s + 1
---------------
5 s^2 + 5 s + 1

>> % time response due to impulse response
>> F = (5*s + 1)/(5*s^2 + 5*s + 1)

F =

(5*s + 1)/(5*s^2 + 5*s + 1)
>> ilaplace(F)

ans =

1/5*exp(-1/2*t)*(  -3*sinh(1/10*t*5^(1/2))*5^(1/2)
+ 5*cosh(1/10*t*5^(1/2)))

>> % time response due to step response
>> F = (5*s + 1)/(s*(5*s^2 + 5*s + 1))

F =

(5*s + 1)/s/(5*s^2 + 5*s + 1)
>> ilaplace(F)

ans =

1 + exp(-1/2*t)*(-cosh(1/10*t*5^(1/2)) +
sinh(1/10*t*5^(1/2))*5^(1/2))
>>
```

Problem 24 A unity feedback control system has open loop transfer function $G(s)=\frac{22}{(s+1)\,(s+4)}$. Determine the characteristic equation of the system, ω_n, ξ, ω_d, t_p, $\%M_p$, the time at which the first undershoot occurs, the time period of oscillations and the number of cycles completed before reaching the steady state.

Solution

```
>> num = [0 22]

num =

   0  22
>> den = [1 5 4]

den =

   1   5   4
>> [p q] = cloop(num, den, -1)
p =
   0   0  22

q =
   1   5  26
>> tf(p, q)

Transfer function:

    22
--------------
s ^ 2 + 5 s + 26
```

```
>> wn = sqrt(22)
wn =
   4.6904
>> zeta = 5/(2*wn)
zeta =
   0.5330
>> wd = wn*sqrt(1 –zeta^2)
wd =
   3.9686
>> tp = pi/wd
tp =
   0.7916
>> mp = exp(–zeta*wn*tp)*100
mp =
   13.8205
>> % time for first undershoot
>> tfirst = 2*pi/wd
tfirst =
   1.5832
>> %period of oscillations
>> period = 2*pi/wd
period =
   1.5832
>> % number of oscillations before steady state
>> oscillations = (wd/(2*pi))*(4/(zeta*wn))
oscillations =
   1.0106
>>
```

Problem 25 A unity feedback control system has the open loop transfer function $G(s) = \dfrac{20}{s(s+4)}$.

(a) Calculate (i) the natural frequency of oscillations, damped frequency of oscillations, damping factor, damping ratio, the maximum overshoot for a unit step input, (ii) e_{ss} for a unit ramp input.

(b) If the damping ratio is to be made 0.55 using a tachometer feedback, calculate the tachometer constant and obtain the maximum overshoot.

Solution

```
>> num = [0 20]
num =
   0  20
>> den = [1 4 0]
den =
   1  4  0
>> [p, q] = cloop(num, den, –1)
p =
   0  0  20
q =
   1  4  20
>> sys = tf(p, q)
Transfer function:
   20
--------------
s ^ 2 + 4 s + 20
>> wn = sqrt(20)
wn =
   4.4721
>> zeta = 4/(2*wn)
zeta =
   0.4472
>> wd = wn*sqrt(1 – zeta ^ 2)
wd =
   4
>> mp = exp((–zeta*pi)/sqrt(1 – zeta ^ 2))*100
mp =
   20.7880
>> %Increase in damping coeffcient from 0.4472 to 0.55
>> zetan = 0.55
zetan =
   0.5500
>> wd = wn*sqrt(1 – zetan ^ 2)
wd =
   3.7350
>> mp = exp((–zetan*pi)/sqrt(1 – zetan ^ 2))*100
mp =
   12.6324
>> kt = (2*zetan*wn – 4.4721)/20
kt =
   0.0224
>> b = 2*zetan*wn
b =
   4.9193
>> den = [1 b 20]
```

```
den =
   1.0000   4.9193   20.0000

>> sys1 = tf(num, den)

Transfer function:

     20
-----------------
s ^ 2 + 4.919 s + 20

>>
```

Problem 26 The open loop transfer function of a unity feedback control system is expressed by $G(s) = \dfrac{12}{s^2 + 7s + 12}$. Calculate (a) K_p, (ii) K_v and (iii) K_a.

Solution

```
>> num = [0 12]

num =

   0   12

>> den = [1 7 12]

den =

   1   7   12

>> G = tf(num, den)

Transfer function:

     12
--------------
s ^ 2 + 7 s + 12

>> kp = dcgain(G)
kp =
   1

>> ess = 1/(1 + kp)
ess =
   0.5000

>> %Ramp input
>> numsg = conv([1 0], num)

numsg =

   0   12   0

>> densg = [1 7 12]

densg =
   1   7   12

>> sG = tf(numsg, densg)

Transfer function:

    12 s
--------------
s ^ 2 + 7 s + 12

>> kv = dcgain(sG)

kv =
   0

>> ess = 1/kv

ess =
  Inf

>> num2sg = conv([1 0 0], num)
num2sg =
   0   12   0   0

>> den2sg = [1 7 12]

den2sg =
   1   7   12

>> s2G = tf(num2sg, den2sg)

Transfer function:

    12 s ^ 2
--------------
s ^ 2 + 7 s + 12

>> ka = dcgain(s2G)

ka =
   0

>> ess = 1/ka

ess =
  Inf

>>
```

Problem 27 Calculate the position, velocity and acceleration error coefficients for a system expressed by

$$G(s)H(s) = \frac{100(s+1)\ (s+35)}{s^3(s^2 + 5s + 250)}$$

Solution

```
>> p = [1 1]
p =
   1   1

>> q = [1 35]
q =
   1   35
```

```
>> p = conv(p, q)
p =
   1   36   35
>> p = 100*p
p =
     100      3600      3500
>> p1 = [ 1 0 0 0]
p1 =
   1    0    0    0
>> q1 = [ 1 5 250]
q1 =
   1    5   250
>> q = conv(p1, q1)
q =
   1    5   250    0    0    0
>> GH = tf(p, q)
Transfer function:
100 s ^ 2 + 3600 s + 3500
-----------------------
 s ^ 5 + 5 s ^ 4 + 250 s ^ 3
>> Kp = dcgain(GH)
Kp =
  Inf
>> p = conv([1 0], p)
p =
     100      3600      3500        0
>> sGH = tf(p, q)
Transfer function:
100 s ^ 3 + 3600 s ^ 2 + 3500 s
---------------------------
  s ^ 5 + 5 s ^ 4 + 250 s ^ 3
>> kv = dcgain(sGH)
kv =
  Inf
>> p = conv([1 0], p)
p =
     100      3600      3500        0        0
>> s2GH = tf(p, q)
Transfer function:
100 s ^ 4 + 3600 s ^ 3 + 3500 s ^ 2
----------------------------
   s ^ 5 + 5 s ^ 4 + 250 s ^ 3
>> ka = dcgain(s2GH)
ka =
  Inf
>>
```

Problem 28 Calculate the position, velocity and acceleration error coefficients for a system expressed by

$$G(s)H(s) = \frac{100(s+1)(s+35)}{s^2+5s+250}$$

Solution

```
>> p = [1 1]
p =
   1   1
>> q = [1 35]
q =
   1   35
>> p = conv(p, q)
p =
   1   36   35
>> p = 100*p
p =
     100      3600      3500
>> q = [1 5 250]
q =
   1    5   250
>> GH = tf(p, q)
Transfer function:
100 s ^ 2 + 3600 s + 3500
------------------------------
   s ^ 2 + 5 s + 250
>> kp = dcgain(GH)
kp =
  14
>> p = conv([1 0], p)
p =
     100      3600      3500        0
>> sGH = tf(p, q)
Transfer function:
100 s ^ 3 + 3600 s ^ 2 + 3500 s
---------------------------------
    s ^ 2 + 5 s + 250
```

```
>> kv = dcgain(sGH)

kv =
   0

>> p = conv ([1 0], p)

p =
     100     3600     3500     0     0

>> s2GH = tf(p, q)

Transfer function:

100 s ^ 4 + 3600 s ^ 3 + 3500 s ^ 2
-----------------------------
     s ^ 2 + 5 s + 250

>> ka = dcgain(s2GH)

ka =
   0

>>
```

Problem 29 The forward path transfer function of a unity feedback control system is given by $G(s) = \dfrac{100}{s(s+3)\,(s^2+10s+8)}$. Determine the roots of the characteristic equation and bode plot of the open loop transfer function.

Solution

```
>> num = [ 0 100]

num =
   0  100

>> p1 = [1 0]
p1 =
   1   0

>> p2 = [1 3]
p2 =
   1   3

>> p3 = [1 10 8]

p3 =
   1  10   8

>> p = conv(p1, p2)
p =
   1   3   0

>> den = conv(p, p3)

den =

   1  13  38  24  0

>> sys = tf(num, den)

Transfer function:

          100
---------------------------
s ^ 4 + 13 s ^ 3 + 38 s ^ 2 + 24 s

>> [n1, d1] = cloop(num, den, -1)

n1 =
   0   0   0   0  100
d1 =
   1  13  38  24  100

>> sys1 = tf(n1, d1)

Transfer function:

          100
---------------------------------
s ^ 4 + 13 s ^ 3 + 38 s ^ 2 + 24 s + 100

>> roots(d1)

ans =

  -8.8842
  -4.3758
   0.1300 + 1.5986i
   0.1300 - 1.5986i

>> pzmap(sys1)
>> bode(num, den)
>> margin(sys)
>>
```

Problem 30 Sketch the bode plot for the open loop transfer function $G(s)H(s) = \dfrac{20(s+0.75)}{s^2(s+1)(s+0.8)}$. Also determine the gain margin, phase margin, gain cross over frequency and phase cross over frequency.

Solution

```
>> p1 = [0 20]
p1 =
   0   20

>> p2 = [1 0.75]
p2 =
   1.0000   0.7500
```

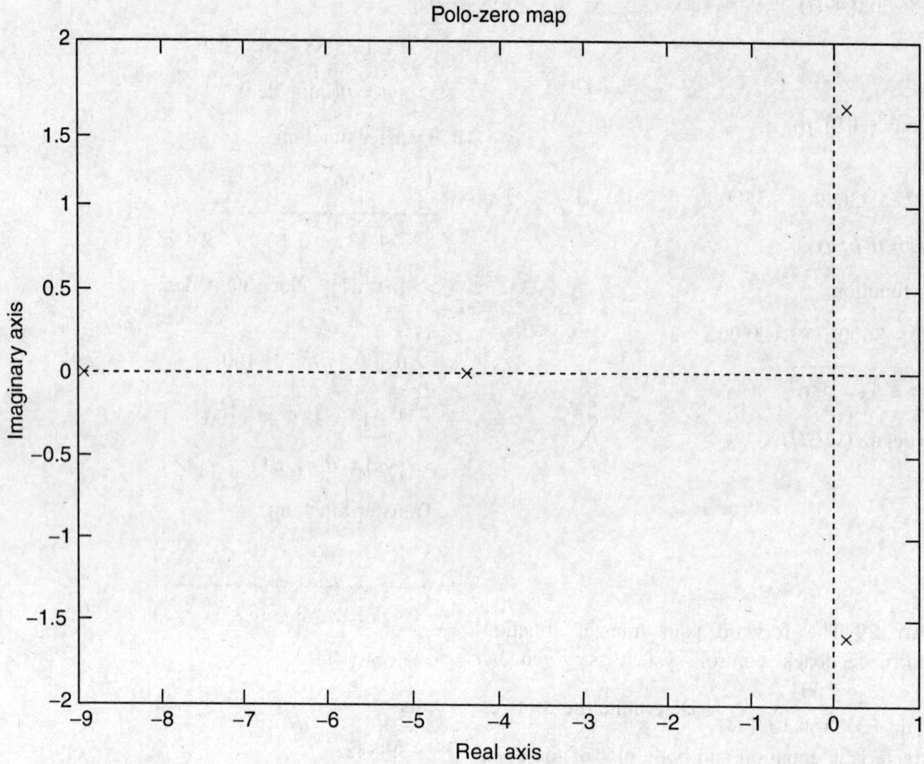

Fig. P29

```
>> num = conv(p1, p2)
num =
   0   20   15
>> q1 = [1 0 0]
q1 =
   1   0   0
>> q2 = [1 1]
q2 =
   1   1
>> q3 = [1 0.8]
q3 =
   1.0000   0.8000
>> q = conv(q1, q2)
q =
   1   1   0   0
>> den = conv(q, q3)
den =
   1.0000   1.8000   0.8000   0   0
>> sys = tf(num, den)
Transfer function:
      20 s + 15
-----------------------
s ^ 4 + 1.8 s ^ 3 + 0.8 s ^ 2
>> bode(num, den)
>> margin(sys)
>>
```

Problem 31 Sketch the bode plot for the open loop transfer function $G(s)H(s) = \dfrac{50(s+2)}{s^2(s+1)(s+5)}$. Also determine the gain margin, phase margin, gain cross over frequency and phase cross over frequency.

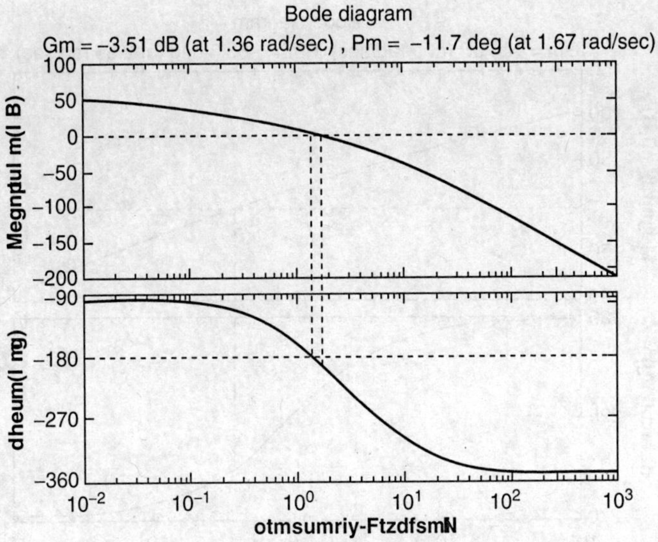

Fig. P29

Solution

```
>> p1 = [ 0 50]
p1 =
   0  50

>> p2 = [1 2]
p2 =
   1   2

>> num = conv(p1, p2)
num =
   0  50  100

>> q1 = [ 1 0 0]
q1 =
   1  0   0

>> q2 = [1 1]
q2 =
   1   1

>> q3 = [1 5]
q3 =
   1   5

>> q = conv(q1, q2)
q =
   1   1   0   0

>> den = conv(q, q3)
den =
   1   6   5   0   0

>> sys = tf(num, den)

Transfer function:

   50 s + 100
-------------------------
s ^ 4 + 6 s ^ 3 + 5 s ^ 2

>> bode(num, den)
>> margin(sys)
>>
```

Problem 32 Sketch the bode plot for the open loop transfer function $G(s)H(s)=\dfrac{50(s+3)}{s(s+2)(s+6)}$.

Also determine the gain margin, phase margin, gain cross over frequency and phase cross over frequency.

Solution

```
>> p1 = [ 0 50]
p1 =
   0  50
>> p2 = [1 3]
p2 =
   1   3
>> num = conv(p1, p2)
```

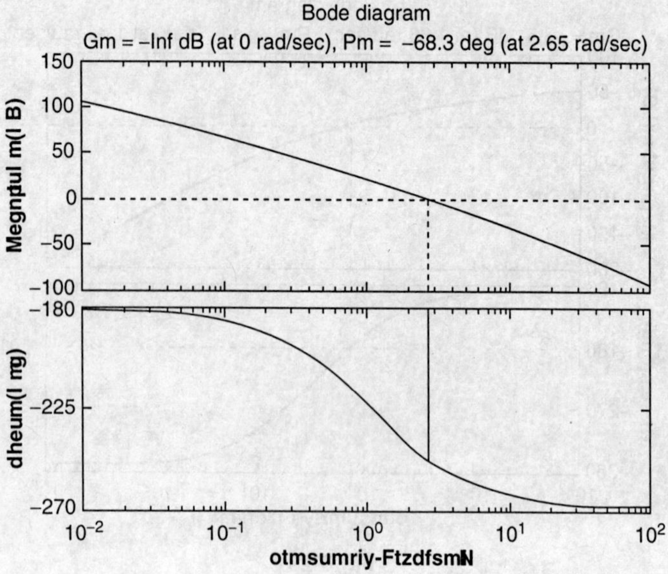

Fig. P30

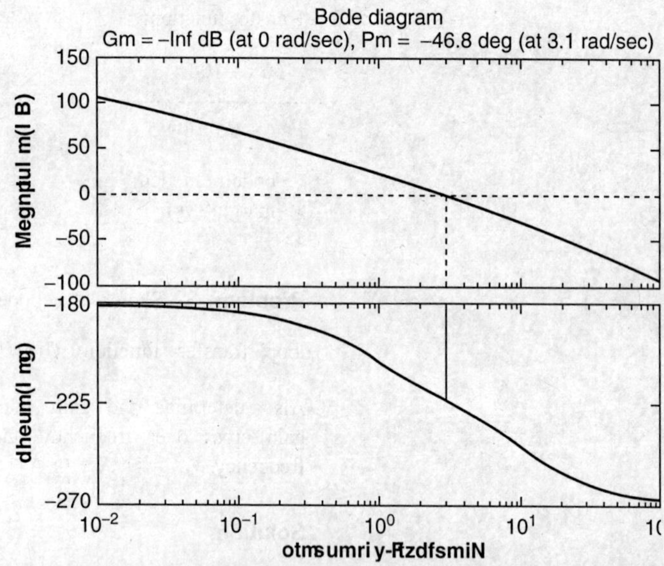

Fig. P31

```
num =
   0   50   150

>> q1 = [ 1 0 ]
q1 =
   1    0

>> q2 = [1 2]
q2 =
   1    2

>> q3 = [1 6]
q3 =
   1    6

>> q = conv(q1, q2)
```

```
q =
   1  2  0
>> den = conv(q, q3)
den =
   1  8  12  0
>> sys = tf(num, den)
Transfer function:
 50 s + 100
--------------------
s ^ 3 + 8 s ^ 2 + 12 s
>> bode(num, den)
>> margin(sys)
>>
```

Problem 33 Sketch the bode plot for the open loop transfer function $G(s)H(s) = \frac{50(s+2)}{s(s+5)}$. Also determine the gain margin, phase margin, gain cross over frequency and phase cross over frequency.

Solution

```
>> p1 = [0 50]
p1 =
   0  50
>> p2 = [1 2]
p2 =
   1  2
>> num = conv(p1, p2)
num =
   0  50  100
>> q1 = [ 1 0 ]
q1 =
   1  0
>> q2 = [1 5]
q2 =
   1  5
>> den = conv(q1, q2)
den =
   1  5  0
>> sys = tf(num, den)
Transfer function:
50 s + 100
--------------
s ^ 2 + 5 s
>> bode(num, den)
>> margin(sys)
>>
```

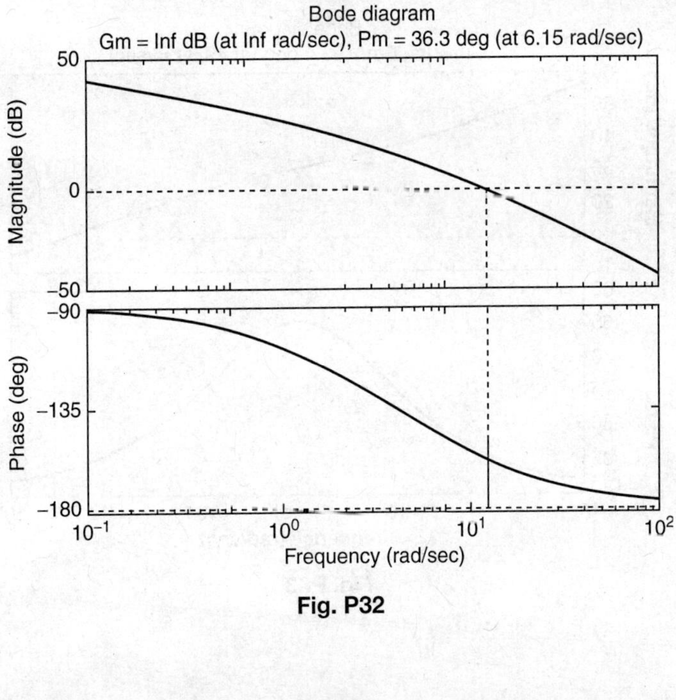

Fig. P32

Problem 34 Sketch the bode plot for the open loop transfer function $G(s)H(s) = \frac{50(s+2)}{(s+1)\ (s+3)}$. Also determine the gain margin, phase margin, gain cross over frequency and phase cross over frequency.

Solution

```
>> p1 = [ 0 50]
p1 =
   0   50

>> p2 = [1 2]
p2 =
   1   2

>> num = conv(p1, p2)
num =
   0   50   100

>> q1 = [ 1 1]
q1 =
   1   1

>> q2 = [1 3]
q2 =
   1   3

>> den = conv(q1, q2)
den =
   1   4   3

>> sys = tf(num, den)
Transfer function:
 50 s + 100
---------------
s ^ 2 + 4 s + 3

>> bode(num, den)
>> margin(sys)
>>
```

Problem 35 Sketch the bode plot for the open loop transfer function $G(s)H(s) = \frac{50}{s(s+1)(s+3)}$. Also determine the gain margin, phase margin, gain cross over frequency and phase cross over frequency.

Solution

```
>> num = [0 50]
num =
   0   50

>> q1 = [1 0]
q1 =
   1   0

>> q2 = [1 1]
```

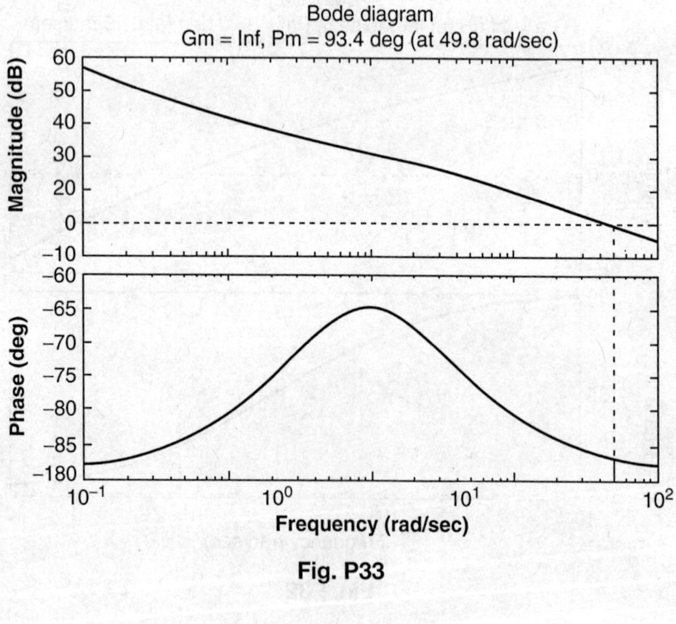

Fig. P33

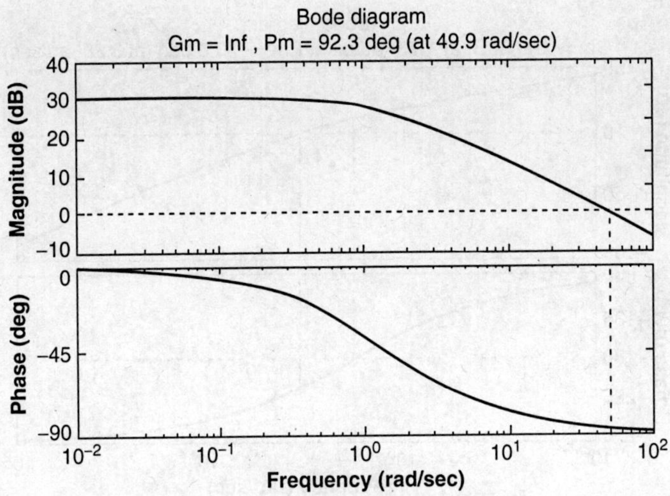

Fig. P34

```
q2 =
   1   1

>> q3 = [1 3]
q3 =
   1   3

>> q = conv(q1, q2)
q =
   1   1   0

>> den = conv(q, q3)

den =
   1   4   3   0

>> sys = tf(num, den)

Transfer function:

      50
-----------------
s ^ 3 + 4 s ^ 2 + 3 s

>> bode(num, den)
>> margin(sys)
>>
```

Problem 36 Sketch the root locus for the open loop transfer function $G(s)H(s) = \dfrac{50}{s(s+1)(s+3)}$.

Solution

```
>> num = [0 50]
num =
   0   50

>> q1 = [1 0]
q1 =
   1   0

>> q2 = [1 1]
q2 =
   1   1

>> q3 = [1 3]
q3 =
   1   3
>> q = conv(q1, q2)
q =
   1   1   0
>> den = conv(q, q3)

den =
   1   4   3   0
>> sys = tf(num, den)

Transfer function:

      50
-----------------
s ^ 3 + 4 s ^ 2 + 3 s
>> rlocus(sys)
>>
```

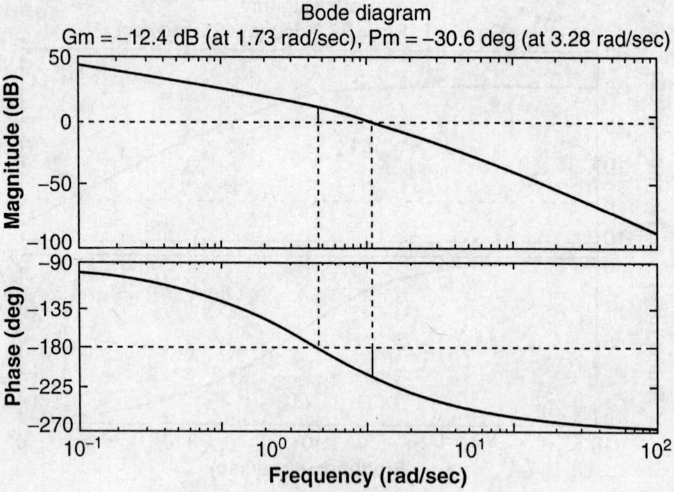

Fig. P35

Problem 37 Sketch the root locus for the open loop transfer function $G(s)H(s) = \dfrac{K}{s^4 + 5s^3 + 8s^2 + 10s + 12}$.

Solution

```
>> p = [0 1]
p =
   0   1
>> q = [1 5 8 10 12]
q =
   1   5   8   10   12
>> sys = tf(p, q)
Transfer function:
              1
-------------------------------
s ^ 4 + 5 s ^ 3 + 8 s ^ 2 + 10 s + 12
>> rlocus(sys)
>>
```

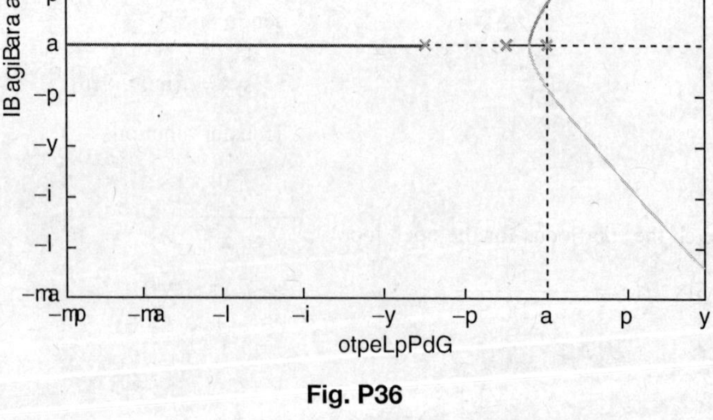

Fig. P36

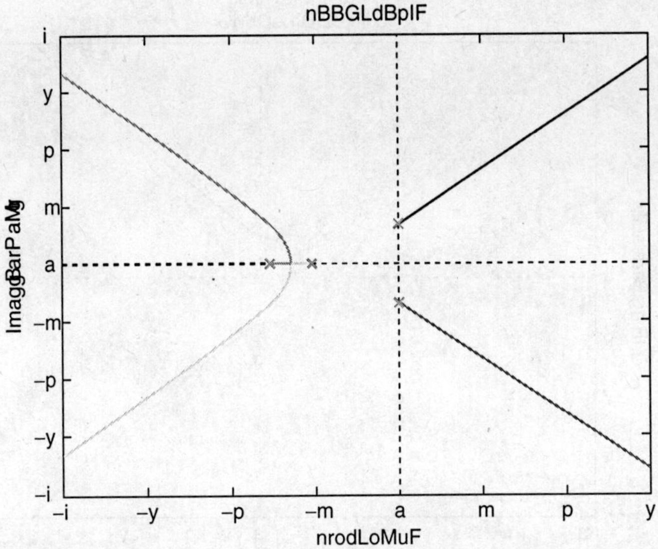

Fig. P37

Problem 38 Sketch the root locus for the open loop transfer function $G(s)H(s)=\dfrac{K}{s^4+5s^3+8s^2+10s}$.

Solution

```
>> p = [0 1]
p =
  0  1
>> q = [1 5 8 10 0]
q =
  1  5  8  10  0
>> sys = tf(p, q)
Transfer function:
      1
--------------------------
s^4 + 5 s^3 + 8 s^2 + 10 s
>> rlocus(sys)
>>
```

Problem 39 Sketch the root locus for the open loop transfer function $G(s)H(s)=\dfrac{K(s-1)}{s(s+1)\,(s+3)}$.

Solution

```
>> p = [1 -1]
p =
  1  -1
>> q1 = [1 0]
q1 =
  1  0
>> q2 = [1 1]
q2 =
  1  1
>> q3 = [1 3]
q3 =
  1  3
>> q = conv(q1, q2)
q =
  1  1  0
>> q = conv(q, q3)
q =
  1  4  3  0
>> sys = tf(p, q)
Transfer function:
     s - 1
----------------
s^3 + 4 s^2 + 3 s
>> rlocus(sys)
>>
```

Problem 40 Sketch the nyquist plot for the open loop transfer function $G(s)H(s)=\dfrac{40}{(s+1)(s+3)}$.

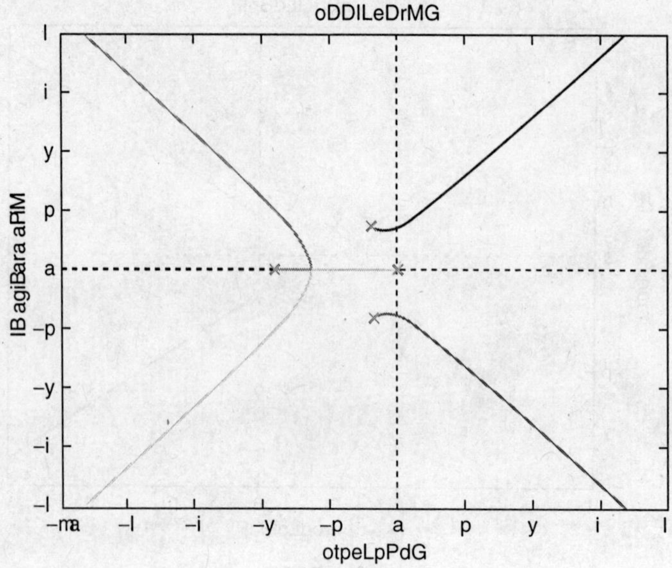

Fig. P38

Solution

```
>> num = [0 40]
num =
    0   40
>> q1 = [1 1]
q1 =
    1    1
>> q2 = [1 3]
q2 =
    1    3
>> den = conv(q1, q2)
den =
    1    4    3
>> sys = tf(num, den)
```

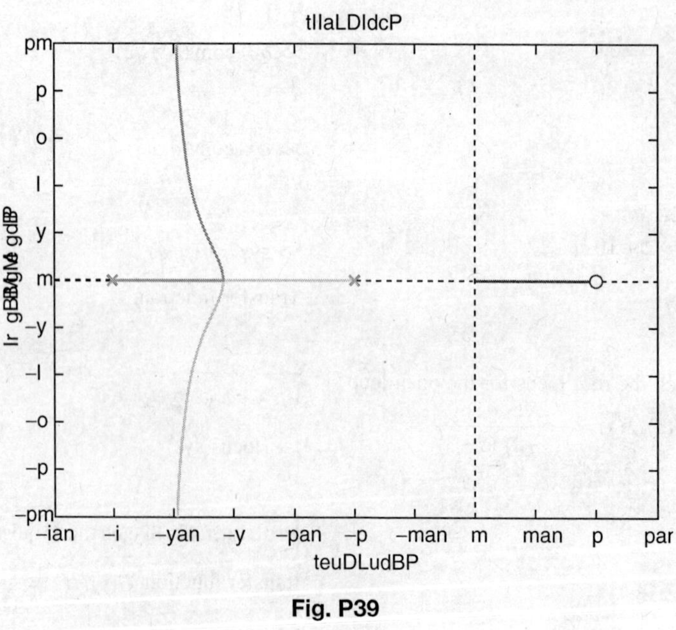

Fig. P39

```
Transfer function:
   40
------------
s ^ 2 + 4 s + 3

>> nyquist(sys)
>>
```

Problem 41 Sketch the nyquist plot for the open loop transfer function $G(s)H(s) = \dfrac{4}{s(s+1)(s+3)}$.

Solution

```
>> num = [0 4]
num =
   0   4
>> q1 = [1 0]
q1 =
   1   0
>> q2 = [1 1]
q2 =
   1   1
>> q3 = [1 3]
q3 =
   1   3
>> q = conv(q1, q2)
q =
   1   1   0
>> den = conv(q, q3)
den =
   1   4   3   0
>> sys = tf(num, den)
Transfer function:
      4
----------------
s ^ 3 + 4 s ^ 2 + 3 s
>> nyquist (sys)
>>
```

Problem 42 Obtain the state model for transfer function shown below:

$$\frac{C(s)}{R(s)} = \frac{s+1}{(s+2)(s+4)}$$

Solution

```
>> num = [ 1 1]
num =
   1   1
>> p1 = [1 2]
p1 =
   1   2
```

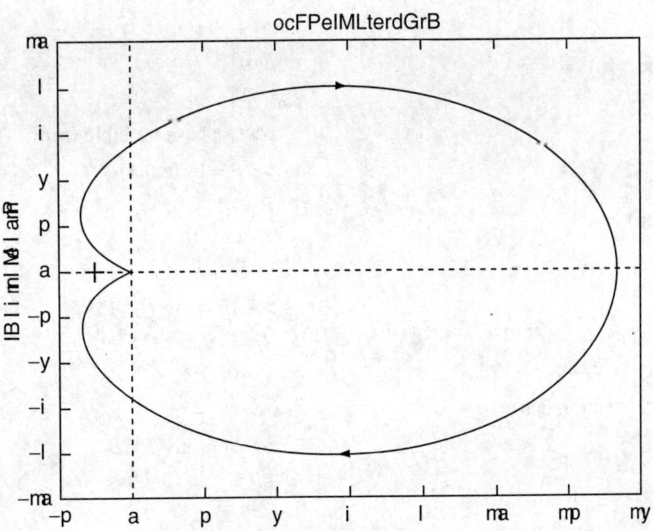

Fig. P40

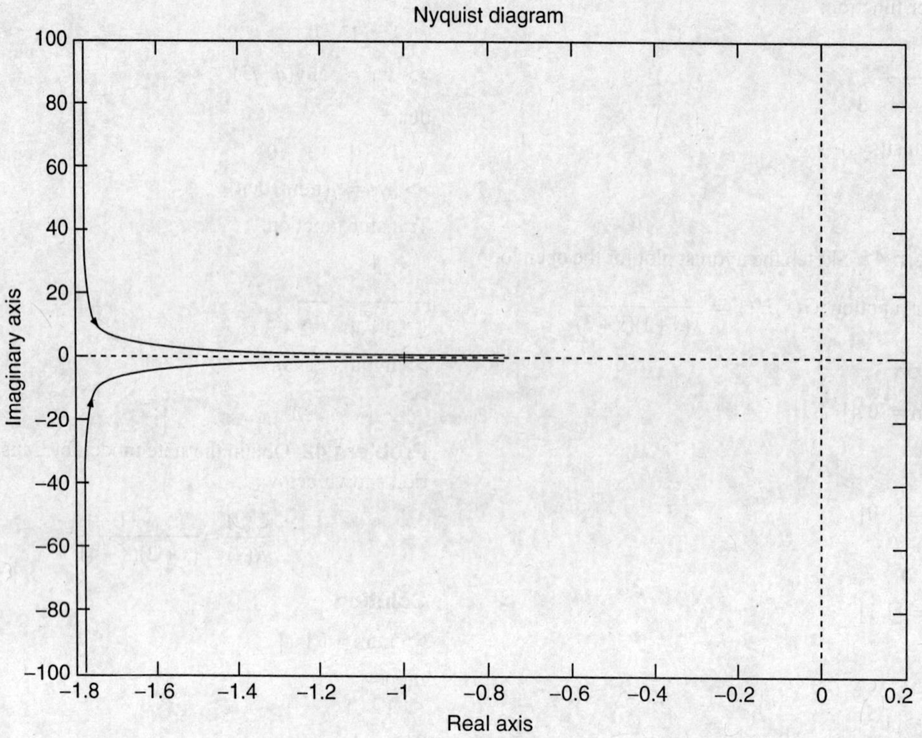

Fig. P41

```
>> p2 = [1 4]
p2 =
   1   4
>> den = conv(p1, p2)
den =
   1   6   8
>> sys = tf(num, den)
Transfer function:
   s + 1
-----------------
s ^ 2 + 6 s + 8
>> [A, B, C, D] = tf2ss (num, den)
A =
  -6  -8
   1   0
B =
   1
   0
C =
   1   1
D =
   0
>> % phase variable form
>> P = [1 0; 0 1]
P =
   1   0
   0   1
>> Ap = inv(P)*A*P
Ap =
  -6  -8
   1   0
>> Bp = inv(P)*B
Bp =
   1
   0
>> Cp = C*P
```

```
Cp =
   1   1
>> Dp = D
Dp =
   0
>>
```

Problem 43 Obtain the state model for transfer function shown below:

$$\frac{C(s)}{R(s)} = \frac{s^2+2s+3}{s^3+11s^2+29s+26}$$

Solution

```
>> num = [1 2 3]
num =
   1   2   3
>> den = [1 11 29 26]
den =
   1   11   29   26
>> sys = tf(num, den)
Transfer function:
   s^2 + 2 s + 3
-----------------------------
s^3 + 11 s^2 + 29 s + 26
>> [A, B, C, D] = tf2ss (num, den)
A =
  -11  -29  -26
    1    0    0
    0    1    0
B =
   1
   0
   0
C =
   1   2   3
D =
   0
>> P = [1 0 0; 0 1 0; 0 0 1]
P =
   1   0   0
   0   1   0
   0   0   1
>> Ap = inv(P)*A*P
Ap =
  -11  -29  -26
    1    0    0
    0    1    0
>> Bp = inv(P)*B
Bp =
   1
   0
   0
>> Cp = C*P
Cp =
   1   2   3
>> Dp = D
Dp =
   0
>>
```

Problem 44 Obtain the state model for transfer function shown below:

$$\frac{C(s)}{R(s)} = \frac{3}{s^3+8s^2+13s+18}$$

Solution

```
>> num = [3]
num =
   3
>> den = [1 8 13 18]
den =
   1   8   13   18
>> sys = tf(num, den)
Transfer function:
      3
-----------------------------
s^3 + 8 s^2 + 13 s + 18
>> [A, B, C, D] = tf2ss(num, den)
A =
   -8  -13  -18
    1    0    0
    0    1    0
B =
   1
   0
   0
C =
   0   0   3
D =
   0
>> P = [1 0 0; 0 1 0; 0 0 1]
P =
   1   0   0
   0   1   0
   0   0   1
```

```
>> Ap = inv(P)*A*P
Ap =
   -8  -13  -18
    1    0    0
    0    1    0
>> Bp = inv(P)*B
Bp =
    1
    0
    0
>> Cp = C*P
Cp =
    0   0   3
>> Dp = D
Dp =
    0
>>
```

Problem 45 Determine the state transition matrix for

$$A = \begin{bmatrix} 1 & 5 \\ -3 & -6 \end{bmatrix}$$

Solution

```
>> A = [ 1 5; -3 -6]
A =
    1    5
   -3   -6
>> B = A*A
B =
  -14  -25
   15   21
>> C = B*A
C =
   61   80
  -48  -51
>> D = [1 0; 0 1]
D =
    1   0
    0   1
>> syms t
>> A = A*t
A =
[   t, 5*t]
[-3*t, -6*t]
>> B = B*t
B =
[-14*t, -25*t]
[ 15*t, 21*t]
>> C = C*t
C =
[ 61*t, 80*t]
[-48*t, -51*t]
>> B = B/2
B =
[ -7*t, -25/2*t]
[ 15/2*t, 21/2*t]
>> C = C/6
C =
[ 61/6*t, 40/3*t]
[   -8*t, -17/2*t]
>> E = D + A + B + C
E =
[ 1 + 25/6*t,   35/6*t]
[  -7/2*t,    1-4*t]
>>
```

Problem 46 The state equations of a system are shown below:

$$\begin{bmatrix} \dot{x}_1 \\ \dot{x}_2 \end{bmatrix} = \begin{bmatrix} 1 & 5 \\ -3 & -6 \end{bmatrix} \begin{bmatrix} x_1 \\ x_2 \end{bmatrix} + \begin{bmatrix} 0 \\ 1 \end{bmatrix} u$$

Check the controllability.

Solution

```
>> a = [1 5; -3 -6]
a =
    1    5
   -3   -6
>> b = [0; 1]
b =
    0
    1
>> P = ctrb(a, b)
P =
    0    5
    1   -6
>> rank(P)
    2
>> % the system is controllable
```

Problem 47 The state equations of a system are shown below:

$$\begin{bmatrix} \dot{x}_1 \\ \dot{x}_2 \end{bmatrix} = \begin{bmatrix} 1 & 5 \\ -3 & -6 \end{bmatrix} \begin{bmatrix} x_1 \\ x_2 \end{bmatrix} + \begin{bmatrix} 0 \\ 1 \end{bmatrix} u$$

$$y = \begin{bmatrix} 1 & 2 \end{bmatrix} \begin{bmatrix} x_1 \\ x_2 \end{bmatrix} + [0]\, u$$

Obtain the transfer function of the system.

Solution

```
>> a = [1 5; -3 -6]
a =
   1   5
  -3  -6
>> b = [0;1]
b =
   0
   1
>> c = [1 2]
c =
   1  2
>> d = [0]
d =
   0
>> [num, den] = ss2tf(a, b, c, d)
num =
       0  2.0000  3.0000
den =
   1.0000  5.0000  9.0000
>> sys = tf(num, den)
Transfer function:
  2 s + 3
---------------
s ^ 2 + 5 s + 9
>>
```

Problem 48 The state equations of a system are shown below:

$$\begin{bmatrix} \dot{x}_1 \\ \dot{x}_2 \end{bmatrix} = \begin{bmatrix} -1 & 2 \\ -3 & 4 \end{bmatrix} \begin{bmatrix} x_1 \\ x_2 \end{bmatrix} + \begin{bmatrix} 1 \\ 2 \end{bmatrix} u$$

$$y = \begin{bmatrix} 2 & 3 \end{bmatrix} \begin{bmatrix} x_1 \\ x_2 \end{bmatrix} + [0]\, u$$

Obtain the transfer function of the system. Check the controllability and observability of the system.

Solution

```
>> a = [-1 2; -3 4]
a =
  -1   2
  -3   4
>> b = [1 ;2]
b =
   1
   2
>> c = [2 3]
c =
   2  3
>> d = [0]
d =
   0
>> [num, den] = ss2tf(a, b, c, d)
num =
       0  8.0000  -3.0000
den =
   1.0000  -3.0000  2.0000
>> sys = tf(num, den)
Transfer function:
  8 s - 3
-------------
s ^ 2 - 3 s + 2
>> P = ctrb(a, b)
P =
   1  3
   2  5
>> rank(P)
   2
>> % the system is controllable
>> Q = obsv(a, c)
Q =
   2   3
 -11  16
>> rank(Q)
   2
>> % the system is observable
>>
```

Multiple-choice Questions

1. The Laplace transform of $e^{-2t}\sin 3t$ will be

   (a) $\frac{9}{s^2+9}$ (b) $\frac{9}{(s+3)^2+9}$ (c) $\frac{4}{(s+2)^2+4}$ (d) $\frac{9}{(s+2)^2+9}$

2. The Laplace transform of $2+3e^{-2t}$ will be

   (a) $2s+\frac{3}{s}$ (b) $2+\frac{3}{s+2}$ (c) $\frac{2}{s}+\frac{2}{s+3}$ (d) $\frac{2}{s}+\frac{3}{s+2}$

3. The initial value of the function $f(t)$ having Laplace transform $F(s)=\frac{s+5}{s^2+3s+2}$ is

   (a) 0 (b) ∞ (c) $\frac{5}{2}$ (d) 1

4. The final value of the function $f(t)$ having Laplace transform $F(s)=\frac{s+5}{s^2+3s+2}$ is

   (a) 0 (b) ∞ (c) $\frac{5}{2}$ (d) 1

5. The following relation is valid

   (a) $(A+B)^T = A^T + B$ (b) $(A+B)^T = A^T + B^T$
   (c) $(A+B)^T = B^T + A$ (d) $(A+B)^T = A + B$

6. The following relation is valid

   (a) $(AB)^T = A^T B$ (b) $(AB)^T = A^T B^T$ (c) $(AB)^T = B^T A$ (d) $(AB)^T = B^T A^T$

7. The transfer function of a system becomes if its impulse response is e^{-2t}

   (a) s (b) $\frac{1}{s}$ (c) $\frac{1}{s+2}$ (d) $\frac{s}{s+2}$

8. The transfer function $\frac{Y(s)}{X(s)}$ of a system represented by the differential equation $\frac{dy}{dt}+2y=\frac{dx}{dt}+4x$ is given by

   (a) $\frac{s+1}{s+2}$ (b) $\frac{s+4}{s+2}$ (c) $\frac{s}{s+2}$ (d) $\frac{s+2}{s+4}$

9. The transfer function $\frac{Y(s)}{X(s)}$ of a system represented by the differential equation $3\frac{dy(t)}{dt}+2y(t)=2x(t-T)$ is given by

   (a) $\frac{2e^{-2T}}{3s+2}$ (b) $\frac{e^{-2T}}{3s+2}$ (c) $\frac{2e^{-2T}}{s+2}$ (d) $\frac{2e^{-2T}}{3s+1}$

10. If the transfer function of a system is $T(s)=\dfrac{3(s+1)}{(s+2)(s+5)}$, the characteristic equation is represented by

(a) $s^2+s+5=0$ (b) $s^2+7s+10=0$ (c) $s^2+3s+5=0$ (d) $s^2+s+7=0$

11. For a gear train the following relation is valid

(a) $r \propto N$ (b) $r \propto \dfrac{1}{N}$ (c) $r \propto N^2$ (d) $r \propto \dfrac{1}{N^2}$

12. For a gear train the following relation is valid

(a) $T \propto \theta$ (b) $T \propto \dfrac{1}{\theta}$ (c) $T \propto \theta^2$ (d) $T \propto \dfrac{1}{\theta^2}$

13. The overall transfer function of a system having individual blocks $G_1(s)$, $G_2(s)$ and $G_3(s)$ in cascade is given by

(a) $G_1(s)G_2(s)+G_3(s)$ (b) $G_1(s)+G_2(s)+G_3(s)$
(c) $G_1(s)G_2(s)-G_3(s)$ (d) $G_1(s)G_2(s)G_3(s)$

14. The overall transfer function of a system having individual blocks $G_1(s)$, $G_2(s)$ and $G_3(s)$ in parallel with positive summation is given by

(a) $G_1(s)G_2(s)+G_3(s)$ (b) $G_1(s)+G_2(s)+G_3(s)$
(c) $G_1(s)G_2(s)-G_3(s)$ (d) $G_1(s)G_2(s)G_3(s)$

15. The inverse Laplace transform of $\dfrac{3}{s(s+2)}$ is given by

(a) $3(1-e^{-2t})$ (b) $2(1-e^{-2t})$ (c) $1.5(1-e^{-2t})$ (d) $3(1-e^{-t})$

16. For the system $\dfrac{C(s)}{R(s)}=\dfrac{9}{s^2+6s+9}$, the nature of the time response will be

(a) overdamped (b) critically damped (c) under damped (d) none of these

17. For the system $\dfrac{C(s)}{R(s)}=\dfrac{9}{s^2+6s+9}$, the natural frequency of oscillation will be

(a) 2 rad/sec (b) 3 rad/sec (c) 1 rad/sec (d) 1.5 rad/sec

18. For the system $\dfrac{C(s)}{R(s)}=\dfrac{16}{s^2+4s+16}$, the damping factor will be

(a) 0.6 (b) 0.5 (c) 0.2 (d) 0.3

19. For the system $\dfrac{C(s)}{R(s)}=\dfrac{16}{s^2+4s+16}$, the damped frequency of oscillation will be

(a) 3.46 (b) 3.85 (c) 4.2 (d) 1.3

20. The velocity error coefficient is given by

(a) $\underset{s\to 0}{Lt}\, sG(s)H(s)$ (b) $\underset{s\to 0}{Lt}\, G(s)H(s)$ (c) $\underset{s\to 0}{Lt}\dfrac{1}{sG(s)H(s)}$ (d) $\underset{s\to 0}{Lt}\dfrac{1}{G(s)H(s)}$

21. The positional error coefficient is given by

(a) $\underset{s\to 0}{Lt}\, sG(s)H(s)$ (b) $\underset{s\to 0}{Lt}\, G(s)H(s)$ (c) $\underset{s\to 0}{Lt}\dfrac{1}{sG(s)H(s)}$ (d) $\underset{s\to 0}{Lt}\dfrac{1}{G(s)H(s)}$

22. The acceleration error coefficient is given by

(a) $\underset{s\to 0}{Lt}\, sG(s)H(s)$ (b) $\underset{s\to 0}{Lt}\, s^2G(s)H(s)$ (c) $\underset{s\to 0}{Lt}\dfrac{1}{sG(s)H(s)}$ (d) $\underset{s\to 0}{Lt}\dfrac{1}{G(s)H(s)}$

23. The steady state error due to a step input Au(t) is given by

(a) $\frac{A}{1+K_p}$ (b) $\frac{A}{K_p}$ (c) $\frac{1}{AK_p}$ (d) $\frac{K_p}{1+A}$

24. The steady state error due to a ramp input Ar(t) is given by

(a) $\frac{A}{1+K_v}$ (b) $\frac{A}{K_v}$ (c) $\frac{1}{AK_v}$ (d) $\frac{K_v}{1+A}$

25. The steady state error due to a parabolic input having magnitude A is given by

(a) $\frac{A}{1+K_a}$ (b) $\frac{A}{K_a}$ (c) $\frac{1}{AK_a}$ (d) $\frac{K_a}{1+A}$

26. The type of the system is having $G(s)=\frac{K(1+2s)}{s^2}$ and $H(s)=\frac{K(1+3s)}{s^2+2s+1}$ is

(a) type 1 (b) type 2 (c) type 0 (d) type 3

27. The transfer function of a zero-order-hold system is

(a) $\frac{1}{s}(1+e^{-st})$ (b) $\frac{1}{s}(1-e^{-st})$ (c) $1+\frac{1}{s}e^{-st}$ (d) $1-\frac{1}{s}e^{-st}$

28. The transfer function of $C(s)/R(s)$ of the system, whose block diagram is shown in Fig. 1 below, is

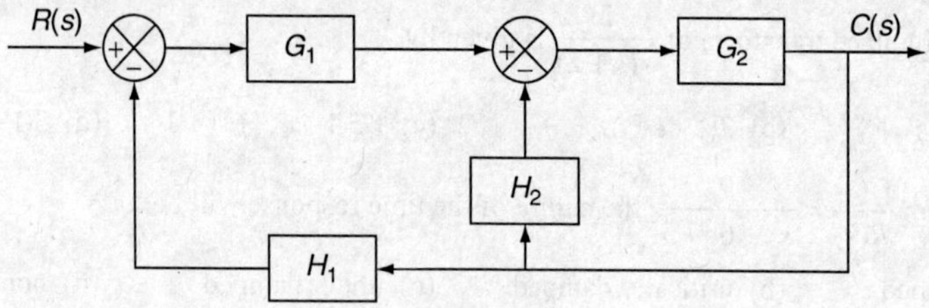

Fig. 1

(a) $\frac{G_1G_2}{1+G_1H_1+G_2H_2-G_1G_2H_1H_2}$ (b) $\frac{G_1G_2}{1+G_1H_1+G_2H_2+G_1G_2H_1H_2}$

(c) $\frac{G_1G_2}{1+G_1H_1+G_2H_2}$ (d) $\frac{G_1(1+G_2H_2)+G_2(1+G_1H_1)}{1+G_1H_1+G_2H_2+G_1G_2H_1H_2}$

29. The system described the mechanical system shown in Fig. 2 is

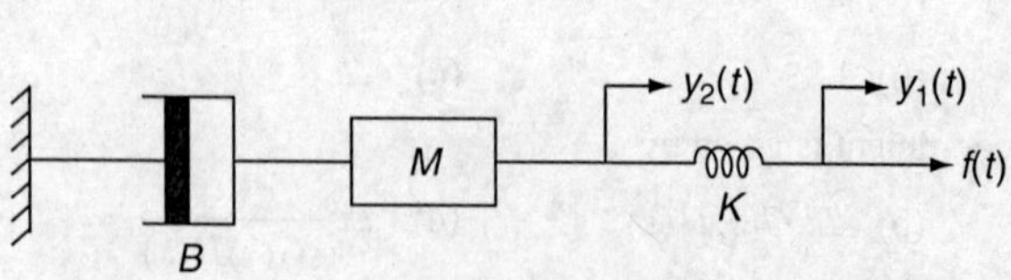

Fig. 2

(a) $M\frac{d^2y_1(t)}{dt^2}+B\frac{dy_1(t)}{dt}=k[y_2(t)-y_1(t)]$ (b) $M\frac{d^2y_2(t)}{dt^2}+B\frac{dy_2(t)}{dt}=k[y_2(t)-y_1(t)]$

(c) $M\frac{d^2y_1(t)}{dt^2}+B\frac{dy_1(t)}{dt}=k[y_1(t)-y_2(t)]$ (d) $M\frac{d^2y_2(t)}{dt^2}+B\frac{dy^2(t)}{dt}=k[y_1(t)-y_2(t)]$

30. The parameter is related to the system shown in Fig. 3 is

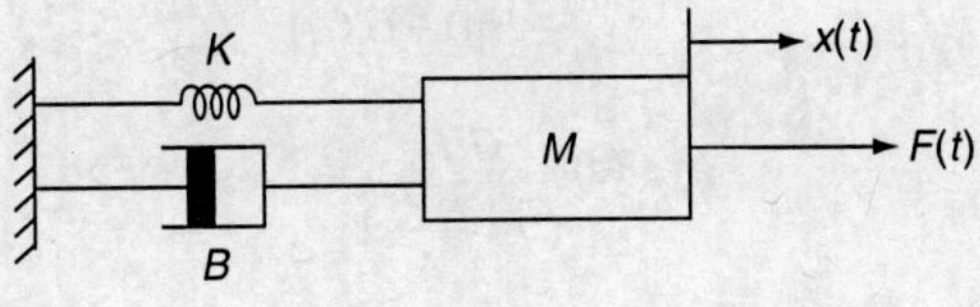

Fig. 3

(a) The time constant is $\frac{1}{M}$

(b) Damping co-efficient is $\frac{B\sqrt{KM}}{2}$

(c) Natural frequency of oscillation is $\sqrt{\frac{K}{M}}$

(d) None of these

31. The signal flow diagram of a system is shown in Fig. 4. The number of forward paths and the number of non-touching loops are respectively

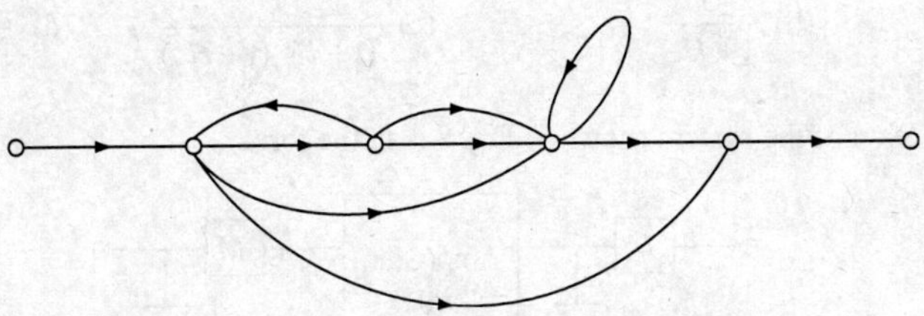

Fig. 4

(a) 3, 1 (b) 3, 2 (c) 4, 2 (d) 2, 4

32. A system is represented by the block diagram is given in the Fig. 5. The transfer function of the system is

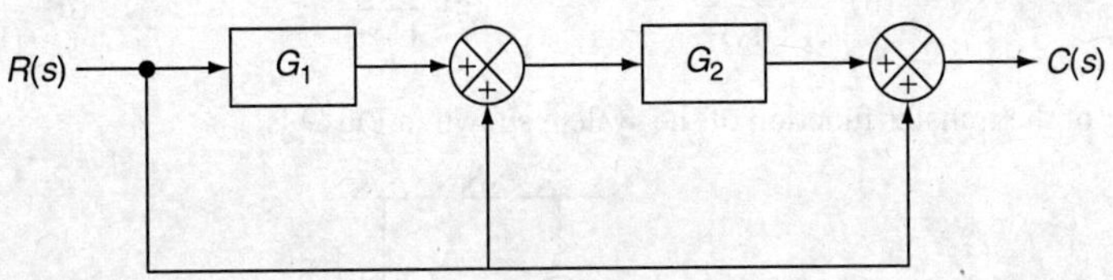

Fig. 5

(a) G_1G_2 (b) G_1+G_2 (c) $1+G_1+G_1G_2$ (d) $1+G_2+G_1G_2$

33. The transfer function of the system shown in Fig. 6 is

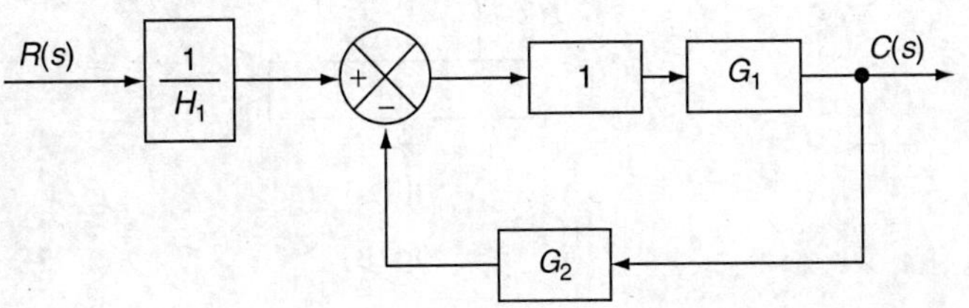

Fig. 6

(a) $\dfrac{G_1G_2}{H_1(1+G_1G_2)}$ (b) $\dfrac{G_1}{H_1(1+G_1G_2)}$

(c) $\dfrac{G_2}{H_1(1+G_1G_2)}$ (d) $\dfrac{G_2H_1}{(1+G_1G_2)}$

34. The overall transfer function $\dfrac{C(s)}{R(s)}$ of the system shown in Fig. 7 is

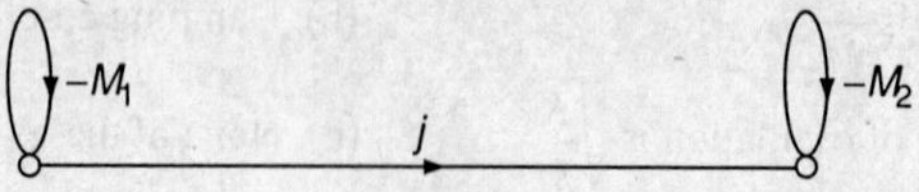

Fig. 7

(a) G (b) $\dfrac{G}{1+H_2}$ (c) $\dfrac{G}{(1+H_1)(1+H_2)}$ (d) $\dfrac{G}{1+H_1+H_2}$

35. The transfer function of the system shown in Fig. 8 is in the form

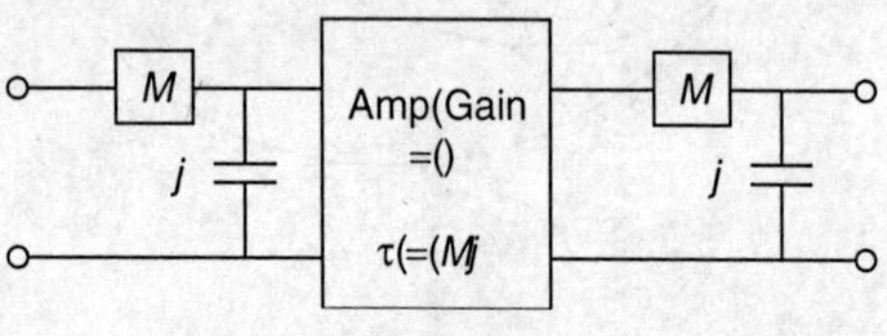

Fig. 8

(a) $\dfrac{1}{1+\tau s}$ (b) $\dfrac{1}{(1+\tau s)^2}$ (c) $\dfrac{\tau s}{1+\tau s}$ (d) $\dfrac{\tau s}{(1+\tau s)^2}$

36. The order of the transfer function of the system shown in Fig. 9 is

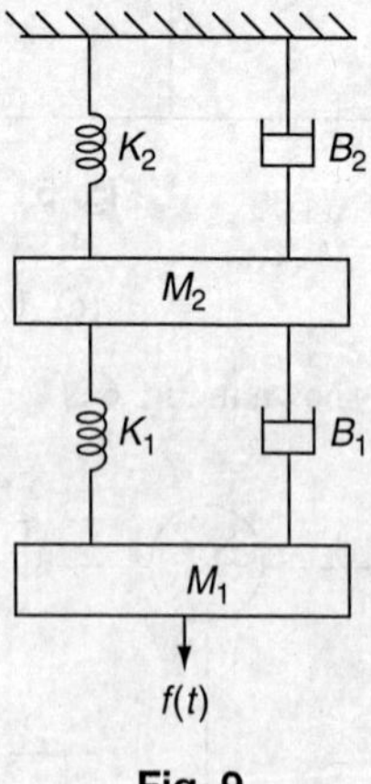

Fig. 9

(a) 1 (b) 2 (c) 3 (d) 4

37. The transfer function $C(s)/N(s)$ of the system shown in Fig. 10 is

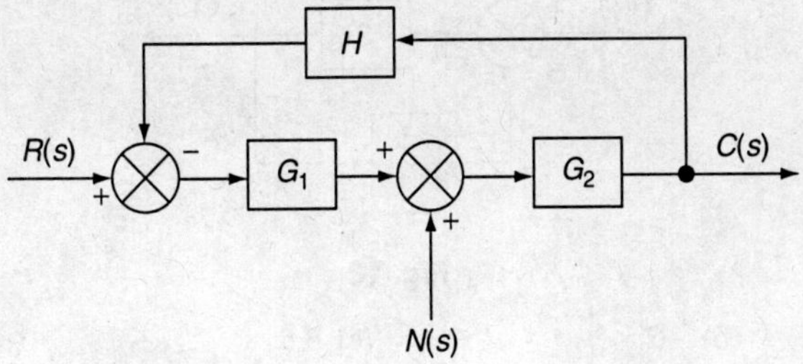

Fig. 10

(a) $\dfrac{G_1G_2}{1+G_1G_2H}$ (b) $\dfrac{G_2}{1+G_2H}$ (c) $\dfrac{G_1}{1+G_1H}$ (d) $\dfrac{G_2}{1+G_1G_2H}$

38. The transfer function $X(s)/F(s)$ of the system shown in Fig. 11 is

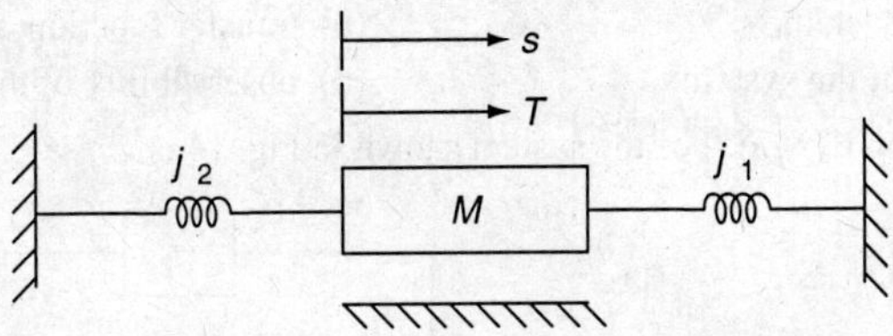

Fig. 11

(a) $\dfrac{1}{Ms^2 + fs + K_1K_2}$ (b) $\dfrac{1}{Ms^2 + fs + K_1 + K_2}$

(c) $\dfrac{1}{Ms^2 + fs + \dfrac{K_1K_2}{K_1 + K_2}}$ (d) $\dfrac{1}{Ms^2 + fs + K_1}$

39. The transfer function $\dfrac{C(s)}{R(s)}$ of the system shown in Fig. 12 is

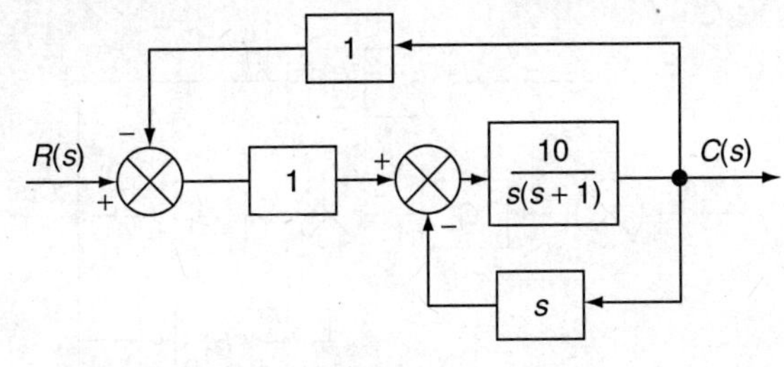

Fig. 12

(a) $\dfrac{10}{s^2+s+10}$ (b) $\dfrac{10}{s^2+11s+10}$ (c) $\dfrac{10}{s^2+9s+10}$ (d) $\dfrac{10}{s^2+2s+10}$

40. The closed loop gain of the system shown in Fig. 13 is

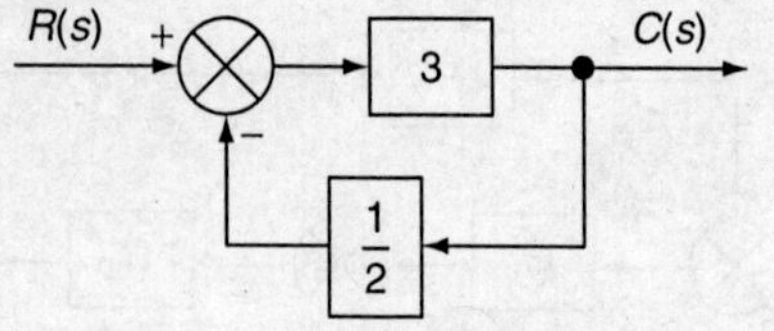

Fig. 13

(a) –9/5 (b) –6/5 (c) 6/5 (d) 9/5

41. The response $c(t)$ of a system due to an input $r(t)$ is given by the following differential equation $\frac{d^2c(t)}{dt^2}+3\frac{dc(t)}{dt}+5c(t)=5r(t)$. The transfer function of the system is

(a) $\frac{5}{s^2+3s+5}$ (b) $\frac{1}{s^2+3s+1}$ (c) $\frac{3s}{s^2+3s+5}$ (d) $\frac{s+3}{s^2+3s+5}$

42. Signal flow graph is used to obtain the

(a) stability of the system (b) transfer function of the system
(c) controllability of the system (d) observability of the system

43. The transfer function $C(s)/R(s)$ of the system shown in Fig. 14 is

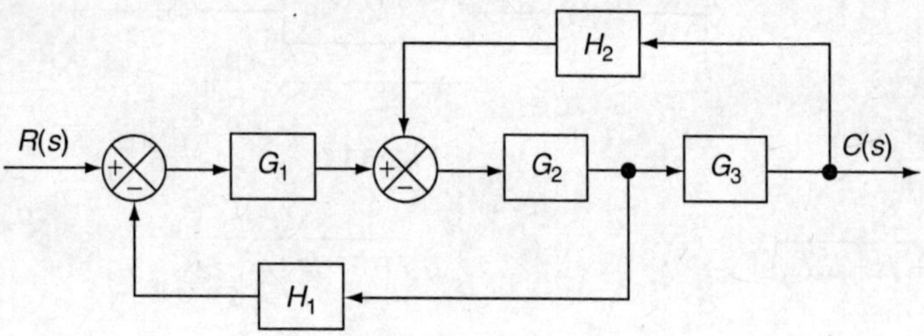

Fig .14

(a) $\frac{G_1G_2G_3}{1+H_2G_2G_3+H_1G_1G_2}$ (b) $\frac{G_1G_2G_3}{1+G_1G_2G_3H_1H_2}$ (c) $\frac{G_1G_2G_3}{1+G_1G_2G_3H_1+G_1G_2G_3H_2}$ (d) $\frac{G_1G_2G_3}{1+G_1G_2G_3H_1}$

44. The overall transfer function $C(s)/R(s)$ of the system shown in Fig. 15 for $G_1=\frac{10}{s}$, $G_2=\frac{10}{s+1}$, $H_1=s+3$ and $H_2=1$ is

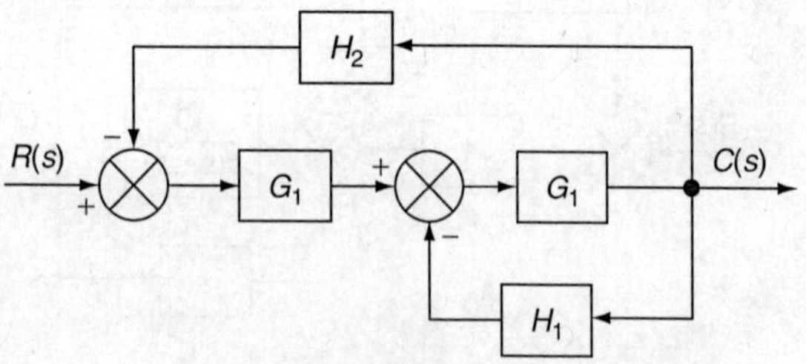

Fig. 15

(a) $\frac{10}{11s^2+31s+10}$ (b) $\frac{100}{11s^2+31s+100}$ (c) $\frac{100}{11s^2+31s+10}$ (d) $\frac{100}{11s^2+31s}$

45. The transfer function $I(s)/V(s)$ of the network shown in Fig. 16 is

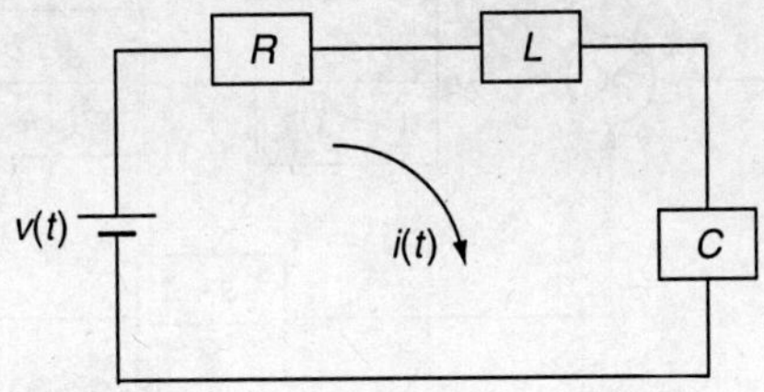

Fig. 16

(a) $\dfrac{Cs}{LCs^2 + RCs + 1}$ (b) $\dfrac{C}{LCs^2 + RCs + 1}$ (c) $\dfrac{Cs}{RCs^2 + LCs + 1}$ (d) $\dfrac{C}{RCs^2 + LCs + 1}$

46. The transfer function of a linear system is the
    (a) ratio of the output $v_0(t)$ and input $v_i(t)$
    (b) ratio of derivative of the output and the input
    (c) ratio of Laplace transform of the output and that of the input with all initial conditions zero
    (d) none of these

47. The value of y/x of the signal flow graph shown in Fig. 17 is

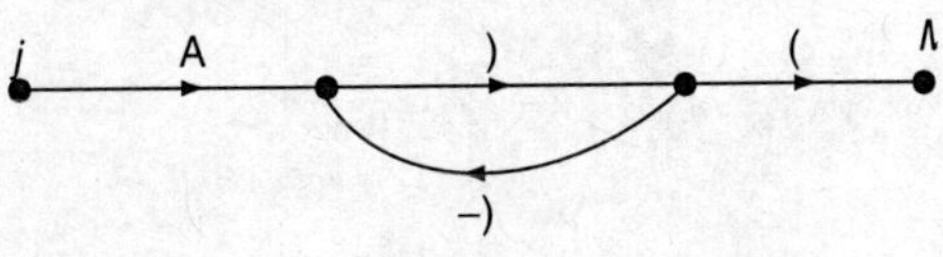

Fig. 17

(a) 3 (b) 5/2 (c) 2 (d) 7

48. The gain $C(s)/R(s)$ of the signal flow graph shown in Fig. 18 is

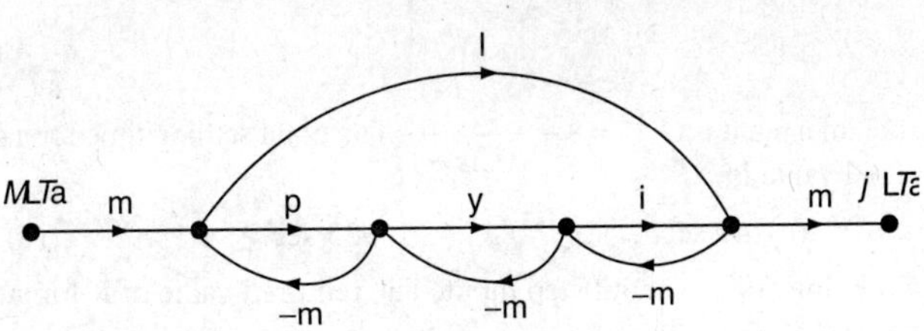

Fig. 18

(a) 11/9 (b) 22/15 (c) 24/23 (d) 22/9

49. The steady state error of a unity feedback control system having open loop transfer function $\dfrac{K}{s(s+2)}$ is due to step input is
    (a) zero (b) K (c) $1/K$ (d) ∞

50. The block diagram of a system shown in Fig. 19 represents a

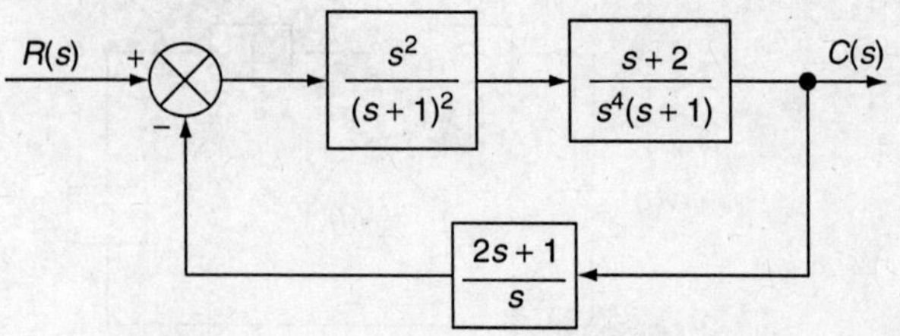

Fig. 19

(a) Type 0 system (b) Type 1 system (c) Type 2 system (d) Type 3 system

51. The forward path transfer function of a unity feedback control system is $\frac{42.25}{s^2(s+6.5)}$. The maximum value of the unity step response starting from rest is at a time equal to

(a) 0 (b) 0.56 (c) 5.6 (d) infinity

52. The correct match is

LIST I (Open Loop Transfer Function)	LIST II (Steady state errors)
A. $\frac{30}{s^2+6s+9}$	1. Zero
B. $\frac{30}{s^2+6s}$	2. 0.2
C. $\frac{30}{s^2+9s}$	3. 0.3
D. $\frac{s+1}{s^2}$	4. Infinity

	A	B	C	D
(a)	1	2	3	4
(b)	4	3	2	1
(c)	1	3	2	4
(d)	4	2	3	1

53. The transfer function of a plant is $G(s) = \frac{1}{s^2+0.2s+1}$. The plant settling time for a step input when it settles to within 2% of its final value is

(a) 20 sec. (b) 40 sec. (c) 35 sec. (d) 45 sec.

54. The system shown in Fig. 20 has a unit step input. The required value of K for steady state error 0.1 is

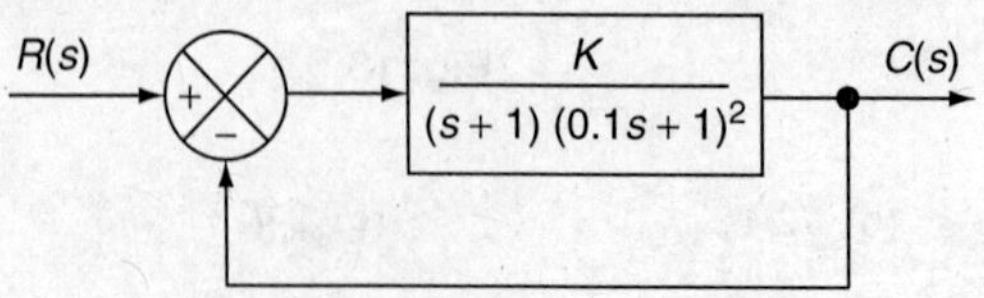

Fig. 20

(a) 0.1 (b) 0.0 (c) 1.0 (d) 9.0

55. If $s^2 + 4s + 16 = 0$ be the characteristic equation of a closed loop control system, the resonant frequency in radians per second of the system is

(a) 2 (b) $2\sqrt{3}$ (c) 4 (d) $2\sqrt{2}$

56. A system has forward path transfer function and feedback path transfer function $\frac{20}{s^2}$ and $(s+5)$ respectively. The steady state error of the system for a unit step input is

(a) 0 (b) 0.5 (c) 0.2 (d) 0.8

57. A system is represented by $\frac{dy}{dt} + 2y = 4tu(t)$. The ramp component in the forced system input is

(a) tu(t) (b) 2tu(t) (c) 3tu(t) (d) 4tu(t)

58. If the unit step response of the network is $(1 - e^{-at})$, its unit impulse response will be

(a) ae^{-at} (b) $ae^{-t/a}$ (c) $\frac{1}{ae^{-at}}$ (d) $(1-a)e^{-at}$

59. The impulse response of an R-L circuit is a
(a) rising exponential function
(b) decaying exponential function
(c) step function
(d) parabolic function

60. A unity feedback second order control system is characterized by $G(s) = \frac{K}{s(Js+B)}$ where J is moment of inertia, K is the system gain and B is the viscous damping co-efficient. The transient response specification which is not affected by the variation of the system gain is

(a) peak overshoot (b) rise time (c) settling time (d) damped frequency of oscillations

61. The performance specifications for a unity feedback control system having an open-loop transfer function $G(s) = \frac{K}{s(s+1)(s+2)}$ are

(a) velocity error co-efficient $K_v > 10\ \text{sec}^{-1}$ (b) stable closed-loop operation.

The value of K, satisfying the above specifications, is

(a) $K > 6$ (b) $6 < K < 10$ (c) $K > 10$ (d) none of these

62. The unit step response of a second order linear system with zero initial states is given by $c(t) = 1+1.25\exp(-6t)\sin(8t-\tan^{-1}1.333);\ t \geq 0$

The damping ratio and the undamped natural frequency of oscillations of the system are respectively

(a) 0.6 and 10 rad/s (b) 0.6 and 13.5 rad/s (c) 0.8 and 10 rad/s (d) 0.8 and 13.5 rad/s

63. For the unity feedback system shown in Fig. 21, the steady state error, when the input is $R = \frac{3}{s} - \frac{1}{s^2} + \frac{1}{2s^3}$, would be

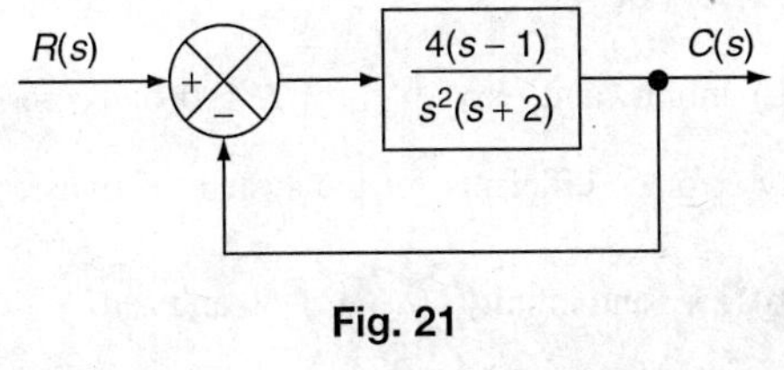

Fig. 21

(a) $\frac{1}{4}$ (b) 0 (c) ∞ (d) $\frac{1}{2}$

64. In the derivative error compensation, damping

(a) decreases and settling time decreases (b) increases and settling time increases

(c) decreases and settling time increases (d) increases and settling time decreases

65. A synchro-transmitter receiver unit is a

(a) two-phase AC device (b) 3-phase AC device

(c) DC device (d) Single-phase AC device

66. Figure 22 shows a system which is distributed so that $C(0) = 1$. The $c(t)$ for a step input will be

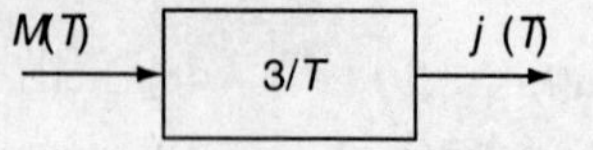

Fig. 22

(a) $1 + t$ (b) $1 - t$ (c) $1 + 3t$ (d) $1 - 3t$

67. The maximum overshoot of a unity feedback control system having open loop transfer function $G(s) = \dfrac{1}{s(s+1)}$ due to the unit step response is

(a) 0.143 (b) 0.153 (c) 0.163 (d) 0.194

68. For a feedback control system of Type-2, the steady state error for a ramp input is

(a) infinite (b) constant (c) zero (d) intermediate

69. The closed loop transfer function of a control system is given by $\dfrac{C(s)}{R(s)} = \dfrac{1}{s+1}$. For the input $r(t) = \sin t$, the steady state value of $c(t)$ is equal to

(a) $\dfrac{1}{\sqrt{2}}\cos t$ (b) 1 (c) $\dfrac{1}{\sqrt{2}}\sin t$ (d) $\dfrac{1}{\sqrt{2}}\sin\left(t - \dfrac{\pi}{4}\right)$

70. The impulse response of an initially relaxed linear system is $e^{-2t}u(t)$. To produce a response of $te^{-2t}u(t)$, the input should be equal to

(a) $2e^{-t}u(t)$ (b) $\dfrac{1}{2}e^{-2t}u(t)$ (c) $e^{-2t}u(t)$ (d) $e^{-t}u(t)$

71. The closed loop transfer function of a control system is $\dfrac{C(s)}{R(s)} = \dfrac{2(s-1)}{(s+1)(s+2)}$. The output for a unit step input is

(a) $-3e^{-2t} + 4e^{-1} - 1$ (b) $-3e^{-2t} + 4e^{-t} - 1$ (c) zero (d) infinity

72. The overshoot of the system for $\dfrac{16K}{s(s^2 + 2s + 16)}$ for a step input will be

(a) 60% (b) 40% (c) 20% (d) 10%

73. The position and acceleration error co-efficients for the open-loop transfer function $G(s) = \dfrac{K}{s^2(s+10)(s+100)}$ respectively are

(a) zero and infinity (b) infinity and zero (c) $K/100$ and zero (d) infinity and $K/1000$

74. The position and velocity error co-efficients for the system of transfer function $G(s) = \dfrac{50}{(1+0.1s)(1+2s)}$ respectively are

(a) zero and zero (b) zero and infinity (c) 50 and zero (d) 50 and infinity

75. The open loop transfer function of a system is $\frac{10}{1+s}$ having unity feedback. The steady state error due to unit step input is

(a) zero (b) $\frac{1}{11}$ (c) 10 (d) infinity

76. The correct statement(s) is (are):

(a) Derivative control improves system's transient performance

(b) Integral control does not improve system's steady state performance

(c) Integral control can convert a second order system into a third order system

77. The principles of homogeneity and super-position are applied to

(a) linear time variant systems

(b) non-linear time variant systems

(c) linear time invariant systems

(d) non-linear time invariant systems

78. In Fig. 23, $r(t) = \sin\omega t$. The steady state response $c(t)$ will exhibit a resonance peak at a frequency of

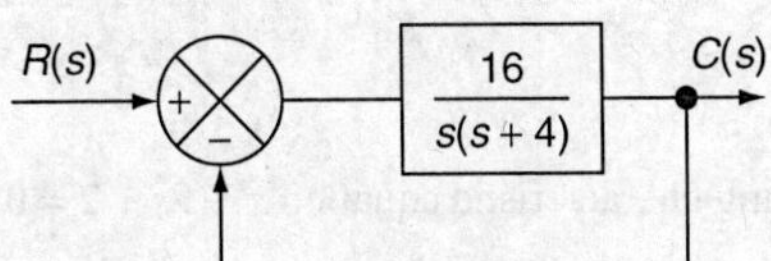

Fig. 23

(a) 4 rad/sec (b) $2\sqrt{2}$ rad/sec (c) 2 rad/sec (d) $\sqrt{2}$ rad/sec

79. If the unit step response of a system is a unit impulse function, the transfer function of such a system will be

(a) 1 (b) $\frac{1}{s}$ (c) s (d) $\frac{1}{s^2}$

80. The correct statement(s) for a type 1 system having unity gain in the forward path and a unity feedback is (are)

(a) Positional error constant K_p is equal to infinity

(b) Acceleration error constant K_a is equal to zero

(c) Steady state error e_{ss} to a unit step displacement input is equal to one

81. The error response of a second order system to a step input is obtained as $e(t) = 1.66e^{-st}\sin(6t + 37°)$. The damping ratio is

(a) 0.4 (b) 0.5 (c) 0.8 (d) 1.0

82. In a linear system, an input of $5\sin\omega t$ produces an output of $10\cos\omega t$. The output corresponding to input $10\cos\omega t$ will be equal to

(a) $5\sin\omega t$ (b) $-5\sin\omega t$ (c) $20\sin\omega t$ (d) $-20\sin\omega t$

83. The step error co-efficient of a system $G(s) = \frac{1}{(s+6)(s+1)}$ with unity feedback is

(a) 1/6 (b) ∞ (c) 0 (d) 1

84. For the system shown in Fig. 24, with a damping ratio ξ of 0.7 and undamped natural frequency ω_n of 4 rad/sec., the values of K and a are

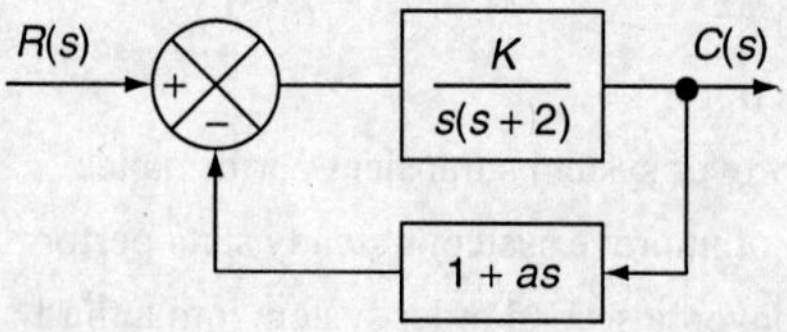

Fig. 24

(a) $K = 4, a = 0.35$ (b) $K = 8, a = 0.455$ (c) $K = 16, a = 0.225$ (d) $K = 64, a = 0.9$

85. The unit impulse response of a system is given as $c(t) = -4e^{-t} + 6e^{-2t}$. The step response of the same system for $t \geq 0$ is equal to

(a) $-3e^{-2t} + 4e^{-t} + 1$ (b) $-3e^{-2t} + 4e^{-t} - 1$ (c) $-3e^{-2t} - 4e^{-t} - 1$ (d) $-3e^{-2t} + 4e^{-t} - 1$

86. The transfer function of a system is $G(s) = \dfrac{s+6}{Ks^2 + s + 6}$. The value of K for which its damping ratio will be 0.5 is

(a) 1/3 (b) 3 (c) 1/6 (d) 6

87. A closed loop system having characteristic equation $s^2 + 2s + 2 = 0$ is

(a) overdamped (b) critically damped (c) underdamped (d) undamped

88. The final value theorem is used to find the

(a) steady state value of the system output

(b) initial value of the system output

(c) transient behaviour of the system output

(d) none of these

89. A process with open-loop model $G(s) = \dfrac{Ke^{-at}}{as+1}$ is controlled by a PID controller. For this process

(a) the integral mode improves transient performance

(b) the integral mode improves steady state performance

(c) the derivative mode improves transient performance

(d) the derivative mode improves steady state performance

90. The steady state error of a stable type 0 unity feedback system for a unit step function is

(a) 0 (b) $\dfrac{1}{1+K_p}$ (c) ∞ (d) $\dfrac{1}{K_p}$

91. The output of a linear time invariant control system is $c(t)$ for a certain input $r(t)$. If $r(t)$ is modified by passing it through a block whose transfer function is e^{-s} and then applied to the system, the modified output of the system would be

(a) $\dfrac{c(t)}{1+e^{t}}$ (b) $\dfrac{c(t)}{1+e^{-t}}$ (c) $c(t-1)u(t-1)$ (d) $c(t)u(t-1)$

92. The Laplace transform of the function $i(t)$ is $I(s) = \dfrac{9s+5}{s(s+1)(s^2+6s+3)}$. Its final value will be

(a) 5 (b) 9 (c) 5/3 (d) 10

93. Consider the unit step response of a unity feedback control whose open loop transfer function is $G(s) = \frac{1}{s(s+1)}$. The maximum overshoot is equal to

(a) 0.143 (b) 0.153 (c) 0.163 (d) 0.173

94. The value of a so that the system shown in Fig. 25 has a zero steady state error for a step input is

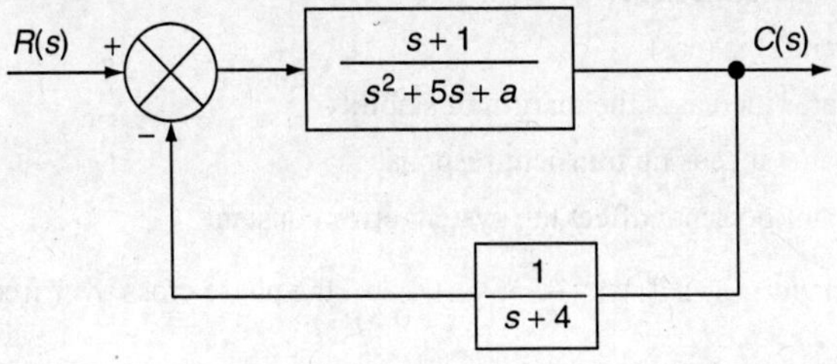

Fig. 25

(a) 0 (b) 2 (c) ≥ 4 (d) none of these

95. The open loop DC gain of a unity negative feedback system with closed loop transfer function $\frac{s+4}{s^2+7s+13}$ is

(a) 4/13 (b) 4/9 (c) 4 (d) 13

96. The phase transfer function of a phase lead controller is $\frac{1+3Ts}{1+Ts}$. The maximum value of phase provided by this controller is

(a) 90° (b) 60° (c) 45° (d) 30°

97. The Nyquist plot of a open loop transfer function $G(j\omega)H(j\omega)$ of a system encloses the $(-1, j0)$ point. The gain margin of the system is

(a) less than zero (b) zero (c) greater than zero (d) infinity

98. By suitable choice of the scalar parameter K, the system shown in Fig. 26 can be made to oscillate continuously at a frequency of

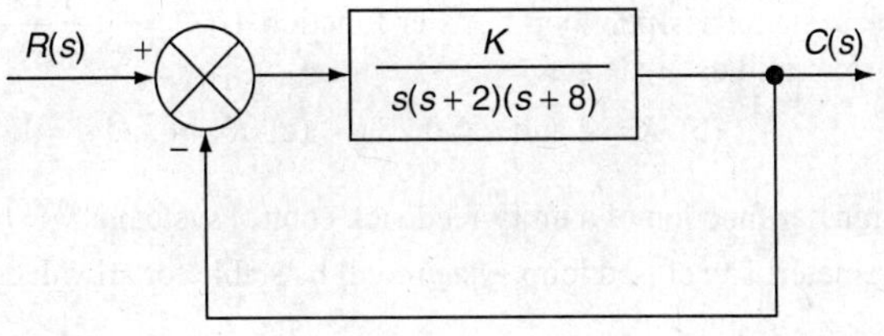

Fig. 26

(a) 1 rad/sec (b) 2 rad/sec (c) 4 rad/sec (d) 8 rad/sec

99. The open loop transfer functions with unity feedback are shown below for different systems:

(a) $G(s) = \frac{2}{s+2}$ (b) $G(s) = \frac{2}{s(s+2)}$ (c) $G(s) = \frac{2}{s^2(s+2)}$ (d) $G(s) = \frac{2(s+1)}{s(s+2)}$

Among these systems, the unstable system is

(a) 1 (b) 2 (c) 3 (d) 4

100. For a gain K, the phase-lead compensator
    (a) reduces the slope of the magnitude curve in the entire range of frequency domain
    (b) decreases the gain cross over frequency
    (c) reduces the phase margin
    (d) reduces the resonance peak M_p
101. The correct statement(s) is (are)
    (a) Lead compensator increases the margin of stability
    (b) Lead compensator speeds up transient response
    (c) Lead compensator does not affect the system error constant
102. For the transfer function $G(s)H(s) = \dfrac{1}{s(s+1)(s+0.5)}$, the phase cross over frequency is
    (a) 0.5 rad/sec. (b) 0.707 rad/sec. (c) 1.732 rad/sec. (d) 2 rad/sec.
103. The maximum phase shift that can be provided by a lead compensator with transfer function $G(s) = \dfrac{1+6s}{1+2s}$ is
    (a) 15° (b) 30° (c) 45° (d) 60°
104. The transfer function of a compensating network is of the form $\dfrac{1+\alpha Ts}{1+Ts}$. If this is a phase-lag network, the value of α should be
    (a) exactly equal to 0 (b) between 0 and 1 (c) exactly equal to 1 (d) greater than 1
105. A phase lag compensation will
    (a) improve relative stability (b) increase the speed of response
    (c) increase bandwidth (d) increase overshoot
106. An effect of phase-lag compensation on servo system performance is that
    (a) for given relative stability, the velocity constant is increased
    (b) for given relative stability, the velocity constant is decreased
    (c) the bandwidth of the system is increased
    (d) the time response is made faster
107. A unity feedback system has open loop transfer function $G(s) = \dfrac{K(s+1)}{s^3 + as^2 + 2s + 1}$. The values of K and a for which the system will oscillate are
    (a) $K = 2$ and a = 1 (b) $K = 2$ and $a = 0.75$ (c) $K = 4$ and $a = 1$ (d) $K = 4$ and $a = 0.75$
108. The open loop transfer function of a unity feedback control system is $G(s)H(s) = \dfrac{30}{s(s+1)(s+T)}$ where T is a variable parameter. The closed loop system will be stable for all values of
    (a) $T > 0$ (b) $0 < T < 3$ (c) $T > 5$ (d) $3 < T < 4$
109. A system has a complex conjugate root pair of multiplicity two or more in its characteristic equation The impulse response of the system will be
    (a) a sinusoidal oscillation which decays exponentially; the system is therefore stable
    (b) a sinusoidal oscillation with time multiplier; the system is therefore unstable
    (c) a sinusoidal oscillation which rises exponentially with time; the system is therefore unstable
    (d) a DC term harmonic oscillation; the system therefore becomes limitingly stable

110. The Nyquist plot of a closed loop system is shown in Fig. 27. The plot indicate that the

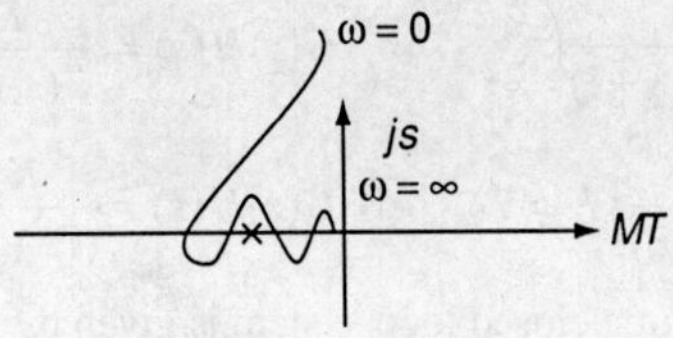

Fig. 27

(a) system is unstable (b) system is stable
(c) system is critically stable (d) stability of the system cannot be determined

111. A control system whose step response is $0.5(1+e^{-2t})$ is cascaded to another block whose impulse response is e^{-t}. The transfer function of the cascaded combination is

(a) $\dfrac{1}{(s+1)(s+2)}$ (b) $\dfrac{1}{s(s+1)}$ (c) $\dfrac{1}{s(s+2)}$ (d) $\dfrac{0.5}{(s+1)(s+2)}$

112. The compensating phase lag network is in Fig. 28.

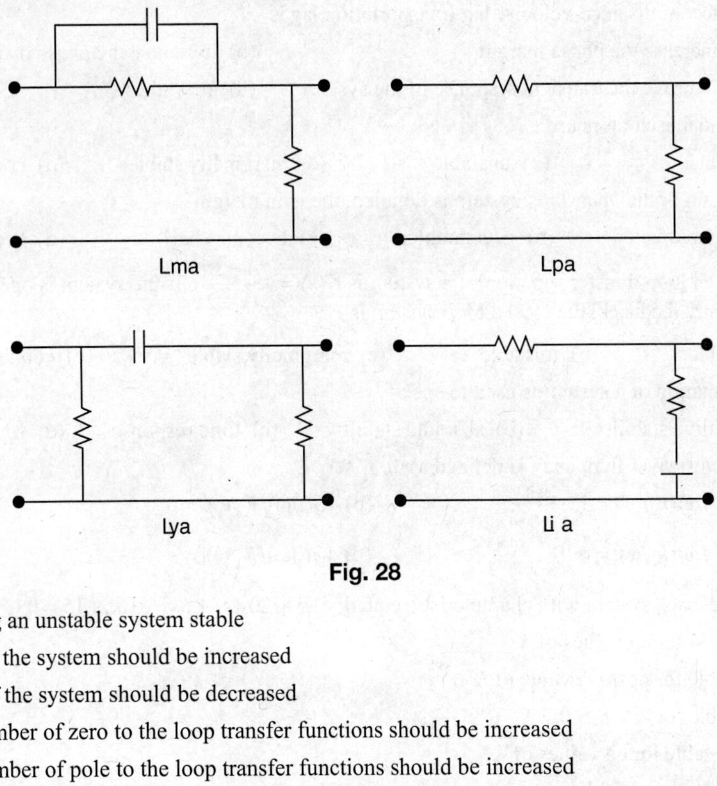

Fig. 28

113. For making an unstable system stable

(a) gain of the system should be increased
(b) gain of the system should be decreased
(c) the number of zero to the loop transfer functions should be increased
(d) the number of pole to the loop transfer functions should be increased

114. Which of the following system is unstable?

(a) $G(s)H(s) = \dfrac{K}{(T_1 s+1)\,(T_2 s+1)}$ (b) $G(s)H(s) = \dfrac{K(s+1)}{s^2(s+4)\,(s+5)}; K > 0.9$

(c) $G(s)H(s) = \dfrac{K(s+2)}{(s+1)\,(s-3)}; K > 2$ (d) $G(s)H(s) = \dfrac{K}{(Ts+1)^3}; -1 < K < 8$

115. The characteristic equation of a closed loop system is given by $s^4 + 6s^3 + 11s^2 + 6s + K = 0$. Stable closed-loop behaviour can be ensured when gain K is such that

(a) $0 < K < 10$ (b) $K > 10$ (c) $-\infty \le K > \infty$ (d) $0 < K \le 20$

116. The maximum phase shift that can be obtained by using a lead compensator with transfer function $G_c(s) = \dfrac{4(1+0.15s)}{1+0.05s}$ is equal to

(a) 15° (b) 30° (c) 45° (d) 60°

117. The transfer function of a system is $F(s) = \dfrac{10(1+0.2s)}{2+0.5s}$. The phase shift at $\omega = 0$ and $\omega = -\infty$ will be

(a) 90° and 0° (b) −180° and 180° (c) −90° and 70° (d) none of these

118. The effect of distance-velocity lag/transportation lag is

(a) to increase the phase margin (b) to reduce the phase margin

(c) to improve the transient response of the system (d) all of these

119. Sinusoidal oscillators are

(a) stable (b) unstable (c) marginally stable (d) conditionally stable

120. If the gain of the open-loop system is doubled, the gain margin

(a) is not affected (b) gets doubled (c) becomes half (d) becomes one-fourth

121. The open loop transfer function of a system is $G(s) = \dfrac{1-s}{s(s+2)}$. If the system is operated in a closed loop with unity feedback, the closed loop system is

(a) stable (b) unstable (c) marginally stable (d) conditionally stable

122. Phase margin of a system is used to specify

(a) relative stability (b) absolute stability (c) time response (d) frequency response

123. Gain cross over frequency is defined as

(a) $|G(j\omega)H(j\omega)| = 1$ (b) $|G(j\omega)H(j\omega)| = 0$

(c) $|G(j\omega)H(j\omega)| = \infty$ (d) $|G(j\omega)H(j\omega)| = \dfrac{1}{2}$

124. The feedback system with characteristic equation $s^4 + 20Ks^3 + 5s^2 + 10s + 15 = 0$ is

(a) stable for all values of K

(b) stable for positive value of K

(c) stable for $\infty > K > 0.7$

(d) unstable for all values of K

125. The gain margin of the given transfer function $G(s) = \dfrac{0.75}{(s+1)(s+2)}$ will be

(a) 4 dB (b) 8 dB (c) 12 dB (d) 16 dB

126. If a unity feedback system having an open loop transfer function

$G(j\omega) = \dfrac{K}{j\omega\,(j0.2\omega+1)\,(j0.05\omega+1)}$ has $GM = 28$ dB for $K = 1$. When $GM = 20$ dB, K will be equal to

(a) 2 (b) 2.5 (c) 4 (d) 5

127. The open loop transfer function of a unity feedback control system is $\dfrac{10}{(s+5)^3}$. The gain margin will be

(a) 20 dB (b) 30 dB (c) 50 dB (d) 80 dB

128. The characteristic equation of a feedback control system is $s^3 + Ks^2 + 5s + 10 = 0$. The value of K so that the system is critically stable will be

(a) 1 (b) 2 (c) 3 (d) 4

129. As compared to a closed loop system, an open loop system is

(a) more stable as well as more accurate

(b) less stable as well as more accurate

(c) more stable but less accurate

(d) less stable but more accurate

130. The transfer function of a simple RC network functioning as a controller is $G_c(s) = \dfrac{s+z_1}{s+p_1}$. The condition for the RC network to act as a phase lead controller is

(a) $p_1 < z_1$ (b) $p_1 > 0$ (c) $p_1 = z_1$ (d) $p_1 > z_1$

131. A unity feedback system has the open loop transfer function $G(s) = \dfrac{1}{(s-1)(s+2)(s+3)}$. The Nyquist plot of G encircles the origin

(a) never (b) once (c) twice (d) thrice

132. The transfer function of a tachometer is of the form

(a) Ks (b) $\dfrac{K}{s}$ (c) $\dfrac{K}{1+s}$ (d) $\dfrac{K}{s(1+s)}$

133. In a synchro error detector, the output voltage is proportional to $[\omega(t)]^n$, where $\omega(t)$ is the rotor velocity and n equals to

(a) −2 (b) −1 (c) 1 (d) 2

134. The AC motor used in servo application is a

(a) single-phase induction motor

(b) two-phase induction motor

(c) three-phase induction motor

(d) synchronous motor

135. In position control systems, the device used for providing rate feedback voltage is called

(a) potentiometer (b) Synchro-transmitter

(c) DC device (d) single-phase AC device

136. A synchro-transmitter receiver unit is a

(a) two-phase AC device (b) 3-phase AC device

(c) DC device (d) single-phase AC device

137. The correct match is

LIST I (Components)	LIST II (Functions)
A. Servomotor	1. Error detector
B. Amplidyne	2. Transducer
C. Potentiometer	3. Actuator
D. Flapper Valve	4. Power amplifier

	A	B	C	D
(a)	3	1	2	4
(b)	2	4	1	3
(c)	4	3	2	1
(d)	3	4	1	2

138. Mechanical impedance is the ratio of
(a) rms force to rms velocity
(b) rms force to rms displacements
(c) rms velocity to rms displacements
(d) none of these

139. Time rate of change in heat energy is analogous to
(a) voltage
(b) charge
(c) current
(d) voltage gradient

140. Temperature is analogous to
(a) Voltage
(b) flux
(c) charge
(d) none of these

141. For the equation $\frac{dQ}{dt} = C_t \frac{d\theta}{dt}$ where Q is heat energy, C_t is thermal capacitance and θ is temperature, the analogous electrical equation is
(a) $i = C\frac{dv}{dt}$
(b) $v = L\frac{di}{dt}$
(c) $e = N\frac{d\phi}{dt}$
(d) none of these

142. For the equation $\frac{dQ}{dt} = G_t(\theta_1 - \theta_2)$ where Q is heat energy, G_t is thermal conductance and θ is temperature, the analogous electrical equation is
(a) $i = G(v_1 - v_2)$
(b) $Q = C\frac{dv}{dt}$
(c) $e = N\frac{d\phi}{dt}$
(d) $v = L\frac{di}{dt}$

143. The resistance R of a liquid flow system is defined as
(a) $\frac{\text{change in flow rate}}{\text{change in level difference}}$
(b) $\frac{\text{change in level difference}}{\text{change in flow rate}}$
(c) change in head × change in flow rate
(d) none of these

144. In liquid flow system the capacitance of a water tank is defined as
(a) change in liquid store/change in head
(b) change in head/change in liquid store
(c) change in flow rate/change in head
(d) none of these

145. A proportional controller is basically
(a) an amplifier with adjustable gain
(b) an integrating amplifier
(c) an amplifier with infinite gain
(d) an amplifier with almost zero gain

146. In an integral controller
(a) the output is proportional to input
(b) the rate of change of output is proportional to input
(c) the output is proportional to rate of change of input
(d) none of these

147. Which control action is called rate control
(a) proportional (b) derivative (c) integral (d) proportional plus integral

148. Derivative control action is effective
(a) only during steady state period
(b) only during transient period
(c) during both steady state and transient periods
(d) either during steady state or during transient periods

149. Which control action is never used alone
(a) proportional (b) derivative (c) integral (d) none of these

150. The frequency at which phase angle is 180° is called
(a) phase cross over frequency
(b) stability limit frequency
(c) frequency of limited stability
(d) gain margin frequency

151. A system has high gain and phase margins. The system is
(a) very stable (b) sluggish (c) very stable and sluggish (d) oscillatory

152. A system is highly oscillatory if
(a) gain margin is high
(b) gain margin is close to 1
(c) gain margin is close to 1 or phase margin is zero
(d) gain margin is high and phase margin is 180°

153. Integral control action
(a) removes offset
(b) may lead to oscillatory response
(c) both (a) and (b)
(d) neither (a) and (b)

154. Whether a linear system is stable or unstable
(a) is a property of the system only
(b) depends on the input function only
(c) both (a) and (b)
(d) either (a) and (b)

155. The correct statement(s) is (are)
(a) Pneumatic system are fire proof
(b) Pneumatic systems are explosion proof
(c) Hydraulic systems are explosion proof
(d) Hydraulic systems are fire proof

156. The correct statement(s) is (are)
(a) Air and gases are compressable
(b) Air, gases and some liquids are compressable
(c) Oil can function as hydraulic fluid as well as lubricator

157. Gas flow resistance is defined as

(a) Change in gas pressure difference/Change in gas flow rate

(b) Change in gas flow rate/Change in gas pressure difference

(c) Change in gas pressure difference × Change in gas flow rate

(d) None of these

158. The capacitance of a gas pressure vessel is defined as

(a) Change in gas stored/Change in gas pressure

(b) Change in gas pressure/Change in gas stored

(c) Change in gas stored × Change in as pressure

(d) None of these

159. The correct statement(s) is (are)

(a) A small size hydraulic actuator can develop good torque

(b) It is easy to maintain a hydraulic system leak proof

(c) Both linear and rotary hydraulic actuators are available

(d) Hydraulic power is more easily available than electric power

160. Transport lag usually exists in

(a) thermal systems (b) hydraulic systems and thermal systems

(c) pneumatic systems (d) thermal, hydraulic and pneumatic systems

161. In a minimum phase system

(a) all poles lie in the left half of s-plane

(b) all zeros lie in the left half of s-plane

(c) all poles and zeros lie n the left half of s-plane

(d) all except one pole or zero lie on the left half of s-plane

162. The addition of a pole to the open loop transfer function pulls the root locus

(a) to the right

(b) to the left and making the system more stable

(c) towards right or left depending on the position of new pole

(d) none of these

163. The addition of a zero to the open loop transfer function

(a) pulls the root locus to the left and makes the system more stable

(b) pulls the root locus to the right and makes the system more stable

(c) either (a) or (b)

(d) neither (a) nor (b)

164. The angular location of poles depends on

(a) undamped natural frequency (b) damping ratio

(c) both (a) and (b) (d) neither (a) nor (b)

165. The distance of poles from the origin depends on

(a) undamped natural frequency (b) damping ratio

(c) both (a) and (b) (d) neither (a) nor (b)

166. If the damping ratio is 0.5, the lines joining complex poles with origin are inclined to negative real axis at
(a) ±90° (b) ±60° (c) ±45° (d) ±30°

167. If the poles lie in first quadrant, the damping ratio is
(a) 1 (b) more than 1 (c) less than 1 (d) zero

168. If the poles lie in first quadrant, the damping ratio is
(a) 1 (b) less than 1 (c) zero (d) more than 1

169. The damping ratio of a pair of complex poles when expressed in terms of angle θ measured from $j\omega$ axis is equal to
(a) sinθ (b) cosθ (c) tanθ (d) cotθ

170. The transfer function $G_c = 5\left(\frac{s+2.9}{s+5.4}\right)$ can be for
(a) lead compensator (b) lag compensator
(c) either (a) or (b) (d) neither (a) nor (b)

171. The transfer function $G_c = 3\left(\frac{s+0.05}{s+0.005}\right)$ can be for
(a) lead compensator (b) lag compensator
(c) neither (a) nor (b) (d) either (a) or (b)

172. The transfer function $G_c = 5\left(\frac{s+2.9}{s+5.4}\right)\left(\frac{s+0.1}{s+0.03}\right)$ can be for
(a) lead compensator (b) lag compensator
(c) lead-lag compensator (d) either (a) or (b)

173. The log magnitude curve for a constant gain K is a
(a) horizontal straight line
(b) horizontal straight line of magnitude $20\log_{10}K$ decibels
(c) an inclined line having slope K
(d) an inclined line having slope $-K$

174. In Bode plot an octave is a frequency band form
(a) ω_1 to $10\omega_1$ (b) ω_1 to $8\omega_1$ (c) ω_1 to $4\omega_1$ (d) ω_1 to $2\omega_1$

175. In Bode diagram the factor $1/j\omega$ in the transfer function gives a line having slope
(a) −20 dB/octave (b) −10 dB/octave (c) −6 dB/octave (d) −2 dB/octave

176. In Bode diagram the factor $(j\omega)^n$ in the transfer function gives a line having slope
(a) 20 dB/decade (b) 20n dB/decade (c) $\frac{20}{n}$ dB/decade (d) −20n dB/decade

177. The open loop transfer function of a system is $\frac{K}{s(Js+F)}$. If the frequency at the intersection of −40 dB/decade line and 0 dB line is ω, then
(a) $\omega = \frac{K}{J}$ (b) $\omega^2 = \frac{K}{J}$ (c) $\omega^3 = \frac{K}{J}$ (d) $\omega^4 = \frac{K}{J}$

178. For the transport lag $G(j\omega) = e^{-j\omega t}$, the magnitude is always equal to
(a) 0 (b) 1 (c) all (d) 2 and 3

179. For the transport lag $G(j\omega) = e^{-j\omega t}$, the phase angle

(a) is constant

(b) varies linearly with frequency

(c) varies linearly with frequency with low frequency only

(d) varies as per square law

180. For the transport lag $G(j\omega) = e^{-j\omega T}$, the polar plot is a

(a) semi circle (b) circle (c) unit circle (d) none of these

181. For the transport lag $G(j\omega) = \dfrac{e^{-j\omega T}}{1+e^{-j\omega T}}$, the polar plot is a

(a) semi circle (b) circle (c) spiral (d) none of these

182. A conditionally stable system is stable for the value of gain between two critical values. It is unstable if

(a) the gain is increased beyond the upper value

(b) the gain is decreased below the lower critical value

(c) both (a) and (b)

(d) none of these

183. The gain margin for a stable system

(a) has a positive decibel value (b) has a negative decibel value

(c) has a large negative decibel value (d) none of these

184. A conditionally stable system has

(a) one phase cross over frequency (b) two phase cross over frequencies

(c) two or more phase cross over frequencies (d) one phase cross over frequencies

185. For a stable system having two or more gain cross over frequencies, the phase margin is measured

(a) at lowest gain cross over frequency (b) at highest gain cross over frequency

(c) at either of the gain cross over frequencies (d) none of these

186. The relative stability of a system is given by

(a) gain margin alone (b) phase margin alone

(c) both phase and gain margin (d) either (a) or (b)

187. For a stable system

(a) both phase and gain margin are negative

(b) both phase and gain margin are positive

(c) phase margin is positive but gain margin is negative

(d) gain margin is positive and phase margin is negative

188. If a system is to follow arbitrary inputs accurately, the bandwidth should be

(a) large (b) small (c) very small (d) neither small nor large

189. From the view point of noise, bandwidth should

(a) be large (b) not be too large (c) be as large as possible (d) be infinite

190. As bandwidth increases the cost of the components

(a) generally decreases (b) generally increases

(c) may increase or decrease (d) does not change

191. The constant magnitude loci on $G(s)$ plane are
(a) a family of straight horizontal lines
(b) a family of straight inclined lines with positive slope
(c) a family of straight inclined lines with negative slope
(d) a family of circles

192. For any transfer function, the constant phase angle loci are
(a) a family of circles (b) a family of straight lines
(c) either (a) or (b) (d) neither (a) nor (b)

193. A lead compensator is basically a
(a) high pass filter (b) low pass filter (c) band stop filter (d) none of these

194. A lag compensator provides attenuation in
(a) very low frequency range (b) low frequency range
(c) high frequency range (d) none of these

195. The primary function of a lag compensator is to provide sufficient
(a) gain margin (b) phase margin (c) both (a) and (b) (d) either (a) or (b)

196. A lag compensator is essentially a
(a) low pass filter (b) high pass filter (c) band pass filter (d) either (b) or (c)

197. The effect of adding poles and zeros can be determined quickly by
(a) Nichols chart (b) Nyquist plot (c) Bode plot (d) Root locus

198. Magnetic amplifier is used for
(a) voltage amplification (b) power amplification
(c) current amplification (d) frequency amplification

199. In position control system the device used for providing rate-feedback voltage is
(a) potentiometer (b) synchro-transmitter (c) synchro receiver (d) tacho generator

200. In terms of ξ and ω_n the settling time of a second order systems for 2% criterion and step input is equal to
(a) $\frac{1}{\xi\omega_n}$ (b) $\frac{2}{\xi\omega_n}$ (c) $\frac{3}{\xi\omega_n}$ (d) $\frac{4}{\xi\omega_n}$

201. In terms of ξ and ω_n the settling time of a second order systems for 5% criterion and step input is equal to
(a) $\frac{1}{\xi\omega_n}$ (b) $\frac{2}{\xi\omega_n}$ (c) $\frac{3}{\xi\omega_n}$ (d) $\frac{4}{\xi\omega_n}$

202. Which of the following is open loop system
(a) wind screen wiper (b) aqualing
(c) respiratory system of animal (d) system for controlling anti rocket missile

203. Which of the following is not correct for two phase servomotor?
(a) Rotor diameter is small (b) Applied voltages are seldom balanced
(c) Rotor resistance is low (d) Torque speed curve is linear

204. Which is the following is used to obtain the output position in a position control system
(a) strain gauge (b) load cell (c) synchro (d) thermistor

205. The correct sequence of events to improve system stability is
(a) insert derivative action, use negative feedback, reduce gain
(b) reduce gain, use negative feedback, insert derivative gain
(c) reduce gain, insert derivative feedback, use negative feedback
(d) use negative feedback, reduce gain, insert derivative action

206. In the field control motor the entire control comes from
(a) armature resistance (b) back emf
(c) motor friction and load (d) field resistance

207. For stable network the real parts of poles and zeros of driving point functions must be
(a) negative (b) negative or zero (c) zero (d) positive

208. The poles with greater displacement from the real axis correspond to
(a) higher frequencies of oscillations (b) stable response
(c) lower frequencies of oscillations (d) unbound output

209. The phase difference $(j\omega - p_1)$ means
(a) phasor drawn from p_1 to $j\omega$ (b) phasor drawn from origin to p_1
(c) phasor directed from origin to jω point (d) phasor directed from $j\omega$ to p_1

210. A reactance function
(a) has a pole at origin
(b) has a zero at origin
(c) may have a pole or zero at origin
(d) has a pole or zero at origin and a pole or zero at infinity

211. The poles of an RC function
(a) are simple and lie on $j\omega$ axis (b) are simple and lie on negative real axis
(c) may have complex conjugate (d) may be anywhere in s plane

212. For a feedback system $\frac{C(s)}{R(s)} = \frac{G(s)}{1+G(s)H(s)}$, the velocity error co-efficient K_v is given by
(a) $K_v = \underset{s\to 0}{Lt}\, G(s)H(s)$ (b) $K_v = \underset{s\to 0}{Lt}\, \frac{sG(s)}{1+G(s)H(s)}$
(c) $K_v = \underset{s\to 0}{Lt}\, sG(s)$ (d) $K_v = \underset{s\to 0}{Lt}\, s[1+G(s)H(s)]$

213. *M* circles are symmetrical with respect to
(a) real and imaginary axis (b) $M = 1$ straight line and real axis
(c) $M = 1$ straight line and imaginary axis (d) $M = 1$ straight line

214. Which of the following is not a necessary advantage of negative feedback system
(a) Feedback improves speed of response
(b) Feedback improves stability
(c) Feedback reduces effect of variations in system parameters
(d) Feedback reduces the effect of external disturbances

215. The correct statement(s) regarding the negative feedback is (are)
(a) it increases system gain (b) it improves sensitivity
(c) it improves system stability (d) it decreases the system gain

216. In control system the word controller includes
(a) error detector and feedback element
(b) feedback element and control elements
(c) error detector, control elements and feedback elements

217. In control systems the output of sensor, usually, is
(a) analog electrical signal (b) digital electrical signal
(c) mechanical signal (d) analog or digital electrical signal

218. DC servomotors are, generally
(a) permanent magnet DC motors (b) shunt field motors
(c) separately excited DC motors (d) any one of above

219. DC tachogenerator can be used in
(a) speed control system (b) position control system
(c) both (a) and (b) (d) neither (a) nor (b)

220. The current rating of DC tachogenerator is uaually
(a) small (b) high (c) medium (d) either (b) or (c)

221. The shape of torque-speed curve of a two phase servo motor mostly depends on
(a) ratio of rotor reactance to rotor resistance
(b) ratio of stator reactance to rotor resistance
(c) ratio of rotor reactance to rotor resistance and ratio of stator reactance to rotor resistance
(d) magnitude of stator voltage

222. An AC tachometer has a drag cup construction so that
(a) rotor inertia is low (b) rotor body is conducting
(c) rotor body is easy to fabricate (d) both (a) and (b)

223. The number of stator field windings in an AC tachometer are
(a) 1 (b) 2 (c) 3 (d) 4

224. The resolution of a wire wound potentiometer is about
(a) 0.001% (b) 0.1% (c) 1% (d) 5%

225. Power handling capacity of wire wound potentiometer used in control system is about
(a) 0.1 W (b) 5 W (c) 25 W (d) 100 W

226. Optical encoders most commonly used in control systems are
(a) absolute encoders (b) secondary encoders
(c) incremental encoders (d) none of these

227. The sensor generally used in incremental optical encoder is
(a) LED (b) LCD (c) photo diode (d) either (a) or (b)

228. A second order differential equation is given by $\frac{d^2x}{dt^2}+3\frac{dx}{dt}+9x=9y$. The frequency and damping ratio are respectively
(a) 2 rad/sec, 3 (b) 6 rad/sec, 9 (c) 1 rad/sec, 3 (d) 3 rad/sec, 0.5

229. Bellow converts
(a) pressure difference into displacement
(b) pressure difference into voltage
(c) displacement into pressure difference
(d) either (a) or (c)

230. The action of bellows is similar to that of
(a) mass
(b) viscous friction
(c) spring
(d) none of these

231. In a pneumatic bellows
(a) the input surface, output surface and side walls are corrugated
(b) the input surface and output surface are flat but the side walls are corrugated
(c) the input surface and side walls are flat but the output surface is corrugated
(d) the input surface and output surface are corrugated but the side walls are flat

232. If ΔP is pressure difference, A is area of each flat surface, K is stiffness and Δx is displacement of movable surface from reference of bellows, then
(a) $\frac{\Delta X(s)}{\Delta P(s)} = \frac{A}{K}$
(b) $\frac{\Delta X(s)}{\Delta P(s)} = \frac{K}{A}$
(c) $\frac{\Delta X(s)}{\Delta P(s)} = AK$
(d) $\frac{\Delta X(s)}{\Delta P(s)} = \sqrt{AK}$

233. In a synchro transmitter
(a) input is angular position of rotor shaft and output is stator voltage
(b) input is stator voltage and output is angular position of shaft
(c) input is rotor voltage and output is angular position of rotor shaft
(d) input is angular position of rotor shaft and output is rotor voltage

234. In a second order system with a unit step input, the speed of response is high if system is
(a) overdamped
(b) undamped system
(c) critically damped system
(d) under damped

235. In a second order undamped system, the poles are on
(a) positive real axis
(b) imaginary axis
(c) negative real axis
(d) either (a) or (c)

236. In a second order system, the peak overshoot is independent of
(a) damping factor
(b) natural frequency
(c) both damping factor and natural frequency
(d) neither damping factor nor natural frequency

237. In a second order system, the time constant of exponential envelopes depends
(a) only on damping factor
(b) only on natural frequency
(c) both on damping factor and natural frequency
(d) neither on damping factor nor on natural frequency

238. If a zero is added to a second order under damped system
(a) the peak overshoot is increased
(b) the peak overshoot is decreased
(c) the peak overshoot remains unchanged
(d) either (a) or (b)

239. In a control system the magnitude of error voltage
(a) is very small
(b) is very large
(c) is neither small nor large
(d) may be any value from 0 to about $50V$

240. The effect of negative feedback on distortion and bandwidth are
(a) both decreased
(b) both increased
(c) distortion reduced and bandwidth increased
(d) distortion increased and bandwidth decreased

241. When a unit step voltage drives a lag network, the output
(a) remains constant at unit step value
(b) increases exponentially from zero to final value
(c) decreases exponentially from 1 to 0
(d) either (b) or (c) depending on values of R and C

242. Integral error compensation changes a second order system to
(a) first order system
(b) third order system
(c) fourth order system
(d) either (a) or (b) depending on the system

243. The pole zero plot shown in Fig. 29 is for

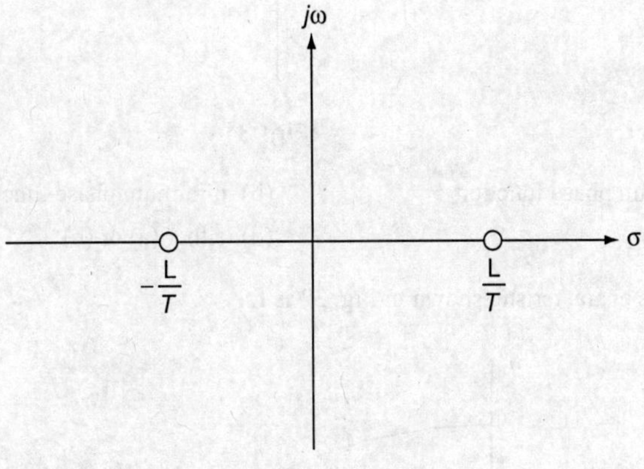

Fig. 29

(a) non minimum phase function
(b) minimum phase function
(c) all pass function
(d) either (a) or (b)

244. The pole zero plot shown in Fig. 30 is for

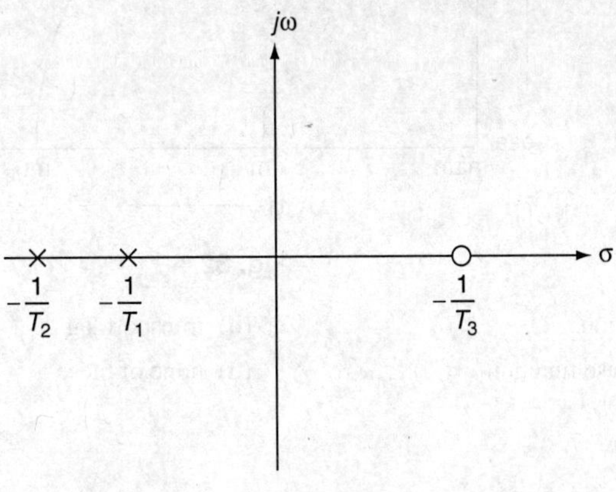

Fig. 30

(a) non minimum phase function
(b) minimum phase function
(c) all pass function
(d) either (a) or (b)

245. The pole zero plot shown in Fig. 31 is for

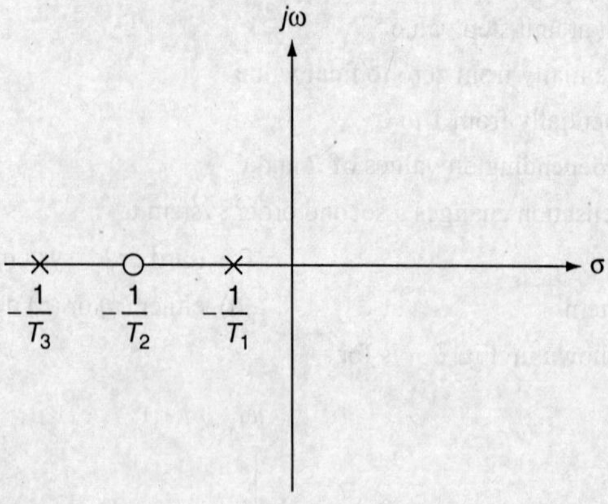

Fig. 31

(a) non minimum phase function
(b) minimum phase function
(c) pass function
(d) either (a) or (b)

246. The phase angle characteristic shown in Fig. 32 is for

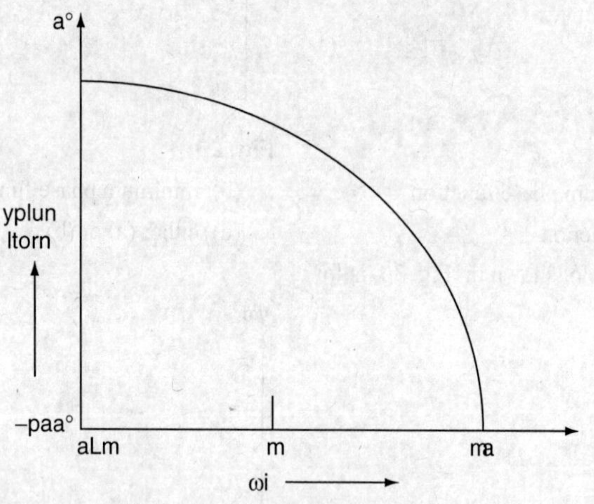

Fig. 32

(a) all pass function
(b) transport lag
(c) minimum phase function
(d) none of these

247. The curves 1, 2 and 3 shown in Fig. 33 are respectively for

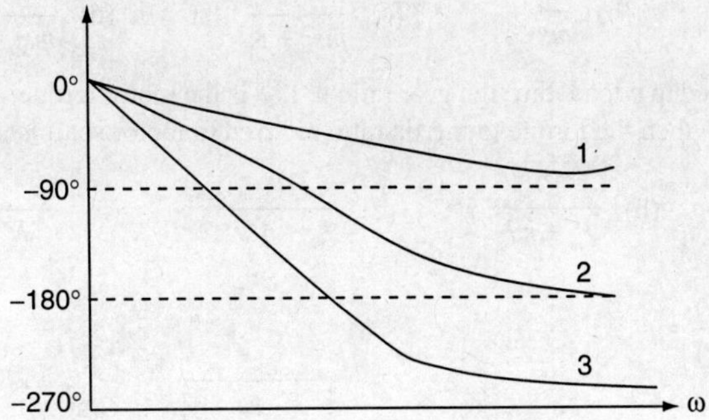

Fig. 33

(a) all pass, minimum phase and non-minimum phase functions
(b) minimum phase, all pass and non-minimum phase functions
(c) non-minimum phase , all pass and minimum phase function
(d) minimum phase, non-minimum phase and all pass functions

248. The polar plot of $G(s) = \dfrac{1}{(1+s)(1+2s)}$

(a) does not cross the real axis
(b) crosses the real axis
(c) crosses real axis at $\omega = 0.7$ rad/sec.
(d) crosses real axis at $\omega = 1.4$ rad/sec.

249. The polar plot of $G(s) = \dfrac{1}{s^2(1+s)(1+2s)}$

(a) does not cross real axis
(b) crosses real axis once
(c) crosses real axis twice
(d) crosses real axis 4 times

250. If $G(s)H(s) = \dfrac{1}{s^2+100}$, the closed loop poles are on

(a) negative real axis (b) $j\omega$ axis (c) right half of s-plane (d) either (a) or (b)

251. A system has $\dfrac{Y(s)}{X(s)} = \dfrac{s+1}{s^2+5s+6}$. If $x(t) = 2t^2$, $Y(s)$ is

(a) $\dfrac{2(s+1)}{s^2+5s+6}$ (b) $\dfrac{2s^2(s+1)}{s^2+5s+6}$ (c) $\dfrac{2(s+1)}{s^2+5s+6}$ (d) $\dfrac{s+1}{2s^3(s^2+5s+6)}$

252. In the system shown in Fig. 34, $x(0) = 0$ and $\dot{x}(0) = 0$. At $t = 0$ the unit impulse $\delta(t)$ is applied. $X(s) =$

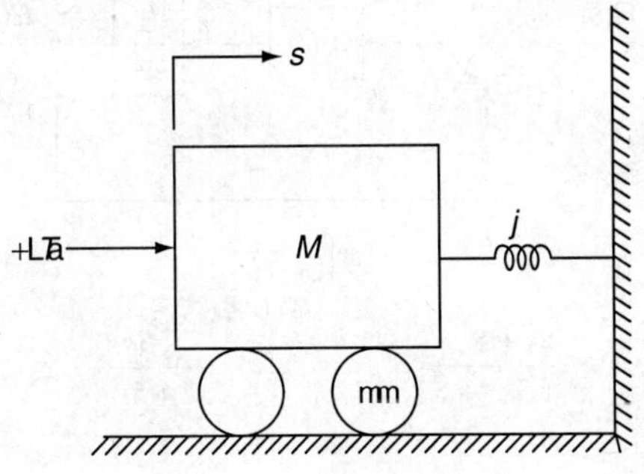

Fig. 34

(a) $\dfrac{1}{ms+K}$ (b) $\dfrac{K}{ms}$ (c) $\dfrac{1}{ms^2+K}$ (d) $\dfrac{1}{ms^3+K}$

253. A motor is coupled to a load through gear ratio n. If T is the motor torque, J_m and J_L are moment of inertia of rotor and load, then the torque to inertia ratio referred to motor shaft is

(a) $\dfrac{T}{J_m+J_L}$ (b) $\dfrac{T}{J_m+nJ_L}$ (c) $\dfrac{T}{J_m+n^2J_L}$ (d) $\dfrac{T}{n^2J_m+J_L}$

254. For Fig. 35, $\dfrac{C(s)}{R(s)}=$

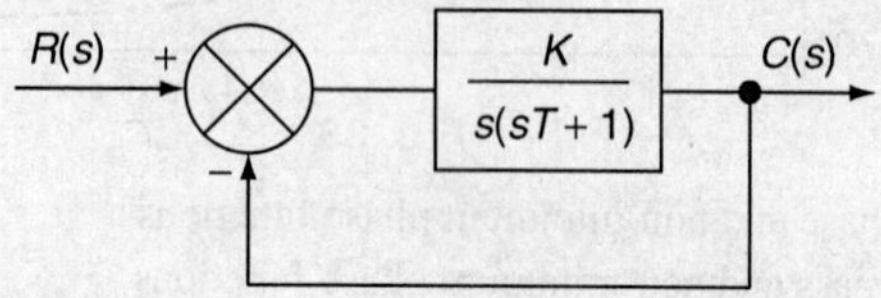

Fig. 35

(a) $\dfrac{Ks}{s(Ts+1)}$ (b) $\dfrac{Ks}{s(Ts+1)}$ (c) $\dfrac{Ks}{Ts^2+s+K}$ (d) $\dfrac{Ks}{Ts^3+s^2+Ks}$

255. For Fig. 36, $\dfrac{C(s)}{R(s)}=$

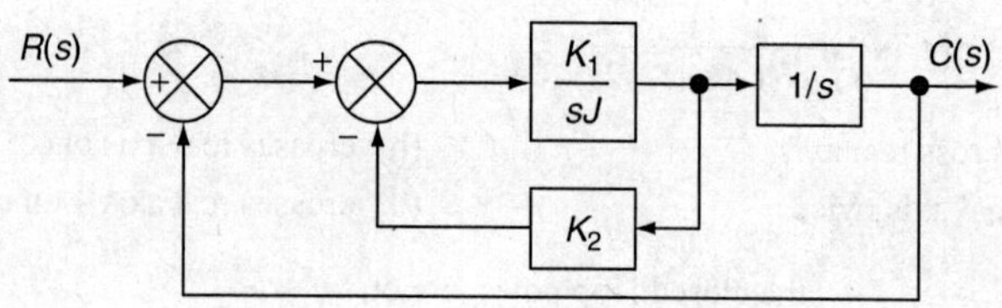

Fig. 36

(a) $\dfrac{K_1}{Js^2+K_1K_2+K_1}$ (b) $\dfrac{K_1}{Js^3+K_1K_2+K_1}$ (c) $\dfrac{K_1K_2}{Js^2+K_1K_2+K_1}$ (d) none of these

256. For Fig. 37, $\dfrac{C(s)}{R(s)}=$

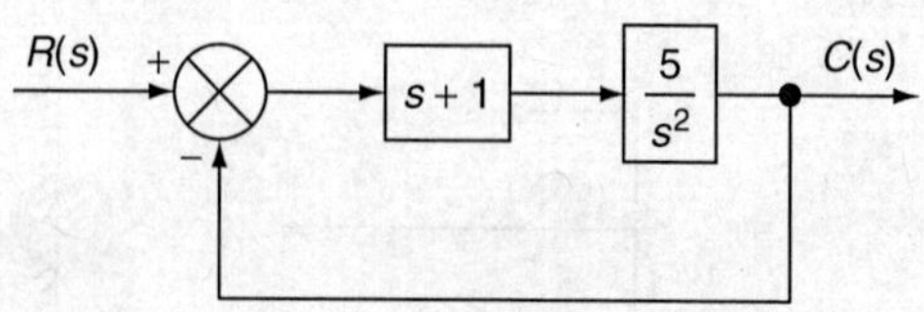

Fig. 37

(a) $\dfrac{5s+5}{s^2+5s+5}$ (b) $\dfrac{5s+5}{s^3+5s^2+5}$ (c) $\dfrac{5s}{s^2+5s+5}$ (d) $\dfrac{5}{s^2+5s+5}$

257. For Fig. 38, $\dfrac{X(s)}{P(s)} =$

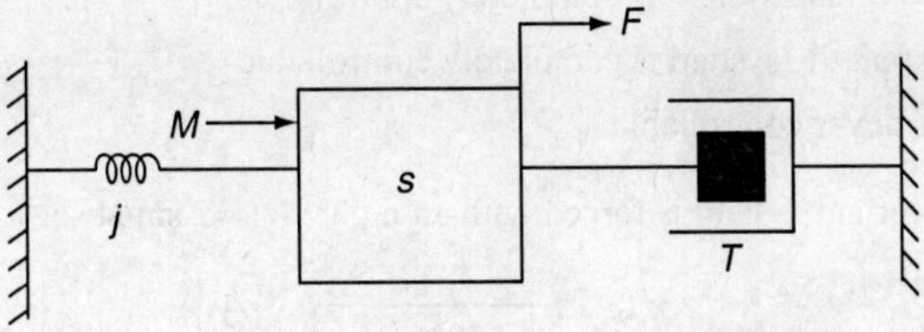

Fig. 38

(a) $\dfrac{1}{ms + b + K}$ (b) $\dfrac{1}{ms^2 + b + K}$ (c) $\dfrac{K}{ms^2 + bs + K}$ (d) $\dfrac{b}{ms^2 + b + K}$

258. If $P = P_m \sin\omega t$ in Fig. 38, then $X(s)$

(a) $\dfrac{P_m\omega}{(s^2+\omega^2)\,(ms^2+b+K)}$ (b) $\dfrac{P_m}{(s^2+\omega^2)\,(ms^2+b+K)}$

(c) $\dfrac{P_m\omega^2}{(s^2+\omega^2)\,(ms^2+b+K)}$ (d) $\dfrac{P_m\omega}{s^2(ms^2+b+K)}$

259. In a second order system $\omega_n = 10$ rad/sec. and $\xi = 0.1$, then $\omega_d =$

(a) 10.1 rad/sec^2 (b) 9.95 rad/sec^2 (c) 9 rad/sec^2 (d) 1 rad/sec^2

260. A unity feedback system has open loop transfer function $\dfrac{1}{s(s+5)}$. The closed loop transfer function is

(a) $\dfrac{1}{s^2+5s}$ (b) $\dfrac{1}{5s^2+s}$ (c) $\dfrac{1}{s^2+5s+1}$ (d) $\dfrac{s}{s^2+5s+1}$

261. A unity feedback system has open loop transfer function $\dfrac{2s+1}{s^2}$. The closed loop transfer function is

(a) $\dfrac{s^2}{2s+1}$ (b) $\dfrac{2s+1}{s^2+2s+1}$ (c) $\dfrac{2s}{s^2+5s+1}$ (d) $\dfrac{1}{s^2+5s+1}$

262. Find the correct statement(s) is (are):

(a) state model is unique

(b) state model can be derived from the system transfer function

(c) state model can be derived from time variant system

(d) state model is not unique

263. The state representation of a second order system is

$\dot{x}_1 = -x_1 + u$

$\dot{x}_2 = x_1 - 2x_2 + u$

Find the correct statement(s) from the following statements regarding the above systems:

(a) The system is completely controllable

(b) If x_1 is the output, the system is completely controllable

(c) If x_2 is the output, the system is completely controllable

(d) The system is never controllable

264. A linear stable invariant system is forced with an input $x(t) = A\sin\omega t$ shown in Fig. 39.

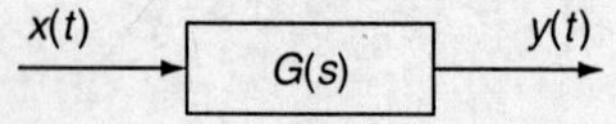

Fig. 39

Under steady state conditions, the output $y(t)$ of the system will be

(a) $A\sin(\omega t+\phi)$ where $\phi = \tan^{-1}|G(j\omega)|$
(b) $|G(j\omega)|\, A\sin[\omega t + \angle G(j\omega)]$
(c) $|G(j\omega)|\, A\sin[2\omega t + \angle G(j\omega)]$
(d) $A\, G(j\omega)\, A\sin[\omega t + G(j\omega)]$

265. The transfer function of a multi input multi output system with the state space representation of $\dot{X} = Ax + Bu$ and $Y = Cx + Du$

where X represents the state and Y the output and u the input vector, will be given by

(a) $C(sI-A)^{-1}B$ (b) $C(sI-A)^{-1}B + D$ (c) $(sI-A)^{-1}B + D$ (d) $C(sI-A)^{-1}B + A$

266. Find the correct statement(s) for a linear system $y = f(x)$ is (are)

(a) $f(x_1 + x_2) = f(x_1) + f(x_2)$
(b) $f[x(t+T)] = f[x(t)] + f[x(T)]$
(c) $f(Kx) = Kf(x)$
(d) $f(x_1 + x_2) = f(x_1) \times f(x_2)$

267. Find the correct statement(s) regarding the state transition matrix is (are)

(a) $\phi(0) = 0$
(b) $\phi^{-1}(t) = \phi(1/t)$
(c) $\phi(t_1 + t_2) = \phi(t_1) + \phi(t_2)$
(d) $\phi(t_2 - t_1)\,\phi(t_1 - t_0) = \phi(t_2 - t_0)$

268. A system is described by the state equation $\dot{X} = Ax + Bu$ and $Y = Cx$ where X represents the state and Y the output and u the input vector and $A = \begin{bmatrix} -4 & -1 \\ 3 & -1 \end{bmatrix}$, $B = \begin{bmatrix} 1 \\ 1 \end{bmatrix}$ and $C = [\,1 \;\; 0\,]$. Transfer function of the system is

(a) $\dfrac{s}{s^2+5s+7}$ (b) $\dfrac{1}{s^2+5s+7}$ (c) $\dfrac{s}{s^2+3s+2}$ (d) $\dfrac{1}{s^2+3s+2}$

269. A set of linear equations is represented by the matrix equation $Ax = B$. The necessary condition for the existence of a solution for this system is

(a) A must be invertible

(b) B must be linearly dependent on the columns of A

(c) B must be linearly independent on the columns of A

(d) None of these

270. A second order system has state space representation of the form $\dot{X} = Ax + Bu$. If $x_1(t) = x_2(t)$, the system is

(a) controllable (b) uncontrollable (c) observable (d) unstable

271. Which of the following quantities under mechanical rotational system and electrical system are not analogous ?
(a) Angular displacement charge
(b) Angular velocity-current
(c) Moment of inertia-conductance
(d) Viscous friction co-efficient resistance

272. Which of the following represents an analogous pair between electrical systems and pneumatic systems?
(a) Charge-air flow
(b) Current-head
(c) Voltage-air flow rate
(d) Resistance-pressure

273. Under force-current analogy, displacement is analogous to
(a) voltage (b) conductance (c) magnetic flux linkage (d) capacitance

274. Under electrical system and pneumatic system analogy, current is considered analogous to
(a) air flow (b) air flow rate (c) pressure (d) velocity

275. Under thermal and electrical system analogy, temperature is considered analogous to
(a) current (b) voltage (c) charge (d) capacitance

276. Mass in mechanical translational system is analogous to which of the following quantities under mechanical rotational system?
(a) Torque (b) Angular velocity (c) Angular displacement (d) Moment of inertia

277. When analogy is drawn between electrical systems and thermal systems, current is considered analogous to
(a) heat flow rate (b) temperature (c) reciprocal of temperature (d) heat flow

278. Laplace transform of 1 is
(a) zero (b) 1 (c) s (d) $1/s$

279. Laplace transform of a unity function is
(a) zero (b) 1 (c) $1/s$ (d) s

280. Laplace transform method of solution is applicable to equations containing
(a) differential terms only
(b) integral terms only
(c) scalar terms only
(d) none of these

281. Under force-voltage analogy, viscous friction co-efficient is analogous to
(a) reciprocal of capacitance
(b) reciprocal of inductance
(c) charge
(d) resistance

282. If a zero appears in the first column of the Routh table, the system is
(a) necessarily stable
(b) necessarily unstable
(c) marginally stable
(d) none of these

283. The damping ratio of a second order system given by $2\frac{d^2y}{dt^2}+4\frac{dy}{dt}+8y=8x$ is
(a) 0.25 (b) 0.50 (c) 0.60 (d) 0.75

284. The undamped natural frequency of Q283 is
(a) 0.1 (b) 5 (c) 8 (d) 0.75

285. The damping co-efficient of Q283 is
(a) 1 (b) 0.8 (c) 0.5 (d) 0.1

286. The time constant of Q283 is

(a) 1 (b) 0.8 (c) 0.5 (d) 0.1

287. Human eye can be considered as

(a) an open loop system (b) a closed loop system
(c) a closed system with multivariable feedback (d) none of these

288. Which system has the tendency to oscillate?

(a) open loop system (b) closed loop system
(c) both open loop as well as closed loop (d) neither open loop nor closed loop

289. Transfer function can be obtained from

(a) analogous table (b) standard block system
(c) output-input ratio (d) signal flow graph

290. The input which is established or varied by some means external to and independent of the feedback control system is known as

(a) signal (b) modulated signal (c) command (d) disturbance

291. Which signal will become zero when the feedback signal and reference signal are equal?

(a) Actuating signal (b) Reference signal
(c) Feedback signal (d) Input signal

292. A control system in which the control action is somehow dependent on the output is known as

(a) open loop system (b) semi-closed system
(c) closed loop system (d) none of the above

293. If the damping factor of a system is unity, the system is

(a) overdamped (b) critically damped (c) underdamped (d) unstable

294. Which of the following must have negative real parts?

(a) The system eigen values (b) The gain margin
(c) The gain factor (d) The zeros of the transfer function

295. Which of the following is not in frequency domain?

(a) Nyquist criterion (b) Bode plot
(c) Root locus plot (d) All of the above

296. The output of a thermocouple is

(a) DC current (b) AC current (c) DC voltage (d) AC voltage

297. Which of the following potentiometer has infinite resolution?

(a) Deposited field potentiometer (b) Gall Tensley potentiometer
(c) Kelvin's double arm potentiometer (d) All of the above

298. For a linear transfer function of the variable, the transfer function of the system

(a) changes (b) does not change
(c) has no relation at all (d) depends on other factor also

299. Which of the following is incorrect?

(a) Routh's criterion is in time domain (b) Bode plot is in frequency domain
(c) Nyquist criterion is in time domain (d) None of these

300. An ideal potentiometer should have
   (a) a fine wire
   (b) an infinite resolution
   (c) resolution proportional to the reference voltage
   (d) zero resolution
301. The effect of adding poles and zeros can be determined quickly by
   (a) Nichols chart (b) Nyquist plot (c) Bode plot (d) Root locus
302. A term not associated with the potentiometer is
   (a) reference voltage (b) resolution (c) least count (d) backlash
303. Which of the following can be extended to system which are non linear?
   (a) Laplace transform (b) Eigen value analysis
   (c) Nyquist criterion (d) State variable analysis
304. For type 2 system, the velocity error is
   (a) low (b) zero (c) infinite (d) variable
305. The system response can be tested better with
   (a) sinusoidal input signal (b) unit impulse input signal
   (c) ramp input signal (d) exponentially decaying signal
306. Well-designed first order control systems have
   (a) a negative time constant (b) a small bandwidth
   (c) a small negative eigen value (d) a large negative transfer function pole
307. When a control signal is required to have a power level higher than the capability of linear electronic amplifiers, which of the following is preferred?
   (a) Amplidyne (b) AC servomotor (c) AC tachometer (d) Any one of the above
308. With feedback which of the following reduces?
   (a) system gain (b) system stability
   (c) Both (a) and (b) above (d) None of the above
309. The normal range of damping ratio is
   (a) 0.04 to 0.1 (b) 0.4 to 1.0 (c) 0.8 to 1.5 (d) 1.2 to 1.8
310. With feedback which of the following increases?
   (a) sensitivity (b) gain
   (c) effects of disturbing signals (d) system stability
311. The system steady state error can be minimized by
   (a) decreasing damped frequency (b) decreasing natural frequency
   (c) increasing damped frequency (d) increasing system gain constant
312. For minimizing the loading in potentiometers
   (a) reference voltage may be increased (b) length of the wire may be increased
   (c) linearity may be introduced (d) non-linearity may be introduced

313. With feedback system

(a) the transient response gets magnified

(b) the transient response decays at a constant rate

(c) the transient response decays slowly

(d) the transient response decays more quickly

314. The number of root loci branches which do not terminate at zeros is given by

(a) the number of zeros

(b) the number of poles

(c) the number of zeros + the number of poles

(d) the number of zeros – the number of poles

315. In case the magnitude of the polar plot at phase cross over is A then the gain margin is

(a) zero (b) A (c) –A (d) 1/A

316. $\frac{(s+3)(s+2)}{s^2(s+4)(s+5)}$ is a

(a) type 0 system (b) type 1 system (c) type 2 system (d) type 3 system

317. The Bode plot is used to analyze

(a) all phase network (b) lag lead network

(c) minimum phase network (d) maximum phase network

318. The loop transfer function of the system shown in Fig. 40 is

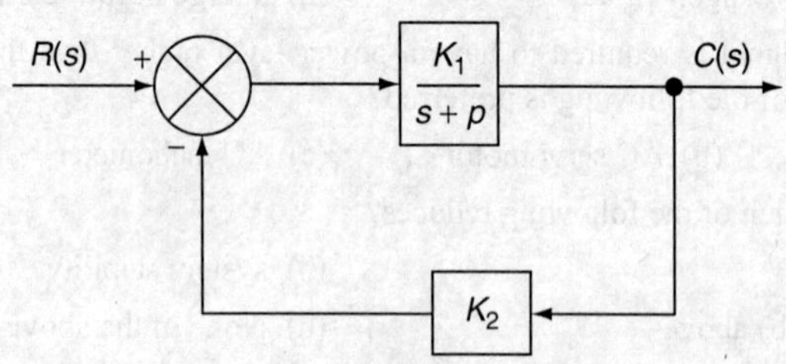

Fig. 40

(a) $\frac{K_1K_2}{s+p}$ (b) $\frac{K_1}{K_2(s+p)}$ (c) $\frac{K_2}{K_1(s+p)}$ (d) $\frac{s+p}{K_1K_2}$

319. The control ratio of Fig. 40 is

(a) $\frac{K_1}{s(s+p-K_1K_2)}$ (b) $\frac{s+p}{K_1K_2}$ (c) $K_1K_2(s+p)$ (d) none of the above

320. The error ratio of Fig. 40 is

(a) K_1K_2 (b) $K_1K_2(s+p)$ (c) $\frac{K_1(s+p)}{K_2}$ (d) $\frac{(s+p)}{(s+p-K_1K_2)}$

321. A 1000 turns potentiometer is excited by 10V source, its percentage resolution will be

(a) 0.1 % (b) 1.0 % (c) 10% (d) 11%

322. The negative real axis of Nyquist plot corresponds to the
(a) negative 180° phase line of Bode plot (b) positive 180° phase line of Bode plot
(c) 0 dB line Bode plot (d) none of the above

323. Which of the following will increase the steady state accuracy?
(a) phase lead compensator (b) phase lag compensator
(c) differentiator (d) integrator

324. Pressure can be measured by
(a) strain gauge (b) selsyn
(c) differential bellows and stain gauge (d) strain gauge and potentiometer

325. Non-linearity caused by servomotor is
(a) saturation (b) backlash (c) static friction (d) none of these

326. Steep cut-off characteristics will have
(a) small peak response (b) low stability margin
(c) accurate cut-off (d) all of the above

327. For the system shown in Fig. 41 below, with a damping ratio ξ of 0.8 and an undamped natural frequency ω_n of $\sqrt{5}$ rad/sec, the values of K are

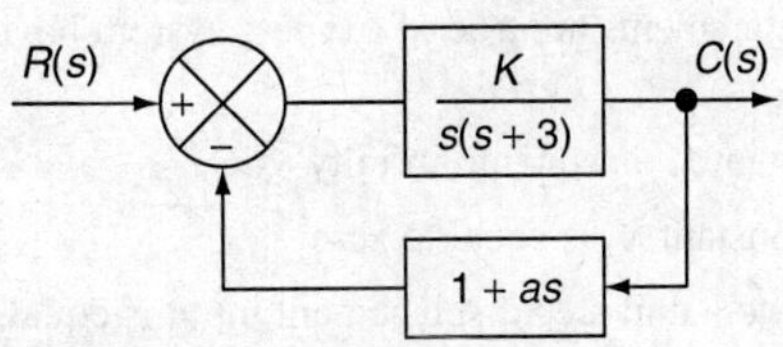

Fig. 41

(a) $K = 14, a = 0.25$ (b) $K = 10, a = 0.455$
(c) $K = 5, a = 0.1155$ (d) $K = 25, a = 0.1155$

328. If the gain of the system is doubled, the gain margin
(a) is not affected (b) gets doubled (c) becomes half (d) becomes one fourth

329. Match List-I with List-II ad select the correct answer using the codes shown below the lists.

List-I (Time Function)	List-II (Laplace Transforms)
A. $e^{-2(t-2)}$	1. $\frac{2}{s^2}$
B. $\frac{d(t^2)}{dt}$	2. $\frac{1}{s+2}e^{-2s}$
C. $\frac{t}{2}$	3. $\frac{1}{2s^2}$
	4. $\frac{1}{2s}$

Codes:

(a)	A	B	C	(b)	A	B	C
	2	1	3		2	3	4
(c)	A	B	C	(d)	A	B	C
	1	2	3		3	1	4

330. Which of the following is an example of an open loop system?
(a) Household refrigerator
(b) Respiratory system of an animal
(c) Stabilization of air pressure entering into a mask
(d) Execution of a program by a computer

331. If a unit step response of a system is a unit impulse function, then the transfer function of such a system will be
(a) 1 (b) $1/s$ (c) s (d) $1/s^2$

332. When a unit step input is applied, a second order underdamped system has a peak overshoot of OP occurring at t_{max}. If another step input equal in magnitude to the peak overshoot OP is applied at $t = t_{max}$, then the system will settle down at
(a) $1 + OP$ (b) $1 - OP$ (c) OP (d) 1.0

333. Which of the following statements are true of a type 1 system having unity gain in the forward path and a unity feedback?
(a) Positional error constant K_p is equal to infinity
(b) Acceleration error constant K_a is equal to zero
(c) Steady-state error e_{ss} to a unit step displacement input is equal to one.
Select the correct answer from the given codes:
Codes:
(a) 1, 2 and 3 (b) 1 and 2 (c) 2 and 3 (d) 1 and 3

334. The bandwidth of a control system is the range of frequency over which $|C(j\omega)/R(j\omega)|$ is greater than or equal to
(a) $\frac{1}{\sqrt{2}}$ and it depends only on the undamped natural frequency of the system
(b) 1 and it depends only on the damping factor of the system
(c) 1 and it depends only on the undamped natural frequency of the system
(d) $\frac{1}{\sqrt{2}}$ and it depends on both the undamped natural frequency and the damping factor of the system

335. The open loop transfer function of a unity feedback control system is given by
$$G(s) = \frac{K(s+2)}{s(s^2+2s+2)}$$
The centroid and angles of root locus asymptotes are respectively
(a) zero and +90°, −90° (b) −2/3 and +60°, −60°
(c) zero and +120°, −120° (d) −2/3 and −90° and +90°, −90°

336. A system with characteristic equation $s^4 + 2s^3 + 11s^2 + 18s + 18 = 0$ will have closed loop poles such that
   (a) All poles lie in the left half of the s-plane
   (b) All poles lie in the right half of the s-plane
   (c) Two poles lie symmetrically on the imaginary axis of the s-plane
   (d) No pole lie on the imaginary axis of the s-plane

337. Which of the following systems is completely state controllable?
   (a) $\begin{bmatrix}\dot{x}_1\\ \dot{x}_2\end{bmatrix} = \begin{bmatrix}-1 & 0\\ 0 & -1\end{bmatrix}\begin{bmatrix}x_1\\ x_2\end{bmatrix} + \begin{bmatrix}2\\ 1\end{bmatrix}u$
   (b) $\begin{bmatrix}\dot{x}_1\\ \dot{x}_2\end{bmatrix} = \begin{bmatrix}-1 & 0\\ 0 & -2\end{bmatrix}\begin{bmatrix}x_1\\ x_2\end{bmatrix} + \begin{bmatrix}2\\ 0\end{bmatrix}u$
   (c) $\begin{bmatrix}\dot{x}_1\\ \dot{x}_2\end{bmatrix} = \begin{bmatrix}-1 & 1\\ 0 & -1\end{bmatrix}\begin{bmatrix}x_1\\ x_2\end{bmatrix} + \begin{bmatrix}1\\ 0\end{bmatrix}u$
   (d) $\begin{bmatrix}\dot{x}_1\\ \dot{x}_2\end{bmatrix} = \begin{bmatrix}0 & 0\\ 2 & -1\end{bmatrix}\begin{bmatrix}x_1\\ x_2\end{bmatrix} + \begin{bmatrix}2\\ 2\end{bmatrix}u$

338. For the field controlled servomotor DC servomotor shown in Fig. 42. The transfer function $\dfrac{\theta(s)}{E(s)}$ is

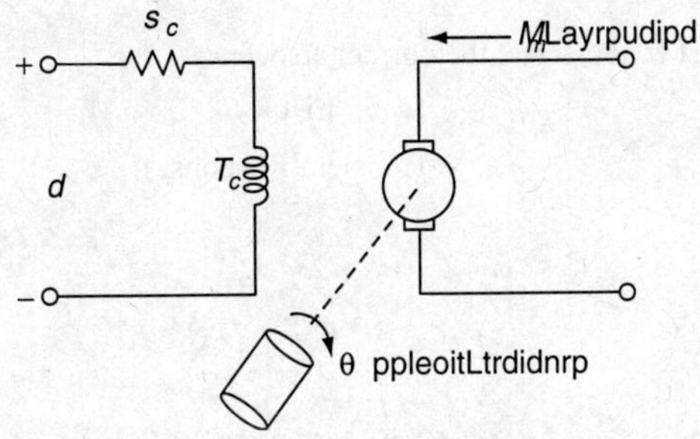

Fig. 42

   (a) two time constants, no integration
   (b) two time constants, one integration
   (c) one time constant, no integration
   (d) one time constant, two integrations

339. Bode plots of an open-loop control system are shown in Fig. 43 below.

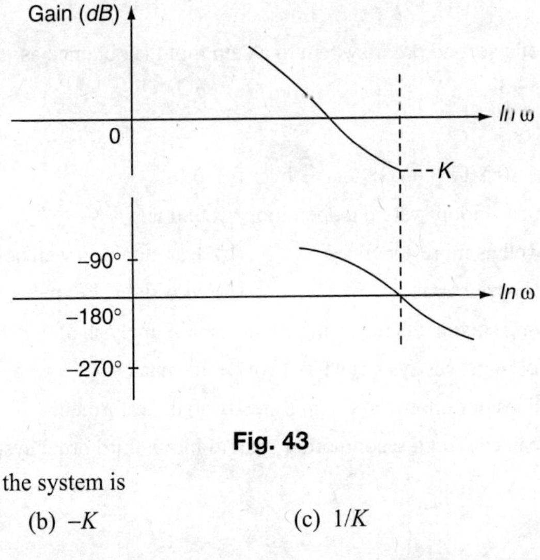

Fig. 43

The gain margin of the system is
   (a) K
   (b) $-K$
   (c) $1/K$
   (d) $-1/K$

340. The first two rows of Rouths tabulation of a fourth-order system are

s^4	1	10	5
s^3	2	20	

The number of poles lying on the right half of s-plane is

(a) 0 (b) 2 (c) 3 (d) 4

341. Match the list I with List II and select the correct answer.

List-I (Signal flow graphs) List II (Simplifications)

Codes:

	A	B	C	D
(a)	2	1	5	4
(b)	3	4	2	1
(c)	2	1	4	3
(d)	1	2	3	5

342. Match the list I with List II and select the correct answer.

List-I (Transfer Function) List II (Transient Response)

Assume $\alpha > 0$ and $\beta > 0$

Codes:

	A	B	C	D
(a)	1	3	2	4
(b)	3	1	2	4
(c)	2	1	4	3
(d)	4	2	1	3

343. A control system whose step response is $-0.5(1+e^{-2t})$ is cascaded to another block whose impulse response is e^{-t}. The transfer function of the cascaded combination is

(a) $\dfrac{1}{(s+1)(s+2)}$ (b) $\dfrac{1}{s(s+1)}$ (c) $\dfrac{1}{s(s+2)}$ (d) $\dfrac{0.5}{(s+1)(s+2)}$

344. The error response of a second order system to a step input is obtained as $e(t) = 1.66e^{-5t}\sin(6t+37°)$

The damping ratio is

(a) 0.4 (b) 0.5 (c) 0.8 (d) 1.0

345. As compared to a closed loop system a open loop system is

(a) more stable as well as more accurate
(b) less stable as well as less accurate
(c) more stable but less accurate
(d) less stable but more accurate

346. Consider the following statements regarding time-domain analysis of a control system:

1. Derivative control improves system's transient performance
2. Integral control does not improve system's steady state performance
3. Integral control can convert a second order system into a third order system

Of these statements:

(a) 1 and 2 are correct (b) 1 and 3 are correct

(c) 2 and 3 are correct (d) 1, 2 and 3 are correct

347. The principles of homogeneity and superposition are applied to

(a) linear time variant systems (b) non-linear time variant systems

(c) linear time invariant systems (d) non-linear time invariant systems

348. Match the list I with List II and select the correct answer.

List-I List II

(Response to a unit step input) (Location of poles in the s-plane)

Assume $\alpha > 0$ and $\beta > 0$

Codes:

(a) A B C D (b) A B C D
4 3 2 1 3 4 1 2

(c) A B C D (d) A B C D
3 4 2 1 4 3 2 1

349. The root locus of a unity feedback system is shown in Fig. 44 below. The open loop transfer function is expressed by

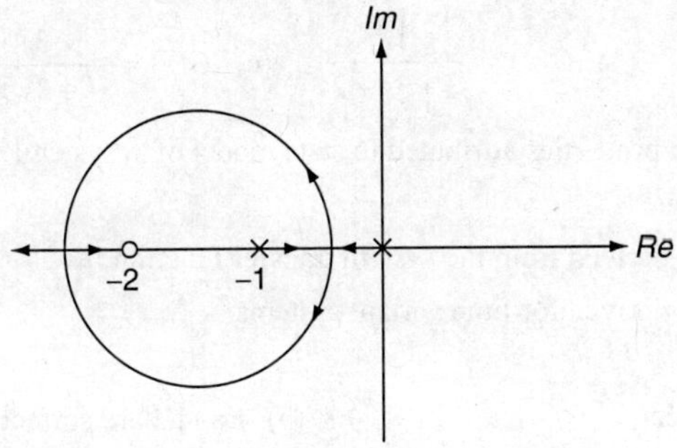

Fig. 44

(a) $\dfrac{K}{s(s+1)(s+2)}$ (b) $\dfrac{K(s+1)}{s(s+2)}$ (c) $\dfrac{K(s+2)}{s(s+1)}$ (d) $\dfrac{Ks}{(s+1)(s+2)}$

350. The open loop transfer function of a unity feedback control system is given by

$G(s) = \dfrac{K}{s(s+1)}$. If the gain K is increased to infinity, the damping ratio will tend to become

(a) $1/\sqrt{2}$ (b) 1 (c) 0 (d) ∞

351. The characteristic equation of a unity feedback control system is given by $s^3 + K_1 s^2 + s + K_2 = 0$. Consider the following statements in this regards:

1. For a given of K_1, all the root-locus branches will terminate at infinity for K_2 in the positive direction.
2. For a given value of K_2, all the root-locus branches will terminate at infinity for variable K_1 in the positive direction
3. For a given value of K_2, only one root locus branch will terminate at infinity for variable K_1 in the positive direction.

Of these statements:

(a) 1 and 2 are correct (b) 3 alone is correct (c) 2 alone is correct (d) 1 and 3 are correct

352. The characteristic equation of a closed loopy stem is given by

$$s^4 + 6s^3 + 11s^2 + 6s + K = 0$$

Stable closed loop behaviour can be ensured when gain K is such that

(a) $0 < K < 10$ (b) $K > 10$ (c) $-\infty \leq K < \infty$ (d) $0 < K \leq 20$

353. For a gain constant K, the phase lead compensator

(a) reduces the slope of the magnitude curve in the entire range of frequency domain

(b) decreases the gain cross over frequency

(c) reduces the phase margin

(d) reduces the resonance peak M_p

Of these statements:

(a) 2 and 3 are correct (b) 1 and 2 are correct

(c) 1 and 3 are correct (d) 1, 2 and 3 are correct

354. Given a system represented by equation

$\dot{x} = -\begin{bmatrix} 0 & 1 \\ -2 & -3 \end{bmatrix} x + \begin{bmatrix} 0 \\ 1 \end{bmatrix} u$ and $y = [1\ 0]x$. The equivalent transfer function representation $G(s)$ of the system is

(a) $G(s) = \dfrac{1}{s^2 + 5s + 2}$ (b) $G(s) = \dfrac{1}{s^2 + 3s + 2}$ (c) $G(s) = \dfrac{3}{s^2 + 3s + 2}$ (d) $G(s) = \dfrac{2}{s^3 + 2s^2 + 2}$

355. Consider the following properties attributed to state model of a system:

1. State model is unique
2. State model can be derived from the system transfer function
3. State model can be derived for time variant systems

Of these statements:

(a) 1, 2 and 3 are correct (b) 1 and 2 are correct

(c) 2 and 3 are correct (d) 1 and 3 are correct

356. In the feedback system shown in Fig. 45 below the noise component of output is given by (assume high loop gain at frequencies of interest)

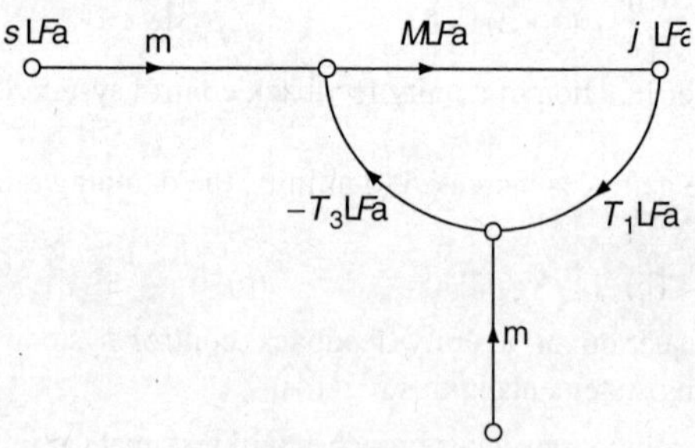

Fig. 45

(a) $-\dfrac{N(s)}{H_1(s)}$ (b) $\dfrac{N(s)}{H_1(s)}$ (c) $\dfrac{N(s)}{H_1(s)H_2(s)}$ (d) $-\dfrac{N(s)}{H_1(s)H_2(s)}$

357. The system shown in Fig. 46 below has a unit step input. I order that the steady state error is 0.1, the value of K required is

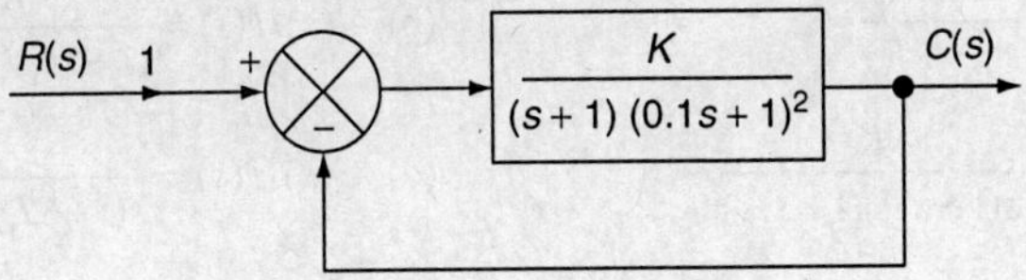

Fig. 46

(a) 0.1 (b) 0.9 (c) 1.0 (d) 9.0

358. Consider the following statements regarding the advantages of closed loop negative feedback control systems over open loop systems:

1. The overall reliability of the closed loop system is more than that of open loop system
2. The transient response in the closed loop system decays more quickly than in the open loop system
3. In the open loop system closing of the loop increases the overall gain of the system
4. In the closed loop system, the effect of variation of component parameters on its performance is reduced.

Of these statements:

(a) 1 and 3 are correct (b) 1 and 2 are correct (c) 2 and 4 are correct (d) 3 and 4 are correct

359. The Nyquist locus of a transfer function $G(s)H(s) = \dfrac{K}{1+sT_1}$ is shown in Fig. 47 below.

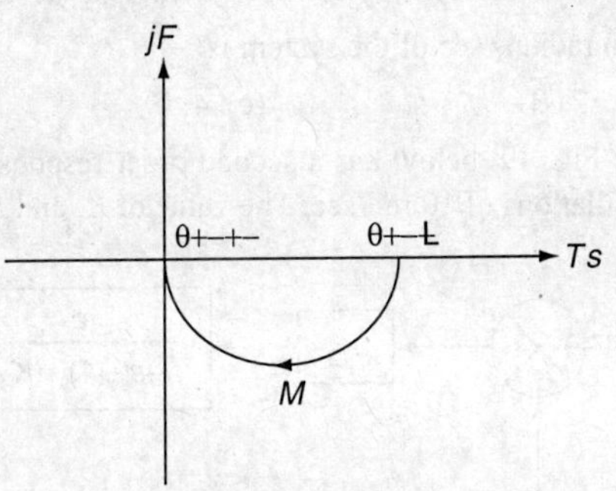

Fig. 47

The locus is modified as shown in Fig. 48 below on addition of poles or poles to the original.

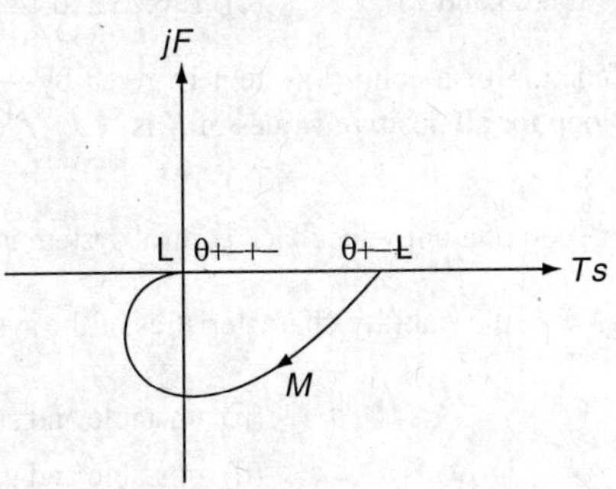

Fig. 48

Then, the modified transfer function of the modified locus is

(a) $G(s)H(s)=\dfrac{K}{s(1+sT_1)}$ (b) $G(s)H(s)=\dfrac{K}{(1+sT_1)(1+sT_2)}$

(c) $G(s)H(s)=\dfrac{K}{s(1+sT_1)(1+sT_2)}$ (d) $G(s)H(s)=\dfrac{K}{s(1+sT_1)(1+sT_2)(1+sT_3)}$

360. Match the list I with List II and select the correct answer.

List-I (Transfer function) List II (Impulse response)

Codes:

	A	*B*	*C*	*D*
(a)	2	1	4	3
(b)	1	2	4	3
(c)	2	1	3	4
(d)	1	2	3	4

361. The constant *M* circles corresponding to the magnitude (*M*) of the closed loop transfer function of a linear system for values of *M* greater than one lie in the *G* plane and to the right of

(a) right of $M = 1$ line (b) left of the $M = 1$ line

(c) upper side of the $M = +j1$ line (d) upper side of the $M = -j1$ line

362. The characteristic equation of a closed loop control system is given as $s^2 + 4s + 16 = 0$

The resonant frequency in radians/sec of the system is

(a) 2 (b) $2\sqrt{3}$ (c) 4 (d) $2\sqrt{2}$

363. The system shown in the Fig. 49 below has a second order response with a damping ratio of 0.6 and a frequency of damped oscillations of 10 rad/sec. The value of K_1 and K_2 are respectively

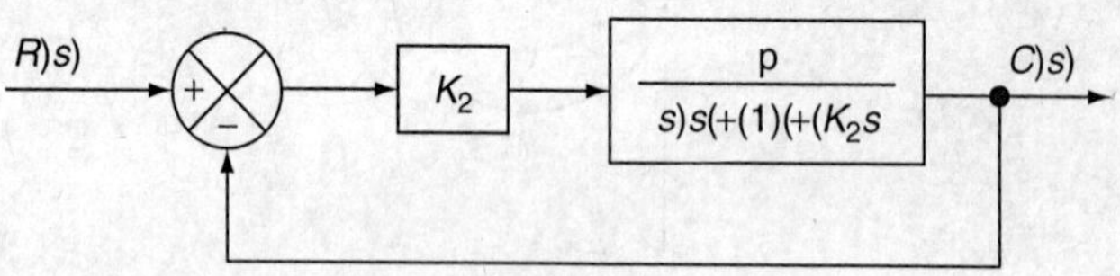

Fig. 49

(a) 12.5 and 15 (b) 156.25 and 15 (c) 156.25 and 14 (d) 12.5 and 4

364. The open loop transfer function of a control system is given by $\dfrac{K(s+10)}{s(s+2)(s+a)}$. The smallest possible value of '*a*' in the closed loop for all positive values of *K* is

(a) 0 (b) 8 (c) 10 (d) 12

365. The open loop transfer function of a unity feedback control system is given by

$G(s)=\dfrac{K(s+2)}{(s+1)(s-7)}$. For $K > 6$, the stability characteristics of the system are respectively

(a) stable and stable (b) unstable and stable

(c) stable and unstable (d) unstable and unstable

366. The phase lag compensation will

(a) improve relative stability (b) increase the speed of response

(c) increase bandwidth (d) increase overshoot

367. The maximum phase shift that can be obtained by using a lead compensator with transfer function $G_c(s) = \dfrac{4(1+0.15s)}{1+0.05s}$ is equal to

(a) 15° (b) 30° (c) 45° (d) 60°

368. Consider the following statements regarding a first order system with a proportional (*P*) controller which exhibits an offset to a step input:

1. increase the gain of the *P* controller
2. add derivative mode
3. add integral mode

Of these statements:

(a) 1, 2 and 3 are correct (b) 1 and 2 are correct

(c) 2 and 3 are correct (d) 1 and 3 are correct

369. Consider the closed loop system shown in the Fig. 50 below.

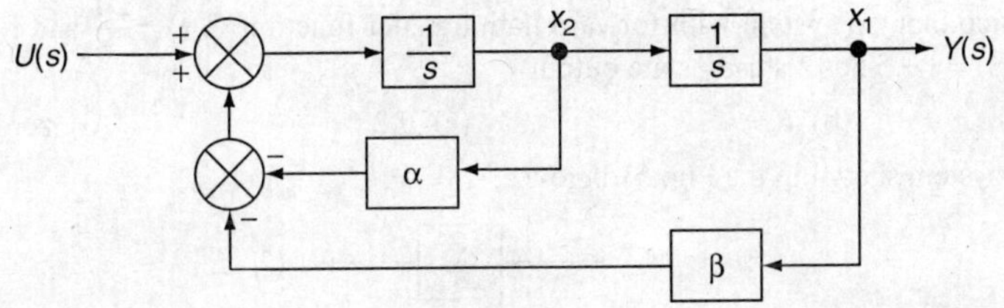

Fig. 50

The state model of the system is

(a) $\begin{bmatrix}\dot{x}_1\\ \dot{x}_2\end{bmatrix} = \begin{bmatrix}1 & 0\\ -\beta & -\alpha\end{bmatrix}\begin{bmatrix}x_1\\ x_2\end{bmatrix} + \begin{bmatrix}0\\ 1\end{bmatrix}u$

$y = [1 \;\; 0]\begin{bmatrix}x_1\\ x_2\end{bmatrix}$

(b) $\begin{bmatrix}\dot{x}_1\\ \dot{x}_2\end{bmatrix} = \begin{bmatrix}0 & 1\\ -\beta & -\alpha\end{bmatrix}\begin{bmatrix}x_1\\ x_2\end{bmatrix} + \begin{bmatrix}0\\ 1\end{bmatrix}u$

$y = [1 \;\; 0]\begin{bmatrix}x_1\\ x_2\end{bmatrix}$

(c) $\begin{bmatrix}\dot{x}_1\\ \dot{x}_2\end{bmatrix} = \begin{bmatrix}1 & 1\\ -\beta & -\alpha\end{bmatrix}\begin{bmatrix}x_1\\ x_2\end{bmatrix} + \begin{bmatrix}0\\ 1\end{bmatrix}u$

$y = [1 \;\; 0]\begin{bmatrix}x_1\\ x_2\end{bmatrix}$

(d) $\begin{bmatrix}\dot{x}_1\\ \dot{x}_2\end{bmatrix} = \begin{bmatrix}1 & 1\\ -\alpha & -\beta\end{bmatrix}\begin{bmatrix}x_1\\ x_2\end{bmatrix} + \begin{bmatrix}0\\ 1\end{bmatrix}u$

$y = [1 \;\; 0]\begin{bmatrix}x_1\\ x_2\end{bmatrix}$

370. A system is described by the state equation

$$\begin{bmatrix}\dot{x}_1\\ \dot{x}_2\end{bmatrix} = \begin{bmatrix}2 & 0\\ 0 & 2\end{bmatrix}\begin{bmatrix}x_1\\ x_2\end{bmatrix} + \begin{bmatrix}1\\ 1\end{bmatrix}u$$

The state matrix of the system is

(a) $\begin{bmatrix} e^{-2t} & 0 \\ 0 & e^{2t} \end{bmatrix}$ (b) $\begin{bmatrix} e^{-2t} & 0 \\ 0 & e^{-t} \end{bmatrix}$ (c) $\begin{bmatrix} e^{2t} & 0 \\ 0 & e^{2t} \end{bmatrix}$ (d) $\begin{bmatrix} e^{-2t} & 1 \\ 1 & e^{2t} \end{bmatrix}$

371. Match List I with List II and select the correct answer using the codes shown below the lists:

List-I (unit)	List-II (Type of rotor)
A. Synchro-transmitter	1. Dumb-bell rotor
B. Control transformer	2. Drag-cup rotor
C. A.C. servo-motor	3. Cylindrical rotor
D. Stepper Motor	4. Toothed rotor
	5. Phase wound rotor

Codes:

	A	*B*	*C*	*D*
(a)	1	3	2	4
(b)	1	5	3	2
(c)	2	4	3	1
(d)	3	2	1	5

372. For a unit step input, a system with forward path transfer function $G(s) = \frac{20}{s^2}$ and feedback path transfer function $H(s) = s + 5$ has a steady state output

(a) 0 (b) 5 (c) 0.2 (d) zero

373. Consider a system shown in the Fig. 51 below.

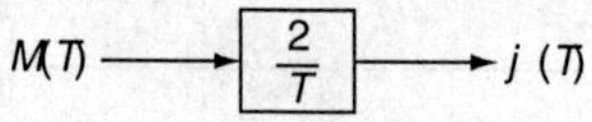

Fig. 51

If the system is disturbed so that $c(0)=1$, then $c(t)$ for a unit step input will be

(a) $1 + t$ (b) $1 - t$ (c) $1 + 2t$ (d) $1 - 2t$

374. Which one of the following statements is true for gain margin and phase margin of two closed loop system having loop transfer function $G(s)H(s)$ and $\exp(-s)G(s)H(s)$?

(a) Both gain and phase margins of two statements will be identical.
(b) Both gain and phase margins of $G(s)H(s)$ will be more.
(c) Gain and phase margins of the two systems are the same but phase margin of $G(s)H(s)$ will be more.
(d) Phase margins of two systems are the same but gain margin of $G(s)H(s)$ will be less.

375. An effect of phase-lag compensation on servo system performance is that

(a) for a given relative stability, the velocity constant is increased
(b) for a given relative stability, the velocity constant is decreased
(c) the bandwidth of the system is increased
(d) the time response is made faster

376. Consider the system

$$\dot{x}(t) = \begin{bmatrix} 1 & 1 \\ 0 & 1 \end{bmatrix} x(t) + \begin{bmatrix} b_1 \\ b_2 \end{bmatrix} u(t)$$

$$c(t) = [d_1 \;\; d_2]\, x(t)$$

The conditions for complete state controllability and complete observability is

(a) $d_1 > 0, b_2 > 0, b_1$ and d_2 can be anything (b) $d_1 > 0, d_2 > 0, b_1$ and b_2 can be anything

(c) $b_1 > 0, b_2 > 0, d_1$ and d_2 can be anything (d) $b_1 > 0, b_2 > 0, b_1$ and d_1 can be anything

377. Canonical decomposition of a linear time invariant system is given by

$$\dot{x} = \begin{bmatrix} -3 & 0 & 8 \\ 0 & -5 & -12 \\ 0 & 0 & -1 \end{bmatrix} x + \begin{bmatrix} 6 & 0 \\ 0 & 6 \\ 0 & 0 \end{bmatrix} u$$

and $$y = \begin{bmatrix} 0 & 0 & -4 \\ 1 & 0 & 4 \end{bmatrix} x$$

It is

(a) completely controllable and observable (b) completely controllable and unobservable

(c) uncontrollable and observable (d) neither completely controllable nor observable

378. Match the List-I and List-II and select the correct answer using the codes given below.

List-I	List-II
A. Transfer function of a lag network	1. $\alpha^{-1}(sI - A)^{-1}$
B. State transition matrix for $\dot{x} = Ax$	2. $\frac{s+z}{s+p}, \frac{z}{p} > 1$
C. DC amplifier transfer function	3. $\frac{K}{1+sT}$
D. Steady state error of type-I system with step-input	4. $\alpha^{-1}(sA - I)^{-1}$
	5. $\frac{s+z}{s+p}, \frac{z}{p} < 1$
	6. $k(1+sT)$
	7. Zero
	8. ∞

Codes:

	A	B	C	D
(a)	1	3	4	5
(b)	2	3	1	4
(c)	5	4	6	8
(d)	2	1	3	7

379. What will be the closed loop transfer function of a unity feedback control system whose step response is given by $c(t) = k[1 - 1.66e^{-8t}\sin(6t + 37°)]$?

(a) $\frac{100k}{s^2 + 16s + 100}$ (b) $\frac{10}{s^2 + 16s + 100}$ (c) $\frac{k}{s^2 + 16s + 100}$ (d) $\frac{10k}{s^2 + 8s + 100}$

380. The state representation of a second order system is

$$\dot{x}_1 = -x_1 + u$$

$$\dot{x}_2 = x_1 - 2x_2 + u$$

Consider the following statements regarding the above system:

1. The system is completely state controllable
2. If x_1 is the output, then the system is completely output controllable
3. If x_2 is the output, then the system is completely output controllable

Of these statements:

(a) 1, 2 and 3 are correct (b) 1 and 2 are correct

(c) 2 and 3 are correct (d) 1 and 3 are correct

381. The AC motor used in servo application is a

(a) single-phase induction motor (b) two-phase induction motor

(c) three-phase induction motor (d) synchronous motor

382. In position control systems, the device used for providing rate-feedback voltage is called

(a) potentiometer (b) synchro-transmitter

(c) synchro-transformer (d) tachogenerator

383. A synchro-transmitter receiver unit is a

(a) two-phase ac device (b) 3-phase ac device

(c) dc device (d) single-phase ac device

384. For the transfer function $G(s)H(s) = \dfrac{1}{s(s+1)(s+0.5)}$, the phase cross over frequency is

(a) 0.5 rad/sec (b) 0.707 rad/sec (c) 1.732 rad/sec (d) 2 rad/sec

385. A unity feedback control system has a forward path transfer function equal to $\dfrac{42.25}{s(s+6.5)}$. The unit step response of this system starting from rest, will have its maximum value at a time equal to

(a) 0 sec (b) 0.56 sec (c) 5.6 sec (d) infinity

386. Match List-I with List-II and select the correct answer using the codes shown below the list:

List-I (Transfer function)	List-II (Controller)
A. $\dfrac{K_1 s + K_2}{K_3}$	1. P-controller
B. $\dfrac{K_1 s^2 + K_2 s + K_3}{K_4 s}$	2. PI-Controller
C. $\dfrac{K_1 s + K_2}{K_3}$	3. PD-controller
D. $\dfrac{K_1 s}{K_2}$	4. PID-controller

Codes:

	A	B	C	D
(a)	3	4	2	1
(b)	4	3	1	2
(c)	3	2	4	1
(d)	4	1	2	3

387. The maximum phase-shift that can be provided by a lead compensator with transfer function $G_c(s) = \dfrac{1+6s}{1+2s}$ is

(a) 15° (b) 30° (c) 45° (d) 60°

388. The correct sequence of steps needed to improve system stability is

(a) insert derivative action, use negative feedback reduce gain

(b) reduce gain, use negative feedback, insert derivative action

(c) reduce gain, insert derivative action, use negative feedback

(d) insert derivative action, reduce gain, insert derivative action

389. A process with open-loop model $G(s) = \dfrac{ke^{-s\tau}4}{s\tau+1}$ is controlled by a PID controller. For this process

(a) the integral mode improves transient performance

(b) the integral mode improves steady-state performance

(c) the derivative mode improves transient performance

(d) the derivative mode improves steady-state performance

390. The 3-dB bandwidth of a typical second-order system with the transfer function $\dfrac{C(s)}{R(s)} = \dfrac{\omega_n^2}{s^2 + 2\xi\omega_n s + \omega_n^2}$ is given by

(a) $\omega_n\sqrt{1-2\xi^2}$ (b) $\omega_n\sqrt{1-\xi^2+\sqrt{\xi^4-\xi^2+1}}$

(c) $\omega_n\sqrt{1-2\xi^2+\sqrt{4\xi^4-4\xi^2+2}}$ (d) $\omega_n\sqrt{1-2\xi^2-\sqrt{4\xi^4-4\xi^2+2}}$

391. For a second order system, damping ratio (ξ) is $0 < \xi < 1$, then the roots of the characteristic polynomials are

(a) real but not equal (b) real and equal (c) complex conjugates (d) imaginary

392. The Laplace transform of the function $i(t)$ is $I(s) = \dfrac{10s+4}{s(s+1)(s^2+4s+5)}$. Its final value will be

(a) 4/5 (b) 5/4 (c) 4 (d) 5

393. A linear second-order system with transfer function $G(s) = \dfrac{49}{s^2+16s+49}$ is initially at rest and is subjected to a step input signal. The response of the system will exhibit a peak overshoot of

(a) 16% (b) 9% (c) 2% (d) zero

394. A system has the following transfer function:

$G(s) = \dfrac{100(s+5)(s+50)}{s^4(s+10)(s^2+3s+10)}$. The closed loop system will be stable if the value of K is

(a) 2 (b) 3 (c) 4 (d) 5

395. Which of the following motors would be suitable for use as servomotors?

1. Two-phase induction motor
2. DC separately excited motor
3. Single-phase capacitance start induction motor
4. Universal motor
5. Single-phase synchronous motor

Select the correct answer using the codes given below:

(a) 1, 2, 3, 4 and 5 are correct (b) 2, 3 and 4 are correct

(c) 1, 2 and 5 are correct (d) 1 and 5 are correct

396. The open loop transfer function of a unity feedback control system is $G(s)H(s)=\dfrac{30}{s(s+1)(s+T)}$ where T is a variable parameter. The closed loop system will be stable for all values of

(a) $T>0$ (b) $0<T<3$ (c) $T>5$ (d) $3<T<5$

397. The Routh-Hurwitz criterion cannot be applied when the characteristic equation of the system contains any co-efficient which is

(a) negative real and exponential and sinusoidal functions of s

(b) negative real, both exponential and sinusoidal functions of s

(c) both exponential and sinusoidal functions of s

(d) complex, both exponential and sinusoidal functions of s

398. A unity feedback system has the plant transfer function $G_p=\dfrac{1}{(s+1)(2s+1)}$. The frequency at which the plant has a phase lag of 90° is

(a) $\dfrac{1}{\sqrt{3}}$ rad/sec (b) $\dfrac{1}{\sqrt{2}}$ rad/sec (c) $\dfrac{1}{\sqrt{6}}$ rad/sec (d) none of the above

399. Find the value of K if a point $-3+j8$ lies on the root locus of the system $G(s)H(s)=\dfrac{K}{s(s+6)}$

(a) 63 (b) 83 (c) 53 (d) none of the above

400. The differential equation of a mass-damper spring system is given by

$$2\frac{d^2x}{dt^2}+\frac{dx}{dt}+0.75x=f(t)$$

where $f(t)$ is the external force acting on the system and x is the displacement of mass. The steady state displacement corresponding to a force is given by

(a) 4 m (b) $\dfrac{12}{11}$ m (c) $\dfrac{4}{5}$ m (d) $\dfrac{9}{4}$ m

ANSWERS

1 (b)	2 (c)	3 (d)	4 (a)	5 (b)	6 (d)	7 (c)	8 (b)	9 (b)	10 (b)
11 (a)	12 (a)	13 (d)	14 (b)	15 (c)	16 (b)	17 (b)	18 (b)	19 (a)	20 (a)
21 (b)	22 (b)	23 (a)	24 (b)	25 (b)	26 (b)	27 (b)	28 (b)	29 (d)	30 (c)
31 (c)	32 (d)	33 (b)	34 (b)	35 (b)	36 (d)	37 (d)	38 (b)	39 (b)	40 (b)
41 (d)	42 (b)	43 (a)	44 (b)	45 (a)	46 (c)	47 (c)	48 (d)	49 (a)	50 (d)
51 (d)	52 (d)	53 (b)	54 (d)	55 (c)	56 (c)	57 (b)	58 (a)	59 (b)	60 (d)
61 (d)	62 (a)	63 (a)	64 (d)	65 (d)	66 (c)	67 (c)	68 (c)	69 (d)	70 (c)
71 (a)	72 (b)	73 (d)	74 (c)	75 (b)	76 (b)	77 (c)	78 (b)	79 (c)	80 (b)
81 (c)	82 (d)	83 (a)	84 (c)	85 (b)	86 (c)	87 (c)	88 (a)	89 (b), (c)	90 (b)
91 (d)	92 (c)	93 (c)	94 (a)	95 (b)	96 (d)	97 (a)	98 (c)	99 (c)	100 (a)
101 (b)	102 (b)	103 (b)	104 (b)	105 (d)	106 (a)	107 (b)	108 (c)	109 (b)	110 (b)

111 (c) 112 (a) 113 (a) 114 (b) 115 (b) 116 (b) 117 (d) 118 (b) 119 (b) 120 (c)
121 (a) 122 (a) 123 (a) 124 (d) 125 (c) 126 (b) 127 (b) 128 (b) 129 (c) 130 (d)
131 (b) 132 (a) 133 (c) 134 (b) 135 (d) 136 (d) 137 (d) 138 (a) 139 (c) 140 (a)
141 (a) 142 (a) 143 (b) 144 (a) 145 (a) 146 (b) 147 (b) 148 (b) 149 (b) 150 (a)
151 (c) 152 (c) 153 (c) 154 (a) 155 (b) 156 (c) 157 (a) 158 (a) 159 (c) 160 (d)
161 (c) 162 (a) 163 (a) 164 (b) 165 (a) 166 (b) 167 (c) 168 (b) 169 (a) 170 (a)
171 (b) 172 (c) 173 (b) 174 (d) 175 (c) 176 (b) 177 (b) 178 (b) 179 (b) 180 (c)
181 (c) 182 (c) 183 (a) 184 (c) 185 (b) 186 (c) 187 (b) 188 (a) 189 (b) 190 (b)
191 (d) 192 (a) 193 (a) 194 (c) 195 (b) 196 (a) 197 (c) 198 (b) 199 (d) 200 (d)
201 (c) 202 (a) 203 (c) 204 (b) 205 (d) 206 (c) 207 (b) 208 (a) 209 (a) 210 (d)
211 (a) 212 (a) 213 (b) 214 (b) 215 (d) 216 (c) 217 (d) 218 (a) 219 (c) 220 (a)
221 (a) 222 (d) 223 (b) 224 (b) 225 (b) 226 (c) 227 (c) 228 (d) 229 (a) 230 (c)
231 (b) 232 (b) 233 (a) 234 (d) 235 (b) 236 (b) 237 (c) 238 (a) 239 (a) 240 (c)
241 (b) 242 (b) 243 (c) 244 (a) 245 (b) 246 (b) 247 (b) 248 (a) 249 (a) 250 (b)
251 (c) 252 (c) 253 (c) 254 (c) 255 (a) 256 (a) 257 (b) 258 (a) 259 (b) 260 (c)
261 (b) 262 (c) 263 (c) 264 (b) 265 (b) 266 (d) 267 (d) 268 (a) 269 (b) 270 (b)
271 (c) 272 (a) 273 (c) 274 (b) 275 (b) 276 (d) 277 (a) 278 (d) 279 (c) 280 (d)
281 (d) 282 (b) 283 (b) 284 (a) 285 (d) 286 (a) 287 (c) 288 (b) 289 (d) 290 (c)
291 (a) 292 (c) 293 (c) 294 (a) 295 (c) 296 (c) 297 (a) 298 (b) 299 (c) 300 (b)
301 (c) 302 (d) 303 (d) 304 (b) 305 (b) 306 (d) 307 (a) 308 (a) 309 (b) 310 (d)
311 (d) 312 (d) 313 (d) 314 (d) 315 (d) 316 (c) 317 (d) 318 (a) 319 (a) 320 (d)
321 (a) 322 (a) 323 (d) 324 (c) 325 (a) 326 (b) 327 (c) 328 (c) 329 (a) 330 (d)
331 (c) 332 (a) 333 (b) 334 (d) 335 (a) 336 (c) 337 (a) 338 (b) 339 (a) 340 (b)
341 (c) 342 (b) 343 (d) 344 (c) 345 (c) 346 (b) 347 (c) 348 (a) 349 (c) 350 (c)
351 (a) 352 (a) 353 (a) 354 (b) 355 (c) 356 (a) 357 (d) 358 (c) 359 (b) 360 (a)
361 (b) 362 (c) 363 (c) 364 (a) 365 (b) 366 (d) 367 (d) 368 (b) 369 (b) 370 (a)
371 (a) 372 (c) 373 (c) 374 (c) 375 (a) 376 (a) 377 (b) 378 (d) 379 (a) 380 (c)
381 (b) 382 (d) 383 (d) 384 (b) 385 (d) 386 (a) 387 (b) 388 (d) 389 (b) 390 (c)
391 (c) 392 (a) 393 (d) 394 (b) 395 (c) 396 (c) 397 (d) 398 (b) 399 (d) 400 (a)

Extra Solved Problems

Example 1 If the transfer function of a system and applied input to it are e^{-5t} and e^{-7t} respectively, find the response of the system.

Solution

Given

$$g(t) = e^{-5t} \text{ and } r(t) = e^{-7t}$$

$$\therefore \quad G(s) = \text{LT}[g(t)] = \frac{1}{s+5}$$

and

$$R(s) = \text{LT}[r(t)] = \frac{1}{s+7}$$

Since

$$C(s) = G(s)\,R(s)$$

$$\therefore \quad C(s) = \frac{1}{(s+5)\,(s+7)}$$

$$\therefore \quad C(s) = \frac{1}{2}\left(\frac{1}{s+5} - \frac{1}{s+7}\right) \tag{1}$$

Taking inverse Laplace transform of Eq.(1), we have

$$c(t) = \frac{1}{2}(e^{-5t} - e^{-7t})$$

Example 2 A system having input $x(t)$ and output $y(t)$ is represented by

$$\frac{dy(t)}{dt} + 9y(t) = \frac{dx(t)}{dt} + 7x(t)$$

Find the transfer function of the system. Find $y(t)$ when $x(t) = u(t)$.

Solution The system is represented by

$$\frac{dy(t)}{dt} + 9y(t) = \frac{dx(t)}{dt} + 7x(t) \tag{1}$$

Taking Laplace transform of Eq. (1), we have

$$sY(s) + 9Y(s) = sX(s) + 7X(s)$$

Initial conditions of the system are assumed to be zero.

$\therefore$ $$Y(s)(s+9) = x(s)(s+7)$$

i.e., $$\frac{Y(s)}{X(s)} = \frac{s+7}{s+9}$$

$\therefore$ The transfer function of the system is $$G(s) = \frac{Y(s)}{X(s)} = \frac{s+7}{s+9}.$$

Here $$x(t) = u(t).$$

$\therefore$ $$X(s) = \frac{1}{s}$$

$\therefore$ $$Y(s) = X(s)\left(\frac{s+7}{s+9}\right)$$

i.e., $$Y(s) = \frac{s+7}{s(s+9)} = \frac{\frac{7}{9}}{s} + \frac{\frac{2}{9}}{s+9}$$

$\therefore$ $$y(t) = \frac{1}{9}(7 + 2e^{-9t})$$

Example 3 Determine the transfer function of a system represented by

$$\frac{d^3y(t)}{dt^3} + 5\frac{d^2y(t)}{dt^2} + 2\frac{dy(t)}{dt} + y(t) = \frac{d^2x(t)}{dt^2} + 7\frac{dx(t)}{dt} + 3x(t)$$

Solution Taking Laplace transform of the given equation and assuming initial condition to be zero, we have

$$s^3Y(s) + 5s^2Y(s) + 2sY(s) + Y(s) = s^2X(s) + 7sX(s) + 3X(s)$$

i.e., $$(s^3 + 5s^2 + 2s + 1)Y(s) = (s^2 + 7s + 3)X(s)$$

$\therefore$ $$\frac{Y(s)}{X(s)} = \frac{s^2 + 7s + 3}{s^3 + 5s^2 + 2s + 1}$$

$\therefore$ the transfer function of the system is $$G(s) = \frac{s^2 + 7s + 3}{s^3 + 5s^2 + 2s + 1}$$

Example 4 The transfer function of a system is given by

$$G(s) = \frac{3s + 5}{s^2 + 5s + 3}$$

Find the differential equation of the system having input x and output y.

Solution

Here $$G(s) = \frac{Y(s)}{X(s)} = \frac{3s+5}{s^2 + 5s + 3}$$

$$s^2Y(s) + 5sY(s) + 3Y(s) = 3sX(s) + 5X(s)$$

Taking inverse Laplace transform, we can write

$$\frac{d^2y(t)}{dt^2} + 5\frac{dy(t)}{dt} + 3y(t) = 3\frac{dx(t)}{dt} + 5x(t)$$

This is the required differential equation of the system.

Example 5 Find the transfer function of a system if its impulse response in e^{-4t} sin3t.

Solution The impulse response of the system is given by

$$c(t) = e^{-4t}\sin 3t$$

$$\therefore \quad \text{LT}[c(t)] = \text{LT}[e^{-4t}\sin 3t] = \frac{3}{(s+4)^2+9} = \frac{3}{s^2+8s+25}$$

Since the Laplace transform of the impulse response of a system represents its transfer function.

∴ the transfer function of the system is

$$G(s) = \frac{3}{s^2+8s+25}$$

Example 6 Obtain the unit step response of a unity feedback (negative) system whose open loop transfer function is $G(s) = \dfrac{5(s+20)}{s(s+4.59)(s^2+3.41s+16.35)}$.

Solution

Here $$G(s) = \frac{5(s+20)}{s(s+4.59)(s^2+3.41s+16.35)} \text{ and } H(s) = 1.$$

$$\therefore \quad \frac{C(s)}{R(s)} = \frac{G(s)}{1 + G(s)H(s)}$$

$$= \frac{5(s+20)}{s^4+3.41s^3+16.35s^2+4.59s^3+15.6519s^2+75.04s+5s+100}$$

$$= \frac{5(s+20)}{s^4+8s^3+32s^2+80s+100} = \frac{5(s+20)}{(s^2+2s+10)(s^2+6s+10)}$$

The unit step response of the system is

$$= \frac{5(s+20)}{s(s^2+2s+10)(s^2+6s+10)} = \frac{1}{s} + \frac{\frac{3}{8}(s+1) - \frac{17}{8}}{(s+1)^2+(3)^2} + \frac{-\frac{11}{8}(s+3) - \frac{13}{8}}{(s+3)^2+(1)^2}$$

$$c(t) = 1 + \frac{3}{8}e^{-t}\cos 3t - \frac{17}{24}e^{-t}\sin 3t - \frac{11}{8}e^{-3t}\cos t - \frac{13}{8}e^{-3t}\sin t$$

Example 7 Find the forced response of $\dfrac{d^2y(t)}{dt^2} + 3\dfrac{dy(t)}{dt} + 2y(t) = x(t)$ where x = sint.

Solution The given differential equation is

$$\frac{d^2y(t)}{dt^2} + 3\frac{dy(t)}{dt} + 2y(t) = x(t) \tag{1}$$

The Laplace transform of Eq.(1) gives

$$(s^2+3s+2)Y(s) = X(s) = \frac{1}{s^2+1}$$

i.e., $$Y(s) = \frac{1}{(s^2+1)(s^2+3s+2)} = \frac{1}{(s^2+1)(s+1)(s+2)} = \frac{-\frac{3s}{10}+\frac{1}{10}}{(s^2+1)} + \frac{\frac{5}{10}}{(s+1)} - \frac{\frac{1}{5}}{(s+2)} \tag{2}$$

The inverse Laplace transform of Eq. (2) gives

$$y(t) = -\frac{3}{10}\cos t + \frac{1}{10}\sin t + \frac{5}{10}e^{-t} - \frac{1}{5}e^{-2t}$$

The forced response $= -\frac{3}{10}\cos t + \frac{1}{10}\sin t$

Example 8 Find the transient and steady state response of the system having differential equation

$$\frac{d^2y(t)}{dt^2} + 3\frac{dy(t)}{dt} + 2y(t) = 1$$

Solution The given differential equation is $\frac{d^2y(t)}{dt^2} + 3\frac{dy(t)}{dt} + 2y(t) = 1$ (1)

The Laplace transform of Eq. (1) gives

$$(s^2 + 3s + 2)Y(s) = \frac{1}{s}$$

i.e.,
$$Y(s) = \frac{1}{s(s^2 + 3s + 2)} = \frac{1}{s(s+1)(s+2)} = \frac{\frac{1}{2}}{s} + \frac{-1}{(s+1)} + \frac{\frac{1}{2}}{(s+2)} \quad (2)$$

The inverse Laplace transform of Eq. (2) gives

$$y(t) = \frac{1}{2} + \frac{1}{2}e^{-2t} - e^{-t}$$

The steady state response $= \frac{1}{2}$ and transient response is $= \frac{1}{2}e^{-2t} - e^{-t}$

Example 9 Find the unit impulse response of the system described by $\frac{d^3y(t)}{dt^3} + \frac{dy(t)}{dt} = x(t)$.

Solution The unit impulse response of the system is the inverse Laplace transform of the transfer function.

The given differential equation is $\frac{d^3y(t)}{dt^3} + \frac{dy(t)}{dt} = x(t)$ (1)

The Laplace transform of Eq. (1) gives

$$(s^3 + s)Y(s) - X(s)$$

i.e.,
$$\frac{Y(s)}{X(s)} = \frac{1}{(s^3 + s)} = \frac{1}{(s^2 + 1)s} = \frac{1}{s} - \frac{s}{s^2 + 1}$$

i.e.,
$$\frac{y(t)}{x(t)} = 1 - \cos t$$

$\therefore$ Impulse response = 1 − cost.

Example 10 Find the response of the differential equation $\frac{d^3y(t)}{dt^3} + 6\frac{d^2y(t)}{dt^2} + 11\frac{dy(t)}{dt} + 6y(t) = 0$ with the initial conditions $y(0) = 1$, $\left.\frac{dy(t)}{dt}\right|_{t=0} = \left.\frac{d^2y(t)}{dt^2}\right|_{t=0} = 0$

Solution

It is given that $$\frac{d^3y(t)}{dt^3}+6\frac{d^2y(t)}{dt^2}+11\frac{dy(t)}{dt}+6y(t)=0 \quad (1)$$

The Laplace transform of Eq. (1) gives

$$s^3Y(s)-s^2y(0)-s\dot{y}(0)-\ddot{y}(0)+6[s^2Y(s)-sy(0)-\dot{y}(0)]+11[sY(s)-y(0)]+6Y(s)=0$$

i.e., $$s^3Y(s)-s^2+6[s^2Y(s)-s]+11[sY(s)-1]+6Y(s)=0$$

i.e., $$Y(s)\ (s^3+6s^2+11s+6)=s^2+6s+11$$

i.e., $$Y(s)=\frac{s^2+6s+11}{s^3+6s^2+11s+6}=\frac{s^2+6s+11}{(s+1)(s+2)(s+3)}=\frac{3}{s+1}-\frac{3}{s+2}+\frac{1}{s+3} \quad (2)$$

The inverse Laplace transform of Eq. (2) gives

$$y=3e^{-t}-3e^{-2t}+e^{-3t}$$

Example 11 Find the weighting function of the differential equation

$\frac{d^2y(t)}{dt^2}-\frac{dy(t)}{dt}-2y(t)=x(t)$. The system is initially relaxed.

Solution Impulse response of a system is known as weighting function and is given by the inverse LT.

It is given that $$\frac{d^2y(t)}{dt^2}-\frac{dy(t)}{dt}-2y(t)=x(t) \quad (1)$$

The Laplace transform of Eq. (1) gives

$$s^2Y(s)-sY(s)-2Y(s)=X(s)$$

i.e., $$(s^2-s-2)Y(s)=X(s)$$

i.e., $$\frac{Y(s)}{X(s)}=\frac{1}{(s^2-s-2)}=\frac{1}{(s-2)(s+1)}=\frac{1}{3}\left[\frac{1}{s-2}-\frac{1}{s+1}\right] \quad (2)$$

The inverse Laplace transform of Eq. (2) gives

∴ $$\frac{y(t)}{x(t)}=\frac{1}{3}[e^{2t}-e^{-t}]$$

Example 12 Find the impulse response of the circuit shown in Fig.1.

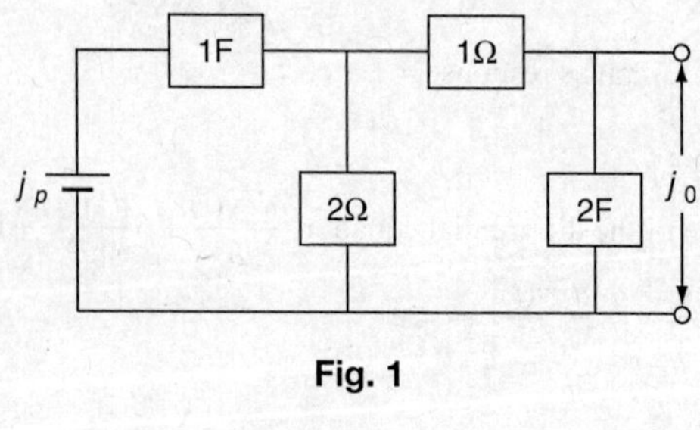

Fig. 1

Solution Figure 2 shows the transformed network of the given circuit.

Using KVL, we have

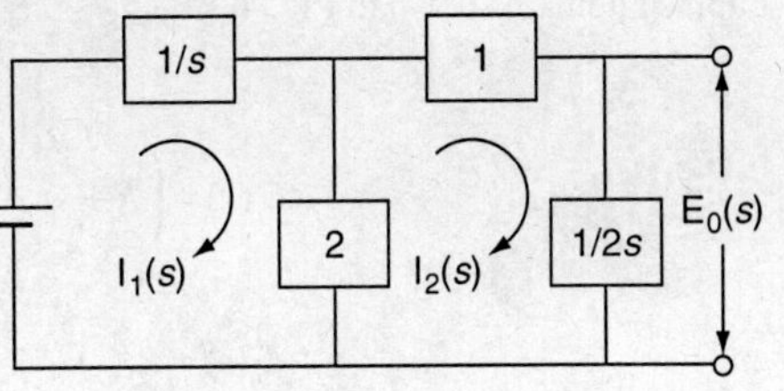

Fig. 2

$$E_i(s) = I_1(s)\left[\frac{1+2s}{s}\right] - 2I_2(s) \quad (1)$$

and $I_2(s)\left[3+\frac{1}{2s}\right] = 2I_i(s)$

i.e., $$I_1(s) = I_2(s)\left[\frac{6s+1}{4s}\right] \quad (2)$$

and $$E_0(s) = \frac{1}{2s}I_2(s) \quad (3)$$

Substituting the value of $I_1(s)$ from Eq. (2) in Eq. (1), we have

$$E_i(s) = I_2(s)\left(\frac{4s^2+8s+1}{4s^2}\right) \quad (4)$$

From Eq. (3) and Eq. (4), we have

$$E_i(s) = 2sE_0(s)\left(\frac{4s^2+8s+1}{4s^2}\right) = E_0(s)\left(\frac{4s^2+8s+1}{2s}\right)$$

i.e., $$\frac{E_0(s)}{E_i(s)} = \frac{2s}{4s^2+8s+1}$$

Here $E_i(s) = 1$

$$\therefore \quad E_0(s) = \frac{2s}{4s^2+8s+1} = \frac{0.5s}{s^2+2s+0.25} = \frac{0.5s}{(s+1)^2 - \left(\frac{\sqrt{3}}{2}\right)^2}$$

i.e., $$E_0(s) = \frac{0.5s}{\left(s+1+\frac{\sqrt{3}}{2}\right)\left(s+1-\frac{\sqrt{3}}{2}\right)} = \frac{0.5386}{s+1+\frac{\sqrt{3}}{2}} - \frac{0.0386}{s+1-\frac{\sqrt{3}}{2}}$$

i.e., $$E_0(s) = \frac{0.5386}{s+1.866} - \frac{0.0386}{s+0.1339} \quad (5)$$

The inverse Laplace transform of Eq. (5) gives

$$e_0(t) = 0.5386e^{-1.866t} - 0.0386e^{-0.1339t}$$

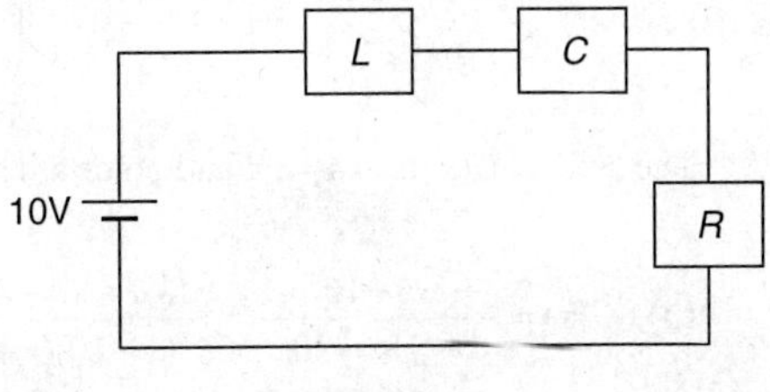

Fig. 3

Example 13 Find the step response of series LCR circuit shown in Fig.3 with $R = 1\Omega$, $L = 1H$ and $C = 1F$. The output is taken across R and input is a step of 10V.

Solution The given network shown in Fig. 3 is transformed in Laplace domain shown in Fig. 4.

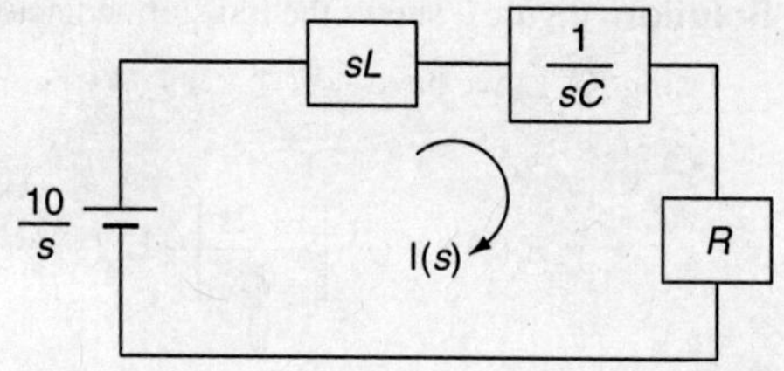

Fig. 4

$$E_i(s) = \left(R + sL + \frac{1}{sC}\right) I(s)$$

and

$$E_0(s) = RI(s)$$

$$\frac{E_0(s)}{E_i(s)} = \frac{RCs}{s^2LC + sRC + 1}$$

It is given that $e_i = 10$ V

Therefore, $E_i(s) = \dfrac{10}{s}$

$$E_0(s) = E_i(s)\left(\frac{RCs}{s^2LC + sRC + 1}\right) = \left(\frac{10}{s}\right)\left(\frac{s}{s^2 + s + 1}\right) = \frac{10}{\left(s + \frac{1}{2}\right)^2 + \left(\frac{\sqrt{3}}{2}\right)^2}$$

i.e.,

$$E_0(s) = \frac{10}{\left(s + \frac{1}{2} + j\frac{\sqrt{3}}{2}\right)\left(s + \frac{1}{2} - j\frac{\sqrt{3}}{2}\right)} = \frac{j\frac{10}{\sqrt{3}}}{s + \frac{1}{2} + j\frac{\sqrt{3}}{2}} + \frac{-j\frac{10}{\sqrt{3}}}{s + \frac{1}{2} - j\frac{\sqrt{3}}{2}}$$

Therefore, $e_0(t) = 11.54e^{-0.5t}\sin 0.866t$

Example 14 Find the poles and zeros of the following functions:

(i) $G(s) = \dfrac{K(s+2)}{s(s^2 + s + 1)}$ (ii) $G(s) = \dfrac{s^2 - 16}{s^5 - 7s^4 - 30s^3}$

Solution

(i) $$G(s) = \frac{K(s+2)}{s(s^2 + s + 1)} = \frac{K(s+2)}{s\left(s + \frac{1 + j\sqrt{3}}{2}\right)\left(s + \frac{1 - j\sqrt{3}}{2}\right)}$$

The zero of $G(s)$ is at $s = -2$ and poles are at $s = 0, \dfrac{-1 \pm j\sqrt{3}}{2}$

(ii) $$G(s) = \frac{s^2 - 16}{s^5 - 7s^4 - 30s^3} = \frac{(s+4)(s-4)}{s^3(s-10)(s+3)}$$

Zeros of $G(s)$ are at $s = 4, -4$ and its poles are at $s = 0, 0, 0, 10, -3$

Example 15 Derive the system equations and find the value of $X_2(s)/F(s)$ for the system shown in Fig. 5 below.

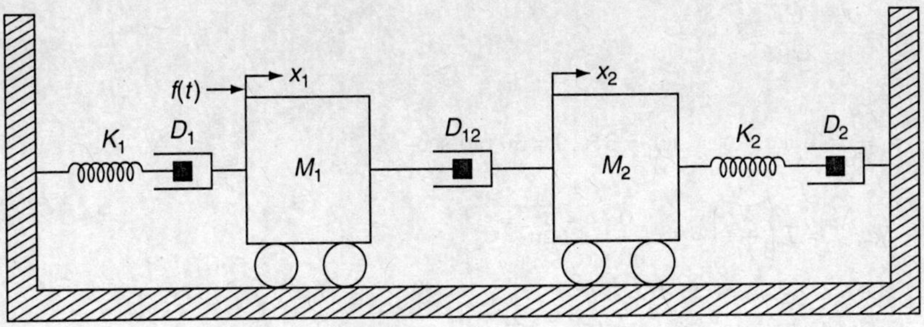

Fig. 5

Solution

For mass M_1:

$$f(t) = M_1\frac{d^2x_1}{dt^2} + D_1\frac{dx_1}{dt} + D_{12}\frac{d}{dt}(x_1 - x_2) + K_1x_1 \tag{1}$$

For mass M_2:

$$D_{12}\frac{d}{dt}(x_1 - x_2) = M_2\frac{d^2x_2}{dt^2} + D_2\frac{dx_2}{dt} + K_2x_2 \tag{2}$$

Laplace transform of Eq. (1),

$$F(s) = X_1(s)[s^2M_1 + sD_{12} + sD_1 + K_1] - sD_{12}X_2(s) \tag{3}$$

Laplace transform of Eq. (2),

$$sD_{12}X_1(s) = X_2(s)[s^2M_1 + sD_{12} + sD_2 + K_2]$$

i.e., $$X_1(s) = \frac{X_2(s)[s^2M_1 + sD_{12} + sD_2 + K_2]}{sD_{12}} \tag{4}$$

Putting the value of $X_1(s)$ in Eq. (3), we have

$$F(s) = \frac{X_2(s)[s^2M_1 + sD_{12} + sD_2 + K_2][s^2M_1 + sD_{12} + sD_1 + K_1]}{sD_{12}} - sD_{12}X_2(s)$$

i.e., $$\frac{X_2(s)}{F(s)} = \frac{sD_{12}}{[s^2M_1 + sD_{12} + sD_2 + K_2][s^2M_1 + sD_{12} + sD_1 + K_1] - s^2D_{12}^2}$$

Example 16

(i) Find the ratio of r_1 and r_2

(ii) Calculate ω_2 if $\omega_1 = 10$ rad/sec

(iii) If the torque of wheel 1 is 9 N-m, what is the torque of wheel 2?

(iv) If the displacement of wheel 1 is 15^0, what is the displacement of wheel 2?

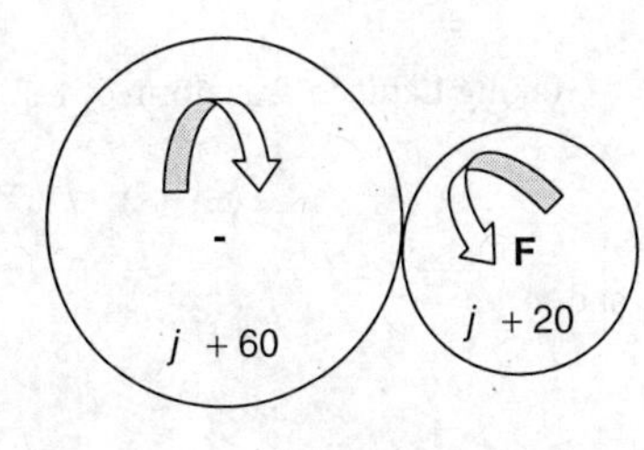

Fig. 6

Solution

(i) $\frac{r_1}{r_2} = \frac{N_1}{N_2}$ i.e., $\frac{r_1}{r_2} = \frac{60}{20} = 3:1$

(ii) $\frac{\omega_2}{\omega_1} = \frac{N_1}{N_2} = \frac{60}{20} = 3$ i.e., $\omega_2 = 3\omega_1 = 3 \times 10 = 30$ rad/sec

(iii) $\frac{T_1}{T_2} = \frac{N_1}{N_2}$ *i.e.*, $T_2 = T_1\left(\frac{N_2}{N_1}\right) = 9 \times \frac{20}{60} = 3$ N-m

(iv) $\frac{\theta_2}{\theta_1} = \frac{N_1}{N_2} = \frac{60}{20} = 3$ *i.e.*, $\theta_2 = 3\theta_1 = 3 \times 15° = 45°$

Example 17 Show that the systems shown in Fig. 7(a) and Fig. 7(b) are analogous system.

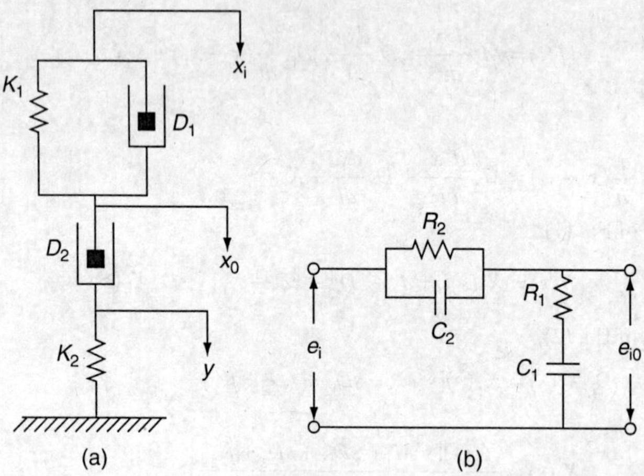

Fig. 7

Solution From Fig. 7(a):

$$D_1 \frac{d}{dt}(x_i - x_0) + K_1(x_i - x_0) = D_2 \frac{d}{dt}(x_0 - y) \tag{1}$$

and
$$D_2 \frac{d}{dt}(x_0 - y) = K_2 y \tag{2}$$

Taking Laplace Transform of Eq. (1) and Eq. (2), we have

$$D_1 sX_i(s) - D_1 sX_0(s) + K_1 X_i(s) - K_1 X_0(s) = D_2 sX_0(s) - D_2 sY(s) \tag{3}$$

and
$$D_2 sX_0(s) - D_2 sY(s) = K_2 Y(s)$$

i.e.,
$$Y(s) = \frac{sD_2 X_0(s)}{K_2 + sD_2} \tag{4}$$

Substituting $Y(s)$ from Eq. (4) in Eq. (3), we have

$$D_1sX_i(s) - D_1sX_0(s) + K_1X_i(s) - K_1X_0(s) = D_2sX_0(s) - D_2s\left[\frac{sD_2X_0(s)}{K_2 + sD_2}\right]$$

i.e.,
$$\frac{X_0(s)}{X_i(s)} = \frac{(sD_1 + K_1)(sD_2 + K_2)}{s^2D_1D_2 + K_1K_2 + sD_1K_2 + sD_2K_1 + sD_2K_2} \tag{5}$$

Using f–v analogy, the Eq. (5) becomes

$$\frac{E_0(s)}{E_i(s)} = \frac{(sR_1C_1 + 1)(sR_2C_2 + 1)}{s^2R_1R_2C_1C_2 + 1 + sR_1C_1 + sR_2C_2 + sR_2C_1}$$

$$= \frac{(1 + sR_1C_1)(1 + sR_2C_2)}{s^2R_1R_2C_1C_2 + sR_1C_1 + sR_2C_2 + sR_2C_1 + 1} \tag{6}$$

From Fig. 7(b), we have $Z_1(s) = \dfrac{R_2}{1 + sR_2C_2}$ and $Z_2(s) = \dfrac{1 + sR_1C_1}{sC_1}$

$$\therefore \quad \frac{e_0(s)}{e_i(s)} = \frac{Z_2}{Z_1 + Z_2} = \frac{\dfrac{1 + sR_1C_1}{sC_1}}{\dfrac{R_2}{1 + sR_2C_2} + \dfrac{1 + sR_1C_1}{sC_1}} = \frac{(1 + sR_1C_1)(1 + sR_2C_2)}{s^2R_1R_2C_1C_2 + sR_1C_1 + sR_2C_2 + sR_2C_1 + 1} \tag{7}$$

Since Eq. (6) and Eq. (7) are identical, Figure (a) and (b) are analogous of each other.

Example 18 Find the transfer function $\dfrac{X(s)}{E(s)}$ for the electromechanical system shown in Fig. 8 below.

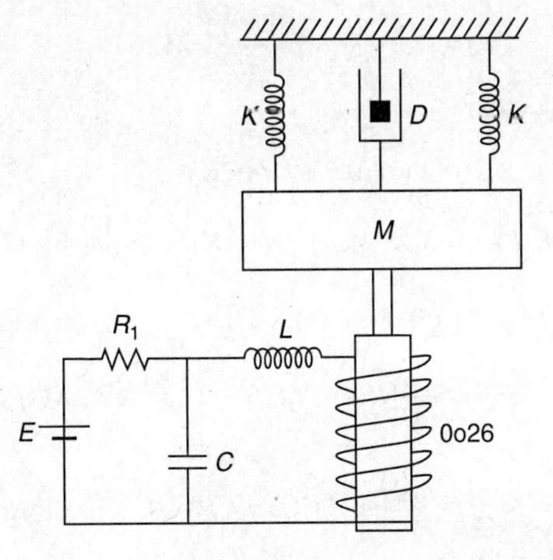

Fig. 8

Solution

There are two following parts in the given system
(i) electrical and (ii) mechanical.

Let us first consider the electrical system shown in Fig. 9. Using KVL, we have

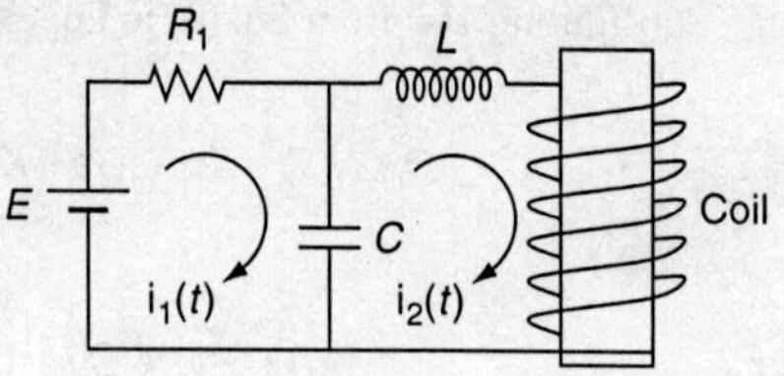

Fig. 9

$$e(t) = R_1 i_1(t) + \frac{1}{C}\int [i_1(t) - i_2(t)]dt \qquad (1)$$

and
$$0 = \frac{1}{C}\int [i_2(t) - i_1(t)]dt + L\frac{di_2(t)}{dt} + e_b \qquad (2)$$

Laplace transform of Eq. (1) and Eq. (2) gives

$$E(s) = R_1 I_1(s) + \frac{1}{sC}[I_1(s) - I_2(s)] \qquad (3)$$

and
$$0 = \frac{1}{sC}[I_2(s) - I_1(s)] + sLI_2(s) + E_b(s) \qquad (4)$$

We know that
$$e_b \propto \frac{dx}{dt} \text{ i.e., } e_b = K_1\frac{dx}{dt} \text{ i.e., } E_b(s) = K_1 sX(s) \qquad (5)$$

where K_1 is the back e.m.f. constant.

Let us now consider the mechanical system shown in Fig.10.

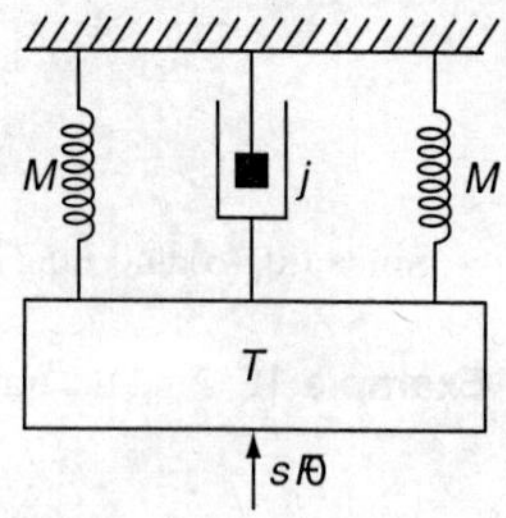

Fig. 10

$$f(t) \propto i_2(t)$$

i.e.,
$$f(t) = K_2 i_2(t) \qquad (6)$$

The Laplace transform of Eq. (6) gives

$$F(s) = K_2 I_2(s) \qquad (7)$$

From the Fig. 10, we have

$$f(t) = M\frac{d^2x(t)}{dt^2} + D\frac{dx(t)}{dt} + 2Kx(t) \qquad (8)$$

The Laplace transform of Eq. (8) gives

$$F(s) = X(s)[Ms^2 + Ds + 2K] \qquad (9)$$

i.e.,
$$K_2 I_2(s) = X(s)[Ms^2 + Ds + 2K] \quad \text{[Using Eq. (7)]} \qquad (10)$$

Substituting $E_b(s)$ from Eq. (5) in Eq. (4), we have

$$0 = \frac{1}{sC}[I_2(s) - I_1(s)] + sLI_2(s) + K_1 sX(s)$$

i.e.,
$$\frac{I_1(s)}{sC} = \frac{I_2(s)}{sC} + sLI_2(s) + K_1 sX(s)$$

i.e., $$I_1(s) = I_2(s) + s^2LCI_2(s) + K_1s^2CX(s) \tag{11}$$

From Eq. (3), we have

$$E(s) = R_1I_1(s) + \frac{I_1(s)}{sC} - \frac{I_2(s)}{sC}$$

i.e., $$E(s) = \left(R_1 + \frac{1}{sC}\right)\left[I_2(s) + s^2LCI_2(s) + K_1s^2CX(s)\right] - \frac{I_2(s)}{sC} \quad \text{[From Eq. (11)]}$$

i.e., $$E(s) = R_1I_2(s) + s^2R_1LCI_2(s) + K_1R_1s^2CX(s) + \frac{I_2(s)}{sC} + sLI_2(s) + K_1sX(s) - \frac{I_2(s)}{sC}$$

i.e., $$E(s) - K_1sX(s) - K_1R_1s^2CX(s) = R_1I_2(s) + s^2R_1LCI_2(s) + sLI_2(s)$$

i.e., $$I_2(s) = \frac{E(s) - K_1s(1 + R_1Cs)X(s)}{s^2R_1LC + sL + R_1} \tag{12}$$

Substituting $I_2(s)$ from Eq. (12) in Eq. (10), we have

$$K_2\left(\frac{E(s) - K_1s(1 + R_1Cs)X(s)}{s^2R_1LC + sL + R_1}\right) = X(s)[Ms^2 + Ds + 2K]$$

After simplification, we have

$$\frac{X(s)}{E(s)} = \frac{K_2}{s^4R_1LCM + s^3L(R_1CD + M) + s^2\{R_1M + LD + R_1C(2LK + K_1K_2)\} + s(R_1D + 2LK + K_1K_2) + 2R_1K}$$

Example 19 Find the transfer function $\frac{E_0(s)}{E_i(s)}$ for the OpAmp shown in Fig. 11 below.

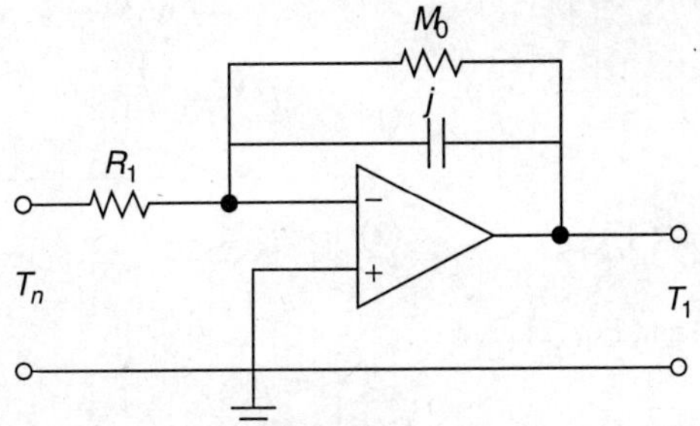

Fig. 11

Solution The transformed network of Fig. 11 is shown in Fig. 12.

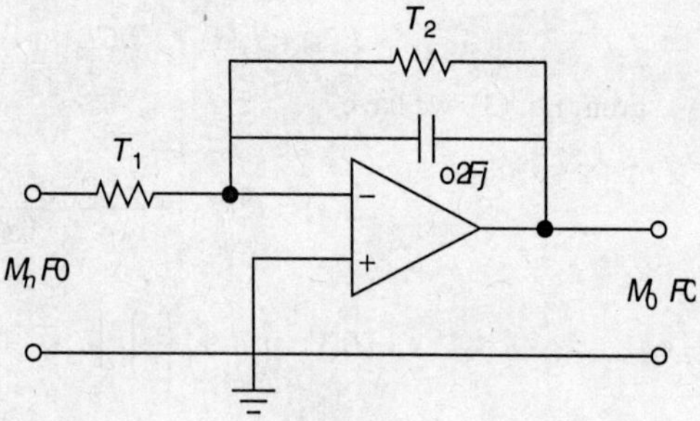

Fig. 12

$$Z_1(s) = R_1$$

$$Z_2(s) = \frac{R_2 \times \frac{1}{sC}}{R_2 + \frac{1}{sC}} = \frac{R_2}{1 + sCR_2}$$

$$\therefore \quad \frac{E_0(s)}{E_i(s)} = -\frac{Z_2(s)}{Z_1(s)} = -\frac{R_2}{R_1(1 + sR_2C)}$$

Example 20 Find the transfer function of the amplifier shown in Fig. 13.

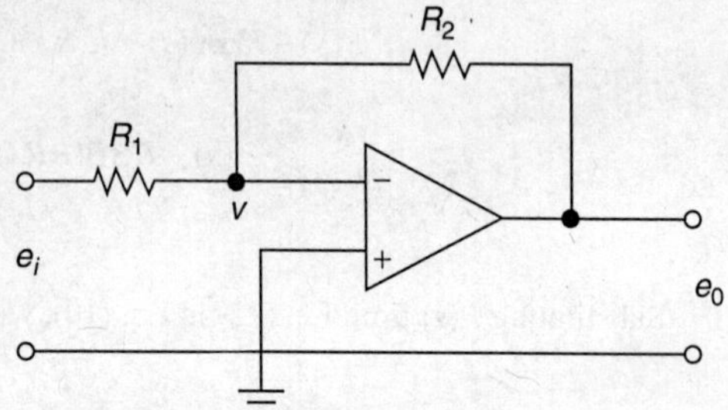

Fig. 13

Solution

$$i_1 = \frac{e_i - v}{R_1} \text{ and } i_2 = \frac{v - e_i}{R_2}$$

Since $\quad v = 0,$

$$i_1 = i_2$$

i.e., $$\frac{e_i}{R_1} = -\frac{e_0}{R_2} \qquad (1)$$

The Laplace transform of Eq.(1) gives

$$\frac{E_i(s)}{R_1} = -\frac{E_0(s)}{R_2} \text{ i.e., } \frac{E_0(s)}{E_i(s)} = -\frac{R_2}{R_1}$$ is the required transfer function.

Example 21 Determine the transfer function of circuit using OP–AMP as shown in Fig. 14.

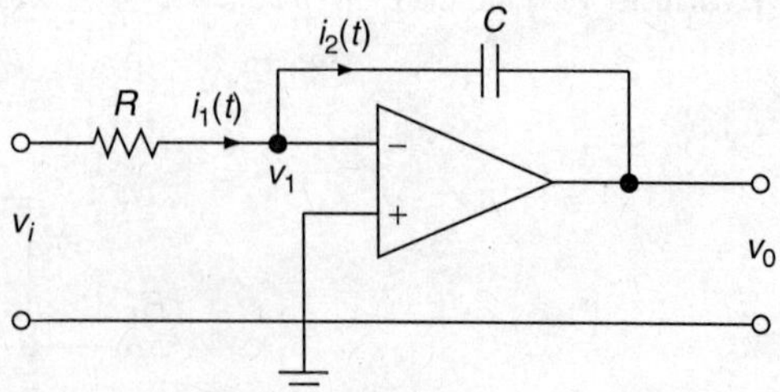

Fig. 14

Solution

$$i_1(t) = \frac{v_i(t) - v_1(t)}{R} \qquad (1)$$

and $$i_2(t) = C\frac{d}{dt}[v_1(t) - v_0(t)] \qquad (2)$$

Laplace transform of Eq. (1) and Eq. (2) give

$$I_1(s) = \frac{1}{R}[V_i(s) - V_1(s)]$$

and $$I_2(s) = sC[V_1(s) - V_0(s)]$$

Since, $$V_1(s) = 0,$$

$$I_1(s) = I_2(s)$$

$$I_1(s) = \frac{V_i(s)}{R} \text{ and } I_2(s) = -sCV_0(s)$$

$\therefore$ $$\frac{V_i(s)}{R} = -sCV_0(s) \quad \text{i.e.,} \quad \frac{V_0(s)}{V_i(s)} = -\frac{1}{sCR}$$

Example 22 Find the transfer function of the circuit shown in Fig. 15 below.

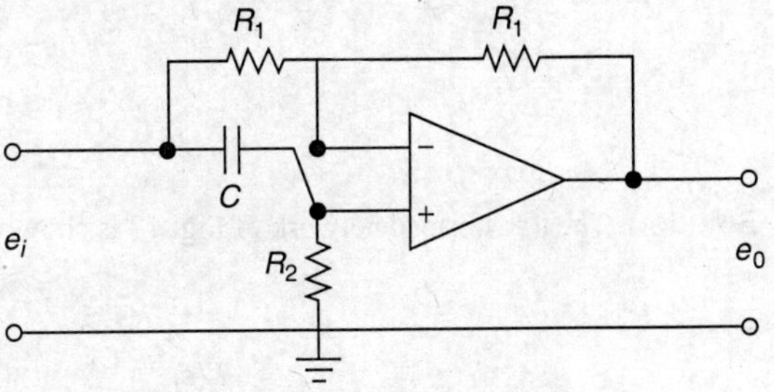

Fig. 15

Solution The transformed network of Fig. 15 is shown in Fig. 16.

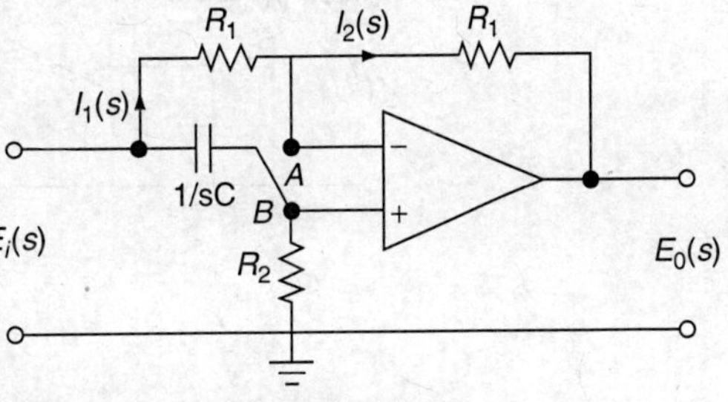

Fig. 16

$$V_B(s) = \frac{R_2 E_i(s)}{R_2 + \frac{1}{sC}}$$

and

$$V_A(s) = V_B(s) = \left(\frac{R_2}{R_2 + \frac{1}{sC}}\right) E_i(s) = \left(\frac{sCR_2}{1 + sCR_2}\right) E_i(s)$$

$$I_1(s) = \frac{E_i(s) - V_A(s)}{R_1} \text{ and } I_2(s) = \frac{V_A(s) - E_0(s)}{R_1}$$

Since $I_1(s) = I_2(s)$

$$\therefore \frac{E_i(s) - V_A(s)}{R_1} = \frac{V_A(s) - E_0(s)}{R_1}$$

i.e., $$2V_A(s) = E_i(s) + E_0(s) \text{ i.e., } E_i(s) = 2V_A(s) - E_0(s)$$

i.e., $$E_i(s) = 2\left(\frac{sCR_2}{1 + sCR_2}\right) E_i(s) - E_0(s) \text{ i.e., } E_0(s) = 2\left(\frac{sCR_2}{1 + sCR_2}\right) E_i(s) - E_i(s)$$

i.e., $$\frac{E_0(s)}{E_i(s)} = \frac{sCR_2 - 1}{sCR_2 + 1}$$ is the required transfer function.

Example 23 Find the transfer function of the circuit shown in Fig. 17 below.

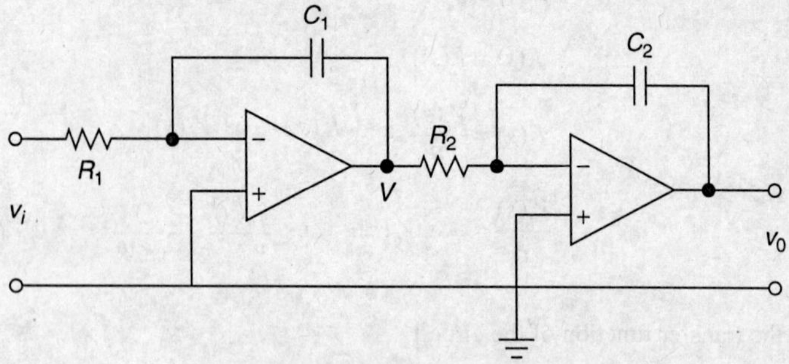

Fig. 17

Solution The transformed network of Fig. 17 is shown in Fig. 18.

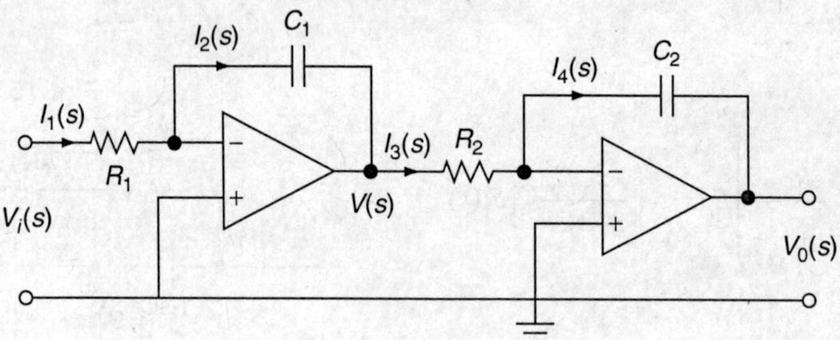

Fig. 18

$$I_1(s) = \frac{V_i(s)}{R_1} \text{ and } I_2(s) = \frac{-V(s)}{\frac{1}{sC_1}} = -sC_1V(s)$$

Since $$I_1(s) = I_2(s)$$

$\therefore$ $$\frac{V(s)}{V_i(s)} = -\frac{1}{sR_1C_1}$$

$$I_3(s) = \frac{V(s)}{R_2}$$

$$I_4(s) = \frac{-V_0(s)}{\frac{1}{sC_2}} = -sC_2V_0(s)$$

Since, $$I_3(s) = I_4(s)\,,$$

$$\frac{V(s)}{R_2} = -V_0(s)\, sC_2$$

i.e., $$\frac{V_0(s)}{V(s)} = -\frac{1}{sC_2R_2}$$

$\therefore$ $$\frac{V_0(s)}{V_i(s)} = \frac{V(s)}{V_i(s)} \times \frac{V_0(s)}{V(s)} = \frac{1}{s^2R_1R_2C_1C_2}$$

Example 24 Find the transfer function of the network shown in Fig. 19 below.

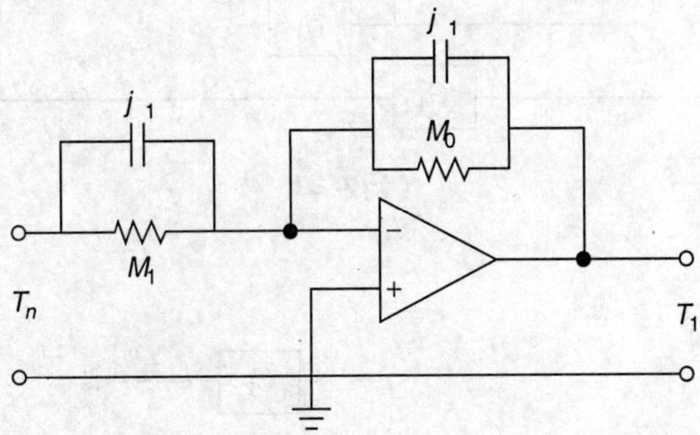

Fig. 19

Solution The transformed network of Fig. 19 is shown in Fig. 20.

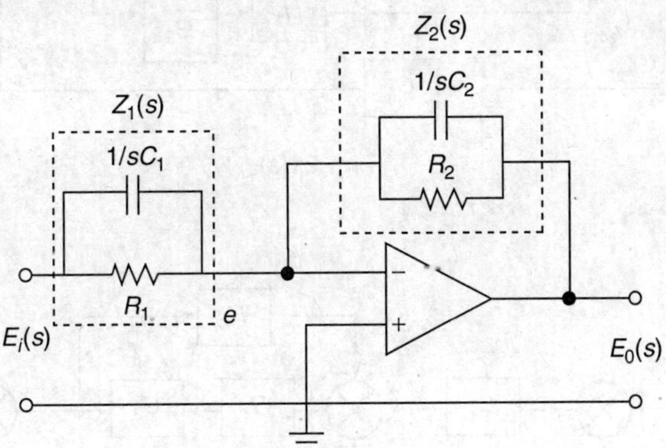

Fig. 20

$$Z_1(s) = \frac{R_1 \times \dfrac{1}{sC_1}}{R_1 + \dfrac{1}{sC_1}} = \frac{R_1}{sR_1C_1 + 1} \quad \text{and} \quad Z_2(s) = \frac{R_2 \times \dfrac{1}{sC_2}}{R_2 + \dfrac{1}{sC_2}} = \frac{R_2}{sR_2C_2 + 1}$$

$$\therefore \qquad \frac{E_0(s)}{E_i(s)} = -\frac{Z_2}{Z_1} = -\frac{R_2(sR_1C_1+1)}{R_1(sR_2C_2+1)}$$ is the required transfer function of the network.

Example 25 Find the transfer function of the block diagram shown in Fig. 21.

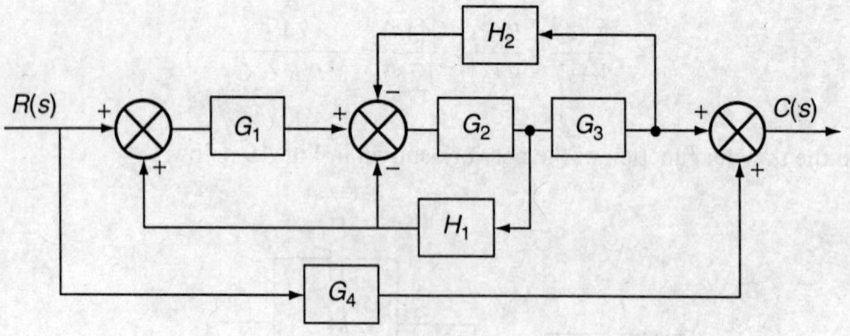

Fig. 21

Solution

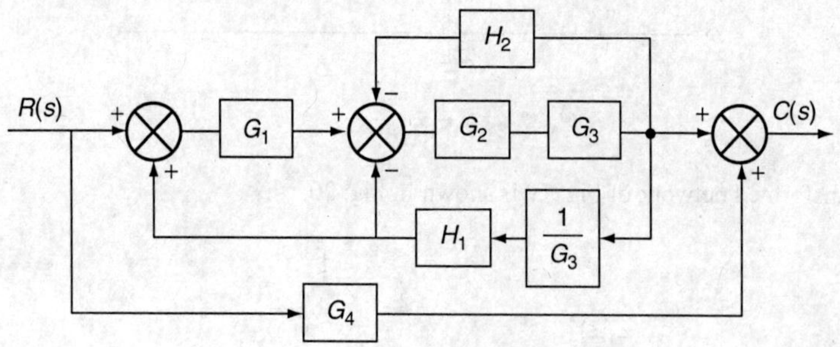

Fig. 21(a)

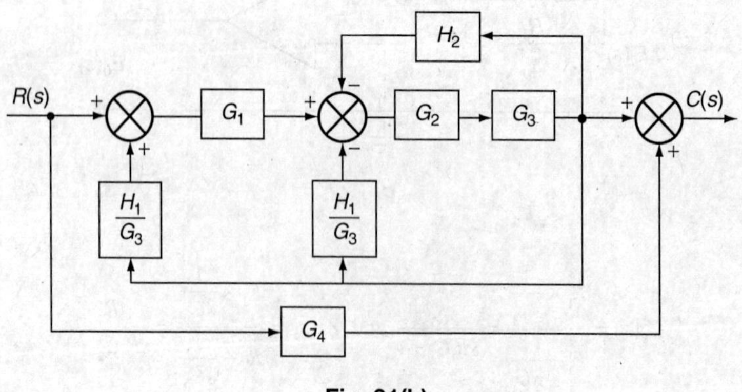

Fig. 21(b)

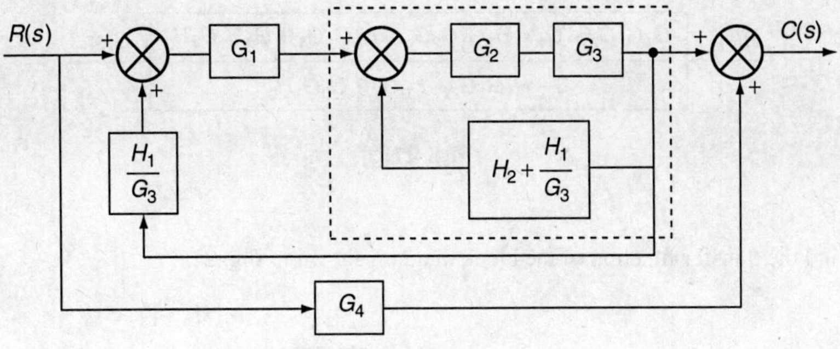

Fig. 21(c)

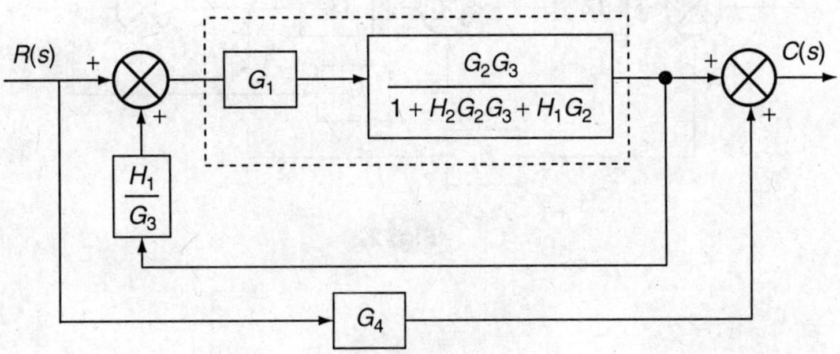

Fig. 21(d)

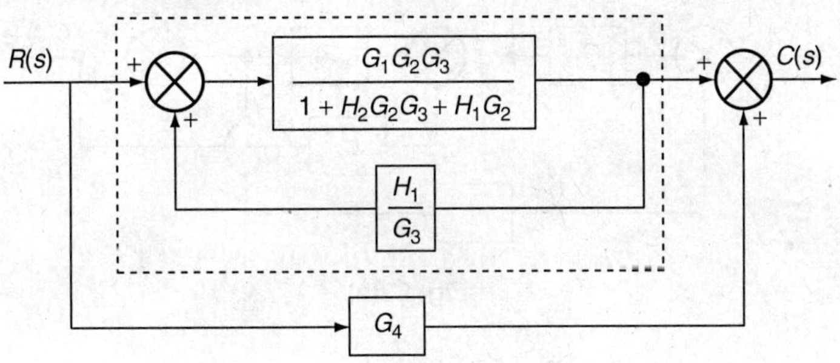

Fig. 21(e)

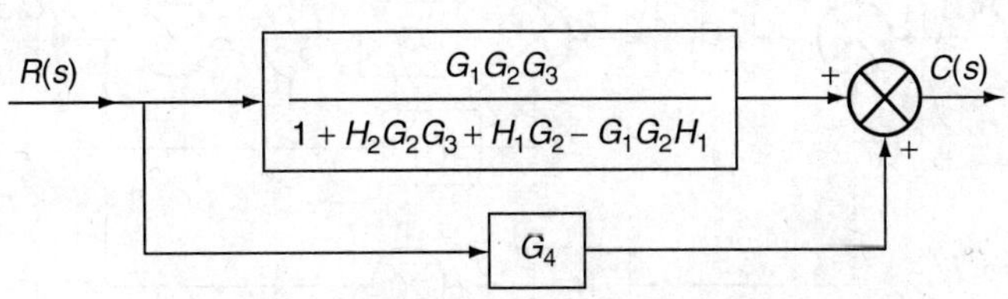

Fig. 21(f)

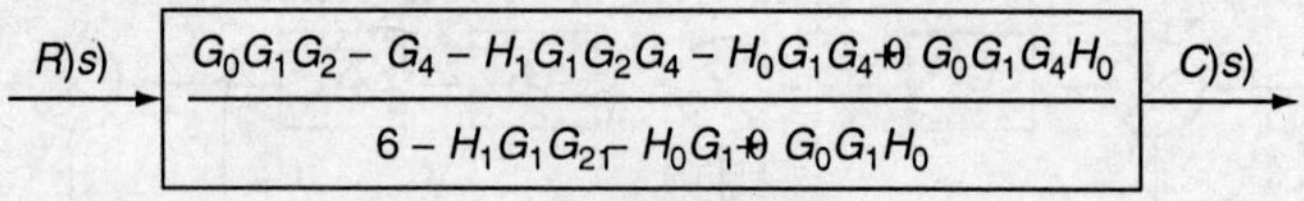

Fig. 21(g)

Example 26 Find the transfer function of the block diagram shown in Fig. 22.

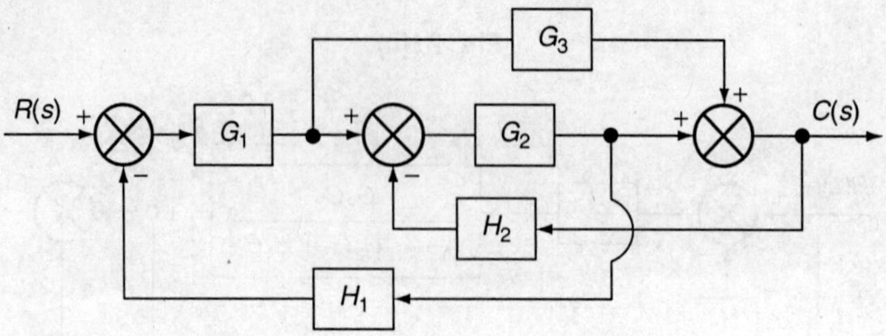

Fig. 22

Solution

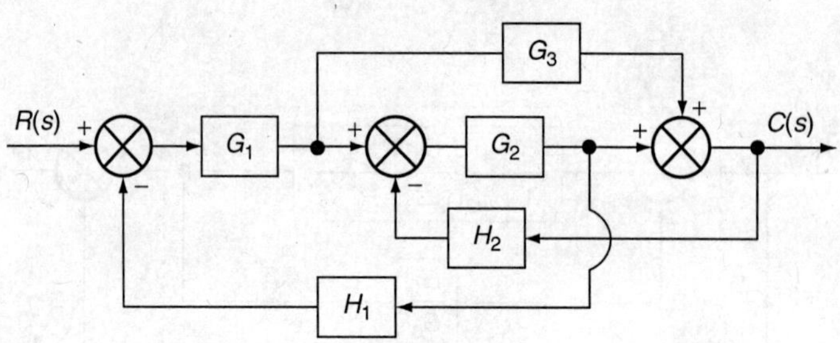

Fig. 22(a)

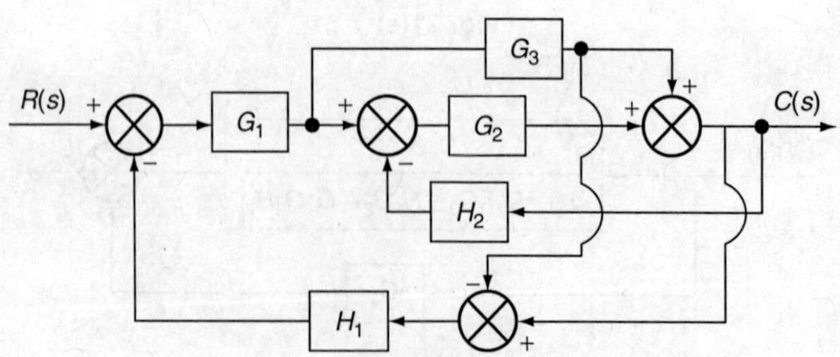

Fig. 22(b)

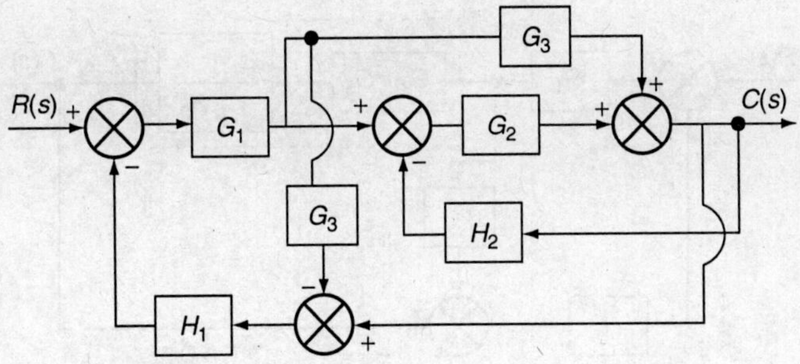

Fig. 22(c)

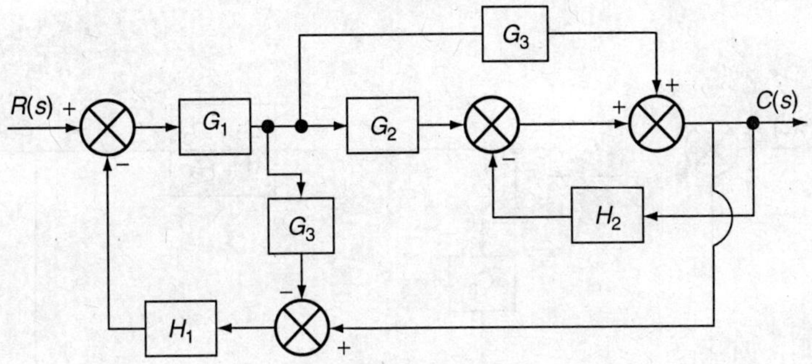

Fig. 22(d)

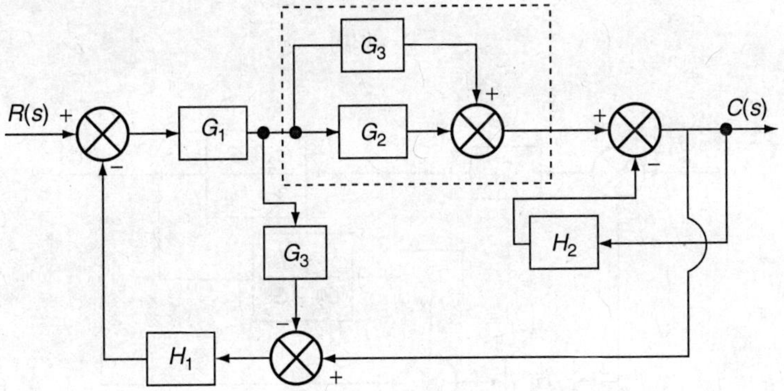

Fig. 22(e)

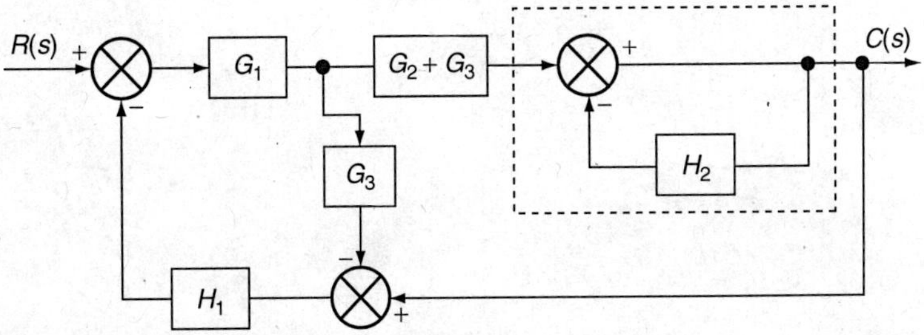

Fig. 22(f)

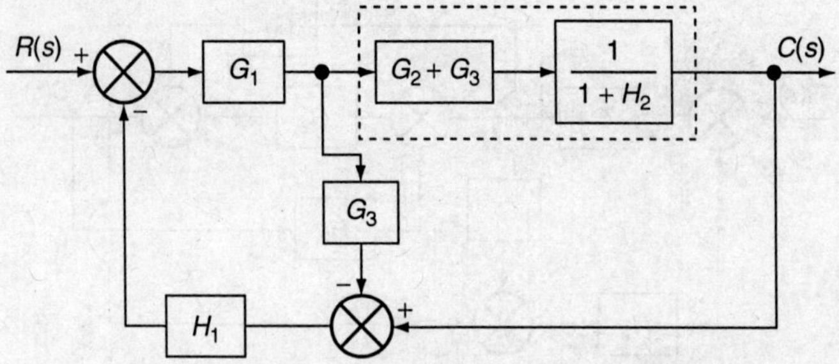

Fig. 22(g)

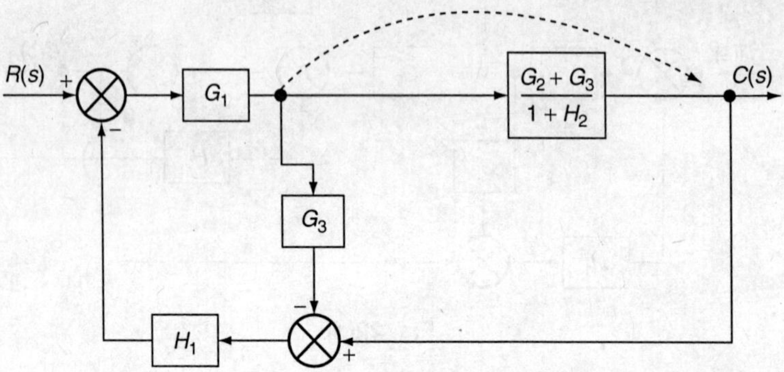

Fig. 22(h)

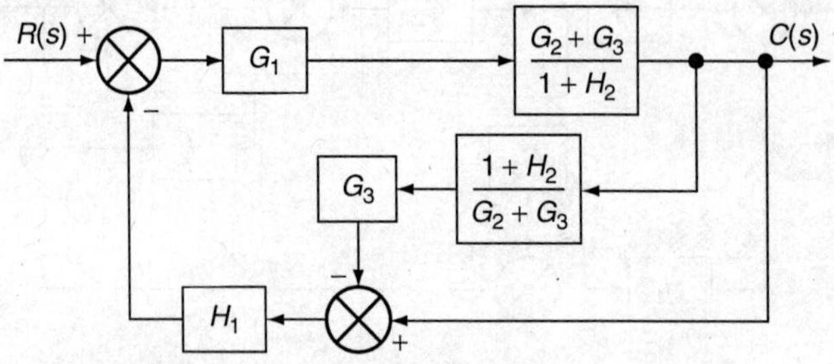

Fig. 22(i)

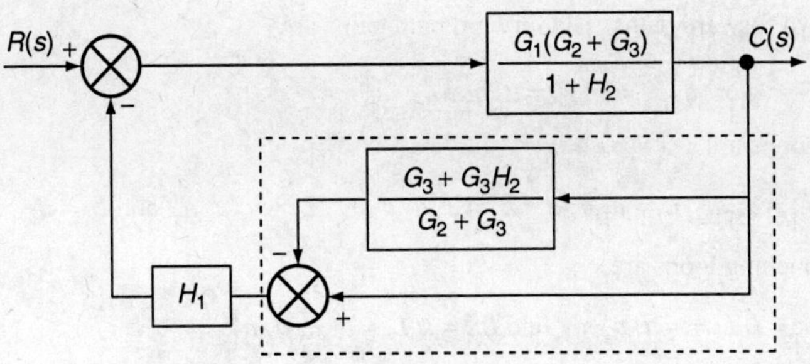

Fig. 22(j)

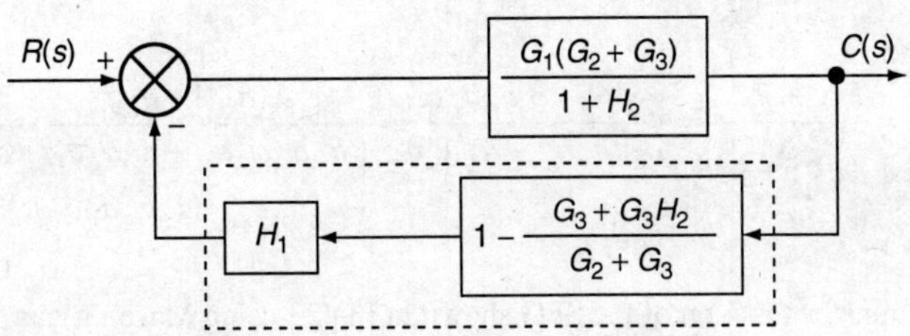

Fig. 22(k)

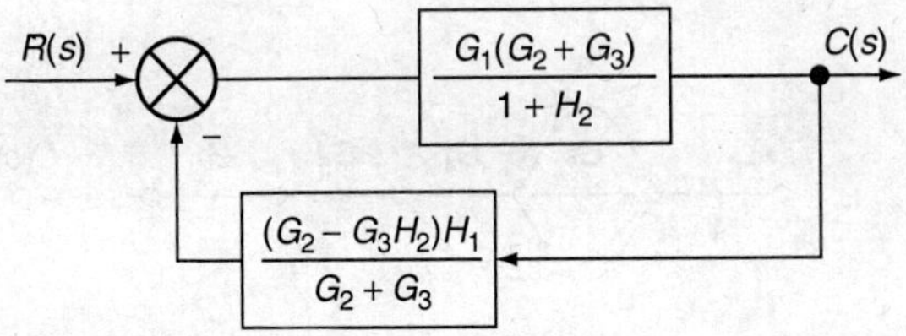

Fig. 22(l)

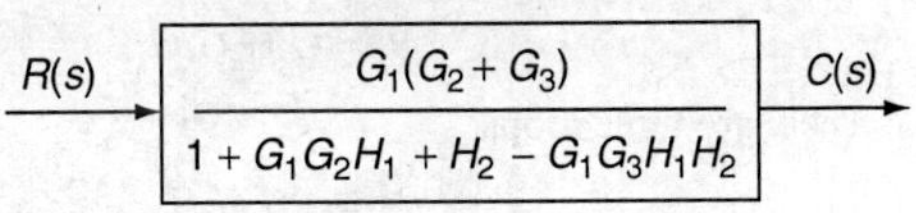

Fig. 22(m)

Example 27 For the system represented by the following differential equations find the transfer function x_5/x_1 using Mason's Gain formula.

$$x_2 = a_{12}x_1 + a_{32}x_3 + a_{42}x_4 + a_{52}x_5$$

$$x_3 = a_{23}x_2$$

$$x_4 = a_{34}x_3 + a_{44}x_4$$

$$x_5 = a_{35}x_3 + a_{45}x_4$$

where x_1 is the input variable and x_5 is the output variable.

Solution There are two forward paths. The forward path gains are

$$P_1 = a_{12}a_{23}a_{34}a_{45} \text{ and } P_2 = a_{12}a_{23}a_{35}$$

Gain of individual loops are

$$L_1 = a_{23}a_{32},\ L_2 = a_{23}a_{34}a_{42},\ L_3 = a_{23}a_{34}a_{45}a_{52},\ L_4 = a_{23}a_{35}a_{52} \text{ and } L_5 = a_{44}$$

Gain of two non−touching loops are

$$L_{21} = L_1L_5 = = a_{23}a_{32}\ a_{44} \text{ and } L_{22} = L_4L_5 = a_{23}a_{35}a_{52}a_{44}$$

Since all the loops touch the forward path 1, $\Delta_1 = 1$
Since L_5 does not touch the forward path 2, $\Delta_2 = 1 - a_{44}$

$$\therefore \quad \Delta = 1-(L_1 + L_2 + L_3 + L_4 + L_5) + (L_{21} + L_{22})$$

$$\therefore \quad \frac{x_5}{x_1} = \frac{P_1\Delta_1 + P_2\Delta_2}{\Delta} = \frac{a_{12}a_{23}a_{34}a_{45} + a_{12}a_{23}a_{35}(1-a_{44})}{1-(a_{23}a_{32} + a_{23}a_{34}a_{42} + a_{23}a_{34}a_{45}a_{52} + a_{23}a_{35}a_{52} + a_{44}) + a_{23}a_{32}a_{44} + a_{23}a_{35}a_{52}a_{44}}$$

Example 28 Find the transfer function of the SFG shown in Fig.23 using Mason's gain formula.

Solution There are three forward paths:

$$P_1 = G_2G_4G_5G_6$$
$$P_2 = G_1G_4G_5G_6$$
$$P_3 = G_3G_4G_5G_6$$

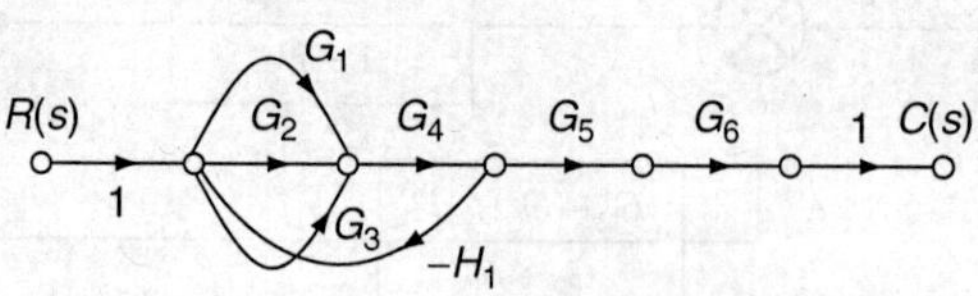

Fig. 23

There are three loops:

$$P_{11} = -G_1G_4H_1$$
$$P_{21} = -G_2G_4H_1$$
$$P_{31} = -G_3G_4H_1$$

Out of these three loops all of them touch each other.

$$\Delta = 1-(P_{11} + P_{21} + P_{31}) = 1 + G_1G_4H_1 + G_2G_4H_1 + G_3G_4H_1$$

For P_1, all loops touch it. Therefore, $\Delta_1 = 1$.
For P_2, all loops touch it. Therefore, $\Delta_2 = 1$.
For P_3, all loops touch it. Therefore, $\Delta_3 = 1$.

$$\therefore \quad T.F. = \frac{P_1\Delta_1 + P_2\Delta_2 + P_3\Delta_3}{\Delta} = \frac{G_1G_4G_5G_6 + G_2G_4G_5G_6 + G_3G_4G_5G_6}{1 + G_1G_4H_1 + G_2G_4H_1 + G_3G_4H_1}$$

Example 29 Find the expression for the outputs C_1 and C_2 for the signal flow graph of the system shown in Fig. 24 below. Also determine the condition that makes C_1 independent of R_2 and C_2 independent of R_1.

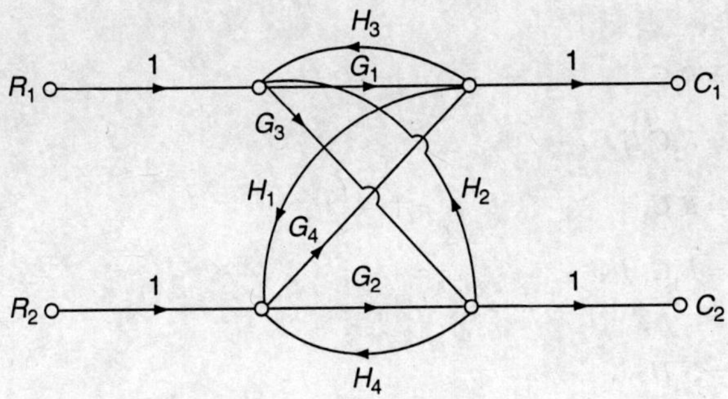

Fig. 24

Solution

Expression for C_1:

$$P_1 = R_1G_1 \qquad \Delta_1 = 1 - G_2H_4$$

$$P_2 = R_2G_4 \qquad \Delta_2 = 1 - G_3H_2$$

$$P_3 = R_1G_3H_4G_4 \qquad \Delta_3 = 1$$

$$P_4 = R_2G_2H_2G_1 \qquad \Delta_4 = 1$$

Individual Loops:

$$L_1 = G_1H_3$$

$$L_2 = G_4H_1$$

$$L_3 = G_3H_2$$

$$L_4 = G_2H_4$$

$$L_5 = G_1H_1G_2H_2$$

$$L_6 = G_4H_3G_3H_4$$

Two non–touching loops:

$$L_{21} = L_1L_4 = G_1H_3G_2H_4$$

$$L_{22} = G_4H_1G_3H_2$$

$$\Delta = 1 - (G_1H_3 + G_4H_1 + G_3H_2 + G_2H_4 + G_1G_2H_1H_2 + G_3G_4H_3H_4) + (G_1G_2H_3H_4 + G_3G_4H_1H_2)$$

$$\therefore \qquad C_1 = \frac{P_1\Delta_1 + P_2\Delta_2 + P_3\Delta_3 + P_4\Delta_4}{\Delta}$$

$$= \frac{R_1G_1(1 - G_2H_4) + R_2G_4(1 - G_3H_2) + R_1G_3G_4H_4 + R_2G_1G_2H_2}{1 - (G_1H_3 + G_4H_1 + G_3H_2 + G_2H_4 + G_1G_2H_1H_2 + G_3G_4H_3H_4) + (G_1G_2H_3H_4 + G_3G_4H_1H_2)} \qquad (1)$$

Expression for C_2:

$$P_1 = R_2G_2 \qquad \Delta_1 = 1 - G_1H_3$$
$$P_2 = R_2G_4H_3G_3 \qquad \Delta_2 = 1$$
$$P_3 = R_1G_3 \qquad \Delta_3 = 1 - G_4H_1$$
$$P_4 = R_1G_1H_1G_2 \qquad \Delta_4 = 1$$

Individual Loops:

$$L_1 = G_1H_3$$
$$L_2 = G_3H_2$$
$$L_3 = G_4H_1$$
$$L_4 = G_2H_4$$
$$L_5 = G_4H_3G_3H_2$$
$$L_6 = G_1H_1G_2H_2$$

Two non-touching loops:

$$L_{21} = L_1L_4 = G_1H_3G_2H_4$$
$$L_{22} = G_3H_2G_4H_1$$

$$\Delta = 1 - (G_1H_3 + G_3H_2 + G_4H_1 + G_2H_4 + G_1G_2H_1H_2 + G_3G_4H_3H_4) + (G_1G_2H_3H_4 + G_3G_4H_1H_2)$$

$$\therefore \quad C_2 = \frac{P_1\Delta_1 + P_2\Delta_2 + P_3\Delta_3 + P_4\Delta_4}{\Delta}$$

$$= \frac{R_2G_2(1 - G_1H_3) + R_2G_3G_4H_3 + R_1G_3(1 - G_4H_1) + R_1G_1G_2H_1}{1 - (G_1H_3 + G_4H_1 + G_3H_2 + G_2H_4 + G_1G_2H_1H_2 + G_3G_4H_3H_4) + (G_1G_2H_3H_4 + G_3G_4H_1H_2)} \quad (2)$$

Condition C_1 independent of R_2:

From the expression (1), we have

$$R_2G_4(1 - G_3H_2) + R_2G_1G_2H_2 = 0$$

i.e.,

$$H_2 = \frac{-G_4}{G_1G_2 - G_3G_4}$$

Condition C_2 independent of R_1:

From the expression (2), we have

$$H_1 = \frac{-G_3}{G_1G_2 - G_3G_4}$$

Example 30 Determine the error coefficients for:

$$G(s)H(s) = \frac{(s + 3)}{s(1 + 0.65)(1 + 0.25)}$$

Solution Here $G(s)H(s)$ is in time constant form.

$$k_p = \underset{s\to 0}{\text{Lt}}\, G(s)H(s) = \underset{s\to 0}{\text{Lt}} \frac{(s+3)}{s(1+0.65)(1+0.25)} = \infty$$

$$K_v = \underset{s\to 0}{\text{Lt}}\, sG(s)H(s) = \underset{s\to 0}{\text{Lt}} \frac{s(s+3)}{s(1+0.65)(1+0.25)} = \frac{3}{(1+0.65)(1+0.25)} = 1.4545$$

$$K_a = \underset{s\to 0}{\text{Lt}}\, s^2G(s)H(s) = \underset{s\to 0}{\text{Lt}} \frac{s^2(s+3)}{s(1+0.65)(1+0.25)} = 0$$

Example 31 A unity feedback system has $G(s) = \dfrac{40(s+2)}{s(s+1)(s+4)}$. Determine (i) Type of the system, (ii) all error coefficients and error for ramp input with magnitude 4.

Solution Here $G(s) = \dfrac{40(s+2)}{s(s+1)(s+4)} = \dfrac{40\times 2\times\left(\dfrac{s}{2}+1\right)}{s(s+1)\left(\dfrac{s}{4}+1\right)\times 4} = \dfrac{20(1+0.5s)}{s(s+1)(1+0.25s)}$

(i) As $j = 1$, it is a type 1 system.

(ii) Error coefficients:

$$K_p = \underset{s\to 0}{\text{Lt}}\, G(s) = \underset{s\to 0}{\text{Lt}} \frac{20(1+0.5s)}{s(s+1)(s+0.25s)} = \infty$$

$$K_v = \underset{s\to 0}{\text{Lt}}\, sG(s) = \underset{s\to 0}{\text{Lt}} \frac{20s(1+0.5s)}{s(s+1)(1+0.25s)} = 20$$

$$K_a = \underset{s\to 0}{\text{Lt}}\, s^2G(s) = \underset{s\to 0}{\text{Lt}} \frac{20s^2(1+0.5s)}{s(s+1)(1+0.25s)} = 0$$

(iii) As ramp input is used K_v is taken into account.

$$e_{ss} = \frac{A}{K_v}$$

Here ramp magnitude $= A = 4$, $K_v = 20$.

$$\therefore \quad e_{ss} = \frac{A}{K_v} = \frac{4}{20} = 0.2$$

Example 32 A second order servo system has poles at $-1 \pm j2$ and a zero at $-1 + j0$. Its steady state output for a unit step input is 2. Determine its transfer function. What is its peak overshoot for a unit step input?

Solution

$$\text{Transfer function} = \frac{K(s+1)}{(s+1-j2)(s+1+j2)} = \frac{K(s+1)}{s^2+2s+5}$$

$\therefore$ Output for unit step $= \dfrac{K(s+1)}{s(s^2+2s+5)}$

$\therefore$ Steady State Output $= \underset{s\to 0}{\text{Lt}}\ \dfrac{sK(s+1)}{s(s^2+2s+5)} = \dfrac{K}{5} = 2$

i.e., $K = 10$

$\therefore$ Transfer function $= \dfrac{10(s+1)}{s^2+2s+5}$

Comparing the denominator with $s^2 + 2\xi\omega_n s + \omega_n^2 = 0$, we have

$$\omega_n = \sqrt{5} \text{ and } 2\xi\,\omega_n = 2 \qquad \therefore \xi = \frac{1}{\sqrt{5}}$$

Maximum overshoot (M_p) $= e^{\frac{-\pi\xi}{\sqrt{1-\xi^2}}} = e^{-\frac{3.14\times\frac{1}{\sqrt{5}}}{\sqrt{1-\frac{1}{5}}}} = 0.2080$

Example 33 A second order system has a transfer function given by $G(s) = \dfrac{25}{s^2+8s+25}$. If the system is initially at rest is subjected to a unit step input at $t = 0$, find the time at which the third peak of response will occur.

Solution

Here $R(s) = \dfrac{1}{s}$

and $C(s) = R(s)G(s) = \dfrac{25}{s(s^2+8s+25)}$

Comparing the denominator with $s^2 + 2\xi\,\omega_n s + \omega_n^2$, we have

$$\omega_n = 5 \text{ rad/sec and } \xi = 8/(2\omega_n) = 0.8$$

The required time is $= \dfrac{3\pi}{\omega_n\sqrt{1-\xi^2}} = \dfrac{3\pi}{5\times\sqrt{1-(0.8)^2}} = \pi$ sec.

Example 34 Figure 25 shows a system. Determine the value of 'a' such that the damping ratio is 0.5. Also obtain the values of rise time and maximum overshoot M_p in its step response.

Solution

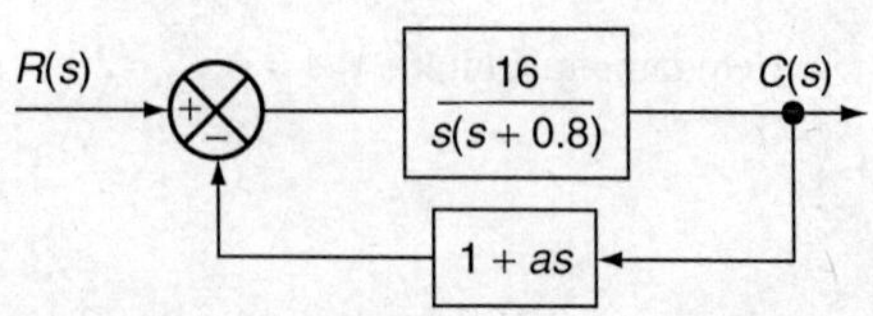

Fig. 25

$$\frac{C(s)}{R(s)} = \frac{\dfrac{16}{s(s+0.8)}}{1+\dfrac{16(1+as)}{s(s+0.8)}} = \frac{16}{s^2+(0.8+16a)s+16}$$

Comparing the denominator with $s^2 + 2\xi\,\omega_n s + \omega_n^2$, we have

$$\omega_n = 4 \text{ rad/sec}$$

and $2\xi\omega_n = 0.8 + 16a$

$\therefore$ $2\times 0.5\times 4 = 0.8 + 16a$

i.e., $a = 0.2$

$$\text{Rise time } (t_r) = \frac{\pi - \tan^{-1}\dfrac{\sqrt{1-\xi^2}}{\xi}}{\omega_n\sqrt{1-\xi^2}} = \frac{\pi - \tan^{-1}\dfrac{\sqrt{1-(0.5)^2}}{0.5}}{\omega_n\sqrt{1-(0.5)^2}} = 0.605 \text{ sec.}$$

$$\text{Maximum overshoot } (M_p) = e^{\frac{-\pi\xi}{\sqrt{1-\xi^2}}} \times 100 = e^{\frac{-\pi\times 0.5}{\sqrt{1-(0.5)^2}}} \times 100 = 16.32\%$$

Example 35 The block diagram of feedback system is shown in Fig. 26(a).

(i) Find closed loop transfer function

(ii) Find the minimum value of G for which the step response of the system would exhibit an overshoot as shown in Fig. 26(b).

(iii) For G equal to twice this minimum value, find the time period 'T' indicated in Figure 26(b).

Assume damping ratio is 0.6.

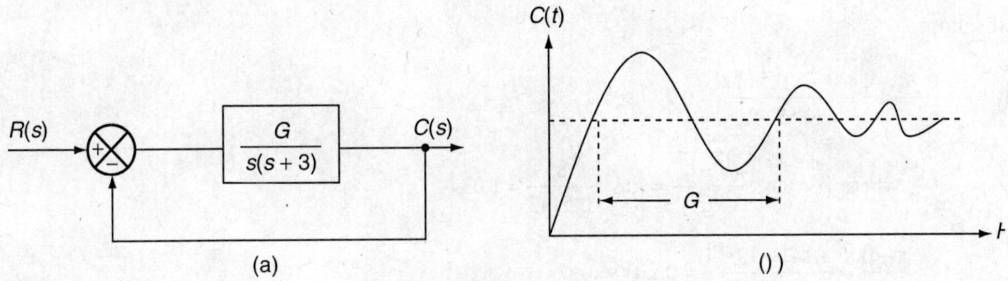

Fig. 26

Solution

(i) The closed loop transfer function of the system is given by

$$\frac{C(s)}{R(s)} = \frac{\dfrac{G}{s(s+3)}}{1+\dfrac{G}{s(s+3)}} = \frac{G}{s^2+3s+G}$$

(ii) Characteristic equation is $s^2 + 3s + G = 0$

Comparing the denominator with $s^2 + 2\xi\,\omega_n s + \omega_n^2$, we have

$$2\xi\omega_n = 3 \text{ and } \omega_n = \sqrt{G}$$

$$\therefore \quad 2\times 0.6\times\sqrt{G} = 3 \quad \text{i.e., } G = 6.25$$

(iii) $G' = 2G = 2\times 6.25 = 12.5$

$$\therefore \quad \omega_n = \sqrt{G'} = \sqrt{12.5} = 3.53 \text{ rad/sec.}$$

$$\xi = \frac{3}{2\omega_n} = \frac{3}{2\times 3.53} = 0.425$$

$$\therefore \quad \therefore\, \omega_d = \omega_n\sqrt{1-\xi^2} = 3.53\times\sqrt{1-0.425^2} = 3.195 \text{ rad/sec.}$$

$$\therefore \quad \frac{2\pi}{T} = 3.195 \qquad \text{i.e., } T = 1.96 \text{ sec.}$$

Example 36 A system has $G(s)=\dfrac{20}{s^2+5s+4}$ with unity feedback. Find the time domain specifications when a unit step input is applied.

Solution

$$\frac{C(s)}{R(s)}=\frac{G(s)}{1+G(s)H(s)}=\frac{\dfrac{20}{s^2+5s+4}}{1+\dfrac{20}{s^2+5s+4}}=\frac{20}{s^2+5s+24}$$

Here $\omega_n^2=24$ i.e., $\omega_n=\sqrt{24}=4.89$ rad/sec.

and $2\xi\omega_n=5$ i.e., $\xi=\dfrac{5}{2\times 4.89}=0.5112$

and $\omega_d=\omega_n\sqrt{1-\xi^2}=4.89\times\sqrt{1-0.5112^2}=4.203$ rad/sec.

Specifications:

(i) $t_d=\dfrac{1+0.7\xi}{\omega_n}=\dfrac{1+0.7\times 0.5112}{4.89}=0.277$ sec

(ii) $t_r=\dfrac{\pi-\beta}{\omega_d}$ where $\beta=\tan^{-1}\dfrac{\sqrt{1-\xi^2}}{\xi}=\tan^{-1}\dfrac{0.859}{0.5112}=1.034$

$\therefore\ t_r=\dfrac{\pi-\beta}{\omega_d}=\dfrac{3.141-1.034}{4.203}=0.501$ sec.

(iii) $\therefore\ t_p=\dfrac{\pi}{\omega_d}=\dfrac{3.14}{4.203}=0.747$ sec

(iv $\therefore\ M_p=100\times e^{-\frac{\xi\pi}{\sqrt{1-\xi}}}=15.44\%$

(v) $t_s=\dfrac{4}{\xi\omega_n}=\dfrac{4}{0.5112\times 4.89}=1.601$ sec

Example 37 The system equation is $\dfrac{d^2y}{dt^2}+4\dfrac{dy}{dt}+8y=7x$ where $y\rightarrow$ output and $x\rightarrow$ input. Find the time domain specification for unit step input.

Solution

Here
$$\frac{d^2y}{dt^2}+4\frac{dy}{dt}+8y=7x \tag{1}$$

The Laplace transform of the above Eq. (1) is given by

$$s^2Y(s)+4sY(s)+8Y(s)=7X(s)$$

i.e.,
$$\frac{Y(s)}{X(s)}=\frac{7}{s^2+4s+8}$$

Here $\omega_n^2=8$ i.e., $\omega_n=\sqrt{8}=2.828$ rad/sec.

and $$\xi = \frac{4}{2\omega_n} = \frac{4}{2\times 2.828} = 0.707 \text{ rad/sec}$$

and $$\omega_d = \omega_n\sqrt{1-\xi^2} = 2.828\times\sqrt{1-0.707^2} = 2 \text{ rad/sec.}$$

Specifications:

(i) $$t_d = \frac{1+0.7\xi}{\omega_n} = \frac{1+0.7\times\left(\frac{1}{\sqrt{2}}\right)}{2.828} = 0.5286 \text{ sec.}$$

(ii) $$t_r = \frac{\pi-\beta}{\omega_d} \quad \text{where} \quad \beta = \tan^{-1}\frac{\sqrt{1-\xi^2}}{\xi} = 0.785$$

$$\therefore \quad t_r = \frac{\pi-\beta}{\omega_d} = \frac{\pi-0.785}{2} = 1.177 \text{ rad/sec}$$

(iii) $$t_p = \frac{\pi}{\omega_d} = 1.57 \text{ sec.}$$

(iv) $$\%M_p = 100\times e^{\frac{-\pi\xi}{\sqrt{1-\xi^2}}} = 4.328\%$$

(v) $$t_s = \frac{4}{\xi\omega_n} = 2 \text{ sec.}$$

and $$c(t) = 1-\frac{e^{-\xi\omega_n t}}{\sqrt{1-\xi^2}}\sin(\omega_d t+\beta) = 1-1.41e^{-2t}\sin(2t+0.785)$$

Example 38 A system has $G(s) = \dfrac{10}{s(s+3)}$ and $H(s) = 1 + \alpha s$. Find the value of damping coefficient, the value of α and the value of steady state error for a step input. The value of peak overshoot is 10%.

Solution

Here $$G(s) = \frac{10}{s(s+3)} \text{ and } H(s) = 1 + \alpha s.$$

$$\therefore \quad \frac{C(s)}{R(s)} = \frac{G(s)}{1+G(s)H(s)} = \frac{\frac{10}{s(s+3)}}{1+\frac{10}{s(s+3)}(1+\alpha s)} = \frac{10}{s^2+s(3+10\alpha)+10} \tag{1}$$

$$\therefore \quad \omega_n^2 = 10 \qquad \text{i.e., } \omega_n = \sqrt{10} \text{ rad/sec.} = 3.162 \text{ rad/sec.}$$

Here $$2\xi\omega_n = (3+10\alpha)$$

$$\therefore \quad \xi = \frac{(3+10\alpha)}{2\omega_n} \tag{2}$$

$$\therefore \quad \%M_p = e^{-\frac{\pi\xi}{\sqrt{1-\xi^2}}}\times 100 = 10$$

i.e.,
$$\ln(0.1) = -\frac{\pi\xi}{\sqrt{1-\xi^2}}$$

i.e.,
$$\xi = 0.5913$$

Substituting the value of ξ in Eq.(2), we have

$\therefore$
$$0.5913 = \frac{(3+10\alpha)}{2\omega_n}$$

i.e.,
$$0.5913 = \frac{(3+10\alpha)}{2\sqrt{10}}$$

i.e.,
$$\alpha = \frac{0.5913 \times 2\sqrt{10} - 3}{10}$$

i.e.,
$$\alpha = 0.074$$

From Eq.(1), we have for step input

$$C(s) = \frac{10R(s)}{s^2 + s(3+10\alpha) + 10}$$

i.e.,
$$C(s) = \frac{\frac{10}{s}}{s^2 + s(3+10\alpha) + 10} = \frac{10}{s[s^2 + s(3+10\alpha) + 10]}$$

$\therefore$
$$E(s) = \frac{R(s)}{1+G(s)H(s)} = \frac{\frac{10}{s(s+3)}}{1+\frac{10(1+\alpha s)}{s(s+3)}} = \frac{10}{s^2 + s(3+10\alpha) + 10}$$

$\therefore$
$$e_{ss} = \underset{s\to 0}{\text{Lt}}\, sE(s) = \underset{s\to 0}{\text{Lt}} \frac{10s}{s^2 + s(3+10\alpha) + 10} = 0$$

Therefore, steady state error is zero for step input.

Example 39 A system has $G(s) = \frac{K}{s(1+sT)}$ with unity feedback where K and T are constants. The overshoot is to be reduced from 75% to 25%. Find the factor by which K should be multiplied. If the system overshoot is to be reduced from 0.6 to 0.2, by change in K only, show that $\frac{TK_1 - 1}{TK_2 - 1} = 43.33$ where K_1 and K_2 are corresponding values of K for 0.6 and 0.2 overshoot.

Solution

Here
$$\frac{C(s)}{R(s)} = \frac{G(s)}{1+G(s)H(s)} = \frac{\frac{K}{s(1+sT)}}{1+\frac{K}{s(1+sT)}} = \frac{K}{s^2T + s + K} = \frac{\frac{K}{T}}{s^2 + \frac{s}{T} + \frac{K}{T}}$$

Here
$$\omega_n^2 = \frac{K}{T} \quad \text{i.e.,} \quad \omega_n = \sqrt{\frac{K}{T}}$$

and
$$\xi = \frac{1}{2\omega_n T} = \frac{1}{2\sqrt{KT}}$$

(i) $M_p = 75\% = 0.75$

Let $$\xi = \xi_1 \text{ for } M_p = 75\% = 0.75$$

$$75 = 100 \times e^{-\frac{\xi_1 \pi}{\sqrt{1-\xi_1^2}}} \text{ i.e., } \xi_1 = 0.0912$$

$$M_p = 25\% = 0.25$$

Let $$\xi = \xi_2 \text{ for } M_p = 25\% = 0.25$$

$$25 = 100 \times e^{-\frac{\xi_2 \pi}{\sqrt{1-\xi_2^2}}} \text{ i.e., } \xi_2 = 0.4037$$

But $$\xi_1 = \frac{1}{2\sqrt{K_1 T}} \text{ and } \xi_2 = \frac{1}{2\sqrt{K_2 T}}.$$

$\therefore$ $$\frac{\xi_2}{\xi_1} = \sqrt{\frac{K_1}{K_2}} \text{ i.e., } 4.431 = \sqrt{\frac{K_1}{K_2}} \text{ i.e., } \frac{K_1}{K_2} = 19.637 \text{ i.e., } K_2 = 0.0509 K_1$$

(ii) For $M_p = 0.6$, $\xi = \xi_1$ and for $M_p = 0.2$, $\xi = \xi_2$.

$$0.6 = e^{-\frac{\xi_1 \pi}{\sqrt{1-\xi_1^2}}} \text{ i.e., } \xi_1 = 0.1604$$

and $$0.2 = e^{-\frac{\xi_2 \pi}{\sqrt{1-\xi_2^2}}} \text{ i.e., } \xi_2 = 0.4553$$

But $$\xi_1 = \frac{1}{2\sqrt{K_1 T}} \text{ and } \xi_2 = \frac{1}{2\sqrt{K_2 T}}.$$

$\therefore$ $$\frac{\xi_1}{\xi_2} = \frac{2\sqrt{K_2 T}}{2\sqrt{K_1 T}} \text{ i.e., } \frac{\xi_1^2}{\xi_2^2} = \frac{K_2 T}{K_1 T}$$

i.e., $$\frac{K_2 T - 1}{K_1 T - 1} = \frac{\xi_1^2 - 1}{\xi_2^2 - 1}$$

i.e., $$\frac{TK_1 - 1}{TK_2 - 1} = \frac{\xi_2^2 - 1}{\xi_1^2 - 1} = 43.33$$

Example 40 Figure 27 shows a feedback system. Find K_1 and K_2 so that peak overshoot of the system is 1.2% and corresponding time is 1 sec.

Find K_1 and K_2 so that peak overshoot of the system is 1.2% and corresponding time is 1 sec.

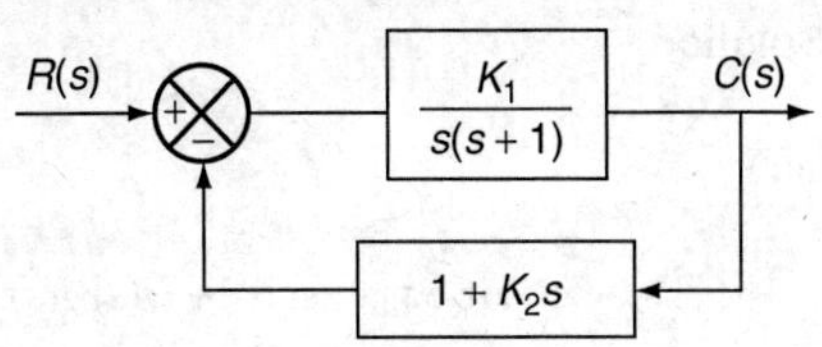

Fig. 27

Solution

Here $$\frac{C(s)}{R(s)} = \frac{G(s)}{1 + G(s)H(s)} = \frac{\frac{K_1}{s(s+1)}}{1 + \frac{K_1(1 + K_2 s)}{s(s+1)}} = \frac{K_1}{s^2 + (K_1 K_2 + 1)s + K_1}$$

Here $\omega_n^2 = K_1$ i.e., $\omega_n = \sqrt{K_1}$ and $\xi = \frac{1+K_1K_2}{2\omega_n} = \frac{1+K_1K_2}{2\sqrt{K_1}}$

$\therefore$ $\omega_d = \omega_n\sqrt{1-\xi^2} = \sqrt{K_1}\sqrt{1-\frac{(1+K_1K_2)^2}{4K_1}}$

$$\%M_p = 100 \times e^{-\frac{\xi\pi}{\sqrt{1-\xi^2}}} = 1.2\%$$

i.e., $\xi = 0.8154$

Also $t_p = \frac{\pi}{\omega_d} = 1$ i.e., $\omega_d = \pi$

But $$\xi = \frac{1+K_1K_2}{2\sqrt{K_1}} = 0.8154 \quad (1)$$

and $$\omega_d = \sqrt{K_1}\sqrt{1-\frac{(1+K_1K_2)^2}{4K_1}} = \pi \quad (2)$$

From Eq. (1), $1+K_1K_2 = 1.63\sqrt{K_1}$ (3)

Using Eq. (3), we have from Eq. (2),

$$\omega_d = \sqrt{K_1}\sqrt{1-\frac{(1.63)^2K_1}{4K_1}} = 0.5794\sqrt{K_1} = \pi$$

i.e., $K_1 = 29.37$

From Eq. (1),

$$\frac{1+K_1K_2}{2\sqrt{K_1}} = 0.8154 \text{ i.e., } K_2 = 0.2668$$

Example 41 The open loop transfer function of a servo system with unity feedback is given by

$$G(s) = \frac{10}{(s+2)\,(s+5)}$$

Determine the damping ratio, undamped natural frequency of oscillation. What is the percentage overshoot of the response to a unit step input?

Solution

Here $G(s) = \frac{10}{(s+2)\,(s+5)}$ and $H(s) = 1$

$\therefore$ The characteristic equation is $1+G(s)H(s) = 0$

i.e., $$1+\frac{10}{(s+2)\,(s+5)} = 0$$

i.e., $$s^2+7s+20 = 0$$

Comparing the characteristic equation with $s^2 + 2\xi\,\omega_n s + \omega_n^2 = 0$, we have

$$\omega_n^2 = 20 \text{ i.e., } \omega_n = 4.472 \text{ rad/sec.}$$

and $2\xi\omega_n = 7$ i.e., $2\times\xi\times 4.472 = 7$ i.e., $\xi = 0.7826$

Maximum overshoot $(M_p) = e^{\frac{-\pi\xi}{\sqrt{1-\xi^2}}}\times 100 = e^{\frac{-\pi\times 0.7826}{\sqrt{1-(0.7826)^2}}}\times 100 = 1.93\%$

Example 42 The open loop transfer function of a unity feedback system is given by

$$G(s) = \frac{K}{s(1+sT)}$$

where 'K' and 'T' are positive constants. By what factor should be the amplifier gain so that the peak overshoot of unit step response of the system reduced form 80% to 30%.

Solution

$$\frac{C(s)}{R(s)} = \frac{G(s)}{1+G(s)H(s)} = \frac{\frac{K}{s(1+sT)}}{1+\frac{K}{s(1+sT)}} = \frac{K}{s^2T+s+K} = \frac{\frac{K}{T}}{s^2+\frac{1}{T}s+\frac{K}{T}}$$

Comparing the denominator with $s^2+2\xi\,\omega_n s+\omega_n^2$, we have

$$2\xi\omega_n = \frac{1}{T} \text{ and } \omega_n = \sqrt{\frac{K}{T}}$$

$$\xi = \frac{1}{2T\omega_n} = \frac{1}{2T\times\sqrt{\frac{K}{T}}} = \frac{1}{2\sqrt{KT}}$$

It is given that

$$M_{p1} = e^{-\frac{\pi\xi_1}{\sqrt{1-\xi_1^2}}} = 0.80 \text{ i.e., } \frac{\pi\xi_1}{\sqrt{1-\xi_1^2}} = 0.223$$

and $$M_{p2} = e^{-\frac{\pi\xi_2}{\sqrt{1-\xi_2^2}}} = 0.30 \text{ i.e., } \frac{\pi\xi_2}{\sqrt{1-\xi_2^2}} = 1.204$$

i.e., $$\xi_1 = 0.0708$$

i.e., $$\xi_2 = 0.358$$

$\therefore$ $$\frac{\xi_1}{\xi_2} = \frac{0.0708}{0.358}$$

i.e., $$\frac{\frac{1}{2\sqrt{K_1T}}}{\frac{1}{2\sqrt{K_2T}}} = \frac{0.0708}{0.358}$$

i.e., $$K_2 = 0.04K_1$$

Example 43 A second order system is represented by a transfer function shown below:

$$\frac{Q_0(s)}{T(s)} = \frac{1}{Js^2 + fs + K}$$

where $Q_0(s)$ is the proportional output and T is the input torque. A step input 10 N-m is applied to the system and test results are shown below.

(a) $M_p = 6\%$, (b) $t_p = 1$ sec., and (c) steady state error of the output is 0.5 N-m. Determine the values of J, K and f.

Solution

Here $$T(s) = \frac{10}{s}$$

Therefore, $$Q_0(s) = \frac{10}{s(Js^2 + fs + K)}$$

Steady state value = $\underset{s\to 0}{\text{Lt}}\ sQ_0(s)$

i.e., $$0.5 = \underset{s\to 0}{\text{Lt}}\ s\left[\frac{10}{s(Js^2 + fs + K)}\right] = \frac{10}{K}$$

i.e., $$K = 20$$

$$M_p = 6\%$$

i.e., $$e^{-\frac{\xi\pi}{\sqrt{1-\xi^2}}} = 0.60 \text{ i.e., } \xi = 0.1605$$

$$t_p = 1 \text{ sec.}$$

i.e., $$\frac{\pi}{\omega_n\sqrt{1-\xi^2}} = 1 \text{ i.e., } \frac{\pi}{\omega_n\sqrt{1-0.1605^2}} = 1 \text{ i.e., } \omega_n = 3.18 \text{ rad/sec.}$$

Characteristic equation is

$$Js^2 + fs + K = 0$$

i.e., $$s^2 + \frac{f}{J}s + \frac{K}{J} = 0$$

Comparing $s^2 + \frac{f}{J}s + \frac{K}{J} = 0$ with $s^2 + 2\xi\,\omega_n s + \omega_n^2 = 0$, we have

$$\omega_n^2 = \frac{K}{J} \text{ i.e., } J = \frac{K}{\omega_n^2} = \frac{20}{3.18^2} = 1.978$$

and $$2\xi\,\omega_n = \frac{f}{J} \text{ i.e., } f = 2\xi\omega_n J = 2 \times 0.1605 \times 3.18 \times 1.978 = 2.02$$

Example 44 A feedback system is described by the following transfer function

$$G(s) = \frac{16}{s^2 + 4s + 16} \quad \text{and} \quad H(s) = Ks$$

The damping factor of the system is 0.8. Determine the overshoot of the system and the value of K.

Solution

$$\frac{C(s)}{R(s)}=\frac{G(s)}{1+G(s)H(s)}=\frac{\dfrac{16}{s^2+4s+16}}{1+\dfrac{16}{s^2+4s+16}\times Ks}=\frac{16}{s^2+(4+16K)s+16}$$

Comparing the denominator with $s^2+2\xi\,\omega_n s+\omega_n^2$, we have

$$\omega_n^2=16 \text{ i.e., } \omega_n=4 \text{ rad/sec.}$$

and $2\xi\omega_n=4+16K$ i.e., $2\times0.8\times4=4+16K$ i.e., $K=0.15$

$$\therefore \quad \frac{C(s)}{R(s)}=\frac{16}{s^2+(4+16\times0.15)s+16}=\frac{16}{s^2+6.4s+16}$$

$$\therefore \quad M_p=e^{-\frac{\pi\xi}{\sqrt{1-\xi^2}}}\times100=e^{-\frac{\pi\times0.8}{\sqrt{1-(0.8)^2}}}\times100=1.5\%$$

Example 45 Figure 28 shows the block diagram model of a position control system. In the absence of feedback ($K_t=0$) determine the damping ratio of the system for amplifier gain $K_A=5$. Also find the steady state error to unit ramp input.

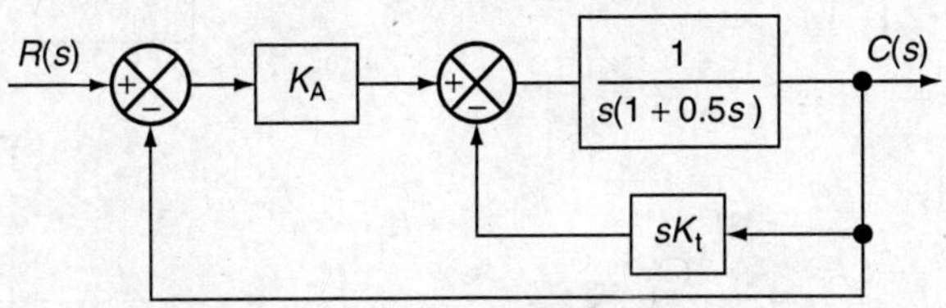

Fig. 28

Find the suitable values of the parameters K_A and K_t so that the damping ratio of the system is increased to 0.7 without affecting the steady state error as obtained in previous part.

Solution

When $$K_t=0,\ G(s)=\frac{10}{s(s+2)}.$$

Characteristic equation is $1+\dfrac{10}{s(s+2)}=0$ i.e., $s^2+2s+10=0$

Comparing the characteristic equation with $s^2+2\xi\,\omega_n s+\omega_n^2=0$, we have

$$\omega_n^2=10 \text{ i.e., } \omega_n=3.16 \text{ rad/sec.}$$

and $$2\xi\omega_n=2 \text{ i.e., } 2\times\xi\times3.16=2 \text{ i.e., } \xi=0.316$$

Steady state error (e_{ss}) $$=\underset{s\to0}{Lt}\frac{sR(s)}{1+G(s)H(s)}=\underset{s\to0}{Lt}\frac{s\times\dfrac{1}{s^2}}{1+\dfrac{10}{s(s+2)}}=\underset{s\to0}{Lt}\frac{s+2}{s^2+2s+10}=0.2$$

The overall transfer function of the system is

$$\frac{C(s)}{R(s)}=\frac{G(s)}{1+G(s)H(S)}=\frac{2K_A}{s^2+(2+2K_t)s+2K_A}$$

Comparing the denominator with $s^2 + 2\xi\,\omega_n s + \omega_n^2$, we have

$$\omega_n^2 = 2K_A \quad \text{i.e., } \omega_n = \sqrt{2K_A}$$

and $$2\xi\,\omega_n = 2 + 2K_t \quad \text{i.e., } 2\times 0.7\times\sqrt{2K_A} = 2 + 2K_t$$

i.e., $$1 + K_t = 0.7\times\sqrt{2K_A} \qquad (1)$$

Steady state error, $$e_{ss} = \underset{s\to 0}{\text{Lt}}\ \frac{sR(s)}{1+G(s)H(s)} = \frac{1+K_t}{K_A} = 0.2$$

i.e., $$1 + K_t = 0.2K_A \qquad (2)$$

Therefore, $$0.7\times\sqrt{2K_A} = 0.2K_A$$

i.e., $$K_A = 24.5$$

From Eq. (2), we have $K_t = 3.9$

Example 46 For a control system shown in Fig. 29 find K_1 and K_2 so that $M_P = 25\%$ and $T_P = 4$ sec.

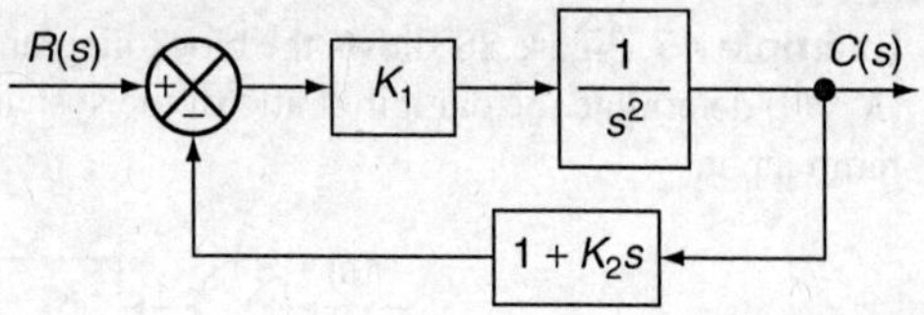

Fig. 29

Solution

Here $G(s) = \dfrac{K_1}{s^2}$ and $H(s) = 1 + K_2 s$

$$\therefore \quad \frac{C(s)}{R(s)} = \frac{G(s)}{1+G(s)H(s)} = \frac{\dfrac{K_1}{s^2}}{1+\dfrac{K_1}{s^2}(1+K_2 s)} = \frac{K_1}{s^2 + K_1K_2 s + K_1}$$

Here $$\omega_n^2 = K_1 \text{ i.e., } \omega_n = \sqrt{K_1} \qquad (1)$$

and $$\xi = \frac{K_1K_2}{2\omega_n} = \frac{K_2\sqrt{K_1}}{2} \qquad (2)$$

Here $$M_P = 100\times e^{-\frac{\xi\pi}{\sqrt{1-\xi^2}}} = 25$$

i.e., $$\xi = 0.4037$$

Also $$t_p = \frac{\pi}{\omega_d} = 4 \text{ i.e., } \omega_d = 0.785$$

Since, $$\omega_d = \omega_n\sqrt{1-\xi^2}$$

i.e., $$\omega_n = \left[\frac{(0.785)^2}{1-(0.4037)^2}\right]^{\frac{1}{2}} = 0.8558 \text{ rad/sec.}$$

Using Eq. (1), we have

$$K_1 = \omega_n^2 = 0.7324$$

Using Eq. (2), we have

$$\xi = \frac{K_2\sqrt{K_1}}{2} \text{ i.e., } K_2 = \frac{2\xi}{\sqrt{K_1}} = \frac{2\times 0.4037}{\sqrt{0.7324}} = 0.9434$$

Example 47 Find K and K_t so that $\xi = 0.6$ and $T_s = 0.1$ for the system shown in Fig. 30 below.

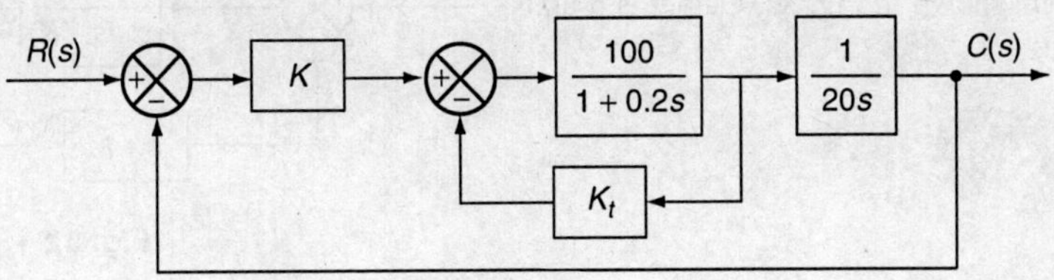

Fig. 30

Solution Using block diagram reduction technique for inner loop the Fig. 30 is reduced shown in Fig. 31.

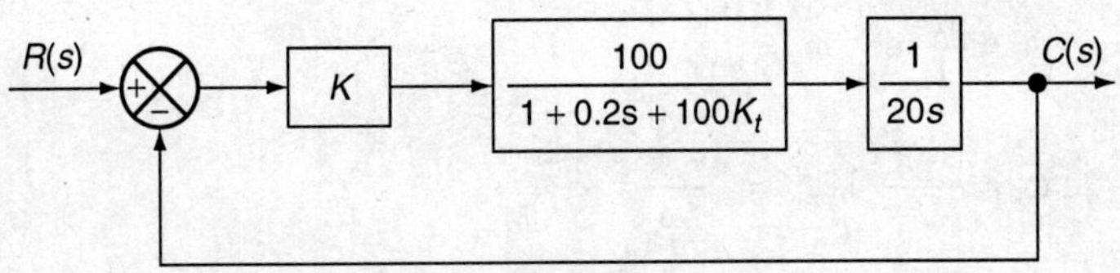

Fig. 31

For inner loop, the transfer function becomes

$$G_1(s) = \frac{\dfrac{100}{1+0.2s}}{1+\left(\dfrac{100}{1+0.2s}\right)K_t} = \frac{100}{1+0.2s+100K_t}$$

Overall $\qquad G(s) = \dfrac{5K}{s(1+0.2s+100K_t)}$

$\therefore \qquad \text{T.F.} = \dfrac{G(s)}{1+G(s)H(s)} = \dfrac{5K}{0.2s^2+s(100K_t+1)+5K} = \dfrac{25K}{s^2+5s(100K_t+1)+25K}$

Here $\qquad \omega_n^2 = 25K$

i.e., $\qquad \omega_n - 5\sqrt{K} \qquad (1)$

and $\qquad \xi = \dfrac{5(1+100K_t)}{10\sqrt{K}} \qquad (2)$

Here $\qquad \xi = 0.6$ and $T_s = 0.1$ sec.

Since $\qquad T_s = \dfrac{4}{\xi\omega_n}$,

i.e., $\qquad 0.1 = \dfrac{4}{0.6\times\omega_n}$ i.e., $\omega_n = 66.67$ rad/sec.

Using Eq. (1), we have

$$K = \frac{\omega_n^2}{25} = 177.8$$

Using Eq. (2), we have

$$\xi = \frac{5(1+100K_t)}{10\sqrt{K}} = 0.6$$

i.e., $\qquad K_t = 0.15$

Example 48 Find values of K, M and J if $M_p = 25\%$ and $T_p = 2$ for the system shown in Fig. 32. Output is 5 units for a unit step input.

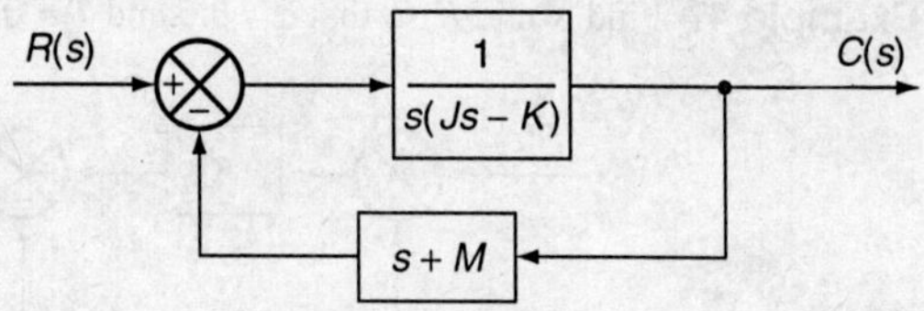

Fig. 32

Solution

Here
$$\frac{C(s)}{R(s)} = \frac{G(s)}{1+G(s)H(s)} = \frac{\frac{1}{s(Js-K)}}{1+\frac{(s+M)}{s(Js-K)}} = \frac{1}{Js^2+s(1-K)+M} = \frac{\frac{1}{J}}{s^2+s\frac{(1-K)}{J}+\frac{M}{J}}$$

Therefore,
$$\omega_n^2 = \frac{M}{J} \text{ i.e., } \omega_n = \sqrt{\frac{M}{J}} \tag{1}$$

and
$$2\xi\,\omega_n = \frac{1-K}{J} \text{ i.e., } \xi = \frac{1-K}{2J\omega_n} = \frac{1-K}{2\sqrt{JM}} \tag{2}$$

(i) $M_p = 25\%$

i.e., $100 \times e^{-\frac{\xi\pi}{\sqrt{1-\xi^2}}} = 25$ i.e., $\xi = 0.4037$

(ii) $t_p = \frac{\pi}{\omega_d} = 2$ i.e., $\omega_d = 1.57$ rad/sec.

We know $\omega_d = \omega_n\sqrt{1-\xi^2}$

i.e., $\omega_n = \frac{\omega_d}{\sqrt{1-\xi^2}} = 1.716$ rad/sec.

(iii) Now $\underset{t\to\infty}{\text{Lt}}\, c(t) =$ final output

By final value theorem,

$$\underset{t\to\infty}{\text{Lt}}\, c(t) = \underset{s\to 0}{\text{Lt}}\, sC(s)$$

Here $R(s) = \frac{1}{s}$

$$\therefore \quad C(s) = \left[\frac{\frac{1}{J}}{s^2+s\frac{(1-K)}{J}+\frac{M}{J}}\right]\times\frac{1}{s}$$

$$\therefore \quad \underset{s\to 0}{\text{Lt}}\, sC(s) = \underset{s\to 0}{\text{Lt}}\, s\times\left[\frac{\frac{1}{J}}{s^2+s\frac{(1-K)}{J}+\frac{M}{J}}\right]\times\frac{1}{s} = 5$$

i.e., $M = 0.2$

From Eq. (1), we have $J = 0.06792$
Again, from Eq. (2), we have $K = 0.8385$

Example 49 For the servomechanism shown in Fig. 33 below, determine the value of K and k so that the maximum overshoot in unit step response is 25% and peak time is 2 sec.

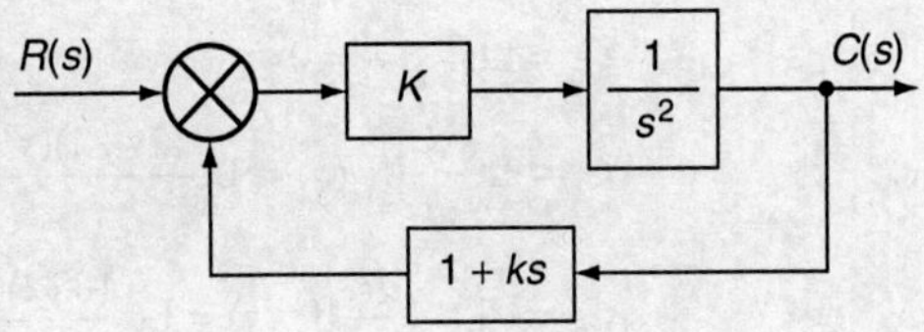

Fig. 33

Solution

Maximum overshoot $(M_p) = e^{-\frac{\xi\pi}{\sqrt{1-\xi^2}}} \times 100 = 25$

i.e., $$e^{-\frac{\xi\pi}{\sqrt{1-\xi^2}}} = 0.25$$

i.e., $$\frac{\pi\xi}{\sqrt{1-\xi^2}} = 1.39 \text{ i.e., } \xi = 0.4$$

The peak time is specified as 2 sec.

$$T_p = \frac{\pi}{\omega_d} = 2 \text{ i.e., } \omega_d = 1.57$$

Again, $$\omega_d = \omega_n\sqrt{1-\xi^2} = 1.57$$

i.e., $$\omega_n = \frac{1.57}{\sqrt{1-\xi^2}} = \frac{1.57}{\sqrt{1-(0.4)^2}} = 1.71$$

$\therefore$ $$\frac{C(s)}{R(s)} = \frac{\frac{K}{s^2}}{1+\frac{K}{s^2}(1+ks)} = \frac{K}{s^2+Kks+K}$$

Comparing the denominator with $s^2 + 2\xi\omega_n s + \omega_n^2 = 0$, we have

$$\omega_n = \sqrt{K} \text{ and } 2\xi\,\omega_n = Kk$$

$\therefore$ $$2\xi\sqrt{K} = Kk$$

i.e., $$k = \frac{2\xi}{\sqrt{K}}$$

$\therefore$ $$K = \omega_n^2 = 2.93 \text{ and } k = 0.47$$

Example 50 Compare the errors of the unity feedback control system with the following feed forward functions

$$G_1(s) = \frac{10}{s(s+1)} \text{ and } G_2(s) = \frac{10}{s(2s+1)}.$$

The input is assumed to be $r = 5 + 2t + 3t^2$.

Solution

For $$G_1(s) = \frac{10}{s(s+1)}$$

$\therefore$ $$W_{1e}(s) = \frac{1}{1+G_1H} = \frac{s^2+s}{s^2+s+10}$$

$$\therefore \quad C_0 = \underset{s\to}{\text{Lt}} W_{1e}(s) = 0$$

$$\therefore \quad C_1 = \underset{s\to 0}{\text{Lt}} \frac{d}{ds} W_{1e}(s) = \underset{s\to 0}{\text{Lt}} \frac{(2s+1)(s^2+s+10)-(s^2+s)(2s+1)}{(s^2+s+10)^2} = \frac{10}{100} = 0.1$$

$$\therefore \quad C_2 = \underset{s\to 0}{\text{Lt}} \frac{d^2}{ds^2} W_{1e}(s) = \underset{s\to 0}{\text{Lt}} \frac{10[(2)(s^2+s+10)^2 - 2(2s+1)(s^2+s+10)(2s+1)]}{(s^2+s+10)^4} = 0.18$$

Here $\quad r = 5 + 2t + 3t^2.$

$$\therefore \quad \frac{dr}{dt} = 2 + 6t \text{ and } \frac{d^2r}{dt^2} = 6.$$

Therefore, the error series is

$$e_{ss1} = 0.1(2+6t) + \frac{0.18}{2}(6) = 0.1(2+6t) + 0.54 = 0.74 + 0.6t$$

For $\quad G_2(s) = \dfrac{10}{s(2s+1)}$

$$\therefore \quad W_{1e}(s) = \frac{1}{1+G_2H} = \frac{2s^2+s}{2s^2+s+10}$$

$$\therefore \quad C_0 = \underset{s\to 0}{\text{Lt}} W_{1e}(s) = 0$$

$$\therefore \quad C_1 = \underset{s\to 0}{\text{Lt}} \frac{d}{ds} W_{1e}(s) = \frac{(4s+1)(2s^2+s+10)-(2s^2+s)(4s+1)}{(2s^2+s+10)^2} = \frac{10}{100} = 0.1$$

$$\therefore \quad C_2 = \underset{s\to 0}{\text{Lt}} \frac{d^2}{ds^2} W_{1e}(s) = \frac{10[(4)(2s^2+s+10)^2 - 2(4s+1)(2s^2+s+10)(4s+1)]}{(2s^2+s+10)^4} = 0.38$$

Here $\quad r = 5 + 2t + 3t^2.$

$$\therefore \quad \frac{dr}{dt} = 2 + 6t \text{ and } \frac{d^2r}{dt^2} = 6.$$

Therefore, the error series is

$$e_{ss1} = 0.1(2+6t) + \frac{0.38}{2}(6) = 0.1(2+6t) + 1.14 = 1.34 + 0.6t$$

Example 51 A closed loop servo system is represented by the differential equation

$$\frac{d^2C}{dt^2} + 8\frac{dC}{dt} = 64e$$

where C is the displacement of the output shaft, r is the displacement input shaft and $e = r - C$. Determine (i) damping ratio, (ii) damped natural frequency and (iii) $\%M_p$.

Solution

$$\frac{d^2C}{dt^2} + 8\frac{dC}{dt} = 64e$$

i.e.,

$$\frac{d^2C}{dt^2} + 8\frac{dC}{dt} = 64(r - C) \quad (1)$$

Taking Laplace transform of Eq.(1), we have

$$s^2C(s) + 8sC(s) = 64\,[R(s) - C(s)]$$

i.e.,

$$(s^2 + 8s + 64)\,C(s) = 64R(s)$$

$$\therefore \quad \frac{C(s)}{R(s)} = \frac{64}{s^2 + 8s + 64}$$

Comparing the denominator with $s^2 + 2\xi\,\omega_n s + \omega_n^2 = 0$, we have

$$\omega_n^2 = 64 \text{ i.e., } \omega_n = 8 \text{ rad/sec.}$$

and

$$2\xi\omega_n = 8 \text{ i.e., } 2 \times \xi \times 8 = 8 \text{ i.e., } \xi = 0.5$$

$$\text{Maximum overshoot } (M_p) = e^{-\frac{\xi\pi}{\sqrt{1-\xi^2}}} \times 100 = e^{-\frac{\pi\times 0.5}{\sqrt{1-(0.5)^2}}} \times 100 = 16.3\%$$

Example 52 The characteristic equation of a system is given by

$s^6 + 3s^5 + 4s^4 + 6s^3 + 5s^2 + 3s + 2 = 0$. Comment on stability.

Solution

s^6	1	4	5	2
s^5	3	6	3	
s^4	2	4	2	
s^3	0	0	0	

Since one row becomes zero, the auxiliary equation becomes

$$A(s) = 2s^4 + 4s^2 + 2 = 0$$

$$\therefore \quad \frac{dA(s)}{ds} = 8s^3 + 8s$$

s^6	1	4	5	2
s^5	3	6	3	
s^4	2	4	2	
s^3	8	8		
s^2	2	2		
s^1	0	0		

Again, one row becomes zero. The auxiliary equation $B(s)$

$$B(s) = 2s^2 + 2 = 0$$

$$\therefore \quad \frac{dB(s)}{ds} = 4s = 0$$

s^6	1	4	5	2
s^5	3	6	3	
s^4	2	4	2	
s^3	4	4		
s^2	2	2		
s^1	4			
s^0	2			

All the elements in the first column are of same sign, therefore, no roots is in right half of s–plane, but there are two rows of zeros, so check the system for marginally stable or unstable.
From the first auxiliary equation,

$$A(s) = 2s^4 + 4s^2 + 2 = 0$$

i.e., $s^4 + 2s^2 + 1 = 0$

i.e., $(s^2 + 1)^2 = 0$

i.e., $s_1 = +j,\ s_2 = -j,\ s_3 = +j,\ s_4 = -j,$

From second auxiliary equation,

$$2s^2 + 2 = 0$$

i.e., $s^2 = -1$

i.e., $s_5 = +j,\ s_6 = -j$

Since on imaginary axis roots are repetitive, hence the system is unstable.

Example 53 Determine number of roots with positive real parts, zero real part and negative real part for the following polynomial equation using Routh criterion.

$$Q(s) = s^4 + 2s^3 + 8s^2 + 10s + 15$$

Solution

s^4	1	8	15
s^3	2	10	
s^2	3	15	
s^1	0		

Auxiliary equation is

$$A(s) = 3s^2 + 15$$

$$\frac{dA(s)}{ds} = 6s$$

s^4	1	8	15
s^3	2	10	
s^2	3	15	
s^1	6		
s^0	15		

There are no changes of sign in the first column, the system may be stable.
Let us solve the auxiliary equation,

$$3s^2 + 15 = 0$$

i.e., $s_1 = j\sqrt{5}$ and $s_2 = -j\sqrt{5}$

After factorization, the given polynomial becomes

$$(s^2 + 5)\ (s^2 + 2s + 3) = 0$$

$$s_1 = j\sqrt{5}\ ,\ s_2 = -j\sqrt{5},\ s_3 = -1 + j1.414 \quad \text{and} \quad s_4 = -1 - j1.414$$

There are no roots with positive real parts. There are two roots with zero real parts. There are two roots with negative real parts.

Example 54 Find the region in which the following closed loop system is stable.

$$G(s)H(s) = \frac{K(s+1)}{s(1+sT)(1+3s)}$$

Solution The characteristic equation is given by

$$1 + G(s)H(s) = 0$$

i.e.,
$$1 + \frac{K(s+1)}{s(1+sT)(1+3s)} = 0$$

i.e.,
$$s(1+sT)(1+3s) + K(s+1) = 0$$

i.e.,
$$3Ts^3 + (3+T)s^2 + (K+1)s + K = 0$$

The Routh's array is shown below:

$$\begin{array}{lll} s^3 & 3T & K+1 \\ s^2 & 3+T & K \\ s^1 & K+1-\dfrac{3KT}{3+T} & \\ s^0 & K & \end{array}$$

For stability,

$$3T > 0 \text{ and } 3 + T > 0$$

This will be satisfied if and only if $T > 0$.
Again, $K > 0$ and

$$K+1-\frac{3KT}{3+T} > 0 \tag{1}$$

i.e.,
$$K\left(1-\frac{3T}{3+T}\right) > -1 \text{ i.e., } K\left(\frac{3-2T}{3+T}\right) > -1 \text{ i.e., } K < \frac{3+T}{2T-3}$$

Also,
$$K+1 > \frac{3KT}{3+T}$$

$$\therefore \quad \frac{K+1}{3K} > \frac{T}{3+T} \text{ i.e., } \frac{3K}{K+1} < \frac{3+T}{T} \text{ i.e., } \frac{2K-1}{K+1} < \frac{3}{T} \text{ i.e., } T > \frac{3(K+1)}{2K-1}$$

Therefore,
$$T > \frac{3}{2} \quad \text{and} \quad K > \frac{1}{2}$$

Example 55 The characteristic equation of a feedback control system is

$$s^4 + 30Ks^3 + 6s^2 + 10s + 20 = 0$$

Find the range of K for which the system is stable.

Solution The Routh's array of the given characteristic equation is shown below:

$$\begin{array}{llll} s^4 & 1 & 6 & 20 \\ s^3 & 30K & 10 & \\ s^2 & 6-\dfrac{10}{30K} & 20 & \\ s^1 & 10-\dfrac{1800K^2}{180\text{K}-10} & & \\ s^0 & 20 & & \end{array}$$

The system will be stable if

$$30K > 0 \text{ i.e., } K > 0,$$

$$6-\frac{10}{30K} > 0 \text{ i.e., } K > \frac{1}{18}$$

Also, $10-\dfrac{1800K^2}{180K-10} > 0$ i.e., $-1800K^2+1800K-100 > 0$. This gives the complex roots. Hence the system is unstable.

Example 56 Find the condition of stability of the following system having characteristic equations

(i) $s^4+15s^3+220s^2+1200s+2000+K=0$

(ii) $s^3+(K+0.8)s^2+5Ks+40=0$

Solution (i) The given characteristic equation is

$$s^4+15s^3+220s^2+1200s+2000+K=0$$

The Routh's array is shown by

$$\begin{array}{llll} s^4 & 1 & 220 & 2000+K \\ s^3 & 15 & 1{,}200 & \\ s^2 & 140 & 2000+K & \\ s^1 & 1200-\dfrac{(2000+K)15}{140} & 0 & \\ s^0 & 2000+K & & \end{array}$$

For stability $2000+K > 0$ i.e., $K > -2000$

Also, $1200-\dfrac{(2000+K)15}{140} > 0$ i.e., $K < \left(\dfrac{1200\times 140}{15}-2000\right)$ i.e., $K < 9200$

The range of K for stability is $-2000 < K < 9200$

If $K = 9200$, the auxiliary equation is given by

$$140s^2 + (2000 + K) = 0$$

i.e., $$140s^2 = -(2000 + 9200)$$

i.e., $$s^2 = -\frac{11200}{140} = -80$$

i.e., $$s = \pm j\,8.94$$

Therefore, the presence of an imaginary pair of roots gives oscillatory response with frequency of oscillations of 8.94 rad/sec.

(ii) The given characteristic equation is

$$s^3 + (K + 0.8)s^2 + 5Ks + 40 = 0$$

The Routh's array is shown by

$$\begin{array}{lll} s^3 & 1 & 5K \\ s^2 & K+0.8 & 40 \\ s^1 & 5K - \dfrac{40}{K+0.8} & 0 \\ s^0 & 40 & \end{array}$$

For stability $0.8 + K > 0$ i.e., $K > -0.8$

Also, $$5K - \frac{40}{K+0.8} > 0$$

i.e., $$5K(K + 0.8) - 40 > 0$$

i.e., $$5K^2 + 4K - 40 > 0$$

i.e., $$K > 2.456 \text{ and } K > -3.254$$

If $K > 2.456$ is satisfied, all condition s will be satisfied.

If $K = 2.456$, the auxiliary equation is given by

$$(K + 0.8)\, s^2 + 40 = 0$$

i.e., $$(K + 0.8)\, s^2 = -40$$

i.e., $$s^2 = -\frac{40}{K+0.8} = -\frac{40}{2.456+0.8} - 12.28$$

i.e., $$s = \pm j\,3.50$$

Therefore, the presence of an imaginary pair of roots gives oscillatory response with frequency of oscillations of 3.50 rad/sec.

Example 57 The open loop transfer function of a unity feedback system is shown by

$$G(s) = \frac{K}{s(s+4)(s^2+s+1)}$$

Determine the values of K that will cause sustained oscillations in the closed loop system. Also find the oscillation frequency.

Solution The characteristic equation is

$$1 + G(s)H(s) = 0$$

i.e.,
$$1+\frac{K}{s(s+4)(s^2+s+1)}=0$$

i.e.,
$$s(s+4)(s^2+s+1)+K=0$$

i.e.,
$$s^4+5s^3+5s^2+4s+K=0$$

The Routh's array is shown below.

$$\begin{array}{llll} s^4 & 1 & 5 & K \\ s^3 & 5 & 4 & 0 \\ s^2 & \frac{21}{5} & K & \\ s^1 & \frac{\frac{84}{5}-5K}{\frac{21}{5}} & 0 & \\ s^0 & K & & \end{array}$$

The condition for stability is $K > 0$ and $\frac{\frac{84}{5}-5K}{\frac{21}{5}} > 0$ i.e., $K < \frac{84}{25}$

Therefore, the range for stability is $0 < K < \frac{84}{25}$.

For $K = \frac{84}{25}$, there will be zero in the fourth row of first column. There will be symmetrical roots which are purely imaginary. Hence $K = \frac{84}{25}$ will cause sustained oscillation.

The auxiliary equation is shown below.

$$\frac{21}{5}s^2+K=0 \text{ i.e., } \frac{21}{5}s^2+\frac{84}{25}=0 \text{ i.e., } s^2=-\frac{4}{5} \text{ i.e., } s=\pm j\frac{2}{\sqrt{5}}$$

The frequency of sustained oscillation is $\pm j0.89$ rad/sec.

Example 58 Determine the values of K and b, so that the system having unity feedback and open loop transfer function $G(s)=\frac{K(s+1)}{s^3+bs^2+4s+1}$ oscillates at a frequency of 3 rad/sec.

Solution The characteristic equation is

$$1 + G(s)H(s) = 0$$

i.e., $$1+\frac{K(s+1)}{s^3+bs^2+4s+1}=0$$

i.e., $$s^3+bs^2+4s+1+K(s+1)=0$$

i.e., $$s^3+bs^2+(K+4)s+1+K=0$$

The Routh's array is shown below.

$$\begin{array}{lll} s^3 & 1 & K+4 \\ s^2 & b & K+1 \\ s^1 & K+4-\frac{K+1}{b} & \\ s^0 & K+1 & \end{array}$$

For sustained oscillation,

$$K+4-\frac{K+1}{b}=0$$

i.e., $$b=\frac{K+1}{K+4} \quad (1)$$

The auxiliary equation is

$$bs^2+(K+1)=0$$

i.e., $$s^2=-\frac{K+1}{b}=(j\omega)^2=(j3)^2=-9$$

∴ $$b=\frac{K+1}{9} \quad (2)$$

From Eq.(1) and Eq.(2), we have

$$\frac{K+1}{K+4}=\frac{K+1}{9} \text{ i.e., } K=5$$

∴ $$b=\frac{5+1}{9}=\frac{2}{3}$$

Example 59 A closed loop control system has the characteristic equation given by

$$s^3+4.5s^2+3.5s+1.5=0$$

Investigate the stability using Routh Hurwitz Criteria

Solution

$$\begin{array}{lll} s^3 & 1 & 3.5 \\ s^2 & 4.5 & 1.5 \\ s^1 & 3.17 & \\ s^0 & 1.5 & \end{array}$$

Since there is not any change of sign in the first column of the Routh Hurwitz array, there is no root in $R\text{–}H$ plane. Hence the system is stable.

Example 60 The open loop transfer function of unity feedback system is $\frac{K}{s(1+0.4s)(1+0.25s)}$. Find the restriction of K so that the closed loop system is absolutely stable.

Solution

$$G(s) = \frac{K}{s(1+0.4s)(1+0.25s)} \text{ and } H(s) = 1$$

The characteristic equation is

$$1 + G(s)H(s) = 0$$

i.e., $$1 + \frac{K}{s(1+0.4s)(1+0.25s)} \times 1 = 0$$

i.e., $$s^3 + 6.5s^2 + 10s + 10K = 0$$

The Routh Hurwitz array is shown below:

$$\begin{array}{lll} s^3 & 1 & 10 \\ s^2 & 6.5 & 10K \\ s^1 & \frac{65-10K}{6.5} & \\ s^0 & 10K & \end{array}$$

For absolute stability,

$$K > 0 \text{ and } 65 - 10\,K > 0 \text{ i.e., } K < 6.5$$

Hence for stability,

$$0 < K < 6.5$$

Example 61 The characteristic equation for certain feedback control system are shown below. In each case, determine the range of values K for the following systems to be stable.

(i) $s^4 + 20Ks^3 + 5s^2 + 10s + 15 = 0$

(ii) $s^3 + 2Ks^2 + (K+2)s + 4 = 0$

Solution

(i) Routh's array of the characteristic equation $s^4 + 20Ks^3 + 5s^2 + 10s + 15 = 0$ is shown by

$$\begin{array}{llll} s^4 & 1 & 5 & 15 \\ s^3 & 20K & 10 & \\ s^2 & \frac{100K-10}{20K} & 15 & \\ s^1 & 10 - \frac{600K^2}{10K-1} & & \\ s^0 & 15 & & \end{array}$$

There must be no root in the R–H plane for stability of the system.

$$20\,K > 0 \text{ i..e., } K > 0$$

$$\frac{100K-10}{20K} > 0 \text{ i.e., } K > \frac{1}{10}$$

and $10 - \dfrac{600K^2}{10K-1} > 0$ i.e., $-600K^2 + 100K - 10 > 0$ This gives the complex roots. Hence the system is not stable.

(ii) Routh's array of the characteristic equation $s^3 + 2Ks^2 + (K+2)s + 4 = 0$ is shown below:

$$\begin{array}{lll} s^3 & 1 & K+2 \\ s^2 & 2K & 4 \\ s^1 & \dfrac{K^2+2K-2}{K} & \\ s^0 & 4 & \end{array}$$

For stability,

$$2K > 0 \text{ i.e., } K > 0$$

and
$$\frac{K^2+2K-2}{K} > 0 \text{ i.e., } K^2 + 2K - 2 > 0 \text{ i.e., } K > 0.73$$

Example 62 The characteristic equation of feedback control system is

$$s^4 + 20s^3 + 15s^2 + 2s + K = 0$$

(i) Determine the range of K for which the system is stable.
(ii) Can the system be marginally stable ? If so, find the required value of K and the frequency of sustained oscillation.

Solution (i) Routh's array of the characteristic equation $s^4 + 20s^3 + 15s^2 + 2s + K = 0$ is shown by

$$\begin{array}{llll} s^4 & 1 & 15 & K \\ s^3 & 20 & 2 & \\ s^2 & 14.9 & K & \\ s^1 & \dfrac{29.8-20K}{14.9} & & \\ s^0 & K & & \end{array}$$

For stability, $K > 0$ and $29.8 - 20K > 0$ i.e., $K < 1.49$

Hence the range is $0 < K < 1.49$

(ii) For marginally stable $K = 1.49$

The auxiliary equation is

$A(s) = 14.9\, s^2 + 1.49 = 0$

i.e., $s = \pm j0.316$ rad/sec.

Therefore, the frequency of sustained oscillation = 0.316 rad/sec.

Example 63 The characteristic equation of a feedback control system is shown by

$$2s^5 + 2s^4 + 5s^3 + 5s^2 + 3s + 5 = 0$$

Use Routh's criteria to determine the stability.

Solution

$$\begin{array}{llll} s^5 & 2 & 5 & 3 \\ s^4 & 2 & 5 & 5 \\ s^3 & \delta & -2 & 0 \\ s^2 & \dfrac{5\delta+4}{\delta} & 5 & 0 \\ s^1 & \dfrac{-2\left(\dfrac{5\delta+4}{\delta}\right)-5\delta}{\dfrac{5\delta+4}{\delta}} & 0 & 0 \\ s^0 & 5 & & \end{array}$$

The fifth term of the first column becomes negative as $\delta \rightarrow 0$. Since there are two changes of sign, two roots lie on the right half of s plane. Hence the system is unstable.

Example 64 The characteristic equation for a feedback control system is shown by

$$s^3 + 20Ks^2 + 5s^2 + 10s + 15 = 0$$

Determine the range of values of K for which the system is stable.

Solution The Routh's array is shown below:

$$\begin{array}{lll} s^3 & 1 & 10 \\ s^2 & 20K+5 & 15 \\ s^1 & \dfrac{200K+50-15}{20K+5} & \\ s^0 & 15 & \end{array}$$

For the system to be stable $\dfrac{200K+50-15}{20K+5} > 0$ i.e., $K > -0.175$

Example 65 Determine the range of K for stability of a unity feedback system whose open loop transfer function is $G(s) = \dfrac{K}{s(s+1)(s+2)}$.

Solution The characteristic equation is $s^3 + 3s^2 + 2s + K = 0$

The Routh's array is shown below:

$$\begin{array}{lll} s^3 & 1 & 2 \\ s^2 & 3 & K \\ s^1 & \dfrac{6-K}{3} & 0 \\ s^0 & K & \end{array}$$

For stability $0 < K < 6$.

Example 66 For what value of k, the polynomial $s^3 + (4 + k)\, s^2 + 6s + 12$ has roots with negative real part?

Solution The Routh's array is shown below:

$$\begin{array}{lll} s^3 & 1 & 6 \\ s^2 & 4+k & 12 \\ s^1 & \dfrac{24+6k-12}{4+k} & 0 \\ s^0 & 12 & \end{array}$$

For the system to be stable

$$\frac{24+6k-12}{4+k} > 0$$

i.e.,
$$12 + 6k > 0$$

i.e.,
$$k > -2$$

Example 67 For what positive value of K does the polynomial $s^4 + 8s^3 + 24s^2 + 32s + K$ have root with zero real part? What are these roots?

Solution The characteristic equation is $s^4 + 8s^3 + 24s^2 + 32s + K = 0$

The Routh's array is

$$\begin{array}{llll} s^4 & 1 & 24 & K \\ s^3 & 8 & 32 & 0 \\ s^2 & 20 & K & \\ s^1 & \dfrac{8K-640}{-20} & 0 & \\ s^0 & K & & \end{array}$$

One condition for the roots to lie on imaginary axis is when $K = 0$,then the auxiliary equation is

$$s^1(32) = 0 \text{ i.e., } s = 0$$

Another condition is when $\dfrac{8K-640}{-20} = 0$ i.e., $K = 80$

The auxiliary equation is $20s^2 + 80 = 0$ i.e., $s = \pm\, 2j$

Hence the polynomial has roots with zero real parts for the following values of K.

$$K = 0 \text{ at } s = 0$$

$$K = 80 \text{ at } s = \pm 2j$$

Example 68 Sketch the root locus plot for the system having $G(s)H(s) = \dfrac{K}{s(s^2 + 4s + 5)}$

Solution

Here $G(s)H(s) = \dfrac{K}{s(s^2 + 4s + 5)} = \dfrac{K}{s(s + 2 + j)\,(s + 2 - j)}$

Step 1: Number of poles (n) = 3 and number for zeros(m) = 0. Therefore, three branches will start from $s = 0$, $s = -2-j$ and $s = -2 + j$ and all three will approach to infinity.

Step 2: The pole zero plot is shown in Fig. 34. One branch is approaching to $-\infty$ from 0 and the real axis loci exists on $-\infty < \sigma < 0$. No breakaway point exists.

Step 3: Since three branches are approaching to ∞, 3 asymptotes are required.

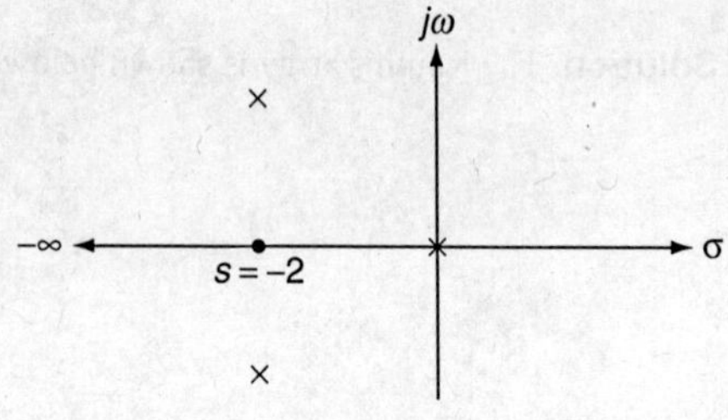

Fig. 34

We know $\theta = \dfrac{(2q+1)180°}{n - m}$, $q = 0, 1, 2$

$$\theta_0 = 60°,\ \theta_1 = 180°,\ \theta_2 = 300°$$

Step 4: Centroid:

$$\text{Centroid } (\sigma_c) = \frac{\text{R.P. of Poles} - \text{R.P. of Zeros}}{n - m} = \frac{0 - 2 - 2 - 0}{3} = -\frac{4}{3} = -1.34$$

Step 5: No breakaway point exists.

Step 6: Intersection with imaginary axis:

The characteristic equation is shown by

$$1 + G(s)H(s) = 0$$

i.e.,

$$1 + \frac{K}{s(s^2 + 4s + 5)} = 0$$

i.e.,

$$s^3 + 4s^2 + 5s + K = 0$$

The Routh's array is shown below:

s^3	1	5
s^2	4	K
s^1	$\dfrac{20 - K}{4}$	0
s^0	K	

Here $K_{mar} = 20$ makes row of $s^1 = 0$.

$\therefore$

$$A(s) = 4s^2 + 20 = 0$$

i.e.,

$$s^2 = -5 \text{ i.e., } s = \pm j\sqrt{5} = \pm j2.236$$

Step 7: Angle of departure:

$$\therefore \quad G(s)H(s)\big|_{s=-2+j} = \frac{K}{(-2+j)\ (-2+j+2+j)} = \frac{K}{2(-2+j)j} = \frac{K}{2\times 2.236\angle 153.43^\circ \angle 90^\circ}$$

$$= \frac{K}{4.47}\angle -243.43^\circ$$

$$\therefore \quad \angle G(s)H(s)\big|_{s=-2+j} = -243.43^\circ$$

At $s=-2+j$, $\phi_d = 180^\circ + (-243.43^\circ) = -63.43^\circ$

Similarly, at $s=-2-j$, $\phi_d = 63.43^\circ$

Step 8: The complete root locus is shown in Fig. 35.

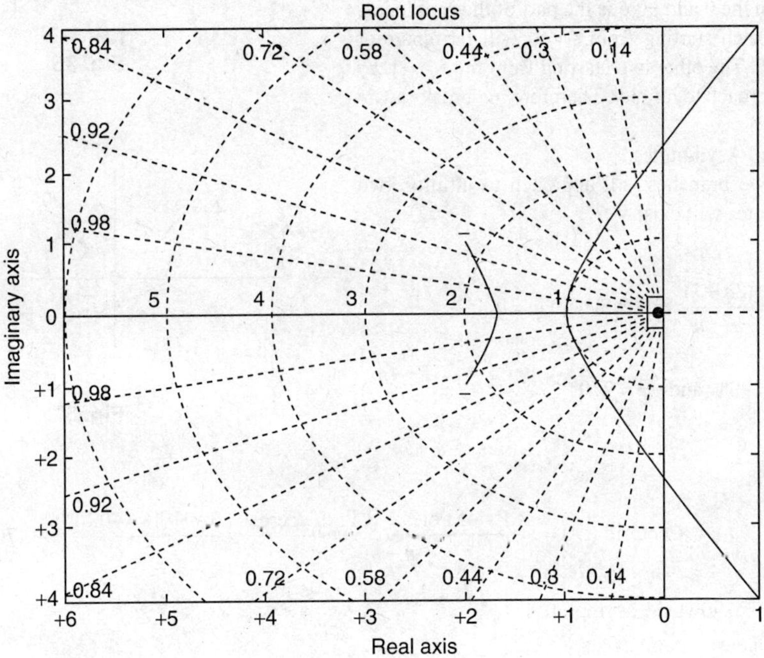

Fig. 35

Step 9: Comment on Stability:

For $0 < K < 20$, all the roots lie on the let half of s–plane. Hence the system is stable.

At $K = 20$, dominant roots are on imaginary axis, system is marginally stable oscillating with 2.236 rad/sec.

For $K > 20$, roots lie on the right half of s–plane and hence the system becomes unstable in nature.

Example 69 Sketch the root locus for the system with $G(s)H(s) = \dfrac{K(s+0.6)}{s(s^2+2s+2)}$. Also find the value of 'K' for damping ratio of 0.6 from the root locus.

Solution

Here $G(s)H(s) = \dfrac{K(s+0.6)}{s(s^2+2s+2)} = \dfrac{K(s+0.6)}{s\ (s+1+j)\ (s+1-j)}$

Step 1: Here the number of poles is $(n) = 3$ and number of zero is $(m) = 1$. The poles are at $s = 0, -1+j$ and $-1-j$ and the zero is at $s = -0.6$. The three branches will start from the poles at the poles and one branch will terminate at the zero $s = -0.6$. The other two will approach to infinity.

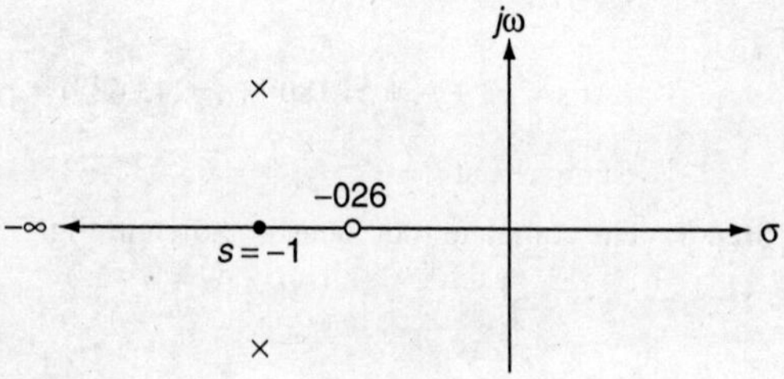

Fig. 36

Step 2: The pole zero plot is shown in Fig. 36. The section between the 0 and −0.6 is the part of the root locus. One branch starting from $s = 0$ will terminate at $s = -0.6$. The other two starting from the $s = -1 \pm j$ will approach to infinity. Therefore, no break−away point exists.

Step 3: Angle of Asymptotes:

Since two branches will approach to infinity, two asymptotes will exist.

$$\theta = \frac{(2q+1)}{n-m} \times 180^\circ \text{ where } q = 0, 1$$

$$\theta = 90^\circ \quad \text{and} \quad \theta = 270^\circ$$

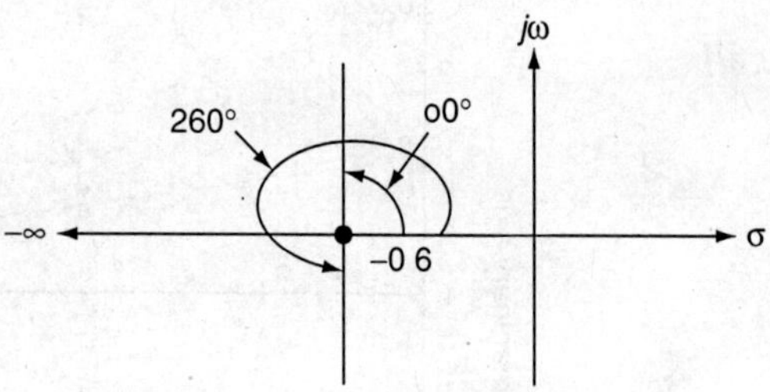

Fig. 37

Step 4: Centroid:

$$\text{Centroid } (\sigma_c) = \frac{\text{R.P. of Poles} - \text{R.P. of Zeros}}{n-m} = \frac{0-1-1-(-0.6)}{2} = -0.70$$

Figure 37 shows the asymptotes.

Step 5: No breakaway point exists.

Step 6: Intersection with the imaginary axis:

The characteristic equation is shown by

$$1 + G(s)\,H(s) = 0$$

i.e.,

$$1 + \frac{K(s+0.6)}{s(s^2+2s+2)} = 0$$

i.e.,

$$s^3 + 2s^2 + (2+K)s + 0.6K = 0$$

The Routh's array is shown below:

$$\begin{array}{lll} s^3 & 1 & 2+K \\ s^2 & 2 & 0.6K \\ s^1 & 2+K-\dfrac{0.6K}{2} & 0 \\ s^0 & 0.6K & \end{array}$$

To have K_{mar},

$$2+K_{mar}-\frac{0.6K_{mar}}{2}=0 \text{ i.e., } K_{mar}=-2.857$$

Since K_{mar} is negative, root locus branches will not intersect the imaginary axis but always located in left half of s–plane.

Step 7: Angle of departure:

$$\therefore \qquad G(s)H(s)\big|_{s=-1+j}=\frac{K(-1+j+0.6)}{(-1+j)(-1+j+1+j)}=\frac{K(-0.4+j)}{2(-1+j)j}=\frac{K}{2\times 2.236\angle 153.43^\circ\angle 90^\circ}$$

$$\therefore \qquad \angle G(s)H(s)\big|_{s=-1+j}=\frac{\angle 111.8^\circ}{\angle 90^\circ\angle 135^\circ}=-113.2^\circ$$

At $s=-1+j$, $\phi_d=180^\circ+(-113.2^\circ)=-66.8^\circ$

Similarly, at $s=-1-j$, $\phi_d=66.8^\circ$

Step 8: The complete root locus is shown in Fig. 38.

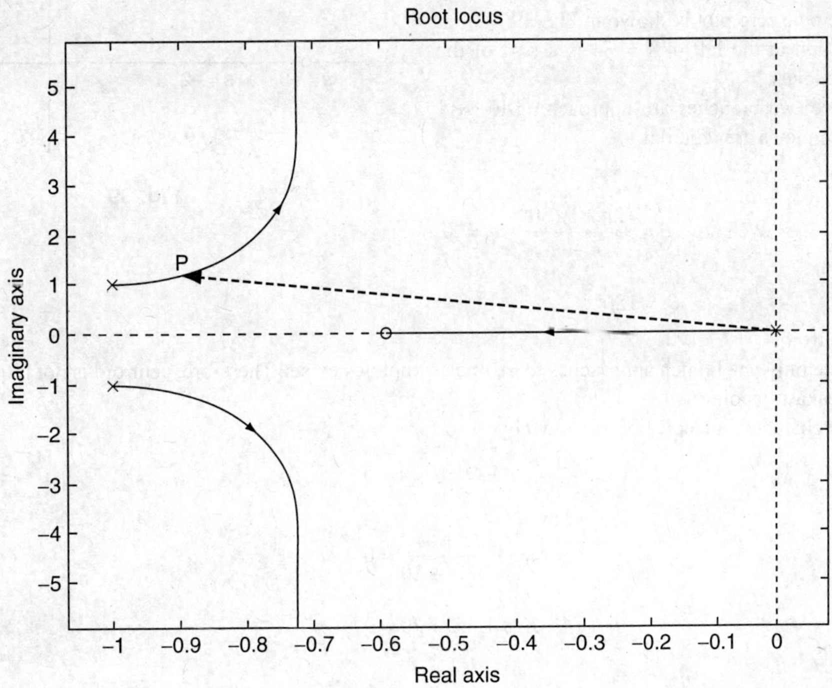

Fig. 38

Step 9: To calculate the value of K for $\xi = 0.6$, a line is drawn from the origin at $\theta = \cos^{-1}(\xi) = 53°$. The point P is shown by $(-0.918, j1.22)$.
Since the point P is on the root locus,

$$|G(s)H(s)|s = -0.918 + j1.22 = \left|\frac{K(-0.918 + j1.22 + 0.6)}{(-0.918 + j1.22)\ \{(-0.918 + j1.22)^2 + 2(-0.918 + j1.22) + 2\}}\right| = 1$$

i.e., $$\frac{1.260K}{1.5268 \times 0.5216} = 1$$

i.e., $$K = 0.632$$

Example 70 Sketch the complete root locus for the system $G(s)H(s) = \dfrac{K(s+4)}{s^2 + 4s + 20}$.

Solution

Here $G(s)H(s) = \dfrac{K(s+4)}{s^2 + 4s + 20} = \dfrac{K(s+4)}{(s+2+j4)\ (s+2-j4)}$

Step 1: Number of poles $(n) = 2$ and number for zeros$(m) = 1$. Therefore, three branches will start from $s = -2-j$ and $s = -2 + j$ and $(n-m)$ i.e., one will approach to infinity.

Step 2: The pole zero plot is shown in Fig. 39. Section to the left of $s = -4$ is a part of the root locus.

Step 3: Since two branches are approaching to ∞, 1 asymptotes are required.

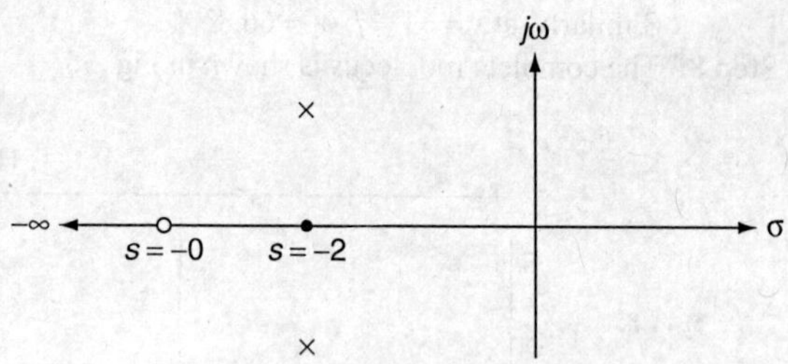

Fig. 39

$$\text{We know } \theta = \frac{(2q+1)180°}{n-m},\ q = 0.$$

$$\theta_1 = 180°$$

Step 4: Centroid:
Since only one branch approaches to ∞, one asymptotes exists. Therefore, centroid is not required.

Step 5: Breakaway point:
The characteristic equation is shown by

$$1 + G(s)H(s) = 0$$

i.e., $$1 + \frac{K(s+4)}{s^2 + 4s + 20} = 0$$

i.e., $$s^2 + (4+K)s + 4K + 20 = 0$$

i.e.,
$$K = -\frac{s^2+4s+20}{s+4}$$

$$\therefore \quad \frac{dK}{ds} = -\frac{(s+4)(2s+4)-(s^2+4s+20)}{(s+4)^2} = 0$$

i.e.,
$$s^2+8s-4=0$$

$$\therefore \quad s = \frac{-8 \pm \sqrt{64+16}}{2} = \frac{-8 \pm 8.94}{2}$$

$$\therefore \quad s = \text{either } -8.94 \text{ or } 0.47$$

For $s = -8.47$, $K = 12.94$ and for $s = -0.47$, $K = -4.94$.
Therefore, $s = -8.47$ is a valid breakaway point.

Step 6: Intersection with imaginary axis:
The characteristic equation is shown by

$$1 + G(s)H(s) = 0$$

i.e.,
$$s^2 + (4+K)s + 4K + 20 = 0$$

The Routh's array is shown below:

$$\begin{array}{lll} s^2 & 1 & 4K+20 \\ s^1 & 4+K & 0 \\ s^0 & 4K+20 & \end{array}$$

Here $K_{mar} = -4$ makes row of $s^1 = 0$. Since K_{mar} has negative value, there is no intersection of root locus with the imaginary axis.

Step 7: Angle of departure:

$$\therefore \quad G(s)H(s)\big|_{s=-2+j4} = \frac{K(-2+j4)}{(-2+j4+2+j4)} = \frac{K(2+j4)}{j8}$$

$$\therefore \quad \angle G(s)H(s)\big|_{s=-2+j} = \frac{\angle 63.43^\circ}{\angle 90^\circ} = -26.57^\circ$$

At
$$s = -2 + j4\ ,\ \phi_d = 180^\circ + (-26.57^\circ) = 153.43^\circ$$

Similarly, at $s = -2 - j$, $\phi_d = -153.43^\circ$

Step 8: The complete root locus is shown in Fig. 40.

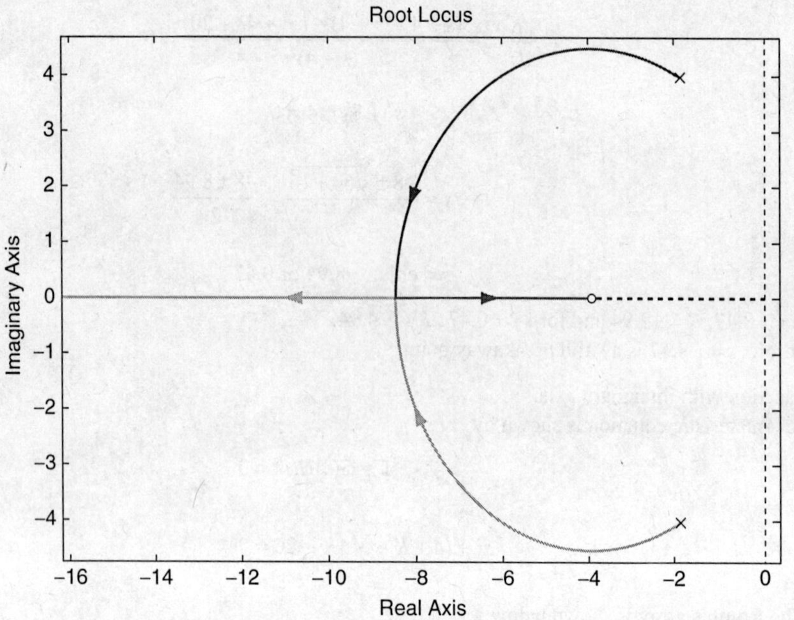

Fig. 40

Step 9: Comment on Stability:
For $0 < K < \infty$, all the roots lie on the left half of s–plane. Hence the system is stable.

Example 71 Sketch the complete root locus for the system $G(s)H(s) = \dfrac{K}{s(s+5)\,(s^2+5s+31.25)}$. Also comment on stability of the system.

Solution

Here $$G(s)H(s) = \frac{K}{s(s+5)\,(s^2+5s+31.25)} = \frac{K}{s(s+5)(s+2.5+j5)\,(s+2.5-j5)}$$

Step 1: Number of poles $(n) = 4$ and number for zeros$(m) = 0$. Therefore, four branches will start from $s = 0$, $s = -5$, $s = -2.5 - j5$ and $s = -2.5 + j5$ and $(n - m)$ i.e., four branches will approach to infinity.

Step 2: The pole zero plot is shown in Fig. 41.
Section $-5 < \sigma < -0$ is a part of the root locus.

Step 3: Since four branches are approaching to ∞, four asymptotes are required.

We know $\theta = \dfrac{(2q+1)180°}{n-m}$, $q = 0, 1, 2, 3$

$\theta_0 = 45°$, $\theta_1 = 135°$, $\theta_2 = 225°$, $\theta_3 = 315°$,

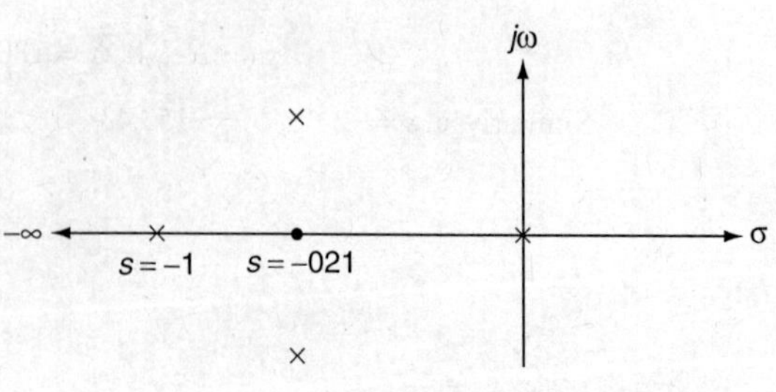

Fig. 41

Step 4: Centroid:

$$\text{Centroid } (\sigma_c) = \frac{\text{R.P. of Poles} - \text{R.P. of Zeros}}{n-m} = \frac{0-5-2.5-2.5-0}{4} = -2.5$$

Step 5: Breakaway point:

The characteristic equation is shown by

$$1 + G(s)\,H(s) = 0$$

i.e., $$1 + \frac{K}{s(s+5)\,(s^2+5s+31.25)} = 0$$

i.e., $$s^4 + 10s^3 + 56.25s^2 + 156.25s + K = 0$$

i.e., $$K = -(s^4 + 10s^3 + 56.25s^2 + 156.25s)$$

$\therefore$ $$\frac{dK}{ds} = -(4s^3 + 30s^2 + 112.5s + 15.25) = 0$$

i.e., $$s^3 + 7.5s^2 + 28.125s + 39.06 = 0$$

Since the poles are located symmetrically, point of symmetry is one of the roots of $\frac{dK}{ds}$. Let us take $s = -2.5$ for the root of the above equation.

−2.5	1	7.5	28.125	39.06
	−2.5	−12.5	−12.5	−39.06
	1	5	15.625	0

$\therefore$ $$s^3 + 7.5s^2 + 28.125s + 39.06 = 0$$

i.e., $$(s+2.5)(s^2+5s+15.625) = 0$$

$\therefore$ $$s = \frac{-5 \pm \sqrt{25 - 4\times 15.625}}{2} = \frac{-5 \pm j6.12}{2} = -2.5 \pm j3.06$$

$\therefore$ $$s = -2.5, \text{ or } -2.5 \pm j3.06$$

For $$s = -2.5,\ K = 156.25.$$

Therefore, $s = -2.5$ is a valid breakaway point.

$$\therefore \angle\{G(s)H(s)\} = \frac{\angle K}{\angle s \angle s+5 \angle s+2.5+j3.06 \angle s+2.5-j3.06}$$

$$(s = -2.5 + j3.06) = \frac{\angle K}{\angle(-2.5+j3.06)\ \angle(2.5+j3.06)\ \angle j6.12 \angle -1.94j}$$

$$= \frac{\angle 0^\circ}{\angle 129.25^\circ \angle 129.25^\circ \angle 90^\circ \angle -90^\circ} = -180^\circ$$

$$= -(2\times 0 + 1)\times 180^\circ$$

It satisfies the angle condition. Hence it is a valid breakaway point.

Step 6: Intersection with imaginary axis:

The characteristic equation is shown by

$$1 + \text{G(s) H(s)} = 0$$

i.e., $$s^4 + 10s^3 + 56.25s^2 + 156.25s + K = 0$$

The Routh's array is shown below:

s^4	1	56.25	K
s^3	10	156.25	0
s^2	40.625	K	
s^1	$156.25 - \frac{10K}{40.625}$	0	
s^0	K		

$$K_{\text{mar}} = \frac{156.25 \times 40.625}{10} = 634.76$$

∴ $$A(s) = 40.625s^2 + 634.76 = 0$$

i.e., $$s = \pm j3.95$$

Step 7: Angle of departure:

$$\therefore \quad G(s)H(s)\big|_{s=-2.5+j5} = \frac{\angle K}{\angle s \angle(s+5)\angle(s+2.5+j5)} = \frac{\angle K}{\angle(-2.5+j5)\angle(2.5+j5)\angle(j10)}$$

$$= \frac{\angle 0^\circ}{\angle 129.25^\circ \angle 50.75^\circ \angle 90^\circ} = -270^\circ$$

At $s = -2.5 + j5$, $\phi_d = 180^\circ + (-270^\circ) = -90^\circ$

Similarly, at $s = -2.5 - j5$, $\phi_d = 90^\circ$

Step 8: The complete root locus is shown in Fig. 42.

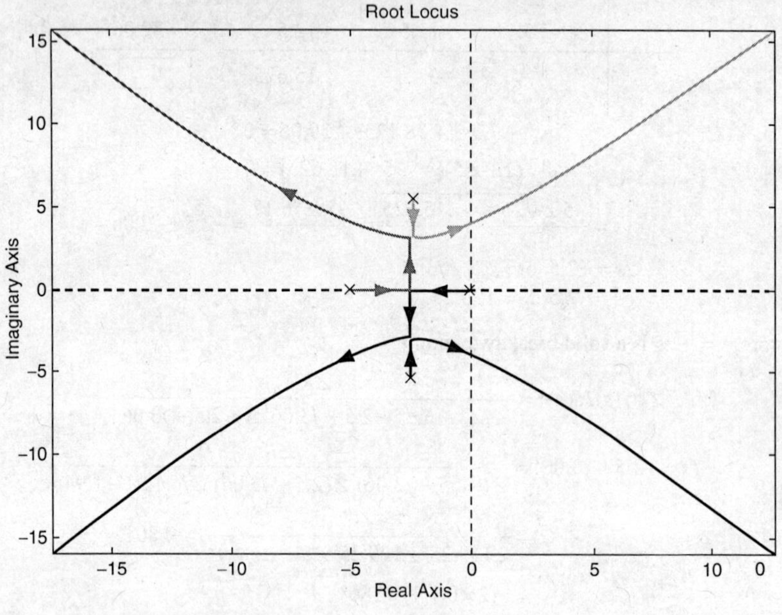

Fig. 42

Step 9: Comment on Stability:

For $0 < K < 634.76$, the system is stable because roots lie on the left hand side of s−plane.

For $K = 634.76$, the system is marginally stable because roots lie on the $j\omega$−axis.

For $K > 634.76$, the system is unstable because roots lie on the right hand side of s−plane.

Example 72 Sketch the complete root locus for the system $G(s)H(s) = \dfrac{K}{s(s+2)(s^2+2s+2)}$. Also comment on stability of the system.

Solution

Here $$G(s)H(s) = \frac{K}{s(s+2)\,(s^2+2s+2)} = \frac{K}{s(s+2)(s+1+j)\,(s+1-j)}$$

Step 1: Number of poles (n) = 4 and number for zeros (m) = 0. Therefore, four branches will start from $s = 0$, $s = -2$, $s = -1-j$ and $s = -1 + j$ and ($n-m$) and these four branches will approach to infinity.

Step 2: The pole zero plot is shown in Fig. 43.
Section $-2 < \sigma < -0$ is a part of the root locus.

Step 3: Since four branches are approaching to ∞, four asymptotes are required.

We know $$\theta = \frac{(2q+1)180°}{n-m},\ q = 0, 1, 2, 3$$

$\theta = 45°, \theta_1 = 135°, \theta_2 = 225°, \theta_3 = 315°$

Step 4: Centroid:

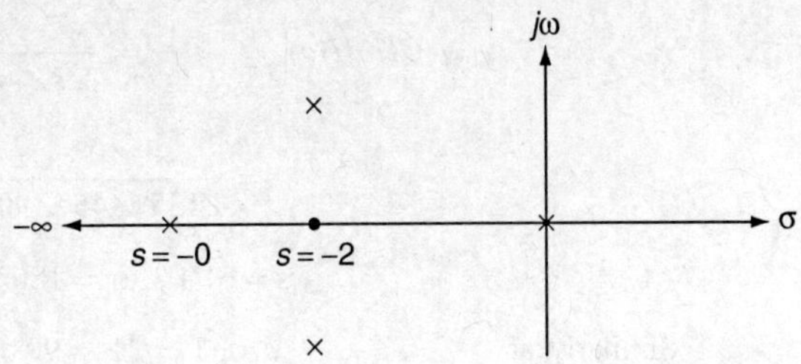

Fig. 43

$$\text{Centroid}\ (\sigma_c) = \frac{\text{R.P. of Poles} - \text{R.P. of Zeros}}{n-m} = \frac{0-2-1-1-0}{4} = -1$$

Step 5: Breakaway point:
The characteristic equation is shown by

$$1 + G(s)H(s) = 0$$

i.e., $$1 + \frac{K}{s(s+2)\,(s^2+2s+2)} = 0$$

i.e., $$s^4 + 4s^3 + 6s^2 + 4s + K = 0$$

i.e., $$K = -(s^4 + 4s^3 + 6s^2 + 4s)$$

∴ $$\frac{dK}{ds} = -(4s^3 + 12s^2 + 12s + 4) = 0$$

i.e., $$s^3 + 3s^2 + 3s + 1 = 0$$

i.e., $$(s+1)^3 = 0$$

i.e., $$s = -1, -1, -1$$

Step 6: Intersection with imaginary axis:
The characteristic equation is shown by

$$1 + G(s)H(s) = 0$$

i.e., $$s^4 + 4s^3 + 6s^2 + 4s + K = 0$$

The Routh's array is shown below:

s^4	1	6	K
s^3	4	4	0
s^2	5	K	
s^1	$\frac{20-4K}{5}$	0	
s^0	K		

$$\frac{20-4K_{\text{mar}}}{5}=0 \text{ i.e., } K_{\text{mar}}=5$$

$\therefore$ $$A(s)=5s^2+K=0 \quad \text{i.e.,} \quad 5s^2+5=0 \qquad \text{i.e., } s=\pm j1.$$

Step 7: Angle of departure:

$\therefore$ $$G(s)H(s)\big|_{s=-1+j}=\frac{\angle K}{\angle s\angle(s+2)\angle(s+1+j)}=\frac{\angle K}{\angle(-1+j)\angle(1+j)\angle(j2)}$$

$$=\frac{\angle 0^\circ}{\angle 135^\circ\angle 45^\circ\angle 90^\circ}=-270^\circ$$

At $s=-1+j,\ \phi_d=180^\circ+(-270^\circ)=-90^\circ$

Similarly, at $s=-1-j,\ \phi_d=90^\circ$

Step 8: The complete root locus is shown in Fig. 44.

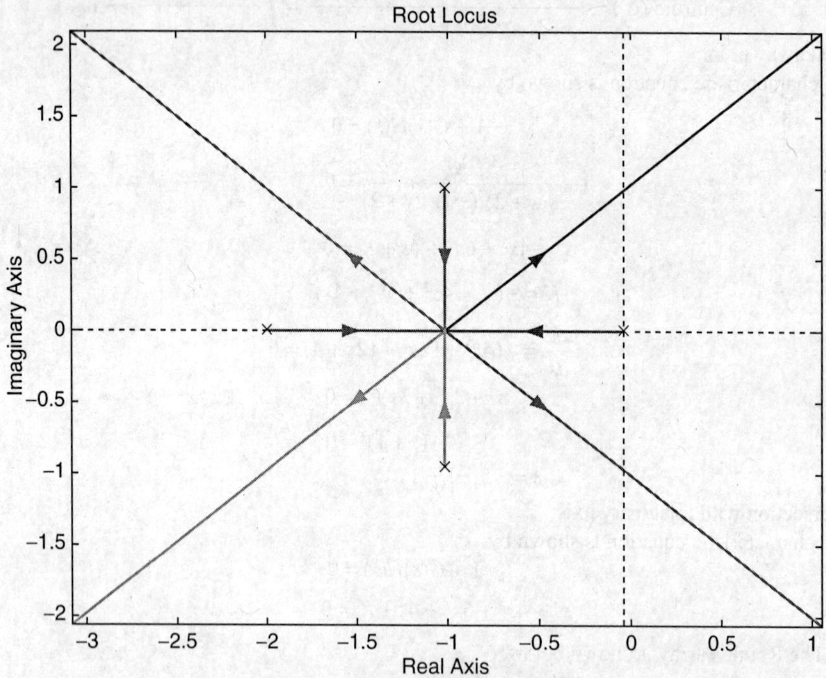

Fig. 44

Step 9: Comment on Stability:

For $0<K<5$, the system is stable because roots lie on the left hand side of s−plane.

For $K=5$, the system is marginally stable because roots lie on the $j\omega$−axis.

For $K>5$, the system is unstable because roots lie on the left hand side of s−plane.

Example 73 Sketch the root locus of a system having $G(s)H(s) = \dfrac{K(s+1)}{s(s-1)(s^2+5s+25)}$

Solution

Here $$G(s)H(s) = \frac{K(s+1)}{s(s-1)(s^2+5s+25)} = \frac{K(s+1)}{s(s-1)(s+2.5+j4.33)(s+2.5-j4.33)}$$

Step 1: Here the number of poles(n) = 4 and number of zeros (m) = 1. Therefore, three branches will approach to infinity.

Step 2: Figure 45 shows the pole zero plot. Two breakaway point exist. One is between 0 and −1 and the other is to the left of $s = -1$.

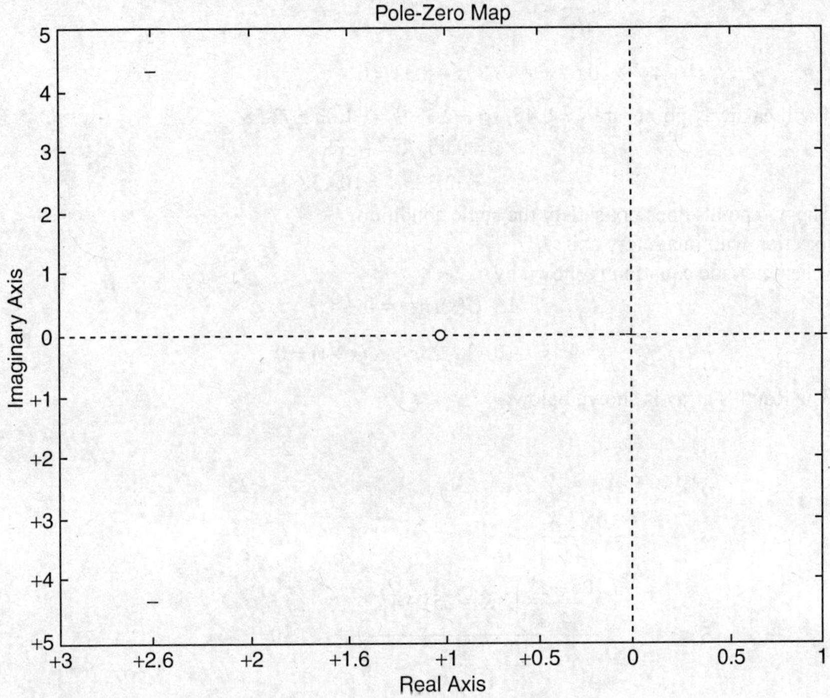

Fig. 45

Step 3: Since three branches are approaching to ∞, three asymptotes are required.

We know $$\theta = \frac{(2q+1)180°}{n-m},\ q = 0, 1, 2$$

$$\theta_0 = 60°,\ \theta_1 = 180°,\ \theta_2 = 300°$$

Step 4: Centroid:

$$\text{Centroid } (\sigma_c) = \frac{\text{R.P. of Poles} - \text{R.P. of Zeros}}{n-m} = \frac{0+1-2.5-2.5+1}{3} = -1$$

Step 5: Breakaway point:

The characteristic equation is shown by

$$1 + G(s)\,H(s) = 0$$

i.e.,
$$1+\frac{K(s+1)}{s(s-1)\,(s^2+5s+25)}=0$$

i.e.,
$$s^4+4s^3+20s^2-25s+K(s+1)=0$$

i.e.,
$$K=-\frac{(s^4+4s^3+20s^2-25s)}{s+1}$$

∴
$$\frac{dK}{ds}=0$$

i.e.,
$$s^4+4s^3+10.67s^2+13.33s-8.33=0$$

The breakaway points are $s = 0.45$, $s = -2.38$, $s = -1.03 \pm j2.58$

For $s = 0.45, K = +4.68$

For $s = -2.38, K = 109.38$

The two points does not satisfy the angle condition.

Step 6: Intersection with imaginary axis:

The characteristic equation is shown by

$$1 + G(s)H(s) = 0$$

i.e.,
$$s^4+4s^3+20s^2-25s+K(s+1)=0$$

The Routh's array is shown below:

$$\begin{array}{llll}
s^4 & 1 & 20 & K \\
s^3 & 4 & K-25 & 0 \\
s^2 & \dfrac{105-K}{4} & K & \\
s^1 & \dfrac{\left(\dfrac{105-K}{4}\right)(K-25)-4K}{\dfrac{105-K}{4}} & s0 & \\
s^0 & K & &
\end{array}$$

For marginally stability, $\frac{105-K}{4} > 0$ i.e., $K < 105$. Therefore, K_{mar} should be less than 105.

For finding K_{mar}, let us put

$$\frac{\left(\dfrac{105-K_{mar}}{4}\right)(K_{mar}-25)-4K_{mar}}{\dfrac{105-K_{mar}}{4}}=0 \text{ i.e., } K_{mar}=32.02 \text{ or } 81.97$$

For $K_{mar} = 32.02$, $s = \pm j1.324$ while for $K_{mar} = 81.97$, $s = \pm j3.77$.

The root locus branches will intersect imaginary axis twice.

Step 7: Angle of departure:

$$\therefore \quad G(s)H(s)\big|_{s=-2.5+j4.33} = -229.8°$$

At $s = -2.5 + j4.33$, $\phi_d = 180° + (-229.8°) = -49.8°$

Similarly, at $s = -2.5 - j4.33$, $\phi_d = +49.8°$

Step 8: The complete root locus is shown in Fig. 46.

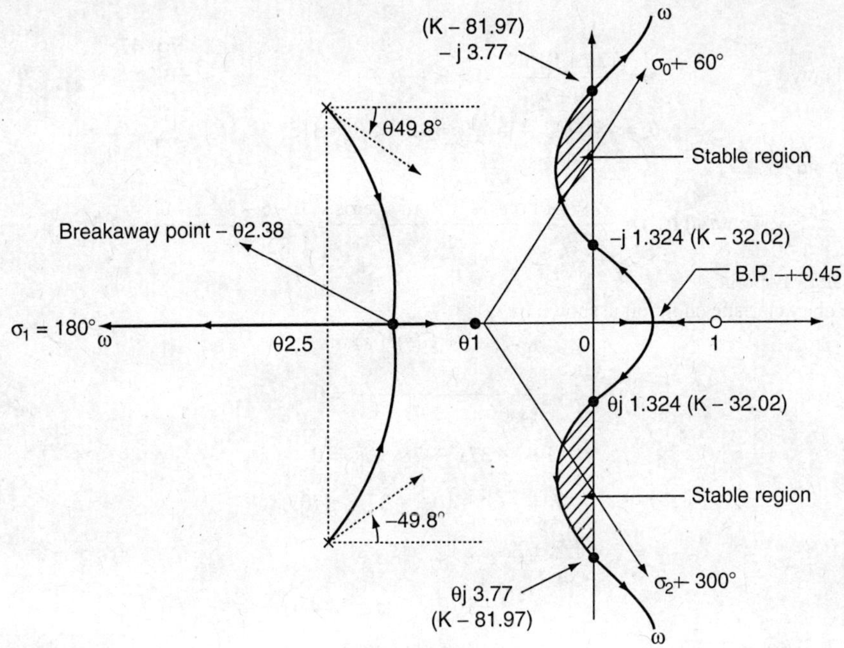

Fig. 46

Step 9: Comment on Stability:
For $0 < K < 32.02$, roots are in the right half of s–plane and the system is unstable.
For $32.02 < K < 81.97$, roots are in the left half of s–plane and the system is stable.
For $K > 81.97$, the system once again becomes unstable.

Example 74 Sketch the root locus of a system having $G(s)H(s) = \dfrac{K}{s(s+6)(s^2+4s+13)}$

Solution

Here $$G(s)H(s) = \frac{K}{s(s+6)(s^2+4s+13)} = \frac{K}{s(s+6)(s+2+j3)(s+2-j3)}$$

Step 1: Here the number of poles(n) = 4 and number of zeros (m) = 0. Therefore, four branches will approach to infinity.

Step 2: Figure 47 shows the pole zero plot. Two breakaway point exist. One is between 0 and −2 and the other is between $s = -2$ and $s = -6$.

Step 3: Since four branches are approaching to ∞, three asymptotes are required.

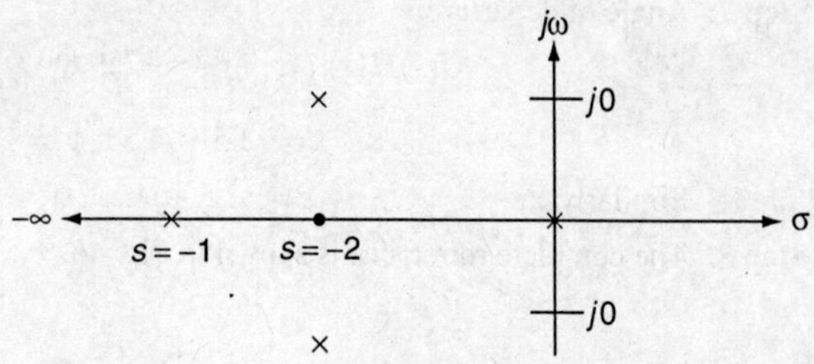

Fig. 47

We know
$$\theta = \frac{(2q+1)180°}{n-m}, q = 0, 1, 2, 3$$
$$\theta_0 = 45°, \theta_1 = 135°, \theta_2 = 225°, \theta_3 = 315°$$

Step 4: Centroid:
$$\text{Centroid } (\sigma_c) = \frac{\text{R.P. of Poles} - \text{R.P. of Zeros}}{n-m} = \frac{0-6-2-2-0}{4} = -2.5$$

Step 5: Breakaway point:
The characteristic equation is shown by
$$1 + G(s)H(s) = 0$$
i.e.,
$$1 + \frac{K}{s(s+6)(s^2+4s+13)} = 0$$
i.e.,
$$s^4 + 10s^3 + 37s^2 + 78s + K = 0$$
i.e.,
$$K = -(s^4 + 10s^3 + 37s^2 + 78s)$$
∴
$$\frac{dK}{ds} = 0$$
i.e.,
$$4s^3 + 30s^2 + 74s + 78 = 0$$
The breakaway points are $s = -4.2$ by trial and error method.

Step 6: Intersection with imaginary axis:
The characteristic equation is shown by
$$1 + G(s)H(s) = 0$$
i.e.,
$$s^4 + 10s^3 + 37s^2 + 78s + K = 0$$
The Routh's array is shown below:

s	1	37	K
s^3	10	78	
s^2	29.2	K	
s^1	$78 - 0.342K$	0	
s^0	K		

For sustained oscillations, $78 - 0.34K = 0$ i.e., $K = 228.07$.
Therefore, auxiliary equation $A(s) = 29.2s^2 + 228.07 = 0$
i.e., $s = \pm 2.8$
Therefore, K_{mar} should be less than 228.07.
The root locus branches will intersect imaginary axis twice.

Step 7: Angle of departure:

$$\therefore \quad G(s)H(s)\big|_{s=-2+j3} = -250.57°$$

At $\quad s = -2.5 + j4.33,\ \phi_d = 180° + (-250.57°) = -70.57°$

Similarly, at $\quad s = -2.5 - j4.33,\ \phi_d = 70.57°$

Step 8: The complete root locus is shown in Fig. 48.

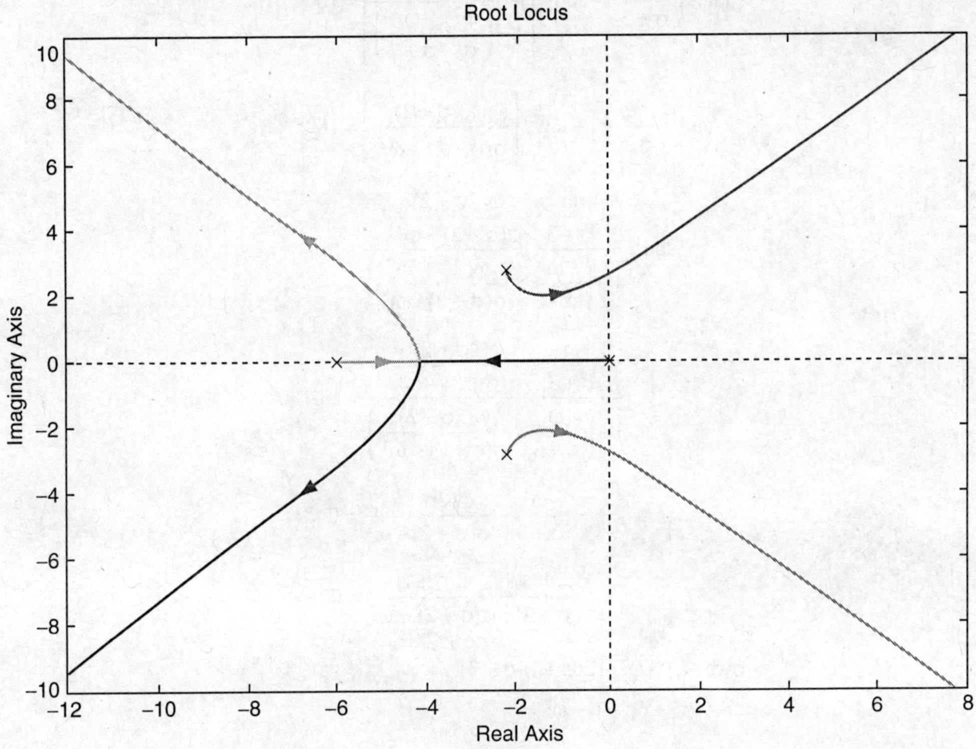

Fig. 48

Example 75 Show that the part of the root locus of a system with $G(s)H(s) = \dfrac{K(s+3)}{s(s+2)}$ is a circle having centre (−3, 0) and radius at $\sqrt{3}$.

Solution

Here
$$G(s)H(s) = \frac{K(s+3)}{s(s+2)}$$

The angle condition must be satisfied for all the points on the root locus. Therefore, $\angle G(s)H(s)$ at any point $s = \sigma + j\omega$ will be 180°.

$$\therefore \quad \angle G(s)H(s)\big|_{s=\sigma+j\omega} = 180°$$

i.e.,
$$\angle\left\{\frac{K(\sigma+j\omega+3)}{(\sigma+j\,\omega)\,(\sigma+j\omega+2)}\right\}=180^\circ$$

i.e.,
$$\tan^{-1}\left(\frac{\omega}{\sigma+3}\right)-\tan^{-1}\left(\frac{\omega}{\sigma}\right)-\tan^{-1}\left(\frac{\omega}{\sigma+2}\right)=180^\circ$$

i.e.,
$$\tan^{-1}\left(\frac{\omega}{\sigma+3}\right)-\tan^{-1}\left[\frac{\frac{\omega}{\sigma}+\frac{\omega}{\sigma+2}}{1-\left(\frac{\omega}{\sigma}\times\frac{\omega}{\sigma+2}\right)}\right]=180^\circ$$

i.e.,
$$\tan^{-1}\left(\frac{\omega}{\sigma+3}\right)-\tan^{-1}\left[\frac{\omega(2\sigma+2)}{\sigma(\sigma+2)-\omega^2}\right]=180^\circ$$

i.e.,
$$\tan^{-1}\frac{\frac{\omega}{\sigma+3}-\frac{\omega(2\sigma+2)}{\sigma(\sigma+2)-\omega^2}}{1+\left(\frac{\omega}{\sigma+3}\times\frac{\omega(2\sigma+2)}{\sigma(\sigma+2)-\omega^2}\right)}=180^\circ$$

i.e.,
$$\frac{\frac{\omega}{\sigma+3}-\frac{\omega(2\sigma+2)}{\sigma(\sigma+2)-\omega^2}}{1+\left(\frac{\omega}{\sigma+3}\times\frac{\omega(2\sigma+2)}{\sigma(\sigma+2)-\omega^2}\right)}=\tan 180^\circ$$

i.e.,
$$\frac{\omega}{\sigma+3}-\frac{\omega(2\sigma+2)}{\sigma(\sigma+2)-\omega^2}=0$$

$$\frac{1}{\sigma+3}-\frac{2(\sigma+1)}{\sigma(\sigma+2)-\omega^2}=0$$

i.e.,
$$\sigma(\sigma+2)-\omega^2-2(\sigma+1)(\sigma+3)$$

i.e.,
$$\sigma^2+6\sigma+6+\omega^2=0$$

i.e.,
$$(\sigma+3)^2+(\omega-0)^2=(\sqrt{3})^2$$

i.e.,
$$\{\sigma-(-3)\}^2+(\omega-0)^2=(\sqrt{3})^2 \qquad (1)$$

Eq. (1) is the equation of a circle having centre (−3, 0) with radius $\sqrt{3}$.

Example 76 Show that the part of the root locus of a system with $G(s)H(s)=\dfrac{K(s+1)}{s(s-1)}$ is a circle having centre (−1, 0) and radius at $\sqrt{2}$.

Solution

Here
$$G(s)H(s)=\frac{K(s+1)}{s(s-1)}$$

The angle condition must be satisfied for all the points on the root locus. Therefore, $\angle G(s)H(s)$ at any point $s=\sigma+j\omega$ will be 180°.

∴
$$\angle G(s)H(s)\big|_{s=\sigma+j\omega}=180^\circ$$

i.e.,
$$\angle\left\{\frac{K(\sigma+j\omega+1)}{(\sigma+j\,\omega)\,(\sigma+j\omega-1)}\right\}=180^\circ$$

i.e.,
$$\tan^{-1}\left(\frac{\omega}{\sigma+1}\right)-\tan^{-1}\left(\frac{\omega}{\sigma}\right)-\tan^{-1}\left(\frac{\omega}{\sigma-1}\right)=180^\circ$$

i.e.,
$$\tan^{-1}\left(\frac{\omega}{\sigma+1}\right)-\tan^{-1}\left[\frac{\frac{\omega}{\sigma}+\frac{\omega}{\sigma-1}}{1-\left(\frac{\omega}{\sigma}\times\frac{\omega}{\sigma-1}\right)}\right]=180^\circ$$

i.e.,
$$\tan^{-1}\left(\frac{\omega}{\sigma+1}\right)-\tan^{-1}\left[\frac{\omega(2\sigma-1)}{\sigma(\sigma-1)-\omega^2}\right]=180^\circ$$

$$\tan^{-1}\left(\frac{\frac{\omega}{\sigma+1}-\frac{\omega(2\sigma-1)}{\sigma(\sigma-1)-\omega^2}}{1+\frac{\omega}{\sigma+1}\times\frac{\omega(2\sigma-1)}{\sigma(\sigma-1)-\omega^2}}\right)=180^\circ$$

i.e.,
$$\frac{\frac{\omega}{\sigma+1}-\frac{\omega(2\sigma-1)}{\sigma(\sigma-1)-\omega^2}}{1+\frac{\omega}{\sigma+1}\times\frac{\omega(2\sigma-1)}{\sigma(\sigma-1)-\omega^2}}=0$$

i.e.,
$$\sigma(\sigma-1)-\omega^2-(\sigma+1)\,(2\sigma-1)=0$$

i.e.,
$$\sigma^2-\sigma-\omega^2-2\sigma^2-\sigma+1=0$$

i.e.,
$$-\sigma^2-2\sigma+1-\omega^2=0$$

i.e.,
$$\sigma^2+2\sigma+1+\omega^2=2$$

i.e.,
$$(\sigma+1)^2+(\omega-0)^2=(\sqrt{2})^2 \qquad (1)$$

Eq. (1) is the equation of a circle having centre (−1, 0) with radius $\sqrt{2}$.

Example 77 Consider the open loop transfer function of a unity feedback system

$$G(s)=\frac{K(s+3)}{s(s^2+2s+2)(s+5)(s+6)}$$

In the root locus diagram for this system determine the following:

(i) The number of branches of this root loci.

(ii) At what location these branches start.

(iii) Where these branches terminate.

(iv) Is $s=-10+j0$, a point on the root locus?

(v) At what points is the imaginary axis crossed by the root loci and what is the corresponding value of gain K?

Solution The given open loop transfer function is

$$G(s) = \frac{K(s+3)}{s(s^2+2s+2)(s+5)(s+6)} = \frac{K(s+3)}{s(s+1+j)(s+1-j)(s+5)(s+6)}$$

Let us determine the important points in the root locus:

(i) The zero is located at $s = -3$.
The poles are located a $s = 0, s = -5, s = -6, s = -1 \pm j1$.
The number of branches of root loci is 5.

(ii) The branches will start from the above poles.

(iii) One branch will terminate at zero and the remaining four will approach to ∞.
Since the four branches of the root locus will approach to infinity, therefore the number of asymptotes is 4 which is equal to $n - m = 5 - 1 = 4$ where n and m are the number of poles and zeros. .
The angles of asymptotes are given by

$$\theta = \frac{(2q+1)180°}{n-m} \quad \text{where} \quad q = 0, 1, 2, 3$$

$$\theta_0 = 45°,\ \theta_1 = 135°,\ \theta_2 = 225°,\ \theta_3 = 315°$$

(iv) Here $0 < \sigma < -3$ and $-6 < \sigma < -5$ are the real part of the root locus. The point $-10 + j0$ is not a point on the root locus because the total number of poles and zeros on the right of it is six.

$$\text{Centroid}(\sigma_c) = \frac{\text{R.P. of Poles} - \text{R.P. of Zeros}}{n-m} = \frac{0-5-6-1-1-(-3)}{4} = -2.5$$

(d) The characteristic equation is given by

$$1 + G(s)H(s) = 1 + \frac{K(s+3)}{s(s^2+2s+2)(s+5)(s+6)} = 0$$

i.e.,

$$s^5 + 13s^4 + 54s^3 + 82s^2 + (60+K)s + 3K = 0$$

The Routh's array is shown below:

s^5	1	54	$60+K$
s^4	13	82	$3K$
s^3	47.7	$\frac{10K+780}{13}$	0
s^2	$\frac{3131.4-10K}{47.7}$	$3K$	0
s^1	d_1	0	
s^0	$3K$		

where

$$d_1 = \frac{47.7\times(3K) - \dfrac{(10K+780)(3131.4-10K)}{47.7\times 13}}{\dfrac{10K-3131.4}{47.7}}$$

$$= \frac{\{143.1K - (0.77K+60)(65.65-0.21K)\}\times 47.7}{10K-3131.4}$$

$$= \frac{(0.1617K^2 + 105.15K - 3939)\times 47.7}{10K-3131.4}$$

One value of root locus with imaginary axis is when the s^0 term $3K = 0$ i.e., $K = 0$.
The auxiliary equation is the s^1 term $d_1 = 0$ or $s = 0$. This is one point.
Also when $d_1 = 0$, we have

$$\frac{(0.1617K^2 + 105.15K - 3939) \times 47.7}{10K - 3131.4} = 0$$

i.e.,
$$0.1617K^2 + 105.15K - 3939 = 0$$

i.e.,
$$K = -685.8, 35.25$$

Since K should be > 0, $K = 35.25$
The auxiliary equation is

$$\frac{3131.4 - 10K}{47.7} s^2 + 3K = 0$$

i.e.,
$$s = \pm j1.35$$

Hence the intersection of the root locus with the imaginary axis is at $s = 0$, when $K = 0$ and

$$s = \pm j1.35 \text{ when } K = 35.25.$$

Figure 49 shows the root locus of the system.

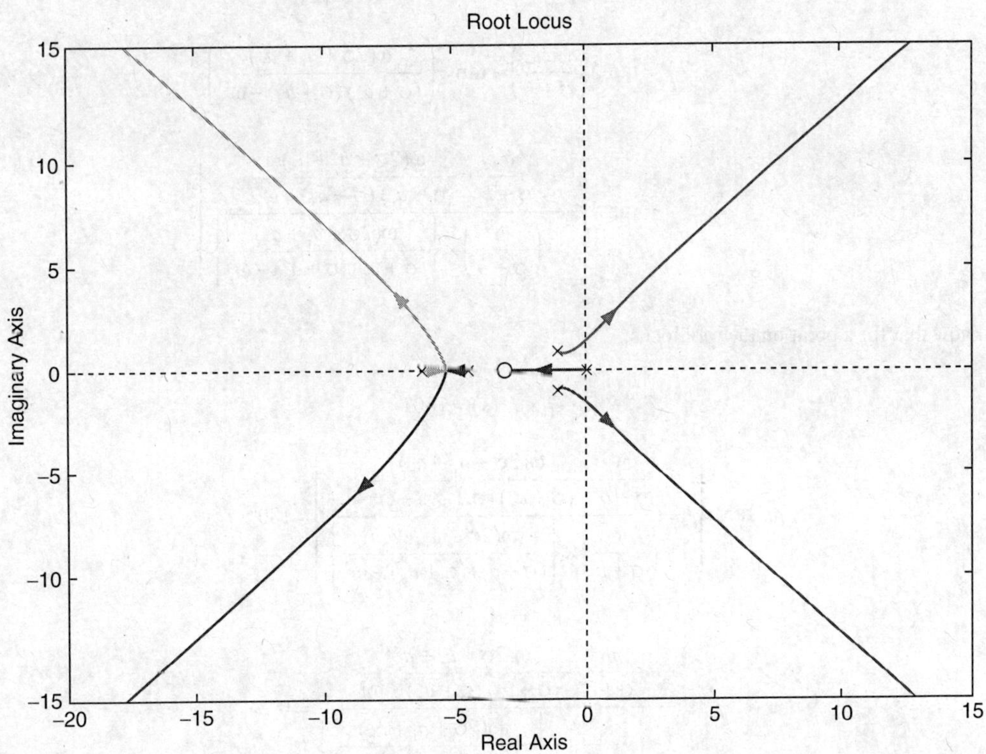

Fig. 49

Example 78 A combination of two poles at $s = -a_1$ and $s = -a_2$ and one zero at $s = -b$ to the left of both the poles will result in root locus whose complex root branches form a circle centred at the zero $s = b$, with a radius given by $\sqrt{(b-a_1)(b-a_2)}$ where the root locus gain varying from 0 to ∞.

Solution

Here
$$G(s)H(s) = \frac{K(s+b)}{(s+a_1)(s+a_2)}$$

Let $s = \sigma + j\omega$ be the complex point on the root locus.

$$\therefore \quad G(s)H(s)\big|_{s=\sigma+j\omega} = \frac{K(\sigma+j\omega+b)}{(\sigma+j\omega+a_1)(\sigma+j\omega+a_2)}$$

$$\therefore \quad \angle G(s)H(s)\big|_{s=\sigma+j\omega} = \tan^{-1}\left(\frac{\omega}{\sigma+b}\right) - \left\{\tan^{-1}\left(\frac{\omega}{\sigma+a_1}\right) + \tan^{-1}\left(\frac{\omega}{\sigma+a_2}\right)\right\}$$

$$= \tan^{-1}\left(\frac{\omega}{\sigma+b}\right) - \tan^{-1}\left(\frac{\dfrac{\omega}{\sigma+a_1} + \dfrac{\omega}{\sigma+a_2}}{1 - \dfrac{\omega^2}{(\sigma+a_1)(\sigma+a_2)}}\right)$$

$$= \tan^{-1}\left(\frac{\omega}{\sigma+b}\right) - \tan^{-1}\left(\frac{\omega(2\sigma+a_1+a_2)}{(\sigma+a_1)(\sigma+a_2)-\omega^2}\right)$$

$$= \tan^{-1}\left(\frac{\dfrac{\omega}{\sigma+b} - \dfrac{\omega(2\sigma+a_1+a_2)}{(\sigma+a_1)(\sigma+a_2)-\omega^2}}{1+\left(\dfrac{\omega}{\sigma+b}\right)\times\left\{\dfrac{\omega(2\sigma+a_1+a_2)}{(\sigma+a_1)(\sigma+a_2)-\omega^2}\right\}}\right)$$

We know that for a point on the root locus,

$$\angle G(s)H(s) = 180°$$

$$\therefore \quad \tan^{-1}\left(\frac{\dfrac{\omega}{\sigma+b} - \dfrac{\omega(2\sigma+a_1+a_2)}{(\sigma+a_1)(\sigma+a_2)-\omega^2}}{1+\left(\dfrac{\omega}{\sigma+b}\right)\times\left\{\dfrac{\omega(2\sigma+a_1+a_2)}{(\sigma+a_1)(\sigma+a_2)-\omega^2}\right\}}\right) = 180°$$

i.e.,
$$\frac{\dfrac{\omega}{\sigma+b} - \dfrac{\omega(2\sigma+a_1+a_2)}{(\sigma+a_1)(\sigma+a_2)-\omega^2}}{1+\left(\dfrac{\omega}{\sigma+b}\right)\times\left\{\dfrac{\omega(2\sigma+a_1+a_2)}{(\sigma+a_1)(\sigma+a_2)-\omega^2}\right\}} = 0$$

i.e.,
$$(\sigma+a_1)(\sigma+a_2)-\omega^2-(2\sigma+a_1+a_2)(\sigma+b)=0$$

i.e.,
$$\sigma^2+\sigma(a_1+a_2)+a_1a_2-\omega^2-2\sigma^2-\sigma(a_1+a_2)-2\sigma b-b(a_1+a_2)=0$$

i.e.,
$$-\sigma^2-2\sigma b-\omega^2-b(a_1+a_2)+a_1a_2=0$$

i.e.,
$$\sigma^2+2\sigma b+b^2+\omega^2=b^2-b(a_1+a_2)+a_1a_2$$

i.e.,
$$(\sigma+b)^2+(\omega-0)^2=(b-a_1)(b-a_2)$$

i.e.,
$$(\sigma+b)^2+(\omega-0)^2=\left\{\sqrt{(b-a_1)(b-a_2)}\right\}^2 \quad (1)$$

Eq. (1) is the equation of a circle. The centre of the circle is at $(-b,0)$ and the radius is $\sqrt{(b-a_1)^2(b-a_2)}$ unit.

Example 79 Show that the root loci for a control system with $G(s)=\dfrac{K(s^2+6s+10)}{s^2+2s+10}$, $H(s)=1$ are the arcs of the circle centered at the origin with radius equal to $\sqrt{10}$.

Solution

Here
$$G(s)H(s)=\frac{K(s^2+6s+10)}{s^2+2s+10}=\frac{K(s+3-j1)(s+3+j1)}{(s+1-j3)(s+1+j3)} \quad (1)$$

Putting $s=\sigma+j\omega$ in Eq. (1), we have

$$G(s)H(s)\big|_{s=\sigma+j\omega}=\frac{K(\sigma+j\omega+3-j1)(\sigma+j\omega+3+j1)}{(\sigma+j\omega+1-j3)(\sigma+j\omega+1+j3)}$$

$$=\frac{K[(\sigma+3)+j\,(\omega-1)]\ [(\sigma+3)+j\,(\omega+1)]}{[(\sigma+1)+j\,(\omega-3)]\ [(\sigma+1)+j\,(\omega+3)]}$$

Here $\angle G(j\omega(j\omega)H)=\left[\tan^{-1}\dfrac{\omega-1}{\sigma+3}+\tan^{-1}\dfrac{\omega+1}{\sigma+3}\right]-\left[\tan^{-1}\dfrac{\omega-3}{\sigma+1}+\tan^{-1}\dfrac{\omega+3}{\sigma+1}\right]$

To satisfy the angle criterion,

$$\left[\tan^{-1}\frac{\omega-1}{\sigma+3}+\tan^{-1}\frac{\omega+1}{\sigma+3}\right]-\left[\tan^{-1}\frac{\omega-3}{\sigma+1}+\tan^{-1}\frac{\omega+3}{\sigma+1}\right]=180^\circ$$

i.e.,
$$\tan^{-1}\left[\frac{\dfrac{\omega-1}{\sigma+3}+\dfrac{\omega+1}{\sigma+3}}{1-\left(\dfrac{\omega-1}{\sigma+3}\right)\left(\dfrac{\omega+1}{\sigma+3}\right)}\right]-\tan^{-1}\left[\frac{\dfrac{\omega-3}{\sigma+1}+\dfrac{\omega+3}{\sigma+1}}{1-\left(\dfrac{\omega-3}{\sigma+1}\right)\left(\dfrac{\omega+3}{\sigma+1}\right)}\right]=180^\circ$$

i.e.,
$$\tan^{-1}\left[\frac{\frac{\omega-1}{\sigma+3}+\frac{\omega+1}{\sigma+3}}{1-\left(\frac{\omega-1}{\sigma+3}\right)\left(\frac{\omega+1}{\sigma+3}\right)}\right]=180°+\tan^{-1}\left[\frac{\frac{\omega-3}{\sigma+1}+\frac{\omega+3}{\sigma+1}}{1-\left(\frac{\omega-3}{\sigma+1}\right)\left(\frac{\omega+3}{\sigma+1}\right)}\right].$$

Taking tangent of both sides we have,

$$\frac{\frac{\omega-1}{\sigma+3}+\frac{\omega+1}{\sigma+3}}{1-\left(\frac{\omega-1}{\sigma+1}\right)\left(\frac{\omega+1}{\sigma+3}\right)}=\frac{\frac{\omega-3}{\sigma+1}+\frac{\omega+3}{\sigma+1}}{1-\left(\frac{\omega-3}{\sigma+1}\right)\left(\frac{\omega+3}{\sigma+1}\right)}$$

i.e.,
$$\frac{\sigma+3}{(\sigma+3)^2-(\omega^2-1)}=\frac{\sigma+1}{(\sigma+1)^2-(\omega^2-9)}$$

i.e.,
$$2\sigma^2+2\omega^2-20=0$$

i.e.,
$$\sigma^2+\omega^2=10$$

i.e.,
$$(\sigma-0)^2+(\omega-0)^2=(\sqrt{10})^2$$

This is the equation of a circle with center at the origin having radius $\sqrt{10}$.

Example 80 Find frequency specifications for a system having $\frac{C(s)}{R(s)}=\frac{64}{s^2+6s+64}$.

Solution Here $\omega_n^2=64$, $\omega_n=8$ rad / sec. and $2\xi\omega_n=6$ or, $\xi=\frac{6}{2\omega_n}=0.375$

Now
$$M_r=\frac{1}{2\xi\sqrt{1-\xi^2}}=\frac{1}{2\times0.375\times\sqrt{1-(0.375)^2}}=1.438$$

and
$$\omega_r=\omega_n\sqrt{1-2\xi^2}=8\times\sqrt{1-2\times(0.375)^2}=6.78 \text{ rad/sec}$$

Again,
$$\omega_c=\omega_n\left\{(1-2\xi^2)+\sqrt{4\xi^4-4\xi^2+2}\right\}^{\frac{1}{2}}$$
$$=8\left[\{1-2\times(0.375)^2\}+\sqrt{4\times(0.375)^4-4\times(0.375)^2+2}\right]^{\frac{1}{2}}=11.17 \text{ rad/ sec.}$$

$\therefore$
$$BW=\omega_c=11.17 \text{ rad/sec.}$$

Example 81 Find frequency specifications for a system having $G(s)=\frac{49}{s(s+14)}$ with unity feedback.

Solution $G(s)=\frac{49}{s(s+14)}=\frac{49}{s^2+14s}$

$\therefore$
$$\frac{C(s)}{R(s)}=\frac{G(s)}{1+G(s)H(s)}=\frac{49}{s^2+14s+49}$$

Now $\omega_n^2=49$ i.e., $\omega_n=7$ rad/ sec. and $2\xi\omega_n=14$ or $\xi=\frac{14}{2\times7}=1$

Now $u_r = \sqrt{1-2\xi^2}$ becomes complex.

The other solution $u_r = 0$. Hence $\omega_r = 0$ and $M_r = 1$ (This is true for $\xi > 0.707$).

Again $\omega_c = \omega_n \left\{1-2\xi^2+\sqrt{4\xi^4-4\xi^2+2}\right\}^{\frac{1}{2}} = 4.50$ rad/sec. $BW = \omega_c = 4.50$ rad/sec.

$BW < \omega_n$ for $\xi > 0.707$.

Example 82 Use Nyquist criterion, determine whether the closed loop system having the following open loop transfer function is stable or not. If not how many closed loop poles lie in the right half s–plane?

$$G(s)H(s) = \frac{1+4s}{s^2(1+s)(1+2s)}$$

Solution

Here
$$G(s)H(s) = \frac{1+4s}{s^2(1+s)(1+2s)} \tag{1}$$

Putting $s = j\omega$, we have

$$G(j\omega)H(j\omega) = \frac{1+4j\omega}{(j\omega)^2(1+j\omega)(1+2j\omega)}$$

Along the $j\omega$–axis, for various values of 'ω', $G(j\omega)\,H(j\omega)$ is plotted.

ω	$G(j\omega)\,H(j\omega)$
0	$\infty\angle-180°$
0.1	$105\angle-175.22°$
1.0	$1.3\angle-212.47°$
2.0	$0.21\angle-236.5°$
4.0	$0.03\angle-252.4°$
10.0	$0.0019\angle-262.8°$
20.0	$0.00024\angle-266.4°$
∞	$0\angle-270°$

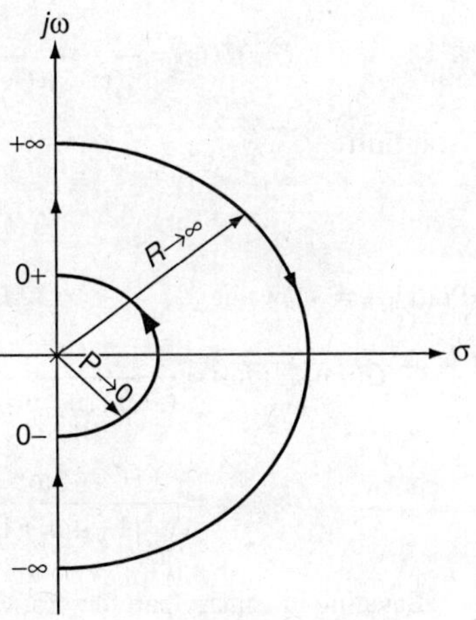

Fig. 50

$$\therefore \quad M = |G(j\omega)H(j\omega)| = \frac{\sqrt{1+16\omega^2}}{\omega^2\sqrt{1+\omega^2}\sqrt{1+4\omega^2}}$$

and $\quad \angle G(j\omega)H(j\omega) = \tan^{-1}4\omega - 180° - \tan^{-1}\omega - \tan^{-1}2\omega$

The point of intersection with $G(j\omega)\,H(j\omega)$ axis is calculated as

$$\angle G((j\omega)\,H(j\omega) = -180°$$

i.e., $\quad \tan^{-1}4\omega - 180° - \tan^{-1}\omega - \tan^{-1}2\omega = -180°$

i.e., $\quad \omega = 0.354$ rad/sec.

$$G(j\omega)\,H(j\omega)\big|_{\omega=0.354} = -10.64$$

Mapping of infinite semicircle :

$$\underset{R\to\infty}{\text{Lt}} \frac{1+4Re^{j\theta}}{(Re^{j\theta})^2\,(1+Re^{j\theta})\,(1+2Re^{j\theta})} = 0\angle-270°$$

Mapping of $j\omega$ axis from $-\infty$ to $0-$, which is the mirror image of $j\omega$ axis from $0+$ to $+\infty$.

Mapping of small semicircle at origin:

$$\underset{\rho \to 0}{\text{Lt}} \frac{1+4\rho e^{j\theta}}{(\rho e^{j\theta})^2\ (1+\rho e^{j\theta})\ (1+2\rho e^{j\theta})} = \infty\ (\angle 180° \to 0° \to 180°)$$

The complete Nyquist plot is shown in Fig. 51 below.

Since $OX = -10.64$, hence $(-1+j0)$ encircles two times in clockwise direction.

$\therefore \quad N = 2, \text{P} = 0$

Since $\quad N = Z - \text{P}$

$\therefore \quad Z = 2$

Therefore, two roots of the characteristic equation lie in the right half of s−plane. Hence the closed loop system is stable.

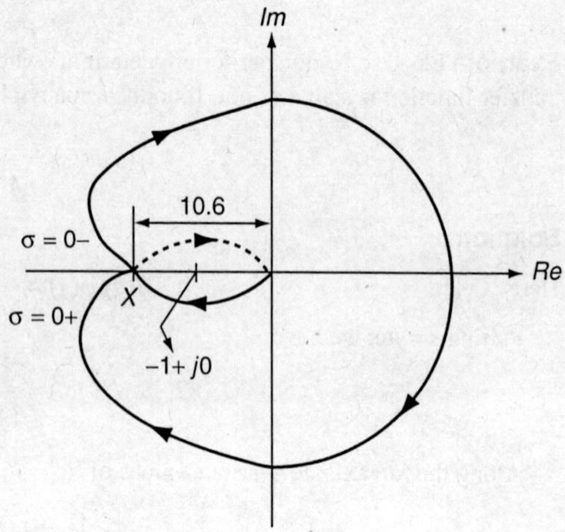

Fig. 51

Example 83 Use Nyquist criterion, determine whether the closed loop system having the following open loop transfer function is stable or not. If not how many closed loop poles lie in the right half *s*−plane?

$$G(s)H(s) = \frac{1}{s(1+s)(1+2s)}$$

Solution

Here $\quad G(s)H(s) = \dfrac{1}{s(1+s)(1+2s)} \quad (1)$

Putting $s = j\omega$, we have

$$G(j\omega)\ H(j\omega) = \frac{1}{(j\omega)\ (1+j\omega)\ (1+2j\omega)} \quad (2)$$

$$= \frac{-3\omega^2}{\omega^2(1+4\omega^2)(1+\omega^2)} - \text{j}\frac{\omega(1-2\omega^2)}{\omega^2(1+4\omega^2)(1+\omega^2)}$$

Equating imaginary part to zero, we have

$$1 - 2\omega^2 = 0$$

i.e., $\quad \omega = \dfrac{1}{\sqrt{2}} = 0.707$ rad/sec.

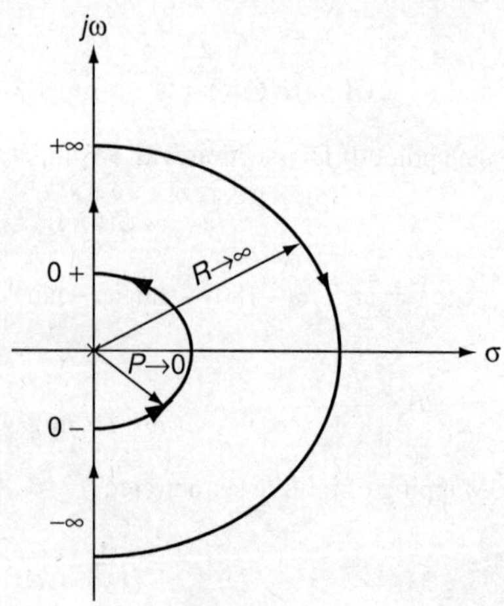

Fig. 52

$G(j\omega)H(j\omega)\big|_{\omega=0.707} = -0.667$ which is the point of intersection on real axis.

From Eq. (1), we have

$$|G(j\omega)H(j\omega)| = \frac{1}{\omega\sqrt{1+4\omega^2}\ \sqrt{1+\omega^2}}$$

and $\quad \angle G(j\omega)H(j\omega) = -90° - \tan^{-1} 2\omega - \tan^{-1}\omega$

The following are the mappings:

(i) For mapping values of ω between 0 + and + ∞.

(ii) Mapping of infinite semicircle.

$$\underset{R\to\infty}{\text{Lt}} \frac{1}{(Re^{j\theta})\,(1+Re^{j\theta})\,(1+2Re^{j\theta})} = 0\angle -270°$$

Mapping of $j\omega$ axis from $-\infty$ to $0-$, which is the mirror image of $j\omega$ axis from $0+$ to $+\infty$.

(iii) Mapping of $j\omega$ axis from $0-$ to $-\infty$, which is the mirror iage of $j\omega$ axis from $0+$ to $+\infty$.

(iv) Mapping of semicircle at origin

$$\underset{\rho\to 0}{\text{Lt}} \frac{1}{(\rho e^{j\theta})\,(1+\rho e^{j\theta})\,(1+2\rho e^{j\theta})} = \infty\angle -90°$$

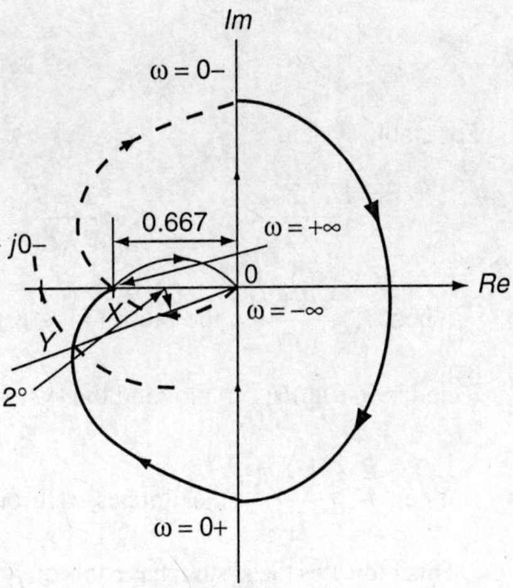

Fig. 53

Figure 52 shows the Nyquist contour and Fig. 53 shows the Nyquist plot.

$$\text{Gain Margin} = 20\log_{10}\left(\frac{1}{0.667}\right) = 3.51 \text{ dB}$$

For phase margin a unit circle is drawn that intersects the Nyquist plot at Y. The points O and Y are joined and the angle is measured.

∴ P.M. = 12°.

Example 84 Using Nyquist criterion investigate the stability of a closed loop control system having the following open loop transfer function

$$G(s)H(s) = \frac{K}{s(1+sT_1)\,(1+sT_2)}$$

Solution

Here
$$G(s)H(s) = \frac{K}{s(1+sT_1)\,(1+sT_2)} \tag{1}$$

Putting $s = j\omega$ in Eq. (1), we have

$$G(j\omega)H(j\omega) = \frac{K}{j\omega(1+j\omega T_1)\,(1+j\omega T_2)}$$

$$= \frac{-K(T_1+T_2)}{\omega(1+\omega^2T_1^2)\,(1+\omega^2T_2^2)} - j\frac{-K(1-\omega^2T_1T_2)}{\omega(1+\omega^2T_1^2)\,(1+\omega^2T_2^2)} \tag{2}$$

To obtain the value of 'ω' at the point of intersection of plot with the real axis , the imaginary part of Eq. (2) must be put to zero.

∴
$$\frac{-K(1-\omega^2T_1T_2)}{\omega(1+\omega^2T_1^2)\,(1+\omega^2T_2^2)} = 0 \quad \text{i.e.,} \quad \omega = \frac{1}{\sqrt{T_1T_2}}$$

$\therefore$ $$G(j\omega)H(j\omega)\Big|_{\omega=\frac{1}{\sqrt{T_1T_2}}} = -\frac{KT_1T_2}{T_1+T_2}$$

For stable system

$$\frac{KT_1T_2}{T_1+T_2} < 1 \quad \text{i.e., } K < \frac{T_1+T_2}{T_1T_2}$$

- When $K < \dfrac{T_1+T_2}{T_1T_2}$, the point $(-1 + j0)$ is not encircled by $G(j\omega)H(j\omega)$ plot and the system is stable.
- When $K = \dfrac{T_1+T_2}{T_1T_2}$, the plot passes through $(-1+j0)$. This indicates the system has roots on $j\omega$–axis. Hence the system is marginally stable.
- When $K > \dfrac{T_1+T_2}{T_1T_2}$, the point $(-1+j0)$ is encircled by $G(j\omega)H(j\omega)$ plot and the system is unstable.

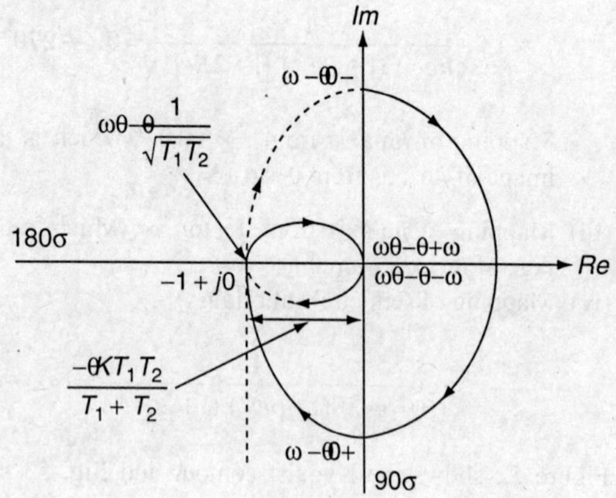

Fig. 54

Figure 54 shows the nyquist plot of the system.

Example 85 Consider the transfer function $G(s)H(s) = \dfrac{60}{(s+1)\,(s+2)\,(s+5)}$. Using Nyquist stability criterion determine whether the closed loop system is stable or not.

Solution

Here $$G(s)H(s) = \frac{60}{(s+1)\,(s+2)\,(s+5)} \qquad (1)$$

Putting $s = j\omega$ in Eq. (1),

$$G(j\omega)H(j\omega) = \frac{60}{(j\omega+1)\,(j\omega+2)\,(j\omega+5)}$$

$$= \frac{60(10-8\omega^2)}{(1+\omega^2)\,(4+\omega^2)\,(25+\omega^2)} - j\frac{60(17\omega-\omega^3)}{(1+\omega^2)\,(4+\omega^2)\,(25+\omega^2)}$$

Equating real part to zero,

$$\frac{60(10-8\omega^2)}{(1+\omega^2)\,(4+\omega^2)\,(25+\omega^2)} = 0$$

i.e., $$\omega = 1.25 \text{ rad/sec.}$$

Therefore, the point of intersection on imaginary axis is

$$G(j\omega)H(j\omega)\Big|_{\omega=1.25} = -3.057$$

Equating imaginary part to zero,

$$\frac{60(17\omega - \omega^3)}{(1+\omega^2)\,(4+\omega^2)\,(25+\omega^2)} = 0$$

i.e.,
$$\omega = 4.123 \text{ rad/sec.}$$

Therefore, the point of intersection on imaginary axis is

$$G(j\omega)H(j\omega)\big|_{\omega=4.123} = -0.47$$

$$\therefore \ \underset{\omega\to 0}{\text{Lt}} |G(j\omega)H(j\omega)| = 6 \quad \text{and} \quad \underset{\omega\to 0}{\text{Lt}} \angle G(j\omega)H(j\omega) = 0^\circ$$

and

$$\underset{\omega\to \infty}{\text{Lt}} |G(j\omega)H(j\omega)| = 0 \quad \text{and} \quad \underset{\omega\to \infty}{\text{Lt}} \angle G(j\omega)H(j\omega) = -270^\circ$$

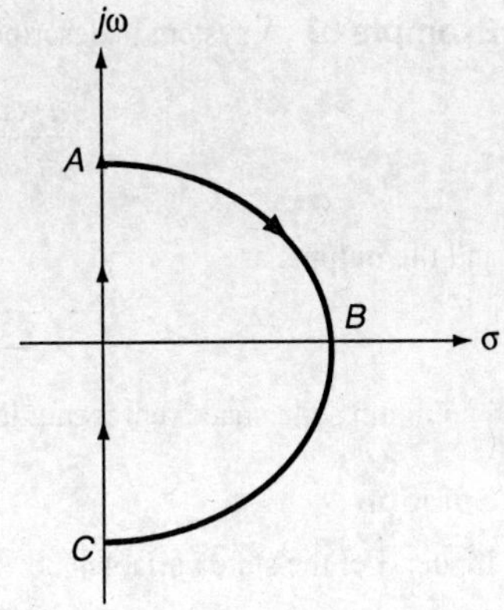

Fig. 55

Figure 55 shows the Nyquist contour and Fig. 56 shows the Nyquist plot.

The point $-1 + j0$ is not encircled by the plot. Therefore, $N = 0$.

There is no pole in the right half. Therefore, $P = 0$. Hence $Z = 0$.

Therefore, there is no zero in right half of s–plane and the system is stable.

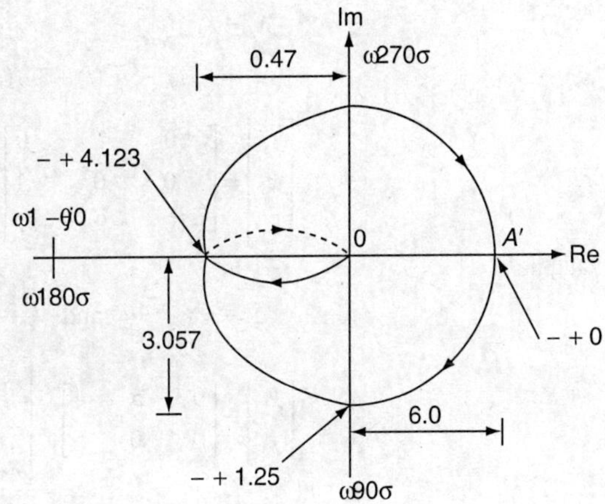

Fig. 56

Example 86 The system is described by the differential equation

$$\frac{d^3y}{dt^3} + 6\frac{d^2y}{dt^2} + 11\frac{dy}{dt} + 10y = 7u(t)$$

where y is the output and u is the input to the system. Obtain state space representation of the system.

Solution

Let the variables be

$$x_1 = y,\ x_2 = \dot{y} \quad \text{and} \quad x_3 = \ddot{y}$$

$$\dot{x}_1 = x_2 \tag{1}$$

$$\dot{x}_2 = x_3 \tag{2}$$

$$\dot{x}_3 = 7u(t) - 10x_1 - 11x_2 - 6x_3 \tag{3}$$

The Eq.(1), Eq.(2) and Eq.(3) is written in matrix form as

$$\begin{bmatrix}\dot{x}_1\\ \dot{x}_2\\ \dot{x}_3\end{bmatrix} = \begin{bmatrix}0 & 1 & 0\\ 0 & 0 & 1\\ -10 & -11 & -6\end{bmatrix}\begin{bmatrix}x_1\\ x_2\\ x_3\end{bmatrix} + \begin{bmatrix}0\\ 0\\ 7\end{bmatrix} u(t)$$

and

$$y = [1 \quad 0 \quad 0]\begin{bmatrix}x_1\\ x_2\\ x_3\end{bmatrix}$$

Example 87 A system is described by the following differential equation

$$\frac{d^3x}{dt^3}+4\frac{d^2x}{dt^2}+9\frac{dx}{dt}+10x=u_1(t)+5u_2(t)+3u_3(t)$$

and the outputs are

$$y_1=5\frac{dx}{dt}+4u_1 \text{ and } y_2=5\frac{d^2x}{dt^2}+5u_2+2u_3$$

Obtain state space representation of the system.

Solution

Inputs: Let the state variables be

$$x_1=x$$

$$\dot{x}_1=\dot{x}=x_2$$

$$\dot{x}_2=x_3$$

$$\dot{x}_3=u_1(t)+5u_2(t)+3u_3(t)-4x_3-9x_2-10x_1$$

$$\therefore \quad \begin{bmatrix}\dot{x}_1\\ \dot{x}_2\\ \dot{x}_3\end{bmatrix}=\begin{bmatrix}0 & 1 & 0\\ 0 & 0 & 1\\ -10 & -9 & -4\end{bmatrix}\begin{bmatrix}x_1\\ x_2\\ x_3\end{bmatrix}+\begin{bmatrix}1\\ 5\\ 3\end{bmatrix}u(t)$$

Outputs:

$$y_1=5x_2+4u_1 \text{ and } y_2=5x_3+5u_2+2u_3$$

$$\begin{bmatrix}y_1\\ y_2\end{bmatrix}=\begin{bmatrix}0 & 5 & 4\\ 0 & 0 & 5\end{bmatrix}\begin{bmatrix}x_1\\ x_2\\ x_3\end{bmatrix}+\begin{bmatrix}4 & 0 & 0\\ 0 & 5 & 2\end{bmatrix}\begin{bmatrix}u_1\\ u_2\\ u_3\end{bmatrix}$$

Example 88 Obtain the state model of the following transfer function:

$$G(s)=\frac{y(s)}{u(s)}=\frac{K(B_2s+B_1)}{s^3+b_1s^2+b_2s+b_3}$$

Solution

$$G(s)=\frac{y(s)}{u(s)}=\frac{x_1(s)}{u(s)}\times\frac{y(s)}{x_1(s)}=\frac{K}{s^3+b_1s^2+b_2s+b_3}\times(B_2s+B_1)$$

Let $\qquad \dfrac{x_1(s)}{u(s)}=\dfrac{K}{s^3+b_1s^2+b_2s+b_3}$

i.e., $\qquad [s^3+b_1s^2+b_2s+b_3]x_1(s)=Ku(s) \qquad (1)$

The inverse Laplace transform of Eq. (1) gives

$$\frac{d^3x_1}{dt^3}+b_1\frac{d^2x_1}{dt^2}+b_2\frac{dx_1}{dt}+b_3x_1=Ku(t)$$

i.e.,
$$\frac{d^3x_1}{dt^3} = Ku(t) - b_1\frac{d^2x_1}{dt^2} - b_2\frac{dx_1}{dt} - b_3x_1$$

Let the state variables be

$$x_1 = x$$

$$\dot{x}_1 = \dot{x} = x_2$$

$$\dot{x}_2 = x_3$$

$$\dot{x}_3 = Ku - b_1x_3 - b_2x_2 - b_3x_1$$

$$\therefore \quad \begin{bmatrix}\dot{x}_1\\ \dot{x}_2\\ \dot{x}_3\end{bmatrix} = \begin{bmatrix}0 & 1 & 0\\ 0 & 0 & 1\\ -b_3 & -b_2 & -b_1\end{bmatrix}\begin{bmatrix}x_1\\ x_2\\ x_3\end{bmatrix} + \begin{bmatrix}0\\ 0\\ K\end{bmatrix}u$$

and
$$\frac{y(s)}{x_1(s)} = B_2s + B_1 \text{ i.e., } Y(s) = (B_2s + B_1)\,x_1(s) \qquad (2)$$

The inverse Laplace transform of Eq. (2) gives $y(t) = B_1x_1 + B_2\dfrac{dx_1}{dt}$

$$y(t) = B_1x_1 + B_2x_2$$

$$y(t) = \begin{bmatrix}B_1 & B_2 & 0\end{bmatrix}\begin{bmatrix}x_1\\ x_2\\ x_3\end{bmatrix}$$

Example 89 A system is shown in Fig. 57. Obtain the state model of the system.

Solution

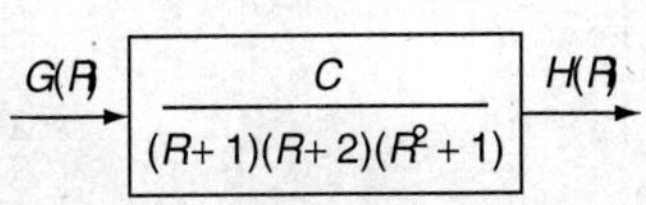

Fig. 57

$$\frac{Y(s)}{U(s)} = G(s) = \frac{K}{(s+1)(s+2)(s^2+1)}$$

i.e.,
$$\frac{Y(s)}{U(s)} = \frac{K}{s^4 + 3s^3 + 3s^2 + 3s + 2}$$

i.e.,
$$Y(s)[s^4 + 3s^3 + 3s^2 + 3s + 2] = KU(s)$$

i.e.,
$$s^4Y(s) + 3s^3Y(s) + 3s^2Y(s) + 3sY(s) + 2Y(s) = KU(s) \qquad (1)$$

The inverse Laplace transform of Eq.(1) gives

$$\frac{d^4x}{dt^4} + 3\frac{d^3y}{dt^3} + 3\frac{d^2y}{dt^2} + 3\frac{dy}{dt} + 2y = Ku$$

Let the state variables be

$$x_1 = y$$

$$\dot{x}_1 = x_2$$

$$\dot{x}_2 = x_3$$

$$\dot{x}_3 = x_4$$

$$\dot{x}_4 = Ku - 3x_4 - 3x_3 - 3x_2 - 2x_1$$

$$\therefore \quad \begin{bmatrix} \dot{x}_1 \\ \dot{x}_2 \\ \dot{x}_3 \\ \dot{x}_4 \end{bmatrix} = \begin{bmatrix} 0 & 1 & 0 & 0 \\ 0 & 0 & 1 & 0 \\ 0 & 0 & 0 & 1 \\ -2 & -3 & -3 & -3 \end{bmatrix} \begin{bmatrix} x_1 \\ x_2 \\ x_3 \\ x_4 \end{bmatrix} + \begin{bmatrix} 0 \\ 0 \\ 0 \\ K \end{bmatrix} u$$

and
$$y = [\,1\;0\;0\;0\,] \begin{bmatrix} x_1 \\ x_2 \\ x_3 \\ x_4 \end{bmatrix}$$

Example 90 A system is described by the following transfer function $G(s) = \dfrac{20(10s+1)}{s^3+3s^2+2s+1}$.

Find the state and output equation of the system.

Solution

Here
$$G(s) = \frac{Y(s)}{U(s)} = \frac{X_1(s)}{U(s)} \times \frac{Y(s)}{X_1(s)} = \frac{20}{s^3+3s^2+2s+1} \times (10s+1)$$

Let
$$\frac{X_1(s)}{U(s)} = \frac{20}{s^3+3s^2+2s+1}$$

i.e.,
$$X_1(s)[s^3+3s^2+2s+1] = 20U(s)$$

i.e.,
$$s^3X_1(s) + 3s^2X_1(s) + 2sX_1(s) + X_1(s) = 20U(s) \tag{1}$$

Taking inverse Laplace transform of Eq. (1), we have

$$\frac{d^3x_1}{dt^3} + 3\frac{d^2x_1}{dt^2} + 3\frac{dx_1}{dt} + x_1 = 20u$$

Let
$$\dot{x}_1 = x_2$$

$$\dot{x}_2 = x_3$$

$$\dot{x}_3 = 20u - 3x_3 - 2x_2 - x_1$$

$$\therefore \quad \begin{bmatrix} \dot{x}_1 \\ \dot{x}_2 \\ \dot{x}_3 \end{bmatrix} = \begin{bmatrix} 0 & 1 & 0 \\ 0 & 0 & 1 \\ -1 & -2 & -3 \end{bmatrix} \begin{bmatrix} x_1 \\ x_2 \\ x_3 \end{bmatrix} + \begin{bmatrix} 0 \\ 0 \\ 20 \end{bmatrix} u$$

and
$$\frac{y(s)}{x_1(s)} = 10s+1 \quad \text{i.e., } y(s) = (10s+1)x_1(s) \tag{2}$$

The inverse Laplace transform of Eq. (2) gives

$$y(t) = x_1 + 10x_2$$

$$y = \begin{bmatrix} 1 & 10 & 0 \end{bmatrix} \begin{bmatrix} x_1 \\ x_2 \\ x_3 \end{bmatrix}$$

Example 91 A system is characterized by the transfer function

$$\frac{Y(s)}{U(s)} = \frac{2}{s^3 + 6s^2 + 11s + 6}$$

Find the state and output equation in matrix form and also test the controllability and observability of the system.

Solution

$$\text{Here } \frac{Y(s)}{U(s)} = \frac{2}{s^3 + 6s^2 + 11s + 6}$$

i.e.,
$$Y(s)[s^3 + 6s^2 + 11s + 6] = 2U(s) \tag{1}$$

Taking inverse Laplace transform of Eq. (1), we have

$$\frac{d^3y}{dt^3} + 6\frac{d^2y}{dt^2} + 11\frac{dy}{dt} + 6y = 2u$$

Let the state variables be

$$x_1 = y$$

$$\dot{x}_1 = x_2$$

$$\dot{x}_2 = x_3$$

$$\dot{x}_3 = 2u - 6x_3 - 11x_2 - 6x_1$$

$$\therefore \quad \begin{bmatrix} \dot{x}_1 \\ \dot{x}_2 \\ \dot{x}_3 \end{bmatrix} = \begin{bmatrix} 0 & 1 & 0 \\ 0 & 0 & 1 \\ -6 & -11 & -6 \end{bmatrix} \begin{bmatrix} x_1 \\ x_2 \\ x_3 \end{bmatrix} + \begin{bmatrix} 0 \\ 0 \\ 2 \end{bmatrix} u$$

and
$$y = \begin{bmatrix} 1 & 0 & 0 \end{bmatrix} \begin{bmatrix} x_1 \\ x_2 \\ x_3 \end{bmatrix}$$

Here
$$A = \begin{bmatrix} 0 & 1 & 0 \\ 0 & 0 & 1 \\ -6 & -11 & -6 \end{bmatrix}, \ B = \begin{bmatrix} 0 \\ 0 \\ 2 \end{bmatrix} \text{ and } C = \begin{bmatrix} 1 & 0 & 0 \end{bmatrix}$$

$$AB = \begin{bmatrix} 0 & 1 & 0 \\ 0 & 0 & 1 \\ -6 & -11 & -6 \end{bmatrix} \begin{bmatrix} 0 \\ 0 \\ 2 \end{bmatrix} = \begin{bmatrix} 0 \\ 2 \\ -12 \end{bmatrix} \text{ and } A^2B = \begin{bmatrix} 0 & 1 & 0 \\ 0 & 0 & 1 \\ -6 & -11 & -6 \end{bmatrix} \begin{bmatrix} 0 \\ 0 \\ 2 \end{bmatrix} = \begin{bmatrix} 2 \\ -12 \\ 50 \end{bmatrix}$$

$$Q = [B : AB : A^2B] = \begin{bmatrix} 0 & 0 & 2 \\ 0 & 2 & -12 \\ 2 & -2 & 50 \end{bmatrix} \quad \therefore |Q| \neq 0$$

Therefore, it is a full rank matrix and hence the system is controllable.

Again, $$C^T = \begin{bmatrix} 1 \\ 0 \\ 0 \end{bmatrix}$$

$$\therefore \quad A^T C^T = \begin{bmatrix} 0 \\ 1 \\ 0 \end{bmatrix} \text{ and } (A^T)^2 C^T = \begin{bmatrix} 0 \\ 1 \\ 0 \end{bmatrix}$$

$$\therefore \quad Q' = [C^T : A^T C^T : (A^T)^2 C^T] = \begin{bmatrix} 1 & 0 & 0 \\ 0 & 1 & 0 \\ 0 & 0 & 1 \end{bmatrix} \text{ and } |Q'| \neq 0$$

Therefore, it is a full rank matrix and hence the system is observable.

Example 92 A system is described by the following equations

$$\dot{x}(t) = \begin{bmatrix} -1 & 1 \\ 0 & -2 \end{bmatrix} x(t) + \begin{bmatrix} 1 & 0 & 1 \\ 0 & 1 & 1 \end{bmatrix} u(t)$$

and $$y(t) = \begin{bmatrix} 1 & 2 \\ 1 & 0 \\ 1 & 1 \end{bmatrix} x(t)$$

Find the transfer function of the system.

Solution The given system is a MIMO system.

$$G(s) = C[sI - A]^{-1}B$$

Here $$A = \begin{bmatrix} -1 & 1 \\ 0 & -2 \end{bmatrix}, \quad B = \begin{bmatrix} 1 & 0 & 1 \\ 0 & 1 & 1 \end{bmatrix} \text{ and } C = \begin{bmatrix} 1 & 2 \\ 1 & 0 \\ 1 & 1 \end{bmatrix}$$

$$\therefore \quad sI - A = \begin{bmatrix} s & 0 \\ 0 & s \end{bmatrix} - \begin{bmatrix} -2 & 1 \\ 0 & -2 \end{bmatrix} = \begin{bmatrix} s+1 & -1 \\ 0 & s+2 \end{bmatrix}$$

$$\therefore (sI - A)^{-1} = \frac{1}{s^2 + 3s + 2} \begin{bmatrix} s+1 & -1 \\ 0 & s+2 \end{bmatrix}$$

$$\therefore \quad G(s) = C(sI - A)^{-1}B = \begin{bmatrix} 1 & 2 \\ 1 & 0 \\ 1 & 1 \end{bmatrix} \left(\frac{1}{s^2 + 3s + 2} \right) \begin{bmatrix} s+1 & -1 \\ 0 & s+2 \end{bmatrix} \begin{bmatrix} 1 & 0 & 1 \\ 0 & 1 & 1 \end{bmatrix} = \frac{s^3 - 3s^2 - 9s + 2}{s^2 + 3s + 2}$$

Example 93 A system is characterised by the equation

$$\frac{Y(s)}{U(s)} = \frac{20(4s+1)}{s^3+5s^2+8s+2}$$

Find its state and output equation and express in matrix form.

Solution

Here $$G(s) = \frac{Y(s)}{U(s)} = \frac{X_1(s)}{U(s)} \times \frac{Y(s)}{X_1(s)} = \frac{20}{s^3+5s^2+8s+2} \times (4s+1)$$

Let $$\frac{X_1(s)}{U(s)} = \frac{20}{s^3+5s^2+8s+2}$$

i.e., $$X_1(s)[s^3+5s^2+8s+2] = 20U(s)$$

i.e., $$s^3X_1(s)+5s^2X_1(s)+8sX_1(s)+2X_1(s) = 20U(s) \qquad (1)$$

Taking inverse Laplace transform of Eq. (1), we have

$$\frac{d^3x_1}{dt^3}+5\frac{d^2x_1}{dt^2}+8\frac{dx_1}{dt}+2x_1 = 20u$$

Let
$$\dot{x}_1 = x_2$$
$$\dot{x}_2 = x_3$$
$$\dot{x}_3 = 20u - 3x_3 - 2x_2 - x_1$$

$$\therefore \quad \begin{bmatrix} \dot{x}_1 \\ \dot{x}_2 \\ \dot{x}_3 \end{bmatrix} = \begin{bmatrix} 0 & 1 & 0 \\ 0 & 0 & 1 \\ -1 & -2 & -3 \end{bmatrix} \begin{bmatrix} x_1 \\ x_2 \\ x_3 \end{bmatrix} + \begin{bmatrix} 0 \\ 0 \\ 20 \end{bmatrix} u$$

and $$\frac{y(s)}{x_1(s)} = 4s+1 \text{ i.e., } y(s) = (4s+1)x_1(s) \qquad (2)$$

The inverse Laplace transform of Eq. (2) gives

$$y(t) = x_1 + 4x_2$$

$$y = \begin{bmatrix} 1 & 4 & 0 \end{bmatrix} \begin{bmatrix} x_1 \\ x_2 \\ x_3 \end{bmatrix}$$

Example 94 A single input single output system is given as

$$\dot{x}(t) = \begin{bmatrix} -1 & 0 & 0 \\ 0 & -2 & 0 \\ 0 & 0 & 3 \end{bmatrix} x(t) + \begin{bmatrix} 1 \\ 1 \\ 0 \end{bmatrix} u$$

and $$y = \begin{bmatrix} 1 & 0 & 2 \end{bmatrix} x(t)$$

Test for controllability and observability.

Solution

Here $$A=\begin{bmatrix}-1 & 0 & 0\\ 0 & -2 & 0\\ 0 & 0 & 3\end{bmatrix},\ B=\begin{bmatrix}1\\1\\0\end{bmatrix} \text{ and } C=\begin{bmatrix}1 & 0 & 2\end{bmatrix}$$

∴ $$AB=\begin{bmatrix}-1\\-2\\0\end{bmatrix} \text{ and } A^2B=\begin{bmatrix}1\\4\\0\end{bmatrix}$$

$$Q=\left[B:AB:A^2B\right]=\begin{bmatrix}1 & -1 & 1\\ 1 & -2 & 4\\ 0 & 0 & 0\end{bmatrix} \quad \therefore |Q|=0$$

The rank of the matrix is < 3 and hence the system is uncontrollable.

Again, $$C^T=\begin{bmatrix}1\\0\\2\end{bmatrix}$$

∴ $$A^TC^T=\begin{bmatrix}-1\\0\\-6\end{bmatrix} \text{ and } (A^T)^2C^T=\begin{bmatrix}1\\0\\18\end{bmatrix}$$

∴ $$Q'=\left[C^T:A^TC^T:(A^T)^2C^T\right]=\begin{bmatrix}1 & -1 & 1\\ 0 & 0 & 0\\ 2 & -6 & 18\end{bmatrix} \text{ and } |Q'|=0.$$

The rank of the matrix is < 3 and hence the system is unobservable.

Example 95 Find the state transition matrix $\phi(t)$ and the characteristic equation of A.

$$A=\begin{bmatrix}-2 & 0\\ 0 & -2\end{bmatrix}$$

Solution The characteristic equation is defined as

$$|sI-A|=0$$

$$sI-A=s\begin{bmatrix}1 & 0\\ 0 & 1\end{bmatrix}-\begin{bmatrix}-2 & 0\\ 0 & -2\end{bmatrix}=\begin{bmatrix}s+2 & 0\\ 0 & s+2\end{bmatrix}$$

∴ $$\phi(s)=(sI-A)^{-1}=\frac{1}{(s+2)^2}\begin{bmatrix}s+2 & 0\\ 0 & s+2\end{bmatrix}=\begin{bmatrix}\dfrac{1}{s+2} & 0\\ 0 & \dfrac{1}{s+2}\end{bmatrix}$$

∴ $$\phi(t)=LT^{-1}[(sI-A)^{-1}]=\begin{bmatrix}e^{-2t} & 0\\ 0 & e^{-2t}\end{bmatrix}$$

The characteristic equation is $s^2+4s+4=0$

Example 96 Find the state transition matrix $\phi(t)$, the characteristic equation and the eigen values of A.

$$A = \begin{bmatrix} \sigma & -\omega \\ \omega & \sigma \end{bmatrix}$$

Solution

$$sI - A = s\begin{bmatrix} 1 & 0 \\ 0 & 1 \end{bmatrix} - \begin{bmatrix} \sigma & -\omega \\ \omega & \sigma \end{bmatrix} = \begin{bmatrix} s-\sigma & \omega \\ -\omega & s-\sigma \end{bmatrix}$$

$$\therefore \qquad \phi(s) = (sI - A)^{-1} = \frac{1}{(s-\sigma)^2 + \omega^2}\begin{bmatrix} s-\sigma & -\omega \\ \omega & s-\sigma \end{bmatrix} = \begin{bmatrix} \dfrac{s-\sigma}{(s-\sigma)^2+\omega^2} & \dfrac{-\omega}{(s-\sigma)^2+\omega^2} \\ \dfrac{\omega}{(s-\sigma)^2+\omega^2} & \dfrac{s-\sigma}{(s-\sigma)^2+\omega^2} \end{bmatrix}$$

$$\therefore \qquad \phi(t) = LT^{-1}[(sI - A)^{-1}] = \begin{bmatrix} e^{\sigma t}\cos\omega t & -e^{-\sigma t}\sin\omega t \\ e^{\sigma t}\sin\omega t & e^{\sigma t}\cos\omega t \end{bmatrix}$$

The roots of the characteristic equation are

$$s = \frac{2\sigma \pm j2\omega}{2} = \sigma \pm j\omega$$

The eigen values of A are $\sigma + j\omega$ and $\sigma - j\omega$ respectively.

Example 97 The following are the system equations of a mechanical system. Obtain the state model of the system taking x_1 and x_2 as the state variables of mass M_1 and M_2 respectively.

$$F = M_1\frac{d^2x_1}{dt^2} + D_1\frac{dx_1}{dt} + D_2\frac{d}{dt}(x_1 - x_2)$$

and

$$0 = M_2\frac{d^2x_2}{dt^2} + Kx_2 + D_2\frac{d}{dt}(x_2 - x_1)$$

Solution

Let the state variables be

$$x_1 = Z_1 \qquad x_2 = Z_3$$

$$\dot{x}_1 = \dot{Z}_1 = Z_2 \qquad \dot{x}_2 = \dot{Z}_3 = Z_4$$

$$\ddot{x}_1 = \dot{Z}_2 \qquad \ddot{x}_2 = \dot{Z}_4$$

The given equations can be written as

$$\therefore \qquad \dot{Z}_2 = \frac{F}{M_1} - \left(\frac{B_1 + B_2}{M_1}\right)Z_2 - \frac{B_2}{M_1}Z_4$$

and

$$\dot{Z}_4 = -\frac{K}{M_2}Z_3 - \frac{B_2}{M_2}Z_4 - \frac{B_2}{M_2}Z_2$$

The state equations are

$$\dot{Z}_1 = Z_2$$

$$\dot{Z}_2 = \frac{F}{M_1} - \left(\frac{B_1 + B_2}{M_1}\right)Z_2 - \frac{B_2}{M_1}Z_4$$

$$\dot{Z}_3 = Z_4$$

$$\dot{Z}_4 = -\frac{K}{M_2}Z_3 - \frac{B_2}{M_2}Z_4 - \frac{B_2}{M_2}Z_2$$

$$\begin{bmatrix}\dot{Z}_1\\ \dot{Z}_2\\ \dot{Z}_3\\ \dot{Z}_4\end{bmatrix} = \begin{bmatrix}0 & 1 & 0 & 0\\ 0 & -\left(\frac{B_1+B_2}{M_1}\right) & 0 & -\frac{B_2}{M_1}\\ 0 & 0 & 0 & 1\\ 0 & -\frac{B_2}{M_2} & -\frac{K}{M_2} & -\frac{B_2}{M_2}\end{bmatrix}\begin{bmatrix}Z_1\\ Z_2\\ Z_3\\ Z_4\end{bmatrix} + \begin{bmatrix}0\\ \frac{1}{M_1}\\ 0\\ 0\end{bmatrix}F$$

and

$$y_1 = x_1 = Z_1$$
$$y_2 = x_2 = Z_2$$

$$\begin{bmatrix}y_1\\ y_2\end{bmatrix} = \begin{bmatrix}1 & 0\\ 0 & 1\end{bmatrix}\begin{bmatrix}Z_1\\ Z_2\end{bmatrix}$$

Example 98 For the state matrix $A = \begin{bmatrix}0 & 1\\ -8 & -6\end{bmatrix}$, find e^{At} by (i) use of diagonal method and (ii) series of summation method.

Solution (i) Use of diagonal:

Here

$$A = \begin{bmatrix}0 & 1\\ -8 & -6\end{bmatrix}$$

$$\lambda I - A = \lambda\begin{bmatrix}1 & 0\\ 0 & 1\end{bmatrix} - \begin{bmatrix}0 & 1\\ -8 & -6\end{bmatrix} = \begin{bmatrix}\lambda & -1\\ 8 & \lambda+6\end{bmatrix} \quad (1)$$

i.e.,

$$\therefore |\lambda I - A| = \lambda^2 + 6\lambda + 8 = 0$$

i.e.,

$$(\lambda + 2)(\lambda + 4) = 0$$

i.e.,

$$\lambda_1 = -2, \lambda_2 = -4$$

Let the eigen vector be

$$p_1 = \begin{bmatrix}p_{11}\\ p_{21}\end{bmatrix} \text{ and } p_2 = \begin{bmatrix}p_{12}\\ p_{22}\end{bmatrix}$$

Putting $\lambda = -2$ in Eq. (1), we have

$$\begin{bmatrix} -2 & -1 \\ 8 & 4 \end{bmatrix} \begin{bmatrix} p_{11} \\ p_{21} \end{bmatrix} = \begin{bmatrix} 0 \\ 0 \end{bmatrix} \quad (2)$$

On solving Eq. (2), we have

$$p_{11} = 1 \text{ and } p_{21} = -2$$

$$\therefore p_1 = \begin{bmatrix} 1 \\ -2 \end{bmatrix}$$

Putting $\lambda = -4$ in Eq. (1), we have

$$\begin{bmatrix} -4 & -1 \\ 8 & 2 \end{bmatrix} \begin{bmatrix} p_{12} \\ p_{22} \end{bmatrix} = \begin{bmatrix} 0 \\ 0 \end{bmatrix} \quad (2)$$

On solving Eq. (2), we have

$$p_{12} = 1 \text{ and } p_{22} = -4$$

$$\therefore \quad p_1 = \begin{bmatrix} 1 \\ -4 \end{bmatrix}$$

$$\therefore \quad M = \begin{bmatrix} p_1 & p_2 \end{bmatrix} = \begin{bmatrix} 1 & 1 \\ -2 & -4 \end{bmatrix}$$

$$\therefore \quad M^{-1} = \begin{bmatrix} 2 & 0.5 \\ -1 & -0.5 \end{bmatrix} \text{ and } \wedge = \begin{bmatrix} -2 & 0 \\ 0 & -4 \end{bmatrix}$$

$$\therefore \quad e^{At} = Me^{\wedge t}M^{-1} = \begin{bmatrix} 1 & 1 \\ -2 & -4 \end{bmatrix} \begin{bmatrix} e^{-2t} & 0 \\ 0 & e^{-4t} \end{bmatrix} \begin{bmatrix} 2 & 0.5 \\ -1 & -0.5 \end{bmatrix}$$

$$= \begin{bmatrix} 2e^{-2t} - e^{-4t} & 0.5e^{-2t} - 0.5e^{-4t} \\ -4e^{-2t} + 4e^{-4t} & e^{-2t} + 2e^{-4t} \end{bmatrix}$$

(ii) Series of Summation Method:

$$A = \begin{bmatrix} 0 & 1 \\ -8 & -6 \end{bmatrix}$$

$$\therefore \quad A^2 = \begin{bmatrix} -8 & -6 \\ 48 & 28 \end{bmatrix}$$

$$\therefore \quad A^3 = \begin{bmatrix} 48 & 28 \\ -224 & -120 \end{bmatrix}$$

$$\therefore \quad A^4 = \begin{bmatrix} -224 & -120 \\ 960 & 496 \end{bmatrix}$$

We know that

$$e^{At} = I + At + \frac{A^2t^2}{2!} + \frac{A^3t^3}{3!} + \cdots\cdots\cdots$$

i.e.,

$$e^{At} = \begin{bmatrix} 1 & 0 \\ 0 & 1 \end{bmatrix} + \begin{bmatrix} 0 & 1 \\ -8 & -6 \end{bmatrix} t + \begin{bmatrix} -8 & -6 \\ 48 & 28 \end{bmatrix} \frac{t^2}{2!} + \begin{bmatrix} 48 & 28 \\ -224 & -120 \end{bmatrix} \frac{t^3}{3!} + \begin{bmatrix} -224 & -120 \\ 960 & 496 \end{bmatrix} \frac{t^4}{4!} + \cdots$$

$$= \begin{bmatrix} t - 4t^2 + 8t^3 - 9.33t^4 \cdots & t - 3t^2 + 4.76t^3 - 5t^4 \cdots \\ -8t + 24t^2 - 37.3t^3 + 40t^4 \cdots & 1 - 6t + 14t^2 - 20t^3 + 20.675t^4 \cdots \end{bmatrix}$$

Example 99 Consider the vector matrix differential equation as

$$\dot{x} = \begin{bmatrix} 0 & 1 \\ -2 & -3 \end{bmatrix} x$$

Obtain the state transition matrix.

Solution

$$sI - A = s\begin{bmatrix} 1 & 0 \\ 0 & 1 \end{bmatrix} - \begin{bmatrix} 0 & 1 \\ -2 & -3 \end{bmatrix} = \begin{bmatrix} s & -1 \\ 2 & s+3 \end{bmatrix}$$

$$\therefore \quad |sI - A| = (s+1)(s+2)$$

$$\therefore \quad \phi(s) = (sI - A)^{-1} = \frac{1}{(s+1)(s+2)} \begin{bmatrix} s+3 & -1 \\ -2 & s \end{bmatrix} = \begin{bmatrix} \frac{s+3}{(s+1)(s+2)} & \frac{1}{(s+1)(s+2)} \\ \frac{-2}{(s+1)(s+2)} & \frac{s}{(s+1)(s+2)} \end{bmatrix}$$

$$\therefore \quad \phi(t) = LT^{-1}[\varphi(s)] = \begin{bmatrix} 2e^{-t} - e^{-2t} & e^{-t} - e^{-2t} \\ -2e^{-t} + 2e^{-2t} & -e^{-t} + 2e^{-t} \end{bmatrix}$$

Example 100 Determine the poles and zeros of the following systems:

$$\dot{x} = \begin{bmatrix} 0 & 1 \\ -20 & -9 \end{bmatrix} x + \begin{bmatrix} 0 \\ 1 \end{bmatrix} r$$

and

$$y = \begin{bmatrix} -17 & -5 \end{bmatrix} x + [1]\, r$$

Solution

$$sI - A = s\begin{bmatrix} 1 & 0 \\ 0 & 1 \end{bmatrix} - \begin{bmatrix} 0 & 1 \\ -20 & -9 \end{bmatrix} = \begin{bmatrix} s & -1 \\ 20 & s+9 \end{bmatrix}$$

$$\therefore \quad |sI - A| = s^2 + 9s + 20 = (s+4)(s+5) = 0$$

Hence the poles are $s = -4$ and $s = -5$.

$$\therefore \quad \phi(s) = (sI - A)^{-1} = \frac{1}{(s+4)(s+5)} \begin{bmatrix} s+9 & 1 \\ -20 & s \end{bmatrix}$$

We know that $$\frac{Y(s)}{U(s)} = C(sI - A)^{-1}B + D \tag{1}$$

Equating the Eq. (1) to zero, we have

$$C(sI - A)^{-1}B + D = 0$$

i.e., $$\begin{bmatrix} -17 & -5 \end{bmatrix} \frac{1}{(s+4)\,(s+5)} \begin{bmatrix} s+9 & 1 \\ -20 & s \end{bmatrix} \begin{bmatrix} 0 \\ 1 \end{bmatrix} [1] = 0$$

i.e., $$-17 - 5s + s^2 + 9s + 20 = 0$$

i.e., $$(s+3)\,(s+1) = 0$$

Therefore, zeros are at $s = -1$ and -3.

Index